INVERTEBRATES

RICHARD C. BRUSCA

CURATOR OF MARINE INVERTEBRATES, SAN DIEGO MUSEUM OF NATURAL HISTORY

INVERTE

WITH ILLUSTRATIONS BY

SINAUER ASSOCIATES, INC. • PUBLISHERS

GARY J. BRUSCA

PROFESSOR OF ZOOLOGY, HUMBOLDT STATE UNIVERSITY

BRATES

NANCY J. HAVER

 SUNDERLAND, MASSACHUSETTS

ABOUT THE BOOK

Book and Cover Design: Joseph J. Vesely
Editor-in-Chief: Andrew D. Sinauer
Copy Editor: Jodi Simpson
Editorial Coordinator: Carol J. Wigg
Production Coordinator: Janice Holabird
Art Director: Joseph J. Vesely
Principal Artist: Nancy Haver
Graphics: Frederic Schoenborn
Photo Assistance: Biological Photo Service
Composition: DEKR Corporation
Prepress: Jay's Publishers Services, Inc.
Book Manufacture: Courier Westford, Inc.

INVERTEBRATES

Library of Congress Cataloging-in-Publication Data
Brusca, Richard C.
    Invertebrates / Richard C. Brusca, Gary J. Brusca.
        p.   cm.
    Includes bibliographical references.
    ISBN 0-87893-098-1 (alk. paper)
    1. Invertebrates.   I. Brusca, Gary J.   II. Title
QL362.B924   1990
592—dc20   90-30061
                                                                   CIP

This book is printed on paper that meets the
guidelines for permanence and durability of the
Committee on Production Guidelines for Book
Longevity of the Council on Library Resources.

Printed in U.S.A.

6   5   4   3   2

Contents in Brief

Contents

Preface

*It is not now, nor will it ever be given to one man
to observe all the things recounted in the following pages.*
Waldo L. Schmitt
Crustaceans, 1965

When we started this project some years ago it was our intention to produce a book of about 400 pages. Obviously, we were unable to do that. The field of invertebrate zoology is so vast, and cuts across so many disciplinary lines, that even in a book of this size we have found it necessary to generalize about some topics and to slight others. As university instructors, we realized early on that the teaching of invertebrate zoology should not be compartmentalized. Thus in preparing a single-volume work on invertebrates we were concerned about two potential dangers. First, the book might become an encyclopedic list of "facts" about one group after another, the sort of "flash-card" approach that we hope you will avoid in your study. Second, the book might be a rambling series of stories or vignettes about the animals and their ways of life. The first book would be dull, would encourage rote memorization instead of understanding, and might reinforce the misconception that there is little left to discover. The second book would be full of interesting "gee whiz" stuff but might seem disorganized and without much continuity or purpose to the serious student. Although our book has some elements of both problems—many new terms and a great deal of factual information, as well as interesting bits of invertebrate natural history—we have organized it in the belief that what we *know* about these animals is not as important as what and how we *think* about them. The primary purpose of this book is to provide enough information and ideas to enable you to understand the broad and dynamic concepts underlying the study of life and some of the ways by which biologists have come to develop these concepts. You should be prepared to assimilate much new material. Be prepared, also, for a great deal of uncertainty; much remains to be discovered.

To establish threads of continuity throughout this mass of information, we have developed our book around three fundamental themes: (1) functional body architecture, or what we call the *Bauplan* concept; (2) developmental patterns and life history strategies, mainly as they relate to adult lifestyles; and (3) evolution and phylogenetic relationships. We feel these themes are basic not only to understanding invertebrates but also to understanding life in general and how it has evolved. The first four chapters explore these themes and provide a foundation on which the rest of the book depends. Please read these chapters carefully and refer back to them throughout your study.

In the remaining chapters, we try to use these three themes as a common vehicle to tie the animal phyla together in a logical and interesting way. Furthermore, we present information in a comparative fashion wherever possible, both within and among major taxa. We also include a considerable amount of information on the general ecology of various groups, but this is generally incorporated into our *Bauplan* discussions.

Zoology texts have traditionally used one or the other of two approaches. One is the strictly taxonomic or "survey" approach, wherein phyla are broken down into their various subordinate taxa and each is discussed separately. This approach usually bears the double burden of redundancy and conceptual fragmentation. The second is the "systems" approach, wherein anatomical or physiological systems are surveyed. This approach can largely ignore the importance of taxonomic diversity, phylogeny, and the relationships of the whole animal to its environment. We have tried to fuse the two approaches by presenting *both* a taxonomic survey *and* a systems survey. We integrate the taxonomic and systems information largely by emphasizing the themes noted above.

The bulk of this book (Chapters 5–23) is devoted to a phylum-by-phylum discussion of invertebrates. Fairly detailed classifications or taxonomic synopses for each phylum are included in separate sections of each chapter to serve as references. We have endeavored to maintain a similar and consistent organization throughout each chapter, although we yield to the important and sometimes different lessons to be learned by investigating the special attributes of each group of animals. In addition, because of their size and diversity, some taxa receive more attention than others—although this does not mean that such groups are more "important" zoologically than smaller or more homogeneous ones. In certain chapters we cover more than one phylum. In some cases the phyla covered together are thought to be closely related to one another; in other cases the phyla merely represent a particular grade of complexity and their inclusion in a single chapter facilitates our comparative approach. Five chapters are devoted to the arthropods and their kin; for all other phyla, we discuss the entire group as a whole. The goals of our approach are to reduce redundancy, illustrate our themes, and underscore both the diversity and unifying attributes of each phylum.

Certain aspects of this book have, of course, been influenced by our own biases; this is especially true of the discussions on phylogeny. We have used a combination of cladistic analyses (cladograms), narrative discussions, and traditional (orthodox) evolutionary trees to discuss existing classifications and derive phylogenetic hypotheses. We have used cladograms whenever feasible because we believe they provide the least ambiguous statements or hypotheses of relationships that can be made about animal groups (see Chapter 2 for a discussion of cladistics). We would have liked to include more cladograms, but the field of cladistics is so new that few invertebrate cladograms have been published; and many of those must be regarded as very preliminary phylogenetic hypotheses in need of further testing and refinement. We know that some of you, professors and students both, will disagree with our methods and ideas to various degrees—at least we *hope* that you will. We implore you never to placidly accept them without question.

Most of the cladograms in this book are new and were derived by cladistic procedures based on the maximum parsimony method (i.e., the use of logical parsimony, or Occam's razor). Simple ones were constructed by hand, but more complex ones were generated by a computer phylogenetics program called PAUP (Phylogenetic Analysis Using Parsimony). PAUP was written and is distributed by David L. Swofford (Illinois Natural History Survey). PAUP analyzes data sets of characteristics to find the most parsimonious tree(s) possible; in other words, it strives to find the tree(s) requiring the fewest character transformations (the shortest tree) and the smallest number of homoplasies (the fewest hypothesized evolutionary convergences, parallelisms, and reversals). Thus, the maximum parsimony method is equivalent to minimizing the total number of evolutionary "steps" needed to explain the evolution of each taxon and character in the tree. Other phylogenetic parsimony programs exist, but our experience has shown that most will arrive at the same (shortest) trees given prudent application. Descriptions of numerical methods of cladistic analysis can be found in the literature (see references in Chapter 2, especially studies by Farris, Felsenstein, Kluge, and Wiley), and documentation for PAUP was, at the time of this writing, available on request from David Swofford.

We believe that it is absolutely essential for an introductory course on invertebrates to be framed within the context of evolutionary biology. To do so, one cannot rely on reductionism but must view organisms as complex systems operating and evolving together and in concert with the environment. We work in an age of specialization, and there is extensive fragmentation within every scientific discipline. This fragmentation has led to a tendency in zoology texts to deemphasize the fundamentally important fields of comparative anatomy and developmental biology. To the extent possible, we have tried to avoid this trend in our book by taking a "whole animal" approach. Most of the major animal body plans that we recognize today originated in a burst of evolutionary radiation during the Precambrian and Cambrian periods. Most of the evolutionary history of life has been a protracted period of adaptive specialization and exploitation of these fundamental body plans, or *Baupläne*. For this reason we emphasize both the phylogenetic origins of animals and the specializations that have evolved within each of the major lineages.

The book's final chapter is a phylogenetic summary, an overview of the major evolutionary ideas presented in the book. In it we reinforce the point that much remains to be explored and learned about invertebrates. Like all scientific knowledge, we are dealing here with provisional, wispy, transient "truths" that always remain open to challenge and revision. This is especially true with regard to the evolutionary relationships among animal phyla—something biologists really know very little about. And, of course, scientists disagree. It is this disagreement and the constant challenging of hypotheses that push the frontiers of knowledge forward.

There are a few other things you should know about this book. We provide a brief historical review of the taxonomy and classification of each major group. We feel this material not only is interesting but also serves to imbue students with a sense of the dynamic nature of taxonomy and the development of our understanding of each group. Unless otherwise indicated, the Classification section in each chapter deals only with extant taxa. Descriptions of taxa in these annotated classifications are written in telegraphic style to save space.

Important new words, when first defined, are set in **boldface** type. These boldfaced terms are also indicated by boldfaced page references in the index; thus the index can also be used as a "glossary." Although we have tried to be consistent in our usage of zoological terminology, the existence of similar terms for entirely different structures in certain groups is notoriously troublesome. We try to point these out where appropriate in the text. We also opt for contemporary spellings and usages where several options are available.

We have tried to be as current as possible with citations listed in the reference section for each chapter, but even as this book goes into production important new publications cross our desks. It has been estimated that the volume of scientific information is doubling about every 10 years. A half-million nonclinical biology papers are published annually. As Professor George Bartholomew noted, "If one equates ignorance with the ratio between what one knows and what is available to be known . . . each biological investigator becomes more ignorant with every passing day." Our goal has been to provide sufficient reference material to lead the interested student quickly into the heart of the literature for all the invertebrate phyla. Most of the references cited in the text will be found at the end of the corresponding chapter. However, to conserve space and eliminate redundancy, in a number of cases (especially in figure legends) references of a general nature may be listed only once, usually in the introductory chapters. In most figure legends, the original sources of the illustrations are cited; but when such sources predate 1940 we cite only our direct source for the illustration. We also include a fair number of references that are quite old, some from the nineteenth century. We have done so not out of whimsy, but because many of these are benchmark research papers or stand out as some of the best available descriptions for the subject in question. It is surprising how many of the illustrations in modern invertebrate texts derive ultimately from nineteenth-century publications.

These things being said, we hope you are now ready to forge ahead in your study of invertebrates. The task may at first seem daunting, and rightly so. We hope that our book will make this seemingly overwhelming task a bit more manageable. If we succeed in enhancing your education, and your enjoyment and appreciation of invertebrates, then our efforts will have been worthwhile.

R.C.B.
G.J.B.

Acknowledgments

We have benefited immeasurably from the careful and professional work of many conscientious reviewers (listed below), most of whom went far beyond simply correcting our factual errors. Many sent extended commentaries on difficult points, reprints of their recent work, and even photographs for our use. We extend to these reviewers our most sincere gratitude; this book is far better for their efforts. Where weaknesses and errors remain, they are of course solely our responsibility. Special thanks go to Michael Ghiselin, Robert Hessler, Robert Higgins, Reinhardt Kristensen, Joel Martin, Todd Newberry, Frederick Schram, and Regina Wetzer for their interest in our book, their thoughtful discussions, and their critical insight on difficult chapters.

We cannot list here all of the colleagues who sent photographs or allowed us to rummage through their files of slides and prints and never complained about how long we kept their pictures. We thank them all and have acknowledged their courtesy in the figure captions throughout the book. We are particularly indebted to Peter Fankboner, Rainer Foelix, and Gary McDonald. Thanks also to Carl May (Biological Photo Service, Moss Beach, California) for his tireless efforts in tracking down hard-to-find photographs. In addition, we are grateful to the authors and publishers who allowed us to reproduce material from their books, journals, and other publications. A special note of appreciation is due Harding Michel and Frederick Bayer, who graciously sent us the original artwork from their wonderful little book *The Free-Living Lower Invertebrates*, and we acknowledge their talented artist, Peter Loewer. We also note that artwork reproduced from *A Naturalist's Seashore Guide* was drawn by Sue Macias.

We extend a special word of thanks to Greg Payne. His faith in us and confidence in our project played a major role in the early development of this book. We wish him well and hope that he enjoys some measure of satisfaction in seeing this text in print. And special thanks also to Ken Rinehart, for all those unexpected phone calls from sundry airports that resulted in rendezvous in remote regions of the world to collect invertebrates.

Most of the new artwork in this text was done from our original sketches or from other sources by Nancy Haver, supported by our publisher, Sinauer Associates, Inc. Ms. Haver has been diligent in her efforts and patient with our nit-picking; we are proud to have her outstanding work in our book. Joseph Vesely was in charge of the art and production programs. He and his staff are magicians with photographs and layouts, and they continually met our "impossible" requests. Joe's expertise at book production is exceeded only by his unwavering good nature and sense of humor.

Most of the final manuscript was word processed by Jan Kastler and Carol Wigg. We particularly appreciate their remarkable ability to read our overlapping layers of editing and produce error-free diskettes. It was our special good fortune that Jodi Simpson agreed to read and edit the manuscript for this book. Her creative insights, editorial talents, and incredible patience are major factors in whatever success this book enjoys.

Andy Sinauer became a part of this project at a time when we wondered if the book would ever be completed. Andy and his staff have treated us fairly and professionally, listened to our concerns, and led us patiently through the complicated maze of publishing a book of this magnitude. To all the folks at Sinauer, and especially to Andy, we extend our lasting gratitude.

Finally, we thank our colleagues, students, friends, and families for encouragement and patience—especially Julie and Anna Mary, who supported us during the conception and development of this project, and Lizzie, whose plane arrived too late for her to be recognized in the seashore guide.

SCIENTIFIC REVIEWERS

(The chapters reviewed by each person are indicated in parentheses.)

GERALD J. BAKUS   University of Southern California, Los Angeles (7)

FREDERICK M. BAYER   Smithsonian Institution, Washington, D.C. (8)

ROBERT BIERI   Antioch College, Yellow Springs, OH (23)

RALPH BRINKHURST   Institute of Ocean Sciences, British Columbia, Canada (13)

DANIEL R. BROOKS   University of Toronto, Ontario, Canada (10)

ANNE COHEN   University of California, Los Angeles (2, 18)

DANIEL COHEN   Los Angeles County Natural History Museum (2)

CLIFTON CONEY   Los Angeles County Natural History Museum (20)

JOHN O. CORLISS   University of Maryland, College Park (5, 12)

EDWARD B. CUTLER   Utica College of Syracuse University, Utica, NY (14)

PAUL M. DELANEY   College of the Virgin Islands, St. Thomas (2)

NILES ELDREDGE   American Museum of Natural History, New York (15, in part)

KRISTIAN FAUCHALD   Smithsonian Institution, Washington, D.C. (13)

RAINER F. FOELIX   Université de Fribourg, Switzerland (16)

MICHAEL T. GHISELIN   California Academy of Sciences, San Francisco (1, 2, 3, 4, 12, 20)

RAY GIBSON   Liverpool Polytechnic, Liverpool, U.K. (1, 2, 3, 4, 11)

JOEL HEDGPETH   Santa Rosa, CA (16)

GORDON HENDLER   Los Angeles County Natural History Museum (22)

ROBERT R. HESSLER   Scripps Institution of Oceanography, La Jolla, CA (15, 18, 19)

ROBERT HIGGINS   Smithsonian Institution, Washington, D.C. (12, 15, 18, 19)

JENS HØEG   University of Copenhagen, Denmark (18, in part)

CHARLES L. HOGUE   Los Angeles County Natural History Museum (17)

DUANE HOPE   Smithsonian Institution, Washington, D.C. (12)

ROY S. HOUSTON   Loyola Marymount University, Los Angeles (20)

EUGENE H. KAPLAN   Hofstra University, Hempstead, NY (1, 2, 3, 4)

EUGENE N. KOZLOFF   University of Washington, Seattle (6)

PATRICIA KREMER   University of Southern California, Los Angeles (9)

REINHARDT M. KRISTENSEN   Institute of Cell Biology and Anatomy, University of Copenhagen, Denmark (15, 18, 19)

CHARLES LAMBERT   California State University, Fullerton (23)

GRETCHEN LAMBERT   California State University, Fullerton (23)

J. G. E. LEWIS   Dover College, Kent, U.K. (17)

DAVID MADDISON   Harvard University, Cambridge, MA (2)

LARRY MADIN   University of California, Santa Barbara (9)

JOEL MARTIN   Los Angeles County Natural History Museum (18)

MICHAEL MIYAMOTO   University of Florida, Gainesville (2)

FRANCOISE MONNIOT   Muséum National d'Histoire Naturelle, Paris, France (23)

JAMES MORIN   University of California, Los Angeles (8)

TODD NEWBERRY   University of California, Santa Cruz (1, 2, 3, 4, 5, 7, 12)

WILLIAM NEWMAN   Scripps Institution of Oceanography, La Jolla, CA (15, 18, 19)

CLAUS NIELSEN   Zoologisk Museum, Copenhagen, Denmark (24)

DAVID PAWSON   Smithsonian Institution, Washington, D.C. (22)

NORMAN PLATNICK   American Museum of Natural History, New York (16, 19)

GARY C. B. POORE   Museum of Victoria, Melbourne, Australia (19)

GREGORY PREGILL   San Diego Natural History Museum (2)

JAY M. SAVAGE   University of Miami, Coral Gables, FL (2)

FREDERICK SCHRAM   San Diego Natural History Museum (1, 2, 3, 4, 15, 16, 17, 18, 19)

TIMOTHY STEBBINS   Point Loma Biology Laboratory, San Diego (1, 2, 3)

CAROL A. STEPIEN   Scripps Institute of Oceanography, La Jolla, CA (2)

ERIK THUESEN   Ocean Research Institute, University of Tokyo, Japan (23)

SETH TYLER   University of Maine, Orono (10)

RUSSEL L. ZIMMER   University of Southern California, Los Angeles (21)

INVERTEBRATES

Chapter One

Introduction

For a gentleman should know something of invertebrate zoology, call it culture or what you will, just as he ought to know something about painting and music and the weeds in his garden.

Martin Wells
Lower Animals, 1968

What are invertebrates?

The vast majority of the species that inhabit this planet belong to one kingdom—the Animalia. These organisms are all **metazoa**, or multicelled animals. Over 1 million different kinds (species) of living metazoan animals have been described, and perhaps as many as 20–50 million more remain to be discovered and named! Among the organisms of Animalia, there are some that possess a backbone (or vertebral column) and many that do not. Those with a backbone constitute a single subphylum of the phylum Chordata, comprising about 47,000 species, or less than 5 percent of all described animals. Those without a backbone constitute the remainder of the phylum Chordata, and over 30 additional phyla; these are the **invertebrates**. We see, then, that the division of animals into invertebrates versus vertebrates is based more on tradition and convenience—reflecting a dichotomy of zoologists' interests—than on a recognition of natural biological groupings.

Courses and texts about invertebrates traditionally also include discussions of certain phyla of single-celled organisms that are members of the kingdom Protista, notably those that are heterotrophic and are referred to as "protozoa" (the "animal-like" protists, which survive by feeding on other organisms). In a broad sense, such creatures are also invertebrates. Thus, this book treats two kingdoms and about 40 phyla, seven of which comprise single-celled organisms (protists) and one (the phylum Chordata) that includes both invertebrates and vertebrates. The rest of the phyla discussed are composed entirely of metazoan invertebrates.

The first five kingdoms of life listed in Box 1 are those formally proposed by R. H. Whittaker in 1959. Whether or not these five kingdoms are themselves really natural (evolutionary) groupings is questionable. For this reason, some biologists prefer to view life on Earth as being of just two fundamental sorts: **prokaryotes**, which have small, simple, anucleate cells lacking mitochondria and chromosomes; and **eukaryotes**, which have larger, more complex, nucleate cells housing mitochondria and chromosomes. The Eukaryota is probably a natural group, defined by the unique trait of chromosomes. The prokaryotes (or microbes) are almost certainly not a natural group. The earliest prokaryote fossils are nearly 3.5 billion years old. Eukaryotes date from about 1.7 billion years ago, and we know very little about their origins or early evolution.

A sixth possible kingdom, Archaebacteria, has been proposed recently for a group of organisms that may represent a third fundamental form of life. Archaebacteria superficially resemble prokaryotic bacteria, but they have genetic and metabolic characteristics that make them unique. They are **methanogens**, anaerobic methane-producing microorganisms that have been found in an odd variety of circumstances (e.g., hot springs, sewage treatment ponds, sediment of natural waters, the gut of humans and other animals). Evidence to date suggests that the Archaebacteria may be no more closely related to prokaryotes than they are to eukaryotes. There is, however, still much to be learned about these odd organisms (see Woese 1981 for a review).

Box One

Six Kingdoms

A current classification, defining six kingdoms of life forms on Earth. Viruses and subviral organisms (viroids and prions) are not included in this classification.

Kingdom Monera
    The prokaryotes (i.e., the bacteria, including the cyanobacteria, or blue-green algae, and the spirochetes).

Kingdom Protista
    The eukaryotic single-celled microorganisms: protozoa, diatoms, and diatom-like algae, slime molds.

Kingdom Fungi
    The fungi: molds, mushrooms, and yeasts. Saprobic, heterotrophic, multicelled organisms.

Kingdom Plantae
    The "true" plants: includes the red, brown, and green algae, bryophytes, and vascular plants. Photosynthetic, autotrophic, multicelled organisms.

Kingdom Animalia
    The multicellular animals. Ingestive, heterotrophic, multicelled organisms.

Kingdom Archaebacteria
    Anaerobic, methane-producing microorganisms

Where did invertebrates come from?

The incredible array of living (extant) invertebrate groups is the product of hundreds of millions of years of evolution; they are the surviving descendants of successful lineages, which today inhabit virtually every environment on Earth. We know little of the history of most groups of invertebrates, however, because they have left us with few fossils. Many were soft-bodied and did not fossilize well under any conditions; many others perished when or where con-

ditions were not suitable for the formation of fossils. We can only speculate about the abundance of members of these groups in times past. Those invertebrates with hard parts, however, offer better, but often incomplete, records of their histories and changes in abundance through time. Groups such as the echinoderms (sea stars, sea urchins), molluscs (clams, snails), arthropods (crustaceans, insects), corals, and vertebrates have left relatively rich fossil records. For some groups, such as the echinoderms, the number of known extinct fossil species far exceeds the number of living forms.

When confronted with the concept of extinct forms of life, many people envision early humans, or perhaps dinosaurs. However, these creatures are newcomers to this planet (dinosaurs appeared just 200–300 million years ago). Representatives of nearly all of the animal phyla present today were present in the late Cambrian period of the Paleozoic era, and some invertebrate groups were probably established during the Precambrian era, perhaps over 1 *billion* years ago. The oldest "proven" metazoan fossils are from the late Precambrian era, 680 million years ago. These earliest animals were almost certainly marine organisms. Between about 670 and 580 million years ago, the Ediacaran fauna appeared (Figure 1). Scientists have yet to decide whether or not these earliest Ediacaran metazoa were precursors of members of modern phyla. Most resemble jellyfish, corals, or worms. However, they seem to have one feature in common—lack of internal plumbing. That is, vascular networks, digestive tracts, and other internal tubelike structures seem to have been missing. This fact has led some paleontologists to speculate that the creatures of the Ediacaran world ocean were not precursors to the great Cambrian marine evolutionary explosion that produced representatives of nearly every animal phylum extant today. Rather, the Ediacaran fauna might have been a short-lived "experiment" wherein animals increased in body size not by the evolution of organ systems, but merely by maintaining a surface-area-to-volume ratio sufficient to allow metabolic processes to proceed by means of simple diffusion. The small body volume relative to external body surface area of these Ediacaran animals was maintained by flat, leaflike, sometimes quilted shapes. Probably few had mineralized skeletons. If this hypothesis is true, the early Cambrian ocean witnessed the mass extinction of one entire kind of animal life and its replacement with another.

Regardless of what the Ediacaran fauna really was, modern metazoan body plans, with true skeletons and internal organ systems, appeared very rap-

Figure 1

The geological time scale.

Era	Period	Epoch	Age (Millions of years from start to present)	Major events in Earth history
Cenozoic	Quaternary	Recent Pleistocene	2	Culmination of mountain-building, followed by erosion and minor invasions of the sea. Drake Passage opens; circumpolar circulation is established, and south polar ice cap forms (north polar ice cap forms in Pleistocene). Early warming trends reversed by middle of Tertiary to cooler and finally glacial conditions; distinct seasonality and latitudinal temperature gradients develop. Series of major glaciations in Pliocene-Pleistocene epochs alter sea levels 300-400 feet. Evolution of humans during last 5-8 million years.
Cenozoic	Tertiary	Pliocene Miocene Oligocene Eocene Paleocene	5 24 38 54 65	
Mesozoic	Cretaceous	Late Early	100 130	Last great spread of epicontinental seas and shoreline swamps. Cool climates in late cretaceous. Angiosperm dominance begins. Extinction of archaic birds and many reptiles by the end of the period. Gondwana begins to split up.
Mesozoic	Jurassic		185	Climate warm and stable with little latitudinal or seasonal variation. Modern genera of many gymnosperms and advanced angiosperms appear. Reptilian diversity high in all habitats. First birds appear. Pangaea splits into northern (Laurasia) and southern (Gondwana) land masses.
Mesozoic	Triassic		225	Continents relatively high with few shallow seas. Climate warm; deserts extensive. Gymnosperms dominate; angiosperms first appear. Mammal-like reptiles replaced by precursors of dinosaurs; earliest true mammals appear. All continents joined as a single land mass (Pangaea).
Paleozoic	Permian		265	Land generally higher than at any previous time. Climate cold at beginning of period but warming progressively toward Triassic. Glossopterid forests develop with decline of coal swamps. Mammal-like reptiles diverse; widespread extinction of amphibians at end of period.
Paleozoic	Carboniferous		355	Generally warm and humid, but some glaciation in Southern Hemisphere. Extensive coal-producing swamps with large arthropod faunas. Many specialized amphibians; first appearance of reptiles. Mountain-building produces locally arid conditions, but extensive lowland forests and swamps were beginnings of the great coal deposits. Extensive radiation of amphibians.
Paleozoic	Devonian		413	Land higher and climates cooler. Freshwater basins develop in addition to shallow seas. First forests appear. First winged insects. Explosive radiation of fishes, followed by disappearance of many jawless forms. Earliest tetrapods appear.
Paleozoic	Silurian		425	Land slowly being uplifted, but shallow seas extensive. Climate warm. Terrestrial plants radiate. Eurypterid arthropods at their maximum abundance in aquatic habitats; first terrestrial arthropods appear. First gnathostomes appear among a diverse group of marine and freshwater jawless fishes.
Paleozoic	Ordovician		475	Maximum recorded extent of shallow seas reached; warming of climate continues. Algae become more complex; vascular plants may have been present; a variety of large invertebrates present. Jawless fish fossils from this period widespread.
Paleozoic	Cambrian		550	Extensive shallow seas in equatorial regions; climate warm. Algae abundant, records of trilobites and brachiopods. First remains of vertebrates appear at the end of this period.
Paleozoic	Ediacaran		670	Multicelled invertebrates flourish, mostly soft-bodied creatures.
Proterozoic (Precambrian)	*Oldest definite fossils*		3500	Changes in lithosphere produce major land masses and areas of shallow seas. Multicellular organisms first appear - algae, fungi, and many invertebrates. First animal fossils appear approximately 700 mya.
Proterozoic (Precambrian)	*Oldest dated rocks*		4000	Formation of earth and slow development of the lithosphere, hydrosphere, and atmosphere. Development of life in the hydrosphere.
Proterozoic (Precambrian)			5000	

idly during the subsequent Cambrian period, in a unique evolutionary outburst. Many paleontologists believe that the rapidity of appearance of animals representing these major lineages, perhaps during only a few million years, precludes their origin by microevolution or species selection and makes the event a prime example of rapid macroevolution outside the modern evolutionary synthesis paradigm. Representatives of several distinct lineages within many phyla first appear in the early Cambrian, with no known fossil intermediates or ancestors (e.g., Porifera, Mollusca, Brachiopoda, Echinodermata, Annelida, and perhaps Arthropoda). It appears as though this period witnessed the rapid evolution of novel morphologies at the level of body plans (*Baupläne*) and subplans. Over 50 phylum-level body plans have been identified so far from the Cambrian period, and it is likely that representatives of additional extinct phyla remain to be found. A great many Cambrian groups are known solely because of the remarkably well-preserved fossils found in the famous Burgess Shale deposits of Canada.

Evaluation of the present-day success of animal groups involves a consideration of diversity (numbers of species and higher taxa) and abundance (numbers of individuals). The former characteristic may be broadly interpreted as a measure of the group's success through adaptation and variation, the latter as a measure of the group's specialization and successful competition, often in particular environmental situations. The predominance of certain kinds of invertebrates is unquestionable. For example, of the 1,000,000+ described species of animals, 90 percent are arthropods, which are important members of virtually every environment on Earth. Most of these arthropods are insects, perhaps the most successful animals on Earth today. Box 2 conveys a general idea of the various levels of diversity achieved within the animal phyla.

Where do invertebrates live?

Marine habitats

The major events leading to early diversification of invertebrates (indeed, of all life) probably occurred

---

Box Two
The Abundance of Animals

The approximate numbers of known extant species in various animal groups. Specialists in certain groups estimate that the known kinds probably represent only a small fraction of actual existing species; this conjecture is especially true of the roundworms (Nematoda), insects, and certain groups of chelicerates (particularly the mites). Note that we have broken down the phyla Arthropoda and Chordata into their respective subphyla, and that we have lumped the protozoan phyla together. See Chapter 5 for a complete classification of the protozoa.

Protozoa (35,000)	Priapula (15)	Mollusca (50,000)
Placozoa (1)	Acanthocephala (700)	Brachiopoda (335)
Mesozoa (100)	Entoprocta (150)	Ectoprocta (4,500)
Porifera (9,000)	Loricifera (9+)	Phoronida (15)
Cnidaria (9,000)	Annelida (15,000)	Chaetognatha (100)
Ctenophora (100)	Echiura (135)	Echinodermata (7,000)
Platyhelminthes (20,000)	Sipuncula (250)	Hemichordata (85)
Nemertea (900)	Pogonophora (135)	Chordata:
Gnathostomulida (80)	Vestimentifera (8)	Urochordata (3,000)
Rotifera (1,800)	Tardigrada (400)	Cephalochordata (23)
Gastrotricha (450)	Onychophora (80)	Vertebrata (47,000)
Kinorhyncha (150)	Arthropoda:	
Nematoda (12,000)	Cheliceriformes (65,000)	
Nematomorpha (230)	Crustacea (32,000)	
	Uniramia (860,000)	

in ancient shallow seas. So it is not surprising to find that the marine environment continues to harbor the greatest variety of these animals. Some groups, such as the echinoderms, have remained exclusively marine. The tremendous variety and abundance of invertebrates in the world's oceans are the result of a number of factors, many of which are related to conditions that reduce the physical and chemical stress on organisms in their day-to-day activities. The oceans are also much bigger volumetrically than inhabitable land surfaces, so they provide much more space in which animals can live and diversify. In addition, productivity in the world's oceans is very high; this rich nutrient source probably contributes to the high diversity of animal life in the sea (net primary productivity of the seas is about 28×10^9 tons of carbon per year). Perhaps most significant, however, is the special nature of sea water itself.

Water is a very efficient thermal buffer. Because of certain unique chemical properties, water has a very high heat capacity and is slow to heat up or cool down. In large bodies of water, such as the oceans (which cover over 71 percent of the Earth's surface), great amounts of heat can be gained or lost with little change in actual water temperature. Indeed, ocean temperatures are very stable in comparison with those of freshwater and terrestrial environments.

The saltiness, or **salinity**, of sea water averages about 3.5 percent (usually expressed as parts per thousand, 35‰), and this property, too, is quite stable, especially in areas away from the shore and the influence of freshwater runoff. The relatively high density of sea water also enhances buoyancy, thereby reducing energy expenditures required for flotation and locomotion. Furthermore, the various ions that contribute to the total salinity occur in fairly constant proportions. These qualities result in a total ionic concentration in sea water that is similar to that in the body fluids of many animals, an arrangement minimizing the problems of osmoregulation and ionic regulation (see Chapter 3). The pH of sea water is also quite stable throughout most of the oceans. Certain naturally occurring compounds (CO_2 and HCO_3^{2-}) participate in a series of chemical reactions that buffer sea water at a slightly alkaline level (pH 7.5–8.5).

In shallow nearshore waters, carbon dioxide, various nutrients, and sunlight are generally available in quantities sufficient to allow high levels of photosynthesis, either seasonally or continuously (depending on latitude and other factors). Dissolved oxygen levels rarely drop below that required for normal respiration.

Because the marine realm is the home for most of the groups of animals discussed in this book, it is useful to introduce some terms that describe the subdivisions of that environment and the categories of animals that inhabit them. Figure 2 illustrates a generalized cross section through an ocean. The ocean shoreline marks the **littoral** region, commonly referred to as the intertidal environment. The littoral region may be described as that area where sea, air, and land meet and interact. Obviously, it is affected by the rise and fall of the tides. We can further subdivide this area into zones or various shore elevations relative to the tides. The **supralittoral** region, or **splash zone**, is rarely if ever covered by water, even at high tide, but it is subjected to frequent spray from waves. The **eulittoral** region, or true intertidal zone, lies between the levels of the highest and lowest tides and can be divided into high, mid, and low intertidal zones. The **sublittoral** region, or **subtidal zone**, is never uncovered, even at very low tides, but it is influenced by tidal action (e.g., by changes in turbulence, turbidity, and light penetration).

Organisms that inhabit the world's littoral regions are subjected to dynamic and often demanding conditions, and yet these areas are exceptionally high in number of species. Many animals and plants are more-or-less restricted to particular elevations along the shore, a condition resulting in the phenomenon of **zonation**. Such zones are visible as distinct bands or assemblages of organisms along the shore (Figure 3A). We may generalize and state that the upper elevation limit of an intertidal organism is often established by its ability to tolerate conditions of exposure to air (desiccation, temperature fluctuations), whereas its lower elevation limit is often determined by biological factors (competition with or predation by other species). There are, however, many exceptions to these "rules."

Extending seaward from the shoreline is the **continental shelf**, a feature of most large land masses. The continental shelf may extend a few kilometers or 1,000 km from shore (50 to 100 km is average in most areas), and it usually reaches a depth of about 150 m. These nearshore shelf environments are among the most productive regions of the sea, being rich in available nutrients and shallow enough to permit plant growth over much of the area. The outer limit of the continental shelf—called the **continental edge**—is usually indicated by a relatively sudden increase in the steepness of the bottom contour. These "steep" parts of the ocean floor—called the **continental slope**—are actually slopes of only 4–6 percent (although the slope is usually much steeper around volcanic islands); note the difference in scale between the vertical and horizontal axes on Figure 2.

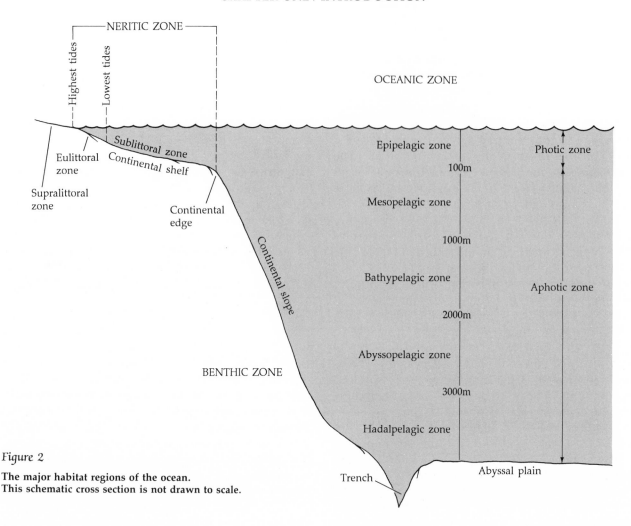

Figure 2

**The major habitat regions of the ocean.
This schematic cross section is not drawn to scale.**

The continental slope extends to the deep ocean floor, where the **abyssal plain** forms the expansive, relatively flat sea bottom. The abyssal plain lies an average of about 4 km below the surface, but it is interrupted by a variety of ridges, sea mounts, trenches, and other formations. Some of the trenches exceed 10 km in depth.

Organisms that inhabit the sea bottom anywhere along the entire contour shown in Figure 2 are collectively referred to as **benthic** organisms, or simply the **benthos**. There is a general decrease in both variety and abundance of benthic organisms with increasing depth, from the rich littoral and shelf environments to the deep abyssal plain. However, an overgeneralization of this relationship can be misleading. For example, shelf and slope habitats in temperate regions are often characterized by low animal density but high diversity or species richness (occasionally as great as that seen in shallow tropical seas). In many areas diversity has been shown to increase

abruptly below the shelf–slope break (100–300 m depth), peak quickly, and then decrease gradually with depth. The first impression of early marine scientists—that the deep sea bed was an environment able to sustain but a few species in impoverished simple communities—is simply not the case.

Depending on the nature of the substratum and the attributes of the organisms in question, benthic animals may live on the surface of the substratum (**epifauna**, or **epibenthic** forms, such as most sea anemones, many snails, barnacles) or burrow within the soft substrata (**infauna**). Infaunal forms include many relatively large invertebrates, such as clams and various worms, and some specialized, very tiny forms that inhabit the spaces between sand grains and are termed **interstitial** organisms.

Benthic animals may also be categorized by their locomotor capabilities, or lack thereof. Animals that are generally quite motile and active are described as being **errant** (e.g., crabs, some worms), whereas those that are firmly attached to the substratum are **sessile** (e.g., sponges, corals, barnacles). Others are unattached or weakly attached but generally do not move around much; they are described as being **sedentary** (e.g., sea stars, anemones, some clams).

The region of water extending from the bottom of the sea to the surface is called the **pelagic region**. It provides an environment for organisms adapted to this distinctly three-dimensional aqueous world. The pelagic region over the continental shelf is called the **neritic zone**, and that over the continental slope and beyond, the **oceanic zone**. The pelagic region may be otherwise subdivided into increments on the basis of depth alone (Figure 2), or of the depth to which light (for photosynthesis) penetrates. The latter factor is, of course, of paramount biological importance. It is only within the **photic zone** (the upper 100 m or so) that photosynthesis can occur, and (except in a few odd circumstances) all life in the **aphotic zone** (the zone beyond the depth to which light penetrates) depends ultimately upon organic input from the overlying sunlit layers of the sea. Notable exceptions are the small and restricted sea-floor thermal vent communities, in which sulfur-fixing microorganisms serve as the basis of the food chain.

Organisms that inhabit the pelagic region are frequently categorized by their relative powers of locomotion. Pelagic animals that are relatively strong swimmers, such as fishes and squids, constitute the **nekton**. Those pelagic forms that simply float and drift, or generally are at the mercy of water movements, are collectively called the **plankton**. Many planktonic animals actually swim very well, but they are so small that they are swept along by prevailing currents even though they are swimming. Both plants (**phytoplankton**) and animals (**zooplankton**) are included in this category, the latter being represented by invertebrates such as jellyfishes, comb jellies, arrow worms, many small crustaceans, and the pelagic larvae of many benthic adults. Planktonic animals that spend their entire lives in the pelagic realm are called **holoplanktonic** animals; those whose adult stage is benthic are called **meroplanktonic** animals.

Freshwater habitats

Because bodies of fresh water are very small in comparison with the oceans, they are relatively unstable environments, being much more readily and drastically influenced by extrinsic environmental factors (Figures 3D and E). Changes in temperature and other conditions may occur quickly and be of a magnitude never experienced in most marine environments. Seasonal conditions are even more extreme and may include freezing of ponds, streams, and lakes during the winter and complete drying in the summer.

The very low salinity of fresh water (rarely more than 1 percent) and the lack of constant relative ion concentrations subject freshwater inhabitants to severe ionic and osmotic stresses. These conditions, along with other factors such as reduced buoyancy, less stable pH, rapid nutrient input and depletion, and flow rates of moving water, produce environments that support far less biological variety than the sea does. Nonetheless, many different invertebrates do live in fresh water and have solved the problems associated with this environment. Special adaptations to life in fresh water are summarized in Chapter 3 and in discussions on appropriate groups of invertebrates in later chapters.

Estuaries and coastal marshlands

Estuaries are special environments; they usually occur along low-lying coasts and are created by the interaction of fresh and marine waters, commonly where rivers enter the sea. Here one finds an unstable blending of fresh and salt water conditions, moving water, tidal influences, and drastic seasonal changes. Estuaries are generally highly productive environments receiving from their freshwater sources high concentrations of nutrients from terrestrial runoff. Temperature and salinity vary greatly with tidal activity and with season. Depending on tides and turbulence, the waters of estuaries may be relatively well mixed and more-or-less homogeneously brackish, or they may be distinctly stratified, with fresh water floating on the more dense and saline water below.

Not only do temperature and salinity vary greatly in an estuary, but amounts of dissolved oxygen also may change markedly throughout a 24-hour cycle as a function of temperature and phytoplankton productivity. Furthermore, vast amounts of silt borne by freshwater runoff are carried into the waters of estuaries; most of this silt settles out and creates extensive tidal flats (Figure 3B). In addition to the natural stresses common to estuarine existence, there are also stresses from human activity—pollution, thermal additions from power plants, dredging and filling, chemicals, and storm drain discharges are some examples.

To further complicate matters, a high degree of variability in environmental conditions exists among different parts of any given estuary. Thus, animals inhabiting a region near the ocean endure an environmental regime very different from that of animals

A

B

C

D

E

F

Figure 3

A few of the Earth's major habitat types. A, Awash rock shows evidence of marine intertidal zonation. B, A tidal flat and bordering salt marsh in northern California. C, A mangrove swamp in Mexico. D, A freshwater stream in a Costa Rican tropical rain forest. E, A mountain lake in Guatemala. F, Sonoran Desert, Mexico. (Photographs by the authors.)

living near the head of the estuary (near the freshwater source).

Estuaries often have distinctly different patches or regions of benthic substrata. So, as a result of these and other variables, different species assemblages inhabit various distinct regions of the estuary. For example, many species of invertebrates and fishes live in one part of the estuary during larval or juvenile stages and in a different part during adulthood.

Some coastal swamps and marshlands, such as **salt marshes** and **mangrove swamps**, are characterized by dense stands of **halophytes** (plants that flourish in saline conditions). These special areas are frequently associated with estuaries (Figures 3B and C). Salt marshes and mangrove swamps are alternately flooded and uncovered by tidal action within the estuary, and are thus subjected to the fluctuating conditions described earlier. The dense halophyte stands and the "instant" mixing of waters of different salinities create an efficient "nutrient trap" in these unique environments. In other words, instead of being swept out to sea, most dissolved nutrients entering an estuary (or generated within it) remain there, a situation producing some of the most productive regions in the world. The net primary productivity of marine halophytes is one of the highest in the world. This great productivity enters the sea in two principal ways: as plant detritus (mainly from halophyte debris) and within the nekton that migrate in and out of the estuary. The contribution of estuaries to general coastal productivity can hardly be exaggerated. For example, it has been shown that the organic matter produced by mangroves forms the base of a major detritus food web that culminates in the rich fisheries of Florida Bay. Furthermore, it has been estimated that 60–80 percent of the world's commercial marine fishes rely on estuaries directly, either as homes for migrating adults or as protective nurseries in which the young grow to adulthood. Estuaries and other coastal wetlands are also of prime importance to both resident and migratory populations of sea birds.

A large number of invertebrates have adapted to life in these stressful and dynamic environments. In general, animals have but two alternatives when encountering stressful conditions. Either they migrate to more favorable environments or they remain and tolerate the conditions. Many animals migrate into estuaries for only a portion of their life cycle, or on a daily basis with the tides. Many others remain there throughout their lives, and these species show a remarkable range of physiological responses to the environmental conditions with which they must cope (see Chapter 3).

Terrestrial habitats

Life on land is in many ways even more rigorous than life in fresh water. Temperature extremes are usually encountered on a daily basis, water balance is a critical problem, and just physically supporting the body requires major expenditures of energy. Water provides a medium for support, for dispersing gametes, larvae, and adults, for diluting waste products, and for retention of dissolved materials (such as oxygen) needed by animals. Animals living in terrestrial environments do not enjoy these numerous benefits of water and must pay the price.

Major invertebrate success on land is found almost entirely among the arthropods, notably the insects and, to a lesser extent, spiders, scorpions and their kin. These arthropod groups include truly terrestrial species that have invaded even the most arid environments (Figure 3F). Except for some snails and nematodes, all other land-dwelling invertebrates are largely restricted to relatively moist areas and include such familiar animals as earthworms and sow bugs. In a very real sense, these and many other smaller land invertebrates survive only by the permanent or periodic presence of water. Special attention is given in later chapters to structural, reproductive, physiological, and behavioral adaptations of terrestrial invertebrates.

A special type of environment: The phenomenon of symbiosis

Many invertebrates have adopted lifestyles involving an intimate association with other animals or plants. This association is termed a symbiotic relationship, or simply **symbiosis**. In most symbiotic situations, a larger organism (called the **host**) provides an environment (its body, burrow, nest, etc.) on or within which a smaller organism (the **symbiont**) lives. Some symbiotic relationships are rather transient—for example, the relationship of ticks or lice and their vertebrate host—whereas others are more-or-less permanent. Some symbionts are opportunistic (**facultative**), whereas others cannot survive unless they are in association with their host (**obligatory**).

The phenomenon of symbiosis can be subdivided into several categories on the basis of the nature of the interactions between the symbiont and its host. Perhaps the most familiar type of relationship is that of **parasitism**, in which the symbiont (a parasite) receives benefits at the host's expense. Parasites may be external (**ectoparasites**), such as lice, ticks, and leeches; or internal (**endoparasites**), such as liver flukes, some roundworms, and tapeworms. Other parasites may be neither strictly internal nor strictly external; rather, they may live in a body cavity or area of the host that communicates with the environment, such as the gill chamber of a fish or the mouth or anus of some animal (**mesoparasites**). A few groups of invertebrates are predominantly or exclusively parasitic, and almost all invertebrate phyla have at least some species that have adopted this lifestyle. Many texts and courses on parasitology pay particular attention to the impact of these animals on humans, crops, livestock, and various economic conditions. However, in keeping with the approach developed in this book, we will treat this form of symbiosis largely from "the parasite's point of view," that is, as a particular lifestyle suited to a specific environment, requiring certain adaptations and endowing certain advantages.

Mutualism is another form of symbiosis, and is generally defined as an association in which both host and symbiont benefit. Such relationships may be extremely intimate and important for the survival of both parties; for example, the bacteria in our own large intestine are important in the production of certain vitamins and in processing material in the gut. Another example is the termite and certain protozoa that inhabit its digestive tract and are responsible for the breakdown of cellulose to compounds that can be assimilated by their insect host. Other mutualistic relationships may be less binding on the organisms involved. For example, there are cleaner shrimps that inhabit coral reef environments, where they establish "cleaning stations" and are visited regularly by a variety of reef-dwelling fishes who present themselves to the shrimps for the removal of parasites. Obviously, even this rather loose association results in benefits for the shrimps (a free meal) as well as for the fishes (free removal of parasites).

A third type of symbiosis is called **commensalism**. This category is somewhat of a catch-all for associations where harm or mutual benefit is not obvious. It is usually described as an association that is advantageous to one party (the symbiont) but leaves the other (the host) unaffected. For instance, among invertebrates, there are numerous examples of one species inhabiting the tube or burrow of another, the former obtaining protection or food or both with little or no apparent impact on the latter.

There is a good deal of overlap among the categories of symbiosis described above, and many animal relationships have elements of two or even of all categories.

Invertebrates and the ecosystem

Because invertebrates play major roles in this planet's ecology, it is worthwhile giving a very brief explanation of the **ecosystem** concept. Figure 4 illustrates a generalized ecosystem. The ultimate source of energy for all ecosystems is the sun (certain strange chemosynthetically based situations excepted). Chlorophyll-containing plants utilize this energy through the process of photosynthesis to combine carbon dioxide, water, and certain inorganic nutrients (notably nitrates and phosphates) into organic compounds. This process is called **primary production**. Thus, plants (the **producers**) form the base of the system's food webs. Animals that feed directly on the producers are **herbivores**, or the **primary consumers** of the system, and may in turn be preyed on by **secondary consumers** (**carnivores**); secondary consumers are themselves preyed on by other carnivores, and so on. Some animals feed at more than one level in the ecosystem, eating both plants and other animals (**omnivores**), the dead bodies of other organisms (**scavengers**), or small bits of organic material (**detritivores**). In one way or another, most of the organic material produced by photosynthesis is eventually incorporated into other levels of the system. A good portion of the organic material in such a system is not eaten by consumers at all, but, along with some waste products, is acted on by **decomposers** (usually bacteria and fungi) whose "role" is to convert these materials to various inorganic compounds, which are cycled back into the system through their eventual reuse in photosynthesis or other processes.

The preceding discussion is a very oversimplified view of a very complex and important topic. Modern-day ecologists spend a good deal of their time investigating and attempting to describe and quantify ecosystems or parts thereof, and there is still a great deal to be learned. Invertebrates—indeed, all organisms—must be adapted to their roles within an ecosystem in order to survive and be successful as species. In

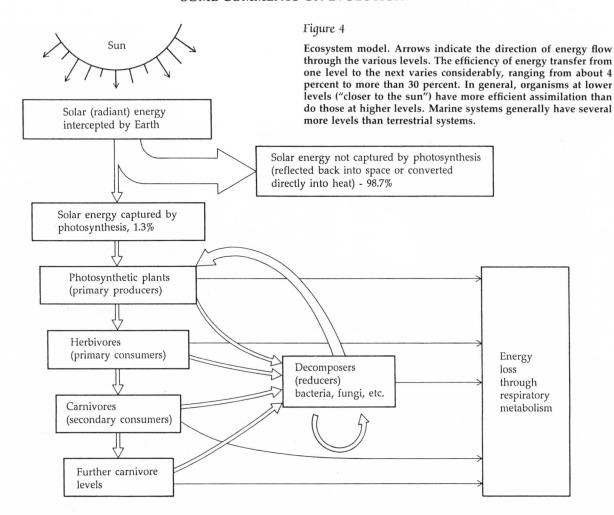

Figure 4

Ecosystem model. Arrows indicate the direction of energy flow through the various levels. The efficiency of energy transfer from one level to the next varies considerably, ranging from about 4 percent to more than 30 percent. In general, organisms at lower levels ("closer to the sun") have more efficient assimilation than do those at higher levels. Marine systems generally have several more levels than terrestrial systems.

this book, we will investigate the invertebrates in this context.

Some comments on evolution

Fitness By Any Other Name
Would Be As Loose

A group inept
Might better opt
To be adept
And so adopt
Ways more apt
To wit, adapt.

John Burns
("Biograffiti," 1975)

Evolutionary biology is in the midst of a major reevaluation of the premises and paradigms by which it has been guided for the past 40 years. This soul-searching has been precipitated by three phenomena.

First is the revolution in molecular genetics, which has produced dramatic discoveries regularly since about 1965 and which will no doubt continue to produce new information for many years to come. Second is the development of a new method of analyzing phylogenetic information, a method called either phylogenetic systematics or cladistics; we discuss this method in Chapter 2. Third is the development of radically new and different hypotheses regarding how evolution itself may operate. We feel it is important to comment briefly on some of the most significant aspects of these new ideas because this text was written specifically within an evolutionary framework.

The neo-Darwinian or so-called modern synthesis that resulted from the integration of Mendelian genetics into Darwinian natural selection theory has dominated evolutionary biology for nearly half a century. Basically, the neo-Darwinian theory holds that most important evolutionary changes result from the action of natural selection on variation within populations. The theory focuses on adaptation, and primarily deals with genes and changes in gene frequencies within populations. These variations are ultimately due to mutations. Although random phenomena such as genetic drift and the founder effect are part of the neo-Darwinian synthesis, they have always been taken to be of minor significance relative to natural selection, which has been viewed as the principal guiding force in the evolution of new species. This view is now being challenged and reexamined in light of new data and new ideas regarding evolutionary change. Whether these new ideas will be incorporated into the existing framework of neo-Darwinism or will give rise to a new evolutionary synthesis remains to be seen.

One of the most fundamental new ways of viewing evolution is through stochastic approaches. **Stochastic** processes are random, or chance (unpredictable), occurrences. **Deterministic** processes, in contrast, are not random but predictable, or directed. Evolution by natural selection, even though largely unpredictable by scientists over geological time, is primarily a deterministic process, although certain elements of chance are accepted within the theory (e.g., mutation, random mating, the founder effect, catastrophic environmental phenomena). Nevertheless, given a complete understanding of the environment and animal genetics, the theory of natural selection implies that one should be able to largely predict evolutionary outcomes. The theory of natural selection implies that essentially all of the characteristics animals possess are products of adaptations leading to increased fitness (increased survival or reproductive success). The dangers of unquestioned acceptance of this adaptationist view manifest themselves in many ways. The most obvious is the tendency to assume that every aspect of an animal's phenotype is the product of adaptations arising by way of natural selection. For example, biologists used to assume that feathers evolved as a selected adaptation for flight in ancestral birds. However, recent studies indicate that feathers probably initially served as a heat insulation layer in small, predatory dinosaurs. "Feathers for flight" are an example of what Stephen Jay Gould and Elizabeth Vrba call an "exadaptation." Similar ideas have been expressed for several decades, and their proponents have called such opportunistic evolutionary events **preadaptations**.

Current critics of the adaptationist view largely agree that any feature an organism possesses may have actually been derived in one of three different ways. It may have been constructed via the process of true adaptation. It may have been co-opted for its current use from a different use, in which case it is an exadaptation. Or, it may have arisen with no particular function at all. In the latter case, if it now serves some specific function it is also classed as an exadaptation, or a preadaptation. However, if a trait arose without specific function and persists today without a specific function, its occurrence is purely a matter of chance, and it may not have been selected for or against—it is evolutionarily "neutral." Exadaptations and chance occurrences of neutral features are largely the products of stochastic evolution. Structures apparently without present function may eventually become important sources of raw material for future evolution. The theory of evolution by natural selection alone does not accommodate the persistence of nonadvantageous or neutral characters, nor does it allow for character evolution "for future use." Herein lies a major element of the selectionist–neutralist debate.

There are a number of new evolutionary hypotheses that range from strongly stochastic, to moderately stochastic, to only partly stochastic. But these new ideas often address different levels of evolution. Eldredge (1985) noted at least 16 different biological entities that can "evolve": base pairs, codons, functional genes, pseudogenes, families of repeated noncoding DNA, chromosomes, cells, mitochondria, chloroplasts, organisms, demes, populations, species, communities, regional biotas, and monophyletic groups. In a more general sense, however, evolution might be viewed as operating on only three levels: microevolution, speciation, and macroevolution.

Study of the microevolution level constitutes population genetics—that is, analysis of changes in gene frequency from generation to generation within a population or deme. It is at this level that natural selection is unquestionably in operation, as are mutation and genetic drift. The modern synthesis approach deals almost exclusively with this level of evolutionary activity.

Speciation is an event by which one species gives rise to one or more new species. This transformation may take place slowly or it may be very rapid. Many of the newer views of speciation suggest that it may or may not involve natural selection. That is to say, some speciation may result from purely stochastic processes.

Macroevolution can be thought of as evolution above the level of speciation. It is the focus of some of the hottest battles among evolutionists today. Phenomena at this level include the origin of taxa above the species level, radiations of lineages, and mass extinctions with subsequent new biotic proliferations. Mass extinction events in Earth's history have obviously played major roles in reshaping the directions of animal and plant evolution. The really large extinction events may have wiped out a majority of life forms, such as the PermoTriassic event in which perhaps 70–90 percent of all species existing may have gone extinct (although probably no phylum went extinct). The world we see around us today is the product of over one billion years of evolutionary history at all three of these phenomenological levels of change.

Recent advances in molecular biology have also spawned a number of novel and interesting theories of genomic changes. The non-Darwinian (neutral) view maintains that a significant amount of change is not under the control of natural selection. This theory has considerable empirical evidence to support it at the molecular genetics level, including unexplained genetic polymorphism and constant rates of genetic change (based on measures of amino acid substitutions) over time. The neutral theory of molecular evolution holds that most new (mutant) genes that appear are adaptively neither more nor less advantageous than the genes they replace, and that at the molecular level most evolutionary changes are caused by random drift of selectively equivalent genes. The molecular drive hypothesis suggests that neutral genes may be fixed in a population by purely intrinsic factors, such as the mechanical–biochemical constraints of mitosis and chromosome replication. Another, extrinsic, view of stochastic evolution points to the recent discovery that genes or groups of genes may be fairly routinely transferred from one species to another, distantly related species through an RNA viral vector. If this random input of new genetic material to the genotype of a species is in fact a common phenomenon, it could contribute to the origin of new evolutionary lineages (macroevolution). Finally, relatively small changes in regulatory gene sequences during development may result in major phenotypic differences in adult stages (see Chapter 4). Regulatory changes of this kind may come about through traditional mechanisms such as mutation, or through more unusual actions such as interspecific RNA viral transfer. The existence of transposable elements that move to different positions on and among chromosomes, of multiple forms of RNA, and of homeotic genes (genes that switch other genes on and off), and an unexpected complexity in nuclear and mitochrondrial DNA suggest that control of genomic organization involves far more than just natural selection and genetic drift.

With the preceding comments in mind, we can understand why some biologists prefer to divide evolution into two processes: **anagenesis** (evolution within a phyletic line) and **cladogenesis** (the splitting of one phyletic line into two or more lines). The major features of anagenetic evolution may be adaptation through natural selection, at least at the phenotypic level, and anagenesis may be important in an evolutionary sense only at the within-species level. Although Darwin titled his book *Origin of Species*, he dealt primarily with the origin of adaptations, or anagenesis. Cladogenesis, on the other hand, is the origin of new species or new evolutionary lineages (**clades**). Natural selection and adaptation may not necessarily be coupled directly to speciation or cladogenesis. Further evidence for this possibly tenuous relationship comes from the fossil record, which shows a history of animal species strongly suggesting that most species do not change significantly throughout their existence; rather, they remain phenotypically very stable for millions of years, then undergo a very rapid change in which they essentially "replace themselves" with one or more new and different species. These new species then, in turn, remain phenotypically static for millions more years. Generally, species of marine invertebrates seem to persist more-or-less unchanged for 5–10 million years, whereas the time required for major anatomical change to occur is a mere 5–50 thousand years. This idea of speciation in rapid bursts, sandwiched between long periods of species stasis, has been explained within the natural selection paradigm, where it is usually referred to as **saltation**. It has also been described outside the natural selection paradigm, in the original **punctuated equilibrium** model of Eldredge and Gould (1972). Most recent thinking on the subject of punctuated equilibrium versus gradual (anagenetic) speciation suggests that both phenomena may occur, and the debate is now centered more on the relative roles of natural selection and stochastic change in the processes of speciation and macroevolution.

Several recent studies have proposed the idea that natural selection may actually be a stabilizing force in evolution. That is, selection may serve to keep a population fit in its changing environment and, in doing so, keep species stable. Speciation, on the other hand, might then be viewed as the result of a breakdown of this stabilizing effect of selection; speciation then is an event precipitated by the inability of selection to keep pace with changing environmental conditions or stochastic biological or environmental events.

The consequence of these debates is that the topic of evolution, and of speciation in particular, is today awash in speculations and reevaluations of basic premises. Biologists are still a long way from understanding all the causes and mechanisms of the evolutionary process. That evolution has occurred and is occurring is well documented and consistent with all of the available data. The debates concern the nature of the evolutionary mechanisms involved and the actual histories of lineages (phylogeny). It seems probable that different processes are at work at different levels, having created the patterns we see in the world today. For these reasons, readers should be cautious in how they view evolutionary discussions and phylogenetic trees presented in this or any other text. Courses in invertebrate zoology place great emphasis on comparative morphology and functional anatomy, and it is easy to slip into a way of thinking in which every aspect of an animal is viewed as an adaptation that is the end product of millions of years of natural selection. While this may be true in a general sense, it should be kept in mind that major new radiations leading to higher taxa may well have been initiated by the chance appearance of new phenotypic attributes or by rare environmental events. In this sense, natural selection may be best viewed as the final casting director, not necessarily the author, of every scene in the evolutionary play.

Despite the many evolutionary questions currently being discussed, biologists are quite able to continue their efforts at reconstructing the evolutionary history of life on Earth, because genetics and the environment meet and interact at the level of the organism. Thus, the processes of evolution (whatever they entail) result in newly evolved organisms that are distinct by virtue of various unique new characters or attributes they have acquired. Their descendants retain these attributes and in time acquire still others, which are retained by their descendants. In this fashion the living world provides us with an analyzable pattern. This pattern consists of hierarchially nested sets of features that are present in living organisms, the "characters" with which we can attempt to reconstruct a history of descent of life. We will have much more to say regarding this reconstruction process in the following chapter, because understanding what characters are and how they are evaluated is fundamental to comparative biology and to an appreciation of the invertebrate world.

A final introductory message to the reader

If you have not already done so, please read the Preface to this text, which explains its limitations and, more important, describes what the book is about and what sort of information we intend to convey. Because of our comparative approach, it is critical that you become familiar with the initial chapters (Chapters 1–4) before attempting to study and comprehend the sections dealing with individual animal groups. These first four chapters are designed to accomplish several goals: (1) to define some basic terminology, (2) to introduce a number of important concepts, and (3) to describe the themes that we use throughout the rest of the book.

The fundamental theme of this book is evolution, and we approach invertebrate evolution primarily through the field of comparative biology (comparative anatomy/morphology, physiology, and ecology). In Chapters 3 and 4 we lay out the fundamental anatomical/morphological designs and life history strategies of invertebrates. Like morphological features of organisms, these designs and strategies are not random but form patterns. Recognition and analysis of these patterns constitute the basic "building blocks" of this book. We then proceed in the "animal chapters" to explore the evolution of the invertebrates as they utilize and build on various combinations of these basic functional body plans and lifestyles. In this fashion, you should be able to follow the evolutionary progressions of the invertebrate phyla, their systems, and their approaches to success on Earth. In Chapter 2 we provide an explanation of how biologists derive evolutionary schemes and classifications in the first place, how theories regarding the evolution of animal groups grow and change, and how the information presented in this text has been used to construct theories on how life evolved on Earth.

Through our approach, we hope to add continuity to the massive subject of invertebrate zoology, which is often covered (in texts and lectures) by a sort of "flash-card" method, where the primary goal is to memorize animal names and characteristics and keep them properly associated, at least until after the examination. Thus, we urge you to look back frequently at these first few chapters as you explore how invertebrates are put together, how they live, and where they came from.

Selected References

General References on Invertebrates

Abbott, D. P. 1987. *Observing Marine Invertebrates.* (Edited by G. H. Hilgard.) Stanford University Press, Stanford, California. [We are delighted Galen Hilgard undertook this project.]

Adiyodi, K. G. and R. G. Adiyodi (eds.). 1983. *Reproductive Biology of Invertebrates.* Vol. 1. Oogenesis, Oviposition, and Oosorption. Wiley, New York.

Alexander, R. M. 1979. *The Invertebrates.* Cambridge University Press, New York.

Ali, M. A. (ed.). 1984. *Photoreception and Vision in Invertebrates.* N.A.T.O. Adv. Sci. Insts., Ser. A: Life Sciences 74.

Barnes, R. D. 1980, 1987. *Invertebrate Zoology,* 4th and 5th Eds. Saunders, Philadelphia. [One of the most popular invertebrate texts available.]

Barnes, R. S. K. 1974. *Estuarine Biology.* The Institute of Biology's Studies in Biology, No. 49. Crane, Rusnak and Co., New York.

Barnes, R. S. K. (ed.). 1984. *A Synoptic Classification of Living Organisms.* Blackwell, Oxford. [An excellent, scaled-down version of Parker's 1982 attempt; treats all living organisms, including prokaryotes.]

Barnes, R. S. K., P. Calow and P. J. W. Olive. 1988. *The Invertebrates: A New Synthesis.* Blackwell, Oxford.

Barrington, E. J. W. 1979. *Invertebrate Structure and Function,* 2nd Ed. Halstead Press, New York. [One of the best available treatments of invertebrate functional morphology.]

Barth, R. H. and R. E. Broshears. 1982. *The Invertebrate World.* Saunders, Philadelphia.

Bayer, F. M. and H. B. Owre. 1968. *The Free-Living Lower Invertebrates.* Macmillan, New York. [Porifera through Nemertea only.]

Beck, D. E. and L. F. Braithwaite. 1968. *Invertebrate Zoology Laboratory Workbook.* Burgess, Minneapolis.

Beklemishev, W. N. 1969. *Principles of Comparative Anatomy of Invertebrates* (2 vols.). University of Chicago Press, Chicago. [Translated from Russian; a very different view of the subject, quite unlike western texts.]

Boardman, R. S., A. H. Cheetham and A. J. Rowell. 1987. *Fossil Invertebrates.* Blackwell, London. [A wonderful distillation of fossil invertebrate zoology.]

Bougis, P. 1976. *Marine Plankton Ecology.* Elsevier, New York.

Briggs, J. C. 1974. *Marine Zoogeography.* McGraw-Hill, New York. [Reviews current information on general distributional patterns of fishes and some marine invertebrates; attempts to update Ekman (1953), with only marginal success.]

Brusca, G. J. 1975. *General Patterns of Invertebrate Development.* Mad River Press, Eureka, California. [A brief introductory account of basic invertebrate embryology.]

Buchsbaum, R., M. Buchsbaum, J. Pearse and V. Pearse. 1987. *Animals Without Backbones,* 3rd Ed. University of Chicago Press, Chicago. [Although somewhat elementary in its coverage, the third edition of this classic introductory text is informative, delightful reading and includes a great many excellent photographs.]

Bullock, T. H. and G. A. Horridge. 1965. *Structure and Function of the Nervous System of Invertebrates.* W. H. Freeman, San Francisco. [An extremely useful two-volume coverage true to the title.]

Calow, P. 1981. *Invertebrate Biology. A Functional Approach.* Wiley, New York.

Carefoot, T. 1977. *Pacific Seashores, A Guide to Intertidal Ecology.* University of Washington Press, Seattle. [A very clear and readable account of ecological concepts as they apply to the seashore; the emphasis is on invertebrates.]

Carthy, J. 1958. *An Introduction to Behavior of Invertebrates.* Allen & Unwin, London. [Somewhat dated but a good review of the subject as understood 30 years ago.]

Carthy, J. and G. Newell (eds.). 1968. *Invertebrate Receptors.* Academic Press, New York. [Excellent.]

Cheng, T. (ed.). 1971. *Aspects of the Biology of Symbiosis.* University Park Press, Baltimore.

Clark, R. B. and A. L. Panchen. 1971. *Synopsis of Animal Classification.* Chapman and Hall, London. [Classifies animals to ordinal level.]

Cloud, P. and M. F. Glaessner. 1982. The Ediacaran period and system: Metazoa inherit the Earth. Science 217:783–792.

Crawford, C. S. 1981. *Biology of Desert Invertebrates.* Springer-Verlag, New York. [A long-overdue text on a subject rich in ideas and information.]

Cushing, D. H. and J. J. Walsh (eds.). 1976. *The Ecology of the Seas.* Saunders, Philadelphia.

Dales, R. P. (ed.). 1981. *Practical Invertebrate Zoology. A Laboratory Manual for the Study of the Major Groups of Invertebrates, Excluding Protochordates,* 2nd Ed. Wiley, New York.

Dawydoff, C. 1928. *Traité d'Embryologie Comparée des Invertébrés.* Masson et Cie, Libraires de l'Academie de Médicine, Paris. [Somewhat out of date, but still useful.]

Dindal, D. L. 1975. Symbiosis: Nomenclature and proposed classification. The Biologist 57(4):129–142.

Dyer, J. C. and F. R. Schram. 1983. *A Manual of Invertebrate Paleontology.* Stipes, Champaign, Illinois. [Apparently the only lab manual available for invertebrate paleontology; current.]

Easton, W. H. 1960. *Invertebrate Paleontology.* Harper & Row, New York.

Ekman, S. 1953. *Zoogeography of the Sea.* Sedgwick and Jackson, London. [Excellent review of marine invertebrate distributions; a classic work.]

Engemann, J. G. and R. W. Hegner. 1981. *Invertebrate Zoology,* 3rd Ed. Macmillan, New York.

Fingerman, M. 1969. *Animal Diversity.* Holt, Rinehart & Winston, New York. [A well-written little paperback on general comparative zoology; emphasizes invertebrates.]

Florkin, M. and B. Sheer (eds.). 1967–1978. *Chemical Zoology.* Academic Press, New York. [Ten volumes, which include articles by numerous authors on animal physiology and biochemistry.]

Freeman, W. H. and B. Bracegirdle. 1971. *An Atlas of Invertebrate Structure.* Heinemann Educational Books, London. [An excellent laboratory aid; includes gross morphology and anatomy as well as histological illustrations and photographs.]

Fretter, V. and A. Graham. 1976. *A Functional Anatomy of Invertebrates.* Academic Press, London.

Gardiner, M. 1972. *The Biology of Invertebrates.* McGraw-Hill, New York. [System-level comparative anatomy.]

Giese, A. and J. S. Pearse (eds.). 1974–1987. *Reproduction of Marine Invertebrates.* Vols. 1–5, 9. Academic Press, New York. [With more on the way; excellent review articles.]

Gotto, R. V. 1969. *Marine Animals, Partnerships and Other Associations.* Elsevier, New York.

Grassé, P. (ed.). 1948– . *Traité de Zoologie.* Masson et Cie, Paris. [Work continues on this multivolume enterprise covering the animal kingdom; perhaps the best single reference source on invertebrates; in French.]

Green, J. 1968. *The Biology of Estuarine Animals.* University of Washington Press, Seattle.

Hardy, A. C. 1956. *The Open Sea*. Houghton Mifflin, Boston. [Still perhaps the best introduction to the world of plankton.]

Harland, W. B., A. V. Cox, P. G. Llewellyn, C. A. G. Pickton, A. G. Smith and R. Walters. 1982. *A Geologic Time Scale*. Cambridge University Press, Cambridge.

Hedgpeth, J. W. (ed.). 1957. *Treatise on Marine Ecology and Paleoecology*. Geol. Soc. Am. Mem. 67: 1–1296. [Still frequently consulted and cited; outstanding reviews of major aspects of marine biology.]

Henry, S. M. 1966. *Symbiosis*. Vol. 1. Associations of Microorganisms, Plants, and Marine Organisms. Academic Press, New York.

Hyman, L. H. 1940–1967. *The Invertebrates* (6 vols.). McGraw-Hill, New York. [This series has probably ended, following the completion of Volume 6 (a partial coverage of the Mollusca) and Libbie Hyman's death. No one seems inclined to attempt comparable coverage of the groups left undone by Dr. Hyman, including the annelids, arthropods, bivalves, scaphopods, cephalopods, and some minor taxa. Naturally, some of the material in early volumes has fallen out of date, but they still remain among the best references available.]

Jackson, J. B. C., L. W. Buss and R. E. Cook (eds.). 1986. *Population Biology and Evolution of Clonal Organisms.*. Yale University Press, New Haven, Connecticut. [Papers from a 1982 symposium.]

Jones, O. A. and R. Endean (eds.). 1973–1977. *Biology and Geology of Coral Reefs*. Academic Press, New York. [A four-volume compilation of excellent review papers, many dealing with invertebrates.]

Kaestner, A. 1967–1970. *Invertebrate Zoology*. Vols. 1–3. Wiley-Interscience, New York. [Translated from German. Volume 3 deals exclusively with the Crustacea and is extremely thorough.]

Kume, M. and K. Dan et al. 1957. *Invertebrate Embryology*. In Japanese. [An English translation has been made available by Prosveta, Belgrade, 1968, published by NOLIT Publishing House, Belgrade, Yugoslavia, for the National Library of Medicine, Public Health Service, U.S. Department of Health, Education, and Welfare, and the National Science Foundation. It is available through the Clearing House for Federal Scientific and Technical Information, Springfield, Virginia. The bulk of this paperback translation presents rather detailed explanations of development of specific invertebrates. It has a good section on eggs, sperm, and fertilization, often neglected in similar books. One of its greatest values is to expose the reader to the incredible variation within taxa. A recommended reference for those interested in embryology.]

Lankester, R. (ed.). 1900–1909. *A Treatise on Zoology*. Adam and Charles Black, London. [A classic multivolume work on invertebrates.]

Larwood, G. and B. Rosen (eds.). 1979. *Biology and Systematics of Colonial Organisms*. Academic Press, New York.

Lauff, G. H. (ed.). 1967. *Estuaries*. Publication No. 83. AAAS, Washington, D.C.

Laverack, M. S. and J. Dando. 1987. *Lecture Notes on Invertebrate Zoology*, 3rd Ed. Blackwell, Palo Alto, California.

Levinton, J. S. 1982. *Marine Ecology*. Prentice-Hall, Englewood Cliffs, New Jersey.

Lewis, J. R. 1964. *The Ecology of Rocky Shores*. The English Universities Press, Ltd., London.

Lincoln, R. J. and J. G. Sheals. 1979. *Invertebrate Animals: Collection and Preservation*. British Museum (Natural History) and Cambridge University Press, London and Cambridge.

Lutz, P. E. 1986. *Invertebrate Zoology*. Addison-Wesley, Reading, Massachusetts.

MacGinitie, G. E. and N. MacGinitie. 1968. *Natural History of Marine Animals*, 2nd Ed. McGraw-Hill, New York. [Emphasis is on the North American Pacific coast.]

Margulis, L. and K. V. Schwartz. 1982. *Five Kingdoms: An Illustrated Guide to the Phyla of Life on Earth*. W. H. Freeman, New York. [Very nicely done.]

Marshall, N. B. 1980. *Deep Sea Biology: Development and Perspective*. Garland STPM Press, New York.

McConnaughey, B. H. and R. Zottoli. 1983. *Introduction to Marine Biology*, 4th Ed. C. V. Mosby, St. Louis. [Possibly the best general text on marine biology available.]

Mechiorri-Santolini, U. and J. W. Hopton (eds.). 1972. Detritus and its role in aquatic ecosystems. Proceedings of an IBP-UNESCO Symposium Mem. Dell'Instituto Italiano di Idrobiologia. Vol. 29 (Suppl.), 1972.

Meglitsch, P. A. 1972. *Invertebrate Zoology*, 2nd Ed. Oxford University Press, New York. [One of the better standard texts, well illustrated, but dated.]

Menzies, R. J., R. Y. George and G. T. Rowe. 1973. *The Abyssal Environment and Ecology of the World Ocean*. Wiley, New York.

Moore, R. C. (ed.). 1952– . *Treatise on Invertebrate Paleontology*. Geological Society of America and University of Kansas Press, Lawrence. [Detailed coverage of fossil forms; many volumes and still incomplete.]

Newell, R. C. 1979. *Biology of Intertidal Animals*, 3rd Ed. Marine Ecological Surveys Ltd., Faversham, Kent, United Kingdom.

Nicol, J. A. C. 1969. *The Biology of Marine Animals*, 2nd Ed. Wiley-Interscience, New York. [One of the finest comparative biology texts available.]

Osman, R. W. and J. A. Haugsness. 1981. Mutualism among sessile invertebrates: A mediator of competition and predation. Science 211: 846–848.

Parker, S. P. (ed.). 1982. *Synopsis and Classification of Living Organisms* (2 vols.). McGraw-Hill, New York. [Encyclopedic and current.]

Pearse, V., J. Pearse, M. Buchsbaum, and R. Buchsbaum. 1987. *Living Invertebrates*. Blackwell, Palo Alto, California. [Very current; easy reading; lots of wonderful photographs of invertebrates.]

Pechenik, J. A. 1985. *Biology of the Invertebrates*. Prindle Weber and Schmidt, Boston.

Prosser, C. L. (ed.). 1973. *Comparative Animal Physiology*, 3rd Ed. Saunders, Philadelphia. [One of the best accounts of comparative physiology.]

Pflugfelder, O. 1962. *Lehrbuch der Entwicklungsgeschichte und Entwicklungsphysiologie der Tiere*. Fischer, Jena.

Reid, G. K. and R. D. Wood. 1976. *Ecology of Inland Waters and Estuaries*, 2nd Ed. Van Nostrand, New York.

Ricketts, E. F., J. Calvin, J. W. Hedgpeth, and D. W. Phillips. 1985. *Between Pacific Tides*, 5th Ed. Stanford University Press. [A standard reference for the natural history of the Pacific coast intertidal region.]

Russell-Hunter, W. D. 1979. *A Life of Invertebrates*. Macmillan, New York. [One biologist's structure and function approach to teaching invertebrate zoology.]

Sherman, I. W. and V. G. Sherman. 1976. *The Invertebrates: Function and Form. A Laboratory Guide*. Macmillan, New York.

Shrock, R. R. and W. H. Twenhofel. 1963. *Principles of Invertebrate Paleontology*. McGraw-Hill, New York.

Stancyk, S. E. 1979. *Reproductive Ecology of Marine Invertebrates*. University of South Carolina Press, Columbia.

Stephensen, T. A. and A. Stephensen. 1972. *Life Between Tide Marks on Rocky Shores*. W. H. Freeman, San Francisco. [A summary of the authors' life work on the subject; primarily deals with algae and invertebrates; global in coverage.]

Sverdrup, H. U., M. W. Johnson and R. H. Fleming. 1942. *The Oceans*. Prentice-Hall, New York. [Much has been learned since the writing of this classic text, so much that no one has attempted such a massive undertaking again!]

Taylor, D. L. 1973. The cellular interactions of algal–invertebrate symbiosis. Adv. Mar. Biol. 11: 1–56.

Thorson, G. 1971. *Life in the Sea.* World University Library, McGraw-Hill, New York and Toronto.

Tombes, A. S. 1970. *An Introduction to Invertebrate Endocrinology.* Academic Press, New York.

Trueman, E. R. 1975. *The Locomotion of Soft-Bodied Animals.* American Elsevier, New York. [A small (200 pp.) but useful book— see our Chapter 3.]

Vernberg, W. B. 1974. *Symbiosis in the Sea.* The Belle W. Baruch Library in Marine Science. No. 2. University of South Carolina Press, Columbia.

Vernberg, W. B. and F. J. Vernberg. 1972. *Environmental Physiology of Marine Animals.* Springer-Verlag, New York.

Welch, P. S. 1952. *Limnology.* McGraw-Hill, New York.

Wells, M. 1968. *Lower Animals.* McGraw-Hill, New York. [Good reading.]

Welsh, J. H., R. I. Smith and A. E. Kammer. 1968. *Laboratory Exercises in Invertebrate Physiology.* Burgess, Minneapolis.

Whittington, H. B. 1985. *The Burgess Shale.* Yale University Press, New Haven, Connecticut. [Modern summary of an important middle-Cambrian deposit in the Rocky Mountains of Canada.]

Winn, H. E. and B. L. Olla. 1972. *Behavior of Marine Animals. Current Perspective in Research.* Vol. 1. Plenum, New York.

Yonge, C. M. 1949. *The Sea Shore.* Collins, London. [This fine book has been reprinted (1963) by Atheneum, New York; the new version lacks color plates.]

Manuals and Field Guides for Identification of Invertebrates

We have included here only a few of the scores of identification guides, booklets, and the like. Some guides to particular taxa are listed in appropriate chapters.

Allen, R. 1969. *Common Intertidal Invertebrates of Southern California.* Peek Publications, Palo Alto, California. [The only available keys to Southern California invertebrates.]

Arnold, A. 1901. *The Sea-Beach at Ebb Tide: A Guide to the Study of the Seaweeds and Lower Animal Life Found Between the Tide-Marks.* Century. [This classic work deals primarily with Atlantic species; has been reprinted (1968) by Dover Publications Co., New York.]

Bright, T.J. and L.H. Pequegnat (eds.) 1974. *Biota of the West Flower Garden Bank.* Gulf Publishing Co., Houston, Texas.

Brusca, G. J. and R. C. Brusca. 1978. *A Naturalist's Seashore Guide: Common Marine Life of the Northern California Coast and Adjacent Shores.* Mad River Press, Eureka, Cal-

ifornia. [Includes introductory remarks on marine ecology and other pertinent matters, followed by descriptions and keys.]

Brusca, R. C. 1980. *Common Intertidal Invertebrates of the Gulf of California,* 2nd. Ed. University of Arizona Press, Tucson. [A fairly exhaustive treatment of the subject, including keys, descriptions and figures for over 1,300 species.]

Colin, P.I. 1978. *Caribbean Reef Invertebrates and Plants.* T.F.H. Publications, Neptune City, New Jersey.

Edmondson, W. T. H. B. Ward, and G. C. Whipple (eds.). 1959. *Freshwater Biology.* Wiley, New York. [Good keys to freshwater invertebrates.]

Fielding, A. 1982. *Hawaiian Reefs and Tidepools.* Oriental, Honolulu.

Gosner, K. L. 1971. *Guide to the Identification of Marine Estuarine Invertebrates.* Wiley-Interscience, New York. [For use on the northeastern coast of the United States.]

Gunson, D. 1983. *Collins Guide to the New Zealand Seashore.* Collins, Auckland.

Hedgpeth, J. W. 1962. *Introduction to Seashore Life of the San Francisco Bay Region and the Coast of Northern California.* Natural History Guide No. 9, University of California Press, Berkeley and Los Angeles.

Johnson, M. and H. Snook. 1927. *Seashore Animals of the Pacific Coast.* Macmillan, New York. [An old classic available as a Dover reprinted edition.]

Kaplan, E. 1982. *A Field Guide to Coral Reefs of the Caribbean and Florida.* Houghton Mifflin, Boston. [Excellent; one of the Peterson Field Guide Series.]

Kerstitch, A. 1989. *Sea of Cortez Marine Invertebrates.* Sea Challengers, Monterey, California.

Kozloff, E. 1973. *Seashore Life of Puget Sound, the Strait of Georgia, and the San Juan Archipelago.* University Washington Press, Seattle.

Kozloff, E. 1974. *Keys to the Marine Invertebrates of Puget Sound, the San Juan Archipelago, and Adjacent Regions.* University of Washington Press, Seattle.

Kozloff, E. 1987. *Marine Invertebrates of the Pacific Northwest.* University of Washington Press, Seattle.

Luther, W. and K. Fiedler. 1976. *A Field Guide to the Mediterranean Sea Shore.* Collins, London.

Morris, R. H., D. P. Abbott and E. C. Haderlie. 1980. *Intertidal Invertebrates of California.* Stanford University Press, Stanford. California. [Although it does not contain keys, this book has a great array of color plates with ecological and other notes, plus excellent literature listings.]

Morton, B. and J. Morton. 1983. *The Sea Shore Ecology of Hong Kong.* Hong Kong University Press.

Newell, G. and R. Newell. 1973. *Marine Plankton: A Practical Guide.* Hutchinson, London.

Pennak, R. W. 1989. *Fresh-water Invertebrates of the United States,* 3rd Ed. Protozoa to Mollusca. Wiley, New York [Excellent keys to most species.]

Riedl, R. (ed.). 1983. *Fauna und Flora des Mittelmeeres.* Verlag Paul Parey, Hamburg. [The best field guide for the Mediterranean.]

Ruppert, E.E. and R.S. Fox. 1988. *Seashore Animals of the Southeast: A Guide to Common Shallow-Water Invertebrates of the Southeastern Atlantic Coast.* University of South Carolina Press, Columbia.

Sefton, N. and S. K. Webster. 1986. *A Field Guide to Caribbean Reef Invertebrates.* Sea Challengers, Monterey, California.

Smith, D. L. 1977. *A Guide to Marine Coastal Plankton and Marine Invertebrate Larvae.* Kendall/Hunt Publishing Co., Dubuque, Iowa. [Despite the abundant errors, still a useful little laboratory guide for beginning invertebrate students.]

Smith, R. I. (ed.). 1964. *Keys to Marine Invertebrates of the Woods Hole Region.* Contribution No. 11, Systematics-Ecology Program, Marine Biological Laboratory, Woods Hole, Massachusetts.

Smith, R. I., and J. Carlton (eds.). 1975. *Light's Manual: Intertidal Invertebrates of the Central California Coast,* 3rd Ed. University of California Press, Berkeley. [A product of the editors' devotion to the task and the contributions of many experts; includes sets of keys to the invertebrates; well referenced.]

Sterrer, W. (ed.). 1986. *Marine Fauna and Flora of Bermuda.* Wiley, New York. [A comprehensive guide.]

Voss, G. L. 1976. *Seashore Life of Florida and the Caribbean.* Seeman, Miami. [Excludes the Mollusca.]

Yamagi, I. 1968. *The Plankton of Japanese Coastal Waters.* Hoikushu Publishing Co., Osaka, Japan. [In Japanese.]

Yamagi, I. 1969. *Illustrations of the Marine Plankton of Japan.* Hoikushu Publishing Co., Osaka, Japan. [In Japanese; a companion to the above work.]

Zeiller, W. 1974. *Tropical Marine Invertebrates of Southern Florida and the Bahama Islands.* Wiley-Interscience, New York.

Recommended References on Evolution

Also see references on evolution at the end of Chapter 2.

Ayala, F. J. 1974. Biological evolution: Natural selection or random walk. Am. Sci. 62:692–701. [Although over 10 years old now, this is still one of the most readable accounts of the subject.]

Ayala, F. J. (ed.). 1976. *Molecular Evolution*. Sinauer Associates, Sunderland, Massachusetts.

Bookstein, F. L., P. D. Gingerich, and A. G. Kluge. 1978. A hierarchical linear modeling of the tempo and mode of evolution. Paleobiology 4:120–134.

Bush, G. L. 1975. Modes of animal speciation. Annu. Rev. Evol. Syst. 6:339–364. [One of the classic review studies on "traditional" speciation models.]

Charlesworth, B., R. Lande and M. Slatkin. 1982. A neo-Darwinian commentary on macroevolution. Evolution 36:474–498. [A current and timely rebuttal to the non-Darwinian arguments; specifically addresses the punctuated equilibrium theory.]

Dobzhansky, Th., F. J. Ayala, G. L. Stebbins and J. W. Valentine. 1977. *Evolution*. W. H. Freeman, San Francisco. [An excellent, current and readable little book.]

Dover, G. A. 1982. Molecular drive: An adhesive mode of species evolution. Nature 299:111–117.

Eldredge, N. 1985. *Time Frames*. Simon & Schuster, New York.

Eldredge, N. 1985. *Unfinished Synthesis. Biological Hierarchies and Modern Evolutionary Thought*. Oxford University Press, New York. [An important, thought-provoking look at the "modern synthesis" of evolution, its shortcomings and some alternative ideas on evolutionary theory.]

Eldredge, N. and S. J. Gould. 1972. Punctuated equilibria: An alternative to phyletic gradulaism. *In* T. J. M. Schopf (ed.), *Models in Paleobiology*. Freeman, Cooper, San Francisco, pp. 82–115.

Fox, G. E. and 18 other authors. 1980. The phylogeny of prokaryotes. Science 209:457–463.

Futuyma, D. J. 1986. *Evolutionary Biology*, 2nd Ed. Sinauer Associates, Sunderland, Massachusetts. [A most enjoyable and interesting approach to the subject; de-emphasizes phylogeny and macroevolution but excellent on the subjects of anagenesis and adaptation.]

Futuyma, D. J. and G. C. Mayer. 1980. Non-allopatric speciation in animals. Syst. Zool. 29:254–271.

Goldschmidt, R. B. 1940. *The Material Basis of Evolution*. Yale University Press, New Haven, Connecticut. [An embryologist's view of evolution.]

Gould, S. J. 1977. *Ontogeny and Phylogeny*. Harvard University Press, Cambridge, Massachusetts. [An intellectual and entertaining synthesis of developmental influences on phylogeny; highly recommended.]

Gould, S. J. 1980. Is a new general theory of evolution emerging? Paleobiology 6:119-130.

Gould, S. J. and N. Eldredge. 1977. Punctuated equilibria: The tempo and mode of evolution reconsidered. Paleobiology 3:115–151.

Hallam, A. 1978. How rare is phyletic gradualism and what is its evolutionary significance? Evidence from Jurassic Bivalvia. Paleobiology 4:16–25.

Hennig, W. 1966. *Phylogenetic Systematics*. University of Illinois Press, Urbana.

King, J. L. and T. H. Jukes. 1969. Non-Darwinian evolution. Science 164:788–798.

Lewontin, R. C. 1974. *The Genetic Basis of Evolutionary Change*. Columbia University Press, New York. [An excellent treatment of evolutionary genetics.]

Mayr, E. 1969. *Principles of Systematic Zoology*. McGraw-Hill, New York.

Milkman, R. 1982. *Perspectives on Evolution*. Sinauer Associates, Sunderland, Massachusetts. [Good reviews of current thinking in evolutionary biology.]

Morris, S. C. 1989. Burgess shale faunas and the Cambrian explosion. Science 246:339–346.

Raff, R. A. and T. C. Kaufman. 1983. *Embryos, Genes, and Evolution*. Macmillan, New York.

Raup, D. M. and J. J. Sepkoski. 1982. Mass extinctions in the marine fossil record. Science 215:1501–1503.

Rensch, B. 1959. *Evolution Above the Species Level*. Columbia University Press, New York. [The standard reference on macroevolution for a decade; one of the most important contributions to appear in the field of evolutionary biology; now getting a bit dated.]

Schopf, J. W. (ed.). 1983. *Earth's Earliest Biosphere: Its Origin and Evolution*. Princeton University Press, Princeton, New Jersey. [An excellent scholarly review of Precambrian biology.]

Schopf, T. M. J. (ed.). 1972. *Models in Paleobiology*. Freeman, Cooper, San Francisco.

Stanley, S. M. 1975. A theory of evolution above the species level. Proc. Natl. Acad. Sci. U.S.A. 72(2):646–650.

Stanley, S. M. 1979. *Macroevolution: Pattern and Process*. W. H. Freeman, San Francisco.

Stanley, S. M. 1982. Macroevolution and the fossil record. Evolution 36:460–473.

Stebbins, G. L. and F. J. Ayala. 1981. Is a new evolutionary synthesis necessary? Science 213:967–971.

Steele, E. J. 1979. *Somatic Selection and Adaptive Evolution: On the Inheritance of Acquired Characteristics*. Williams and Wallace International, Toronto. [A controversial book on an obviously controversial subject.]

Templeton, A. R. 1981. Mechanisms of speciation: A population genetic approach. Annu. Rev. Syst. 12:23–48.

Vrba, E. S. 1980. Evolution, species and fossils: How does life evolve? So. Afr. J. Sci. 76:61–84.

White, M. J. D. 1978. *Modes of Speciation*. W. H. Freeman, San Francisco. [One of the best general accounts of orthodox views on the subject.]

Wright, S. 1968–1978. *Evolution and the Genetics of Populations*. Vols. 1–4. University of Chicago Press, Chicago. [One of the classic treatises on population genetics and anagenesis.]

References on the Kingdoms of Organisms

Copeland, H. F. 1938. The kingdoms of organisms. Q. Rev. Biol. 13:383–420.

Copeland, H. F. 1947. Progress report on basic classification. Am. Nat. 81:340–361.

Copeland, H. F. 1956. *The Classification of Lower Organisms*. Pacific Books, Palo Alto, California.

Diener, T. O. 1983. The viroid—A subviral pathogen. Am. Sci. 71:481–488.

Dodson, E. O. 1971. The kingdoms of organisms. Syst. Zool. 20:265–281.

Hadzi, J. 1953. An attempt to reconstruct the system of animal classification. Syst. Zool. 2:145–154.

Leedale, G. F. 1974. How many are the kingdoms of organisms? Taxon 23:261–270.

Marcus, E. 1958. On the evolution of the animal phyla. Q. Rev. Biol. 33:24–58.

Margulis, L. 1981. *Symbiosis in Cell Evolution. Life and Its Environment on the Early Earth*. W. H. Freeman, San Francisco.

Moore, R. C. 1954. Kingdom of organisms named Protista. J. Paleontol. 28:588–598.

Whittaker, R. H. 1957. The kingdoms of the living world. Ecology 38:536–538.

Whittaker, R. H. 1959. On the broad classification of organisms. Q. Rev. Biol. 34:210–226.

Whittaker, R. H. 1969. New concepts of kingdoms of organisms. Science 163:150–160.

Woese, C. R. 1981. Archaebacteria. Sci. Am. 244(6):98–122.

Chapter Two

Classification, Systematics, and Phylogeny

Our classifications will come to be,
as far as they can be so made, genealogies.
   Charles Darwin
      The Origin of Species, 1859

Prodromos

This book deals with the field of **comparative biology**, or what may be called the science of the diversity of life (Figure 1). To understand invertebrate zoology, one must understand comparative biology, the tasks of which are to describe the patterns inherent in living systems and to explain those patterns by the scientific method. When the patterns have resulted from evolutionary processes, such studies illuminate the history of life on Earth. Because one cannot directly observe that history, one must rely on the strength of the scientific method to reconstruct it. This chapter provides an overview of both the operational and philosophical approaches to some of these tasks. Comparative biology, in its attempt to understand the living world, deals with three distinguishable elements: (1) descriptions of organisms, and similarities and differences in their characteristics, (2) the history of organisms in time, and (3) the distributional history of organisms in space. Although the present text emphasizes the first of these elements, it addresses, to a limited extent, the other two as well. Our view of comparative biology is an evolutionary one, which we believe is the only view that realistically explains natural patterns and diversity.

Until the advent of modern ecological theory in the late 1940s, virtually all aspects of comparative biology were within the purview of biological systematists, who in earlier days were referred to simply as naturalists. Systematics today includes many areas of investigation other than biological classification (e.g., biogeography, genetics, population biology). The field of biological systematics is currently experiencing a revolution in its theory and practice, and considerable controversy exists. The philosophical aspects of this exciting controversy are described in this chapter ("Current Schools of Thought in Classification Theory"), and some practical examples appear in later chapters of this book. It is absolutely essential that biology students have a basic grasp of current theories of classification and phylogenetic reconstruction. The latter part of this chapter provides intellectually challenging reading in this regard. It is not merely a listing of facts, but a discussion of concepts, and we urge you to reflect carefully on the ideas discussed.

Biological classification

The term **biological classification** has two meanings. First, it means the *process* of classifying, which consists of the delimiting, ordering, and ranking of organisms into groups. Second, it means the *product* of this process itself, or the classificatory scheme. The natural world has an objective structure that can be empirically documented and described. One goal of science is to describe this structure, and classifications are one way of doing this. Carrying out the process of classification constitutes one of the principal tasks of the systematist or taxonomist.

The construction of a classification (the product) may at first appear straightforward; basically, the process consists of analyzing patterns in the distribution of characters among organisms. By such analyses, specimens are grouped into **species** (the word *species* is both singular and plural); related species are

Figure 1

The relationship of comparative biology to descriptive/functional biology and to evolutionary biology. The flow of information (arrows) is primarily from descriptive/functional biology to evolutionary biology. Roughly paralleling this flow of information is a corresponding transition in levels at which animal/plant evolution is studied, as depicted at the left.

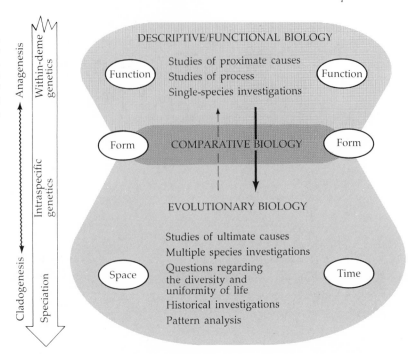

grouped into **genera** (singular, **genus**); related genera are grouped to form **families**; and so forth. The grouping process creates a system of subordinated, or nested, **taxa** (singular, **taxon**) arranged in a **hierarchical** fashion. If the taxa are properly grouped according to their degree of shared similarity, the hierarchy will reflect patterns of descent—the "descent with modification" of Darwin.

The concept of similarity is fundamental to taxonomy, the classificatory process, and comparative biology as a whole. **Similarity**, as evaluated by shared characteristics, is generally accepted by biologists to be a measure of biological (evolutionary) relatedness among taxa. The concept of **relatedness,** or genealogical kinship, is also fundamental to systematics and evolutionary biology. Patterns of relatedness are usually depicted by biologists in various kinds of branching diagrams, called **trees (phylogenetic, genealogical,** or **evolutionary trees)**. Once constructed, such trees can then be converted into classification schemes, which are a dynamic way to represent our understanding of the history of life on Earth. They are theories of natural order and the evolution of life.

Classifications are necessary for several reasons, not the least of which is to efficiently catalog the enormous number of species of organisms on Earth. About a million different animal species have so far been named and described! The insects alone comprise over 800,000 named species—about 300,000 of these are beetles! Thus, classifications provide sci-

entists and people in general with a detailed system for storage and retrieval of names. Second, and most important to evolutionary biologists, classifications serve a descriptive function. By this we refer not only to the descriptions that define each taxon, but also to detailed hypotheses of evolutionary relationships of the animals and plants that inhabit Earth. In other words, classifications are (or should be) constructed on evolutionary relationships, that is, on patterns of ancestry and descent depicted in phylogenetic trees of one sort or another. This second, interpretive function of classification is the focal point of both agreement and argument among many biologists today. While the majority of scientists agree that classifications should attempt to reflect the evolutionary history of life, not all agree on how best to construct such classifications. The philosophical premise in classifications and phylogenetic trees is that descendants genealogically or phylogenetically close to their ancestors resemble those ancestors more closely than they resemble more distantly related ancestors, this resemblance in turn being a reflection of genetic similarity.

A biologically sound classification scheme is in reality a set of hypotheses defined and summarized by a phylogenetic tree. The tree and the classification describe the phylogeny (evolutionary history) of the taxa they contain. Thus, classifications, like other hypotheses and theories in science, have a third function, that of prediction. The more accurate and less ambiguous the classification, the greater its predictive value. Predictability is another way of saying testability, and it is testability that places an endeavor in the realm of science rather than in the realm of art or rhetoric. Like other theories, classifications are always subject to change, refinement, and growth as new data become available. These new data may be in the form of newly discovered species, new tools for the analysis of characters, or new ideas regarding how characters may be evaluated. Changes in classifications reflect changes in our views and understanding of the natural world. Stability in classification is, of course, desirable and sought after, but it can only be attained by giving way to new advances in science. Throughout this book we will discuss the evolution of various invertebrate groups; as you read these discussions, keep in mind that the evolutionary hypotheses we discuss are depicted by the classifications that accompany the text.

Nomenclature

The names employed within classifications are governed by rules and recommendations that may be thought of as being analogous to the rules of grammar that govern the use of the English language. Adherence to these nomenclatural codes is strictly voluntary; there is no "enforcement" power. Thus, adherence constitutes an outstanding example of agreement among concerned individuals all over the world. The goal of **biological nomenclature** is the creation of a classification in which (1) any single kind of organism has one and only one correct name and (2) no two kinds of organisms bear the same name. Nomenclature is a tool of the biologist that facilitates communication.

Prior to the mid-1700s, animal and plant names consisted of one to several words or often simply a descriptive phrase. In 1758 the great Swedish naturalist Carl von Linné (Carolus Linnaeus, in the latinized form he preferred) established a system of naming animals now referred to as **binomial nomenclature**. Linnaeus' system required that every organism have a two-part name—a **binomen**. The two parts of a binomen are the generic, or genus, name and the **specific epithet** (= **trivial name**). For example, the scientific name for one of the common Pacific coast sea stars is *Pisaster giganteus*. These two names together constitute the binomen; *Pisaster* is the animal's generic (genus) name, and *giganteus* is its specific epithet. The specific epithet is never used alone but must be preceded by the generic name, and an animal's "species name" thus refers to the complete binomen. Use of the first letter of a genus name preceding the specific epithet is also acceptable once the name has appeared spelled out on the page or in a short article (e.g., *P. giganteus*).

The 1758 version of Linnaeus' system is actually the tenth edition of his *Systema Naturae*, in which he listed all animals known to him at that time and included critical guidelines for classifying organisms. Linnaeus' *Species Plantarum* had done the same for the plants in 1753. Linnaeus was one of the first naturalists to emphasize the use of the *similarities* among species or other taxa in constructing a classification, rather than using *differences* between them. In doing so, he unknowingly began classifying organisms by virtue of their genetic, and hence evolutionary, relatedness. Linnaeus produced his *Systema Naturae* about 100 years prior to the appearance of Darwin and Wallace's theory of evolution by natural selection (1859), and thus his use of similarities in classification foreshadowed the subsequent emphasis by biologists on evolutionary relationships among taxa.

Binomens are Latin (or latinized) because of the custom followed prior to the eighteenth century of publishing scientific papers in Latin, the universal language of educated people of the time. For several decades after Linnaeus, names for animals and plants proliferated, and there were often several names for any given species (different names for the same organism are called **synonyms**). The name in common use was usually the most descriptive one, or often it was simply the one used by the most eminent authority of the time. This lack of nomenclatural·uniformity led, in 1842, to the adoption of a code of rules formulated under the auspices of the British Association for the Advancement of Science, called the Strickland Code. In 1901 the newly formed International Commission on Zoological Nomenclature adopted a revised version of the Strickland Code, called the International Code of Zoological Nomenclature (**I.C.Z.N.**). Botanists had adopted a·similar code for plants in 1813, the Théorie Élémentaire de la Botanique, which became in 1930 the International Code of Botanical Nomenclature. Similarly, bacteriologists have a separate code of nomenclature.

The I.C.Z.N. established January 1, 1758 (the year the tenth edition of Linnaeus' *Systema Naturae* appeared) as the starting date for modern zoological nomenclature. Any names published the same year, or in subsequent years, are regarded as having appeared after the *Systema*. The I.C.Z.N. also changed slightly the description of Linnaeus' naming system, from binomial nomenclature (names of two parts) to

binominal nomenclature (names of two names). One still sees the former designation in common use. This subtle change implies that the system must be truly binary, that is, both the generic and trivial names must be of one word only. Although the system is binary, it also accepts the use of **subspecies** names, constituting a **trinomen** (three names) within which is contained the mandatory binomen. For example, the sea star *Pisaster giganteus* is known to have a distinct form occurring in the southern part of its range; this southern form is designated as a subspecies, *Pisaster giganteus capitatus*.

All codes of nomenclature share the following six basic principles.

1. Botanical, zoological, and bacteriological nomenclature are independent of one another. It is therefore permissible, although not recommended, for a plant genus and an animal genus to bear the same name.
2. A taxon can bear one and only one correct name.
3. No two genera within a given code can bear the same name (i.e., generic names are unique); and no two species within one genus can bear the same name (i.e., binomens are unique).
4. Scientific names are treated as Latin, regardless of their linguistic origin, and hence subject to Latin rules of grammar.
5. The correct or valid name of a taxon is based on priority of publication (first usage).
6. For the category of superfamily in animals, and order in plants, and for all categories below these, taxa names must be based on **type specimens**, **type species**, or **type genera**.*

When strict application of a code results in confusion or ambiguity, problems are referred to the appropriate commission for a "legal" decision. Rulings of the International Commission on Zoological Nomenclature are published regularly in their journal, the Bulletin of Zoological Nomenclature. Note that the international commissions rule only on nomenclature or "legal" matters, not on questions of scientific or biological interpretation; these latter problems are the business of systematists. The hierarchical categories established and recognized by the

*When a biologist first names and describes a new species, he takes a typical specimen, declares it a type specimen, and deposits it in a safe repository such as a large natural history museum. If later workers are ever uncertain about whether they are working with the same species described by the original author, they can compare their material to the type specimen. Designation of a "typical" type species for a genus, or a "typical" genus for a family serves a somewhat similar purpose in establishing typical groups of species upon which a genus or family is based.

I.C.Z.N. are as follows:

> Kingdom
> Phylum
> Subphylum
> Superclass
> Class
> Subclass
> Cohort
> Superorder
> Order
> Suborder
> Superfamily
> Family
> Subfamily
> Tribe
> Genus
> Subgenus
> Species
> Subspecies

The above names are **categories**; the actual animal group that is placed at any particular categorical level forms the **taxon**. Thus, the taxon Echinodermata is placed at the hierarchical level corresponding to the category Phylum—Echinodermata is the taxon; phylum is the category. The common Pacific sea star *Pisaster giganteus* is classified as follows:

CATEGORY	TAXON
Phylum	Echinodermata
Subphylum	Eleutherozoa
Class	Asteroidea
Order	Forcipulatida
Family	Asteriidae
Genus	*Pisaster*
Species	*Pisaster giganteus* (Stimpson, 1857)

Notice the person's name following the species name in this classification. This is the name of the "author" of that species—the person who first described the organism (*Pisaster giganteus*) and gave the animal its specific name. In this particular case the author's name is in parentheses, to indicate that the species is now placed in a different genus than that originally assigned by Professor Stimpson. Authors' names usually follow the first usage of a species name in the primary literature (i.e., articles published in scientific journals). In the secondary literature, such as textbooks, authors' names are rarely used.

The names given to animals and plants are usually descriptive in some way, or perhaps indicative of the geographical area in which the species occurs. Others are named in honor of people, for one reason or another. Occasionally one runs across purely whimsical names, or even names that seem to have been formulated for purely diabolical reasons.*

All categories (and taxa) at the level of subgenus and higher are referred to as the **higher categories** (or **higher taxa**), as distinguished from the **species group categories** (species and subspecies). This distinction reflects a profound difference in the way evolutionists view these groups. At this point, suffice it to say that species are real biological entities, evolutionary units whose boundaries are defined in space and time by nature. The higher taxa are clusters of species grouped by biologists to reflect the state of knowledge regarding their evolutionary relationships, and as such they may be viewed as evolutionary hypotheses. However, higher taxa do, if correctly constructed, represent natural groups of species having a common ancestry and descent, and thus they are also evolutionary units whose boundaries are defined by nature. Higher taxa are phyletic lineages that share a common genetic history, and in this sense they may be thought of as "historical entities." Interestingly, family-level taxa often tend to be stable taxonomic groupings, usually even recognizable to lay persons; for example, humans (Hominidae), cats (Felidae), dogs (Canidae), abalone (Haliotidae), ladybird beetles (Coccinellidae), mosquitoes (Culicidae), octopuses (Octopodidae), shore crabs (Grapsidae),

swimming crabs (Portunidae), and porcelain crabs (Porcellanidae). This generalization makes family groups very convenient taxa to deal with and to write about.

The science of **systematics** (or **taxonomy**, as it is sometimes called) is perhaps the most encompassing of all fields of biology. The eminent biologist George Gaylord Simpson referred to systematics as "the study of the kinds and diversity of life on Earth, and of any and all relationships between them." The modern systematist is a natural historian of the first order. A systematist's training is broad, cutting across the fields of zoology and botany, genetics, paleontology, biogeography, geology, historical biology, ecology, and even ethology, chemistry, philosophy, and cellular biology.

Ernst Mayr has said that the field of systematics can be thought of as a continuum, from the routine naming and describing of species, through the compilation of large faunal compendia and monographs, to more sophisticated levels of study, such as the fitting of these species into classifications that depict evolutionary relationships, biogeographical analyses, population biology and genetics, evolutionary and speciation studies, and paleoecology. Mayr designated three stages of study within this continuum: alpha, beta, and gamma, corresponding to three general levels of complexity he perceived in taxonomy. When an animal group is first discovered or is in a poorly known state, work on that group is of necessity at the alpha level (e.g., the describing of new species). It is only when most, or at least many, species in a taxon become known that the systematist is able to work at the beta or gamma levels within that group (e.g., evolutionary studies). Some biologists choose to refer to those people working at the alpha level as taxonomists, reserving the term *systematist* for those engaging in studies at the beta or gamma level. Although this may be an instructive way to scrutinize the spectrum of endeavors systematists engage in, it is an oversimplification. These stages in systematic study overlap and cycle back on themselves to a great extent, and it is often through studies of population genetics, ethology, or biogeography that classifications are revised or brought up to date.*

Modern systematists use a great variety of tools to classify species and study the relationships among taxa. These tools include not only the traditional and

---

*Some clever names for animals are *Agra vation* (a tropical beetle that was extremely difficult for Dr. T. Erwin to collect); *Lightiella serendipida* (a small crustacean; the generic name honors the famous Pacific naturalist S. F. Light [1886–1947], while the trivial name is taken from "serendipity," a word coined by Walpole in allusion to a tale, "The Three Princes of Serendip," who in their travels were always discovering, by chance or sagacity, things they did not seek—the term is said to aptly describe the circumstances of the initial discovery of this species). The nineteenth-century naturalist W. E. Leach erected numerous genera of isopod crustaceans whose spellings were anagrams of Carolina (or Caroline), the name of either his wife or his mistress (just what the relationship was between Professor Leach and Carolina is not clear). These include *Cirolana, Lanocira, Rocinela, Nerocila, Anilocra, Conilera, Olincera,* and others. A light-hearted attitude toward naming organisms has not always been without Freudian overtones, as there also exist *Thetys vagina* (a large, hollow, tubular pelagic salp), *Succinea vaginacontorta* (a hermaphroditic snail whose vagina twists in corkscrew fashion), *Phallus impudicus* (a slime-covered mushroom), and *Amanita phalloides* and *Amanita vaginata* (two species of highly toxic mushrooms around which numerous aboriginal ceremonies and legends exist). A few biologists have gone overboard in erecting names for new animals, and many binomens exceed 30 letters in length, including the common north Pacific sea urchin *Strongylocentrotus drobachiensis* (31 letters). Amphipod crustaceans seem to be popular in the "longest name" category—e.g., *Polichinellobizarrocomic burlescomagicaraneus* (44 letters) and *Siemienkiewicziechinogammarus siemenkiewitschii* (46 letters). The weedy chaparral bush, *Eriophyllum* (*Eriophyllum*) *staechadifolium artemisiaefolium,* is 53 letters long including both subgeneric and subspecific designation. And, in a stroke of whimsy, the entomologist G. W. Kirkaldy created the bug genera *Pollychisme* ("Polly kiss me"), *Peggichisme, Marichisme, Dolychisme,* and *Florichisme.*

*Europeans tend to use the terms *systematics* and *biosystematics,* whereas North Americans tend to use *taxonomy* more frequently. In this text, we use the terms taxonomy, systematics, and biosystematics interchangeably.

highly informative techniques of comparative and functional anatomy but also the methods of embryology, serology, immunology, biochemistry, population and molecular genetics, and behavioral and physiological ecology. Systematists are often ecologists in the broad sense of the word, but the reverse is less often the case. Ecologists today are rarely trained in the fundamentals of systematics. In contrast, many of the last generation's most eminent ecologists began their careers as systematists on one group of animals or another, later to find their "bugs" ideally suited to investigations of general ecological problems. A sound classification lies at the root of any study of evolutionary significance, as does a thorough appreciation for the enormous diversity of life. Without systematics, the science of biology would grind to a slow and painful halt, or worse yet, would drift off into pockets of isolated reductionist or deterministic schools with no conceptual framework or continuity.

The field of systematics is currently experiencing a welcome revival in popularity. Within the worldwide literature, there are now about 150 scientific journals publishing specifically in the fields of systematics and evolution, and another 1,000 or so cover the general field of natural history. The causes for this recent revival appear to be fourfold.

First is the growing awareness that too few systematists have been trained over the past 30 years. As the previous generation's systematists retire, few are left to continue work on important taxa and evolutionary problems. The laboratory biologist often feels this crisis with great urgency when there is no one capable of providing an accurate identification of his experimental animal.

Second is the recent discovery of a great many potential anticancer, antibiotic, and other pharmacologically important compounds in native species of animals and plants. This newly popularized field of study has been dubbed "natural products chemistry," and many of the most highly "active" plants and animals that the chemists are discovering come from the most poorly known regions of the world, such as the tropics, where an estimated 90 percent of the native species are yet to be named and described. Even today, about one-half of the prescriptions written in this country contain ingredients of natural origin.

Third is the state of affairs in the tropics, which are said to harbor 80–90 percent of the total animal and plant species on Earth. These regions are rapidly being destroyed by humans, at the rate of 50 million acres per year (an area slightly larger than the state of Kansas). A committee of the National Academy of Sciences has reported that the tropical forests of the world are being altered so rapidly that most will not exist in their present form by the end of this century. Estimates of the loss of species of terrestrial life alone in the tropics range as high as 40 percent (millions of species) of the total world fauna and flora between now and the year 2000—if present trends of human exploitation continue. The extirpation of millions of animal and plant species not only represents an enormous loss of potential food, drug, and other product sources for mankind, but also is an outrageous insult to the natural environment. Lately, several national and international organizations have urged educational institutions to train new systematists as quickly as possible, and to survey and inventory the threatened natural "gene reservoirs" of the tropics.

Last, but certainly not least, is the recent formulation of new quantitative approaches or methods of classification and phylogeny reconstruction, most notably cladistic taxonomy. These new concepts are discussed in the latter half of this chapter. The emergence of cladistic techniques and philosophy and their use in conjunction with new empirical approaches to biogeographical analysis are two of the most exciting events in biological systematics since the development of modern population genetics in the 1940s.

Some terminology and concepts

One of the concepts most crucial to our understanding of biological systematics and evolutionary theory in general is monophyly. A **monophyletic group** is a group of species that includes an ancestral species and all of its descendants, that is, a *natural group*. In other words, a monophyletic taxon is a group of species whose members are related to one another through a unique history of descent (with modification) from a common ancestor—a single evolutionary line. A group in which member species are all descendants of a common ancestor, but which does not contain *all* the species descended from that ancestor, is a **paraphyletic group**. Paraphyly implies that for some reason (e.g., lack of knowledge, purposeful manipulation of the classification, accidents) some members of a natural lineage have been removed and classified in a group separate from the group comprising other members of that lineage. Many paraphyletic taxa exist within animal classifications today, to the consternation of those who prefer to recognize only monophyletic taxa.

A third possibility is polyphyly. A **polyphyletic taxon** is a group comprising species that arose from two (or more) different immediate ancestors. Such composite taxa have been established primarily because of insufficient knowledge concerning the species in question. One of the principal goals of systematists is to discover such polyphyletic or "artificial" taxa and, through careful study, reclassify their members into appropriate monophyletic taxa. These three kinds of taxa or species groups are illustrated diagrammatically in Figure 2.

There are many examples of known or suspected polyphyletic taxa in the zoological literature. For example, the old phylum Gephyrea contained what we now recognize as three distinct phyla—Sipuncula, Echiura, and Priapula. Another example is the old group Radiata, which included any animals possessing radial symmetry (e.g. cnidarians and echinoderms). Still another example is the former phylum Protozoa, whose members now make up seven phyla. Polyphyletic taxa usually are established because the features or characters used to recognize and diagnose them are the result of evolutionary convergence in different lineages, as discussed below. Convergence can only be discovered by careful comparative morphological, biochemical, embryological, or anatomical studies, sometimes requiring the efforts of several generations of specialists.

Characters are the attributes, or features, of organisms or groups of organisms (taxa) that biologists rely on to indicate relatedness to other similar organisms (or other taxa) and to distinguish them from different groups. They are the observable products of the genotype. We are just beginning to learn how to measure genetic relatedness directly, through gene sequencing. But we can estimate it indirectly through examination of the various kinds of characters produced by an organism's genotype (i.e., through examination of the phenotype). A character can be just about anything taxonomists are able to examine and measure in some fashion; it can be a morphological or anatomical feature of an organism, its chromosome make-up (karyotype) or biochemical "fingerprint," or even an ethological (behavioral) attribute. Most recently a variety of biochemical techniques designed to indirectly measure "genetic distances" between organisms have become popular among systematists; these include DNA hybridization, starch-gel electrophoresis of proteins and amino acids, and immunological similarity indices. These and other kinds of data provide systematists with sets of characters to define and characterize species and higher taxa.

The fundamental basis for comparative anatomy

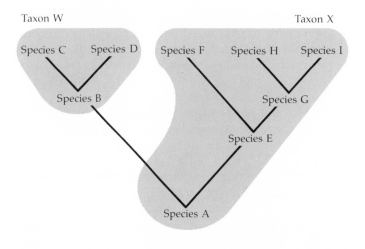

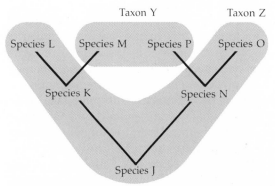

Figure 2

Two dendrograms, illustrating three kinds of taxa. Taxon W, comprising three species, is monophyletic because it contains all descendants (C and D) and their immediate common ancestor (B). Taxon X is paraphyletic because it includes an ancestor (A) but only *some* of the descendants (E through I). Taxon Y is polyphyletic because it contains taxa not derived from an immediate common ancestor; species M and P may look very much alike, as a result of evolutionary convergence or of parallelism, and therefore mistakenly have been placed together in a single taxon. Taxon Z is paraphyletic; in this case, further work on these species (J through P) should eventually reveal the correct relationships among these taxa, resulting in species M being classified with species K and L, and species P with species N and O.

(or comparative biochemistry, or comparative ethology, etc.) is the concept of homology. In a general sense, similarities due to descent from a common ancestor are called **homologies**. More specifically, in order to compare features (characters) among different organisms or groups of organisms, it must be established that the features being compared are homologous. **Homologous characters** are features that have the same evolutionary origin—the same genetic basis and the same developmental sequence. Another way of saying it is, homologues are characters that are present in two or more taxa and are traceable phylogenetically and ontogenetically to the same character in some common ancestor of those taxa. Our ability to recognize homologues usually depends upon embryological evidence and on the relative structural position of the feature in adults (see Chapter 4). Attempts to relate two taxa by comparison of nonhomologous characters results in error. For example, the hands of chimpanzees and humans are homologous characters because they have the same evolutionary and developmental origin; the wings of bats and butterflies, although similar in some ways, are not homologous characters because they had completely different origins. The process of evolutionary descent with modification has produced a hierarchical pattern of appearances of homologies (often referred to as adaptations) that can be traced through lineages of living organisms.

It should be noted that the concept of homology has nothing to do, in the strict sense, with similarity or degree of resemblance; it is associated only with whether or not specific characters share a common genetic and phylogenetic origin. Some homologous features look very different in different taxa (e.g., pectoral fins of whales and arms of humans; hair of mammals and feathers of birds; the forewings of beetles and flies). Furthermore, through the phenomenon of **convergent evolution**, similar-appearing structures may evolve in entirely unrelated groups of organisms; for example, the vertebrate eye and the cephalopod eye; the bivalve shells of molluscs and of brachiopods; and the sucking mouthparts of true bugs (Hemiptera) and of mosquitoes (Diptera). Structures such as these, which appear superficially similar but have arisen more than once and have separate genetic and phylogenetic origins, are called convergent characters. Failure to recognize convergences among different groups of organisms has led to the creation of many "unnatural," or polyphyletic, taxa in the past.

Convergence is often confused with **parallelism**. Parallel characters are similar structures that have arisen more than once in different species or species groups within a single extended lineage, and that share a common genetic and developmental basis (Figure 3). Parallel evolution is the result of "distant" or "underlying" homology; for parallel evolution to occur, the genetic potential for certain features must persist within a group, thus allowing the feature to appear and reappear in various taxa. Parallelism is commonly encountered in characters of morphological "reduction," such as reduction in the number of segments, spines, fin rays, and so on in many different kinds of animals.

When comparing homologues among species,

one quickly sees that variation in expression of the character is the rule rather than the exception. The various conditions of a homologous character are often referred to as the **character states**.* A character may have only two contrasting states, or it may have several different states within a taxon. **Polymorphic species** show a significant range in phenotypic variability as a result of the presence of numerous character states for the various features being examined. A simple example is hair color in humans: black, brown, red, and blond are all states of the character "hair color."

Not only can characters vary within a species, but they typically can have several states among groups of species within a higher taxon, such as patterns of body hair among various primates or the spine patterns on the legs of crustaceans. It is important to understand that a character may be thought of as a hypothesis—that two attributes that appear different in different organisms are nonetheless simply alternative states of the same feature (i.e., they are homologues). Recognition and selection of proper characters in biological systematics is clearly of primary importance, and a great deal has been written on this subject (see the references at the end of this chapter). Systematics is, to a very great extent, a search for the defining homologues of various evolutionary lineages.

Another important tool in systematics and comparative biology is the dendrogram. A **dendrogram** is a branching diagram, or tree, depicting relationships among groups of organisms. It is a visual means of expressing one view of the relationships among species or other taxa. Most dendrograms are intended to depict evolutionary relationships, with the trunk representing the oldest (earliest) ancestors and the branches indicating successively more recent divisions of various evolutionary lineages. But dendrograms can be constructed with different goals in mind. The traditional dendrograms drawn by biologists are called **evolutionary trees**, and are meant to depict all ideas concerning the evolution of the organisms in question. Such trees usually have (at least implied) a time component as the vertical axis, and often some measurement of divergence as the horizontal axis. Three examples of evolutionary trees are given in Figure 4. Recall from our earlier discussion that classification schemes are ultimately derived

---

*In practical usage the terms *character* and *character state* are often used interchangeably when comparing two or more species. This practice can be a bit confusing. When the term *character* is used in a discussion of two or more homologues, it is being used in the same sense as *character states*.

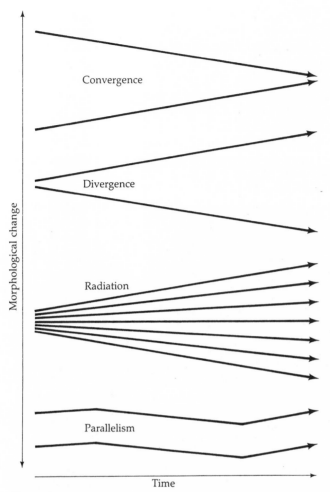

Figure 3

Some common patterns of evolution displayed by independent lineages. Convergence occurs when two or more lineages (or characters) evolve toward a similar state. Divergence occurs when two lineages (or characters) evolve independently and become less similar. Convergence generally refers to unrelated (or very distantly related) taxa and to characters sharing no common phylogenetic or ontogenetic basis. Radiations are multiple divergences from a common ancestor (involving more than two lineages). Parallel evolution occurs when two or more lineages change similarly so that despite evolutionary activity they remain similar in some ways, or become more similar over time. Parallelism usually refers to closely related taxa, usually species, within which the characters or structures in question share a common genetic (phylogenetic or ontogenetic) basis. (After Ayala and Valentine 1979.)

from phylogenetic trees or dendrograms of some sort. Various kinds of dendrograms are discussed in further detail in succeeding pages, and they also appear throughout this book to provide the reader with current theories on the evolution of various invertebrate taxa.

When examining dendrograms and classifica-

tions derived from them, it is important to understand the concept of grades and clades. As depicted in Figure 5, a **clade** is a monophyletic group or branch, which may undergo very little or a great deal of diversification. A clade is a group of species related by direct descent. A **grade**, on the other hand, is a group of species defined by somewhat more abstract measures. In fact, it is a group of species defined by the level of functional and morphological complexity achieved by all the species in the group. Thus, a grade can be polyphyletic, paraphyletic, or monophyletic (in the latter case, it is also a clade). A good example of a grade is the large group of gastropod taxa that achieved shellessness. These slugs, however, do not constitute a clade, because this anatomical stage was achieved in several different lineages; thus slugs are a polyphyletic group. The pseudocoelomates probably are another example of a polyphyletic grade. An example of a monophyletic grade is the subphylum Vertebrata.

One last concept important to our understanding of systematics and evolutionary biology is that of primitive versus derived character states. **Primitive character states** are attributes of a taxon that are relatively "old" and have been retained from some remote ancestor; in other words, that have been around for a long time, geologically speaking. Character states of this kind are often referred to as "ancestral." **Derived character states**, on the other hand, are those attributes of a taxon that are of relatively recent origin—often called "advanced" character states. As mentioned in Chapter 1, the gradual, or not so gradual, accumulation of changes that takes place in the evolution of a group or lineage (an ancestor–descendant line) is called **anagenesis**. Hence, primitive character states of a taxon represent those features that evolved early in an anagenetic series; derived (or advanced) character states are those features that evolved more recently in an anagenetic series. Within the phylum Chordata, for example, the possession of hair, milk glands, three middle-ear bones, and a placenta are derived character states whose evolutionary appearance marked the origin of the mammals (thus distinguishing them from all other chordates). Within a subset of the Mammalia, however, such as the primates, these same features represent retained primitive features, whereas possession of an opposable thumb is a defining derived trait.

It should be apparent from the preceding paragraph that the designations *primitive* and *derived* are relative, and that any given character state or attribute can be viewed *either* as primitive or as derived, depending on the level of the phylogenetic tree or

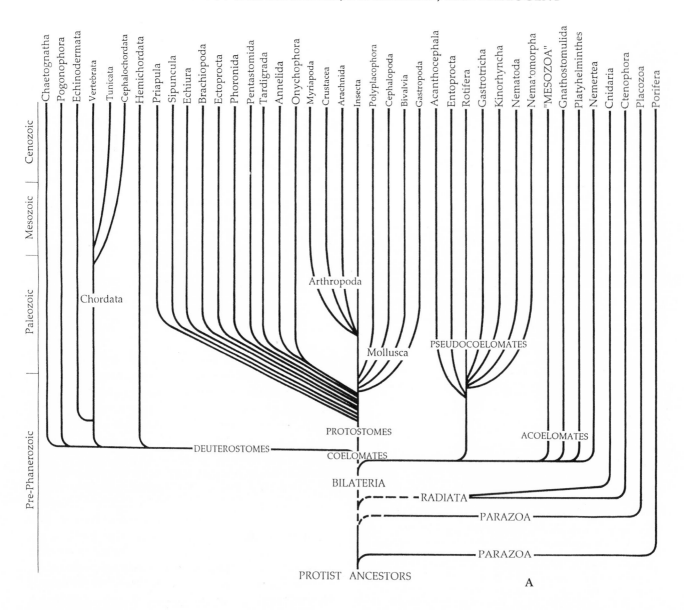

A

classification being examined. The opposable thumb may be a derived trait defining primates within the mammal lineage, but it is not a derived character state if one refers only to the genus *Homo*, in which case certain features of the nervous system that distinguish humans from the "lower apes" would be considered derived (such as Broca's center in the human brain). The most precise and unambiguous way to describe and use this important concept of primitive and derived character states is to define the exact place in the history of a group of organisms at which a character actually undergoes the evolutionary (anagenetic) transformation from one state to another. At the specific point on a phylogenetic tree where such a transformation takes place the new (derived)

character state is said to be an **apomorphy** and the former (primitive) state, a **plesiomorphy**. Use of these terms thus implies a precise phylogenetic placement of the character in question, and this placement constitutes a testable phylogenetic hypothesis in itself.

Current schools of thought in classification methodology

From what you have read so far in this chapter, it should be evident that comparative biologists spend a great deal of their time seeking to identify and unambiguously define monophyletic groups.

Figure 4

Three types of dendrograms, which are called evolutionary trees and depict phylogeny among the invertebrates. (A after Margulis 1981; B from Barnes 1969, *Invertebrate Zoology*, 2nd Ed., Saunders, as derived from Hadzi 1963; C after Hyman 1940.)

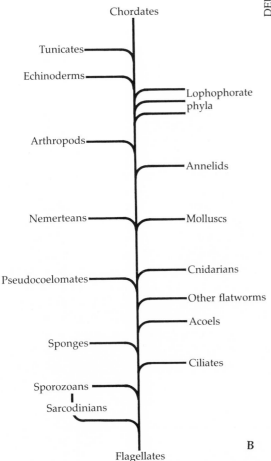

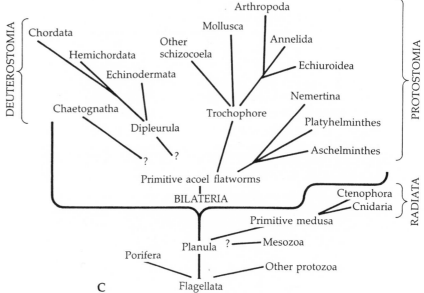

These biologists may present their ideas on such matters of relationship either in trees, classifications, or narrative discussions (evolutionary scenarios). In all cases, these presentations represent sets of evolutionary hypotheses—hypotheses of common ancestry (or ancestor–descendant relationships). The least ambiguous (most testable) way to present evolutionary hypotheses is in the form of a dendrogram, or branching tree. Although classification schemes are ultimately derived from such dendrograms, they do not always reflect precisely the arrangement of natural groups in the tree. There are three basic kinds of dendrograms that can be constructed: phenograms, cladograms, and evolutionary trees. These different kinds of trees epitomize the three prevailing

Figure 5

Clades and grades. Clades are monophyletic branches that may undergo various degrees of diversification. Grades are groups of animals classified together on the basis of levels of functional or morphological complexity. Grades may be monophyletic, paraphyletic, or polyphyletic. In this figure, grade I is monophyletic, encompassing only a single clade (clade 3); grade II is polyphyletic, because the associated level of complexity has been achieved independently by two separate lineages, clades 1 and 2. (After Ayala and Valentine 1979.)

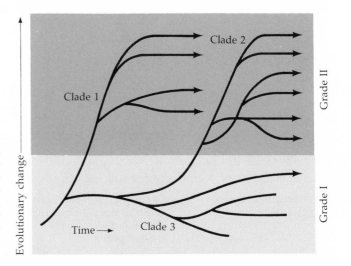

philosophies of biological systematics recognized today: **numerical phenetics** (= numerical taxonomy), **cladistics** (= phylogenetic taxonomy), and **orthodox**, or **"evolutionary" taxonomy**.

Many systematists would dislike being categorized as a member of only one of these three schools, as many modern workers utilize an eclectic approach in their studies. But because this categorization provides an excellent framework for a discussion of modern classification theory, these three schools are characterized separately below. Most of you encountered the general philosophy of "orthodox systematics" in freshman biology or general zoology, but the other two approaches are newer and perhaps less familiar to you. Table 1 gives a concise comparison of the three systematic schools.

Numerical phenetics

Phenetics, also referred to as numerical taxonomy, developed as an outgrowth of the increased availability of computers in the late 1950s. The philosophical basis for numerical phenetics is the argument that it is never possible to know with certainty which of various competing phylogenies is the correct one—the real description of what happened in the evolution of any particular group. Hence, rather than attempt to establish classifications based on hypothetical reconstructions of the phylogenetic history of a group of animals or plants, organisms should be classified strictly for the sake of convenience, like books in a library. Pheneticists claim that if one avoids all considerations of the evolution of taxa and simply measures as many characters as possible within the taxa under consideration, one can generate classifications based on overall similarity that will be the most utilitarian systems devisable.

Reasoning further, they claim that because phenetic similarity (whether it be based on morphological features, protein chemistry, or whatever) is a reflection of genetic similarity, then a large sample of randomly chosen phenotypic characters should represent a large sample of the genome. If each character is given equal weight, the sums of the differences and similarities between taxa should serve as the best possible measure of genetic (= evolutionary) distance. In an attempt to be wholly objective, pheneticists avoid the subjective evaluations that systematists have traditionally relied on to classify organisms. By analyzing as many phenotypic characters as possible (the more the better) and giving them all equal weight, one can secondarily construct a scheme depicting theoretical genetic relatedness. To some biologists, this reasoning appears contradictory, for it begins with the premise that one cannot hope to discover the evolutionary history of taxa but ends by stating that numerically derived classifications can actually depict patterns of genetic (thus, evolutionary) relatedness.

Table One

Comparison of classification methods

ATTRIBUTE	PHENETICS	CLADISTICS	ORTHODOX SYSTEMATICS
Relationships depicted by tree or classification	Overall similarity or dissimilarity	Genealogy	Genealogy + overall similarity/dissimilarity (genetic relatedness)
Evolutionary similarity[a]	All kinds used	Apomorphies only	All kinds used
Character weighting	Not used	Generally not used	Used
Homology	Not considered	Of primary importance	Important
Fossils	Not used	May be considered, but of no greater importance than living species	May be very important
Ecological and "evolutionary" data	Not used	Rarely used	May be very important
Rates of evolution	Not considered	Not considered	Very important
Transformation of tree into classification	No overall rules; arbitrary levels of overall similarity/dissimilarity chosen to delimit taxa	Classification precisely depicts branching pattern on cladogram	Classification reflects both branching pattern and degree of difference between taxa

[a]Apomorphies, plesiomorphies, convergences, parallelisms, evolutionary reversals.

Pheneticists call their groups of similar-appearing organisms—regardless of the group size (a group can even consist of a single individual)—operational taxonomic units, or OTUs. Their procedures tend to cluster taxa by absences (negative occurrences) or by "degree of difference" rather than by the presence of shared similarities. Critics view these techniques of grouping taxa as one of the principal weaknesses of the phenetic approach. Pheneticists do not adhere to the phylogenetic definition of homology discussed earlier in this chapter, nor do they assess characters as being primitive or derived. They generally claim that any such weighting or evolutionary assessment introduces bias. Critics argue that evolutionary studies have shown that characters do have highly different values at different hierarchical levels, depending on whether they are derived or primitive, are homologous or nonhomologous, represent convergent or parallel evolution, are evolving rapidly or slowly, and so on.

Pheneticists analyze data by one or more computer programs that cluster OTUs on indices of overall similarity or difference and produce a branching diagram (dendrogram) called a **phenogram** (Figure 6). The phenogram can then be transformed into a classification. This transformation requires considerable subjectivity, because there are no clear guidelines for determining what level of similarity (or "differentness") is sufficient to establish categorical levels in a classification.

During the 1960s, many scientists were attracted to numerical phenetics; but by the mid-1970s its popularity had greatly declined. First, hundreds of computer programs have been devised by pheneticists, but when the same set of data is analyzed by different programs, different phenograms may be produced. Second, so much manipulation of the data takes place in many programs that the stated goal of producing an unbiased analysis may be completely abandoned. Third, phenograms tend to generate polyphyletic taxa, because they do not account for character convergence or consider evolutionary homologies. Fourth, many working systematists, even some of those using these numerical techniques, failed to understand exactly what it was that the computer was doing with their data (the "black box syndrome"). During the 1960s and 1970s, the number of programs available to numerical taxonomists multiplied enormously, but many were designed and written by people whose real areas of expertise were mathematics and computer programming, not evolutionary biology.

Finally, pheneticists have attempted to convert the totality of the similarity/difference values of individual characters into a single "phenetic distance" or "overall similarity" value; this process not only leads to an enormous loss of information as a result of the oversimplification of biological reality, but also ignores the fundamental basis of homology. Overall similarity or percentage difference, of course, have nothing to do with homology per se and mean nothing in a genealogical context; only specific homologies indicate genealogical relationships. And, homology either does or does not exist—it does not come in degrees. Phenetic biochemists are especially fond of using the term "sequence homology" for sequences of base pairs (in a gene) that are similar between two species; but what they are really referring to is "overall sequence similarity," which has nothing to do with homology or genealogical relationships. Most gene sequence trees are generated by computer algorithms that literally count the number of *differences* between taxa, placing those with the most differences the greatest distance apart, those with the least differences closest together, and so on. There seems to be no logical reason to assume these kinds of trees might reflect genealogical relationships. If "numbers of differences" were a measure of genealogical relatedness, crocodiles and birds would probably be classified in different phyla, when in fact they are very closely related!

Perhaps the best summary statement we have seen describing the weaknesses of phenetic taxonomy is that of G. G. Simpson. He stated that the basic fallacy of phenetic taxonomy is the attempt to equate living organisms with inanimate objects. The real fact is that "members of a taxon are similar because they share a common heritage; they do not belong to the same taxon because they are similar," just as "two brothers are not identical twins because they look alike, but because they are both derived from the same zygote." Simpson's argument is really a plea for genealogical classification. Another illustrative example is that of cousins and brothers. Two cousins may look more alike (be more "similar") than either resembles his own brother. But knowledge of their genealogies tells us that the brothers are more closely related to one another than either is to his cousin.

Cladism

Cladistic systematics (= cladism; = phylogenetic systematics) had its origin in 1950 in a German textbook by Willi Hennig; the English translation (with revisions) appeared in 1966. Its popularity has grown steadily since that time. For a thorough discussion of current ideas in cladistic systematics see Nelson and Platnick (1981), Eldredge and Cracraft (1980), and

Figure 6

Examples of dendrograms (phenograms) derived by the technique of numerical phenetics. A, Phenogram of selected beetle genera used to illuminate relationships among two families, the ground beetles (Carabidae) and tiger beetles (Cicindelidae). The phenogram is based on correlation coefficient values of immunological data for 11 species of ground and tiger beetles, as well as 9 species from 6 other families. Abscissa values are correlation coefficients. B, Phenogram of 54 species of stony corals known from the Caribbean and Hawaii, derived from coefficient of correlation analysis of 60 characters. In this somewhat unconventional analysis, ecological attributes of the various species were included among the 60 characters analyzed (e.g., bathymetric range, relative dominance on reefs, found mostly on windward or leeward side of reefs). (A after Basford et al. 1968, Syst. Zool. 17: 388–406; B after Powers and Rohlf 1972, Syst. Zool. 21: 53–64.)

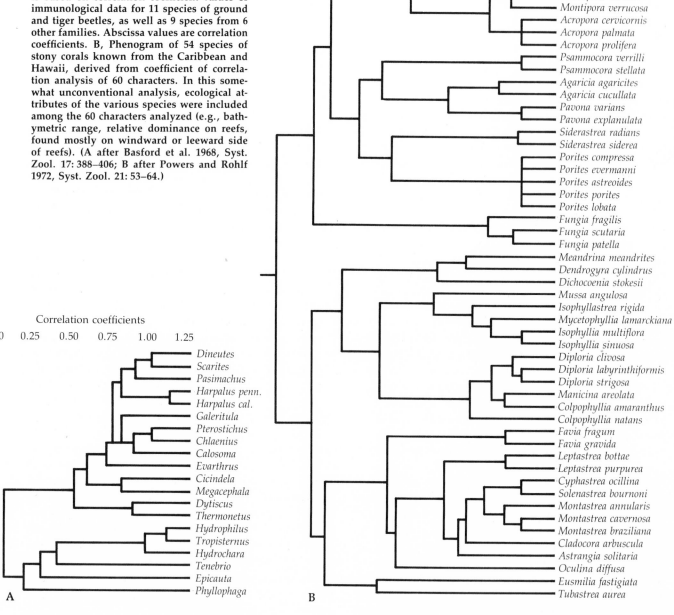

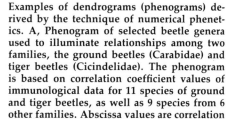

especially Wiley (1981). Through the years, cladism has evolved well beyond the framework Hennig originally proposed. Its detailed methodology has been expanded and formalized and will probably continue to be elaborated for some time to come.

The goal of cladistics is to produce *testable* hypotheses of genealogical relationships among mon-ophyletic groups of organisms. As a systematic methodology, cladism is based entirely on *recency of common descent* (i.e., strict genealogy). The dendrograms used by cladists are called **cladograms**, and they are constructed to depict only genealogy, or ancestor–descendant relationships. The term **clado-genesis** refers to splitting; in the case of biology, this means the splitting of one species into two or more species. It is this splitting process that produces genealogical (ancestor–descendant) relationships.

Cladists rely heavily on the concept of primitive versus derived character states (discussed earlier in this chapter). They identify these character homologies in the strict sense as plesiomorphies and apomorphies. Apomorphies restricted to a single species may be referred to as **autapomorphies**, whereas apomorphic character states that are shared between two or more species (or other taxa) are called **synapomorphies**. Identifying synapomorphies, or shared derived characters, constitutes the cladists' most powerful means of recognizing close evolutionary (genealogical) relationship. Because synapomorphies are shared homologies inherited from an immediate common ancestor, all homologues may be considered synapomorphies at one (but only one) level of phylogenetic relationship, and they therefore constitute **symplesiomorphies** at all lower levels. For example, the appearance of hair, milk glands, and so on are synapomorphies uniquely defining the appearance of the mammals, but these are symplesiomorphies *within* the group Mammalia.

Perhaps a more relevant example of the plesiomorphy/apomorphy concept is the excretory system of many invertebrates. The protonephridial excretory system of acoelomate and "pseudocoelomate" animals is generally considered to be homologous with the metanephridial system (and its derivatives) of at least some coelomate animals. But it is hypothesized that the metanephridial system arose from the protonephridial system. Thus, the former is the synapomorphic condition, the latter the symplesiomorphic condition. When no synapomorphy is shared with any other taxa, no cladistic (genealogical) hypotheses can exist. For example, a new group of animals, the Loricifera, was recognized in 1983 based on a single species and had to be given a separate phylum status, because it appears to share no *unique* derived characters (synapomorphies) with any other known phylum of animals (see Chapter 12).

Numerous techniques or criteria have been used to determine which is the apomorphic and which is the plesiomorphic form of two character states—a process referred to as **character state polarity analysis**. No method is foolproof but some may be better than others under specific circumstances. Only three of the dozen or so techniques appear to have both a strong evolutionary basis *and* a reasonably powerful means for recognizing the relative place of origin of a synapomorphy on a tree: out-group analysis (seeking clues to ancestral character states in groups "known" to be more primitive than the study group), developmental studies (ontogenetic analysis, seeking clues to ancestral conditions in the embryogeny or ontogeny of the group), and study of the fossil record. Out-group analysis seeks to identify the states of the characters in question in closely related taxa outside the study taxon. Ontogenetic analysis seeks to identify character changes that occur during the development of a species (see the discussion of ontogeny and phylogeny in Chapter 4). The use of fossils and associated dating and stratigraphic techniques provides some direct historical information. However, the fossil record is very incomplete, and such fragmentary data can be misleading. These techniques of polarity analysis are not discussed here; we refer those of you with a serious interest in systematics, evolution, and comparative biology to the references at end of this chapter. We must point out that for some groups, determination of which are primitive and which are advanced character states is very difficult indeed.

A cladistic analysis consists of (1) identifying homologous characters among the organisms being studied; (2) assessing the direction of character change or character "evolution" (character state polarity analysis); and (3) constructing a cladogram of the taxa possessing the characters analyzed. The cladogram depicts only one kind of event—the origins or sequence of appearances of unique derived character states (synapomorphies). Hence, cladograms may be thought of in the most fundamental way simply as synapomorphy patterns or, more specifically, as dendrograms depicting the pattern of shared similarities thought to be evolutionary novelties among taxa. However, biologists generally recognize and define taxa by the traits they possess. Thus, in a larger sense, the sequential branching of nested sets of evolutionary novelties (the derived character states or synapomorphies) in a cladogram creates a "family tree"—an evolutionary pattern of hypothesized monophyletic lineages.

Construction of a cladogram can be a time-consuming process. The number of mathematically possible cladograms for more than a few species is enormous—for 3 taxa there are only 4 possible cladograms, but for 10 taxa there are about 280 million possible cladograms. Needless to say, a thorough analysis of a family of several dozen species is an overwhelming prospect unless a computer is used. Algorithms for computer-assisted cladogram construction began appearing in the late 1970s. These programs generate trees (cladograms) by clustering taxa on the basis of nested sets of synapomorphies.

Cladograms, like other dendrograms, represent sets of hypotheses or predictions regarding evolutionary relationships (in this case, common ancestry relationships). Unlike other kinds of dendrograms, however, the cladogram is based on only one specific kind of characteristic, synapomorphies. In identifying the precise place at which a character state transformation occurs (that is, where a synapomorphy appears), cladograms unambiguously define monophyletic lineages. The synapomorphies are "markers" that identify specific places in the tree where new taxa arise. This restriction results in less ambiguity, greater precision, and hence greater testability than that of traditional or orthodox evolutionary trees.

As new homologues or new species are identified and their character states elucidated, they are compared with character hierarchies in existing cladograms. In this fashion, the various hypotheses in a cladogram are tested. New data can be used to generate new cladograms to compare with previous ones. Hypotheses (trees) consistently resisting refutation are said to be highly corroborated. Obviously, the more "testable" a hypothesis is, the better it is (in a scientific sense). Cladists recognize that all homologous characters are relevant to defining monophyletic groups at some level. The problem, of course, is recognizing the correct level at which each character state should be employed, that is, at what level the character becomes a synapomorphy.

The final step in a cladistic analysis may be the conversion of the cladogram into a classification scheme. Here lies another fundamental difference between cladists and traditional (orthodox) systematists. For the cladists, phylogeny consists of a genealogical branching pattern expressed in a cladogram. As envisioned on the cladogram, each split or dichotomy produces a pair of newly derived taxa called **sister-taxa**, or **sister-groups** (for example, sister-species) (Figure 7). Cladists generally convert their cladograms *directly* into classifications strictly on the basis of the branching sequence depicted. They use only as much information for the construction of the classification as is contained in the cladogram. Thus, strict cladists would erect classifications based solely on genealogy. Cladists give no taxonomic consideration to the *degree of difference* between taxa (i.e., the number and kinds of characters used to separate taxa), or to differential rates of change in various groups, or to evolutionary events other than those involving the origin of new apomorphies. Figure 8 shows a dendrogram (an evolutionary tree *or* a cladogram) that is widely accepted by both cladists and orthodox taxonomists. In the two different clas-

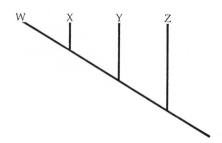

Figure 7

A cladogram of four taxa, illustrating the concept of sister-groups. Taxon W is the sister-group of taxon X; taxon W + X is the sister-group of taxon Y; taxon W + X + Y is the sister-group of taxon Z. See text for further discussion.

sifications derived from this dendrogram, the orthodox systematists have recognized a hierarchical arrangement of the taxa different from that of the cladists. This difference is due entirely to the fact that the orthodox view considers the overall degree of difference between taxa (viewed as a result of differing evolutionary rates), whereas the cladist has considered only the branching sequence.

As depicted on a cladogram, the product of cladogenesis (or the splitting of a taxon) is two (or more) new lineages that constitute sister-groups. Another way of stating this is to say that the two subsets of any set defined by a synapomorphy constitute sister-groups. In Figure 7, set W is the sister-group of set X; set W + X is the sister-group of set Y; and W + X + Y is the sister-group of Z.

A good example of the sister-group concept can be seen in a series of four families of flabelliferan isopod crustaceans. These four families show an evolutionary trend from free-living (the Cirolanidae) to parasitic lifestyles (the Cymothoidae). In the cladogram in Figure 9, the Cymothoidae (a family of isopods that are obligatory parasites on fishes) is the sister-group of Aegidae (a family of "temporary" fish parasites); together they constitute a sister-group of the Corallanidae ("micropredators" on fishes); and all three constitute a sister-group of the Cirolanidae (predatory scavenging isopods). Each of these sister-group pairs shares one or more unique synapomorphies that define them. In Figure 9, the synapomorphies that define the sister-groups Cirolanidae and Corallanidae + Aegidae + Cymothoidae become symplesiomorphies at lower levels in the cladogram (i.e., for each of the separate families). Sister-groups are monophyletic by definition.

As illustrated in Figure 9, some cladists early on suggested that every sister-group must be designated by a formal name and categorical rank; and each

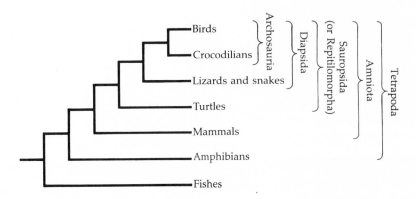

Classification scheme A

    Phylum Chordata
        Subphylum Vertebrata
            Class Pisces (fishes)
            Class Amphibia (amphibians)
            Class Reptilia (turtles, crocodilians, snakes,
                    lizards, *Sphenodon*)
            Class Aves (birds)
            Class Mammalia (mammals)

Classification scheme B

    Phylum Chordata
        Vertebrata
            Tetrapoda
                Lissamphibia (amphibians)
                Amniota
                    Mammalia (mammals)
                  Reptilomorpha
                    Anapsida (turtles)
                    Diapsida
                      Lepidosaura (snakes, lizards, *Sphenodon*)
                    Archosauria
                        Crocodilia (crocodilians)
                      Aves (birds)

Figure 8

Sometimes genealogy and overall similarity/dissimilarity seem to lead to conflicting conclusions about relationships. The conflict between cladists (genealogical priority) and orthodox systematists (phenetic similarity/dissimilarity + genealogy) is well exemplified in the case of the birds and reptiles. The cladogram in this figure depicts the generally accepted relationships of the major groups of living vertebrates. Classification scheme A depicts a traditional, or "orthodox," classification of the tetrapods (four-limbed vertebrates) in which crocodilians are classified with lizards, snakes, and turtles in the taxon Reptilia, while birds are retained as a separate taxon Aves. Orthodox systematists, in their desire to express both branching patterns and degree of dissimilarity in classifications, are willing to accept such paraphyletic taxa in order to classify birds and reptiles in separate groups. Classification scheme B, which omits categorical names for simplicity's sake, strictly reflects the branching pattern of the cladogram; thus, the "reptiles" are broken into separate taxa in recognition of their genealogical relationships, and the birds and crocodilians are classified together as a sister-group (called "Archosauria" in this scheme). In scheme B all taxa are monophyletic. The hierarchical cladistic grouping of the tetrapods (and their associated names) are indicated on the cladogram.

member of a sister-group pair must be of the *same* categorical rank, no matter how different they may appear phenotypically. A moment's thought reveals that giving names to every branching point in a cladogram would result in an enormous proliferation of names and ranks. Other cladists have proposed a method of avoiding such name proliferation, called the phylogenetic sequencing convention. When the sequencing convention is used, *linear* sequences of taxa branching off a main stem or other line can all be given equal categorical designation (i.e., they can all be classified as genera, or all as families, and so on) so long as they are listed in the classification scheme in the precise sequence in which the branches appear on the cladogram. Thus, either method of creating a classification scheme allows one to convert the classification directly back into a cladogram, that is, to visualize the genealogical branching pattern (e.g., Figure 9).

One of the most illustrative examples of the difference of opinion between cladists and orthodox taxonomists regarding the categorical ranking of sister-groups is the case of the crocodilians and the birds, which are probably more recently descended from a common ancestor than either is from an ancestor held in common with any of the other reptiles. Because of this relationship, crocodilians and birds form a sister-group to most other reptiles (see cladogram in Figure 8). That is, birds originated from the branch of reptiles that also gave rise to the crocodilians. By cladistic methodology (on genealogical grounds), birds and crocodilians should therefore be ranked together, separate from the remaining reptiles (cladists recognize such a group, calling it the Archo-

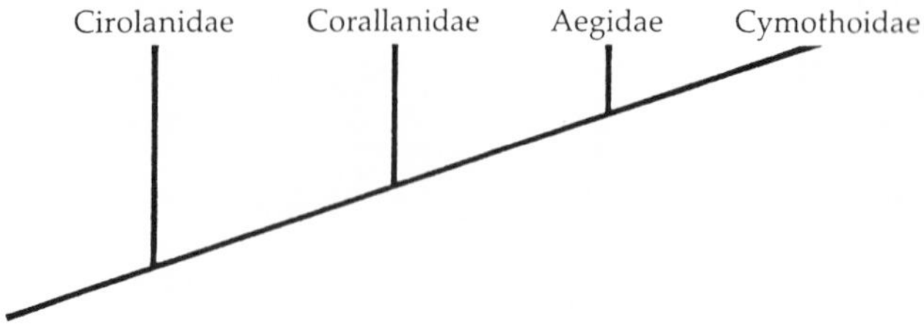

Classification scheme A

Family Aegidae
Family Cirolanidae
Family Corallanidae
Family Cymothoidae

Classification scheme B

Family Cirolanidae
Family Corallanidae
Family Aegidae
Family Cymothoidae

Classification scheme C

Family Cirolanidae
Family Corallanidae
  Subfamily Corallaninae
  Subfamily Aeginae
    Tribe Aegini
    Tribe Cymothoini

Figure 9

A dendrogram of four closely related groups (families) of isopod crustaceans (marine "pillbugs;" see Chapter 18). The dendrogram can be viewed as either a cladogram or an evolutionary tree. In this particular example, the four taxa listed constitute an interesting "evolutionary series," from the free-living predacious Cirolanidae, through facultative fish parasites (Corallanidae and Aegidae), to obligatory fish parasites (Cymothoidae). Classification scheme A views the dendrogram as an evolutionary tree and is the "traditional" classification developed by orthodox systematists and currently in use. The families can be listed in any order (here, alphabetically) in the classification, so the order of listing does not necessarily mirror their arrangement in the tree. Scheme B views the dendrogram as a cladogram and utilizes the convention of phylogenetic sequencing to arrange the taxa in the exact sequential order in which they appear on the tree. Scheme C also views the tree as a cladogram, arranging the taxa in a *subordinated (hierarchial) classification* but again depicting precisely the arrangement of the cladogram. There is no way to convert the orthodox classification of scheme A directly into the tree from which it was derived; hence phylogenetic relationships cannot be ascertained from the classification. Schemes B and C can be directly transformed into the tree from which they were derived, in that they precisely reflect the genealogical (phylogenetic) scheme. Cladists claim that the ability of cladistic classifications to reflect strict genealogical relationships makes them preferable to orthodox classifications.

sauria); or birds should be classified with "reptiles" and the definition of that group expanded to include birds (cladists also recognize this grouping, often referring to it as the Sauropsida, or Reptilomorpha). Orthodox systematists argue that even though birds and crocodilians may be "most closely related" on a genealogical basis (on a tree or cladogram), birds are genetically very different from reptiles (as reflected in their considerable morphological difference) and hence the two groups should be classified in entirely different taxa. Furthermore, orthodox systematists argue that taking all attributes into consideration, the crocodilians are clearly members of the reptilian grade, whereas the birds have evolved many new attributes. In other words, the crocodilians have retained more primitive reptilian features (symplesiomorphies) than have the birds, and for this reason crocodilians should be classified with the other reptiles, not with the birds.

One oft-heard criticism of cladism is that it depicts the speciation process only as the splitting of an ancestral species into two sister-species, despite the probability that numerous other speciation models exist. Several of these alternative speciation models are illustrated in Figure 10. From the cladistic viewpoint, once a new species appears, a "split" must be placed on a cladogram; and the two (or more) lines represent sister-groups whether or not the original species has in fact "changed" at all. Acceptance of this working assumption is necessary to facilitate the methods of cladogram construction. Many evolutionary systematists claim that this is an unrealistic evolutionary assumption and therefore an inappropriate methodology. This criticism is unfounded, however, and derives more from simple lack of understanding than from anything else. Cladograms are not always completely dichotomous; they can have branching points that are trichotomous or even polytomous. Furthermore, terminal taxa on a cladogram may lack any defining synapomorphies (in fact, they often do), thus indicating that such a taxon not only may be the sister-group of its adjacent lineage, but also may contain the actual ancestor of that lineage. The cladogram of annelids (see Figure 38A in Chapter 13) is an example of this. The oligochaetes (earthworms and their kin) lack any unique defining synapomorphies, so they are the direct ancestors of the hirudinidans (the leeches and their kin)—that is, leeches probably evolved from an oligochaetous ancestor. A cladogram can express any kind of speciation event; it simply does so in a restricted way (by way of branches depicting a synapomorphy pattern).

Cladists claim that their technique is superior because it is the only method that is scientific in the strict sense, namely, that it utilizes the general principles of prediction, analysis, and testing. The cladistic method forces the systematist to be explicit about groups and character analysis. The method is also largely independent of the biases of the discipline in which it is applied. In its fundamental principles, it

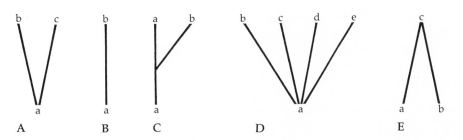

Figure 10

Common models of speciation. A, One species splits into two new species. B, One species is transformed into another. This type of speciation may be viewed as either gradual or rapid. C, One species remains unchanged while an isolated peripheral population evolves into a distinct new species. This model probably represents for most evolutionists the most common mode of speciation. D, A model of "explosive radiation," in which one species suddenly splits into many new species. The speciation events represented by this model are predicted to occur when a species is suddenly confronted with a vast new array of habitats or "unfilled niches" to exploit, a circumstance resulting in rapid specialization and reproductive isolation as the new niches are filled. Explosive radiation might also occur when the range of a widespread species is fragmented into numerous smaller, isolated populations. E, A new species is "created" by hybridization of two other species; this type of speciation probably occurs almost exclusively in plants.

is not restricted to biology but is applicable to a variety of fields in which the relations that characterize groups are comparable to the homology concept and possess a hierarchical nature. Thus, cladistic analyses have been applied to linguistics and textual criticism (where the "homologues" are shared tongues or texts), and even to the classification of musical instruments. It is also used in biogeographical analyses (in this case the "homologues" are sister-groups shared by areas). Although the information stored in a cladogram is restricted to genealogy, such trees are often used to test other kinds of hypotheses, such as modes of speciation, the historical relationships among geographical areas, and coevolution in host–parasite lineages.

As mentioned earlier, the concept of "similarity" plays a central role in much of the debate between cladists and noncladists. However, there are really only three kinds of evolutionary similarities expressed among organisms: (1) shared evolutionary novelties inherited from an immediate common ancestor; (2) similarity inherited from some more remote ancestor (any number of descendant taxa may retain the similarity); and (3) similarities due to evolutionary convergence. Cladists accept only the first kind of similarity (synapomorphy) as valid evidence of close affinity (common ancestry) between two taxa. Orthodox systematists also rely heavily on synapomorphies, but consider the second kind of similarity (symplesiomorphy) in their analyses as well. They also use "degree of difference" (i.e., the number and kinds of synapomorphies distinguishing a lineage) to classify organisms. Pheneticists give no particular emphasis to (1) or (2), and may even include the third (convergence) in their analyses simply by not considering evolutionary processes and homology during character analysis prior to tree construction.

Orthodox (= "evolutionary") taxonomy

The term *evolutionary taxonomist* is a bit unfortunate because it suggests that the use of evolutionary theory is unique to this particular philosophy of classification. Hence, perhaps the phrase *orthodox taxonomy* is more appropriate. Orthodox taxonomists are generally regarded as "traditional" taxonomists, and their views have been championed over the years by a great many followers (for a good overview of evolutionary systematics, see Mayr 1969). Orthodox classification theory developed slowly. Beginning with Linnaeus, it went through several periods of rapid growth and change, such as Darwin and Wallace's theory of evolution by natural selection in the 1850s, and the emergence of the "new synthesis" in the 1940s. The latter approach is based largely upon the development of population genetics and the subsequent shift away from typological approaches in taxonomy toward recognition of important population-level phenomena (e.g., species polymorphism). Most existing classifications recognized today are the product of 150 years of gradual refinement by orthodox systematists.

The basic hypothesis or premise by which orthodox systematists carry out their work is the general theory of evolution by natural selection. The fundamental assumptions are that speciation involves two independent processes: (1) the acquisition of reproductive isolation, and (2) the acquisition of niche differences, which leads to character divergence. Orthodox systematists point out that character gaps exist between all taxa, and differences in the sizes of the gaps are the result of differential rates of evolution and/or the extinction of intermediate forms. Orthodox systematists attempt to "measure" character gaps in lineages, thus emphasizing overall genetic distance to produce their classifications. Perceived differential rates in evolution are emphasized in such classifications. But cladists and pheneticists pay little or no attention to evolutionary rates; so identical dendrograms (the phenograms of pheneticists, the cladograms of cladists, the evolutionary trees of orthodox systematists) may be transformed into quite different classifications by practitioners of these three competing schools. Unlike pheneticists, orthodox systematists routinely weight characters from an "evolutionary point of view." This implies some kind of value judgment by the individual specialist regarding how characters are to be used (or not used). For example, use of features that are "conservative" (i.e., appear to evolve slowly) may be favored over characters thought to be more evolutionarily labile.

Orthodox taxonomists emphasize that reproductive isolation and character divergence do not necessarily take place at the same time, and these two processes may not be linked in any strong or general way. For example, when massive rapid splitting of a lineage occurs, phenotypic character divergence in a clade may be relatively small. Note, for example, the 75,000 species of weevils (Curculionidae), or the thousands of species of *Drosophila* that all look very much alike. In these cases, there has been little impact on the basic morphology despite the enormous amount of new reproductive isolation. There has also been no extensive shift into new adaptive zones in these cases. To the orthodox systematist, such examples imply that much more than genealogy must be considered when attempting to reconstruct the evolutionary history of a taxon, and it is for this reason that proponents of this school consider all aspects of taxa in construction of their classifications (e.g., ecology and distribution, as well as morphology).

Orthodox taxonomists claim that extensive shifts in character and niche warrant recognition of new and distinct categorical rankings (e.g., the birds). Thus, they construct classifications that express both genealogy *and* degree of divergence, whereas cladists rely solely on genealogy in classification construction. The classification schemes of orthodox systematists depict as closely as possible all aspects of the evolution of the group in question as envisioned by their authors; thus they usually do not precisely reflect genealogy.

Orthodox systematics developed gradually, out of the Linnaean tradition, but specific methodologies or rules that all proponents of the system utilize for creating classifications were never formulated. Rather than being based on a clearly expressed methodological underpinning, much of the process is steeped in intuition, or based on ambiguously expressed hypotheses of evolutionary processes. For these reasons, trees produced by this method do not lend themselves well to testing or falsification.

Biological classification: A new synthesis?

We have discussed the three principal philosophies of classification, pointing out the major attributes of each, areas where they are in conflict, and some of the most often heard criticisms of each. It is important to keep in mind that these are philosophies of classification methodology, not of evolution. In fact, many cladistic and phenetic taxonomists will argue that the process of constructing a cladogram is one of pattern analysis alone—although the pattern is presumed to be the product of the evolutionary process, reconstructing the pattern requires no prior knowledge of evolutionary mechanisms, only acceptance that evolution has occurred and is expressed in the characters that organisms possess. However, this claim ignores the fact that one cannot identify homologies or determine character state polarity without a framework of evolutionary hypotheses or assumptions.

Although only a handful of systematists apply themselves wholly to just one of the above philosophies, it is important to recognize the attributes each has to offer and to reflect on the arguments for and against each philosophy. Many systematists utilize an eclectic combination of two or three methodologies in their work. Were we to inquire of taxonomists what particular combination or recipe they favored in their work, most would very likely tell us, "The combination that best fits the taxon being studied at the particular time." This is a reasonable answer. As Mayr (1981) pointed out, "the more moderate representatives [of the three competing schools] have quietly incorporated some of the criteria of the opposing schools, so that the differences among them have been partially obliterated."

For the past 20 years or so, taxonomists have indeed been incorporating for the most part what they felt was the best (or most suitable) of each of these methodologies. For example, while the strict school of numerical taxonomy has few followers left among the ranks of today's systematists, many of the analytical "packages" designed by the adherents of that school have found their way into fairly common use among biologists. Most workers now using these computer programs, however, do so with a sense of evolutionary perspective (e.g., applying characters prudently, utilizing only homologous characters, and using alternative analyses for comparison and corroboration). Numerical analyses based on overall similarity (i.e., cluster algorithms) have become especially popular with taxonomists using biochemical and gene sequencing techniques. It is important to keep in mind, however, that phenetic methodologies do not attempt to distinguish apomorphy from plesiomorphy, and as such they are fundamentally incompatible with the cladistic philosophy.

The concept of shared derived characters has been around for many decades, and a careful review of the work produced by the most critical systematists reveals that most were striving to delimit monophyletic taxa and construct phylogenetic trees based, as the cladists prescribe, on nested sets of synapomorphies. In a majority of taxa, however, character polarity analyses have not yet been accomplished, and in these cases taxonomists have simply done the best they could, but always with the understanding that as the group became better understood their classifications would be refined or improved. Absence of reliable data on character polarity has led to the construction of many phylogenies based, at least in part, on autapomorphies and symplesiomorphies rather than solely on synapomorphies. All of these types of data have some kind of information content and certainly should not be wholly ignored during evolutionary analyses, especially if distinct synapomorphies are few or wanting.

There has been a clear trend over the past 30 years toward redefining taxa so that they are based strictly upon "positive characters," or the possession of distinct recognizable features. Formerly recognized taxa based on "negative characters" (the absence of features) have largely been redefined and reorganized, or are simply no longer considered valid. The most obvious example of this is, of course, the "Invertebrata." The invertebrates are a group of convenience, useful for didactic purposes but no more. They are not evolutionarily related by their *lack* of a backbone—whereas vertebrates *are* related by their possession of a backbone (a vertebrate synapomorphy).

Probably the most important contribution that both the pheneticists and cladists have made to the field of systematic biology has been to force conscientious taxonomists of all persuasions to reevaluate their methods and premises, thereby creating an atmosphere of healthy self-criticism from which has emerged a new generation of systematists that takes greater care in designing classifications that are testable (and actually tested). There can be no doubt that the future of biological systematics will be an exciting one. As our present methodologies and philosophies are refined and as new tools are discovered, they will interact with our view of evolution and stimulate continued growth and improvement in the understanding of diversity and the history of life.

Selected References

Interestingly, there are no texts that present a completely unbiased treatment of the three competing schools of classification theory. For this reason, the books listed here are arranged by schools (corresponding to the discussion in this chapter), and we caution you that most are heavily biased in favor of a particular approach.

Texts and Key Papers on Cladistic Systematics

Duncan, T. and T. Stuessy (eds.). 1984. *Perspectives on the Reconstruction of Evolutionary History*. Columbia University Press, New York.

Eldredge, N. and J. Cracraft. 1980. *Phylogenetic Pattern and the Evolutionary Process: Method and Theory in Comparative Biology*. Columbia University Press, New York. [An excellent text on the theory of cladistic classification, with a good review of different kinds of trees; also presents the authors' personal views on evolutionary change and speciation—through an entirely nongenetic approach!]

Farris, J. S., A. G. Kluge and M. J. Eckardt. 1970. A numerical approach to phylogenetic systematics. Syst. Zool. 19:172–189.

Felsenstein, J. 1983. Parsimony in systematics: Biological and statistical issues. Annu. Rev. Ecol. Syst. 14:313–333.

Felsenstein, J. 1985. Phylogenies and the comparative method. Am. Nat. 126:1–25.

Funk, V. A. and D. R. Brooks (eds.). 1981. *Advances in Cladistics: Proceedings of the First Meeting of the Willi Hennig Society*. The New York Botanical Garden, Bronx, New York. [Thirteen papers presented at the above conference; for advanced students of cladistics.]

Hennig, W. 1979. *Phylogenetic Systematics*. University of Illinois Press, Urbana. [The "third edition" of Hennig's original 1950 text on cladistic classification theory; the philosophy and methodology of cladistics has changed/grown a great deal since Hennig's original ideas.]

Kluge, A. G. and J. S. Farris. 1969. Quantitative phyletics and the evolution of anurans. Syst. Zool. 18:1-32.

Meacham, C. A. and G. F. Estabrook. 1985. Character compatibility analysis. Annu. Rev. Ecol. Syst. 16:431–446. [Excellent review of cladistic procedure known as "character-compatibility analysis."]

Nelson, G. and N. Platnick. 1981. *Systematics and Biogeography. Cladistics and Vicariance*. Columbia University Press, New York. [Excellent review of the history and development of systematics and biogeography, as well as a thorough, although theoretical, treatment of cladistics and vicariance biogeography.]

Platnick, N. and V. Funk (eds.). 1983. *Advances in Cladistics: Proceedings of the Second Meeting of the Willi Hennig Society*. Columbia University Press, New York. [For the advanced student.]

Wiley, E. O. 1981. *Phylogenetics: The Theory and Practice of Phylogenetic Systematics*. Wiley, New York. [A thorough and detailed analysis of cladistic methodology; less theoretical and more operational in its approach than the Nelson/Platnick and Eldredge/Cracraft volumes on the same subject; also includes a good discussion of species and speciation concepts.]

Wiley, E. O. 1988. Vicariance biogeography. Annu. Rev. Ecol. Syst. 19:513–542.

Texts on Numerical Taxonomy

Clifford, H. T. and W. Stephenson. 1975. *An Introduction to Numerical Classification*. Academic Press, New York. [Some of the "nuts and bolts" of numerical methods of classification; deals more with statistical analyses (e.g., multivariate analysis) and computer programs to classify ecological data than with the philosophy of numerical taxonomy per se.]

Dunn, G. and B. S. Everitt. 1982. *An Introduction to Mathematical Taxonomy*. Cambridge University Press, Cambridge. [The only introductory text on the subject; covers all important techniques of phenetics; weak on philosophy of taxonomy and evolutionary trees.]

Sneath, P. H. A. and R. R. Sokal. 1973. *Numerical Taxonomy: The Principles and Practice of Numerical Classification*. W. H. Freeman, San Francisco. [This "second edition" of Sokal and Sneath's (1963) *Principles of Numerical Taxonomy* provides detailed discussion of both philosophy and techniques of phenetics; 1963 edition provides excellent review of concepts and literature of phenetic classification.]

Texts on Orthodox ("Evolutionary") Systematics

Blackwelder, R. E. 1967. *Taxonomy. A Text and Reference Book*. Wiley, New York. [A pleasant text on practical taxonomy (e.g., identification of specimens, curatorial practices, the use of names, the use of taxonomic literature, publication); a considerable portion of the text is devoted to the intricacies of nomenclature and the rules for name publication, changes, etc.]

Mayr, E. 1969. *Principles of Systematic Zoology*. McGraw-Hill, New York. [For many years this was the "bible" of systematics; good discussion of the philosophy of orthodox systematists; largely an update of the classic text by Mayr, Linsley, and Usinger, *Methods and Principles of Systematic Zoology* (McGraw-Hill, 1953).]

Ross, H. H. 1974. *Biological Systematics*. Addison-Wesley, Reading, Massachusetts. [Good treatment of orthodox systematics; more practical information and less theory than Mayr's book; the two complement one another well.]

Simpson, G. G. 1961. *Principles of Animal Taxonomy*. Columbia University Press, New York. [Aging fast but still provides a sound look at "traditional" thinking on evolution, the species concept, and orthodox classification methods.]

Systematic Biology: Proceedings of an International Conference Conducted at the University of Michigan (June 14–16, 1967). 1969. Publ. 1962, National Academy of Science, Washington, D.C. [A collection of 19 papers that are primarily "orthodox" in orientation, although two are on numerical techniques; includes papers on the uses of molecular, cytological, behavioral, and ecological data in systematic research; one of the articles by W. H. Wagner presents the theory and "ground-plan method" upon which many cladistic computer algorithms are ultimately based.]

Related Works

Cracraft, J. and N. Eldredge (eds.). 1979. *Phylogenetic Analysis and Paleontology*. Columbia University Press, New York. [A collection of papers presented at a symposium titled "Phylogenetic Models" at the North American Paleontological Convention II, 1977. Some excellent discussions of phylogenetic reconstruction, not entirely from the paleontologist's point of view; an eclectic overview of systematics and evolution; good reading.]

Gould, S. J. 1977. *Ontogeny and Phylogeny*. Harvard University Press, Cambridge, Massachusetts. [A scholarly treatment of the myriad relationships postulated, over the past 100 years, to exist between ontogeny and phylogeny; state-of-the-art view on important, but little understood, phenomena such as recapitulation, paedomorphosis, and neoteny; highly recommended. Also see G. J. Nelson's 1978 article, "Ontogeny, phylogeny, paleontology and the biogenetic law" (Syst. Zool. 27: 324–345).]

Lincoln, R. J., G. A. Boxshall, and P. F. Clark. 1982. *A Dictionary of Ecology, Evolution and Systematics*. Cambridge University Press, New York. [Excellent.]

Mayr, E. 1981. Biological classification: Toward a synthesis of opposing methodologies. Science 214:510–516.

Milkman, R. (ed.). 1982. *Perspectives on Evolution*. Sinauer Associates, Sunderland, Massachusetts. [Although this chapter did not discuss evolution per se, the theory and practice of biological systematics is intimately related to evolutionary theory, which has itself undergone major revolutions in method and philosophy recently. We include this reference because we feel it is one of the best concise overviews on current thinking in evolutionary theory. For an eye-opener, compare the ideas presented in this text to those expressed in Mayr's *Animal Species and Evolution* (1963, Belknap/Harvard University Press), the latter being the orthodox view of the Synthetic Theory of evolution that dominated the field for the past 40 years. Another good modern treatment of evolution (lying somewhere between the above two volumes) is *Evolution* by Dobzhansky, Ayala, Stebbins, and Valentine (1975, W. H. Freeman.)]

Osterbrock, D. E. and P. H. Raven (eds.). 1988. *Origins and Extinctions*. Yale University Press, New Haven, Connecticut.

Otte, D. and J. A. Endler (eds.). 1989. *Speciation and Its Consequences*. Sinauer Associates, Sunderland, Massachusetts.

Patterson, C. 1978. *Evolution*. British Museum (Natural History) and Cornell University Press, Ithaca, New York. [One of the best modern, concise descriptions of evolution and classification.]

Rensch, B. 1959. *Evolution Above the Species Level*. Columbia University Press, New York. [The English translation of Rensch's classic treatment of a conceptually perplexing subject. The original text (1947; 1954) had considerable influence on the development of modern evolutionary theory.]

Stanley, S. M. 1979. *Macroevolution: Pattern and Process*. W. H. Freeman, San Francisco.

Also see references on evolution at the end of Chapter 1. In addition, the Systematics Association (of London) publishes at irregular intervals volumes on various topics. Some of the titles are *Function and Taxonomic Importance* (1959), *Taxonomy and Geography* (1962), *Speciation in the Sea* (1963), *Phenetic and Phylogenetic Classification* (1964), *Organisms and Continents Through Time* (1973), *The New Systematics* (1940), *Chemotaxonomy and Serotaxonomy* (1968), *Taxonomy and Ecology* (1973), *Problems of Phylogenetic Reconstruction* (1982).

Biological Nomenclature

Ayers, D. M. 1972. *Bioscientific Terminology. Words from Latin and Greek stems*. University Arizona Press, Tucson. [A wonderful little workbook to learn the basics of biological terminology.]

International Code of Zoological Nomenclature, 3rd Ed. Adopted by XX General Assembly of the International Union of Biological Sciences. 1985. International Trust for Zoological Nomenclature in association with the British Museum (Natural History) and University of California Press, Berkeley. [The rule book.]

Jeffrey, C. 1977. *Biological Nomenclature*, 2nd Ed. Crane, Russak & Co., New York.

Savory, T. 1962. *Naming the Living World*. The English Universities Press, Ltd., London.

Smith, H. M. and O. Williams. 1970. The salient provisions of the International Code of Zoological Nomenclature: A summary for nontaxonomists. BioScience 20:553–557. [A highly readable summary of the ICZN.]

The Primary Literature

Annual Reviews of Ecology and Systematics. Annual Reviews, Palo Alto, California. [Yearly volumes with contributed chapters on a great variety of subjects in the indicated fields; state-of-the-art views by specialists.]

Cladistics: The International Journal of the Willi Hennig Society. Meckler, Westport, Connecticut.

Evolution: The International Journal of Organic Evolution. The quarterly journal of the Society for the Study of Evolution, Lancaster, Pennsylvania. [The title says it all.]

Evolutionary Biology. Plenum, New York. [An annual volume containing contributed review papers on subjects of current interest in evolutionary biology.]

Systematic Zoology. The journal of the Society of Systematic Zoology. [A quarterly journal emphasizing current topics of discussion in systematics; usually two or three articles presenting "real" data appear in each issue, but for the most part, the journal has become theoretical in its coverage (to the delight of some and the dismay of others); it was largely in the pages of this journal that the "taxonomy wars" raged between the advocates of the three schools of systematic methodology described in this chapter.]

Zeitschrift für Zoologische Systematik und Evolutionsforschung (Journal of Systematic Zoology and Evolution Research). Paul Parey, Hamburg. [An excellent quarterly journal on aspects of evolution and systematics; a European counterpart to the American journal *Systematic Zoology*.]

Discussions of Character Analysis

deJong, R. 1980. Some tools for evolutionary and phylogenetic studies. J. Syst. Zool. Evol. Res. 18:1–23.

Hecht, M. K. 1976. Phylogenetic inference and methodology applied to the vertebrate record. Evol. Biol. 9:353–363.

Maddison, W. P., M. J. Donoghue and D. R. Maddison. 1984. Outgroup analysis and parsimony. Syst. Zool. 33(1):83–103.

Marx, H. and G. B. Rabb. 1970. Character analysis: An empirical approach to advanced snakes. J. Zool. Lond. 161:525–548.

Maslin, T. P. 1952. Morphological criteria of phyletic relationships. Syst. Zool. 1:49–70.

Nelson, G. J. 1973. The higher-level phylogeny of vertebrates. Syst. Zool. 22:87–91.

Stevens, P. F. 1980. Evolutionary polarity of character states. Annu. Rev. Ecol. Syst. 11:333–358.

Watrous, L. E. and Q. D. Wheeler. 1981. The out-group comparison method of character analysis. Syst. Zool. 30(1):1–11.

Chapter Three

Animal Architecture and the *Bauplan* Concept

The business of animals is to stay alive until they reproduce themselves, and . . . the business of zoologists is to try to understand how they do it.

    E. J. W. Barrington
       Invertebrate Structure and Function, 1967

This chapter must be read carefully in order to appreciate the remainder of the book. In it, we reduce the incredible diversity of invertebrate architecture, structure, and function to a number of easily understandable basic themes. The German language includes a wonderful word that expresses the essence of these architectural themes: **Bauplan** (plural, **Baupläne**). The word means, literally, "a structural plan or design," but a direct translation is not entirely adequate. An animal's *Bauplan* is, in part, its "body plan"—but it is more than that. The concept of a *Bauplan* really captures in a single word the essence of both structural range and architectural limits, as well as the functional aspects of a design. If an organism is to "work," *all* of its body components must be both structurally *and* functionally compatible. The entire organism encompasses a definable *Bauplan,* and the specific organ systems themselves also encompass describable *Baupläne;* in both cases the structural and functional components of the particular plan establish its capabilities and limits. Thus *Baupläne* determine the major constraints that operate at both the organismic and the organ system levels.

All animals must accomplish certain basic tasks to survive and reproduce successfully. They must acquire, digest, and metabolize food and distribute its usable products throughout their bodies. They must obtain oxygen for various chemical processes, while at the same time ridding themselves of metabolic wastes and undigested materials. The strategies employed by animals to maintain life are extremely varied, but they rest upon relatively few biological, physical, and chemical principles. Within the constraints imposed by particular *Baupläne,* groups of animals have a limited number of options available to accomplish life's tasks. For this reason a few recurring fundamental themes become apparent. This chapter is a general review of the structural–functional aspects of invertebrate *Baupläne* and the basic themes or survival strategies employed within each. It is a description of how invertebrates are put together and how they manage to survive. Each of the various subjects discussed in this chapter also reflects fundamental principles of animal mechanics, physiology, and evolution. Thus the *Bauplan* approach has the additional advantage of serving to instruct the student in these basic biological concepts.

Keep in mind that even though this chapter is organized on the basis of what might be called the "components" of animal structure, whole animals are integrated functional combinations of these components. Furthermore, there is a strong element of predictability in the concepts discussed here. For example, given a particular type of symmetry, one can make reasonable guesses about other aspects of an animal's structure that are compatible with that symmetry—some combinations work, others do not. Herein are explained many of the concepts and terms used throughout this book, and we encourage you to become familiar with this material now as a basis for understanding the remainder of the text.

Body symmetry

A fundamental aspect of an animal's *Bauplan* is its overall shape or geometry. So in order to discuss invertebrate architecture and function, we must first

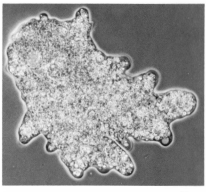

B

Figure 1

Examples of asymmetrical invertebrates. A, An assortment of sponges. B, An ameba. (A courtesy of J. DeMartini; B photo by P. Sieburth/BPS.)

A

acquaint ourselves with a basic aspect of body form: symmetry. **Symmetry** refers to the arrangement of body structures relative to some axis of the body. Animals that can be bisected or split along at least one plane, so that the resulting halves are similar to one another, are said to be **symmetrical**. For example, a shrimp can be bisected vertically through its midline, head to tail, to produce right and left halves that are mirror images of one another. Animals that have no **plane of symmetry** are said to be **asymmetrical**. Many sponges, for example, have an irregular growth form and so lack any clear plane of symmetry. Similarly, many protozoa, particularly the ameboid forms, are asymmetrical (Figure 1).

Most animals possess some kind of symmetry, but within this context a great variety of body design has evolved. The simplest form of symmetry is **spherical symmetry**; it is seen in an animal that assumes the form of a sphere, with its parts arranged concentrically around, or radiating from, a central point (Figure 2). A sphere has an infinite number of planes of symmetry that can pass through its center to divide the organism into like halves. Spherical symmetry is rare in nature and, in the strictest sense, is found only in certain protozoa. Animals with spherical symmetry share an important functional attribute with asymmetrical organisms, in that both groups lack **polarity**. That is, there exists no clear differentiation along an axis, other than from the center of the animal toward its surface. In all other forms of symmetry, some level of polarity has been achieved; and with polarity comes specialization of body regions and structures.

A body displaying **radial symmetry** has the general form of a cylinder, with one main axis around which the various body parts are arranged (Figure 3). In a body displaying perfect radial symmetry, the body parts are arranged equally around the axis, and any plane of sectioning that passes along that axis results in similar halves (rather like a cake being divided and subdivided into equal halves and quar-

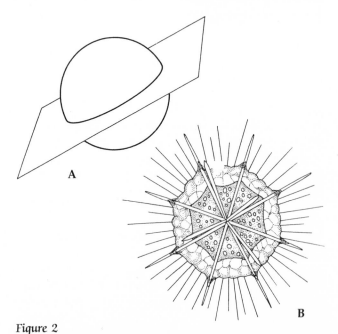

Figure 2

Spherical symmetry in animals. A, An example of spherical symmetry; any plane passing through the center divides the organism into like halves. B, A radiolarian (protozoon).

Figure 3

Radial symmetry in invertebrates. The body parts are arranged radially around a central oral–aboral axis. A, Representation of perfect radial symmetry. B, The sponge *Xetospongia*. C, The sea anemone *Epiactis*, whose mouth alignment and internal organization produce biradial symmetry. D, The hydromedusa *Scrippsia*, with quadriradial symmetry. E, The sea star *Patiria*, with pentaradial symmetry. F, The sea star *Solaster*, with multiradial symmetry. (B photo by S. K. Webster, Monterey Bay Aquarium/BPS; C–E courtesy of G. McDonald; F photo by C. E. Mills, Univ. of Washington/BPS.)

ters). Nearly perfect radial symmetry occurs in the simplest sponges and in many cnidarian polyps (Figures 3A and B). Perfect radial symmetry is relatively rare, however, and most radially symmetrical animals have evolved modifications on this theme. **Biradial symmetry**, for example, occurs where portions of the body are specialized and only two planes of sectioning can divide the animal into similar halves. Common examples of biradial organisms are sea anemones and ctenophores (Figure 3C). Further specializations of the basic radial body plan can produce nearly any combination of multiradiality. For example, many jellyfishes possess **quadriradial symmetry** (Figure 3D), and most sea stars display **pentaradial symmetry** (Figure 3E), although many multiarmed sea stars are also known. The common Pacific sun star *Solaster stimpsoni* (Figure 3F), for example, has many arms (multiradial symmetry).

One adaptively significant feature of radial symmetry is that such animals can confront their environment in numerous directions. A radially symmetrical animal has no front or back end; rather it is organized about an axis that passes through the center of its body, like an axle through a wheel. When a gut is present, this axis passes through the mouth-bearing (**oral**) surface to the opposite (**aboral**) surface. Radial symmetry is most common in sessile species (e.g., sponges and sea anemones) and drifting pelagic species (e.g., jellyfishes and ctenophores).

Given these lifestyles, it is clearly advantageous to be able to confront the environment equally in a variety of directions. In such creatures one generally finds feeding structures and sensory receptors placed in such a way that they contact the environment more-or-less equally in all directions from the body axis. Furthermore, many fundamentally bilaterally symmetrical animals have become "functionally" radial in certain ways associated with sessile lifestyles. For example, their feeding structures may be in the form of a whorl of radially arranged tentacles, an arrangement allowing more efficient contact with their surroundings.

The body parts of **bilaterally symmetrical** animals are oriented about an axis that passes from the front (**anterior**) to the back (**posterior**) end. There is but a single plane of symmetry, which passes along the axis of the body to separate right and left sides,

the **midsagittal plane** (or median sagittal plane). A longitudinal plane passing perpendicular to the sagittal plane and separating the backside (**dorsal**) from the underside (**ventral**) is called a **frontal plane**. Any plane that cuts across the body, from side to side, is called a **transverse plane** (or simply, a cross section) (Figure 4). In bilaterally symmetrical animals the term **lateral** refers to the sides of the body, or to structures away from and to the right and left of the midsagittal plane. The term **medial** refers to the midline of the body, or to structures on or near the midsagittal plane.

Whereas spherical and radial symmetry are typically associated with sessile or drifting animals, bilaterality is generally found in animals with controlled mobility; in these animals, the anterior end of the body confronts the environment first. With the evolution of bilateral symmetry and unidirectional

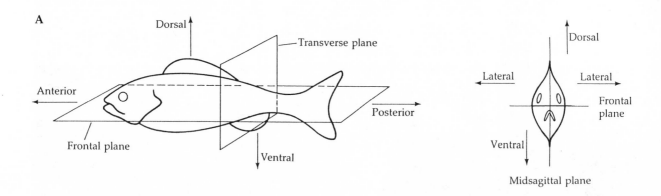

Figure 4

Bilateral symmetry in animals; a single plane passing through the body divides it into equal halves. A, Diagrammatic illustration of bilateral symmetry, with terms of orientation and planes of sectioning. B, The crab *Petrolisthes*. C, A nereid polychaete. (B–C courtesy of G. McDonald.)

movement, one finds an associated concentration of feeding and sensory structures at the anterior end. The formation of a head end is called **cephalization**. Furthermore, with the differentiation of dorsal and ventral surfaces, the ventrum often becomes loco-motor and the dorsum generally specialized for pro-tection. A variety of secondary asymmetrical modi-fications of bilateral (and radial) symmetry have occurred, for example, the spiral coiling of gastro-pods and hermit crabs. Most asymmetrical modifi-cations result from the displacement of certain body parts because of the habit of being fastened on one side rather than on the ventral surface, or from un-equal development of certain paired structures.

Cellularity, body size, germ layers, and body cavities

One of the main characteristics used in defining grades of animal complexity is the presence or ab-sence of true **tissues**. Tissues are aggregations of morphologically similar cells organized to perform a specific function. The protozoa (Chapter 5), of course, do not possess tissues, but occur only as single cells or as simple colonies of cells. They may be said to have a **unicellular grade** of construction. Beyond the unicellular phyla is the vast array of mul-ticellular life, the metazoa. Within the metazoa is a hierarchy of increasing complexity of body form that can be divided into three major levels, or grades: mesozoa, parazoa, and eumetazoa. These names do not represent formal taxa, but may be used to group the metazoa by their level of overall structural com-plexity. The mesozoa and parazoa are not generally considered to possess true tissues (or possess only incipient tissue formation), and for this reason they are separated from the rest of the metazoa. The eu-metazoa pass through distinct embryonic stages dur-ing which tissue layers form (see Chapter 4). Box 1 has an outline of these general grades of body archi-tecture.

Each of the grades of body complexity noted above is associated with inherent constraints and ca-pabilities, and within each grade there are obvious limits to size. As the British biologist D'Arcy Thomp-son (1942) wrote, "Everything has its proper size . . . men and trees, birds and fishes, stars and star-sys-tems, have . . . more or less narrow ranges of abso-lute magnitudes." As a cell (or organism) increases in size, its volume increases at a rate faster than the rate of increase of its surface area (surface area in-creases as the square of linear dimensions; volume

increases as the cube of linear dimensions). Because a cell ultimately relies on transport of material across its plasma membrane for survival, this disparity quickly reaches a point at which the cytoplasm can no longer be adequately serviced by simple cellular diffusion. Some unicellular forms compensate for this problem by developing complexly folded surfaces or by assuming flattened or threadlike shapes. In this way they can become quite large; but eventually a limit is reached. Thus, we have no two-meter-long protozoa.

Ultimately, the only way around the surface-to-volume dilemma is to increase the number of cells constituting a single organism; hence the metazoa. But, for similar reasons, size increase in the metazoa is also limited. Some metazoa lack complex speciali-zations of tissues and organs; therefore diffusion into and out of the body cannot occur unless the ma-jority of the body's cells are near or in contact with the external environment. For example, diffusion is an effective method of oxygenation only when the diffusion path is less than about 1.0 mm. Here, too, there are limits. An animal simply cannot increase indefinitely in volume when most of its cells must lie in close proximity to the body surface. Primitive met-azoa solve this problem to some degree by arranging their cellular material so that diffusion distances from cell to environment are comfortably short. One method of accomplishing this is to pack the internal bulk of the body with nonliving material, such as the jelly-like mesoglea of medusae and ctenophores. An-other is to assume a body geometry that increases the surface area as much as possible. Increase in one dimension leads to a vermiform body plan, like that of ribbon worms. Increase in two dimensions results in a flat, sheetlike body like that of the flatworms. In both cases the diffusion distances are kept short. It has been shown experimentally that the very flatness of flatworms is necessary to maintain adequate partial pressures of oxygen throughout the body (special dorsoventral muscles that have evolved enable flat-worms to actively maintain their distinctive shape). Sponges effectively increase their surface area by a process of complex branching and folding of the body, both internally and externally.

If these were the only solutions to the surface-to-volume dilemma, the natural world would be filled with tiny, thin, and flat animals, or convoluted spongelike creatures. However, another solution was discovered during the course of animal evolution, and that is to bring the environment functionally closer to each cell in the body by the use of internal transport systems. A three-dimensional increase in

Box One
Body Construction and Phyla Organization

Organization of the animal phyla on the basis of their grade of body construction.*

I. Unicellular animals: the protozoa [Greek *proto*, "first"; *zoa*, "animals"]
   Phyla **Ciliophora, Sarcomastigophora, Labyrinthomorpha, Apicomplexa, Microspora, Ascetospora, Myxozoa**

II. Multicellular animals: the metazoa [Greek *meta*, "later"; *zoa*, "animals"]
   A. Without true tissues
      1. The mesozoa [Greek *meso*, "middle"; *zoa*, "animals"]
         Phyla **Orthonectida, Rhombozoa, Placozoa** (?), and **Monoblastozoa** (?)
      2. The parazoa [Greek *para*, "alongside"; *zoa*, "animals"
         Phylum **Porifera** (sponges)
   B. With true tissues: the eumetazoa [Greek *eu*, "true"; *meta*, "later"; *zoa*, "animals"]
      1. The diploblastic eumetazoa (lacking true mesoderm): the "radiata"
         Phylum **Cnidaria** (sea anemones, medusae)
         Phylum **Ctenophora** (comb jellies)
      2. The triploblastic eumetazoa (with true mesoderm): the "bilateria"
         a. Acoelomates: without a body space other than the digestive tract; mesenchyme and muscle fills region between gut and epidermis.
            Phylum **Platyhelminthes** (flatworms)
            Phylum **Nemertea** (ribbon worms) (?)
         b. Pseudocoelomates: with a persistent blastocoel (= pseudocoel) between gut and body wall
            Phylum **Nematoda** (roundworms)
            Phylum **Nematomorpha** (horsehair worms)
            Phylum **Entoprocta** (entoprocts) (?)
            Phylum **Rotifera** (rotifers)
            Phylum **Gastrotricha** (gastrotrichs) (?)
            Phylum **Kinorhyncha** (kinorhynchs) (?)
         c. Coelomates (or Eucoelomates). With a true coelom (= mesodermal cavity).
            All other phyla: **Mollusca, Annelida, Arthropoda, Echinodermata, Chordata**, etc.

*Only those names set in boldface type are recognized taxa; other names are simply designations used to group various taxa by the level of body complexity they have achieved. Taxa of uncertain placement are indicated by a question mark. The enigmatic phyla Priapula, Gnathostomulida, and Loricifera are not included in the table. See text for discussion.

body size thus necessitated the development of sophisticated transport mechanisms for nutrients, oxygen, waste products, and so on. In time these transport structures became the organs and the organ systems of higher metazoa. For example, the body volume in humans is so large that we require an elaborately branched network of gas exchange surfaces (called lungs) to provide an adequate surface area for oxygen diffusion. This network has about 1,000 square feet of surface—as much area as half a tennis court! The same constraints apply to food ab-

sorption surfaces; hence the evolution of highly folded or branched guts.

The embryonic tissue layers of eumetazoa are called **germ layers** (from the Latin *germen*, "a sprout, bud, or embryonic primordium"), and it is from the embryonic germ layers that all adult structures develop. Chapter 4 presents the details of germ layer formation and other aspects of metazoan developmental patterns. At this point we need only point out that the germ layers initially form as outer and inner sheets or masses of embryonic tissue, termed ectoderm and entoderm (or endoderm), respectively. In the embryogeny of the radiate phyla, Cnidaria and Ctenophora, only these two germ layers develop (or if a middle layer develops, it appears to be produced by the ectoderm and is not considered a true germ layer). Hence these animals are regarded as **diploblastic** (Greek, *diplo*, "two"; *blast*, "bud" or "sprout," in reference to the embryonic germ layer). In the embryogeny of most metazoa, however, a true third germ layer, the cellular **mesoderm**, arises between the ectoderm and the entoderm; these metazoan groups are said to be **triploblastic**. This "true" mesoderm arises from embryonic cells initially associated with the entoderm. The evolution of a mesoderm greatly expanded the evolutionary potential for animal complexity. As we shall see, the triploblastic phyla have achieved many unique and more highly sophisticated *Baupläne* than are possible within the confines of a diploblastic body plan. Simply put, a developing triploblastic embryo has more kinds of potential "building material" than does a diploblastic embryo.

One of the major trends in the evolution of the triploblastic metazoa has been the development of a fluid-filled cavity between the outer body wall and the digestive tube; that is, between the derivatives of the ectoderm and the entoderm. The evolution of this structural device created a radically new architecture, a tube-within-a-tube design in which the inner tube (the gut and its associated organs) was freed from the constraint of being attached to the outer tube (the body wall), except at the ends. The fluid-filled cavity not only served as a mechanical buffer between these two largely independent tubes but also allowed for the development and expansion of new structures within the body, served as a storage chamber for various body products (e.g., gametes), provided a medium for circulation, and was in itself an incipient hydrostatic skeleton. The nature of this cavity (or the absence of it) is associated with the formation and subsequent development of the mesoderm, as discussed in detail in Chapter 4.

Three major grades of construction are recognizable among the triploblastic metazoa: **acoelomate**, **pseudocoelomate**, and **eucoelomate**. The acoelomate grade (Greek *a*, "without"; *coel*, "hollow, cavity") occurs in two triploblastic phyla, Platyhelminthes and Nemertea. In the animals in these taxa the mesoderm forms a more-or-less solid mass of tissue between the gut and body wall (Figure 5A). In animals in most other triploblastic phyla, an actual space develops as a fluid-filled cavity between the body wall and the gut. In many phyla (e.g., annelids, molluscs, and echi-

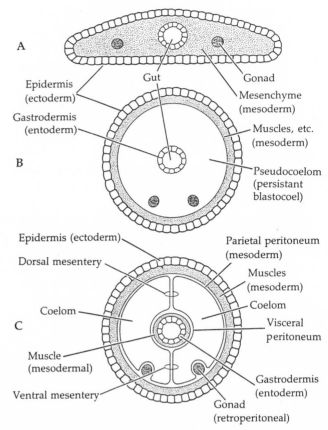

Figure 5

Principal body plans of metazoa (diagrammatic cross sections). A, The acoelomate body plan. B, The pseudocoelomate body plan. C, The eucoelomate body plan.

noderms) this cavity arises within the mesoderm itself and is completely enclosed within a thin cellular lining called the **peritoneum**, which is derived from the mesoderm. Such a cavity is called a true **coelom** (or eucoelom). Notice that the organs of the body are not actually free within the coelomic space itself, but are separated from it by the peritoneum (Figure 5C).

Several phyla of triploblastic metazoa (rotifers, roundworms, and others) possess a body cavity that is neither formed from the mesoderm nor fully lined by peritoneum or any other form of mesodermally derived tissue. Such a cavity is called a **pseudocoelom**, or simply a pseudocoel (Greek *pseudo*, "false"; *coel*, "hollow, cavity") (Figure 5B). Unlike the enclosed organs of true coelomates, the organs of pseudocoelomate animals actually lie free within the body cavity and are bathed directly in its fluid.

Within the constraints inherent in each of the basic body organizations discussed above, the various animal phyla have, over time, evolved a multitude of variations on these themes. Each additional level of complexity that evolved opened new avenues for potential variation and adaptation. Throughout the remainder of this chapter we describe the fundamental organizational plans of major body systems as they have evolved within these basic *Baupläne*. In subsequent chapters on each animal phylum, we describe how the various groups have modified these basic plans through their own particular evolutionary program or direction.

Locomotion and support

As life progressed from the single-celled stage to multicellularity and then to bilaterality, body size tended to increase dramatically. This increase in body size, coupled with directed movement, was accompanied by the evolution of a variety of support structures and associated locomotor mechanisms. Because these two body systems evolved mutually and work in a complementary fashion in living animals, the basic patterns of their functional morphology are best discussed together.

There are four fundamental locomotor patterns in animals: ameboid movement, ciliary and flagellar movement, hydrostatic propulsion, and locomotor limb movement. There are three fundamental kinds of support systems: structural endoskeletons, structural exoskeletons, and hydrostatic skeletons. In this section we briefly describe the basic architecture and mechanics of the various combinations of these two systems.

Most invertebrates live in water, and aquatic environments present obstacles and advantages to support and locomotion quite different from those of terrestrial environments. Animals moving through water (or moving water over their bodies—the effect is the same) face certain problems of fluid dynamics created by the interaction between a solid body and a surrounding liquid. What happens during this interaction is tied to the concept of **Reynolds number**, a unitless value based on the experiments of Osborne Reynolds (1842–1912). Reynolds number represents a ratio of inertial force to viscous force. At higher Reynolds numbers, inertial force predominates and determines the behavior of water flow around an object. At lower Reynolds numbers, viscous force predominates and determines the behavior of the water flow. The importance of this concept is being increasingly recognized and applied to biological systems (see for examples Vogel 1981; LaBarbera 1984; Yates 1986; Jorgensen et al. 1984). Although there is

still a great deal to be done in this area, some interesting generalizations can be made about locomotion of aquatic animals and, as we discuss later, aquatic suspension feeding. Reynolds number is expressed by the following equation:

$$\mathrm{Re} = \frac{plU}{v}$$

where p equals the density of fluid, l is some measurement of the size of the solid body, U equals the relative velocity of the fluid over the body surface, and v is the viscosity of the fluid. The formula was derived by Reynolds to describe the behavior of cylinders in water. Of course, since animals' bodies are not perfect cylinders, the "size variable" (l) is difficult to standardize. Nonetheless, meaningful relative estimates can be derived and applied to living creatures in water.

Without belaboring this issue beyond its importance here, it turns out that the problems of a large animal swimming through water are very different from those of a small animal. Large animals such as fishes, whales, or even humans, by virtue of their size or high velocity or both, move in a world of high Reynolds numbers. Under such conditions, fluid viscosity becomes less and less significant as far as the animal's energy output during locomotion is concerned. At the same time, however, inertia becomes more and more important. With increasing size, an animal must expend more energy to put its body in motion. But, by the same token, inertia works in favor of the moving animal by carrying it forward when the animal stops swimming. The effect of inertia when large animals move at high Reynolds numbers also imparts motion to the water around the animal's body. Thus, as the Reynolds number increases, a point is reached at which the flow of water changes from laminar to turbulent, thus decreasing swimming efficiency.

Small animals generally move in a world of low Reynolds numbers. For them, inertia and turbulence are virtually nonexistent. But viscosity becomes important, increasingly so as body size and velocity decrease (i.e., as the Reynolds number decreases). Small animals swimming through water have been likened to a human swimming through liquid tar, or thick molasses. The effect of this situation is that tiny animals, such as ciliate and flagellate protozoa and many small metazoa, start and stop instantaneously, and the motion of the water set up by their swimming also ceases immediately if the animal stops moving. Thus, small creatures neither pay the price nor reap the benefits of the effects of inertia. The organism only moves forward when it is expending energy to swim; as soon as it stops moving its cilia, or flagella, or appendages, it stops—and so does the fluid surrounding it. Tiny animals swimming at low Reynolds numbers must expend an incredible amount of energy to propel themselves through their "viscous" surroundings.

Ameboid locomotion

In the animal kingdom, ameboid movement is used principally by certain protozoa and by numerous kinds of ameboid cells that occur internally, *within* the bodies of most metazoa. Ameboid cells possess a gel-like **ectoplasm**, which surrounds a more fluid **endoplasm** (Figure 6). Movement is facilitated by changes in the states of these regions of the cell. At one (or several) points on the cell surface, **pseudopodia*** develop; and as endoplasm flows into a growing pseudopodium, the cell creeps in that direction. This seemingly simple process actually involves complex changes in cell fine structure, chemistry, and behavior. The innermost endoplasm moves "forward" while the outermost endoplasm takes on a granular appearance and remains fairly stable. The advancing portion of endoplasm pushes forward and then becomes semirigid ectoplasm at the tip of the advancing pseudopodium. Concurrently, endoplasm is recruited from the trailing end of the cell, from whence it streams forward to join in the "growing" pseudopodium.

Although biologists have been studying ameboid locomotion for over 100 years, the precise mechanism is not yet fully understood. The physiological basis of ameboid movement is probably essentially the same as that of vertebrate muscle contraction, involving actin, myosin, and ATP. Two principal theories exist to explain the process. Perhaps the more popular of the two ideas has the actin molecules floating freely in the endoplasm, polymerizing into their filamentous form at the point of active pseudopod growth, where they interact with myosin molecules. The resultant contraction literally pulls the streaming endoplasm forward, while in the same stroke converting it to the ectoplasm that rings the forward-streaming pseudopodium. The second theory suggests that the actin–myosin interaction takes place at the rear of the cell, where it produces a contraction of the ectoplasm. This contraction squeezes the cell like a tube of toothpaste and causes the endoplasm to stream forward and create a pseu-

*Singular, pseudopodium; plural, pseudopodia. The diminutive is often used: singular, pseudopod; plural, pseudopods.

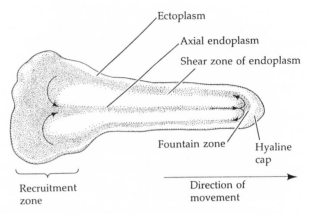

Figure 6

Ameboid locomotion: pseudopod formation in an ameba.

dopodium directly opposite the point of ectoplasmic contraction. Even though several other theories have been proposed recently to explain pseudopodial movement, the definitive answer to the question of how a simple single-celled ameba moves remains elusive. Some modifications of pseudopodial movement are discussed in Chapter 5.

Cilia and flagella

Cilia or **flagella** or both occur in virtually every animal phylum (with the qualified exception of the Arthropoda; see Chapter 15). Structurally, cilia and flagella are very similar, but the former are shorter and tend to occur in relatively larger numbers (in patches or tracts), whereas the latter are long and generally occur singly or in pairs.

The movement of cilia and flagella creates a propulsive force that either moves the organism through a liquid medium or, if the animal (or cell) is anchored, creates a movement of fluid over it. Such action virtually always occurs at very low Reynolds numbers. The general structure of a cilium or flagellum consists of a long flexible rod, the outer covering of which is simply an extension of the **plasma membrane** of the cell (Figure 7A). Inside is a circle of nine paired **microtubules** (often called doublets) that runs the length of the cilium or flagellum. Down the center of the doublet circle is an additional pair of microtubules. This familiar 9 + 2 pattern is characteristic of nearly all flagella and cilia (Figure 7B). During cell development, each new flagellum arises from an organelle called the **basal body** (or **kinetosome**), to which it remains anchored. Flagellar microtubules are modified proteinaceous, hollow tubules similar to those present in the matrix of most cells. The principal function of the tubules appears to be support.

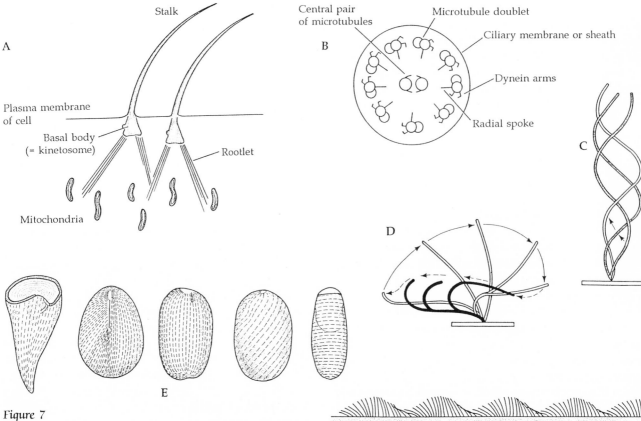

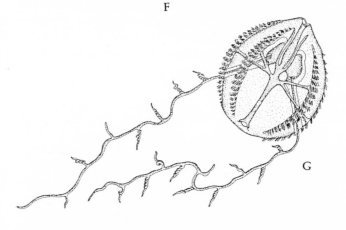

Figure 7

Cilia and flagella. A, Structures of two adjacent cilia. B, Cross section of a cilium. C, Three successive stages in the undulatory movement of a flagellum. D, Successive stages in the oarlike action of a cilium. The power stroke is shown in white, the recovery stroke in black. E, Examples of ciliary tract patterns in various ciliate protozoa (tracts indicated by dashed lines). F, Appearance of metachronal waves of a line of cilia. G, The comb jellies (ctenophores) are the largest animals known to rely primarily on cilia for locomotion. Shown here is the rather small ctenophore, *Pleurobrachia bachei* (about an inch or less in diameter). (G from Brusca and Brusca 1978.)

Just as the ectoplasm helps retain the shape and integrity of a protozoan cell (acting as a type of rudimentary "exoskeleton"), so the cytoplasmic microtubules act as a sort of simple "endoskeleton" to help protozoa (and other cells) retain their shape. Microtubules are also responsible for formation of the spindle and distribution of the chromosomes during cell division.

In addition to the locomotor function seen in some protozoa and small metazoa, cilia and flagella actually have an enormous variety of functions in many other animals. For example, they create feeding and gas exchange currents; they line digestive tracts and facilitate food movement; they propel sex cells and larvae. Special uses and occurrences of these structures are discussed in subsequent chapters. Here we focus on their use as locomotor structures.

Analysis by high-speed photography reveals that the movement of these structures is complex and differs somewhat among various groups. Some flagella beat back and forth, while others beat in a helical rotary pattern that drives flagellate protozoan cells like the propeller of an outboard motor (Figure 7C). Depending on whether the undulation moves from base-to-tip or from tip-to-base, the effect will be respectively to push or pull the cell along. Some flagella possess tiny hairlike side branches called **mastigo-nemes** that increase the surface area and thus improve the propulsive capability. The beat of a cilium is generally simpler, consisting of a forward stroke, or **power stroke**, with a relaxed **recovery stroke**, or **backstroke** (Figure 7D). When many cilia are present on a cell, they often occur in distinct tracts, and their action is integrated, with beats usually moving in **metachronal waves** over the cell surface (Figure 7F). Since at any one time some cilia are always performing a power stroke, metachronal coordination ensures a uniform and continuous propulsive force.

It was once thought that ciliary tracts on individual cells were coordinated by a primitive sort of cellular "nervous system." However, evidence for this hypothesis has not been forthcoming and current thinking suggests that the coordinated beating of cilia is probably due to hydrodynamic constraints imposed on them by the interference effects of the surrounding water layers and by the simple mechanical stimulation of moving, adjacent cilia. Nevertheless, some ciliary responses in metazoa may be under neural control, for example, reversal of power stroke direction.

Ciliated protozoa are the swiftest of the single-celled organisms. Flagellated protozoa are the next most rapid, and amebas are the slowest. Most amebas move at rates around 5 μm/sec (about 2 cm/hr), or about 100 times slower than most ciliates. Cilia are also used for locomotion by several metazoan groups, such as the mesozoa, ctenophores, platyhelminthes, many pseudocoelomates, some gastropods, and by the larval stages of many others.

Muscles and skeletons

Almost all animals have some sort of a skeleton, the major functions of which are to maintain body shape and provide support. These functions may be attained either by hard tissues or secretions, or by the turgidity of body fluids under pressure. Muscles, skeletons, and body form are closely integrated, both developmentally and functionally. When rigid skeletal elements are present, they can serve as fixed points for muscle attachment. For example, the rigid and jointed exoskeleton of arthropods allows for a complex system of levers that results in very precise and restricted limb movements. Many invertebrates lack hard skeletons and can change their body shape by alternate contraction and relaxation of various muscle groups attached to tough connective tissues or to the inside of the general body surface. These "soft-bodied" invertebrates usually have a **hydrostatic skeleton**.

The hydrostatic skeleton. The performance of a hydrostatic skeleton is based on two fundamental properties of liquids: their incompressibility and their ability to assume any shape. Because of these features, body fluids transmit pressure changes rapidly and equally in all directions.

It is important to realize a basic physical limitation concerning the action of muscles—they can only actively exert a force (and thus perform direct work) by contracting, that is, by shortening. Muscles cannot directly "push," they can only "pull." To facilitate the extension or protrusion of a body region, the contractile force of a muscle can be imparted to body fluids to create a hydrostatic pressure that accomplishes the action. Such indirect muscle actions can be compared to the action of squeezing a rubber glove filled with water in order to force the fingers of the glove to become erect. The enclosure of a fluid-filled chamber (e.g., a coelom) within sets of opposing muscle layers establishes a system in which muscles in one part of the body can contract to force body fluids into another region of the body (where the muscles are relaxed) and thus extend or otherwise change the shape of the body.

In the most common plan, two muscle layers surround a fluid-filled body cavity, and the fibers of the layers run in different directions (i.e., a circular muscle layer and a longitudinal muscle layer). A soft-bodied invertebrate can move forward by using its hydrostatic skeleton in the following way. The circular muscles at the posterior end of the animal contract, so the hydrostatic pressure generated there pushes anteriorly to extend the front of the animal's body. Then, contraction of the longitudinal muscles at the posterior end pulls that end of the body forward. This sequence of muscle contractions results in a directed and controlled movement forward. Such movement requires that the posterior end be anchored when the anterior end is extended, and the anterior end be anchored when the posterior end is pulled forward. This system is commonly used for locomotion by many worms. The same sort of hydrostatic system can be used to temporarily or intermittently extend selected parts of the body, such as the feeding proboscis of various worms, suckered tube feet of echinoderms, and siphons of clams.

A little thought reveals that the contraction of circular muscles at one end of a vermiform animal may actually have four possible effects: the contracting end may elongate, the opposite end may elongate, or it may thicken, or both ends may elongate. Which event transpires depends not on the contraction of the circular muscles of the contracting end, but on the state of contraction of the longitudinal and circular muscles in other parts of the body (Figure 8).

It is essential that fluid not leak out of the hydrostatic skeleton; or, more realistically, the rate of leakage should not be greater than the rate at which the fluid can be replaced. Body fluids must be retained despite the existence of "holes" in the body wall, such as the excretory pores of many coelomate animals or the mouth opening of cnidarians. Of course, animals with hydrostatic skeletons have solved this problem. Structures such as the excretory pores of

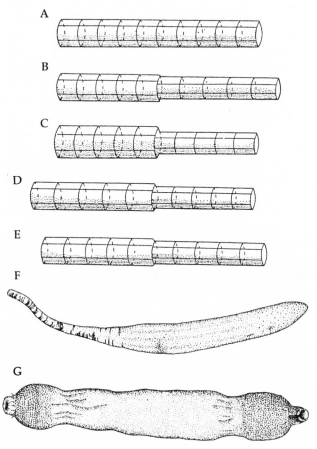

Figure 8

The hydrostatic skeleton. A–E, The initial state and the four possible results of contraction of the circular muscles at one end of a cylindrical animal with a hydrostatic skeleton. In A the muscles are all relaxed. In B the circular muscles of the right-hand end have contracted and this end has elongated; the left-hand end has remained unaltered. In C the length of the right-hand end has remained the same but the diameter of the left-hand end has increased. In D the length of the right-hand end and the diameter of the left-hand end have remained the same, but the length of the left-hand end has increased. In E the lengths of both ends have increased, but their respective diameters have remained the same. F–G, Two animals that rely on hydrostatic skeletons for support and locomotion. F, The sipunculan worm *Phascolosoma*. G, The echiuran worm *Urechis caupo*. (F–G from Brusca and Brusca 1978.)

annelids and the mouths of sea anemones generally have their openings encircled by **sphincter muscles** that can regulate loss of body fluids. When the circular muscles contract, the sphincter muscles also contract and close the body openings.

One way in which movement by a hydrostatic skeleton can be made more precise is to divide an animal up into a series of separate compartments. For example, in annelid worms partitioning of the coelom and body muscles into segments with separate neural control enables body expansions and contractions to be confined to a few segments at a time. By manipulations of selected sets of segmental muscles, most annelids not only can move forward and backward but can turn and twist in complex maneuvers.

The rigid skeleton. In "hard-bodied" invertebrates, a fixed or rigid skeletal system prevents the gross changes in body form seen in soft-bodied invertebrates. This trade-off in flexibility gives hard-bodied animals several advantages: the capacity to support a larger body (an advantage that is especially useful in terrestrial environments, which lack the buoyancy provided by aquatic environments), more precise or controlled body movements, better defense against predators, and often greater speed of movement.

Hard skeletons can be broadly classed as either **endoskeletons** or **exoskeletons**. Endoskeletons are generally derived from mesoderm, whereas exoskeletons are derived from ectoderm; both usually have organic and inorganic components. It has been hypothesized that rigid skeletons may have originated by chance, as by-products of general metabolic pathways. By sheer accident (preadaptation, or exadaptation), for example, the accumulation of nitrogenous wastes and their incorporation into complex organic molecules may have resulted in the evolution of a **chitinous** exoskeleton like that of arthropods. Similarly, a metabolic system originally designed to eliminate excess calcium from the body might have produced the calcareous shell of molluscs. Marine invertebrates are now known to be capable of forming, through their various biological activities, a vast array of minerals, some of which cannot be formed inorganically in the biosphere. In fact, it is probable that the ever-increasing amounts of these **biominerals** have radically altered the character of the biosphere over the past 600 million years. Most common among these biominerals are various carbonates, phosphates, halides, sulfates, and iron oxides.

Invertebrate skeletons may be of the articulating

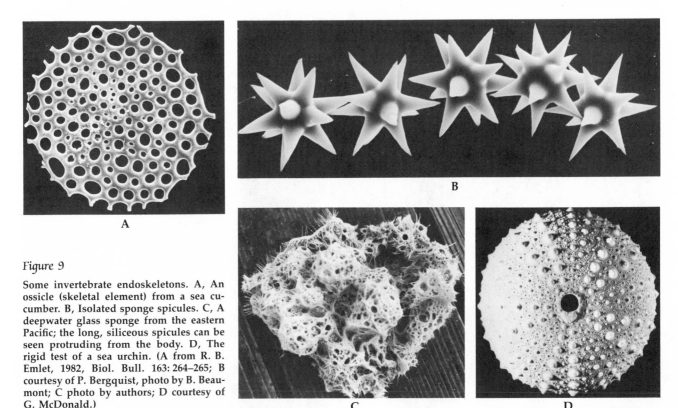

Figure 9

Some invertebrate endoskeletons. A, An ossicle (skeletal element) from a sea cucumber. B, Isolated sponge spicules. C, A deepwater glass sponge from the eastern Pacific; the long, siliceous spicules can be seen protruding from the body. D, The rigid test of a sea urchin. (A from R. B. Emlet, 1982, Biol. Bull. 163: 264–265; B courtesy of P. Bergquist, photo by B. Beaumont; C photo by authors; D courtesy of G. McDonald.)

type (e.g., the exoskeleton of arthropods, bivalves, and brachiopods and the endoskeleton of some echinoderms), or they may be of the nonarticulating type, as seen in the simple one-piece exoskeletons of snails and the rigid endoskeletons composed of interlocking and fused plates of sea urchins and sand dollars. Animal endoskeletons may be as simple as the microscopic calcareous or siliceous spicules embedded in the body of a sponge, cnidarian, or sea cucumber, or they may be as complex as the bony skeleton of vertebrates (Figure 9).

In the broadest sense, all invertebrates possess an exoskeleton of sorts, even soft-bodied types (Figure 10). Protozoan cells all possess a semirigid ectoplasm, and some have surrounded themselves with a **test** composed of bits of sand or other foreign matter glued together. Other protozoa have evolved the capacity to build a test made from chemicals that they either extract from sea water or produce themselves.

From their epidermis, many metazoa secrete an external, nonliving layer called the **cuticle**, which serves as an exoskeleton for these animals. The cuticle varies in thickness and complexity, but is often composed of several definable layers of differing structure and composition. In the arthropods, for example, the cuticle comprises a complex combination of chitin and a protein called glycoprotein. This complex compound may be strengthened by the formation of internal cross-linkages (a process called tanning) and by the addition of calcium. In most insects, the outermost layer is impregnated with wax to decrease its permeability to water. The cuticle is often ornamented with various spines, tubercles, scales, or striations; and it is frequently divided into rings or segments, a feature lending flexibility to the body. Other examples of exoskeletons are the calcareous shells of many molluscs and the casings of coral animals (Figure 10).

The term **chitin** refers to a family of closely related chemical compounds that, in various forms, are produced by and incorporated into the cuticles of many invertebrates. Certain types of chitin are also produced by some fungi and diatoms. The chitins are high-molecular-weight, nitrogenous polysaccharides that are tough yet flexible (Figure 10F). In addition to its supportive and protective functions in the formation of exoskeletons, chitin is also a major component of the teeth, jaws, and grasping and grinding structures of a wide variety of invertebrates. Increased use of chitin in fields as diverse as medical research and wastewater treatment has led to an emerging field of study—chitin chemistry. Much of

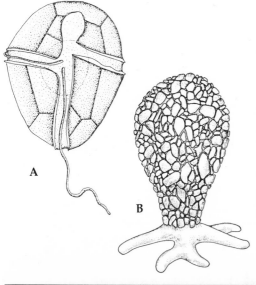

Figure 10

Some invertebrate exoskeletons. A, The dinoflagellate protozoon *Gonyaulax*, encased in cellulose plates. B, The ameba *Difflugia*, with a test of minute sand grains. C, The foraminiferan *Cyclorbiculina*, with a calcareous, multichambered shell. D, An assassin bug, with a jointed, chitinous exoskeleton. E, The giant clam, *Tridacna*, among corals. These two very different animals both have calcareous exoskeletons. F, Chemical structure of the polysaccharide chitin. (A from Brusca and Brusca 1978; B after Sherman and Sherman 1976; C from G. F. Lutze and G. Wefer, 1980, J. Foram. Res. 10(4): 251–260; D courtesy G. McDonald; E courtesy P. Fankboner.)

cles, hydrostatic forces, or elastic structures. In animals possessing rigid but articulated skeletons, antagonistic muscles often appear in pairs, for example, **flexors** and **extensors**. These muscles extend across a joint and are used to move a limb or other body part (Figure 11). Most muscles have a discrete **origin**, or point of little body movement, and an **insertion**, or point of major body or limb movement. A classic vertebrate example of this system is the biceps muscle of the human arm, in which the origin is on the scapula and the insertion on the radius bone of the forearm; contraction of the biceps causes flexion of the arm by decreasing the angle between the upper arm and the forearm. Movement of a limb toward the body is brought about by flexor muscles, of which the biceps is an example (Figures 11A and B). The

this recent chemical research has been reviewed in a text by R. A. A. Muzzarelli (1977).

Most skeletons act as the body elements against which muscles operate and by which muscle action is converted to body movement. Because muscles cannot elongate by their own volition, they must be stretched by antagonistic forces—usually other mus-

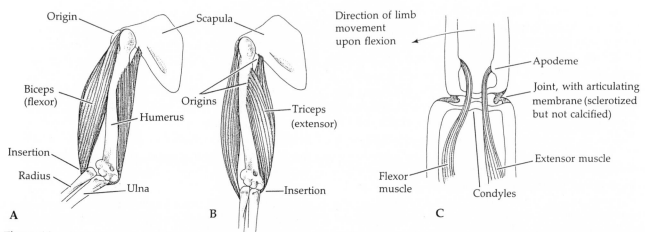

Figure 11

How antagonistic muscles work. A, The biceps is contracted and the triceps relaxed; this combination flexes the forearm. B, The biceps is relaxed and the triceps is contracted; this combination extends the forearm. C, A diagrammatic representation of an arthropod joint, illustrating a similar relationship between flexor and extensor. In this animal, however, the attachment is within an exoskeleton.

muscle antagonistic to the biceps is the triceps, an extensor muscle whose contraction extends the forearm away from the body. Other common sets of antagonistic muscles and actions are **protractors** and **retractors**, which respectively cause anterior and posterior movement of entire limbs at their place of juncture with the body; and **adductors** and **abductors**, which move a body part toward or away from a particular point of reference. Although vertebrates have endoskeletons and arthropods have exoskeletons, most arthropod muscles are arranged in antagonistic sets similar to those seen in vertebrates (Figure 11C). The muscles of arthropods attach to the inside of the skeletal parts, whereas those of vertebrates attach to the outside; but they both operate systems of levers.

Not all muscles attach to rigid endo- or exoskeletons. Some form masses of interlacing muscle fibers, like those in the foot of a snail, or form layers in the walls of "hollow" organs, like those surrounding a gut tube or embedded in the body wall of worms. In these cases the muscles have no definite origin and insertion but act on each other and the surrounding tissues and body fluids to effect changes in the shape of the body or body parts.

The basic physiology and biochemistry of muscle contraction is the same in all animals (vertebrates and invertebrates), although a variety of specialized variations on the basic model have evolved in various taxa. For example, the adductor muscles of clams (the muscle that holds the shell closed) is divided into two parts. One part is heavily striated and used for rapid shell closure (the "quick" muscle); the other is smooth, or tonic, and is used to hold the shell closed for hours or even days at a time (the "catch" muscle). Brachiopods have a similar adductor muscle specialization. Other specializations are found in crustacean muscle innervation, which differs from that typically seen in invertebrates, and in certain insect flight muscles that are capable of contracting at frequencies far higher than any that can be induced by nerve impulses.

Feeding mechanisms

Intracellular and extracellular digestion

Every animal has to locate, select, capture, ingest, and finally digest and assimilate food. Although the physiology of digestion is more-or-less uniform at the biochemical level, considerable variation exists in the mechanisms of capture and digestion as a result of constraints placed on animals by their overall *Baupläne*. For example, unicellular animals do not manipulate and process food in the same fashion as multicelled animals do; nor do diploblastic organisms behave in the same manner as the triploblastic animals do. Digestion is the process of breaking down food by hydrolysis into units suitable to the nutrition of cells. When this breakdown occurs outside the body altogether, it is called **extracorporeal digestion**; when it occurs in a gut chamber of some sort, it is referred to as **extracellular digestion**; and when the process occurs *within* a cell, it is called **intracellular digestion**. Regardless of the site of digestion, all organisms are ultimately faced with the fundamental challenge of cellular capture of nutritional products (food, digested or not). This cellular challenge is met by the processes of **phagocytosis** (literally, "eating by cells") and **pinocytosis** ("drinking by cells"). These processes are mechanically simple and involve the engulfment of food "particles" at the cell surface.

In phagocytosis, extensions of the plasma membrane encircle the food particle, form a pocket in the cell surface, then pinch off the pocket inside the cell (Figure 12A). The resultant intracellular structure is called a **food vacuole**. Because the food particle is inside a chamber formed and bounded by a piece of the original plasma membrane of the cell, some biologists consider that it is not actually "inside" the cell. This point is irrelevant. The plasma membrane surrounding the food vacuole is, of course, no longer part of the cell's outer membrane and in this sense it and whatever is in the vacuole are now "inside" the cell, and the subsequent digestive processes that take place should be considered intracellular (not extracellular). However, food inside the food vacuole is not actually incorporated into the cell's cytoplasm until digestion is completed and the resultant molecules released. Protozoa and sponges rely on phagocytosis as a feeding mechanism, and the digestive cells of metazoan guts take up food particles in the same fashion. Once a cell has phagocytized a food particle and intracellular digestion has been com-

pleted, the remaining waste particles may be carried back to the cell surface by what remains of the old food vacuole, which fuses with the plasma membrane to discharge its wastes in a sort of reverse phagocytosis.

Pinocytosis can be thought of as a highly specialized form of phagocytosis, in which molecule-sized particles are taken up by the cell. Such molecules are always dissolved in some fluid (e.g., a body fluid, or sea water). During pinocytosis, minute invaginations (**pinocytotic channels**) form at specific sites on the cell surface, fill with liquid from the surrounding medium (which includes the dissolved nutritional molecules), and then pinch off to enter the cytoplasm as **pinocytotic vesicles** (Figure 12B). Pinocytosis generally occurs in cells lining some body cavity (e.g., the gut) in which considerable extracellular digestion has already taken place and nutritive molecules have been released from the original food source. In some cases, however, nutritional molecules may be taken up directly from sea water, and there is growing evidence that many invertebrates rely heavily on the direct uptake of dissolved organic matter (DOM) from their environment.

Metazoa generally possess some sort of an internal digestive tract into which food passes. In some there is but one opening, the mouth, through which food is ingested and undigested materials are eliminated (e.g., cnidarians and flatworms). These animals are said to have an **incomplete** or **blind gut**. Most other metazoa have both mouth and anus (a complete gut). This latter arrangement allows a one-way flow of food, as well as specialization of gut regions for various functions such as grinding, secretion, storage, digestion, and absorption. As the noted biologist Libbie Hyman so aptly put it, "The advantages of an anus are obvious."

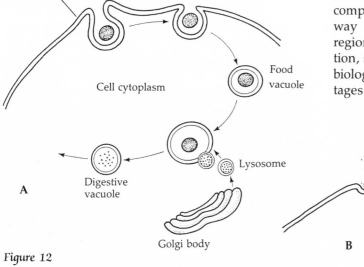

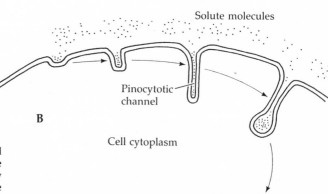

Figure 12

A, Phagocytosis. This diagram illustrates the formation of a food vacuole, the fusion of a lysosome from the Golgi body and the food vacuole, and the remaining digestive vacuole that will carry wastes back to the cell surface. B, Pinocytosis. Nutritive solute molecules attach to binding sites on the plasma membrane of the cell. These then form pinocytotic channels and finally pinch off as pinocytotic vesicles.

Feeding strategies

Just as body architecture influences and limits the digestive modes of invertebrates, it is also intimately associated with the processes of food location, selection, and ingestion. Animals are generally defined as **heterotrophic** organisms (as opposed to autotrophs and saprophytes)—they ingest organic material in the form of other organisms, or parts thereof. However, one group of protozoa utilizes photosynthesis as a primary nutritional strategy—the class Phytomastigophorea (phylum Sarcomastigophora). This protist group is, as one might expect, also classified and discussed in botany texts as single-celled "algae." In addition, many nonphotosynthetic invertebrate groups have developed intimate symbiotic relationships with single-celled algae, especially with certain species of dinoflagellates. These invertebrates use photosynthetic by-products as an accessory (or occasionally as the primary) food source. Most notable in this regard are reef-building corals, giant clams (tridacnids), and certain species of *Paramecium*, flatworms, sea slugs, hydroids, ascidians, sea anemones, and sponges (the freshwater Spongillidae). However, the overwhelming majority of invertebrates are strictly heterotrophic.

Biologists classify heterotrophic animal feeding strategies in a number of ways. For example, animals can be considered **herbivores, carnivores**, or **omnivores**; or they can be classed as **grazers, predators**, or **scavengers**. Animals can be classified by the comparative size of their food or prey (**microphages** or **macrophages**), or they can be classified by the environmental source of their food—**suspension feeders, deposit feeders**, or **detritivores**. These classification schemes and others are instructive, and they are important in the field of ecology, particularly evolutionary ecology.

The forces of evolution that shape animal morphology generally work in concert with those directing animal behavior, and despite myriad "exceptions to the rule," a number of clear relationships between morphology and feeding behavior can be identified. The study of the relationship between structure and function and behavior is called **functional morphology** (or functional anatomy). Studies of this sort have revealed that lineages of animals have radiated within the constraints of restricted anatomical themes. In the remainder of this section we define some important feeding-strategy terms and explain some common themes of feeding.

We assume you are already familiar with the basic concepts of herbivory and carnivory and that you are aware that few animals are strictly one or the other, even though most show a clear preference for either a vegetable or a meat diet. A typical example is the Atlantic purple sea urchin *Arbacia punctulata*, which usually feeds, as do most urchins, on micro- and macroalgae. However, in selected portions of its range, where algae may become seasonally scarce, epifaunal animals constitute the bulk of this urchin's diet. Omnivores, of course, must have the anatomical and physiological capability to capture, handle and digest both plant and animal material. Among invertebrates there are two large categories of feeding strategies in which true omnivory largely prevails: suspension feeding and deposit feeding.

Suspension feeding. **Suspension feeding** is the removal of suspended food particles from the surrounding medium by some sort of capture, trapping, or filtration mechanism. It comprises three basic steps: transport of water past the feeding structures, removal of particles from the water, and transport of the captured particles to the mouth. It is a major mode of feeding in sponges, ascidians, appendicularians, brachiopods, ectoprocts, entoprocts, phoronids, most bivalves, and many crustaceans, polychaetes, and gastropods. The main food selection criterion is particle size, and the size limits of food are determined by the nature of the particle-capturing device. In some cases potential food particles may also be "sorted" on the basis of their size or specific gravity.

Suspension feeding invertebrates generally consume bacteria, phytoplankton, zooplankton, and some detritus. All suspension feeders probably have optimal ranges of particle size; but some are capable, experimentally, of preferentially selecting "enriched" artificial food capsules over "nonenriched" (nonfood) capsules, an observation suggesting that chemosensory selectivity may occur in situ as well. In order to capture food particles from their environment, suspension feeders either must move part or all of their body through the water, or water must be moved past their feeding structures. As with locomotion in water, the relative motion between a solid and liquid during suspension feeding creates a system that behaves according to the concept of Reynolds numbers. Because the flow rates in such systems are almost always very low and the feeding structures generally small (e.g., cilia, flagella, setae), particle capture by suspension feeders typically takes place at very low Reynolds numbers. Furthermore, most of the feeding structures of such animals are more-or-less cylindrical, making the application of Reynolds' experimentally derived principles particularly appropriate.

Recall that at low Reynolds numbers viscous forces dominate, and water flow over small feeding structures is laminar and nonturbulent and ceases instantaneously when energy input stops. Thus, in the absence of inertial influence, suspension feeders that generate their own feeding currents expend a great deal of energy in the process. Some suspension feeders conserve energy by depending to various degrees on prevailing water movements to continually replenish their food supplies (e.g., barnacles on wave-swept shores and mole crabs in the wash zone on sandy beaches). For those living in relatively still waters, however, the effort expended for feeding is a major part of their energy budget.

Only a relatively few suspension feeders are truly filterers. Because of the principles outlined above, it is energetically costly to drive water through a fine-meshed filtering device. For small animals, this is somewhat analogous to moving a fine-mesh filter through molasses. Such actual sieving does occur, most notably in some ciliate protozoa, certain bivalve molluscs, many tunicates, some larger crustaceans, and some worms that produce mucous nets. However, most suspension feeders employ some less expensive method of capturing particles from the water, one that does not involve continuous filtration. Many invertebrates expose a sticky surface, such as a coating of mucus, to flowing water. Suspended particles contact and adhere to the surface and then are moved to the mouth by ciliary tracts (used by crinoids), setal brushes (used by certain crustaceans), or by some other means of transport. Other "contact" suspension feeders living in still water may simply expose a sticky surface to the rain of particulate material settling down from the water above, thus letting gravity do much of the work of food-getting. Some species of oysters are suspected of this feeding strategy, at least on a part-time basis. Several other "contact" methods of suspension feeding may occur, but all eliminate the costly activity of actual sieving at low Reynolds numbers.

The trick of removing food particles from the local environment has been achieved by the use of five fundamentally different mechanisms. Because there are a limited number of ways in which animals can suspension feed, it is not surprising that a great deal of evolutionary convergence has appeared in this regard.

Among some aquatic arthropods, certain limbs are modified with rows of feather-like setae adapted for generating water currents across parts of the body and removing selected particulate matter from it (Figure 13). The size of the particles captured is often directly proportional to the "mesh" size of the interlaced setae on the actual food-capture structure. In sessile arthropods, like barnacles, the feeding appendages are swept through the water or held taut against moving water. In either case, such sessile animals are dependent upon local currents to continually replenish their food supply and particle capture is by simple filtration. Motile setal-net feeders, like many larger planktonic crustaceans, aquatic insect larvae, or certain benthic crustaceans (e.g., porcelain crabs), may have particular appendages modified to generate a current across the feeding appendages that bear the capture setae. Sometimes these same appendages serve simultaneously for locomotion. In the cephalocarid crustaceans, for example, complex coordinated movements of the highly setose thoracic legs produce a constant current of water (Figure 13E). These appendages simultaneously capture food particles from the water and collect them in a median ventral food groove at the leg bases, where they are passed forward to the mouth region.

A second, nonfiltering, suspension feeding method has been called "scan-and-trap" (LaBarbera 1984). The general strategy here is to move water over part or all of the body, detect suspended food particles, isolate the particle(s) in a small parcel of water, and process only that water by some method of particle extraction. The animal thus avoids the energetic expense of continuously driving water over the feeding surface at low Reynolds numbers and paying the price of viscous forces. The precise method of particle detection, isolation, and capture varies among different invertebrates that use the scan-and-trap technique; but this basic strategy is probably employed by many crustaceans (e.g., planktonic copepods), many ectoprocts, and a variety of larval forms. The details of suspension feeding mechanisms and their relationships to the principles of fluid dynamics are only beginning to be understood. Papers cited at the end of this chapter summarize much of what is known and provide additional references.

A third suspension feeding device is the **mucous net**, or **mucous trap**, wherein patches or a sheet of mucus may be used to trap suspended food particles. Most mucous-net feeders consume their net along with the food, and thus recycle the chemicals used to produce it. Again, sessile and sedentary species must rely largely on local currents to keep a fresh supply of food coming their way. Many such forms, especially benthic burrowers, actively pump water through their burrow or tube, where it passes across

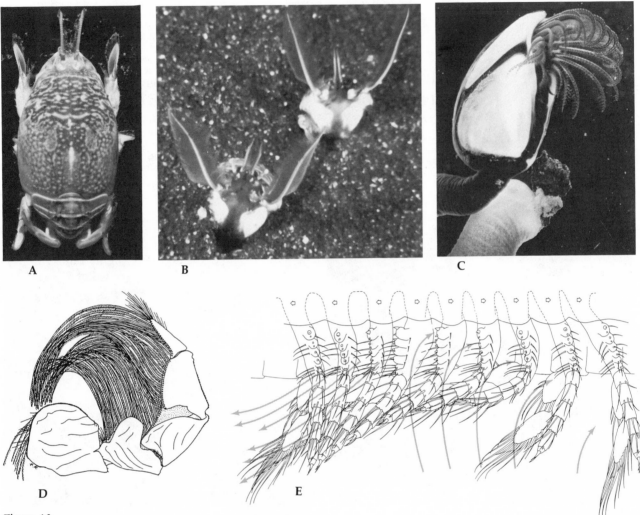

Figure 13

A variety of setal-net suspension-feeding invertebrates. A, The sand crab *Emerita*. B, *Emerita* feeding on a beach, with its feeding antennae extended out of the sand. C, Goose barnacles, *Pollicipes*, with feeding appendages extended. D, The third maxilliped of the porcelain crab *Petrolisthes elegans*. Note the long, dense setae used in feeding. E, A portion of the trunk (sagittal view) of a cephalocarid crustacean during the metachronal cycle of the feeding limbs. The arrows indicate the direction of water currents; the arrow direction above each trunk limb indicates the limb's direction of movement. (A,C courtesy G. McDonald; B courtesy of A. Wenner; D courtesy of J. Haig; E from Sanders 1963.)

or through the mucous sheet. A classic example of mucous-trap feeding is seen in the annelid worm *Chaetopterus* (Figure 14A). This animal lives in a U-shaped tube in the sediment and pumps water through the tube and through a mucous "net." As the "net" fills with trapped food particles, it is periodically manipulated and rolled into a ball, which is then passed to the mouth and swallowed. Another good example of mucous-trap feeding is seen the tube-building gastropods (family Vermetidae). These wormlike snails construct colonies of meandering calcareous tubes in the intertidal zone. Each animal secretes a mucous trap that is deployed just outside the opening of the tube, until nearly the entire colony surface is covered with mucus. Suspended particulate matter settles and becomes trapped in the mucus. At periodic intervals each animal withdraws its mucous sheet and swallows it, whereupon a new sheet is immediately constructed.

A fourth type of suspension feeding is the **ciliary-mucous mechanism**, in which rows of cilia carry a mucous sheet across some structure while water is passed through or across it. Ascidians (sea squirts) are ciliary-mucous feeders (Figure 14B). They move a more-or-less continuous mucous sheet across their sievelike pharynx, while at the same time pumping

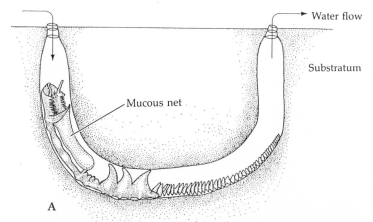

Water flow

Substratum

Mucous net

A

Figure 14

Some mucous-net and ciliary-mucus suspension feeders. A, The annelid worm *Chaetopterus* in its burrow. Note the direction of water flow through its mucous net. B, The solitary ascidian *Styela* has incurrent and excurrent siphons through which water enters and leaves the body. Inside, the water passes through a sheet of mucus covering holes in the wall of the pharynx. C–D, A maldanid polychaete, *Praxillura maculata*. This animal constructs a membranous tube that bears 6–12 stiff radial spokes. A mucous web hangs from these spokes and passively traps passing food particles. In D, the worm's head is seen sweeping around the radial spokes to retrieve the mucous web and its trapped food particles. (A after Fauchald and Jumars 1979; B courtesy G. McDonald; C, D from N. McDaniel and K. Banse, 1979, Mar. Biol. 55: 129–132.)

B   C   D

water through it. Fresh mucus is secreted at one side of the pharynx while the food-laden mucus is ingested at the other side, an arrangement not unlike a conveyer belt. Several polychaete groups also make primary use of the ciliary-mucous feeding technique (Figures 14C and D). For example, some species of tube-dwelling fan worms feed with a crown of tentacles that are covered with cilia and mucus, or may be grooved and have ciliary-mucous tracts in the grooves that slowly move captured food particles to the mouth. In this case, size selection may result in the larger particles being rejected, the smallest particles being ingested, and medium-sized particles being carried to a special storage area where they may eventually be used in tube construction. Many sand dollars capture suspended particles, especially diatoms, on the mucus-covered spines of the test surface, where they are transported by the tube feet

and ciliary currents to **food tracts**, and then to the mouth.

A fifth general kind of suspension feeding is **tentacle** or **tube feet suspension feeding**. In this strategy, some sort of tentacle-like structure captures larger food particles, with or without the aid of mucus. Food particles captured by this mechanism are generally larger than those captured by setal or mucous traps and sieves. Examples of tentacle or tube feet suspension feeding are most commonly encountered in the echinoderms (e.g., many brittle stars and crinoids) and cnidarians (e.g., certain sea anemones and corals) (Figure 15).

Much research has been done on suspension feeding in the past 15 years, and we now know what size range of particles many animals feed on and what kinds of capture rates they have. In general, feeding rates increase with food particle concentra-

Figure 15

Tube feet suspension feeding. Food-particle capture in the brittle star *Ophiothrix fragilis*. The photographs show two views of a captured food particle being transported by the arm podia to the mouth. (From G. F. Warner and J. D. Woodley, 1975, J. Mar. Biol. Assoc. U.K. 55: 199-211.)

tion to a plateau, above which the rate levels off. At still higher particle concentrations, entrapment mechanisms may become overtaxed or clogged and feeding is inhibited or simply ceases. In sessile and sedentary suspension feeders, for example, pumping rates decrease quickly as the amount of suspended inorganic sediment (mud, silt, and sand) increases beyond a given concentration. For this reason, the amount of sediment in coastal waters limits the distribution and abundance of certain invertebrates such as clams, corals, sponges, and ascidians. Many tropical coral reefs are dying as a result of increased coastal sediment loads generated by deforestation and run-off.

Deposit feeding. The deposit feeders make up another major group of omnivores. These animals obtain nutrients from the sediments of soft-bottom habitats (muds and sands) or terrestrial soils, but their techniques for feeding are diverse. **Direct deposit feeders** simply swallow large quantities of sediment—mud, sand, soil, organic matter, everything. The usable organics are digested and the unusable materials passed out the anus. The resultant fecal material is essentially "cleaned dirt." This kind of deposit feeding is seen in many polychaete annelids, some snails, some sea urchins, and also in most earthworms (Figure 16A). Some deposit feeders utilize tentacle-like structures to consume sediment, such as some sea cucumbers, most sipunculans, certain clams, and several types of polychaetes (Figures 16B and C). Tentacle-utilizing deposit feeders pref-

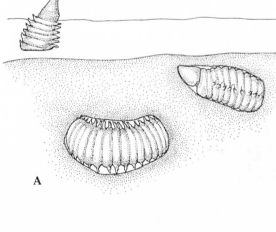

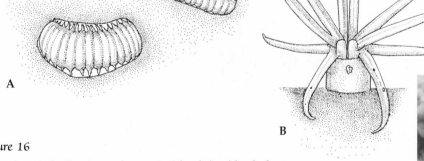

Figure 16

Some deposit-feeding invertebrates. A, A lumbrinerid polychaete burrowing in the sediment. This worm is a subsurface deposit feeder. B, The sabellid polychaete *Manayunkia aestuarina* in its feeding posture. A pair of branchial filaments are being used to feed. The large particle falling in front of the tube has just been expelled from the branchial crown by the ejection current. C, A surface deposit feeding holothurian. (A,B after Fauchald and Jumars 1979; C courtesy of P. Fankboner.)

erentially remove only the uppermost deposits from the sediment surface and thus consume a far greater percentage of living (especially bacteria and protozoa) and detrital organic material that accumulates there than do the burrowing deposit feeders. These animals are generally called **selective deposit feeders**. Aquatic deposit feeders may also rely to a significant extent on fecal material that accumulates on the bottom, and many will actively consume their own fecal pellets (**coprophagy**), which may contain some undigested or incompletely digested organic material as well as microorganisms.

The ecological role of deposit feeding in sediment turnover is a critical one. When burrowing deposit feeders are removed from an area, organic debris quickly accumulates, subsurface oxygen is depleted by bacterial decomposition, and anaerobic sulfur bacteria eventually bloom. On land, earthworms and other burrowers are important in maintaining the health of agricultural and garden soils.

Herbivory. The following discussion deals with **macroherbivory**, that is, the consumption of macroscopic plants. Herbivory is common throughout the animal kingdom. It is most dramatically illustrated when certain invertebrate herbivore species undergo a temporary population explosion. Famous examples are outbreaks of locust, which can destroy virtually all plant material in their path of migration. In a similar fashion, extremely high numbers of the Pacific sea urchin *Strongylocentrotus franciscanus* result in the wholesale destruction of kelp beds (Figure 18). Unlike

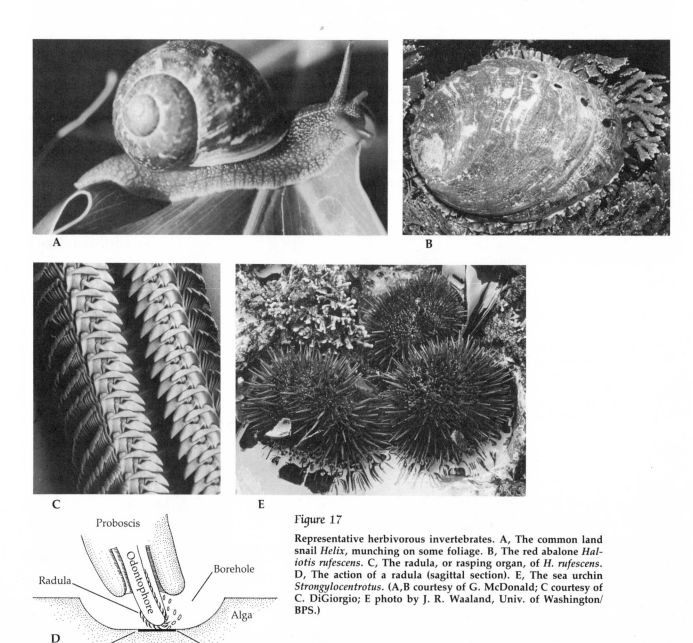

Figure 17

Representative herbivorous invertebrates. A, The common land snail *Helix*, munching on some foliage. B, The red abalone *Haliotis rufescens*. C, The radula, or rasping organ, of *H. rufescens*. D, The action of a radula (sagittal section). E, The sea urchin *Strongylocentrotus*. (A,B courtesy of G. McDonald; C courtesy of C. DiGiorgio; E photo by J. R. Waaland, Univ. of Washington/BPS.)

suspension and deposit feeding herbivory, in which mostly single-celled and microscopic plant matter is consumed, macroherbivory requires the ability to "bite and chew" large pieces of vegetable matter. Although the evolution of biting and chewing mechanisms took place within the architectural framework of a number of different invertebrate lineages, it is always characterized by the development of hard (usually calcified or chitinous) "teeth," which are manipulated by powerful muscles. Members of a number of major invertebrate taxa have evolved macroherbivorous lifestyles, including molluscs, polychaetes, arthropods, and sea urchins (Figures 17 and 18).

Many molluscs have a unique rasping structure called a **radula**, which is a muscularized beltlike rasp armed with chitinous teeth. Herbivorous molluscs use the radula to scrape algae off rocks or tear pieces of algal fronds or the leaves of terrestrial plants. The radula acts like a curved file that is drawn across the feeding surface (Figures 17C and D). Herbivorous polychaetes such as nereids (family Nereidae) have sets of large chitinous teeth on an eversible pharynx or proboscis. The proboscis is protracted by hydrostatic pressure, exposing the teeth, which by muscular action tear or scrape off pieces of algae that are swallowed when the proboscis is retracted. As might be expected, the toothed pharynx of polychaetes is also reasonably suited for carnivory, and some primarily herbivorous polychaetes can switch to meat-eating when algae are scarce.

Macroherbivory in arthropods can be found in nearly every major group, but it is best studied in certain insects and crustaceans. Both of these large groups have powerful mandibles capable of biting off pieces of plant material and subsequently grinding or chewing them before passing them on to the mouth. Macroherbivorous arthropods usually are able to temporarily switch to carnivory when necessary. This switching is rarely seen in the terrestrial herbivores because it is almost never necessary; terrestrial plant matter can almost always be found. This is not the case in marine environments, however, where algal supplies may at times be very limited. Some herbivorous invertebrates cause serious damage to wooden man-made structures (homes, pier pilings, boats, etc.) by burrowing through and consuming the wood (Figure 19).

Carnivory and scavenging. The most sophisticated methods of feeding are those that require the capture of live food by pursuit, or **predation**. Most predators will, however, consume dead or dying animal matter when live food is scarce. Only a few generalizations about the many kinds of predation are presented here; detailed discussions of various

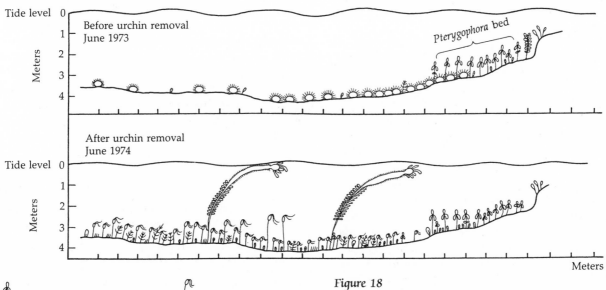

Pterygophora bed

Before urchin removal June 1973

After urchin removal June 1974

Tide level
Meters

Meters

ꙮ	*Pterygophora californica*
ꙮ	*Laminaria groenlandica*
ꙮ	*Laminaria setchellii*
ꙮ	*Costaria costata*
ꙮ	Young plants
ꙮ	*Desmarestia ligulata*

ꙮ	*Nereocystis leutkeana*
ꙮ	*Polysiphonia pacifica*
ꙮ	*Macrocystis integrifolia*
ꙮ	*Strongylocentrotus franciscanus*

Figure 18

In a study by D. Pace, the sea urchin *Strongylocentrotus franciscanus* was removed and continuously excluded from an experimental plot on the coast of western Canada. The exclusion experiment lasted a full year. This illustration gives profile views of the experimental site before and after urchin removal and illustrates the effect of this herbivore in controlling numerous algae, including the giant kelps *Macrocystis* and *Nereocystis*. (After D. Pace, 1974, Proc. 8th Internat. Seaweed Symp., Wales.)

Figure 19

Wood from the submerged part of an old dock piling, split open to show the work of the wood-boring bivalve shipworm, *Teredo navalis.* **The shell valves are so reduced that they can no longer enclose the animal; instead they are used as "auger-blades" in boring. The walls of the burrow are lined with a smooth, calcareous, "shelly" material. (Photo courtesy of P. Fankboner.)**

2. **Arrestant**   a stimulus that causes the prey animal to cease locomotion when in close contact with the apparent source.
3. **Repellent**   a stimulus that causes the prey animal to orient away from or become nonreceptive to the apparent source.
4. **Incitant**   a stimulus that evokes initiation of feeding (tasting).
5. **Suppressant**   a stimulus that inhibits or prevents initiation of feeding.
6. **Stimulant**   a stimulus that promotes ingestion and continuation of feeding.
7. **Deterrent**   a stimulus that prevents continuation of feeding or hastens termination of feeding.

Feeding attractants, arrestants, and repellents are borne in the medium (air or water) and are characteristically detectable in very low concentrations. Incitants, suppressants, and deterrents may also be borne in the medium; but, more often, they are detected only after direct contact between chemoreceptor and food source.

Chemoreceptors tend to be equally distributed around the bodies of radially symmetrical carnivores (e.g., jellyfish) but, coincidentally with cephalization, most invertebrates have their "tasters" concentrated in the head region. Chemosensation, of course, is not restricted to carnivores, but is found in species exhibiting all kinds of feeding. Critical thresholds of certain water-borne chemicals are sometimes necessary to initiate feeding activities of suspension feeders. Even certain bacteria are known to migrate along a concentration gradient in response to certain substances usable in their metabolism.

Predators may be classified by how they capture their prey—as **motile stalkers, lurking predators** (ambushers), **sessile opportunists,** or **grazers.** Some examples are shown in Figure 20. Stalkers, who actively pursue their prey, include members of dispar-

taxa are presented in their appropriate chapters.

Active predation involves several distinct and recognizable steps: prey location (predator orientation), pursuit, capture, handling, and, finally, ingestion. Prey location requires a certain level of nervous system sophistication in which specialized sense organs are present (see later in this chapter). Most invertebrates rely primarily on **chemosensory** location of prey, although many also use visual orientation, touch, and vibration detection. In most metazoa, chemosensation is used first to "home in" on or track the prey. The visual/tactile sensory modes come into play once the prey is in close proximity. A complex array of chemically initiated behaviors exists. Lindstedt (1971) has classified prey-produced chemicals into seven categories:

1. **Attractant**   a stimulus to which a prey animal responds by orientation toward or becoming receptive to the apparent source; may operate over long distances.

Figure 20

Some predatory invertebrates. A, Most octopuses are active hunting predators; this one is a member of the genus *Eledone.* **B, The crown-of-thorns sea star,** *Acanthaster,* **feeds on corals. C, The moon snail,** *Polinices,* **drills holes in the shells of bivalve molluscs to feed on the soft parts. D, A mantis shrimp (stomatopod); the two drawings (E) depict its raptorial strike to capture a passing fish. F, The predatory flatworm** *Mesostoma* **attacking a mosquito larva. G, A three-photo sequence showing the cone snail** *Conus purpurescens* **eating a fish. Tiny, harpoon-like radular teeth are fired into the prey, poisoning it prior to ingestion. H,** *Acanthina,* **a predatory gastropod feeding on a barnacle. (A–C courtesy of P. Fankboner; D courtesy of A. Kerstitch; E after R. Caldwell and H. Dingle, 1976, Sci. Am. (Jan.): 81-89; F courtesy T. Case, photo by J. K. Clark; G courtesy J. Nykbakken; H courtesy of D. Perry.)**

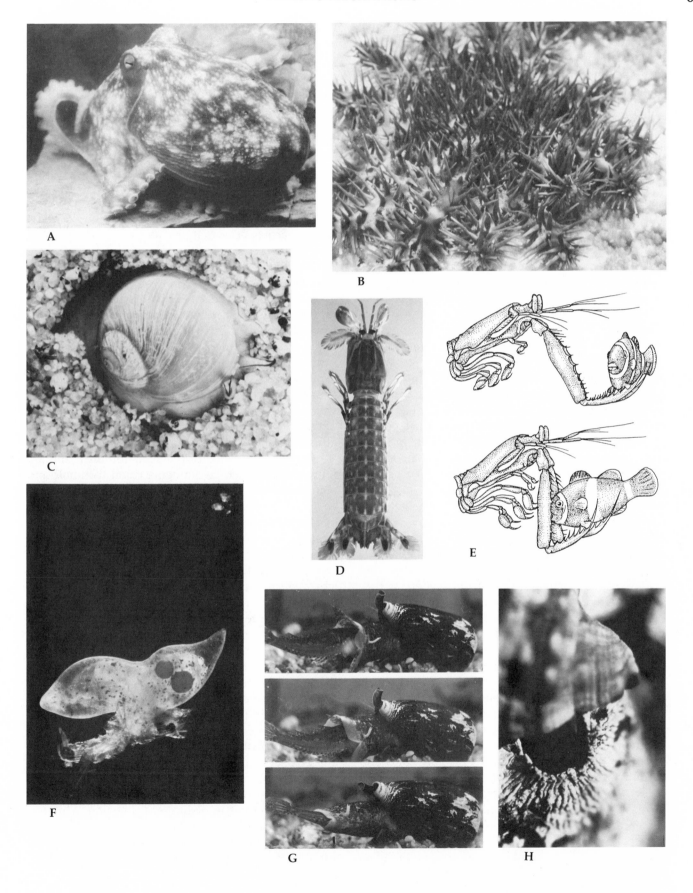

A

B

C

D

E

F

G

H

ate invertebrate groups such as ciliate protozoa; polyclad, nemertean, and polychaete worms; gastropods; octopuses and squids; crabs; and sea stars. In all these groups, chemosensation is very important in locating potential prey, although cephalopods are known to be the most highly visual of all the invertebrate predators. One does not usually think of flatworms as particularly voracious predators. However, most are; and certain species are known to be substantial predators on corals, barnacles, and aquatic insect larvae. Fluctuations in mosquito population density in some areas have been attributed to predation on the mosquito larvae by flatworms.

Lurking predators are those that sit and wait for their prey to come within capture distance, whereupon they quickly seize the victim. Many lurking predators, such as certain species of mantis shrimps (stomatopods), crabs, spiders, and polychaetes, live in burrows from which they emerge to capture passing prey. Perhaps the best known of the lurking predators are certain spiders, whose webs do most of the work for them. There are even ambushing planarians, which produce mucous patches that form sticky traps for their prey. The cost of building traps (e.g., webs, mucous patches, and burrows) is significant. Ant lions, for example, may increase their energy consumption as much as eightfold when building their sand capture pits, and energy lost in mucus secretion by planarians may account for 20 percent of the worm's energy.

Sessile opportunists operate in much the same fashion as lurking predators do, but they lack the mobility of the latter. The same may be said for drifting opportunists, such as jellyfishes. Many sessile predators, such as some protozoa, barnacles, and cnidarians, are actually suspension feeders with a strong preference for live prey.

Grazing carnivores move about the substratum picking at the epifauna. Grazers may be fairly indiscriminate, consuming whatever happens to be present, or they may be fairly choosy about what they eat. In either case, their diet consists largely of sessile and slow-moving animals, such as sponges, ectoprocts, tunicates, snails, small crustaceans, and worms. As might be suspected, most grazers are omnivorous to various degrees, consuming plant material along with their animal prey. Many crabs and shrimps are excellent grazers, continuously moving across the benthos and picking through the epibiota for tasty morsels. Sea spiders (pycnogonids) and some carnivorous sea slugs can also be classed as grazers on hydroids, ectoprocts, sponges, tunicates, and other sessile epifauna. Ovulid snails (Ovulidae)

inhabit, and usually mimic, the gorgonians and corals upon which they slowly crawl about, nipping off polyps as they go.

A popular new field of ecological research is called **optimal foraging theory**. Basically, this theory suggests that animals evolve toward the most economical or efficient feeding strategies possible within the constraints of their morphological–physiological capabilities and local environmental regime. For example, say a predatory snail population is specialized to feed on barnacles. If several kinds of barnacles are available in the snail's local environment, optimal foraging theory suggests that the snail should prefer the barnacle species that provides it with the greatest nutritional return for its feeding-energy output. While a large barnacle may provide more nutrition per animal, it may take so long to open and consume it (handling time) that the snail would get a better return for its energy by consuming several smaller barnacles instead. Prey handling and capture time do influence feeding behavior in many different animal groups, both invertebrate and vertebrate. Several recent studies have suggested that, generally speaking, predators prefer medium-sized prey individuals over very small or very large individuals; that is, food obtained per unit handling time is often maximized at intermediate prey sizes.

On the other side of the predator–prey coin are prey adaptations to avoid being eaten. Animals and plants have evolved an impressive variety of mechanisms that help deter would-be predators. These mechanisms include structural protection, cryptic coloration, chemical repellents and deterrents, and avoidance behavior. One of the most intriguing areas of research in this regard is that of alarm and repellent pheromones. A **pheromone** is any chemical released by an organism into the environment for the purpose of inter- or intraspecific communication. In the terrestrial environment, alarm pheromones have been studied extensively in social insects. In the marine environment, they are less well understood. One recent study (Sleeper et al. 1980) showed that the predatory sea slug *Chelidonura* (= *Navanax*), which produces a typical gastropod mucopolysaccharide slime trail as it stalks its prey or mate, releases a potent alarm pheromone onto its trail when threatened. A second *Chelidonura*, encountering the trail of the first one, will react violently when the alarm substance is confronted, contracting and turning to beat a fast retreat in a different direction. This has been referred to as "trail-breaking behavior"; it is a well-known strategy of social insects. Ants are quite adept at "smelling out" differences between themselves and ants of different colonies or different species that might cross their path. Predatory species take advantage of this to locate the trails of other ant species, which they may prey upon or take for slaves by a process that involves the release of other pheromones that throw the enemy colony into disorganized panic.

One special category of carnivory is **cannibalism**, or intraspecific predation. G. A. Polis (1981) examined over 900 published reports describing cannibalism in about 1,300 different species of animals. In general, Polis found that larger species (and also larger individuals in any given species) are the most likely to be cannibals. By far, the majority of the victims are juveniles. Interestingly, in a number of invertebrate groups cannibalism occurs when smaller individuals band together to attack and consume a larger individual. Furthermore, females tend generally to be more cannibalistic than males, and males tend to be eaten far more often than females. Polis concluded that cannibalism is a major factor in the biology of many species and may influence population structure, life history, behavior, and competition for mates and resources. In many species, filial cannibalism is common, in which a parent will eat its dying, deformed, weak, or sick offspring. Polis goes so far as to point out that *Homo sapiens* may be "the only species capable of worrying whether its food is intra- or extraspecific."

Dissolved organic matter. The possibility that dissolved organic compounds may contribute significantly to the nutrition of marine organisms is not a new idea (e.g., see the excellent review by Krogh 1931). However, the *relative* role of dissolved organic matter (DOM) in the nutrition of aquatic animals is still a controversial problem. The total living biomass of the world's oceans is estimated at about 2×10^9 tons of organic carbon (this is 500 times the amount of organic carbon in the terrestrial environment). Furthermore, an additional 20×10^9 tons of particulate organic matter is estimated to occur in the seas, and another 200×10^9 tons of organic carbon (C) may occur in the seas as DOM. Thus, only a small fraction of the organic carbon in the world's seas actually exists in living organisms. Amino acids and carbohydrates are the most commonly utilized dissolved organics. Typical oceanic values of DOM range from 0.4 to 1.0 mg C/liter, but may reach 8.0 mg C/liter near shore. Pelagic and benthic algae release copious amounts of DOM into the environment, as do certain invertebrates. Coral mucus, for example, has been shown to be an important fraction of reef detrital material, as it contains significant amounts of nitrogen-rich and energy-rich compounds, including mono- and polysaccharides and amino acids. Other sources of DOM include decomposing tissue, detritus, fecal material, and metabolic by-products discharged into the environment.

It appears as though many members of almost all aquatic invertebrate phyla are capable of absorbing DOM to some extent, and virtually all marine ciliary-mucous suspension feeders tested have shown the ability to take up dissolved free amino acids from a dilute external medium. Furthermore, inorganic particles of colloidal dimensions provide a surface on which small organic molecules are concentrated by adsorption, thereon to be captured and utilized by suspension feeding invertebrates. Reliance on DOM as a substantial nutritive resource is probably greatest in invertebrate larvae and soft-bodied animals, and especially in benthic invertebrates living in sediments rich in organic matter. One study (Reid and Bernard 1980) showed that certain species of bivalves of the genus *Solemya* have highly reduced guts and apparently survive entirely by active absorption through the gill lamellae of DOM from sea water. For many years, similar suspicions have held for pogonophorans, which lack a functional alimentary canal altogether. Evidence from numerous studies indicates absorption of DOM occurs directly across the body wall of invertebrates, as well as via the gills and gut lining. Interestingly, most freshwater organisms seem incapable of removing small organic molecules from solution at anything like the rate characteristic of marine invertebrates. It has been suggested that the uptake of DOM is retarded by the process of osmoregulation. Also, with the probable exception of the aberrant hagfish, marine vertebrates seem not to utilize DOM to any significant extent.

Excretion and osmoregulation

We define excretion broadly as the elimination from the body of metabolic waste products, including carbon dioxide and water, which are produced primarily by cellular respiration, and excess nitrogen, which is produced as ammonia from deamination of amino acids. The excretion of respiratory carbon dioxide is generally accomplished by structures other than those associated with these other waste products and is discussed in the following section of this chapter. The excretion of nitrogenous wastes is usually intimately associated with the regulation of water and ion balance (**osmoregulation**) within the body fluids, so these processes are considered together in this section. The various aspects of excretion, osmoregulation, and ion regulation serve not only to rid the body of potentially toxic wastes, but also to maintain concentrations of the various components of body fluids at levels appropriate for metabolic activities. As we shall see, these processes are structurally and functionally tied to the overall level of body complexity and construction, the nature of other physiological systems, and the environment in which the animal lives. We again emphasize the necessity of looking at whole animals, the integration of all aspects of their biology and ecology, and the possible evolutionary histories that could have produced compatible and successful combinations of functional systems.

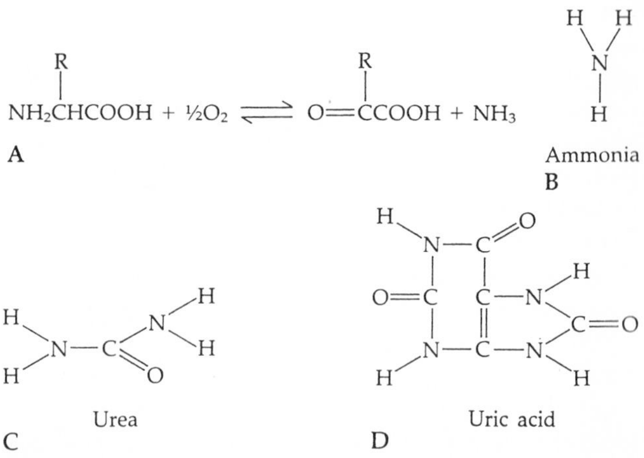

Nitrogenous wastes and water conservation

The source of most of the nitrogen in an animal's system is amino acids produced from the digestion of proteins in the diet. Once absorbed, these amino acids may be utilized for building new proteins, or they may be deaminated and the residues used to form other compounds (Figure 21). The excess nitrogen released during deamination is typically liberated from the amino acid in the form of **ammonia** (NH_3), a highly soluble but quite toxic substance that must be eliminated quickly or converted to a less toxic form. The excretory products of vertebrates have been studied much more extensively than those of invertebrates, but the available data on the latter allow some generalizations. Probably no animal excretes excess nitrogen in a single chemical form. However, one nitrogenous waste form tends to predominate in a given species, and the nature of that chemical is related to the availability of environmental water.

The major excretory product in most marine and freshwater invertebrates is simply ammonia, since their environment provides an abundance of water as a medium for rapid dilution of this toxic substance. Such animals are said to be **ammonotelic**. Being highly soluble, ammonia diffuses easily through fluids and tissues, and much of it is lost straight across the body walls of ammonotelic animals. Animals that do not possess definite excretory organs (e.g., sponges, cnidarians, and echinoderms) are more-or-less limited to the production of ammonia and thus are restricted to aquatic habitats. Furthermore, ammonia tolerance is known to be much higher in lower eukaryotes (and prokaryotes) than in higher eukaryotes.

Terrestrial invertebrates (indeed, all land ani-

Figure 21

Nitrogenous waste products. A, The general reaction for deamination of an amino acid, producing a keto acid and ammonia. B–D, The structures of three common excretory compounds. B, Ammonia. C, Urea. D, Uric acid.

mals) have water conservation problems. They simply cannot afford to lose much body water in the process of diluting their wastes. These animals have sophisticated excretory organs or other structures that process the nitrogenous wastes to produce more complicated but far less toxic substances. These compounds are energetically expensive to produce, but they often require relatively little or no dilution by water, and they can be stored within the body prior to excretion.

There are two major metabolic pathways for the detoxification of ammonia: the urea pathway and the uric acid pathway. The products of these pathways, **urea** and **uric acid**, are illustrated in Figure 21 along with ammonia for comparison. **Ureotelic** animals include amphibians, mammals, and cartilaginous fishes (sharks and rays). Urea is a relatively rare and insignificant excretory compound among invertebrates. On the other hand, the ability to produce uric acid is critically associated with the success of invertebrates on land. **Uricotelic** animals have capitalized on the relative insolubility (and very low toxicity) of uric acid, which is generally precipitated and excreted in a solid or semisolid form with little water loss. Most of the land-dwelling arthropods and snails are uricotelic, and all have evolved structural and physiological mechanisms for the incorporation of excess nitrogen into molecules of uric acid. We emphasize that various combinations of these and other forms of nitrogen excretion are found in most animals. In some cases, individual animals can vary the combination depending on short-term environmental changes affecting water loss problems.

Osmoregulation and habitat

In addition to its relationship to excretion, osmoregulation is directly associated with environmental conditions. As mentioned in Chapter 1, the composition of sea water and that of the body fluids of most invertebrates is very similar, both in terms of total concentration and the concentrations of many particular ions. Thus, the body fluids of many marine invertebrates and their fluid habitats are close to being **isotonic**. But we hasten to add that probably no animal has body fluid that is exactly isotonic with sea water, and therefore all are faced with the need for some degree of osmoregulation. Nonetheless, marine invertebrates certainly do not face the extreme osmoregulatory problems encountered by land and freshwater forms.

As shown in Figure 22, the body fluids of freshwater animals are strongly **hypertonic** with respect to their environment and thus face serious problems of water influx as well as potential loss of precious body salts. Terrestrial animals are exposed to air and thus also to problems of water loss. The evolutionary invasion of land and fresh water must have been accompanied by the development of mechanisms to solve these problems, and only a relatively small number of invertebrate groups have managed to do this. Animals inhabiting freshwater and terrestrial habitats generally have excretory structures that are responsible for eliminating or retaining water as needed, and they frequently possess modifications of the body wall to reduce overall permeability. The most successful invertebrate *Baupläne* on land, and in some ways for all environments, are those of the arthropods and gastropods. Their effective excretory structures and thickened exoskeletons provide them with physiological osmoregulatory capabilities plus protection against desiccation.

Osmoregulatory problems of aquatic animals are, of course, determined by the salinity of the environmental water relative to the body fluids (Figure 22). Invertebrates respond physiologically to changes in environmental salinities in one of two basic ways. Some, such as most freshwater forms (certain crustaceans, protozoa, oligochaetes, and other worms), maintain their internal body fluid concentrations regardless of external conditions and are thus called **osmoregulators**. Others, including a number of intertidal and estuarine forms (mussels and some other bivalves, and a variety of soft-bodied animals), allow their body fluids to vary with changes in environmental salinities; they are appropriately called **osmoconformers**. Again, even the body fluids of marine, so-called osmoconformers are generally not exactly isotonic with respect to their surroundings; thus these animals must osmoregulate slightly. Neither of these strategies is without limits, and tolerance to various environmental salinities varies among different species. Those that are restricted to a very narrow range of salinities are said to be **stenohaline**, and those that tolerate relatively extensive variations, such as many estuarine animals, are **euryhaline**.

While the preceding discussion may seem clearcut, we hasten to add that it is an oversimplification. Research indicates that experimental data from whole animals tell only part of the story of osmoregulation. When a whole marine animal is placed in a hypotonic medium it tends to swell (if it is an osmoconformer) or to maintain its normal body volume (if it is an osmoregulator). Even at this gross level, most inver-

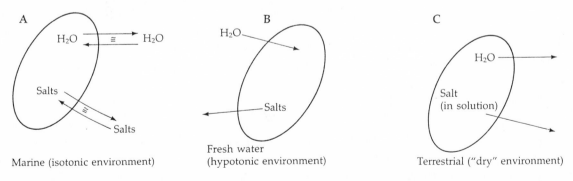

Marine (isotonic environment)

Fresh water (hypotonic environment)

Terrestrial ("dry" environment)

Figure 22

Relative osmotic and ionic conditions existing between marine, freshwater, and terrestrial invertebrates and their environments. The arrows indicate the directions in which water and salts move passively in response to concentration gradients. Remember that in each of these cases movement occurs in both directions, but it is the potential *net* movement along the gradient that is important and against which freshwater and terrestrial animals must constantly battle. For marine invertebrates, the body fluids and the environment are nearly isotonic to one another and there is little net movement in either direction.

tebrates usually show evidence of both conforming and regulating to various degrees. For instance, an osmoconformer generally swells for a period of time in a lowered salinity environment and then begins to regulate. Its swollen volume will decrease, although probably not to its original size. The same is true of most osmoregulators when faced with a decrease in environmental salinity, but the degree of original swelling is much reduced. In both cases, the swelling of the body is a result of an influx of environmental water into the extracellular body fluids (blood, coelomic fluids, and intercellular fluids). Within limits, this excess water is handled by excretory organs and various surface epithelia of the gut and body wall. However, the second part of the osmoregulatory phenomenon takes place at the cellular level. Obviously, as the tonicity of the body fluids drops with the entrance of water, the cells in contact with those fluids are placed in conditions of stress—they are now in hypotonic environments. To be sure, these stressed cells swell to some degree because of the diffusion of water into their cytoplasm, but not to the degree one might expect given the magnitude of the osmotic gradient to which they are subjected. Cellular-level osmoregulation is accomplished by a loss of dissolved materials from the cell into the surrounding intercellular fluids. The solutes released from these cells include both inorganic ions and free amino acids. The actual mechanisms involved remain somewhat elusive. The point here is that osmoconformers are not passive animals that inactively tolerate extremes of salinities. Nor are marine invertebrates free from osmotic problems just because we read statements that they are "98 percent water" or other such comments.

Excretory and osmoregulatory structures

The form and function of organs or systems associated with excretion and osmoregulation are related not only to environmental conditions, but also to body size (especially the surface-to-volume ratio) and other basic features of an animal's *Bauplan*. In very small animals, notably the protozoa, most metabolic wastes diffuse easily across the body covering, because these animals have sufficient body surface (environmental contact) relative to their volume. However, this high surface-to-volume ratio presents a distinct osmoregulatory problem, particularly for freshwater forms. Freshwater protozoa (and even some marine species) typically possess specialized organelles called "**contractile vacuoles**," or **water expulsion vesicles (WEV)**, which actively excrete excess water (Figure 23). The water expulsion vesicles ac-

cumulate cytoplasmic water and expel it from the cell body. Both of these activities apparently require energy, as suggested in part by the large numbers of mitochondria typically associated with WEV. The idea that WEV are primarily osmoregulatory in function is supported by a good deal of evidence. Most convincing is the fact that their rates of filling and emptying change dramatically when the animal is exposed to different salinities. Interestingly, WEV also occur in freshwater sponges, where they probably perform similar osmoregulatory functions.

Although a few groups of metazoan invertebrates possess no known excretory structures, most have some sort of **nephridia** that serve for excretion or osmoregulation, or both. The evolution of various types of invertebrate nephridia and their relationships to other structures were discussed by E. S. Goodrich in 1945 in a classic paper, "The Study of Nephridia and Genital Ducts since 1895." Probably the earliest type of nephridium to appear in the evolution of animals was the **protonephridium** (Figure 24A). Protonephridial systems are characterized by a tubular arrangement opening to the outside of the body via one or more **nephridiopores** and terminating internally in closed unicellular units. These units are the **cap cells** (or **terminal cells**) and may occur singly or in clusters. Each cell leads to an excretory duct (**nephridioduct**) and eventually to the nephridiopore. Two generally recognized types of protonephridia are **flame bulbs**, bearing a tuft of numerous cilia within the cavity, and **solenocytes**, usually with only one or two flagella. Although both are referred to as protonephridia, the precise nature of the homology between the two structural types is still uncertain. There is some evidence that several different types of flame bulb nephridia have been independently derived from solenocyte precursors.

The cilia or flagella drive fluids down the nephridioduct, thereby creating a lowered pressure within the tubule lumen. This lowered pressure draws body fluids, carrying wastes, across the thin cell membranes and into the duct. Selectivity is primarily based on molecular size. Protonephridia are common in adult acoelomates (flatworms and ribbon worms), many pseudocoelomates (rotifers), and some annelids, but rare among adult eucoelomates (although they occur frequently in various larval types). Protonephridia are probably more important in osmoregulation than in excretion. In most of these animals, nitrogenous wastes are expelled primarily across the general body surface by diffusion.

A second and probably more advanced type of excretory structure among invertebrates is the **meta-**

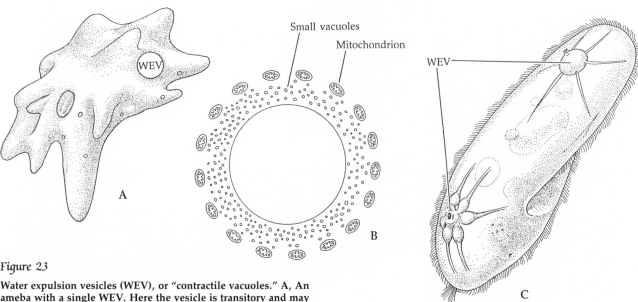

Figure 23

Water expulsion vesicles (WEV), or "contractile vacuoles." A, An ameba with a single WEV. Here the vesicle is transitory and may form anywhere within the cell. B, The WEV of an ameba, and its association with mitochrondria. The numerous small vacuoles accumulate water and then contribute their contents to the main WEV. C, *Paramecium*. Note the positions of two fixed WEV surrounded by an arrangement of collecting canals that pass water to the vesicle. D, The WEV of *Paramecium*, in filled (bottom) and emptied (top) conditions. Enlarged areas show details of a collecting canal surrounded by cytoplasmic tubules that accumulate cell water. The water is passed into the main vesicle, which is collapsed by the action of contractile fibrils, thereby expelling the water through a discharge channel to the outside. (B after Mercer, 1959, Proc. R. Soc. Lond. B 150: 216-237; D from Jurand and Selman 1969.)

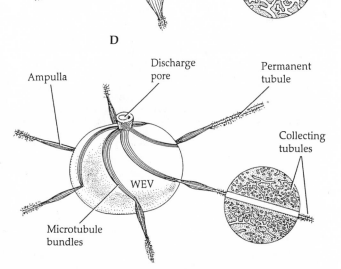

nephridium (Figure 24B). There is a critical structural difference between protonephridia and metanephridia. Both open to the outside, but metanephridia are open internally to the body fluids as well. Metanephridia are also multicellular. The inner end typically bears a ciliated funnel (**nephrostome**), and the duct is often elongated and convoluted and may include a bladder-like storage region. Metanephridia function by taking in large amounts of coelomic fluid through the open nephrostome and then selectively absorbing most of the nonwaste components back into the body fluids through the walls of the bladder or the excretory duct. In very general terms, we can relate the structural and functional differences between proto- and metanephridia to the types of *Baupläne* with which they are commonly associated. Whereas protonephridia can adequately serve animals that have solid bodies (acoelomates), body cavities of small volume (pseudocoelomates), or very small bodies (e.g., larvae), metanephridia cannot. Open funnels would be ineffective in acoelomates, and would quickly drain small pseudocoelomates of their body fluids.

Conversely, except in a few cases, protonephridia are not capable of handling the relatively large body and fluid volumes typical of eucoelomate invertebrates. Thus, in many large coelomate animals (e.g., annelids, molluscs, sipunculans, echiurans) one or more pairs of metanephridia are found.

We have very broadly interpreted the use of the

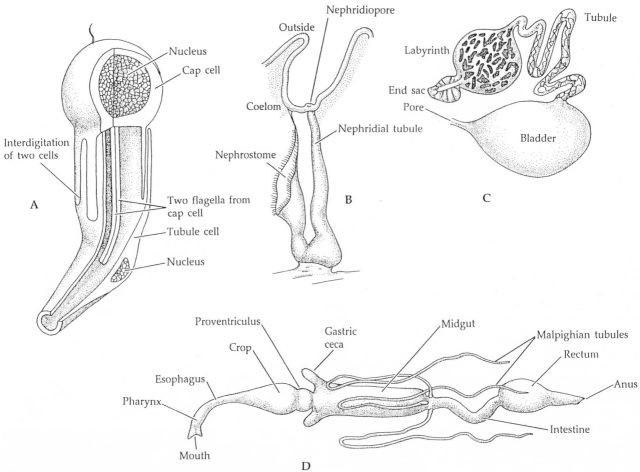

Figure 24

Some invertebrate excretory structures. A, A single protonephridium, with the cap cell and tubule cell (cutaway view). B, A simple metanephridium from a polychaete worm. The nephrostome opens to the coelom, and the pore opens to the exterior. C, The internally closed nephridium (antennal gland) of a crustacean. D, An insect's digestive tract. Excretory Malpighian tubules extract wastes from the hemocoel and empty them into the gut. (A after Wilson and Webster 1974; B after Goodrich 1945; C after various sources; D after Snodgrass 1952.)

terms protonephridia and metanephridia in the above discussion, and we use them that way throughout this text unless specified otherwise. Furthermore, there is a frequent association of nephridia, especially metanephridia, with structures called coelomoducts. **Coelomoducts** are tubular connections arising from the coelomic lining and extending to the outside via special pores in the body wall. Their inner ends are frequently funnel-like and ciliated, resembling the nephrostomes of metanephridia. Coelomoducts may have arisen evolutionarily as a means of allowing the escape of gametes to the outside; they are, in fact, considered homologous to reproductive ducts of many invertebrates. Primitively, the coelom-

oducts and nephridia were separate units; however, through evolution they have in many cases fused in various fashions to become what are called **nephromixia**.

Generally speaking, there are three types of nephromixia. When the coelomoducts are joined with protonephridia and share a common duct, the structure is called a **protonephromixium**. When coelomoducts are united with metanephridia, the result is either a **metanephromixium** or **mixonephridium**, depending upon the structural nature of the union. Whereas coelomoducts originate from the coelomic lining, the nephridial components arise from the outer body wall, so nephromixia are a combination of mesodermally and ectodermally derived parts. Obviously there is some confusion at times about which term applies to a particular "nephridial" type if the precise developmental origin is not clear. We do not wish to belabor this point, so we leave it here to be resurrected periodically in later chapters (see especially Chapter 13).

Not all metazoa possess excretory organs that are clearly proto- or metanephridia as described above.

In some taxa (e.g., sponges, echinoderms, chaetognaths, and cnidarians) no definite excretory structures are known. In such cases wastes are eliminated across surface epithelia (skin or gut lining), perhaps with the aid of ameboid phagocytic cells that collect and transport these products. Other groups possess excretory organs that may represent highly modified nephridia or secondarily derived ("new") structures. The **antennal** and **maxillary glands** of crustaceans appear to be derived from metanephridia, whereas the **Malpighian tubules** of insects arose independently (Figures 24C and D). The details of these structures are discussed in appropriate later chapters.

Circulation and gas exchange

Internal transport

The transport of materials from one place to another within an organism's body depends on the movement and diffusion of substances in body fluids. Various nutrients, gases, and metabolic waste products are generally carried in solution or bound to other soluble compounds within the body fluid itself or sometimes in loose cells, such as blood cells, suspended in fluid. Any system of moving fluids that reduces the functional diffusion distance that these products must transverse may be referred to as a **circulatory system**, regardless of its embryological origin or its ultimate design. The nature of the circulatory system is directly related to the size, complexity, and lifestyle of the organism in question. Usually, the circulatory fluid is an internal, extracellular, aqueous medium produced by the animal. There are, however, a few instances in which circulatory functions are accomplished at least partly by other means. For instance, in most protozoa the protoplasm itself serves as the medium through which materials diffuse to various parts of the cell body, or between the organism and the environment. Sponges and most cnidarians utilize water from the environment as a circulatory fluid, the sponges by passing the water through a series of channels in their bodies and the cnidarians by circulating water through the gut (Figures 25A and B).

In all animals except protozoa, the intercellular tissue fluids play a critical role as a transport mechanism. Even where complicated circulatory plumbing exists, tissue fluids are still necessary to bring dissolved materials in contact with cells, a vital process for life support. In some animals (such as flatworms), there are no special chambers or vessels for body fluids other than the gut and intercellular spaces through which materials diffuse on a cell-to-cell level.

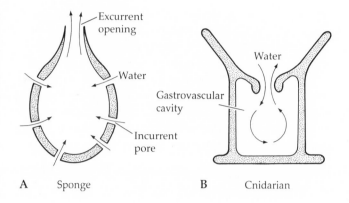

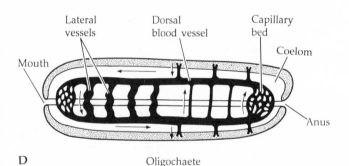

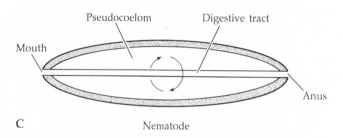

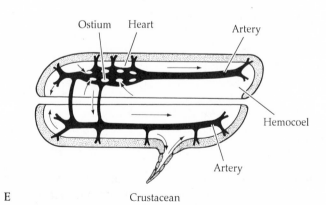

Figure 25

Invertebrate circulatory systems. Sponges (A) and cnidarians (B) utilize environmental water as their circulatory fluid. C, Pseudocoelomates (e.g., rotifers and nematodes) use their body cavity fluid for internal transport. D, The closed circulatory system of an earthworm contains blood that is kept separate from the coelomic fluid. E, Arthropods are characterized by an open circulatory system, in which the blood and body cavity (hemocoelic) fluid are one and the same.

This condition limits these animals to relatively small sizes or to shapes that maintain low diffusion distances. Most animals, however, have some specialized structure to facilitate the transport of various body fluids and their contents. This structure may include the body cavities themselves or actual circulatory systems of vessels, chambers, sinuses, and pumping organs. Actually, most animals employ both their body cavity and a circulatory system for internal transport.

Pseudocoelomate invertebrates use the fluids of the body cavity for circulation (Figure 25C). Most of these animals (e.g., rotifers and roundworms) are quite small, and adequate circulation is accomplished by the movements of the body against the pseudocoelomic fluids, which are in direct contact with internal tissues and organs. Several types of cells are generally present in the body fluids of pseudocoelomates. These cells may serve in various activities such as transport and waste accumulation, but their functions have not been well studied. A few eucoelomate invertebrates (e.g., sipunculans, ectoprocts, and most echinoderms) also depend largely on the body cavity as a coelomic circulatory chamber.

Circulatory systems

Beyond the relatively rudimentary circulatory mechanisms discussed above, there are two principal designs or structural plans for accomplishing internal transport (exceptions and variations are discussed under specific taxa). These two organizational plans are **closed** and **open circulatory systems**, both of which contain a circulatory fluid, or **blood**. In closed systems the blood stays in distinct vessels and perhaps lined chambers; exchange of circulated material with parts of the body occurs in special areas of the system such as capillary beds (Figure 25D). Since the blood itself is physically separated from the intercellular fluids, the exchange sites must offer minimal resistance to diffusion; thus one finds capillaries that typically have very thin membranous walls a single cell-layer thick. Closed circulatory systems are common in animals with well developed or spacious coelomic compartments (e.g., annelids, echiurans, phoronids, and vertebrates). Such arrangements facilitate the transport from one body area to another of materials that might otherwise be isolated by mesenteries of the body cavity. In such situations the blood and coelomic fluid may be quite different from one another both in composition and in function. For example, the blood may transport nutrients and gases, while the coelomic fluid accumulates metabolic wastes for removal by nephridia and also acts as a hydrostatic skeleton.

It takes power to keep a fluid moving through a plumbing system. Many invertebrates with closed systems rely on body movements and the exertion of coelomic pressure on vessels to move their blood. These activities are frequently supplemented by muscles of the blood vessel walls that contract in peristaltic waves. In addition, there may be special heavily muscled pumping areas along certain vessels. These regions are sometimes referred to as "hearts," but most are more appropriately called **contractile vessels**.

Open circulatory systems are associated with a reduction of the adult coelom, including a secondary loss of most of the peritoneal lining around the various organs and inner surface of the body wall. The circulatory system itself usually includes a distinct heart as the primary pumping organ and various vessels, chambers, or ill-defined sinuses (Figure 25E). The degree of elaboration of such systems depends primarily on the size, complexity, and to some extent the activity level of the animal. This kind of system, however, is "open" in that the blood, often called the **hemolymph**, empties from vessels into the body cavity and directly bathes the organs. The body cavity is appropriately called a **hemocoel**. Open circulatory systems are typical of arthropods and noncephalopod molluscs.

Just because the open circulatory system seems "sloppier" in its organization and operates under lower blood velocities and pressures, it should not be presumed to be poorly designed or inefficient. In fact, in many groups this type of system has assumed a variety of functions beyond circulation. For example, in bivalves and gastropods, the hemocoel functions as a hydrostatic skeleton for locomotion and certain types of burrowing activities. In aquatic arthropods, it also serves a hydrostatic function when the animal molts and temporarily loses its exoskeletal support. In large terrestrial insects, the transport of respiratory gases has been largely assumed by the tracheal system, and one of the primary responsibilities taken on by the open circulatory system appears to be thermal regulation. In most spiders extension of the limbs is accomplished by forcing hemolymph into the appendages.

Hearts and other pumping mechanisms

Circulatory systems, open or closed, generally have structural mechanisms for pumping the blood and maintaining adequate blood pressures. Beyond the influence of general body movements, most of these structures fall into the following categories: **contractile vessels** (found in annelids), **tubular hearts** (found in most arthropods), and **chambered hearts** (found in molluscs and vertebrates). Furthermore, the method of initiating contraction of these various pumps (the pacemaker mechanism) may be intrinsic (originating within the musculature of the structure itself) or extrinsic (originating from motor nerves arising outside the structure). The first case describes the **myogenic hearts** of molluscs and vertebrates; the second describes the **neurogenic hearts** of most arthropods and, at least in part, the contractile vessels of annelids.

Blood pressure and flow velocities are intimately associated not only with the activity of the pumping mechanism but also with vessel diameters. Energetically, it costs a good deal more to maintain flow through a narrow pipe than through a wide pipe. This cost is minimized in animals with closed circulatory systems by keeping the narrow vessels short and using them only at sites of exchange (i.e., capillary beds), and by using the larger vessels for long-distance transport from one exchange site to another. Of course, for a system of vessels, reducing the diameter of a single vessel increases flow velocity, which poses problems at an exchange site. This problem is solved by the presence of large numbers of small vessels, the total cross-sectional area of which exceeds that of the larger vessel from which they arise. The result is that blood pressure and total flow velocity actually decrease at capillary exchange sites. A drop in blood pressure and a relative rise in blood osmotic pressure along the capillary bed facilitate exchanges between the blood and surrounding tissue fluids. In open systems, of course, both pressure and velocity drop dramatically once the blood leaves the heart and vessels and enters the hemocoel.

Gas exchange and transport

One of the principal functions of most circulatory fluids is to carry oxygen and carbon dioxide through the body and exchange these gases with the environment. With few exceptions, oxygen is necessary for the process of cellular respiration. Although a number of invertebrates can survive periods of environmental oxygen depletion—either by dramatically reducing their metabolic rate or by switching to anaerobic respiration—most cannot; thus they depend upon a relatively constant oxygen supply. All animals can take in oxygen from their surroundings while at the same time releasing carbon dioxide, a metabolic waste product of respiration. We define the uptake of oxygen and the loss of carbon dioxide at the surface of the organism as **gas exchange** and reserve the term **respiration** for the energy-producing metabolic activities within cells. Some authors distinguish these two processes with the terms external respiration and cellular (or internal) respiration.

Gas exchange in nearly all animals operates according to certain common principles regardless of any structural modifications that serve to enhance the process under different conditions. The basic strategy is to bring the environmental medium (water or air) close to the appropriate body fluid (blood or body cavity fluid) so that the two are separated only by a wet membrane across which the gases can diffuse. The system must be moist because the gases must be in solution to diffuse across the membrane. The diffusion process depends, of course, on the concentration gradients of the gases at the exchange site; the gradients are maintained by the circulation of internal fluids to and away from these areas (Figure 26).

Gas exchange structures. A number of invertebrates lack special gas exchange structures. In such animals gas exchange is said to be **integumentary** or **cutaneous**, and occurs over the whole body surface. Such is the case in many tiny animals with very high surface-to-volume ratios and in some larger soft-bodied forms (e.g., cnidarians and flatworms). Most an-

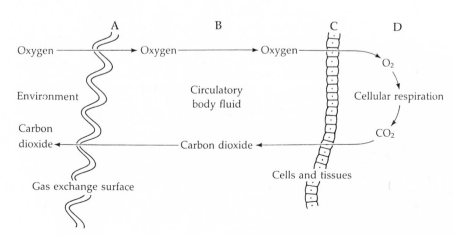

Figure 26

Gas exchange in animals. Oxygen is obtained from the environment at a gas exchange surface (A), and is transported by a circulatory body fluid (B) to the body's cells and tissues (C), where cellular respiration occurs (D). Carbon dioxide follows the reverse path. See text for details.

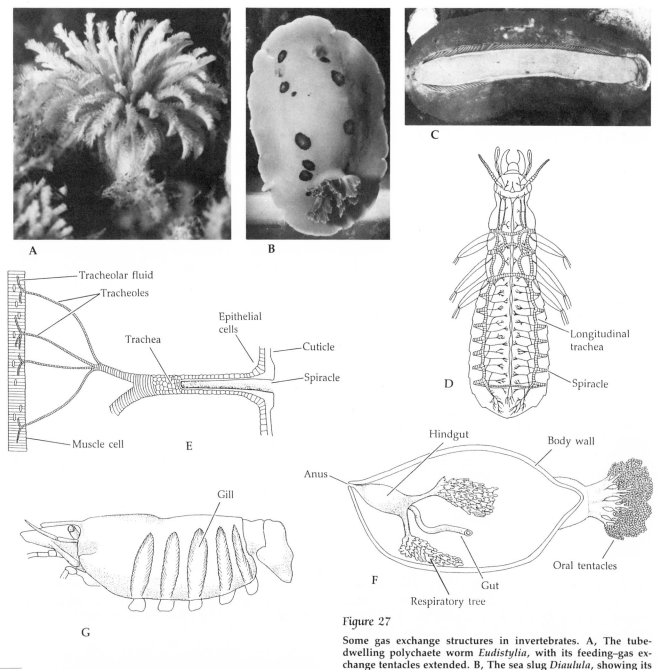

Figure 27

Some gas exchange structures in invertebrates. A, The tube-dwelling polychaete worm *Eudistylia*, with its feeding–gas exchange tentacles extended. B, The sea slug *Diaulula*, showing its branchial plume. C, The chiton *Cryptochiton stelleri*. This ventral view shows the gills. These highly folded gas exchange structures lie along each side of the foot in a channel through which water is passed from front to back. D, A general plan of the tracheal system of an insect. E, A single insect trachea and its branches (tracheoles), which lead directly to a muscle cell. F, A sea cucumber dissected to expose the paired respiratory trees, which are flushed with water by hindgut irrigation. G, The placement of gills beneath the flaps (carapace) of the thorax in a crustacean (lateral view). H, A terrestrial banana slug has a pneumostome that opens to the air sac, or "lung." (A courtesy G. McDonald; B photo by C. E. Mills, Univ. of Washington/BPS; C photo by the authors; D–G after Barnes 1980; H photo by P. J. Bryant/BPS.)

imals with integumentary gas exchange are restricted to aquatic or damp terrestrial environments where the body surface is kept moist. Integumentary gas exchange also supplements other methods in many animals, even certain vertebrates (e.g., amphibians).

Most marine and many freshwater invertebrates possess gills (Figures 27A,B,C,G), which are external organs or restricted areas of the body surface specialized for gas exchange. Basically, gills are thin-walled processes, well supplied with blood or other body fluids, that promote diffusion between this fluid and the animal's environment. Gills are frequently highly folded or digitate, increasing the diffusive surface area. A number of nonhomologous structures have evolved as gills in different taxa, and they often serve other functions in addition to exchange (sensory input and feeding). By their very nature, gills are permeable surfaces that must be protected during times of osmotic stress (e.g., as occur in estuaries and intertidal environments). In such instances, the gills may be housed within chambers or be retractable.

A few marine invertebrates employ the lining of the gut as the gas exchange surface. Water is pumped in and out of the hindgut, or a special evagination thereof, in a process called **hindgut irrigation** (or cloacal irrigation). Many sea cucumbers and certain echiuran worms are known to use this method of gas exchange (Figure 27F).

As we have defined them, protruding gills will not work on dry land. Here the gas exchange surfaces must be internalized to keep them moist and protected and to prevent body water loss through the wet surfaces. The lungs of terrestrial vertebrates offer the most familiar example of such an arrangement. Among the invertebrates, the arthropods have managed to solve the problems of "air breathing" in two basic ways. Spiders and their kin possess **book lungs**, whereas most insects, centipedes, and millipedes, possess **tracheae** (Figures 27D and E). Book lungs are inpocketings with highly folded inner linings across which gases diffuse in the usual manner. Tracheae, however, are branched, usually anastomosed invaginations of the outer body wall and are open both internally and externally.

The tracheae of most insects allow diffusion of gases from air directly to the tissues of the body, and the blood plays little or no role in gas transport. Rather, intercellular fluids extend part way into the tracheal tubes as a solvent for gases. Atmospheric pressure tends to prevent these fluids from being drawn too close to the external body surface where evaporation is a potential problem. In addition, the outside openings (the **spiracles**) of the tracheae are

often equipped with some mechanism of closure. In many insects, especially large ones, special muscles act to ventilate the tracheae by actually pumping air in and out.

The only other major group of terrestrial invertebrates whose members have evolved distinct air-breathing structures is the molluscan subclass Pulmonata—the land snails and slugs (Figure 27H). The gas exchange structure here is a **lung** that opens to the outside via a pore called the **pneumostome**. This lung is derived from a feature common to molluscs in general, the **mantle cavity**, which in other molluscs houses the gills and other organs.

Gas transport. As illustrated in Figure 26, oxygen must be transported from the sites of environmental gas exchange to the various cells of the body, and carbon dioxide must get from the cells where it is produced to the gas exchange surface for release. Generally, groups displaying marked cephalization circulate freshly oxygenated blood first through the "head" region and secondarily to the rest of the body. Invertebrates vary considerably in their oxygen requirements. In general, more active animals consume more oxygen. In slow-moving and sedentary invertebrates, oxygen consumption and utilization is quite low; for example, no more than 20 percent O_2 withdrawal from the gas exchange water current has ever been demonstrated in sessile sponges, bivalves, or tunicates. The amount of oxygen available to an organism also varies greatly in different environments. For example, the concentration of oxygen in dry air at sea level is uniformly about 210 ml/liter, whereas in fresh and salt water it ranges from near zero to nearly 10 ml/liter. This variation in aquatic environments is due to factors such as depth, surface turbulence, photosynthetic activity, temperature, and salinity (oxygen concentrations drop as temperature and salinity increase). With the exception of certain areas prone to oxygen depletion (e.g., muds rich in organic detritus), most habitats provide adequate sources of oxygen to sustain animal life. However, given the relatively low capacity of aqueous mediums to carry oxygen in solution, animals usually bind oxygen with complex organic compounds called **respiratory pigments**, which greatly increase their carrying capacity for oxygen.

Respiratory pigments differ in molecular architecture and in their affinities for oxygen, but all have a metal ion (iron or copper) with which the oxygen combines. These pigments may be in solution within the blood or other body fluid, or they may be in specific cells (blood corpuscles). In general, the pigments respond to high oxygen concentrations by "loading" (combining with oxygen) and to low oxygen concentrations by "unloading" or dissociating with oxygen (releasing oxygen). The loading and unloading qualities are different for various pigments in terms of their relative saturations at different levels of oxygen in their immediate surroundings, and are generally expressed in the form of dissociation curves. Respiratory pigments load at the site of gas exchange, where environmental oxygen levels are high relative to the body fluid, and unload at the cells and tissues, where surrounding oxygen levels are low relative to the body fluid. In addition to simply carrying oxygen from the loading to the unloading sites, some pigments may carry reserves of oxygen that are released only when tissue levels of oxygen are unusually low. Other factors such as temperature and carbon dioxide concentrations also influence the carrying capacities of respiratory pigments.

Hemoglobin is among the most common respiratory pigments in animals. Actually there are several different hemoglobins. Some function primarily for transport, whereas others serve to store oxygen and then release it during times of low environmental oxygen availability. Hemoglobins are reddish pigments containing iron as the oxygen-binding metal. They are found in a variety of invertebrates and, with the exception of a few fishes, in all vertebrates. Among the major groups of invertebrates, hemoglobin occurs in some annelids, the more primitive crustaceans, and a few molluscs. Interestingly, hemoglobin is not restricted to metazoa; it is also produced by some protozoa, certain fungi, and in the root nodules of leguminous plants, a distribution suggesting that it is a chemical that evolved very early in the history of life. Among animals, hemoglobin may be carried within red blood corpuscles (**erythrocytes**) or simply be dissolved in the blood or coelomic fluid.

Hemocyanins are the most commonly occurring respiratory pigments in molluscs and higher crustaceans. Although hemocyanins, like hemoglobins, are proteins, they display significant structural differences, contain copper rather than iron, and tend to have a bluish color when oxygenated. The oxygen-binding site on a hemocyanin molecules is a pair of copper atoms that are linked to amino acid side chains. Unlike most hemoglobin, hemocyanins tend to release oxygen easily and provide a ready source of oxygen to the tissues as long as there are relatively high concentrations of available environmental oxygen. Hemocyanins are always found in solution, never in corpuscles, a characteristic probably related to the necessity for rapid oxygen unloading.

Two other types of respiratory pigments occur incidentally in certain invertebrates; these are **hemerythrins** and **chlorocruorins**, both of which contain iron. The former is violet to pink when oxygenated; the latter is green in dilute concentrations but red in high concentrations. Chlorocruorins generally function as efficient oxygen carriers when environmental levels are relatively high; hemerythrins function more in oxygen storage. Chlorocruorin is structurally similar to hemoglobin in a number of respects and may have evolved from it. Chlorocruorin occurs in several families of polychaete worms; hemerythrin is known from sipunculans, at least one genus of polychaetes, and some priapulans and brachiopods.

Table 1 gives some of the basic properties of these pigments. There seems to be no obvious phylogenetic rhyme or reason to the occurrence of these pigments among various taxa. Their sporadic and inconsistent distribution suggests that some of them may have evolved more than once, through parallel evolution. Interestingly, respiratory pigments are rare among insects and are known only from the occurrence of hemoglobin in chironomid midges and certain parasitic flies of the genus *Gastrophilus*. The general absence of respiratory pigments among the insects reflects the fact that most of them do not use the blood as a medium for gas transport but employ extensive tracheal systems to carry gases directly to the tissues. In those insects without well developed tracheae, oxygen is simply carried in solution in the hemolymph.

Respiratory pigments raise the oxygen-carrying capacity of body fluids far above what would be achieved by transport in solution. Similarly, carbon dioxide levels in body fluids (and in sea water as well) are much higher than would be expected strictly on the basis of its solubility. The enzyme **carbonic anhydrase** greatly accelerates the reaction between carbon dioxide and water to form carbonic acid:

$$CO_2 + H_2O \rightleftharpoons H_2CO_3$$

Furthermore, the carbonic acid ionizes to hydrogen and bicarbonate ions so that a series of reversible reactions takes place:

$$CO_2 + H_2O \rightleftharpoons H_2CO_3 \rightleftharpoons H^+ + HCO_3^-$$

By "tying up" CO_2 in other forms, the concentration of CO_2 in solution is lowered, thus raising the overall carrying capacity of the blood. This set of reactions responds to changes in pH, and in the presence of appropriate cations (e.g., Ca^{2+} and Na^+) it shifts back and forth, serving as a buffering agent by regulating hydrogen ion concentration.

Table One

Properties of oxygen-carrying respiratory pigments

PIGMENT	MOLECULAR WEIGHT	METAL	RATIO OF METAL TO O_2	METAL ASSOCIATE
Hemoglobin	65,000	Fe	1:1	Porphyrin
Hemerythrin	40,000–108,000	Fe	2:1	Protein chains
Hemocyanin	40,000–9,000,000	Cu	2:1	Protein chains
Chlorocruorin	3,000,000	Fe	1:1	Porphyrin

Nervous systems and sense organs

Invertebrate nervous systems

All living cells respond to some stimuli and conduct some sort of "information," at least for short distances. Thus, even when no real nervous "system" is present—the condition found in protozoa and sponges—coordination and reaction to external stimulation do occur. The regular metachronal beating of cilia in ciliate protozoa or the responses of certain flagellates to varying light intensities cannot be attributed to chance. But the integration and coordination of bodily activities in metazoa are in large part due to the processing of information by a discrete or true nervous system. The functional units of true nervous systems are cells called **neurons**, which are specialized for relatively high-velocity impulse conduction. We prefer the definition of a nervous system offered by Bullock and Horridge (1965): "A nervous system may be defined as an organized constellation of cells (neurons) specialized for the repeated conduction of an excited state from receptor sites or from other neurons to effectors or to other neurons." This definition certainly applies to those animals that have achieved the tissue or organ level of complexity (the diploblastic and triploblastic

metazoa), but it clearly excludes the protozoa and sponges.

The generation of an impulse within a true nervous system usually results from a stimulus imposed upon the nervous elements. The source of stimulation may be external or internal. A typical pathway of events occurring in a nervous system is shown in Figure 28. The stimulus is received by some **receptor** (e.g., a sense organ) and turned into an impulse, which is conducted along a **sensory nerve (afferent nerve)** via a series of adjacent neurons to some coordinating center or region of the system. The information is processed and an appropriate response is "selected." A **motor nerve (efferent nerve)** then conducts a second impulse from the central processing center to an **effector** (e.g., a muscle) where the response occurs. Once an impulse is initiated within the system, the mechanism of conduction is essentially the same in all neurons, regardless of the stimulation. The wave of depolarization along the length of each neuron and the chemical neurotransmitters crossing the synaptic gaps between neurons are common to virtually all nervous conduction. How then is the information interpreted within the system for response selection? The answer to this question involves two basic considerations.

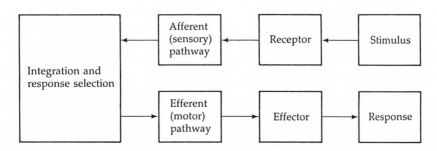

Figure 28

A generalized pathway within the nervous system. A stimulus initiates an impulse within some sensory structure (the receptor); the impulse is then transferred to some integrative portion of the nervous system via sensory nerves. Following response selection, an impulse is generated and transferred along motor nerves to an effector (e.g., muscle), where the appropriate response is elicited.

First is the occurrence of a point called a **threshold** that corresponds to the minimum intensity of stimulation necessary to generate an impulse. Receptor sites consist basically of specialized neurons whose thresholds for various kinds of stimuli are drastically different from one another because of structural or physiological qualities. For example, a sense organ whose threshold for light stimulation is very low (compared with other potential stimuli) functions as a light sensor or **photoreceptor**. In any such specialized sensory receptor, the condition of differential thresholds essentially "screens" incoming stimuli so that an impulse normally is generated by only one kind of information (e.g., light, sound, heat, pressure).

Second, the overall "wiring" or circuitry of the entire nervous system is such that impulses received by the integrative (response-selecting) areas of the system from any particular nerve will be interpreted according to the kind of stimulus for which that sensory pathway is specialized. For example, all impulses coming from a photoreceptor are "understood" as being light-induced. Threshold and circuitry can be demonstrated simply by introducing "false" information into the system by stimulating a specialized sense organ in an inappropriate manner: if photoreceptors in the eye are stimulated by electricity or pressure, the nervous system will interpret this input as light. Remember that an impulse can be generated in any receptor by nearly any form of stimulation if the stimulus is of sufficient intensity to exceed the relevant threshold. A blow to the eye often results in "seeing stars," or flashes of light, even when the eye is closed. The photoreceptor's threshold to mechanical stimulation has been reached in such a situation. By the same token, the application of extreme cold to a heat receptor may "feel" hot.

Nervous systems in general operate on the principles outlined above. However, this description applies largely to nervous systems that have structural centralized regions. Following a discussion below of the basic types of sense organs (receptor units), we discuss centralized and noncentralized nervous systems and their relationships to general body architecture.

Sense organs

Invertebrates possess an impressive array of receptor structures through which they receive information about their environment. An animal's behavior is in large part a function of its responses to that information. These responses often take the form of some sort of movement relative to the source of a particular stimulus. A response of this nature is called a **taxis** and may be positive or negative depending on the reaction of the animal to the stimulus. For example, animals that tend to move away from bright light are said to be negatively phototactic.

Since the activities of receptor units represent the initial step in the usual functioning of the nervous system, they are a critical link between the organism and its surroundings. Consequently, the kinds of sense organs present and their placement on the body are intimately related to the overall complexity, mode of life, and general *Bauplan* of any animal. The following general review provides some concepts and terminology that serve as a basis for more detailed coverage in later chapters. The first five categories of sense organs may all be viewed as "mechanoreceptors," in that they respond to various mechanical stimuli (e.g., touch, vibrations, and pressure). The last three are sensitive to nonmechanical input (e.g., chemicals, light, and temperature).

Tactile receptors. Touch or **tactile** receptors are generally derived from modifications of epithelial cells associated with sensory neurons. The nature of the epithelial modifications depends a great deal on the structure of the body wall itself. For instance, the form of a touch receptor in an arthropod with a rigid exoskeleton must be different from that in a soft-bodied cnidarian. Most, however, involve projections from the body surface, such as bristles, spines, setae, tubercles, and assorted bumps and pimples (Figure 29). Objects in the environment with which the animal makes contact move these receptors, thereby creating mechanical deformations that are imposed upon the underlying sensory neurons to initiate an impulse.

Virtually all animals are touch-sensitive, but their responses are many and varied and are often integrated with other sorts of sensory input. For example, the gregarious nature of many animals may involve a positive response to touch (positive thigmotaxis) combined with a chemical recognition of members of the same species. Some touch receptors are highly sensitive to mechanically induced vibrations propagated in water or through solid substrata. Such vibration sensors are common in certain tube-dwelling polychaetes that retract quickly into their tubes in response to movements in their surroundings. Certain crustacean ambush-predators are able to detect the vibrations induced by nearby potential prey animals. Similarly, web-building spiders sense struggling prey in their webs through vibrations of the threads.

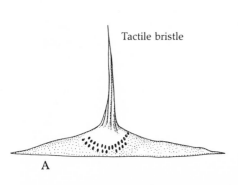

Tactile bristle

A

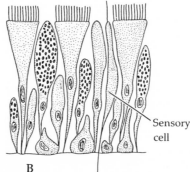

Sensory cell

B

Figure 29

Some invertebrate tactile receptors. A, Tactile organ of *Sagitta bipunctata* (an arrow worm, phylum Chaetognatha). B, A sensory epithelial cell of a nemertean worm. C, Touch-sensitive bristles on the leg of a young lobster (scanning electron micrograph). (A after Kuhl 1938; B after Gibson 1972; C from C. Derby, 1982, J. Crust. Biol. 2: 1-21.)

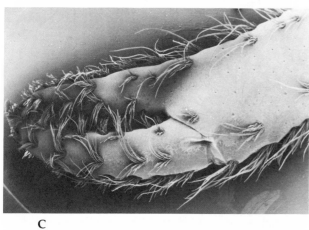

C

Georeceptors. **Georeceptors** respond to the pull of gravity, and thus give animals information about their orientation relative to "up and down." Most georeceptors are structures called **statocysts** (Figure 30). Statocysts usually consist of a fluid-filled chamber containing a solid granule or pellet called the **statolith**. The inner lining of the chamber includes a touch-sensitive epithelium from which project bristles or "hairs" associated with underlying sensory neurons. In aquatic invertebrates, some statocysts are open to the environment and thus are filled with water. In some of these the statolith is a sand grain obtained from the animal's surroundings. Most statoliths, however, are secreted by the organisms themselves within closed capsules.

Because of the resting inertia of the statolith within the fluid, any movement of the animal results in a change in the pattern or intensity of stimulation of the sensory epithelium by the statolith. Additionally, when the animal is stationary, the position of the statolith within the chamber provides information about the organism's orientation to gravity. The fluid within statocysts of at least some invertebrates (especially certain crustaceans) also acts something like

the fluid of the semicircular canals in vertebrates. When the animal moves, the fluid tends to remain stationary—the relative "flow" of the fluid over the sensory epithelium provides the animal with information about its linear and rotational acceleration relative to its environment.

Whether stationary or in motion, animals utilize the input from georeceptors in different ways, depending on their habitat and lifestyle. The information from these statocysts is especially important under conditions where other sensory reception is rendered inadequate. For example, burrowing invertebrates cannot rely on photoreceptors for orientation when moving through the substratum, and some employ statocysts for that purpose. Similarly, planktonic animals face orientation problems in their three-dimensional aqueous environment, especially in deep water and at night, and many such creatures possess statocysts.

There are a few exceptions to the standard statocyst arrangements described above. For example, a number of aquatic insects detect gravity by using air bubbles trapped in certain passageways (e.g., tracheal tubes). The bubbles move according to their

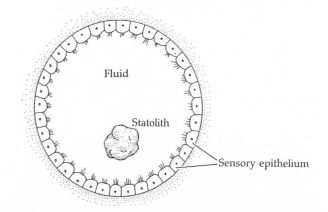

Fluid

Statolith

Sensory epithelium

Figure 30

A generalized statocyst, or georeceptor (section).

orientation to the vertical, much like the air bubble in a carpenter's level, and stimulate sensory bristles lining the tube in which they are located.

Proprioceptors. Internal sensory organs that respond to mechanically induced changes caused by stretching, compression, bending, and tension are called **proprioceptors**, or sometimes simply "stretch receptors." These receptors give the animal information about the movement of its body parts and their positions relative to one another. Proprioceptors have been most thoroughly studied in vertebrates and arthropods, where they are associated with appendage joints and certain body extensor muscles. The sensory neurons involved in proprioception are associated with and attached to some part of the body that is stretched or otherwise mechanically affected by movement or muscle tension. These parts may be specialized muscle cells, elastic connective tissue fibers, or membranes that span joints. As these structures are stretched, relaxed, and compressed, the sensory endings of the attached neurons are distorted accordingly and thus stimulated. Some of these receptor arrangements can detect not only changes in position but also static tension.

Phonoreceptors. General sensitivity to sound (**phonoreception**) has been demonstrated in a number of invertebrates (certain annelid worms and a variety of crustaceans), but true auditory receptors are known only in a few groups of insects and perhaps some arachnids and centipedes. Crickets, grasshoppers, and cicadas possess phonoreceptors called **tympanic organs** (Figure 31). A rather tough but flexible **tympanum** covers an internal air sac that allows the tympanum to vibrate when struck by sound waves. Sensory neurons attached to the tympanum are stimulated directly by the vibrations. Most arachnids possess structures called **slit sense organs** which, although poorly studied, are suspected to perform auditory functions; at least they appear to be capable of sensing sound-induced vibrations. Certain centipedes bear so-called **Organs of Tömösvary**, which some workers believe may be sensitive to sound.

Baroreceptors. Sensitivity of invertebrates to pressure changes (**baroreception**) is not well understood, and no structures for this purpose have been positively identified. However, a number of invertebrates do respond to pressure in a variety of ways. Behavioral responses to pressure changes have been demonstrated in several pelagic invertebrates includ-

Figure 31

An arthropod phonoreceptor, or auditory organ, on the fork-tailed katydid, *Scudderia furcata*. Note the position of the tympanum on the tibia of the first walking leg. (Photo by P. J. Bryant/BPS.)

ing medusae, ctenophores, squids, and copepod crustaceans, and in various planktonic larvae. Aquatic insects also sense changes in pressure, and may use a variety of methods to do so. Some intertidal crustaceans coordinate daily migratory activities with tidal movements, perhaps partly in response to pressure as water depth changes.

Chemoreceptors. Many animals have a **general chemical sensitivity**, which is not a function of any definable sensory structure but is due to the general irritability of protoplasm itself. Chemicals that are in and of themselves noxious or generally irritating can induce responses via this general chemical sensitivity when they occur in sufficiently high concentrations. In addition, most animals have specific chemoreceptors. Chemoreception is a rather direct sense in that the molecules stimulate sensory neurons by contact after diffusing in solution across a thin epithelial covering. The chemoreceptors of many aquatic invertebrates are located in pits or depressions, through which water may be circulated by ciliary action. In arthropods, the chemoreceptors are usually in the form of hollow "hairs" or other projections, within which are chemosensory neurons. While chemosensitivity is a universal phenomenon among invertebrates, a wide range of specificities and capabilities exists.

The types of chemicals to which particular animals respond are closely associated with their lifestyles. Chemoreceptors may be specialized for certain tasks such as general water analysis, humidity detection, sensitivity to pH, prey tracking, mate location, substratum analysis, and food recognition. Probably all aquatic organisms "leak" certain amounts of amino acids into their environment through the skin and gills as well as in urine and feces. These released amino acids form an organism's "body odor," which can create a chemical picture of the animal that others detect to identify such characteristics as species, sex, stress level, population density, and perhaps size and individuality. Amino acids are widely distributed in the aquatic environment, where they serve as general indicators of biological activity. Many aquatic animals are capable of detecting amino acids with much greater sensitivity than our most sophisticated laboratory equipment.

Photoreceptors. Nearly all animals are sensitive to light, and most have some kind of identifiable photoreceptors. These receptors share the common quality of possessing **light-sensitive pigments**. These pigment molecules are capable of absorbing light energy in the form of photons, a process necessary for the initiation of any light-induced, or **photic**, reaction. The energy thus absorbed is ultimately responsible for stimulating the sensory neurons of the photoreceptor unit. Beyond this basic commonality, however, there is an incredible range of variation in complexity and capability of light-sensitive structures. Arthropods, molluscs, and some polychaete annelids possess eyes having extreme sensitivity, good spatial resolution, and, in some cases, multiple spectral channels. Most classifications of photoreceptors are based upon grades of complexity, and the same categorical term may be applied to a variety of nonhomologous structures, from simple pigment spots (found in protozoa) to extremely complicated lensed eyes (found in squids and octopuses). Functionally, the capabilities of these receptors range from simply perceiving light intensity and direction to forming images with a high degree of visual discrimination and resolution.

Certain protozoa, particularly flagellates, possess subcellular organelles called **stigmata**, which are simple spots of light-sensitive pigment (Figure 32A). The simplest photoreceptors in metazoa are unicellular structures scattered over the epidermis or concentrated in some area of the body. These are usually called **eye spots**. Multicellular photoreceptors may be classified into three general types, with some subdivisions. These types include **ocelli** (sometimes called "simple eyes" or "eye spots"), **compound eyes** (found in many arthropods), and **complex eyes** (the so-called camera eyes of cephalopod molluscs and

vertebrates). In multicellular ocelli, the light-sensitive (**retinular**) cells may face outward; these ocelli are then said to be **direct**. Or the light-sensitive cells may be **inverted**. The inverted type is common among flatworms and nemerteans and is made up of a cup of reflective pigment and retinular cells (Figure 32B). The light-sensitive ends of these neurons face into the cup. Light entering the opening of the pigment cup is reflected back onto the retinular cells. Because light can enter only through the cup opening, this sort of ocellus gives the animal a good deal of information about light direction as well as variations in light intensity.

Compound eyes are composed of a few to many distinct units called **ommatidia** (Figure 32C). Although eyes of multiple units occur in certain annelid worms and some bivalve molluscs, they are best developed and best understood among the arthropods. Each ommatidium is supplied with its own nerve tract leading to a large optic nerve, and apparently each has its own discrete field of vision. The visual fields of neighboring ommatidia overlap to some degree, with the result that a shift in position of an object within the total visual field causes changes in the impulses reaching several ommatidial units; based in part on this phenomenon, compound eyes are especially suitable for detecting movement. Compound eyes are described in more detail in Chapter 15.

The complex (camera) eyes of squids and octopuses (Figure 32D) are probably the best image-forming eyes among the invertebrates. Cephalopod eyes are frequently compared with those of vertebrates, but they differ in many respects. The eye is covered by a transparent protective **cornea**. The amount of light that enters the eye is controlled by the **iris**, which regulates the size of the slitlike **pupil**. The lens is held by a ring of **ciliary muscles** and serves to focus light on the **retina**, a layer of densely packed photosensitive cells from which the neurons arise. The receptor sites of the retinal layer face in the direction of the light entering the eye. This **direct eye** arrangement is quite different from the **indirect eye** condition in vertebrates, where the retinal layer is inverted. Another difference between cephalopod eyes and those of most vertebrates is in the mechanism by which focusing occurs. In most vertebrates focusing is accomplished by the action of muscles that change the shape of the lens, whereas in cephalopods it is achieved by moving the lens back and forth with the ciliary muscles and by compressing the eyeball.

A good deal of work suggests that metazoan pho-

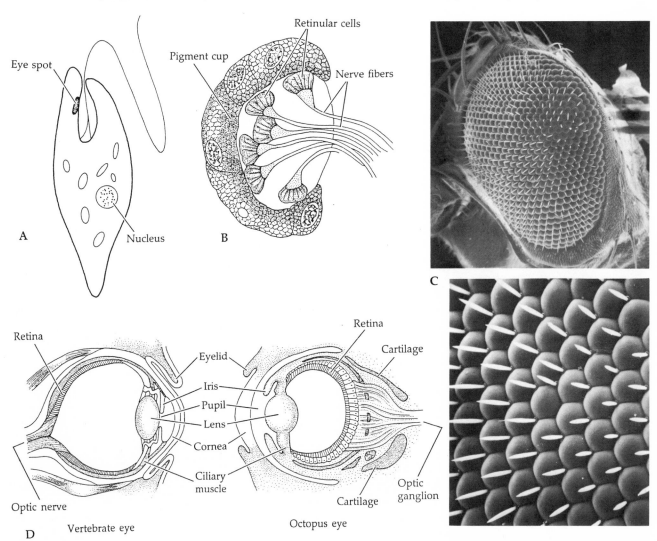

Figure 32

Some invertebrate photoreceptors. A, A protozoon, *Euglena*. Note the position of the eye spot, or stigma. B, An inverted pigment-cup ocellus of a flatworm (section). C, An insect's compound eye. A single unit is called an ommatidium. D, A vertebrate eye (left) and a cephalopod eye (right) (vertical sections). (B after Prosser and Brown 1961; C photos by P. J. Bryant/BPS; D after various sources.)

toreceptors have evolved along two lines (see Eakin 1963). On one hand are photoreceptor units derived from or closely associated with cilia (found in, e.g., cnidarians, echinoderms, and chordates). These types of eyes are called **ciliary eyes**. On the other hand are photoreceptors derived from microvilli or microtubules and referred to as **rhabdomeric eyes** (found in, e.g., flatworms, annelids, arthropods, and molluscs). It is not yet known if these different categories represent actual lineages of homologous structures, but it is interesting to note that the two groups of taxa roughly parallel the two distinct lines of evolutionarily related taxa known as deuterostomes and protostomes (see Chapters 4 and 24).

Thermoreceptors. The influence of temperature changes on all levels of biological activity is well documented. Certainly every student of general biology has learned about the basic relationships between temperature and rates of metabolic reactions. Furthermore, even the casual observer must have noticed that many organisms' activity levels range from lethargy at low temperatures to hyperactivity at elevated temperatures, and that thermal extremes can result in death. The problem is determining whether the organism is simply responding to the effects of temperature at a general physiological level, or whether discrete thermoreceptor organs are also involved.

There is considerable circumstantial evidence

that at least some invertebrates are capable of directly sensing differences in environmental temperature, but actual receptor units are for the most part unidentified. A number of insects, some crustaceans, and the horseshoe crab (*Limulus*) apparently can sense thermal variation. The only nonarthropod invertebrates that have received much attention in this regard are certain leeches, which apparently are drawn to warm-blooded hosts by some heat-sensing mechanism. Other ectoparasites (e.g., ticks) of warm-blooded vertebrates may also be able to sense the "warmth of a nearby meal," but very little work has been done on this matter.

Independent effectors

Independent effectors are specialized sensory response structures that not only receive information from the environment but also elicit a response to the stimulus directly, without the intervention of the nervous system per se. In this sense, independent effectors are like closed circuits. The stinging capsules (nematocysts) of cnidarians and the adhesive cells of ctenophores are, at least under some circumstances, independent effectors. More details on these and other independent effectors are discussed in later chapters.

Nervous systems and body *Baupläne*

The nervous system proper is constantly receiving information via its associated receptors, process-ing this information, and eliciting appropriate responses. We limit our discussion at this point to those conditions in which distinct systems of identifiable neurons exist, leaving the special situations in protozoa and sponges for later chapters. The structure of the nervous system of any organism is related to its *Bauplan* and mode of life. Some examples will demonstrate this relationship. Consider first a radially symmetrical animal with limited powers of locomotion, such as a planktonic jellyfish or a sessile sea anemone. In such animals the major receptor organs are more or less regularly (and radially) distributed around the body; the nervous system itself is a noncentralized, diffuse meshwork generally called a **nerve net** (Figure 33A). Interestingly, at least in cnidarians, there are both polarized and nonpolarized synapses within the nerve net. Impulses can travel in either direction across the nonpolarized synapses because the neuronal processes on both sides are capable of releasing synaptic transmitter chemicals. This capability, coupled with the gridlike form of the nerve net, enables impulses to travel in all directions from a point of stimulation. From this brief description, it might be assumed that such a simple and "unorganized" nervous system would not provide enough integrated information to allow complex behaviors and coordination. But many cnidarians are in fact capable of fairly intricate behavior. In the absence of a structurally recognizable integrating center, the nerve net does not fit well with our earlier

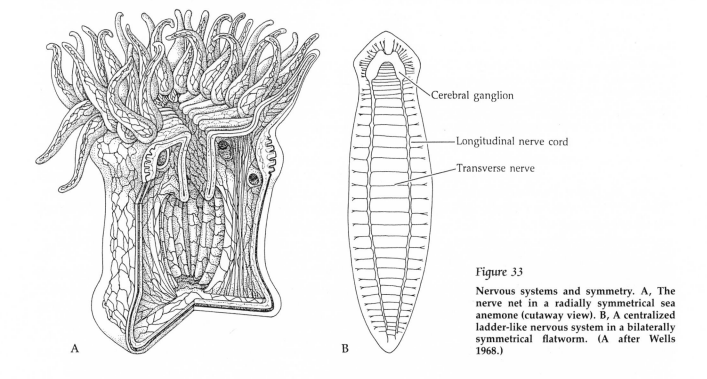

Cerebral ganglion

Longitudinal nerve cord

Transverse nerve

A   B

Figure 33

Nervous systems and symmetry. A, The nerve net in a radially symmetrical sea anemone (cutaway view). B, A centralized ladder-like nervous system in a bilaterally symmetrical flatworm. (A after Wells 1968.)

description of the sequence of events from stimulus to response. The system does work, however, and often in ways that are as yet unexplained. In any case, symmetry, sense organ distribution, nervous system organization, and lifestyles are clearly correlated to one another. Radially symmetrical animals tend to be able to respond equally well to stimuli coming from any direction—a useful ability for creatures with either sessile or free-floating (planktonic) lifestyles.

The tremendous evolutionary success of bilateral symmetry and unidirectional locomotion must have depended in large part on associated changes in the organization of the nervous system and the distribution of sense organs. The basic trend has been to centralize or concentrate the major coordinating elements of the nervous system. This **central nervous system** is generally made up of an anteriorly located neuronal mass (ganglion) from which arise one or more longitudinal nerve cords that often bear additional ganglia (Figure 33B). The anterior ganglion is referred to by a variety of names. We largely abandon the term *brain* for such organs because of the multifaceted implications of that word and adopt the more neutral term **cerebral ganglion** for the general case. In many instances, a term of relative position to some other organ is applied. For example, the cerebral ganglion commonly lies dorsal to the anterior portion of the gut and is thus a **supraenteric** (or supraesophageal, or suprapharyngeal) ganglion.

In addition to the cerebral ganglion, most bilaterally symmetrical animals have many of the major sense organs placed anteriorly. The concentration of these organs at the front end of an animal is called **cephalization**—the formation of a "head region." Even though cephalization may seem an obvious and predictable outcome of bilaterality and mobility, it is nonetheless extremely important. It simply would not do to have information about the environment gathered by the trailing end of a motile animal, lest it enter adverse and potentially dangerous conditions unawares. Hunting, tracking, or other forms of food location are greatly facilitated by having the appropriate receptors placed anteriorly—toward the direction of movement.

Longitudinal nerve cords receive information through peripheral nerves from whatever sense organs are placed along the body, and they carry motor impulses from the cerebral ganglion to peripheral nerves to effector sites. Additionally, nerve cords and peripheral nerves often serve animals in reflex actions and in some highly coordinated activities that do not depend on the cerebral ganglion. The most primitive centralized nervous system may have been similar to that seen today in some free-living flatworms, with several pairs of longitudinal cords attached to one another by a series of transverse connectives (Figure 33B). This arrangement is commonly referred to as a **ladder-like nervous system**. Among metazoa that have developed active lifestyles (e.g., errant polychaetes, most arthropods, cephalopod molluscs, and vertebrates), the nervous system has become more and more centralized through a reduction in the number of longitudinal nerve cords. However, a number of bilaterally symmetrical invertebrates (e.g., ectoprocts, tunicates, and echinoderms) have secondarily taken up sedentary or sessile modes of existence. Within these taxa there has been a corresponding decentralization of the nervous system and a general reduction in and dispersal of sense organs.

Hormones and pheromones

We have stressed the significance of the integrated nature of the parts and processes of living organisms and have discussed the general role of the nervous system in this regard. Organisms also produce and distribute within their bodies a variety of chemicals that control and coordinate biological activities. This very broad description of what may be called **chemical coordinators** obviously includes nearly any substance that has some effect on bodily functions. One special category of chemical coordinators comprises **hormones**; this term is not easy to define so that it fits all cases. It generally refers to any chemicals that are produced and secreted by some organ or tissue and are then carried by the blood or other body fluid to exert their influence elsewhere in the body. In vertebrates we associate this sort of phenomenon with the **endocrine system**, which includes well known glands as production sites. For our purposes we may subdivide hormones into two types. First are **endocrine hormones**, which are produced by more-or-less isolated glands and released into the circulatory fluid. Second are **neurohormones**, which are produced by special neurons called **neurosecretory cells**.

Much remains unknown concerning hormones in invertebrates. Nearly all our information comes from studies on insects and crustaceans, although hormonal activity has been demonstrated in a few other taxa and is suspected in many others. Among the arthropods, hormones are involved in the control of growth, molting, reproduction, eye pigment migration, and probably other phenomena; in at least some other taxa (e.g., annelid worms), hormones influence growth, regeneration, and sexual maturation.

Notice that the definition of hormones is based upon the sites of production and action. Hormones do not belong to any particular class of chemical compounds, nor do they all produce the same effects at the action sites; some are excitatory, some are inhibitory. Since endocrine hormones are carried in the circulatory fluid, they reach all parts of an animal's body. The site of action, or **target site**, must be able to recognize the appropriate hormone(s) among all the other myriad chemicals in its surroundings. This recognition usually involves an interaction between the hormone and the cell surface at the target site. Thus, under normal circumstances, even though a particular hormone is contacting many parts of the body, it will elicit activity only from the appropriate target organ or tissue that recognizes it.

In a general sense, **pheromones** are substances that act as "interorganismal hormones." These chemicals are produced by organisms and released into the environment, where they have an effect on other organisms. Most pheromone research has been on intraspecific actions, especially in insects, where activities such as mate attraction are frequently related to these airborne chemicals. We may view intraspecific pheromones as coordinating the activities of populations, just as hormones help coordinate the activities of individual organisms. There is also a great deal of evidence for the existence of interspecific pheromones, if we interpret the word in a broad sense. Some predatory species (e.g., some sea stars) release chemicals into the water that elicit extraordinary behavioral responses on the part of potential prey species, generally in the form of escape behavior. We will be discussing examples of various pheromone phenomena for specific animal taxa later in the book.

Reproduction

As noted in the passage from Barrington that introduces this chapter, the biological success of any

species depends upon its members staying alive long enough to reproduce themselves. The following account includes a discussion of the basic methods of reproduction among invertebrates and leads to the account of embryology and developmental strategies provided in Chapter 4.

Asexual reproduction

Asexual reproductive processes do not involve the production and subsequent fusion of haploid cells but rely solely on vegetative growth through mitosis. Cell division itself is a common form of asexual reproduction among protozoa, and many other invertebrates engage in various types of body fission, budding, or fragmentation, followed by growth to new individuals (Figure 34). These asexual processes depend largely on the animal's "reproductive exploitation" of its ability to **regenerate** (regrow lost parts). Even wound healing is a form of regeneration, but many animals have much more dramatic capabilities. The replacement of lost appendages in familiar animals such as sea stars and crabs is a common example of regeneration. However, these particular regenerative abilities are not "reproduction" because no new individuals result, and their presence does not imply that an animal capable of replacing a lost leg can necessarily reproduce asexually. Examples of animals that possess regenerative abilities of a magnitude permitting asexual reproduction include protozoa,

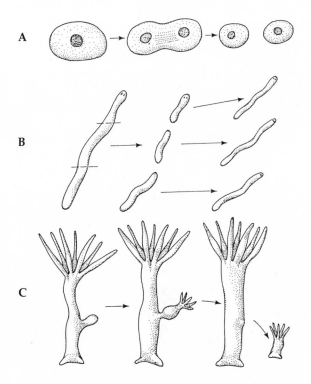

Figure 34

Some common asexual reproductive processes among invertebrates. A, Simple mitotic binary fission; this process occurs in most protozoa. B, Fragmentation, followed by regeneration of lost parts. This process occurs in a number of vermiform invertebrates. C, Budding may produce separate solitary individuals, as it does in *Hydra* (shown here); or it may produce colonies (see Figure 35). (Adapted from various sources.)

sponges, many cnidarians (corals, hydroids), certain types of worms, and sea squirts.

In many cases asexual reproduction is a relatively incidental process and rather insignificant to a species' overall survival strategy. In others, however, it is an integral and even necessary step in the life cycle. There are important evolutionary and adaptive aspects to asexual reproduction. Animals capable of rapid asexual reproduction can quickly take advantage of favorable environmental conditions by exploiting temporarily abundant food supplies, newly available living space, or other resources. This competitive edge is frequently evidenced by extremely high numbers of asexually produced individuals in certain disturbed or unique habitats, or in other unusual conditions. In addition, asexual processes are often employed in the production of resistant cysts or overwintering bodies, which are capable of surviving through periods of harsh environmental conditions. When favorable conditions return, these structures grow to new individuals.

A word about colonies. A frequent result of asexual reproduction, particularly various forms of budding, is the formation of **colonies** (Figure 35). This phenomenon is especially common in certain taxa (e.g., cnidarians, ascidians, and ectoprocts). The

term *colony* is not always easy to define. It may initially bring to mind ant or bee colonies, or even groups of people; but these examples are more appropriately viewed as social units rather than as colonies, at least in the context of our discussions. We accept Barrington's (1967) definition that "True colonies can be defined as . . . associations in which the constituent individuals are not completely separated from each other, but are organically connected together, either by living extensions of their bodies, or by material that they have secreted." This definition will suffice for now; we describe the nature of particular examples of colonial life in later chapters.

The formation of colonies not only may enhance the benefits of asexual reproduction in general, but also produces overall functional units of a much

A

Figure 35

Representative invertebrate colonies. A, *Botryllus*, a colonial ascidian. B, Massive growth of coral colonies (on either side of diver). C, *Aglaophenia*, a colonial hydroid. (A photo by H. W. Pratt/BPS; B courtesy P. Fankboner; C courtesy G. McDonald.)

B

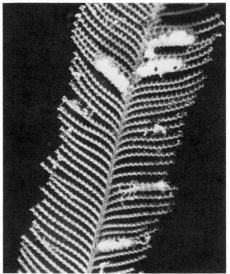

C

greater size than mere individuals; thus this growth habit may be viewed as a partial solution to the surface-to-volume dilemma. Increased functional size through colonialism can result in a number of advantages for animals; it can increase feeding efficiency, facilitate the handling of larger food items, reduce chances of predation, increase their competitive edge (for food, space, and other resources), and allow for specialization of individuals within the colony. In the latter instance specialization may result in individuals assuming distinct, organ-like roles in a colony of an otherwise simple animal.

Sexual reproduction

Among nature's many riddles is the fact that, although reproduction is critical to a species' survival, it is also the one major physiological activity that is not essential to an individual organism's survival. In fact, when animals are stressed, reproduction is usually the first activity that ceases. Sexual reproduction is especially energy costly. Given the advantages of asexual reproduction, one might wonder why all animals do not employ it and give up sexual activities entirely. The most frequently given explanation for the popularity of sexual reproduction (aside from anthropomorphic views, of course) focuses on the long-term benefits of genetic variation. Through regular meiosis and recombination, a level of genetic variation is maintained generation after generation, within and between populations; thus species are thought to be more "genetically prepared" for environmental changes. Although this advantage must surely be real, it does not seem to satisfactorily account for any short-term selection (i.e., generation by generation) for sexual reproduction. However, perhaps even in the short term the advantage may lie in the maintenance of variability. That is, apart from social implications such as pair bonding in many vertebrate species, there may be some short-term selective advantage(s) to the genetic variation that results from recombination. For example, a number of hypotheses center on the idea that a genetically varied and variable population may have a better chance of resisting predation, disease, and parasitism by virtue of their more heterogenous genotype, and also by "staying one step ahead" (evolutionarily) of predators or disease-causing organisms and parasites.

Sexual reproduction involves the formation of haploid cells through meiosis and the subsequent fusion of pairs of those cells to produce a diploid **zygote** (Figure 36). The haploid cells are **gametes**—sperm and eggs—and their fusion is the process of **fertilization** (exceptions to these general terms and

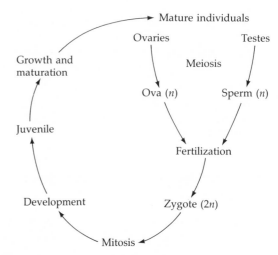

Figure 36

A generalized metazoan life cycle.

processes are common among protozoa, as discussed in Chapter 5). The production of gametes is accomplished by the **gonads**—**ovaries** (female) and **testes** (male)—or their functional equivalents. The gonads are frequently associated with reproductive systems of differing complexity, which may include various arrangements of ducts and tubes, accessory organs such as yolk glands or shell glands, and structures for copulation. The levels of complexity of these systems are related in part to the kinds of developmental strategies used by the organisms in question, as discussed in Chapter 4 and described in the coverage on each phylum. Reproductive structures also differ in the methods used to bring sperm and eggs together for fertilization. The variation in such matters is immense, but at this point we introduce some basic terminology of structure and function.

Many invertebrates simply release their gametes into the water in which they live (**broadcast spawning**), where external fertilization occurs. In such animals the gonads are usually simple, often transiently occurring structures associated with some means of getting the eggs and sperm out of the body. This release is achieved through a discrete plumbing arrangement (coelomoducts, metanephridia, or gonoducts—sperm ducts and oviducts), or by temporary pores in or rupture of the body wall.

On the other hand, invertebrates that pass sperm directly from the male to the female, where fertilization occurs internally, must have structural features to facilitate such activities. Figure 37 illustrates stylized male and female reproductive systems. A general scenario leading to internal fertilization in such

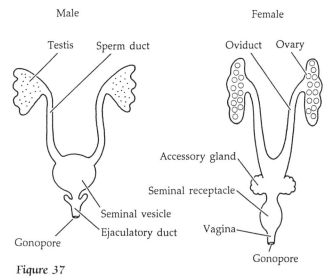

Figure 37

Schematic and generalized male and female reproductive systems. See text for explanation.

systems might be as follows. Sperm are produced in the testes and transported via the **sperm duct** to a precopulatory storage area called the **seminal vesicle**. Prior to mating, many invertebrates incorporate groups of sperm cells into sperm packets, or **spermatophores**. Spermatophores provide a protective casing for the sperm and facilitate transfer with minimal loss. In addition, many spermatophores are motile, acting as independent sperm carriers. Some sort of male copulatory or intromittent organ (e.g., **penis, cirrus, gonopod**) is inserted through the female's **gonopore** and into the **vagina**. Sperm are passed through the male's **ejaculatory duct** directly, or by way of a copulatory organ, into the female system, where they are received and often stored by a **seminal receptacle**.

In the female, eggs are produced in the ovaries and then transported into the **oviducts**. The sperm eventually travel up the oviducts, where they encounter the eggs; fertilization often takes place in the oviducts. Among invertebrates, the sperm may move by flagellar or ameboid action, or by locomotor structures on the spermatophore packet; they may be aided by ciliary action of the lining of the female reproductive tract. Various **accessory glands** may be present both in males (such as those that produce spermatophores or seminal fluids) and in females (such as those that produce egg capsules or shells).

This simple sequence is typical (although with many elaborations) of most invertebrates that rely on internal fertilization. Animals in which the sexes are separate, each individual being either male or female,

are termed **gonochoristic** (or **dioecious**). However, many invertebrates are **hermaphroditic*** (or **monoecious**), each animal containing both ovaries and testes and thus capable of producing both eggs and sperm (though not necessarily at the same time). Although self-fertilization may seem to be a natural advantage in this condition, such is not the case. In fact, with few exceptions, self-fertilization in hermaphrodites is prevented by various means. Fertilizing one's self would be the ultimate form of inbreeding, and would presumably result in a dramatic decrease in potential genetic variation and heterozygosity. The general rule for many hermaphroditic invertebrates is **mutual cross fertilization**, wherein two individuals function alternately or simultaneously as males and females, and exchange sperm, and then use the mate's sperm to fertilize their own eggs. The real advantage of hermaphroditism now becomes clear: the sexual encounter of two individuals can result in the impregnation of two individuals, rather than only one as in the gonochoristic condition. A frequently noted phenomenon associated with hermaphroditic invertebrates is that of **protandric hermaphroditism**, or simply **protandry** (Greek *proto*, "first"; *andro*, "male"), where individuals are first functional males, but later in life reverse sex to become functional females. The reverse situation, female first and then male, is called **protogynic hermaphroditism**, or simply **protogyny** (Greek *gynos*, "female"). At least some invertebrates alternate regularly between being functional males and females, as explained by Jerome Tichenor:

> Consider the case of the oyster,
> Which passes its time in the moisture;
>   Of sex alternate,
>   It chases no mate,
> But lives in a self-contained cloister.

> (*Poems in Contempt of Progress*, 1974)

In addition to the clever oysters immortalized by Professor Tichenor, other taxa in which the hermaphroditic condition is common include barnacles and certain other arthropods, advanced gastropods, arrow worms (phylum Chaetognatha), flatworms, oligochaetous annelids, leeches, and tunicates.

Parthenogenesis

Parthenogenesis (Greek *partheno*, "virgin"; *genesis*, "birth") is a special reproductive strategy in which unfertilized eggs undergo division to develop into viable adult individuals. Parthenogenetic species

---

*Hermaphroditus, the beautiful son of Hermes and Aphrodite, was united with a water nymph at the Carian fountain. Thus his body became both male and female.

have been reported from a number of invertebrate (and vertebrate) groups, including gastrotrichs, rotifers, tardigrades, nematodes, gastropods, certain insects and crustaceans, and several species of fishes, amphibians, and lizards. Among the invertebrates, parthenogenesis usually occurs in small-bodied species that are parasites, or are free-living but inhabit extreme or highly variable habitats such as temporary freshwater ponds. In "parthenogenetic taxa" inhabiting both fresh and marine waters (e.g., cladocerans, rotifers, and ostracods) parthenogenesis is far more prevalent among the freshwater species. There is also a general trend for parthenogenesis to become more prevalent in any group as one moves toward higher latitudes or into harsher environments.

Ova that develop parthenogenetically are often diploid, and the young are genetically identical to the mother. Thus, long-term parthenogenetic reproduction results in all-female clones. In most species that have been studied, parthenogenetic periods alternate with periods of sexual reproduction. In temperate freshwater habitats, parthenogenesis often occurs during summer months, with the population switching to sexual reproduction as winter approaches. In some species, parthenogenesis has been shown to take place for many generations, or several years, eventually to be punctuated by a brief period of sexual reproduction. In some rotifers, parthenogenesis predominates until the population attains a certain critical size, at which time males appear and a period of sexual reproduction ensues. Cladocerans have been shown to switch from parthenogenesis to sexual reproduction under a variety of conditions, such as overcrowding, adverse temperature, food scarcity, or even when the nature of the food changes. Many parasitic species alternate between a free-living sexual stage and a parasitic parthenogenetic one; this arrangement is seen in nematodes, thrips (Thysanoptera), gall wasps, aphids, and other homopterans.

One of the most interesting examples of parthenogenesis occurs in honeybees; in these animals the queen is fertilized by one or more males (drones) at only one period of her lifetime, in her "nuptial flight." The sperm are stored in her seminal receptacles. If sperm are released when the queen lays eggs, fertilization occurs and the eggs develop into females (queens or workers). If the eggs are not fertilized, they develop parthenogenetically into males (drones).

The question of the existence or prevalence of purely parthenogenetic species has been debated for decades. Many species once thought to be entirely parthenogenetic have proved, upon closer inspection, to alternate between parthenogenesis and brief periods of sexual reproduction. In some species purely parthenogenetic populations apparently exist only in some localities. In other species, parthenogenetic lineages have been traced to sexual ancestral populations occupying relictual habitats. Nevertheless, for some parthenogenetic animals, males have yet to be found in any population, and these may indeed be purely clonal species. One cannot help but wonder how long such species can exist in the face of natural selection without the benefits of any genetic exchange. One might imagine, as with any form of asexual reproduction, that obligatory parthenogenesis would eventually lead to genetic stagnation. There may, however, be some as yet unexplained genetic mechanisms to avoid this, because some parthenogenetic animals (e.g., some earthworms and insects) are capable of inhabiting a wide range of habitats. Thus presumably they either have a significant level of genetic adaptability or possess "general purpose genotypes."

Selected References

General References on Form and Function

Alexander, R. M. 1979. *The Invertebrates*. Cambridge University Press, London.

Barrington, E. J. W. 1967. *Invertebrate Structure and Function*, 2nd Ed. Halstead Press, New York.

Beklemishev, V. N. 1969. *Principles of Comparative Anatomy of Invertebrates* (2 vols.). Trans. J. M. MacLennan; ed. Z. Kabata. University of Chicago Press, Chicago. [One of the best coverages of form and function, although written from a very non-Western perspective.]

Brock, T. D. Life at high temperatures. Science 230:132–138.

Calow, P. 1977. Ecology, evolution and energetics: A study in metabolic adaptation. Adv. Ecol. Res. 10:1–62.

Calow, P. 1981. *Invertebrate Biology: A Functional Approach*. Halstead Press, New York.

Clark, R. B. 1964. *Dynamics in Metazoan Evolution*. Clarendon Press, Oxford. [A functional approach to phylogeny of metazoans; also see Chapter 4.]

Clark, R. B. and J. B. Cowey. 1958. Factors controlling the change in shape in certain nemertean and turbellarian worms. J. Exp. Biol. 35:731–748.

Eckert, R. and D. Randall. 1978. *Animal Physiology*. W. H. Freeman, San Francisco.

Fretter, V. and A. Graham. 1976. *A Functional Anatomy of Invertebrates*. Academic Press, New York. [An excellent reference for the serious student.]

Giese, A. C. 1979. *Cell Physiology*, 5th Ed. Saunders, Philadelphia. [A standard classic.]

Gordon, M. S. 1982. *Animal Physiology: Principles and Adaptations*. Macmillan, New York.

Hyman, L. H. 1940, 1951. *The Invertebrates*. Vol. 1, Protozoa through Ctenophora; Vol. 2, Platyhelminthes and Rhynchocoela: The Acoelomate Bilateria. McGraw-Hill, New York. [These two volumes of Hyman's series are especially useful in their discussion of body architecture in lower metazoa.]

Jackson, J. B. C., L. W. Buss and R. E. Cook (eds.). 1985. *Population Biology and Evolution of Clonal Organisms*. Yale University Press, New Haven, Connecticut.

Jacobs, M. H. 1967. *Diffusion Processes*. Springer-Verlag, New York.

Keegan, B. F., P. O. Ceidigh and P. J. S. Boaden (eds.). 1971. *Biology of Benthic Organisms*. Pergamon Press, New York.

Nicol, J. A. C. 1967. *The Biology of Marine Animals*, 2nd Ed. Wiley Interscience, New York. [One of the finest summaries of ecological physiology in print.]

Prosser, C. L. (ed.). 1973. *Comparative Animal Physiology*, 3rd. Ed. Saunders, Philadelphia. [Another classic and excellent reference, perhaps the best of its kind.]

Ramsay, J. A. 1952. *Physiological Approach to the Lower Animals*. Cambridge University Press, London.

Rockstein, M. (ed.). 1965–1974. *Physiology of Insecta*. Academic Press, New York.

Schmidt-Nielsen, K. 1979. *Animal Physiology*, 2nd Ed. Cambridge University Press, New York.

Schmidt-Nielsen, K., L. Bolis and S. H. P. Maddrell (eds.). 1978. *Comparative Physiology: Water, Ions and Fluid Mechanics*. Cambridge University Press, New York.

Swanson, C. P. and P. L. Webster. 1977. *The Cell*, 4th Ed. Prentice-Hall, Englewood Cliffs, New Jersey.

Thompson, D'Arcy. 1942. *On Growth and Form*, Rev. Ed. Macmillan, New York.

Waterman, T. H. (ed.). 1960. *Physiology of Crustacea*. Academic Press, New York.

Wigglesworth, V. B. 1974. *Insect Physiology*. Chapman and Hall, London.

Wilbur, K. M. and C. M. Yonge (eds.). 1964, 1967. *Physiology of Mollusca*. Academic Press, New York.

Locomotion and Support

Alexander, R. M. and G. Goldspink (eds.). 1977. *Mechanics and Energetics of Animal Locomotion*. Chapman and Hall, London.

Allen, R. D. 1962. Amoeboid movement. Sci. Am. 206(2):112–122. [The controversy on the mechanism of amoeboid movement continues today; also see Science (1972) 177:636–638, "Capillary suction test of the pressure gradient theory of amoeboid movement."]

Allen, R. D., D. Francis and R. Zek. 1971. Direct test of the positive pressure gradient theory of pseudopod extension and retraction in amoebae. Science 174:1237–1240.

Bereiter-Hahn, J., A. G. Matoltsy and K. S. Richards (eds.). 1984. *Biology of the Integument*. Vol. 1, Invertebrates. Springer-Verlag, New York.

Blake, J. R. and M. A. Sleigh. 1974. Mechanics of ciliary locomotion. Biol. Rev. 49:85–125.

Chapman, G. 1958. The hydrostatic skeleton in the invertebrates. Biol. Rev. 33:338–371.

Denny, M. W., T. L. Daniel and M. A. R. Koehl. 1985. Mechanical limits to size in wave-swept organisms. Ecol. Monogr. 55(1):69–102.

Herreid, C. T. II and C. R. Rourtner (eds.). 1981. *Locomotion and Energetics in Arthropods*. Plenum, New York.

Huxley, T. 1965. The mechanism of muscle contraction. Sci. Am. 213:18–27.

Jahn, T. L. and E. C. Bovee. 1967. Motile behavior of Protozoa. *In* T. Chen (ed.), *Research in Protozoology*. Pergamon Press, New York, pp. 41–200.

Jahn, T. L. and E. C. Bovee. 1969. Protoplasmic movements within cells. Phys. Rev. 49(4):830–862.

Jeffrey, D. J. and J. D. Sherwood. 1980. Streamline patterns and eddies in low Reynolds number flow. J. Fluid Mech. 96:315–334.

Jones, A. R. 1974. *The Ciliates*. St. Martin's Press, New York.

Koehl, M. A. R. 1984. How do benthic organisms withstand moving water? Am. Zool. 24:57–70.

Lowenstam, H.A. 1981. Minerals formed by organisms. Science 211:1126–1131.

Muzzarelli, R. 1977. *Chitin*. Pergamon Press, New York.

Satir, P. 1974. How cilia move. Sci. Am. 231:44–54.

Sleigh, M. A. (ed.). 1974. *Cilia and Flagella*. Academic Press, New York.

Trueman, E. R. 1975. *The Locomotion of Soft-Bodied Animals*. American Elsevier, New York.

Vogel, S. 1981. *Life in Moving Fluids: The Physical Biology of Flow*. Willard Grant, Boston.

Vogel, S. 1988. How organisms use flow-induced pressures. Am. Sci. 76:28–34.

Warner, F. D. and P. Satir. 1974. The structural basis of ciliary bend formation. Radial spoke positional changes accompanying microtubule sliding. J. Cell Biol. 63:35–63.

Wilkie, D. R. 1968. *Studies in Biology 11: Muscle*. The Camelot Press, London.

Yates, G. T. 1986. How microorganisms move through water. Am. Sci. 74:358–365.

Feeding Mechanisms

Anderson, J. M. and A. MacFayden (eds.). 1976. *The Role of Terrestrial and Aquatic Organisms in Decomposition Processes*. Blackwell Scientific, Oxford.

American Zoologist 22(3). 1982. The role of uptake of organic solutes in nutrition of marine organisms. pp. 611–733. [A collection of ten papers that offer an introduction to the literature on dissolved organic matter and its role in invertebrate nutrition.]

Blake, J. R., N. Liron and G. K. Aldis. 1982. Flow patterns around ciliated microorganisms and in ciliated ducts. J. Theor. Biol. 98:127–141.

Case, T. J. and R. K. Washino. 1979. Flatworm control of mosquito larvae in rice fields. Science 206:1412–1414.

Deibel, D., M.-L. Dickson and C. V. L. Powell. 1985. Ultrastructure of the mucous feeding filter of the house of the appendicularian *Oikopleura vanhoeffeni*. Mar. Ecol. Prog. Ser. 27:79–86.

Emlet, R. B. and R. R. Strathmann. 1985. Gravity, drag, and feeding currents of small zooplankton. Science 228:1016–1017.

Fauchald, K. and P. A. Jumars. 1979. The diet of worms: A study of polychaete feeding guilds. Oceanogr. Mar. Biol. Annu. Rev. 17:193–284.

Ford, M. J. 1977. Energy costs of the predatory strategy of the web-spinning spider *Lepthyphantes zimmermanni* Bertkau (Linyphiidae). Oecologia (Berlin) 28:341–349.

Gerritsen, J. and K. G. Porter. 1982. The role of surface chemistry in filter feeding by zooplankton. Science 216:1225–1227.

Jennings, J. B. 1973. *Feeding, Digestion and Assimilation in Animals*, 2nd Ed. Pergamon Press, New York.

Jorgensen, C. B. 1966. *Biology of Suspension Feeding*. Pergamon Press, New York.

Jorgensen, C. B. 1976. August Putter, August Krogh, and modern ideas on the use of dissolved organic matter in aquatic environments. Biol. Rev. 51:291–329.

Jorgensen, C. B. 1982. Uptake of dissolved amino acids from natural sea water in the mussel *Mytilus edulis* L. Ophelia 21:215–221.

Jorgensen, C. B., T. Kiorboe, J. Mohlenberg and H. U. Riisgard. 1984. Ciliary and mucus-net filter feeding, with special reference to fluid mechanical characteristics. Mar. Ecol. Prog. Ser. 15:283–292.

Koehl, M.A.R. and J.R. Strickler. 1981. Copepod feeding currents: Food capture at low Reynolds number. Limnol. Oceanogr. 26:1062–1093.

Krogh, A. 1931. Dissolved substances as food of aquatic organisms. Biol. Rev. 6:412–442.

LaBarbera, M. 1984. Feeding and particle capture mechanisms in suspension feeding animals. Am. Zool. 24:71–84.

Lehman, J. T. 1976. The filter-feeder as an optimal forager and the predicted shapes of feeding curves. Limnol. Oceanogr. 21:501–516.

Lindstedt, K. J. 1971. Chemical control of feeding behavior. Comp. Biochem. Physiol. 39A:553–581.

Manahan, D. T., S. H. Wright, G. C. Stephens and M. A. Rice. 1982. Transport of dissolved amino acids by the mussel, *Mytilus edulis*: Demonstration of net uptake from natural sea water. Science 215:1253–1255.

Menge, B. 1972. Foraging strategies in starfish in relation to actual prey availability and environmental predictability. Ecol. Monogr. 42:25–50.

Meyers, D. G. and J. R. Stickler (eds.). 1984. *Trophic Interactions within Aquatic Ecosystems*. AAAS Selected Symp., Vol 85. Westview Press, Boulder, Colorado.

Meyhofer, E. 1985. Comparative pumping rates in suspension-feeding bivalves. Mar. Biol. 85:137–142.

Nicol, E. A. 1930. The feeding mechanism, formation of the tube and physiology of digestion in *Sabella pavonina*. Trans. Royal Soc. Edinburgh 56:537–596.

O'Brien, W. J., D. Keetle and H. Riessen. 1979. Helmets and invisible armour: Structures reducing predation from tactile and visual planktivores. Ecology 60:287–294.

Polis, G. A. 1981. The evolution and dynamics of intraspecific predation. Annu. Rev. Ecol. Syst. 12:225–251.

Poulet, S.A. and P. Marsot. 1975. Chemosensory grazing by marine calanoid copepods (Arthropoda: Crustacea). Science 200:1403–1405.

Reid, R. G. B. and F. R. Bernard. 1980. Gutless bivalves. Science 208:609–610.

Roth, L. E. 1960. Electron microscopy of pinocytosis and food vacuoles in *Pelomyxa*. J. Protozool. 7:176–185.

Sammarco, P. W. 1980. *Diadema* and its relationship to coral spat mortality: Grazing, competition, and biological disturbance. J. Exp. Mar. Biol. Ecol. 45:245–272.

Schembri, P. J. 1982. Feeding behavior of fifteen species of hermit crabs (Crustacea: Decapoda: Anomura) from the Otago Region, southeastern New Zealand. J. Nat. Hist. 16:859–878.

Schoener, T. W. 1971. Theory of feeding strategies. Annu. Rev. Ecol. Syst. 2:369–404.

Sleeper, H. L., V. J. Paul and W. Fenical. 1980. Alarm pheromones from the marine opisthobranch *Navanax inermis*. J. Chem. Ecol. 6:57–70.

Smith, D., L. Muscatine, and O. Lewis. 1969. Carbohydrate movement from autotrophs to heterotrophs in parasitic and mutualistic symbioses. Biol. Rev. 44:17–90.

Smith, D. C. and Y. Tiffon (eds.). 1980. *Nutrition in the Lower Metazoa*. Pergamon Press, Elmsford, New York.

Spielman, L. A. 1977. Particle capture from low-speed laminar flows. Annu. Rev. Fluid Mech. 9:297–319.

Strathmann, R. R., T. L. Jahn and J. R. C. Fonseca. 1972. Suspension feeding by marine invertebrate larvae: Clearance of particles by ciliated bands of a rotifer, pluteus, and trochophore. Biol. Bull. 142:505–519.

Strathmann, R. R. 1978. The evolution and loss of feeding larval stages of marine invertebrates. Evolution 32:894–906.

Wharton, W. G. and K. H. Mann. 1981. Relationship between destructive grazing by the sea urchin *Strongylocentrotus droebachiensis*, and the abundance of American lobster, *Homarus americanus*, on the Atlantic coast of Nova Scotia. Can. J. Fish. Aquatic Sci. 38:1339–1349.

Yonge, C. M. 1928. Feeding mechanisms in the invertebrates. Biol. Rev. 3:21–76.

Yonge, C. M. 1937. Evolution and adaptation in the digestive system of the metazoa. Biol. Rev. 12:87–115.

Zaret, T. M. 1980. *Predation in Freshwater Communities*. Yale University Press, New Haven, Connecticut.

Excretion and Osmoregulation

Beadle, L. C. 1957. Osmotic and ionic regulation in aquatic animals. Annu. Rev. Physiol. 19:329–358.

Cohen, P. P. and G. W. Brown. 1960. Ammonia metabolism and urea biosynthesis. *In* M. Florkin and H. S. Mason (eds.), *Comparative Biochemistry* 21:161–244. Academic Press, New York.

Fisher, R. S., B. E. Persson and K. R. Spring. 1981. Epithelial cell volume regulation: Bicarbonate dependence. Science 214:1357–1359.

Giles, R. and A. Pequeux. 1981. Cell volume regulation in crustaceans: Relationship between mechanisms for controlling extracellular and intracellular fluids. J. Exp. Zool. 215:351–362.

Goodrich, E. S. 1945. The study of nephridia and genital ducts since 1895. Q. J. Microsc. Sci. 86:113–392. [Goodrich's classic paper on the evolution of coelomoducts, gonoducts, and nephridia.]

Oglesby, L. C. 1981. Volume regulation in aquatic invertebrates. J. Exp. Zool. 215:289–301. [Excellent summary paper.]

Pierce, S. K. 1982. Invertebrate cell volume control mechanisms: A coordinated use of intracellular amino acids and inorganic ions as osmotic solute. Biol. Bull. 163:405–419. [Osmoregulation at the cellular level; an important paper.]

Potts, W. T. W. and G. Parry. 1964. *Osmotic and Ionic Regulation in Animals*. Pergamon Press, New York.

Ramsay, J. A. 1954. Movements of water and electrolytes in invertebrates. Symp. Soc. Exp. Biol. 8:1–15.

Ramsay, J. A. 1961. The comparative physiology of renal function in invertebrates. *In*

J. A. Ramsay and V. B. Wigglesworth (eds.), *The Cell and the Organism*, Cambridge University Press, London, pp. 158–174.

Remmert, H. 1969. Über Poikilosmotic and Isoosmotic. Z. Vergl. Physiol. 65:424–427.

Ruppert, E. E. and P. R. Smith. 1988. The fundamental organization of filtration nephridia. Biol. Rev. 6:231–258.

Wilson, R. A. and L. A. Webster. 1974. Protonephridia. Biol. Rev. 49:127–160.

Yancey, P. H., M. E. Clark, S. C. Hand, R. D. Bowlus and G. N. Somero. 1982. Living with water stress: Evolution of osmolyte systems. Science 217:1214–1222.

Circulation and Gas Exchange

Bird, R. B., W. E. Stewart and N. Lightfoot. 1960. *Transport Phenomena*. Wiley, New York.

Krogh, A. 1941. *The Comparative Physiology of Respiratory Mechanisms*. University of Pennsylvania Press, Philadelphia.

Mill, P. J. 1972. *Respiration in the Invertebrates*. St. Martin's Press, London.

Ratcliffe, N. A. and A. F. Rowley (eds.). 1981. *Invertebrate Blood Cells*. Vols. 1–2. Academic Press, New York. [For a good introduction to the literature on circulation and gas exchange in invertebrates, see *American Zoologist* 19(1) (1979), "Comparative Physiology of Invertebrate Hearts;" for a good introduction to the literature on respiratory pigments, see *American Zoologist* 20(1) (1980), "Respiratory Pigments."]

Nervous Systems, Endocrines, and Behavior

Ali, M. A. (ed.). 1982. *Photoreception and Vision in Invertebrates*. Plenum, New York.

Atwood, H. L. and D. C. Sandeman (eds.). 1982. *The Biology of Crustacea*. Vol. 3, Neurobiology: Structure and Function. Academic Press, New York.

Autrum, H., R. Jung, W. R. Loewenstein, D. M. Mackay and H. L. Teuber (eds.). 1972–1981. *Handbook of Sensory Physiology*. Vols. 1–7. Springer-Verlag, New York. [This multivolume work includes the efforts of over 400 authors and contains excellent coverage.]

Barrington, E. J. W. 1963. *An Introduction to General and Comparative Endocrinology*. Clarendon Press, Oxford.

Bullock, T. and G. Horridge. 1969. *Structure and Function of the Nervous System of Invertebrates*. W. H. Freeman, San Francisco. [This two-volume work is still one of the finest summaries available on the subject.]

Bullock, T., R. Ork and A. Grinnel. 1977. *Introduction to Nervous Systems*. W. H. Freeman, San Francisco.

Carthy, J. D. 1968. *The Behavior of Arthropods*. Academic Press, New York.

Corning, W. C., J. A. Dyal and A. O. D. Willows (eds.). 1973. *Invertebrate Learning* (2 vols.). Plenum, New York.

Cronin, T. W. 1986. Photoreception in marine invertebrates. Am. Zool. 26:403–415.

Dumont, J. P. C. and R. M. Robertson. 1986. Neuronal circuits: An evolutionary perspective. Science 233:849–853.

Eakin, R. M. 1963. Lines of evolution of photoreceptors. *In* D. Mazia and A. Tyler (eds.), *General Physiology of Cell Specialization*. McGraw-Hill, New York, pp. 393–425.

Fein, A. and E. Z. Szuts. 1982. *Photoreceptors: Their Role in Vision*. Cambridge University Press, New York.

Hanstrom, B. 1928. *Vergleichende Anatomie des Nervensystems der wirbellosen Tiere*. Springer-Verlag, Berlin. Reprinted in 1968 by A. Asher and Co., Amsterdam.

Highnam, K. C. and L. Hill. 1977. *The Comparative Endocrinology of the Invertebrates*, 2nd Ed. University Park Press, Baltimore.

Jennings, H. S. 1976. *Behavior of the Lower Organisms*. Indiana University Press, Bloomington.

Laverack, M. S. 1968. On the receptors of marine invertebrates. Oceanogr. Mar. Biol. Annu. Rev. 6:249–324.

Lentz, T. L. 1968. *Primitive Nervous Systems*. Yale University Press, New Haven, Connecticut. [Interesting reading concerning the origin and evolution of nervous systems among the "lower" metazoa.]

Mill, P. J. 1976. *Structure and Function of Proprioceptors in the Invertebrates*. Halsted Press, New York. [The author assumes a very broad definition of proprioceptors, hence this large book includes fine descriptions of a number of mechanoreceptors.]

Salanki, J. (ed.). 1973. *Neurobiology of Invertebrates*. Academiai Riado, Budapest.

Sandeman, D. C. and H. L. Atwood (eds.). 1982. *The Biology of Crustacea*. Vol. 4, Neural Integration and Behavior. Academic Press, New York.

Wells, M. J. 1965. Learning by marine invertebrates. Adv. Mar. Biol. 3:1–62.

Wolken, J. J. 1971. *Invertebrate Photoreceptors*. Academic Press, New York.

Also see: *Oceanus* 23(3), 1980, for a number of introductory review articles on senses of marine animals.

Reproduction

Also see references for Chapter 4.

Boardman, R. S., A. M. Cheetham and W. A. Oliver (eds.). 1973. *Animal Colonies: Development and Function Through Time*. Dowden, Hutchinson and Ross, Stroudsburg, Pennsylvania.

Braverman, M. H. and R. G. Schrandt. 1966. Colony formation of a polymorphic hydroid as a problem in pattern formation. *In* W. J. Rees (ed.), *The Cnidaria and their Evolution*. Symp. Zool. Soc. Lond. 16:169–198.

Cohen, J. 1977. *Reproduction*. Butterworth, London.

Giese, A. C. and J. A. Pearse (eds.). 1974–1987. *Reproduction of Marine Invertebrates*. Vols. 1–5, 9. Academic Press, New York. [Volumes 6–8 are not yet published as we write; Volume 9 has outstanding reviews of most aspects of invertebrate gametogenesis and development.]

Lockwood, G. and B. R. Rosen (eds.). 1979. *Biology and Systematics of Colonial Animals*. Academic Press, New York. Published for the Systematics Association, Special Vol. No. 11.

Maynard Smith, J. 1978. *The Evolution of Sex*. Cambridge University Press, Cambridge.

Policansky, D. 1982. Sex change in plants and animals. Annu. Rev. Ecol. Syst. 13: 471–496.

Rose, S. M. 1970. *Regeneration*. Appleton-Century-Crofts, New York.

Stancyk, S. E. 1979. *Reproductive Ecology of Marine Invertebrates*. University of South Carolina Press, Columbia.

Thorson, G. 1950. Reproduction and larval ecology of marine bottom invertebrates. Biol. Rev. 25:1–45.

For an introduction to the literature on the evolutionary implications of asexual reproduction, see *American Zoologist* 19(3) (1979), "Ecology of Asexual Reproduction."

Chapter Four

The Metazoa: Development, Life Histories, and Phylogeny

He who sees things grow from their beginning will have the finest view of them.
    Aristotle

Metazoa are multicellular, as opposed to the protozoa, which are usually viewed as being unicellular (see Chapter 5). This distinction, however, is sometimes blurred, for there are a number of protozoa that form rather complex colonies with some division of labor among different cell types. The metazoa possess certain qualities that must be considered in concert with the basic idea of multicellularity. The cells of metazoa are organized into functional units, generally as tissues and organ structures, with specific roles for supporting the life of the whole organism. These cell types are interdependent and their activities are coordinated into predictable patterns and relationships. Structurally, the cells of metazoa are organized as layers, which develop through a series of events early in an organism's embryogeny. These embryonic tissues, or **germ layers**, form the framework upon which the metazoan body is constructed (see Chapter 3). *Thus, the cells of metazoa are specialized, interdependent, coordinated in function, and develop through layering during embryogeny. This combination of features is absent from protozoa.*

Eggs and embryos

The attributes that distinguish the metazoa are the result of metazoan embryonic development. To put it another way, adult phenotypes result from specific sequences of developmental stages. Indeed, embryogenesis is unique to multicelled life. Bearing this in mind, one is not surprised to discover that both animal unity and diversity are as evident in patterns of development as they are in the architecture of adults. The patterns of development discussed below reflect unity and diversity, and serve as a basis for understanding the sections on embryology in later chapters.

Eggs

Biological processes in general are cyclical. The production of one generation after another through reproduction exemplifies this generality, as the term *life cycle* implies. At what point one begins describing such things is more-or-less a matter of convenience. For our purposes in this chapter we choose to begin with the **egg**, or **ovum**, that remarkable cell capable of developing to a new individual. Once the egg is fertilized, all of the different cell types of adult metazoa ultimately are derived during embryogenesis from this single totipotent cell. A fertilized egg contains not only the information necessary to direct development but also some quantity of nutrient material called **yolk**, which sustains the early stages of life.

Eggs tend to be polarized along what is called their **animal–vegetal axis**. This polarity may be apparent in the egg itself, or it may be recognizable only as development proceeds. The vegetal pole is commonly associated with the formation of nutritive organs (e.g., the digestive system), whereas the animal pole tends to produce other regions of the embryo. These and many other manifestations of the egg's polarity will be more completely explored throughout this chapter.

Metazoan ova are categorized primarily by the amount and location of yolk within the cell (Figure 1), two factors that greatly influence certain aspects of development. **Isolecithal eggs** contain a relatively small amount of yolk that is more-or-less evenly dis-

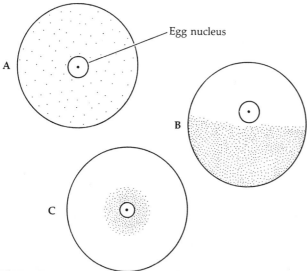

Figure 1

Types of ova. The stippling denotes the distribution and relative concentration of yolk within the cytoplasm. A, An isolecithal ovum has a small amount of yolk distributed evenly. B, The yolk in a telolecithal ovum is concentrated toward the vegetal pole. The amount of yolk in such eggs varies greatly. C, A centrolecithal ovum has yolk concentrated at the center of the cell.

tributed throughout the cell. Ova in which the yolk is concentrated at one end (the vegetal pole) are termed **telolecithal eggs**; those in which the yolk is concentrated in the center are called **centrolecithal eggs**. The actual amount of yolk in telolecithal and centrolecithal eggs is highly variable.

Cleavage

The stimulus that initiates development in an egg is generally provided by the penetration of a sperm cell and the subsequent fusion of the male and female nuclei to produce a fertilized egg, or **zygote**. The

initial process of cell division of a zygote is called **cleavage**, and the resulting cells are called **blastomeres**. Certain aspects of the patterns of early cleavage are determined by the amount and placement of yolk, whereas other features are apparently inherent in the genetic programming of the particular organism. Isolecithal and weakly to moderately telolecithal ova generally undergo **holoblastic cleavage**. That is, the cleavage planes pass completely through the cell, producing blastomeres that are separated from one another by thin cell membranes (Figure 2A). Whenever very large amounts of yolk are present (e.g., as in strongly telolecithal eggs), the cleavage planes do not pass readily through the dense yolk, so the blastomeres are not fully separated from one another by cell membranes. This pattern of early cell division is called **meroblastic cleavage** (Figure 2B). The pattern of cleavage in centrolecithal eggs is dependent on the amount of yolk and varies from holoblastic to various modifications of meroblastic (e.g., see the descriptions of arthropod development in Chapter 15).

Orientation of cleavage planes. A number of terms are used to describe the relationship of the planes of cleavage to the animal–vegetal axis of the egg and the relationships of the resulting blastomeres to each other. Figure 3 illustrates the patterns described below. Cell divisions during cleavage are often referred to as either **equal** or **unequal**, the terms indicating the comparative sizes of groups of blastomeres. The term **subequal** is used when blastomeres are only slightly different in size. When cleavage is distinctly unequal, the larger cells are called **macromeres** and usually lie at the vegetal pole. The smaller cells are called **micromeres** and are usually located at the animal pole.

Cleavage planes that pass parallel to the animal–

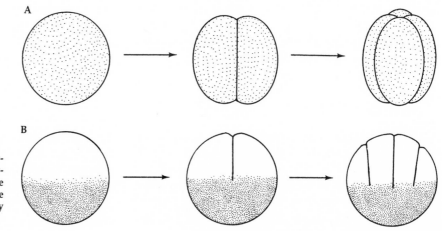

Figure 2

Types of early cleavage in developing zygotes. A, Holoblastic cleavage. The cleavage planes pass completely through the cytoplasm. B, Meroblastic cleavage. The cleavage planes do not pass completely through the yolky cytoplasm.

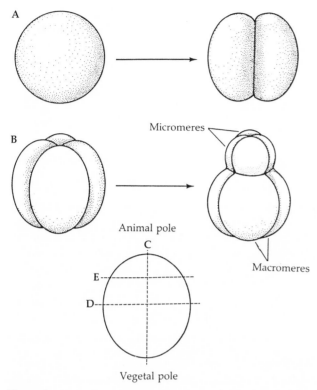

Figure 3

Planes of holoblastic cleavage. A, Equal cleavage. B, Unequal cleavage produces micromeres and macromeres. C–E, Planes of cleavage relative to the animal–vegetal axis of the egg or zygote. C, Longitudinal (= meridional) cleavage parallel to the animal–vegetal axis. D, Equatorial cleavage perpendicular to the animal–vegetal axis and bisecting the zygote into equal animal and vegetal halves. E, Latitudinal cleavage perpendicular to the animal–vegetal axis but not passing along the equatorial plane.

vegetal axis produce **longitudinal** or **meridional** divisions; those that pass at right angles to the axis produce **transverse** divisions. Transverse divisions may be either **equatorial**, when the embryo is separated equally into animal and vegetal halves, or simply **latitudinal**, when the division plane does not pass through the "equator" of the embryo.

Radial and spiral cleavage. Most invertebrates display one of two basic cleavage patterns defined on the basis of the orientation of the blastomeres

about the animal–vegetal axis. These patterns are called **radial cleavage** and **spiral cleavage** and are illustrated in Figure 4. Radial cleavage involves strictly meridional and transverse divisions. Thus, the blastomeres are arranged in rows either parallel or perpendicular to the animal–vegetal axis. The placement of the blastomeres shows a radially symmetrical pattern in polar view.

Spiral cleavage is quite another matter. Although not inherently complex, it can be difficult to describe. The first two divisions are meridional, generally equal or subequal. Subsequent divisions, however,

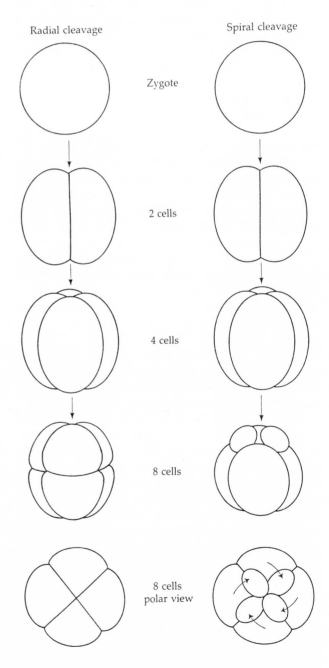

Figure 4

Comparison of radial versus spiral cleavage through the 8-cell stage. During radial cleavage, the cleavage planes all pass either perpendicular or parallel to the animal–vegetal axis of the embryo. Spiral cleavage involves a tilting of the mitotic spindles, commencing with the division from four to eight cells. The resulting cleavage planes are neither perpendicular nor parallel to the axis. The polar views of the resulting 8-cell stages illustrate the differences in blastomere orientation.

result in the displacement of blastomeres in such a way that they lie in the furrows between one another. This condition is a result of the formation of the mitotic spindles at acute angles rather than parallel or perpendicular to the axis of the embryo; hence the cleavage planes are neither perfectly meridional nor perfectly transverse. The division from four to eight cells involves a displacement of the cells near the animal pole in a clockwise (**dextrotropic**) direction (viewed from the animal pole). The next division, from eight to sixteen cells, occurs with a displacement in a counterclockwise (**levotropic**) direction; the next is clockwise, and so on—"twisting" back and forth until approximately the 64-cell stage. We hasten to add that in reality divisions are frequently nonsynchronous; not all of the cells divide at the same rates. Thus, a particular embryo may not proceed from four cells to eight, to sixteen, and so on, as neatly as in our generalized example.

An elaborate coding system for spiral cleavage was developed by E. B. Wilson (1892) during his extensive studies on the polychaete worm *Neanthes succinea* conducted at Woods Hole Biological Station. Wilson's system is usually applied to spiral cleavage in order to trace cell fates and compare development among various taxa. The following account of spiral cleavage is a general one, but it will provide a point of reference for later consideration of the patterns in different groups of animals. Although this coding system may at first seem a bit confusing, it will quickly become evident that it is a rather simple and elegant means by which one may follow the developmental lineage of each and every cell in a developing embryo.

At the 4-cell stage, following the initial meridional divisions, the cells are given the codes of A, B, C, and D, and are labeled clockwise in that order when viewed from the animal pole (Figure 5A). These four cells are referred to as a quartet of macromeres, and they may be collectively coded as simply Q. The next division is more-or-less unequal, with the four cells nearest the animal pole being displaced in a dextrotropic fashion, as explained above. These four smaller cells are called the first quartet of micromeres (collectively the lq cells) and are given the individual codes of 1a, 1b, 1c, and 1d. The numeral "1" indicates that they are members of the first micromere quartet to be produced; the letters correspond to their respective macromere origins. The capital letters designating the macromeres are now preceded with the numeral "1" to indicate that they have divided once and produced a first micromere set (Figure 5B). We may view this 8-celled em-

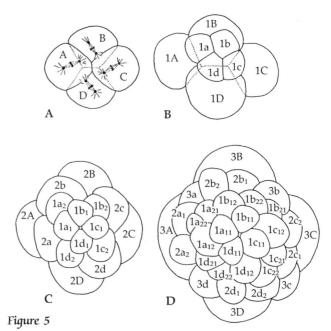

Figure 5

Spiral cleavage from 4 to 32 cells (assumed synchronous) labeled with E. B. Wilson's coding system. All diagrams are animal-pole surface views. See text for explanation.

bryo as four pairs of daughter cells that have been produced by the divisions of the four original macromeres as follows:

$$A \nearrow \begin{matrix} 1a \\ 1A \end{matrix} \qquad B \nearrow \begin{matrix} 1b \\ 1B \end{matrix} \qquad C \nearrow \begin{matrix} 1c \\ 1C \end{matrix} \qquad D \nearrow \begin{matrix} 1d \\ 1D \end{matrix}$$

One "rule" that will aid you in tracing the cells and their codes through spiral cleavage is that the *only* code numbers that are changed through subsequent divisions are the prefix numbers of the macromeres. These are changed to indicate the number of times these individual macromeres have divided, and to correspond to the number of micromere quartets thus produced. So, at the 8-cell stage, we can designate the existing blastomeres as the 1Q (= 1A, 1B, 1C, 1D) and the lq (= 1a, 1b, 1c, 1d).

It should be mentioned that although the macromeres and micromeres are sometimes similar in size, these terms are nonetheless always used in describing spiral cleavage. Much of the size discrepancy depends upon the amount of yolk present at the vegetal pole in the original egg; this yolk tends to be retained primarily in the larger macromeres.

The division from eight to sixteen cells occurs levotropically and involves cleavage of each macromere and micromere. The macromeres (1Q) divide to produce a second quartet of micromeres (2q = 2a, 2b, 2c, 2d), and the prefix numeral of the daughter macromeres is changed to "2." The first micromere quartet also divides and now comprises eight cells, each of which is identifiable not only by the letter corresponding to its parent macromere but now by the addition of subscript numerals. For example, the 1a micromere (of the 8-cell embryo) divides to produce two daughter cells coded the $1a_1$ and the $1a_2$ cells. The cell that is physically nearer the animal pole of the embryo receives the subscript "1," the other cell the subscript "2." Thus, the 16-cell stage (Figure 5C) includes the following cells:

$$\text{derivatives of the 1q} \begin{cases} 1a_1 & 1b_1 & 1c_1 & 1d_1 \\ 1a_2 & 1b_2 & 1c_2 & 1d_2 \end{cases}$$

$$\text{derivatives of the 1Q} \begin{cases} 2q = 2a & 2b & 2c & 2d \\ 2Q = 2A & 2B & 2C & 2D \end{cases}$$

The next division (from 16 to 32 cells) involves dextrotropic displacement. The third micromere quartet (3q) is formed, and the daughter macromeres are now given the prefix "3" (3Q), and all of the 12 existing micromeres divide. Subscripts are added to the derivatives of the first and second micromere quartets according to the rule of position as stated above. Thus, the $1b_1$ cell divides to yield the $1b_{11}$ and $1b_{12}$ cells; the $1a_2$ cell yields the $1a_{21}$ and $1a_{22}$ cells; the 2c yields the $2c_1$ and $2c_2$, and so on. Do not think of these subscripts as double digit numbers (e.g., "twenty-one" and "twenty-two") but rather as two-digit sequences reflecting the precise lineage of each cell ("two-one" and "two-two").

The elegance of Wilson's system is that each code tells the history as well as the position of the cell in the embryo. For instance, the code $1b_{11}$ indicates that the cell is a member (derivative) of the first quartet of micromeres, that its parent macromere is the B cell, that the original 1b micromere has divided twice since its formation, and that this particular cell rests uppermost in the embryo relative to its sister cells. The 32-cell state (Figure 5D) is composed of the following:

$$\text{derivatives of the 1q} \begin{cases} 1a_{11} & 1b_{11} & 1c_{11} & 1d_{11} \\ 1a_{12} & 1b_{12} & 1c_{12} & 1d_{12} \\ 1a_{21} & 1b_{21} & 1c_{21} & 1d_{21} \\ 1a_{22} & 1b_{22} & 1c_{22} & 1d_{22} \end{cases}$$

$$\text{derivatives of the 2q} \begin{cases} 2a_1 & 2b_1 & 2c_1 & 2d_1 \\ 2a_2 & 2b_2 & 2c_2 & 2d_2 \end{cases}$$

$$\text{derivatives of the 2Q} \begin{cases} 3q = 3a & 3b & 3c & 3d \\ 3Q = 3A & 3B & 3C & 3D \end{cases}$$

The division to 64 cells follows the same pattern, with appropriate coding changes and additions of sub-scripts. The displacement is levotropic and results in the following cells:

$$\text{derivatives of the 1q} \begin{cases} 1a_{111} & 1b_{111} & 1c_{111} & 1d_{111} \\ 1a_{112} & 1b_{112} & 1c_{112} & 1d_{112} \\ 1a_{121} & 1b_{121} & 1c_{121} & 1d_{121} \\ 1a_{122} & 1b_{122} & 1c_{122} & 1d_{122} \\ 1a_{211} & 1b_{211} & 1c_{211} & 1d_{211} \\ 1a_{212} & 1b_{212} & 1c_{212} & 1d_{212} \\ 1a_{221} & 1b_{221} & 1c_{221} & 1d_{221} \\ 1a_{222} & 1b_{222} & 1c_{222} & 1d_{222} \end{cases}$$

$$\text{derivatives of the 2q} \begin{cases} 2a_{11} & 2b_{11} & 2c_{11} & 2d_{11} \\ 2a_{12} & 2b_{12} & 2c_{12} & 2d_{12} \\ 2a_{21} & 2b_{21} & 2c_{21} & 2d_{21} \\ 2a_{22} & 2b_{22} & 2c_{22} & 2d_{22} \end{cases}$$

$$\text{derivatives of the 3q} \begin{cases} 3a_1 & 3b_1 & 3c_1 & 3d_1 \\ 3a_2 & 3b_2 & 3c_2 & 3d_2 \end{cases}$$

$$\text{derivatives of the 3Q} \begin{cases} 4q = 4a & 4b & 4c & 4d \\ 4Q = 4A & 4B & 4C & 4D \end{cases}$$

Notice that no two cells share precisely the same code, so exact identification of individual blastomeres and their lineages is always possible.

The problem of cell fates. Tracing the fates of cells through development has been a popular and productive endeavor of embryologists for over a century. Such studies have played a major role in enabling researchers not only to describe development but also to establish homologies among attributes in different animals. Although the cells of embryos eventually become established as functional parts of tissues or organs, there is a great deal of variation in the timing of the establishment of cell fates and in how "firmly fixed" the fates eventually become. Even in the adult stages of some animals (e.g., sponges) the cells retain the ability to change their structure and function, although under normal conditions they are relatively specialized. Furthermore, many groups of animals have remarkable powers of regeneration of lost parts, wherein cells may dedifferentiate and then generate new tissues and organs. In other cases cell fates are relatively firmly fixed and are able only to produce more of their own kind.

By carefully watching the development of any animal, it becomes clear that certain cells are going to predictably form certain structures. In some cases the cell fates are determined very early during cleavage—as early as the 2- or 4-cell stage. When one experimentally removes a blastomere from the early embryo of such an animal, that embryo will fail to develop normally; the fates of the cells have already become fixed, and the missing cell cannot be replaced. Animals whose cell fates are established very early are said to have **determinate cleavage**. On the other hand, the blastomeres of some animals can be separated at the 2-cell, 4-cell, or even later stages, and each separate cell will develop normally; in such cases the fates of the cells are not fixed until relatively late in development. Such animals are said to have **indeterminate cleavage**. Eggs that undergo determinate cleavage are often called **mosaic ova**, because the fates of regions of undivided cells can be mapped. Eggs that undergo indeterminate cleavage are often called **regulative ova**, in that they can "regulate" to accommodate lost blastomeres and thus cannot easily be predictably mapped prior to division.

In any case, formation of the basic body plan is generally complete by the time the embryo comprises about 10^4 cells (usually after one or two days). By this time, all available embryonic material has been apportioned into specific cell groups, or "founder regions." These specific regions are few in number, and large, each becoming a territory within which still more intricate developmental patterns occur. As these zones of undifferentiated tissue are established, the unfolding genetic code drives them to develop into their "preassigned" body tissues, organs, or other structures.

In the past it has been a general practice to equate mosaic eggs and determinate cleavage with spirally cleaving embryos, and to equate regulative ova and indeterminate cleavage with radially cleaving embryos. These relationships appeared to be so important to some workers that W. Schleip (1929) coined and defined the term **Spiralia** as a formal taxon including all animals with spirally cleaving embryos (e.g., platyhelminths, annelids, molluscs, arthropods, and a few others). The tendency has been to observe the cleavage pattern of an embryo and note whether it is spiral or radial, and then to assume that it is determinate or indeterminate on the basis of the physical placement of the blastomeres. Surprisingly few actual tests for determinacy have been performed, and what evidence is available suggests that there are many exceptions to this generalization. That is, some embryos with spiral cleavage appear indeterminate, and some with radial cleavage appear determinate. Much more work remains to be done on these matters, but for the present the relationships among these features of early development are questionable. (For additional information see Costello and Henley 1976; Siewing 1980; and Ivanova-Kazas 1982.)

In spite of the variations and exceptions, there is a remarkable underlying consistency in the fates of blastomeres among embryos that develop by typical spiral cleavage. Many examples of these similarities are discussed in later chapters, but we illustrate the point by noting that the germ layers of spirally cleaving embryos tend to arise from the same groups of cells. The first three quartets of micromeres and their derivatives give rise to ectoderm (the outer germ layer), the 4a, 4b, 4c, and 4Q cells to entoderm (the inner germ layer), and the 4d cell to mesoderm (the middle germ layer). Many students of embryology view this uniformity of cell fates as strong evidence that taxa sharing this pattern are related to one another in some fundamental way and that they share a common evolutionary heritage. We will have more to say about this idea throughout this book.

Blastula types

The product of early cleavage is called the **blastula**, which may be defined developmentally as the embryonic form preceding the formation of embryonic germ layers. Several types of blastulae are recognized among invertebrates. A **coeloblastula** frequently results from radial cleavage of ova with relatively small amounts of yolk (Figure 6A). This blastula is a hollow ball of cells, the wall of which is usually one cell-layer thick. The space within the sphere of cells is the **blastocoel**, or **primary body cavity**. Spiral cleavage often results in a solid ball of cells called a **stereoblastula** (Figure 6B); obviously there is no blastocoel at this stage. Meroblastic cleavage typically results in a cap or disc of cells at the animal pole over an uncleaved mass of yolk. This arrangement is appropriately termed a **discoblastula** (Figure 6C). Some centrolecithal ova undergo odd cleavage patterns to form a **periblastula**, similar in some respects to a coeloblastula that is centrally filled with noncellular yolk (Figure 6D).

Gastrulation and germ layer formation

Through one or more of several methods the blastula develops toward a multilayered form, a process called **gastrulation** (Figure 7). The structure of the blastula dictates to some degree the nature of the process and the form of the resulting embryo, the **gastrula**. Gastrulation is the formation of the embryonic germ layers, the tissues on which all subsequent development and body plans eventually depend. In fact, we may view gastrulation as the embryonic analogue of the transition from protozoan to metazoan grades of complexity. It achieves separation of those cells that must interact directly with the environment (i.e., locomotor, sensory, and protective functions)

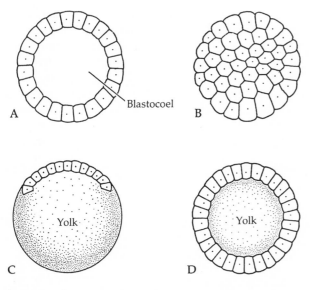

Figure 6

Types of blastulae. These diagrams represent sections along the animal–vegetal axis. A, Coeloblastula. The blastomeres form a hollow sphere with a wall one cell layer thick. B, Stereoblastula. Cleavage results in a solid ball of blastomeres. C, Discoblastula. Cleavage has produced a cap of blastomeres that lies at the animal pole, above a solid mass of yolk. D, Periblastula. Blastomeres form a single cell layer enclosing an inner yolky mass.

from those that process materials ingested from the environment (i.e., nutritive functions).

The names of the germ layers have already been introduced, and we concentrate here on their formation and fates. The initial inner and outer sheets of cells are the entoderm (= endoderm) and ectoderm, respectively; a third germ layer, the mesoderm, is produced in most animals between the ectoderm and the entoderm. One of the principal examples of the unity among the metazoa is the consistency of the fates of these germ layers. For example, ectoderm always forms the nervous system and the outer skin and its derivatives; entoderm the main portion of the gut and associated structures; and mesoderm the coelomic lining, the circulatory system, most of the internal support structures, and the musculature. The process of gastrulation, then, is a critical one in establishing the basic materials and their locations for the building of the whole organism.

Coeloblastulae often gastrulate by **invagination**, a process commonly used to illustrate gastrulation in general zoology classes. The cells in one area of the surface of the blastula (frequently at or near the vegetal pole) grow inward as a sac within the blastocoel (Figure 7A). These invaginated cells are now called the entoderm, the sac thus formed is the embryonic

gut, or **archenteron**, and the opening to the outside is the **blastopore**. The outer cells are now called ectoderm, and a double-layered hollow **coelogastrula** has been formed.

The coeloblastulae of many cnidarians undergo gastrulation processes that result in solid gastrulae (**stereogastrulae**). Usually the cells of the blastula divide in such a way that the cleavage planes are perpendicular to the surface of the embryo. Some of the cells detach from the wall and migrate into the blastocoel, eventually filling it with a solid mass of entoderm. This process is called **ingression** (Figure 7B) and may occur only at the vegetal pole (unipolar ingression) or more-or-less over the whole blastula (multipolar ingression). In a few instances (e.g., certain hydroids) the cells of the blastula divide with cleavage planes that are parallel to the surface, a process called **delamination** (Figure 7C). This process also produces a layer or mass of entoderm surrounded by a layer of ectoderm.

Stereoblastulae that result from holoblastic cleavage generally undergo gastrulation by **epiboly**. Because there is no blastocoel into which the presumptive entoderm can migrate by any of the above methods, gastrulation involves a rapid growth of presumptive ectoderm around the presumptive entoderm (Figure 7D). Cells of the animal pole proliferate rapidly, growing down and over the vegetal cells to enclose them as entoderm. The archenteron typically forms secondarily as a space within the developed entoderm.

Figure 7E illustrates gastrulation by **involution**, a process that usually follows the formation of a discoblastula. The cells around the edge of the disc divide rapidly and grow beneath the disc, thus forming a double-layered gastrula with ectoderm on the surface and entoderm below. There are several other types of gastrulation, mostly variations or combinations of the above processes. These gastrulation methods occur in particular taxa and are discussed in later chapters.

Mesoderm and body cavities

Some time following gastrulation, a middle layer forms between the ectoderm and the entoderm. This middle layer may be derived from ectoderm, as it is in members of the diploblastic phyla Cnidaria and Ctenophora, or from entoderm, as it is in members of the triploblastic phyla. In the first case the middle layer is said to be **ectomesoderm**, and in the latter case **entomesoderm** (or "true mesoderm"). Thus, the triploblastic condition, by definition, includes entomesoderm. In this text, and most others, the term

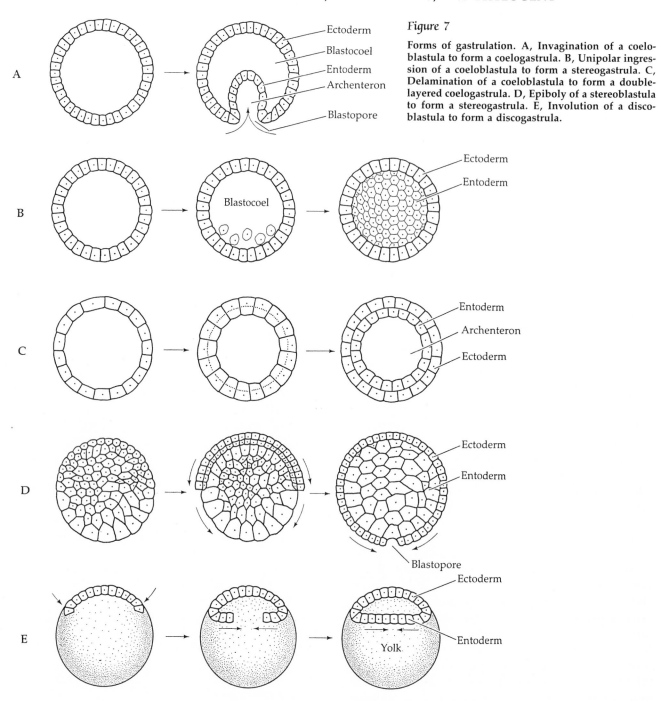

Figure 7

Forms of gastrulation. A, Invagination of a coelo-blastula to form a coelogastrula. B, Unipolar ingression of a coeloblastula to form a stereogastrula. C, Delamination of a coeloblastula to form a double-layered coelogastrula. D, Epiboly of a stereoblastula to form a stereogastrula. E, Involution of a disco-blastula to form a discogastrula.

mesoderm in a general sense refers to entomesoderm rather than ectomesoderm.

In the diploblastic and certain triploblastic phyla (the acoelomates), the middle layer does not form thin sheets of cells, rather it produces a more-or-less solid but loosely organized **mesenchyme** consisting of a gel matrix (the **mesoglea**) containing various cellular and fibrous inclusions. In a few cases (the

hydrozoans) a virtually noncellular mesoglea lies between the ectoderm and entoderm (see Chapter 8).

In most metazoa the area between the inner and outer body layers includes a fluid-filled space. As discussed in Chapter 3, this space may be either a pseudocoelom, which is a cavity not completely lined by mesoderm, or a true coelom, which is a cavity fully enclosed within thin sheets of mesodermally

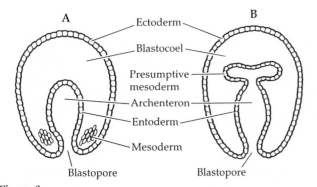

Figure 8

Methods of mesoderm formation in late gastrulae (frontal sections). A, Mesoderm formed from derivatives of a mesentoblast. B, Mesoderm formed by archenteric pouching.

derived tissues. Mesoderm generally originates in one of two basic ways as described below (Figure 8); modifications of these processes are discussed in later chapters. In most phyla that undergo spiral cleavage (e.g., flatworms, annelids, molluscs), a single micromere—the 4d cell, called the **mesentoblast**—proliferates as mesoderm between the walls of the developing archenteron (entoderm) and the body wall (ectoderm) (Figure 8A). The other cells of the 4q (the 4a, 4b, and 4c cells) contribute to entoderm; thus the 4d derivatives are true entomesoderm. In some other taxa (e.g., echinoderms and chordates) the mesoderm arises from the wall of the archenteron itself (that is, from preformed entoderm), either as a solid sheet or as pouches (Figure 8B).

The formation and subsequent development of mesoderm is intimately associated with the formation of the body cavity in coelomate metazoa. In those instances where mesoderm has been produced as solid masses derived from a mesentoblast, the body cavity arises through a process called **schizocoely**. Normally in such cases, bilaterally paired packets of mesoderm gradually enlarge and hollow, eventually becoming thin-walled coelomic spaces (Figures 9A and B). The number of such paired coeloms varies among different animals and is frequently associated with segmentation, as it is in annelid worms (Figure 9C).

The other general method of coelom formation is called **enterocoely**; it accompanies the process of mesoderm formation from the archenteron. In the most direct sort of enterocoely, mesoderm production and coelom formation are one and the same process. Figure 10A illustrates this process, which is called **archenteric pouching**. A pouch or pouches form in the gut wall. Each pouch eventually pinches off from the

gut as a complete coelomic compartment. The walls of these pouches are defined as mesoderm. In some cases the mesoderm arises from the walls of the archenteron as a solid sheet or plate that later becomes bilayered and hollow (Figure 10B). Some authors consider this process to be a form of schizocoely (because of the "splitting" of the mesodermal plate), but it is in fact a modified form of enterocoely. Enterocoely frequently results in a tripartite arrangement of the body cavities, which are designated **protocoel**, **mesocoel**, and **metacoel** (Figure 10C).

As mentioned earlier, the nature of the egg and early cleavage are often predictive of the pattern of subsequent development. As a way of summarizing our account of embryology thus far, we illustrate some generalized developmental sequences in Figure 11.

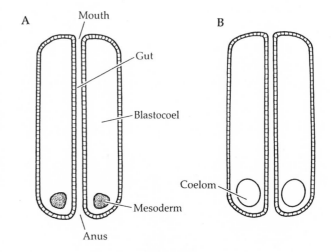

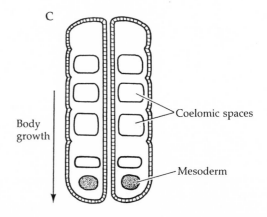

Figure 9

Coelom formation by schizocoely (frontal sections). A, Precoelomic conditions with paired packets of mesoderm. B, Hollowing of the mesodermal packets to produce a pair of coelomic spaces. C, Progressive proliferation of serially arranged pairs of coelomic spaces. This process occurs in metameric annelids.

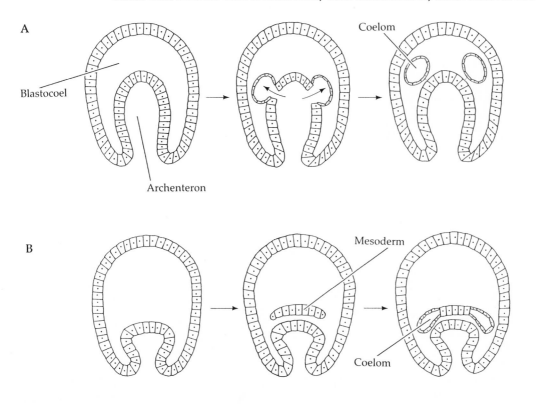

A

Coelom

Blastocoel

Archenteron

B

Mesoderm

Coelom

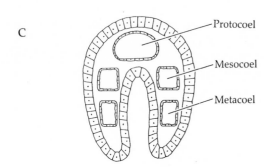

C

Protocoel

Mesocoel

Metacoel

Figure 10

Coelom formation by enterocoely (frontal sections). A, Archenteric pouching. B, Proliferation and subsequent hollowing of a plate of mesoderm from the archenteron. C, The typical tripartite arrangement of coeloms in a deuterostome embryo (frontal section).

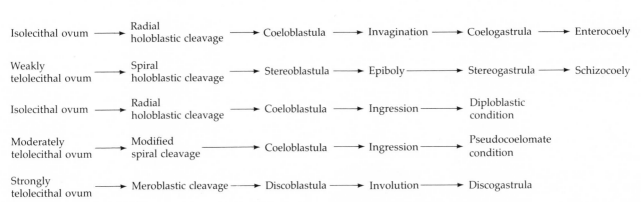

Isolecithal ovum	→	Radial holoblastic cleavage	→	Coeloblastula	→	Invagination	→	Coelogastrula	→	Enterocoely
Weakly telolecithal ovum	→	Spiral holoblastic cleavage	→	Stereoblastula	→	Epiboly	→	Stereogastrula	→	Schizocoely
Isolecithal ovum	→	Radial holoblastic cleavage	→	Coeloblastula	→	Ingression	→	Diploblastic condition		
Moderately telolecithal ovum	→	Modified spiral cleavage	→	Coeloblastula	→	Ingression	→	Pseudocoelomate condition		
Strongly telolecithal ovum	→	Meroblastic cleavage	→	Discoblastula	→	Involution	→	Discogastrula		

Figure 11

Several generalized patterns of development, illustrating some predictable relationships among various early stages and processes.

Life cycles: Sequences and strategies

The patterns of early development described above are not merely isolated sequences of events but are related to the mode of sexual reproduction, the presence or absence of larval stages in the life cycle, and the ecology of the adults. Efforts to classify various invertebrate life cycles and to explain their respective advantages and disadvantages have produced a host of publications and a lot of controversy. Most of these studies concern marine invertebrates, on which we center our attention first. We then present some comments on the special adaptations of terrestrial and freshwater forms.

Classification of life cycles

A number of classification schemes for life cycles have been proposed over the past four decades (see papers by Thorson, Mileikovsky, Vance, Chia, Strathmann, Jablonsky and Lutz, Day and McEdward). We have generalized from the works of various authors and suggest that most animals display one of the three following patterns (Figure 12).

1. **Indirect development.** The life cycle includes early release of gametes or zygotes, followed by some sort of free larva (usually a swimming form) that is distinctly different from the adult and which must undergo a more-or-less drastic metamorphosis to reach the juvenile or young adult stage. In aquatic groups, two basic larval types can be recognized.
   a. Indirect development with **planktotrophic larvae**: The larva survives primarily by feeding, usually on plankton.
   b. Indirect development with **lecithotrophic larvae**: The larva survives primarily on yolk supplied to the egg by the mother.
2. **Direct development.** The life cycle does not include a free larva. In these cases the embryos are cared for by the parents in one way or another (generally by brooding or encapsulation).
3. **Mixed development.** The life cycle involves brooding or encapsulation of the embryos at early stages of development and subsequent release of free planktotrophic or lecithotrophic larvae. The initial source of nutrition and protection is the adult.

Not every species can be conveniently categorized into just one of the above developmental patterns. For example, some species have free larvae that depend upon yolk for a time, and then feed once they develop the capability to do so. Some species actually display different developmental strategies under different environmental conditions—convincing evidence that embryogenies are adaptable, evolutionarily plastic, and responsive to selection pressures (as are adults).

These life cycle patterns provoke two basic ques-

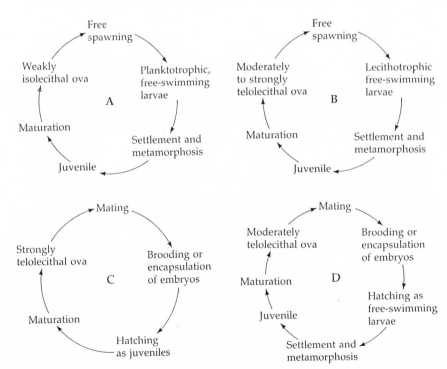

Figure 12

Some generalized invertebrate life cycle strategies. A, Indirect development with planktotrophic larvae. B, Indirect development with lecithotrophic larvae. C, Direct development. D, Mixed life history. See text for further explanation.

tions. First, how do these different sequences relate to other aspects of reproduction and development such as ova types, cleavage patterns, and mating or spawning activities? Second, how do the overall developmental sequences relate to the survival strategies of the adults? Given the large number of interacting factors to be considered, these are very complex questions, and our understanding of these matters is still quite incomplete. However, by examining a few clear cases of direct and indirect development, we can illustrate some of the principles that underlie their relationships to different ecological situations.

Indirect development

Consider first a life cycle with planktotrophic larvae (Figure 12A). The metabolic expense incurred on the part of the adults involves only the production and release of gametes. Animals with fully indirect development generally do not mate; instead, they shed their eggs and sperm directly into the water, thus divorcing the adults from any further responsibility of parental care. Such animals typically undergo synchronous (**epidemic**) broadcast spawning of very large numbers of gametes, thereby ensuring some level of successful fertilization.

The eggs are usually isolecithal and individually inexpensive to produce. The cost to the parent is in the production of high numbers of eggs. Being supplied with little yolk, the embryos must develop quickly into a feeding larva to survive. Mortalities among the embryos and larvae are extremely high and can result from a variety of factors including lack of food, predation, or adverse environmental conditions. Each successful larva must accumulate enough nutrients from feeding to provide the energy necessary for immediate survival and for the processes of settling and metamorphosis from larva to juvenile or subadult. That is, they must feed to excess as they prepare for a new lifestyle as a juvenile. Frequently survival rates from zygote to settled juvenile are less than 1 percent; such mortalities are compensated for by the initial high production of gametes. By the same token, the mortalities compensate for this high production of gametes—if all of these zygotes survived, this planet would quickly be covered by the offspring of animals with indirect development.

What are the advantages and limitations of such a life history, and under what circumstances might it be successful? This sort of planktotrophic development is most common and largely predominates among benthic marine invertebrates in relatively shallow water and the intertidal zones of tropical and warm temperate seas. Here the planktonic food

sources are more consistently available (although often in low concentration) than they are in colder or deeper waters, thus reducing the danger of starvation of the larvae. Such meroplanktonic life cycles allow animals to take advantage of two distinct resources (upper-water-column plankton as larvae, benthos and bottom plankton as adults). This arrangement reduces or eliminates competition between larvae and adults. Indirect development also provides a mechanism for dispersal, a particularly significant benefit to species that are sessile or sedentary as adults. There is good evidence to suggest that animals with free-swimming larvae are likely to recover more quickly from damage to the adult population than those engaging in direct development. A successful "set" of larvae is a ready-made new population to replace lost adults.

The disadvantages of planktotrophic development result from the unpredictability of larval success. Massive larval deaths can result in poor recruitment and the possibility of invasion of suitable habitats by competitors. Conversely, unusually high survival rates of larvae can lead to overcrowding and intraspecific competition upon settling.

Animals that produce fully lecithotrophic larvae (Figure 12B) must produce yolky and thus more metabolically expensive eggs. This built-in nutrient supply releases the larvae from a dependence on environmental food supplies and generally results in reduced mortalities; consequently, these animals produce fewer ova than those with planktotrophic larvae. The eggs are either spawned directly into the water or are fertilized internally and then released as zygotes. Again, the adults' parental responsibility ends with the release of gametes or zygotes into the environment. While survival rates of lecithotrophic larvae are generally higher than those of planktotrophic types, they are low compared with those of embryos that undergo direct development. There is a tendency for marine invertebrates that live in relatively deep benthic environments to produce lecithotrophic larvae. Here some of the advantages of indirect development are realized, but without dependency on environmental food supplies and without subjecting the larvae to the intense predation commonly encountered in surface water. The trade-off is clear: in deeper water the trend is to produce fewer (more expensive) zygotes with a higher survival rate than planktotrophic larvae and an ability to survive where planktotrophic larvae could not.

Settling and metamorphosis

Of particular importance to the successful completion of life cycles with indirect development, and thus to the perpetuation of such species, are the processes of settlement and metamorphosis. Settling and metamorphosis represent crucial and dangerous times in an animal's life cycle, when the organism is changing habitats and lifestyles. Typically, the free-swimming larva metamorphoses into a benthic juvenile and must survive this transformation in form and function as it adopts a new mode of life. Throughout its free-swimming life the larva has been "preparing" for these events, until it reaches a condition in which it is physiologically capable of metamorphosis; such a larva is then termed "competent." The duration of the free-swimming period varies greatly among invertebrate larvae and is dependent on many factors such as original egg size, yolk content, availability of food for planktotrophic forms, and various physical factors (e.g., water temperature). There is a good deal of controversy about many of these factors and how they influence larval life (see Perron and Carrier 1981; Steele 1975, 1977; Strathman 1977; Underwood 1974; Chia 1978; Chia and Rice 1978).

Once a larva becomes competent, it generally begins to respond to certain environmental cues that induce settling behavior. Metamorphosis is often preceded by settling, but some species metamorphose prior to settling and still others engage in both processes simultaneously. In any case, larvae typically become negatively phototactic and/or positively geotactic and swim toward the bottom. Once contact with a substratum is made, a larva tests it to determine its suitability as a habitat. This act of **substratum selection** may involve processing physical, chemical, and biological information about the immediate environment. A number of studies suggest that important factors include substratum texture, composition, and particle size; presence of conspecific adults; presence of certain food organisms; the nature of currents; and turbulence. Assuming an appropriate situation is encountered, metamorphosis is induced and proceeds to completion. Interestingly, some types of larvae are able to resume planktonic life and postpone metamorphosis if they initially encounter an unsuitable substratum. In such cases, however, the larvae become gradually less selective, and eventually metamorphosis ensues regardless of the availability of a proper substratum.

Direct development

Direct development avoids some of the disadvantages but misses some of the advantages of indirect development. A typical scenario involves the production of relatively few, very yolky eggs, followed by some sort of mating activity and internal fertiliza-

tion (Figure 12C). The embryos then receive prolonged parental care either directly by brooding in or on the parent's body, or indirectly by encapsulation in egg cases provided by the parent. Animals that simply deposit their fertilized eggs, either freely or in capsules, are said to be **oviparous**. A great number of invertebrates as well as some vertebrates (fishes, amphibians, reptiles, and birds) display oviparity. Animals that brood their embryos internally and nourish them directly, such as placental mammals, are described as **viviparous**. **Ovoviviparous** animals brood their embryos internally but rely on the yolk within the eggs to nourish their developing young. Most internally-brooding invertebrates are ovoviviparous.

The large, yolky eggs of most invertebrates with direct development are metabolically expensive to produce. Only a few can be afforded; but the investment is protected and survival rates are relatively high. The dangers of planktonic larval life and metamorphosis are avoided and the embryos eventually hatch as juveniles. What sorts of environments and lifestyles might result in selection for such a developmental sequence? One such situation occurs when the adults have no dispersal problems. We find that holoplanktonic species with pelagic adults often undergo direct development either by brooding or by producing floating egg cases. A second situation is one in which critical environmental factors (e.g., food and temperature) are highly variable or unpredictable. There is a trend toward direct development among benthic invertebrates at increasingly higher latitudes. The relatively harsh conditions and strongly seasonal occurrence of planktonic food sources in polar and subpolar areas partially explain this phenomenon.

In addition to avoiding some of the dangers of larval life, direct development has another distinct advantage. The juveniles hatch in suitable habitats where the adults brooded them or deposited the eggs in capsules. Thus, there is a reasonable assurance of appropriate food sources and other environmental factors for the young.

Adaptations to land and fresh water

The foregoing account of life cycle strategies applies largely to marine invertebrates. Many invertebrates, however, have invaded land or fresh water, and their success in these habitats requires not only an adaptation of the adults to special problems, but also adaptation of the developmental forms. As discussed in Chapter 1, terrestrial and freshwater environments are more rigorous and unstable than the sea, and they are generally unsuitable for reproductive strategies that involve free spawning of gametes or the production of delicate, separate larval forms. Most groups of terrestrial and freshwater invertebrates have adopted internal fertilization followed by direct development, while their marine counterparts often produce free-swimming larvae. A notable exception is the insects, many of which have evolved elaborate indirect developmental life histories. In these cases, the larvae are highly adapted to their freshwater or aerial environment and probably evolved secondarily rather than from any marine larval ancestor. A review of insect life history patterns is given in Chapter 17.

Parasite life cycles

There is no doubting the success of parasitism as a lifestyle. Most parasites have rather complicated life cycles, and specific examples are given in later chapters. At this point, however, we may view the situation in a general way and examine the strategies of parasitism in terms of their life cycles, and at the same time introduce some basic terminology.

As outlined in Chapter 1, parasites may be classed as **ectoparasites** (living upon the host), **endoparasites** (living internally, within the host), or **mesoparasites** (living in some cavity of the host that opens directly to the outside such as the oral, nasal, anal, and gill cavities). While associated with a host, an adult parasite engages in sexual reproduction; but the eggs or embryos are usually released to the outside via some avenue through the host's body. The problems at this point are very similar to those encountered during indirect development: some mechanism(s) must be provided to ensure adequate survival through the developmental stages, and the sequence of events must bring the parasite back to an appropriate host (the proper "substratum") for maturation and reproduction.

Parasites exploit at least two different habitats in their life cycles. This practice is essential because their hosts eventually die. Thus, the developmental period from zygote to adult parasite involves either the invasion of another host species or a free-living period. When more than one host species is utilized for the completion of the life cycle, the organism harboring the adult parasite is called the **primary** or **definitive host**, and those hosts in which any developmental or larval forms reside are called **intermediate hosts**. The completion of such a complex life cycle often requires elaborate methods of transfer from one host to the other, and of surviving the changes from one habitat to another. Losses are high, and it is common to find life cycle stages that compensate by periods

of rapid asexual reproduction in addition to the sexual activities of the adult. Many parasites are also parthenogenetic.

Thus, we find that many parasites enjoy some of the benefits of indirect development (e.g., dispersal and exploitation of multiple resources) while being subjected to accompanying high mortalities and the dangers of very specialized lifestyles.

We hasten to emphasize again that the above discussions of life cycles are generalities to which there are many exceptions. But given these basic patterns, you should recognize the adaptive significance of life history patterns of the different invertebrate groups discussed later. You might also be able to predict the sorts of sequences that would be likely to occur under different conditions. For example, given a situation in which a particular species is known to produce very high numbers of free-spawned isolecithal ova, what might you predict about cleavage pattern, blastula and gastrula type, presence or absence of a larval stage, type of larva, adult lifestyle, and ecological settings in which such a sequence would be advantageous? We hope you will develop the habit of asking these kinds of questions and thinking this way about all aspects of your study of invertebrates.

The relationships between ontogeny and phylogeny

Of the many fields of study from which we draw information used in phylogenetic investigations, embryology has been one of the most important. The construction of phylogenies may be accomplished and subsequently tested by a variety of methods (see Chapter 2). But regardless of method, one of the principal problems of phylogeny reconstruction—in fact, central to the process—is separating true homologies from similar characters resulting from evolutionary parallelism and convergence. Even when these problems involve comparative adult morphology, one must often seek answers in studies of the development of the organisms and structures in question. The search is for developmental processes or structures that are homologues and thus demonstrate relationships between ancestors and descendants (i.e., synapomorphies and symplesiomorphies). Changes that take place in developmental stages are not trivial evolutionary events. It has been effectively argued that developmental phenomena may in and of themselves provide the evolutionary mechanisms by which entire new lineages originated.

Although few workers would argue against a significant relationship between ontogeny and phylogeny, the exact nature and extent of the relationship have historically been subjects of considerable controversy, a good deal of which continues today. Stephen Jay Gould has presented a fine analysis of these debates in his book *Ontogeny and Phylogeny* (1977). Central to much of the controversy is the concept of recapitulation.

The concept of recapitulation

In 1866 Ernst Haeckel, a physician who found a greater calling in zoology and never practiced medicine, introduced his **law of recapitulation** (or **the biogenetic law**), most commonly stated as "ontogeny recapitulates phylogeny." This idea suggested that a species' embryonic development (ontogeny) reflects the adult forms of that species' evolutionary history (phylogeny). According to Haeckel, this was no accident, but a result of the mechanistic relationship between the two processes: phylogenesis is the actual *cause* of embryogenesis. Restated, animals have an embryogeny *because of* their evolutionary history. Evolutionary change over time has resulted in a continual "adding on" of morphological stages to the developmental process of organisms. The implications of Haeckel's proposal are immense. Among other things, it means that to trace the phylogeny of an animal one need only examine its development to find therein a sequential or "chronological" parade of the animal's adult ancestors.

Ideas and disagreement concerning the relationship between ontogeny and phylogeny were by no means new even at Haeckel's time. Over 2,000 years ago Aristotle described a sequence of "souls" or "essences" of increasing quality and complexity through which animals pass in their development. He related these conditions to the adult "souls" of various lower and higher organisms, a notion suggestive of a type of recapitulation.

Descriptive embryology began to flourish in the nineteenth century, stimulating vigorous controversy regarding the relationship between development and evolution. Many of the leading developmental biologists of the time were in the thick of things, each proposing his own explanation (Meckel 1811a,b; Serres 1824; von Baer 1828; and others). It was Haeckel, however, who really stirred the pot with his eloquent discourse on the law of recapitulation. He offered a focal point around which biologists argued pro or con for 50 years; sporadic skirmishes still erupt periodically. Walter Garstang critically examined the biogenetic law and gave us a different line of thinking. His ideas were presented in 1922 and are reflected in many of his poems (published posthumously in 1951). Garstang made clear what a number of other workers had suggested, that evolution must be viewed not as a succession of ancestral adult forms but as a succession of ontogenies; each animal is a result of its own developmental processes and any change in an adult must represent a change in its ontogeny. So what we see in the embryogeny of a particular species are not tiny replicas of its adult ancestors, but rather an evolved pattern of development in which clues or traces of ancestral ontogenies and thus phylogenetic relationships to other organisms may be found.

Arguments over these matters did not end with Garstang and they continue today in many quarters. In general, we tend to agree with the approach (if not the details) of Gosta Jägersten in his book *Evolution of the Metazoan Life Cycle* (1972). Recapitulation per se should not categorically be accepted or dismissed as an "always" or "never" phenomenon. The term must be clearly defined in each case investigated, not locked in to Haeckel's original definition and implications. For instance, similar, distinctive, homologous larval types within a group of animals reflect some degree of shared ancestry (e.g., crustacean nauplii or molluscan veligers). And, we may speculate on such matters at various taxonomic levels, even when the adults are quite different from one another (e.g., similar trochophore larvae of polychaetes, molluscs, and sipunculans). These phenomena may be viewed as developmental evidence of relatedness through shared ancestry, and thus they are examples of "recapitulation" in a broad sense.

Jägersten's example of vertebrate gill slits is particularly appropriate because, to him, it provides a case in which Haeckel's strict concept of recapitulation is manifest. In writing of this feature Jägersten (1972) states,

> The fact remains . . . that a character which once existed in the adults of the ancestors but was lost in the adults of the descendants is retained in an easily recognizable shape in the embryogenesis of the latter. This is my interpretation of recapitulation (the biogenetic 'law').

Hyman (1940) perhaps put it most reasonably when she wrote,

> Recapitulation in its narrow Haeckelian sense, as repetition of adult ancestors, is not generally applicable; but ancestral resemblance during ontogeny is a general biological principle. There is no need to quibble over

the word recapitulation; either the usage of the word should be altered to include any type of ancestral reminiscence during ontogeny, or some new term should be invented.

Other authors, however, are not comfortable with such flexibility and have made great efforts to categorize and define the various possible relationships between ontogeny and phylogeny, of which strict recapitulation is considered only one (see especially Chapter 7 of Gould 1977). Although much of this material is beyond the scope of this book, we discuss a few commonly used terms here because they bear on topics in later chapters. We have drawn on a number of sources cited in this chapter to mix freely with our own ideas in explaining these concepts.

Heterochrony and paedomorphosis

When comparing two ontogenies, one often finds that some feature or set of features appears either earlier or later in one sequence than in the other. Such temporal displacement is called **heterochrony**. When comparing suspected ancestral and descendant embryogenies, for example, one may find the very rapid (accelerated) development of a particular feature and thus its relatively early appearance in the descendant species or lineage. Conversely, the development of some trait may be slower (retarded) in the descendant than in the ancestor and thus appear later in the descendant's ontogeny. This retardation may be so pronounced that a structure may never develop to more than a rudiment of its ancestral condition.

Particular types of heterochrony result in a condition known as **paedomorphosis**, where sexually mature adults possess features characteristically found in early developmental stages of related forms (e.g., juvenile or larval features). Paedomorphosis results when the reproductive structures develop before completion of the development of all the nonreproductive (somatic) structures. Thus, we find a reproductively functional animal retaining what in the ancestor were certain embryonic, larval, or juvenile characteristics. According to Gould and others, this condition can result from two different heterochronic processes. These are **neoteny**, in which somatic development is retarded, and **paedogenesis**, in which reproductive development is accelerated. These two terms are frequently used interchangeably in much of the literature; certainly it is not always possible to know which process has given rise to a particular paedomorphic condition. Recognition of paedomorphosis may play a significant role in ex-

amining phylogenetic hypotheses concerning the origins of certain taxa. For example, the evolution of precocious sexual maturation of a planktonic larval stage (that would "normally" continue developing to a benthic adult) might result in a new diverging lineage in which the descendants pursue a fully pelagic existence. This particular scenario, for example, is widely believed to have been responsible for the origin of the crustacean subclass Maxillopoda (see Chapter 18).

Myriad questions about the role of embryogenesis in evolution and the usefulness of embryology in constructing and testing phylogenies persist. As the following accounts show, different authors continue to hold a variety of opinions about these matters.

Origins of major groups of metazoa

One thread of continuity we develop throughout this book is the evolutionary relationship within and among various invertebrate taxa. Life has probably existed on this planet for at least 3.5 billion years; man has been observing it scientifically for only a few thousand years, and evolutionarily for only about 125 years. Thus, the thread of evolutionary continuity we actually see around us today looks much like frazzled ends, representing the legions of successful animals that have survived to today. It is only through conjecture, study, and inference that we are able to trace phylogenetic strands back in time, joining them at various points to produce hypothetical pathways of evolution. We do not operate blindly, however, but use rigorous scientific methodology to draw upon information from many disciplines in attempts to make our evolutionary hypotheses meaningful and (we hope) increasingly closer to the truth—to the actual history of life on Earth.

Discussed below are some ideas concerning the origins and evolution of certain major conditions or *Baupläne* within the metazoa. Consult the references at the end of this chapter for more detailed and comprehensive treatments of these topics. Our goal in this section is to introduce the reader only to the major ideas concerning the origins of metazoan grades. We reserve our discussion of the origins of specific taxa for later chapters.

Origin of the metazoa

The origin of the metazoan condition has received considerable attention for more than a century. There is little doubt that the metazoa arose from protozoa, perhaps about 700 million years ago; the debates concern which protozoan group was ancestral to the first metazoa, what the first metazoa were like and what their habitat was, whether the metazoa are monophyletic or polyphyletic (that is, whether there was one or more than one protozoan ancestor to the metazoa), and how the changes from unicellularity to multicellularity took place. The two monophyletic hypotheses that have enjoyed most support are usually referred to as the **syncytial theory** and the **colonial theory**.

The syncytial theory was supported in this century primarily by J. Hadži (1953, 1963) and E. D. Hanson (1958). They suggested that the metazoan ancestor was a multinucleate, bilaterally symmetrical, ciliated protozoon that assumed a benthic lifestyle, crawling about on the bottom with its oral groove directed toward the substratum. In a major evolutionary step, the surface nuclei became partitioned off from one another by the formation of cell membranes to produce a cellular epidermis surrounding an inner syncytial mass. The result of this and other changes was an acoel flatworm-like creature (phylum Platyhelminthes). Figure 13 illustrates the basic steps in this ciliate-to-acoel hypothesis.

The principal arguments in support of the syncytial theory rest upon certain similarities between modern ciliates and acoel flatworms: size (large ciliates can actually exceed small acoels in size), shape, symmetry, mouth location, and surface ciliation. Most of the objections to this hypothesis concern developmental matters and differences in general levels of adult complexity. Like all flatworms, acoels undergo a complex embryonic development; nothing of this sort occurs in ciliates. Also of major importance is the recently discovered fact that the interior of acoel flatworms is generally cellular, not syncytial (see Chapter 10). The syncytial theory suggests that the flatworms (acoelomate triploblastic bilateria) were the first and thus the ancestral metazoa, leaving us to somehow derive the seemingly more primitive cnidarians and ctenophores (diploblastic radiates) as well as more advanced groups. We hope this whole idea is finally laid to rest.

The colonial theory is based upon ideas first expressed by E. Haeckel (1874), who suggested that a colonial flagellated protozoon gave rise to a planuloid metazoan ancestor (the planula is the basic larval type of cnidarians; see Chapter 8). The ancestral protozoon in this theory was a hollow sphere of flagellated cells that developed anterior and posterior locomotor orientation, and also evolved some level of specialization of cells into separate somatic and reproductive functions. As we explain in Chapter 5, such conditions are common in living colonial protozoa. Haeckel

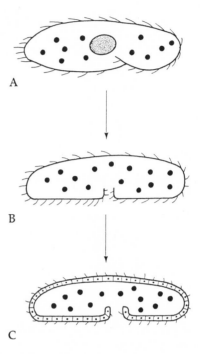

Figure 13

The syncytial theory. Schematic representation of the hypothetical transition from a multinucleate, ciliated protozoon to an acoel-like flatworm. A, A ciliated protozoon. B, The hypothetical metazoan precursor (sagittal section) as it assumed a benthic, crawling lifestyle and developed a ventral mouth and simple pharynx. C, The hypothetical metazoan precursor (sagittal section) after it achieved the acoel grade via cellularization of the epidermis, which surrounded a syncytial endodermis.

called this hypothetical protometazoan ancestor the **blastea** and supported its validity by noting the widespread occurrence of coeloblastulae among modern metazoa. Thus, the first metazoa arose by invagination of the blastea; the resulting animals had a double-layered, gastrula-like body (Haeckel's **gastrea**) with a blastopore-like opening to the outside (Figure 14B) similar to the gastrulae of many modern animals. Haeckel believed that these ancestral creatures (the blastea and gastrea) were recapitulated in the ontogeny of modern metazoa. Thus, the gastrea was the metazoan precursor to the cnidarians. In addition to the above, the colonial theory has been supported by the argument that the body walls of many lower metazoa (e.g., members of the phyla Porifera and Cnidaria) bear flagellated ("monociliated") cells.

Haeckel's original ideas have been modified over the years by various authors investigating a colonial protozoan ancestry to the metazoan condition (Metschnikoff 1883; Hyman 1940). Some have argued that the transition to a layered construction occurred by

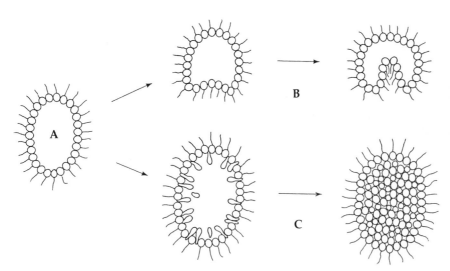

Figure 14

Two versions of the colonial theory of the origin of the metazoa. A, The hypothetical colonial flagellate ancestor, Haeckel's "blastea" (section). B, According to Haeckel, the transition to a multicellular condition occurred by invagination, a developmental process that resulted in a hollow "gastrea." C, According to Metschnikoff, the formation of a solid "gastrea" occurred by ingression. (C after Hyman 1940.)

ingression rather than by invagination, and that the original metazoa were solid, not hollow (Figure 14C). This idea is based in large part on the conviction that ingression is the primitive form of gastrulation among cnidarians.

For some reason, people tend to look among living creatures for possible ancestral types and "missing links." When attempting to support the colonial theory, particularly Haeckel's blastea, in this way, investigators long considered the volvocines (freshwater, colonial, photosynthetic flagellates, such as *Volvox*; Figure 16A) to be the most likely candidates. Consequently most of the arguments against the colonial theory have dealt with the implied plant-like nature and freshwater habitat of the presumed ancestor.

An interesting offshoot of the colonial theory was presented by Otto Bütschli in 1883. He suggested that the primitive metazoon was a bilaterally symmetrical flattened creature of two layers of cells; he called this hypothetical animal a **plakula** (Figure 15). According to Bütschli, the plakula crawled about ingesting food through its "ventral" cell layer. Eventually the animal hollowed somewhat by separation of the dorsal and ventral cell layers, and this development allowed an invagination of the nutritive cells (Figure 15). The formation of a "gut" chamber increased the digestive surface area and at the same time produced inner and outer cell layers, an arrangement approaching the metazoan grade of complexity. As we will soon see, this old idea has been revitalized during the past decade and may have some merit. In any case, the evidence is strong that the metazoa had their origin in flagella-bearing protozoa.

Many evolutionary theorists have tended to em-

phasize monophyletic hypotheses. But in fact there have been some interesting suggestions that the metazoa are polyphyletic, and that different major lineages arose from different protozoan ancestors. Most of these ideas, however, still focus on ciliates and flagellates as probable ancestral groups (see Greenberg 1959; Anderson and Ingils 1985).

Among all of the ideas concerning the evolutionary origin of the metazoan condition, there exists a common problem: the search for protozoan–metazoan intermediates. Some authors have chosen to design logical but hypothetical forms of life for this

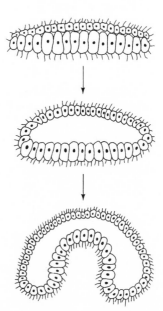

Figure 15

Otto Bütschli's hypothetical plakula and its transformation to a metazoan "gastrea" by invagination of a digestive chamber. (After Bütschli 1883.)

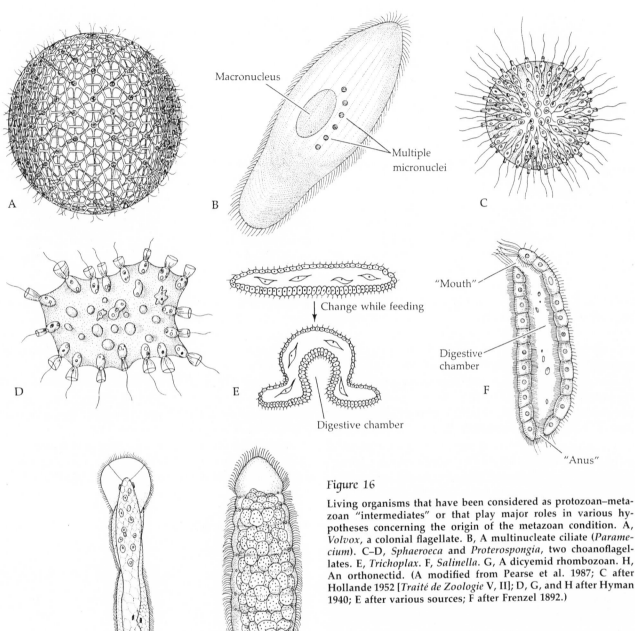

Figure 16

Living organisms that have been considered as protozoan–metazoan "intermediates" or that play major roles in various hypotheses concerning the origin of the metazoan condition. A, *Volvox*, a colonial flagellate. B, A multinucleate ciliate (*Paramecium*). C–D, *Sphaeroeca* and *Proterospongia*, two choanoflagellates. E, *Trichoplax*. F, *Salinella*. G, A dicyemid rhombozoan. H, An orthonectid. (A modified from Pearse et al. 1987; C after Hollande 1952 [*Traité de Zoologie* V, II]; D, G, and H after Hyman 1940; E after various sources; F after Frenzel 1892.)

purpose, while others rummage among extant types, arguing the advantages of using "real" organisms. Even though it is very probable that the actual precursor of the metazoa long ago joined the ranks of the extinct, the presence of modern-day forms that somewhat combine protozoan and metazoan traits keeps debate alive. These organisms or groups of organisms include not only various multinucleate cil-

iates and the colonial *Volvox* but also several other protozoa and some enigmatic little multicellular animals of uncertain position. Figure 16 illustrates these odd animals for comparative purposes, along with a *Paramecium*-like ciliate and *Volvox*. *Proterospongia*, *Sphaeroeca*, and other choanoflagellates are animal-like colonial protozoa (Figures 16C and D). These organisms possess collar cells similar to those found in sponges (Chapter 7), and some workers have suggested that such protozoan colonies may have been the precursors of the first sponges.

About a century ago a tiny, flagellated, yet multicellular creature was discovered in a marine aquarium. This animal has been named *Trichoplax adhaerens* (Figure 16E) and placed in its own phylum, the Placozoa (see papers by Grell). For many years following its discovery, *Trichoplax* was considered to be a larval stage of some invertebrate, but workers are now convinced that it is an adult of uncertain affinity. The body of *Trichoplax* consists of an outer, partly flagellated epithelium surrounding an inner mesenchymal cell mass. The body margins are irregular, and it changes shape like an ameba. There is definitely some division of labor among various cells and areas of the body, but our information still lacks many details. When feeding, *Trichoplax* "hunches" up to form a temporary digestive chamber on its underside, thereby producing a form strikingly similar to Bütschli's hypothetical plakula. It is through discoveries of such real animals that hypothetical creatures gain credence. *Trichoplax* may represent the most primitive of all living metazoa and perhaps a conservative descendant line of the ancestral metazoan type.

In 1892, J. Frenzel described a tiny organism reportedly collected from salt beds in Argentina. He named this animal *Salinella* (Figure 16F). Although *Salinella* does not possess the layered construction of metazoa, it appears to display a higher level of functional organization than colonial protozoa. A single layer of cells forms the entire body wall and separates the digestive cavity from the outside. The digestive cavity is open at both ends as a mouth and anus; the animal feeds on organic detritus. The phylum name Monoblastozoa has been proposed for this odd animal. Sadly, *Salinella* has not been seen again, and many zoologists suspect that poor Frenzel seriously misinterpreted whatever creature he saw.

Finally, we mention briefly two other so-called mesozoan phyla, Rhombozoa and Orthonectida (Figures 16G and H). These animals are structurally rather simple but display complicated life cycles; they are all endoparasites of various invertebrates. Some workers consider them to be derived from another group (or groups) of metazoa, perhaps from parasitic trematodes (flukes), which undergo similarly complex life cycles. Other authors suggest that they are primitively simple and thus may have been derived from early metazoan or premetazoan stock. We have more to say about these and other controversies regarding the mesozoa in Chapter 6.

Origin of the bilateral condition

We discussed the functional significance of bilaterality to some extent in Chapter 3. The evolution of an anterior–posterior body axis, unidirectional movement, and cephalization almost certainly coevolved

to some degree, and probably coincided with the invasion of benthic environments and the development of creeping locomotion. Furthermore, it is likely that the origin of the triploblastic condition took place soon after the appearance of the first bilateral forms. At least bilaterality and triploblasty generally occur together in modern-day invertebrates.

There are several ideas concerning the origin of bilateral symmetry within the metazoa. If one accepts Hadži's syncytial theory or Bütschli's plakula, then the problem is already solved, since bilaterality would characterize the first metazoa. On the other hand, supporters of the colonial theory generally assume that the first bilateral metazoa arose from a gastrula-like ancestor, presumably one with spherical or radial symmetry, or its planuloid descendant. Jägersten suggests that the blastea took up a benthic lifestyle, invaginated, and then assumed bilaterality. In this scenario, the first metazoon was a bilaterally symmetrical gastrea (Jägersten's **bilaterogastrea**) from which all other major groups arose. This idea, and many others, can be found in Jägersten's works (1955, 1959, 1972).

Figures 17 and 18 illustrate two evolutionary schemes that show the implications of some of the ideas discussed above. Most animal phylogenies that are proposed are based on the widely held view that two major phyletic lines arose during the evolution of bilateral symmetry and the triploblastic condition, particularly among the coelomate animals (Figure 19). All coelomate animals (including those that have secondarily lost the coelom) can be placed along one or the other of these lineages, although some fit much

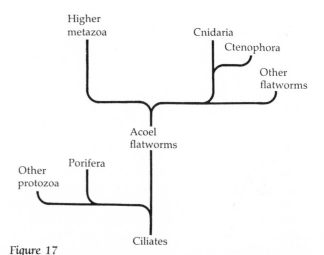

Figure 17

Scheme of possible metazoan relationships based on a ciliate ancestor and the "syncitial theory."

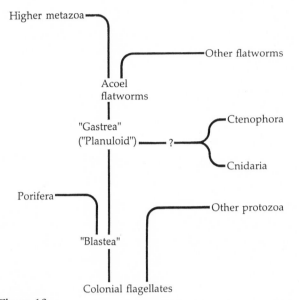

Figure 18

Scheme of possible metazoan relationships based on a colonial flagellate ancestor and the "colonial theory."

coelomate metazoa. But as additional information on the details of development has become available for more and more animals, exceptions to these broad descriptions have surfaced. Some of these problems are discussed in later chapters and summarized in Chapter 24. In fact, the whole idea of a protostome–deuterostome dichotomy has been challenged by certain authors (see, for example, Løvtrup 1975), although most zoologists still accept that these supraphylum groups represent actual clades.

Origin of the coelomate condition

No less controversial than the origin of the metazoan condition is the evolutionary appearance of the coelom. The various hypotheses concerning this matter were summarized by R. B. Clark in his fine book *Dynamics in Metazoan Evolution* (1964). Clark's approach is a functional one and emphasizes the adaptive significance of the coelom as the central criterion for evaluating ideas concerning its origin.

When early soft-bodied, bilaterally symmetrical animals larger than a few millimeters or so assumed a benthic, crawling, or burrowing lifestyle, a fluid (hydrostatic) skeleton was essential for certain types of movement. Hence, the evolution of a body cavity filled with fluid against which muscles could operate would have offered a tremendous locomotor advantage in addition to providing a circulatory medium and space for organ development. How might such spaces have originated?

more conveniently than others. These two evolutionary lines are generally called the Protostomia and the Deuterostomia and are distinguished from one another on the basis of several relatively consistent differences in development (Box 1).

The protostome–deuterostome distinction provides a basis for comparison when discussing the

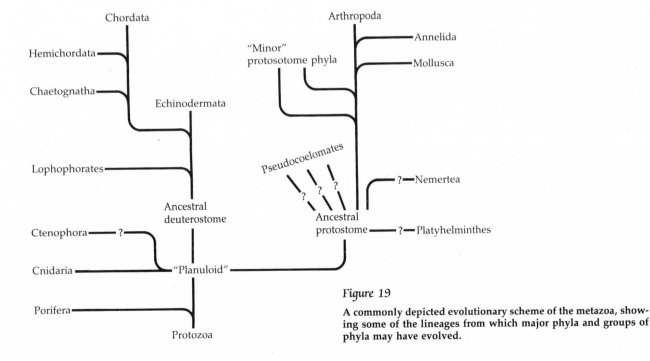

Figure 19

A commonly depicted evolutionary scheme of the metazoa, showing some of the lineages from which major phyla and groups of phyla may have evolved.

Box One

Protostomes and Deuterostomes

Developmental differences between protostomes and deuterostomes
and some representative eucoelomate taxa

PROTOSTOMIA	DEUTEROSTOMIA
Spiral cleavage	Radial cleavage
Blastopore becomes the mouth	Blastopore does not become the mouth (often becomes the anus)
Mesoderm derived from mesentoblast (usually the 4d cell)	Mesoderm arises from wall of archenteron
Schizocoelous coelom formation	Enterocoelous coelom formation
Annelida, Mollusca, Arthropoda, Nemertea, Sipuncula, Echiura	Echinodermata, Hemichordata, Chordata

---

Note: Some authors include certain noncoelomate taxa in these listings; for example, the flatworms demonstrate all the features of the Protostomia except the formation of the coelom.

Most of the ideas concerning the evolutionary origin of the coelom were developed from the mid-nineteenth to early twentieth century, during the heyday of comparative embryology. Most of these hypotheses shared the premise of monophyly—that the coelomic condition arose but once. The inherent problem with a monophyletic approach is the difficulty of relating all existing coelomate animals to a single common coelomate ancestor. Considering the advantages of possessing a coelom, the very different methods of embryonic development (e.g., schizocoely and various forms of enterocoely), and the variety of adult coelomic *Baupläne*, it may be more biologically reasonable to suggest that the coelomic condition arose at least twice. That is, perhaps the coelom represents a nonhomologous, convergent feature in the metazoa. There are several currently debated ideas about how this might have happened, and a number of others have mostly been discarded in recent years as being incompatible with existing evidence or our standard definition of the coelom.

The coelom may have originated by the pinching off and isolation of gut diverticula as occurs in the embryogeny of extant enterocoelous animals (Figure 20). This so-called **enterocoel theory** (in several versions) has enjoyed relatively strong support by many authors since it was originally proposed by Lankester in 1877 (e.g., Lang 1881; Sedgwick 1884; Masterman 1897; Hubrecht 1904; Jägersten 1955). An obvious point in favor of this general idea is that enterocoely

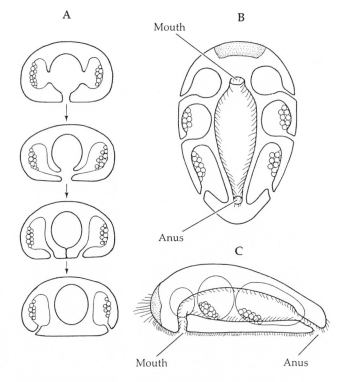

Figure 20

Jägersten's bilaterogastrea theory, according to which the coelomic compartments arise by enterocoelic pouching. A, The formation of paired coeloms from the wall of the archenteron. The slitlike blastopore of the bilaterogastrea closes midventrally, leaving mouth and anus at opposite ends (B). B–C, The tripartite coelomic condition in Jägersten's hypothetical early coelomate animal (ventral and lateral views). (After Jägersten 1955.)

does occur in many living animals, thus paralleling the hypothetical ancestral process. In addition, various authors cite examples of noncoelomate animals (anthozoans and turbellarian flatworms) in which gut diverticula exist in arrangements that resemble possible ancestral patterns.

Another popular idea concerning coelom origin is the **gonocoel theory** (Bergh 1885; Hatschek 1877, 1878; Meyer 1890, 1901; Goodrich 1946). This hypothesis suggests that the first coelomic spaces arose by way of mesodermally derived gonadal cavities that persisted subsequent to the release of gametes (Figure 21). The serial arrangement of gonads in animals such as flatworms and nemerteans could have resulted in serially arranged coelomic spaces and linings in annelids, where, at least primitively, they still produce and store gametes. The major argument against this hypothesis is the fact that nowhere do gonads develop *before* coelomic spaces in modern-day coelomate animals. As we have seen, however, heterochrony can account for such turnabouts.

Clark (1964) speculated that schizocoely as we know it today could have evolved by the formation of spaces within the solid mesoderm of acoelomate animals and then have been retained in response to the positive selection for the resulting hydrostatic skeleton. These matters are complicated by suggestions that modern acoelomate animals (platyhelminths and nemerteans) may have arisen from eucoelomate ancestors (see Chapters 10 and 11).

Another idea on coelom origin is called the **nephrocoel theory** (Lankester 1874; Ziegler 1898, 1912; Faussek 1899, 1911; Snodgrass 1938). The association between the coelom and excretion has prompted different versions of this hypothesis through about 75 years of moderate support. One idea is that the protonephridia of flatworms expanded to coelomic cavities, arguing that the coelom first arose from ectodermally derived structures. Another view is that coelomic spaces arose as cavities within the mesoderm and served as storage areas for waste products. Certainly the coelomic cavities of many animals are related to excretory functions, but there is no convincing evidence that this relationship was the primary selective force in the origin of the coelomate condition.

As we mentioned earlier, these hypotheses tend to share the fundamental constraint of arguing a monophyletic origin to all coelomate animals. The basic developmental differences between what may be two clades of coelomate metazoa (the protostomes and the deuterostomes) suggest to us that the coelom may have arisen separately in these two lineages; we

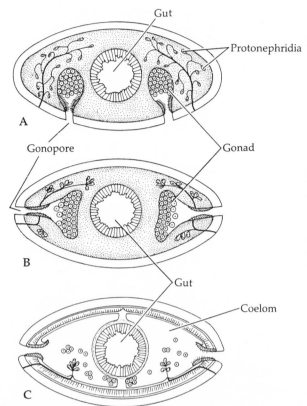

Figure 21

A version of the gonocoel theory (schematic cross sections). **A,** The condition in flatworms, which have mesodermally derived gonads leading to ventral gonopores. **B,** The condition in nemerteans, which have serially arranged gonadal masses leading to laterally placed gonopores. **C,** The condition in polychaetes, in which the linings of the gonads have expanded to produce coelomic spaces with coelomoducts to the outside. (After Goodrich 1946.)

explore this further in Chapter 24. Given the strong similarities between the coelomate protostomes and acoelomate flatworms and nemerteans, it is easy to envision the protostome clade arising from a triploblastic acoelomate ancestor. Hollowing of the mesoderm in such a precursor to produce fluid-filled hydrostatic spaces can be easily explained both developmentally (modern-day schizocoely) and functionally (peristaltic burrowing, increased size, and so on). On the other hand, schizocoely and the origin of mesoderm in the ancestral protostome may have arisen simultaneously. As mentioned above, the acoelomate conditon may be secondarily derived from coelomate precursors (see Rieger 1985, 1986).

To derive deuterostome and protostome clades from a common immediate coelomate ancestor creates a complicated scenario. The most parsimonious hypothesis might be to view the deuterostome ancestor as a diploblastic animal, perhaps a planuloid form, in which enterocoely occurred. Deriving the deuterostome lineage *before* the evolution of spiral cleavage and the other features of protostomes avoids many of the complications inherent in a monophyletic view of coelom origin. Imagine a hollow, invaginated gastrula-like metazoon, swimming with its blastopore trailing, as in the planula larvae of some cnidarians. Enterocoely may have accompanied a tendency toward benthic life, giving the animal a peristaltic burrowing ability. The archenteron may then have opened anteriorly as a mouth, and the new coelomate creature adopted a deposit-feeding lifestyle. If such a story began at the level of diploblastic metazoa (e.g., cnidarians), then the radial cleavage seen today in deuterostomes may have also been present in the ancestor to this group. A diphyletic origin of the coelomic condition from larval-like ancestors has been presented by Nielsen and Nørrevang (1985). In their hypothesis, a pelagic gastrea gave rise to the cnidarians and to a second lineage—another larval-like creature they call a **trochea**. This ancestor was the precursor to the protostomes and the deuterostomes, but the coelom arose separately in each group.

As you can see, phylogenetic analysis at the level of phyla is highly problematical. We trust, however, that you have gained some basic insights not only into the particular hypotheses discussed here, but also into phylogenetic speculation. A fundamental caveat should be kept in mind: any number of evolutionary pathways can be proposed and made to appear convincing on paper by imagining appropriate hypothetical ancestors or intermediates, but one must always ask whether these hypothetical creatures would have worked. Do they possess a realistic *Bauplan*? Clark (1964) spends a good deal of time on this point and emphasizes it in his conclusion with the following passage (p. 258):

> The most important and least considered of these [principles] is that hypothetical constructs which represent ancestral, generalized forms of modern groups, or stem forms from which several modern phyla diverge, must be possible animals. In other words, they must be conceived as living organisms, obeying the same principles that we have discovered in existing animals.

It is in such terms that phylogenetic hypotheses can be evaluated. Recently it has become fashionable to avoid initial speculation on what a hypothetical ancestor might have looked like at all, and instead rely on phylogenetic trees (cladograms) of known taxa to establish genealogical relationships or branching patterns. Once a cladogram has been constructed, the pattern of features that are associated with the taxa on the tree will themselves predict the nature (character combination) of the ancestor for each branch. Such a method attempts to avoid the potential problem of circular reasoning, in which a hypothetical ancestor is established first, and hence constrains (predicts) the nature of the taxa descended from it. In either case, for the hypotheses to be truly scientific, they must be testable with new data gathered outside the framework of that used to construct the initial hypotheses themselves.

Selected References

General Invertebrate Embryology

Adiyodi, K. G. and R. G. Adiyodi (eds.). 1983. *Reproductive Biology of Invertebrates.* Vol. 1, Oogenesis, Oviposition, and Oosorption. Wiley, New York.

Brusca, G. J. 1975. *General Patterns of Invertebrate Development.* Mad River Press, Eureka, California.

Cooke, J. 1988. The early embryo and the formation of body pattern. Am. Sci. 76:35–41.

Costello, D. P. and C. Henley. 1976. Spiralian development: A perspective. Am. Zool. 16:277–291. [The introduction to an entire issue of *American Zoologist* (Spring 1976) devoted to an examination of those creatures showing spiral cleavage.]

Dawydoff, C. 1928. *Traité d'Embryologie Comparée des Invertébrés.* Masson et Cie, Libraires de l'Academie de Medicine, Paris.

Giese, A. C. and J. S. Pearse (and V. B. Pearse, Vol. 9) (eds.). 1974–1987. *Reproduction of Marine Invertebrates.* Vols. 1–5, 9. Blackwell, Palo Alto, California. [An outstanding series of volumes containing reviews of every invertebrate phylum; Volumes 6 (lophophorates and echinoderms), 7 (nonmalacostracan arthropods), and 8 (malacostracans) are in preparation. Volume 9 provides general overviews of invertebrate gametogenesis and development.]

Harrison, F. W. and R. R. Cowden (eds.). 1982. *Developmental Biology of Freshwater Invertebrates.* Alan R. Liss, New York.

Korschelt, E. and K. Heider. 1900. *Lehrbuch der vergleichenden Entwicklungsgeschichte der wirbellosen Tiere.* Fischer, Jena.

Kume, M. and K. Dan. 1957. *Invertebrate Embryology.* [English translation published by NOLIT Publishing House, Belgrade, Yugoslavia (see references in Chapter 1).]

MacBride, E. 1914. *Textbook of Embryology.* Macmillan, London.

MacBride, E. 1917. Recapitulation as a proof of the inheritance of acquired characters. Scientia 22:425–434. [Interesting historical perspective.]

Malacinski, G. M. (ed.). 1984. *Pattern Formation: A Primer in Developmental Biology.* Macmillan, New York.

Masterman, A. 1897. On the Diplochordata. 1. The structure of *Actinotrocha.* 2. The structure of *Cephalodiscus.* Q. J. Microsc. Sci. 40:281–366.

Pflugfelder, O. 1962. *Lehrbuch der Entwicklungsgeschichte und Entwicklungsphysiologie der Tiere.* Fischer, Jena.

Strathmann, M. F. 1987. *Reproduction and Development of Marine Invertebrates of the Northern Pacific Coast.* University of Washington Press, Seattle.

Wilson, E. B. 1892. The cell lineage of *Nereis*. J. Morphol. 6:361–480. [Wilson's classic work establishing the coding system for spiral cleavage.]

Life Histories

Ayal, Y. and U. Safriel. 1982. r-curves and the cost of the planktonic stage. Am. Nat. 119:391–401.

Cameron, R. A. 1986. Proceedings of the Invertebrate Larval Biology Workshop held at Friday Harbor Laboratories, University of Washington, 26–30 Mar. 1985. Bull. Mar. Sci. 39:145–622. [Thirty-seven papers on larval biology.]

Caswell, H. 1978. Optimal life histories and the age–specific cost of reproduction. Bull. Ecol. Soc. Am. 59:99. [This paper and the two below include interesting discussions of the adaptive qualities of various life cycle patterns, especially those that do not fit the typical direct or indirect definitions.]

Caswell, H. 1980. On the equivalence of maximizing fitness and maximizing reproductive value. Ecology 61:19–24.

Caswell, H. 1981. The evolution of "mixed" life histories in marine invertebrates and elsewhere. Am. Nat. 117(4):529–536.

Chia, F. S. 1976. Classification and adaptive significance of developmental patterns in marine invertebrates. Thalassia Jugosl. 10:121–130.

Chia, F. S. 1978. Perspective: Settlement and metamorphosis of marine invertebrate larvae. *In* Chia and Rice (1978).

Chia, F. S. and M. Rice (eds.). 1978. *Settlement and Metamorphosis of Marine Invertebrate Larvae*. [Proceedings of the Symposium on Settlement and Metamorphosis of Marine Invertebrate Larvae, Am. Zool. Soc. Meeting, Toronto, Ontario, Canada, Dec. 27–28, 1977.] Elsevier/North-Holland, New York.

Christiansen, F. and T. Fenchel. 1979. Evolution of marine invertebrate reproductive patterns. Theor. Pop. Biol. 16:267–282.

Crisp, D. 1974. Energy relations of marine invertebrate larvae. Thalassia Jugosl. 10:103–120.

Crisp, D. 1974. Factors influencing the settlement of marine intertidal larvae. *In* P. T. Grant and A. M. Macie (eds.), *Chemoreception in Marine Organisms*. Academic Press, New York, pp. 177–265. [Donald Crisp continues today to attempt to unravel the problems of chemical cues in larval settlement.]

Day, R. and L. McEdward. 1984. Aspects of the physiology and ecology of pelagic larvae of marine benthic invertebrates. *In* K. A. Steidinger and L. M. Walker (eds.). *Marine Plankton Life Cycle Strategies*. C.R.C. Press, Boca Raton, Florida, pp. 93–120.

Gilbert, L. I. and E. Frieden (eds.). 1981. *Metamorphosis: A Problem in Developmental Biology*, 2nd Ed. Plenum Press, New York. [Mostly biochemical.]

Jablonsky, D. and R. A. Lutz. 1983. Larval ecology of marine benthic invertebrates: Paleobiological implications. Biol. Rev. 58:21–89. [An excellent, up-to-date review of larval ecology from an evolutionary perspective.]

Jeffrey, W. R. and R. A. Raff (eds.). 1982. *Time, Space, and Pattern in Embryonic Development*. Alan R. Liss, New York.

Mileikovsky, S. 1971. Types of larval development in marine bottom invertebrates, their distribution and ecological significance: A re-evaluation. Mar. Biol. 10:193–213. [An excellent treatment of the subject.]

Pechnick, J. A. 1979. Role of encapsulation in invertebrate life histories. Am. Nat. 114:859–870. [Ideas concerning mixed life cycles.]

Perron, F. and R. Carrier. 1981. Egg size distribution among closely related marine invertebrate species: Are they bimodal or unimodal? Am. Nat. 118:749–755. [Explains, in part, how egg size varies with other aspects of the developmental pattern.]

Stancyk, S. E. (ed.). 1979. *Reproductive Ecology of Marine Invertebrates*. University of South Carolina Press, Columbia.

Steele, D. H. 1975. Egg size and duration of embryonic development in Crustacea. Int. Rev. Ges. Hydrobiol. 60(5):711–715.

Steele, D. H. 1977. Correlation between egg size and developmental period. Am. Nat. 111:371–372.

Strathmann R. 1977. Egg size, larval development and juvenile size in benthic marine invertebrates. Am. Nat. 111:373–376.

Strathmann, R. 1978. The evolution and loss of feeding larval stages of marine invertebrates. Evolution 32(4):894–906.

Strathmann, R. and M. Strathmann. 1982. The relationship between adult size and brooding in marine invertebrates. Am. Nat. 119:91–101.

Steidinger, K. A. and L. M. Walker (eds.). 1984. *Marine Plankton Life Cycle Strategies*. C.R.C. Press, Boca Raton, Florida.

Todd, C. D. and R. W. Doyle. 1981. Reproductive strategies of marine benthic invertebrates: A settlement-timing hypothesis. Mar. Ecol. Prog. Ser. 4:75–83.

Thorson, G. 1946. Reproduction and larval development of Danish marine bottom invertebrates with special reference to the planktonic larvae in the South (Øresund). Medd. Danm. Fisk., Havunders., Ser. Plankton, 4.

Thorson, G. 1950. Reproduction and larval ecology of marine bottom invertebrates. Biol. Rev. 25:1–45. [These two works by G. Thorson laid the foundation for more recent studies concerning the classification of invertebrate life cycles and their significance.]

Underwood, A. 1974. On models for reproductive strategy in marine benthic invertebrates. Am. Nat. 108:874–878.

Vance, R. 1973. On reproductive strategies in marine benthic invertebrates. Am. Nat. 107:339–352.

Vance, R. 1973. More on reproductive strategies in marine benthic invertebrates. Am. Nat. 107:353–361. [Vance's papers form a modern mathematical synthesis concerning life history classifications, and have prompted many additional investigations and theoretical speculations.]

On the Origins of Major Invertebrate Lines

Alberch, P., S. J. Gould, G. F. Osta, and D. B. Wake. 1979. Size and shape in ontogeny and phylogeny. Paleobiology 5(3):296–317.

Anderson, D. T. 1982. Origins and relationships among the animal phyla. Proc. Linn. Soc. N. S. W. 106(2):151–166.

Baer, K. E. von. 1828. *Entwicklungsgeschichte der Thiere: Beobachtung und Reflexion*. Borntrager, Konigsberg.

Bergh, R. S. 1885. Die Exkretionsorgane der Würmer. Kosmos, Lwow 17:97–122.

Bütschli, O. 1883. Bemerkungen zur Gastrea Theorie. Morph. Jahrb. 9.

Carter, G. S. 1954. On Hadži's interpretations of animal phylogeny. Syst. Zool. 3:163–167. [An analysis, sometimes quite pointed, of Hadži's views.]

Clark, R. B. 1964. *Dynamics in Metazoan Evolution*. Oxford University Press, New York. [A fine functional approach to metazoan evolution, especially concerning the origin of the coelom and metamerism.]

Dougherty, E. C. (ed.). 1963. *The Lower Metazoa: Comparative Biology and Phylogeny*. University of California Press, Berkeley.

Eaton, T. H. 1953. Paedomorphosis: An approach to the chordate–echinoderm problem. Syst. Zool. 2:1–6.

Faussek, V. 1899. Über die physiologische Bedeutung des Cöloms. Trav. Soc. Nat. St. Petersberg 30:40–57.

Faussek, V. 1911. Vergleichend–embryologische Studien. (Zur Frage über die Bedeutung der Cölom-hölen). Z. Wiss. Zool. 98:529–625. [Faussek's works include his views on the nephrocoel theory.]

Field, K. G. et al. 1988. Molecular phylogeny of the animal kingdom. Science 239:748–753. [Very preliminary, with some odd results.]

Frenzel, J. 1892. *Salinella*. Arch. Naturgesch. 58, Pt. 1.

Garstang, W. 1922. The theory of recapitulation. J. Linn. Soc. Lond. Zool. 35:81–101. [Garstang's revolutionary ideas on Haeckel's recapitulation concept.]

Garstang, W. 1985. *Larval Forms and Other Zoological Verses.* University of Chicago Press, Chicago. [A wonderful collection of prose by Garstang, published after his death. The biographical sketch by Sir Alister Hardy and the Foreword by Michael LaBarbera chronicle many of Garstang's contributions to our understanding of the relationships between ontogeny and phylogeny and serve as a delightful introduction to the 26 poems in this little volume. This new edition of the original (1951) version also includes Garstang's famous address on "The Origin and Evolution of Larval Forms."]

Goodrich, E. S. 1946. The study of nephridia and genital ducts since 1895. Q. J. Microsc. Sci. 86:113–392. [One of the great classics concerning the origin of the coelom and related evolutionary matters.]

Gould. S. J. 1977. *Ontogeny and Phylogeny.* Harvard University Press, Cambridge, Massachusetts. [A scholarly and often controversial coverage of ideas concerning recapitulation and other interactions between development and evolution.]

Greenberg, M. J. 1959. Ancestors, embryos, and symmetry. Syst. Zool. 8:212–221. [On the polyphyletic origin of the Metazoa.]

Grell, K. G. 1971a. *Trichoplax adhaerens* F. E. Schulze, und die Entstehung der Metazoen. Naturwiss. Rundsch. 24(4):160–161.

Grell, K. G. 1971b. Embryonalentwicklung bei *Trichoplax adhaerens* F. E. Schulze. Naturwiss. 58:570.

Grell, K. G. 1972. Formation of eggs and cleavage in *Trichoplax adhaerens*. Z. Morphol. Tiere 73(4):297–314.

Grell, K. G. 1973. *Trichoplax adhaerens* and the origin of the Metazoa. Actualite's Protozooligiques. IVe. Cong. Int. Protozoologie. Paul Couty, Clermont–Ferrand.

Grell, K. G. and G. Benwitz. 1971. Die Ultrastruktur von *Trichoplax adhaerens* F. E. Schulze. Cytobiologie 4(2):216–240.

Gutman, W. F. 1981. Relationships between invertebrate phyla based on functional–mechanical analysis of the hydrostatic skeleton. Am. Zool. 21:63–81.

Hadži, J. 1953. An attempt to reconstruct the system of animal classification. Syst. Zool. 2:145–154. [Odd ramblings about lumping all animals into a few phyla, the author's views on the origins of the metazoan condition, and other things.]

Hadži, J. 1963. *The Evolution of the Metazoa.* Macmillan, New York. [Overkill. But then, any book that begins with the sentence, "It was in 1903, 58 years ago, that I, then a young man who had just left the classical grammar school at Zagreb, went to Vienna to study natural sciences and above all my beloved Zoology at Vienna University," can't be all bad!]

Haeckel, E. 1866. *Generelle Morphologie der Organismen: Allgemeine Grundzüge der organischen Formen-Wissenschaft mechansch begründet durch die von Charles Darwin reformierte Descendenz-Theorie.* Vols. 1–2. George Reimer, Berlin.

Haeckel, E. 1874. The gastrea-theory, the phylogenetic classification of the animal kingdom and the homology of the germlamellae. Q. J. Microsc. Sci. 14:142–165; 223–247. [Haeckel's concepts of recapitulation and blastea-gastrea idea of metazoan origin. A translation of the original German paper that introduced the colonial theory of metazoan origin (Jena. Z. Naturwiss. 8: 1–55).]

Hanson, E. D. 1958. On the origin of the Eumetazoa. Syst. Zool. 7:16–47. [Support for Hadži's views.]

Hanson, E. D. 1977. *The Origin and Early Evolution of Animals.* Wesleyan University Press, Middletown, Connecticut.

Hatschek, B. 1877. Embryonalentwicklung und Knospung der *Pedicellina echinata*. Z. Wiss. Zool. 29:502–549. [Some early thoughts on the gonocoel theory.]

Hatschek, B. 1878. Studien über Entwicklungsgeschichte der Anneliden. Ein Beitrag zur Morphologie der Bilaterien. Arb. Zool. Inst. Wien 1:277–404.

House, M. R. (ed.). 1979. *The Origin of Major Invertebrate Groups.* Academic Press, New York. Published for The Systematics Association, Special Vol. No. 12.

Hyman, L. H. 1940–1967. *The Invertebrates.* Vols. 1–6. McGraw-Hill, New York. [All volumes include especially fine discussions on embryology of the included taxa. Volumes 1 and 2 include the author's views on the origin of the metazoa, bilaterality, and coelom, and other related matters.]

Hubrecht, A. 1904. Die Abstammung der Anneliden und Chordaten und die Stellung der Ctenophoren und Platyhelminthen im System. Jena. Z. Naturwiss. 39:152–176. [Odd ideas about a cnidarian ancestry for the annelids and chordates relating to the enterocoel theory.]

Inglis, W. G. 1985. Evolutionary waves: Patterns in the origins of animal phyla. Aust. J. Zool. 33:153–178.

Ivanova-Kazas, O. M. 1982. Phylogenetic significance of spiral cleavage. Soviet J. Mar. Biol. 7(5):275–283.

Jägersten, G. 1955. On the early phylogeny of the Metazoa. The bilaterogastrea theory. Zool. Bidr. Uppsala 30:321–354.

Jägersten, G. 1959. Further remarks on the early phylogeny of the Metazoa. Zool. Bidr. Uppsala 33:79–108.

Jägersten, G. 1972. *Evolution of the Metazoan Life Cycle.* Academic Press, London. [The phylogeny of the metazoa according to Jägersten, based in part on his bilaterogastrea hypothesis. Included are some of the author's thoughts on recapitulation.]

Jefferies, R. P. S. 1986. *The Ancestry of the Vertebrates.* British Museum (Natural History), London.

Lang, A. 1881. Der Bau von *Gunda segmentata* und die Verwandtschaft der Platyhelminthen mit Coelenteraten und Hirudineen. Mitt. Zool. Sta. Neapel. 3:187–251.

Lang, A. 1903. Beiträge zu einer Trophocoltheorie. Jena. Z. Naturwiss. 38:1–373. [Lang's 1881 paper was in support of the enterocoel theory, suggesting that the coelom arose from pinched-off gut diverticula in flatworms; this opinion was based upon his study of the turbellarian *Gunda* (now *Procerodes*). However, Lang eventually switched his allegiance to the gonocoel theory (1903).]

Lankester, E. R. 1874. Observations on the development of the pond snail (*Lymnaea stagnalis*), and on the early stages of other Mollusca. Q. J. Microsc. Sci. 14:365–391. [Some of the author's ideas about coelom origin.]

Lankester, E. R. 1877. Notes on the embryology and classification of the animal kingdom; comprising a revision of speculations relative to the origin and significance of the germ layers. Q. J. Microsc. Sci. 17:399–454. [In addition to the ambitious title, this work includes thoughts about the gonocoel theory.]

Løvtrup, S. 1975. Validity of the Protostomia–Deuterostomia theory. Syst. Zool 24: 96–108. [Arguments against the concept of protostomes and deuterostomes.]

Marcus, E. 1958. On the evolution of the animal phyla. Q. Rev. Biol. 33:24–58.

Margulis, L. 1981. *Symbiosis in Cell Evolution: Life and Its Environment on the Early Earth.* W. H. Freeman, San Francisco.

Masterman, A. 1897. On the theory of archimeric segmentation and its bearing upon the phyletic classification of the Coelomata. Proc. R. Soc. Edinburgh 22:270–310. [Masterman was generally a proponent of the enterocoel theory.]

Meckel, J. 1811a. Entwurf einer Darstellung der zwischen dem Embryozustande der höheren Tiere und dem Permanenten der niedere stattfindenden Parallele: Beiträge zur vergleichenden Anatomie, Vol. 2. Carl Heinrich Reclam., Leipzig, pp. 1–60.

Meckel, J. 1811b. Über den Charakter der allmähligen Vervollkommung der Organisation, oder den Unterschied zwischen den höheren und niederen Bildungen: Beyträge zur vergleichenden Anatomie, Vol. 2. Carl Heinrich Reclam., Leipzig, pp. 61–123. [Works by Meckel contain interesting pre-Haeckelian concepts of relationships between development and evolution as it was understood before Darwin.]

Metschnikoff, E. 1883. Untersuchungen über die intracellulare Verdauung bei wirbellosen Thieren. Arb. Zool. Inst. Wien. 5:141–168. [Translated into English and published as "Researches on the Intracellular Digestion of Invertebrates," Q. J. Microsc. Sci. (1884) 24:89–111. This paper includes some of the studies that led Metschnikoff and eventually others to conclude that ingression was the original form of gastrulation.]

Meyer, E. 1890. Die Abstimmung der Anneliden. Der Ursprung der Metamerie und die Bedeutung des Mesoderms. Biol. Cbl. 10:296–308. [An English translation appeared in Am. Natur. 24:1143–1165.]

Meyer, E. 1901. Studien über den Körperbau der Anneliden. V. Das Mesoderm der Ringelwürmer. Mitt. Zool. Sta. Neapel. 14:247–585. [The two papers by Meyer include coverage of the gonocoel theory.]

Morris, S. C., J. D. George, R. Gibson and H. M. Platt (eds.). 1985. *The Origins and Relationships of Lower Invertebrates*. Clarendon Press, Oxford. Published for the Systematics Association, Special Vol. 28.

Nielsen, C. 1985. Animal phylogeny in light of the trochaea theory. Biol. J. Linn. Soc. London 25:243–299. [One of the most carefully done modern analyses of invertebrate phyla.]

Nielsen, C. and A. Nørrevang. 1985. The trochea theory: An example of life cycle phylogeny. *In* Morris et al. (1985), pp. 28–41.

Raff, R. A. and T. C. Kaufman. 1983. *Embryos, Genes, and Evolution*. Macmillan, New York.

Remane, A. 1963. The evolution of the Metazoa from colonial flagellates *vs.* plasmodial ciliates. *In* E. C. Dougherty et al. (eds.), *The Lower Metazoa: Comparative Biology and Phylogeny*. University of California Press, Berkeley, pp. 78–90.

Rieger, R. M. 1985. The phylogenetic status of the acoelomate organization within the Bilateria: A histological perspective. *In* Morris et al. (eds.), pp 101–122.

Rieger, R. M. 1986. Über der Bilateria: Die Bedeutung der Ultrastrukturforschung für ein neues Verstehen der Metazoenevolution. Verh. Dtsch. Zool. Ges. 79:31–50.

Salvini-Plawen, L. V. 1982. A paedomorphic origin of the oligomerous animals? Zool. Scr. 11:77–81. [Dubious and difficult reading.]

Sarvaas, A. E. du Marchie. 1933. La theorie du coelome. Thesis, University of Utrecht. [Some ideas on the schizocoel theory that never quite took hold.]

Schleip, W. 1929. *Die Determination der Primitiventwicklung*. Akad. Verlags, Leipzig. [The origin of the concept of the "Spiralia."]

Sedgwick, A. 1884. On the nature of metameric segmentation and some other morphological questions. Q. J. Microsc. Sci. 24:43–82. [This work provided the main driving force behind the idea that the coelom arose (via enterocoely) from cnidarian gut pouches rather than by a pinching off of the diverticula in flatworm digestive tracts.]

Serres, E. R. A. 1824. Explication de système nerveux des animaux invertébrés. Ann. Sci. Nat. 3:377–380.

Serres, E. R. A. 1830. Anatomie transcendante—Quatrieme mémoire: Loi de symétrie et de conjugaison du système sanguin. Ann. Sci. Nat. 21:5–49.

Siewing, R. 1980. Das Archichelomatenkonzept. Zool. Jahrb. Abt. Anat. Ontog. Tiere 8,103:439–482.

Smith, J. III and S. Tyler. 1985. The acoel turbellarians: Kingpins of metazoan evolution or a specialized offshoot? *In* Morris et al. (eds.), pp. 123–142.

Snodgrass, R. E. 1938. Evolution of the Annelida, Onychophora and Arthropoda. Smithson. Misc. Collect. 97(6)1–159. [A proponent of the nephrocoel theory explains his views.]

Thiele, J. 1902. Zur Cölomfrage. Zool. Anz. 25:82–84.

Thiele, J. 1919. Über die Auffassung der Leibeshöhle von Mollusken und Anneliden. Zool. Anz. 35:682–695.

Valentine, J. 1973. Coelomate superphyla. Syst. Zool. 22:97–102.

Valentine, J. 1975. Adaptive strategy and the origin of grades and ground-plans. Am. Zool. 15:391–404.

Wilson, E. B. 1898. Considerations in cell-lineage and ancestral reminiscence. Ann. N. Y. Acad. Sci. 11:1–27.

Ziegler, H. E. 1898. Über den derzeitigen Stand der Cölomfrage. Verh. Dtsch. Zool. Ges. 8:14–78.

Ziegler, H. E. 1912. Leibeshöhle. Handwörterbuch Naturwiss. 6:148–165.

Chapter Five

The Protozoa

Cut down the light. Search the slide for an irregular granular sharply outlined object, apparently motionless. Ask the instructor whether or not you have an Amoeba or have him help you find one.

Libbie Hyman
A Laboratory Manual for Elementary Zoology, 1926

Most zoologists today would agree that the organisms called protozoa constitute a grade of complexity but not a monophyletic assemblage meriting single-phylum status. We present them here in a single chapter partly out of tradition and convenience, but primarily because together they demonstrate the survival strategies of nonmetazoan *Baupläne*. As we noted in Chapters 3 and 4, the protozoa are most easily defined by their lack of metazoan attributes, notably the absence of a layered, cellular construction. They are traditionally and most frequently described as unicellular eukaryotic organisms, although some workers (see Hyman 1940) have preferred to think of them as acellular. This latter view contends that the protozoa are too complex to be thought of as single cells; rather they should be considered simply as small organisms whose bodies are not subdivided into cellular units. We view their structure as decidedly cell-like, because they possess all of the organelles found in "typical" eukaryotic cells of metazoa; therefore, we retain the traditional terminology of unicellularity.

The bodies of protozoa consist of but a single cell, although many are colonial. The roughly 35,000 described extant species include an awesome array of shapes and functional types, and there are certainly many thousands of species yet to be discovered. Figure 1 illustrates some of this variety. Another 40,000 or so species are known from the fossil record.

Most protozoa are microscopic, ranging in size from about 5 to 300 μm. A few, such as the foraminiferans, are much larger and are visible to the naked eye (some fossil forms exceed 15 cm in diameter). Many protozoa are actually larger than the smallest of the metazoa (e.g., some gastrotrichs; kinorynchs). The protozoa include marine, freshwater, terrestrial, and symbiotic species, the last category including many serious pathogens.

Taxonomic history and classification

— Antony van Leeuwenhoek is generally credited with being the first person to report seeing protozoa, in about 1675. In fact, Leeuwenhoek was the first to describe a number of microscopic aquatic life forms (protozoa, rotifers, and others), referring to them as "animalcules." For nearly 200 years the protozoa were classified along with a great variety of other microscopic life forms under various names (e.g., Infusoria). The name protozoon (Greek, *proto*, "first"; *zoon*, "animal") was coined by Goldfuss in 1818 as a subgrouping of a huge assemblage of animals known at that time as the Zoophyta (protozoa, sponges, cnidarians, rotifers, and others). Following the discovery of cells in 1839, the distinctive nature of the protozoa became apparent. On the basis of this distinction, von Siebold (1845) coined the name Protozoa to apply to all unicellular forms of animal life.

Throughout the early part of the twentieth century, a relatively standard classification scheme developed and was widely accepted, undergoing only minor periodic modifications. This scheme was based on the idea that all of the protozoa were descended from a common ancestor and thus merited recognition as a single phylum that could be subdivided into various subphyla and classes. However, extensive

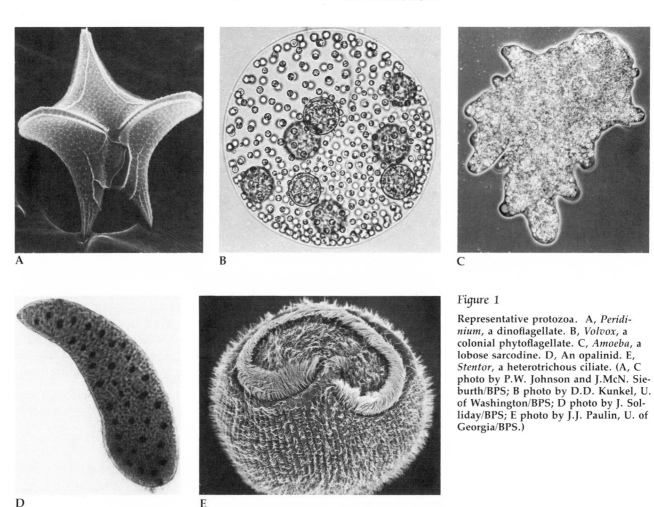

Figure 1

Representative protozoa. A, *Peridinium*, a dinoflagellate. B, *Volvox*, a colonial phytoflagellate. C, *Amoeba*, a lobose sarcodine. D, An opalinid. E, *Stentor*, a heterotrichous ciliate. (A, C photo by P.W. Johnson and J.McN. Sieburth/BPS; B photo by D.D. Kunkel, U. of Washington/BPS; D photo by J. Solliday/BPS; E photo by J.J. Paulin, U. of Georgia/BPS.)

research over the past two decades has led most specialists to recognize a new classification scheme. The Committee on Systematics and Evolution of the Society of Protozoologists (1980, N. D. Levine, Chairman), proposed the establishment of seven separate phyla. But the proposition includes the disclaimer that the classification does not entirely reflect evolutionary relationships and is only meant to serve as a starting point for further investigation. Box 1 provides a brief outline of this protozoan classification. These taxa are discussed in detail in subsequent sections of this chapter.

The protozoan *Bauplan*

General synopsis

While realizing that the protozoa probably do not represent an actual monophyletic clade, it is still advantageous to examine them together from the standpoint of the strategies and constraints of a unicellular *Bauplan*. Remember that within the limitations imposed by unicellularity, these creatures still must accomplish all of the basic life functions common to metazoa.

As we discussed in Chapter 3, most life processes are dependent on activities associated with surfaces, notably with cell membranes. Even in the largest multicellular organisms, the regulation of exchanges across cell membranes and the metabolic reactions along the surfaces of various cell organelles are the phenomena on which all life ultimately depends. Consequently, the total area of these important surfaces must be great enough relative to the volume of the organism to provide adequate exchange and reaction sites. Nowhere is the "lesson" of the surface area:volume ratio made more clear than among the protozoa, where it demonstrates the impossibility of bulky-bodied, 100-kg amebas and other such 1950s horror-movie beasts. Without either an efficient

---

Box One

Summary of protozoan classification

PHYLUM SARCOMASTIGOPHORA. The amebas, flagellates, and opalinids
   SUBPHYLUM MASTIGOPHORA. The flagellates
      CLASS PHYTOMASTIGOPHOREA. The photosynthetic flagellates
      CLASS ZOOMASTIGOPHOREA. The nonphotosynthetic flagellates
   SUBPHYLUM SARCODINA. The amebas
      SUPERCLASS RHIZOPODA. The naked amebas, foraminiferans
      SUPERCLASS ACTINOPODA. Radiolarians, heliozoans
   SUBPHYLUM OPALINATA, CLASS OPALINATEA. The opalinids
PHYLUM LABYRINTHOMORPHA. The labyrinthomorphans
*PHYLUM APICOMPLEXA. The gregarines, coccidians
*PHYLUM MICROSPORA. The microsporans
*PHYLUM ASCETOSPORA. The ascetosporans
*PHYLUM MYXOZOA. The myxozoans
PHYLUM CILIOPHORA. The ciliates
      CLASS KINETOFRAGMINOPHOREA. The suctorians, hypostomes, gymnostomes, etc.
      CLASS OLIGOHYMENOPHOREA. The paramecia and other hymenostomes, peritrichs, etc.
      CLASS POLYHYMENOPHOREA. The spirotrichs (tintinnids, heterotrichs, hypotrichs, etc.)

*The former taxon Sporozoa has been abandoned and its members divided among the phyla Apicomplexa, Microspora, Myxozoa, and Ascetospora.

---

mechanism of circulation within the body or the presence of membrane partitions (multicellularity) to enhance and regulate exchanges of materials, protozoa must remain relatively small; the diffusion distance between their cell membranes (body surface) and the innermost parts of their bodies can never be so great that it prevents adequate movement of materials from one place to another within the cell. Certainly there are structural elements (e.g., microtubules, endoplasmic reticula) and various processes (e.g., protoplasmic streaming, active transport) that supplement passive circulatory phenomena. But the fact remains that unicellularity mandates that a high surface area :volume ratio be maintained by restricting shape or size. This principle is behind the fact that the largest protozoa (other than certain colonies) have assumed shapes—elongate, thin, or flattened—that maintain small diffusion distances.

The formation of membrane-bounded pockets or vesicles is common in protozoa, and helps maintain a high surface area for internal reactions and exchanges. The elimination of metabolic wastes and excess water, especially in freshwater forms living in hypotonic environments, is facilitated by **water expulsion vesicles** (Figures 4, 15, and others). As explained in Chapter 3, these vesicles (frequently called **contractile vacuoles**) release their contents to the outside in a more-or-less controlled fashion, often counteracting the normal diffusion gradients between the cell and the environment.

Various types of nutrition are seen among the protozoa. Indeed, they may be either autotrophic or heterotrophic (some may be both). Photosynthetic flagellates depend upon membrane-bounded organelles (chloroplasts) to provide the surface area needed for photosynthetic reactions. But most protozoa are heterotrophic, requiring organic chemicals from their environment. Heterotrophic forms may be **saprobic**, taking in dissolved organics by diffusion, active transport, or pinocytosis. Or they may be **holozoic**, taking in solid foods such as organic detritus or whole prey by phagocytosis. Many heterotrophic protozoa are symbiotic on or within other organisms. All heterotrophic protozoa acquire food through some cell surface–environment interaction. Those that engage in pinocytosis or that engulf solid food rely on the formation of membrane-bounded capsules or vesicles called **food vacuoles** (Figure 2). These structures may

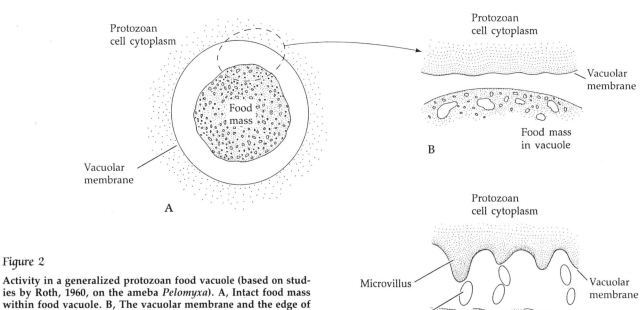

Figure 2

Activity in a generalized protozoan food vacuole (based on studies by Roth, 1960, on the ameba *Pelomyxa*). A, Intact food mass within food vacuole. B, The vacuolar membrane and the edge of the food mass (magnified view). C, Formation of microvilli and vesicles of vacuolar membrane. D, Uptake of vesicles containing products of digestion into the cytoplasm. See text for additional explanation. (B,C,D after Meglitsch 1972, adapted from electron micrographs in Roth 1960.)

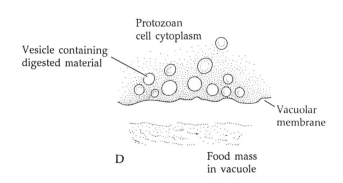

form at nearly any site on the cell surface, as they do in the amebas, or at particular sites associated with some sort of "cell mouth," or **cytostome**, as they do in most protozoa with more-or-less fixed shapes. The cytostome may be associated with further elaborations of the cell surface that form permanent invaginations or feeding structures (discussed under specific taxa).

Numerous studies over the past 20 years have added greatly to our understanding of the complex activities of digestion within food vacuoles in protozoa, and they further demonstrate the significance of membrane surfaces in biology. Once a food vacuole has been formed and has moved into the cytoplasm, it begins to swell as various chemicals and enzymes are secreted into it. The vacuole first becomes acidic, and the vacuolar membrane develops numerous inwardly directed microvilli (Figure 2). As digestion proceeds, the vacuolar fluid becomes increasingly alkaline. The cytoplasm just inside the vacuolar membrane takes on a distinctive appearance, presumably from the products of digestion. Then the vacuolar membrane forms tiny vesicles that pinch off and carry these products into the cytoplasm. Much of this activity resembles surface pinocytosis. The result is numerous tiny nutrient-carrying "vacuoles" offering a greatly increased surface area for absorption of the digested products into the cell's cytoplasm. During

this period of activity, the original vacuole gradually shrinks and undigested materials eventually are expelled from the cell. In naked protozoa (e.g., many amebas), the spent vacuole may discharge anywhere on the cell surface; but in cases where a relatively impermeable covering exists around the cell (see below), the covering bears a permanent pore (**cytoproct**) through which the vacuole releases material to the outside.

The cell surface is critical not only in providing a means of exchange of materials with the environment but also in providing protection and structural integrity to the cell. The plasma membrane itself serves as a mechanical and chemical boundary to the protozoan "body," and when present alone (as in the asymmetrical naked amebas), it allows great flexibility and plasticity of shape. However, many protozoa maintain a more-or-less constant shape (spherical, radial, or bilateral symmetry) by thickening the cell membrane to form a **pellicle**, by secreting a shell-like covering called a **test**, by accumulating particles from the environment, or by other skeletal arrangements. Furthermore, locomotor capabilities are also ultimately provided by interactions between the cell surface and the surrounding medium. Pseudopodia, cilia, and flagella provide the means by which protozoa push or pull themselves along.

Thus, the plasma membrane of protozoa provides on the one hand a large surface area for contact between body and environment, and on the other hand a selective barrier whose structural and functional integrity must be maintained so that the nature of the internal milieu does not exceed the limits of tolerance of the cell.

Many protozoa display remarkable degrees of sensitivity to environmental stimuli and are capable of some fairly complex behaviors. But, unlike those of metazoa, the entire stimulus–response circuit lies within the confines of the single cell. Response behavior may be a function of the general sensitivity and conductivity of protoplasm, or it may involve special organelles. Sensitivity to touch often involves distinctive locomotor reactions in motile protozoa and avoidance responses in many sessile forms. Cilia and flagella are touch-sensitive organelles; when mechanically stimulated, they typically stop beating or beat in a pattern that moves the organism away from the point of stimulus. These types of responses are dramatically expressed by most sessile stalked ciliates, which display very rapid reactions when the cilia of the cell body are touched. Contractile elements within the stalk shorten, pulling the animal's body away from the source of the stimulus.

Thermoreception is known among many protozoa, but it is not well understood. Under experimental conditions, most motile protozoa will "seek" optimal temperatures when given a choice of environments. This behavior probably is a function of the general sensitivity of the organism and not of special receptors. Analogous chemotactic responses are probably similarly induced. Most protozoa react positively or negatively to various chemicals or concentrations of chemicals. For example, amebas are able to distinguish food from nonfood items and quickly egest the latter from their vacuoles.

Many flagellates, some ciliates, and some sarcodines respond to various intensities of light. The phytoflagellates typically show a positive taxis to low or moderate light intensities, an obviously advanta-

geous response for these photosynthetic creatures. They usually become negatively phototactic in very strong light. Ciliates and amebas that respond to light are generally negatively phototactic. Specialized light-sensitive organelles are known among many flagellates, especially the phytoflagellates. These **eye spots** or **stigmata** are frequently located at or near the anterior end, but some are found associated with the chloroplasts. Some are very simple pigment spots; others are complex, with lenslike structures.

A major aspect of protozoan success is their surprising range of reproductive strategies. Most protozoa have been able to capitalize on the advantages of both asexual and sexual reproduction, although some apparently reproduce only asexually. Many of the complex cycles seen in certain protozoa (especially parasitic forms) involve alternation between sexual and asexual processes, and often these cycles include a series of asexual divisions between sexual phases.

Protozoa undergo a variety of strictly asexual reproductive processes including **binary fission**, **multiple fission**, and **budding**. Some engage in a process called **plasmotomy**, in which a multinucleate adult simply divides into two multinucleate daughter cells.

Sexual reproduction produces genetic variation. The protozoa have evolved a variety of methods that achieve this end, not all of which result in the immediate production of additional individuals. If we expand our traditional definition of meiosis to include any nuclear process that results in a haploid condition, then meiosis can be considered a protozoan as well as a metazoan phenomenon. This disclaimer is necessary because protozoan "meiosis" appears to be somewhat different from that seen in metazoa, and it is certainly less well understood. Nonetheless, reduction division does occur, and haploid cells or nuclei of one kind or another are produced and then fuse to restore the diploid condition. This production and subsequent fusion of gametes is called **syngamy**. It may involve gametes that are all similar in size and shape (**isogamy**), or the more familiar condition of gametes of two distinct types (**anisogamy**). Thus, as in metazoa, both haploid and diploid phases are produced in the life histories of sexual protozoa. The meiotic process may immediately precede the formation and union of gametes (prezygotic reduction division), or it may occur immediately after fertilization (postzygotic reduction division), as it does in many lower plants. Postzygotic reduction division occurs in the photosynthetic flagellates (Phytomastigophorea). Other sexual processes that result in genetic mixing by the exchange of nuclear material between "mates" (**conjugation**) or by the reformation of a genetically "new" nucleus within a single individual (**autogamy**) are best known among the ciliates and are discussed under that taxon.

In summary, protozoan success reflects the tremendous potential for variation among single cells. The following accounts of protozoan phyla explore this diversity and potential.

Phylum Sarcomastigophora

The phylum Sarcomastigophora includes the amebas, opalinids, and flagellates. These groups are combined in one phylum because they are all homokaryotic (possess a single type of nucleus), have similar methods of sexual reproduction, and use flagella or pseudopodia or both for locomotion (Box 2). Certain flagellates—the ameboflagellates—possess both flagella and pseudopodia, a resemblance further suggesting a close affinity between the groups. Certainly this phylum is a mixed assemblage of forms, and it remains to be shown whether the Sarcomastigophora is truly a monophyletic taxon.

Subphylum Mastigophora (the flagellates)

The great array of life histories among the flagellates makes them a study in evolutionary diversity just in themselves. It has long been thought that the flagellates may have been the first eukaryotic life on this planet and the ancestral pool from which the metaphyta and metazoa arose. The approximately 8,500 described species of extant flagellates are ex-

---

Box Two

Characteristics of the Phylum Sarcomastigophora

1. Unicellular or colonial protozoa
2. Locomotion by flagella or pseudopodia or both
3. Autotrophic or heterotrophic
4. Homokaryotic (possess a single type of nucleus)

---

traordinarily diverse. They are usually divided into two classes on the basis of the presence or absence of chloroplasts. Those with chloroplasts make up the class Phytomastigophorea, or plantlike flagellates. Those that lack chloroplasts are called the Zoomastigophorea, or animal-like flagellates.

Flagellates are among the most common free-living protozoa in all aquatic environments. There are also a number of endosymbiotic forms. The free-living species include both solitary and colonial forms, and although most are motile, many sessile types are known. Figures 3 and 4 illustrate a variety of flagellates and introduce some anatomical terms.

The most prominent unifying feature of the flagellates is the possession of one or more flagella, which are used as locomotor organelles. Most specialists agree, however, that this character is a rather

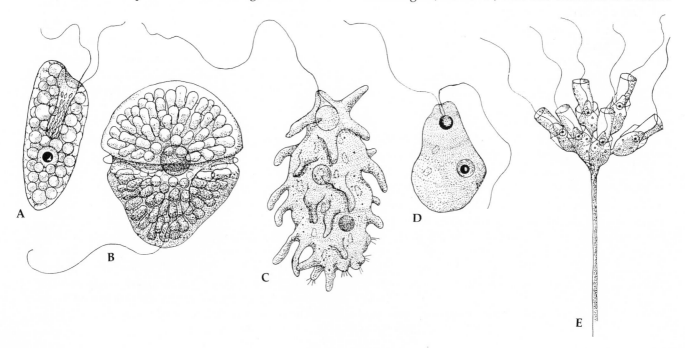

superficial one, given the widespread occurrence of flagella throughout the living world. Consequently, the flagellates are considered by many workers to represent a polyphyletic assemblage. The systematic resumé that follows includes only the more common orders.

SUBPHYLUM MASTIGOPHORA

CLASS PHYTOMASTIGOPHOREA: Photosynthetic flagellates (Figures 1A,B; 3A,B,F,G; 4A). Possess chloroplasts (occasionally secondarily lost); always with chlorophyll *a*, and often with other pigments (such as chlorophylls *b*, c_1, c_2; carotenes; xanthophylls).

ORDER CRYPTOMONADIDA: Cryptomonads. Biflagellate, free-living in most aquatic habitats; most are red, brown, or yellow (*Chilomonas* is colorless); all have a deep depression near base of flagellum through which food is ingested; sexual reproduction unknown. (e.g., *Chilomonas*, *Cryptochrysis*, *Cryptomonas*)

ORDER DINOFLAGELLIDA: Dinoflagellates. Biflagellate, free-living or symbiotic, in marine and fresh water; most are yellow-brown to golden; cell walls composed of cellulose plates; flagella lie in grooves in cell wall; sexual and asexual reproduction. Some dinoflagellates (e.g., *Noctiluca*) are phosphorescent, and when present in high densities near shore produce the familiar blue-white glow in the breaking surf. Others (e.g., *Gonyaulax*) are responsible for "red tide" associated with paralytic shellfish poisoning. (e.g., *Ceratium*, *Gonyaulax*, *Gymnodinium*, *Noctiluca*)

ORDER EUGLENIDA: Euglenids. Most are biflagellate with one short and one long flagellum; free-living in marine and freshwater habitats; green, red, or colorless; flagella arise from anterior depression or vestibule; sexual reproduction unknown. (e.g., *Euglena*, *Peranema*, *Phacus*)

ORDER CHRYSOMONADIDA: Chrysomonads. Most are biflagellate with one short and one long flagellum, free-living in all aquatic environments; yellow or brown; some capable of engulfing solid food by pseudopodia; solitary or colonial, sessile or motile; many species secrete small siliceous plates or vase-shaped casings (loricae); sexual reproduction occurs in some groups. (e.g., *Chromulina*, *Dinobryon*, *Synura*)

ORDER VOLVOCIDA: Volvocids. Unicellular or colonial, usually with two flagella per cell; all are green with distinct cellulose cell walls; mostly in fresh water; asexual and sexual reproduction common. (e.g., *Chlamydomonas*, *Eudorina*, *Gonium*, *Pandorina*, *Pleodorina*, *Volvox*). In small colonial volvocids with only a few cells (e.g., *Gonium*, *Pandorina*), there is little or no differentiation of cells into different functional types; and where sexual reproduction occurs, it is by isogamy. But in those colonial forms with high numbers of cells (e.g., *Volvox*), a definite polarity occurs and differentiation between somatic and reproductive cells within each colony is evident; sexual reproduction is anisogamous. Furthermore, these large colonies move with a particular pole consistently directed forward. The cells of the "anterior end" bear well-developed eye spots and have significantly larger flagella than the trailing cells.

ORDER SILICOFLAGELLIDA: Silicoflagellates. Uniflagellate, marine plankters; sometimes placed within the Chrysomonadida; yellow to brown with siliceous tests; sexual reproduction unknown. (e.g., *Dictyocha*)

CLASS ZOOMASTIGOPHOREA: "Animal-like" flagellates (Figures 3C,D,E; 4B,C). Lack chloroplasts; number of flagella varies from none to many.

ORDER CHOANOFLAGELLIDA: Choanoflagellates. Uniflagellate, freshwater protozoa; solitary or colonial, mostly sessile; cells of colonial forms embedded in gelatinous mass or surrounded by siliceous lorica; with unique protoplasmic collar extending from cell around proximal portion of flagellum. (e.g., *Codosiga*, *Monosiga*, *Proterospongia*)

ORDER KINETOPLASTIDA: Kinetoplastids. Uni- or biflagellate; free-living or parasitic mastigophorans; possess a unique DNA-housing organelle called the kinetoplast; includes many agents of serious diseases (leishmaniasis, sleeping sickness, Chagas disease). (e.g., *Bodo*, *Cryptobia*, *Leishmania*, *Trypanosoma*)

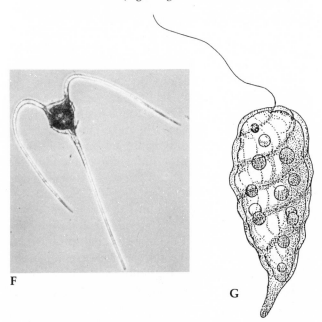

F   G

Figure 3

Representative mastigophorans. A, *Chilomonas*, **a cryptomonad phytoflagellate that has secondarily lost its pigments. B,** *Gymnodinium*, **a dinoflagellate. C,** *Mastigamoeba*, **a zooflagellate with pseudopodia. D,** *Bodo*, **a free-living member of the order Kinetoplastida. E,** *Codosiga*, **a colonial choanoflagellate. F,** *Ceratium*, **a dinoflagellate. G,** *Phacus*, **a euglenoid phytoflagellate. (A–G after various sources; F photo by P.W. Johnson and J.McN. Sieburth/BPS.)**

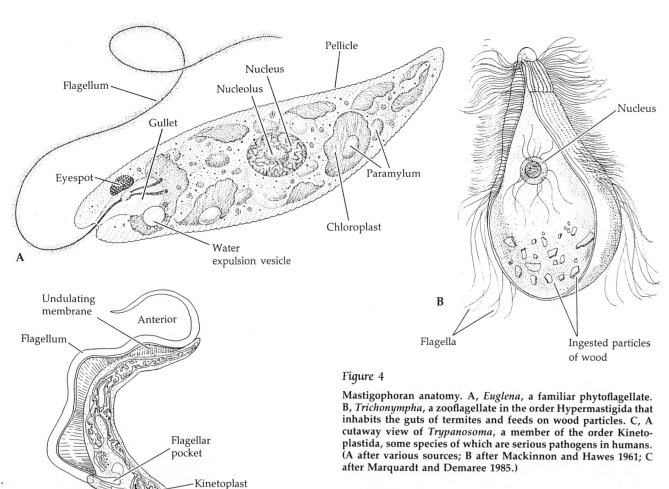

Figure 4

Mastigophoran anatomy. A, *Euglena*, **a familiar phytoflagellate. B,** *Trichonympha*, **a zooflagellate in the order Hypermastigida that inhabits the guts of termites and feeds on wood particles. C, A cutaway view of** *Trypanosoma*, **a member of the order Kinetoplastida, some species of which are serious pathogens in humans. (A after various sources; B after Mackinnon and Hawes 1961; C after Marquardt and Demaree 1985.)**

ORDER RETORTAMONADIDA: Retortamonads. Exclusively parasitic, mostly in the digestive tracts of insects; with two or four flagella and a distinct cytostome. (e.g., *Chilomastix*.)

ORDER DIPLOMONADIDA: Diplomonads. Called dikaryomastigonts because they possess two nuclei; each nucleus is associated with a bundle of one to four flagella; includes free-living and parasitic species. (e.g., *Enteromonas*, *Giardia*, *Trepomonas*)

ORDER TRICHOMONADIDA: Trichomonads. Commensal or parasitic endosymbionts; most bear four to six flagella (rarely one or none). (e.g., *Dientamoeba*, *Histomonas*, *Trichomonas*)

ORDER HYPERMASTIGIDA: Hypermastigids. Multiflagellate endosymbionts in insects, especially termites and roaches; flagella typically occur in distinctive patterns; produce cellulase and feed primarily on wood ingested by the hosts. (e.g., *Eucomonympha*, *Trichonympha*)

Stephanopogon

Before leaving our resumé of mastigophorans, we must briefly mention the enigmatic genus *Stephanopogon*. These animals have played an important role in phylogenetic speculations regarding protozoan evolution. Several theories have implicated these organisms not only in the origin of the ciliates from a flagellate ancestor and of the ciliate binuclear condition, but also in the origin of the metazoa from a ciliate protozoan line. Until recently, *Stephanopogon* was classified in the phylum Ciliophora (as a gymnostomatid). In 1982, however, D. L. Lipscomb and J. O. Corliss provided evidence based on ultrastructural studies that these protists are actually highly specialized flagellates. Lipscomb and Corliss found that the "cilia" of *Stephanopogon* are actually short,

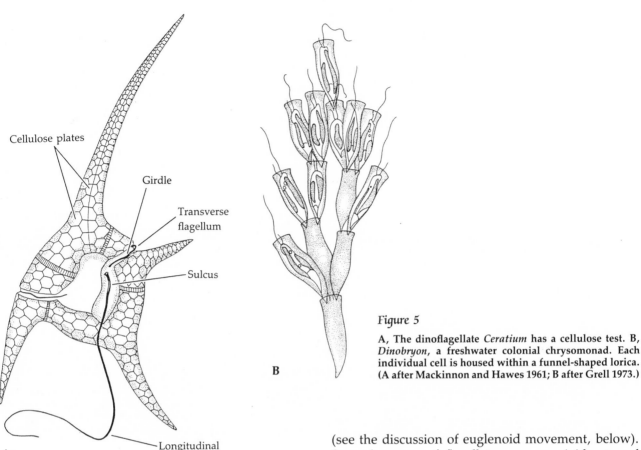

Figure 5

A, The dinoflagellate *Ceratium* has a cellulose test. B, *Dinobryon*, a freshwater colonial chrysomonad. Each individual cell is housed within a funnel-shaped lorica. (A after Mackinnon and Hawes 1961; B after Grell 1973.)

modified flagella. The two (or up to 16) nuclei of *Stephanopogon* are identical rather than differentiated into macro- and micronuclei as they are in the ciliates (this long-ignored fact was actually discovered in the 1920s). Nuclear division is very much like that seen in zooflagellates, and the cells lack a kinetidal system. While *Stephanopogon* cells possess an unusually short kinetosome at the base of each flagellum, these are not linked by kinetodesmata. Lipscomb and Corliss have proposed that the genus *Stephanopogon* be considered as a new order of zooflagellates. They point out that it appears to be far from the main trunk on any phylogenetic tree that depicts the origin of ciliates from flagellates. This proposal means that the use of *Stephanopogon* to derive the metazoa from a ciliate ancestry is also no longer plausible.

Support and locomotion. Most mastigophorans possess a pellicle consisting of spirally arranged strips of protein. The pellicle of flagellates is flexible enough to allow modest changes in shape, but resilient enough to bring the body back to its normal form

(see the discussion of euglenoid movement, below). Several groups of flagellates possess rigid external tests or **loricae**, which are secreted by the cell as an exoskeleton. These tests may be composed of a variety of substances, including cellulose (dinoflagellates) and siliceous compounds (silicoflagellates and choanoflagellates) (Figure 5).

Nearly all motile mastigophorans employ one or more patterns of **flagellar movement** (see Chapter 3). Much of our understanding of flagellar motion, and of other forms of protozoan locomotion, came from the laboratory of T. L. Jahn and his co-workers during the 1960s. The beating of a single flagellum can usually be described as being either **uniplanar** or **helicoidal**. In the first case the flagellum moves within a single plane in space; in the second, the movement follows a helical pattern. The flagellum acts much like an oar in water (e.g., paddling, sculling fore or aft) by pushing against the environmental medium, thereby resulting in movement of the protozoon's body (Figure 7). Many protozoan flagella possess tiny, hairlike side branches called **mastigonemes**, which increase the surface area of the "paddle" and which may be oriented in various ways to provide a means of steerage (Figure 6). The particular pattern of flagellar movement depends upon the position in which the flagellum is held as it moves, the

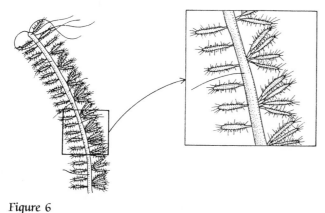

Figure 6

Branched and unbranched mastigonemes on a flagellum of the chrysomonad flagellate *Ochromonas*. (Redrawn from Grell 1973, after Bouck 1971.)

portion of the flagellum undergoing movement, the pattern and direction of wave motion along the flagellum, and the number and location of flagella possessed by the animal. Several of these patterns are illustrated and explained further in Figure 7.

Some flagellates lack a rigid test and can engage in a type of locomotion called **euglenoid movement**. The flexibility of the pellicle allows various sorts of squirming, wormlike undulations of the body, which augment flagellar movement and allow turns and twists through cramped areas (Figure 8).

Nutrition. Members of the class Phytomastigophorea are usually either facultative or obligate photosynthetic autotrophs, whereas the zoomastigophorans all exhibit some type of heterotrophy. All

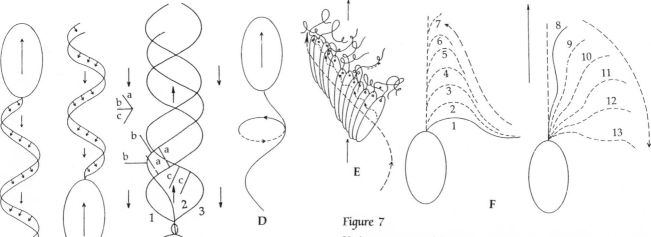

Figure 7

Various patterns of flagellar movement in mastigophorans. A–C, Locomotion resulting from uniplanar flagellar beating. The large arrows within the cells indicate the direction of movement. A, Sine wave moves from the proximal to the distal end along the trailing flagellum. This motion pushes the animal forward. B, Sine wave moves from the distal to the proximal end along the leading flagellum. This motion pulls the animal forward. C, Sine wave moves from the distal to the proximal end along the leading flagellum but mastigonemes produce a posteriorly directed force to propel the animal forward. Three pairs of mastigonemes are illustrated here (represented by the lowercase letters). As the flagellum beats (1 to 2 to 3), each mastigoneme is carried through a beat stroke (a to b to c). Thus, each mastigoneme acts as a tiny oar to move the animal forward. D, Helicoidal pattern of beating of a trailing flagellum produces forward movement. E, Locomotion in *Euglena*. The complex looping action of the flagellum proceeds from proximal to distal. Overall movement of the whole cell is forward. The dotted line indicates the path of the organism. F, Relatively simple pattern of flagellar beating in *Monas*. 1–7 represent recovery stroke; 8–13 represent power stroke, which drives the animal forward. G, *Euglena* can employ the flagellar tip alone as a tiny propeller to pull itself forward. H, Locomotor currents in the dinoflagellate *Ceratium*. I, Pattern of flagellar beating in the phytoflagellate *Polytomella*. J, Undulatory flagellar movement in *Trypanosoma*. (A–C redrawn from Grell 1973, after Jahn, Landman, and Fonseca 1964; D redrawn from Grell 1973, adapted from several sources; E, H–J after Jahn and Bovee 1967; F redrawn from several sources.)

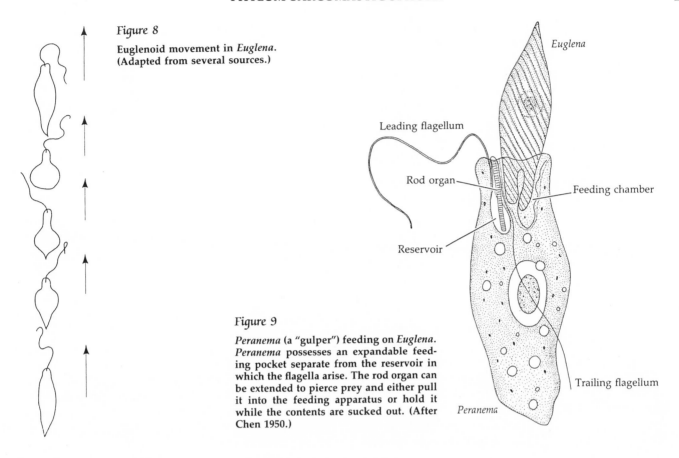

Figure 8

Euglenoid movement in *Euglena*.
(Adapted from several sources.)

Euglena

Leading flagellum

Rod organ

Reservoir

Feeding chamber

Trailing flagellum

Peranema

Figure 9

Peranema (a "gulper") feeding on *Euglena*. *Peranema* possesses an expandable feeding pocket separate from the reservoir in which the flagella arise. The rod organ can be extended to pierce prey and either pull it into the feeding apparatus or hold it while the contents are sucked out. (After Chen 1950.)

photosynthetic flagellates possess chlorophyll *a*, which produces the characteristic green color of familiar genera such as *Euglena*, *Chlamydomonas*, and *Volvox*. Many species possess additional pigments, especially various xanthophylls, which tend to mask the green pigments and impart a yellow, brown, or golden color (e.g., in various dinoflagellates). Some of the phytoflagellates (e.g., *Chlamydomonas*) are capable of complete and typical photosynthesis, requiring only various inorganic chemicals and sunlight to produce carbohydrate end products. Many others, although capable of photosynthesis, require certain organic molecules for the production of starch or related compounds. Photosynthetic flagellates with these additional nutrient requirements are termed **mixotrophic**. The storage products vary among photosynthetic flagellates. Many species store food reserves as typical plant starch encased within proteinaceous organelles called **pyrenoids**. These pyrenoids are usually located close to, or even within, the chloroplasts or chromoplasts.* Some flagellates store food

as nonstarch polysaccharides (including **paramylum** and **leucosin**) and some as fats and oils. A great number of phytoflagellates are **amphitrophic**: they are facultative autotrophs or mixotrophs but can switch to heterotrophy in the absence of sufficient light. Many of these species lose their chloroplasts altogether if kept long enough in the dark.

Virtually all of the cell-feeding mechanisms described earlier in this chapter occur among heterotrophic mastigophorans. Some species, notably the mixotrophic forms, obtain organic molecules by simple diffusion or active transport across the cell membrane. Although not well documented, it is suspected that many flagellates obtain larger molecules by pinocytosis. This activity has been demonstrated in certain species of zoomastigophorans of the genus *Trypanosoma*. These blood parasites possess a gullet-like depression near the base of the flagellum through which host fluids are ingested by pinocytosis.

Many mastigophorans feed by phagocytosis of relatively large (sometimes comparatively huge) food materials. Most of these holozoic forms have a cytostome located near the base of a flagellum where food vacuoles form. In some species, a depression called

*The term chromoplast is often used in the literature to refer to plastids in which both accessory pigments (e.g., xanthophylls) and chlorophyll are present.

the **vestibule** or **reservoir** occurs in the same general area as the cytostome, but its involvement in feeding is not well understood. In others (e.g., *Peranema*; Figure 9), the cytostome and feeding apparatus are separate from the vestibule. Most holozoic flagellates obtain food by creating water currents that bring potential food items in contact with the area of the cytostome. Prey or other food materials are ingested by the formation of food vacuoles. Figure 10 illustrates feeding currents and ingestion activities in motile and sessile flagellates.

A large number of mastigophorans are endosymbiotic, and some cause serious diseases in humans. Species of *Leishmania* cause a variety of ailments collectively called **leishmaniases**, including kala-azar (a visceral infection that particularly affects the spleen), oriental sore (characterized by boils resulting from blockage of the skin capillaries by the parasite), and several other skin and mucous membrane infections. Perhaps more familiar are certain diseases caused by members of the genus *Trypanosoma*. Some species infect wild and domestic animals, and two species (*T. gambiense* and *T. rhodesiense*) cause sleeping sickness in humans. These parasites are introduced into the blood of humans from the salivary glands of the tsetse fly. From the blood, the trypanosomes can enter the lymphatic system and ultimately the cerebrospinal fluid of the host. The disease is usually fatal. *Trypanosoma cruzi* causes Chagas disease and is transmitted to humans from hemipteran insects of the genus *Triatoma*. The parasites leave the insect

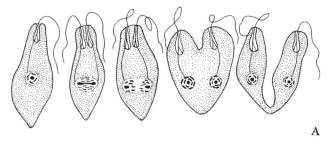

Figure 11

Binary fission in mastigophorans. A, Longitudinal fission in *Euglena*; flagella and reservoir duplicate prior to cell division. B, Oblique fission in the dinoflagellate *Ceratium*; each daughter cell produces a replacement for the missing portion of the test. C, Longitudinal fission in *Trypanosoma*. D, Binary fission in *Devescovina*. (A after various sources; B after Grell 1973; C after Mackinnon and Hawes 1961; D redrawn from Grell 1973, after Kirby 1944.)

vector's body by way of its feces and enter the human host through mucous membranes. The parasites migrate to the host's bloodstream, where they cause destruction of erythrocytes.

Various other mastigophorans inhabit humans as commensals or as agents of less serious diseases. The diplomonad *Giardia* is a very common intestinal parasite that causes enteritis. Various species of *Trichomonas* occur in the gut, mouth, vagina, and upper urinary tract of humans and cause inflammation of the mucous membranes. A number of other trichomonads inhabit the human digestive tract, where they feed on bacteria.

Certain wood-eating insects, notably termites and roaches, house mutualistic flagellates of the order Hypermastigida. These protozoa produce cellulase, which breaks down the wood that is ingested by the host insect. The breakdown products nourish both the host and the protozoa.

Reproduction and some life cycles. All mastigophorans reproduce asexually, usually by longitudinal binary fission (Figure 11). In uniflagellate forms, the flagellum typically duplicates prior to cell division; in those with many flagella, the flagella are sometimes lost and reformed by each daughter cell following division, or they may be retained and divided equally or unequally between the daughter cells. Many dinoflagellates (e.g., *Ceratium*) undergo oblique binary fission, and each daughter cell retains a portion of the parent's test (Figure 11B). *Devescovina lemniscata*, a multiflagellate trichomonad from the gut of termites, first undergoes nuclear and flagellar division. The daughter nuclei then migrate to opposite ends of the cell, and the cytoplasm subsequently divides (Figure 11D).

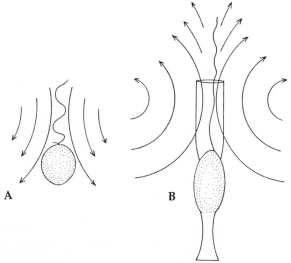

Figure 10

Feeding currents produced by two "swirlers." A, *Ochromonas*, a free-swimming chrysomonad flagellate. B, *Codosiga*, a sessile choanoflagellate. (Adapted from several sources.)

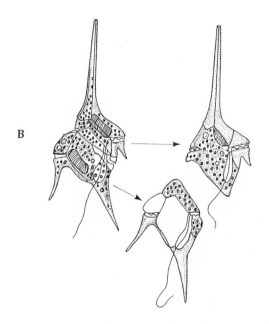

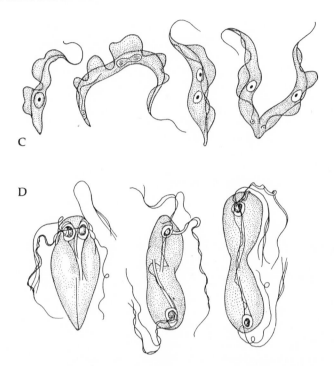

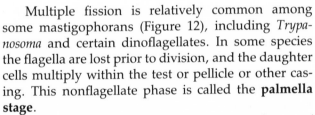

Multiple fission is relatively common among some mastigophorans (Figure 12), including *Trypanosoma* and certain dinoflagellates. In some species the flagella are lost prior to division, and the daughter cells multiply within the test or pellicle or other casing. This nonflagellate phase is called the **palmella stage**.

Some special and singularly curious forms of multiple fission occur in some of the volvocids (e.g., *Volvox, Chlamydomonas, Eudorina*). Repeated cell divisions produce arrangements of cells in patterns reminiscent of those seen among the metazoa (Figure

13). In the colonial volvocids, this type of multiple fission is involved in the formation of daughter colonies.

Sexual reproduction is known among many phytoflagellates, certain endosymbiotic zooflagellates (e.g., trichomonads and trichonymphids), and opalinids. Both isogamy and anisogamy (heterogamy) occur among the flagellates (Figure 14). Autogamy is also known in certain groups. In most of the phytoflagellates meiosis is postzygotic (as it is in the lower

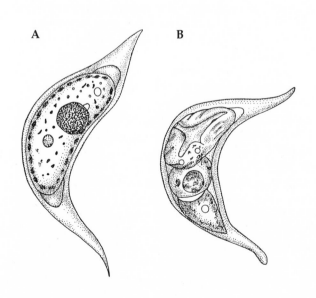

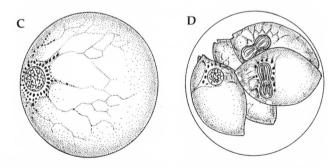

Figure 12

Multiple fission in some phytoflagellate protozoa. The palmella stage (A), followed by multiple fission (B) in the dinoflagellate *Cystodinium*. C, The palmella stage of the dinoflagellate *Dissodinium*. Following loss of its flagella, the cell divides to form four oval individuals (D), each of which subsequently elongates and divides several more times. (A,B redrawn from Barnes 1980, after other sources; C,D redrawn from Grell 1973, after other sources.)

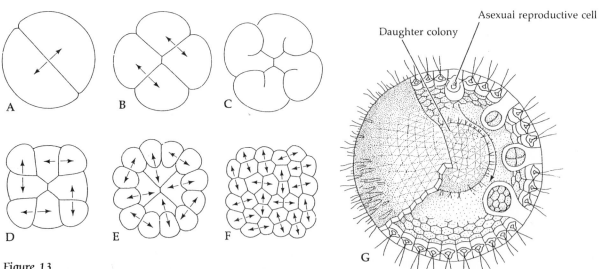

Daughter colony

Asexual reproductive cell

G

Figure 13

Multiple fission in two colonial volvocid phytoflagellates, *Eudorina* and *Volvox*. A–F, Sequence of cell divisions during colony formation in *Eudorina*. Double-headed arrows indicate pairs of daughter cells. This pattern of division is as regular and predictable as the cleavage-stage divisions of many metazoa. G, Composite, cutaway drawing of *Volvox*, showing the sequence of stages in the asexual production of a daughter colony from a single generative cell of the parent colony. Note the cleavage-like progression of the successive stages. (A–F redrawn from Grell 1973, after Gerisch 1959; G after various sources.)

plants), whereas with few exceptions (e.g., *Trichonympha*) prezygotic reduction division occurs in most of the zooflagellates.

As might be anticipated, life cycles vary widely among the mastigophorans. Many alternate between sexual and asexual phases, frequently with a period of encystment such as the palmella stage, whereas others are fully asexual in nature (e.g., *Trypanosoma*). In many cases the alternation between sexual and asexual activities is irregular, or facultative, and is timed to environmental changes rather than set to some prescribed pattern. For example, sexual activity in the hypermastigids is induced by the molting hormones of their insect hosts. Many of the parasitic mastigophorans require intermediate hosts and complex transmission mechanisms for the completion of their life cycles.

The phenomenon of "red tide" is associated with periodic bursts of population growth among certain planktonic dinoflagellates. Red tides have nothing to do with actual tides, nor are they truly red. A red tide is simply a streak or patch of ocean water discolored (generally a pinkish orange) by the presence of billions of single-celled dinoflagellate protozoa. Exactly why single-species population explosions of these specific organisms occur is not clear, and red tides still are not predictable events. During a red tide, densities of these dinoflagellates may be as high as 50–60 million cells per liter of sea water (it takes about a million cells per liter to discolor the water).

Red tide events have been recorded from many parts of the world, including the coast of California and the southeastern United States. California red tides are caused by *Gonyaulax polyhedra* (south of Point Conception) and *G. catenella* (north of Point Conception). Off the west coast of South America, red tides are usually produced by a species of *Gymnodinium*. Most of the red tides in Florida are caused by *Gymnodinium breve* or *Pyrodinium bahamense*. Although red tides appear gradually, they usually disappear rather abruptly, often in association with strong winds and changing sea surface conditions. Apparently this rapid decrease in abundance comes about when the cells lose their flagella and settle to the bottom as resting cysts.

Most red tide organisms manufacture highly toxic substances. When suspension feeders such as mussels and clams eat these protozoa, they store the toxins in their bodies. These shellfish then become toxic to animals that eat them, including humans. Extremely high concentrations of toxic dinoflagellates will even kill the suspension feeders. The dinoflagellate poison blocks the sodium–potassium pump of nerve cells and prevents normal impulse transmission, producing a disease known as paralytic shellfish poisoning (PSP). Extreme cases of PSP result in muscular paralysis and respiratory failure. Over 300 human deaths worldwide have been documented from PSP. *Gymnodinium breve* apparently releases several toxins directly into sea water, and red tides off Florida

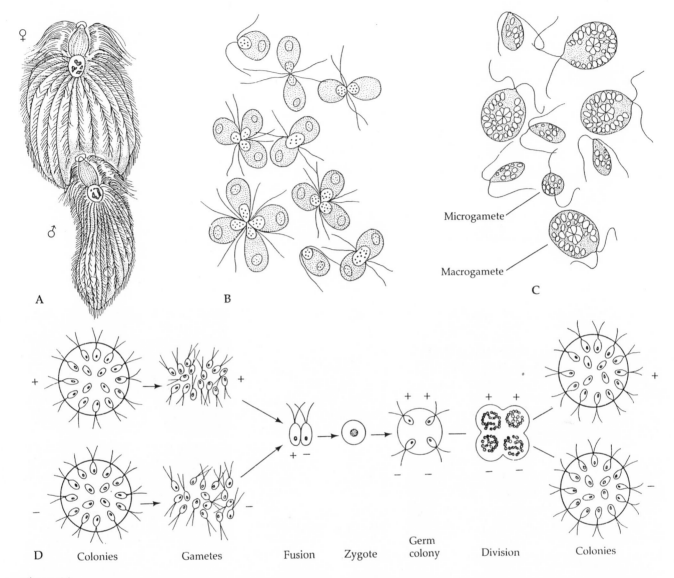

Microgamete

Macrogamete

D   Colonies   Gametes   Fusion   Zygote   Germ   Division   Colonies
                                                                  colony

Figure 14

Sexual reproduction and life cycles of flagellates. A, Mating activity in *Eucomonympha* (Hypermastigida), in which entire individuals act as gametes. A male pierces the cell of a female in order to transfer its nucleus. B, Fusion of biflagellated isogametes of the phytomastigophoran *Dunaliella*. C, Anisogametes of *Chlamydomonas*. D, Sexual reproduction in the volvocid phytoflagellate *Gonium*. The + and − signs indicate a gene-level sexual difference between mating colonies, even though the gametes are anatomically similar (isogamy). Gametes from two colonies of different sex type fuse to produce a zygote that divides twice, thereby yielding a four-celled germ colony. Each of the four cells divides to produce a new daughter colony. As indicated, segregation of the cells by sex type occurs during the division of the zygote, so each daughter colony is either all + or all −. (A redrawn from Grell 1973, after Cleveland 1950; B redrawn from Grell 1973; C redrawn from Grell 1973, after Tschermak-Woess 1959; D after Sleigh 1973b, adapted from Grell in Chen 1967.)

and along the Gulf Coast have resulted in massive fish kills. Ocean spray containing *G. breve* toxins can blow ashore and cause temporary health problems for seaside residents and visitors (skin, eye, and throat problems). Dinoflagellate toxins are among the strongest known poisons. In crystalline form, an aspirin-sized tablet of saxotoxin from *G. catenella* would be enough to kill 35 people or produce moderate poisoning in 350 people.

Subphylum Sarcodina (the amebas)

There are approximately 13,500 described living species of sarcodines. The subphylum is divided into two superclasses, each with a number of classes and orders. Much of the higher classification of sarcodines is based upon the nature of the pseudopodia, presence or absence of a test or other skeletal elements, and various characters associated with repro-

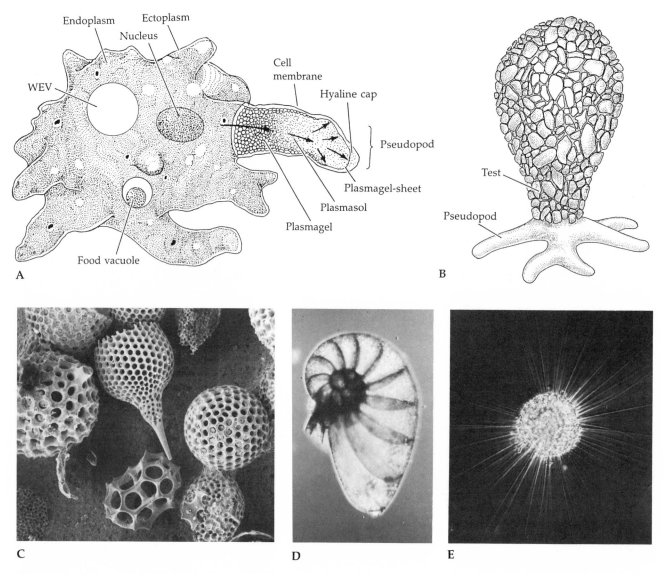

Figure 15

Representative examples of the subphylum Sarcodina. A, Naked ameba (order Amoebida); arrows indicate the direction of cytoplasmic flow during locomotion. B, *Difflugia*, an ameba of the order Arcellinida. C, Skeletons of radiolarians. D, A foraminiferan. E, The heliozoan *Echinosphaerium*. (A–B after Sherman and Sherman 1976; C photo by D.D. Kunkel, U. of Washington/BPS; D photo by J.R. Porter, U. of Georgia/BPS; E photo by L.E. Roth, U. of Tennessee/BPS.)

duction. Several sarcodines are illustrated in Figure 15.

The sarcodines are characterized by an absence of permanent locomotor organelles. They move by the production of temporary pseudopodia or by simple protoplasmic streaming. Flagella do occur in some amebas but are generally restricted to certain developmental or reproductive stages. Sarcodines are typically single-celled, although some are multinucleate.

Others are plasmodial and can fragment into uninucleate cytoplasmic parts that may fuse again into a multinucleate cytoplasmic mass. Amebas can be found in nearly any moist or aquatic habitat: in moist soil or sand, on aquatic vegetation, on wet rocks, in lakes, streams, glacial meltwater, tidepools, bays, estuaries, on the ocean floor, and afloat in the open ocean. Many are ectocommensals on aquatic animals and some are parasites of diatoms, fishes, molluscs, arthropods, and mammals. Most specialists agree that the sarcodines are related to the flagellates and are probably a polyphyletic taxon.

Of critical importance in the classification of amebas are the various types of pseudopodia that occur in this group (Figure 16). Most familiar to the general zoology student are **lobopodia**, which are blunt,

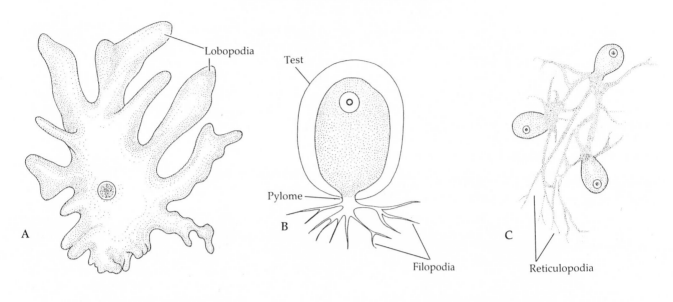

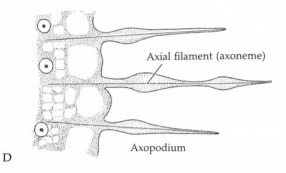

Figure 16

Representative pseudopodia in sarcodines. A, Lobopodia of *Amoeba*. B, Filopodia of *Chlamydophorus*, a shelled ameba. C, Anastomosing rhizopodia (reticulopodia) of *Microgromia*. D, Axopodia on the surface of the heliozoan *Actinosphaerium*. (A–D redrawn from Meglitsch 1972 after several sources.)

rather thick extensions of the cell surface. Long thin pseudopodia are called **filopodia**. Some amebas produce complex branching or anastomosing masses of thin pseudopodia called **reticulopodia** or **rhizopodia**. Lobopodia, filopodia, and reticulopodia are characteristic of various members of the superclass Rhizopoda, whereas members of the superclass Actinopoda usually possess pseudopodia that radiate from the body and are typically held rigid by an internal **axoneme**, or axial filament (Figure 16D). These **axopodia** (or **actinopodia**) assist in food capture and provide sufficient surface area for flotation. The type of pseudopodia, along with other characteristics, are used to describe some of the more common taxa within the Sarcodina in the following classification.

SUBPHYLUM SARCODINA

SUPERCLASS RHIZOPODA: Rhizopodans (Figures 1C; 15A,B,D). With lobopodia, filopodia, or reticulopodia.

CLASS LOBOSEA: With lobopodia or broad filopodia.

ORDER AMOEBIDA: Naked lobosid amebas. Most with lobopodia; lack test or shell; uninucleate; without any flagellated stage; fresh water, marine, and endosymbiotic. (e.g., *Amoeba*, *Chaos*, *Entamoeba*, *Pelomyxa*)

ORDER ARCELLINIDA: Shelled lobosid amebas. Most with lobopodia; with single-chambered shell or test of secreted material or cemented particles; pseudopodia emerge through single opening called the pylome; flagellated gametes occur in some species; fresh water, some associated with bryophytes. (e.g., *Arcella*, *Difflugia*)

CLASS FILOSEA: Filose amebas. With simple, branching, or anastomosing filopodia; with shells (order Gromiida) or naked (order Aconchulinida); shelled forms with one or more apertures; without flagellated stages; marine and fresh water. (e.g., *Gromia*)

CLASS GRANULORETICULOSEA: Reticulose amebas. With reticulopodia; with organic or calcareous test (absent in order Athalamida).

ORDER FORAMINIFERIDA: Foraminiferans. A large and ecologically important group; test siliceous, membranous, particulate, or (most commonly) calcareous; test usually a series of chambers of increasing size with a main aperture in the largest chamber and numerous small perforations through which cytoplasm extends as a thin external layer; complex network of reticulopodia produced from external cytoplasmic layer; marine, benthic, planktonic, and ectocommensal; many with complex life cycles. (e.g., *Cyclorbiculina*, *Elphidium*, *Fusilina*, *Globigerina*, *Homotrema*, *Iridia*, *Rotaliella*, *Tinoporus*, *Tretomphalus*)

SUPERCLASS ACTINOPODA: Actinopodans. (Figures 15C,E). With axopodia and delicately formed skeletons.

CLASS ACANTHAREA: Acanthareans. With a central capsule lacking perforations; skeletal spines of strontium sulfate; mostly planktonic. (e.g., *Acanthocolla*, *Acanthometra*, *Lithoptera*)

CLASSES POLYCYSTINEA AND PHAEODAREA (= RADIOLARIA): Radiolarians. Central capsule perforated; skeleton siliceous and spherically symmetrical; extracapsular cytoplasm with characteristic foamy appearance, called the calymma; marine, planktonic. (e.g., *Coccodiscus*, *Collosphaera*, *Eucornis*)

CLASS HELIOZOA: Heliozoans. Without a central capsule; skeleton siliceous or rarely organic, usually spherical; most are planktonic; some are benthic and sessile; nearly all live in fresh water. (e.g., *Actinosphaerium*, *Clathrulina*, *Gymnosphaera*)

Support and locomotion. Members of the subphylum Sarcodina may be supported only by the cell membrane (Amoebida), or they may possess some sort of test or skeletal framework (Arcellinida, Foraminiferida, Radiolaria). The naked forms are typically asymmetrical, because they change shape as they form pseudopodia. Those with tests or skeletons generally have fixed shapes determined by the architecture of the supporting elements. Many approach perfect spherical symmetry; others are radially or bilaterally symmetrical.

The skeletons of amebas may be composed of particulate material either gathered from the environment (as in *Difflugia* and some foraminiferans) or secreted by the cell (as in most foraminiferans, *Arcella*, and radiolarians). A few species secrete organic skeletons, frequently of chitin, but most produce skeletons of calcium carbonate (foraminiferans) or siliceous compounds (radiolarians). The skeletons of arcellinids are external and single-chambered. They have a single opening called the **pylome** through which the pseudopodia extend (Figure 15B).

Foraminiferan tests are usually constructed as a series of chambers of increasing size, with a main aperture in the largest chamber. Much of the rest of the shell typically bears numerous tiny perforations through which the cytoplasm extends as a thin sheet around the outside of the test. Certain planktonic foraminiferans occur in such high numbers that the tests of dead individuals provide a major contribution to the sediments of ocean basins. In some parts of the world, these sediments—called foraminiferan oozes—are many meters thick. Such sediments are restricted to depths shallower than about 3000–4000 m, however, because the $CaCO_3$ dissolves under high pressure. Foraminiferan tests not only are very abundant in recent and fossil deposits but also are extremely durable. On the island of Bali, the tests of one species are mined and used as gravel in walks and roads. Much of the world's chalk, limestone, and marble is composed largely of foraminiferan tests or the residual calcareous material derived from the tests. Most of the stones used to build the great pyramids of Egypt are foraminiferan in origin.

Most radiolarians secrete skeletons of siliceous compounds. These skeletons vary greatly in construction and ornamentation, and frequently bear radiating spines that aid in flotation (Figure 15C). The core of the skeletal framework is a perforated **central capsule**. The living part of the animal is housed partly within this capsule as capsular cytoplasm and partly outside the capsule as extracapsular cytoplasm. The extracapsular cytoplasm has a rather foamy appearance and is called the **calymma**. This region of the cell contains abundant vacuoles, which house oil droplets and other low-density fluids and aid in flotation. When sea surface conditions become rough and potentially dangerous to these delicate protozoa, the calymma expels some of its contents and the animal sinks to calmer depths. Eventually the cell replaces the oils and other fluids, and the organism rises toward the surface again. Like the foraminiferan tests, radiolarian skeletons persist after the death of the animals, and they accumulate as thick deposits on the floors of deep ocean basins (usually between 3,500 and 10,000 m in depth). These siliceous skeletons do not dissolve under great pressure.

We generally think of the pseudopodia as locomotor structures. While such is frequently the case, the pseudopodia of many sacrodines (e.g., radiolarians and heliozoans) serve primarily as flotation devices and food-gathering structures. Furthermore, some of the naked amebas move by wavelike cytoplasmic flowing without the formation of distinct pseudopodia. Such amebas are referred to as "limax" (sluglike) forms (Figure 17B). Others (e.g., *Thecamoeba*) employ yet another method of nonpseudopodial movement. In these amebas, the cell surface and outer cytoplasm roll much like the tread on a tractor or tank, the leading surface adhering temporarily to the substratum as the animal progresses (Figure 17C).

Nonetheless, most of the amebas that live in contact with a substratum do employ pseudopodia to accomplish locomotion. As described in Chapter 3, the physical and chemical processes involved in pseudopodial or **ameboid movement** are still not fully understood. It is likely that more than one method of pseudopod formation occurs among the

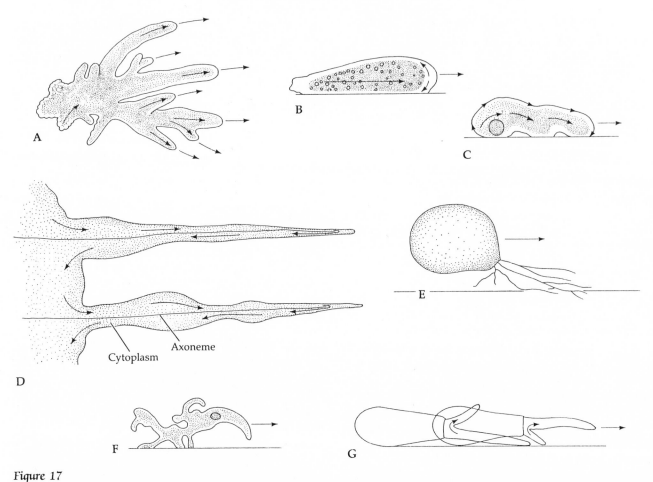

Figure 17

Locomotion in sarcodines. A, "Typical" ameboid movement by lobopodia in *Amoeba proteus*. B, Creeping, "limax" form of movement. C, Rolling, treadlike movement. D, Cytoplasmic streaming within axopodia. E, Filopodial creeping in *Chlamydophorus*. F, "Walking" locomotion of certain naked amebas. G, "Bipedal-stepping" in *Difflugia*. See text for additional details. (A,B,C,F,G after Grell 1973, adapted from various sources.)

different amebas; certainly the mechanics involved in the use of pseudopodia vary greatly (Figure 17).

Recall the typical differentiation of the cytoplasm into **ectoplasm** (plasmagel) and **endoplasm** (plasmasol), the latter being much more fluid than the former. The formation of broad lobopodia as well as the "limax" form of movement results from the streaming of the inner plasmasol into areas where the constraints of the plasmagel have been temporarily relieved. In contrast, the formation of filopodia and thin axopodia and reticulopodia typically does not involve sol–gel interactions, but simply the rapid streaming of the cytoplasm to form single or branching pseudopodial extensions. In many amebas that form axopodia (e.g., heliozoans) the cytoplasm streams outward from the core of the cell along the

rigid supportive axonemes (Figure 17D). The axonemes are made up of a group of microtubules that originate deep in the cell's cytoplasm.

While many amebas move by "flowing" into their pseudopodia or by "creeping" with numerous filopodia or reticulopodia, some engage in more bizarre methods of getting from one place to another. Some hold their bodies off the substratum by extending pseudopodia downward; leading pseudopodia are then produced and extended forward sequentially, pulling the animal along in a sort of "multilegged" walking fashion. Some of the shelled amebas (e.g., *Difflugia*) that possess a single pylome extend two pseudopodia through the aperture. By alternately extending and retracting these pseudopodia, the animal "steps" forward. The reticulopodia of benthic foraminiferans are known to form and retract rapidly. During locomotion, one pseudopodium is extended and used to "pull" the animal along, trailing the other pseudopodium behind the cell. When pelagic foraminiferans are placed in contact with a substratum the pseudopodia (normally used for feeding) perform like those of benthic species.

Nutrition. Like all other heterotrophic creatures, sarcodines ingest organic materials from their environment. Much of our knowledge of ingestion and digestion in amebas comes from the careful work of Cicily Chapman-Andresen, who has contributed a wealth of information on the nutritional biology of this group of protozoa (see especially Chapman-Andresen 1973).

While there is little doubt that amebas take up dissolved organics directly across the cell membrane, the most common mechanisms of ingestion are pinocytosis and phagocytosis (Figure 18). The size of the resulting food vacuoles varies greatly, depending primarily on the size of the food material ingested. Generally, ingestion can occur anywhere on the surface of the body, there being no distinct cytostome. Most sarcodines are carnivores and are frequently predaceous. Some, such as *Pelomyxa*, inhabit soil or mud and are predominantly herbivorous, but they are known to ingest nearly any sort of organic matter in their environment. As explained earlier, a food vacuole forms from an invagination in the cell surface—sometimes called a food cup—that pinches off and drops inward. This process—sometimes called **endocytosis**—occurs in response to some stimulus at the interface between the cell membrane and the environmental medium. Vacuole formation in amebas may be induced by either mechanical or chemical

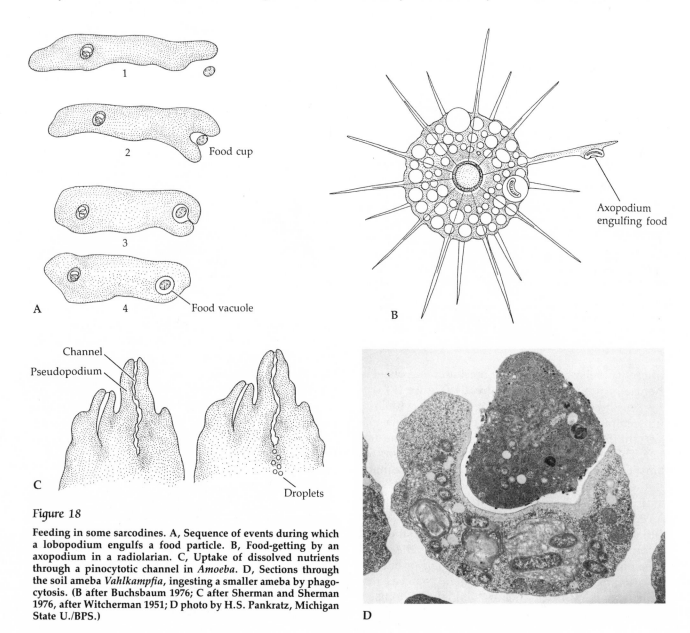

Figure 18

Feeding in some sarcodines. A, Sequence of events during which a lobopodium engulfs a food particle. B, Food-getting by an axopodium in a radiolarian. C, Uptake of dissolved nutrients through a pinocytotic channel in *Amoeba*. D, Sections through the soil ameba *Vahlkampfia*, ingesting a smaller ameba by phagocytosis. (B after Buchsbaum 1976; C after Sherman and Sherman 1976, after Witcherman 1951; D photo by H.S. Pankratz, Michigan State U./BPS.)

stimuli; even nonfood items may be incorporated into food vacuoles, but they are soon egested.

Not only the size of a food item, but also the amount of water taken in during feeding determine the size of the food vacuole. Frequently the pseudopodia that form the food cup do not actually contact the food item; thus, a packet of the environmental medium is taken in with the food. In other cases, the walls of the forming vacuole press closely against the food material; thus, little water is included in the vacuole.

Food vacuoles move about the cytoplasm and sometimes coalesce. If live prey have been ingested, they generally die within a few minutes, presumably from suffocation. Undigested material that remains within the vacuole eventually is expelled from the cell when the vacuole wall reincorporates into the cell membrane. In most amebas this process of cell defecation may occur anywhere on the body, but in some active forms it tends to take place at or near the trailing end of the moving cell.

Feeding in sarcodines with skeletal elements varies with the form of the skeleton and the type of pseudopodia. Those amebas with a relatively large single pylome, such as *Arcella* and *Difflugia*, feed much as described above. By extending lobopodia through the pylome, they engulf food in typical vacuoles. Sarcodines that possess reticulopodia or axopodia (foraminiferans, radiolarians, and heliozoans) use the pseudopodia as traps. Potential prey items adhere to the mucous covering of these extended pseudopodia. Small prey are engulfed in food vacuoles directly, whereas large prey may be partially digested extracellularly by the action of secretory lysosomes in the mucous coating. The extracellular food is drawn toward the cell body by cytoplasmic streaming, finally enclosed within the food vacuoles, and completely digested in the central portion of the cell, the exact location varying among different groups. All major categories of digestive enzymes have been identified among the amebas, including cellulase in certain forms.

Numerous members of the order Amoebida are endosymbiotic, and many of these are considered to be parasites. They occur most commonly in arthropods, annelids, and vertebrates (including people). Under severe stress, certain symbiotic gut amebas (e.g., *Entamoeba coli*) can increase to abnormally high numbers and cause temporary mild gastrointestinal distress. Of the numerous amebas known from the human gut, however, only *Entamoeba histolytica* is considered a true pathogen. This species causes amoebic dysentery, an intestinal disorder resulting

from destruction of cells lining the gut. The parasite is usually ingested in its cyst stage. Emergence of individuals in the active (motile) stage, called ***trophozoites***, takes place quickly, and it is these individuals that release the histolytic enzymes that break down the epithelium of the large intestine and rectum. In severe cases, motile amebas penetrate the intestinal wall and are carried in the bloodstream to the liver; there they can produce massive lesions. Hepatic abscesses are occasionally fatal; and in rare cases *E. histolytica* may even invade the lungs and brain. The severity of the disease varies with the level of infection, the strain of *E. histolytica* acquired, and the overall physical condition of the host. Furthermore, partial or complete resistance to amebiasis is acquired by individuals who recover from the disease (due to antibody production and cellular immune responses). Most animals are largely immune to amebiasis, and humans seem to be the main reservoir of this pathogen.

Reproduction and some life cycles. Several asexual and sexual processes are known among the Sarcodina. Simple binary fission is the most common form of asexual reproduction, differing only in minor details among the different groups of amebas (Figure 19). In naked amebas, nuclear division occurs first and cytoplasmic division follows. During cytoplasmic division, the two potential daughter cells form locomotor pseudopodia and pull away from each other. In the heliozoans, binary fission occurs along any plane through the body; in the radiolarians and various shelled forms, however, division occurs along planes predetermined by body symmetry and skeletal arrangement. In species with an external test, the shell itself may divide more-or-less equally in conjunction with the formation of daughter cells (e.g., *Pamphagus*); or, as occurs more frequently, it may be retained by one daughter cell, the other producing a new shell (e.g., *Arcella*). The relative density and rigidity of the test determine which process occurs.

Multiple fission is also known among sarcodines (Figure 19D). Certain endosymbiotic naked amebas, including *Entamoeba histolytica*, produce cysts in which multiple fission takes place; and heliozoans, radiolarians, and foraminiferans commonly engage in various forms of multiple fission.

Different life-cycle patterns are known among the sarcodines, but some are still poorly understood. Sexual reproduction is fully documented only in the foraminiferans, but related processes are known in other groups. Some naked amebas undergo a phenomenon known as **hologamy**, wherein two entire individuals

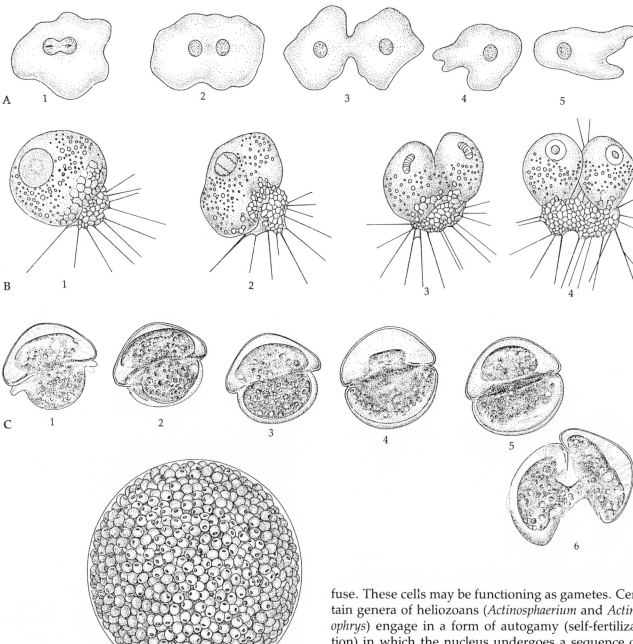

Figure 19

Asexual reproduction in the sarcodines. A, Simple binary fission in *Amoeba*. B, Binary fission in the shelled ameba *Pamphagus*; the test is partitioned more-or-less equally between the two daughter cells. C, Binary fission in *Arcella*; the parent test is retained by one daughter cell, and a new test is produced by the other daughter cell. D, Mass of swarmers produced by multiple fission within the central capsule of the radiolarian *Thalassophysa*. (B redrawn from Grell 1973; C redrawn from a series of photographs in Grell 1973, from film E-1643 by Netzel and Heunert 1971; D redrawn from Grell 1973, after Hollande and Enjumet 1953.)

fuse. These cells may be functioning as gametes. Certain genera of heliozoans (*Actinosphaerium* and *Actinophrys*) engage in a form of autogamy (self-fertilization) in which the nucleus undergoes a sequence of meiotic divisions followed by fusion of two of the haploid daughter nuclei (Figure 20). Sexual processes are suspected in the radiolarians, because many produce free-swimming flagellated cells; but, again, most of the details are yet to be described.

The life cycles of foraminiferans are frequently complex, and many remain incompletely understood. These cycles often involve an alternation of sexual and asexual phases (Figure 21). It is not uncommon to find individuals of the same foraminiferan species differing greatly in size and shape at different phases of the life cycle. The size difference is generally determined by the size of the initial shell

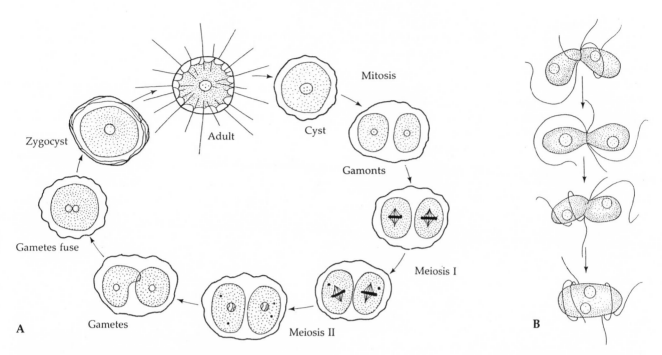

Figure 20

A, Autogamy in the heliozoan *Actinophrys*. The adult enters a cyst stage and undergoes mitosis to produce a pair of gamonts. The nucleus of each gamont undergoes meiosis, but only one haploid nucleus survives in each cell. The gametes (and their haploid nuclei) fuse to produce an encysted zygote, which eventually grows into a new individual. B, Fusion of isogametes of the foraminiferan *Iridia lucida*. (A after Sleigh 1973b; B redrawn from Grell 1973.)

chamber (the **proloculum**) produced following a particular life-cycle event. Often the proloculum formed following asexual processes is significantly larger than one formed after syngamy. Individuals with large prolocula are called the **macro-** or **megaspheric** generation, individuals with small prolocula, the **microspheric** generation. During the sexual phase of the life cycle, the haploid individuals are known as **gamonts**; they undergo repeated divisions to produce and release bi- or triflagellated isogametes, which pair and fuse to form the asexual individuals. Each asexual diploid individual (called an **agamont**) undergoes meiosis and produces haploid gamonts—the

Figure 21

Life cycle of the foraminiferan *Tretomphalus bulloides*. 1, The settled zygote is shell-less and ameboid. 2, The animal grows and matures as an agamont, which (3) asexually produces young gamonts. Each mature gamont (4) accumulates particles of detritus (5–6) to produce a flotation chamber (7). 8, The gamont floats to the surface and produces and releases gametes (9), which fuse to produce a swimming zygote (10, 11, 12). (Redrawn from Grell 1973, after Myers 1943.)

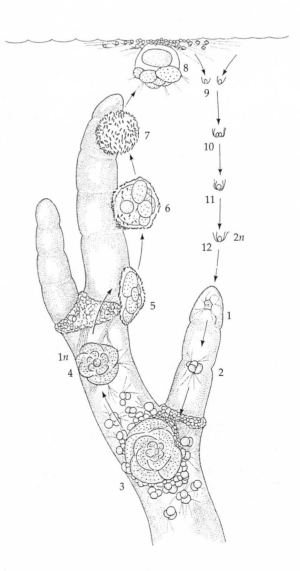

sexual individuals. The means of return to the diploid condition varies. In many foraminiferans (*Elphidium*, *Tretomphalus*, *Iridia*, and others), flagellated gametes are produced and released; fertilization occurs free in the sea water to produce a young agamont. In others, such as *Glabratella*, two or more gamonts come together and temporarily attach to one another. The gametes, which may be flagellated or ameboid, fuse within the chambers of the paired tests. The shells eventually separate, releasing the newly formed agamonts. True autogamy occurs in *Rotaliella*: each gamont produces gametes that pair and fuse within a single test; the zygote is then released as an agamont.

Subphylum Opalinata

While no longer considered mastigophorans, we mention the opalinids at this point for purposes of comparison (Figure 1D). Once classified as protociliates, then as an order within the zoomastigophorans, their placement here as a separate subphylum should not be taken as the final word on the matter; it is meant merely to draw attention to the enigmatic nature of these creatures. The numerous longitudinal rows of cilia-like locomotor structures appear to be secondarily shortened flagella rather than true cilia. During asexual reproduction, the fission plane par-

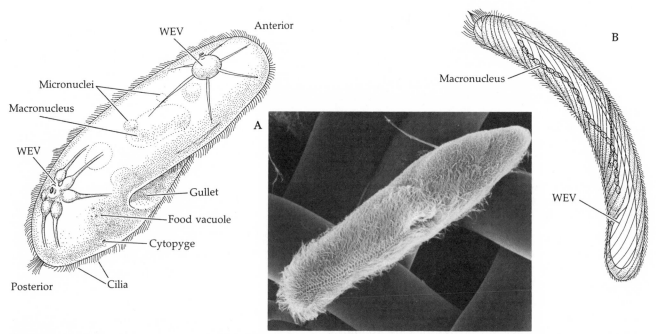

Figure 22

Representative ciliates and their anatomy. A, Photo and drawing of the familiar hymenostome, *Paramecium*. B, *Spirostomum*, a large (up to 1.5 mm) freshwater spirotrich. C, Photo and drawing of *Stentor*, another spirotrich, this one with an attachment structure, or holdfast. D, The peritrich *Cothurnia*, a sessile colonial ciliate with a lorica. E, The hypotrich *Euplotes*, in which groups of cilia fuse to form thick cirri. F, *Eudiplodinium*, a rumen symbiont in bovines; note the greatly reduced somatic ciliature. G, The tintinnid *Tintinnopsis* within its test (lorica); note the loss of somatic ciliature and the well developed adoral zone membranelles and paroral membrane. H, Photo of *Tintinnopsis*. I, *Paracineta*, a suctorian with numerous capitate tentacles and no cilia. (A drawing after Jurand and Selman 1969; photo by D.D. Kunkel, U. of Washington/BPS; B after various sources; C photo by J.J. Paulin, U. of Georgia/BPS; D photo by P.W. Johnson and J. McN. Sieburth/BPS; E,F after various sources; G after Sherman and Sherman 1976; H,I photos by P.W. Johnson and J. McN. Sieburth/BPS.)

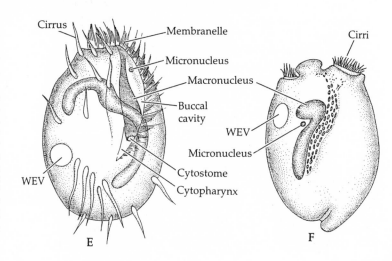

allels these rows; thus it is longitudinal (as it is in flagellates) rather than transverse (as it is in ciliates). Some opalinids are binucleate, others multinucleate, but all are homokaryotic (i.e., the nuclei are all identical).

Opalinids are all endosymbiotic in the hindgut of anurans (frogs and toads), where they ingest food anywhere on their body surface. As with other sarcomastigophorans, sexual reproduction is by syngamy and asexual reproduction is by binary fission and **plasmotomy**, the latter involving cytoplasmic divisions that produce multinucleate offspring. *Opalina* and *Protopalina* are two common genera.

Phylum Ciliophora (the ciliates)

There are nearly 8,000 described species of ciliates. Ciliates are extremely common in all aquatic environments. Both sessile and errant types are known; many are ecto- or endosymbionts, including a number of parasitic forms. The ciliates shown in Figure 22 illustrate the variety of body forms within this large and complex group of protozoa. The ciliates are characterized by several features (Box 3), one of which is the presence of cilia. The phylum is divided into three classes on the basis of the nature of the

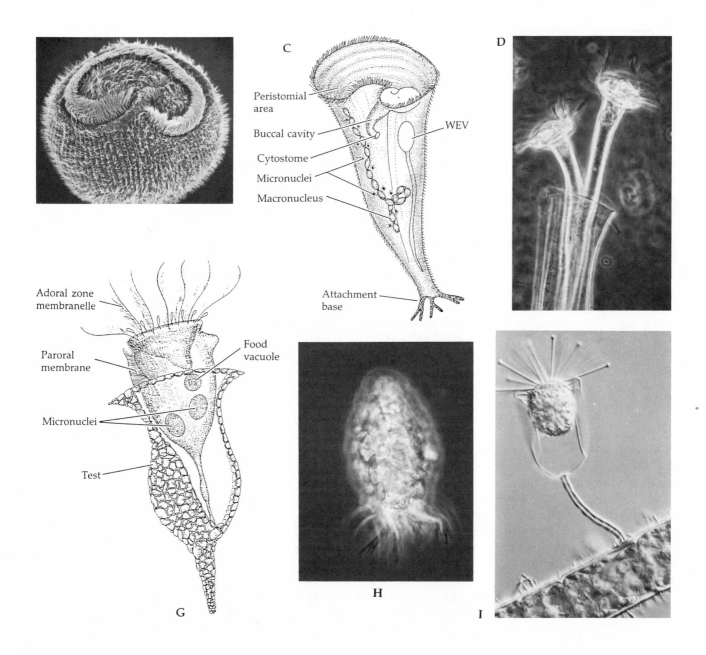

> **Box Three**
> ## Characteristics of the Phylum Ciliophora
>
> 1. Possess cilia for locomotion and for the generation of feeding currents
> 2. With a relatively rigid pellicle and more-or-less fixed cell shape
> 3. With distinct cytostome
> 4. With dimorphic nuclei, typically a larger macronucleus and one or more smaller micronuclei

ciliature and the feeding apparatus. Because ciliates possess a relatively rigid pellicle, they have a more-or-less fixed shape. The pellicle of ciliates consists of tightly packed, double-walled chambers called **alveoli**, whose arrangement helps maintain the structural integrity of the body wall (Figure 23C). Most species occur as single cells, but branching and linear colonies are known in several groups. Ciliates are further characterized by the presence of a distinct cytostome, which is often associated with some sort of gullet or other depression to facilitate feeding. Ciliates also have two types of nuclei in each cell. The larger type—the **macronucleus**—has a largely vegetative function, controlling the general operation of the cell. The macronucleus is usually hyperpolyploid (containing many sets of chromosomes) and may be compact, ribbon-like, beaded, or branched. The smaller type—the **micronucleus**—has a reproductive function, synthesizing the DNA associated with various sexual processes. It is usually diploid and less than 5 μm in diameter.

The details of the ciliature of these protozoa are of utmost importance in their classification and, hence, in phylogenetic speculation. The ciliature of the cell can be grouped into two functional and two structural categories. Cilia associated with the cytostome and the surrounding feeding area constitute the **oral ciliature**, whereas cilia of the general body surface constitute the **somatic ciliature**. The cilia in both of these categories may be single, or **simple cilia**, or they may be fused together to form **compound ciliature** (Figures 23A and B). All of the cilia of these protozoa are interconnected by a complicated latticework of microorganelles that lies beneath the pellicle; these organelles constitute the **infraciliature** (Figure 23C). Ciliatologists have developed a detailed and complicated terminology for talking about their favorite creatures. This special language reaches almost overwhelming proportions in matters of ciliature, and we present here only a necessary minimum of new words in order to adequately describe these animals. Beyond this, we offer the reference list at the end of this chapter, especially the extensive and illustrated glossary in J. O. Corliss's *The Ciliated Protozoa* (1979).

The infraciliature serves as a sort of anchoring network for the cilia themselves. It is tempting to compare the infraciliature to a kind of "nervous system" coordinating the action of the cilia, but evidence to date suggests this is not the case. Each cilium arises from and is subsequently associated with a **kinetosome** (basal body). The cilia and their kinetosomes are generally arranged in regular longitudinal rows, the kinetosomes being connected by longitudinal microfibrils called interciliary fibrils or **kinetodesmata** (singular, kinetodesma). Each row of kinetosomes and their associated kinetodesmata make up a **kinety**. In addition to the longitudinally oriented kineties, most ciliates possess a variety of transverse interciliary fibrils between kinetosomes; these fibrils apparently anchor the cilia and the rest of the infraciliature complex. Although the infraciliature is most well developed and clearly depicted in those ciliates that bear more-or-less uniform somatic ciliature, it persists to varying degrees even in those with modified or reduced surface ciliation.

The **adoral zone membranelles**, or simply the **AZM** (see Figure 26) typically comprise a series of membrane-like organelles formed of fused cilia and associated with the area of the cytostome or surrounding buccal cavity. In addition, certain ciliates possess another membrane-like structure of fused cilia, called the **paroral membrane**.

The classification given below includes a brief characterization of some of the common taxa within the Ciliophora.

PHYLUM CILIOPHORA

CLASS KINETOFRAGMINOPHOREA: Kinetofragminophoreans (Figures 22I; 26A,B,C). Oral and body cilia similar; cytostome generally at the cell tip (or on the underside); compound ciliature usually absent.

SUBCLASS GYMNOSTOMATIA: Gymnostome ciliates. An apical or slightly subapical cytostome at or near the cell surface, not within an obvious depression; somatic ciliature generally quite uniform over the whole body; com-

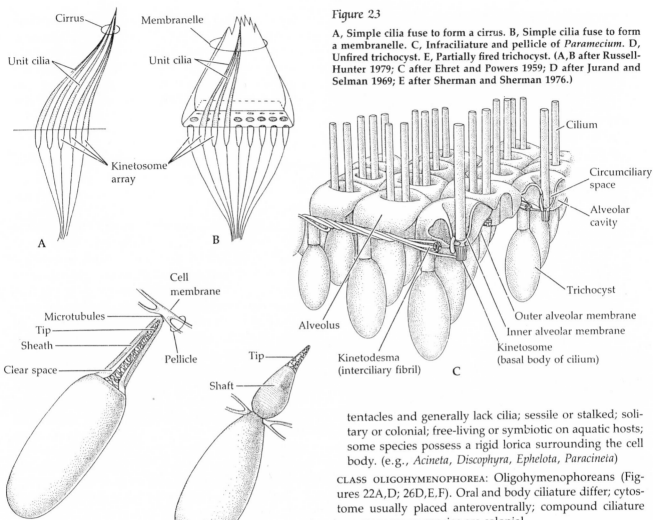

Figure 23

A, Simple cilia fuse to form a cirrus. B, Simple cilia fuse to form a membranelle. C, Infraciliature and pellicle of *Paramecium*. D, Unfired trichocyst. E, Partially fired trichocyst. (A,B after Russell-Hunter 1979; C after Ehret and Powers 1959; D after Jurand and Selman 1969; E after Sherman and Sherman 1976.)

monly possess toxicysts. Four orders diagnosed in part by location and shape of cytostome (oval, round, slitlike, and so on). (e.g., *Amphileptus, Coleps, Didinium, Geleia, Loxodes*)

SUBCLASS VESTIBULIFERIA: Vestibuliferians. Usually with apical or subapical cytostome, set in a depression or vestibulum, lined with relatively complex ciliature derived from somatic rather than oral cilia. Three orders, distinguished in part on the nature (including reduction) of the somatic ciliature. (e.g., *Balantidium, Colpoda, Entodinium, Isotricha, Raabena, Woodruffia*)

SUBCLASS HYPOSTOMATIA: Hypostome ciliates. Midventral cytostome frequently depressed into an atrium lined with specialized cilia. Free-living and many ecto- or endosymbionts of invertebrates. Some symbiotic forms possess a distinctive adhesive organelle. (e.g., *Chlamydodon, Collinia, Dysteria, Gymnodinioides, Heliochona, Isochona, Microthorax, Nassula*)

SUBCLASS SUCTORIA: Suctorians. Adults bear suctorial tentacles and generally lack cilia; sessile or stalked; solitary or colonial; free-living or symbiotic on aquatic hosts; some species possess a rigid lorica surrounding the cell body. (e.g., *Acineta, Discophyra, Ephelota, Paracineta*)

CLASS OLIGOHYMENOPHOREA: Oligohymenophoreans (Figures 22A,D; 26D,E,F). Oral and body ciliature differ; cytostome usually placed anteroventrally; compound ciliature common; many species are colonial.

SUBCLASS HYMENOSTOMATIA: Hymenostome ciliates. Includes many of the familiar, relatively large, freshwater ciliates; most are solitary and motile; free-living or symbiotic; oral region usually well developed, with cytostome located ventrally either near the body midpoint or somewhat anteriorly; buccal region frequently bears distinct adoral membranelles and a paroral membrane. (e.g., *Colpidium, Paramecium, Tetrahymena*)

SUBCLASS PERITRICHA: Peritrichs. With well developed oral ciliature and generally reduced somatic ciliature; with distinct adoral membranelles and paroral membrane; many are sessile or sedentary, with bell-shaped bodies, attached to substratum by contractile stalk; others are motile parasites with conical or cylindrical bodies and a single band of somatic cilia. (e.g., *Trichodina, Urceolaria, Vaginicola, Vorticella*)

CLASS POLYHYMENOPHOREA: (With one subclass, SPIROTRICHA.) Spirotrichs (Figures 22B,C,E,G,H). With well developed adoral zone membranelles (AZM); somatic ciliature highly variable, modified in some as fused bundles of cilia called cirri; cytostome set within a deep buccal cavity.

ORDER HETEROTRICHIDA: Heterotrichs. With strong somatic ciliature and elongate multilobed macronucleus; body form varies (conical, trumpet-shaped, vermiform, and other shapes); free-living or parasitic. (e.g., *Blepharisma*, *Spirostomum*, *Stentor*)

ORDER ODONTOSTOMATIDA: Odontostomatids. Somatic ciliature reduced; body usually small and often laterally flattened; most live in fresh water among decaying organic matter; a few are marine. (e.g., *Mylestoma*, *Saprodinium*)

ORDER OLIGOTRICHA: Oligotrichs. With reduced somatic ciliature and well developed AZM; mostly marine, pelagic; includes the tintinnids, which bear a conical or cylindrical lorica. (e.g., *Dictyocysta*, *Tintinnopsis*)

ORDER HYPOTRICHIDA: Hypotrichs. Somatic and oral ciliature (including the AZM) largely of stiff compound cilia, with ventral locomotor cirri; mostly benthic, motile; present in all aquatic environments; a few are ectosymbiotic on invertebrates. (e.g., *Euplotes*, *Hypotrichidium*, *Oxytricha*, *Urostyla*)

Support and locomotion. The generally fixed cell shape of ciliates is maintained by the semirigid pellicle. A few types (e.g., tintinnids) possess external skeletons, or **loricae**. The locomotor organelles of ciliates are, of course, cilia. Their structural similarities to flagella are well known, and many workers treat cilia simply as specialized flagella; but ciliates do not move like flagellates. The differences are due in large part to the facts that cilia are much shorter than flagella, that cilia are much more numerous and densely distributed than flagella, and that the patterns of ciliation on the body are extremely varied and thus allow a range of diverse locomotor strategies not possible with just one or a few flagella.

As discussed in Chapter 3, each individual cilium undergoes an effective (power) stroke and a recovery stroke as it beats. The cilium does not move on a single plane, but describes something like a distorted cone as it beats (Figures 24A and B). This counterclockwise rotation (when viewed from the outside) is an intrinsic feature of cilia and occurs even in isolated cilia. The beating of a ciliary field occurs in metachronal waves that pass over the body surface (Figure 24C). The coordination of these waves is apparently due largely to hydrodynamic effects generated as each cilium moves. Microdisturbances created in the water by the action of one cilium stimulate movement in the neighboring cilium, and so on over the cell surface. As mentioned earlier, the subpellicular infraciliature probably serves as an anchoring device, not as a network for coordinating ciliary movement.

The direction of the effective stroke of the cilia, coupled with the direction of the metachronal waves, results in various movement patterns characteristic of different ciliates (see Figure 24D–G). In a few types of ciliates, both the effective strokes and the metachronal waves move parallel to the longitudinal axis of the body, from anterior to posterior. This condition is called **symplectic metachrony**. In some cases the effective stroke is from anterior to posterior but the metachronal waves move from posterior to anterior, a condition called **antiplectic metachrony**. Most ciliates, however, exhibit **diaplectic metachrony**, in which the effective strokes of the cilia occur at a right angle to the metachronal waves, so the animal spirals on its axis as it swims.

A good deal of variation on these basic patterns is known among ciliates. Members of the familiar genus *Paramecium* have been well studied and engage in what might be considered an intermediate pattern of locomotion (**oblique metachrony**). As a paramecium moves forward, the metachronal waves pass generally from posterior to anterior but sweep at an angle toward the right of the body axis. The effective stroke, which is anterior to posterior, is also displaced at a similar angle, so each cilium beats from anterior-left to posterior-right. As a result of this complex situation, the animal progresses along a line describing a left-handed helix, while the body's rotation on its axis keeps the oral area directed toward the center of the helix (Figure 24H).

Many hypotrichs with cirri have assumed a crawling or nearly walking mode of locomotion (Figure 24I). Sometimes the cirri move in metachronal waves; at other times, they move singly or in independent groups. This variability suggests a high level of coordination. Many ciliates (e.g., *Paramecium*, *Didinium*) can vary direction of ciliary beating and metachronal waves. In such forms, complete reversal of the body's direction of movement is possible by simply reversing the ciliary beat and wave directions.

Perhaps more than any other protozoan group, the ciliates have been studied for their complex locomotor behavior. *Paramecium*, a popular laboratory animal, has received most of the attention of protozoological behaviorists. When a swimming paramecium encounters a mechanical or chemical environmental stimulus of sufficient intensity, it begins a series of rather intricate response activities. The animal first initiates a reversal of movement, effectively backing away from the source of the stimulus. Then, while the posterior end of the body remains more-or-less stationary, the anterior end swings around in a circle. This action is appropriately called the cone-swinging phase. The paramecium then proceeds forward again, typically along a new pathway. Much of

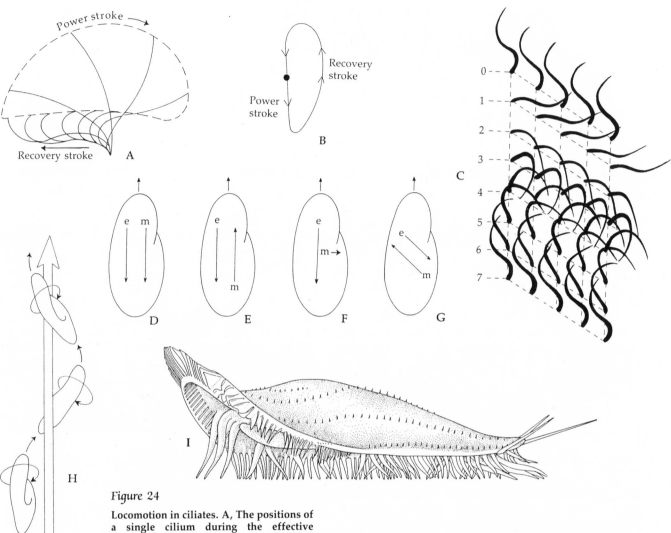

Figure 24

Locomotion in ciliates. A, The positions of a single cilium during the effective (power) and the recovery strokes. B, Flattened oval described by the tip of a beating cilium. C, A ciliary field fixed during metachronal beating. Rows 0–2 are engaged in the power stroke, whereas rows 3–7 are at various stages of the recovery stroke. D, Symplectic metachrony. (e, direction of effective stroke; m, direction of metachronal wave). E, Antiplectic metachrony. F, Diaplectic metachrony. G, Oblique metachrony seen in *Paramecium*. H, Helical pattern of forward movement of *Paramecium*. I, *Stylonychia* uses cirri for "walking." (A–B after Sleigh 1973a; C redrawn from Grell 1973, after Parducz 1954; I redrawn from Grell 1973, original courtesy of H. Machemer to Grell.)

the literature refers to this behavior in terms of simple "trial and error," but the situation cannot be so easily explained. The response pattern is not constant, because the cone-swinging phase may not always occur; sometimes the animal simply changes direction in one movement and swims forward again. Furthermore, the cone-swinging phase occurs even in the absence of recognizable stimuli and thus may be regarded as a phenomenon of "normal" locomotion. Recent studies suggest that paramecium swimming behavior is governed by the cell's membrane potential. When the membrane is "at rest," the cilia beat posteriorly and the cell swims forward. When the membrane becomes depolarized, the cilia beat in a reverse direction (ciliary reversal) and the cell backs up. In *Paramecium*, certain genetic mutations are known to result in abnormal behavior that is characterized by prolonged periods of continuous ciliary reversal (the so-called "paranoiac paramecium").

Many sessile ciliates are capable of movement in response to stimuli. The attachment stalk of many peritrichs (e.g., *Vorticella*) contains contractile myonemes that serve to pull the cell body against the substratum.

Nutrition. Most ciliates are holozoic heterotrophs, but as a group they display a great variety of feeding methods. Most of these strategies may be characterized by one or the other of Grell's (1973) colorful terms, "gulpers" (primarily motile stalkers) and "swirlers" (primarily lurking predators). Some, however, do not fit either of these descriptive terms, as we will discuss.

The particular method of food-getting by any ciliate is largely dependent on the anatomy of the feeding area. Although most ciliates have a cytostome, this "cell mouth" has been secondarily lost in some groups (e.g., astomatids). Food vacuoles are formed at the cytostome and then are circulated through the cytoplasm as digestion occurs (Figure 25). The cytostome and its surrounding structures effectively localize the activity of feeding to a particular region of the cell. The position of the cytostome and the relative complexity of the oral ciliature vary and are considered to be of phylogenetic significance. The presumed primitive condition is seen in some gymnostomes; in these animals the cytostome is lo-

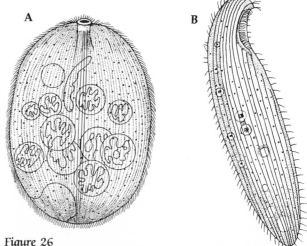

Figure 26

Trends and variations in the elaboration of the oral area among ciliates. A, *Prorodon teres* (Gymnostomatia), presumably has retained the ancestral condition of an anterior cytostome. The orifice is flush with body surface, and there is no differentiation between oral and somatic ciliature. B, *Loxodes rostrum* (Gymnostomatia) has a subterminal cytostome located in a superficial depression. C, *Colpoda cucullus* (Vestibuliferia) has a cytostome almost at the midventral position and set within a vestibule (see inset); the distinct cilia of the oral area are derived directly from somatic ciliature rather than from special oral ciliature. D, *Pleuronema marinum* (Hymenostomatia) has a midventral cytostome and well developed AZM and a paroral membrane. E, *Paramecium caudatum* (Hymenostomatia) has a broad peristomial depression and cytopharynx. F, *Vorticella nebulifera* (Peritricha), a sessile colonial ciliate with a deep buccal cavity, has well developed oral ciliature, but it has lost its somatic ciliature completely. (All redrawn from Grell 1973, after various sources.)

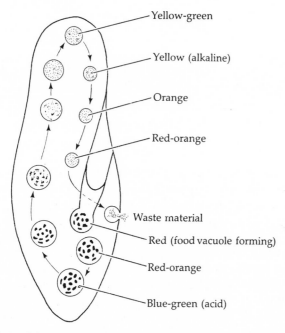

Figure 25

Formation of and digestion within a food vacuole in *Paramecium caudatum* (schematic). The sequence of digestive events may be followed by staining yeast cells with Congo red and allowing the stained cells to be ingested by the organism. The changes in color from red to red-orange to blue-green reflect the change to an acid condition within the food vacuole and thus the initial stages of the digestive process. The change back to red-orange occurs as the vacuole subsequently becomes more alkaline. The pattern of movement of the food vacuole (arrows) is typical of this animal and is often termed cyclosis. (After various sources.)

cated at or near the anterior end of the body, usually almost flush with the cell surface and without distinct oral infraciliature. The evolutionary trends from this ancestral state among other ciliates have been to move the cytostome toward a midventral position, to form various sorts of depressions in the cell surface in which the cytostome is located, and to develop a distinct oral ciliature associated with feeding. Figures 26 and 27 illustrate some of these variations.

Most ciliates with a distinct cytostome possess a nonciliated duct or tube—called the **cytopharynx**—that extends from the cytostome deep into the cytoplasm. The walls of the cytopharynx are often reinforced with rodlike structures—**nematodesmata**—

Figure 27

A–F, The oral regions of various ciliates (compare with Figure 26). A, Gymnostomatia. B, Vestibuliferia. C–D, Hymenostomatia. E, Peritricha. F, Hypotrichida. G, *Blepharisma*, showing relationship of the AZM and the paroral undulating membrane to the cytostome. (A–F after Russell-Hunter 1979 after other authors, especially Corliss 1979; G after Sherman and Sherman 1976.)

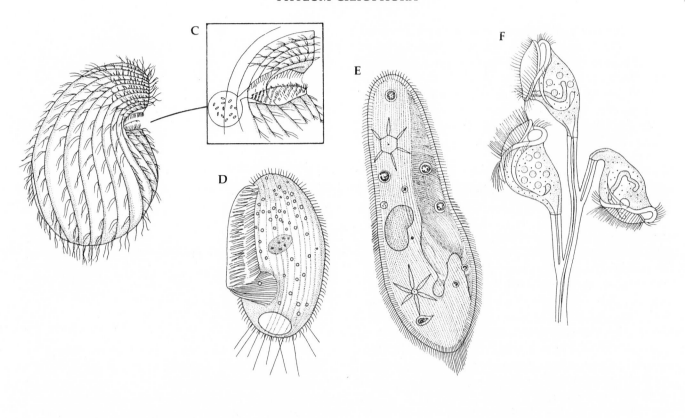

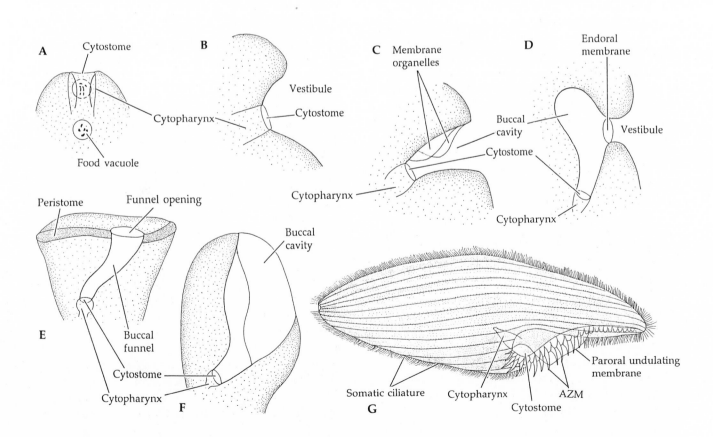

that may incidentally occur beneath the pellicle in other parts of the body as well. In all but the most primitive and some very specialized ciliates, the cytostome is set within some sort of depression or **preoral chamber** (Figure 27). The preoral chamber may contain simple cilia that are merely an extension of the somatic ciliature, or it may contain specialized ciliary organelles (AZM and paroral membrane) distinct from the rest of the body cilia. In the first case the preoral chamber is called a **vestibule** or **prebuccal area** and in the latter case a **buccal cavity** (Figures 26 and 27).

Ciliates that feed as "gulpers" take in prey through the cytostome, forming food vacuoles at the inner end of the cytopharynx. Most free-swimming forms attack and eat various microscopic and nearly microscopic organisms, including other protozoa and rotifers. Some are specialized for feeding upon particular prey species, sometimes organisms larger than themselves. For example, *Didinium nasutum* feeds exclusively on *Paramecium*, but it must stretch its oral area to engulf its relatively gigantic prey. A few ciliates are herbivorous, feeding primarily on unicellular or filamentous algae.

Many of the most familiar ciliates may be categorized as lurking "swirlers." They use the oral ciliature to carry food material to the cytostome, where it is eventually passed to food vacuoles. The size of the food eaten by such ciliates depends primarily upon the nature of the feeding current and, when present, the size of the vestibule or buccal cavity. Such ciliates typically possess a well developed AZM and paroral membrane, the latter frequently called an **undulating membrane**. "Swirlers" include such common genera as *Euplotes*, *Stentor*, and *Vorticella* (Figure 28). Many hypotrichs (e.g., *Euplotes*) that move about the substratum with their oral region oriented ventrally use their specialized oral ciliature to swirl settled material into suspension and then into the buccal cavity for ingestion.

Among the most specialized ciliate feeding methods are those used by the suctorians, which lack cilia as adults but possess instead knobbed feeding tentacles (Figure 29). A few suctorians have two types of tentacles, one form for food capture and another for ingestion. The swellings at the tips of the tentacles contain organelles called **haptocysts**, which are discharged upon contact with a potential prey. Portions of the haptocyst penetrate the victim and hold it to the tentacle. Sometimes prey are actually paralyzed after contact with haptocysts, presumably by some secretion released during discharge. Following attachment to the prey, a temporary "tube" forms

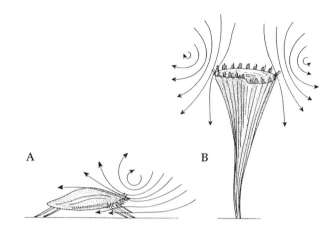

Figure 28

Feeding currents in two ciliates. A, *Euplotes*. B, *Stentor*. The ciliary currents produced by these organisms bring suspended food to the cell, where it can be ingested. (After Sleigh 1973a.)

within the tentacle and the contents of the prey are sucked into the tentacle and incorporated into food vacuoles (Figure 29B,C,D).

Some predatory ciliates (e.g., gymnostomes) have tubular extrusible organelles—called **toxicysts** —in the oral region of the cell. During feeding, the toxicysts are extruded and release their contents, which apparently include both paralytic and digestive components. Active prey are first immobilized and then partially digested by the discharged chemicals; this partially digested food is later taken into food vacuoles by the ciliate.

Some ciliates, including a number of hymenostomes, possess organelles that are called **mucocysts** and **trichocysts** and are located just beneath the pellicle. Mucocysts discharge mucus onto the surface of the cell as a protective coating; they may also play a role in cyst formation. Trichocysts (Figure 23C,D,E) contain nail-shaped structures that can be discharged through the pellicle. Most specialists suggest that these structures are not used in prey capture, but serve a defensive function.

A number of ciliates are ecto- or endosymbionts associated with a variety of vertebrate and invertebrate hosts. In some cases these symbionts depend entirely upon their hosts for food. Some suctorians, for example, are true parasites, often living within the cytoplasm of other ciliates! A number of hypostome ciliates are ectoparasites on freshwater fishes and may cause significant damage to their hosts' gills. *Balantidium coli*, a large vestibuliferan ciliate, is common in pigs and is known to cause intestinal lesions in people. Certain spirotrichs (*Entodinium*) reside in

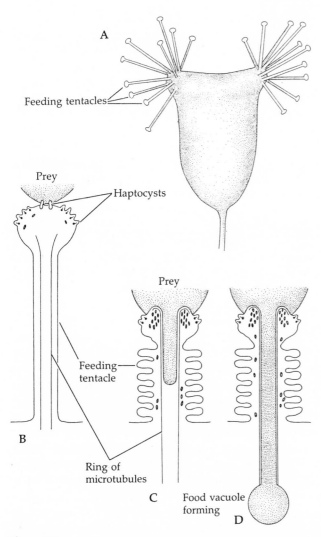

Figure 29

Feeding in the suctorian ciliate *Acineta*. **A,** *Acineta* has capitate feeding tentacles; note absence of cilia. **B–D,** Schematic drawings of enlarged feeding tentacles, showing the sequence of events in prey capture and ingestion. **B,** Contact with prey and firing of haptocysts into prey. **C,** Shortening of tentacle and formation of a temporary feeding duct within a ring of microtubules. **D,** Drawing of contents of prey into duct and formation of food vacuole. (A after Hyman 1940; B,C,D redrawn from Grell 1983, after Bardele and Grell 1967.)

the rumen of ungulates and can digest cellulose. Members of the order Chonotrichida (Hypostomatia) are mostly ectosymbiotic on crustaceans. Chonotrichs are sessile, attaching to their hosts by a stalk produced from a special adhesive organelle. Many other hypostome ciliates are symbiotic on a variety of hosts, including bivalve and cephalopod molluscs, crustaceans, polychaete worms, and perhaps mites.

Reproduction. Asexual reproduction in ciliates is usually by binary fission, although multiple fission and budding are known (Figure 30). Binary fission in ciliates is usually transverse—except among the peritrichs, where it appears to be longitudinal. The micronucleus is the reservoir of genetic material in ciliates. As such, each micronucleus within the cell (even when there are many) forms a mitotic spindle during fission, thus contributing daughter micronuclei equally to the progeny of division. No spindle forms in the macronucleus, which typically divides by simple constriction. In some species multiple macronuclei may fuse to produce a single macronucleus prior to fission. Similarly, elongate or multilobed macronuclei may shorten and form a compact mass before dividing. Since many ciliates are anatomically complex and frequently bear structures that are not centrally or symmetrically placed on the body (especially structures associated with the area of the cytostome), a significant amount of reconstruction must occur following fission. Such reformation of parts or special ciliary fields does not take place haphazardly; it apparently is controlled, at least partly by the macronucleus and certain portions of the infraciliature.

Binary fission in the peritrichs differs between solitary and colonial forms. In colonial forms the division is equal, and both daughter cells remain attached to the growing colony. But, in the solitary forms fission is unequal; and the smaller, stalkless daughter cell bears a posterior locomotor ciliary ring that it uses to swim about, later to settle and produce a stalk of its own. Another kind of unequal division is frequently referred to as budding, and it occurs in a variety of sessile ciliates, including chonotrichs and suctorians. In each case the bud is released as a so-called **swarmer** that swims about before adopting the adult morphology and lifestyle. In some cases, several buds are formed and released simultaneously.

True multiple fission is known in a few groups of ciliates and typically follows the production of a cyst by the prospective parent organism. Repeated divisions within the cyst produce numerous offspring, which are eventually released with breakdown of the cyst coating.

Sexual reproduction (using the broad connotation of the term) by ciliates is usually by conjugation, less commonly by autogamy. Conjugation is perhaps most easily understood by first describing it in *Paramecium*. As with any sexual process, the biological "goal" of the activity is genetic mixing or recombination, and it is accomplished during conjugation by an exchange of micronuclear material. The following

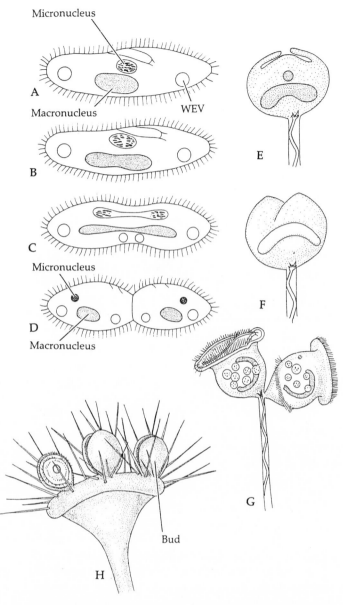

Figure 30

Asexual reproduction in ciliates. A–D, Transverse binary fission in *Paramecium*; the micronucleus divides mitotically, whereas the macronucleus simply splits. E–G, Binary fission in *Vorticella*. H, Budding in the suctorian *Ephelota gigantea*. (E–H after Hyman 1940.)

account (Figure 31) is of *Paramecium caudatum*—details vary in other species of the genus.

As paramecia move about and encounter one another, they apparently recognize compatible "mates." After making contact at their anterior ends, the "mates"—called **conjugants**—orient themselves side by side and attach to each other at their oral areas. In each conjugant, the micronucleus enlarges some-

what and then undergoes two divisions that are equivalent to meiosis and reduce the chromosome number to the haploid condition. Three of the daughter micronuclei in each conjugant disintegrate and are incorporated into the cytoplasm; the remaining haploid micronucleus in each cell divides once more. The products of this postmeiotic micronuclear division are called gametic nuclei. One gametic nucleus in each conjugant remains in the "parent" conjugant while the other (the smaller of the two) is transferred to the other conjugant via a cytoplasmic connection formed at the point of joining. Thus, each conjugant sends a haploid micronucleus to the other, thereby accomplishing the exchange of genetic material. Each migratory gametic nucleus then fuses with the stationary micronucleus of the recipient, thereby producing a diploid nucleus, or **synkaryon**, in each conjugant. This process is analogous to mutual cross-fertilization in metazoan invertebrates.

Following cross-fertilization, the cells separate from each other and are now called **exconjugants**. The process is far from complete, however, for the new genetic combination must be incorporated into the macronucleus if it is to impart any phenotypic changes in the organism. The macronucleus of each exconjugant has disintegrated during the meiotic and cross-fertilization process, and the diploid synkaryon divides mitotically three times, producing eight small nuclei (all, remember, containing the combined genetic information from the two conjugants). Four of the eight nuclei in each cell enlarge to become macronuclei. Three of the remaining four small nuclei break down and are absorbed into the cytoplasm. The single remaining micronucleus divides twice mitotically as the entire organism undergoes two binary fissions to produce four daughter cells, each of which receives one of the four macronuclei and one micronucleus. Thus, the ultimate product of conjugation and the subsequent fissions is four diploid daughter organisms.

Variations on the sequence of events described above for *Paramecium caudatum* include differences in the number of divisions, which seem to be determined in part by the normal number of micronuclei present in the cell. Even when two or more micronuclei are present, they typically all undergo meiotic divisions. All but one disintegrate, however, and the remaining micronucleus divides again to produce the stationary and migratory gametic nuclei.

In most ciliates the members of the conjugating pair are indistinguishable from each other in terms of size, shape, and other morphological details. However, some species, especially in the Peritricha, dis-

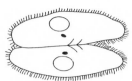

1. Two conjugants unite at oral areas.

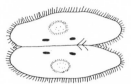

2. Macronuclei begin to disintegrate; micronuclei divide twice (meiosis).

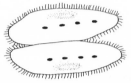

3. Three (of four) micronuclei disintegrate in each conjugant.

8. Micronucleus and animal divide to produce four individuals, each receiving one macronucleus.

7. Four micronuclei become macronuclei; three of the others disintegrate and one remains as the new micronucleus.

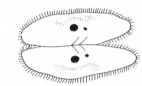

4. Remaining micronuclei divide unequally to produce two "gamete nuclei", the smaller of which is exchanged between conjugants.

6. Nucleus divides three times.

Animals separate (only one is shown in remaining drawings).

5. Micronuclei fuse within each conjugant to form a synkaryon.

Figure 31

Conjugation in *Paramecium*.

play distinct and predictable differences between the two conjugants, particularly in size. In such cases, we refer to the members of the mating pair as the **microconjugant** and the **macroconjugant** (Figure 32). The formation of the microconjugant generally occurs by one or a series of unequal divisions, which may occur in a variety of ways. The critical difference between conjugation of similar mates and that of dissimilar mates is that in the latter case there is often a one-way transfer of genetic material. The microconjugant alone contributes a haploid micronucleus to the macroconjugant; thus only the larger individual is "fertilized." Following this activity, the entire microconjugant usually is absorbed into the cytoplasm of the macroconjugant (Figure 32E). A similar process occurs in most chonotrichs; one conjugant appears

to be swallowed into the cytostome of another, after which nuclear fusion and reorganization take place. There are a number of other modifications on this complex sexual process in ciliates, but all have the same fundamental result of introducing genetic variation into the populations.

One other aspect of conjugation that deserves mention is that of **mating types**. Individuals of the same genetic mating type (e.g., members of a clone produced by binary fission) cannot successfully conjugate with one another. In other words, conjugation is not a random event but can occur only between members of different mating types, the number of which varies among different species of ciliates. For more details, we refer you to the discussion of this matter by Grell (1973, pp. 212–219).

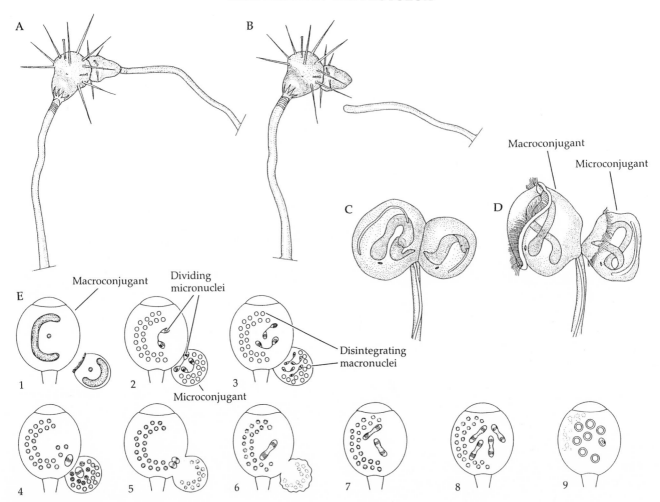

Figure 32

Sexual processes in ciliates. A–B, *Ephelota gemmipara* (a suctorian); two mating partners of unequal sizes are attached to each other, apparently following chemical recognition. Both have undergone nuclear meiosis. The smaller mate detaches from its stalk and is absorbed by the larger one, then the gametic nuclei fuse. Subsequent nuclear divisions produce the multimicronuclear and macronuclear components of the normal individual. C–D, Unequal division of *Vorticella campanula* results in macro- and microconjugants; conjugation follows. E, Schematic diagrams of sexual activities in certain peritrichs. Unequal divisions result in macro- and microgamonts; the latter detach from their stalks and become free-swimming organisms, eventually the free-swimming microgamont attaches itself to a sessile macrogamont (1–2). The macronuclei begin to disintegrate (2) and ultimately disappear (9). The micronucleus of the macrogamont divides twice (2-3) and the micronucleus of the microgamont divides three times (2–3). All but one of the micronuclei in each gamont disintegrate, and the remaining micronucleus of the microgamont moves to fuse with the micronucleus of the macrogamont (4–5). As the zygotic nucleus (synkaryon) begins to divide, the microgamont is absorbed into the cytoplasm of the macrogamont. The synkaryon divides three times (6–8); one of the daughter nuclei becomes the micronucleus and the others eventually form the new macronucleus (9). It should be noted that the sequence of nuclear activities and numbers of divisions vary among different peritrichs. (A,B redrawn from Grell 1973, after Grell 1953; C,D redrawn from Grell 1973, after Mugge 1957; E after Grell 1973.)

The second basic sexual process in ciliates is that of autogamy, which has been mentioned in passing in our discussions of other protozoa (certain members of the Sarcodina). Among ciliates in which autogamy occurs (e.g., certain species of *Euplotes* and *Paramecium*), the nuclear phenomena are similar if not identical to those occurring in conjugation. However, only a single individual is involved. When the point is reached at which the cell contains two haploid micronuclei, these two nuclei fuse with one another, rather than one being transferred to a mate. This is a case of true hermaphroditic "self-fertilization." Autogamy is known in relatively few ciliates, although it may actually be much more common than demonstrated thus far.

Other protozoa

We have discussed in some detail the three groups of protozoa that you are most likely to encounter in your general studies—the flagellates, the amebas, and the ciliates. In this section we offer a briefer review of the other protozoan phyla currently recognized. The five remaining phyla are Labyrinthomorpha, Apicomplexa, Microspora, Ascetospora, and Myxospora. All of the protozoa in these phyla are parasitic except a few labyrinthomorphs that are saprobic on algae. For the most part, these five groups previously made up the no-longer-recognized phyla Sporozoa and Cnidospora.

Phylum Labyrinthomorpha

This group contains a single class (Labyrinthulea) and order (Labyrinthulida) of still poorly understood protozoa (Figure 33). They are distinguished by the formation of complex networks of ectoplasmic extensions that connect the cells to one another. In some forms, the cells are ameboid and glide about within the ectoplasmic network. They reproduce by binary fission and by the production of motile spores. All known species are found on algae, in marine or estuarine environments, and are either saprobic or parasitic.

Phylum Apicomplexa

The phylum Apicomplexa includes about 4,600 species of parasitic protozoa (e.g., gregarines and coccidians) that formed a portion of the now-abandoned phylum Sporozoa. They are characterized by the presence of a unique **apical complex** (Figure 34). Locomotor organelles are absent from adult forms, which glide by simple body undulations; flagella are present during some reproductive stages. These protozoa typically alternate between sexual and asexual phases and often have complex life cycles.

The class Sporozoea includes the most common

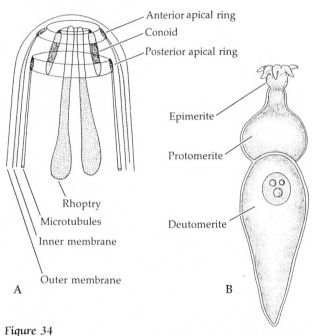

Figure 34

The Apicomplexa. A, The apical complex. B, The body of a gregarine commonly is divided into three recognizable regions. (A after Sleigh 1973b; B after Marquardt and Demaree 1985.)

members of this phylum. The gregarines are relatively large (up to 1 mm long) sporozoean parasites in the guts and body cavities of several kinds of invertebrates, including annelids, sipuculans, tunicates, and arthropods. The anterior end is frequently equipped with hooks or suckers for attachment to the host's epithelia (Figure 34B). The best studied gregarines are those that inhabit insect hosts, and the life cycles of these forms are well understood; an example is diagrammed in Figure 35. Sexual reproduction usually involves the enclosure of a mating pair of gamonts within a cyst or capsule (a phenomenon known as **syzygy**). Each gamont undergoes multiple fission to produce many gametes. Both isogamy and anisogamy are known among different gregarines. Each zygote formed by the fusion of two gametes becomes a spore, which divides to produce as many as eight **sporozoites**. Each sporozoite enters a period of growth, to mature as a trophozoite, which eventually becomes a sexual gamont, completing the cycle.

The coccidian sporozoeans are parasites of several groups of animals, mostly vertebrates. They typically reside within the cells of their hosts, at least during some stages, and many are pathogenic. Some coccidians pass their entire life cycle within a single host; many others require an intermediate host that

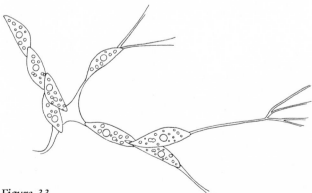

Figure 33

Labyrinthula coenocytis **has spindle-shaped cells within an anastomosing reticulum. The cells are motile along these threads. (Drawn from a photograph in Grell 1973.)**

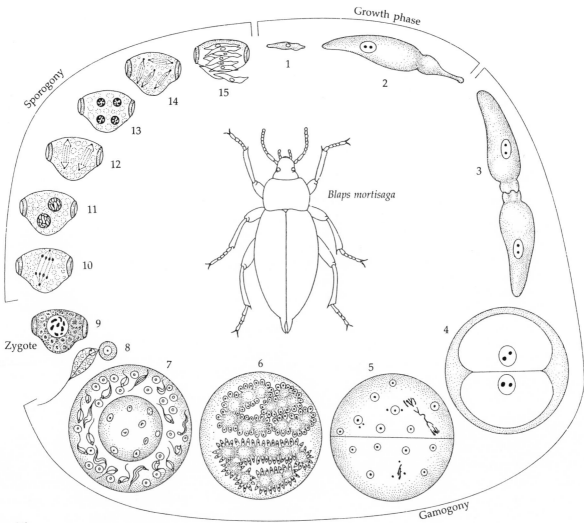

Figure 35

Life cycle of the gregarine *Stylocephalus longicollis*, a gut parasite of the coleopteran *Blaps mortisaga*. Stages 1–4 take place within the host, 5–15 outside the host. The spores (15) are ingested by the beetle and release sporozoites within the gut lumen. Each sporozoite grows into a gamont (2); the gamonts subsequently mate (3–4), becoming enclosed within a mating cyst, which leaves the host with the feces. Repeated mitotic divisions within the cyst produce anisogametes (5–7); these ultimately fuse (8) to produce a zygote (9), which eventually becomes a spore. The first divisions of the spore cell are meiotic (10), so all subsequent stages leading back to gamete fusion are haploid. (After Grell 1973.)

serves as a vector. Coccidians are responsible for a variety of diseases, including coccidiosis in rabbits and birds, and toxoplasmosis and malaria in people.

Malaria is caused by several species of the genus *Plasmodium* (Figure 36). Although malaria was nearly wiped out worldwide during the 1960s, it is making an alarming comeback today, and is one of the most prevalent and severe health problems in the Third

World. In 1977, 30–50 million people in India alone fell victim to malaria parasites. One of the principal causes of this resurgence is the dramatic rise in the numbers of pesticide-resistant strains of *Anopheles* mosquitoes, the insect vector for *Plasmodium*. By 1968, 38 strains or species of *Anopheles* in India had been identified as largely pesticide resistant. By 1975 the incidence of malaria in Central America was three times what it had been a decade earlier.

Phylum Microspora

The 800 or so species of the phylum Microspora (Figure 37) are intracellular parasites in nearly every group of animals; some are even found in other protozoa, including gregarines. Much remains to be learned about the various stages in the life histories of microsporans; there is even disagreement about whether any sexual phases are present. The common

Figure 36

Life cycle of *Plasmodium*, the causative agent of malaria in humans. When a female anopheline mosquito takes a blood meal, she releases sporozoites into the victim's bloodstream (1). These sporozoites enter the host's liver cells and undergo multiple fission, producing many merozoites (2); each sporozoite may produce as many as 20,000 merozoites in a single liver cell. The infected liver cells rupture, releasing the merozoites into the blood where they invade red blood cells (3). Through continued multiple fission, more merozoites are produced. The red blood cells eventually burst, releasing merozoites, which enter other RBCs. Some merozoites differentiate to become gametocytes (4), which are picked up by mosquitoes. The female gametocyte forms a single macrogamete; the male gametocyte typically undergoes multiple fission to produce several motile, flagellated, microgametes within the gut of the mosquito (5). After fertilization occurs, the zygote migrates to the mosquito's salivary glands and divides to form numerous sporozoites, thereby completing the life cycle. (After Miller et al. 1986.)

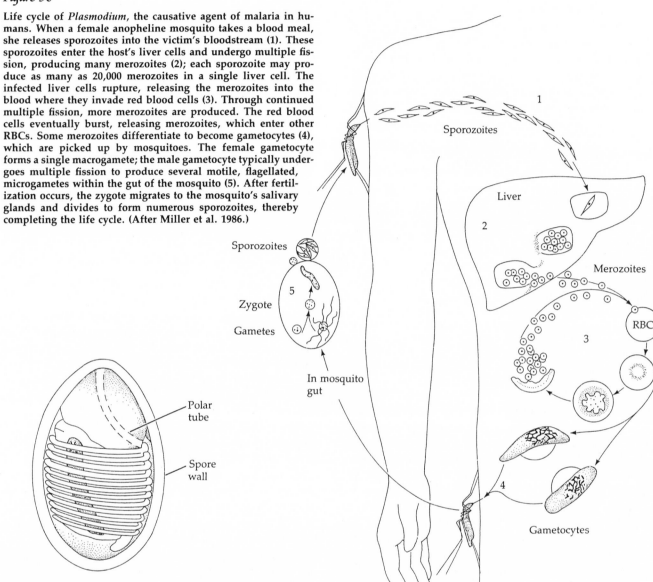

Figure 37

The spore of *Thelohania californica* has a coiled polar tube. (Redrawn from Grell 1973, after Kudo and Daniels 1963.)

visible form that occurs in host tissues is the spore, and it is on the basis of the details of spore structure that the phylum is defined and the species characterized. The microsporans have unicellular, uni- or binucleate spores within a typically multilayered cyst. The spore bears a **polar cap** and a single **polar tube** (the polar tube has also commonly been referred to as the polar filament). When a spore is ingested by a potential host, the polar tube extends, carrying with it the spore cytoplasm, which attaches to and penetrates cells of the host's gut wall. These organisms cause damage to host tissues. Species that infect silkworms and honeybees are of considerable economic importance.

Phylum Ascetospora

The ascetosporans are another exclusively parasitic group of protozoa. We know little of their life cycles and they, like microsporans, are characterized on the basis of spore structure. The spores of most forms are multicellular or distinctly bicellular; some may be unicellular. The spores lack polar filaments or tubes.

Phylum Myxozoa

The Myxozoa (or Myxospora, as they are also known) includes about 1,200 species. They are parasitic in annelids and poikilothermic vertebrates. The spores, which are enclosed within one or more plate-like valves, are uni- or multicellular, containing from one to several polar tubules (filaments) (Figure 38). The polar filaments of these organisms (like the polar tubes of microsporans) can be everted; and they are presumably used for attachment to the host. The action of these organelles is somewhat similar to that of the nematocysts (see Chapter 8) of cnidarians.

Hosts become infected by ingesting spores. The polar filaments evert, thereby causing drag and slowing the passage of the spores through the gut. Then the valves open, releasing the spore cytoplasm as a motile **amebula**, which penetrates the host's gut epithelium and migrates to a final site of infection. Some myxozoa include both a vertebrate and an invertebrate host in their life cycle (see Wolf and Markiw 1984).

Protozoan phylogeny

We can do no more than touch upon the myriad questions and interesting points of view concerning the origin and evolution of the protozoa. Beyond the problems of relationships among the various protozoa themselves, we are faced on one hand with questions about the very origin of eukaryotic life on this planet, and on the other with interpreting the ancestral forms of the rest of the living world. The origin of eukaryotic cells probably took place about 2.5 to 3.5 billion years ago, and we may consider this event as the origin of the protistan grade. Although there are over 30,000 known fossil species of protozoa, they are of little use in establishing the origin or subsequent evolution of the protozoan groups. Only those with hard parts have left us much of a fossil record, and only the foraminiferans and radiolarians have well established records in Precambrian rocks (and there is some debate even about this). The origin of the eukaryotic condition was, of course, a momentous event in the biological history of the Earth, for it enabled animals to escape from the limitations of the prokaryotic *Bauplan* by providing the various subcellular units that have formed the basis of specialization among the Protista.

The first living organisms on Earth probably fed upon dissolved organic molecules abundant in ancient seas, perhaps by some form of active uptake

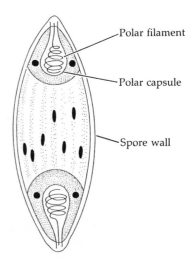

Figure 38

Myxozoa. Stylized drawing of a myxozoan spore. (After Russell-Hunter 1979.)

across their cell membranes or by pinocytosis. Oxygen became available in the primeval environment after photosynthesis became the dominant form of nutrition; and eventually, organisms that might now be classified as phytoflagellates originated. This scenario, one among several, suggests that the first eukaryotes were autotrophs, most closely resembled today by the Phytomastigophorea. If we assume for the moment that the first eukaryotes were photosynthetic and that they had a monophyletic origin, a phylogenetic tree such as the one shown in Figure 39 can be constructed and viewed as a reasonable outline (hypothesis) of evolutionary history. In it, the photosynthetic flagellates serve as an ancestral pool from which radiated the various other protists, metazoa, and higher plants.

Acceptance of this tree requires acceptance of several assumptions, two of which are quite fundamental. First, this theory, which has been called **direct filiation**, implies that all the cellular organelles of eukaryotes evolved by the accumulation of "usual" evolutionary events (e.g., mutations, deletions, duplications). Second, direct filiation assumes that autotrophy is a symplesiomorphy shared by all photosynthetic organisms. Heterotrophy, then, may be viewed as an apomorphic character, derived once or several times (convergence) and shared by all those organisms that have become feeders of various sorts. The placement of the line to the red algae (division Rhodophyta) in this scheme is problematic, because these organisms are multicellular but do not possess any flagella with the 9 + 2 filament arrangement typical of the other eukaryotes.

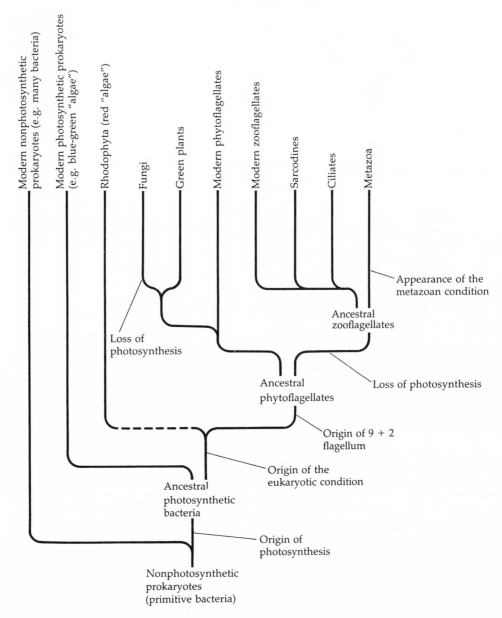

Figure 39

A generalized evolutionary tree depicting the major events and lines of descent by the direct-filiation theory (see text). This particular tree depicts Eukaryota as monophyletic and the protozoa as paraphyletic, with separate origins for the modern phytoflagellates and the remaining protozoan groups. Both the phytofla-gellates and the remaining protozoa could have had multiple origins; the point of this model, however, is that these two large groups had their ultimate origins in ancestral photosynthetic protozoa (the ancestral phytoflagellates). This model hypothesizes autotrophy as the plesiomorphic eukaryotic condition.

Alternatively, we might view the eukaryotes originating as nonautotrophs, giving rise later to photosynthetic forms. Of the number of models explaining how this might have happened, one of the most interesting is the **serial endosymbiotic theory** (SET), which is strongly supported by Margulis (1970, 1975, 1981) and others. Figure 40 outlines the basic steps in the transition from a prokaryotic to a eukaryotic condition according to this theory. The premise in this scenario is that the eukaryotes arose through intimate symbiotic relationships between various prokaryotic cells. Unlike prokaryotic cells, all eukaryotic cells contain several kinds of membrane-bounded organelles that harbor distinct genetic systems. Three classes of organelles—mitochondria, cilia/flagella, and photosynthetic plastids—are hy-

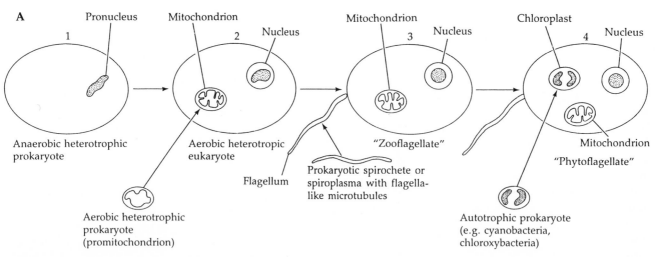

A

1 Pronucleus   Mitochondrion   2 Nucleus   Mitochondrion   3   Nucleus   Chloroplast   4   Nucleus

Anaerobic heterotrophic prokaryote

Aerobic heterotropic eukaryote

"Zooflagellate"

Flagellum

Prokaryotic spirochete or spiroplasma with flagella-like microtubules

Aerobic heterotrophic prokaryote (promitochondrion)

Autotrophic prokaryote (e.g. cyanobacteria, chloroxybacteria)

Mitochondrion

"Phytoflagellate"

Figure 40

A, A simple model of the origin of eukaryotic cells by symbiosis (the serial endosymbiotic theory). The three major events depicted are: acquisition of an aerobic heterotrophic prokaryote (origin of mitochondrion); acquisition of a spirochete or spiroplasma-like prokaryote (origin of flagellum); and acquisition of an autotrophic prokaryote (origin of chloroplast). See text for discussion. B, One view of life, depicting the major lines of descent by the serial endosymbiotic theory. This particular tree depicts the protozoa as polyphyletic; in fact, various protozoan taxa could have had numerous unrelated origins by a variety of possible symbiotic associations. This model, in contrast to the direct-filiation model (see Figure 39), views heterotrophy as the plesiomorphic eukaryotic (protozoan) condition; the advent of autotrophy is seen as a synapomorphy distinguishing the phytoflagellates and green plants. (A after Sleigh in House 1979, modified from Margulis 1970.)

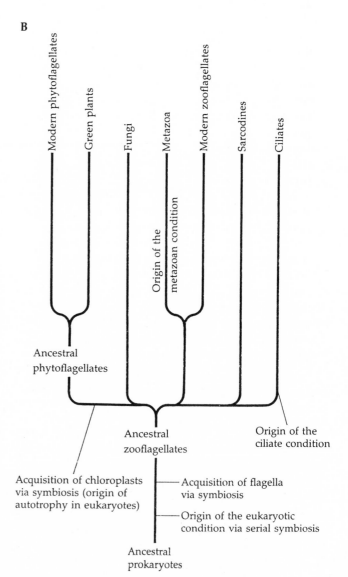

B

Modern phytoflagellates

Green plants

Fungi

Metazoa

Modern zooflagellates

Sarcodines

Ciliates

Origin of the metazoan condition

Ancestral phytoflagellates

Ancestral zooflagellates

Origin of the ciliate condition

Acquisition of chloroplasts via symbiosis (origin of autotrophy in eukaryotes)

Acquisition of flagella via symbiosis

Origin of the eukaryotic condition via serial symbiosis

Ancestral prokaryotes

pothesized to have once been free-living prokaryotes that were acquired symbiotically and in a certain sequence by another (host) prokaryote. Hence, the functions now performed by these organelles are postulated to have evolved long before the eukaryotic cell itself evolved. The SET theory suggests that a prokaryotic heterotroph ingested other, mitochondrion-like prokaryotes and, roughly at the same time, began forming an organized nucleus. Subsequently, this nonmotile cell established a symbiotic relationship with yet another prokaryote in the form of a spirochete or spiroplasma bacterium attached to the outside of the cell. Such bacteria contain protein microtubules and are capable of something like flagellar activity; thus a protozooflagellate evolved. Eventually a photosynthetic prokaryote was engulfed by this now-eukaryotic organism, the "prey" representing a chloroplast precursor—thus, the origin of the phytoflagellates. Margulis believes this photosynthetic prokaryote might have been a bacterium that evolved in an anaerobic environment, very early in the history of life. The type of photosynthesis that produces oxygen would have evolved later, and oxygen-respiring organisms still later, after photosynthetically produced oxygen began accumulating in the environment. While this account may initially seem rather bizarre, there is a great deal of evidence in its favor, and many protistologists support it. For example, there are in fact prokaryotic organisms that are very similar to those viewed as symbionts in this story (see Margulis 1978, 1981; John and Whatley 1975). From this symbiotic basis the modern protistan groups may be derived in a variety of ways (e.g., see Figure 40B).

As we pointed out early in this chapter, most modern workers do not hold the protozoa to be monophyletic. Because the origin of the various groups of protozoa is so intimately associated with the origin of the eukaryotic condition, we may consider polyphyly at different levels. For example, did the protozoa as we know them originate as two or more distinct clades from separate prokaryotic ancestors, or from separate eukaryotic ancestors? In other words, did the eukaryotic condition arise as a synapomorphy common to all protozoa (and hence all metazoan life on Earth) or does it represent a convergent feature in two or more lineages (Figure 41)? Biologists are still a long way from answering these questions. Careful reading of the protozoological literature reveals that dozens of different hypotheses have been offered describing the multiple origins of photosynthesis in the phytoflagellates. For example, two popular and competing hypotheses differ on the origin of autotrophy in euglenoids: from symbiosis with a cyanobacterium versus from symbiosis with a green alga. Similarly, Steidinger and Cox (1980) presented a phylogenetic hypothesis in which the dinoflagellates evolved through several symbiotic events; consequently, various dinoflagellates have different carotenoid pigment patterns. One fairly popular and conservative view of polyphyly is that

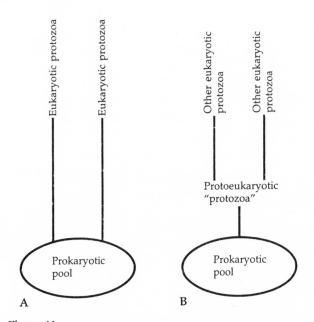

Figure 41

Two possible views of protozoan origin. Scheme A depicts a polyphyletic origin of the eukaryotic unicellular condition. Scheme B depicts a monophyletic origin of the eukaryotic unicellular condition, with a subsequent polyphyletic origin of various modern protozoan groups.

the ciliates are alone on one evolutionary line, and the remainder of the protozoa constitute another.

We realize that we have slighted the protozoa, and especially their evolution, in our necessarily limited treatment here. We have not, for example, considered the many intriguing questions about the evolution of reproductive strategies among these groups. However, we hope that you have gained enough information about their complexity and diversity to appreciate their success and general relationships to other invertebrates, and that you now have an appreciation of the success of unicellular *Baupläne*.

Selected References

General Works

Baker, J. R. 1969. *Parasitic Protozoa.* Hutchinson, London.

Bamforth, S. S. 1980. Terrestrial protozoa. J. Protozool. 27:33–36.

Cavalier-Smith, T. 1981. Eukaryote kingdoms: Seven or nine? BioSystems 14:461–481.

Chen, T.-T. (ed.). 1967–1972. *Research in Protozoology.* Pergamon Press, Oxford. [Four volumes on recent advances in protozoan biology.]

Corliss, J. O. 1978–79. A salute to 54 great microscopists of the past: A pictorial footnote to the history of protozoology. Parts I and II. Trans. Amer. Microsc. Soc. 97:419–458; 98:26–58.

Corliss, J. O. 1982. Protozoa. *In* S. Parker (ed.), *Synopsis and Classification of Living Organisms.* Vol. 1. McGraw-Hill, New York. pp. 491–637. [Includes coverage of all protozoan taxa by several specialists.]

Fenchel, T. 1987. *Ecology of Protozoa: The Biology of Free-Living Phagotrophic Protists.* Springer-Verlag, New York.

Grassé, P. (ed.) 1952. *Traité de Zoologie.* Vol. 1, pts. 1–2. Masson et Cie, Paris.

Grell, K. B. 1973. *Protozoology.* Springer-Verlag, New York. [This translation from the original German text is a classic and among the very best of protozoology texts. Although falling slightly out of date on certain subjects, it remains a most valuable resource.]

Hanson, E. D. 1967. Protozoan development. *In* M. Florkin and B. J. Sheer (eds.), *Chemical Zoology,* Vol. 1. Academic Press, New York.

Hanson, E. D. 1977. *The Origin and Early Evolution of Animals.* Wesleyan University Press, Middletown, Connecticut, and Pitman Publishing Ltd., London. [Contains nearly 300 pages on the protozoa.]

Hollande, A. 1972. Le déroulement de la cryptomitose et les modalités de la ségrégation des chromatides dans quelques groupes de protozoaires. Ann. Biol. 11: 427–466.

House, M. R. (ed.) 1979. *The Origin of Major Invertebrate Groups.* Academic Press, London. Published for the Systematics Association, Special Vol. 12. [This fine book includes 18 articles concerning many groups of invertebrates and the origin of chordates. The work presented by M. A. Sleigh is especially important as regards the protozoa.]

Hyman, L. H. 1940. *The Invertebrates.* Vol. 1, Protozoa through Ctenophora. McGraw-Hill, New York.

Jahn, T. L. and J. J. Votta. 1964. Locomotion of Protozoa. Annu. Rev. Fluid Mech. 4:93–116.

Jahn, T. L. and E. C. Bovee. 1964. Protoplasmic movements and locomotion of Protozoa. *In* S. H. Hutner (ed.) *Biochemistry and Physiology of Protozoa,* Vol. 3. Academic Press, New York.

Jahn, T. L. and E. C. Bovee. 1965. Movement and locomotion of microorganisms. Annu. Rev. Microbiol. 19:21–58.

Jahn, T. L. and E. C. Bovee. 1967. Motile behavior of Protozoa. *In* T.-T. Chen (ed.) *Research in Protozoology,* Vol. 1. Pergamon Press, Oxford.

Jahn, T. L., E. C. Bovee and F. F. Jahn. 1979. *How to Know the Protozoa,* 2nd Ed. W. C. Brown, Dubuque, Iowa.

Krylov, M. V. and Y. I. Starobogatov (eds.) 1980. *Principles of the Construction of the Macrosystem of the Unicellular Animals.* (Proc. Zool. Inst. 94), Academy of Sciences of the USSR, Leningrad.

Kudo, R. R. 1966. *Protozoology,* 5th Ed. Charles C. Thomas, Springfield, Illinois. [One of the most popular texts for courses in protozoology a few years ago; still a good treatment.]

Laybourn-Parry, J. 1985. *A Functional Biology of Free-Living Protozoa.* University of California Press, Berkeley.

Lee, J. J., S. H. Hunter and E. C. Bovee (eds.) 1985. *An Illustrated Guide to the Protozoa.* Allen Press, Lawrence, Kansas. Published for the Society of Protozoologists.

Levandowsky, M. and S. H. Hutner (eds.) 1979–1981. *Biochemistry and Physiology of Protozoa.* Vols. 1–4. Academic Press, New York. [A scholarly series of state-of-the-art reviews on various aspects of protozoan physiology and chemistry.]

Levine, N. D., J. O. Corliss, F. E. G. Cox, G. Deroux, J. Grain, B. M. Honigberg, G. F. Leedale, A. R. Loeblish III, J. Lom, D. Lynn, E. G. Merinfeld, F. C. Page, G. Poljansky, V. Sprague, J. Vavra and F. G. Wallace. 1980. A newly revised classification of the Protozoa. J. Protozool. 276(1):37–58. [The long list of authors includes the members of The Committee on Systematics and Evolution of the Society of Protozoologists, chaired by Dr. N. D. Levine. We have adopted their scheme in this book as it represents the most recent major treatment of protozoan classification. This is not the final word, however; as the authors caution, ". . . probably none of them [members of the committee] agrees with it [the scheme] completely; however, unanimity can hardly be expected. What we have produced is something with which we can live and which we can modify as suggested by our differing needs and ideas."]

Mackinnon, D. L. and R. S. Hawes. 1961. *An Introduction to the Study of Protozoa.* Clarendon Press, Oxford.

Manwell, R. D. 1968. *Introduction to Protozoology.* Dover Publications, New York.

Marquardt, W. C. and R. S. Demaree, Jr. 1985. *Parasitology.* Macmillan, New York.

Nisbet, B. 1983. *Nutrition and Feeding Strategies in Protozoa.* Croom Helm, London.

Noble, E. R. and G. A. Noble. 1982. *Parasitology: The Biology of Animal Parasites.* 5th Ed. Lea & Febiger, Philadelphia. [Includes extensive coverage of pathogenic protozoa found in humans and other hosts. We have fond memories of our undergraduate course in parasitology from G. Noble so many years ago.]

Patterson, D. J. 1980. Contractile vacuoles and associated structures: Their organization and function. Biol. Rev. 55:1–46.

Raikov, I. B. 1982. *The Protozoan Nucleus: Morphology and Evolution.* Cell Biol. Monogr. 9, Springer-Verlag, Vienna and New York.

Sandon, H. 1963. *Essays on Protozoology.* Hutchinson Educational Ltd., London. [We like the way this little book is written; it is lively and entertaining reading and provides a nice introduction to the protozoa.]

Seravin, L. N. 1971. Mechanisms and co-ordination of cellular locomotion. Comp. Physiol. Biochem. 4:37–111.

Seravin, L. N. and E. E. Orlovskaja. 1978. Feeding behavior of unicellular animals. I. The main role of chemoreception in the food choice of carnivorous protozoa. Acta Protozool. 16:309–332.

Sleigh, M. A. 1971. Cilia. Endeavour 30: 11–17.

Sleigh, M. A. (ed.) 1973a. *Cilia and Flagella.* Academic Press, London.

Sleigh, M. A. 1973b. *The Biology of Protozoa.* Edward Arnold Ltd., London. [A good treatment for those with some basic background in protozoology.]

Sonneborn, T. M. 1974. Breeding systems, reproductive methods and species problems in Protozoa. *In* E. Mayr (ed.), *The Species Problem.* Orno, New York.

Sarcomastigophorans

Allen, R. D. 1970. Comparative aspects of amoeboid movement. Acta Protozool. 7: 291–299.

Allen, R. D., D. Francis and R. Zek. 1971. Direct test of the positive pressure gradient theory of pseudopod extension and retraction in amoebae. Science 174:1237–1240.

Anderson, O. R. 1983. *Radiolaria.* Springer-Verlag, New York.

Be, A. 1982. Biology of planktonic Foraminifera. University of Tennessee Studies in Geology 6:51–92.

Bovee, E. C. and T. L. Jahn. 1966. Mechanisms of movement in taxonomy of Sarcodina. III. Orders, suborders, families, and subfamilies in the superorder Lobida. Syst. Zool. 25:229–240.

Bovee, E. C. and T. K. Sawyer. 1979. Marine flora and fauna of the northeastern United States. Protozoa: Sarcodina: Amoebae. NOAA Tech. Rpt. NMFS Circular No. 419, U.S. Department of Commerce.

Buetow, D. E. (ed.) 1968. *The Biology of Euglena.* Vols. 1–2. Academic Press, New York.

Buetow, D. E., (ed.) 1982. *The Biology of Euglena.* Vol. 3, Physiology. Academic Press, New York and London.

Chapman-Andresen, C. 1973. Endocytic processes. *In* K. Jeon (ed.), *The Biology of Amoeba.* Academic Press, New York, pp. 319–348.

Chen, T. 1950. Investigations of the biology of *Peranema trichophorum.* J. Microsc. Sci. 9:279–308.

Cox, E. R. (ed.) 1980. *Phytoflagellates.* Elsevier/North-Holland, New York. [An excellent current review of the major groups, by appropriate specialists; includes a number of articles addressing the origin and phylogeny of various phytoflagellate taxa.]

Cullen, K. J. and R. D. Allen. 1980. A laser microbeam study of amoeboid movement. Exp. Cell Res. 128:353–362.

Donelson, J. E. and M. J. Turner. 1985. How the trypanosome changes its coat. Sci. Am. 252(2):44–51.

Gibbons, I. R. and A. V. Grimstone. 1960. On flagellar structure in certain flagellates. J. Biophys. Biochem. Cytol. 7:697–716.

Gojdics, M. 1953. *The Genus Euglena.* University of Wisconsin Press, Madison. [Includes keys, descriptions and figures of all species known at the time.]

Groto, K., D. L. Laval-Martin and L. N. Edmunds Jr. 1985. Biochemical modeling of an autonomously oscillatory circadian clock in *Euglena.* Science 228:1284–1288.

Hedley, R. H. and C. G. Adams (eds.) 1974. *Foraminifera.* Academic Press, New York.

Jahn, T. L., W. M. Harmon and M. Landman. 1963. Mechanisms of locomotion in flagellates. I. *Ceratium.* J. Protozool. 10:358–363.

Jahn, T. L., M. Landman and J. R. Fonesca. 1964. The mechanism of locomotion of flagellates. II. Function of the mastigonemes of *Ochromonas.* J. Protozool. 11:291–296.

Jeon, K. W. (ed.) 1973. *The Biology of Amoeba.* Academic Press, New York and London. [Twenty-one of the world's ameba specialists have contributed to this fine book dealing with virtually all aspects of the biology of free-living amebas.]

Jepps, M. W. 1956. *The Protozoa: Sarcodina.* Oliver and Boyd, Edinburgh.

Leedale, G. F. 1967. *Euglenoid Flagellates.* Prentice-Hall, Englewood Cliffs, New Jersey.

Leith, A. 1980. Variability of the limited life span state in amoeba. Exp. Cell Res. 127:261–268.

Lumsden, W. H. R. and D. A. Evans (eds.) 1976, 1979. *Biology of the Kinetoplastida.* Vols. 1 and 2. Academic Press, New York.

Murray, J. W. 1973. *Distribution and Ecology of Living Benthic Foraminiferida.* Crane, Russak, New York.

Nigrini, C. and T. C. Moore. 1979. *A Guide to Modern Radiolaria.* Special Publ. No. 16, Cushman Foundation for Foraminiferal Research, Washington, DC.

Ogden, C. G. and R. H. Hedley. 1980. *An Atlas of Freshwater Testate Amoebae.* Oxford University Press, Oxford. [A book of magnificent SEM photographs of ameba shells accompanied by descriptions of the pictured species.]

Roth, L. E. 1960. Electron microscopy of pinocytosis and food vacuoles in *Pelomyxa.* J. Protozool. 7:176–185.

Seliger, H. H. (ed.) 1979. *Toxic Dinoflagellate Blooms.* Elsevier/North Holland, New York.

Steidinger, K. A. and E. R. Cox. 1980. Free–living dinoflagellates. *In* E. R. Cox (ed.), *Phytoflagellates.* Elsevier/North-Holland, New York.

Steidinger, K. A. and K. Haddad. 1981. Biologic and hydrographic aspects of red tides. BioScience 31(11):814–819.

Vickerman, K. and T. M. Preston. 1976. Comparative cell biology of the kinetoplastid flagellates. *In* W. H. R. Lumsden and D. A. Evans (eds.), *Biology of the Kinetoplastida,* Vol. 1. Academic Press, New York and London, pp. 35–130.

Wefer, G. and W. H. Berger. 1980. Stable isotopes in benthic Foraminifera: Seasonal variation in large tropical species. Science 209:803–805.

Ciliates

Bardele, C. F. 1972. A microtubule model for ingestion and transport in the suctorian tentacle. Z. Zellforsch. Mikrosk. Anat. 130:219–242. [One view on how suctorians may eat.]

Borror, A. C. 1973. Marine flora and fauna of the northeastern United States. Protozoa: Ciliophora. NOAA Tech. Rpt. NMFS Circular No. 378, U.S. Deptartment of Commerce.

Brehm, P. and R. Eckert. 1978. An electrophysiological study of the regulation of ciliary beating frequency in *Paramecium.* J. Physiol. 282:557–568.

Byrne, B. J. and B. C. Byrne. 1978. An ultrastructural correlate of the membrane mutant "Paranoiac" in *Paramecium.* Science 199:1091–1093.

Canella, M. F. 1964. Contributi alla conoscenza dei Ciliati. IV. Structure buccali, infraciliature, filogenesi e sistematica dei ciliofori. Ann. Univ. Ferrara 2:119–188.

Corliss, J. O. 1979. The impact of electron microscopy on ciliate systematics. Am. Zool. 19(2):573–587.

Corliss, J. O. 1979. *The Ciliated Protozoa: Characterization, Classification and Guide to the Literature,* 2nd Ed. Pergamon Press, New York. [A landmark effort by one of the most famous ciliatologists in the world. The extensive illustrated glossary (Chapter 2) is especially useful to those of us not so well versed in ciliate biology.]

Eckert, R. 1972. Bioelectric control of ciliary activity. Science 176:473–481.

Eckert, C. F. and E. L. Powers. 1959. The cell surface of *Paràmecium.* Int. Rev. Cytol. 8:97–133.

Eckert, C. F. and E. W. McArdle. 1974. The structure of *Paramecium* as viewed from its constituent levels of organization. *In* W. J. van Wagtendonk (ed.), *Paramecium: A Current Survey.* Elsevier, New York, pp. 263–338.

Elliott, A. M. (ed.) 1973. *The Biology of Tetrahymena.* Dowden, Hutchinson, and Ross, Stroudsburg, Pennsylvania.

Fenchel, T. 1980. Suspension feeding in ciliated protozoa: Structure and function of feeding organelles. Arch. Protistenkd. 123:239–260. [A fine review of the subject.]

Gall, J. G. (ed.). 1986. *The Molecular Biology of Ciliated Protozoa.* Academic Press, Orlando, Florida.

Gates, M. A. 1978. An essay on the principles of ciliate systematics. Trans. Am. Microsc. Soc. 97:221–235.

Grain, J., P. de Puytorac and J. Bohatier. 1973. Essai de systématique des ciliés gymnostomes fondée sur les caractéristiques de l'infraciliature circumorales. C. R. Acad. Sci. Paris 277:69–74.

Grimes, G. W. 1982. Pattern determination in hypotrich ciliates. Am. Zool. 22:35–46.

Jahn, T. L. 1961. The mechanism of ciliary movement. I. Ciliary reversal and activation by electric current; the Ludloff phenomenon in terms of core and volume conductors. J. Protozool. 8:369–380.

Jahn, T. L. 1962. The mechanism of ciliary movement. II. Ion antagonism and ciliary reversal. J. Cell. Comp. Physiol. 60:217–228.

Jahn, T. L. 1967. The mechanism of ciliary movement. III. Theory of suppression of reversal by electric potential of cilia reversed by barium ions. J. Cell. Physiol. 70:79–90.

Jones, A. R. 1974. *The Ciliates.* St. Martin's Press, New York.

Jurand, A. and G. G. Selman. 1969. *The Anatomy of Paramecium aurelia.* Macmillan, London, and St. Martin's Press, New York. [All you ever wanted to know about the anatomy of *Paramecium.* Many fine illustrations and micrographs.]

Kuhlmann, H.-W. and K. Heckmann. 1985. Interspecific morphogens regulating prey–predator relationships in protozoa. Science 227:1347–1349.

Mohr, J. L., H. Matsudo and Y.-M. Leung. 1970. The ciliate taxon Chonotricha. Oceanogr. Mar. Biol. Annu. Rev. 8:415–456.

Parducz, B. 1967. Ciliary movement and coordination in ciliates. Int. Rev. Cytol. 21:91–128.

Pitelka, D. R. 1970. Ciliate ultrastructure: Some problems in cell biology. J. Protozool. 17:1–10.

de Puytorac, P. and J. Grain. 1976. Ultrastructure du cortex buccal et évolution chez les Ciliés. Protistologica 12:49–67.

Spoon, D. M., G. B. Chapman, R. S. Cheng and S. F. Zane. 1976. Observations on the behavior and feeding mechanisms of the suctorian *Heliophyra erhardi* (Reider) Matthes preying on *Paramecium.* Trans. Am. Microsc. Soc. 95:443–462.

Tartar, V. 1961. *The Biology of Stentor*. Pergamon Press, New York.

Wichterman, R. 1986. *The Biology of Paramecium*, 2nd Ed. Plenum, New York. [A major reference.]

Other protozoa

Godsen, G. N. 1985. Molecular approach to malarial vaccines. Sci. Am. 252(2):52–59.

Levine, N. D. 1988. *The Protozoan Phylum Apicomplexa*. Vols. 1–2. CRC Press, Boca Raton, Florida. [Provides diagnoses of all genera and species.]

Mazier, D., R. Beaudoin, S. Mellour, P. Druithe, B. Texier, J. Trosper, F. Miltgen, I. Landau, C. Paul, O. Brandicourt, C. Guguen-Guillouzo and P. Langlois. 1985. Complete development of hepatic stages of *Plasmodium falciparum* in vitro. Science 227:440–442.

Miller, L. H. R. J. Howard, R. Carter, M. F. Good, V. Nussenzweig and R. S. Nussenzweig. 1986. Research toward malaria vaccines. Science 234:1349–1356.

Olive, L. S. 1975. *The Mycetozoans*. Academic Press, New York.

Schulman, S. S. and V. N. Semenovich. 1973. The life cycle in the Myxosporidia and the taxonomic position of the Cnidosporidia in the animal kingdom. Proc. IV Int. Cong. Protozool., Clermont Ferrand.

Sprague, V. 1977. Classification and phylogeny of the Microsporidia. *In* L. A. Bulla and T. Cheng (eds.), *Comparative Pathobiology*, Vol. 2, Systematics of the Microsporida. Plenum, New York.

Wolf, K. and M. E. Markiw. 1984. Biology contravenes taxonomy in the Myxozoa: New discoveries show alternation of invertebrate and vertebrate hosts. Science 225:1449–1452.

Phylogeny

Cavalier-Smith, T. 1975. The origin of nuclei and of eukaryotic cells. Nature 256:463–468.

Corliss, J. O. 1972. The ciliate Protozoa and other organisms: Some unresolved questions of major phylogenetic significance. Am. Zool. 12:739–753.

Corliss, J. O. 1974. Time for evolutionary biologists to take more interest in protozoan phylogenetics? Taxon 23:497–522.

Corliss, J. O. 1975. Nuclear characteristics and phylogeny in the protistan phylum Ciliophora. BioSystems 7:338–349.

Corliss, J. O. 1981. What are the taxonomic and evolutionary relationships of the protozoa to the Protista? BioSystems 14:445–459.

Corliss, J. O. and E. Hartwig. 1977. The "primitive" interstitial ciliates: Their ecology, nuclear uniquenesses, and postulated place in the evolution and systematics of the phylum Ciliophora. Akad. Wiss. Lit. (Mainz) Math.-Naturwiss. Kl. Mikrofauna Meeresboden 61:65–88.

Gray, M. W. and W. F. Doolittle. 1982. Has the endosymbiont hypothesis been proven? Microbiol. Rev. 46:1–42.

Jeon, K. W. (ed.) 1983. *Intracellular Symbiosis*. Int. Rev. Cytol. (Suppl.) 14:1–379.

John, P. and F. W. Whatley. 1975. *Paracoccus dentrificans*: A present-day bacterium resembling the hypothetical free-living ancestor of the mitochondrion. Symp. Soc. Exp. Biol. 29:39–40.

Lipscomb, D. L. and J. O. Corliss. 1982. *Stephanopogon*, a phylogenetically important "ciliate," shown by ultrastructural studies to be a flagellate. Science 215:303–304.

Margulis, L. 1970. *Origin of Eukaryotic Cells*. Yale University Press, New Haven, Connecticut.

Margulis, L. 1975. Symbiotic theory of the origin of eukaryotic organelles: Criteria for proof. Symp. Soc. Exp. Biol. 29:21–38.

Margulis, L. 1976. The genetic and evolutionary consequences of symbiosis. Exp. Parsitol. Rev. 39:277–349.

Margulis, L. 1978. Microtubules in prokaryotes. Science 200:1118–1124.

Margulis, L. 1980. Flagella, cilia, and undulipodia. BioSystems 12:105–108.

Margulis, L. 1981. *Symbiosis in Cell Evolution*. W. H. Freeman, San Francisco. [A current reassessment of the serial endoysymbiotic theory (SET), and an excellent review of the evolution of life on earth.]

Mahler, H. A. and R. A. Raff. 1975. The evolutionary origin of the mitochondrion: A non-symbiotic model. Int. Rev. Cytol. 43:1–124. [An argument against the serial symbiosis theory.]

Pohley, H. J., R. Dornhaus and B. Thomas. 1978. The amoebo-flagellate transformation: A system-theoretical approach. BioSystems 10:349–360.

Raff, R. A. and H. A. Mahler. 1975. The symbiont that never was: An inquiry into the evolutionary origin of the mitochondrion. Symp. Soc. Exp. Biol. 29:41–92.

Ragan, M. A. and D. J. Chapman. 1978. *A Biochemical Phylogeny of the Protists*. Academic Press, New York. [An exhaustive treatment of the biochemical data available on protistans and their evolutionary implications.]

Raikov, I. B. 1976. Evolution of macronuclear organization. Annu. Rev. Genet. 10:413–440.

Schwartz, R. M. and M. Dayhoff. 1978. Origins of prokaryotes, eukaryotes, mitochondria, and chloroplasts. Science 199:395–403.

Taylor, F. J. R. 1976. Flagellate phylogeny: A study in conflicts. J. Protozool. 23:28–40.

Taylor, F. J. R. 1978. Problems in the development of an explicit hypothetical phylogeny of the lower eukaryotes. BioSystems 10:67–89.

Chapter Six

Four Phyla of Uncertain Affinity

The only solid piece of scientific truth about which I feel totally confident is that we are profoundly ignorant about nature.

Lewis Thomas
The Medusa and the Snail, 1979

As we discussed in Chapter 4, there are a number of extant organisms that cannot conveniently be categorized at either the protozoan or metazoan grade of complexity. Over the years, three phylum names have emerged to include these animals: Placozoa (for *Trichoplax adhaerens*); Monoblastozoa (for *Salinella*); and Mesozoa, usually divided into the orders Rhombozoa or Dicyemida (e.g., *Dicyema, Pseudicyema*) and Orthonectida (e.g., *Rhopalura*). Altogether these creatures make up only 100 or so species. *Trichoplax* is marine (first discovered in marine aquaria); *Salinella* lives in salt beds; the rhombozoans are symbionts in the nephridia of cephalopod molluscs; and the orthonectids are parasitic in a variety of invertebrates (e.g., echinoderms, molluscs, nemerteans, polyclads, and polychaete worms). All of these animals have been the subjects of a great deal of taxonomic and phylogenetic controversy, much of which will no doubt remain unresolved for some time to come.

Taxonomic history

The first group of these enigmatic animals to be discovered was the Rhombozoa, described and named by A. Krohn (1839) in Germany. But it was not until 1876 that a careful study of dicyemids was published by the Belgian zoologist Edouard van Beneden. He was convinced that these odd parasites represented a true link between the protozoa and the metazoa, and coined the name Mesozoa (= middle animals) to emphasize his point of view. By the end of the nineteenth century, *Trichoplax, Salinella,* and the orthonectids had also been discovered. The phylum Mesozoa soon became a dumping ground for a host of multicellular but presumed nonmetazoan organisms. Over time, all of these animals except the rhombozoans and orthonectids were removed to other taxa as further studies revealed them to be protozoa, larval stages, or simply unrelated to one another or to other established groups. In the past, *Salinella* has been treated by some as a protozoon and by others as a larval stage, and *Trichoplax* has been considered to be a hydrozoan planula larva. These suggestions, however, have been rejected by most workers. Based on our current understanding of these organisms, it is probably most appropriate to leave them as sole members of two separate phyla until and unless more information dictates otherwise.

The rhombozoans and orthonectids were for many years treated as closely related groups composing a monophyletic taxon of phylum rank—the Mesozoa. These organisms have typically been assigned to two orders, sometimes in a single class called the Moruloidea to characterize their "ball-of-cells" grade of construction. More recently, some authors have assigned these organisms to two classes rather than to two orders. However, some workers (Kozloff 1969; Lapan and Morowitz 1972) are convinced that Rhombozoa and Orthonectida are not at all closely related to each other and should not even be classified together in a single phylum. We agree with this point of view and abandon the name Mesozoa as a formal taxon. We use the term mesozoa to represent organisms at a particular grade of complexity—just as the terms protozoa and metazoa are used in this book—without implying any particular evolutionary relationships. Thus, Rhombozoa,

Orthonectida, Placozoa, and Monoblastozoa are treated as separate phyla, each containing animals belonging to the mesozoan grade of complexity. These and other phylogenetic matters are discussed at the end of this chapter.

Mesozoan *Baupläne*

You will recall that multicellularity is only one of the criteria by which the metazoan grade of complexity is established. While mesozoa certainly satisfy this requirement, they do not have the typical layered construction of higher animals, and they do not obviously pass through any developmental stage that may be equated unequivocally with gastrulation. The survival strategies among mesozoa are varied and are reflected partly in the body plans of the adults (Figure 1). In each case, certain advantages of regional specialization are realized by division of labor among their component cells, but there has been no development of true tissues or complex organ systems.

Phylum Placozoa

Trichoplax adhaerens was discovered in 1883 in a seawater aquarium at the Graz Zoological Institute in Austria. Specimens have subsequently been found in various marine situations around the world. In recent years this organism has been studied extensively by Grell and Ruthmann (see references). The body of *Trichoplax* consists of several thousand cells arranged as a simple double-layered plate (Figure 1A). It lacks anterior–posterior polarity and symmetry. The upper and lower cell layers are distinct from one another; the cells of the two layers differ in shape, and there is a consistent dorsal–ventral orientation of the body relative to the substratum. The dorsal cells are flattened, bear a single flagellum, and contain lipid droplets. The ventral cells are more columnar, some bear a flagellum, and they lack distinct oil droplets. Furthermore, as described in Chapter 4 (see Figure 16E in Chapter 4), the ventral epithelium can be temporarily invaginated, presumably for feeding. This observation supports the notion that there are functional as well as structural differences between the two cell layers. Between these two epithelial sheets is a mesenchymal layer of stellate ameboid cells embedded in a supportive gel matrix. Grell (1982) considers *Trichoplax* to be a true diploblastic metazoon and suggests that the upper and lower epithelia are homologous to ectoderm and entoderm, respec-

tively. However, a basal membrane (basal lamina) has not yet been identified beneath the epithelia, which suggests that *Trichoplax* is closer in organization to the Porifera than to the true diploblastic metazoa.

Trichoplax moves by flagellar gliding along a solid surface, aided by irregular shape changes along the body edges. Very small individuals—probably just recently developed from eggs—can swim, while larger individuals crawl. Most evidence suggests that *Trichoplax* feeds by phagocytosis of organic detritus. The phagocytosis occurs only in the invaginated cells of the ventral epithelium. Although there is no evidence for extracellular digestion, the speculation has been made that the *Trichoplax* may secrete digestive enzymes onto its food within the ventral "digestive pocket."

Trichoplax reproduces asexually by fission of the entire body into two new individuals, and by a budding process that yields numerous multicellular flagellated "swarmers," each of which forms a new individual. Sexual reproduction is also known, followed by a developmental period of holoblastic cell division and growth. Eggs have been observed within the mesenchyme, but their origin is unknown. A *Trichoplax* has very little DNA, about as much as a bacterium or protist, and its chromosomes are very small.

Phylum Monoblastozoa

Salinella has apparently not been studied at all since its discovery in 1892 by Frenzel, who found it in cultures of Argentine salt-bed material. There is serious question about the accuracy of the original description, and we caution you that *Salinella* may have existed more in Frenzel's imagination than in Argentina. According to Frenzel, the body wall of *Salinella* consists of but a single layer of cells. The inner cell borders line a cavity, which is open at both ends (Figure 1B). The openings function as an anterior "mouth" and posterior "anus," both of which are ringed by bristles. The rest of the body, inside and out, is densely ciliated.

The animal is said to move by ciliary gliding, much like ciliated protozoa and small flatworms. *Salinella* is thought to feed by ingesting organic detritus through the "mouth" and digesting it in the internal cavity. Undigested material is carried to the "anus" by ciliary action. Asexual reproduction takes place by transverse fission of the body, and sexual reproduction is suspected. The real nature of this animal, including its very existence, remains elusive.

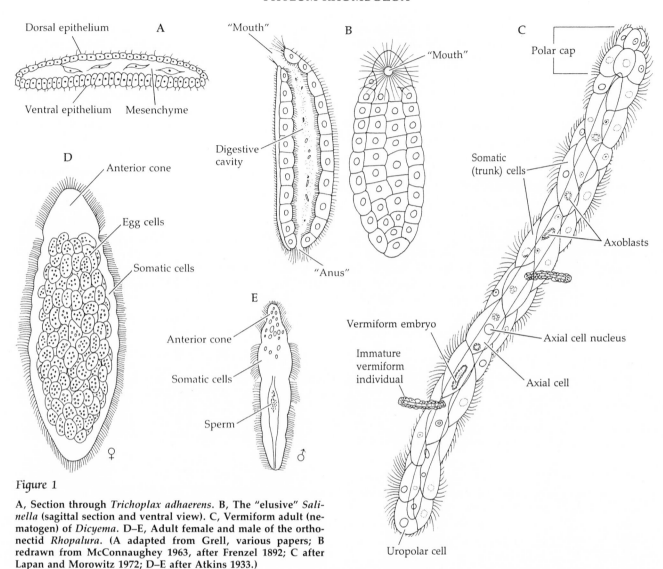

Figure 1

A, Section through *Trichoplax adhaerens*. B, The "elusive" *Salinella* (sagittal section and ventral view). C, Vermiform adult (nematogen) of *Dicyema*. D–E, Adult female and male of the orthonectid *Rhopalura*. (A adapted from Grell, various papers; B redrawn from McConnaughey 1963, after Frenzel 1892; C after Lapan and Morowitz 1972; D–E after Atkins 1933.)

Phylum Rhombozoa

Stunkard (1982) considers the taxon Rhombozoa as a class comprising the orders Dicyemida and Heterocyemida. While we treat the Rhombozoa as a phylum, we retain the ordinal level assignment of the two subtaxa. The rhombozoan *Bauplan* includes a solid body construction. An outer layer of somatic/nutritive cells surrounds an inner core of reproductive cells (or often a single cell). This general architecture is also seen in the Orthonectida, and that similarity has provided the major argument in favor of a close relationship between the two groups.

Dicyemids are more common and better understood than heterocyemids. Members of both groups are obligate symbionts in the nephridia of cephalopod molluscs. Various terminologies have been applied to the rhombozoans, especially to the different stages in their life cycles. We have drawn from several sources in an attempt to use the most descriptive terms. Some of the frequently used alternative terms are also included.

The Dicyemida

An adult dicyemid is called a **vermiform adult** or a **nematogen**. The body of a nematogen consists of an outer sheath of ciliated **somatic cells**, the number of which is constant for any particular species.* Within the covering composed of somatic cells lies a single, long **axial cell** (Figure 1C). Eight or nine so-

*The constancy of the number of cells is called **eutely** and is a common feature of many microscopic and near-microscopic organisms.

matic cells at the anterior end form a distinctive **polar cap**. Immediately behind the polar cap are two **parapolar cells**. The rest of the 10 to 15 somatic cells are sometimes called **trunk cells**; the two most posterior cells are called the **uropolar cells**.

Young dicyemids are motile and swim about in the host's urine by ciliary action. The adults, however, attach to the inner linings of the nephridia by their polar caps. There is no conclusive evidence that these animals cause any damage to their hosts, but when present in very high numbers they may interfere with the normal flow of fluids through the nephridia. Nematogens take in particulate and molecular nutrients from the host's urine by phagocytotic and pinocytotic action of the somatic cells. Once the adult has attached to the host, the somatic cilia probably serve to keep fluids moving over the body to bring nutrients in contact with the surface cells. Although in nature dicyemids appear to be obligately associated with cephalopods, they have been suc-

cessfully maintained in experimental nutrient media (see Lapan and Morowitz 1972).

The stages of the dicyemid life cycle that occur outside the host are still incompletely known. Lapan and Morowitz (1972) have posed some likely inferences, however. This portion of the life cycle includes both asexual and sexual processes, but without a regular alternation between them. The cytoplasm of the axial cell of the vermiform adult contains numerous tiny cells called **axoblasts**. Immature vermiform organisms are produced asexually by a sort of embryogeny of individual axoblasts within the parent axial cell (Figure 2). The first division of an axoblast is unequal and produces a large presumptive axial cell and a small presumptive somatic cell. The presumptive somatic cell divides repeatedly, and its daughters move by epiboly to enclose the presumptive axial cell, which has not yet divided. When this inner cell finally divides, it divides unequally, and the smaller daughter cell is then engulfed by the

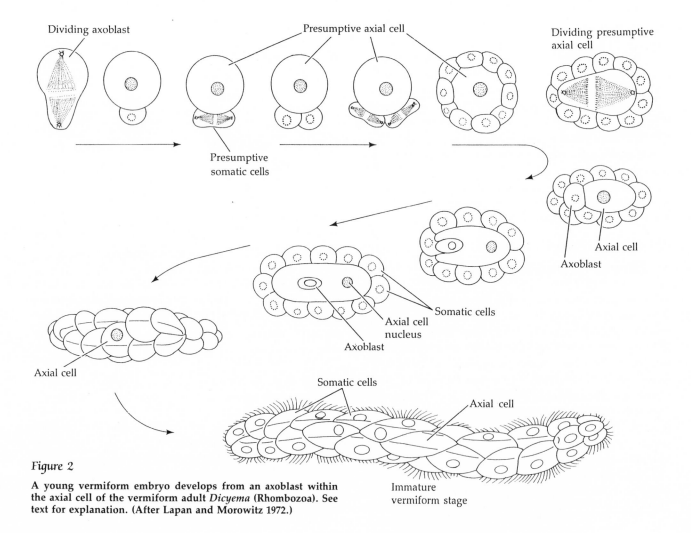

Figure 2

A young vermiform embryo develops from an axoblast within the axial cell of the vermiform adult *Dicyema* (Rhombozoa). See text for explanation. (After Lapan and Morowitz 1972.)

larger one. The larger one becomes the progeny's axial cell proper with its single nucleus, and the smaller engulfed cell becomes the progenitor of all future axoblasts within that axial cell. The "embryo," which now consists of its own central axial cell surrounded by somatic cells, elongates and the somatic cells acquire cilia. The resulting structure is a miniature vermiform organism. The immature vermiform organism leaves the parent vermiform adult and swims about in the nephridial fluids. Eventually, it attaches to the host and enters the adult stage of the life cycle.

The initiation of sexual reproduction in dicyemids may be a density-dependent phenomenon associated with high numbers of vermiform individuals within the host's nephridia. The switch from asexual to sexual processes might be a response to some chemical factor that accumulates in the urine of the host (Lapan and Morowitz 1972). Other workers suggest that sexual reproduction in dicyemids is brought on by the sexual maturation of the host (e.g., Hyman 1940; Stunkard 1982; Hochberg 1983). In any event, as the vermiform adults become sexually "motivated," their somatic cells usually enlarge as they become filled with a yolky material; the name **rhombogen** is often applied to the individuals in this stage (Figure 3). Because the reported differences between rhombogens and nematogens are not consistently found, it is better to call them simply sexual and asexual vermiform adults, respectively.

The axoblasts of sexual vermiform adults develop into multicellular structures called **infusorigens**. These forms consist of an outer layer of ova and an inner mass of sperm (Figure 3B). The infusorigens are retained within the parent's axial cell. They have been likened to separate hermaphroditic individuals or to transient double-sexed gonads. The centrally located sperm fertilize the peripherally arranged ova, and each zygote develops into a ciliated **infusoriform larva** (Figure 3C). This larva has a fixed number of cells; the two anteriormost cells—called **apical cells**—contain very high density substances within their cytoplasm. The rest of the surface cells are ciliated and form a sheath around a ring of **capsule cells**, which in turn enclose four central cells. The infusoriform larvae escape from the parent vermiform adult and pass out of the host's body with the urine.

The events of the dicyemid life cycle that occur outside the cephalopod host remain a mystery. Some workers have held to the belief that the infusoriform larva enters an intermediate host (presumably some benthic invertebrate), but the best evidence to date suggests that this is not the case. While much re-

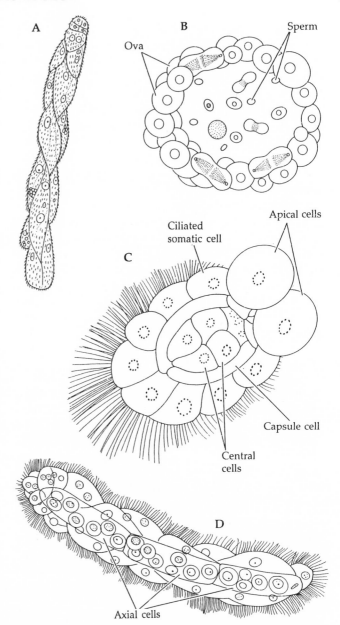

Figure 3

Sexual reproduction in dicyemid rhombozoans A, Sexual (rhombogen) form of vermiform adult. B, Infusorigen of sperm and ova formed within the axial cell of the vermiform adult. C, Infusoriform larva produced by fertilization. D, Stem nematogen with three axial cells. See text for explanation. (A redrawn from Hyman 1940; B–C after Lapan and Morowitz 1972; D redrawn from McConnaughey 1963, after Nouvel 1948.)

mains to be learned, the following scenario seems most plausible. After leaving the host, the larva sinks to the bottom—the dense contents of the apical cells serving as ballast. The larva, or some persisting part of the larva (perhaps the innermost four cells), enters another cephalopod host. This infectious individual

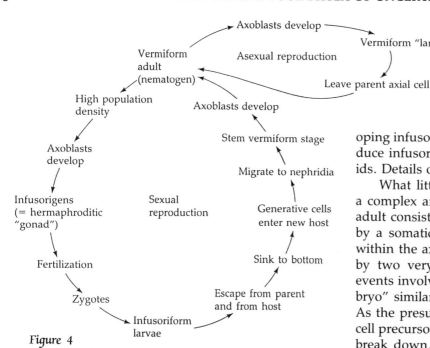

Figure 4

Life cycle of a dicyemid.

travels through the host, probably via the circulatory system, and enters the nephridia, where it becomes a so-called **stem nematogen** (Figure 3D). The stem nematogen is similar to the vermiform adult except that the former has three axial cells rather than one. Axoblasts within the axial cells of the stem nematogen give rise to more vermiform adults, just as the axoblasts within the adults described earlier did. The vermiform adults produce more individuals like themselves until the onset of sexual reproduction is triggered again, presumably by the high population density. This life cycle is schematically represented in Figure 4.

The Heterocyemida

Only two species are included in this group of rhombozoans. *Conocyema polymorpha* lives in the nephridia of octopuses, and *Microcyema gracile* in *Sepia* (the cuttlefish). These two heterocyemids differ from each other in certain respects.

The vermiform adult of *Conocyema* bears a polar cap of four enlarged cells and has a trunk of somatic cells around an inner axial cell; all the cells of the body lack cilia (Figure 5A). The axial cell contains axoblasts, which give rise to ciliated "larvae" that escape from the parent, lose their cilia, and grow into more vermiform adults within the host. The individuals that produce the infusorigens lack a polar cap. They have only a very thin layer of somatic cells surrounding the axial cell, which contains the devel-

oping infusorigens (Figure 5B). The infusorigens produce infusoriform larvae similar to those of dicyemids. Details of the life cycle of *Conocyema* are lacking.

What little is known about *Microcyema* suggests a complex and distinctive life cycle. The vermiform adult consists of a single inner axial cell surrounded by a somatic syncytium (Figure 6A). The axoblasts within the axial cell produce more vermiform adults by two very different methods. One sequence of events involves the formation of a multicellular "embryo" similar to that seen in dicyemids (Figure 6B). As the presumptive somatic cells surround the axial cell precursor, the cell boundaries of the somatic cells break down, thereby resulting in an ameboid individual in which a syncytial mass surrounds the growing axial cell (Figures 6C and D). This individual apparently develops into a new vermiform adult. The other asexual process involves the formation of ciliated **Wagener's larvae** from the axoblasts (Figure 6E). These larvae leave the parent, swim about in the

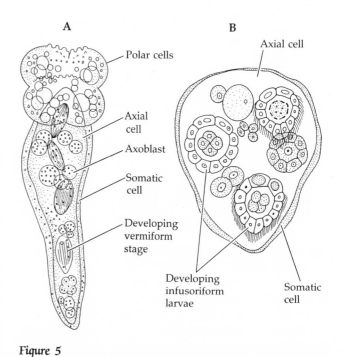

Figure 5

The heterocyemid *Conocyema*. A, Vermiform adult. B, During the reproductive phase, infusoriform larvae are formed within the adult's axial cell (cross section). (After Hyman 1940.)

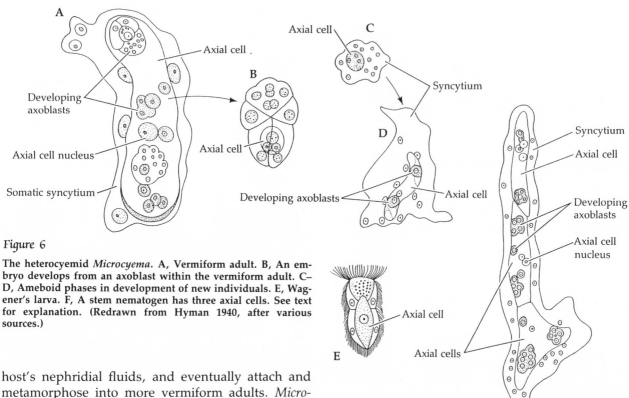

Figure 6

The heterocyemid *Microcyema*. A, Vermiform adult. B, An embryo develops from an axoblast within the vermiform adult. C–D, Ameboid phases in development of new individuals. E, Wagener's larva. F, A stem nematogen has three axial cells. See text for explanation. (Redrawn from Hyman 1940, after various sources.)

host's nephridial fluids, and eventually attach and metamorphose into more vermiform adults. *Microcyema* adults also produce infusorigens and infusoriform larvae that are much like those of the dicyemids. The infusoriform larvae apparently leave the host via the urine, but nothing is known about the stages of the life cycle outside the host. It is assumed that the infusoriform larva enters a host and matures into a ciliated nematogen, which has three axial cells. Stem nematogens have been observed in host animals. Eventually, the cilia and the cell boundaries between adjacent somatic cells are lost (Figure 6F), and the animal develops into another vermiform adult.

Phylum Orthonectida

The life cycles of some orthonectids are well known and in certain respects differ markedly from those of the rhombozoans. Asexual individuals dominate the life cycle; they are ameboid syncytial forms, commonly called **plasmodial stages** (Figure 7A). Some plasmodia grow and spread to such an extent that they cause severe damage to the host. For example, *Rhopalura ophiocomae* (in the brittle star *Amphipholis squamata*) and *R. granosa* (in the bivalve mollusc *Heteranomia squamula*) destroy the gonads of their hosts.

The plasmodium produces more syncytial masses by fragmentation, and it also gives rise to the sexual individuals. Certain nuclei within the plasmodium (frequently called **agametes**) accumulate and partition off a small amount of cytoplasm (Figure 7A). These cellular units within the plasmodium undergo cleavage (Figures 7B–D), and eventually form the sexual individuals. In most species, a single plasmodium produces only males or females. In a few species, however, both sexes are produced from one plasmodium. The sexual organism consists of an outer layer of ciliated somatic cells and an inner mass of gametes (Figures 1D,E and 7D,E). Between the gametes and the outer somatic layer are what appear to be contractile cells. Upon maturation, the sexual forms leave the parent plasmodium; then they leave the host and swim about in the environment. In the gonochoristic species, the male attaches to the female and deposits sperm through a small genital pore located near the posterior end of her body (Figure 7E). The sperm fuse with ova to form zygotes. Each zygote develops into a ciliated larva (Figure 7F), which escapes from the female's body and eventually enters another host animal. Once inside the host, the larva loses its somatic cells and releases the mass of inner

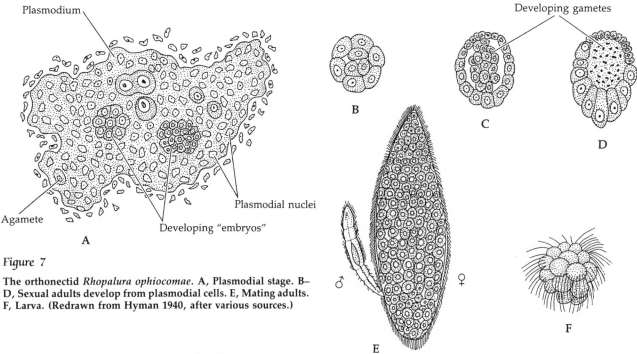

Figure 7

The orthonectid *Rhopalura ophiocomae*. A, Plasmodial stage. B–D, Sexual adults develop from plasmodial cells. E, Mating adults. F, Larva. (Redrawn from Hyman 1940, after various sources.)

cells, each of which apparently develops into a new plasmodium. The life cycle is shown in Figure 8.

Mesozoan phylogeny

Some of the controversies about the origins and evolution of mesozoan creatures have already been mentioned. We cannot, of course, solve these problems here, but we can offer some hypotheses that provide a basis for discussion, speculation, and further testing.

If *Salinella* is real, it is difficult to relate it to any known mesozoan or metazoan taxon. Structurally and functionally, it seems to belong to the protozoan grade. Perhaps *Salinella* arose from some colonial flagellate ancestor that took up benthic life and assumed a somewhat bilateral form. By this hypothesis, the most striking apomorphic feature of *Salinella* is its complete digestive cavity with inwardly directed cilia. The animal's nutritive activities are assumed by the inside surfaces of the single layer of body cells, while its locomotor functions are performed by the outer cell surfaces.

It is tempting to view *Trichoplax* as a surviving descendant of some premetazoan ancestor. The most acceptable hypotheses concerning the origin of the metazoan condition depend upon the evolution of a layered construction through some form of gastrulation (see Chapter 4). Perhaps a number of premetazoa "experimented" evolutionarily with various at-

tempts at layered construction, one or more of which eventually led to the origin of the metazoa. One of these events could have been the formation of a temporary digestive pouch by invagination or simple inpocketing, thus increasing the surface area for feeding. *Trichoplax* may be an extant remnant of that ancient event. We know that certain colonial flagellates tend to have groups of cells that are somewhat specialized for various functions. It would seem to be a relatively small evolutionary step to turn the nutritive cells inward during feeding. Such an event would have been particularly advantageous to benthic animals.

The rhombozoans and orthonectids present phylogenetic problems that have been argued enthusiastically for decades. Some authors suggest that the relatively simple construction of these animals is primitive rather than the result of degeneration from some more complex metazoan ancestor. The rhombozoans and orthonectids may have arisen as a side branch from some protozoan–metazoan lineage, from some ancient planuloid, or from ciliate protozoa. Many other specialists have championed the idea that these animals are descended from an established metazoan stock, and that their simple construction represents a degeneration associated with parasitic habits. Lameere (1922) suggested that the orthonectids may have arisen from echiuran worms, apparently because some echiurans (e.g., *Bonellia*) show extreme sexual dimorphism, with tiny, reduced males. Lameere likened this feature to the dimorphic nature of some orthonectids, but his hypothesis never gained much favor. The most likely candidates for such an ancestral group are probably found among the parasitic trematodes (Platyhelminthes), an idea most strongly supported by Stunkard (1954, 1972). This contention is based largely on general morphological features and on the complex life cycles of these animals. Space does not permit a complete examination of Stunkard's views, but arguments against them have been presented by several authors (Dodson 1956; Kozloff 1969; Lapan and Morowitz 1972; Hochberg 1983).

In our opinion, one can view the complex life cycles of these organisms and the trematodes as examples of convergence associated with their parasitic lifestyles. Such life cycles are common among many parasitic organisms, metazoa and protozoa, and we thus view the mesozoan grade of construction as primitive.

Our treatment of the orthonectids and rhombozoans as unrelated taxa is based upon several considerations. Their similarities appear to be superficial results of convergence related to their similar lifestyles and general level of complexity. In terms of body construction, the rhombozoa and orthonectids are similar only in their basic strategy of relegating

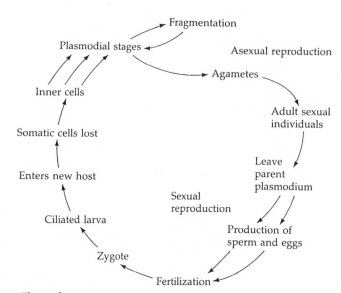

Figure 8
Generalized life cycle of an orthonectid.

reproductive cells to the inside while maintaining an outer covering of somatic cells that function in nutrition, protection, and locomotion. Many workers have compared the sexual adult stages of orthonectids to the asexual vermiform adults of dicyemids. However, the sexual stages are drastically different from one another, and this lack of similarity is a major criterion for separating the two groups.

Selected References

Atkins, D. 1933. *Rhopalura granosa* sp. nov. an orthonectid parasite of a lamellibranch *Heteranomia squamula* L. with a note on its swimming behavior. J. Mar. Biol. Assn. U.K. 19:233–252.

Caullery, M. 1961. Classe des Orthonectides. *In* P. Grassé (ed.), *Traité de Zoologie.* Masson et Cie, Paris, pp. 695–706.

Caullery, M. and A. Lavellée. 1908. La fécondation et le développement de l'oeuf des Orthonectides. I. *Rhopalura ophiocomae.* Arch. Zool. Exp. Gén., Series 4(8):421–469.

Caullery, M. and F. Mesnil. 1901. Recherches sur les Orthonectides. Arch. Anat. Microsc. 4:381–470.

Dodson, E. O. 1956. A note on the systematic position of the Mesozoa. Syst. Zool. 5(1):37–40.

Frenzel, J. 1892. Untersuchungen über die mikroskopische Fauna Argentiniens. Arch. Naturgesch. 58:66–96.

Grassé, P. 1961. Classe des Dicyémides. *In* P. Grassé (ed.), *Traité de Zoologie.* Masson et Cie, Paris, pp. 707–729.

Grell, K. 1971a. *Trichoplax adhaerens* F.E. Schulze und die Entstehung der Metazoen. Naturwiss. Rundsch. 24:160–161.

Grell, K. 1971b. Embryonalentwicklung bei *Trichoplax adhaerens* F.E. Schulze. Naturwiss. 58:570.

Grell, K. 1972. Eibildung und Furchung von *Trichoplax adhaerens* F.E. Schulze (Placozoa). Z. Morphol. Tiere 73:297–314.

Grell, K. 1973. *Trichoplax adhaerens* and the origin of the Metazoa. Actualites Protozoologiques. I^ve Cong. Int. Protozoologie. Paul Couty, Clermont-Ferrand.

Grell, K. 1982. Placozoa. *In* S. Parker (ed.), *Synopsis and Classification of Living Organisms.* McGraw-Hill, New York, p. 639.

Grell, K. and G. Benwitz. 1971. Die Ultrastruktur von *Trichoplax adhaerens* F.E. Schulze. Cytobiologie 4:216–270.

Grell, K. and G. Benwitz. 1974. Elektronenmikroskopische Beobachtungen über das Wachstum der Eizelle und die Bildung der "Befruchtungsmembran" von *Trichoplax adhaerens* F.E. Schulze (Placozoa). Z. Morphol. Tiere 79:295–310.

Hatschek, B. 1888. *Lehrbuch der Zoologie.* Jena.

Hochberg, F. G. 1983. The parasites of cephalopods: A review. Mem. Natl. Mus. Victoria 44:109–145.

Hyman, L. 1940. *The Invertebrates.* Vol. 1, Protozoa through Ctenophora. McGraw-Hill, New York.

Kozloff, E. 1965. *Ciliocincta sabellariae* gen. and sp. n., an orthonectid mesozoan from the polychaete *Sabellaria cementarium* Moore. J. Parasitol. 51(1):37–44.

Kozloff, E. 1969. Morphology of the orthonectid *Rhopalura ophiocomae.* J. Parasitol. 55(1):171–195.

Kozloff, E. 1971. Morphology of the orthonectid *Ciliocincta sabellariae.* J. Parasitol. 57(3):585–597.

Lameere, A. 1916. Contributions à la connaissance des Dicyémides. Première partie. Bull. Sci. Fr. Belg. 59:1–35.

Lameere, A. 1918. Contributions à la connaissance des Dicyémides. Deuxieme partie. Bull. Biol. 51:347–390.

Lameere, A. 1922. L'histoire naturelle des Dicyémides. Brussels Acad. R. Belg. Bul. Cl. Sci. Ser. 5(8):779–792.

Lapan, E. and H. Morowitz. 1972. The Mesozoa. Sci. Am. 227(6):94–101.

McConnaughey, B. H. 1963. The Mesozoa. *In* E. C. Dougherty (ed.), *The Lower Metazoa*. University of California Press, Berkeley, pp. 151–168.

Nouvel, H. 1948. Les Dicyémides. 2° partie: Infusoriforme, teratologie, spécificité du parasitisme, affinitiés. Arch. Biol. Paris 59:147–223.

Ruthmann, A. 1977. Cell differentiation, DNA content, and chromosomes of *Trichoplax adhaerens* F.E. Schulze. Cytobiologie 15:58–64.

Ruthmann, A. and H. Wenderoth. 1975. Der DNA-Gehalt der Zellen bei dem primitiven Metazoon *Trichoplax adhaerens* F.E. Schulze. Cytobiologie 10:421–431.

Stunkard, H. 1954. The life-history and systematic relations of the Mesozoa. Q. Rev. Biol. 29:220–244.

Stunkard, H. 1972. Clarification of taxonomy in Mesozoa. Syst. Zool. 21(2):210–214.

Stunkard, H. 1982. Mesozoa. *In* S. Parker (ed.), *Synopsis and Classification of Living Organisms*. McGraw-Hill, New York, pp. 853–855.

van Beneden, É. 1876. Recherches sur les Dicyémides. Bull. Acad. Belg. Cl. Sci. Series 2(41):1160–1205; 2(42):35–97.

van Beneden, É. 1882. Contribution a l'histoire des Dicyémides. Arch. Biol. Paris 3:195–228.

Whitman, C. O. 1882. A contribution to the embryology, life history, and classification of the dicyemides. Mitt. Zool. Sta. Neapel. 4:1–89.

Chapter Seven

Phylum Porifera: The Sponges

*Concerning the relationship of the three classes
of sponges, practically nothing can be said.*
    Libbie Hyman
    The Invertebrates, Vol. 1, 1940

The phylum Porifera (Latin *porus*, "pore"; *ferre*, "to bear") contains those animals commonly called sponges. Figures 1 and 2 illustrate a variety of sponge body forms and some poriferan anatomy. Box 1 lists the major characteristics of sponges. Poriferans are sessile, suspension-feeding metazoa that utilize flagellated cells called **choanocytes** to circulate water through a unique system of water canals. Porifera is the only phylum within the **parazoan** grade (i.e., metazoa lacking true embryological germ layering). Not only are true tissues wanting, but most of the body cells are totipotent—they retain a high degree of mobility and are capable of changing form and function. Despite the fact that sponges are large-bodied multicellular animals, they function largely like organisms at the unicellular grade of complexity. As you will discover in this chapter, their strategies of nutrition, cellular organization, gas exchange, reproduction, and response to environmental stimuli are all very protozoa-like.

About 9,000 living species of sponges have been described, nearly all of which are restricted to benthic marine environments. They occur at all depths, but unpolluted littoral habitats harbor especially rich sponge faunas. Most littoral sponges are **encrusting**, forming thick or thin layers on hard surfaces. Benthic sponges that live on soft substrata are often upright and tall, thus avoiding burial by shifting sediments in their environment. Sponges on tropical reefs may reach considerable size (a meter or more in height in the Caribbean), and in such areas they may constitute a significant portion of reef biomass. Subtidal and deeper-water species that do not confront strong tidal currents or surge are usually large and exhibit a stable symmetrical external form. The deeper water hexactinellid sponges often assume unusual shapes, many being delicate glasslike structures, others round and massive, and still others ropelike (Figure 1).

Sponges come in just about every color imaginable, including bright lavenders, blues, yellows, crimsons, and white. Many species harbor symbiotic bacteria or unicellular algae that can color their bodies.

Taxonomic history and classification

The sessile nature of sponges and their generally amorphous (asymmetrical) growth form convinced the earliest naturalists that they were plants. It was not until 1765, when the nature of the internal water currents was described, that sponges were recognized as animals. The great naturalists of the late eighteenth and early nineteenth centuries (Lamarck, Linnaeus, Cuvier) classified the sponges under Zoophytes or Polypes, regarding them as allied to anthozoan cnidarians. Throughout much of the nineteenth century they were placed with cnidarians under the name Coelenterata or Radiata. The morphology and physiology of sponges were first adequately understood by R. E. Grant. Grant created for them the name Porifera, although other names were frequently used (e.g., Spongida, Spongiae, Spongiaria). Huxley (1875) and Sollas (1884) first proposed the complete separation of sponges from other metazoa, but this view was not generally accepted until the early twentieth century.

Historically, the classes of Porifera have been defined by the nature of their internal skeletons. Until recently, four classes were recognized: Calcarea, Hex-

Box One

Characteristics of the Phylum Porifera

1. Metazoa at the cellular grade of construction; without true tissues; adults asymmetrical or radially symmetrical

2. Cells tend to be totipotent

3. With unique flagellated cells—choanocytes—that drive water through canals and chambers constituting the aquiferous system

4. Adults are sessile suspension-feeders; larval stages are motile

5. Outer and inner cell layers lack a basement membrane

6. Middle layer—the mesohyl—variable, but always includes motile cells and usually some skeletal material

7. Skeletal elements, when present, composed of calcium carbonate, silicon dioxide, and/or collagen fibers

sponge biochemical taxonomy). The great variability in sponge morphology and the difficulty in precisely setting the limits of a sponge species have probably driven many potential poriferologists to frustration (and to other taxa) early in their careers. Even the great sponge taxonomist Arthur Dendy was known to frequently end a species diagnosis with a question mark. This state of affairs was summarized by one student regarding California sponge studies (Ristau 1978):

> The study of California sponges has not generated a fervor of activity over the years, nor has the literature been saturated with information about this little-studied . . . phylum. Probably the greatest interest generated by sponges occurred recently, when several news agencies reported that giant, and presumably mutant, sponges were found growing on undersea nuclear waste storage containers (San Francisco Chronicle, September 14, 1976). It has been rumored that the Japanese are now planning a motion picture in which a sleeze* of giant sponges rises from the depths of the Farallon Islands and phagocytizes the North Beach area of San Francisco. Undoubtedly, when this epic materializes, research and interest in California sponges will increase. Until that time, however, those interested in the sponge fauna of this area must be content with the paucity of scientific literature on this subject.

Recently a host of important bioactive compounds has been discovered in sponges, many of which appear to be of pharmacological significance. This discovery has led to a renewed interest in sponges and a call for the training of more sponge taxonomists.

actinellida, Demospongiae, and Sclerospongiae. The Sclerospongiae included those species that produce a solid, calcareous, rocklike matrix on which the living animal grows. These poriferans are also known as coralline sponges. Demospongiae is the largest class, comprising about 95 percent of the living species. Because of its size and variability, the Demospongiae presents the most problems to sponge taxonomists. In a series of papers published between 1953 and 1957, Levi proposed an important reappraisal of the Demospongiae, incorporating reproductive characteristics for the first time. The class Sclerospongiae was recently abandoned, and its members relegated to the Calcarea and Demospongiae (Vacelet 1985). It has also been suggested that the Hexactinellida should be removed from the Porifera and assigned to a separate phylum, Symplasma (Bergquist 1985).

More and more, specialists are resorting to embryological, biochemical, histological, and cytological methods to positively identify or diagnose sponge taxa (see papers by Minale et al. 1976, Faulkner 1977, and Bergquist and Wells 1983 for recent reviews of

PHYLUM PORIFERA

CLASS CALCAREA: Calcareous sponges (Figure 1A). Spicules of mineral skeleton composed entirely of calcium carbonate laid down as calcite; skeletal elements often not differentiated into megascleres and microscleres; body with asconoid, synconoid, or leuconoid construction; all marine.

SUBCLASS CALCINIA: Free-living larvae are hollow and flagellated, and can become solid parenchymula-like structures by cellular ingression; choanocyte nuclei located basally; flagellum arises independent of nucleus. (e.g., *Clathrina, Dendya, Leucaltis, Leucetta*)

SUBCLASS CALCARONIA: Free-living larvae are partly flagellated amphiblastulae; choanocyte nuclei apical; flagellum arises directly from nucleus. (e.g., *Amphoriscus, Grantia, Lelapia, Leucandra, Leucilla, Leucosolenia, Scypha* [= *Sycon*])

*"Sleeze" is a term coined by Ristau for an aggregation of sponges; the usage is comparable to other such nouns that define animal groups (e.g., flock, herd, gaggle).

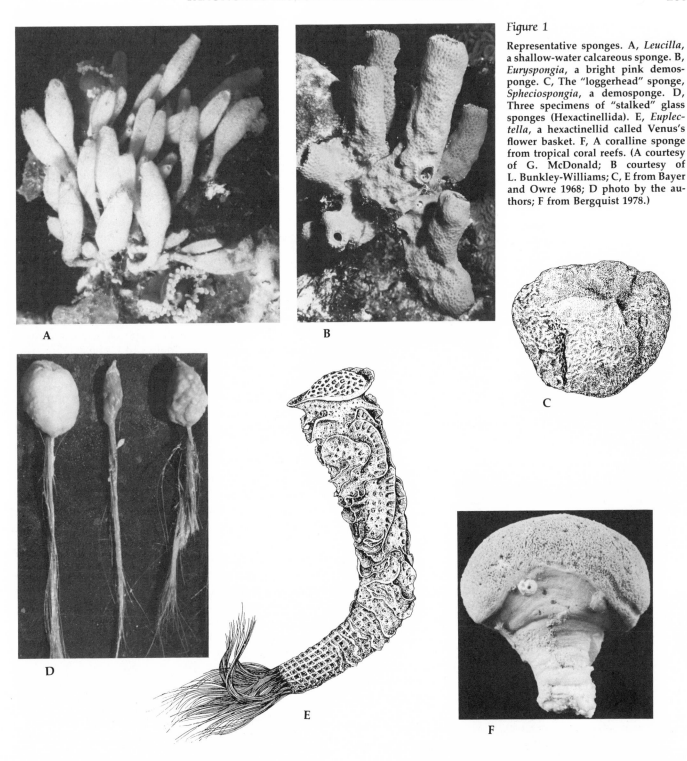

Figure 1

Representative sponges. A, *Leucilla*, a shallow-water calcareous sponge. B, *Euryspongia*, a bright pink demosponge. C, The "loggerhead" sponge, *Spheciospongia*, a demosponge. D, Three specimens of "stalked" glass sponges (Hexactinellida). E, *Euplectella*, a hexactinellid called Venus's flower basket. F, A coralline sponge from tropical coral reefs. (A courtesy of G. McDonald; B courtesy of L. Bunkley-Williams; C, E from Bayer and Owre 1968; D photo by the authors; F from Bergquist 1978.)

SUBCLASS PHARETRONIDIA: Larvae and choanocytes variable; massive reinforcement of calcite added to skeleton as discrete spicules (spicules aligned in tracts, welded into a framework or consolidated as an external or internal aspicular calcareous matrix); oscula with tetractinal (= quadriradiate) spicules. (e.g., *Murrayona*, *Petrobiona*)

SUBCLASS SPHINCTOZOA: Represented by only a single living species (*Neocoelia crypta*) but numerous extinct forms. Spicules absent, but body contains a solid cortical aragonite skeleton that consists of a series of chambers added one before another as the sponge grows; parenchymula larvae develop from "coeloblastulae." Classification of this group is still unresolved; the histology of the mesohyl, the arrangement of the choanocyte chambers, and the occurrence of a parenchymula larva all suggest a possible relationship to the Demospongiae.

CLASS HEXACTINELLIDA: Glass sponges (Figures 1D and E). Spicules siliceous and basically six-rayed (hexactinal); both megascleres and microscleres always present; body wall cavernous, with trabecular network; exclusively marine; primarily deep-water.

SUBCLASS AMPHIDISCOPHORA: Body never attached to a hard substratum but anchored in soft sediments by a basal tuft or tufts of spicules; megascleres discrete spicules, never fused into a rigid network; microscleres never hexasters; mostly deep-water. (e.g., *Hyalonema*, *Monorhaphis*, *Pheronema*)

SUBCLASS HEXASTEROPHORA: Usually attached to hard substrata, but sometimes attached to sediments by a basal spicule tuft or mat; microscleres are hexasters; megascleres sometimes free, but usually fused into a rigid skeletal framework, in which case sponge may assume large and elaborate morphology. (e.g., *Aphrocallistes*, *Caulophacus*, *Euplectella*, *Hexactinella*, *Leptophragmella*, *Lophocalyx*, *Rosella*, *Sympagella*)

CLASS DEMOSPONGIAE: Demosponges (Figures 1B, C, and F). With siliceous spicules; spicule skeleton may be supplemented or replaced by an organic collagenous network ("spongin"); marine, brackish, or freshwater sponges, occurring at all depths.

SUBCLASS HOMOSCLEROMORPHA: Embryos incubated, larvae amphiblastulae; differentiation of spicules into mega- and microscleres not evident; all spicules very small (usually less than 100 μm) and distributed in large numbers throughout body, with little regional organization. (e.g., *Oscarella*, *Plakina*, *Plakortis*)

SUBCLASS TETRACTINOMORPHA: Reproduction typically oviparous, but incubation with direct development occurs in one order; larvae, when present, typically parenchymulae; with distinct megascleres and microscleres; megascleres organized into distinct patterns, either axial or radial; numerous orders and families (e.g., *Agelas*, *Asteropus*, *Axinella*, *Calthropella*, *Chondrilla*, *Chondrosia*, *Cliona*, *Corallistes*, *Geodia*, *Polymastia*, *Rhabderemia*, *Spheciospongia*, *Stelletta*, *Suberites*, *Tethya*, *Tetilla*). This subclass now contains some sponges from the recently abandoned Sclerospongiae: the merliids (*Merlia*) and at least some tabulates (*Tabulospongia*).

SUBCLASS CERACTINOMORPHA: Mostly viviparous, with incubation of parenchymula larvae; distinct microscleres and megascleres present; spongin present in all but one family (Halisarcidae); includes the freshwater families Spongillidae and Potamelepidae (e.g., *Adocia*, *Aplysilla*, *Axociella*, *Callyspongia*, *Clathria*, *Coelosphaera*, *Ephydatia*, *Euryspongia*, *Halichondria*, *Haliclona*, *Halisarca*, *Halispongia*, *Hymeniacidon*, *Ircinia*, *Lissodendoryx*, *Microciona*, *Mycale*, *Myxilla*, *Neofibularia*, *Siphonodictyon*, *Spongia*, *Spongilla*, *Tedania*, *Verongia*). The Ceractinomorpha is now judged to include some sponges previously assigned to the Scle-

rospongiae. These include the stromatoporids (e.g., *Astrosclera*, *Calcifibrospongia*), and the ceratoporellids (e.g., *Ceratoporella*, *Stromatospongia*, *Hispidopetra*, *Goreauiella*).

The poriferan *Bauplan*

In Chapter 3 we discussed some of the limitations of the parazoan grade of construction, in which true tissues and organs are wanting. Now we discuss the various ways in which sponges have overcome the handicaps imposed by their rather primitive level of organization. You will notice a striking resemblance to protozoa in many regards. Two unique organizational attributes define sponges and have played major roles in poriferan success. These two features are the water current channels, or the **aquiferous system** (and its **choanocytes**), and the highly totipotent nature of sponge cells. The tremendous diversity among sponges in size and shape has occurred both evolutionarily and individually, and is largely derived from these two unique characteristics. Increases in size and surface area are accomplished by folding of the body wall into a variety of patterns. Furthermore, variation in the overall shapes of sponges results from different growth patterns in various environments. This general plasticity in form and the fact that most individual sponge cells are capable of radically altering their form and function as needed compensate in part for the absence of tissues and organs. The aquiferous system brings water through the sponge and close to the cells responsible for food gathering and gas exchange. At the same time, excretory and digestive wastes and reproductive products are expelled. The volume of water moving through a sponge's aquiferous system is remarkable. A 1 × 10-cm individual of the complex sponge *Leucandra* pumps about 22.5 liters of water through its body daily. Researchers have recorded sponge pumping rates that range from 0.002 to 0.84 milliliters of water per second per cubic centimeter of sponge body.

Early workers viewed sponges as essentially colonial animals. One view was that the "individual" was synonymous with a single cell and that all sponges were colonial. Another idea was that each choanocyte chamber represented an individual, and that all but the simplest Calcarea were colonies. Two later theories, however, have gained more popularity. One idea is that each excurrent opening (**osculum**) and its attendant flagellated surfaces, mesohyl, and surface cells constitute an "individual." By this reasoning, a sheet of encrusting sponge that grows on a rock and possesses ten oscula would be a colony of ten individuals. The other theory, and the one that we prefer, holds that the entirety of a sponge—that is, any and all sponge material bounded by a continuous outer cellular covering—constitutes a single individual. The fact is, a whole sponge grows as a whole body, dictated largely by environmental factors (e.g., water flow dynamics, substratum contours), and changes in body form can arise anywhere in or on the organism in response to these environmental pressures. Sponges grow by continually adding new cells that differentiate as needed. This is not usually viewed as colonial asexual reproduction. The existence of some coordinated behavior in sponges (e.g., water current reversal, cessation of choanocyte pumping, synchronous oscular contractions) further supports the view that each sponge is, in its entirety, an "individual."

Body structure and the aquiferous system

The outer squamous surface cells of a sponge make up the **pinacoderm** and are called **pinacocytes**. The inner surface is called the **choanoderm**; it is composed of flagellated cells called **choanocytes**. Both of these epithelial layers are a single cell thick. Between these two thin cellular sheets is the **mesohyl**, which often makes up the bulk of the sponge body (Figures

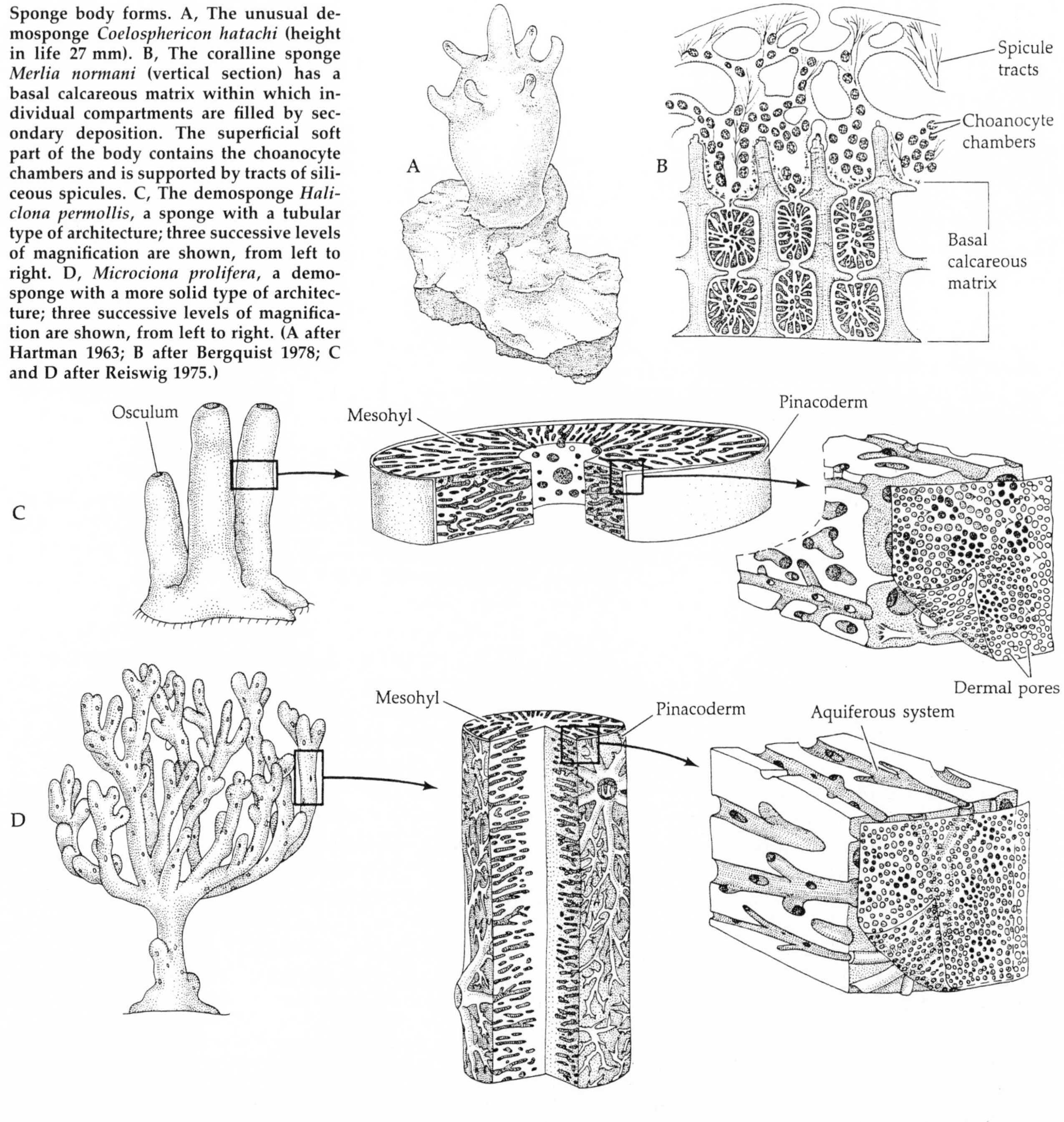

Figure 2

Sponge body forms. A, The unusual demosponge *Coelosphericon hatachi* (height in life 27 mm). B, The coralline sponge *Merlia normani* (vertical section) has a basal calcareous matrix within which individual compartments are filled by secondary deposition. The superficial soft part of the body contains the choanocyte chambers and is supported by tracts of siliceous spicules. C, The demosponge *Haliclona permollis*, a sponge with a tubular type of architecture; three successive levels of magnification are shown, from left to right. D, *Microciona prolifera*, a demosponge with a more solid type of architecture; three successive levels of magnification are shown, from left to right. (A after Hartman 1963; B after Bergquist 1978; C and D after Reiswig 1975.)

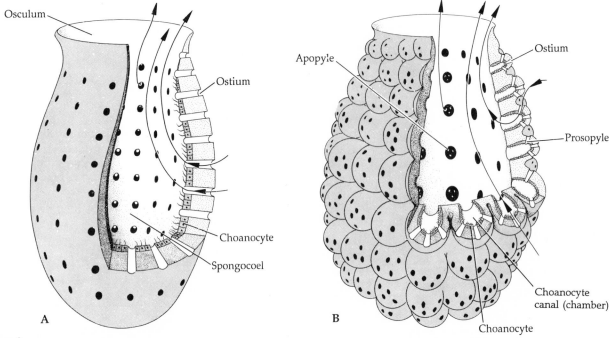

Figure 3

Body complexity in sponges. A, The asconoid condition. B, A simple syconoid condition. C, A complex syconoid condition with cortical growth. D, A leuconoid condition. (From Bayer and Owre 1968.)

2 and 3). The pinacoderm is perforated by small holes or **incurrent pores** called **dermal pores** or **ostia** (singular, ostium), depending on whether the opening is surrounded by several or one cell, respectively (Figure 3). Water is pulled through these openings and is driven across the choanoderm by the beating of the choanocyte flagella. The choanocytes pump large volumes of water through the sponge body at very low pressures, establishing a water current (aquiferous) system.

The pinacoderm can be a simple external sheet, but typically it also lines some of the internal cavities of the aquiferous system where choanocytes do not occur. These pinacoderm cells that line internal canals are called **endopinacocytes**. The choanoderm also can be simple and continuous, or folded and subdivided in various ways. The mesohyl ranges from thin to quite thick, and it plays vital roles in digestion, gamete production, transport of nutrients and waste products by special ameboid cells, and secretion of the skeleton. The mesohyl includes a noncellular colloidal mesoglea in which are embedded collagen fibers, spicules, and various ameboid cells; as such, it is really a type of mesenchyme. A great number of cell types may be found in the mesohyl. Most of these

cells are able to change from one type to another as required; some differentiate irreversibly, however, such as those that commit themselves to reproduction or to skeleton formation. The mobility of all cells, including pinacocytes and choanocytes, has been demonstrated recently by dramatic time-lapse cinematography. The cells of the pinacoderm and choanoderm are more stable than those of the mesohyl, but in general the whole structure may be thought of as a continuously mobile system.

During growth, the pinacoderm and choanoderm are only one cell thick. By increasing their folding as mesohyl volume increases, these layers maintain a surface area:volume ratio sufficient to sustain adequate nutrient and waste exchange throughout the whole individual. The one-cell thick choanoderm may remain simple and continuous (the **asconoid** condition), or it may become folded (the **syconoid** condition), or it may become greatly subdivided into separate flagellated chambers (the **leuconoid** condition) (Figure 3).

The asconoid condition is found in some adult, radially symmetrical calcareous sponges (e.g., *Leucosolenia*, *Clathrina*) and in the early growth stage (**olynthus**) of all newly settled calcareous sponges (Figure 4A). Asconoid sponges rarely exceed 10 cm in height and remain as simple, vase-shaped, tubular units. The thin walls enclose a central cavity called the **atrium** or **spongocoel**, which opens to the outside

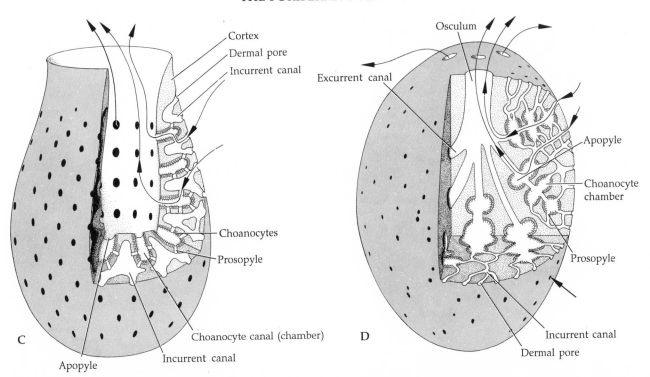

via a single **osculum**. The pinacoderm of asconoid sponges has specialized cells called **porocytes**. During embryogeny, each porocyte elongates and rolls to form a cylindrical canal. The porocyte extends all the way through the pinacoderm, the thin mesohyl, and the choanoderm into the atrium, emerging between adjacent choanocytes (Figure 4B and 7D). The external opening of the porocyte canal is called an ostium or incurrent pore. The choanoderm is a simple, unfolded layer of choanocytes lining the entire spongocoel. Water moving through an asconoid sponge flows through the following structures: ostium → spongocoel (over the choanoderm) → osculum.

Simple folding of the pinacoderm and choanoderm produces the syconoid condition, within which several levels of complexity can be recognized (Figures 3B and C). As complexity increases, the mesohyl may thicken and appear to have two layers. The outer "cortical region," or **cortex**, often contains skeletal elements that are distinct from those found in the interior portion of the mesohyl. In those sponges with a cortex, the incurrent openings are lined by several cells and are referred to as dermal pores. In the syconoid condition, choanocytes are restricted to specific chambers or diverticula of the atrium called **flagellated chambers**, **choanocyte chambers**, or **radial canals**. Each choanocyte chamber opens to the spongocoel by a wide aperture called an **apopyle**.

Syconoid sponges with a thick cortex possess a system of channels or **incurrent canals** that lead from the dermal pores through the mesohyl to the choanocyte chambers. The openings from these channels to the choanocyte chambers are called **prosopyles**. In such syconoid sponges, water moving from the sponge surface into the body flows through the following structures: incurrent pore → incurrent canal → prosopyle → choanocyte chamber → apopyle → spongocoel → osculum. Syconoid construction is found in many calcareous sponges (e.g., *Sycon*, also known as *Scypha*). Although symmetry in some syconoid sponges may appear radial, this appearance is only superficial, and the complex internal organization must be considered largely asymmetrical.

The leuconoid condition is produced by still further folding of the choanoderm and further thickening of the mesohyl by cortical growth. These modifications are accompanied by subdivision of the flagellated surfaces into discrete oval choanocyte chambers (Figure 3D). In the leuconoid condition, one finds an increase in number and a decrease in size of the choanocyte chambers, which typically cluster in groups in the thickened mesohyl. The spongocoel is reduced to a series of **excurrent canals** (or **exhalant canals**) that carry water from the choanocyte chambers to the oscula (Figure 5). The flow of water through a leuconoid sponge is: dermal pore → incurrent canal → prosopyle → choanocyte chamber

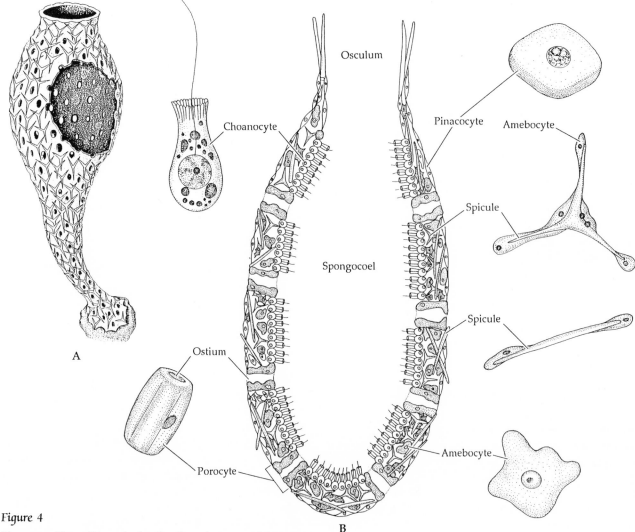

Figure 4

The asconoid condition. A, An olynthus, the asconoid form that follows larval settlement in calcareous sponges. B, Major cell types in an asconoid sponge. (A from Bayer and Owre 1968; B after Sherman and Sherman 1976.)

→ apopyle → excurrent canal → osculum. Leuconoid organization is typical of most calcareous sponges and all members of the class Demospongiae.

It is important to realize that the flow rate is not uniform through the various parts of the aquiferous system. Functionally, it is critical that water be moved very slowly over the choanoderm, allowing time for exchanges of nutrients, gases, and wastes between the water and the choanocytes. The changes in water flow velocity through this plumbing are a function of the effective accumulated cross-sectional diameters of the channels through which the water moves. Water flow velocity decreases as the cross-sectional diameter increases, and thus, in a sponge, velocities are

lowest over the choanoderm. Furthermore, water leaving the oscules must be carried far enough away to prevent recycling by the sponge. In environments of relatively high turbulence, currents, or wave action, this potential recycling of wastes is not a problem. However, sponges that reside in relatively calm water rely on the maintenance of high water-flow velocities through the oscula (or on modified body shapes) to push the excurrent water far enough away from the sponge to avoid the incoming currents. In an irregularly shaped leuconoid sponge living in quiet water, the combined cross-sectional diameter of all the incurrent pores is far less than that of all the choanocyte chambers. But the total oscular diameter is even less than that of the incurrent pores. Simply put, the water enters at some velocity x, slows to a small fraction of x as it passes over the choanoderm, then exits the sponge at a velocity much greater than

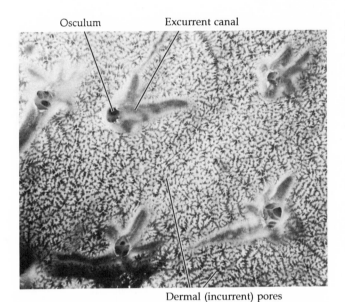

Osculum   Excurrent canal

Dermal (incurrent) pores

Figure 5

The surface of a living poecilosclerid sponge (Demospongiae). The complex system of ostia open into underlying inhalant canals, and large oscules receive several exhalant canals. (From Bergquist 1978.)

x. In complex sponges, the differences in velocity are dramatic. Flow rate regulation is also facilitated in part by the activity of ameboid cells (called **central cells**) that reside near the apopyles of the choanocyte chambers. These cells can slow or speed the exit of water from the chambers by changing shape and position across the apopyle (Figure 7I).

The recognition of the three levels of organization among poriferans allows one to quickly and simply describe a sponge's basic anatomical plan. There is very little evidence, however, that the asconoid plan is necessarily the most primitive, or that all sponge lineages have moved through these three levels of complexity during their evolution. Nor do all sponges pass through three such developmental stages. In addition, gradations of and intermediates between the three basic plans are common. Nonetheless, the simplest organizations (asconoid and syconoid) occur in adults of only the class Calcarea, which is thought to be the most primitive class of living poriferans. Furthermore, calcareous sponges of the leuconoid condition do pass through asconoid and syconoid stages as they grow.

The hexactinellids differ considerably from calcareous sponges and demosponges (Figure 6). The bodies of hexactinellid sponges display a greater degree of radial, or superficial radial, symmetry than any other group. There is no pinacoderm or its equiv-

alent in hexactinellids. A **dermal membrane** is present, but it is extremely thin; no discrete or continuous cellular structure supports it. Incurrent pores are simple holes in this dermal membrane. Cellular material is sparsely distributed and forms a **trabecular network** stretching across interconnecting internal cavities called **subdermal lacunae** (Figure 6A). The thimble-shaped flagellated chambers are arranged in a single layer and are supported within the trabecular network. Both the trabecular network and the walls of the flagellated chambers are syncytial (i.e., discrete choanocytes do not exist). Water enters the incurrent pores, passes into the subdermal lacunae, and from there enters the flagellated chambers via the prosopyles.

The unique structure of hexactinellids is so striking that some workers (e.g., Bergquist 1985) have

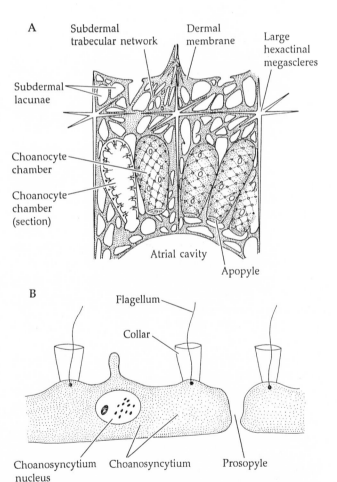

Figure 6

Internal anatomy of Hexactinellida. A, The body wall of *Euplectella* (transverse section). A dermal layer covers the trabecular network. B, The choanosyncytium of *Aphrocallistes vastus* (vertical section). (A after Bergquist 1978; B after Reiswig 1979.)

even suggested the hexactinellids might be regarded as a separate phylum (the Symplasma). However, as explained in Chapter 3, phylogenetic relationships are best sought in similarities, not in differences, among groups, and by this reasoning we treat the hexactinellids as poriferans.

Cell types

Because of the nontissue nature of sponges and because cellular totipotency plays a major role in poriferan biology, considerable effort has gone into describing and classifying sponge cell types. Pre-1970s texts generally recognized only a few basic kinds of poriferan cells. However, more recent detailed histochemical and ultrastructural studies have revealed ever increasing numbers of cell types. This discovery, combined with the dynamic and totipotent nature of sponge cells, makes succinct classification of their cell types difficult. We utilize an abbreviated version of Bergquist's (1978) cell classification below.

Cells that line surfaces. Pinacoderm forms a continuous layer on the external surface of sponges and also lines all inhalant and exhalant canals. The pinacocytes that make up this layer are usually flattened and often overlapping (Figures 7A and B). Internal, canal-lining pinacocytes (endopinacocytes) are usually more fusiform in shape and have less overlap than outer exopinacocytes. Furthermore, ciliated endopinacoderm occurs in the large exhalant canals of some leuconoid sponges. Although the endopinacoderm is epithelial in function, and probably phagocytic as well, the absence of a basal membrane distinguishes sponge pinacoderm from the epithelia of other metazoa. External cells of the basal or attaching region of a sponge surface are called **basopinacocytes**. These flattened, T-shaped cells are responsible for secreting a fibrillar collagen–polysaccharide complex called the "basal lamina," which is the actual attachment structure. In freshwater sponges, the basopinacocytes are active in feeding and extend ameba-like "filopodia" to engulf bacteria. Freshwater sponge basopinacocytes also play an active role in osmoregulation and contain large numbers of water expulsion vesicles, or "contractile vacuoles."

Porocytes are cylindrical, tubelike cells of the pinacoderm that form the ostia (Figures 7C and D). They are contractile and can open and close the pore to regulate the ostial diameter; however, no microfilaments have been observed in them. Some can produce across the ostial opening a diaphragm-like cytoplasmic membrane that regulates pore size.

Figure 7

Cells that line sponge surfaces. A, A pinacocyte from the surface of the demosponge *Halisarca* (drawn from an electron micrograph). The outer surface is covered with a polysaccharide-rich coat. The cell is fusiform and overlaps adjacent pinacocytes. B, Pinacoderm from a calcareous sponge (section). T-shaped pinacocytes alternate with fusiform pinacocytes. C–D, A porocyte from the calcareous asconoid sponge *Leucosolenia*. (C, cross section; D, side view.) E, Myocytes surrounding a prosopyle. F, A section of choanoderm, showing three choanocytes; arrows indicate direction of water current. G, A choanocyte. H, Ultrastructure of a choanocyte (longitudinal section, drawn from an electron micrograph). I, A choanocyte chamber opening into an exhalant canal in a demosponge. (A–D after Connes et al. 1971; E, G from Bayer and Owre 1968; F after Barnes 1980; H after Brill 1973; I after Connes et al. 1971.)

Choanocytes are the flagellated cells that make up the choanoderm and create the currents that drive water through the aquiferous system (Figures 7F–H). Choanocytes are not coordinated in their beating, not even within a given chamber. They are aligned in such a way that the flagella are directed toward the apopyle and beat from base to tip. Water is thus drawn into the chamber through the prosopyles, driven across the choanoderm, and then out the apopyle into the spongocoel or an excurrent canal. The long flagellum is always surrounded by a so-called **collar**, which is made up of about 20 cytoplasmic microvilli, or "tentacles." The "tentacles" have microfilament cores and are connected to one another by a mucous reticulum. Choanocytes rest on the mesohyl, held in place by interdigitation of adjacent basal surfaces. Because of their central role in phagocytosis and pinocytosis, choanocytes are highly vacuolated.

Cells that secrete the skeleton. There are several types of ameboid cells in the mesohyl, some of which serve to secrete the various elements of sponge skeletons.

In almost all sponges, the entire supportive matrix is built upon a framework of fibrillar collagen. The cells that secrete this material are called **collencytes**, **lophocytes**, and **spongocytes**. Collencytes are morphologically nearly indistinguishable from pinacocytes, whereas lophocytes are large, highly motile cells that can be recognized by a collagen band they typically trail behind them (Figure 8C). The primary function of both cell types is the secretion of the dispersed fibrillar collagen found intercellularly in virtually all sponges. Spongocytes produce the fibrous supportive collagen referred to as **spongin** (Figure 10A). Spongocytes operate in groups and are always found wrapped around a spicule or collagen fiber (Figure 8D).

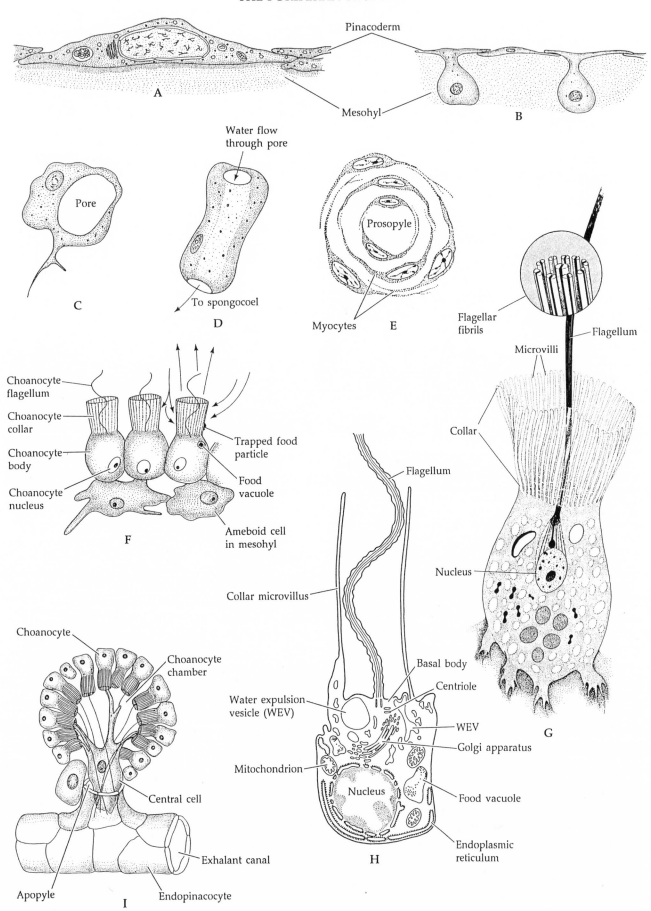

Pinacoderm

Mesohyl

A

B

Water flow through pore

Pore

C

To spongocoel

D

Prosopyle

Myocytes

E

Flagellar fibrils

Flagellum

Microvilli

Choanocyte flagellum

Choanocyte collar

Choanocyte body

Choanocyte nucleus

Collar

Trapped food particle

Food vacuole

Ameboid cell in mesohyl

F

Nucleus

Flagellum

Collar microvillus

Choanocyte

Choanocyte chamber

Central cell

Basal body

Centriole

Water expulsion vesicle (WEV)

WEV

Golgi apparatus

Mitochondrion

Nucleus

Food vacuole

Apopyle

Exhalant canal

Endopinacocyte

I

Endoplasmic reticulum

H

G

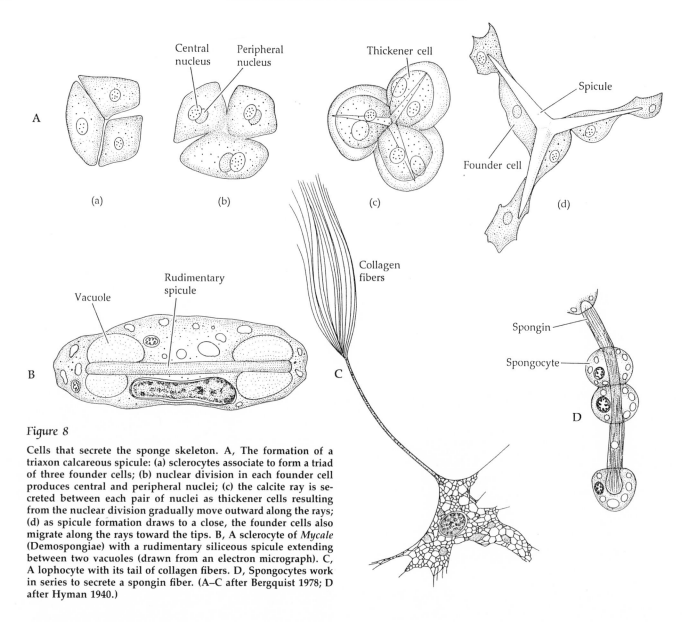

Figure 8

Cells that secrete the sponge skeleton. A, The formation of a triaxon calcareous spicule: (a) sclerocytes associate to form a triad of three founder cells; (b) nuclear division in each founder cell produces central and peripheral nuclei; (c) the calcite ray is secreted between each pair of nuclei as thickener cells resulting from the nuclear division gradually move outward along the rays; (d) as spicule formation draws to a close, the founder cells also migrate along the rays toward the tips. B, A sclerocyte of *Mycale* (Demospongiae) with a rudimentary siliceous spicule extending between two vacuoles (drawn from an electron micrograph). C, A lophocyte with its tail of collagen fibers. D, Spongocytes work in series to secrete a spongin fiber. (A–C after Bergquist 1978; D after Hyman 1940.)

Sclerocytes are responsible for the production of calcareous and siliceous sponge spicules (Figures 8A and B). They are active cells that possess abundant mitochondria, cytoplasmic microfilaments, and small vacuoles. Numerous types of sclerocytes have been described; these cells always disintegrate after spicule secretion is complete.

Contractile cells. Contractile cells in sponges, called **myocytes**, are found in the mesohyl (Figure 7E). They are usually fusiform and grouped concentrically around oscula and major canals. Myocytes are distinguished by the great numbers of microtubules and microfilaments contained in their cyto-

plasm. Because of the nature of their filament arrangement, it has been suggested that myocytes are homologous with the smooth muscle cells of higher invertebrates. Myocytes are independent effectors with a slow response time, and, unlike true neurons and muscle fibers, they are insensitive to electrical stimuli.

Some other cell types. **Archaeocytes** are ameboid cells that are capable of differentiating and giving rise to virtually any other cell type. Archaeocytes are large, highly motile cells that play a major role in digestion and food transport (Figure 9). These cells possess a variety of digestive enzymes (e.g., acid

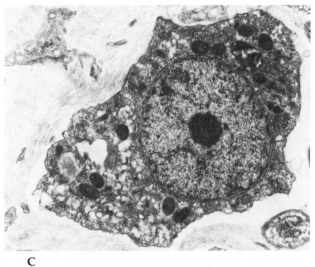

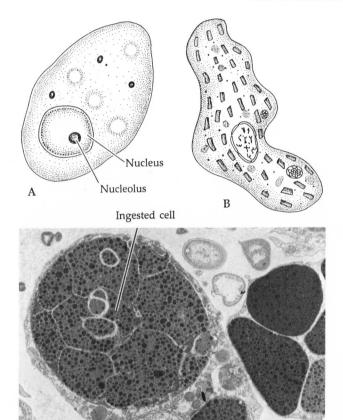

Figure 9

Archaeocytes and rhabdiferous cells. A, A typical archaeocyte with a large nucleus and a prominent nucleolus. B, A rhabdiferous cell from *Microciona*, with dispersed rodlike inclusions of uncertain function. C, Photo of a typical archaeocyte. D, An archaeocyte engages in phagocytosis. (B after Bergquist 1978; C, D from Bergquist 1978.)

phosphatase, protease, amylase, lipase) and can accept phagocytized material from the choanocytes. They also phagocytize material directly through the pinacoderm of water canals. As the principal macrophage of a sponge, archaeocytes carry out much of the digestive, transport, and excretory activities. As cells of maximum totipotency, archaeocytes are essential to the developmental program of sponges and to various asexual processes (e.g., gemmule formation).

Rhabdiferous cells are large mesohyl cells. They contain rodlike mucopolysaccharide inclusions that are aligned parallel to the long axis of the cell (Figure 9B). Rhabdiferous cells periodically discharge their contents into the mesohyl and apparently contribute to the polysaccharide ground substance of the mesohyl matrix.

Several other cell types have been identified in sponges, but most of these have been characterized only morphologically and their functions remain unknown.

Cell aggregation. Around the turn of this century, H. V. Wilson first demonstrated the remarkable ability of sponge cells to reaggregate after being mechanically dissociated. Although this discovery was interesting in itself, lending further insight into the plasticity and cellular organization of sponges, it also foreshadowed more far-reaching cytological research. Recent studies on sponges have shed light on basic questions of how cells adhere, segregate, and specialize. Almost any sponge dissociated and maintained under proper conditions will form aggregates, and many will eventually reconstitute their aquiferous system. For example, when pieces of the Atlantic "red beard sponge" (*Microciona prolifera*) are pressed through fine cloth, the separated cells immediately begin to reorganize themselves by active cell migration. Within two to three weeks, a functional sponge reforms and the original cells mostly return to their respective functions. Furthermore, if cell suspensions of two different species are mixed, the cells sort themselves out and reconstitute individuals of each separate species.

Support

The skeletal elements of sponges are of two types, organic and inorganic. The former is always

Figure 10

Sponge skeletal systems. A, Photomicrograph of the superficial dermal spongin–fiber skeleton typical of the demosponge family Callyspongiidae. B, Arrangement of calcareous triaxon spicules near the oscular opening in *Leucosolenia*. C, Arrangement of monaxon and triaxon calcareous spicules near the oscular opening of *Scypha*. D, Cross section of a simple syconoid calcareous sponge (spongocoel on right) illustrating placement of triaxon spicules. E, Some common types of siliceous spicules from demosponges. F, Some siliceous spicules from sclerosponges. G, Various siliceous spicule types (SEM micrographs). (A, G, in part from Bergquist 1978; photo of asterose microscleres by B. Beaumont, courtesy of P. Bergquist; B–D after Hyman 1940; E from various sources; F after Hartman 1969.)

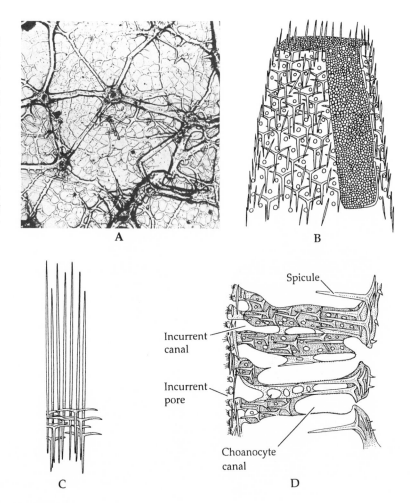

A

B

C

D

Spicule

Incurrent canal

Incurrent pore

Choanocyte canal

collagenous and the latter either siliceous (hydrated silicon dioxide) or calcareous (calcium carbonate in the form of calcite or aragonite). Sponges are the only metazoa that utilize hydrated silica as a skeletal material.

Collagen is the major structural protein in invertebrates; it is found in virtually all metazoan connective tissues. In sponges, it is either dispersed as thin fibers in the intercellular matrix or organized as a framework called **spongin** in the mesohyl. True spongin is found only in members of the class Demospongiae; dispersed collagen fibers are found in all sponges. The amount of this fibrillar collagen varies greatly from species to species. In hexactinellids it is quite sparse, whereas in demosponges it is abundant and may form dense bands in the cortex.

Traditionally, the organic skeleton has been termed spongin. This term, however, should be restricted to the form of collagen that constitutes a distinct organized network in the mesohyl of demosponges (Figure 10A). The network often contains very thick fibers, and may incorporate siliceous spicules into its structure. Spongin often cements siliceous spicules together at their points of intersection. The encysting coat of the asexual gemmules of freshwater (and some marine) sponges is also composed largely of spongin fibers.

Mineral skeletons of silica or calcium are found in almost all sponges, except certain members of the class Demospongiae. Sponges lacking mineral skeletons possess only fibrous collagen networks and are still used as bath sponges despite the prevalence nowadays of synthetic "sponges." Sponges have been harvested for millennia; Homer and other ancient Greek writers mention an active Mediterranean sponge trade. Prior to the 1950s active natural sponge fisheries thrived in south Florida, the Bahamas, and the Mediterranean. The industry peaked in 1938, when the world's annual sponge catch (including cultivated sponges) exceeded 2.6 million pounds, 700,000 pounds of which came from the United States and the Bahamas. The majority of the commercial

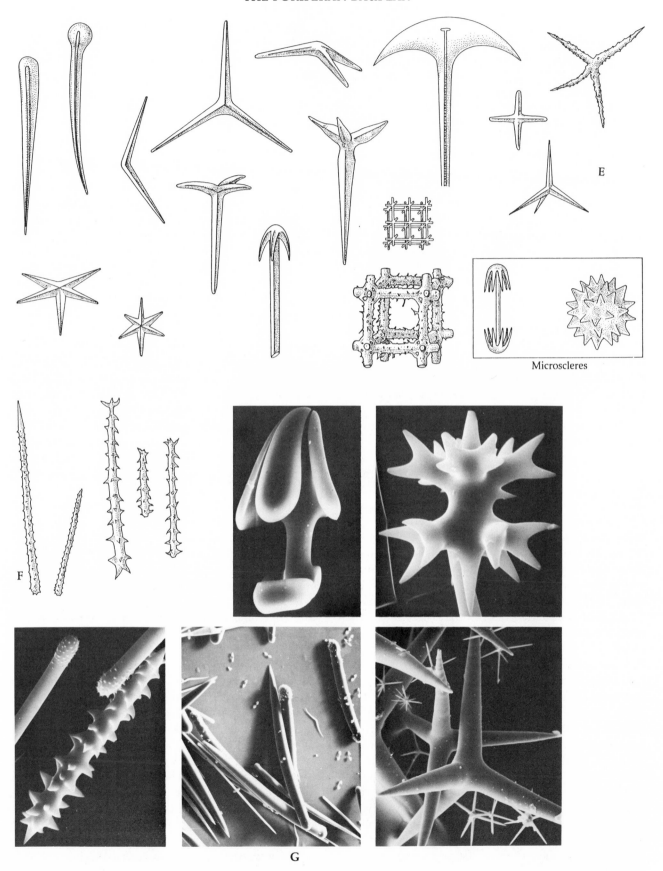

E

Microscleres

F

G

sponges belong to the genera *Euspongia*, *Hippiospongia*, and *Spongia*.

Sponge spicules (Figure 10) are produced by special mesohyl cells called sclerocytes, which are capable of accumulating calcium or silicate and depositing it in an organized way. In some cases, one sclerocyte produces one spicule; in others, several sclerocytes work together to produce a single spicule, often two cells per spicule ray (Figure 8A). The construction of a siliceous spicule begins with the secretion of an organic axial filament within an elongated vacuole in a sclerocyte. As the axial filament elongates at both ends, hydrated silica is secreted into the vacuole and deposited around the filament. Hexactinellid siliceous spicules are secreted in syncytial scleroblast masses. Whereas the rays of siliceous spicules of demosponges are triangular or hexagonal in cross section, those of hexactinellids are always square in cross section. Unlike siliceous spicules, calcium carbonate spicules do not have an organic axial structure, although a central core of uncertain nature is often discernible. Calcareous spicules are produced extracellularly, in intercellular spaces bounded by a number of sclerocytes. Each spicule is essentially a single crystal of calcite or aragonite.

Considerable taxonomic weight (perhaps too much) has been given to spicule morphology, and an elaborate nomenclature has been developed to classify these skeletal structures. According to their morphology, spicules are termed either microscleres or megascleres. The former are small to minute reinforcing or packing spicules; the latter are large structural spicules. The demosponges and hexactinellids have both types; calcareous sponges often have only megascleres. Descriptive terms that designate the number of axes in a spicule end in the suffix *axon* (e.g., monaxon, triaxon). Terms that designate the number of rays end in the suffixes *actine* or *actinal* (e.g., monactinal, hexactinal, tetractinal). In addition, there is a detailed nomenclature specifying shape and ornamentation of various spicules (Figure 10).

A spicular skeleton may be viewed as a supplemental supporting structure. If the amount of inorganic material is increased in relation to organic material, the sponge becomes increasingly solid until the texture approaches that of a rock, as it does in members of the demosponge orders Choristida and Lithistida. In contrast to discrete spicules, the massive calcareous skeletons of some species (the coralline sponges and "sclerosponges") have a polycrystalline microstructure; they are composed of needles ("fibers") of either calcite or aragonite embedded in an organic fibrillar matrix. The advantage of incorporating organic matter into the calcareous framework has been compared to lathe-and-plaster, or reinforced concrete. The mix of organic and inorganic materials probably yields fibrous calcites and aragonites that are less prone to fracture while also producing substances that are more easily molded by the organism.

Nutrition, excretion, and gas exchange

Although sponges lack the complex organs and organ systems seen in the eumetazoa, they are nevertheless a highly successful group of animals. Their success appears to be due largely to three attributes: cell totipotency, the aquiferous system, and the general plasticity of their body form.

Unlike most metazoa, sponges rely solely on intracellular digestion, and thus on phagocytosis and pinocytosis as means of food capture. The aquiferous system has already been described; sponges more-or-less continuously circulate water through their bodies, bringing with it the microscopic food particles upon which they feed. They are size-selective particle feeders, and the arrangement of the aquiferous system creates a series of "sieves" of decreasing mesh size (e.g., inhalant ostia or dermal pores → canals → prosopyles → choanocyte collar tentacles → intertentacular mucous reticulum). The upper limit of the diameter of incurrent openings is usually around 50 μm, so larger particles do not enter the aquiferous system. A few species have larger incurrent pores, reaching diameters of 150–175 μm. Internal particle capture in the 2-to-5 μm range (e.g., certain protozoa, unicellular algae, organic detritus) is by phagocytotic motile archaeocytes that move to the lining of the inhalant canals. Then, as water passes over the choanoderm, eddies are formed around the choanocyte collars. This water passes between the tentacles, into the collar, and is driven out the collar opening. Particles in the 0.1-to-1.5 μm range (e.g., bacteria, large free organic molecules) are trapped between the collar tentacles. The distance between adjacent tentacles is consistently 0.1–0.2 μm. Undulations of the collar move the trapped food particles down to the choanocyte cell body, where they are ingested by phagocytosis or pinocytosis.

In the case of archaeocyte phagocytosis, digestion takes place in the food vacuole formed at the time of capture. In the case of choanocyte capture, food particles are partly digested in the choanocytes and then quickly passed on to a mesohyl archaeocyte (or other wandering "amebocyte") for final digestion. In both cases, mobility of the mesohyl cells assures transport of nutrients throughout the sponge body.

The efficiency of food capture and digestion was dramatically shown in a study by Schmidt (1970) using fluorescence-tagged bacteria fed to the freshwater sponge *Ephydatia fluviatilis*. By monitoring the movement of the fluorescent material, Schmidt determined that 30 minutes elapsed from the onset of feeding until the bacteria had been captured by choanocytes and moved to the base of the cells. Transfer of the fluorescent material to the mesohyl commenced 30 minutes later. Twenty-four hours later, fluorescent wastes began to be discharged into the water, and no fluorescent material remained in the sponges after 48 hours. Additional studies on this same species led to an estimate of 7,600 choanocyte chambers per cubic millimeter of sponge body, each chamber pumping approximately 1,200 times its own volume of water daily. More complex leuconoid sponges have as many as 18,000 choanocyte chambers per cubic millimeter. In some thin-walled asconoid and simple syconoid sponges a distinctive mesohyl is hardly present. In these sponges, the choanocytes assume both capture and digestive/assimilative functions.

Sponges also take up significant amounts of dissolved organic matter (DOM) by pinocytosis from the water within the aquiferous system. Studies by H. M. Reiswig on three Jamaican sponge species showed that 80 percent of the organic matter taken in by these sponges was of a size below that resolvable by light microscopy. The other 20 percent comprised primarily bacteria and dinoflagellates.

Excretion (primarily ammonia) and gas exchange are by simple diffusion, much of which occurs across the choanoderm. We have already seen how folding of the body, combined with the presence of an aquiferous system, overcomes the surface-to-volume dilemma posed by an increase in size. The efficiency of the poriferan *Bauplan* is such that diffusion distances never exceed about 1.0 mm, the distance at which gas exchange by diffusion becomes inefficient. In addition, water expulsion vesicles occur in freshwater sponges, and serve to rid the body of excess water.

Activity and sensitivity

There is no conclusive evidence that sponges possess neurons or discrete sense organs. Furthermore, action potentials have never been recorded in sponges, and nothing resembling eumetazoan synaptic connections are known in these animals. However, they are capable of responding to a variety of environmental stimuli by closure of the ostia or oscula, canal constriction, backflow, and reconstruction of flagellated chambers. The usual effect of these

actions is to reduce or stop the flow of water through the aquiferous system. For example, when suspended particulates become too large or too concentrated, sponges typically respond by closing the ostia and immobilizing the choanocyte flagella. Direct physical stimulation will also elicit this reaction, which is easily observed by simply running one's finger across a sponge surface and observing the dermal pore or oscular contractions with a hand lens or low-power microscope.

Activity also varies with certain endogenous factors. For example, during a major growth phase, such as canal or chamber reorganization, activity levels typically fall and pumping rates go down. Periods of reproductive activity also cause substantial drops in water pumping, because many choanocytes are expended in the reproduction process (see below). Even under "normal" conditions, variations in pumping rates occur. Important studies by Reiswig (see references) on Caribbean sponges have documented a number of endogenous activity patterns. Some sponges cease pumping activity periodically, for a few minutes or for hours at a time; others cease activity for several days at a time.

The switch from full pumping activity to complete cessation requires at least several minutes; considering the organism, however, this is a fairly short response time. The spread of stimulation and response in sponges appears to be by simple mechanical stimulation from one cell to the adjacent cells, and perhaps also occurs by diffusion of certain chemical messenger substances associated with the irritability of cytoplasm in general. The contractile myocytes of sponges have already been described. They act as independent effectors and are organized into a network formed by contacts between filopodial extensions of adjacent myocytes and pinacocytes. Response time of myocytes is very slow. Latency periods average 0.01–0.04 seconds, and conduction velocities are typically less than 0.04 cm/sec (except in the hexactinellids, where velocities of 0.30 cm/sec have been recorded). Conduction is always unpolarized and diffuse. Considerable research once focused on the myocytes in attempts to shed light on the possible presence of a sponge nervous system analogous or homologous to that in higher metazoa. But, in spite of these efforts, there has been no verification of such a system.

One recent study on sponge activity hypothesized a diffuse "conduction system" in the hexactinellid sponge *Rhabdocalyptus* (Lawn et al. 1981). Both mechanical and electrical stimulation elicited a diffuse all-or-none response wherein pumping activity ceased within 20–50 seconds. Conduction velocities of 0.17 to 0.30 cm/sec were estimated. Although the authors agreed that this is too slow for a true neuronal system, they felt it was too fast for conduction by simple chemical diffusion. We await additional research on this problem.

Reproduction and development

All sponges appear to be capable of sexual reproduction, and several types of asexual processes are also common. Many of the details of these processes are unknown, however, largely because sponges lack distinct or localized gonads; gametes and embryos occur throughout the mesohyl. Furthermore, within any population there is generally a marked asynchrony among individuals in terms of reproductive activity; that is, at any given moment reproductive activity may be taking place in only a small number of individuals in any area.

Asexual reproduction. Probably all sponges are capable of regenerating viable adults from fragments. Some branching species "pinch" off branch ends by a process of cellular reorganization. The dislocated pieces fall off and regenerate into new individuals. This regenerative ability is used by Florida commercial sponge farmers, who propagate their sponges by attaching "cuttings" to submerged cement blocks. Additional asexual processes of poriferans include formation of gemmules, budding, and possibly formation of asexual larvae.

In freshwater sponges of the family Spongillidae, small spherical structures called **gemmules** are produced at the onset of winter (Figure 11). These overwintering bodies are invested with a thick spongin coat in which supportive microscleres called **amphidiscs** are embedded. Gemmules are highly resistant to both freezing and drying. Their formation and eventual growth are remarkable examples of poriferan cell totipotency. As winter approaches, archaeocytes aggregate in the mesohyl and undergo rapid mitosis. "Nurse cells" called **trophocytes** stream to the archaeocyte mass and are engulfed by phagocytosis. The result is packed archaeocytes containing food reserves stored in elaborate **vitelline platelets**. This entire mass eventually becomes surrounded by a three-layered spongin covering. Developing amphidisc spicules are transported by their parent cells to the growing gemmule and incorporated into the spongin envelope. The final bit of the gemmule to be enclosed by the spongin case is covered only by a single layer of spongin that is devoid of spicules; this single-layered patch is the **micropyle**. Thus formed, hibernation of the gemmule commences, while the parent sponge usually dies and disintegrates.

When environmental conditions are again favorable, the micropyle opens and the first archaeocytes begin to flow out (Figure 11C). They immediately flow over the gemmule and onto the substratum, whereupon they begin to construct a framework of new pinacoderm and choanoderm. The second wave of archaeocytes to leave the gemmule colonizes this framework. In the course of gemmule "hatching," archaeocytes give rise to every cell type of the adult sponge.

No other sponge group produces gemmules as complex as those of the Spongillidae. However, many marine species produce asexual reproductive bodies (called **reduction bodies**) that are basically similar to freshwater gemmules but incorporate a variety of amebocytes and have a less complex wall structure.

Many marine sponges produce buds of various types. They appear as squat or elongate club-shaped protrusions arising on the sponge surface. The buds fall from the parent sponge surface and may be carried about by water currents for a brief period before attaching to the substratum to form a new individual. Some of the Clionidae (Tetractinomorpha) produce unique armored buds that are rich in stored foods and can drift in the plankton for extended periods of time.

Some sponges are reported to be capable of producing larvae by asexual means. This little-studied process may prove to be a common strategy in littoral Demospongiae, where it has been interpreted as a process assuring production of a free dispersal stage even when fertilization has failed.

Sexual processes. When it comes to sex, sponges probably win the prize for variety. Most sponges are hermaphroditic, but they produce eggs and sperm at different times. This sequential hermaphroditism may take the form of protogyny or protandry, and the sex change may occur only once, or an individual may repeatedly alternate between male and female. In some species individuals appear to be permanently male or female. In still other species, some individuals may be permanently gonochoristic, whereas others in the same population are hermaphroditic. However, in all cases cross-fertilization probably takes place.

Sperm appear to arise primarily from choanocytes; eggs arise from choanocytes or archaeocytes. Spermatogenesis usually occurs in distinct **spermatic cysts** (= **sperm follicles**), which form either when all the cells of a choanocyte chamber are transformed

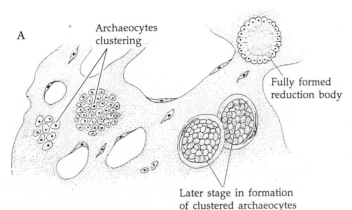

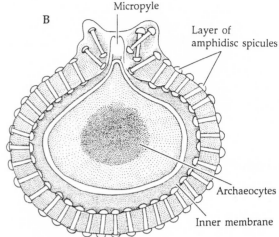

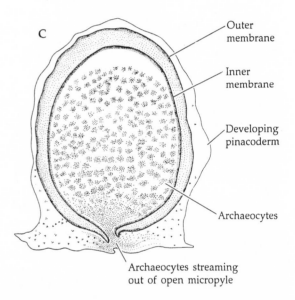

Figure 11

A, Reduction bodies (= gemmules) forming in a marine sponge. B, A gemmule (section) of a freshwater sponge (Spongillidae). C, A gemmule (section) of the freshwater sponge *Spongilla* in the process of hatching. Note the absence of amphidiscs at this point in time. (A from Bayer and Owre 1968; B, C after Hyman 1940.)

into spermatogonia or when transformed choanocytes migrate into the mesohyl and aggregate there (Figure 12A). Little is known about oogenesis, although available information suggests that solitary oocytes develop within "cysts" surrounded by a layer of follicle cells and nurse cells (trophocytes). Meiosis commences after an oogonium has accumulated a sufficient quantity of food reserves, presumably supplied by feeding on the trophocytes (Figure 12B).

Mature sperm are released into the environment through the aquiferous system. The rapid release of sperm from sponge oscula is dramatic, and such individuals are often referred to as "smoking sponges" (Figure 12C). Sperm release may be synchronized in a local population or restricted to certain individuals. Once in the water, sperm are taken into the aquiferous system of neighboring oocyte-containing individuals. They must then cross the cellular barrier of the choanoderm, enter the mesohyl, locate the oocytes, penetrate the follicular barrier, and finally fertilize the egg. This rather impressive feat initially involves sperm capture by choanocytes and enclosure in an intracellular vesicle (much like the formation of a food vacuole during feeding). The choanocyte then loses its collar and flagellum and migrates through the mesohyl as an ameboid cell, transporting the sperm to the oocyte (Figure 13). The migratory choanocyte is called a **carrier cell**, or **transfer choanocyte**. Choanocytes no doubt regularly consume the unlucky sperm of different species of sponges and other benthic invertebrates but, by some as yet undiscovered recognition mechanism, they respond with a remarkably different behavior to sperm of their own kind.

There is only one somewhat detailed account of the embryogeny of a hexactinellid (Okada 1928). Therefore, our discussion is restricted to generalities about the Demospongiae and Calcarea, virtually all of which undergo mixed life histories.

The zygote undergoes equal or unequal holoblastic cleavage and develops into a blastula-like stage. The developing embryo may be released shortly after fertilization, or considerable development may take place in the mesohyl of the parent sponge. A form of oviparity occurs in some demosponges (subclass Tetractinomorpha) in which eggs and early embryos are extruded in gelatinous strings that cling to the outer surface of the maternal parent sponge; development to the parenchymula larva stage ensues in this mucus envelope (Figure 12D). Once the embryo is released, it is typically a motile

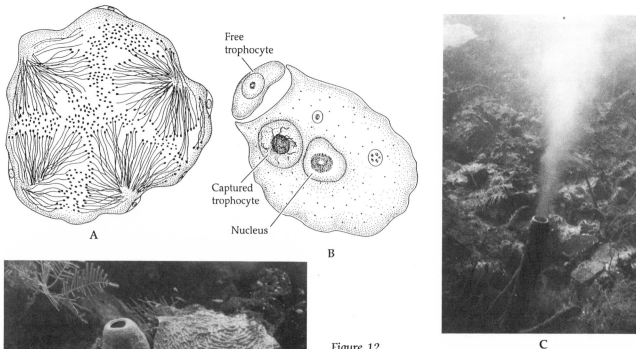

Free
trophocyte

A

Captured
trophocyte

Nucleus

B

Figure 12   C

Sexual reproduction in sponges. A, Sperm follicle (section) containing mature spermatozoa. B, An oocyte (section) of *Ephydatia fluviatilis* (Demospongiae) is phagocytizing a trophocyte. Inside the oocyte is a trophocyte that was recently ingested. C, Sperm release from a tubular West Indian sponge, *Verongia archeri* (Demospongiae). The sponge is about 1.5 m tall. D, Egg and embryo release in the oviparous sponge *Agelas* (Demospongiae). The individual in the foreground is covered by cords of yellow mucus that surround the embryos during their early development; two specimens in the center show no sign of egg release. (A and B after Brien and Meewis 1938; C from Reiswig 1970; D from Reiswig 1976; photo by R. Kinzie.)

D

larva. Release of the larva is through either the excurrent plumbing of the aquiferous system or a rupture in the body wall. Larvae may settle directly, they may swim about for several hours or a few days before settling, or they may simply "crawl" about the substratum until ready to attach. In all known cases, the larvae are lecithotrophic. In general, littoral sponges tend to produce planktonic larvae, whereas subtidal species' larvae tend to settle directly or move about on the ocean floor for a few days before beginning growth into a new adult individual.

Three basic larval types have been described in sponges: **"coeloblastula" larvae** (= "blastula" larvae), **parenchymula larvae** (= parenchymella larvae), and **amphiblastula larvae**. Most demosponges incubate the embryos until a late stage, producing a solid parenchymula larva with an outer surface of flagellated cells and an inner mesohyl-like core of matrix and cells (Figure 14). Parenchymula larvae have a short planktonic life, usually just a few days. Following settling, the external flagellated cells disappear and flagellated choanocytes appear internally, as choanoderm. This process has long been attributed to a unique embryological "inversion" process wherein external cells drop their flagella, migrate to the inner cell layer, then re-form the flagella. Recent work has challenged the existence of this inversion process and suggests that the external flagellated cells are simply shed or phagocytized during larval metamorphosis, the internal choanocytes subsequently forming anew from archaeocytes. In any case, the result of this postsettlement metamorphosis is a tiny leuconoid form called a **rhagon**.

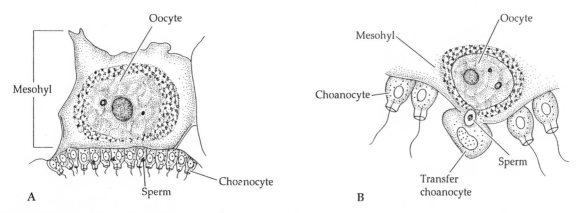

Figure 13

Fertilization in the calcareous sponge *Grantia*. A, A sperm is trapped by choanocyte; an egg is lying in the mesohyl adjacent to the choanoderm. B, A transfer choanocyte gives up the sperm to the egg; note that the egg lies next to the choanoderm and that the choanocyte has lost its flagellum. (After Hyman 1940.)

Calcareous sponges (and a few demosponges) often release their embryos early, as free-swimming "coeloblastula" larvae (Figure 15A). These larvae may undergo one of two developmental processes. In the simplest case, transformation of the larva involves an inward migration of surface cells that have lost their flagella; these same cells subsequently regain their flagella as they metamorphose into choanocytes.

A more complex embryonic development produces two distinct cell types resembling the macro-

meres and micromeres of some true metazoa. At the 16-cell stage, eight large round cells ("macromeres") rest at one pole, and eight smaller cells ("micromeres") form the rest of the hollow embryo. The larger cells are destined to be future pinacoderm and mesohyl, and the smaller cells become the choanoderm. The "micromeres" divide rapidly and develop flagella that extend into the embryo's cavity. The "macromeres" remain undivided for some time and never develop flagella; in the center of the "macromere" cell cluster is a pore to the outside. This stage is called the **stomoblastula**. While still within the mesohyl of the adult sponge, the stomoblastula ingests nutrient-rich amebocytes. As development proceeds, a remarkable process of inversion takes place, in which the stomoblastula turns inside out through the pore, moving the flagella from the inside to the outside and producing a hollow, flagellated, amphiblastula larva (Figures 15B–E). This larva subse-

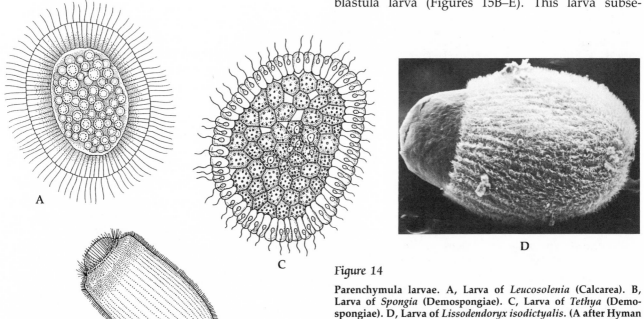

Figure 14

Parenchymula larvae. A, Larva of *Leucosolenia* (Calcarea). B, Larva of *Spongia* (Demospongiae). C, Larva of *Tethya* (Demospongiae). D, Larva of *Lissodendoryx isodictyalis*. (A after Hyman 1940; B after Bergquist 1978; C after Levi 1956; D from Bergquist 1978.)

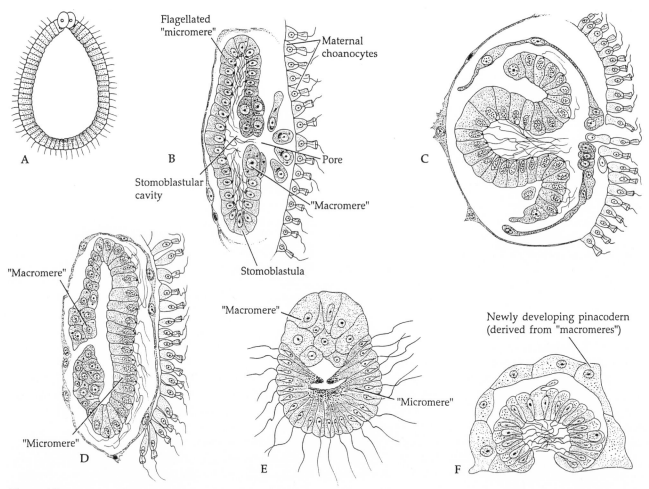

Figure 15

"Coeloblastula" and amphiblastula larvae (in section). A, Typical "coeloblastula" larva with its posterior "macromeres." B–D, During the remarkable process of inversion in *Scypha*, the stomoblastula turns itself inside out to form an amphiblastula larva with externally directed flagella. E, A typical amphiblastula larva (*Scypha*). F, Settled young sponge (*Scypha*) after invagination of flagellated cells. (A adapted from several sources; B–F from Bayer and Owre 1968, after various authors.)

quently is released from the parent sponge. There is no known counterpart to this process in any other sponge group, or in the eumetazoa.

After a free-swimming period, the amphiblastula larva settles on its flagellated end. Metamorphosis involves a rapid proliferation of the "macromeres" to form pinacoderm that overgrows the flagellated hemisphere. The flagellated cells pocket inward to form a chamber lined with cells destined to become choanocytes (Figure 15F). An osculum breaks through, and the tiny asconoid-like sponge becomes capable of circulating water and feeding. This initial functional stage is called an **olynthus** (Figure 4A).

After further growth, it will become an asconoid, syconoid, or leuconoid adult.

The preceding account of development and larval types is drastically simplified. In fact, sponges show more variation in embryological development than many eumetazoan groups. We recommend Bergquist (1978) as a good starting place if you wish to learn more details.

Some additional aspects of sponge biology

Some basic sponge ecology has been presented in the previous sections of this chapter. However, because sponges play such important roles in so many marine habitats, we add here some special aspects of their natural history.

Although sponges have invaded virtually all aquatic habitats, certain ecological–distributional patterns are evident among the three classes. Calcareous sponges (and coralline demosponges) are largely limited to shallow waters (less than 100 m), in part by the nature of calcium carbonate chemistry—secretion of calcareous skeletons becomes more difficult at greater depths, where the solubility of $CaCO_3$ increases. In contrast, the solubility of silica generally decreases with decreasing temperatures (and hence, increasing depth). Hexactinellids, which were common in shallow seas of past eras, are now largely restricted to depths below 200 m, except in cold Antarctic environments, where they occur in shallow waters. The demosponges live at all depths. Calcareous sponges are further restricted to shallow waters by their need for a firm substratum for attachment. On the other hand, many demosponges and hexactinellids grow on soft sediments, attaching by means of rootlike spicule tufts or mats. The coralline sponges, once a predominant group on shallow tropical reefs, are now largely restricted to shaded crevices and caves, or subreef depths, where their potential competitors (the hermatypic corals) cannot grow. They are believed to be relicts of major reef-constructing groups of Mesozoic and Paleozoic seas.

Sponges are the dominant animals in a great many benthic marine habitats. Most rocky littoral regions harbor enormous numbers of sponges, and recent work indicates that they even occur in large numbers (and larger size) around Antarctica. Although many animals prey on sponges, the amount of serious damage they do is usually slight. Some tropical fishes and turtles crop certain kinds of sponges, and small predators (mainly gastropods) consume limited amounts of sponge "tissue" in both warm and temperate seas. Overall, however, sponges appear to be very stable and long-lived animals.

Little is known regarding growth rates in sponges, but available data suggest that rates vary widely among species. Some species are annuals (especially Calcarea of colder waters); hence they grow from larvae or gemmules to reproductive adulthood in a matter of months. Others are perennials and grow so slowly that almost no change can be seen from one year to the next; this growth pattern is especially true of tropical and polar demosponges. Age estimates of perennial species range from 20 to 100 years.

Even a casual seashore explorer or SCUBA diver will quickly realize that sponges are just about everywhere. Most grow on open rock or occasionally sand/mud surfaces, where they are obviously exposed to potential predation. Clearly, some mechanism(s) must be working to prevent these animals from being cropped excessively by various predators. Many biologists suggest that the primary defense mechanism in sponges is biochemical (see Burkholder 1973 and Bergquist and Bedford 1978 for good reviews). Studies over the past two decades have shown that sponges manufacture a surprisingly broad spectrum of biotoxins, some of which are quite potent. A few, such as those of *Tedania* and *Neofibularia*, can cause painful skin rashes in humans. Research in sponge biochemistry has also revealed the widespread occurrence of various antimicrobial agents in sponges. Sponges appear to use "chemical warfare" not only to reduce predation and prevent infection by microbes but also to compete for space with other sessile invertebrates such as ectoprocts, ascidians, and even other sponges. For example, the coral-inhabiting sponge *Siphonodictyon* releases a toxic chemical into the mucus exuded from the oscula, thus preventing potential crowding by maintaining a zone of dead coral polyps around each osculum (Figure 16). Sponges are apparently strong competitors for space in the benthic environment, and different species have evolved various chemicals (**allelochemicals**) that may be species-specific deterrents or actually lethal weapons for use against competing sessile and encrusting invertebrates. In diverse habitats such as coral reefs, there is evidence to suggest that complex networks or competitive hierarchies of various sessile sponges may exist, each sponge competing for space with its own arsenal of chemicals.

Many of the chemicals produced by sponges and

Figure 16

Siphonodictyon coralliphagum infests the hermatypic coral *Montastrea cavernosa* on a Caribbean reef. Note the "dead zone" between the oscular chimneys of the sponge and the coral polyps. (From Sullivan et al. 1983.)

other marine invertebrates are being closely studied by "natural products" chemists and biologists interested in their potential as pharmaceutical agents. Compounds with respiratory, cardiovascular, gastrointestinal, anti-inflammatory, antitumor, and antibiotic activities have been identified from many marine sponges. One New Zealand sponge (*Halichondria moorei*) has long been used by native Maoris to promote wound healing, and was recently discovered to contain remarkably high concentrations (10 percent of the sponge dry weight) of the potent anti-inflammatory agent potassium fluorosilicate. The coming decades will undoubtedly witness the emergence of many new pharmacological compounds of poriferan origin.

Some sponges are capable of very rapid growth and regularly overgrow neighboring flora and fauna. For example, the tropical encrusting sponge *Terpios* grows over both living and nonliving substrata. In Guam this sponge grows at rates averaging 23 mm per month over almost every live coral in the area as well as over hydrocorals, molluscs, and many algae. Experiments have shown that *Terpios* is toxic to living corals and presumably to many other animals. Still another physiological "trick" of some sponges is the ability to rapidly produce copious amounts of mucus when disturbed. On the west coast of North America, the beautiful red-to-orange *Plocamia karykina* covers itself with a thick layer of mucus when injured or disturbed. Yet the little red sea slug *Rostanga pulchra* has evolved the ability to live and feed inconspicuously on this and other sponges, and even lays its camouflaged red egg masses on the sponge's exposed surface without eliciting the mucous reaction.

Commensalism is common among sponges of all kinds. It would be difficult to find a sponge that was not utilized by at least some smaller invertebrates and often by fishes (e.g., gobies and blennies) as refuge. The porous nature of sponges makes them ideally suited for habitation by opportunistic crustaceans, ophiuroids, and various worms. In one study, a single specimen of *Spheciospongia vesparia* from Florida was found to have over 16,000 alphaeid shrimps living in it, and another study from the Gulf of California reported over 100 different species of plants and animals from a 15 × 15-cm piece of *Geodia mesotriaena*. Most symbionts of sponges use their hosts only for space and protection, but some rely on the sponge's water current for a supply of suspended food particles. A classic example of this phenomenon is the male–female pair of shrimp (*Spongicola*) that inhabit hexactinellid sponges known as Venus's flower basket (*Euplectella*; Figure 1E). The shrimp enter the sponge when they are young, only to become entrapped in their host's glasslike case as they grow too large to leave. Here they spend their lives in bonded bliss—or perhaps as prisoners of love. Appropriately, this sponge (with its guests) is a traditional wedding gift in Japan—a symbol of the lifetime bond between two partners.

Other, even more intimate, symbiotic relationships with sponges are common. Some snails and clams characteristically have specific sponges encrusting their shells, and many species of crabs (hermits and brachyurans) effectively collect certain sponges and cultivate them on their shell or carapace. Demosponges, such as *Suberites*, are commonly involved in these commensalistic relationships. The sponge serves primarily as protective camouflage for its host, and it benefits by being carried about to new areas. And the sponge no doubt feeds off small bits of animal matter dislodged during the feeding activities of its host.

Another spectacular example of poriferan symbiosis is certain sponge–bacteria and sponge–algae associations that appear to be mutualistic. For example, a typical member of the demosponge order Verongida contains a mesohyl bacterial population accounting for some 38 percent of its body's volume, far exceeding the actual sponge-cell volume of only 21 percent. Most of these bacteria belong to the genera *Pseudomonas* and *Aeromonas*. Presumably, the sponge matrix provides a rich medium for bacterial growth, and the host benefits by being able to conveniently phagocytize the bacteria for food. Similar relationships are common between poriferans and various blue-green algae. Sponges are the only metazoa known to maintain such symbiotic relationships with cyanobacteria. Recent evidence suggests that some products of normal cyanobacterial metabolism (e.g., glycerol and certain organic phosphates) are translocated directly to the sponge for nutrition. In many sponges, both regular bacteria and cyanobacteria occur, the former in deeper cellular regions, the latter closer to the surface where light is available. In a remarkable study, C. R. Wilkinson (1983) showed that 6 of the 10 most common sponge species on the fore-reef slope of Davies Reef (Great Barrier Reef) are net primary producers, with three times more oxygen produced by photosynthesis than consumed by respiration. In some areas of the Great Barrier Reef, sponges are second only to corals in overall biomass, and they appear to owe their rapid growth and net primary productivity to the presence of large populations of symbiotic cyanobacteria. Most freshwater spongillids (and a few marine sponges as well) maintain similar relationships with zoochlorellae (symbiotic dinoflagellates). These sponges grow larger and more rapidly than specimens of the same species that are kept in dark conditions. Commensalistic relationships have also been reported between sponges and red algae, filamentous green algae, and diatoms.

Not all sponge symbioses are commensal or mutualistic. In fact, some are clearly harmful, such as the boring demosponges that excavate complex galleries in calcareous material such as corals and mollusc shells (Figure 17). The phenomenon of boring, known as **bioerosion**, causes significant damage to commercial oysters as well as to natural coral, clam, and scallop populations. The active boring process involves a chemical removal of fragments or chips of the calcareous material by specialized archaeocytes called **etching cells**. One study (Hatch 1980) has implicated the use of carbonic anhydrase in this process. The chips are expelled in the excurrent canal system and can contribute significantly to local sediments.

In some areas such sponge bioerosion has a significant impact on coral reefs. Perhaps even more important than actual erosion is the weakening of attachment regions of large corals. This action may result in much coral loss during heavy tropical storms. Boring sponges do not appear to gain any direct nourishment from their host coral; rather they use it as a protective casing in which they reside. If you carefully examine the shells of dead bivalves along any beach, you will discover that many of them are perforated with minute holes and galleries of boring sponges. These poriferans are responsible for a major portion of the initial breakdown of such calcareous structures, setting the stage for their eventual decomposition and recycling through one of the Earth's geochemical cycles.

Poriferan phylogeny

The origin of sponges

Sponges are an ancient group, and the important events in their origin and early evolution lie hidden in Precambrian time. The unique nature of the poriferan *Bauplan* is clear, however, and it is strikingly revealed by their cellular totipotency, their lack of true tissues or reproductive organs, their reproductive flexibility, and their lack of body polarity and a basal membrane. These features, in combination with the prevalence and importance of flagellated (i.e., monociliated) cells in sponges, strongly suggest a direct protozoan ancestry. It would seem that the poriferans share as many, if not more, similarities with protozoa than they do with other metazoa (e.g., cellular totipotency; excretory, respiratory and osmoregulatory strategies; reliance on flagellated cells in a variety of ways, including feeding; entirely intracellular digestion). At the same time, sponges appear to stand apart from all other metazoa in their possession of the unique aquiferous system, which represents a key synapomorphy defining this phylum. In addition, sponges have been shown to differ substantially from virtually all other animals in the chemical nature of their various lipoid compounds (sterols in particular).

Current opinion favors the origin of the Porifera from flagellated protistan ancestors—either a simple, hollow, free-swimming colonial form or a colonial choanoflagellate. The choanoflagellates possess cer-

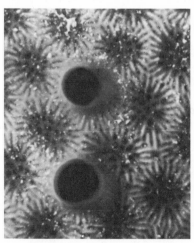

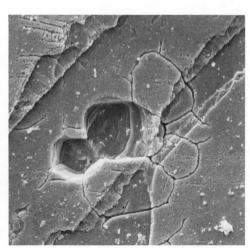

A   B

Figure 17

Boring sponges. A, Surface of a coral (stellate openings) infected by the sponge *Cliona* (circular oscula); B, A close view (SEM) of the surface of a clam shell, showing several eroded "chips," two of which have been entirely removed and others that are only partly etched by *Cliona*. (From Rutzler and Reiger 1973.)

tain features that seem to ally them strongly to sponges. For example, the collar cells of sponges are strikingly similar to the collared cells of colonial choanoflagellates. Similar collar cells have also been found in some widely divergent metazoan animals (certain echinoderm larvae, in the oviducts of some sea cucumbers, and in certain corals). However, these discoveries do little to diminish the force of the argument for a choanoflagellate ancestry to the sponges. In either case, the mesohyl is viewed as originating by simple ingression of surface cells, as seen in the embryogeny of many living sponges. Adoption of a benthic lifestyle could have fostered increased body size. Increase in size led to surface-to-volume problems that were rapidly overcome by the evolution of the syconoid and leuconoid *Baupläne*, increasing the surface area of the choanoderm-lined areas and maintaining small diffusion distances as the aquiferous system became more complex.

The solutions that poriferans evolved to problems of survival created a group of animals unlike any other. Sponges achieved multicellularity and large body size without such typically metazoan traits as embryological tissue layering, neuronal coordination, extracellular digestion, excretory structures, or fixed reproductive structures. Taken together, these and other poriferan attributes suggest that the sponges either arose very early in metazoan cladogenesis, or perhaps even from a separate protozoan line.

Evolution within the Porifera

Sponges are such an ancient and enigmatic phylum that their phylogeny has largely eluded biologists. There is no generally agreed-upon phylogenetic hypothesis of relationships among the classes. We do know that sponges probably evolved during the Precambrian period (Figure 19). Their hard skeletal components have left good fossil records for all three extant classes, beginning in the Cambrian and extending to the present. Well over 1,000 fossil genera have been described, about 20 percent of which are still extant. The early Paleozoic witnessed the growth of massive tropical reefs composed largely of two spongelike groups, the archaeocyathans and stromatoporoids. The archaeocyathans probably were not true sponges (Figure 18). Current thinking on this group views them as a unique, long-extinct, separate phylum or subkingdom called Archaeozoa. They were all extinct by the late Cambrian.

Stromatoporoids have a long geological history, extending from late Cambrian times to the present. The nature of these reef-building invertebrates has

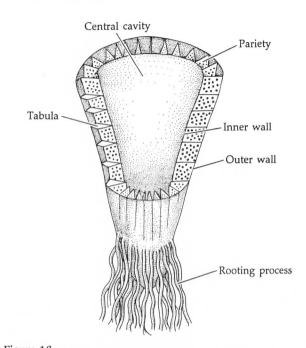

Figure 18

A typical archaeocyathan. A vertical section has been partly cut away to show the structure between the inner and outer walls (i.e., vertical parieties and horizontal tabula). (After Bergquist 1978.)

been hotly debated. That the fossil stromatoporoids were true sponges is most strongly suggested by the apparent homology of structures called **astrorhizae**, found in their calcareous skeletons, to similar stellate impressions in the skeletons of recent coralline sponges. In living sponges these stellate marks are the traces of the converging exhalant canal systems (beneath the oscula). The absence of siliceous spicules in the calcareous skeletons of fossil stromatoporoids has been cited as evidence against a true sponge relationship. However, not all Recent stromatoporoids have siliceous spicules; and in some that do, these spicules are never incorporated into the calcareous basal skeleton that would be fossilized anyway. Nevertheless, there is still disagreement about the relationships of fossil stromatoporoids, especially those of Paleozoic age. Affinities with the Cyanobacteria and Cnidaria have often been postulated, as well as with the coralline red algae, foraminiferans, and ectoprocts. Some zoologists who accept them as true sponges place stromatoporoids in a separate class. However, a coralline sponge affinity is suggested by studies on two Recent relict genera of Stromatoporoida, in which an astrorhizae-bearing basal layer of calcium carbonate (aragonite) is present.

Figure 19

History of the three sponge classes, the coralline sponges, and Archaeocyatha through geological time. Dashed lines indicate suggested occurrence, even though fossils have not yet been found. "R" indicates the times when the group in question is known to have been an important marine reef builder.

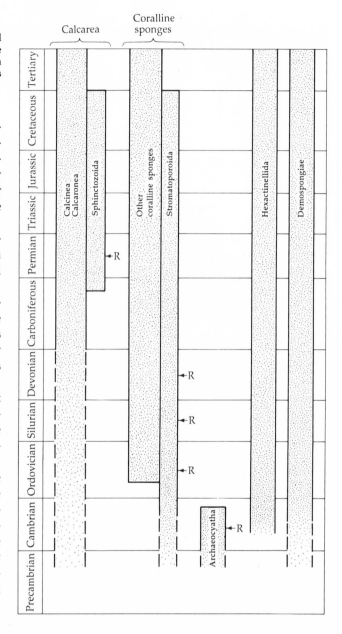

Unlike the coralline sponges, which have decreased in abundance and diversity since the Mesozoic, the remaining calcareous sponges and demosponges appear to have increased in diversity throughout their history. Hexactinellids, on the other hand, were most diverse and abundant during the Cretaceous.

The oldest hexactinellid fossils, of the early Cambrian, were all thin-walled, saclike sponges with a dispersed surface spicule layer that probably could not support a thick body wall. During the Paleozoic, hexactinellids were common in shallow-water environments. Since then, however, they have become restricted largely to the deep oceans. The calcareous sponges have also had a long history, and many distinctive early lines have died out since their origins in Cambrian and Precambrian times. Interestingly, some of these early calcareous forms were important reef builders during the Permian, but like stromatoporoids, the few surviving genera of those lines have all retreated to marine cave habitats in modern tropical seas.

The demosponges were well established by the mid-Cambrian, from which the earliest fossils are known, and all known orders of modern demosponges are found in Cretaceous rocks. But only recently have the complex relationships among the dozen or so orders of Demospongiae been critically examined. The challenge of unraveling the relationships in this ponderous group of sponges is still in an early phase.

Selected References

General

Bakus, G. J. 1964. The effects of fish-grazing on invertebrate evolution in shallow tropical waters. Occ. Papers Allan Hancock Fnd. 27:1–29.

Bakus, G. J. 1969. Energetics and feeding in shallow marine waters. Int. Rev. Gen. Exp. Zool. 4:275–369.

Bayer, F. M. and H. B. Owre. 1968. *The Free-Living Lower Invertebrates.* Macmillan, New York.

Bergquist, P. R. 1978. *Sponges.* University of California Press, Berkeley.

Bergquist, P. 1985. Poriferan relationships. *In* S. C. Morris et al. (eds.), *The Origin and Relationships of Lower Invertebrates.* Syst. Assoc. Spec. Vol. No. 28, Oxford, pp. 14–27.

Bergquist, P. R. and R. J. Wells. 1983. Chemotaxonomy of the Porifera: The development and current status of the field. *In Marine Natural Products, III.* Academic Press, New York, pp. 1–50.

Borojevic, R. 1970. Différentiation cellulaire dans l'embryogenèse et la morphogenèse chez les spongiaires. Symp. Zool. Soc. Lond. 25:467–490.

Borojevic, R., W. G. Fry, W. C. Jones, C. Levi, R. Rasmont, M. Sara and J. Vacelet. 1967. Mise au point actuelle de la terminologie des Èponges. (A reassessment of the terminology for sponges). Bull. Mus. Hist. Nat. Paris (2) 39:1224–1235.

Bowerbank, J. S. 1861, 1862. On the anatomy and physiology of the Spongiidae. Part 1: On the spicula. Philos. Trans. R. Soc. Lond. 148:279–332. Part 2: Proc. R. Soc. Lond. 11:372–375.

Bowerbank, J. S. 1861. On the anatomy and physiology of the Spongiidae. Philos. Trans. R. Soc. Lond. 152:747–829, 1087–1138.

Bowerbank, J. S. 1864, 1866, 1874. A monograph of the British Spongiidae. Vols. 1, 2, and 3. Ray Society, London.

Brien, P. 1968. The sponges, or Porifera. In M. Florkin and B. T. Scheer (eds.), Chemical Zoology, Vol. 2. Academic Press, New York, pp. 1–30.

Brien, P., C. Levi, M. Sara, O. Tuzet and J. Vacelet. 1973. Spongiaires. In P. Grassé (ed.), Traité de Zoologie 3(1):716. Masson et Cie, Paris.

Brill, B. 1973. Untersuchungen zur Ultastruktur der Choanocyte von Ephydatia fluviatilis, L. Z. Zellforsch. 144:231–245.

Burkholder, P. R. 1973. The ecology of marine antibiotics and coral reefs. In O. A. Jones and R. Endean (eds.), Biology and Geology of Coral Reefs, Vol. 1. Biology, 1. Academic Press, New York, pp. 117–182.

Burkholder, P. R. and K. Ruetzler. 1969. Antimicrobial activity of some marine sponges. Nature 222:983–984.

Burton, M. 1932. Sponges. Discovery Rep. 6:327–392.

Cox, G. and A. W. D. Larkum. 1983. A diatom apparently living in symbiosis with a sponge. Bull. Mar. Sci. 33(4):943–945.

Curtis, A. 1979. Individuality and graft rejection in sponges: A cellular basis for individuality in sponges. In G. Larwood and B. Rosen (eds.), Biology and Systematics of Colonial Organisms. Academic Press, New York, pp. 39–48.

Dayton, P. K. 1979. Observations of growth, dispersal and population dynamics of some sponges in McMurdo Sound, Antarctica. In C. Levi and N. Boury-Esnault (eds.), Biologie des Spongiaires, Centre Nat. Recherche Scient., Paris, pp. 271–282.

Dayton, P. K., G. A. Robilliard and R. T. Paine. 1970. Benthic faunal zonation as a result of anchor ice at McMurdo Sound, Antarctica. In M. W. Holdgate (ed.), Antarct. Ecol. 1:244–258.

Dayton, P. K., G. A. Robilliard, R. T. Paine and L. B. Dayton. 1974. Biological accommodation in the benthic community at McMurdo Sound, Antarctica. Ecol. Monogr. 44:105–128.

de Laubenfels, M. W. 1932. The marine and fresh-water sponges of California. Proc. U.S. Nat. Mus. 81:1–140.

de Laubenfels, M. W. 1935. Some sponges of Lower California (Mexico). Am. Mus. Novit. 779:1–14.

de Laubenfels, M. W. 1936. A discussion of the sponge fauna of the Dry Tortugas in particular and West Indies in general, with material for a revision of the families and orders of the Porifera. Carnegie Inst. Washington Publ. 467, Tortugas Lab. Papers 30:1–225.

de Laubenfels, M. W. 1954. The sponges of the West-Central Pacific. Oregon State College, Corvallis.

de Laubenfels, M. W. 1955. Porifera. In R. C. Moore (ed.), Treatise on Invertebrate Paleontology. Archaeocyatha and Porifera, E21–E112. Geol. Soc. Am. and University of Kansas Press.

Dickinson, M. G. 1945. Sponges of the Gulf of California. Allan Hancock Pacific Exped. 11:1–252.

Faulkner, D. J. 1973. Variabilin, an antibiotic from the sponge, Ircinia variabilis. Tetrahedron Let. 29:3821–3822.

Faulkner, D. J. 1977. Interesting aspects of marine natural products chemistry. Tetrahedron Let. 33:1421–1443.

Fell, P. E. 1974. Porifera. In A. C. Giese and J. S. Pearse (eds.), Reproduction of Marine Invertebrates, Vol. 1. Academic Press, New York, pp. 51–132.

Fell, P. E. 1976. Analysis of reproduction in sponge populations: An overview with specific information on the reproduction of Haliclona loosanoffi. In F. W. Harrison and R. R. Cowden (eds.), Aspects of Sponge Biology. Academic Press, New York, pp. 51–67.

Finks, R. M. 1970. The evolution and ecologic history of sponges during Palaeozoic times. Symp. Zool. Soc. Lond. 25:3–22.

Frost, T. M. 1976. Sponge feeding: A review with a discussion of continuing research. In F. W. Harrison and R. R. Cowden (eds.), Aspects of Sponge Biology. Academic Press, New York, pp. 283–298.

Fry, W. G. (ed.). 1970. The Biology of the Porifera. Academic Press, New York.

Garrone, R. and J. Pottu. 1973. Collagen biosynthesis in sponges: Elaboration of spongin by spongocytes. J. Submicrosc. Cytol. 5:199–218.

Goodwin, T. W. 1968. Pigments of Porifera. In M. Florkin and B. T. Scheer (eds), Chemical Zoology, Vol. 2. Academic Press, New York, pp. 53–64.

Harrison, F. W. and R. R. Cowden (eds.) 1976. Aspects of Sponge Biology. Academic Press, New York.

Hartman, W. D. 1958. Natural history of the marine sponges of southern New England. Bull. Peabody Mus. Nat. Hist. 12:1–155.

Hartman, W. D. 1982. Porifera. In S. P. Parker (ed.), Synopsis and Classification of Living Organisms, Vol. 1. McGraw-Hill, New York, pp. 641–666.

Hartman, W. D. and H. M. Reiswig. 1973. The individuality of sponges. In R. S. Boardman, A. H. Cheetham and W. A. Oliver, Animal Colonies. Dowden, Hutchinson & Ross, Inc., Stroudsburg, Pennsylvania, pp. 567–584.

Hartman, W. D., J. W. Wendt and F. Wiedenmayer. 1980. Living and Fossil Sponges. (Compiled by R. N. Ginsburg and P. Reid.) Sedimenta VIII (Comparative Sedimentology Lab., Div. Mar. Geol. and Geophysics, University of Miami).

Hildemann, W. H., I. S. Johnson and P. L. Jobiel. 1979. Immunocompetence in the lowest metazoan phylum: Transplantation immunity in sponges. Science 204:420–422.

Hill, D. 1972. Archaeocyatha. In R. C. Moore (ed.), Treatise on Invertebrate Paleontology 1:158. Geol. Soc. Am. and University of Kansas Press, Lawrence.

Humphreys, T. 1963. Chemical dissolution and in vitro reconstruction of sponge cell adhesions. I. Isolation and functional demonstration of components involved. Dev. Biol. 8:27–47.

Humphreys, T. 1970. Species specific aggregation of dissociated sponge cells. Nature 228:685–686.

Hyman, L. H. 1940. The Invertebrates, Vol. 1, Protozoa through Ctenophora. McGraw-Hill, New York, pp. 284–364.

Jackson, J. B. C. 1977. Competition on marine hard substances: The adaptive significance of solitary and colonial strategies. Am. Nat. 111:743–767.

Jackson, J. B. C. and L. Buss. 1975. Allelopathy and spatial competition among coral reef invertebrates. Proc. Natl. Acad. Sci. USA 72:5160–5163.

Jackson, J. B. C., T. F. Goreau and W. D. Hartman. 1971. Recent brachiopod–coralline sponge communities and their paleoecological significance. Science 173:623–625.

Jones, W. C. 1962. Is there a nervous system in sponges? Biol. Rev. 37:1–50. [Apparently not.]

Kaye, H. and T. Ortiz. 1981. Strain specificity in a tropical marine sponge. Mar. Biol. 63:165–173.

Koltun, V. M. 1968. Spicules of sponges as an element of the bottom sediments of the Antarctic. In Symposium on Antarctic Oceanography, Scott Polar Res. Inst., Cambridge, pp. 121–123.

Kuhns, W., G. Weinbaum, R. Turner and M. Burger. 1974. Sponge cell aggregation: A model for studies on cell–cell interactions. Ann. N.Y. Acad. Sci. 234:58–74.

Lawn, I. D., G. O. Mackie, and G. Silver. 1981. Conduction system in a sponge. Science 211:1169–1171.

Lecompte, M. 1956. Stromatoporoidea. In R. C. Moore (ed.), Treatise on Invertebrate Paleontology, F:F107–F114. Geol. Soc. Am. and University of Kansas Press, Lawrence.

Levi, C. 1957. Ontogeny and systematics in sponges. Syst. Zool. 6:174–183.

Levi, C. and N. Boury-Esnault (eds.). 1979. Sponge Biology. Colloques Internationaux du Centre National de la Recherche Scientifique. Ed. Cen. Nat. Resch. Sci. No. 291.

Mackie, G. O. 1979. Is there a conduction system in sponges? In C. Levi and N. Boury-Esnault (eds.), Sponge Biology, Colloques Internat. C.N.R.S. 291:145–151.

Minale, L., G. Cimino and S. DeStefano. 1976. Natural products from Porifera. Forts. Chemie Org. Naturst. (B) 33:1–72.

Minchin, E. A. 1900. Sponges. *In* E. R. Lankester (ed.), *A Treatise in Zoology*, Pt. 2. Adam and Charles Black, London, pp. 1–178.

Neigel, J. E. and G. P. Schmahl. 1984. Phenotypic variation within histocompatibility-defined clones of marine sponges. Science 224:413–415.

Paine, R. T. 1964. Ash and calorie determinations of sponge and opisthobranch tissue. Ecology 45:384–387.

Palumbi, S. R. 1984. Tactics of acclimation: Morphological changes of sponges in an unpredictable environment. Science 225:1478–1480.

Randall, J. E. and W. D. Hartman. 1968. Sponge-feeding fishes of the West Indies. Mar. Biol. 1:216–225.

Reiswig, H. 1975. Bacteria as food for temperate–water marine sponges. Can. J. Zool. 53:582–589.

Rezvoi, P. D., I. T. Zhuravleva and V. M. Koltun. 1971. Phylum Porifera. *In* Y. A. Orlov and B. S. Sokolov, *Fundamentals of Paleontology*, Vol. 1, Pt. II. Porifera, Archaeocyatha, Coelenterata, Vermes. Israel Program for Scientific Translations, Jerusalem, pp. 5–97.

Rinehart, K. L., Jr., and 25 other authors. 1981. Marine natural products as sources of antiviral, antimicrobial, and antineoplastic agents. Pure Appl. Chem. 53:795–817.

Rützler, K. 1970. Spatial competition among Porifera: Solution by epizoism. Oecologia 5:85–95.

Sara, M. 1970. Competition and cooperation in sponge populations. Symp. Zool. Soc. Lond. 25:273–285.

Sara, M. 1974. Sexuality in the Porifera. Bull. Zool. 41:327–348.

Sara, M. and J. Vacelet. 1973. Écologie des Demosponges. *In* P. Grassé (ed.), *Traité de Zoologie* 3(1):462–576. Masson et Cie, Paris.

Schwab, D. W. and R. E. Shore. 1971. Fine structure and composition of a siliceous sponge spicule. Biol. Bull. 140:125–136.

Sharma, G. H. and B. Vig. 1972. Studies on the antimicrobial substances of sponges. VI. Structure of two antibacterial substances isolated from the marine sponge *Dysidea herbacea*. Tetrahedron Lett. 28:1715–1718.

Shore, R. E. 1972. Axial filament of siliceous sponge spicules, its organic components and synthesis. Biol. Bull. 143:689–698.

Simpson, T. L. 1984. *The Cell Biology of Sponges*. Springer-Verlag, New York.

Stearn, C. W. 1975. The stromatoporoid animal. Lethaia 8:89–100.

Stearn, C. W. 1977. Studies of stromatoporoids by scanning electron microscopy. Mem. Bur. Rech. Geol. Min. 89:33–40.

Sullivan, B., D. J. Faulkner and L. Webb. 1983. Siphonodictidine, a metabolite of the burrowing sponge *Siphonodictyon* sp. that inhibits coral growth. Science 221:1175–1176.

Tuzet, O. 1963. The phylogeny of sponges according to embryological, histological and serological data, and their affinities with the Protozoa and Cnidaria. *In* E. C. Dougherty, Z. N. Brown, E. D. Hanson and W. D. Hartman (eds.), *The Lower Metazoa: Comparative Biology and Phylogeny*. University of California Press, Berkeley, pp. 129–148.

Vacelet, J. 1985. Coralline sponge, and the evolution of the Porifera. *In* S. C. Morris et al. (eds.), *The Origins and Relationships of Lower Invertebrates*. Syst. Assoc. Spec. Vol. No. 28, Oxford, pp. 1–13.

Van de Vyver, G. 1975. Phenomena of cellular recognition in sponges. *In* A. Moscona and A. Monroy (eds.), *Current Topics in Developmental Biology*, Vol. 4. Academic Press, New York, pp. 123–140.

Warburton, F. E. 1966. The behaviour of sponge larvae. Ecology 47:672–674.

Westinga, E. and P. C. Hoetjes. 1981. The intrasponge fauna of *Speciospongia vesparia* at Curacao and Bonaire. Mar. Biol. 26:139.

Wilkinson, C. R. 1983. Net primary productivity in coral reef sponges. Science 219:410–412.

Wilkinson, C. R. and J. Vacelet. 1979. Transplantation of marine sponges to different conditions of light and current. J. Exp. Mar. Biol. Ecol. 37:91–104.

Wilson, H. V. 1891. Notes on the development of some sponges. J. Morphol. 5:511–519.

Calcarea

Borojevic, R. 1979. Evolution des spongiaires Calcarea. *In* C. Levi and N. Boury-Esnault (eds.), *Sponge Biology*, Colloques Internat. C.N.R.S. 291:527–530.

Burton, M. 1963. A revision of the classification of the calcareous sponges. Brit. Mus. Nat. Hist., London.

Hartman, W. D. 1958. A re-examination of Bidder's classification of the Calcarea. Syst. Zool. 7:97–110.

Jones, W. C. 1965. The structure of the porocytes in the calcareous sponge *Leucosolenia complicata* (Montagu). J. R. Microsc. Soc. 85:53–62.

Jones, W. C. 1970. The composition, development, form and orientation of calcareous sponge spicules. Symp. Zool. Soc. Lond. 25:91–123.

Ledger, P. W. and W. C. Jones. 1978. Spicule formation in the calcareous sponge *Sycon ciliatum*. Cell Tiss. Res. 181:553–567.

Tanita, S. 1942. Key to all the described species of the genus *Leucosolenia* and their distribution. Sci. Rep. Tohoku University (4) Biol. 17:71–93.

Tuzet, O. 1973. Éponges Calcaires. *In* P. Grassé (ed.), *Traité de Zoologie* 3(1):27–132. Masson et Cie, Paris.

Ziegler, B. and S. Rietschel. 1970. Phylogenetic relationships of fossil calcisponges. Symp. Zool. Soc. Lond. 25:23–40.

Hexactinellida

Okada, Y. 1928. On the development of a hexactinellid sponge *Farrea sollasii*. Tokyo University Fac. Sci. J., Sect. IV, 2:1–27.

Reiswig, H. M. 1979. Histology of Hexactinellida (Porifera). *In* C. Levi and N. Boury-Esnault (eds.), *Sponge Biology*, Colloques Internat. C.N.R.S. 291:173–180.

Sclerospongiae (the coralline sponges; now in Calcarea and Demospongiae)

Hartman, W. D. and T. F. Goreau. 1975. A Pacific tabulate sponge, living representative of a new order of sclerosponges. Postilla 167:1–21.

Hill, D. and E. C. Strumm. 1956. Tabulata. *In* R. C. Moore (ed.), *Treatise on Invertebrate Paleontology*, F:444–477. Geol. Soc. Am. and University of Kansas Press, Lawrence.

Kazmierczak, J. 1984. Favositid tabulates: Evidence for poriferan affinity. Science 225:835–837.

Lang, J. D., W. D. Hartman and L. S. Land. 1975. Sclerosponges: Primary framework constructors on the Jamaican deep forereef. J. Mar. Res. 33:223–231.

Stearn, C. W. 1972. The relationship of the stromatoporoids to the sclerosponges. Lethaia 5:369–388.

Demospongiae

Ayling, A. L. 1980. Patterns of sexuality, asexual reproduction and recruitment in some subtidal marine Demospongiae. Biol. Bull. 158:271–282.

Ayling, A. L. 1983. Growth and regeneration rates in thinly encrusting Demospongiae from temperate waters. Biol. Bull. 165:343–352.

Bergquist, P. R. and J. H. Bedford. 1978. The incidence of antibacterial activity in marine Demospongiae: Systematic and geographic considerations. Mar. Biol. 46:215–221.

Bergquist, P. R. and W. D. Hartman. 1969. Free amino acid patterns and the classification of the Demospongiae. Mar. Biol. 3:247–268.

Bergquist, P. R. and J. J. Hogg. 1969. Free amino acid patterns in Demospongiae: A biochemical approach to sponge classification. Cah. Biol. Mar. 10:205–220.

Bergquist, P. R., M. E. Sinclair and J. J. Hogg. 1970. Adaptation to intertidal existence: Reproductive cycles and larval behaviour in Demospongiae. Symp. Zool. Soc. Lond. 25:247–271.

Brien, P. and H. Meewis. 1938. Contribution à l'étude de l'embryogenése des Spongillidas. Arch. Biol. 49:177–250.

Bryan, P. G. 1973. Growth rate, toxicity and distribution of the encrusting sponge Terpios sp. (Hadromerida: Suberitidae) in Guam, Mariana Islands. Micronesica 9:237–242.

Cimino, G., S. DeStefano, L. Minale and G. Sodano. 1975. Metabolism in Porifera. III. Chemical patterns and the classification of the Demospongiae. Comp. Biochem. Physiol. 50B:279–285.

Connes, R., J.-P. Diaz and J. Paris. 1971. Choanocytes et cellule centrale chez la Demosponge Suberites massa Nardo. C. R. Hebd. Séanc. Acad. Sci. 273:1590–1593.

de Laubenfels, M. W. 1948. The order Keratosa of the phylum Porifera—A monographic study. Occ. Pap. Allan Hancock Fnd. 3:1–217.

Elvin, D. W. 1976. Seasonal growth and reproduction of an intertidal sponge, Haliclona permoilis (Bowerbank). Biol. Bull. 151:108–125.

Fell, P. E. 1969. The involvement of nurse cells in oogenesis and embryonic development in the marine sponge Haliclona ecobasis. J. Morphol. 127:133–149.

Fell, P. E. 1976. The reproduction of Haliclona loosanoffi and its apparent relationship to water temperature. Biol. Bull. 150:200–210.

Fell, P. E. and K. B. Lewandrowski. 1981. Population dynamics of the estuarine sponge Halichondria sp., within a New England eelgrass community. J. Exp. Mar. Biol. Ecol. 55:49–63.

Finks, R. M. 1967. The structure of Saccospongia laxata Bassler (Ordovician) and the phylogeny of the Demospongiae. J. Paleontol. 41:1137–1149.

Frost, T. M. and C. E. Williamson. 1980. In situ determination of the effect of symbiotic algae on the growth of the freshwater sponge Spongia lacustris. Ecology 61:1361–1370.

Gerrodette, T. and A. O. Fleschig. 1979. Sediment–induced reduction in the pumping rate of the tropical sponge Verongia lacunosa. Mar. Biol. 55:103–110.

Guida, V. G. 1976. Sponge predation in the oyster reef community as demonstrated with Cliona celata Grant. J. Exp. Mar. Biol. Ecol. 25:109–122.

Hatch, W. I. 1980. The implication of carbonic anhydrase in the physiological mechanism of penetration of carbonate substrata by the marine burrowing sponge Cliona celata (Demospongiae). Biol. Bull. 159:135–147.

Levi, C. 1956. Étude des Halisarca de Roscoff. Embryologie et systématiques des Demosponges. Arch. Zool. Exp. Gen. 93:1–181.

Levi, C. 1973. Systematique de las classe des Demospongiaria (Demosponges). In P. Grassé (ed.) Traité de Zoologie 3(1):577–631. Masson et Cie, Paris.

Penney, J. T. 1960. Distribution and bibliography (1892–1957) of the fresh water sponges. University of South Carolina Publ. Biol. 3:1–97.

Penney, J. T. and A. A. Racek. 1968. Comprehensive revision of a worldwide collection of freshwater sponges (Porifera: Spongillidae). Bull. U.S. Nat. Mus. No. 272:1–184.

Rasmont, R. 1962. The physiology of gemmulation in fresh water sponges. In D. Rudnick (ed.), Regeneration, 20th Growth Symposium. Ronald Press, New York, pp. 1–25.

Reiswig, H. M. 1970. Porifera: Sudden sperm release by tropical Demospongiae. Science 170:538–539.

Reiswig, H. M. 1971a. Particle feeding in natural populations of three marine Demosponges. Biol. Bull. 141:568–591.

Reiswig, H. M. 1971b. In situ pumping activities of tropical Demospongiae. Mar. Biol. 9:38–50.

Reiswig, H. M. 1973. Population dynamics of three Jamaican Demospongiae. Bull. Mar. Sci. 23:191–226.

Reiswig, H. M. 1974. Water transport, respiration and energetics of three tropical sponges. J. Exp. Mar. Biol. Ecol. 14:231–249.

Reiswig, H. M. 1975. The aquiferous systems of three marine Demospongiae. J. Morphol. 145(4):493–502.

Reiswig, H. M. 1976. Natural gamete release and oviparity in Caribbean Demospongiae. In F. W. Harrison and R. R. Cowden (eds.), Aspects of Sponge Biology. Academic Press, New York, pp. 99–112.

Ristau, D. A. 1978. Six new species of shallow–water marine demosponges from California. Proc. Biol. Soc. Wash. 91:569–589.

Rützler, K. 1975. The role of burrowing sponges in bioerosion. Oecologia 19:203–216.

Rützler, K. and G. Reiger. 1973. Sponge burrowing: Fine structure of Cliona lampa penetrating calcareous substrata. Mar. Biol. 21:144–162.

Schmidt, I. 1970. Phagocytose et pinocytose chez les Spongillidae. Z. Vgl. Physiol. 66:398–420.

Simpson, T. L. and J. J. Gilbert. 1973. Gemmulation, gemmule hatching and sexual reproduction in freshwater sponges. I. The life cycle of Spongilla lacustris and Tubella pennsylvanica. Trans. Am. Microsc. Soc. 92:422–433.

Sollas, W. J. 1884. On the origin of freshwater faunas: A study in evolution. Trans. R. Soc. Dublin 2(3):87–118.

Chapter Eight

Phylum Cnidaria

"Cyanea!" I cried. "Cyanea! Behold the Lion's Mane!"
Sherlock Holmes
The Adventure of the Lion's Mane

The phylum Cnidaria is a highly diverse assemblage containing jellyfish, sea anemones, corals, and the common laboratory *Hydra*, as well as many less familiar forms such as hydroids, sea fans, siphonophores, and zoanthids (Figure 1). There are about 9,000 extant species of cnidarians. Much of the striking diversity seen in cnidarians results from two fundamental aspects of their lifestyle. First is the tendency to form colonies by asexual reproduction; the colony can achieve dimensions and forms unattainable by single individuals. Second, many cnidarians are **dimorphic**, so called because they assume two entirely different adult morphologies during their life histories: a **polypoid form** and a **medusoid form**. This phenomenon is known as **alternation of generations** and has major evolutionary implications touching on nearly every aspect of cnidarian biology.

Cnidarians are diploblastic metazoa at a tissue grade of construction. They possess primary radial symmetry, tentacles, stinging or adhesive structures called **cnidae**, an entodermally derived, incomplete gastrovascular cavity as their only "body cavity," and a middle layer (called mesenchyme* or mesoglea)

derived largely from ectoderm. They lack cephalization, a centralized nervous system, and discrete respiratory, circulatory, and excretory organs (Box 1). This basic *Bauplan* is retained in both the polypoid and medusoid forms (Figure 2).

Cnidarians are mostly marine, but a few groups have successfully invaded fresh waters. Most are sessile (polyps) or planktonic (medusae) carnivores, although some employ suspension feeding and many species harbor symbiotic intracellular algae from which they may derive nutrients. Cnidarians range

*There exists a suite of terms in zoological literature that is frequently confused, misused, and generally messy. These terms include mesenchyme, mesoglea, collenchyme, parenchyme, and coenenchyme. In this book, these terms are defined as follows. **Mesenchyme** (literally meaning "middle juices") refers to a primitive "connective tissue" derived at least in part from ectoderm and located between the epidermis and the gastrodermis (endodermis). Mesenchyme generally consists of two components: a noncellular, jelly-like matrix called **mesoglea**, and various cells and cell products (e.g., fibers). When no cellular material is present, this layer may be properly called mesoglea. Mesenchyme is the typical middle layer of sponges (where it is called the mesohyl),

and of members of the phyla Cnidaria and Ctenophora. In these diploblastic groups, where no "true" (ento-) mesoderm exists, the mesenchyme is fully ectomesodermally derived. Mesenchyme may sometimes be more specifically designated as **collenchyme** or **parenchyme**, when cellular material is sparse or densely packed, respectively. The term parenchyme is sometimes used for the middle (mesenchymal) layer of triploblastic acoelomate animals (e.g., flatworms), in which the dense layer includes tissues derived from both ecto- and entomesoderm.

In some colonial cnidarians, particularly anthozoan polyps, the individuals are embedded in and arise from a mass of mesenchyme that is perforated with gastrovascular channels that are continuous among the members of the colony. The term **coenenchyme** refers to this entire matrix of common basal material, which is itself covered by a layer of epidermis.

Adding to the potential confusion, the term mesenchyme is used in a second, very different way by some biologists. Embryologists use the term to refer to that part of entomesoderm from which all connective tissues, blood vessels, blood cells, the lymphatic system, and the heart are derived. Thus, to an embryologist, the term "mesenchymal cell" often denotes any undifferentiated cell found in the embryonic mesoderm that is capable of differentiating into such tissues. Because of this confusion, some authors prefer to use the term mesoglea, in lieu of mesenchyme, when referring to the middle layers of sponges and diploblastic metazoa. However, we adhere to the former definition and hope that this note will lessen rather than add to the muddle.

A word of caution regarding spelling: the meanings of some of these terms can be altered by changing the terminal "e" to an "a." The termination "-chyme" is preferred for animals, "-chyma" for plants. **Mesenchyma** refers to tissue lying between the xylem and phloem in plant roots; **collenchyma** refers to certain primordial leaf tissues. **Parenchyma** is a very general botanical term used in reference to various supportive tissues. Unfortunately, the same spelling is sometimes used by zoologists.

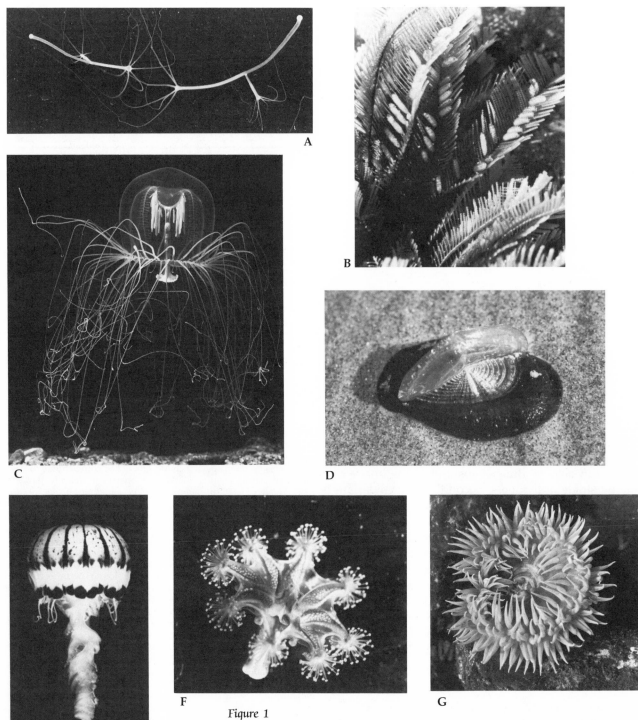

Figure 1

Some cnidarians. A–D, Hydrozoa. A, *Hydra*, an aberrant freshwater anthomedusan. B, A colony of the leptomedusan *Aglaophenia*. C, The medusa of *Polyorchis*, an anthomedusan. D, A chondrophoran, *Velella*, stranded on the beach. E–F, Scyphozoa. E, *Pelagia noctiluca*, a large semaeostoman medusa. F, *Haliclystis*, the strange, sessile stauromedusan. G–K, Anthozoa. G, An actinarian, the giant green anemone *Anthopleura xanthogrammica*. H, An expanse of scleractinian corals. I, Diver over a bed of sea pens, *Ptilosarcus* (order Pennatulacea). J, A gorgonacean from the Gulf of California. K, *Renilla*, the sea pansy (Pennatulacea). (A, C photos by C. E. Mills, Univ. of Washington/ BPS; B, G, J photos by the authors; E, F, K courtesy of G. McDonald; D courtesy of G. Staley; H photo by H. W. Pratt/Biological Photo Service; I courtesy of C. Birkeland.)

in size from nearly microscopic polyps and medusae to individual jellyfish 2 m across with tentacles 25 m long. Colonies, such as corals, may be many meters across. The phylum dates from the Precambrian, and its members have played important roles in various ecological settings throughout their long history, just as modern coral reefs are important today. Although we discuss corals throughout this chapter, we do not devote a separate section to them. The references at the end of this chapter list many excellent books and review articles on corals and reefs, and we particularly recommend Stoddart (1969) and papers by C. M. Yonge and P. Glynn.

Taxonomic history and classification

As is the case with sponges, the nature of cnidarians was long debated. In reference to their sting-ing tentacles, Aristotle called the medusae Acalephae (*akalephe*) and the polyps Cnidae (*knide*), both names derived from terms meaning "nettle." Renaissance scholars considered them plants, and it was not until the eighteenth century that the animal nature of the cnidarians was recognized. Nineteenth-century naturalists classified them along with the sponges and a few other groups under Linnaeus's Zoophytes, a category for organisms deemed somewhat between plants and animals. Lamarck instituted the group Radiata (or "Radiaires") for medusoid cnidarians, ctenophores, and echinoderms. In the early nineteenth century the great naturalist Michael Sars demonstrated that medusae and polyps were merely different forms of the same group of organisms. Sars also demonstrated that the genera *Scyphistoma*, *Strobila*, and *Ephyra* actually represented stages in the life history of certain jellyfish (scyphozoans). The names have been retained and are now used to identify

H

I

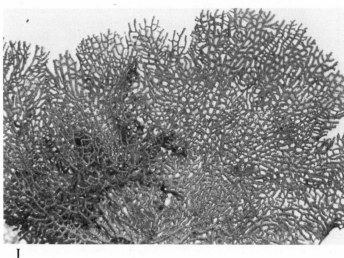

J

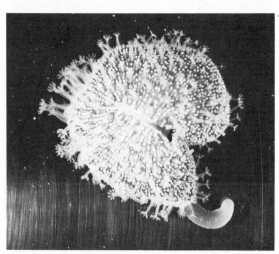

K

Box One
Characteristics of the Phylum Cnidaria

1. Diploblastic metazoa with ectoderm and entoderm separated by a (primarily) ectodermally derived acellular mesoglea or partly cellular mesenchyme

2. Possess primary radial symmetry, often modified as biradial or quadriradial; primary body axis is oral–aboral

3. Possess unique stinging or adhesive structures called cnidae; each cnida resides in and is produced by one cell, a cnidocyte. The most common cnidae are called nematocysts

4. The entodermally derived gastrovascular cavity (coelenteron) is the only "body cavity"

5. The digestive cavity (coelenteron) is saclike or branched, but has only a single opening, which serves as both mouth and anus

6. With no head, no centralized nervous system, and no discrete gas exchange, excretory, or circulatory systems

7. Nervous system is a simple nerve net(s), composed of naked and largely nonpolar neurons

8. The musculature is formed of epitheliomuscular cells, derived from ectoderm and entoderm (epidermis and gastrodermis); the muscle cells are the most primitive in the eumetazoa

9. Exhibit alternation of asexual polypoid and sexual medusoid generations; but there are many variations on this basic theme

10. Typically have planula larvae (ciliated, motile, gastrula larvae)

enteron, "intestine") for the former group in his recognition of the "intestine" as the sole body cavity. In 1888 Hatschek split Leuckart's Coelenterata into the three phyla recognized today: Porifera, Cnidaria, and Ctenophora. Although some workers have been inclined to retain the cnidarians and ctenophores together in the Coelenterata (or even the Radiata), these two groups are almost universally recognized as distinct phyla. The older term "Coelenterata" is still preferred by some specialists, who regard it as a synonym of Cnidaria. Four classes of Cnidaria are recognized: Hydrozoa, Anthozoa, Cubozoa, and Scyphozoa

PHYLUM CNIDARIA

CLASS HYDROZOA: Hydroids and hydromedusae (Figures 1A–D). Alternation of generations occurs in most genera (typically asexual benthic polyps alternate with sexual planktonic medusae), although one or the other generation may be suppressed or lacking; medusoids often retained on the polyp; polyps usually colonial, with interconnected coelenterons; often polymorphic, individual polyps modified for various functions (e.g., gastrozooids are nutritive, gonozooids are reproductive, dactylozooids are for defense and prey capture); exoskeleton usually of chitin or occasionally calcium carbonate; coelenteron of polyps and medusae lacks a pharynx and septa; mesoglea acellular; tentacles solid or hollow; cnidae appear only in epidermis; gonads epidermal; medusae mostly small and transparent, nearly always craspedote (with a velum), and with a ring canal; mouth typically borne on pendant manubrium; medusae lack rhopalia. About 2,700 species in 7 extant orders; includes some freshwater groups.

ORDER HYDROIDA: Hydroids and their medusae. Polypoid generation often predominant; polyps may have a chitinous exoskeleton; oral tentacles filiform or capitate, rarely branched or absent; colonies often polymorphic; many do not release free medusae but release gametes from sporosacs or sessile attached medusoids (gonophores) on colony; colonies dioecious. A large group; over 55 described families. Hydroids occur at all depths, but polypoid forms are very common in the littoral zone.

SUBORDER ANTHOMEDUSAE (= GYMNOBLASTEA OR ATHECATA): Polyps solitary or colonial; hydranths and gonozooids lack exoskeleton; free medusae tall and bell-shaped, with or without ocelli, but without statocysts; medusae bear gonads on subumbrella or manubrium; free medusae absent in many species. (e.g., *Bougainvilla, Calycopsis, Eleutheria, Eudendrium, Hydra, Hydractinia, Hydrocoryne, Janaria, Lar* [= *Proboscidactyla*], *Pennaria, Podocoryne, Polyorchis, Rathkea, Sarsia, Staurocladia, Stylactis, Syncoryne, Tubularia*)

SUBORDER LEPTOMEDUSAE (= CALYPTOBLASTEA OR THECATA): Polyps always colonial; hydranths and gonozooids encased in exoskeleton; free medusae usually

these stages in the life cycle. Leuckart eventually recognized the fundamental differences between the two great "radiate" groups, the Porifera/Cnidaria/Ctenophora and the Echinodermata, and in 1847 created the name Coelenterata (Greek *koilos*, "cavity'"

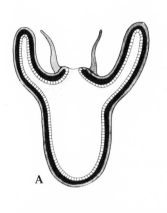
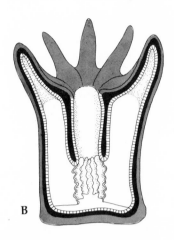
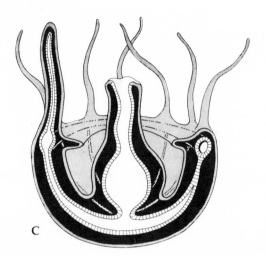

Figure 2

Tissue layer homologies in cnidarians. A, A hydrozoan polyp. B, An anthozoan polyp. C, A hydrozoan medusa, shown upside down for similar orientation. The outer tissue layer is ectodermal (= epidermis); the inner tissue layer is entodermal (= gastrodermis); and the middle layer is the mesenchyme/mesoglea. (From Bayer and Owre 1968.)

absent, but when present flattened and with statocysts; medusae with gonads on subumbrella beneath radial canals. (e.g., *Abietinaria, Aequorea, Aglaophenia, Bonneviella, Campanularia, Cuvieria, Eucheilota, Lovenella, Lytocarpus, Obelia, Plumularia, Sertularia*)

ORDER MILLEPORINA: "Fire corals" or millepore hydrocorals. Polyps form massive or encrusting calcareous coral-like skeleton; calcareous matrix covered by thin epidermal layer; gastrozooids with 3–7 short capitate tentacles on column as well as around mouth; each gastrozooid is surrounded by 4–8 discrete dactylozooid-like tentacles, each tentacle in a separate skeletal cup; gonophores housed in pits (ampullae) in skeleton; small, free medusae lack mouth, tentacles, and velum. Like the true (stony) corals, milleporines rely on a commensal relationship with zooxanthellae and are thus restricted to the photic zone. One extant genus, *Millepora*.

ORDER STYLASTERINA: Stylasterine hydrocorals. Polyps form erect or encrusting calcareous skeleton, often brightly colored (purple, red, yellow); skeleton secreted within epidermis and covered by thick epidermal layer; calcareous style rises from base of polyp cup; polyps may have tentacles; free medusae not produced but sessile medusoid gonophores retained in shallow chambers (ampullae) of colony; several dactylozooids surround each gastrozooid, although the polyp pits are joined. (e.g., *Allopora, Stylaster*)

ORDER TRACHYLINA: Trachyline medusae. Polypoid generation greatly reduced or absent; medusae produce planula larvae that usually develop directly into actinula larvae, which metamorphose into adult medusae; medusae craspedote, with tentacles often arising from exumbrellar surface, well above bell margin; medusae mostly gonochoristic. The trachylines are probably a polyphyletic group, currently including three suborders: Laingiomedusae, Narcomedusae and Trachymedusae. (e.g., *Aegina, Botrynema, Craspedacusta, Cunina, Gonionemus, Hydroctena, Liriope, Polypodium, Rhopalonema, Solmissus*)

ORDER SIPHONOPHORA: Siphonophorans. Polymorphic swimming or floating colonies of polyps and medusae; siphonophorans exhibit the highest degree of polymorphism among all cnidarians, with a number of distinct types of polyps and attached modified medusae; most have a gas-filled flotation zooid. (e.g., *Agalma, Apolemia, Eudoxoides, Nectocarmen, Physalia, Rhizophysa, Sphaeronectes, Stephalia*)

ORDER CHONDROPHORA: Chondrophorans. Enigmatic group viewed either as colonies comprising gastrozooids, gonozooids, and dactylozooids, or as a solitary but highly specialized polypoid individual; "zooids" are attached to a chitinous, multichambered, dislike float, that may or may not have an oblique sail; "gonozooids" bear medusiform gonophores that are released and shed gametes; most are richly supplied with zooxanthellae. Once considered a highly modified group of siphonophorans, their position within the Hydrozoa is still debated. (e.g., *Porpita, Velella*)

ORDER ACTINULIDA: Actinulidans. Free-living, solitary, minute (to 1.5 mm), motile, polypoid, hydrozoans; no medusa stage; use cilia to swim and crawl among sand grains; interstitial; no sexual reproduction has been recorded. (e.g., *Halammohydra, Otohydra*)

CLASS ANTHOZOA: Sea anemones, corals, sea pens. (Figures 1G–K). Exclusively marine; solitary or colonial; without a medusoid stage (mostly benthic polyps); cnidae epidermal and gastrodermal; divided by longitudinal (oral–aboral) mesenteries (= septa), the free edges of which form thick, cordlike mesenterial filaments; mesenchyme thick; tentacles usually number eight or occur in multiples of six and contain extensions of the coelenteron; stomodeal pharynx (= actinopharynx) extends from the mouth into the coelenteron and bears one or more ciliated grooves (siphonoglyphs); polyps usually reproduce both sexually and asexually; gonads gastrodermal. About 6,000 species divided into three subclasses.

SUBCLASS OCTOCORALLIA (= ALCYONARIA): Octocorals. Colonial; polyps with eight hollow, marginal, usually pinnate tentacles, and eight complete (perfect) mesenteries, each with retractor muscle on sulcal side, facing the single siphonoglyph; with free or fused calcareous sclerites embedded in mesenchyme; stolons, or coenenchyme connect the polyps; new polyps are usually budded from stolons. Six representative orders are listed below.

ORDER GORGONACEA: Sea fans and sea whips. Colonies typically brightly colored and arborescent and may be several meters across; firm internal axial skeleton composed of horny proteinaceous material (gorgonin); occasionally the skeleton is calcareous; colonies always covered with a thin layer of sclerite-filled mesoglea; one family (Isidae) has calcareous "segments" that alternate with thin, horny intercalary plates, giving flexibility to the otherwise rigid colony; polyps interconnected by gastrodermal solenia. A large and diverse group, with 18 recognized families. (e.g., *Acanthogorgia, Briareum, Corallium, Eugorgia, Eunicella, Gorgonia, Isis, Leptogorgia, Lophogorgia, Muricea, Parisis, Psammogorgia, Subergorgia, Swiftia*)

ORDER TELESTACEA: Telestaceans. Colonies usually branched; polyps simple, cylindrical, very tall, and typically bud off lateral polyps; polyps are connected at the base and grow from a creeping stolon; axis never solid, although axial spicules may be somewhat fused, providing rigidity (e.g., *Coelogorgia, Paratelesto, Telesto, Telestula*)

ORDER PENNATULACEA: Sea pens and sea pansies. Colonies complex and polymorphic; adapted for life on soft benthic substrata; often luminescent; elongate primary axial polyp extends length of the colony (to 1 m) and consists of a basal bulb or peduncle for anchorage and a distal stalk, the latter giving rise to dimorphic secondary polyps; coelenteron of axial polyp with skeletal axes of calcified horny material in canals. (e.g., *Anthoptilum, Balticina, Cavernularia, Funiculina, Pennatula, Ptilosarcus, Renilla, Stylatula, Umbellula, Virgularia*)

ORDER ALCYONACEA: Soft corals. Colonies encrusting or erect, often massive; usually fleshy and flexible, although the coenenchyme is sclerite-filled; fleshy distal portions of polyps retractable into more compact basal portion. (e.g., *Alcyonium, Anthomastus, Ceratocaulon, Gersemia, Parerythropodium*)

ORDER HELIOPORACEA: Helioporaceans. Colonies produce rigid calcareous skeletons of aragonite crystals (not fused sclerites) similar to those of millepores and stony corals; polyps monomorphic. Two genera: *Epiphaxum* and *Heliopora* ("blue coral").

ORDER STOLONIFERA: Stoloniferans. Simple polyps arise separately from ribbonlike stolon that forms an encrusting sheet or network; oral disc and tentacles retractable into anthostele (stiff proximal portion of polyp); mesenchyme with sclerites; horny external skeleton covers polyps and stolons. Three families; in one, the organ-pipe "corals" (*Tubipora*), sclerites fuse to form a calcareous covering. (e.g., *Clavularia, Cornularia, Sarcodictyon, Tubipora*)

SUBCLASS HEXACORALLIA (= ZOANTHARIA): Sea anemones and true corals. Solitary or colonial; naked, or with calcareous skeleton or chitinous cuticle, but never with isolated sclerites; mesenteries paired, usually in multiples of six; septa bear longitudinal retractor muscles arranged so that those of each pair either face toward each other or away from each other; mesenterial filaments typically trilobed, with two ciliated bands flanking a central one bearing cnidae and gland cells; one to several circles of hollow tentacles arise from endocoels (the spaces between the members of each mesentery pair) and exocoels (the spaces between adjacent mesentery pairs); pharynx may have 0, 1, 2, or many siphonoglyphs; cnidae very diverse; endodermal zooxanthellae may be profuse.

ORDER ACTINIARIA: Sea anemones. Solitary or clonal, but never colonial; calcareous skeleton lacking, although some species secrete a chitinous cuticle; most harbor zooxanthellae; usually with cinclides (perforations in column that permit extrusion of water or acontia); column often with specialized structures, such as warts or verrucae, acrorhagi, pseudotentacles, or vesicles; oral tentacles conical, digitiform, or branched, usually hexamerously arrayed in one or more circles; typically with two siphonoglyphs. About 41 families; Actiniidae contains most of the true sea anemones. (e.g., *Actinia, Actinostola, Adamsia, Aiptasia, Alicia, Anthopleura, Anthothoe, Bartholomea, Bunodactis, Calliactis, Condylanthus, Diadumene, Edwardsia, Epiactis, Halicampa, Haliplanella, Heteractis* (= *Radianthus*), *Lebrunia, Liponema, Metridium, Minyas, Peachia, Phyllodiscus, Sagartiomorphe, Stomphia, Tealia, Triactis, Zaolutus*)

ORDER SCLERACTINIA (= MADREPORARIA): True or stony corals. Mostly colonial; polyp morphology almost identical to that of Actiniaria, except corals lack siphonoglyphs and ciliated lobes on the mesenterial filaments; zooxanthellae usually abundant; colony forms delicate to massive calcareous (aragonite) exoskeleton, with platelike extensions (septa) inside the mesenteries. Over 2,500 extant species, in about 23 families. (e.g., *Acropora* [over 200 species], *Agaricia, Astrangia, Balanophyllia, Dendrogyra, Flabellum, Fungia, Goniopora, Meandrina, Montipora, Oculina, Oxypora, Pachyseris, Porites, Psammocora, Siderastraea, Stylophora*)

ORDER ZOANTHIDEA: Zoanthids. Resemble true sea anemones but typically are colonial, with polyps arising from a basal mat or stolon containing gastrodermal solenia or canals; new polyps budded from gastrodermal solenia of stolons; pharynx flattened, with one siphonoglyph; mesenteries numerous; mesentery musculature poorly developed; tentacles never pinnate; without intrinsic skeleton, but many species incorporate sand, sponge spicules, or other debris into the thick body wall; most have a very thick cuticle; zooxanthellae are generally abundant; many species are epizoic. (e.g., *Epizoanthus, Isaurus, Isozoanthus, Palythoa, Parazoanthus, Thoracactus, Zoanthus*)

ORDER CORALLIMORPHARIA: Coral-like anemones. Solitary or colonial polyps, without a skeleton; lack siphonoglyphs and ciliated bands on mesenterial filaments. (e.g., *Amplexidiscus, Corynactis, Rhodactis, Ricordea*)

SUBCLASS CERIANTIPATHARIA: Ceriantipatharians. Mesenteries complete, but with feeble musculature; with six primary mesenteries, but others added immediately opposite the single siphonoglyph.

ORDER ANTIPATHARIA: Black or thorny corals. Gorgonian-like colonies up to 6 m tall; hard axial skeleton usually brown or black and covered by a thin coenosarc bearing small polyps, usually with six (but up to 24) nonretractable tentacles; with feeble mesenteries and a single siphonoglyph; skeleton produces thorns on its surface; primarily in deep water in tropical seas. (e.g., *Antipathes*)

ORDER CERIANTHARIA: Cerianthids or tube anemones. Large, solitary, elongate anemones living in vertical tubes in soft sediments; tube constructed of interwoven specialized cnidae (ptychocysts) and mucus; aboral end lacks a pedal disc and possesses a terminal pore; long thin tentacles arise from margin of oral disc, fewer shorter labial tentacles encircle mouth; one siphonoglyph; mesenteries complete; gonads occur only on alternate mesenteries; protandric hermaphrodites. (e.g., *Arachnanthus, Botruanthus, Ceriantheomorphe, Ceriantheopsis, Cerianthus, Pachycerianthus*)

CLASS CUBOZOA: Sea wasps and box jellyfish (Figure 15). Medusae small, 15–25 cm tall, largely colorless; polyps each produce a single medusa by complete metamorphosis (strobilation does not occur); medusa bell square in cross section, with four flat sides; hollow interradial tentacles hang from bladelike pedalia, one at each corner of umbrella; unfrilled bell margin drawn inward to form a velum-like structure (the velarium). The single order Cubomedusae (= Carybdeida) was formerly placed in the class Scyphozoa. Cubozoans occur in all tropical seas but are especially abundant in the Indo-West Pacific region. Their sting is very toxic, in some cases fatal to humans, hence the name "sea wasps." (e.g., *Carybdea, Chironex, Tamoya, Tripedalia*)

CLASS SCYPHOZOA: Jellyfish (Figures 1E–F). Medusoid stage predominates, and polypoid individual (scyphistoma) is small and inconspicuous, even lacking in some groups; polyps produce medusae by asexual budding (strobilation); coelenteron divided by four longitudinal (oral–aboral) septa; medusae acraspedote (without a velum), typically with a thick mesogleal (or collenchy-mal) layer, distinct pigmentation, filiform or capitate tentacles, and marginal notches producing lappets; sense organs occur in notches and alternate with tentacles; gonads gastrodermal; cnidae present in epidermis and gastrodermis; mouth may or may not be on a manubrium; usually without a ring canal. Scyphozoans are exclusively marine; planktonic, demersal, or attached; About 200 species are divided into four orders.

ORDER STAUROMEDUSAE: Small, sessile medusae that develop directly from benthic planula larvae; no polyp stage; with aboral stalked adhesive disc in center of exumbrella by which individuals attach "upside down" to substratum; with eight tentacle-bearing "arms"; sexual reproduction only. Occur in shallow water at high latitudes. (e.g., *Haliclystis, Lucernaria*)

ORDER CORONATAE: High bell divided into upper and lower regions by a coronal groove encircling exumbrella; margin of bell deeply scalloped by gelatinous thickenings termed pedalia, which give rise to tentacles, rhopalia, and marginal lappets; gonads present on the four gastrovascular septa. Primarily bathypelagic. (e.g., *Atolla, Linuche, Nausithoe, Periphylla, Stephanoscyphus, Tetraplatia*)

ORDER SEMAEOSTOMAE: Corners of mouth drawn out into four broad, gelatinous, frilly lobes; stomach with gastric filaments; hollow marginal tentacles contain extensions of radial canals; without coronal furrow or pedalia; gonads on folds of gastrodermis. This order contains most of the typical jellyfish of temperate and tropical seas; moderate to very large forms. (e.g., *Aurelia, Chrysaora, Cyanea, Pelagia, Stygiomedusa*)

ORDER RHIZOSTOMAE: Lack a central mouth; the frilled edges of the four oral lobes are fused over the mouth so that many suctorial "mouths" (ostioles) open from a complicated canal system on eight branching armlike appendages; bell without marginal tentacles or pedalia; stomach without gastric filaments; gonads on folds of gastrodermis. Small to large jellyfish that swim vigorously with a well developed subumbrellar musculature; they primarily occur in low latitudes. (e.g., *Cassiopeia, Cephea, Eupilema, Mastigias, Rhizostoma, Stomolophus*)

The cnidarian *Bauplan*

Cnidarians are clearly metazoa (Box 1), and, as such, they show marked advances over the groups covered thus far. However, they possess only two embryonic germ layers—the ectoderm and the entoderm—which become the adult epidermis and gastrodermis, respectively. The middle mesoglea or mesenchyme in adults is derived largely from ectoderm and never produces the complex organs seen in triploblastic metazoa.

Figure 3

Cnidarian radial symmetries. A, Quadriradial symmetry of a hydromedusa. B, Radial symmetry of a hydrozoan polyp. C, Biradial symmetry of an actiniarian polyp (a sea anemone). D, Biradial symmetry of an octocoral polyp (class Anthozoa). (From Bayer and Owre 1968.)

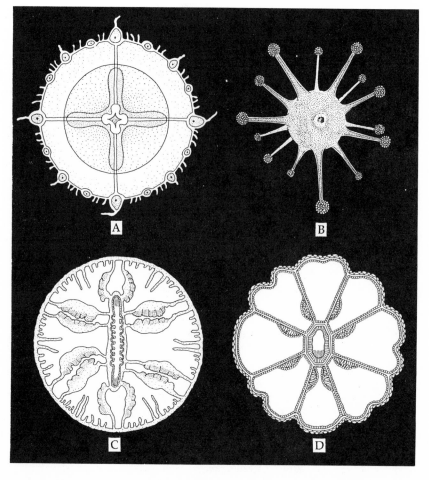

The essence of the cnidarian *Bauplan* is radiality (Figure 3). As discussed in Chapter 3, radial symmetry is associated with various architectural and strategic constraints. The cnidarians are either sessile, sedentary, or pelagic, and do not engage in the active unidirectional movement seen in bilateral, cephalized creatures. Furthermore, radial symmetry demands certain arrangements, particularly of those parts that interact directly with the environment, such as feeding structures and sensory receptors. Thus, we typically find a ring of tentacles, arranged to collect food from any direction, and a diffuse, noncentralized nerve net with radially distributed sense organs. These and other implications of radial symmetry will be explored further throughout this chapter.

In spite of the limitations of a diploblastic, radially symmetrical *Bauplan*, cnidarians are a very successful and diverse group. Much of their success has resulted from the apparent evolutionary plasticity of their dimorphic life histories, a trait allowing myriad variations on the theme of alternation between polypoid and medusoid phases. Although polyps and medusae are very different in appearance, they are really just variations on the basic cnidarian *Bauplan*. Nonetheless, the two stages are so different ecologically that their presence in a single life history allows an individual species to exploit different environments and resources by leading a double life. This sort of dimorphism is unique to the Cnidaria. There is no known counterpart of the cnidarian polyp elsewhere in the animal kingdom, and the medusa is approximated only in the phylum Ctenophora (Chapter 9).

Body structure of the polypoid and medusoid forms

The polypoid form. Polyps are much more diverse than medusae, largely as a result of their capacities for asexual reproduction and colony formation (Figures 4 through 12). The polypoid stage occurs in all four classes of cnidarians, although it is greatly reduced in the Scyphozoa and Cubozoa. Polyps are tubular structures with an outer epidermis, an inner gut sac (**coelenteron**) lined with gastrodermis, and a layer of jelly-like mesoglea or mesenchyme in between (Figure 4). The epidermis is composed largely of **epitheliomuscular cells**. These cells have a columnar cell body and flattened, contractile, basal extensions called **myonemes** (Figure 19). The epidermis also contains sensory cells, cnidocytes, gland cells and **interstitial cells**. The latter are undifferentiated cells capable of developing into germ cells, cnidocytes, epitheliomuscular cells, and so on. In hydrozoans the mesoglea is very thin and contains few or no living cells; in anthozoans, it is a thick and richly cellular mesenchyme. Scyphozoans and cubozoans have few living cells in their thick middle layers.

The basic polypoid symmetry is radial, although as a result of various modifications some species possess a biradial or quadriradial symmetry (Figure 3). The main body axis runs longitudinally through the mouth (oral end) to the base (aboral end) of the polyp. The aboral end may be a **pedal disc** for attaching to hard substrata (as in most common sea anemones); it may be a rounded digging and anchoring structure—called a **physa**—adapted to soft substrata (as in cerianthids); or it may attach to a common mat, stalk, or stolon in colonial forms.

The mouth may be on a flat **oral disc** as it is in the anthozoans, or it may be set on an elevated **hypostome** or **manubrium** as in the hydrozoans (Figures 5 and 7). In anthozoans the mouth is often slitlike and leads to a muscular pharynx that extends into the coelenteron. The pharynx usually bears from one to several ciliated grooves called **siphonoglyphs**,

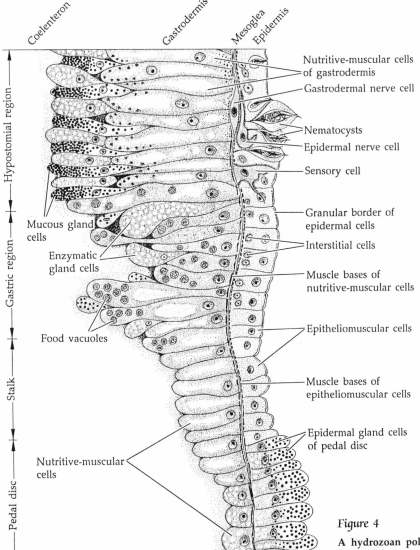

Figure 4

A hydrozoan polyp column wall (cross section) illustrates the basic cnidarian cell and tissue types. (From Bayer and Owre 1968.)

which function to drive water into the gut cavity (Figure 7). It is in part the presence of siphonoglyphs that gives these polyps a secondary biradial symmetry. The side of an anthozoan polyp that bears a single siphonoglyph is called the **sulcal** side, and the opposite side is called the **asulcal** side. In many anthozoans, the column has pores (**cinclides**) that connect to the coelenteron, through which water is expelled during rapid contraction.

The coelenteron, or **gastrovascular cavity**, serves for circulation as well as digestion and distribution of food. In hydrozoan polyps, the coelenteron is a single, uncompartmented chamber. In scyphozoan polyps (the **scyphistomae**), it is partially subdivided by four longitudinal, ridgelike **mesenteries**; and in anthozoan polyps, it is extensively compartmentalized by mesenteries. Anthozoan mesenteries are projections of the inner body wall and thus are lined with gastrodermis and filled with mesenchyme. They extend from the inner body wall toward the pharynx, some or all of them fusing with it as **complete mesenteries**. Those that do not connect to the pharynx are called **incomplete mesenteries**. In anthozoan polyps, the free inner edges of the mesenteries below the pharynx have a thickened, cordlike margin armed with cnidae, cilia, and gland cells and called the **mesenterial filament** (Figures 7 and 20). In some anemones these filaments give rise to long threads, called

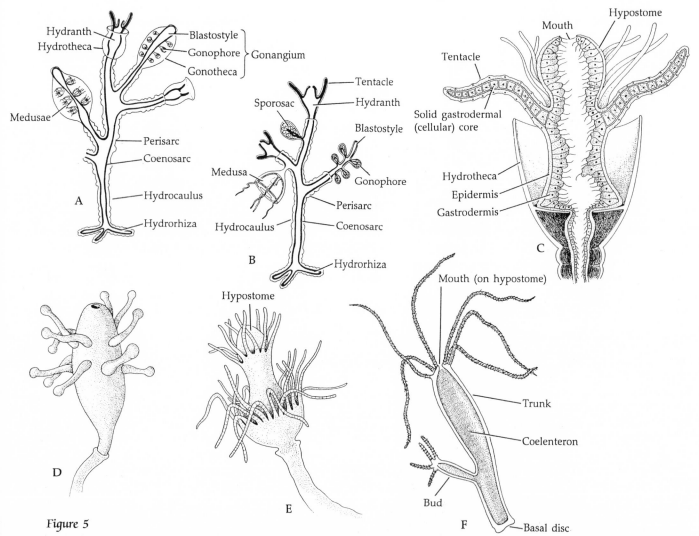

Figure 5

Hydrozoan polyps. A, A thecate hydroid colony. The blastostyles produce either sporosacs or medusae. B, An athecate hydroid colony, illustrating various types of reproductive structures. Note that A and B are *composite diagrams*; a given species produces *either* sporosacs or medusae, never both. C, A thecate hydranth (= gastrozooid) (longitudinal section). D, A hydranth with capitate tentacles. E, A hydranth with two whorls of filiform tentacles. F, The freshwater *Hydra* (body is shown in longitudinal section). (A–E from Bayer and Owre 1968.)

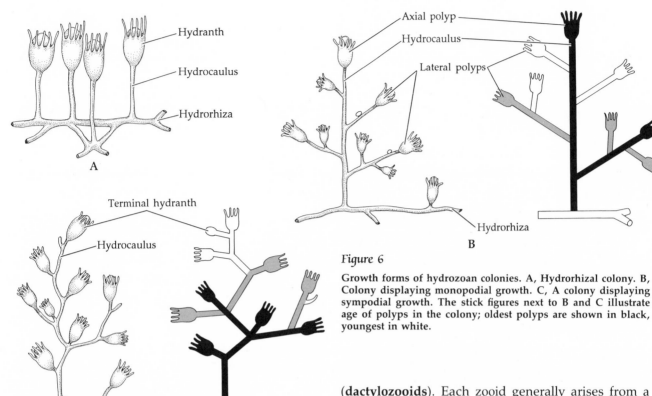

Figure 6

Growth forms of hydrozoan colonies. A, Hydrorhizal colony. B, Colony displaying monopodial growth. C, A colony displaying sympodial growth. The stick figures next to B and C illustrate age of polyps in the colony; oldest polyps are shown in black, youngest in white.

acontia, that hang free in the gastrovascular cavity. They function in defense and feeding, as discussed later. In most colonial anthozoans the cellular mesenchyme surrounds and unites the individual zooids (Figure 8). The gastrovascular cavities of such polyps are connected to one another by canals called **solenia**.

The **tentacles** that surround the mouth may contain hollow extensions of the coelenteron, as they do in most anthozoans, or they may contain a solid core of packed gastrodermal cells, as in most hydrozoans. Tentacles may taper to a point (**filiform tentacles**) or may terminate in a conspicuous knob of cnidae (**capitate tentacles**). In some polyps the tentacles are branched, often as pinnately arranged **pinnules** (e.g., in the octocorals).

A complex terminology has been developed to describe hydrozoan polyps, or **hydroids** (Figures 5 and 6). One reason for this special nomenclature is that hydroid colonies are usually polymorphic, comprising more than one kind of polyp, or zooid. The term **hydranth**, or **gastrozooid**, refers to feeding zooids, which typically bear tentacles and a mouth. Other commonly occurring polyp types include reproductive polyps (**gonozooids**) and defense polyps

(**dactylozooids**). Each zooid generally arises from a stalk, called a **hydrocaulus**. In most colonial hydrozoans, the individual polyps are anchored in a rootlike stolon called a **hydrorhiza**, which grows over the substratum. From the hydrorhiza arise hydrocauli bearing polyps singly or in clusters.

Branched hydrozoan colonies grow in two basic patterns (Figure 6). In **monopodial growth**, the first polyp elongates continuously from a growth zone at the distal end of the hydrocaulus. This primary (axial) polyp may even lose its hydranth and persist merely as a stalk. The primary hydrocaulus gives rise to secondary polyps by lateral budding. These secondary polyps grow and may give rise to lateral tertiary polyps in the same fashion. In hydrozoan colonies developing by **sympodial growth**, the primary polyp does not continue to elongate but produces one or more lateral polyps by budding and then stops growing. The new polyps extend the colony upward some distance, then stop growing and bud off more new polyps. In these colonies the main stem or axis actually represents the combined hydrocauli of many polyps and the age of the polyps decreases from base to tip along each branch.

Most marine hydroids are surrounded, at least in part, by a nonliving protein–chitin exoskeleton secreted by the epidermis and called the **perisarc** (Figure 5). However, this outer covering is absent in freshwater hydroids. The living tissue inside the perisarc is termed the **coenosarc**. The perisarc may ex-

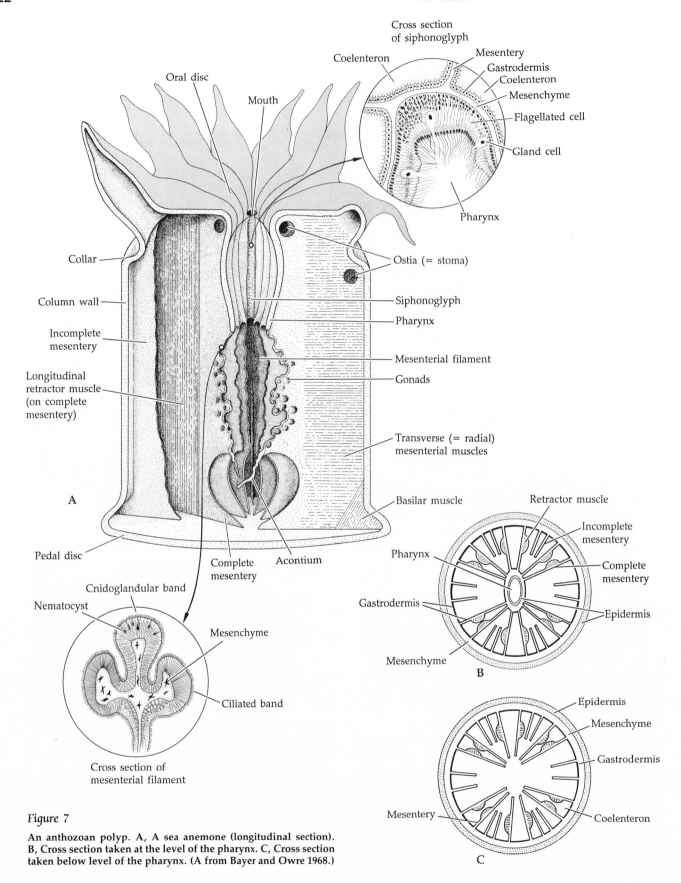

Figure 7

An anthozoan polyp. A, A sea anemone (longitudinal section). B, Cross section taken at the level of the pharynx. C, Cross section taken below level of the pharynx. (A from Bayer and Owre 1968.)

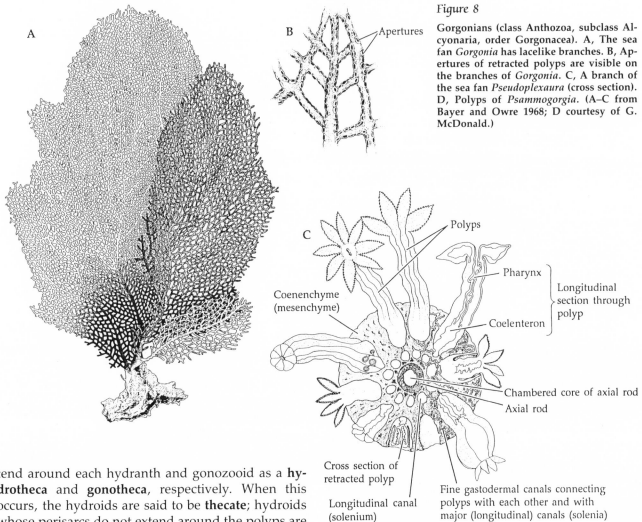

Figure 8

Gorgonians (class Anthozoa, subclass Alcyonaria, order Gorgonacea). A, The sea fan *Gorgonia* has lacelike branches. B, Apertures of retracted polyps are visible on the branches of *Gorgonia*. C, A branch of the sea fan *Pseudoplexaura* (cross section). D, Polyps of *Psammogorgia*. (A–C from Bayer and Owre 1968; D courtesy of G. McDonald.)

tend around each hydranth and gonozooid as a **hydrotheca** and **gonotheca**, respectively. When this occurs, the hydroids are said to be **thecate**; hydroids whose perisarcs do not extend around the polyps are **athecate**.

The gastrozooids capture and ingest prey and provide nutrients to the rest of the colony, including all the nonfeeding polyps. The dactylozooids, which occur in a variety of sizes and shapes, are heavily armed with cnidae. Often several dactylozooids surround each individual gastrozooid and serve for both defense and food capture. Gonozooids produce medusa buds, called **gonophores**, that either are released or are retained on the colony. Whether released as medusae or retained as gonophores, they are responsible for producing gametes for the sexual phase of the hydrozoan life cycle. The living tissue (coenosarc) of the gonozooid is called the **blastostyle**; the gonophores arise from this tissue. When a gonotheca surrounds the blastostyle, the zooid is called a gonangium.

The most dramatic examples of polymorphism among polyps are seen in the hydrozoan order Siphonophora and the anthozoan order Pennatulacea

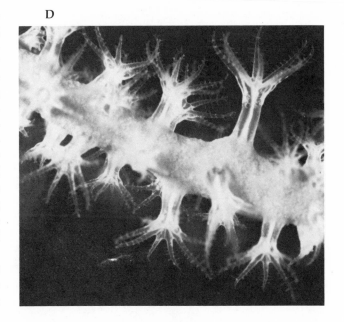

Figure 9

Diversity of form among the colonial Hydrozoa. A, *Proboscidactyla*, a two-tentacled hydroid that lives around the open end of polychaete worm tubes. B, *Monobrachium*, a one-tentacled hydroid that lives on clam shells. C, *Hydractinia*, a colonial hydroid commensal on shells inhabited by hermit crabs. D, The chondrophoran *Velella* ("by-the-wind-sailor"). E, The chondrophoran *Porpita* (aboral view). F, A siphonophoran, *Physalia* ("man-of-war"). G, A colony of the calcareous milleporinid hydrocoral *Millepora*. H, A colony of the calcareous stylasterine hydrocoral *Allopora*. I, A colony of the calcareous stylasterine hydrocoral *Distichopora*. J, *Nectocarmen antonioi*, a colonial calycophoran siphonophore from California. (A–E, G–I from Bayer and Owre 1968; F after Bayer and Owre 1968; J from Alvariño 1983.)

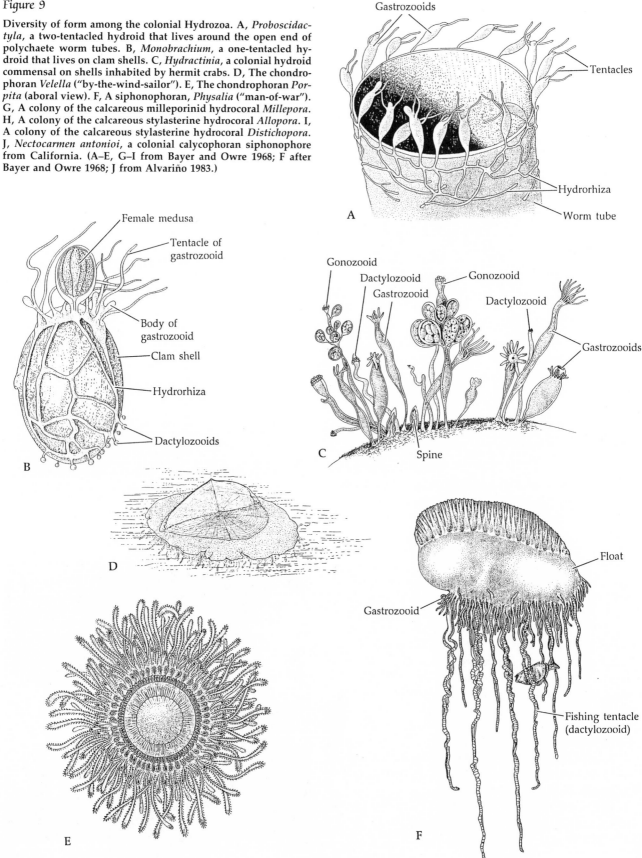

(Figures 1K, 9F and J, 10, 12D and F). Siphonophorans are hydrozoan colonies composed of both polypoid and medusoid individuals, with as many as 1,000 zooids in a single colony. This large order contains a great variety of unusual and poorly understood forms, including the famous man-of-war, *Physalia* (Figures 9F and 10). The gastrozooids of siphonophorans are highly modified polyps bearing a large mouth and only one long, hollow tentacle, which has branches called **tentilla** that bear many cnidae. This long feeding tentacle reaches lengths of 13 m in the Atlantic species *Physalia physalis*. The nonfeeding dactylozooids also bear one long (unbranched) tentacle. The gonozooids are usually branched; they produce sessile gonophores that are never released as free medusae.

Siphonophorans utilize a swimming bell (**nectophore**) or a gas-filled float (**pneumatophore**), or both, to help maintain their position in the water column. The nectophore is a true medusoid individual with many of the structures common to free-swimming medusae, although it has lost its mouth, tentacles, and sense organs. The pneumatophore, once also thought to be a modified medusa, is now known to be derived directly from the larval stage and probably represents a highly modified polyp. Pneumatophores are double-walled chambers lined with chitin. Each float houses a gas gland, which consists of a mitochondrion-laden glandular epithelium lining a pit or chamber. The gland secretes a gas usually similar to air in composition, although in *Physalia* it includes a surprisingly large proportion of carbon monoxide. Many siphonophorans have mechanisms by which they regulate gas in their floats to keep the colony at a particular depth, much like the swim bladders of many fishes.

Siphonophorans are grouped into three suborders on the basis of colony structure: the suborder Calycophorae includes colonies with swimming bells but no float; members of the suborder Physonectae have a small float and a long train of swimming bells; and those of the suborder Cystonectae have a large float and no bell. Calycophorans have a long tubular **stem** extending from the swimming bell, from which various types of zooids are budded in groups called **cormidia** (Figure 9J). Each cormidium acts as a colony-within-a-colony, and is usually composed of a shieldlike **bract**, a gastrozooid, and one or more gonophores that may function as swimming bells. The cormidia commonly break loose from the parent col-

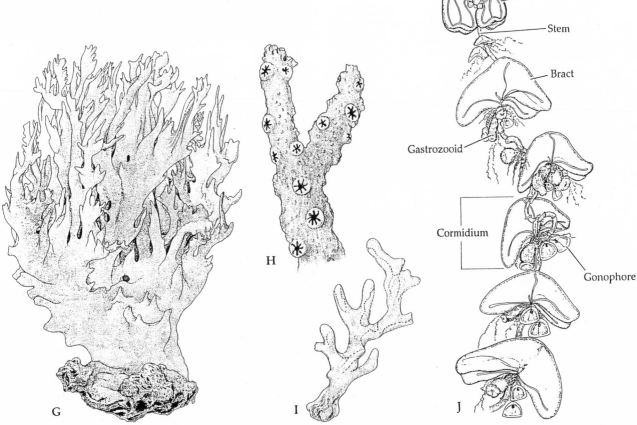

G   H   I   J

Swimming bell

Stem

Bract

Gastrozooid

Cormidium

Gonophore

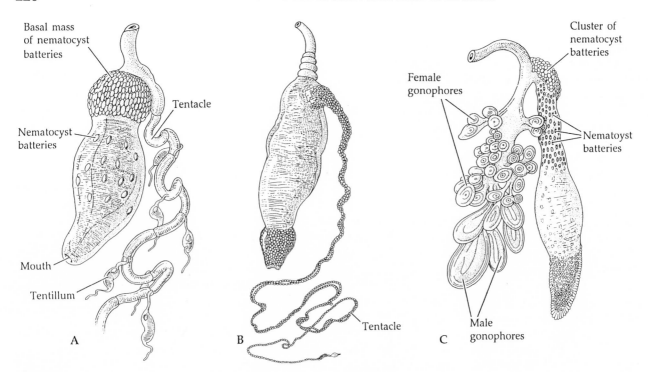

Figure 10

Siphonophoran zooids. A, A gastrozooid. B, A dactylozooid. C, A gonozooid. (From Bayer and Owre 1968.)

ony to live an independent existence, at which time they are termed **eudoxids**. The physonectans have an apical float with a long stem bearing a series of nectophores followed by a long train of cormidia. The cystonectans, such as *Physalia*, usually have a large pneumatophore with a prominent budding zone at its base, which produces the various polyps and medusoids (Figure 9F).

The hydrozoan order Chondrophora is still somewhat of an enigma. These colorful oceanic organisms drift about on the sea surface in enormous flotillas, occasionally washing ashore to coat the beach with their bluish-purple bodies (Figures 1D and 9D). Various specialists have considered them to be modified medusae, colonial hydroids related to siphonophorans, or individual zooids. Most recent opinions hold these animals to be large, solitary, athecate hydranth polyps floating upside down at the surface instead of sitting on a stalk attached to the bottom. Figure 11 compares a chondrophoran and a sessile

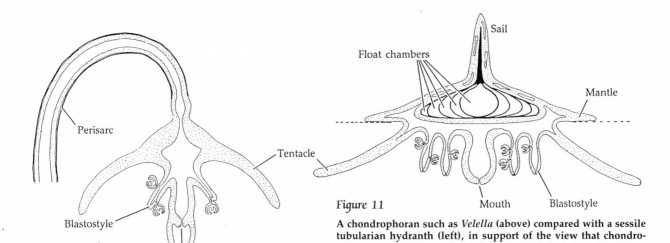

Figure 11

A chondrophoran such as *Velella* (above) compared with a sessile tubularian hydranth (left), in support of the view that chondrophorans are highly specialized, solitary tubularian zooids. (From Fields and Mackie 1971.)

hydroid, such as *Tubularia* or *Corymorpha*. The aberrant medusae of chondrophorans are short-lived and do not possess a functional mouth or gut, probably relying instead on their symbiotic zooxanthellae for nutrition. The aboral sail in *Velella* (the "by-the-wind-sailor") has no counterpart in sessile hydroids. In its ability to sail at an angle to the wind, *Velella* resembles the siphonophoran *Physalia*, but this similarity is now attributed to convergent evolution.

The pennatulaceans are the most complex and polymorphic members of the class Anthozoa (Figure 12D). The colony is built around a main supportive

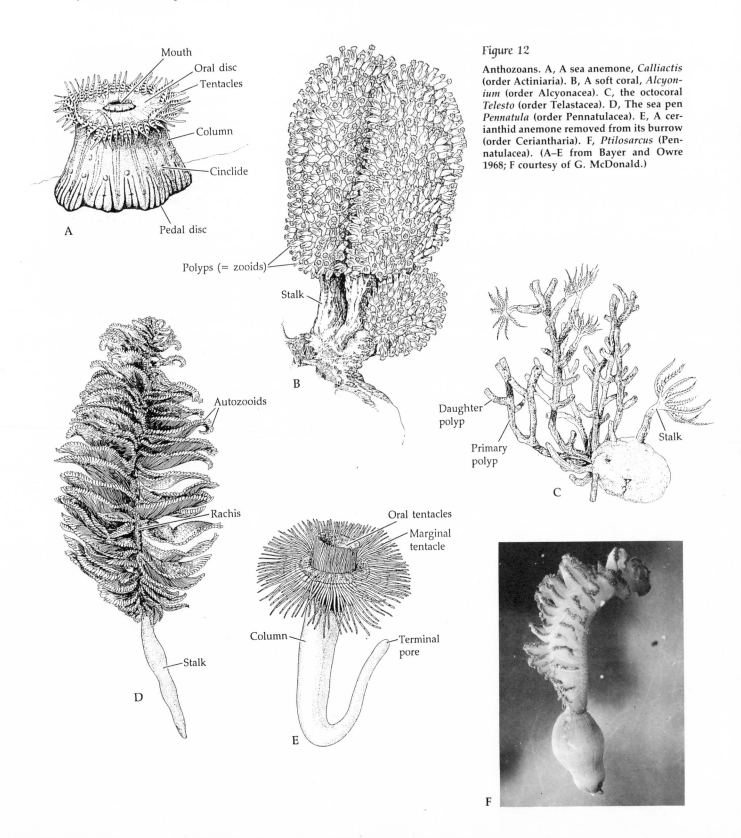

Figure 12

Anthozoans. A, A sea anemone, *Calliactis* (order Actiniaria). B, A soft coral, *Alcyonium* (order Alcyonacea). C, the octocoral *Telesto* (order Telastacea). D, The sea pen *Pennatula* (order Pennatulacea). E, A cerianthid anemone removed from its burrow (order Ceriantharia). F, *Ptilosarcus* (Pennatulacea). (A–E from Bayer and Owre 1968; F courtesy of G. McDonald.)

stem, which is actually the primary polyp and buds off lateral polyps in a regular fashion. The base of the primary polyp is anchored in sediment, but the upper (exposed) portion (the **rachis**) produces polyps in whorls or rows, or sometimes united in crescent-shaped "leaves." Often these polyps are of two distinct types. **Autozooids** bear tentacles and function in feeding; **siphonozooids** are small, have reduced tentacles, and serve to create water currents through the colony. In sea pens the rachis is elongated and cylindrical; in sea pansies it is flattened and shaped like a large leaf (Figure 1K). In the odd deep-sea genus *Umbellula*, the secondary polyps radiate outward to give the colony the appearance of a pinwheel set on the end of a tall narrow stalk. The first deep benthic photos of *Umbellula* had biologists scratching their heads for years, wondering to which phylum this preposterous creature might belong.

Gorgonians are also colonial anthozoans (Figures 1J and 8). Some grow in bushy shapes, whereas others are planar; size and shape of the colony are often mediated by the hydrodynamics of the local surge and current regime. Where prevailing currents are more-or-less in one plane (although they may move back and forth, in two directions), the branches of the colony tend to grow largely in one plane also, perpendicular to the flow. In regions of mixed currents the same species tends to grow in two planes.

The medusoid form. Free medusae occur in all cnidarian classes except Anthozoa. Although variation in form does exist, medusae are far less diverse than polyps, and it is much easier to generalize about their anatomy. The relative uniformity of the medusoid form is a result in part of their usually similar lifestyles in open water, and of their inability to form colonies by asexual reproduction. They are participants in colonial life only insofar as some remain attached to hydrozoan colonies as sessile gonophores. Sessile benthic forms are rare (see Figure 1F).

Even though the general structure of the body walls of medusae and polyps are similar and both adhere to the general cnidarian *Bauplan* outlined earlier, their gross morphologies are adapted to their very different lifestyles. Medusae are bell- or umbrella-shaped, and generally embued with a thick, jelly-like mesogleal layer—hence the name "jellyfish." The convex upper (aboral) surface is called the **exumbrella**; the concave lower (oral) surface is the **subumbrella** (Figures 13, 14, and 15). The mouth is located in the center of the subumbrella, often suspended on a pendant, tubular extension called the **manubrium**, which is almost always present on hydromedusae, but usually reduced or absent in scyphomedusae.

The coelenteron or gastrovascular cavity occupies the central region of the umbrella and extends radially via **radial canals**. In most hydromedusae, a marginal **ring canal** within the rim of the bell connects the ends of the radial canals. The presence of four radial canals and of tentacles in multiples of four (in

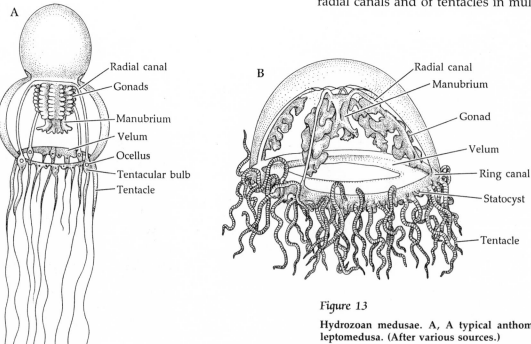

Figure 13

Hydrozoan medusae. A, A typical anthomedusa. B, A typical leptomedusa. (After various sources.)

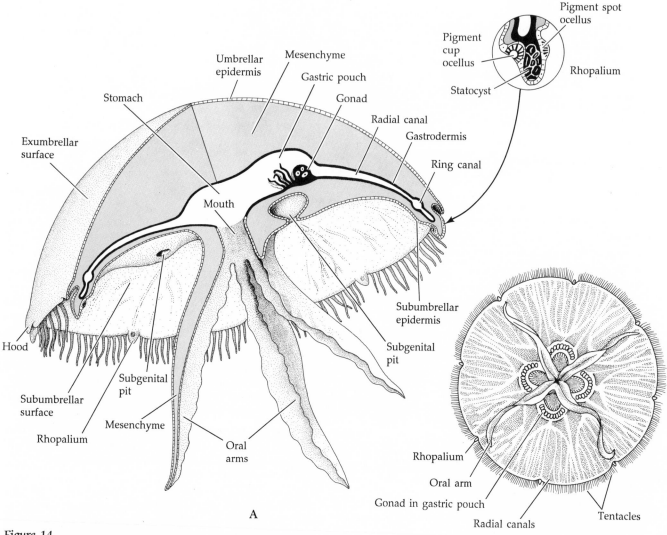

Figure 14

A typical scyphozoan medusa. A, Cutaway side view. B, Oral view. (A from Bayer and Owre 1968; B after Barnes 1987.)

hydromedusae) and the division of the stomach by septa into four gastric pouches (in scyphomedusae) give most jellyfish a **quadriradial** (= **tetramerous**) symmetry (Figure 3A). Many hydromedusae have a thin circular flap of tissue, the **velum**, within the margin of the bell (Figure 13). Such medusae are termed **craspedote**. Those lacking a velum, such as scyphomedusae, are said to be **acraspedote** (Figures 14 and 15). Like that of polyps, the entire external surface of medusae is covered with epidermis, and the internal surfaces (coelenteron and canals) are lined with gastrodermis. The bulky gelatinous middle layer is either an acellular mesoglea or a partly cellular mesenchyme.

Support and movement

Support. Cnidarians employ a wide range of support mechanisms. Polypoid forms often rely substantially upon a hydrostatic skeleton, which consists of the water-filled coelenteron constrained by the muscular walls of the gut and body. In addition, the mesenchyme may be stiffened with fibers of various sorts, particularly in the anthozoans. Colonial anthozoans may incorporate bits of sediment and shell fragments into the column wall for further support. Many colonial hydrozoans produce a flexible, horny perisarc, composed largely of chitin secreted by the epidermis. In medusae, the principal support mech-

Figure 15

Comparison of cubomedusae and scypho-medusae. A, A cubomedusa. B, A coronate scyphomedusa (order Coronatae). C, A se-maeostome scyphomedusa (order Se-maeostomae). D, A rhizostome scypho-medusa (order Rhizostomae). E, A sessile stauromedusa (order Stauromedusae). (After Larson 1976.)

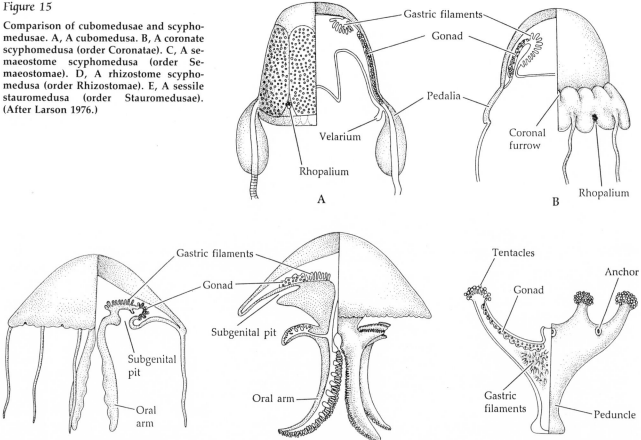

anism is the middle layer, which ranges from a fairly thin and flexible mesoglea to an extremely thick and stiffened fibrous mesenchyme, which may be almost cartilaginous in consistency.

In addition to these soft or flexible support structures, there is an impressive array of hard skeletal structures. These are of three fundamental types: horny or woodlike axial skeletal structures, calcareous sclerites, and massive calcareous frameworks. Horny **axial skeletons** occur in several groups of anthozoan colonies such as gorgonians, sea pens and antipatharian corals (Figure 8). Amebocytes in the coenenchyme secrete a flexible or stiff internal axial rod as a supportive base embedded in the coenenchymal mass; the polyps are spread on this base. Axial rods are protein–mucopolysaccharide complexes (called gorgonin in the order Gorgonacea), but little is known of the chemistry of these structures. In the antipatharians (black coral), the axial skeleton is so hard and dense that it is ground and polished to make jewelry.

In most octocorals, mesenchymal cells called scle-roblasts secrete a variety of calcareous **sclerites**. In octocorals with an axial skeleton, such as many gorgonians, the coenenchyme contains sclerites of various shapes and colors (Figure 17). It is usually the sclerites that give gorgonians their characteristic color. In many species, the sclerites become quite dense and may even fuse to form a more-or-less solid calcareous framework. The precious red coral *Corallium* is actually a gorgonian with fused red coenenchymal sclerites. In the stoloniferan organ-pipe corals (*Tubipora*) the sclerites of the body walls of the individual polyps are fused into rigid tubes.

Massive calcareous skeletons are found in two anthozoan orders and two hydrozoan orders. The best known are the true, or stony, anthozoan corals (order Scleractinia), in which epidermal cells on the lower half of the column secrete a calcium carbonate skeleton (Figure 18). The skeleton is covered by the thin layer of living epidermis that secretes it, and thus it might technically be considered to be an internal skeleton. However, because the coral colony generally sits atop a large nonliving calcareous frame-

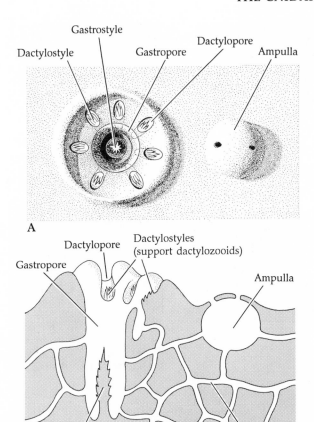

Gastrostyle
Dactylostyle
Gastropore
Dactylopore
Ampulla

A

Dactylopore
Gastropore
Dactylostyles
(support dactylozooids)
Ampulla

B
Gastrostyle
(supports gastrozooid)
Canal

Figure 16

Hydrozoan skeletons. The stylasterine hydrocoral *Allopora*, with its calcareous skeleton. Plane view, from above (A) and section through the skeleton (B). (From Bayer and Owre 1968.)

Members of the hydrozoan orders Milleporina and Stylasterina produce calcareous exoskeletons, and they are often referred to as the hydrocorals. Like true corals, milleporinid colonies may assume a great variety of shapes, and some grow as stony encrustations. The milleporinid exoskeleton, termed a **coenosteum**, is perforated by pores of two sizes; these pores accommodate two different kinds of polyps. The gastrozooids live in large holes, or **gastropores**, and are surrounded by a circle of smaller **dactylopores**, which house the dactylozooids. Canals lead downward from the pores into the coenosteum and are closed off below by a transverse calcareous tabula. As growth proceeds and the colony thickens, new tabulae are formed; keeping the polyp pores at a more-or-less fixed depth. The stylasterine skeleton is similar to the milleporinid skeleton, but the margins of the gastropores often bear notches that serve as dactylopores. The zooids are supported by spine-like **styles**, called **gastrostyles** and **dactylostyles**, respectively. Gonophores arise in chambers called **ampullae**, which connect to the feeding zooids through the coenosteum (Figure 16).

Movement. The muscle cells of cnidarians are the most primitive in the animal kingdom. Unlike higher metazoa, the musculature in cnidarians is derived exclusively from the relatively thin ectoderm and entoderm layers. Rather than being discrete cells and tissues, cnidarian muscle cells are merely modified epidermal and gastrodermal cells, whose basal regions develop contractile processes (Figures 4 and 19). Although the derivations of the musculature in polyps and medusae are the same, comprising sets of longitudinal and circular muscles in both epidermis and gastrodermis, the modes of movement in these two forms are quite different, and are reflected in the increased development of gastrodermal musculature in polyps and an emphasis on epidermal musculature in medusae.

The cells of the epidermis that are specialized for contraction are termed **epitheliomuscular cells** (= **myoepithelial cells**). Most of the polyp epidermis is composed of these unique columnar cells. The bases rest against the inner mesenchyme and the opposite ends form the outer body surface. Unlike true columnar epithelium, cnidarian epitheliomus-

work, most authors speak of the skeleton as being "external." The entire skeleton of a scleractinian coral is termed the **corallum**, regardless of whether the animal is solitary or colonial; the skeleton of a single polyp, however, is called a **corallite**. The outer wall of the corallite is the **theca**; the floor is the **basal plate** (Figure 18). Rising from the center of the basal plate is often a supportive skeletal process called the **columella**. The basal plate and inner thecal walls give rise to numerous radially arranged calcareous partitions, the **septa**, which project inward and support the mesenteries of the polyp. Polyps occupy only the uppermost surface of the corallum. Skeletal thickness increases as polyps grow, and the bottoms of the corallites are sealed off by transverse calcareous partitions called **tabulae**, which become the new basal portion of the polyp. The corallum can assume a great variety of shapes and sizes, from simple cup-shaped structures in solitary corals to large branching forms in colonial species. Colonial forms may be upright, low and massive, or even encrusted on other hard substrata.

Figure 17

The skeleton of gorgonians, illustrated by scanning electron micrographs at successively greater magnification of the gorgonian *Muricea fruticosa*. A, A complete colony. B, Colony branches bear whorls of polyps. C, Sclerites from the tissues of a single polyp. (Courtesy of F. Bayer and W. R. Brown, Smithsonian Institution.)

A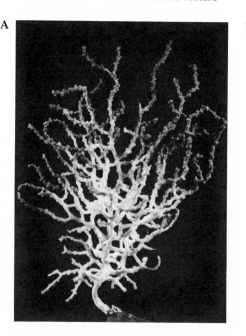

B

cular cells possess two or more basal extensions or processes running parallel to body surface and containing the contractile myofibrils or **myonemes** (Figure 19). These processes interconnect with the processes of neighboring epitheliomuscular cells to form a contractile sheet between the outer epidermis and the middle body layer. In polyps, the processes are largely arranged longitudinally (in an oral–aboral direction). The cnidarian epidermis is thus both a protective and supportive layer and a contractile layer.

The gastrodermis is histologically very similar to the epidermis and also includes myoneme-containing **nutritive–muscular cells**. These are similar in design to the epitheliomuscular cells of the epidermis, and also form both longitudinal and circular contractile layers. In polyps, these two muscle systems work in conjunction with the gastrovascular cavity as an efficient hydrostatic skeleton, as well as providing a means of movement.

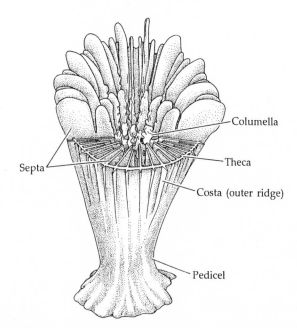

Figure 18

The corallite of a solitary scleractinian coral (diagrammatic), illustrating morphological features. (After Cairns 1981.)

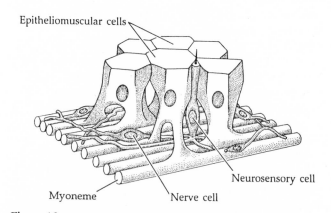

Figure 19

Epitheliomuscular cells and the nerve net of the cnidarian epithelium. (After Mackie and Passano 1968.)

C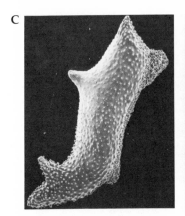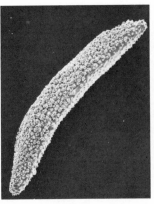

Body musculature is most highly specialized and well developed in the anthozoans, particularly the sea anemones. In anemones the muscles of the column wall are largely gastrodermal, although epitheliomuscular cells occur in the tentacles and oral disc. Bundles of longitudinal fibers lie along the sides of the mesenteries and act as **retractor muscles** for shortening the column (Figure 20). **Circular muscles** derived from the gastrodermis of the column wall are also well developed. In most anemones, the circular muscles form a distinct sphincter at the junction of the column and the oral disc; this sphincter can pinch the mouth closed. Circular fibers also occur in the tentacles and the oral disc. When a sea anemone contracts, the upper rim of the column is pulled over to cover the oral disc. In many anemones, a circular fold—the **collar**, or **parapet**—occurs near the sphincter to further cover and protect the delicate oral surface upon contraction.

Most polyps are sedentary or sessile. Their movements consist mainly of food-capturing actions and the withdrawal of the upper portion of the polyp during body contractions. These activities are accomplished primarily by the epidermal muscles of the tentacles and oral disc, and by the strong gastrodermal muscles of the column. Circular muscles work in conjunction with the hydrostatic skeleton to distend the tentacles and body.

A variety of locomotory methods has evolved among the anemones (Figure 21). Most can creep about slowly by use of pedal disc musculature. In many species, the column can bend far enough to allow the tentacles to contact the substratum, whereupon the pedal disc releases its hold and the animal somersaults or moves like an inchworm. Some solitary hydrozoan polyps (e.g., *Hydra*) also move in this manner. Adhesive cnidae may aid in anchoring the tentacles during this behavior. A few anemones (e.g.,

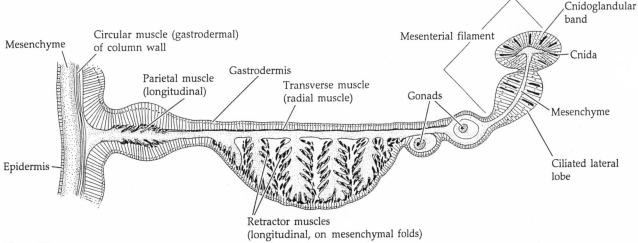

Figure 20

Mesentery (cross section) of a sea anemone (order Actiniaria). (From Bayer and Owre 1968.)

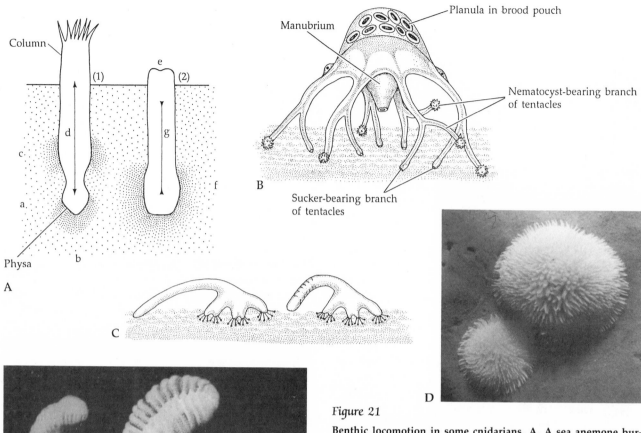

Figure 21

Benthic locomotion in some cnidarians. A, A sea anemone burrowing: (1) Eversion of the physa with displacement of sand (a) and further penetration (b) into substratum; the anemone is held by a column "anchor" (c) as the lower portion of the column is extended (d). (2) With the tentacles folded inward (e), the physa is swollen to form an anchor (f), which allows retractor muscles (g) to pull the anemone into the sand. B, The hydromedusan *Eleutheria*, which creeps about on its tentacles. C, The scyphomedusan *Lucernaria*, which also creeps about on its tentacles. D, *Liponema brevicornis*, a sea anemone that folds itself into a "ball" and rolls about on the sea floor with the bottom currents. E, The anemone *Actinostola* "swimming" off the substratum (between two sea pens) by undulatory back-and-forth contractions of the column—an escape response to the predatory sea star *Gephyriaster swiftti*, visible in this photo (Puget Sound, Washington). (B after Hyman 1940; D from Dunn and Bakus 1977; E courtesy of C. Birkeland.)

Actinostola, Stomphia, Tealia) can detach from the substratum and actually swim away by "rapid" flexing or bending of the column; others swim by thrashing the tentacles. These swimming activities are temporary behaviors, generally elicited by the approach or contact of a predator. In a few species of anemones, the basal disc may detach and secrete a gas bubble, thereby permitting the polyp to float to a new location. Anemones of one family (Minyadidae) are wholly pelagic and float upside down in the sea by means of a gas bubble enclosed within the folded pedal disc. *Hydra* also is known to float upside down by means of a mucus-coated gas bubble on the bottom of its pedal disc. One of the oddest forms of polyp locomotion is that of the anemone *Liponema brevicornis* of the Bering Sea, which is capable of drawing itself into a spherical form that can be rolled freely about the sea floor by the bottom currents (Figure 21D). Even colonial sea pansies (Pennatulacea) are motile, using their muscular peduncle to migrate to different depths.

Cerianthid anemones (Ceriantharia) are burrowing, tube-building organisms (Figure 12E). They differ from the "true" anemones (Actiniaria) in several important ways. They have no sphincter muscle and only weak longitudinal gastrodermal muscles that do' not form distinct retractors in the mesenteries. As a result, cerianthids cannot contract the oral disc and tentacles as they withdraw into their tubes. In contrast to other anemones, however, they possess a complete layer of longitudinal epidermal muscles in the column, which allows a very rapid withdrawal response. Merely the shadow of a passing hand will cause a cerianthid to rapidly pull itself deep into its long buried tube.

In medusae, epidermal musculature predominates, and the gastrodermal muscles that are so important in polyps are reduced or lacking altogether. The epidermal musculature is best developed around the bell margin and over the subumbrellar surface. Here the muscle fibers usually form circular sheets —called **coronal muscles**—that are partly embedded in the mesenchyme or mesoglea. Contractions of the coronal muscles produce rhythmic pulsations of the bell, thereby driving water out from beneath the subumbrella and moving the animal by jet propulsion. The stiffened cellular collenchyme of scyphomedusae

and cubomedusae includes elastic fibers that provide the antagonistic force to restore the bell shape between contractions. Many medusae also possess radial muscles that aid in opening the bell between pulses. In craspedote forms, the velum serves to reduce the size of the subumbrellar aperture, thus increasing the force of the water jet (Figure 13). The velarium of the fast-swimming cubomedusae has the same effect (Figure 15A), and the evolutionary forces that produced these two convergent features were probably similar.

Most medusae spend their time swimming upward in the water column, then sinking slowly down to capture prey by chance encounter, thereafter to pulsate upward once again. Some have the ability to change direction as they swim, however, and many are strongly attracted to light, especially those harboring symbiotic zooxanthellae. Medusae are sometimes very abundant (Figure 22), and some are known to aggregate at temperature or salinity discontinuity layers in the sea, where they feed on zoo-

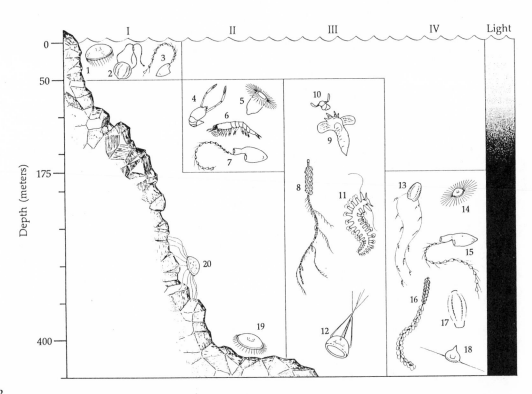

Figure 22

Hydromedusae and other cnidarians are major contributors to the pelagic biomass in the world's temperate seas. This figure, drawn by George Mackie, illustrates the distribution of pelagic cnidarians (and other macroplankton) off the coast of British Columbia, Canada. Epipelagic species (group I) are found in an upper mixed layer, roughly the top 50 m. Species in group II live deeper, where there is not enough light for photosynthesis but enough to affect animal behavior. Most of these species migrate to the surface at night, presumably responding to changes in light intensity. Be-

low 175 m there is not enough light to affect behavior, and species living here do not migrate (group IV). Group III includes species with rather wide vertical range; some migrate, others do not. Cnidarian species indicated on this figure include (1) *Phialidium gregarium*, (3) *Muggiaea atlantica*, (5) *Aglantha digitale*, (7) *Dimophyes arctica*, (8) *Nanomia cara*, (12) *Aegina citrea*, (14) *Pantachogon haeckeli*, (15) *Lensia baryi*, (16) *Cordagalma cordiformis*. Also shown are the deep benthic species (19) *Foersteria purpurea* and (20) *Ptychogastria polaris*. (Courtesy of G. Mackie.)

plankton, which also concentrate at these boundaries. A few unusual groups of medusae are benthic. Some hydromedusae (e.g., *Eleutheria*, *Gonionemus*) crawl about on algae or sea grasses by adhesive discs on their tentacles (Figure 21B). Members of the scyphozoan order Stauromedusae (e.g., *Haliclystis*) develop directly from the polypoid stage and affix to algae and other substrata by an aboral adhesive disc (Figure 1F).

Cnidae

Before considering feeding and other aspects of cnidarian biology, it is necessary to present some information on the structure and function of cnidae. Cnidae, which are often referred to collectively as nematocysts in many works, are the most unique and characteristic structures in cnidarians. They serve a variety of functions, including prey capture, defense, locomotion, and attachment. They are produced by cells called **cnidoblasts**, which develop from interstitial cells in the epidermis and, to a limited extent, in the gastrodermis. Once fully formed, the cell is properly called a **cnidocyte**, in which resides the cnida. During formation of a cnida, the cnidoblast produces a large internal vacuole in which a complex, poorly understood, intracellular reorganization takes place. Evidence suggests that the cnida is a complex secretory product of the Golgi apparatus of the cnidoblasts.

Cnidae are among the largest and most complex intracellular structures known. When fully formed, they are cigar- or flask-shaped capsules, 5–100 or more μm long, with thin walls composed of a collagen-like protein (Figures 23 and 24). One end of the capsule is turned inward as a long, hollow, coiled thread or tube. A complex system of what may be contractile fibers surrounds the body of the cnidocyte, and the entire structure is anchored to adjacent epithelial cells (**supporting cells**) or to the underlying mesenchyme. When sufficiently stimulated, the tube everts to the exterior. In members of the classes Hydrozoa, Scyphozoa, and perhaps Cubozoa, the capsule is covered by a hinged lid, or **operculum**, which is thrown open when the cnida discharges. In members of these three classes, each cnida bears a long cilium-like bristle called a **cnidocil**, a mechanoreceptor that elicits discharge when stimulated. Anthozoan cnidae lack a cnidocil and have a tripartite apical flap in lieu of an operculum. They often bear a cilium that acts as a mechanoreceptor. Recent studies indicate that the cilium responds to specific water-borne vi-

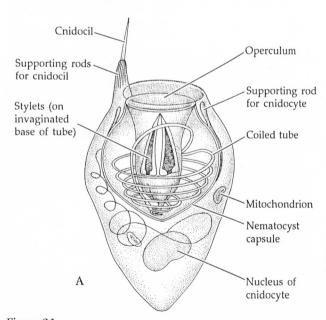

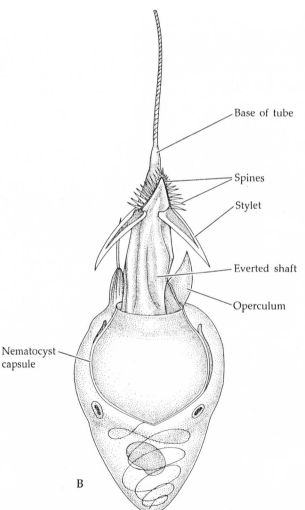

Figure 23

Nematocyst. A, Before discharge. B, After discharge. (After Sherman and Sherman 1976.)

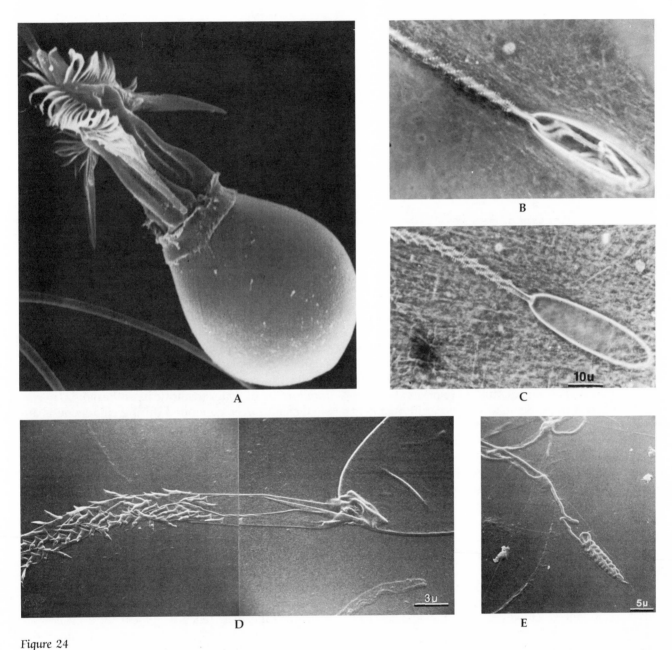

Figure 24

Discharged nematocysts. A, The base of a discharged nematocyst from the hydrozoan *Hydra* (scanning electron micrograph). B, A nematocyst of the anthozoan *Corynactis californica* (order Corallimorpharia). The nematocyst has been "stopped" when partially everted; the everting thread can be seen passing up through the already external region (light micrograph). C, A fully everted nematocyst of *C. californica* (light micrograph). D, A fully everted nematocyst of *C. californica* (scanning electron micrograph of the base of the everted thread and the tip of capsule). E, Everting nematocyst of the anthozoan coral *Balanophyllia elegans*. (A from Holstein and Tardent 1983; B–E from Mariscal 1974.)

bration frequencies. Chemoreceptors that are found on the adjacent supporting cells may actually "tune" the cilium to the proper reception frequency for available prey (Watson and Hessinger 1989). Cnidocytes are most abundant in the epidermis of the oral region and the tentacles, where they often occur in clusters of wartlike structures called "nematocyst batteries."

About three dozen kinds of cnidae have been described (Figures 24 and 25), but they can be assigned to three basic types. **True nematocysts** have double-walled capsules containing a toxic mixture of phenols and proteins. The tube is usually armed with spines or barbs that aid in penetration of and anchorage in the victim's flesh. The toxin is injected

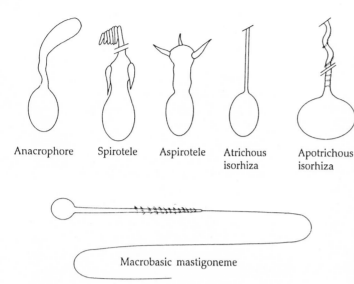

Anacrophore   Spirotele   Aspirotele   Atrichous   Apotrichous   Holotrichous
                                          isorhiza   isorhiza   isorhiza

Figure 25

Some types of cnidae. (After Mariscal, in Muscatine and Lenhoff 1974.)

Macrobasic mastigoneme

into the victim, often through a terminal pore in the thread. **Spirocysts** have single-walled capsules containing mucoprotein or glycoprotein. Their adhesive tubes wrap around and stick to the victim rather than penetrating it. The capsule tubes of spirocysts never have an apical pore. Nematocysts occur in members of all four cnidarian classes; spirocysts occur only in certain anthozoans. The third kind of cnidae, the **ptychocysts**, differs morphologically and functionally from both nematocysts and spirocysts. The capsule tube of ptychocysts lacks spines and an apical pore and is strictly adhesive in nature. In addition, the tube is folded into pleats rather than coiled within the capsule. Ptychocysts occur only in the ceriantharians and function in forming the unique tube in which these animals reside.

Cnidae have usually been viewed as independent effectors, and, indeed, they often discharge upon direct stimulation. However, recent evidence suggests that the animal does have at least some control over the action of its cnidae. For example, starved anemones seem to have a lower firing threshold than satiated animals. It has also been demonstrated that stimulating discharge of cnidae in one area of the body results in discharge in surrounding areas. In addition, the discharge of cnidae is inhibited in some cnidarians during certain activities (e.g., during locomotion in some anemones). Both chemical and mechanical stimuli apparently are necessary for most cnidae to fire. Cnidarians are known to discharge their cnidae in the presence of various sugars, low-molecular-weight amino compounds, and glutathione—the latter being a chemical that is liberated when animals are injured or when tissue breaks

down. Glutathione also causes feeding tentacles and gastrozooids to become active, writhe about, and prepare for feeding.

The ejection of the tube from a cnida is called **exocytosis**, and an individual cnida can be fired only once. Three hypotheses have been proposed to explain the mechanism of firing: (1) the discharge is the result of increased hydrostatic pressure caused by a rapid influx of water (**osmotic hypothesis**); (2) intrinsic tension forces generated during cnidogenesis are released at discharge (**tension hypothesis**); and (3) contractile units enveloping the cnida cause the discharge by "squeezing" the capsule (**contractile hypothesis**). Because of the small size of cnidae and the extreme rapidity of the exocytosis process, these hypotheses have been difficult to test. Recent work using ultrahigh-speed microcinematography suggests that both the osmotic and tension models may be at work. The coiled capsular tube is forcibly everted and thrown out of the bursting cell to penetrate or wrap around a portion of the unwary victim. It takes only a few milliseconds for the cnida to fire, and the everting tube may reach a velocity of 2 m/sec—an acceleration of about 40,000 *g*—making it one of the fastest cellular processes in nature.

Although nematocyst toxins vary in strength, as a class of chemicals they are potent biological poisons capable of subduing large active prey, including fishes. Most appear to be neurotoxins. The toxins of some cnidarians are powerful enough to seriously affect humans (e.g., some jellyfish; certain colonial hydroids, such as *Lytocarpus*; many hydrocorals, such as *Millepora*; many siphonophores, such as *Physalia*). The toxin of most cubomedusans is more potent than cobra venom, and numerous human deaths have been attributed to these box-jellies. Stings by *Chironex* and *Chiropsalmus* usually result in severe pain at the least, and respiratory or cardiac failure at worst.

Feeding and digestion

All cnidarians are carnivores. Typically, cnidocyte-laden feeding tentacles capture small animal prey and carry it to the mouth region where it is ingested whole (Figure 26E). Digestion is initially extracellular in the coelenteron. The gastrodermis is abundantly supplied with **mucous gland cells** and **enzymatic gland cells**, which facilitate digestion (Figure 4). In many groups gastrodermal cilia (or flagella) aid in mixing. The product of this preliminary breakdown is a soupy broth, from which polypeptides, fats, and carbohydrates are taken into the nutritive cells by phagocytosis and pinocytosis. Digestion is completed intracellularly within food vacuoles. Undigested wastes in the coelenteron are expelled through the mouth.

In anthozoan guts the free edges of most mesenteries are thickened to form three-lobed mesenterial filaments (Figure 7 and 20). The lateral lobes are ciliated and aid in circulating the digestive juices in the coelenteron. The middle lobe, called the **cnidoglandular band**, bears cnidae and gland cells. In some sea anemones (e.g., *Aiptasia*, *Anthothoe*, *Calliactis*, *Diadumene*, *Metridium*, *Sagartia*), the cnidoglandular band continues beyond the base of the septum as a free thread called an **acontium**, which hangs freely in the coelenteron. The cnida-bearing acontia not only subdue live prey within the gut, but may be shot out through the mouth or pores in the body wall (**cinclides**) when the animal contracts violently; when this occurs, the acontia presumably play a defensive role.

Several groups of cnidarians have adopted feeding methods other than the direct use of cnidocyte-laden tentacles. One group of large tropical anemones in the order Corallimorpharia (e.g., *Amplexidiscus*) lacks cnidocytes on the external surfaces of most tentacles. These remarkable anemones capture prey directly with the oral disc, which can be expanded and thrown upward to envelop crustaceans and small fishes, rather like a fisherman's cast-net (Figures 26A–D).

In addition to tentacular feeding on various small plankters, many corals are capable of mucous-net suspension feeding. This process is accomplished by spreading thin mucous strands or sheets over the colony surface and collecting fine particulate matter that rains down from the water. The food-laden mucus is driven by cilia to the mouth. In a few corals

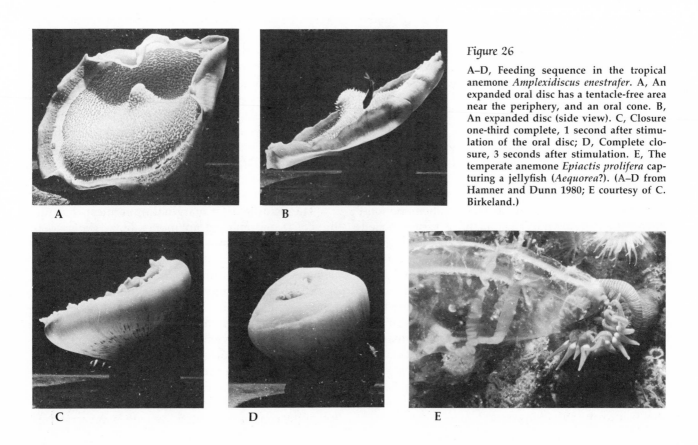

A   **B**   **C**   **D**   **E**

Figure 26

A–D, Feeding sequence in the tropical anemone *Amplexidiscus enestrafer*. A, An expanded oral disc has a tentacle-free area near the periphery, and an oral cone. B, An expanded disc (side view). C, Closure one-third complete, 1 second after stimulation of the oral disc; D, Complete closure, 3 seconds after stimulation. E, The temperate anemone *Epiactis prolifera* capturing a jellyfish (*Aequorea*?). (A–D from Hamner and Dunn 1980; E courtesy of C. Birkeland.)

(e.g., members of the family Agariciidae), the tentacles are greatly reduced or absent, and all direct feeding is by the mucous-net suspension method. The amount of mucus produced by corals is so great that it serves as an important food source for certain fishes and other reef organisms, which feed directly off the coral and also recover mucus sloughed into the surrounding sea water. Coral mucus released into the sea contains a variable mixture of macromolecular components (glycoproteins, lipids, and mucopolysaccharides) or a mucous lipoglycoprotein of specific character for a given organism. These loose mucous webs, or flocs, are usually enriched by bacterial colonies and entrapped detrital materials, further enhancing their nutritional value.

The role of cnidarians as potentially significant members of food webs depends largely on location and circumstance. Corals obviously hold critical trophic positions in tropical reef environments, as do zoanthids and gorgonians in some tropical and subtropical habitats. In many warm and temperate areas sea pens and sea pansies dominate benthic sandy habitats (Figure 1I). Large scyphomedusae often occur in great swarms, and they may consume high numbers of larvae of commercially important fishes in some areas, as well as competing with fishes for food. Swarms of jellyfish may be so dense that they clog and damage fishing nets and power plant intake systems. Hydromedusae are also known to be major components of temperate pelagic food webs (Figure 22). Members of several genera of hydrozoans occur in huge congregations in tropical seas, where they are important carnivores in the epipelagic food web. Best known among these are *Porpita* (which actively feeds on motile crustaceans, such as copepods), *Velella* (which feeds on relatively passive prey, such as fish eggs and crustacean larvae), and *Physalia* (which actively catches and consumes fishes).

Cnidarians have played numerous roles in folklore around the world. In Samoa, the anemone *Rhodactis howesii*, known as mata malu, is served boiled as a festive holiday dish. Eaten raw, however, mata malu causes death and is a traditional device in Samoan suicide. Hawaiians refer to the zoanthid *Palythoa toxica* as "limu-make-o-Hana" (the sacred deadly "seaweed" of Hana). Hawaiians used to smear their spear tips with this cnidarian, the toxin from which is called **palytoxin**. Interestingly, the toxin may be produced by an unidentified symbiotic bacterium, not by the cnidarian itself. It is one of the most powerful toxins known, being more deadly than that of poison arrow frogs (batrachotoxin) and paralytic shellfish toxin (saxitoxin).

Defense, interactions, and symbiosis

There are so many other interesting aspects of cnidarian biology that do not fall neatly into our usual coverage of each group that we present this special section in this chapter. The following discussion also points out the surprising level of sophistication possible at the relatively primitive diploblastic–radiate grade of complexity.

In most cnidarians defense and feeding are intimately related. The tentacles of most anemones and jellyfish usually serve both purposes; the defense polyps (dactylozooids) of hydroid colonies often aid in feeding; and so on. In some cases, however, the two functions are performed by distinctly separate structures (as in many siphonophorans).

Some species of acontiate anemones (e.g., *Metridium*) bear two different types of tentacles: feeding tentacles and defense tentacles. Whereas the former usually move in concert to capture and handle prey, the defense tentacles move singly, in a so-called searching behavior in which they extend to three or four times their resting length, gently touch the substratum, retract, and then reextend. Defense tentacles are used in aggressive interactions with other anemones, either those of a different species or those that are nonclonemates of the same species. The aggressive behavior consists of an initial contact with the opponent, followed by autotomous separation of the defense tentacle tip; this action leaves the tip behind, attached to the other anemone. Severe necrosis develops at the site of the attached tentacle tip, occasionally leading to death of the victim. Defense tentacles develop from feeding tentacles and tend to increase under crowded, nonclonemate conditions. The development involves loss of typical feeding tentacle cnidae (largely spirocysts) and the acquisition of true nematocysts and gland cells, which dominate in defense tentacles.

The **acrorhagi** (= **marginal spherules**) that ring the collar of some sea anemones (e.g., *Anthopleura*) also serve a defensive function. These normally inconspicuous vesicles at the base of the tentacles bear nematocysts and usually spirocysts. Contact of an acrorhagi-bearing anemone with nonclonemates or other species causes the acrorhagi in the area of contact to swell and elongate. The expanded acrorhagi are placed on the victim and withdrawn; the application may be repeated. Pieces of acrorhagial epidermis that remain on the victim result in localized necrosis. In addition to this behavior, the acrorhagi are exposed as a ring of nematocyst batteries around the top of the constricted column whenever an acrorhagi-bearing anemone contracts in response to violent stimulation. Similar competitive interactions are known among stony corals (Figure 27).

There are a number of examples of associations between cnidarians and other organisms, some of which are truly symbiotic, others of which are less intimate. Very few cnidarians are truly parasitic, although several species of hydroids infest marine fishes. In some of these, the polyps lack feeding tentacles and occasionally even nematocysts. The basal portion of the polyp erodes the fish's epidermis and underlying tissues, and nutrients are absorbed directly from the host. One species invades the ovaries of Russian sturgeons (a caviar feeder!). Mutualism is common among cnidarians. Many species of hydroids live on the shells of various gastropods, hermit crabs, and other crustaceans, the hydroid getting a free ride and the host perhaps getting some protection and camouflage. Hydroids of the genus *Zanclea* are epizoic on ectoprocts, where they sting small predators and adjacent competitors, helping the ectoproct to survive and overgrow competing species. The ectoproct lends protection to the hydroid with its coarse skeleton, and the mutualism allows both taxa to cover a larger area than either could individually. The aberrant hydroid *Proboscidactyla* lives on the rim of polychaete worm tubes (Figure 9A) and dines on food particles dislodged by the host's activities.

Some sea anemones attach to gastropod shells inhabited by hermit crabs. Some of these partnerships appear to be mutualistic in nature; the anemone

gains motility while protecting its associate from predation. Perhaps the most extreme case of such mutualism is that of the cloak anemones (e.g., *Adamsia*, *Stylobates*), which wrap themselves around the hermit crab's gastropod shell and grow as the crab does (Figure 28). Initially, the anemone's pedal disc secretes a chitinous cuticle over the small gastropod shell occupied by the hermit. Such fortunate crabs need not seek new, larger shells as they grow, for the cloak anemone simply grows to provide the hermit with a living protective cnidarian "shell," often dissolving the original gastropod shell over time. As if it were itself a gastropod, the anemone grows to produce a flexible coiled house called a **carcinoecium**. These odd anemone "shells" have been mistakenly described and classified as flexible gastropod shells. A similar relationship exists between *Parapagurus* and certain species of *Epizoanthus*. The hydroid *Janaria mirabilis* secretes a shell-like casing that is inhabited by hermit crabs and, in an odd case of evolutionary convergence, so does the ectoproct (see Chapter 21) *Hippoporida calcarea* (Figure 29).

Several other groups of animals have evolved ways to utilize the cnidae of cnidarians for their own defense, a phenomenon known as **kleptocnidae**. Several aeolid sea slugs (see nudibranchs, Chapter 20) consume cnidarian prey, ingesting their unfired nematocysts and storing them in processes on their dorsal surfaces. Once in place, the nudibranchs may use the nematocysts for defense. The ctenophore *Haeckelia rubra* (Chapter 9) feeds on certain hydromedusae and incorporates their nematocysts into its tentacles. The freshwater turbellarian flatworm (Chapter 10) *Microstoma caudatum* feeds on *Hydra*, risking being eaten itself, and then uses the stored nematocysts to capture other prey. Several species of hermit crabs and true crabs carry anemones (e.g., *Calliactis*, *Sagartiomorphe*) on their shells or claws and use them as living weapons to deter would-be predators. Some hermit crabs of the genus *Pagurus* often have their shell covered by a mat of symbiotic colonial hydroids (e.g., *Hydractinia*, *Podocoryne*). The presence of the hydroid coat deters more aggressive hermits (*Clibinarius*) from commanding the pagurid's shell.

Several cases of fish–cnidarian symbiosis have been documented. The well-known association of anemone fishes and their host anemones serves an obvious protective function for the fishes. A fish's ability to live among the anemone's tentacles is still not fully understood. However, the available evidence suggests that the anemone does not voluntarily fail to spend its nematocysts on its fish partner; rather, the fish alters the chemical nature of its mu-

Figure 27

Competition between true corals (order Scleractinia) in the Virgin Islands. The coral *Eusmilia fastigiata* is seen extruding its mesenterial filaments and externally digesting the edge of a colony of *Porites astereoides*. (Courtesy of C. Birkeland.)

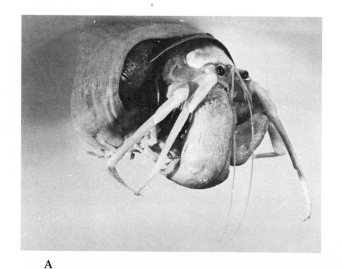

A

B

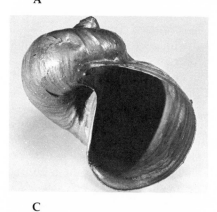

C

D

Figure 28

The "cloak anemone" (class Anthozoa, order Actiniaria) *Stylobates aenus*. A–B, The anemone is forming a "shell," or carcinoecium, around the hermit crab *Parapagurus dofleini*. C–D, The empty carcinoecium of *S. aenus*. (From Dunn et al. 1980.)

cous coating, perhaps by accumulating mucus from the anemone, thereby masking the normal chemical stimulus to which the anemone's cnidae would respond. *Neomus* is a small fish that lives symbiotically among the tentacles of *Physalia* and appears to survive by simply avoiding direct contact with the beast. When stung accidentally, however, it shows a much higher survival rate than other fishes of the same size. *Neomus* feeds on prey captured by its host.

One of the most noteworthy evolutionary achievements of cnidarians is their close relationship with unicellular photosynthetic partners. The relationship is widespread and occurs in many shallow-water cnidarians. The symbionts of freshwater hydrozoans (e.g., *Chlorhydra*) are single-celled species of green algae called "zoochlorellae" (Chlorophyta). In marine cnidarians, the protists are unicellular cryptomonads and dinoflagellates called "zooxanthellae" (e.g., *Symbiodinium*) (Figure 30A). These algae may be capable of living free from their hosts, and perhaps do so normally, but very little is known about their natural history. The algae typically reside in the host's gastrodermis or epidermis, although some cnidarians harbor extracellular zooxanthellae in the mesoglea. It is often the algal symbionts that give some cnidarians their green, blue-green, or brownish color. Corals (including all reef-builders) that harbor zooxanthellae are called **hermatypic corals**. Resident populations of zooxanthellae in corals may reach a density of 30,000 algal cells per cubic millimeter of host tissue.

It has only been in the last two decades that the physiological and adaptive nature of this relationship has been detailed to any extent. Some of this information comes from studies on the scyphomedusa *Mastigias* (Figure 30B), which lives in marine lakes on the islands of Palau, where it may occur in densities exceeding 1,000 per m^3. In these lakes, *Mastigias* makes daily vertical migrations between the oxygenated, nutrient-poor upper layers and the anoxic, nutrient-rich lower layers. This behavior appears to be related to the light and nutrient requirements of its symbiotic zooxanthellae. Unlike the zooxanthellae in benthic cnidarians, which tend to reproduce more-

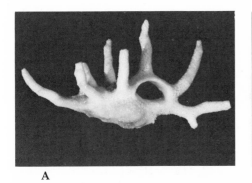
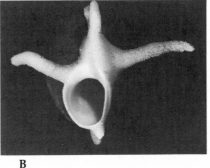
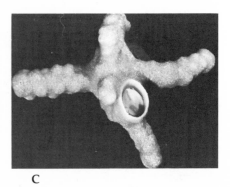

A

B

C

Figure 29

A case of remarkable evolutionary convergence. A–B, The hydrozoan colony *Janaria mirabilis* (suborder Anthomedusae) forms a shell-like corallum inhabited by hermit crabs. C, The ectoproct *Hippoporida calcarea*, which forms a similar structure, is also inhabited by hermit crabs. (From Cairns and Barnard 1984.)

or-less evenly over a 24-hour period, the zooxanthellae of *Mastigias* show a distinct reproductive peak during the hours when their host occupies a position in the deeper nitrogen-rich layers of the lakes. This reproductive peak may be a result of the alga's use of free ammonia as a nutrient source.

Many cnidarians derive only modest nutritional benefit from their algal symbionts, or they utilize them primarily to help eliminate metabolic wastes. But in many others a significant amount of the hosts' nutritional needs is provided by the algae. In such cases, a large portion of the organic compounds pro-

duced by photosynthesis of the symbiont is passed on to the cnidarian host, mainly as glycerol but also as glucose and the amino acid alanine. In return, metabolic wastes produced by the cnidarian provide the alga with nitrogen and phosphorus. In corals, the symbiosis is especially important for rapid growth and for efficient deposition of the calcareous skeleton, and many of them can only form reefs when they maintain a viable dinoflagellate population in their tissues (Figure 31). Algal photosynthesis probably increases the rate of calcium carbonate production (precipitation) by utilizing CO_2. Corals and other cnidarians can be deprived of their algal symbionts by experimentally placing the hosts in dark environments. In such cases the algae may simply die, they may be expelled from the host, or they may (to a limited extent) actually be consumed directly by the host. Because they are dependent on light, herma-

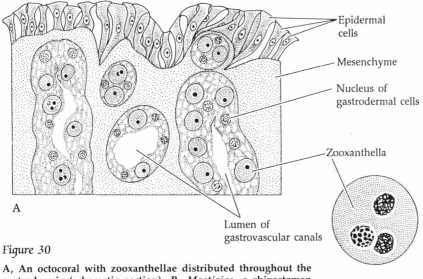

Epidermal cells

Mesenchyme

Nucleus of gastrodermal cells

Zooxanthella

A

Lumen of gastrovascular canals

Figure 30

A, An octocoral with zooxanthellae distributed throughout the gastrodermis (schematic section). B, *Mastigias*, a rhizostoman medusa, harbors zooxanthellae within its cells. (A after Bayer and Owre 1968; B courtesy of J. King.)

B

Figure 31

Possible pathways of calcium and carbonate during calcification in a hermatypic coral. In this diagrammatic cross section of the calicoblastic body wall at the base of a polyp, the parts are not drawn to scale. The coelenteron and the flagellated gastrodermis containing a zooxanthella are shown at the top of the figure, the calicoblastic epidermis is in the middle, and the organic membrane with crystals of calcareous matter is at the bottom. The direction of growth is upward (i.e., calcium deposition is in a downward direction). C.A., carbonic anhydrase. (After Goreau 1959.)

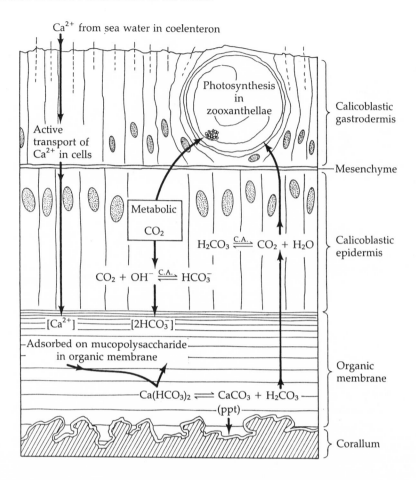

Circulation, gas exchange, excretion, and osmoregulation

There is no independent circulatory system in cnidarians. The coelenteron serves in this role to a limited extent by circulating nutrients through the interior of the body, absorbing metabolic wastes from the gastrodermis, and eventually expelling waste products of all types through the mouth. But large anemones and large medusae confront a serious surface area:volume dilemma. In such cases, the efficiency of the gastrovascular system as a transport device is enhanced by the presence of septa in the anemones and the radially arranged canal system in the medusae. Cnidarians also lack special organs for gas exchange or excretion. The body wall of most polyps is either fairly thin or has a large internal surface area, and the thickness of many medusae is due largely to the gellike mesoglea or mesenchyme. Thus, diffusion distances are kept to a minimum. Gas exchange occurs across the internal and external body surfaces. Facultative anaerobiasis occurs in some species, such as anemones that are routinely buried in soft sediments. Nitrogenous wastes are in the form of ammonia, which diffuses through the general body surface to the exterior or into the coelenteron. In freshwater species there is a continual influx of water into the body. Osmotic stress in such cases is relieved by periodic expulsion of fluids from the gastrovascular cavity, which is kept hypoosmotic to the tissue fluids.

Nervous system and sense organs

Consistent with their radially symmetrical *Bauplan*, cnidarians have a diffuse, noncentralized nervous system. The neurosensory cells of the system are the most primitive in the animal kingdom, being naked and largely nonpolar. The neurons are arranged in two reticular arrays, called **nerve nets**, one between the epidermis and the mesenchyme and another between the gastrodermis and the mesenchyme (Figure 32). The subgastrodermal net is generally less well developed than the subepidermal net, and it may be absent altogether. Some hydrozoans possess one or two additional nerve nets.

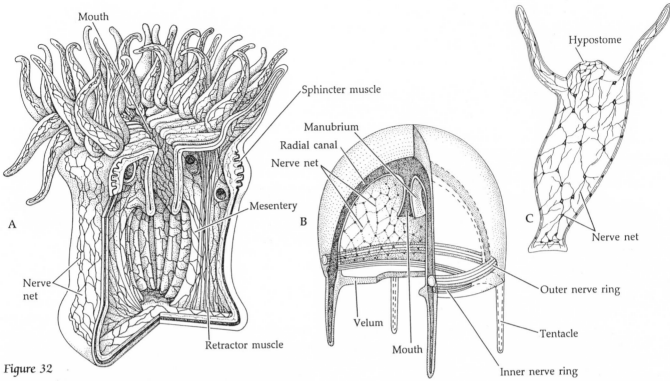

Figure 32

Cnidarian nerve nets. A, Nerve net of a typical sea anemone (class Anthozoa). B, Nerve net in a hydromedusa (class Hydrozoa). C, Nerve net of *Hydra* (Hydrozoa). (A after Wells 1968; B after Barnes 1987, from various sources; C from Bayer and Owre 1968.)

Most of the neurons and synapses are nonpolar—that is, impulses can travel in either direction along the cell or across the synapse. Thus, sufficient stimulus sends an impulse spreading in every direction. In some cnidarians where both nerve nets are well developed, one net serves as a diffuse slow-conducting system of nonpolar neurons, and the other as a rapid through-conducting system of bipolar neurons. A few nerve cells and synapses are polarized (bipolar) and allow for transmission in only one direction.

Polyps generally have very few sensory structures. The general body surface has various minute hairlike projections developed from individual cells. These serve as mechanoreceptors, and perhaps as chemoreceptors, and are most abundant on the tentacles and other regions where cnidae are concentrated. They are involved in behavior such as tentacle movement toward a prey or predator and in general body movements. Some appear to be associated specifically with discharged cnidae, such as the **ciliary cone apparatus** of anthozoan polyps, which is believed to function like the cnidocil in hydrozoan and scyphozoan nematocysts (Figure 33). Oddly, these structures do not appear to be connected directly to

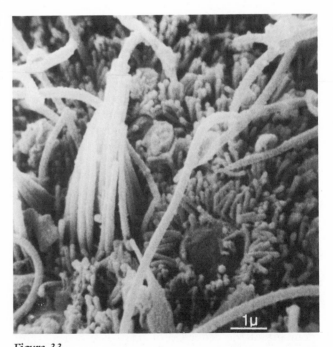

Figure 33

A ciliary cone on the tentacle of the corallimorpharian "anemone" *Corynactis californica* lies adjacent to cnidocyte (the circle of microvilli). (From Mariscal 1974.)

the nerve nets. In addition, most polyps show a general sensitivity to light; it is not mediated by any known receptor, but is presumably associated with neurons concentrated in or just beneath the translucent surface of epidermal cells.

As might be expected, motile medusae have more sophisticated nervous systems and sense organs than the sessile polyps do (Figure 34). In many groups, especially the hydromedusae, the epidermal nerve net of the bell is concentrated into two **nerve rings** near the bell margin. These nerve rings connect with fibers innervating the tentacles, muscles, and sense organs. The lower ring stimulates rhythmic pulsations of the bell. This ring is also connected to statocysts on the bell margin, which is supplied with **general sensory cells** and with radially distributed ocelli, statocysts, and (probably) chemoreceptors. The general sensory cells are neurons whose receptor processes are exposed at the epidermal surface. The ocelli are usually simple patches of pigment and photoreceptor cells organized as a disc or a pit. Statocysts may be in the form of pits or closed vesicles, the latter housing a calcareous statolith adjacent to a sensory cilium. Statocyst stimulation inhibits muscular contraction on the stimulated side of the bell; the medusa subsequently rights itself by contracting muscles on the opposite side. Many medusae maintain themselves in a particular photo-regime by directed swimming behaviors. This action is seen especially in those medusae that harbor abundant populations of zooxanthellae, such as the medusa *Cassiopeia*, which lies upside down on the shallow sea floor, exposing to light the dense zooxanthellae population residing in its tissues.

Cubomedusae possess as many as 24 well-developed eyes located near the bell margin. The most complex of these have a true epidermal cornea, a spherical cellular lens, and an upright retina (Figure 34F). The retina is multilayered, containing a sensory layer, a pigmented layer, a nuclear layer, and a region of nerve fibers. There are roughly 11,000 sensory cells in each of these remarkable eyes. Cubomedusae only 3 cm tall swim at speeds up to 6 m/min and can orient accurately to the light of a match as far away as 1.5 m. This combination of speed and sensitivity to light may enable them to locate and feed on luminescent prey at night.

Cubomedusae and scyphomedusae lack well-developed nerve rings (although they are present in members of the order Coronatae). The bell margins of the cubomedusae and scyphomedusae usually bear club-shaped structures—called **rhopalia**—that are situated between a pair of flaps, or **lappets** (Figure 34). The rhopalia are sensory centers, each containing a concentration of epidermal neurons, a pair of sensory pits, a statocyst, and often an ocellus. One sensory pit is located on the exumbrellar side of the hood of the rhopalium, the other on the subumbrellar side; they may be chemosensory in function.

In addition to the neuronal system just described, cnidarians are said to also possess a "cytoplasmic conducting system" similar in nature to that of the sponges. Epidermal cells and muscle elements appear to be the principal components of the system. Impulse conduction, which travels very slowly and in a highly diffuse fashion, is not well understood but probably relies primarily on physical contact and stimulation of adjacent cells.

Bioluminescence is common in cnidarians and has been documented in all classes except the Cubozoa. In some forms (e.g., many hydromedusae), luminescence consists of single flashes in response to a local stimulus. In others, bursts of flashes propagate as waves across the body or colony surface (e.g., sea pens and sea pansies). The most complicated luminescent behaviors occur in hydropolyps, where a series of multiple flashes is propagated. Propagated luminescence is probably controlled by the nervous system, although this phenomenon is not well understood.

Reproduction and development

Reproductive processes in cnidarians are intimately tied to the alternation of generations that characterizes this phylum. As you have already learned, the basic cnidarian life cycle involves an asexually reproducing polyp stage, alternating with a sexually reproducing medusoid stage. Although numerous variations on this general life history occur, the basic theme underlies all cases. Many cnidarians enjoy the benefits of asexual reproduction. Sexually, most employ a dispersal phase in the form of a characteristic **planula larva**. Thus, we generally find a complex indirect life history, although brooding occurs to various degrees in some groups. The life cycles of each of the four classes are discussed separately, beginning with that of the Hydrozoa, which may be used as a prototype or model for discussing the life cycles of the other three.

Hydrozoan reproduction. Hydrozoan polyps reproduce asexually by budding. This is a rather simple process wherein the body wall evaginates as a bud, incorporating an extension of the gastrovascular cavity with it. A mouth and tentacles arise at the distal end, and eventually the bud either detaches from the

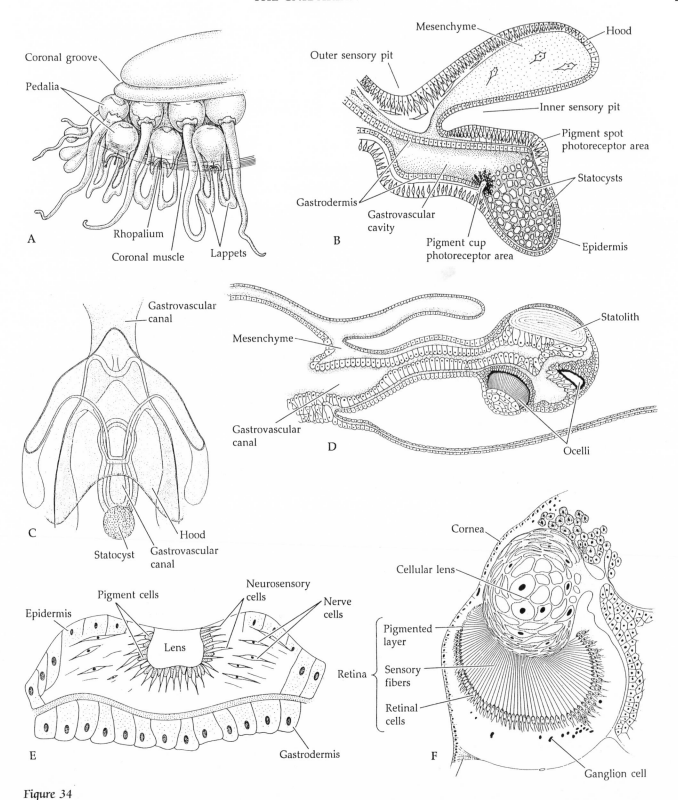

Figure 34

Sensory structures in medusae. A, The rhopalia of the scypho-medusa *Atolla* are situated between the marginal lappets. B, A rhopalium (section) has various sensory regions. C, A rhopalium of *Aurelia* (diagrammatic). A portion of the gastrovascular canal has been cut away. D, A cubozoan rhopalium (section). E, A pigment-cup ocellus (cross section) of a hydrozoan medusa. F, The eye of a cubozoan (*Carybdea*) (cross section). (A, B after Hyman 1940; C, D, F from Bayer and Owre 1968; E after Conant 1900.)

parent and becomes an independent polyp, or, in the case of colonial forms, remains attached. Medusa buds, or gonophores, are also produced by polyps in a similar fashion, although the process is sometimes quite complex. A rather special kind of budding occurs in many siphonophorans, in which the floating colonies produce chains of individuals called **cormidia**, which may break free to begin a new colony.

Certain hydromedusae also undergo asexual reproduction, either by the direct budding of young medusae (Figure 35), or by longitudinal fission. The latter process often involves the formation of multiple gastric pouches (**polygastry**), followed by longitudinal splitting, which produces two daughter medusae. In some species (e.g., *Aequorea macrodactyla*), **direct fission** may take place. Polygastry does not occur during this process; instead, the entire bell folds in half, severing the stomach, ring canal, and velum (Figure 36). Eventually the entire medusa splits in half and each part regenerates the missing portions.

Cnidarians in general have a great capacity for regeneration, as exemplified by experiments on *Hydra*. The eighteenth-century naturalist Abraham Trembley had the clever idea of turning a *Hydra* inside out, which he did. To his delight, the animal survived quite well, with the gastrodermal cells functioning as the "new epidermis" and vice versa. Cells removed from the body of a *Hydra* also have a modest degree of reaggregative ability, like that seen in sponges. In some cases, entire animals can be reconstituted from cells taken only from the gastrodermis or only from the epidermis. Although *Hydra* is an unusual and somewhat atypical cnidarian, this great capacity for cellular reorganization is a reflection of the primitive state of tissue development in the animals belonging to this phylum.

A typical *Hydra* consists of only about 100,000 cells of roughly a dozen different types. Although distinct epidermis and gastrodermis exist, these tissues are very similar to one another, comprising mainly epitheliomuscular cells. The nervous system is, of course, also very simple. It takes only a few weeks for all the cells in a *Hydra* to reproduce themselves, or "turn over." These attributes make *Hydra* an ideal creature for studies of developmental biology, histogenesis, and morphogenesis. This work has shed light on many fundamental biological phenomena, and is summarized in Lenhoff and Loomis (1961), Gierer (1974), and Lenhoff (1983).

All hydrozoan cnidarians have a sexual phase in their life cycle (Figure 37). In solitary species (e.g., *Hydra*) and some colonial forms, the medusoid phase is suppressed or lost. The polyps develop simple, transient, epidermal gamete-producing structures called **sporosacs** (Figures 5A,B). Most colonial hydroids, however, produce medusa buds (**gonophores**) either from the walls of the hydranths or from separate gonozooids. The gonophores may grow to medusae that are released as free-living, sexually reproducing individuals, or they may remain

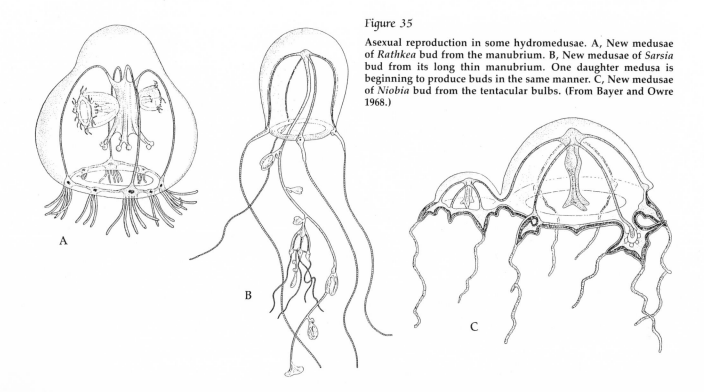

Figure 35

Asexual reproduction in some hydromedusae. A, New medusae of *Rathkea* bud from the manubrium. B, New medusae of *Sarsia* bud from its long thin manubrium. One daughter medusa is beginning to produce buds in the same manner. C, New medusae of *Niobia* bud from the tentacular bulbs. (From Bayer and Owre 1968.)

A

B

C

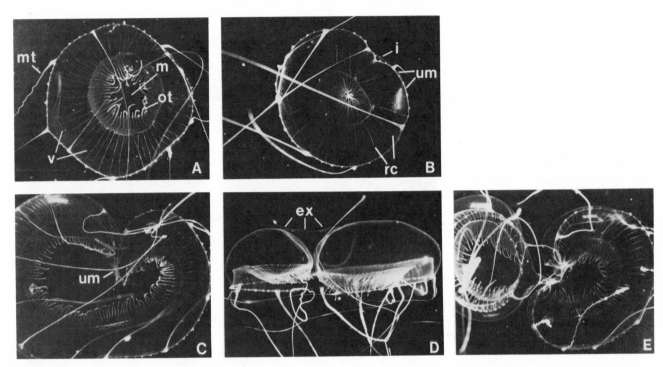

Figure 36

Asexual reproduction in the hydromedusa *Aequorea*. The sequence of photographs shows the direct fission of *A. macrodactyla*. A, This oral view shows a nondividing medusa with its marginal fishing tentacles (mt) deployed. B, Initiation of invagination (i). C–E, A progression of the direct fission process. The oral (C) and marginal (D) views illustrate the severing of the umbrellar margin (um) and the separation of exumbrellar halves; E shows the exumbrellar surface (ex) beginning to pull apart, producing free-swimming daughter medusae; healing is nearly complete in the smaller daughter medusa on the left. ot, oral tentacles; m, mouth; v, velum; rc, radial canals. (From Stretch and King 1980.)

attached to the polyps as incipient medusae and produce gametes.

In free-living hydromedusae, germinal cells arise from either epidermal or gastrodermal interstitial cells that migrate to specific sites on the bell surface, where they consolidate into temporary gonadal masses on the surface of the manubrium, beneath the radial canals, or on the general subumbrellar surface. Hydromedusae are usually gonochoristic, with either sperm or eggs usually being released directly into the water, where fertilization occurs. In some, only sperm are released and fertilization occurs on or in the female medusa's body.

Although several cleavage patterns occur in the hydrozoa, it is generally radial and holoblastic. A coeloblastula forms, which gastrulates by uni- or multipolar ingression to a stereogastrula. The interior cell mass is entoderm; the exterior cell layer is ectoderm (Figure 38). The stereogastrula elongates to form a solid or hollow nonfeeding, ciliated, free-swimming **planula larva** (Figure 39). The planula larva is radially symmetrical, but it possesses distinct "anterior" and "posterior" ends. The trailing posterior end of the larva (of all cnidarians) becomes the oral end of the adult. Hydrozoan planula larvae swim about for a few hours or for days before settling by attaching at the anterior end; during this time the entoderm hollows out to become the coelenteron. The mouth opens at the unattached oral end and tentacles develop as the larva metamorphoses into a young solitary polyp.

This overview of the hydrozoan reproductive cycle includes some minor variations on the basic theme. However, far more variety actually exists than we have space to discuss in detail (Figure 37). For example, in some trachylines, the polypoid stage is apparently lost altogether; the medusae produce planula larvae that develop into **actinula larvae**, which metamorphose into adult medusae, bypassing any sessile polypoid phase. Some trachylines and some siphonophorans undergo direct development, bypassing the larval stage altogether. The order Actinulida includes minute interstitial polyps that lack a medusoid stage and have suppressed the larval phase. The adult polyp is ciliated and resembles an actinula larva (hence the name).

Gonangium with developing medusae

Free hydromedusa

Fertilized egg

Planula larva

A

B

(a)

Planula (b)

Actinula

(c)

(d)

(e)

D

(a)

(b)

(c)

(d)

(e)

E

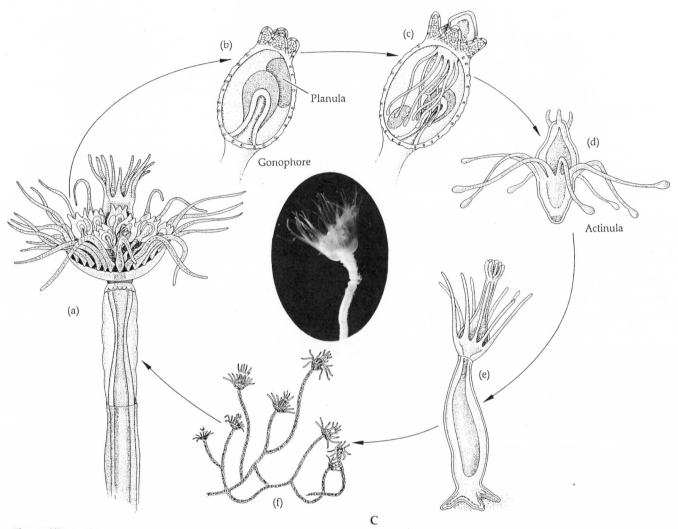

Planula

Gonophore

(d)

Actinula

(e)

(f)

C

Figure 37

Some hydrozoan life cycles. A, Life cycle of *Hydra*. Sperm produced by the male polyp (a) fertilizes the eggs of the female polyp (b). During cleavage, the eggs secrete a chitinous theca about themselves. After hatching, the embryos (c) grow into polyps that reproduce asexually by budding (d–e), until environmental conditions again trigger sexual reproduction. B, Life cycle of *Obelia*, a thecate hydroid with free medusae. C, Life cycle of *Tubularia*, an athecate hydroid that does not release free medusae. The polyp (a) bears many gonophores, whose eggs develop *in situ* into planulae (b) and then into actinula larvae (c) before release; the liberated actinula larvae (d) settle and transform directly into new polyps (e), each of which proliferates to form a new colony (f). D, Life cycle of a trachyline hydrozoan medusa without a polypoid stage (*Aglaura*). After fertilization, a gonochoristic adult (a) releases a planula larva (b), which adds a mouth and tentacles (c) to become an actinula larva (d). Subsequently the actinula larva becomes a young medusa (e). E, Life cycle of a trachyline hydrozoan with a polypoid stage, the freshwater *Limnocnida*. Gonochoristic medusae (a) release fertilized eggs (b) that grow into planula larvae (c). Planula larvae settle to form small hydroid colonies (d), which bud off new medusae (e). (From Bayer and Owre 1968; photo courtesy of G. McDonald.)

Scyphozoan reproduction. The asexual form of scyphozoan cnidarians is a small polyp called the **scyphistoma** (= **scyphopolyp**; Figures 40 and 41). It may produce new scyphistomae by budding from the column wall or from stolons. At certain times of the year, generally in the spring, medusae are produced by repeated transverse fission of the scyphistoma, a

Ectoderm   Entoderm

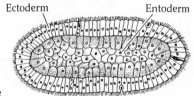

Figure 38

A typical solid hydrozoan planula larva resulting from ingression.

Figure 39

The hollow planula larva of the hydroid *Gonothyraea* (longitudinal section). (From Bayer and Owre 1968.)

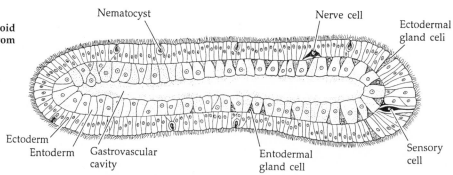

Nematocyst

Nerve cell

Ectodermal gland ceII

Ectoderm

Entoderm

Gastrovascular cavity

Entodermal gland cell

Sensory cell

process called **strobilation**. During this process the polyp is known as a **strobila**. Medusae may be produced one at a time (**monodisc strobilation**), or numerous immature medusae may stack up like soup bowls and then be released singly as they mature (**polydisc strobilation**). Immature and newly released medusae are called **ephyrae**. An individual scyphistoma may survive only one strobilation event, or it may persist for several years, releasing ephyrae annually.

Ephyrae are very small animals with characteristically incised bell margins (Figure 40B). The ephyral arms, or primary tentacles, mark the position of the adult lappets and rhopalia. Maturation involves growth between these arms to complete the bell. Development into sexually mature adult scyphomedusae takes a few months to a few years, depending on the species.

The gonads in adult scyphomedusae are always borne on the gastrodermis, usually on the floor of the gastric pouches, and gametes are generally released through the mouth. Most species are gono-

choristic. Fertilization takes place in the open sea or in the gastric pouches of the female. Cleavage and blastula formation are similar to the processes in hydrozoans. Gastrulation is by ingression or invagination, and results in a mouthless, double-layered planula larva; when invagination occurs, the blastopore closes. The planula larva eventually settles and grows into a new scyphistoma.

The medusa phase clearly dominates the life cycles of most scyphozoans. The polyp stage is often significantly suppressed or dispensed with altogether. For example, many pelagic scyphomedusae have eliminated the scyphistoma, and the planula larva transforms directly into a young medusa (e.g., *Atolla*, *Pelagia*, *Periphylla*). In others, the larvae are brooded, developing in cysts on the parent medusa's body (e.g., *Chrysaora*, *Cyanea*). A few genera have branching colonial scyphistomae with a supportive skeletal tube and abbreviated medusoid stage (e.g., *Nausithoe*, *Stephanoscyphus*). In none, however, has the medusoid stage been lost altogether. Some scyphozoan life cycles are shown in Figure 41.

A

B

Figure 40

Scyphistoma (A), and strobila releasing ephyra (B) of *Aurelia* (order Semaeostomae, class Scyphozoa). (Photos by C. E. Mills, Univ. of Washington/BPS.)

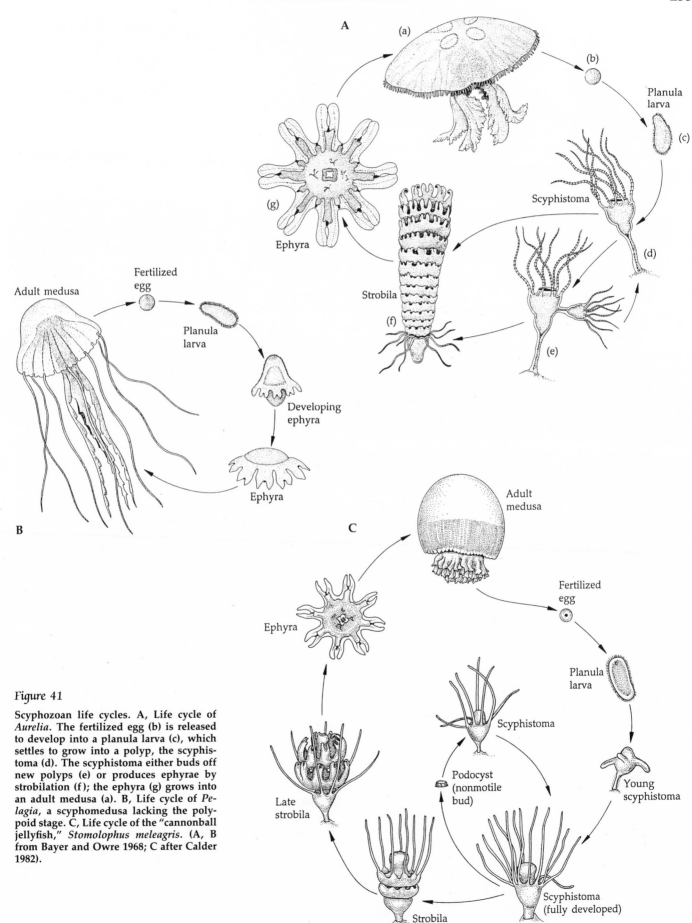

Figure 41

Scyphozoan life cycles. A, Life cycle of *Aurelia*. The fertilized egg (b) is released to develop into a planula larva (c), which settles to grow into a polyp, the scyphistoma (d). The scyphistoma either buds off new polyps (e) or produces ephyrae by strobilation (f); the ephyra (g) grows into an adult medusa (a). B, Life cycle of *Pelagia*, a scyphomedusa lacking the polypoid stage. C, Life cycle of the "cannonball jellyfish," *Stomolophus meleagris*. (A, B from Bayer and Owre 1968; C after Calder 1982).

Anthozoan reproduction. Evolution in this class has so emphasized the polyp form that the medusoid stage has disappeared entirely (if it was ever present in this group). Asexual reproduction is common in anthozoan polyps. **Longitudinal fission** can result in large groups, or clones, of genetically identical individuals (e.g., seen in some species of *Anthopleura*, *Diadumene*, and *Metridium*), as can the less common process of **pedal laceration** (e.g., seen in some acontiate anemones: *Diadumene*, *Haliplanella*, *Metridium*). In the latter phenomenon, the pedal disc spreads, and the anemone simply moves away, leaving behind small fragments from the disc, each of which develops into a young anemone. In addition to these methods of asexual reproduction, at least one scleractinian coral, *Pocillopora damicornis*, produces planula larvae parthenogenetically and broods them until release (Stoddart 1983).

Little is known about the reproductive biology of most alcyonarians and ceriantipatharians. Zoantharians (anemones and corals) may be gonochoristic or hermaphroditic. The gonads are borne on the gastrodermis of all or only some mesenteries. Eggs are fertilized either in the coelenteron, followed by early development in the gut chambers, or more commonly outside the body, in the sea. The northeast Pacific anemone *Aulactinia incubans* can release its brooded young through a pore at the tip of each tentacle! Some corals undergo internal fertilization and early development, and then release planula larvae. One soft coral, *Parerythropodium fulvum*, broods its embryos in a mucous coat on the body surface; then the planula larvae escape. Others shed their gametes and rely on external fertilization. Some coral planula larvae are long-lived, spending several weeks in the plankton, an obvious means of dispersal.

Other corals release benthic planulae that crawl away from the parent and settle nearby. Recently, corals have been shown to undergo synchronous spawning over large areas on reefs. In some cases this synchrony is restricted to colonies of a single species, but at least one study has revealed the synchronous spawning of 105 *different* coral species on the Great Barrier Reef (Babcock et al. 1986).

Sagartia troglodytes is the only anemone known to copulate. The coupling starts when a female glides up to a receptive male, whereupon their pedal discs are pressed together in such a way as to create a chamber into which the gametes are shed and fertilization occurs. The copulatory position is maintained for several days, presumably until planula larvae have developed. It has been suggested that this behavior is an adaptation to areas of great water movement that might otherwise scatter gametes and reduce the probability of successful fertilization.

Cleavage in anthozoan embryos is radial and holoblastic, resulting in a coeloblastula that undergoes gastrulation by ingression or, more frequently, by invagination to form a ciliated planula larva. When invagination occurs, the blastopore remains open and sinks inward, drawing with it a tube of ectoderm that becomes the pharynx. Most anthozoan planula larvae are planktotrophic, unlike the nonfeeding lecithotrophic larvae of other cnidarians. The ability of these larvae to feed allows a potentially longer larval life, enhancing dispersal for anthozoans. As development proceeds, septa begin to grow in couples on opposite sides. A planula larva bearing eight complete septa is sometimes called an **edwardsia stage**, named after the octaseptate genus *Edwardsia*. The larva eventually settles on its aboral end and tentacles grow around the upwardly directed mouth and oral disc. A typical anthozoan life cycle is shown in Figure 42.

Reproduction in cubomedusae. The biology of this class is not yet well known and the polyps of only a few species have been described. Apparently, each polyp metamorphoses directly into a single medusa, rather than undergoing strobilation as scyphozoan polyps do. Some cubozoan medusae engage in a form of copulation, in which sperm are transferred directly from the male to an adjacent female in the water column.

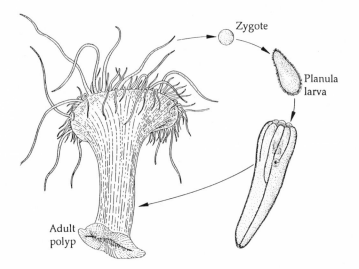

Figure 42

Anthozoan sexual life cycle. An adult polyp releases gametes, which fuse, or fertilized eggs, each of which develops into a planula larva; the larva settles and transforms directly into a young polyp. (From Bayer and Owre 1968.)

Cnidarian phylogeny

The origin of the Cnidaria

The cnidarians have one of the longest fossil histories among the metazoa. The first documented cnidarian fossils are from the famous Ediacara Hills of South Australia, which contain several kinds of medusae and sea pens that lived about 700 million years ago. The only suggestions of metazoan life preceding the appearance of cnidarians are questionable trace fossils. Indeed, the origin of the cnidarians is intimately tied to the origin of the metazoa themselves. (Theories on the origin of the metazoa are discussed in Chapter 4.) The colonial theory depicts a colonial flagellated protozoon giving rise to a hollow metazoan ancestor, termed a blastea, which in turn gave rise to a diploblastic planuloid animal called a gastrea. On the other hand, the syncytial theory of the origin of the metazoa implies that the ancestors of the cnidarians were triploblastic, acoelomate organisms, perhaps something like rhabdocoel turbellarians, that underwent "degenerative evolution" to produce what we recognize today as the cnidarians. This view, sometimes called the **turbellarian theory**, holds the Anthozoa to be the most primitive cnidarian class and cites the "remnants" of bilateral symmetry in that class as evidence of a bilateral ancestry.

The turbellarian theory is not generally promoted by most contemporary zoologists, and we find it weak on several counts. "Degenerative evolution" (a poor choice of words perhaps) is a phenomenon primarily associated with the evolution of parasites and of the exploitation of smallness (e.g., interstitial forms), wherein it may result in the loss of certain systems and the specialized adaptive development of others. General loss of fundamental body architecture in other kinds of free-living animals seems to be an unlikely event. Thus, we view the idea of a free-living, triploblastic, bilateral, motile flatworm taking up a sessile existence and transforming into a radially symmetrical, diploblastic, anthozoan polyp to be an unlikely evolutionary scenario. The adoption of radiality (or at least "functional" radiality) of bilaterally symmetrical animals is well documented in some taxa (e.g., Echinodermata), but does not involve the kinds of "degeneration" required by the turbellarian theory. To us the transformation suggested by the turbellarian theory simply involves the loss or drastic simplification of too many complex systems, and major changes in fundamental design. The case against an anthozoan ancestral form is also supported by arguments favoring radial symmetry as the primitive

feature of the phylum Cnidaria. Both larvae and adults of extant cnidarians maintain a basic radial symmetry. The so-called remnant bilaterality of anthozoan polyps is not true bilaterality at all, but biradiality about an oral–aboral axis, which develops late in the ontogeny of these animals. Such features are thus best viewed as derived attributes within the Cnidaria. The turbellarian theory is also weak on other embryological grounds, such as differences in cleavage patterns and germ layer formation. Finally, analysis *within* the phylum Cnidaria (see below) provides strong evidence that the class Anthozoa is a derived, not a primitive, group.

There are several competing theories concerning the nature of the ancestral cnidarian; some of these ideas focus on whether the first cnidarian was polypoid or medusoid in form. One popular view is called the **medusa theory**. If the Cnidaria did arise from a swimming or creeping, flagellated or ciliated, planuloid ancestor, then the development of tentacles could have produced an animal resembling an actinula larva (Figure 37C). The transition from planula to actinula to the modern medusa form can be seen today in the life cycle of certain hydrozoans (e.g. order Trachylina). Asexual reproduction, such as budding, by a benthic actinula larva could have led to the establishment of a distinct polypoid stage. If so, the polyp can be viewed as an extended larval form specialized for asexual reproduction and benthic existence. A likely scenario is that once the polypoid form became established, some cnidarians began to suppress the medusoid phase of their life cycles, various degrees of which can be seen among the hydrozoans. The epitome of this trend is seen in the class Anthozoa, the members of which have lost the medusa stage altogether.

Some zoologists hold the polyp to be the original cnidarian body form; they view the medusa as a derived dispersal stage that could have evolved independently among the hydrozoans and the scyphozoans. We view this idea as an unnecessarily complicated hypothesis with little supporting evidence. The fact that the medusa produces the gametes when both forms are present supports the concept that the medusoid form is ancestral.

Several lines of evidence suggest that the Hydrozoa is the oldest living cnidarian class. They are structurally less complex, and less specialized overall than other cnidarians. If the medusa theory is valid, then the hydrozoan order Trachylina appears to be the most primitive group. The life cycle is dominated by a relatively simple medusoid form and the polyp stage is absent. Gastrulation in this group is by ingression, yielding solid stereogastrula. The precnidarian may have been a solid-bodied, ciliated, planuloid form.

Evolution within the Cnidaria

The lack of unique synapomorphies for the hydrozoan line (Figure 43) implies that the ancestral cnidarian may have been what we would today classify as a member of the class Hydrozoa. Subsequent to the appearance of the first medusoid cnidarian (perhaps resembling modern trachylines), a polyp phase evolved, and the ancestral form began to experience an alternation between the two life history stages. From this ancestral line evolved a group of cnidarians with increased specializations of the mesenchyme (cellularity), gastrodermis and gastrovascular system, and nervous system. Among these events were the movement of the gonadal tissue to the gastrodermis, formation of septa subdividing the coelenteron, and the origin of internal nematocyst-bearing filaments. In addition to allowing greater specialization of the digestive cavity, these changes set the stage for an increased size in individuals rather than for colonies of small zooids. These and other evolutionary processes resulted in an animal that we envision as ancestral to the two major lineages we recognize today as the Anthozoa and the Scyphozoa/Cubozoa. The resemblance of the scyphistoma to anthozoan polyps also supports the idea of a common ancestry for the scyphozoans and anthozoans (Figure 43).

As the two lines diverged, one emphasized the medusoid form and the polyp stage was greatly reduced or lost. Emphasis on the medusoid form favored the development of complex sensory units, the rhopalia. The main clade of this medusa-dominated lineage resulted in the Scyphozoa, the members of which tend towards large body size and a fully pelagic life style. The retention of sessile polyps (scyphistomae) by many species was eventually accom-

Figure 43

Evolution of the Cnidaria. A, Cladogram depicting the geneology, or sequence of orgin of the cnidarian classes (see text for discussion). The unique cnidarian attributes (synapomorphies) indicated at the base of the cladogram include evolution of the radial, medusoid body form; cnidae; planula larvae; and cnidarian coelenteron with mouth surrounded by tentacles. Structures such as the cnidarian simple nerve net system and simple epitheliomuscular cells may be cnidarian synapomorphies, or may be primitive features retained from an earlier ancestor (i.e., symplesiomorphies). Numbered synapomorphies on the cladogram are as follow: (1) evolution of the polypoid body form and "alternation of generations;" (2) gonads relocated in gastrodermis; (3) septa appear, subdividing the coelenteron; (4) the polyp stage is secondarily reduced or lost; (5) evolution of the rhopalium; (6) evolution of strobilation; (7) acquisition of a boxlike medusa body; (8) evolution of complex lensed rhopalial eyes; (9) invention of velarium; (10) complete suppression of the medusoid stage; (11) development of hexaradial and octaradial symmetry; (12) evolution of the anthozoan actinopharynx; (13) evolution of the siphonoglyph; (14) coelenteron acquires mesenterial filaments; (15) loss of the cnidal operculum; (16) loss of the cnidocil; (17) evolution of tripartite series of flaps on cnidae; (18) evolution of special ciliary cones associated with cnidae. B, A generalized evolutionary tree depicting the origin and subsequent evolution of the Cnidaria. This evolutionary tree is currently the one most widely accepted; it is in agreement with the cladogram.

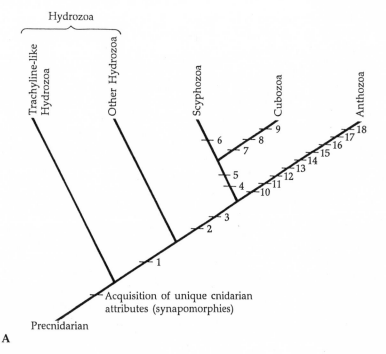

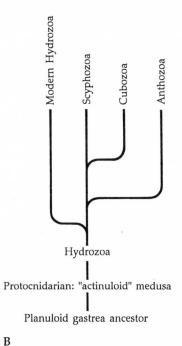

panied by the evolution of a unique form of budding, strobilation, as a means of asexually producing new medusae. A side branch from this medusoid line led to the Cubozoa, characterized by the evolution of a suite of distinguishing features, including the boxlike shape, complex lensed rhopalial eyes, and the velarium (Figure 43), the latter a convergence to the velum of certain hydromedusae.

The other main line diverging from the early hydrozoans led to the anthozoans through reduction and eventual loss of the medusoid form. The anthozoans are highly specialized and demarcated in our cladogram (Figure 43) by several unique synapomorphies: complete loss of the medusoid stage; development of hexaradial or octaradial symmetry (transformed into biradial symmetry in most); evolution of an actinopharynx, siphonoglyphs, and unique mesenterial filaments in the coelenteron; loss of the cnidal operculum and cnidocil; evolution of the tripartite flaps on the cnidae; and evolution of special ciliary cones associated with the cnidocytes. Other noteworthy trends from the ancestral hydrozoan condition to the anthozoans are progressively more complex gastrovascular systems, and a marked increase in the degree of cellularity of the mesenchyme.

Certainly other characters exist that can be used to describe the genealogy and evolution of the cnidarian classes. But even these obvious features provide a tenable theory of phylogeny in this phylum, as depicted in the cladogram and conventional evolutionary tree (Figure 43). Phylogeny within each of the cnidarian classes is equally interesting but largely beyond the scope of this text. However, a few generalizations can be made about some important events. Colonialism in the hydrozoans probably arose by retention of young polyps during asexual reproduction, and this development ultimately led to the highly specialized colonial orders such as the Siphonophora, Chondrophora, Milleporina and Stylasterina. In the class Scyphozoa, evolution has clearly favored increasing specialization of the pelagic medusoid form and diminishing importance of the polypoid stage in their life cycle. Scyphomedusae have evolved large size, special musculature, a highly cellular or fibrous mesenchyme, a complex gastrovascular system, and a sophisticated sensory system. Among members of the class Anthozoa, evolution has eliminated the medusa stage and produced a grand series of experiments in colonial polypoid living, resulting in such "super-organisms" as corals, gorgonians, pennatulaceans, and zoanthidians. An increase in polyp size has occurred with the evolution of complex structural components of the mesenchyme and with efficient musculature. These cnidarians have also exploited the commensal relationship with zooxanthellae to a greater degree than members of the other classes. Convergent evolution has occurred frequently throughout the Cnidaria, as witnessed by such features as colonialism, calcareous skeletons, the velum-velarium structures, and various means of suppressing the medusoid or polypoid stage in the life cycle.

Selected References

General

Bayer, F. M. and H. B. Owre. 1968. *The Free-Living Lower Invertebrates*. Macmillan, New York.

Benson, A. A. and R. F. Lee. 1975. The role of wax in oceanic food chains. Sci. Am. 232(3):76–86.

Blank, R. J. and R. K. Trench. 1985. Speciation and symbiotic dinoflagellates. Science 229:656–658.

Boardman, R. S., A. H. Cheetham and W. A. Oliver (eds.). 1973. *Animal Colonies*. Dowden, Hutchinson and Ross, Stroudsburg, Pennsylvania.

Bullock, T. H. 1965. Coelenterata and Ctenophora. In T. H. Bullock and G. A. Horridge (eds.), *Structure and Function in the Nervous System of Invertebrates*, Vol. 1. W. H. Freeman, San Francisco, pp. 459–534.

Campbell, R. D. 1974. Cnidaria. In A. C. Giese and J. S. Pearse (eds.), *Reproduction of Marine Invertebrates*, Vol. 1. Academic Press, New York, pp. 133–200.

Carre, D. 1980. Hypothesis on the mechanism of cnidocyst discharge. J. Cell Biol. 20:265–271.

Cheng, L. 1975. Marine pleuston—animals at the sea–air interface. Oceanogr. Mar. Biol. Annu. Rev. 13:181–212.

Chia, F. S. and L. R. Bickell. 1978. Mechanisms of larval attachment and the induction of settlement and metamorphosis in coelenterates: A review. In F. S. Chia and M. Rice (eds.), *Settlement and Metamorphosis of Marine Invertebrate Larvae*. Elsevier/North-Holland Biomedical Press, New York.

Conklin, E. J. and R. N. Mariscal. 1977. Feeding behavior, ceras structure, and nematocyst storage in the aeolid nudibranch, *Spurilla neapolitana* (Mollusca). Bull. Mar. Sci. 27(4):658–667.

Crowell, S. (ed.). 1965. Behavioral physiology of coelenterates. Am. Zool. 5:335–589.

Devaney, D. M. and L. G. Eldredge (eds.). 1977. *Reef and Shore Fauna of Hawaii*. Section 1: Protozoa through Ctenophora. Bernice P. Bishop Museum Special Publ. 64(1). (Bishop Museum Press, Honolulu, Hawaii.)

Dunn, D. F. 1982. Cnidaria. In S. P. Parker (ed.), *Synopsis and Classification of Living Organisms*, Vol. 1. McGraw-Hill, New York, pp. 669–706.

Elder, H. Y. 1973. Distribution and functions of elastic fibers in the invertebrates. Biol. Bull. 144:43–63.

Fautin, D. G. 1988. Importance of nematocysts to actinian taxonomy. In D. A. Hessinger and H. M. Lenhoff (eds.), *The Biology of Nematocysts*. Academic Press, New York, pp. 487–500.

Florkin, M. and B. T. Scheer (eds.). 1968. *Chemical Zoology*, Vol. 2, Porifera, Coelenterata, and Platyhelminthes. Academic Press, New York, pp. 81–284.

Gladfelter, W. B. 1973. A comparative analysis of locomotor system of medusoid Cnidaria. Helgol. Wiss. Meeresunters. 25:228–272.

Hand, C. 1959. On the origin and phylogeny of the coelenterates. Syst. Zool. 8:191–202.

Hessinger, D. A. and H. M. Lenhoff (eds.). 1988. *The Biology of Nematocysts.* Academic Press, London. [Cutting-edge stuff on many aspects of nematocyst structure and function.]

Hinsch, G. W. 1974. Comparative ultrastructure of cnidarian sperm. Am. Zool. 14:457–465.

Holstein, T. and P. Tardent. 1983. An ultrahigh-speed analysis of exocytosis: Nematocyst discharge. Science 223:830–833.

Hyman, L. H. 1940. *The Invertebrates,* Vol. 1, Protozoa through Ctenophora. McGraw-Hill, New York.

Hyman, L. H. 1959. Coelenterata. *In* W. T. Edmondson, H. B. Ward and G. C. Whipple (eds.), *Freshwater Biology,* 2nd Ed. Wiley, New York, pp. 313–344.

Kramp, P. L. 1957. On development through alternating generations, especially in Coelenterata. Vidensk. Medd. Dan. Naturhist. Foren. Khobenhavn. 107:13–32.

Kramp, P. L. 1961. Synopsis of the medusae of the world. J. Mar. Biol. Assoc. U.K. 40:1–469. [No keys or figures, but a landmark taxonomic listing; excellent bibliography.]

Lenhoff, H. M. and L. Muscatine (eds.). 1971. *Experimental Coelenterate Biology.* University of Hawaii Press, Honolulu.

Mackie, G. O. (ed.). 1976. *Coelenterate Ecology and Behavior.* Plenum Press, New York. [Although over 10 years old, still one of the best treatments of the subject.]

Mackie, G. O. and L. M. Passano. 1968. Epithelial conduction in hydromedusae. J. Gen. Physiol. 52;600–608.

Mariscal, R. N., C. H. Bigger and R. B. McLean. 1976. The form and function of cnidarian spirocysts. 1. Ultrastructure of the capsule exterior and relationship to the tentacle sensory surface. Cell Tissue Res. 168:465–474.

Mariscal, R. N. and R. B. McLean. 1976. The form and function of cnidarian spirocysts. 2. Ultrastructure of the capsule tip and wall and mechanism of discharge. Cell Tissue Res. 169:313–321.

Mariscal, R. N., R. B. McLean and C. Hand. 1977. The form and function of cnidarian spirocysts. 3. Ultrastructure of the thread and function of spirocysts. Cell Tissue Res. 178:427–433.

Miller, R. L. and C. R. Wyttenbach (eds.). 1974. The developmental biology of the Cnidaria. Am. Zool. 14:440–866.

Moore, R. C. (ed.). 1956. *Treatise on Invertebrate Paleontology. Coelenterata,* Vol. F. Geol. Soc. Am. and University of Kansas Press, Lawrence.

Moore, R. E. and P. J. Scheuer. 1971. Palytoxin: A new marine toxin from a coelenterate. Science 172:495–498.

Muscatine, L. and H. M. Lenhoff (eds.). 1974. *Coelenterate Biology. Reviews and New Perspectives.* Academic Press, New York. [A highly useful volume with excellent reviews of histology, skeletal systems, cnidae, development, symbiosis, and bioluminescence.]

Muscatine, L., R. R. Pool and R. K. Trench. 1975. Symbiosis of algae and invertebrates: Aspects of the symbiont surface and the host–symbiont interface. Trans. Am. Microsc. Soc. 94(4):450–469.

Nielsen, C. 1987. *Haeckelia* (= *Euchlora*) and *Hydroctena* and the phylogenetic interrelationships of Cnidaria and Ctenophora. Z. Zool. Syst. Evolutionsforsch. 25:9–12.

Rees, W. J. (ed.). 1966. *The Cnidaria and Their Evolution.* Zool. Soc. Lond. Symp. No. 16, Academic Press, London. [A benchmark publication with major papers by most specialists on the subject of cnidarian evolution.]

Russell, F. S. 1954, 1970. *Medusae of the British Isles,* Vols. 1 and 2. Cambridge University Press, London.

Sebens, K. P. 1979. The energetics of asexual reproduction and colony formation in benthic marine invertebrates. Am. Zool. 19:683–697.

Shimomura, O. and F. H. Johnson. 1975. Chemical nature of bioluminescence systems in coelenterates. Proc. Natl. Acad. Sci. USA 72:1546–1549.

Tardent, P. and R. Tardent (eds.). 1980. *Developmental and Cellular Biology of Coelenterates.* Elsevier/North-Holland Biomedical Press, New York.

Taylor, D. L. 1973. The cellular interactions of algal–invertebrate symbioses. Adv. Mar. Biol. 11:1–56.

Trench, R. K. 1979. The cell biology of plant–animal symbioses. Annu. Rev. Plant Physiol. 30:485–531.

Tursch, B., J. C. Braeckman, D. Daloze and M. Kaisin. 1978. Terpenoids from coelenterates. *In* P. Scheuer (ed.), *Marine Natural Products,* Vol. 2. Academic Press, New York, pp. 347–396.

Watson, G. M. and D. A. Hessinger. 1989. Cnidocyte mechanoreceptors are tuned to the movements of swimming prey by chemoreceptors. Science 243:1589–1591.

Westfall, J. A. 1973. Ultrastructural evidence for neuromuscular systems in coelenterates. Am. Zool. 13:237–246.

Wyttenbach, C. R. (ed.). 1974. The developmental biology of the Cnidaria. Am. Zool. 14(2):540–866. [Papers presented at a 1972 symposium.]

Hydrozoa

Alvariño, A. 1978. *Nectocarmen antonioi,* a new Prayinae, Calycophorae, Siphonophorae, from California. Proc. Biol. Soc. Washington 96:339–348.

Arai, M. N. and A. Brinckmann-Voxx. 1980. The hydromedusae of British Columbia and Puget Sound, Canada. Can. Bull. Fish. Aquat. Sci. 204:1–192.

Bieri, R. 1970. The food of *Porpita* and niche separation in three neuston coelenterates. Publ. Seto Mar. Biol. Lab. 27:305–307.

Biggs, D. C. 1977. Field studies of fishing, feeding and digestion in siphonophores. Mar. Behav. Physiol. 4:261–274.

Bellamy, N. and M. J. Risk. 1982. Coral gas: Oxygen production in *Millepora* on the Great Barrier Reef. Science 215:1618–1619.

Benos, D. J. and R. D. Prusch. 1972. Osmoregulation in fresh-water hydra. Comp. Biochem. Physiol. [A], 43:165–171.

Bonnet, D. D. 1946. The portuguese man-of-war as a food source for the sand crab (*Emerita pacifica*). Science 103:148–149.

Burnett, A. L. (ed.). 1973. *Biology of Hydra.* Academic Press, New York.

Cairns, S. D. and J. L. Barnard. 1984. Redescription of *Janaria mirabilis,* a calcified hydroid from the eastern Pacific. Bull. South. Calif. Acad. Sci. 83:1–11.

Cornelius, P. F. S. 1977. The linking of polyp and medusa stages in *Obelia* and other coelenterates. Biol. J. Linn. Soc. 9:45–57.

Cornelius, P. F. S. 1975. The hydroid species of *Obelia,* with notes on the medusa stage. Bull. Br. Mus. Nat. Hist. (Zool.) 28(6):249–294.

Cornelius, P. F. S. 1982. Hydroids and medusae of the family Campanulariidae recorded from the eastern North Atlantic, with a world synopsis of genera. Bull. Br. Mus. Nat. Hist. (Zool.) 42(2):37–148.

Eakin, R. M. and J. A. Westfall. 1962. Fine structure of photoreceptors in the hydromedusan, *Polyorchis penicillatus.* Proc. Natl. Acad. Sci. USA 48:826–833.

Edwards, C. 1966. *Velella velella* (L.): The distribution of its dimorphic forms in the Atlantic Ocean and the Mediterranean, with comments on its nature and affinities. *In* H. Barnes (ed.), *Some Contemporary Studies in Marine Science.* Allen and Unwin, London, pp. 283–296.

Fields, W. G. and G. O. Mackie. 1971. Evolution of the Chondrophora: Evidence from behavioural studies on *Velella.* J. Fish. Res. Bd. Can. 28:1595–1602.

Francis, L. 1985. Design of a small cantilevered sheet: The sail of *Velella velella*. Pac. Sci. 39(1):1–15.

Fraser, C. 1954. *Hydroids of the Atlantic Coast of North America*. University of Toronto Press, Toronto.

Freeman, G. 1983. Experimental studies on embryogenesis in hydrozoans (Trachylina and Siphonophora) with direct development Biol. Bull. 165:591–618.

Gierer, Z. 1974. *Hydra* as a model for the development of biological form. Sci. Am. 231(6):44–54.

Lane, C. E. 1960. The Portuguese man-of-war. Sci. Am. 202:158–168.

Lenhoff, H. M. 1983. *Hydra: Research Methods*. Plenum, New York.

Lenhoff, H. M. and W. F. Loomis (eds.). 1961. *The Biology of Hydra and of Some Other Coelenterates: 1961*. University of Miami Press, Coral Gables, Florida.

Lentz, T. L. 1966. *The Cell Biology of Hydra*. Wiley, New York.

Mackie, G. O. 1959. The evolution of the Chondrophora (Siphonophora: Disconanthae): New evidence from behavioral studies. Trans. R. Soc. Can. 53:7–20.

Mackie, G. O. 1960. The structure of the nervous system in *Velella*. Q. J. Microsc. Sci. 101:119–133.

Martin, W. E. 1975. *Hydrichthys pietschi*, new species (Coelenterata) parasitic on the fish, *Ceratias holboelli*. Bull. South. Calif. Acad. Sci. 74:1–6. [A parasitic hydroid from California.]

Naumov, D. V. 1960. Hydroids and hydromedusae of the USSR. Opred. Faune SSSR, No. 70:1–785. Translated from the Russian text (1960) by Israel Program for Scientific Translations, Jerusalem. [Keys to the fauna of the USSR published by the Zoological Institute of the Academy of Sciences of the USSR, No. 70.]

Ostarello, G. L. 1973. Natural history of the hydrocoral *Allopora californica* Verrill (1866). Biol. Bull. 145:548–564.

Pardy, R. L. and B. N. White. 1977. Metabolic relationships between green hydra and its symbiotic algae. Biol. Bull. 153:228–236.

Purcell, J. E. 1980. Influence of siphonophore behavior upon their natural diets: Evidence for aggressive mimicry. Science 209:1045–1047.

Purcell, J. E. 1984. The functions of nematocysts in prey capture by epipelagic siphonophores (Coelenterata, Hydrozoa). Biol. Bull. 166:310–327.

Rees, W. J. 1957. Evolutionary trends in the classification of capitate hydroids and medusae. Bull. Br. Mus. Nat. Hist. (Zool.) 4:453–534.

Singla, C. L. 1975. Statocysts of hydromedusae. Cell Tissue Res. 158:391–407.

Stretch, J. J. and J. M. King. 1980. Direct fission: An undescribed reproductive method in hydromedusae. Bull. Mar. Sci. 30:522–526.

Totton, A. K. 1960. Studies on *Physalia physalis* (L.). Part 1, natural history and morphology. Discovery Rpt. 30:301–367.

Vervoort, W. 1968. Report on a collection of Hydroida from the Caribbean region, including an annotated checklist of Caribbean hydroids. Zool. Verh. Rijksmus. Nat. Hist. Leiden 92:3–124.

Wahle, C. M. 1980. Detection, pursuit, and overgrowth of tropical gorgonians by milleporid hydrocorals: Perseus and Medusa revisited. Science 209:689–691.

West, D. A. 1978. The epithelio-muscular cell of hydra: Its fine structure, three-dimensional architecture and relationship to morphogenesis. Tissue Cell 10:629–646.

Scyphozoa

Alexander, R. M. 1964. Visco-elastic properties of the mesoglea of jellyfish. J. Exp. Biol. 41:363–369.

Anderson, P. A. V. and G. O. Mackie. 1977. Electrically coupled photosensitive neurons control swimming in jellyfish. Science 197:186–188.

Berrill, M. 1963. Comparative functional morphology of the Stauromedusae. Can. J. Zool. 41:741–752.

Calder, D. R. 1971. Nematocysts of *Aurelia*, *Chrysaora* and *Cyanea* and their utility in identification. Trans. Am. Microscop. Soc. 90:269–274.

Calder, D. R. 1982. Life history of the cannonball jellyfish, *Stomolophus meleagris* L. Agassiz, 1860 (Scyphozoa, Rhizostomida). Biol. Bull. 162:149–162.

Horridge, A. 1954. Observations on the nerve fibers of *Aurelia aurita*. Q. J. Microsc. Sci. 95:85–92.

Horridge, A. 1956. The nervous system of the ephyra larva of *Aurelia aurita*. Q. J. Microsc. Sci. 97:59–73.

Larson, R. J. 1976. Marine flora and fauna of the northeastern United States. Cnidaria: Scyphozoa. NOAA Tech. Rpt. NMFS Circ. No. 397.

Mayer, A. G. 1910. *The Medusae of the World. I,II, Hydromedusae. III, Scyphomedusae*. Carnegie Inst. Washington Publ. 109. [Reprinted in 1977 by A. Asher, Amsterdam.]

Moller, H. 1984. Reduction of a larval herring population by jellyfish predator. Science 224:621–622.

Uchida, T. 1973. The systematic position of the Stauromedusae. Publ. Seto Mar. Biol. Lab. 20:133–139.

Cubozoa

Conant, F. S. 1900. The Cubomedusae. Mem. Biol. Lab. Johns Hopkins University 4(1):1–52.

Pearse, J. S. and B. B. Pearse. 1978. Vision in cubomedusan jellyfishes. Science 199:458.

Anthozoa

Adey, W. H. 1978. Coral reef morphogenesis: A multidimensional model. Science 202:831:837.

Anderson, P. A. V. and J. F. Case. 1975. Electrical activity associated with luminescence and other colonial behavior in the pennatulid *Renilla kollikeri*. Biol. Bull. 149:80–95.

Arai, M. N. 1965. The ceriantharian nervous system. Am. Zool. 5:424–429.

Babcock, R., G. Bull, P. Harrison, A. Heyward, J. Oliver, C. Wallace and B. Willis. 1986. Synchronous spawnings of 105 scleractinian coral species on the Great Barrier Reef. Mar. Biol. 90:379–394.

Barham, E. G. and I. E. Davies. 1968. Gorgonians and water motion studies in Gulf of California. Underwater Nat. [Bull. Am. Littoral Soc.] 5(3):24–28, 42.

Batham, E. J. 1960. The fine structure of epithelium and mesoglea in a sea anemone. Quart. J. Microscop. Sci. 101:481–485.

Batham, E. J., C. F. A. Pantin and E. A. Robson. 1960. The nerve-net of the sea-anemone *Metridium senile*: The mesenteries and the column. Q. J. Microsc. Sci. 101:487–510.

Bayer, F. M. 1961. *The Shallow-Water Octocorallia of the West Indian Region: A Manual for Marine Biologists*. Martinus Nijhoff, The Hague.

Bayer, F. M. 1981. Key to the genera of Octocorallia exclusive of Pennatulacea (Coelenterata: Anthozoa), with diagnoses of new taxa. Proc. Biol. Soc. Wash. 94:902–947.

Bayer, F. M., M. Grasshoff and J. Verseveldt (eds.). 1983. *Illustrated Trilingual Glossary of Morphological and Anatomical Terms Applied to Octocorallia*. E. J. Brill, Leiden.

Benayahu, Y. and Y. Loya. 1983. Surface brooding in the Red Sea soft coral *Parerythropodium fulvum fulvum* (Forskål, 1775). Biol. Bull. 165:353–369.

Bigger, C. H. 1980. Interspecific and intraspecific acrorhagial aggressive behavior among sea anemones: A recognition of self and not-self. Biol. Bull. 159:117–134.

Birkeland, C. 1974. Interactions between a sea pen and seven of its predators. Ecol. Monogr. 44(2):211–232.

Burnett, J. W. and J. S. Sutton. 1969. The fine structural organization of the sea nettle fishing tentacles. J. Exp. Zool. 172:335–348.

Cairns, S. D. 1977. *Guide to the Commoner Shallow–Water Gorgonians (Sea Whips, Sea Feathers and Sea Fans) of Florida, The Gulf of Mexico and The Caribbean Region*. University of Miami Sea Grant Program, Field Guide Ser., No. 6.

Cairns, S. D. 1981. *Marine Flora and Fauna of the Northeastern United States. Scleractinia*. NOAA Tech. Rept. NMFS Circ. No. 438.

Carlgren, O. 1949. A survey of the Ptychodactiaria, Corallimorpharia, and Actiniaria. K. Sven. Vetenskapsakad. Handl. 1(4):1–121. [Lists and keys to the species of sea anemones known at the time.]

Coffroth, M. A. 1984. Ingestion and incorporation of coral mucus aggregates by a gorgonian soft coral. Mar. Ecol. Prog. Ser. 17:193–199.

Conklin, E. J., C. H. Bigger and R. N. Mariscal. 1977. The formation and taxonomic status of the microbasic Q-mastigophore nematocyst of sea anemones. Biol. Bull. 152:159–168.

Dana, T. F. 1975. Development of contemporary eastern Pacific coral reefs. Mar. Biol. 33:355–374.

Darwin, C. 1842. *The Structure and Distribution of Coral Reefs*. Reprinted in 1984 by the University of Arizona, Tucson.

Ducklow, H. W. and R. Mitchell. 1979. Composition of mucus released by coral reef coelenterates. Limnol. Oceanogr. 24(4):706–714.

Dunn, D. F. 1975. Reproduction of the externally brooding sea anemone *Epiactis prolifera* Verrill, 1869. Biol. Bull. 148:199–218.

Dunn, D. F. 1977. Dynamics of external brooding in the sea anemone *Epiactis prolifera*. Mar. Biol. 39:41–49.

Dunn, D. F. 1981. The clownfish sea anemones: Stichodactylidae (Coelenterata: Actinaria) and other sea anemones symbiotic with pomacentrid fishes. Trans. Am. Philos. Soc. 71:1–115.

Dunn, D. F. and G. J. Bakus. 1977. Redescription and ecology of *Liponema brevicornis* (McMurrich, 1893), with definition of the family Liponematidae (Coelenterata, Actiniaria). Astarte 10:77–85.

Dunn, D. F., D. M. Devaney and B. Roth. 1980. *Stylobates*: A shell-forming sea anemone (Coelenterata, Anthozoa, Actiniidae). Pac. Sci. 34:379–388.

Durham, J. W. 1947. Corals from the Gulf of California and the North Pacific Coast of America. Geol. Soc. Am. Mem. 20:1–68.

Durham, J. W. and J. L. Barnard. 1952. Stony corals of the eastern Pacific collected by the *Velero III* and *Velero IV*. Allan Hancock Pacific Expeds. 16(1):1–110. [Still one of the best reviews of west American corals.]

Fautin, D. G. 1986. Why do anemonefishes inhabit only some actinians? Environ. Biol. Fishes, 15(3):171–180.

Fenical, W., R. K. Okuda, M. M. Bandurraga, P. Culver and R. S. Jacobs. 1981. Lophotoxin: A novel neuromuscular toxin from Pacific sea whips of the genus *Lophogorgia*. Science 212:1512–1514.

Francis, L. 1973. Clone specific segregation in the sea anemone *Anthopleura elegantissima*. Biol. Bull. 144:64–72.

Francis, L. 1973. Intraspecific aggression and its effect on the distribution of *Anthopleura elegantissima* and some related sea anemones. Biol. Bull. 144:73–92.

Francis, L. 1976. Social organization within clones of the sea anemone *Anthopleura elegantissima*. Biol. Bull. 150:361–376.

Francis, L. 1979. Contrast between solitary and colonial lifestyles in the sea anemone *Anthopleura elegantissima*. Am. Zool. 19:669–681.

Fricke, H. and L. Hottinger. 1983. Coral bioherms below the euphotic zone in the Red Sea. Mar. Ecol. Prog. Ser. 11:113–117.

Gladfelter, E. H. 1983. Circulation of fluids in the gastrovascular system of the reef coral *Acropora cervicornia*. Biol. Bull. 165:619–636.

Glynn, P. W. 1976. Some physical and biological determinants of coral community structure in the eastern Pacific. Ecol. Monogr. 46:431–456.

Glynn, P. W. 1980. Defense by symbiotic Crustacea of host corals elicited by chemical cues from predators. Oecologia 47:287–290.

Glynn, P. W. 1982. Coral communities and their modifications relative to past and prospective Central American seaways. Adv. Mar. Biol. 19:91–132.

Glynn, P. W., G. M. Wellington and C. Birkeland. 1978. Coral reef growth in the Galapagos: Limitation by sea urchins. Science 203:47–49.

Glynn, P. W., G. M. Wellington and J. W. Wells. 1983. *Corals and Coral Reefs of the Galapagos Islands*. University of California Press, Berkeley.

Goreau, T. F. 1959. The physiology of skeleton formation in corals. Biol. Bull. 116:59–75.

Goreau, T. F. 1963. Calcium carbonate deposition by coralline algae and corals in relation to their roles as reef builders. Ann. N.Y. Acad. Sci. 109:127–167.

Grigg, R. W. 1965. Ecological studies of black coral in Hawaii. Pac. Sci. 19(2):244–260.

Grigg, R. W. 1972. Orientation and growth form of the sea fans. Limnol. Oceanogr. 17:185–192.

Grigg, R. W. 1977. Population dynamics of two gorgonian corals. Ecology 58:279–290.

Grimstone, A. V., R. W. Horne, C. F. A. Pantin and E. A. Robson. 1958. The fine structure of the mesenteries of the sea-anemone *Metridium senile*. Q. J. Microscop. Sci. 99:523–540.

Hamner, W. M. and D. F. Dunn. 1980. Tropical Corallimorpharia (Coelenterata: Anthozoa): Feeding by envelopment. Micronesica 16:37–41.

Hand, C. 1954. The sea anemones of central California. I. The corallimorpharian and athenarian anemones. Wasmann J. Biol. 12(3):345–375.

Hand, C. 1955a. The sea anemones of central California. II. The endomyarian and mesomyarian anemones. Wasmann J. Biol. 13(1):37–99.

Hand, C. 1955b. The sea anemones of central California. III. The acontiarian anemones. Wasmann J. Biol. 13(2):189–251.

Howe, N. R. and Y. M. Sheikh. 1975. Anthopleurine: A sea anemone alarm pheromone. Science 189:386–388.

Jennison, B. L. 1979. Gametogenesis and reproductive cycles in the sea anemone *Anthopleura elegantissima*. Can. J. Zool. 57:403–411.

Jones, O. A. and R. Endean (eds.). 1973. *Biology and Geology of Coral Reefs. I. Geology 1*. Academic Press, New York.

Jones, O. A. and R. Endean (eds.). 1974. *Biology and Geology of Coral Reefs. II. Biology 1*. Academic Press, New York.

Jones, O. A. and R. Endean (eds.). 1976. *Biology and Geology of Coral Reefs. III. Biology 2*. Academic Press, New York.

Josephson, R. K. 1974. The strategies of behavioral control in a coelenterate. Am. Zool. 14:905–915.

Josephson, R. K. and S. C. March. 1966. The swimming performance of the sea-anemone *Boloceroides*. J. Exp. Biol. 44:493–506.

Kastendiek, J. 1976. Behavior of the sea pansy *Renilla kollikeri* Pfeffer and its influence on the distribution and biological interactions of the species. Biol. Bull. 151:518–537.

Lang, J. 1973. Interspecific aggression by scleractinian corals: Why the race is not only to the swift. Bull. Mar. Sci. 23:269–279.

Lewis, D. H. and D. C. Smithe. 1971. The autotrophic nutrition of symbiotic marine coelenterates with special reference to hermatypic corals. Proc. R. Soc. Lond. Ser. B 178:11–129.

Lewis, J. B. 1977. Processes of organic production on coral reefs. Biol. Rev. 52:305–347.

Lewis, J. B. and W. S. Price. 1975. Feeding mechanisms and feeding strategies of Atlantic reef corals. J. Zool. 176:527–544.

Lewis, J. B. and W. S. Price. 1976. Patterns of ciliary currents in Atlantic reef corals and their functional significance. J. Zool. 178: 77–89.

Manuel, R. L. 1981. *British Anthozoa*. Synopses of the British Fauna, No. 18. Academic Press, New York.

Mariscal, R. N. 1970. Nature of symbiosis between Indo–Pacific anemone fishes and sea anemones. Mar. Biol. 6: 58.

Mariscal, R. N. 1974. Scanning electron microscopy of the sensory epithelia and nematocysts of corals and a corallimorpharian sea anemone. Proc. Second Int. Coral Reef Symp. 1: 519–532.

Mariscal, R. N., E. J. Conklin and C. H. Bigger. 1977. The ptychocyst, a major new category of cnida used in tube construction by a cerianthid anemone. Biol. Bull. 152: 392–405.

Mauzey, K. P., C. Birkeland and P. Dayton. 1968. Feeding behavior of asteroids and escape responses of their prey in the Puget Sound region. Ecology 49: 603–619.

Meyer, J. L., E. T. Schultz and G. S. Helfman. 1983. Fish schools: An asset to corals. Science 220: 1047–1049.

Micronesica, Vol. 12, No. 1. 1976. Symposium issue. "International Symposium on Indo-Pacific Tropical Reef Biology" held at Guam and Palau, 6/23–7/5, 1974.

Muscatine, L. and C. Hand. 1958. Direct evidence for the transfer of materials from symbiotic algae to the tissues of a coelenterate. Proc. Natl. Acad. Sci. USA 44: 1259–1263.

Muscatine, L. and J. W. Porter. 1977. Reef corals: Mutualistic symbioses adapted to nutrient-poor environments. Biol. Sci. 27: 454–459.

Newell, N. D. 1972. The evolution of reefs. Sci. Am. 226(6): 54–65.

Patton, J. S., S. Abraham and A. A. Benson. 1977. Lipogenesis in the intact coral *Pocillopora capitata* and its isolated zooxanthellae: Evidence for a light-driven carbon cycle between symbiont and host. Mar. Biol. 44: 235–247.

Pearson, R. G. 1981. Recovery and recolonization of coral reefs. Mar. Ecol. Prog. Ser. 4: 105–122.

Proceedings of the Symposium on Coral Reefs. 1974. Published by The Great Barrier Reef Committee, Brisbane, Australia. [In 3 vols.]

Proceedings of the Third International Coral Reef Symposium. 1977. Published by the Rosenstiel School of Marine and Atmospheric Science, University of Miami, Coral Gables, Florida.

Proceedings of the Fourth International Coral Reef Symposium, Vol. 1. 1980. Marine Sciences Center, University of the Philippines. [This volume, and the two symposium sets cited above, are major sources of information on corals and coral reefs.]

Purcell, J. E. 1977. Aggressive function and induced development of catch tentacles in the sea anemone *Metridium senile* (Coelenterata: Actiniaria). Biol. Bull. 153: 355–368.

Purcell, J. E. and C. L. Kitting. 1982. Intraspecific aggression and population distributions of the sea anemone *Metridium senile*. Biol. Bull. 162: 345–359.

Reese, E. S. 1981. Predation on corals by fishes of the family Chaetodontidae: Implications for conservation and management of coral reef ecosystems. Bull. Mar. Sci. 31(3): 594–604.

Reimer, A. A. 1971. Feeding behavior in the Hawaiian zoanthids *Palythoa* and *Zoanthus*. Pac. Sci. 25(4): 257–260.

Richmond, R. H. 1985. Reversible metamorphosis in coral planula larvae. Mar. Ecol. Prog. Ser. 22: 181–185.

Rinkevich, B. and Y. Lola. 1983. Short-term fate of photosynthetic products in a hermatypic coral. J. Exp. Mar. Biol. Ecol. 73: 175–184.

Ross, D. M. 1970. Behavioral and ecological relationships between sea anemones and other invertebrates. Annu. Rev. Mar. Biol. Oceanogr. 5: 291–316.

Sandberg, D. M., P. Kanciruk and R. N. Mariscal. 1971. Inhibition of nematocyst discharge correlated with feeding in a sea anemone, *Calliactis tricolor* (Leseur). Nature 232: 263–264.

Schmidt, H. 1972. Die Nesselkapseln der Anthozoa und ihre Bedeutung für die Phylogenetische systematik. Helgol. Wiss. Meeresunters. 23: 422–458.

Schmidt, H. 1974. On the evolution of the Anthozoa. Proc. Second Internat. Coral Reef Symp. 1: 533–560.

Schuhmacher, H. and H. Zibrowius. 1985. What is hermatypic? A redefinition of ecological groups in corals and other organisms. Coral Reefs 4: 1–9.

Sebens, K. P. 1980. The regulation of asexual reproduction and indeterminate body size in the sea anemone *Anthopleura elegantissima* (Brandt). Biol. Bull. 158: 370–382.

Sebens, K. P. 1981a. Recruitment in a sea anemone population: Juvenile substrate becomes adult prey. Science 213: 785–787.

Sebens, K. P. 1981b. Reproductive ecology of the intertidal sea anemones *Anthopleura xanthogrammica* (Brandt) and *A. elegantissima* (Brandt): Body size, habitat and sexual reproduction. J. Exp. Mar. Biol. Ecol. 54: 225–250.

Sebens, K. P. 1984. Agonistic behavior in the intertidal sea anemone *Anthopleura xanthogrammica*. Biol. Bull. 166: 457–472.

Sebens, K. P. and K. DeRiemer. 1977. Diel cycles of expansion and contraction in coral reef anthozoans. Mar. Biol. 43: 247–256.

Shick, J. M. 1981. Heat production and oxygen uptake in intertidal sea anemones from different shore heights during exposure to air. Mar. Biol. Lett. 2: 225–236.

Shick, J. M. and J. A. Dykens. 1984. Photobiology of the symbiotic sea anemone *Anthopleura elegantissima*: Photosynthesis, respiration, and behavior under intertidal conditions. Biol. Bull. 166: 608–619.

Smith, F. G. W. 1971. *Atlantic Reef Corals*. University of Miami Press, Coral Gables, Florida.

Spaulding, J. G. 1974. Embryonic and larval development in sea anemones (Anthozoa: Actiniaria). Am. Zool. 14: 511–520.

Stoddart, D. R. 1969. Ecology and morphology of recent coral reefs. Biol. Rev. 44: 433–498.

Stoddart, J. A. 1983. Asexual production of planulae in the coral *Pocillopora damicornis*. Mar. Biol. 76: 279–284.

Walsh, G. E. 1967. An annotated bibliography of the families Zoanthidae, Epizoanthidae, and Parazoanthidae (Colenterata, Zoantharia). University of Hawaii, Hawaii Inst. Mar. Biol. Tech. Rpt. 13: 1–77.

Watson, G. M. and R. N. Mariscal. 1983a. Comparative ultrastructure of catch tentacles and feeding tentacles in the sea anemone *Haliplanella*. Tissue Cell, 15(9): 939–953.

Watson, G. M. and R. N. Mariscal. 1983b. The development of a sea anemone tentacle specialized for aggression: Morphogenesis and regression of the catch tentacle of *Haliplanella luciae* (Cnidaria, Anthozoa). Biol. Bull. 164: 506–517.

Wells, J. W. 1957. Coral reefs. Geol. Soc. Am. Mem. 67 1: 609–631.

Westfall, J. A. 1965. Nematocysts of the sea anemone *Metridium*. Am. Zool. 5: 377–393.

Williams, R. B. 1975. Catch-tentacles in sea anemones: Occurrence in *Haliplanella luciae* (Verrill) and a review of current knowledge. J. Nat. Hist. 9: 241–248.

Yonge, C. M. 1963. The biology of coral reefs. Adv. Mar. Biol. 1: 209–260. [Although somewhat dated, still one of the most readable reviews of coral reefs available.]

Yonge, C. M. 1968. Living corals. Proc. R. Soc. Lond. Biol. 169: 329–344.

Yonge, C. M. 1973. The nature of reef-building (hermatypic) corals. Bull. Mar. Sci. 23(1): 1–15.

Yonge, C. M. 1974. Coral reefs and molluscs. Trans. R. Soc. Edinburgh 69(7): 147–166.

Chapter Nine

Phylum Ctenophora: The Comb Jellies

Their power of destruction is not surprising once we see their method of obtaining food.

Sir Alister Hardy (1965)

(speaking of ctenophores)

Ctenophores (Greek *cten*, "comb"; *phero*, "to bear"), commonly called comb jellies, sea gooseberries, or sea walnuts, are transparent, gelatinous animals that drift in open-ocean planktonic environments from surface waters to depths of at least 3,000 m. Their transparency and fragile nature make them difficult to capture or observe by traditional sampling methods such as towing or trawling with nets, and until the recent advent of special "blue water" SCUBA techniques they were thought to be only modestly abundant animals. However, they are now known to form a major portion of the planktonic biomass in many areas of the world, and they may periodically be the predominant zooplankters in some areas. Over 100 species have been described, but there are probably many deep-sea forms yet to be discovered. Figure 1 includes photographs of a few ctenophores.

The ctenophores are radially symmetrical, diploblastic (or perhaps triploblastic) metazoa, and their *Bauplan* includes many features similar to the cnidarians. This similarity is immediately obvious, for example, in features such as symmetry, a gelatinous mesenchyme or collenchyme (formed of ectomesoderm), absence of a body cavity between the gut and the body wall, and a relatively simple, netlike nervous system. Some zoologists, however, view these similarities as convergent features resulting from adaptations to pelagic lifestyles. Ctenophores are significantly different from cnidarians in regard to their more extensively organized digestive system, their mesenchymal (perhaps mesodermal) musculature, and certain other features (see Box 1).

Unlike cnidarians, comb jellies are monomorphic throughout their life histories, and they are never colonial. Most species are pelagic, and all lack any trace of an attached sessile stage. Ctenophores lack a hard skeleton, an excretory system, and a special gas exchange system. Most ctenophores are simultaneous hermaphrodites capable of self-fertilization. A distinctive larval stage, the **cydippid larva**, is usually produced.

Ctenophores are exclusively marine. They display a variety of shapes and range in size from less than 1 cm in height to ribbon-shaped forms 1 m long. Some have evolved rather bizarre body forms, and a few have secondarily taken up a benthic creeping existence. They occur in all the world's seas, at all latitudes.

Taxonomic history and classification

Perhaps because most well-known ctenophores are brilliantly luminescent and are commonly seen from ships, the group has been known since ancient times. In fact, the first recognizable figures of ctenophores were drawn by a ship's doctor and naturalist in 1671. Linnaeus placed them in his group Zoophyta, along with various other "primitive" invertebrates. Cuvier classified them with medusae and anemones in Zoophytes. In the early nineteenth century, Eschscholtz designed the first rational classification of pelagic medusae and ctenophores by creating the orders Ctenophorae (for comb jellies), Discophorae (for all the solitary cnidarian medusae), and Siphonophorae (for the colonial siphonophorans and chondrophorans). Eschscholtz viewed these orders as subdivisions of the class Acalepha, regarding them as intermediate between Zoophytes and Echinoderms (on the basis of the common presence of radial

A

B

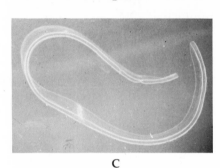

C

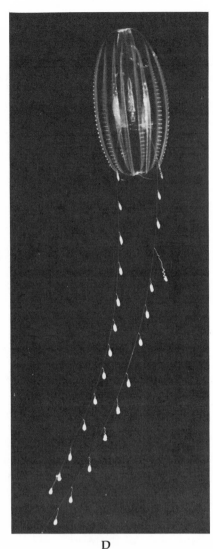

D

Figure 1

Representative ctenophores. A, *Pleurobrachia* (order Cydippida), with its tentacles retracted. B, *Beröe* (order Beroida). C, *Cestum* (order Cestida). D, *Euplokamis* (order Cydippida), with tentacles extended. (A courtesy of G. McDonald; B, D photos by C. E. Mills, Univ. of Washington/BPS; C courtesy of P. Fankboner.)

symmetry). Recall that it was Leuckart who, in 1847, first separated the Coelenterata from the echinoderms, although his Coelenterata also included sponges and ctenophores. Vosmaer (1877) was responsible for removing the sponges and Hatschek (1889) for removing the ctenophores as separate groups.

Following Harbison and Madin (1983), ctenophores are divided among seven orders containing 19 families. Figure 2 illustrates some of the general anatomy of most major groups.

PHYLUM CTENOPHORA

ORDER CYDIPPIDA: (Figures 1A,D; 2A,B) Pelagic; with well-developed comb rows; tentacles long and retractable into sheaths; body globular or ovoid, occasionally flattened in the stomodeal plane; meridional canals end blindly, para-

gastric canals (when present) end blindly at mouth. (e.g., *Aulococtena, Bathyctena, Callianira, Euplokamis, Haeckelia, Hormiphora, Lampea, Mertensia, Pleurobrachia, Tinerfe*)

ORDER PLATYCTENIDA: (Figures 2F,G,H) Planktonic or benthic; most species greatly flattened, with part of stomodeum everted as a creeping sole; often with tentacular sheaths; tentacular canals bifid; gastrovascular system complexly anastomosing; most species possess anal pores; many are ectocommensals on other organisms. Unlike most ctenophores, fertilization is often internal, and many platyctenids brood their embryos to the larval stage; asexual reproduction is common. (e.g., *Coeloplana, Ctenoplana, Lyrocteis, Tjalfiella*)

ORDER GANESHIDA: (Figure 2C) Pelagic; body form somewhat intermediate between Cydippida and Lobata, compressed in tentacular plane; tentacles branched and with sheaths; interradial canals arise from infundibulum and divide into adradial canals, which join the aboral ends of the meridional canals; meridional canals and paragastric canals join and form a circumoral canal (as in Beroida); mouth large and expanded in tentacular plane; without auricles or oral lobes. One genus, *Ganesha*, with two known species.

ORDER THALASSOCALYCIDA: Pelagic; body extremely fragile, expanded orally into medusa-like bell, to 15 cm along tentacular axis; body slightly compressed in stomodeal plane; tentacular sheaths absent; tentacles arise near mouth and bear lateral filaments; comb rows short; mouth and pharynx borne on central conical peduncle; meridional canals long, describing complex patterns in bell; all meridional canals end blindly aborally. Monotypic: *Thalassocalyce inconstans*.

ORDER LOBATA: (Figures 2J–L) Pelagic; body compressed in tentacular plane; with a pair of characteristic oral lobes and four flap-like auricles; a ciliated auricular groove extends to base of auricles from each side of each tentacular base; paragastric and subtentacular meridional canals unite orally. (e.g., *Bolinopsis, Deiopea, Leucothea, Mnemiopsis, Ocyropsis*)

ORDER CESTIDA: (Figures 1C, 2I) Pelagic; body extremely compressed in tentacular plane, and greatly elongate in stomodeal plane, producing a ribbon-like form to 1 m long in some species; substomodeal comb rows elongated, extending along entire aboral edge; subtentacular meridional canals arise under subtentacular comb rows (*Cestum*) or equatorially from interradial canals (*Velamen*); paragastric canals extend along oral edge and fuse with meridional canals; tentacles and tentacle sheaths present. Two genera: *Cestum* and *Velamen*.

ORDER BEROIDA: (Figures 1B, 2D,E) Pelagic; body cylindrical or thimble-shaped and strongly flattened in tentacular plane; tentacles and sheaths absent; aboral end rounded (*Beröe*) or with two prominent keels (*Neis*); stomodeum greatly enlarged; aboral sense organ well developed; comb rows present; meridional canals with numerous side branches; paragastric canals simple or with side branches. Two genera: *Beröe* and *Neis*.

The ctenophoran *Bauplan*

Ctenophores are usually regarded as among the most primitive living metazoa. They possess true tissues, but there is disagreement over the presence of true mesoderm. Between the epidermis and the gastrodermis is a well-developed middle layer, which is always a cellular mesenchyme. However, the ctenophoran mesenchyme differs from that of the cnidarians in several respects. Recall that the musculature of cnidarians is composed largely of simple, modified, epithelial cells of the epidermis and the gastrodermis. In ctenophores, however, true muscle cells develop within the mesenchyme itself, a condition characteristic of triploblastic metazoa.

Box One

Characteristics of the Phylum Ctenophora

1. Diploblastic (or triploblastic?) metazoa, with ectoderm and entoderm separated by a cellular mesenchyme

2. Biradial symmetry; body axis oral–aboral

3. With adhesive structures called colloblasts

4. Gastrovascular cavity (gut) is the only "body cavity"; gut with stomodeum and canals that branch complexly throughout body; gut ends in two small anal pores

5. Without discrete gas exchange, excretory, or circulatory systems (other than the gut)

6. Nervous system in the form of a nerve net or plexus, but more specialized than that of cnidarians

7. Musculature always formed of true mesenchymal cells

8. Monomorphic, without alternation of generations and without any kind of an attached sessile life stage

9. With eight rows of ciliary plates (combs or ctenes) at some stage in their life history; comb rows controlled by unique apical sense organ

10. Some adults and most juveniles with a pair of long tentacles, often retractable into sheaths

11. Most are hermaphroditic; typically with a characteristic cydippid larval stage

---

As we noted in the preceding chapter, the essence of the cnidarian–ctenophoran *Bauplan* is radiality (in the ctenophores, biradiality); and we have explained some of the architectural and structural constraints and advantages that derive from this symmetry. Thus, the nervous system of ctenophores is in the form of a simple, noncentralized nerve net, the locomotor structures are arranged radially about the body, and so on. The ctenophores possess several unique features that define the phylum. These in-

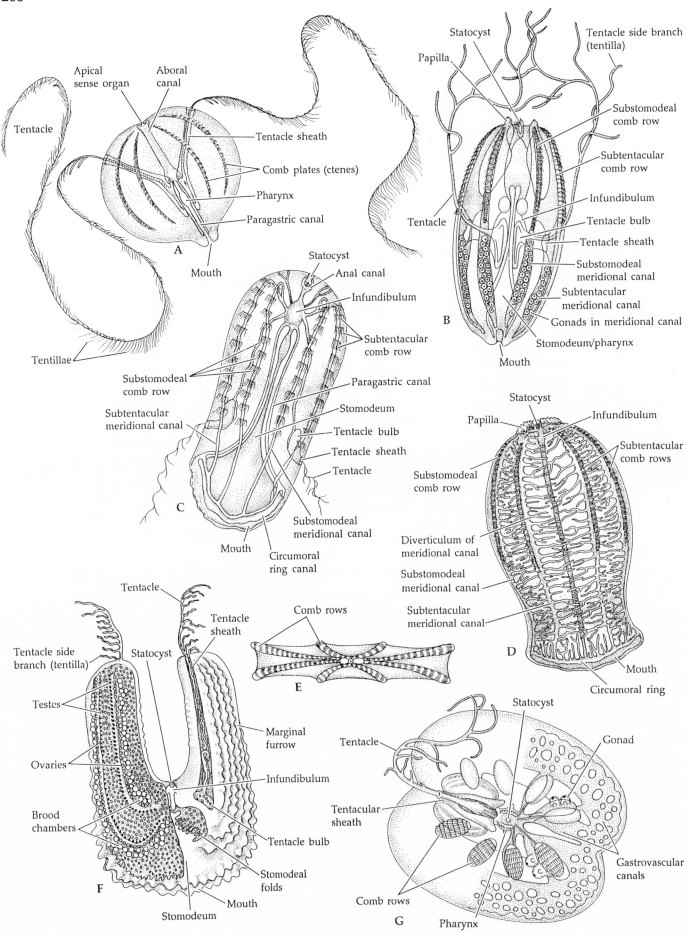

A

Tentacle

Apical sense organ

Aboral canal

Tentacle sheath

Comb plates (ctenes)

Pharynx

Paragastric canal

Mouth

Tentillae

B

Statocyst

Papilla

Tentacle side branch (tentilla)

Substomodeal comb row

Subtentacular comb row

Infundibulum

Tentacle bulb

Tentacle sheath

Substomodeal meridional canal

Subtentacular meridional canal

Gonads in meridional canal

Stomodeum/pharynx

Mouth

Tentacle

C

Statocyst

Anal canal

Infundibulum

Subtentacular comb row

Substomodeal comb row

Subtentacular meridional canal

Paragastric canal

Stomodeum

Tentacle bulb

Tentacle sheath

Tentacle

Substomodeal meridional canal

Mouth

Circumoral ring canal

D

Statocyst

Papilla

Infundibulum

Subtentacular comb rows

Substomodeal comb row

Diverticulum of meridional canal

Substomodeal meridional canal

Subtentacular meridional canal

Mouth

Circumoral ring

E

Comb rows

F

Tentacle

Tentacle sheath

Statocyst

Tentacle side branch (tentilla)

Testes

Ovaries

Brood chambers

Marginal furrow

Infundibulum

Tentacle bulb

Stomodeal folds

Mouth

Stomodeum

G

Statocyst

Tentacle

Gonad

Tentacular sheath

Gastrovascular canals

Comb rows

Pharynx

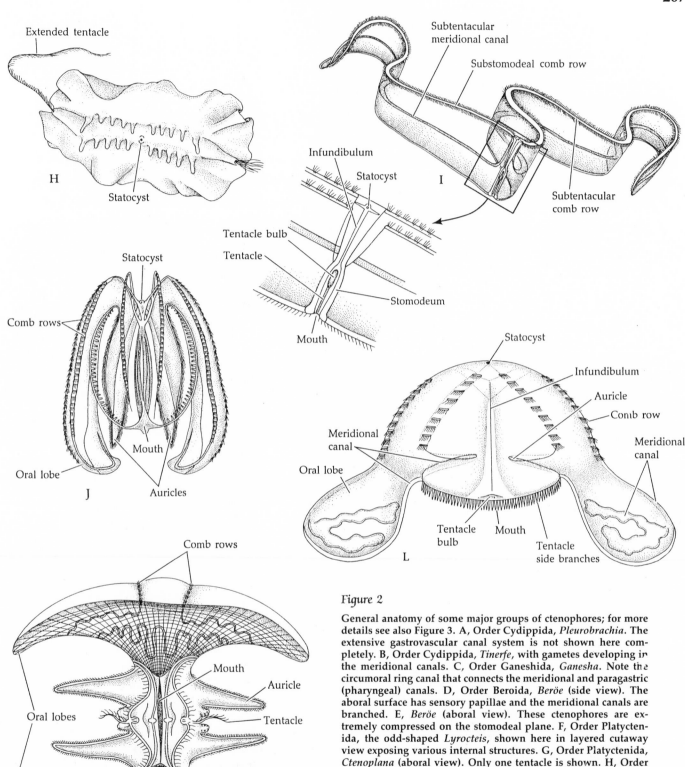

Figure 2

General anatomy of some major groups of ctenophores; for more details see also Figure 3. A, Order Cydippida, *Pleurobrachia*. The extensive gastrovascular canal system is not shown here completely. B, Order Cydippida, *Tinerfe*, with gametes developing in the meridional canals. C, Order Ganeshida, *Ganesha*. Note the circumoral ring canal that connects the meridional and paragastric (pharyngeal) canals. D, Order Beroida, *Beröe* (side view). The aboral surface has sensory papillae and the meridional canals are branched. E, *Beröe* (aboral view). These ctenophores are extremely compressed on the stomodeal plane. F, Order Platyctenida, the odd-shaped *Lyrocteis*, shown here in layered cutaway view exposing various internal structures. G, Order Platyctenida, *Ctenoplana* (aboral view). Only one tentacle is shown. H, Order Platyctenida, *Coeloplana*, a benthic form. I, Order Cestida, *Cestum*. This ctenophore exhibits an extreme modification of body form. J, Order Lobata, *Mnemiopsis* (side view). This ctenophore has oral lobes and auricles. K, *Mnemiopsis* (oral view). Note the greatly expanded oral lobes with their distinctive pattern of muscle fibers. L, Order Lobata, *Deiopea*. (B, C, F, L after Harbison and Madin 1983; E, K after Mayer 1912; G after Komai 1934; H after Bayer and Owre 1968.)

A

Key

1 Anal canal
2 Anal pore
3 Apical sense organ (statocyst)
4 Aboral canal
5 Tentacle
6 Infundibulum
7 Transverse canal
8 Interradial canal
9 Tentacle sheath
10 Tentilla
11 Ctenes of comb row
12 Mouth
13 Pharynx
14 Pharyngeal canal
15 Tentacular canal
16 Meridional canal
17 Adradial canal

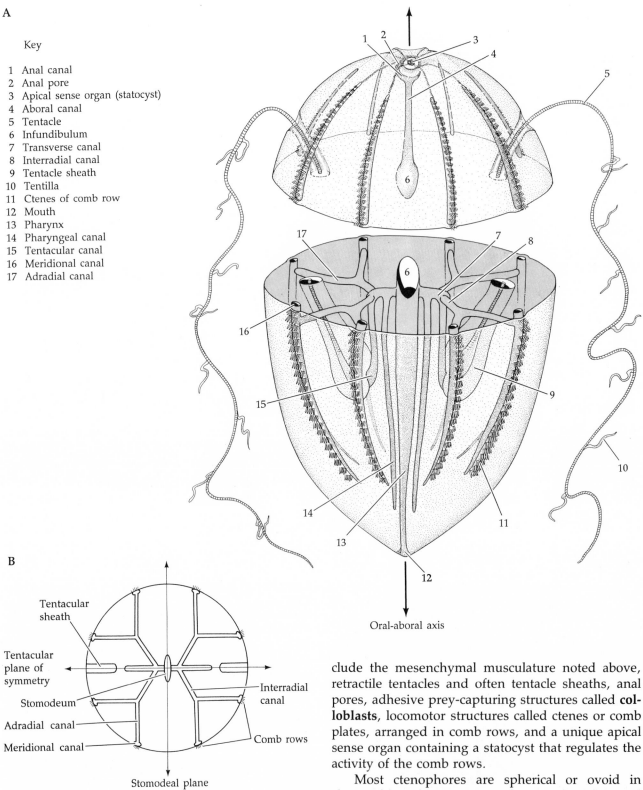

Oral-aboral axis

B

Tentacular
sheath

Tentacular
plane of
symmetry

Stomodeum

Adradial canal

Meridional canal

Interradial
canal

Comb rows

Stomodeal plane
of symmetry

Figure 3

The ctenophoran *Bauplan*. A, A cydippid ctenophore (cross section). B, Ctenophoran biradiality and the planes of symmetry (oral view). (A from Bayer and Owre 1968.)

clude the mesenchymal musculature noted above, retractile tentacles and often tentacle sheaths, anal pores, adhesive prey-capturing structures called **colloblasts**, locomotor structures called ctenes or comb plates, arranged in comb rows, and a unique apical sense organ containing a statocyst that regulates the activity of the comb rows.

Most ctenophores are spherical or ovoid in shape, although some species have evolved flattened shapes through compression and elongation in one of the two planes of body symmetry (Figures 1 and 2). The general body plan can best be understood by first examining a generalized cydippid ctenophore (Figure 3). Most specialists consider the cydippids to be primitive within the phylum. As in cnidarians, the principal axis is oral–aboral. The mouth is at the oral pole; the aboral pole bears the apical sense organ. On the surface of the body are eight equally spaced meridional rows of **comb plates**. Each comb plate, or **ctene**, is composed of a transverse band of long fused cilia. On each side of the body of many species is a deep, ciliated epidermal pouch, the **tentacle** (or tentacular) **sheath**, from whose inner wall a tentacle arises. The tentacles are typically very long and contractile, and bear lateral branches called filaments, or **tentillae** (Figures 2 and 3). The epidermis of both the tentacle and the lateral tentillae is richly armed with colloblasts. Most species can retract the tentacles into the sheaths by muscles. It is the tentacles and certain aspects of the internal anatomy that give ctenophores a biradial symmetry. The elongate stomodeum lies on the oral–aboral axis of the body. It is distinctly flattened in one plane of body symmetry, the **stomodeal plane** (Figure 3B). Bisecting the animal along the stomodeal plane separates the two tentacular halves of the body. The second plane of body symmetry, called the **tentacular plane**, is defined by the position of the tentacular sheaths (Figure 3B).

Some variations of the basic ctenophoran body plan are illustrated in Figures 1 and 2. In members of the order Lobata, the body is compressed in the tentacular plane and the oral end is expanded on each side into rounded, contractile **oral lobes**. The mouth sits on an elongate manubrium, the base of which bears four long flaps called **auricles**. The tentacles are reduced and lack sheaths. From either side of each tentacle base, a ciliated **auricular groove** arises and extends to the auricles. Members of the order Cestida are also compressed in the tentacular plane and extremely elongated in the stomodeal plane, giving these ctenophores a striking snake-like or ribbon-like appearance. The sheathed tentacles are reduced and shifted alongside the mouth. Beroids are thimble-shaped and also flattened in the tentacular plane. They lack tentacles and sheaths. In the single species of Thalassocalycida (*Thalassocalyce inconstans*), the body is expanded around the mouth to form a medusa-like bell.

The oddest ctenophores are members of the order Platyctenida (Figure 2F–H). Platyctenids are primarily benthic and small, often less than 1 cm in length, and, in contrast to most pelagic ctenophores, they are pigmented rather than transparent. The body is oval and markedly flattened. Despite these unusual features, early naturalists recognized them

as ctenophores by the presence of an apical sense organ, comb rows, and a pair of tentacles. Detailed studies have shown that the flattened oral surface is actually an everted portion of the pharynx. The platyctenid pharynx was, in a sense, preadapted to conversion to a creeping foot or sole by its intrinsic musculature. Most of these animals crawl about on the sea bottom, but some are ectocommensals on alcyonarian cnidarians, echinoderms, or pelagic salps.

Support and locomotion

Ctenophores rely primarily on their elastic mesenchyme for structural support. The watery gelatinous mesenchyme makes up most of the body mass, and ctenophore dry weights are only about 4 percent of the live wet weight. The mesenchyme contains both elastic supportive cells and muscle cells, the general tonus of the latter being primarily responsible for maintaining body shape. Figure 4 shows a highly stylized cutaway section of a cydippid ctenophore and illustrates the arrangement of the supportive mesenchymal muscle fibers. Tension in the looped muscles tends to maintain the spherical geometry. Action of the radial muscles diminishes the radius and hence the circumference, and also serves to open the pharynx. These two muscle sets work antagonistically to one another.

Most ctenophores are pelagic, and the gelatinous body and low specific gravity maintain a relatively neutral buoyancy, allowing these creatures to float about with the ocean currents. Neutral buoyancy appears to be maintained by passive osmotic accommodation. Because these passive buoyancy adjust-

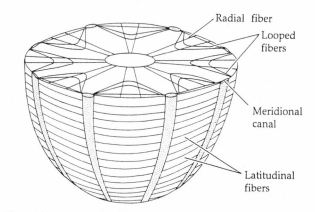

Figure 4

Stereogram of the arrangement of muscle fibers in *Pleurobrachia*, a cydippid ctenophore. The diagram depicts a transverse section through the region of the pharynx; the gastrovascular system and tentacle sheaths have been omitted for clarity. (After Chapman 1958.)

ments take time, ctenophores may accumulate at discontinuity layers in the sea, where a water mass of one density overlies a water mass of a slightly different density.

The beating of the ctenes provides most of the modest locomotor power that allows ctenophores to move up and down in the water column and to locate richer feeding grounds or preferred environmental conditions. Each comb row comprises many ctenes. Each ctene consists of a transverse band of hundreds of very long, partly fused cilia (to 2 mm in length) that beat together as a unit. Ctenophores are the largest animals known to use cilia for locomotion.

The mesenchymal musculature that sets ctenophores apart from cnidarians is usually used to maintain body shape and assist in feeding; it is involved in behaviors such as prey swallowing, pharyngeal contractions, and tentacular movements. Usually both longitudinal and circular muscles are present just beneath the epidermis. In the benthic and epifaunal platyctenids, stomodeal musculature facilitates a creeping locomotion. In the snakelike cestids, body muscles may generate graceful swimming undulations, and the lobate ctenophore *Ocyropsis* swims by muscular flapping of its oral lobes. Giant smooth muscle fibers, the first to be discovered, were recently found in *Beröe*.

Feeding and digestion

Comb jellies are, so far as is known, entirely predatory in their habits. The long tentacles of cydippids (and of the larvae of most other forms) have a muscular core with a colloblast-laden epidermal covering (Figure 5). The tentacles trail passively or are "fished" by various swirling movements of the body. Upon contact with zooplankton prey, the colloblasts (sometimes called lasso cells) burst and discharge a strong adhesive material. Each colloblast develops from a single cell and consists of a hemispherical mass of secretory granules attached to the muscular core of the tentacle by a spiral filament coiled around a straight filament (Figure 6). The straight filament is actually the highly modified nucleus of the colloblast cell. The spiral filament, which uncoils upon discharge, adheres to the prey by the sticky material produced in the secretory granules. As the tentacles accumulate prey, they are periodically wiped across the mouth by muscular contractions, occasionally combined with a coordinated somersaulting action of the animal that brings the mouth to the trailing tentacle. In members of the orders Lobata and Cestida, which bear very short tentacles, small zooplankton are trapped in mucus on the body surface and then

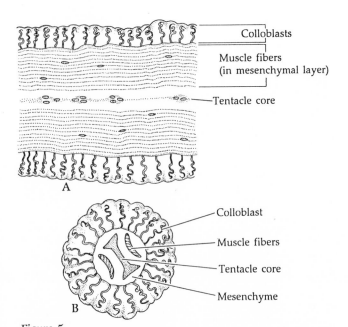

Figure 5

Ctenophore tentacle structure. A, Longitudinal section of tentacle. B, Cross section of a lateral filament (tentilla) of a tentacle. (After Hyman 1940.)

carried to the mouth by ciliary currents (along the ciliated **auricular grooves** in lobate forms and ciliated **oral grooves** in cestids). Most of the benthic platyctenids also feed by capturing zooplankton in a somewhat similar fashion.

Some ctenophores prey upon larger animals, especially gelatinous forms. The cydippid *Lampea* (formerly *Gastrodes*), for example, lives embedded in the body of pelagic tunicates (*Salpa*), upon which it feeds. Figure 7 is a series of remarkable photographs showing the cydippid ctenophore *Haeckelia* eating the tentacles of the trachyline hydromedusa *Aegina*. After consuming the tentacles, one by one, *Haeckelia* retains the prey's unfired nematocysts, incorporates them into its epidermis, and then uses them for its own defense. This phenomenon, known as kleptocnidae, occurs in several unrelated groups of predators on cnidarians (see Chapter 8).

The mouth opens into an elongate, highly folded, and flattened muscular pharynx, or stomodeum; the pharynx is richly endowed with gland cells that produce the digestive juices. Large food items are tumbled within the pharynx by ciliary action. Digestion takes place extracellularly, mostly in the pharynx. The largely digested food passes via a small chamber (the **infundibulum**, **funnel**, or **stomach**) from the pharynx into a complex system of radiating gastrovascular canals (Figures 2 and 3). The details of the

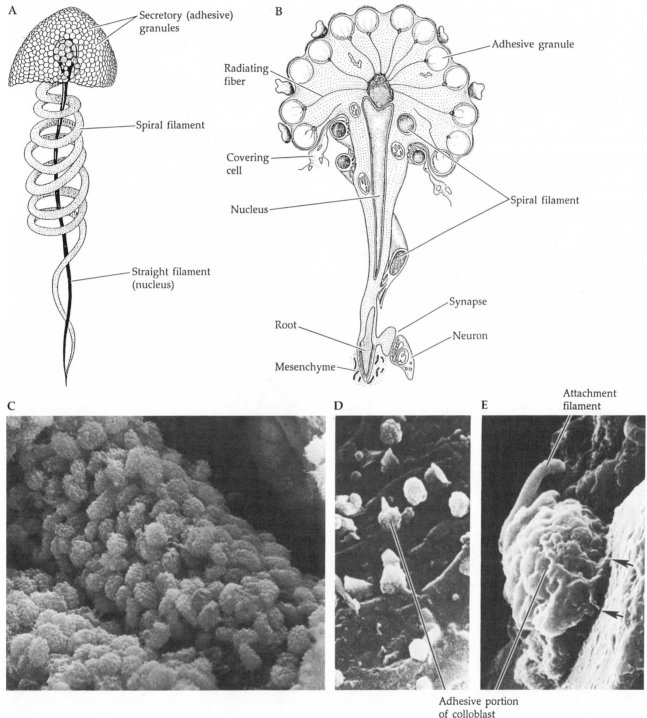

A, Secretory (adhesive) granules

Spiral filament

Straight filament (nucleus)

B, Radiating fiber

Adhesive granule

Covering cell

Nucleus

Spiral filament

Root

Synapse

Neuron

Mesenchyme

C

D

E, Attachment filament

Adhesive portion of colloblast

Figure 6

Colloblasts. A, The functional parts of a colloblast. B, Longitudinal section. C, Colloblasts on the lateral tentacular filaments (tentillae) of *Pleurobrachia* (scanning electron micrograph). D, Fired colloblasts of *Pleurobrachia*, showing adhesive spherules attached to fragments of a copepod (small crustacean). E, Fired colloblasts are still attached to the tentacular filament. The adhesive ends of the coiled filaments are stuck (arrows) to a bit of copepod. (A after Bayer and Owre 1968; B after Franc 1978; C–E courtesy of P. Fankboner.)

arrangement of the canals vary among different groups; the following description applies to the arrangement in a cydippid. Two **paragastric** or **pharyngeal canals** recurve and lie parallel to the pharynx. Two **transverse canals** depart at right angles to the stomodeal plane and divide into three more branches. The middle branch of each triplet, the **ten-**

Figure 7

The cydippid ctenophore *Haeckelia rubra* (= *Euchlora rubra*) is feeding on the trachyline hydromedusa *Aegina citrea*. A, Intact specimen of *Aegina*, with all four tentacles present. B, *Haeckelia* begins to consume one of *Aegina's* tentacles. C, Most of the first tentacle of the medusa has been ingested. D, Same animals, two minutes after feeding began. E, *Aegina* has lost all four of its tentacles to a hungry *Haeckelia*. (From Mills and Miller 1984.)

tacular canal, leads to the base of the tentacular sheath. Each of the other two branches (the **interradial canals**) bifurcates to form a total of four **adradial canals** on each side of the animal. These in turn connect to the eight **meridional canals**, one beneath each comb row. Finally, an **aboral canal** passes from the infundibulum to the aboral pole, where it divides beneath the apical sense organ into four short canals, two ending blindly and two (the **anal canals**) opening to the outside via small **anal pores**. The anal pores serve as a primitive anus, assisting the mouth in the voiding of indigestible wastes. They may also serve as an exit for metabolic wastes.

Within this complicated gastrovascular canal system, digestion is completed, nutrients are distributed through the body, and absorption takes place. Minute pores lead from the various canals into the mesenchyme (Figure 8). Surrounding these pores are circlets of ciliated gastrodermal cells called **cell rosettes**, which appear to regulate the flow of the digestive soup and perhaps also play a role in excretion. Except for the stomodeal pharynx, the gastrovascular system is lined by a simple epithelium of entodermal origin.

Circulation, excretion, gas exchange, and osmoregulation

There is no independent circulatory system in ctenophores; the gastrovascular canal system serves in this role by distributing nutrients to most parts of the body. The gastrovascular system probably also picks up metabolic wastes from the mesenchyme for eventual expulsion out of the mouth or anal pores. The cell rosettes may transport wastes to the gut. Gas exchange occurs across the general body surface and across the walls of the gastrovascular system. All of these activities are augmented by diffusion through the gelatinous mesenchyme. Movement of water over the body surface is enhanced by the beating of the comb plates. Thus, the extensive canal system and the ciliary bands help to overcome the problem of long diffusion distances.

Nervous system and sense organs

Although the nervous systems of both ctenophores and cnidarians are noncentralized nerve nets,

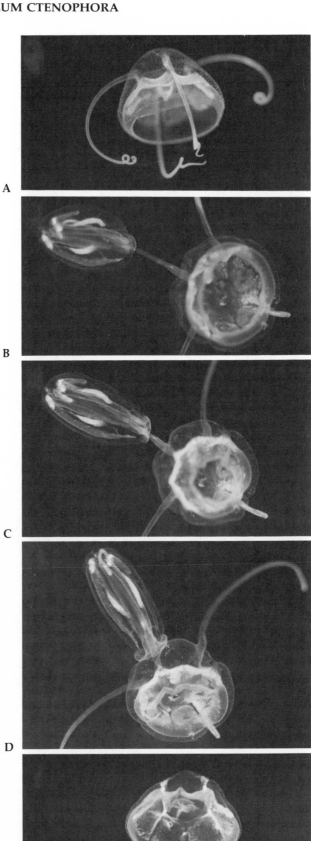

A

B

C

D

E

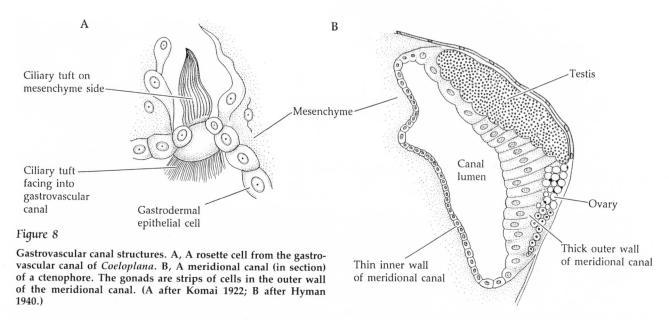

Figure 8

Gastrovascular canal structures. A, A rosette cell from the gastro-vascular canal of *Coeloplana*. B, A meridional canal (in section) of a ctenophore. The gonads are strips of cells in the outer wall of the meridional canal. (A after Komai 1922; B after Hyman 1940.)

there are certain important differences. In a cteno-phore, nonpolar neurons form a diffuse subepider-mal plexus. Beneath the comb rows, the nerve cells form elongate plexes or meshes such that the neurons form nervelike strands. The bases of the ctenes are thus in contact with a rich plexus of nerve cells. A similar concentrated plexus surrounds the mouth. However, as in cnidarians, no true ganglia occur. Their absence contrasts markedly with the presence of a centralized nervous system in bilateral metazoa.

The apical sense organ is a statolith that functions in balance and orientation. The calcareous statolith is supported by four, long, springlike tufts of cilia called **balancers** (Figure 9). The whole structure is enclosed in a transparent dome that is apparently derived from cilia. From each balancer there arises a pair of **ciliated furrows** , or **ciliated grooves**, each of which connects with one comb row. Thus, each balancer innervates the two comb rows of its particular quadrant. Tilting the animal causes the statolith to press more heavily on one of the balancers, and the resulting stimulus elicits a vigorous beating of the appropriate comb rows to right the body.

The two comb rows in each quadrant innervated by a single ciliated furrow beat synchronously. If a ciliated furrow is cut, the beating of the two corre-sponding comb rows becomes asynchronous. The normal direction of ciliary power strokes is toward the aboral pole, so that the animal is driven forward oral end first. The beat in each row, however, begins at the aboral end of the comb row and proceeds in metachronal waves toward the oral end. Stimulation

of the oral end reverses the direction of both the wave and the power stroke. Removal of the apical sense organ or statolith results in an overall lack of coor-dination of the comb rows, and the injured cteno-phore loses its ability to maintain a vertical position. The comb rows are very sensitive to contact; when a comb row is touched, many species retract it into a groove formed in the jelly-like body.

In cydippids and beroids, the stimulation for any given ctene to beat is triggered mechanically, by hy-drodynamic forces arising from the movements of the preceding plate. However, in the lobate cteno-phores, the ctenes are not coordinated in the same mechanical fashion. In these animals a narrow tract of shorter cilia, the **interplate ciliated groove**, runs between successive ctenes and is responsible for co-ordinating their activity. It is not known how the cilia of the groove are coordinated or how the grooves stimulate the appropriate comb row, so these actions may also be mechanical. The interplate ciliated grooves develop only as the lobate ctenophores ma-ture into adulthood; the free-swimming larvae resem-ble cydippids and lack the grooves.

In some ctenophores, a pair of oval tracts of cilia called **polar fields** lies on the stomodeal plane of the aboral surface (Figure 9). These structures are pre-sumed to be sensory in function.

Reproduction and development

Asexual reproduction and regeneration. Cteno-phores can regenerate virtually any lost part, includ-ing the apical sense organ. Entire quadrants and even

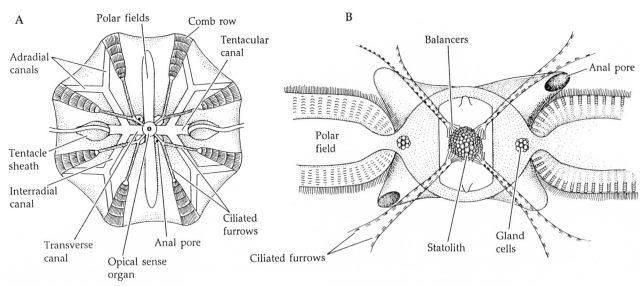

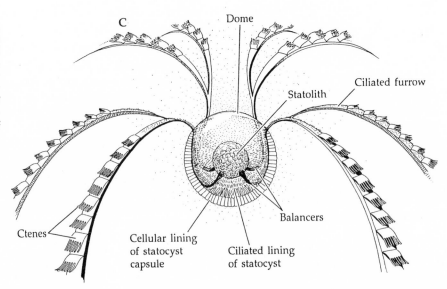

Figure 9

The ctenophore apical sense organ. A, (aboral view). Note the relationship of the statocyst to the comb rows, the ciliated furrows, and the gastrovascular canals. B, Apical sense organ of the cydippid *Hormiphora*. C, Apical sense organ and its relationship to the eight comb rows. (A after Hyman 1940; B after G. C. Bourne, in E. R. Lankester (ed.), 1900, *Treatise on Zoology*; C from Bayer and Owre 1968.)

whole halves will regenerate. Speculation that ctenophores may reproduce by fission has not been documented. Platyctenids reproduce asexually by a process that resembles pedal laceration in sea anemones; small fragments break free as the animal crawls, and each piece can regenerate into a complete adult.

Sexual reproduction and development. Most ctenophores are hermaphroditic, but a few gonochoristic species are known. The gonads are entodermal in origin, arising on the walls of the meridional canals (Figure 8). Pelagic ctenophores generally shed their gametes, via the mouth, into the surrounding sea water where fertilization takes place. Special sperm ducts occur in at least some platyctenid species. Those that free-spawn produce embryos that grow

quickly to planktotrophic **cydippid larvae** (Figure 10). Development is thus indirect, although growth to the adult is gradual rather than metamorphic. In the benthic *Coeloplana* and *Tjalfiella*, fertilization is internal and embryos are brooded until a cydippid larva is formed and released. This mixed life history provides a means of dispersal for these benthic, sedentary animals.

During early cleavage, the first four blastomeres arise by the usual two meridional cleavages, which mark the adult planes of symmetry. The third division is also nearly vertical and results in a curved plate of eight cells (macromeres). The next division is latitudinal and unequal, and gives rise to micromeres on the concave side of the macromere plate. The micromeres continue to divide and spread by

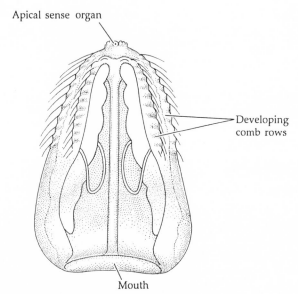

Apical sense organ

Developing comb rows

Mouth

Figure 10

A typical young cydippid larva. (After Hyman 1940.)

epiboly over the aboral pole and eventually over the macromeres. The latter also invaginate into the interior, so the gastrula arises through a combination of epiboly and invagination. Thus, the micromeres become ectoderm and the macromeres become entoderm. Just prior to gastrulation, the macromeres divide and produce additional micromeres on the oral side of the embryo. Whereas the aboral micromeres become ectoderm, these oral micromeres are incorporated into the entoderm and, in at least some species, give rise to photoreceptor cells. There is some question about the fate of all of these oral micromeres. Metschnikoff (1885) suggested that these cells may contribute to the mesenchyme and may thus be viewed as true entomesoderm. More recently, Harbison (1985) has also made a case for a triploblastic condition in ctenophores.

As the micromeres cover the embryo to form the epidermis, four interradial bands of small, rapidly dividing cells become apparent. Each of these thickened ectodermal bands differentiates into two of the comb rows. The aboral ectoderm differentiates into the apical sense organ and its related parts; the oral ectoderm invaginates to form the stomodeum. The gastrovascular system develops from entodermal outgrowths and the tentacle sheaths arise as ectodermal invaginations from the points where the tentacles sprout. The embryo eventually develops into a free-swimming cydippid larva (Figure 10), which closely resembles adult ctenophores of the order Cy-

dippida. Some authors have taken this as evidence that the Cydippida is the most primitive of the extant ctenophore orders.

The development of ctenophores differs markedly from that of cnidarians. In the latter group, early cleavage results in an irregular mass of cells whose fates are not clearly predictable until later development. In the ctenophores, on the other hand, a very precise cleavage pattern unfolds, in which the ultimate morphology is definitely mapped. Furthermore, ctenophores lack the planula larva that characterizes cnidarians; instead, they produce a cydippid larval type having no obvious counterpart among the cnidarians.

Ctenophoran phylogeny

Although the ctenophores and cnidarians are generally regarded as belonging to the same general grade of construction, it is difficult to derive ctenophores from any existing cnidarian group. Some zoologists suggest that ctenophores arose from the hydrozoans, by way of an intermediate medusa possessing an aboral statocyst and two tentacle sheaths, such as is seen today in the aberrant trachyline medusa *Hydroctena* (Figure 11). In this medusa, the number of tentacles has been reduced to two; and these are set high on the bell, like the tentacles of trachylines in general. The tentacles also arise from deep epidermal pockets that resemble the tentacular sheaths of ctenophores. Furthermore, *Hydroctena* has a single apical sense organ, although its construction differs from that in ctenophores. Several other trachyline medusae also have solitary aboral sense organs. Although these similarities may suggest a relationship to the trachylines, ctenophores also show certain similarities to the scyphozoans and anthozoans, such as the stomodeum and the highly cellular mesenchyme, and the four-lobed gastrovascular cavity of the cydippid larva.

As we have seen, however, ctenophores are really quite different from cnidarians in many fundamental ways. These differences are evident both in adult morphology and in patterns of development. Many of the similarities between ctenophores and medusae may well be convergences reflecting adaptations to their similar lifestyles; many gelatinous zooplankters show superficial similarities in body form and construction.

The presence of true mesenchymal muscle cells and gonoducts in some species, and certain features of early cleavage, have led some zoologists to suggest

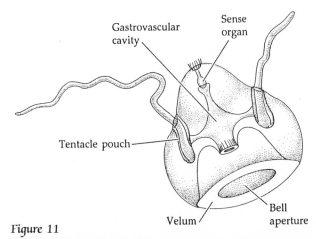

Figure 11

The aberrant trachyline medusa *Hydroctena*, which superficially resembles a ctenophore in its possession of an apical sensory structure and tentacular pouches. (After Hyman 1940.)

a relationship between the ctenophores and flatworms (Platyhelminthes; Chapter 10). Some workers view the ctenophores as ancestral to the flatworms; but a reverse scenario has also been suggested, wherein the ctenophores arose from the flatworms. The presence of benthic, crawling ctenophores (e.g., *Ctenoplana* and *Coeloplana*) is used as evidence that ecological and anatomical intermediates between the two groups are plausible. Harbison (1985) reviews these matters and concludes that there is no more evidence to link the ctenophores to the cnidarians than to the flatworms.

There is also disagreement about evolution within the Ctenophora, centering largely on whether the tentaculate or atentaculate condition is primitive. Without a clearer picture of the origin of the group, it is difficult at best to resolve such questions. The known fossil record offers virtually no help in these matters, as it consists of but a single cydippid-like specimen discovered in 1983 in Devonian deposits. On this sparse and controversial information, we can only hypothesize that the ctenophores are a monophyletic group that probably arose early in the evolution of the metazoan condition. Harbison (1985) indicates that the present classification of ctenophores is not phylogenetic, and more information is needed to understand the intraphylum relationships.

Selected References

Abbott, J. F. 1907. Morphology of *Coeloplana*. Zool. Jahrb. Abt. Anat. Ontog. Tiere 24.

Arai, M. N. 1976. Behavior of planktonic coelenterates in temperature and salinity discontinuity layers. *In* G. O. Mackie (ed.), *Coelenterate Ecology and Behavior*. Plenum, New York, pp. 211–218.

Bigelow, H. B. 1912. Reports on the scientific results of the expedition to the eastern tropical Pacific, in charge of Alexander Agassiz, by the U.S. Fish Commission Steamer *Albatross*, from October 1904, to March 1905, Lieutenant Commander L. M. Garrett, U.S.N., commanding. XXVI. The Ctenophores. Bull. Mus. Comp. Zool. Harvard College 54(12):369–404.

Carré, C. and D. Carré. 1980. Les cnidocysts du ctenophore *Euchlora rubra* (Kolliker 1853). Cah. Biol. Mar. 21:221–226.

Coonfield, B. R. 1936. Regeneration in *Mnemiopsis*. Biol. Bull. 71.

Dawydoff, C. 1963. Morphologie et biologie des *Ctenoplana*. Arch. Zool. Exp. Gen. 75.

Dunlap, H. L. 1966. Oogenesis in the Ctenophora. Ph.D dissertation, University of Washington, Seattle.

Farfaglio, G. 1963. Experiments on the formation of the ciliated plates in ctenophores. Acta Embryol. Morphol. Exp. 6:191–203.

Franc, J.-M. 1978. Organization and function of ctenophore colloblasts: An ultrastructural study. Biol. Bull. 155:527–541.

Freeman, G. 1976. The effects of altering the position of cleavage planes on the process of localization of developmental potential in ctenophores. Dev. Biol. 51:332–337.

Freeman, G. 1977. The establishment of the oral–aboral axis in the ctenophore embryo. J. Embryol. Exp. Morphol 42:237–260.

Harbison, G. R. 1985. On the classification and evolution of the Ctenophora. *In* Morris et al. (eds.), *The Origins and Relationships of Lower Invertebrates*. Syst. Assoc. Spec. Vol. No. 28, pp. 78–100.

Harbison, G. R. and L. P Madin. 1979. A new view of plankton biology. Oceanus 22(2):18–27.

Harbison, G. R. and L. P. Madin. 1983. Ctenophora. *In* S. P. Parker (ed.), *Synopsis and Classification of Living Organisms*, Vol. 1. McGraw-Hill, New York, pp. 707–715.

Harbison, G. R., L. P. Madin and N. R. Swanberg. 1984. On the natural history and distribution of oceanic ctenophores. Deep-Sea Res. 25:233–256.

Harbison, G. R. and R. L. Miller. 1986. Not all ctenophores are hermaphrodites. Studies on the systematics, distribution, sexuality and development of two species of *Ocyropsis*. Mar. Biol. 90:413–424.

Hirota, J. 1974. Quantitative natural history of *Pleurobrachia bachei* in La Jolla Bight. Fish. Bull. U.S. 72:295–335.

Horridge, G. A. 1965. Relations between nerves and cilia in ctenophores. Am. Zool. 5:357–375.

Horridge, G. A. 1974. Recent studies on the Ctenophora. *In* L. Muscatine and H. M. Lenhoff (eds.), *Coelenterate Biology*. Academic Press, New York, pp. 439–468.

Hyman, L. H. 1940. *The Invertebrates*, Vol. 1, Protozoa through Ctenophora. McGraw-Hill, New York, pp. 662–696.

Komai, T. 1922. Studies on two aberrant ctenophores—*Coeloplana* and *Gastrodes*. Kyoto. [Published by the author].

Komai, T. 1934. On the structure of *Ctenoplana*. Kyoto Univ. Col. Sci. Mem. (Ser. B) 9.

Komai, T. 1936. Nervous system, *Coeloplana*. Kyoto Univ. Col. Sci. Mem. (Ser. B) 11.

Komai, T. and T. Tokioka. 1940. *Kiyohimea aurita* n. gn., n. sp., type of a new family of lobate Ctenophora. Annot. Zool. Jpn. 19:43–46.

Komai, T. and T. Tokioka. 1942. Three remarkable ctenophores from the Japanese seas. Annot. Zool. Jpn. 21:144–151.

Kremer, P. 1977. Respiration and excretion by the ctenophore *Mnemiopsis leidyi*. Mar. Biol. 44:43–50.

Krumbach, T. 1925. Ctenophora. *In* W. Kukenthal and T. Krumbach (eds.), *Handbuch des Zoologie* 1:905–995.

Madin, L. P. and G. R. Harbison. 1978. *Bathocyroe fosteri* gen. nov., sp. nov.: A mesopelagic ctenophore observed and collected from a submersible. J. Mar. Bio. Assoc. U.K. 58:559–564.

Madin, L. P. and G. R. Harbison. 1978. *Thalassocalyce inconstans*, new genus and species, an enigmatic ctenophore representing a new family and order. Bull. Mar. Sci. 28(4):680–687.

Main, R. J. 1928. Observations on the feeding mechanism of a ctenophore, *Mnemiopsis leidyi*. Biol. Bull. 55:69–78.

Martindale, M. Q. 1987. Larval reproduction in the ctenophore *Mnemiopsis meeradyi* (order Lobata). Mar. Biol. 94:409–414.

Matsumoto, G. I. and W. M. Hamner. 1988. Modes of water manipulation by the lobate ctenophore *Leucothea* sp. Mar. Biol. 97:551–558.

Mayer, A. G. 1912. *Ctenophores of the Atlantic Coast of North America*. Carnegie Inst. Washington Publ. No. 162.

Metschnikoff, E. 1885. Gastrulation und mesodermbildung der Ctenophoren. Z. Wiss. Zool. 42.

Mills, C. E. 1984. Density is altered in hydromedusae and ctenophores in response to changes in salinity. Biol. Bull. 166:206–215.

Mills, C. E. and R. L. Miller. 1984. Ingestion of a medusa (*Aegina citrea*) by the nematocyst-containing ctenophore *Haeckelia rubra* (formerly *Euchlora rubra*): Phylogenetic implication. Mar. Biol. 78:215–221.

Mills, C. E. and R. G. Vogt. 1984. Evidence that ion regulation in hydromedusae and ctenophores does not facilitate vertical migration. Biol. Bull. 166:216–227.

Mortensen, T. 1912. Ctenophora. Danish Ingolf-Expedition Vol. 5, Pt. 2:1–95. Zoological Museum, University of Copenhagen.

Mortensen, T. 1913. Regeneration in ctenophores. Vidensk. Medd. Dan. Naturhist. Foren. Khobenhavn 66.

Nielsen, C. 1987. *Haeckelia* (= *Euchlora*) and *Hydroctena* and the phylogenetic interrelationships of the Cnidaria and Ctenophora. Z. Zool. Syst. Evolutionsforsch. 25:9–12.

Pianka, H. D. 1974. Ctenophora. *In* A. C. Giese and J. S. Pearse (eds.), *Reproduction of Marine Invertebrates*, Vol. 1. Academic Press, New York, pp. 201–265.

Picard, J. 1955. Les nematocystes du ctenaire *Euchlora rubra* (Kolliker, 1953). Recl. Trav. Stn. Mar. Endoume-Marseille Fasc. Hors. Ser. Suppl. 15:99–103.

Rankin, J. J. 1956. The structure and biology of *Vallicula multiformis* gen. et sp. nov. a platyctenid ctenophore. Zool. J. Linn. Soc. 43:55–71.

Reeve, M. R. and L. D. Baker. 1974. Production of two planktonic carnivores (chaetognath and ctenophore) in South Florida inshore waters. Fish. Bull. U.S. 73:238–248.

Reeve, M. R. and M. A. Walter. 1976. A large-scale experiment on the growth and predation of ctenophore populations. *In* G. Mackie (ed.), *Coelenterate Ecology and Behavior*. Plenum, New York, pp. 187–199.

Reeve, M. R. and M. A. Walter. 1978. Nutritional ecology of ctenophores–a review of recent research. *In* F. S. Russell and M. Yonge, *Advances in Marine Ecology*, Vol. 15. Academic Press, New York, pp. 249–289.

Reeve, M. R., M. A. Walter and T. Ikeda. 1978. Laboratory studies of ingestion and food utilization in lobate and tentaculate ctenophores. Limnol. Oceanogr. 23:740–751.

Robilliard, G. A. and P. K. Dayton. 1972. A new species of platyctenean ctenophore, *Lyrocteis flavopallidus* sp. nov., from McMurdo Sound, Antarctica. Can. J. Zool. 50:47–52.

Stanlaw, K. A., M. R. Reeve and M. A. Walter. 1981. Growth rates, growth variability, daily rations, food size selection and vulnerability to damage by copepods of the early life history stages of the ctenophore *Mnemiopsis mccradyi*. Limnol. Oceanogr. 26:224–234.

Stanley, G. D., Jr. and A. Sturmer. 1983. The first fossil ctenophore from the Lower Devonian of West Germany. Nature 303:518–520.

Stretch, J. J. 1982. Observations on the abundance and feeding behavior of the cestid ctenophore, *Velamen parallelum*. Bull. Mar. Sci. 32(3):796–799.

Sullivan, B. K. and M. R. Reeve. 1982. Comparison of estimates of the predatory impact of ctenophores by two independent techniques. Mar. Biol. 68:61–65.

Tamm, S. L. 1973. Mechanisms of ciliary coordination in ctenophores. J. Exp. Biol. 59:231–245.

Tamm, S. L. 1980. Ctenophores. *In* G. A. B. Shelton (ed.), *Electrical Conduction and Behavior in Invertebrates*. Oxford University Press, New York.

Totton, A. 1954. Egg-laying in Ctenophora. Nature 174:360.

Wiley, A. 1896. *Ctenoplana*. Q. J. Microsc. Sci. 39.

Chapter Ten

Phylum Platyhelminthes

Planarians are hermaphroditic with complicated reproductive systems, unsuitable for beginners to study.
Libbie Hyman
A Laboratory Manual for Elementary Zoology, 1926

The phylum Platyhelminthes (Greek *platy*, "flat"; *helminth*, "worm") includes about 20,000 extant species of free-living and parasitic worms. These animals are at a grade of complexity that may be called the triploblastic acoelomate bilateria. Platyhelminths display a variety of body forms (Figure 1) and are successful inhabitants of a wide range of environments. The majority of flatworms are parasitic members of the classes Trematoda and Monogenea (the flukes) and Cestoda (the tapeworms). The class Turbellaria includes primarily free-living forms in marine and freshwater benthic habitats; a few are terrestrial and some are symbiotic in or on other invertebrates. As their name suggests, most of these animals are strikingly flattened dorsoventrally, although the body shape varies from broadly oval to elongate and ribbon-like; a few bear tentacles at the anterior end or have other elaborations of the body surface. The free-living forms range from less than 1 mm to about 30 cm long. The largest of all flatworms are certain tapeworms that attain lengths of several meters.

The combined features of the platyhelminths may represent a set of derived traits marking major advancements in the evolution of metazoa (Box 1), although some recent work suggests that these animals might have had a coelomate ancestry (see later section on phylogeny). Coupled with a third germ layer (mesoderm), bilateral symmetry, and cephalization are some sophisticated organs and organ systems and a trend toward centralization of the nervous system. The solid (acoelomate) *Bauplan* usually includes a relatively dense mesenchyme between the gut and the body wall. Within this mesenchyme we find discrete excretory/osmoregulatory structures. These structures, **protonephridia**, are found in a number of invertebrate taxa, especially among protostomes and pseudocoelomates. Most flatworms possess complex reproductive systems and an incomplete yet complex gut with a single opening serving for both ingestion and egestion. The mouth leads to a pharynx of varying complexity and thence to a blind intestine. The digestive area contains no permanent cavity in the turbellarian order Acoela, and the gut is entirely lacking in tapeworms.

Taxonomic history and classification

In his first edition of *Systema Naturae* (1735), Linnaeus established two phyla to encompass all of the known invertebrates. To one he assigned the insects and to the other the rest of the invertebrates. Linnaeus called this latter taxon Vermes (Greek, "worms"). By the thirteenth edition of *Systema Naturae* (1788), the various groups of flatworms were placed together in the order Intestina. During the early 1800s, several biologists, including Lamarck and Cuvier, questioned and rejected the concept of the phylum Vermes, although the taxon continued to surface from time to time and actually persisted into the twentieth century as a dumping ground for almost any creatures with wormlike bodies (and many that were not so wormlike).

During the nineteenth century, the unique features of flatworms were gradually recognized, and they were eventually separated from most other groups of worms and wormlike creatures. In 1851, Vogt isolated the flatworms and the nemerteans as a single taxon, which he called the Platyelmia, a name changed to Platyelminthes by Gegenbaur in 1859.

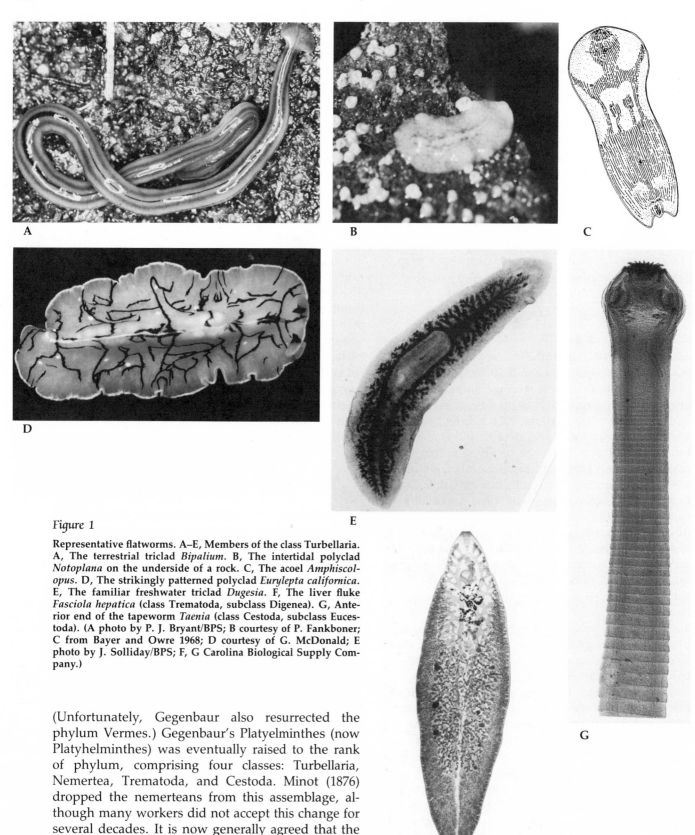

Figure 1

Representative flatworms. A–E, Members of the class Turbellaria. A, The terrestrial triclad *Bipalium*. B, The intertidal polyclad *Notoplana* on the underside of a rock. C, The acoel *Amphiscol-opus*. D, The strikingly patterned polyclad *Eurylepta californica*. E, The familiar freshwater triclad *Dugesia*. F, The liver fluke *Fasciola hepatica* (class Trematoda, subclass Digenea). G, Anterior end of the tapeworm *Taenia* (class Cestoda, subclass Eucestoda). (A photo by P. J. Bryant/BPS; B courtesy of P. Fankboner; C from Bayer and Owre 1968; D courtesy of G. McDonald; E photo by J. Solliday/BPS; F, G Carolina Biological Supply Company.)

(Unfortunately, Gegenbaur also resurrected the phylum Vermes.) Gegenbaur's Platyelminthes (now Platyhelminthes) was eventually raised to the rank of phylum, comprising four classes: Turbellaria, Nemertea, Trematoda, and Cestoda. Minot (1876) dropped the nemerteans from this assemblage, although many workers did not accept this change for several decades. It is now generally agreed that the flukes represent two classes, the Trematoda (digenetic flukes) and the Monogenea (monogenetic

---

Box One
Characteristics of the Phylum Platyhelminthes

1. Parasitic or free-living, unsegmented worms (class Cestoda is strobilated, or segmented)
2. Triploblastic, acoelomate, bilaterally symmetrical; usually flattened dorsoventrally
3. Complex, though incomplete, gut usually present; gut absent in some parasitic forms (Cestoda)
4. Cephalized, with a central nervous system comprising an anterior cerebral ganglion and (usually) longitudinal nerve cords connected by transverse commissures (ladder-like nervous system)
5. With protonephridia as excretory/osmoregulatory structures
6. Hermaphroditic, with complex reproductive systems

---

flukes). However, the classification of flatworms is still a matter of considerable controversy, and consequently is subjected to frequent revisions. No single scheme is likely to be acceptable to all workers. There are about 4,500 species of turbellarians, 9,000 species of flukes, and 5,000 species of tapeworms.

PHYLUM PLATYHELMINTHES

CLASS TURBELLARIA: Free-living flatworms (Figures 1A–E). Predominately free living and aquatic; unsegmented; mouth leads to a stomodeal pharynx, the structure of which differs among orders; epidermis cellular and usually ciliated. The orders of turbellarians were previously grouped into two superorders on the basis of whether yolk is deposited within the cytoplasm of the ova (entolecithal ova) or separately, outside the ova (ectolecithal ova). Those with entolecithal ova were placed in the superorder Archoöphora and those with ectolecithal ova in the Neoöphora. Even though these names have been largely abandoned as formal taxa, the placement of yolk still provides an additional character for describing the orders and has important implications in the early development of these animals. There are 12 recognized orders; those that are relatively common or of particular importance are characterized below.

ORDER NEMERTODERMATIDA: Nemertodermatids. Mouth and pharynx present or absent; pharynx when present simple; gut cavity with interdigitating processes from intestinal lining; uniflagellate sperm (all other flatworms possess sperm with 0 or 2 flagella); with entolecithal ova. These small turbellarians inhabit subtidal marine muds and sands. One genus (*Meara*) is parasitic in sea cucumbers. (e.g., *Flagellophora, Meara, Nemertoderma*)

ORDER ACOELA: Acoels. Pharynx when present is simple; no permanent gut cavity; mouth or pharynx leads to a solid or cellular endodermal mass; entolecithal ova. These small (1–5 mm), common flatworms inhabit marine and brackish water sediments; a few are planktonic or symbiotic. (e.g., *Amphiscolops, Convoluta, Haplogonaria, Polychoerus*)

ORDER CATENULIDA: Catenulids. Simple pharynx; simple, saclike gut; mesenchyme sometimes reduced to a fluid matrix; with entolecithal ova. Catelunids are elongate freshwater and marine forms. (e.g., *Catenula, Paracatenula, Stenostomum*)

ORDER MACROSTOMIDA: Macrostomids. Simple pharynx; simple, saclike gut; with entolecithal ova. These turbellarians are small and predominately interstitial, inhabiting marine and freshwater environments. (e.g., *Macrostomum, Microstomum*)

ORDER LECITHOEPITHELIATA: Pharynx variable; gut simple. Small group of about 30 species united on the basis of an intermediate condition between entolecithal and ectolecithal ova. (e.g., *Gnosonesima, Prorhynchus*)

ORDER RHABDOCOELA: Rhabdocoels. Bulbous pharynx; simple, saclike gut without diverticula; ectolecithal ova produced by ovaries that are usually fully separate from the yolk glands. This extremely large and diverse group is divided into four suborders.

SUBORDER DALYELLIOIDA: Dalyellioids. Mouth anterior; free living or are ecto- or endosymbionts of various marine and freshwater invertebrates. (e.g., *Callastoma, Graffilla, Pterastricola*)

SUBORDER TYPHLOPLANOIDA: Typhloplanoids. Mouth not anterior; free living, marine and fresh water. (e.g., *Kytorhynchus, Mesostoma, Typhlorhynchus*)

SUBORDER KALYPTORHYNCHIA: Kalyptorhynchs. Mouth not anterior; with a complex eversible proboscis at anterior end that is separate from the mouth and pharynx; free living, marine and fresh water. (e.g., *Cystiplex, Gnathorhynchus, Gyratrix*)

SUBORDER TEMNOCEPHALIDA: Temnocephalids. Small symbionts on freshwater decapod crustaceans (a few live on other invertebrates or on turtles); with posterior sucker and anterior tentacles used for attachment and inchworm-like locomotion. (e.g., *Temnocephala*)

ORDER PROLECITHOPHORA: Prolecithophorans. Pharynx plicate or bulbous; gut simple; with ectolecithal ova. Small, free living, marine and fresh water. (e.g., *Plagiostomum, Urostoma*)

ORDER PROSERIATA: Proseriatans. Cylindrical plicate pharynx; simple gut; with ectolecithal ova. Most are free living, marine. (e.g., *Nemertoplana, Octoplana, Taboata*)

ORDER TRICLADIDA: Triclads. Cylindrical plicate pharynx; gut three-branched with numerous diverticula; ectolecithal ova. Marine or fresh water; some terrestrial. Most are free living, including the familiar planarians. (e.g., *Bdelloura, Bipalium, Crenobia, Dugesia, Geoplana, Polycelis, Procotyla*)

ORDER POLYCLADIDA: Polyclads. Most with ruffled plicate pharynx; gut multibranched with diverticula; with entolecithal ova. Polyclads are a diverse group of relatively large turbellarians. Nearly all are marine; they are common in the littoral zone throughout the world, especially in the tropics; predominately benthic and free living; a few are pelagic or symbiotic. (e.g., *Eurylepta, Hoploplana, Leptoplana, Notoplana, Planocera, Pseudoceros, Stylochus, Thysanozoon*)

CLASS MONOGENEA: Monogenetic flukes (Figure 3C). Body covered by a tegument; oral sucker reduced or absent; acetabulum absent; with anterior prohaptor and posterior, hooked opisthaptor; life cycle involves only one host. Most are ectoparasitic, usually on fishes (occasionally on turtles, frogs, hippos, copepods, or squids); a few are endoparasitic in ectothermic vertebrates.

SUBCLASS MONOPISTHOCOTYLEA: Opisthaptor simple and single, but sometimes divided by septa; oral sucker reduced or absent. (e.g., *Gyrodactylus, Polystoma*)

SUBCLASS POLYOPISTHOCOTYLEA: Opisthaptor complex, with multiple suckers; oral sucker absent. (e.g., *Diplozoon*)

CLASS TREMATODA: Digenetic and aspidogastrean flukes (Figures 1F, 3A,B,D). Body covered by a tegument; with one or more suckers; lacking prohaptor and opisthaptor. Most have 2 or 3 hosts during the life cycle; most are endoparasitic.

SUBCLASS DIGENEA: With 2–3 hosts during life cycle; first intermediate host a mollusc, final host a vertebrate; with oral and usually a ventral (acetabulum) sucker. (e.g., *Echinostoma, Fasciola, Microphallus, Opisthorchis* [= *Clonorchis*], *Sanguinicola, Schistosoma*)

SUBCLASS ASPIDOGASTREA: Most with a single host (a mollusc) in life cycle; second host, when present, is a fish or turtle; oral sucker absent, ventral sucker large, divided by septa as a row of suckers. (e.g., *Aspidogaster, Cotylaspis, Multicotyl*)

CLASS CESTODA: Tapeworms (Figures 1G, 4). Exclusively endoparasitic; body covered by a tegument; body usually consisting of an anterior scolex, followed by a short neck, and then a strobila composed of a series of segments or proglottids; digestive tract absent. Cestodes are divided into two subclasses.

SUBCLASS CESTODARIA: Cestodarians. Small group of flattened tapeworms lacking scolex and proglottids; some with suckers. They are endoparasitic in the guts or coelomic cavities of cartilaginous and certain primitive bony fishes, less commonly in turtles. Their life cycles are poorly known. (e.g., *Gyrocotyle, Gyrometra*)

SUBCLASS EUCESTODA: Eucestodes. Often very large (some over 10 m long), segmented tapeworms; almost all with well developed scolex, neck, and strobila. They are endoparasitic in the guts of various vertebrates; most require one or more intermediate hosts during the life cycle. (e.g., *Diphyllobothrium, Dipylidium, Echinococcus, Hymenolepis, Taenia*)

The flatworm *Bauplan*

Compared with taxa discussed in the preceding chapters, the flatworms display some of the most important advances found in the animal kingdom. They represent the acoelomate bilateria and, according to some hypotheses, the basic *Bauplan* from which many of the remaining triploblastic animals may have ultimately been derived.

The evolution of the triploblastic condition and bilateral symmetry almost certainly occurred in concert with the evolution of sophisticated internal "plumbing" (organs and organ systems) and the tendencies to cephalize, to centralize the nervous system, and to develop specialized units within the nervous system for sensory, integrative, and motor activities. With these features came unidirectional movement and a lifestyle more active than that of radially symmetrical animals. The primary evolutionary advantages of these coincidental changes derive chiefly from the ability of these "new" creatures to move around more-or-less freely and thus exploit survival strategies theretofore impossible.

These strategies can be appreciated by examining the rather complex structural features displayed by the free-living turbellarians (Figure 2). The presence of mesoderm facilitates the formation of a fibrous and muscular mesenchyme that provides structural support and allows patterns of locomotion not possible in diploblastic radiates. Osmoregulatory structures in the form of protonephridia were instrumental in the invasion of fresh water. Furthermore, elaborate reproductive systems evolved in the platyhelminths, providing for internal fertilization and enhancing the production of yolky and encapsulated eggs. Most flatworms have abandoned indirect development.

This *Bauplan* is not without constraints, however. Higher energy demands accompany an active lifestyle. The major limiting factor for flatworms, functionally, is the absence of an efficient circulatory mechanism to move materials throughout the body.

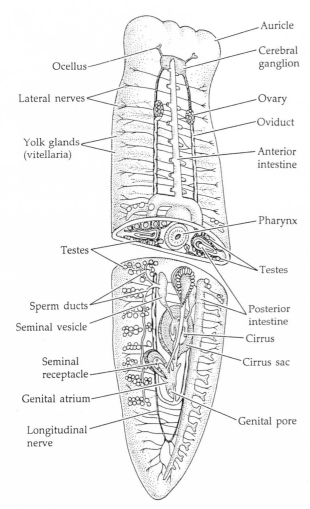

Figure 2

A generalized freshwater turbellarian (order Tricladida). (After various sources.)

areas. They have, however, exploited a variety of marine and freshwater habitats, and are particularly successful as parasites and commensals, enjoying the benefits of living on or in their hosts.

Below we present a comparative discussion of the classes and explore their survival strategies and their diversity within the general platyhelminth *Bauplan*. It is generally assumed that the ancestral flatworm was a free-living form from which the present-day turbellarians evolved and diversified. The flukes and tapeworms were undoubtedly derived from within this varied turbellarian assemblage, as discussed in more detail later. Thus, in each of the following sections we examine the basic features of the turbellarians and set the stage for understanding not only the diversity within that class but the derivation of the specialized parasitic taxa as well.

The basic anatomy of turbellarians, flukes, and tapeworms is shown in Figures 2, 3, and 4. Turbellarians vary from broadly oval to ribbon-like, and are usually flattened dorsoventrally; very small ones are more-or-less cylindrical. The head is usually ill-defined, except for the presence of sense organs. The mouth is located ventrally, either near the middle of the body or more anteriorly. Flukes vary in shape (Figure 3), but most are oval or leaf-shaped. Most species bear various external attachment organs such as hooks and suckers. As their common name suggests, the tapeworms are typically elongate and ribbon-like (Figure 4). The anterior end is a tiny **scolex**, modified for attachment within the host, and the rest of the body is essentially a reproductive machine. They live in the guts of various vertebrates. Most species belong to the subclass Eucestoda and possess three distinguishable regions of the body. The scolex serves for attachment and is usually armed with hooks and suckers. Immediately behind the scolex is a short, unsegmented region called the **neck**, followed by an elongated, segmented trunk, or **strobila**, consisting of individual **proglottids**. The proglottids are produced just behind the neck, an area of strobilation, so the oldest and largest proglottids are at the posterior end of the worm.

Members of the subclass Cestodaria are somewhat flukelike in appearance. They lack a scolex, and the body is not divided into proglottids. They are placed within the Cestoda because of the absence of a digestive tract and because of certain features of the life cycle.

Body wall

The body wall of turbellarians is a layer of various functional cell types (Figure 5). The epidermis is com-

This problem is compounded by the lack of any special structures for gas exchange. These problems relate, of course, to the surface-to-volume dilemma that we discussed earlier (see Chapter 3). In the absence of circulatory and gas exchange structures, the flatworms (particularly the free-living ones) have been limited in terms of size and shape. They have remained relatively small, or have assumed shapes that maintain short diffusion distances. The largest free-living flatworms have highly branched guts that assume much of the responsibility of internal transport.

Having a high surface-to-volume ratio and using the entire body surface for gas exchange create potential problems of ionic balance and osmoregulation, and of desiccation in intertidal and terrestrial habitats. The permeable body surface must be kept moist; thus, flatworms have invaded land only in very damp

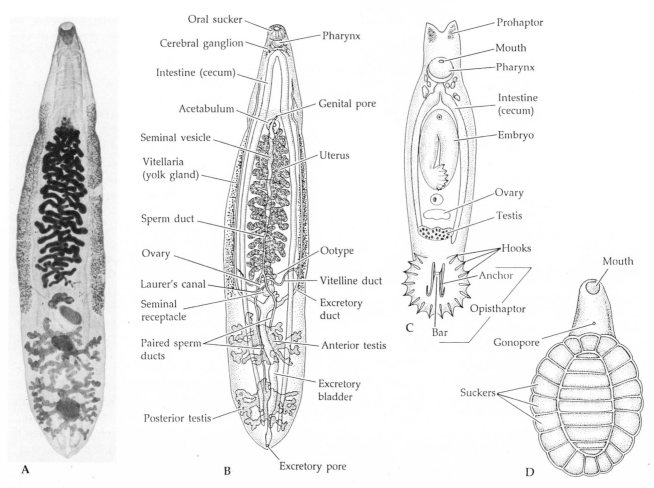

Figure 3

Representative flukes. A–B, *Opisthorchis (Clonorchis) sinensis*, a digenetic fluke that inhabits human livers. C, *Gyrodactylus* (class Monogenea), an ectoparasite on fishes. D, The trematode *Cotylaspis* (subclass Aspidogastrea). (A Carolina Biological Supply Company; B,C after various sources; D after Hyman 1951.)

posed of a wholly or partially ciliated, syncytial or cellular epithelium, with gland cells and sensory nerve endings distributed in various patterns. Beneath the epidermis is a fibrous basement membrane, which is often thick enough to lend some structural support to the body. In the orders Acoela, Catenulida, and Macrostomida, the basement membrane is absent. Internal to the basement membrane are smooth muscle cells, frequently arranged in rather loosely organized outer circular, middle diagonal, and inner longitudinal layers. The area between the body wall and the internal organs is usually filled with a mesenchyme (often called a parenchyme) that includes a variety of loose and fixed cells, muscle fibers, and connective tissue. Most acoels, and perhaps the macrostomids, lack a cellular mesenchyme.

The gland cells of the body wall are generally derived from ectoderm. When mature, many of these cells lie in the mesenchyme, with a "neck" extending between epidermal cells to the body surface. These cells produce mucous secretions that serve a variety of functions. In semiterrestrial and intertidal turbellarians, the mucus provides protection from desiccation and serves as a moist covering that aids in gas exchange. Most benthic flatworms possess a ventral concentration of mucous gland cells that secrete a slime that aids in locomotion. Mucus secretion around the mouth aids in prey capture and swallowing. Other gland cells or complexes of cells provide adhesives for temporary attachment. In some ectocommensal forms (e.g., *Bdelloura* and various temnocephalids; Figure 6) these adhesive glands are associated with special plates or suckers for attachment to the host.

Most turbellarians possess epidermal structures called **rhabdoids** (Figure 5B). These unique rod-

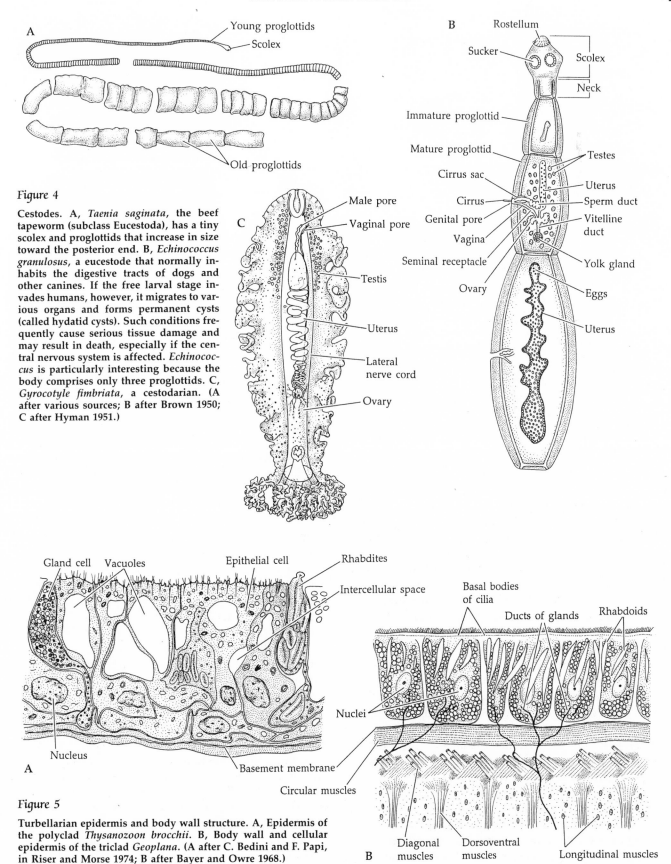

A   Young proglottids
Scolex

Old proglottids

B   Rostellum
Sucker
Scolex
Neck
Immature proglottid
Mature proglottid
Cirrus sac
Cirrus
Genital pore
Vagina
Seminal receptacle
Ovary
Testes
Uterus
Sperm duct
Vitelline duct
Yolk gland
Eggs
Uterus

C   Male pore
Vaginal pore
Testis
Uterus
Lateral nerve cord
Ovary

Figure 4

Cestodes. A, *Taenia saginata*, the beef tapeworm (subclass Eucestoda), has a tiny scolex and proglottids that increase in size toward the posterior end. B, *Echinococcus granulosus*, a eucestode that normally inhabits the digestive tracts of dogs and other canines. If the free larval stage invades humans, however, it migrates to various organs and forms permanent cysts (called hydatid cysts). Such conditions frequently cause serious tissue damage and may result in death, especially if the central nervous system is affected. *Echinococcus* is particularly interesting because the body comprises only three proglottids. C, *Gyrocotyle fimbriata*, a cestodarian. (A after various sources; B after Brown 1950; C after Hyman 1951.)

Gland cell   Vacuoles   Epithelial cell   Rhabdites
Intercellular space
Basal bodies of cilia
Ducts of glands
Rhabdoids
Nuclei
Nucleus
Basement membrane
Circular muscles
Diagonal muscles
Dorsoventral muscles
Longitudinal muscles
A
B

Figure 5

Turbellarian epidermis and body wall structure. A, Epidermis of the polyclad *Thysanozoon brocchii*. B, Body wall and cellular epidermis of the triclad *Geoplana*. (A after C. Bedini and F. Papi, in Riser and Morse 1974; B after Bayer and Owre 1968.)

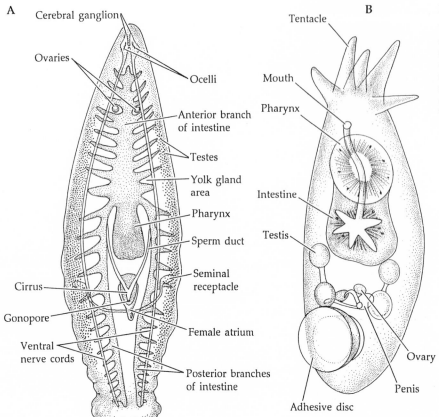

A
Cerebral ganglion
Ovaries
Ocelli
Anterior branch of intestine
Testes
Yolk gland area
Pharynx
Sperm duct
Seminal receptacle
Cirrus
Gonopore
Female atrium
Ventral nerve cords
Posterior branches of intestine
Caudal adhesive disc

B
Tentacle
Mouth
Pharynx
Intestine
Testis
Ovary
Penis
Adhesive disc

Figure 6

Two symbiotic turbellarians with adhesive attachment organs. A, *Bdelloura candida,* **a triclad ectocommensal on horseshoe crabs (***Limulus***). B,** *Temnocephala caeca,* **a rhabdocoel ectocommensal on** *Phreatoicopis terricola* **(a tunneling isopod). (A after Sherman and Sherman 1976; B after various sources.)**

shaped inclusions generally are produced by epithelial cells and then stored in packets within the epidermis. Upon release, rhabdoids produce copious amounts of mucus that may help protect the animal from desiccation and from possible predators. Rhabdoids that are produced by gland cells in the mesenchyme are called **rhabdites**. These structures can reach the body surface through intercellular spaces in the epidermis (Figure 5A) and contribute to mucus production. They may be responsible for the release of noxious defense chemicals by some turbellarians. Some turbellarians have prominent tubercles covering the dorsal surface; these structures probably serve a defensive role. In some species, unfired nematocysts from hydroid prey are transported to the tubercles. In others, such as *Thysanozoon*, the dorsal papillae appear capable of releasing a powerful acid that may deter would-be predators.

Modifications of the outer body covering are common among parasites. Unlike the turbellarians, the flukes and tapeworms possess an external covering called a **tegument**, formed of nonciliated cytoplasmic extensions of large cells whose cell bodies lie

in the mesenchyme (Figure 7). The tegument not only provides some protection but is an important site of exchange between the body and the environment. Gases and nitrogenous wastes are moved across this surface by diffusion, and some nutrients, especially amino acids, are taken in by pinocytosis. In tapeworms, the uptake of nutrients occurs solely across the body wall, and the surface area of the tegument is greatly increased by many tiny folds called **microtriches** (Figure 7B). These folds may interdigitate with the intestinal microvilli of the host and aid in the absorption of nutrients.

The nature of the tegument in flukes and tapeworms is viewed by some zoologists as unique and of major phylogenetic importance (see papers by Ehlers). As discussed later, the larvae of these parasitic worms have a ciliated epidermis over at least part of their bodies. However, this epidermis is shed, and postlarval stages develop a "new," syncytial body covering—the tegument. According to Ehlers (1985) and others, this phenomenon occurs in no other animals and should be regarded as a valid synapomorphy uniting the Trematoda, Monogenea, and Cestoda as a monophyletic taxon that he calls the Neodermata (in reference to the "new skin" of these animals).

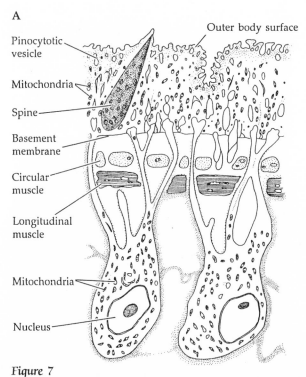

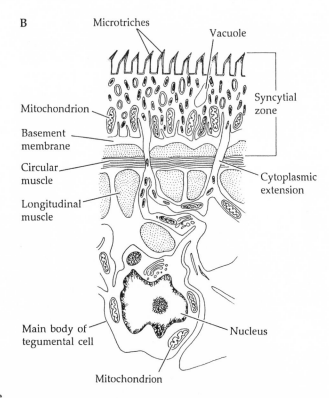

Figure 7

A, The tegument and underlying body wall of a digenetic fluke (*Fasciola hepatica*; longitudinal section). B, The tegument and body wall of a cestode (cross section). (A after L. T. Threadgold, 1963, Q. J. Microsc. Sci. 104; B after Barth and Broshears 1982.)

The tegument is underlaid by a basement membrane, beneath which is the mesenchyme. Within this latter layer, most flukes and tapeworms have circular and longitudinal muscles, and sometimes diagonal, transverse, and dorsoventral muscles as well. The mesenchyme varies from masses of densely packed cells to syncytial and fibrous networks with fluid-filled spaces. In some digenetic flukes, spaces form vessels through the mesenchyme, called lymphatic channels, which contain free cells that have been likened to lymphocytes. The mesenchyme also contains gland cells with connections to the surface of the body through the tegument. These gland cells are few in number compared with those of turbellarians, and they are primarily adhesive in nature and associated with certain organs of attachment.

One of the least explored yet most interesting attributes of tapeworms, and indeed of all intestinal parasites, is their ability to thrive in an environment of hydrolytic enzymes without being digested. One popular hypothesis is that gut parasites produce enzyme inhibitors (sometimes called "antienzymes"). One recent study showed that *Hymenolepis diminuta* (a common tapeworm in rats and mice) releases pro-

teins that appear to inhibit trypsin activity. This tapeworm can also regulate the pH of its immediate environment to about 5.0 by excreting organic acids; this acidic output also may inhibit the activity of trypsin (Uglem and Just 1983).

Support, locomotion, and attachment

Only a very few flatworms possess any sort of special skeletal elements. In a few turbellarians, tiny calcareous plates or spicules are embedded in the body wall (see Rieger and Sterrer 1975). Body support in all other flatworms is provided by the hydrostatic qualities of the mesenchyme, the elasticity of the body wall, and the general body musculature.

Most benthic turbellarians move on their ventral surface by ciliary-powered gliding. Mucus provides lubrication as the animal moves, and serves as a viscous medium against which the cilia act. Some of the larger or more elongate forms also use muscular contractions. The ventral surface of the body is thrown into a series of alternating transverse furrows and ridges that move as waves along the animal, propelling it forward. Muscular undulations of the lateral body margins allow some large polyclads to swim for short periods of time. Very small turbellarians (e.g., acoels) swim or glide by virtue of the action of cilia that cover the entire body surface. Muscular action allows the body to twist and turn, providing steerage and movement among debris and interstitial spaces. Some interstitial forms are highly elongate and use the body wall muscles to slither between sand grains. Many of these types of flatworms possess adhesive glands, the secretions of which provide temporary stickiness and enable the animals to gain purchase and leverage as they move.

Adult flukes lack external cilia, and movement depends on the flukes' own body wall muscles or on the body fluids of their host. Some move about slowly on or within their host by muscle action, and a few (e.g., blood flukes) are carried in the host's circulatory system. Certain larval stages are highly motile and do swim using ciliary action.

Once established within or on a host, it is advantageous for a fluke to stay more-or-less in one place. In that regard, nearly all of them are equipped with external organs for temporary or permanent attachment (Figures 3C and 8). Monogenetic flukes typically have an anterior and a posterior adhesive organ, called the **prohaptor** and the **opisthaptor**, respectively. The prohaptor consists of a pair of adhesive structures, one on each side of the mouth, bearing suckers or simple adhesive pads. The opisthaptor is usually the major organ of attachment, and includes one or more well developed suckers with hooks or claws.

The digenetic flukes possess two hookless suckers. One, the **oral sucker**, surrounds the mouth, and the other, the **acetabulum**, is located on the ventral surface (Figure 3B). These suckers are usually supplied with adhesive gland cells, although the well developed ones operate mainly on suction produced by muscle action. The aspidogastrean flukes lack an oral sucker but have a large, subdivided ventral sucker (Figure 3D).

Adult tapeworms do not move around much, but they are capable of muscular undulations of the body. They remain fixed to the host's intestinal wall by the scolex (or, in the case of members of the subclass Cestodaria, an anterior adhesive organ) and by the microtriches.

The details of scolex anatomy are extremely variable and of critical importance in the taxonomy of the Eucestoda (Figure 9). The tip of the scolex in many cestodes (e.g., *Taenia*) is equipped with a movable hook-bearing **rostellum**, which is sometimes retractable into a cavity in the scolex. In others, such as *Cephalobothrium*, the anterior end bears a protrusible sucker, or adhesive pad, called a **myzorhynchus**. The rest of the scolex bears various sorts of suckers or sucker-like structures and sometimes hooks or spines.

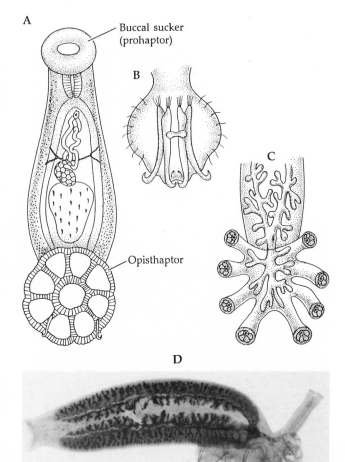

Figure 8

Some attachment organs of monogenetic flukes. A, *Anoplocotyloides papillata*. B–C, Opisthaptors from monogenetic flukes. D, An unidentified fluke with suckered prohaptor and elaborate opisthaptor. (A after Schell 1982; B,C after Marquardt and Demaree 1985; D courtesy of J. DeMartini.)

There are three basic categories of adhesive suckers upon which ordinal and subordinal classification of cestodes is partially based. **Bothria** are elongate, longitudinal grooves on the scolex. They possess weak muscles but are capable of some sucking action. Bothria occur as a single pair and are typical of the order Pseudophyllidea (e.g., *Diphyllobothrium*). Members of the order Tetraphyllidea (e.g., *Acanthobothrium*, *Phyllobothrium*) bear four symmetrically placed **bothridia** around the scolex. These foliose structures are often equipped with suckers at their anterior ends. The third and most familiar type of attachment structures on the scolex are true suckers, or **acetabula**. They are identical in structure and probably homologous to the acetabula of digenetic trematodes.

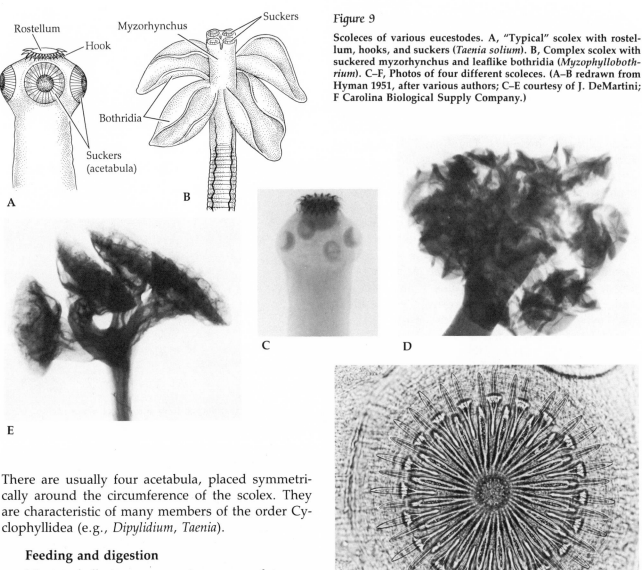

Figure 9
Scoleces of various eucestodes. A, "Typical" scolex with rostellum, hooks, and suckers (*Taenia solium*). B, Complex scolex with suckered myzorhynchus and leaflike bothridia (*Myzophyllobothrium*). C–F, Photos of four different scoleces. (A–B redrawn from Hyman 1951, after various authors; C–E courtesy of J. DeMartini; F Carolina Biological Supply Company.)

There are usually four acetabula, placed symmetrically around the circumference of the scolex. They are characteristic of many members of the order Cyclophyllidea (e.g., *Dipylidium, Taenia*).

Feeding and digestion

Most turbellarians are carnivorous predators or scavengers, feeding on nearly any available animal matter. A few are herbivorous on microalgae; some species switch from herbivory to carnivory as they mature. Their prey includes almost any invertebrate small enough to be captured and ingested (e.g., protozoa, small crustaceans, worms, tiny gastropods). Some species graze on sponges, ectoprocts, and tunicates, while others consume the flesh of barnacles, leaving behind the empty shell. Most turbellarians locate food by chemoreception.

We should mention here that over 100 species of turbellarians are known to be symbiotic with other invertebrates. Some of these are simply commensals that derive some protection from their associations, showing only physical modifications for temporary attachment. Others, however, feed upon their hosts, causing various degrees of damage and displaying true physiological dependency on the relationship. While we can devote space to mentioning only a few examples of symbiotic turbellarians, recognition of these situations is of considerable importance. First, it emphasizes the evolutionary plasticity of the turbellarian *Bauplan*; and second, it provides some essential foundation for our later discussion of the origins of the flukes and tapeworms. (For an excellent survey of the symbiotic turbellarians, see Jennings 1980.)

Most of the symbiotic turbellarians belong to the

order Rhabdocoela (suborders Dalyellioida and Temnocephalida). One notable exception is the triclad *Bdelloura* (Figure 6A), an ectocommensal on the gills of *Limulus*, the horseshoe crab. The temnocephalids (Figure 6B) are ectocommensals within the branchial chambers of freshwater decapod crustaceans, where they feed on microorganisms in the host's gas exchange currents. Several families of dalyellioids include symbiotic members. The umagillids (e.g., *Syndesmis*) live within the gut and coelomic fluid of echinoids (Figure 10). These tiny flatworms feed on protozoa and bacteria, and some may devour cells of their hosts. Graffillid dalyellioids (*Graffilla* and *Paravortex*) include seven species of parasites in the digestive tracts of gastropod and bivalve molluscs. These worms derive their nutrients from the host tissues. Members of the family Fecampiidae (*Fecampia, Kronborgia, Glanduloderma*) are parasites in marine crustaceans and certain polychaete worms. They reside in the host's body fluids and absorb soluble organic nutrients. One dalyellioid, *Oekiocolax* (family Provorticidae), is a parasite of other marine turbellarians. An undescribed species of *Prosthiostomum* (Polycladida) is a parasite on the Hawaiian coral *Montipora* (Jokiel and Townsley 1974).

The feeding methods of free-living turbellarians vary with the size of the animal and the complexity of their food-getting apparatus, especially the pharynx. As noted in the classification scheme, the nature of the pharynx varies greatly among taxa. The details of pharynx structure are discussed below. As far as feeding is concerned, it is important to realize that pharyngeal types range from simple, noneversible, ciliated tubes to complexly folded (**plicate**) eversible structures of different shapes.

Those turbellarians with a simple tubular pharynx are generally quite small, with the mouth frequently located more-or-less midventrally. These forms usually feed by sweeping small organic particles and tiny prey into the pharynx by ciliary action. Those with an eversible pharynx usually fold the body around the prey or other potential food and cover it with mucus from the epidermal glands. Then the pharynx is everted over or into the food item. Some turbellarians, especially triclads, secrete digestive enzymes externally via special glands that empty through the pharyngeal lumen or from the tip of the pharynx. The food is partially digested and reduced to a soupy consistency prior to swallowing. Many other turbellarians swallow their food whole by the action of powerful pharyngeal muscles.

Active prey can be subdued in several ways. Some turbellarians produce mucus, which, in addi-

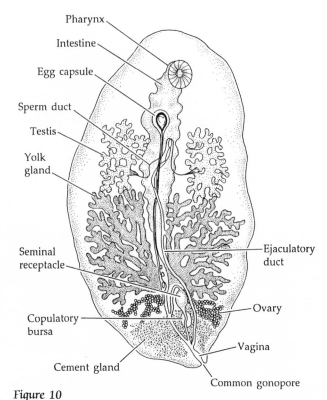

Figure 10

Syndesmis, a rhabdocoel from the gut of a sea urchin. (After Hyman 1951.)

tion to entangling the prey, may contain poisonous or narcotizing chemicals. A few flatworms use the sharp stylet of the copulatory organ to stab prey. One cannot help but concede the remarkable adaptive capacity of the flatworms. Members of the suborder Kalyptorhynchia (order Rhabdocoela) are unique among turbellarians in their possession of a muscular proboscis that is situated at the anterior end of the body and is separate from the mouth (Figure 11E); the proboscis, which in some species is armed with hooks, can be everted to grab prey.

The general plan of the turbellarian digestive system includes a mouth and a pharynx, which lead to a blind intestine, or enteron. The mouth varies in position from midventral to anterior. The pharynx is derived from embryonic ectoderm, so it is really a stomodeum lined with epidermis. Various epithelial **pharyngeal glands** are associated with the lumen of the pharynx; they produce mucus that aids in feeding and swallowing, and (in some) proteolytic enzymes that initiate digestion outside the body. There are three basic pharynx types among the turbellarians: **simple, bulbous,** and **plicate** (Figure 11). A simple pharynx (or **pharynx simplex**) is a short ciliated tube connecting the mouth and intestine (Figure 12). This

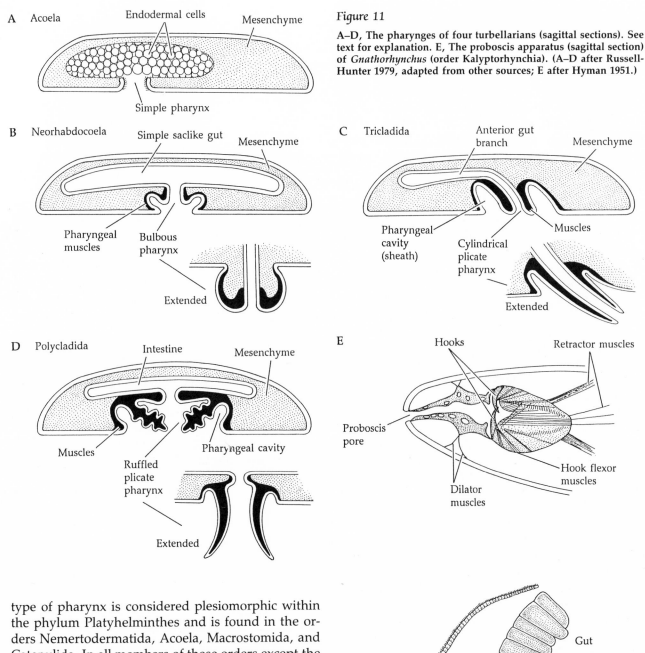

A Acoela

Endodermal cells

Mesenchyme

Simple pharynx

B Neorhabdocoela

Simple saclike gut

Mesenchyme

Pharyngeal muscles

Bulbous pharynx

Extended

C Tricladida

Anterior gut branch

Mesenchyme

Pharyngeal cavity (sheath)

Cylindrical plicate pharynx

Muscles

Extended

D Polycladida

Intestine

Mesenchyme

Muscles

Pharyngeal cavity

Ruffled plicate pharynx

Extended

E

Hooks

Retractor muscles

Proboscis pore

Hook flexor muscles

Dilator muscles

Figure 11

A–D, The pharynges of four turbellarians (sagittal sections). See text for explanation. E, The proboscis apparatus (sagittal section) of *Gnathorhynchus* (order Kalyptorhynchia). (A–D after Russell-Hunter 1979, adapted from other sources; E after Hyman 1951.)

type of pharynx is considered plesiomorphic within the phylum Platyhelminthes and is found in the orders Nemertodermatida, Acoela, Macrostomida, and Catenulida. In all members of these orders except the Acoela, the pharynx leads to a simple saclike or elongate intestine generally lacking diverticula. Members of the order Acoela lack any permanent digestive cavity; instead the pharynx leads to a solid cellular mass of internal digestive tissue (Figure 13A). It was long thought that the digestive mass of acoel turbellarians was a sign of their primitiveness, but newer evidence suggests that this is not the case. This discovery bears significantly on certain phylogenetic hypotheses of flatworm origin and corresponding ideas regarding the origin of the metazoa (see Chapter 4).

Rhabdocoels typically possess a slightly protrac-

Gut

Mouth

Simple pharynx

Figure 12

Sagittal section through anterior end of *Macrostomum* (class Turbellaria, order Macrostomida), which has a simple tubular pharynx. (After Bayer and Owre 1968.)

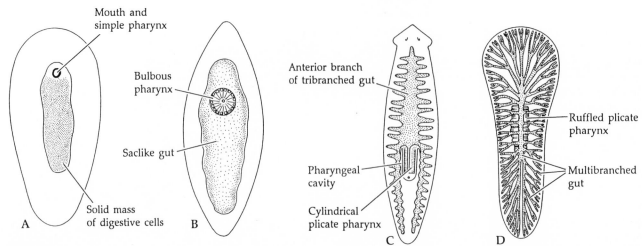

Figure 13

Pharynx type and gut shape combinations among turbellarians. A, Acoela. B, Rhabdocoela. C, Tricladida. D, Polycladida. (B–D after Hyman 1951.)

tile, muscular, bulbous pharynx and a simple saclike gut (Figure 13B). Members of the orders Proseriata, Tricladida, and Polycladida have eversible plicate pharynges. The eversible portion of a plicate pharynx lies within a space called the **pharyngeal cavity**, which is produced by a muscular fold of the body wall (Figures 11 and 13). Proseriates and triclads possess cylindrical plicate pharynges oriented along the body axis. Most polyclads have a ruffled, skirtlike, plicate pharynx attached dorsally within the pharyngeal cavity. During feeding, a plicate pharynx is protruded through the mouth by a squeezing action of extrinsic pharyngeal muscles. Once extended, the pharynx tube can be moved about by intrinsic muscles of the pharyngeal wall. Retractor muscles pull the pharynx back inside the cavity.

Most of the turbellarians that possess a plicate pharynx are relatively large, especially the triclads and the polyclads. Associated with this large body size is an elaboration of the intestine. The triclad intestine comprises three main branches, one anterior and two posterior, each with numerous diverticula; the intestine of polyclads is multibranched with diverticula (Figures 13C and D). These ramifications of the intestine provide not only an increased surface area for digestion and absorption, but also are a means of distributing the products of digestion in the absence of a circulatory system. The lining of the intestine is a single cell layer of phagocytic nutritive cells and enzymatic gland cells (Figure 14). In some groups, the gastrodermis is ciliated.

Digestive physiology has been studied to various degrees in different turbellarian taxa (see Jennings

1974). In most of these animals digestion begins extracellularly, with the action of endopeptidases secreted by the pharyngeal glands or by the enzymatic gland cells of the intestine. The partially digested material is distributed throughout the gut, then phagocytized by the intestinal cells, wherein final digestion occurs. There are, however, some notable exceptions to this basic sequence of digestive events. In some acoels, temporary spaces form within the gastrodermal mass. Primary digestion occurs within these spaces, and the products are phagocytized by the surrounding cells. Bowen (1980) has described an interesting phagocytic process in the small freshwater triclad *Polycelis tenuis*. Following the ingestion of tiny

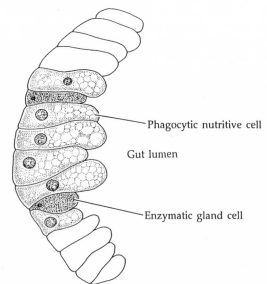

Figure 14

The gut lining (cross section) of a freshwater triclad contains enzymatic gland cells and phagocytic nutritive cells. (After Hyman 1951.)

food particles or the preliminary extracellular digestion of larger food, the intestinal phagocytic cells extend processes into the gut lumen, nearly occluding the digestive cavity. These processes interdigitate to form a complex web, forcing food material into the phagocytes where digestion is completed. Certain polyclads in the suborder Cotylea apparently digest their food entirely extracellularly; phagocytosis of particulate matter is unknown in this group.

Since the gut is generally incomplete, any undigested material must be expelled through the mouth. As discussed in Chapter 3, the major limitation of single-opening guts is the restriction on regional specialization. One macrostomid, *Haplopharynx rostratus*, possesses a minute anal pore, and some polyclads have pores at the ends of gut branches; some proseriates (e.g., *Taboata*) may form a temporary anus.

Adult flukes feed on host tissues and fluids or, in some cases, material within the host's gut. Most of the food is taken in through the mouth by a pumping action of the muscular pharynx, but some organic molecules are picked up across the tegument by pinocytosis. The anterior part of the digestive system includes a mouth, a muscular pharynx, and a short esophagus. The esophagus leads to a pair of intestinal ceca (occasionally, a single cecum), which extend(s) posteriorly in the body (Figures 3 and 15). The lining of the ceca includes absorptive nutritive cells and enzymatic gland cells. Digestion is at least partly extracellular. Some flukes secrete enzymes from the gut out of the mouth, or from the suckers, to partially digest host tissue prior to ingestion.

Cestodes lack any vestige of a mouth or digestive tract, and all nutrients must be taken into the body across the tegument. Uptake probably occurs by pinocytosis and by diffusion across the increased surface area of the microtriches. Some work suggests that tapeworms are unable to take in large molecules and thus rely to a considerable extent on the digestive processes of their hosts and the secretion of enzymes outside their bodies to chemically reduce the size of potential nutrient material. It has also been proposed that the surface of the scolex may absorb host tissue fluids through the site of attachment to the gut wall.

Circulation and gas exchange

As mentioned earlier in this chapter, except for the lymphatic channels in some flukes, flatworms lack special circulatory or gas exchange structures. This condition imposes restrictions on size and shape. The key to survival with such limitations and a generally solid mesenchyme is the maintenance of small diffusion distances. Thus, the flatness of their

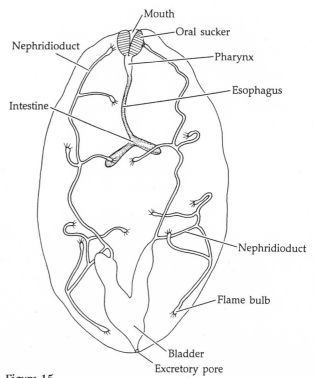

Figure 15

Gut and protonephridial system of *Microphallus* (subclass Digenea). See also Figure 3. In most monogenetic flukes, the protonephridial ducts are separate and terminate anteriorly in separate pores. (After Hyman 1951.)

bodies facilitates gas exchange across the body wall, between the tissues and the environment; nutrients are distributed internally by the digestive system and by diffusion, which is aided by general body movements.

The endoparasitic flatworms are capable of surviving in areas of their hosts where oxygen is absent. In such cases, they rely on anaerobic metabolism, producing a variety of reduced end products (e.g., lactate, succinate, alanine, long-chain fatty acids). Curiously, these anaerobic animals also possess the appropriate enzymes for and are capable of aerobic respiration in the presence of oxygen.

Excretion and osmoregulation

One of the major advances of flatworms over diploblastic animals is the development of protonephridia. These structures occur in all turbellarians except members of the orders Nemertodermatida and Acoela, and some marine catenulids. Turbellarian protonephridia are flame bulbs and may occur singly (as they do in some catenulids) or in pairs (from one to a few pairs in different taxa). The protonephridia are connected to variously complex networks of col-

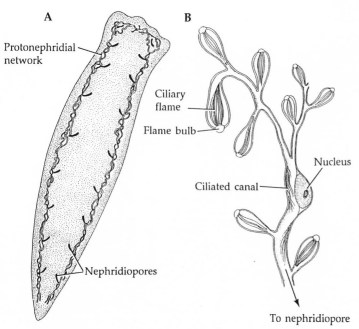

Figure 16

A, The protonephridial system in a freshwater triclad. B, The nephridial arrangement in a turbellarian that has anucleate flame bulbs attached to collecting tubules. (From Bayer and Owre 1968.)

lecting tubules that lead to one or more nephridiopores (Figure 16).

Turbellarian protonephridia function primarily as osmoregulatory structures. Freshwater turbellarians tend to have more protonephridia and more complex tubule systems than their marine counterparts do. Although a small amount of ammonia is released via the protonephridia, most metabolic wastes are lost by diffusion across the body wall.

Flukes also possess variable numbers of flame bulb protonephridia. Two nephridioducts drain the nephridia and lead to a storage area, or **bladder**, which in turn connects with a single posterior nephridiopore in the digenetic flukes or a pair of anterior pores in the monogenetic types (Figure 15). Nitrogenous waste in the form of ammonia is excreted largely across the tegument, and the protonephridia are primarily osmoregulatory, at least in the class Monogenea.

Tapeworms possess numerous flame bulb protonephridia throughout the body. The flame bulbs drain to pairs of dorsolateral and ventrolateral nephridioducts that run the length of the body (Figure 17). Although some variation in plumbing occurs, the ventral ducts are typically connected to one another by transverse tubules near the posterior end of each proglottid. In relatively young worms that have not

lost any proglottids (see the section on reproduction, below), the excretory ducts lead to a collecting bladder in the most posterior proglottid. Once this terminal proglottid is lost, the nephridioducts open separately to the outside on the posterior margin of the remaining hindmost proglottid.

There is still much to be learned about the functioning of the protonephridia in cestodes. They probably function both in excretion and osmoregulation. They may also serve to eliminate certain organic acid products of their anaerobic cellular metabolism. Some experimental work indicates that tapeworms are capable of precipitating and storing some wastes within the body.

Nervous system and sense organs

The nervous system of turbellarians varies from a simple netlike nerve plexus with only a minor concentration of neurons in the head (e.g., acoels) to a distinctly bilateral arrangement with a well developed cerebral ganglion and longitudinal nerve cords connected by transverse commissures (Figure 18). The more advanced condition is referred to as a **ladder-like** nervous system. Even many of those turbellarians that possess distinctly centralized nervous systems have a plexus formed by the repeated branching of nerve endings (e.g., polyclads). In general, larger flatworms show an increasing concentration of the peripheral nerves into fewer and fewer

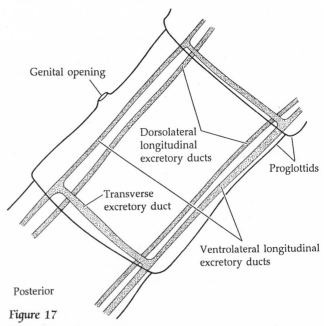

Figure 17

The arrangement of major protonephridial ducts in a eucestode proglottid (ventral view).

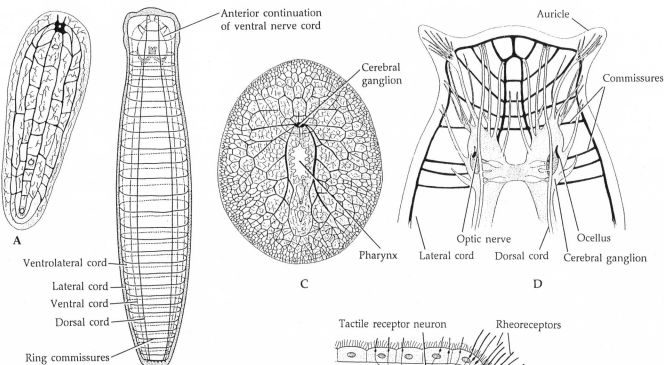

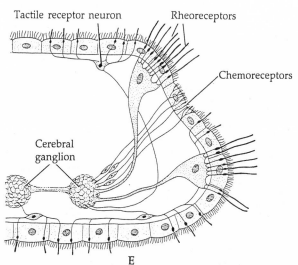

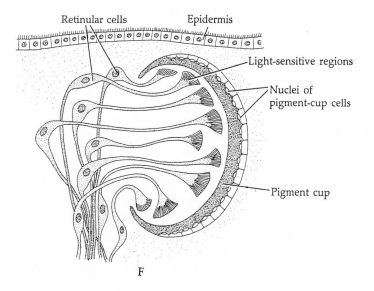

Figure 18

Turbellarian nervous systems and sense organs. A, The netlike nervous system of the acoel *Convoluta*. Note the weak concentration of anterior neurons. B, The ladder-like nervous system of the rhabdocoel *Bothrioplana*. C, The nervous system of the polyclad *Planocera*. D, The cerebral ganglion and associated nerves in the triclad *Crenobia*. E, The anterior end (cross section) of the rhabdocoel *Mesostoma*, showing tactile, chemo-, and rheoreceptors. (Rheoreceptors detect water movements over the surface of the animal.) F, A typical inverted pigment-cup ocellus (section). (From Bayer and Owre 1968.)

longitudinal cords and an accumulation of neurons in the head as an associative center or cerebral ganglion. Furthermore, they show a tendency to separate the elements of the nervous system into distinct sensory and motor pathways and to develop a circuitry that operates primarily on unidirectional impulse transmission.

The turbellarian nervous system and sense organs probably evolved in association with bilateral symmetry and unidirectional movement. The result is a general concentration of sense organs at the anterior end of the body and an elaboration of those receptor types that are compatible with the turbellarian lifestyle. Tactile receptors are abundant over much of the body surface as sensory bristles projecting from the epidermis. These receptors tend to be concentrated at the anterior end and around the pharynx. Benthic turbellarians orient to the substratum by touch; they are positively thigmotactic ventrally and negatively thigmotactic dorsally.

Most turbellarians are equipped with chemoreceptors that aid in location of food. Although sensitive over most of the body, turbellarians have distinct concentrations of chemoreceptors anteriorly, particularly on the sides of the head. Some forms, such as the familiar freshwater planarians, have the chemoreceptors located in flaplike processes called **auricles** on the head (Figures 2 and 18D), whereas others have these sense organs in ciliated pits, on tentacles, or distributed over more-or-less the whole anterior end of the body. The epithelium bearing the chemoreceptors is often ciliated and frequently forms depressions or grooves. The cilia serve to circulate water, thus facilitating sensory input from the environment.

The utilization of chemoreception in locating food has been demonstrated in many turbellarians. Some are known to home in on concentrations of dissolved chemicals associated with potential food. Others, such as *Dugesia*, can be seen to "hunt" by constantly waving the head back and forth, exposing the auricles to any chemical stimulus in their path. When exposed to diffuse chemical attractants, some turbellarians begin a trial-and-error behavior pattern. If unable to determine the direction of the attractant, the worm begins moving in a straight line. If the stimulus weakens, the animal makes apparently random turns until it encounters sufficient stimulus, then moves toward it in a straight line. This behavior can eventually bring the animal near enough to the food source to home in on it directly.

Statocysts are common in certain turbellarians, notably in members of the Nemertodermatida, Acoela, Catenulida, and Proseriata. These orders include mostly swimming and interstitial forms in which orientation to gravity could not be accomplished by touch. When present, the statocyst is usually located on or near the cerebral ganglion. Some turbellarians orient to water movements by rheoreceptors located on the sides of the head (Figure 18E).

Most turbellarians possess photoreceptors in the form of inverted pigment-cup ocelli (Figure 18F). A few types of acoels and macrostomids possess simple pigment-spot ocelli, which are presumed to be primitive within the flatworms. Many turbellarians bear a single pair of ocelli on the head; but some, such as certain polyclads and terrestrial triclads, may have many pairs of eyes. In a few terrestrial forms (e.g., *Geoplana mexicana*) and many of the large tropical polyclads, numerous eyes extend along the edges of the body. Most free-living turbellarians are negatively phototactic. The dorsal placement of the eyes and the

orientation of the pigment cups facilitate the detection of light direction as well as intensity. Larvae of the flatworm *Pseudoceros canadensis* possess two dissimilar kinds of eyes. The right eye appears to be microvillar (i.e., rhabdomeric), but the left one has components of both microvillar and ciliary origin (Eakin and Brandenberger 1980). The phylogenetic lineages of these two eye types were noted in Chapter 3. The discovery of both types of eyes in a flatworm larva suggests to some researchers the possibility that this animal stands at a major point of evolutionary divergence.

Neurosecretory cells have been known in various turbellarians for over three decades, and work continues on exploring their functions. These special cells are generally located in the cerebral ganglion, but they also occur along major nerve cords in at least some species. Neurosecretions probably play important roles in regeneration, asexual reproduction, and gonad maturation.

The nervous system of flukes is distinctly ladder-like and very similar to that in many turbellarians (Figure 19). The cerebral ganglion comprises two well-defined lobes connected by a dorsal transverse commissure. Nerves from the cerebral ganglion extend anteriorly to supply the area of the mouth, adhesive organs, and any cephalic sense organs. Ex-

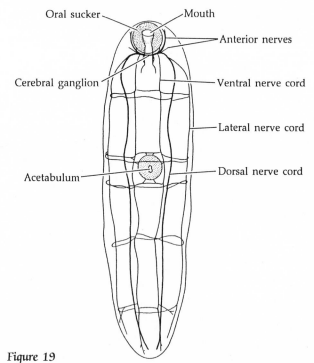

Figure 19

The ladder-like nervous system of a fluke (ventral view). (After Hyman 1951.)

tending posteriorly from the cerebral ganglion are up to three pairs of longitudinal nerve cords with transverse connectives. A pair of ventral cords is most well developed, and dorsal cords are present in the digenetic flukes. Most flukes also have a pair of lateral nerve cords.

The suckers of flukes bear tactile receptors in the form of bristles and small spines. There is also some evidence of reduced chemoreceptors. Nearly all monogenetic flukes possess a pair of rudimentary pigment-cup ocelli near the cerebral ganglion.

The cerebral ganglion of cestodes is usually a complex nerve ring located in the scolex (Figure 20). The ring bears ganglionic swellings and gives rise to a number of nerves. Anterior nerves, in the form of a ring or plexus, serve the rostellum (when present) and other attachment organs. Lateral cerebral ganglionic swellings give rise to a pair of major lateral longitudinal nerves, which extend the length of the animal. In each proglottid, these nerves bear additional ganglia from which transverse commissures arise and connect the two longitudinal cords. Additional longitudinal nerves are often present; the most typical pattern includes two pairs of accessory lateral cords—a pair of dorsal cords and a pair of ventral cords.

As might be expected, sense organs are greatly reduced in cestodes and are limited to abundant tactile receptors in the scolex. These tactile receptors are associated with the organs of attachment.

Reproduction and development

Asexual processes. Asexual reproduction is common among freshwater and terrestrial turbellarians, and it generally occurs by transverse fission. In the catenulids and macrostomids, an odd sort of multiple transverse fission occurs wherein the individuals thus produced remain attached to one another in a chain until they mature enough to survive as individuals (Figure 21A). Some freshwater triclads (e.g., *Dugesia*) split in half behind the pharynx, and each half goes its own way, eventually regenerating the lost parts. A few (e.g., *Phagocata*) reproduce by fragmentation, each part encysting until the new worm forms.

The remarkable regenerative abilities of most turbellarians have been studied intensely for a number of years. Much of the experimental work has been conducted on the common triclad *Dugesia*, a familiar animal to beginning zoology students. Underlying all of the bizarre results of various surgeries performed on these animals (see Figures 21B and C) is the fact that the cells of organisms like *Dugesia* are not totipotent. There exists an anterior–posterior body polarity in terms of the regenerative capabilities of the cells. The cells in the midbody region are less fixed in their potential to produce other parts of the body than are those toward the anterior or posterior ends. Thus, if the flatworm is cut through the middle of the body (as it is in normal transverse fission), each half will regenerate the corresponding lost part. However, if the animal is cut transversely near one end, say separating a small piece of the tail from the rest of the body, the larger piece will grow a new posterior end, but the piece of tail lacks the capability to produce an entire new anterior end. This gradient of cell potency has intrigued cell biologists and medical researchers because of its relevance to healing and regeneration potential in higher animals.

Asexual reproduction is common and is an important feature of the life cycle of digenetic flukes. As we have seen in other examples of parasitic strategies (e.g., certain protozoa), the ability to reproduce asexually helps ensure survival.

Sexual reproduction. Turbellarians are hermaphroditic and possess complex and highly diverse reproductive systems (Figure 22). The following generalizations will help you understand how the systems function.

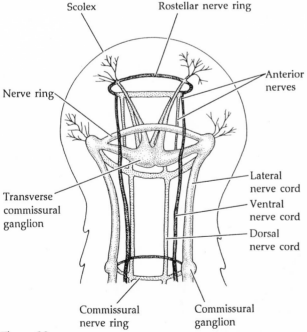

Figure 20

Cestode nervous system. The anterior end of *Moniezia*. The longitudinal cords extend the length of the animal. (After Hyman 1951.)

Labels in figure: Scolex · Rostellar nerve ring · Nerve ring · Anterior nerves · Transverse commissural ganglion · Lateral nerve cord · Ventral nerve cord · Dorsal nerve cord · Commissural nerve ring · Commissural ganglion

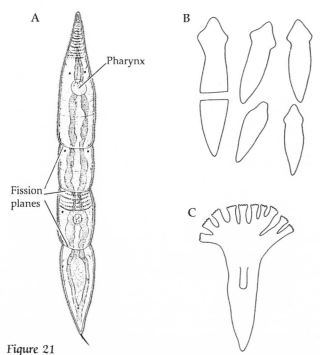

The male system includes single (e.g., macrostomids), paired (e.g., many rhabdocoels), or multiple (e.g., polyclads) testes. The testes are generally drained by collecting tubules that unite to form one or two sperm ducts, which often lead to a precopulatory storage area or seminal vesicle. **Prostatic glands**, which supply seminal fluid to the sperm, are often associated with and empty into the seminal vesicle. The seminal vesicle is typically part of a muscular chamber called the **male atrium**, which houses the copulatory organ. The actual organ of sperm transfer may be a papilla-like penis or an eversible **cirrus**, through which sperm are forced by muscular action of the atrium.

The female reproductive system is more variable than that of the male. Much of the variation is related

Figure 21

A, Asexual reproduction by transverse fission in the catenulid rhabdocoel *Alaurina*. B–C, Regeneration after injury in planarians. (A from Bayer and Owre 1968, after Ivanov 1955; B after various sources; C after Hyman 1951.)

Figure 22

Turbellarian reproductive systems. A, Generalized acoel condition, without separate yolk glands (archoöphoran condition). B, Generalized triclad condition with separate ovaries and yolk glands (neoöphoran condition). C, The copulatory structures of a triclad (sagittal section). (A after von Graaf 1904–1908; B,C from Bayer and Owre 1968.)

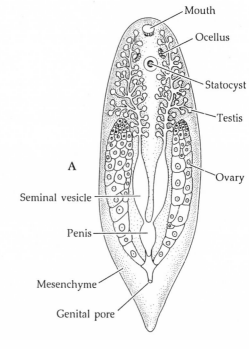

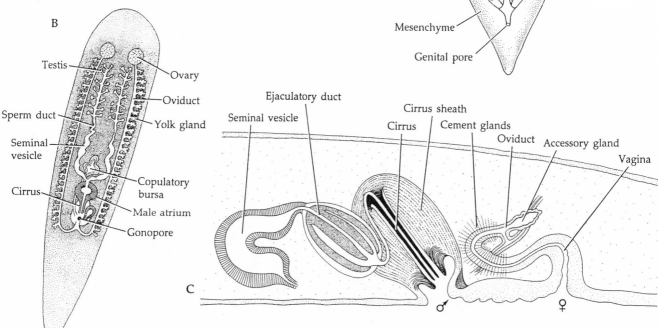

to whether the flatworm in question produces entolecithal or ectolecithal ova—that is, whether the worm is described as archoöphoran or neoöphoran (see the classification scheme). The archoöphorans (e.g., members of the orders Nemertodermatida, Acoela, Macrostomida, and Polycladida) typically possess an organ that produces both eggs and yolk. The final product is entolecithal ova. Such an organ is called a **germovitellarium**, and may occur either singly or paired. In the neoöphorans (e.g., members of the orders Rhabdocoela, Prolecithophora, and Tricladida), the ovary (**germarium**) is separate from the yolk gland (**vitellarium**). Yolk-free eggs are produced by the ovary and then yolk is transported through a vitelline duct and deposited alongside the ova inside the eggshell, a process resulting in ectolecithal ova.

In either case, the eggs are typically moved via an oviduct toward the **female atrium**, which often bears special chambers for receipt and storage of sperm (i.e., **copulatory bursa** and **seminal receptacle**). Associated with this arrangement may be a variety of accessory glands, such as cement glands and glands for the production of shells and egg cases.

The male and female gonopores are often separate, the female opening usually located posterior to the male pore. In some species, however, the two systems share a common genital opening, and in a few the male atrium opens just inside the mouth. In the latter cases, the mouth is referred to as an **orogenital pore**.

Mating is usually by mutual cross-fertilization. The two mates align themselves so that the male gonopore of each is pressed against the female gonopore of the mate (Figure 23A). The male copulatory organ (penis or cirrus) is everted by hydrostatic pressure caused by the muscles surrounding the atrium and then is inserted into the mate's female atrium, where sperm are deposited. The mates then separate, each going its own way and carrying foreign sperm. Fertilization usually occurs as the eggs pass into the female atrium or within the oviduct itself. The zygotes are frequently stored for a period of time in special parts of the female system or in enlarged oviducts; any such storage area is called a **uterus**. A few turbellarians exhibit **hypodermic impregnation**, whereby the male copulatory organ is thrust through the body wall of the mate; the sperm are forcibly injected into the mesenchyme. By some method not yet understood, the sperm find their way to the female system and fertilize the eggs.

Once fertilization is accomplished, the zygotes are either retained by the parent within the uterus of the female reproductive tract or laid in various sorts of gelatinous or encapsulated egg masses (Figure 23B). Thus, most maternal turbellarians are obliged to contribute substantially toward the care of their embryos; they may be described as oviparous or ovoviviparous. Some freshwater triclads produce special overwintering zygotes, which are encapsulated and retained within the female reproductive tract until spring.

The general strategy of the vast majority of turbellarians is to produce relatively few zygotes, which are protected by brooding or encapsulation and undergo direct development. The only notable exceptions to this scheme occur in a few polyclads that

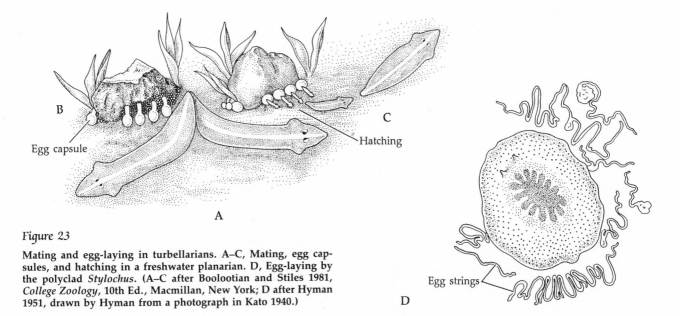

Figure 23

Mating and egg-laying in turbellarians. A–C, Mating, egg capsules, and hatching in a freshwater planarian. D, Egg-laying by the polyclad *Stylochus*. (A–C after Boolootian and Stiles 1981, *College Zoology*, 10th Ed., Macmillan, New York; D after Hyman 1951, drawn by Hyman from a photograph in Kato 1940.)

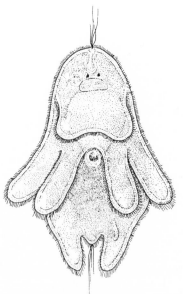

Figure 24

Face-on view of Müller's larva of a polyclad (*Planocera*). (From Bayer and Owre 1968, after Kato 1940.)

produce a **Müller's larva**, which swims about for a few days prior to settling and metamorphosing (Figure 24). This larva is equipped with eight ventrally directed ciliated lobes, by means of which it swims. A few species of parasitic polyclads of the genus *Stylochus* produce a **Götte's larva**, which bears four rather than eight lobes; and members of the freshwater catenulid genus *Rhynchoscolex* pass through a vermiform stage in their development that has been referred to as a larval form (Ruppert 1978).

Early embryogeny differs greatly between the archoöphoran and neoöphoran turbellarians. The entolecithal ova of the archoöphoran worms (e.g., acoels, polyclads, macrostomids) undergo some form of spiral cleavage, the details of which we have described in Chapter 4. The pattern and cell fates in many of these archoöphoran embryos are distinctly protostome-like, and appear to represent an evolutionary step or precursor to the protostome line. (We examine this relationship more fully at the end of this chapter.)

The typical spiral cleavage in polyclads has been well studied. Quartets of cells are produced, the fates of which may be described using Wilson's coding system (Chapter 4). By the end of spiral cleavage, the embryo is considered a stereoblastula, oriented with the derivatives of the first micromere quartet at the animal pole and the macromeres at the vegetal pole. The $1q$ cells become the anterior ectoderm, cerebral ganglion, and most of the rest of the nervous system.

The $2q$ derivatives contribute to ectoderm and ectomesoderm, particularly that of the pharyngeal apparatus and its associated musculature. The remainder of the ectoderm and probably some ectomesoderm are formed from the derivatives of the third micromere quartet. The $4d$ cell, normally associated solely with entomesoderm in typical protostomes, divides to produce $4d_1$ and $4d_2$ cells in polyclads. The $4d_1$ gives rise to entoderm and thus to the intestine; the $4d_2$ produces the entomesoderm from which the body wall muscles, mesenchymal muscles, much of the mesenchymal mass, and most of the reproductive system are derived. The remaining cells ($4a$, $4b$, $4c$, and the $4Q$) include most of the yolk and are incorporated into the developing archenteron as embryonic food.

Gastrulation is by epiboly of the presumptive ectoderm derived from some of the cells of the first three micromere quartets. The ectoderm grows from the animal pole to the vegetal pole, surrounding the $4q$ and $4Q$ cells. At the vegetal pole the ectoderm turns inward as a stomodeal invagination, which later elaborates as the pharynx and connects with the developing intestine (Figure 25). As development proceeds, the embryo flattens, with the mouth directed ventrally, and hatches as a tiny polyclad. If development is mixed, the larva emerges about the time the intestine is hollowing.

Members of the order Acoela are unique in that spiral cleavage occurs by the production of duets rather than quartets of cells (Figure 26). Thus, spiral displacement begins during the division from two to four cells, and a 32-cell stereoblastula is formed. The $4A$, $4B$, $4a$, and $4b$ cells move inward from the vegetal pole and contribute to the majority of the inner cellular mass, particularly the digestive cells. The rest of the interior of the animal apparently derives from ectomesoderm arising from the second and third micromere duets, while the first duet divides to form an outer ectoderm and the nervous system. The inner cellular mass is never readily divisible into entoderm and mesoderm; and, because no $4d$ cell is produced, no "true" entomesoderm forms.

Because of the deposition of separate yolk cells on the surface of ectolecithal ova, the development of neoöphoran turbellarians is highly modified from the plan described above. Certain species of rhabdocoels and triclads have been most extensively studied, and the two groups differ—especially in the early stages. In both cases cleavage is so distorted that cell fates and germ layer formation cannot be easily compared with the typical spiralian pattern. In the rhabdocoels, early cleavage leads to the formation

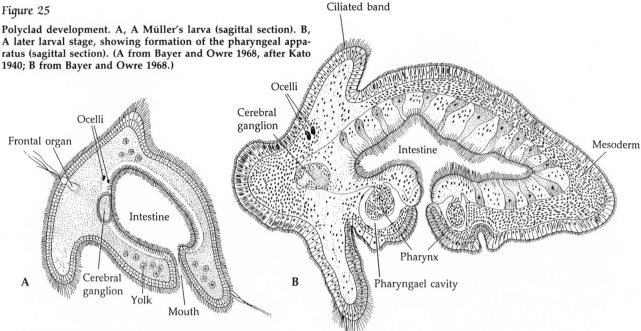

Figure 25

Polyclad development. A, A Müller's larva (sagittal section). B, A later larval stage, showing formation of the pharyngeal apparatus (sagittal section). (A from Bayer and Owre 1968, after Kato 1940; B from Bayer and Owre 1968.)

of three masses of cells positioned along the presumptive ventral surface of the embryo beneath the mass of yolk (Figure 27A). The cell masses then produce a layer of cells that extends around to enclose the yolk. This covering thickens to several cell layers, the innermost eventually becoming the intestinal lining (and enclosing the yolk), the outermost becoming the epidermis. The anterior cell mass produces the nervous system, the middle cell mass the pharynx and associated muscles. The posterior cell mass forms the rear portion of the worm and the reproductive system.

Early development in triclads differs from that of other neoöphorans. During early cleavage, the blastomeres are loose within a surrounding mass of fluid yolk. A few of the blastomeres migrate away from the others and flatten to produce a thin membrane enclosing a packet of the yolk, including the remaining blastomeres (Figures 27B and C). Additional yolk cells are produced as a syncytial mass around a group of developing embryos and encapsulated, as many as 40 per capsule. Through migration and differentiation of various blastomeres, each embryo forms a temporary intestine, pharynx, and mouth, through which it ingests the yolky syncytium. The embryonic mouth eventually closes, and the wall of the embryo thickens to form anterior, middle, and posterior cell masses, whose fates are similar to those in rhabdocoels.

Like the turbellarians, flukes are hermaphroditic and typically engage in mutual cross-fertilization. Self

fertilization only occurs in rare cases. There is a great deal of variation in the details of the reproductive systems among flukes, but most are built around a common plan similar to that in certain turbellarians (Figure 28). The male system includes a variable number of testes (usually many in the monogenetic flukes and two in the digenetic flukes), all of which drain to a common sperm duct that leads to a **copulatory apparatus**, usually an eversible cirrus. The lumen of the cirrus is continuous with that of the sperm duct, and their junction is frequently enlarged as a seminal vesicle. Prostatic glands are typically present, opening into the cirrus lumen near the seminal vesicle. All of these terminal structures are housed within a muscular **cirrus sac**, the contraction of which causes eversion of the cirrus as an intromittent organ (Figure 28C). The common genital pore opens ventrally near the anterior end of the animal and leads to a shallow atrium, usually shared by both the male

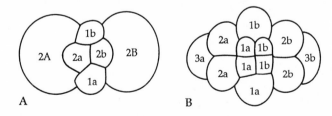

Figure 26

Spiral cleavage in the acoel *Polychoerus*. Animal pole views of 6-cell (A) and 16-cell (B) stages, illustrating the formation of duets rather than quartets of micromeres. (After Hyman 1951.)

Figure 27

Neoöphoran development. A, The embryo of a typical rhabdocoel
has three cell masses with large, vacuolated, external yolk cells.
B, A triclad egg capsule containing three embryos surrounded by
yolk syncytium. C, This single triclad embryo has ingested the
yolk through the temporary embryonic pharynx, and shows the
three cell masses. See text for additional explanation. (Redrawn
and stylized from several sources.)

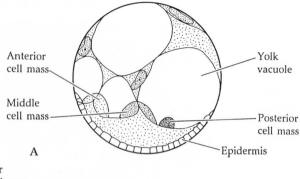

Anterior cell mass

Middle cell mass

Yolk vacuole

Posterior cell mass

Epidermis

A

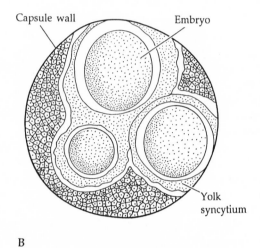

Capsule wall

Embryo

Yolk syncytium

B

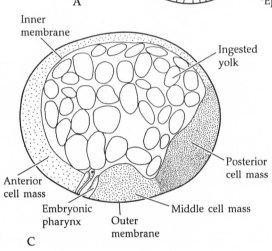

Inner membrane

Ingested yolk

Posterior cell mass

Anterior cell mass

Embryonic pharynx

Outer membrane

Middle cell mass

C

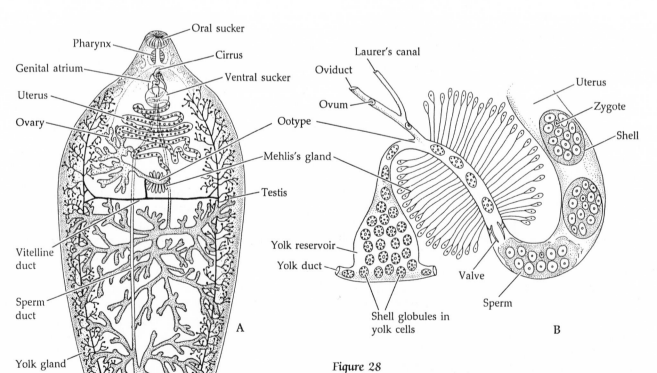

Oral sucker

Pharynx

Cirrus

Genital atrium

Ventral sucker

Uterus

Ovary

Ootype

Mehlis's gland

Testis

Vitelline duct

Sperm duct

Yolk gland

Testis

A

Laurer's canal

Oviduct

Ovum

Uterus

Zygote

Shell

Yolk reservoir

Yolk duct

Valve

Sperm

Shell globules in yolk cells

B

Figure 28

Fluke reproductive system. A, Reproductive structures of *Fasciola
hepatica*. B, The region of the ootype in *F. hepatica*. C, Male
copulatory apparatus with cirrus extended. (A after various
sources; B after Barnes 1980, from Smyth and Clegg 1959; C after
Noble and Noble 1982.)

and female systems. Many monogenetic flukes have simpler male systems than that just described, often lacking much elaboration of the terminal structures and possessing a simple penis papilla rather than an eversible cirrus.

The female reproductive system (Figure 28B) usually bears a single ovary connected by a short oviduct to a hollow structure called the **ootype**. The oviduct is joined by a **yolk duct** (= vitelline duct) formed by the union of paired ducts, which carry yolk from the multiple laterally placed yolk glands. A seminal receptacle is usually present as a blind pouch off the oviduct. Extending anteriorly to the genital atrium is a single uterus, which is sometimes modified as a **vagina** near the female gonopore.

Sperm are produced in the testes and stored prior to copulation in the seminal vesicle. During mating, two flukes align themselves such that the cirrus of each can be inserted into the female orifice of the other. Sperm, along with semen from the prostatic glands, are ejaculated into the female system by muscular contractions. The sperm move to, and are stored within, the seminal receptacle, and the mates separate. As eggs pass through the oviduct to the ootype, they are fertilized by sperm released from the seminal receptacle into the oviduct. Flukes produce ectolecithal ova. The yolk glands produce yolk, which is deposited outside the eggs along with secretions that form a tough shell around the zygote. Thus encapsulated, the zygotes move from the ootype into the uterus, probably aided by secretions from clusters of unicellular **Mehlis's glands**. The zygotes may be stored within the uterus for various lengths of time prior to release through the female gonopore.

Some flukes possess an additional canal that arises from the oviduct and serves as a special copulatory duct. This duct, called **Laurer's canal**, opens on the dorsal body surface and receives the male cirrus during mating. A few polyclad and triclad turbellarians also possess a Laurer's canal.

High fecundity is a general rule among parasites, and the flukes are no exception. The dangers of complex life cycles and host location result in extremely high mortality rates that must be offset by increased zygote production or asexual processes. Flukes may produce as many as 100,000 times as many eggs as free-living turbellarians.

The early stages of development in flukes are usually highly modified, as they are in neoöphoran turbellarians, because of the ectolecithal nature of the ova. In species where little yolk is present, cleavage is holoblastic, and cell fates and germ-layer formation have been traced accurately. Development is always indirect, involving one or more independent larval stages.

The life cycles of monogenetic flukes are relatively simple and involve only a single host. Most of the adults are ectoparasites on fishes, although some attach to turtles, various amphibians, and even some invertebrates. A few members of the Monogenea have taken up mesoparasitic life and reside in host body chambers that open to the environment (e.g., gill chambers, mouth, bladder, cloacal cavity). When the embryos are released from the uterus they often attach to the host tissue by means of special adhesive threads on the shell. Upon hatching, a larval stage called an **oncomiracidium** is released to the environment (Figure 29). The oncomiracidium is densely ciliated and swims about until it encounters another appropriate host. The prohaptor and especially the opisthaptor develop during the larval stage and facilitate attachment to the new host, whereupon the larva metamorphoses to a juvenile trematode. It is about this time that the ciliated larval skin is shed and the tegument (= **neodermis**) is formed. There are many variations on this basic life cycle among members of the class Monogenea, and we present only two in outline form in Figure 30.

The trematode subclass Digenea includes some of the most successful parasites known. A good deal of variation exists not only in adult morphology but also in the life cycles (Figure 31). In general, eggs are

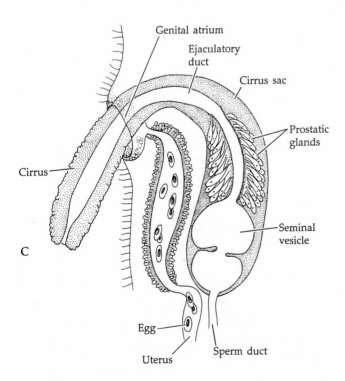

C

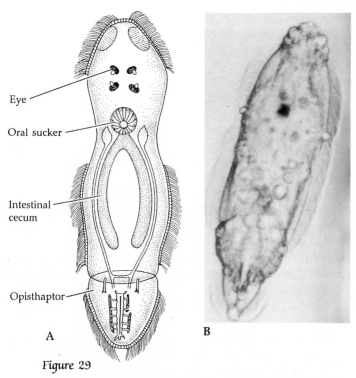

Figure 29

A, An oncomiracidium larva of a monogenetic fluke. B, An oncomiracidium. (A redrawn from several sources; B courtesy of J. DeMartini.)

produced by adult worms in their definitive host and are eventually discharged via the host's feces, urine, or sputum. Eggs reaching the water either are eaten by an intermediate host or hatch as free-swimming ciliated larvae called **miracidia**, which actively penetrate an intermediate host. Several asexual generations of larval forms occur in the intermediate host, eventually producing a free-swimming form called a **cercaria** (Figure 31). The cercaria usually encysts within a second intermediate host as a **metacercaria**. Infection of the definitive host occurs when the metacercaria are eaten, or, when there is no second intermediate host, when the cercaria penetrate directly. The larval skin is lost in the definitive host, and the syncytial tegument then develops.

In their adult stages, nearly all of the digenetic flukes are endoparasites of vertebrates. They are known to inhabit nearly every organ of the body, and many are serious pathogens of humans and livestock. The intermediate hosts of most digenetic flukes are gastropods, although some are known to use other invertebrates or even certain vertebrates. Most species of snails host one or more species of digenetic flukes; the common California tidal flat gastropod *Cerithidea californica* serves as intermediate host to

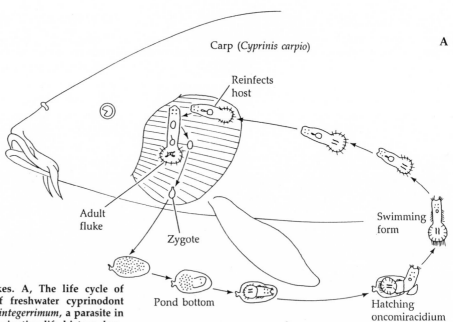

Figure 30

Life cycles of two monogenetic flukes. A, The life cycle of *Dactylogyrus vastator*, a parasite of freshwater cyprinodont fishes. B, The life cycle of *Polystoma integerrimum*, a parasite in the urinary bladders of frogs. This fascinating life history demonstrates the rather dramatic influences exerted by the developmental stage of the host on the development of the parasite. Under normal conditions, the adult fluke resides in the bladder of adult frogs. The fluke releases fertilized eggs into the water, where they hatch as oncomiracidia. These in turn become so-called gyrodactylid larvae, which attack the tadpole larval stages of the host. If the tadpole is very young, the fluke larvae attach to the external gills of the host and undergo precocious sexual maturation to produce more zygotes; these flukes die upon metamorphosis of the host. However, if the fluke larvae encounter

more advanced tadpoles, they enter the branchial chambers and attach to the host's internal gills, where they reside until the host undergoes metamorphosis. At that time, the flukes leave the branchial chamber, migrate to the cloacal pore, and enter the host's bladder. Here the flukes live and grow, but they do not become sexually active until they are influenced by the host's sex hormones. Thus, sexual reproduction of the host and its parasites are synchronized, a pattern guaranteeing the availability of larval hosts for larval parasites! **(A after Olsen 1974; B adapted from Smyth 1977.)**

nearly two dozen fluke species, most of which ultimately infect shore birds. Space does not permit an account of more than a few of the life cycles of these worms. We begin below with a general case, using the Chinese liver fluke (*Opisthorchis sinensis*) as an example. This trematode may be familiar to you from general zoology courses, and it includes all of the common stages found in the life cycles of most digenetic flukes.

The adult liver fluke usually lives within the branches of the bile duct in humans. This animal may reach several centimeters in length and in high numbers causes serious problems. While still in the uterus of the female reproductive tract, the zygotes develop to miracidia, each housed within its original egg case. Once released from the female system and passed out of the host with the feces, the miracidia are eaten by the first intermediate host, a snail of the genus *Bythinia*. The ciliated, swimming miracidium hatches from its egg case in the gut of the snail and then migrates into the digestive gland. Here, each miracidium becomes an asexually active form called a

sporocyst, within which germinal cells become yet another larval form called a **redia**. Subsequently, germinal cells within the redia produce larvae called **cercariae**. This double sequence of rapid asexual reproduction results in perhaps 250,000 cercariae from each original miracidium!

The cercariae leave the snail, swim about, and enter the second intermediate host, the Chinese golden carp (*Macropodus opercularis*). The cercariae of *Opisthorchis* burrow through the skin of the fish and encyst in the muscle tissue as **metacercariae**. If the fish is insufficiently cooked and then eaten by a human, the metacercariae survive and are released from their cysts by the action of the host's digestive enzymes. Once freed, they migrate into the bile duct, metamorphose into juvenile worms, mature, and complete the cycle. With this general life cycle in mind, we refer you to the illustrations and caption in Figure 31 for a brief overview of two additional examples.

The critical point here is the strategy for survival displayed by these parasites. The advantages of such

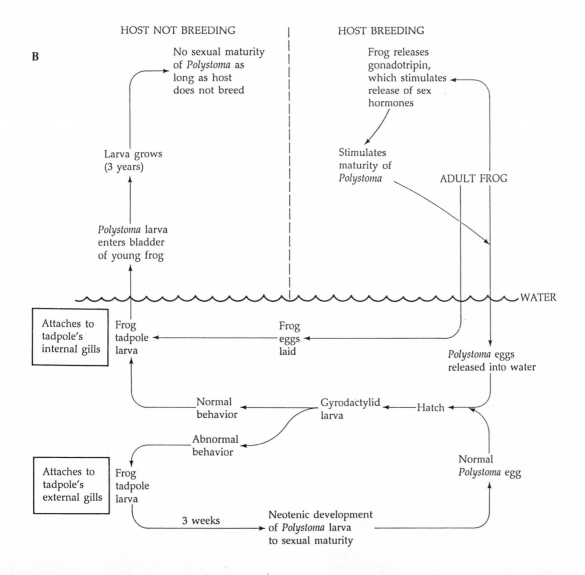

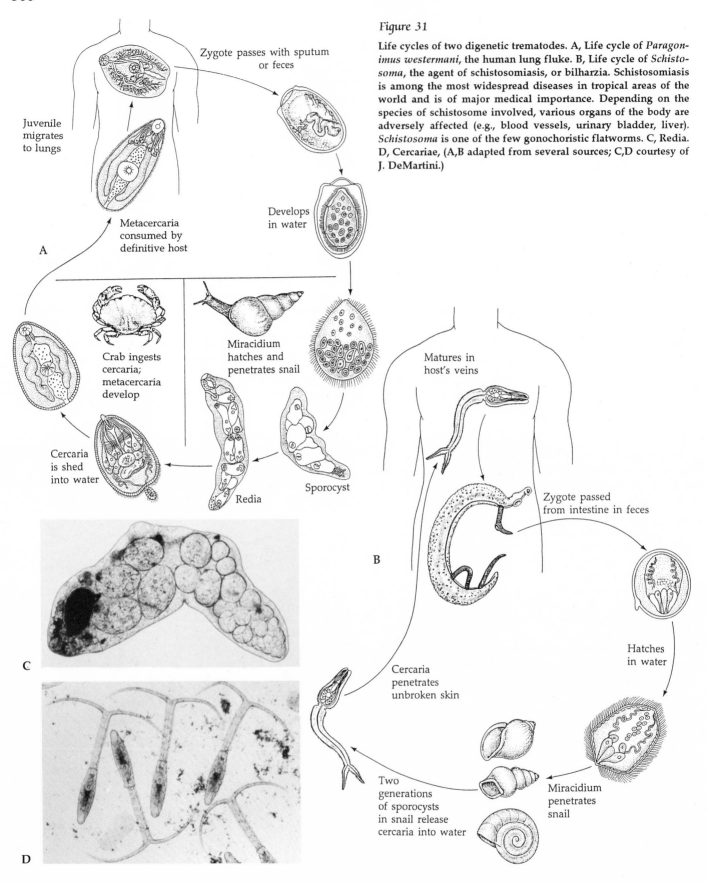

Zygote passes with sputum or feces

Juvenile migrates to lungs

Metacercaria consumed by definitive host

A

Develops in water

Figure 31

Life cycles of two digenetic trematodes. A, Life cycle of *Paragonimus westermani*, the human lung fluke. B, Life cycle of *Schistosoma*, the agent of schistosomiasis, or bilharzia. Schistosomiasis is among the most widespread diseases in tropical areas of the world and is of major medical importance. Depending on the species of schistosome involved, various organs of the body are adversely affected (e.g., blood vessels, urinary bladder, liver). *Schistosoma* is one of the few gonochoristic flatworms. C, Redia. D, Cercariae, (A,B adapted from several sources; C,D courtesy of J. DeMartini.)

Crab ingests cercaria; metacercaria develop

Miracidium hatches and penetrates snail

Cercaria is shed into water

Redia

Sporocyst

C

D

Matures in host's veins

B

Zygote passed from intestine in feces

Hatches in water

Cercaria penetrates unbroken skin

Two generations of sporocysts in snail release cercaria into water

Miracidium penetrates snail

high specialization are efficiency and reduced competition with other species (once established in a proper host). It is, however, an expensive strategy. There is, of course, no assurance of finding the proper hosts at the proper times, and mortalities are incredibly high. As we have emphasized earlier, the compensation for these mortalities is high fecundity coupled with asexual reproduction—and therein lies the expense.

That the business of animals is to reproduce themselves (to paraphrase Barrington) is a lesson well learned by the cestodes. Most of their time, energy, and body mass is devoted to the production of more tapeworms. Cestodes are hermaphroditic and practice mutual cross-fertilization when mates are available. However, many eucestodes are known to self-fertilize.

The cestodarians possess a single male and a single female reproductive system, whereas the eucestodes contain complete systems repeated in each proglottid. There is a good deal of variation in the details of these systems, but they are basically constructed along the plan described for trematodes. The following is a generalized description of the male and female systems as they occur in a single proglottid of a eucestode (Figure 32).

The testes are numerous. Some are scattered throughout the mesenchyme, but most are concentrated along the lateral body margins. Collecting tubules lead from the testes to a single coiled sperm duct, which extends laterally to a cirrus housed within a muscular **cirrus sac**. The male system empties into a common genital atrium.

The female system usually includes two ovaries from which an oviduct extends to an ootype surrounded by shell glands. The ootype is typically situated at the anterior end of the proglottid. The uterus is a branched blind sac extending from the ootype. A duct extends from the female gonopore in the genital atrium to the oviduct; its junction with the oviduct, near the tube, is swollen to serve as a seminal receptacle. The portion of the duct near the genital atrium is called the vagina. A diffuse yolk gland empties via a vitelline duct into the oviduct.

Mating is typically by mutual cross-fertilization, wherein the cirrus of each mate is inserted into the vagina of the other. Some tapeworms double back on themselves and two proglottids of the same worm cross-fertilize; in a few cases self-fertilization occurs within a single proglottid. Sperm are injected into the vaginal duct and are stored in the seminal receptacle. Eggs are fertilized as they move through the oviduct from the ovaries to the ootype. Capsule material and yolk cells are deposited around each zygote, and the zygotes are moved into the uterus for temporary storage. The reproductive systems mature and become functional with age as they are moved more posteriorly along the body by the production of new proglottids. If mating has occurred, the proglottids toward the posterior end become filled with an expanded uterus engorged with developing embryos (Figure 32B). These proglottids eventually break free from the body and are lost from the host with the feces.

Early development of tapeworm embryos is drastically modified from the turbellarian pattern and var-

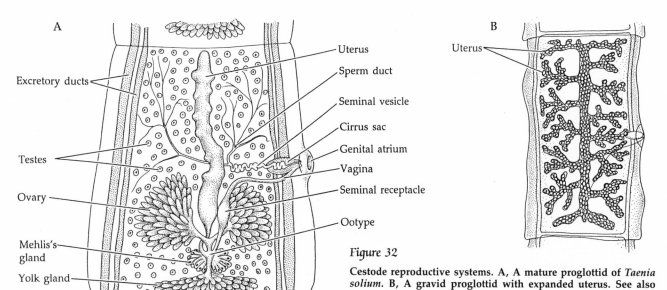

Figure 32

Cestode reproductive systems. **A,** A mature proglottid of *Taenia solium*. **B,** A gravid proglottid with expanded uterus. See also Figure 4. (A after Marquardt and Demaree 1985; B after Noble and Noble 1982.)

ies somewhat among different groups. The ectolecithal ova have lost most vestiges of spiral cleavage, and even germ-layer formation is often difficult or impossible to trace.

The life cycles of cestodes are usually quite complex, involving two or more hosts. A few have become secondarily simplified through the elimination of one or more stages, and the life cycles of many groups are still poorly understood. Described below is the life cycle of *Diphyllobothrium latum*, the so-called broad fish tapeworm (order Pseudophyllidea), which includes two intermediate hosts (Figure 33).

Nearly any fish-eating mammal, including humans, can serve as the definitive host for this tapeworm. The encapsulated embryos are shed with the host's feces and hatch in water to release a ciliated free-swimming larval stage called the **coracidium** (Figure 33). The coracidium is eaten by a copepod (Crustacea), which is the first intermediate host. The ciliated epithelium is lost and replaced by the tegu-

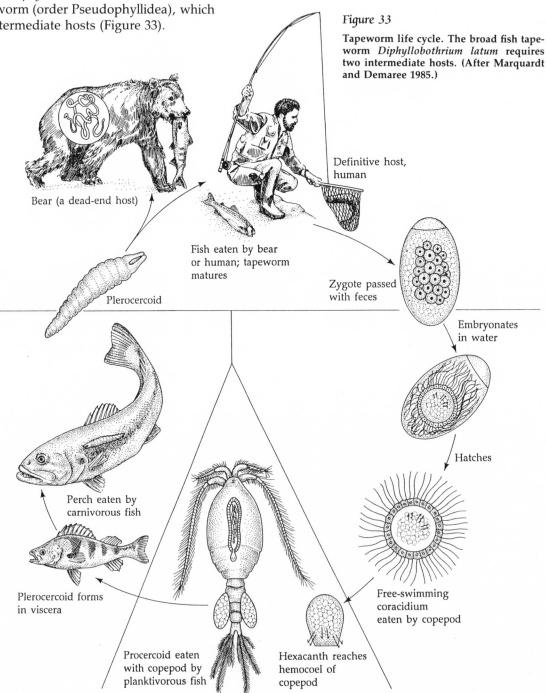

Figure 33

Tapeworm life cycle. The broad fish tapeworm *Diphyllobothrium latum* requires two intermediate hosts. (After Marquardt and Demaree 1985.)

Definitive host, human

Bear (a dead-end host)

Fish eaten by bear or human; tapeworm matures

Plerocercoid

Zygote passed with feces

Embryonates in water

Perch eaten by carnivorous fish

Hatches

Plerocercoid forms in viscera

Procercoid eaten with copepod by planktivorous fish

Hexacanth reaches hemocoel of copepod

Free-swimming coracidium eaten by copepod

ment. Within the copepod, the coracidium becomes a second larval stage, the **procercoid**, which remains inside the copepod until both are eaten by a fish, which serves as the second intermediate host. Within the fish the procercoid migrates from the gut to the body muscles where it encysts as a **plerocercoid** stage, which has the basic form of an unsegmented juvenile. If the fish is eaten by a mammal, the plerocercoid attaches to the gut, begins strobilation, and matures.

It should be understood that many tapeworms (e.g., order Cyclophyllidea) require only one intermediate host; some require none, inhabiting only one definitive host. In various cases certain larval stages are modified or bypassed completely.

Platyhelminth phylogeny

Ideas about the origin of flatworms, their relationship to other taxa, and evolution within the group have been hotly debated for decades. We hinted at some of this controversy earlier in this chapter, and we discussed some of its implications in Chapter 4. In this section we first explore some hypotheses about flatworm origin and then examine some views on the relationships of taxa within the phylum. As you will see, opinions on these matters differ greatly and reflect some extremely diverse views about the position of flatworms in animal evolution.

There have been several hypotheses concerning the origin of flatworms that have received much support. The ciliate-to-acoel hypothesis (discussed in Chapter 4 as part of the syncytial theory of Hadži and others) has been abandoned by most modern zoologists. It is no longer tenable in its original form, due in part to the discovery that the endodermis of acoels is, in fact, cellular. Another hypothesis has been called the ctenophore–polyclad theory. Some workers have suggested that the ctenophores (Chapter 9) gave rise to polyclad turbellarians. This scenario envisions a flattened ctenophore that assumed a benthic, crawling lifestyle, with the mouth directed against the substratum. By reducing the tentacles and moving them forward along with the apical sense organ, a bilateral condition was achieved. Couple these events with increased gut branching and the formation of a plicate pharynx, and a polyclad *Bauplan* is approximated—at least on paper. This hypothesis, too, no longer has much popular support.

Having dispensed with the above proposals, at least temporarily, we can examine the major persist-

ing ideas on the origin of the flatworm body plan. We can safely assume that the original flatworm was turbellarian-like, although not necessarily assignable to any extant order. Peter Ax (1963) and Tor Karling (1974) have derived versions of the turbellarian archetype. Both versions have a simple pharynx and a saclike gut without diverticula (Figures 34A and C). Another popular version of the archetype is an acoel form, lacking any gut cavity (Figure 34B), although both Ax and Karling hold that the solid acoel condition was secondarily derived early in the evolution of the Turbellaria. Furthermore, both agree that the

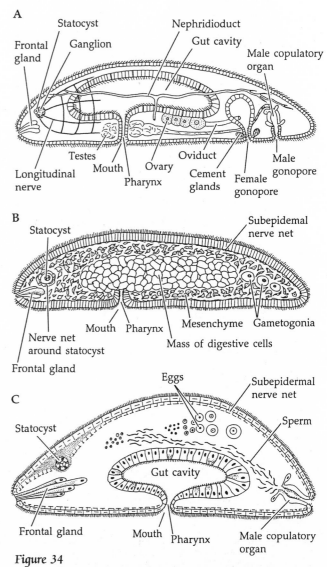

Figure 34

Three examples of hypothetical turbellarian archetypes. A, Macrostomid-like archetype suggested by Ax (1963). B, Acoel-like archetype proposed by several authors. C, Archetype proposed by Karling (1974). (A after Ax 1963; B after Barnes 1980; C after Karling 1974.)

primitive turbellarian also included an archoöphoran condition with spiral cleavage and a single-layered, completely ciliated epidermis.

We are left today with two very different ideas on the origin of the Platyhelminthes, regardless of which of the above archetypes seems most probable. First, as discussed in Chapter 4, is the hypothesis that turbellarian flatworms arose from some diploblastic, radially symmetrical, planuloid ancestor. Various versions of this scenario enjoy some support, but all share a common theme that the very first triploblastic, bilaterally symmetrical organisms were like flatworms. This view implies that the acoelomate condition is plesiomorphic within triploblastic phyla, at least within the lineage leading to the protostomes. Taking this idea further, the flatworms represent the

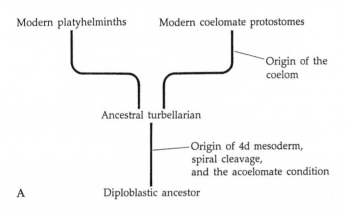

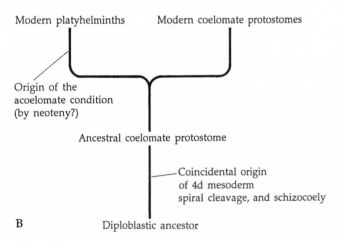

A

B

Figure 35

Two views on the origin of flatworms and the acoelomate condition. A, An evolutionary tree based on the assumption that the acoelomate condition is primitive within the triploplastic spiralians, making Platyhelminthes the "first" descendant group on this lineage. B, An evolutionary tree based on the coincidental origin of the coelom, spiral cleavage, and mesoderm; here the flatworms are viewed as arising (by neoteny?) from a coelomate ancestor.

"first" spiralian taxon, and the stem group from which the protostome clade arose. An evolutionary tree depicting these events is shown in Figure 35A.

In 1963, Peter Ax summarized the work of other zoologists in support of the so-called spiralian theory and explained its implications for the origin of the Platyhelminthes. Ax suggested that the flatworms represent a series of reductions from a vermiform coelomate ancestor. Since that time, other authors (see Rieger 1986, and papers in Morris et al. 1985, especially Smith and Tyler) have presented mounting evidence that the platyhelminth acoelomate condition is secondarily derived. The proposals that the acoelomate condition arose from a coelomate ancestor implies that the origin of the protostome mesoderm probably occurred simultaneously with the origin of the coelom, and that acoelomate animals (and probably pseudocoelomate ones as well) branched from that coelomate clade (Figure 35B). Some workers (Rieger 1986) suggest that these lineages may have arisen through neoteny from developmental stages of protostomes prior to the embryonic appearance of the coelomic cavities—that is, from larval or other stages that had solid bodies (were "acoelomate") or still contained the blastocoel (were "pseudocoelomate"). Controversies like this illustrate why it is so difficult to derive formal cladograms at the phylum level.

Being uncertain about the origin of the flatworms obviously creates difficulties in analyzing phylogeny within the phylum. These problems are compounded by the absence of clear synapomorphies defining the Platyhelminthes. Any cladogram, therefore, must be taken as tentative at the present time. Figure 36A is a cladogram that emphasizes the uniqueness of the tegument as a synapomorphy uniting the Monogenea, Trematoda, and Cestoda, as discussed earlier in the chapter. Our cladogram is admittedly a simple one, based on very few characters (some of which are questionable), but does reflect current understanding of this phylum. However, several authors have presented alternative hypotheses or refinements of the one shown here. For example, based on similarities in larval armature, Ehlers (1985) suggested that the Cestoda and Monogenea are sister taxa.

Figure 36B is a traditional (orthodox) evolutionary tree illustrating one set of hypotheses about overall platyhelminth relationships, including the proposition that the parasitic taxa arose from a dalyellioid-like turbellarian ancestor. Again, much of this is highly speculative. The situation is in such a state of flux that we anticipate substantial changes in the next

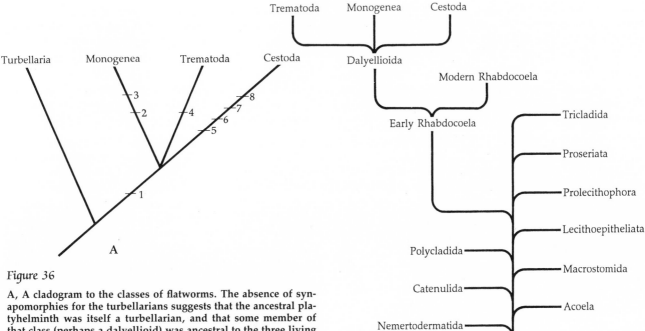

Figure 36

A, A cladogram to the classes of flatworms. The absence of synapomorphies for the turbellarians suggests that the ancestral platyhelminth was itself a turbellarian, and that some member of that class (perhaps a dalyellioid) was ancestral to the three living parasitic groups. The classes Monogenea, Trematoda, and Cestoda share the unique synapomorphy of a tegument that arises during development through replacement of the larval epidermis by extensions of mesenchymal cells (1). This arrangement is strengthened by the exclusively parasitic nature of members of these three classes. We cannot convincingly resolve the trichotomy leading to these three taxa, but each line is defined by synapomorphies as follows: prohaptor (2); opisthaptor (3) (Monogenea); ventral sucker or acetabulum in most (4) (Trematoda); tegumental microtriches (5); scolex (6); loss of digestive tract (7); and segmentation of the body (8) (Cestoda). B, A traditional evolutionary tree depicting general ideas about the relationships among the orders of Turbellaria and the parasitic classes.

few years, perhaps leading to a fundamental restructuring of flatworm classification. The interested reader should stay alert for new publications by Daniel Brooks and Ulrich Ehlers, among others.

Selected References

We have included only a few major references on the trematodes and cestodes, and refer you to texts on general parasitology and helminthology for additional literature lists. A few such texts are listed at the end of our reference compilation.

General

Ax, P. 1987. *The Phylogenetic System.* Wiley, New York. [Quite difficult reading but with an interesting chapter on platyhelminth phylogeny.]

Brooks, D. R. 1982. Higher classification of parasitic Platyhelminthes and fundamentals of cestode classification. *In* D. F. Mettrick and S. S. Dresser (eds.), *Parasites— Their World and Ours.* Elsevier Biomedical Press, Amsterdam.

Beauchamp, P. de, M. Caullery, L. Euzet, P. Grassé and C. Joyeux. 1961. Plathelminthes et Mesozoaires. *In* P. Grassé (ed.), *Traité de Zoologie,* Vol. 4, Pt. 1. Masson et Cie, Paris, pp. 1–729.

Dougherty, E. C. 1963. *The Lower Metazoa: Comparative Biology and Phylogeny.* University of California Press, Berkeley and Los Angeles. [See in particular Sections I (3, 4, 8, 14, 15, 16), II (26), and III (33).]

Ehlers, U. 1984. *Das phylogenetische System der Platyhelminthes.* Akademie der Wissenschaften und der Literatur, Mainz, and Gustav Fischer Verlag, Stuttgart.

Ehlers, U. 1986. Comments on a phylogenetic system of the Platyhelminthes. Hydrobiologia 132:1–12.

Florkin, M. and B. T. Scheer. 1968. *Chemical Zoology,* Vol. 2. Academic Press, New York.

Gegenbaur, C. 1859. *Grundzuge der vergleichenden Anatomie.*

Hyman, L. H. 1951. *The Invertebrates,* Vol. 2, Platyhelminthes and Rhynchocoela: The Acoelomate Bilateria. McGraw-Hill, New York.

Llewellyn, J. 1965. The evolution of parasitic platyhelminths. *In* A. E. R. Taylor (ed.), *Evolution of Parasites.* Symp. Brit. Soc. Parasitol. Vol. 3. Blackwell Scientific Publications, Oxford.

Llewellyn, J. 1986. Phylogenetic inference from platyhelminth life-cycle stages. Int. J. Parasitol. 17:281–289.

Martin, G. G. 1978. Ciliary gliding in lower invertebrates. Zoomorphologie 91:249–262.

Morris, S. C., J. D. George, R. Gibson and H. M. Platt (eds.). 1985. *The Origins and Relationships of Lower Invertebrates.* The Systematics Association, Spec. Vol. No. 28. Oxford. [Includes several papers addressing recent views on flatworm phylogeny; see especially papers by Ax, Ehlers, and Smith and Tyler.]

Rieger, R. M. 1986. Über den Ursprung der Bilateria: Die Bedeutung der Ultrastrukturforschung für ein neues Verstehen der Metazoenevolution. Verh. Dtsch. Zool. Ges. 79:31–50.

Rose, S. M. 1970. *Regeneration*. Appleton-Century-Crofts, New York.

Turbellaria

Ax, P. 1963. Relationships and phylogeny of the Turbellaria. *In* E. C. Dougherty (ed.), *The Lower Metazoa: Comparative Biology and Phylogeny*. University of California Press, Berkeley and Los Angeles, pp. 191–224.

Ax, P. and G. Apelt. 1965. Die "Zooxanthellen" von *Convoluta convoluta* (Turbellaria, Acoela) entstehen aus Diatomen. Naturwissenschaften 52(15):444–446.

Bowen, I. D. 1980. Phagocytosis in *Polycelis tenuis*. *In* D. C. Smith and Y. Tiffon (eds.), *Nutrition in the Lower Metazoa*. Pergamon Press, Oxford, pp. 1–14.

Boyer, B. C. 1971. Regulative development in a spiralian embryo as shown by cell deletion experiments on the acoel *Childia*. J. Exp. Zool. 176(1):97–105.

Crezee, M. 1982. Turbellaria. *In* S. P. Parker (ed.), *Synopsis and Classification of Living Organisms*, Vol. 1. McGraw Hill, New York, pp. 718–740.

Eakin, R. M. and J. L. Brandenberger. 1980. Unique eye of probable evolutionary significance. Science 211:1189–1190.

Heitkamp, C. 1977. The reproductive biology of *Mesostoma ehrenbergii*. Hydrobiologia 55(1):21–32.

Henley, C. 1974. Platyhelminthes (Turbellaria). *In* A. C. Giese and J. S. Pearse (eds.), *Reproduction of Marine Invertebrates*, Vol. 1, Acoelomate and Pseudocoelomate Metazoans. Academic Press, New York.

Holleman, J. J. 1972. Marine turbellarians of the Pacific coast. I. Proc. Biol. Soc. Wash. 85(34):405–412.

Hurley, A. C. 1975. The establishment of populations of *Balanus pacificus* Pilsbry (Cirripedia) and their elimination by predatory Turbellaria. J. Anim. Ecol. 44:521–532.

Hurley, A. C. 1976. The polyclad flatworm *Stylochus tripartitus* Hyman as a barnacle predator. Crustaceana 3(1):110–111.

Jennings, J. B. 1974. Digestive physiology of the Turbellaria. *In* N. W. Riser and M. P. Morse (eds.), *Biology of the Turbellaria*. McGraw-Hill, New York, pp. 173–197.

Jennings, J. B. 1980. Nutrition in symbiotic Turbellaria. *In* D. C. Smith and Y. Tiffon (eds.), *Nutrition in the Lower Metazoa*. Pergamon Press, Oxford, pp. 45–56.

Jokiel, P. L. and S. J. Townsley. 1974. Biology of the polyclad *Prosthiostomum* sp., a new coral parasite from Hawaii. Pacific Sci. 28(4):361–375.

Karling, T. G. 1966. On nematocysts and similar structures in turbellarians. Acta Zool. Fenn. 116:1–21.

Karling, T. G. 1974. On the anatomy and affinities of the turbellarian orders. *In* N. W. Riser and M. P. Morse (eds.), *Biology of the Turbellaria*. McGraw-Hill, New York, pp. 1–16.

Karling, T. G. and M. Meinander (eds.). 1978. The Alex Luther Centennial Symposium on Turbellaria. Acta Zool. Fenn. 154: 193–207.

Kato, K. 1940. On the development of some Japanese polyclads. Jpn. J. Zool. 8:537–573.

Lauer, D. M. and B. Fried. 1977. Observations on nutrition of *Bdelloura candida*, an ectocommensal of *Limulus polyphemus*. Am. Midl. Nat. 97(1):240–247.

Lender, T. 1974. The role of neurosecretion in freshwater planarians. *In* N. W. Riser and M. P. Morse (eds.), *Biology of the Turbellaria*. 475, McGraw-Hill, New York, pp. 460–475.

Lus, J. 1924. Some consideration on polarity and heteromorphosis in fresh-water planarians. Bull. Soc. Nat. Moscow Sect. Exp. Biol. 1. [In Russian.]

McKanna, J. A. 1968. Fine structure of the protonephridial system in planaria. I. Female cells. Z. Zellforsch. Mikrosk. Anat. 92:509–523.

Moraczewski, J. 1977. Asexual reproduction and regeneration of *Catenula*. Zoomorphologie 88:65–80.

Moraczewski, J., A. Czubaj and J. Bakowska. 1977. Organization and ultrastructure of the nervous system in Catenulida. Zoomorphologie 87(1):87–95.

Muscatine, L., J. E. Boyle and D. C. Smith. 1974. Symbiosis of the acoel flatworm *Convoluta roscoffensis* with the alga *Platymonas convolutae*. Proc. R. Soc. Lond. 187(1087): 221–234.

Nentwig, M. R. 1978. Comparative morphological studies after decapitation and after fission in the planarian *Dugesia dorotocephala*. Trans. Am. Microsc. Soc. 97(3): 297–310.

Prudhoe, S. 1985. *A Monograph on the Polyclad Turbellaria*. Oxford University Press, New York.

Prusch, R. D. 1976. Osmotic and ionic relationships in the freshwater flatworm *Dugesia dorotocephala*. Comp. Biochem. Physiol. 54A:287–290.

Rieger, R. M. and W. Sterrer. 1975. New spicular skeletons in Turbellaria, and the occurrence of spicules in marine meiofauna. Z. Zool. Syst. Evolutionsforsch. 13(3):207–248.

Riser, N. W. and M. P. Morse (eds.). 1974. *Biology of the Turbellaria*. Libbie H. Hyman Memorial Volume. McGraw-Hill, New York.

Ruppert, E. E. 1978. A review of metamorphosis of turbellarian larvae. *In* F. S. Chia and M. Rise (eds.), *Settlement and Metamorphosis of Marine Invertebrate Larvae*. Elsevier/North-Holland Biomedical Press, Amsterdam, pp. 65–81.

Smith, J., S. Tyler, M. B. Thomas and R. M. Rieger. 1982. The morphology of turbellarian rhabdites: Phylogenetic implications. Trans. Am. Microsc. Soc. 101:209–228.

Smith, J. III and S. Tyler. 1985. The acoel turbellarians: Kingpins of metazoan evolution or a specialized offshoot. *In* S. C. Morris et al. (eds.), *The Origins and Relationships of Lower Invertebrates*. Syst. Assoc. Spec. Vol. No. 28, Oxford, pp. 123–142.

Steinbock, O. 1966. Die Hofsteniiden (Turbellaria Acoela): Grundsätzliches zur Evolution der Turbellarien. Z. Zool. Syst. Evolutionsforsch. 4:58–195.

Tyler, S. 1976. Comparative ultrastructure of adhesive systems in the Turbellaria. Zoomorphologie 84:1–76.

Tyler, S. and R. M. Rieger. 1975. Uniflagellate spermatozoa in *Nemertoderma* and their phylogenetic significance. Science 188:730–732.

Tyler, S. and R. M. Rieger. 1978. Ultrastructural evidence for the systematic position of the Nemertodermatida (Turbellaria). *In* T. G. Karling and M. Meinander (eds.), The Alex Luther Centennial Symposium on Turbellaria. Acta Zool. Fenn. 154:193–207.

Trematoda and Monogenea

Bychowsky, B. E. 1957. *Monogenetic Trematodes: Their Systematics and Phylogeny*. American Institute of Biological Sciences, Washington, D.C. [Translated from the Russian.]

Combes, C. et al. 1980. The world atlas of Cercariae. Mem. Mus. Natl. Hist. Nat. Ser. A Zool. 115:1–235.

Crane, J. W. 1972. Systematics and new species of marine Monogenea from California. Wasmann J. Biol. 30(1/2):109–166.

Dawes, D. 1956. *The Trematoda*. Cambridge University Press, New York.

Erasmus, D. A. 1972. *The Biology of Trematodes*. Crane, Russak, New York.

Lebedev, B. I. 1978. Some aspects of mongenean existence. Folia Parasitol. 25:131–136.

Martin, W. E. 1972. An annotated key to the cercariae that develop in the snail *Cerithidea californica*. Bull. South. Calif. Acad. Sci. 71(1):39–43.

Pearson, J. C. 1972. A phylogeny of life-cycle patterns of the Digenea. Adv. Parasitol. 10:153–189.

Schell, S. C. 1982. Trematoda. *In* S. P. Parker (ed.), *Synopsis and Classification of Living Organisms*. Vol. 1. McGraw-Hill, New York, pp. 740–807.

Smyth, J. D. and D. W. Halton. 1985. *The Physiology of Trematodes*, 2nd Ed. W. H. Freeman, San Francisco.

Sproston, N. G. 1946. A synopsis of the monogenetic trematodes. Trans. Zool. Soc. Lond. 25:185–600.

Yamaguti, S. 1963. *Systema Helminthum*, Vol. 4, Monogenea and Aspidocatylea. Interscience, New York.

Yamaguti, S. 1971. *Synopsis of Digenetic Trematodes of Vertebrates*, Vols. 1–2. Keigaku, Tokyo.

Yamaguti, S. 1975. *A Synoptical Review of Life Histories of Digenetic Trematodes of Vertebrates*. Keigaku, Tokyo.

Cestoda

Aral, H. P. (ed.). 1980. *Biology of the Tapeworm Hymenolepsis diminuta*. Academic Press, New York.

Schmidt, G. D. 1982. Cestoda. *In* S. P. Parker (ed.), *Synopsis and Classification of Living Organisms*, Vol. 1. McGraw-Hill, New York, pp. 807–822.

Smyth, J. D. 1969. *The Physiology of Cestodes*. W. H. Freeman, San Francisco.

Uglem, G. L. and J. J. Just. 1983. Trypsin inhibition by tapeworms: Antienzyme secretion or pH adjustment. Science 220:79–81.

Wardle, R., J. McLeod and S. Radinovsky. 1974. *Advances in the Zoology of Tapeworms*. University of Minnesota Press, Minneapolis.

Xylander, W. E. R. 1987. Das Protonephridialsystem der Cestoda: Evolutive Veränderungen und ihre mögliche funktionelle Bedeutung. Verh. Dtsch. Zool. Ges. 80:257–258.

Yamaguti, S. 1959. *Systema Helminthum*, Vol. 2, The Cestodes of Vertebrates. Interscience, New York.

General Parasitology

Baer, J. G. 1952. *Ecology of Animal Parasites*. University of Illinois Press, Urbana.

Burt, D. R. R. 1970. *Platyhelminthes and Parasitism*. American Elsevier, New York.

Llewellyn, J. 1965. The evolution of parasitic platyhelminthes. *In* A. E. R. Taylor (ed.), *Evolution of Parasites*, 3rd Ed. Symp. Soc. Parasitol. Blackwell, Oxford, pp. 47–78.

Marquardt, W. C. and R. S. Demaree, Jr. 1985. *Parasitology*. Macmillan, New York.

Mueller, J. F. 1965. Helminth life cycles. Am. Zool. 5:131–139.

Noble, E. and G. Noble. 1976. *Parasitology*, 4th Ed. Lea and Febiger, Philadelphia.

Olsen, O. W. 1974. *Animal Parasites: Their Life Cycles and Ecology*, 3rd Ed. University Park Press, Baltimore.

Read, C. P. 1970. *Parasitism and Symbiology*. Ronald Press, New York.

Schmidt, G. D. and L. S. Roberts. 1977. *Foundations of Parasitology*. Mosby, St. Louis.

Smyth, J. D. 1977. *Introduction to Animal Parasitology*. 2nd Ed. Wiley, New York.

Smyth, J. D. and J. A. Clegg. 1959. Egg shell formation in trematodes and cestodes. Exp. Parasitol. 8:286–323.

Chapter Eleven

Phylum Nemertea:
The Ribbon Worms

Nemerteans are rewarding and satisfying animals to work with, and there is still so very much that we do not know about them.

  Ray Gibson
  Nemerteans, 1972

Members of the phylum Nemertea (Greek, "a sea nymph") or Rhynchocoela (Greek *rhynchos*, "snout"; *coel*, "cavity") are commonly called ribbon worms. Figure 1 illustrates a variety of body forms within this taxon and the major features of their anatomy. These unsegmented vermiform animals are usually flattened dorsoventrally and are moderately cephalized; they possess highly extensible bodies. Many ribbon worms are rather drab in appearance, but many are also brightly colored in various shades of red, green, purple, and orange, and some bear distinctive markings.

About 900 species of nemerteans have been described. They range in length from less than 1 cm to several meters. Many can stretch easily to several times their contracted lengths. One specimen of *Lineus longissimus* was nearly 60 m long! They are predominately benthic marine animals. A few, however, are planktonic, and some are symbiotic in molluscs or other marine invertebrates. A few freshwater and terrestrial species are known, the latter commonly found in greenhouses.

Some features of the nemertean body plan (Box 1) are similar to the conditions seen in flatworms, and the two taxa are often viewed together as the triploblastic acoelomate bilateria. There are similarities in the overall architecture of the nervous systems, the types of sense organs, and the protonephridial excretory structures. However, in other respects ribbon worms differ greatly from flatworms: nemerteans possess an anus (a complete, one-way digestive tract), a closed circulatory system, and an eversible **proboscis** surrounded by a hydrostatic cavity called the **rhynchocoel**. The structure of the proboscis apparatus is unique to nemerteans and represents a novel synapomorphy that distinguishes the phylum Nemertea from all other invertebrate taxa. Furthermore, recent work suggests strongly that the nemerteans are coelomate (Turbeville and Ruppert 1985); we discuss this matter later.

Taxonomic history and classification

The earliest report of a nemertean was that of Borlase (1758), who described his specimen as "the sea long-worm" and categorized it "among the less perfect kind of sea animals." For nearly a century, most authors placed the ribbon worms with the turbellarian flatworms, although others suggested that they were allied to the annelids, the sipunculans, the roundworms, and even the molluscs and the insects. It was during this period that Cuvier (1817) described a particular ribbon worm and called it *Nemertes*, from which the phylum name was eventually derived. It was not until 1851 that some substantial evidence for the distinctive nature of the ribbon worms was published by Max Schultze, who described the functional morphology of the proboscis, established the presence of nephridia and an anus, and discussed many other features of these animals. Schultze even proposed the basis for the classification of these worms, which is still employed today by most authorities. Interestingly, he persisted in considering them to be turbellarians, although he coined the names Nemertina and Rhynchocoela. Minot separated the nemerteans from the flatworms in 1876, but it was not until the mid-twentieth century (see Coe 1943 and Hyman 1951) that the unique combination of characters displayed by the ribbon worms was fully accepted. Since that time they have been treated as a valid phylum.

Classification

Since the publication of Schultze's classical accounts, the primary effort of nemertean taxonomists has been to refine his scheme, with only relatively minor controversies periodically surfacing. The classification scheme used here is a traditional one established by Coe (1943) and followed by most specialists (see Gibson 1972). In it the phylum is subdivided into two classes, each with two clearly differentiated orders. The principal features used to distinguish between the classes and among the orders of nemerteans include proboscis armature, mouth location relative to the position of the cerebral ganglion, gut shape, layering of the body wall muscles, and position of the longitudinal nerve cords.*

PHYLUM NEMERTEA (= RHYNCHOCOELA)

CLASS ANOPLA: Unarmed nemerteans (Figures 1B and C). Proboscis not armed with stylets and not morphologically specialized into three regions. Mouth separate from proboscis pore and located directly below or somewhat posterior to cerebral ganglion; longitudinal nerve cords located within epidermis, dermis, or muscle layers of body wall (not within the mesenchymal mass internal to the body wall muscles).

ORDER PALAEONEMERTEA: Two or three layers of body wall muscles, from external to internal either circular-longitudinal or circular-longitudinal-circular; dermis thin and gelatinous, or absent; longitudinal nerve cords epidermal, dermal, or intramuscular within the longitudinal layer; cerebral organs and ocelli frequently lacking. Palaeonemerteans are marine, primarily littoral forms. (e.g., *Carinoma, Cephalothrix, Tubulanus*)

ORDER HETERONEMERTEA: Three layers of body wall muscles, from external to internal longitudinal-circular-longitudinal; dermis usually thick, partly fibrous; longitudinal nerve cords intramuscular, between outer longitudinal and middle circular layers; cerebral organs and ocelli usually present. These nemerteans are primarily marine littoral forms. (e.g., *Baseodiscus, Cerebratulus, Lineus, Micrura, Paralineus*)

CLASS ENOPLA: Armed nemerteans (Figures 1D and E). Proboscis usually armed with distinct stylets and morphologically specialized into three regions (except in Bdellonemertea); mouth and proboscis pore usually united into a common aperture; mouth located anterior to cerebral gan-

*Iwata (1960) proposed an alternative classification scheme based on certain embryological and morphological features, but his proposal has not been widely accepted. Notice that, according to our cladogram, the class Anopla is not a monophyletic group, but a paraphyletic one.

> ### Box One
> # Characteristics of the Phylum Nemertea
>
> 1. Triploblastic, acoelomate or coelomate, bilaterally symmetrical unsegmented worms
> 2. Digestive tract complete, with an anus
> 3. With protonephridia
> 4. With lobed, supraenteric cerebral ganglion, and two or more longitudinal nerve cords connected by transverse commissures
> 5. With two or three layers of body wall muscles arranged in various ways
> 6. With a unique proboscis apparatus lying dorsal to the gut and surrounded by a coelom-like hydrostatic chamber called the rhynchocoel
> 7. With a closed circulatory system
> 8. Most are gonochoristic; early development typically spiralian, either direct or indirect

glion; longitudinal nerve cords within mesenchyme, internal to body wall muscles.

ORDER HOPLONEMERTEA: Proboscis armed; trunk lacks a posterior sucker; gut more-or-less straight but nearly always with numerous lateral diverticula; cerebral organs and ocelli usually present.

SUBORDER MONOSTILIFERA: Stylet apparatus consists of a single main stylet and two or more sacs housing accessory (replacement) stylets. Most species are marine and benthic, but freshwater, terrestrial, and ectoparasitic forms are also known. (e.g., *Amphiporus, Annulonemertes, Carcinonemertes, Emplectonema, Geonemertes, Paranemertes*)

SUBORDER POLYSTILIFERA: Stylet apparatus consists of many small stylets borne on a basal shield. All species are marine, either benthic (Reptantia) or pelagic (Pelagica). (e.g., *Hubrechtonemertes, Nectonemertes, Pelagonemertes*)

ORDER BDELLONEMERTEA: Proboscis unarmed; trunk with large posterior sucker; proboscis apparatus opens into foregut; gut convoluted and lacks lateral diverticula. Bdellonemerteans are commensal in the mantle cavities of marine bivalves and, in one species, a freshwater gastropod. Monogeneric: *Malacobdella*.

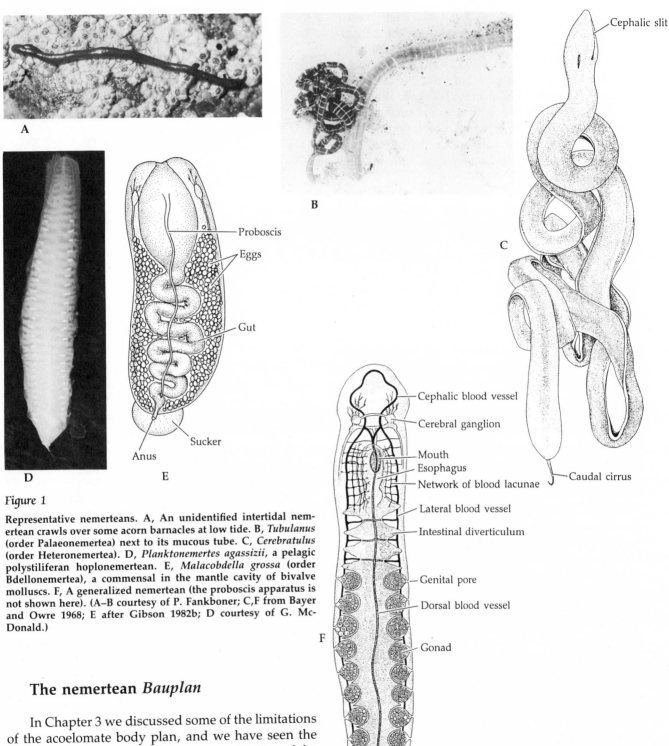

Figure 1

Representative nemerteans. A, An unidentified intertidal nemertean crawls over some acorn barnacles at low tide. B, *Tubulanus* (order Palaeonemertea) next to its mucous tube. C, *Cerebratulus* (order Heteronemertea). D, *Planktonemertes agassizii*, a pelagic polystiliferan hoplonemertean. E, *Malacobdella grossa* (order Bdellonemertea), a commensal in the mantle cavity of bivalve molluscs. F, A generalized nemertean (the proboscis apparatus is not shown here). (A–B courtesy of P. Fankboner; C,F from Bayer and Owre 1968; E after Gibson 1982b; D courtesy of G. Mc-Donald.)

The nemertean *Bauplan*

In Chapter 3 we discussed some of the limitations of the acoelomate body plan, and we have seen the results of these constraints in our examination of the flatworms. It might be said that the ribbon worms have made the best of a rather difficult situation. Even though there is the possibility that nemerteans have coelomic cavities (the rhynchocoel and certain blood vessels), these worms have relatively solid bodies. Thus they are at least "functionally," if not evolutionarily, acoelomate. Recall that many of the problems inherent in the acoelomate architecture are related to restricted internal transport capabilities. The evolution of a circulatory system in nemerteans has largely eased this problem, and the functional anatomy of many other systems is related directly or indirectly to the presence of this circulatory mechanism. For example, the nemertean protonephridia are usually intimately associated with the blood, from which wastes are drawn, rather than with the mesenchymal tissues as in flatworms.

The increased capabilities for internal circulation and transport have allowed a number of developments that would otherwise be impossible. First, the circulatory system provides a solution to the surface-to-volume dilemma, and as a result nemerteans tend to be much larger and more robust than flatworms, having been largely relieved of the constraints of relying on diffusion for internal transport and exchange. Second, the digestive tract is complete and somewhat regionally specialized. With a one-way movement of food materials through the gut, and a circulatory system to absorb and distribute digested products, the anterior region of the gut has been freed for feeding and ingestion. Third, since the animal does not have to rely on diffusion for transport through a loosely organized mesenchyme, that general body area is freed for the development of other structures, notably the well developed layers of muscles. Thus, in summary, the development of a circulatory system in concert with these other changes has resulted in relatively large, robust, active animals, capable of more complex feeding and digestive activities than platyhelminths are.

This general *Bauplan* is enhanced by the presence of the unique proboscis apparatus (which usually functions in prey capture), the distinctly anterior location of the mouth, and well developed cephalic sensory organs for prey location. Thus, while variation exists, the "typical" nemertean may be viewed as an active benthic hunter/tracker that moves among nooks and crannies preying on other invertebrates.

Body wall

The body wall of nemerteans comprises an epidermis, a dermis, relatively thick muscle layers, and a mesenchyme of varying thickness (Figure 2). The epidermis is basically a ciliated columnar epithelium (Figure 2C). Mixed among the columnar cells are sensory cells (probably tactile), mucous gland cells, and interstitial replacement cells that may extend beneath the epidermis. Below the epidermis is the dermis, which varies greatly in both thickness and

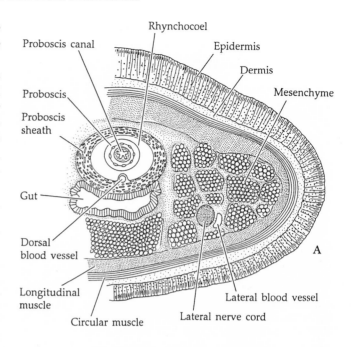

composition. In some ribbon worms (e.g., the palaeonemerteans) the dermis is extremely thin or composed of only a homogeneous gel-like layer, whereas in others (e.g., the heteronemerteans) it is typically quite thick and densely fibrous and usually includes a variety of gland cells. Beneath the dermis are well developed layers of circular and longitudinal muscles. As indicated in the classification outline, the organization of these muscles varies among taxa and may occur in either a two- or three-layered plan (Figure 3). The layering arrangement may also vary to some degree along the body length of individual animals. Internal to the muscle layers is a dense, more-or-less solid mesenchyme, although in some nemerteans the muscle layers are so thick that they nearly obliterate this inner mass. The mesenchyme includes a gel matrix and often a variety of loose cells, fibers, and dorsoventrally oriented muscles. Figure 3 depicts cross-sectional views of the four orders, showing mesenchyme thickness, muscles, placement of longitudinal nerve cords, major longitudinal blood vessels, and other features.

Support and locomotion

In the absence of any rigid skeletal elements, the support system of nemerteans is provided by the muscles and other tissues of the body wall and by the hydrostatic qualities of the mesenchyme. These features permit dramatic changes in both length and cross-sectional shape and diameter, characteristics that are closely associated with various sorts of lo-

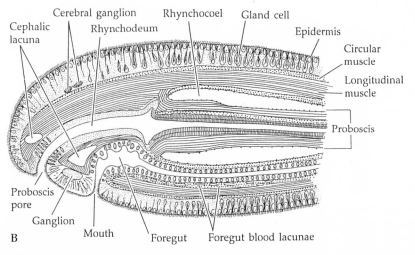

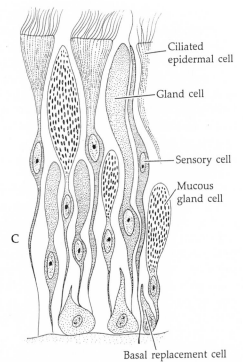

Figure 2

Organization of the nemertean body wall. A, A hoplonemertean (cross-section). B, The anterior end of a palaeonemertean (longitudinal section). C, Epidermal cells. (A,B after Hyman 1951; C from Bayer and Owre 1968, after Coe 1943.)

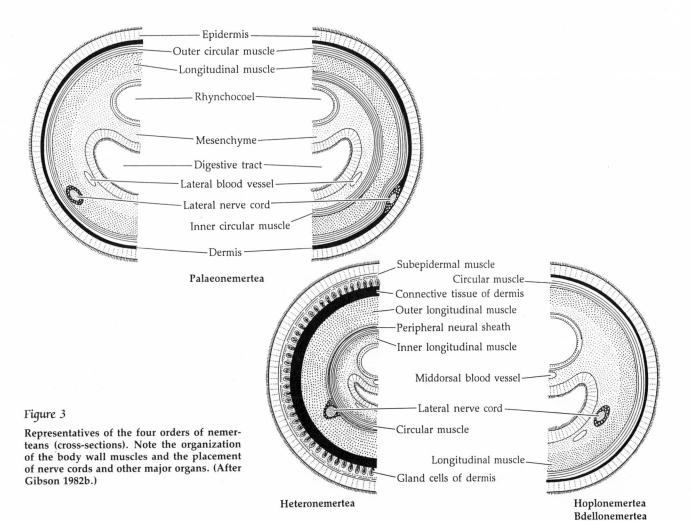

Figure 3

Representatives of the four orders of nemerteans (cross-sections). Note the organization of the body wall muscles and the placement of nerve cords and other major organs. (After Gibson 1982b.)

comotion and accommodation to cramped quarters. Most very small benthic ribbon worms are propelled by the action of their epidermal cilia. A slime trail is produced by the body wall mucous glands and provides a lubricated surface over which the worm slowly glides. Little nemerteans commonly live among the interstices of filamentous algae or in the spaces of other irregular surfaces such as those found in mussel beds and sand, mud, or pebble bottoms. Larger epibenthic ribbon worms and most of the burrowing forms employ peristaltic waves of the body wall muscles to propel them over moist surfaces or through soft substrata. Some of the larger forms (e.g., *Cerebratulus*) use undulatory swimming as a means of locomotion, and perhaps as an escape reaction to benthic predators. Fully pelagic nemerteans (certain hoplonemerteans) generally drift or swim slowly. Some of the terrestrial forms produce a slime sheath through which they glide by ciliary action, and some use their proboscis for rapid escape responses.

Feeding and digestion

Feeding behavior. Most ribbon worms are active predators on small invertebrates, but some are scavengers and others feed on plant material (at least under laboratory conditions). There is evidence to suggest that species of the commensal genus *Malacobdella*, which inhabit the mantle cavity of bivalve molluscs, feed largely on phytoplankton captured from their host's feeding and gas exchange currents. Field observations indicate that predatory forms may have either extremely varied or quite restricted diets, depending on the species involved. Some are capable of tracking prey over long distances, whereas others must locate food by direct contact. Distant prey location and assessment of food acceptability are almost certainly chemotactic responses. Ribbon worms that actually hunt and track can recognize the trails left by potential prey and may fire their proboscis along the trail ahead of them to capture food (Figure 4). Similar reactions are elicited when infaunal nemerteans encounter burrows in which potential prey may be located. Surface hunters that live in intertidal areas generally forage during high tides or at night and thus avoid the threats of desiccation and visual predators. Members of some genera (e.g., *Tubulanus, Paranemertes, Amphiporus*), however, may frequently be seen during low tides, on foggy mornings, gliding over the substratum in search of prey.

The behavior involved in the capture and ingestion of live prey is significantly different from that associated with scavenging on dead material. In predation, the proboscis is employed both in captur-

Figure 4

Paranemertes peregrina (order Hoplonemertea) capturing a nereid polychaete. The proboscis is coiled around the polychaete. (Courtesy of S. Stricker.)

ing prey and in moving it to the mouth for ingestion. The proboscis is everted and wrapped around the victim (Figure 4). The prey is not only physically "held down" by the proboscis but may be subdued or killed by toxic secretions from the proboscis. In the Pacific species *Paranemertes peregrina*, which feeds primarily on nereid polychaetes, the glandular epithelium of the everted proboscis secretes a potent neurotoxin. Nemerteans with an armed proboscis (Hoplonemertea) actually use the stylets to pierce the prey's body (often numerous times) to introduce the toxin. Once captured, the prey is drawn to the mouth by retraction and manipulation of the proboscis; it is generally swallowed whole. The mouth is expanded and pressed against the food, and swallowing is accomplished by peristaltic action of the body wall muscles aided by ciliary currents in the anterior region of the gut. Scavenging, in contrast, usually does not involve the proboscis. The worm simply ingests the food directly by muscular action of body wall and foregut. In some predatory hoplonemerteans (those in which the lumen of the proboscis is connected with the anterior gut lumen), the foregut itself may be everted for feeding on animals too large to be swallowed whole. In such cases, fluids and soft tissues are generally sucked out of the prey's body. Polychaete worms and amphipod crustaceans seem to be favorite food items for many predatory nemerteans.

Species of the hoplonemertean genus *Carcinonemertes* are egg predators on brachyuran crabs. Different species inhabit different regions of the host's body, but all migrate to the egg masses on gravid female crabs and feed on the yolky eggs. In high

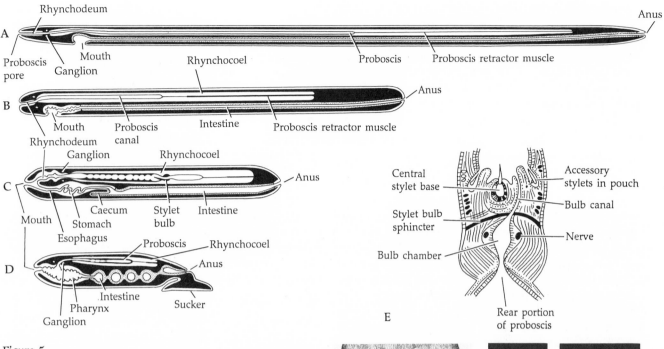

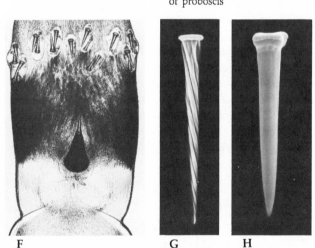

Figure 5

A–D, The arrangements of the proboscis apparatus and digestive tube in the orders Palaeonemertea (A), Heteronemertea (B), Hoplonemertea (C), and Bdellonemertea (D). E, Stylet apparatus in the proboscis of *Prostoma graecense*. F, Stylet apparatus of *Amphiporus formidabilis*. G, Stylet from *Paranemertes peregrina* (scanning electron micrograph). H, Stylet from *Amphiporus bimaculatus* (scanning electron micrograph) (A–D after Russell-Hunter 1979, based on papers by R. Gibson; E after Gibson 1982b; F–H courtesy of S. Stricker.)

numbers, these egg predators can kill all of the embryos in the host's clutch. Recent studies (Wickham 1979, 1980) report 99 percent infestation rates of *Carcinonemertes errans* on the commercially important Dungeness crab (*Cancer magister*), with up to 100,000 worms per host. This predator has been implicated in the general collapse of the central California Dungeness crab fishery.

The proboscis apparatus. The proboscis apparatus is a complex arrangement of tubes, muscles, and hydraulic systems (Figure 5). The proboscis itself is an elongate, eversible blind tube, which either is associated with the foregut or opens through a separate **proboscis pore**. The proboscis may be regionally specialized and bear **stylets** in various arrangements (Figures 5E–H). Nemertean stylets are nail-shaped structures that typically reach lengths of 50–200 μm. Each stylet is composed of a central organic matrix surrounded by an inorganic cortex that contains crystalline calcium and phosphorus. The stylets are

formed within large epithelial cells called **styletocytes**. Because growing ribbon worms must replace their stylets (with new larger ones), and because they often lose the stylet during prey capture, new stylets are continuously produced in **reserve stylet sacs** and stored until needed.

The basic structure and action of the proboscis are most easily described where the apparatus is entirely separate from the gut. As shown in Figures 5A and B, the proboscis pore leads from the outside directly into the anterior proboscis lumen, called the **rhynchodeum**, the lining of which is continuous with the epidermis. Posterior to the rhynchodeum, the lumen continues as the **proboscis canal**, which is surrounded by the muscular wall of the proboscis itself; these muscles are derived from the muscles of the body wall. The proboscis is surrounded by a closed, fluid-filled, perhaps coelomic space called the **rhynchocoel**, which in turn is surrounded by additional muscle layers. The inner blind end of the proboscis is connected to the posterior wall of the rhynchocoel by a **proboscis retractor muscle**. In a few taxa (e.g., *Gorgonorhynchus*) there is no retractor muscle; and eversion *and* retraction are accomplished hydrostatically.

Eversion of the proboscis (Figure 6) is accomplished by contraction of the muscles around the rhynchocoel; this increases the hydrostatic pressure within the rhynchocoel itself, squeezing on the proboscis and causing its eversion. The everted proboscis moves with the muscles in its wall; the proboscis is retracted back inside the body by the coincidental relaxation of the muscles around the rhynchocoel and contraction of the proboscis retractor muscle. The retracted proboscis may extend nearly to the posterior end of the worm, and usually only a portion of it is extended during eversion.

Digestive system. In contrast to the flatworms, nemerteans possess an anus (Figure 7). Associated with the one-way movement of food from mouth to anus we find various degrees of regional specialization (both structural and functional) in the guts of ribbon worms. The mouth leads inward to an ectodermally derived foregut (stomodeum) consisting of

a bulbous **buccal cavity**, sometimes a short **esophagus**, and a **stomach**. The stomach leads to an elongate **intestine** or midgut, which is more-or-less straight and usually bears numerous lateral diverticula. In *Malacobdella*, the intestine is loosely coiled and lacks diverticula; diverticula are also lacking in the strange, "segmented" *Annulonemertes* (Berg 1985). At the posterior end of the intestine is a short ectodermally derived hindgut (proctodeum) or **rectum**, which terminates in the anus. Elaborations on this basic plan are common in certain taxa and may include various ceca arising from the stomach or from the intestine at its junction with the foregut.

The entire digestive tube is ciliated, the foregut more densely than the midgut. The gut epithelium is basically columnar, variably mixed with gland cells. The foregut contains a variety of mucus-producing cells, sometimes multicellular mucous glands, and

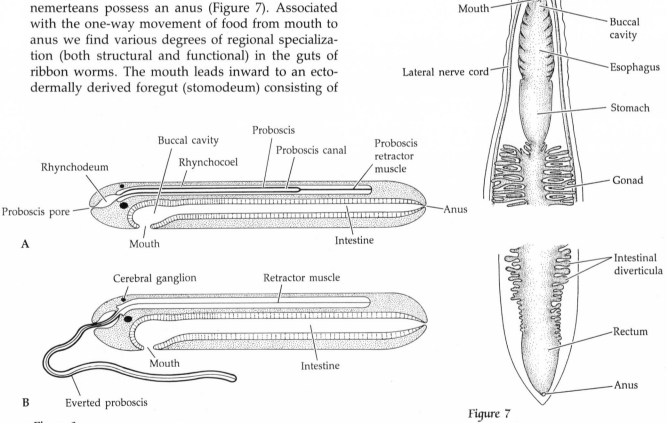

Figure 6

A retracted (A) and an extended (B) proboscis of a hoplonemertean. (After Gibson 1982b.)

Figure 7

A nemertean digestive system. Anterior and posterior regions of the gut of *Carinoma* (ventral view). (After Hyman 1951.)

occasionally enzymatic gland cells in the stomach region. The midgut is lined with vacuolated ciliated columnar cells; these are phagocytic and bear microvilli between their cilia, greatly increasing their surface area. Enzymatic gland cells are abundantly mixed with the ciliated cells of the midgut. The hindgut typically lacks gland cells. Food is moved through the digestive tract by cilia and by the action of the body wall muscles; there are usually no muscles in the gut wall itself, except in the foregut of some heteronemerteans.

The process of digestion in carnivorous nemerteans is a two-phase sequence of protein breakdown. The first step involves the action of endopeptidases released from gland cells into the gut lumen. This extracellular digestion is quite rapid and is followed by phagocytosis (and probably pinocytosis) of the partially digested material by the ciliated columnar cells of the midgut. Protein digestion is completed intracellularly by exopeptidases within the food vacuoles of the midgut epithelium. Lipases have been discovered in at least one species (*Lineus ruber*), and carbohydrases are known in the omnivorous commensal, *Malacobdella*. Food is stored primarily in the form of fats, and to a much lesser extent as glycogen, in the wall of the midgut. Transportation of digested materials throughout the body is accomplished by the circulatory system, which absorbs these products from the cells lining the intestine. Undigestible materials are moved through the gut and out the anus.

Circulation and gas exchange

We have mentioned briefly the evolutionary and adaptive significance of the circulatory system in nemerteans and its general relationship to other systems and functions. This closed system consists of vessels and thin-walled spaces called **lacunae** (Figure 2B). At least some of these vessels are thought to be homologous to coelomic cavities. There is a good deal of variation in the architecture of nemertean circulatory systems (Figure 8). The simplest arrangement occurs in certain palaeonemerteans in which a single pair of longitudinal vessels extends the length of the body, connecting anteriorly by a **cephalic lacuna** and posteriorly by an **anal lacuna**. Elaboration on this basic scheme may include transverse vessels between the longitudinal vessels, enlargement and compartmentalization of the lacunar spaces, and the addition of a middorsal vessel. The walls of the blood vessels are only slightly contractile, and general body movements generate most of the blood flow. There is no consistent pattern to the movement of blood through the system; it may flow either anteriorly or posteri-

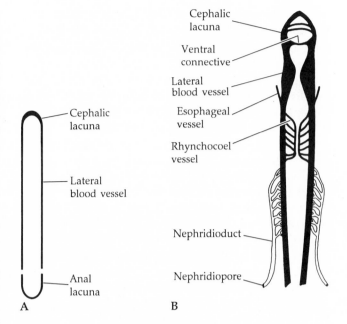

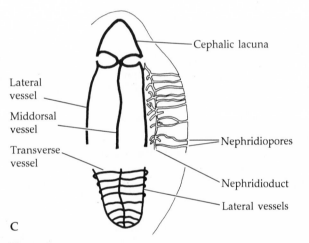

Figure 8

Nemertean circulatory systems. A, The simple circulatory loop of *Cephalothrix* consists of a pair of lateral blood vessels connected by cephalic and anal lacunae. B, The complex circulatory system of *Tubulanus*. Note the intimate association of the nephridial system with the lateral blood vessels. C, The circulatory system of *Amphiporus* includes a middorsal vessel and numerous transverse vessels. (After Hyman 1951.)

orly in the longitudinal vessels, and currents often reverse directions.

The blood consists of a colorless fluid in which are suspended various cells, which can include pigmented corpuscles (yellow, orange, green, red), at least some of which contain hemoglobin, and a variety of so-called lymphocytes and leukocytes of uncertain function. The anatomical association of the circulatory system with other structures, as well as the composition of the blood, suggest several circulatory functions. Although conclusive evidence is lacking in some areas, the circulatory system appears to be involved with the transport of nutrients, gases, neurosecretions, and excretory products. Some intermediary metabolism probably occurs in the blood, as several appropriate enzymes have been identified in solution. The blood may also serve as an aid to body support through changes in hydrostatic pressure within the vessels and lacunar spaces. There is some evidence to support the idea that the blood may also function in osmoregulation.

Gas exchange in nemerteans is epidermal, and no special structures are involved. Oxygen and carbon dioxide diffuse readily across the moist body surface, which is usually covered with mucous secretions. Some robust forms (e.g., *Cerebratulus*) augment this passive exchange of gases across the skin with regular irrigation of the foregut, where there is an extensive system of blood vessels. In those species in which hemoglobin occurs, this pigment probably aids in oxygen transport or storage within the blood.

Excretion and osmoregulation

The excretory system of most nemerteans consists of from two to thousands of flame bulb protonephridia (Figures 8 and 9) similar to those found in turbellarian flatworms. However, apparently none occurs in the deep-sea pelagic hoplonemerteans. The flame bulbs are usually intimately associated with the lateral blood vessels or less commonly with other parts of the circulatory system. The nephridial units

are often pressed into the blood vessel walls, and in some instances the walls are actually broken down so that the nephridia are bathed directly in blood. In the simplest case, a single pair of flame bulbs leads to two nephridioducts, each with its own laterally placed nephridiopore. More complex conditions include rows of single flame bulbs or clusters of flame bulbs with multiple ducts. In some species the walls of the nephridioducts are syncytial and lead to hundreds or even thousands of pores on the epidermis. The most elaborate conditions occur in certain terrestrial nemerteans where approximately 70,000 clusters of flame bulbs (six to eight in each cluster) lead to as many surface pores. In some heteronemerteans (e.g., *Baseodiscus*), the excretory system discharges into the foregut.

The functioning of nemertean protonephridia in the excretion of metabolic wastes has not been well studied. The close association of the flame bulbs with the circulatory system suggests that nitrogenous wastes (probably ammonia), excess salts, and other metabolic products are removed from the blood as well as from the surrounding mesenchyme by the nephridia. If such is the case, it explains again the significance of the circulatory system in overcoming the surface-to-volume problems and the constraints of simple diffusion on body size. Relatively active animals produce large amounts of metabolic wastes. Dependence on diffusion alone would seriously limit any increase in body bulk, but the transport of these wastes from the tissues to the protonephridial system by the circulatory vessels greatly eases this limitation.

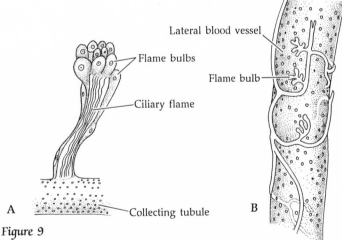

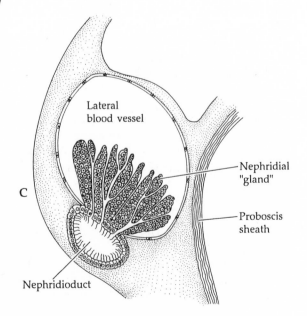

Figure 9

Nemertean excretory systems (see also Figure 8). A, A protonephridial cluster of *Drepanophorus*. B, Nephridial ducts associated with a lateral blood vessel in *Amphiporus*. C, Excretory system of *Carinina* in which the excretory units (so-called nephridial gland) project into the lumen of the lateral blood vessel. (After Hyman 1951.)

One of the most remarkable evolutionary achievements of the nemerteans has been their ability to grow to great size, particularly in length, without segmentation or the development of a large body cavity.

There is some morphological and experimental evidence that the protonephridia also play an important role in osmoregulation, especially in freshwater and terrestrial ribbon worms. It is in some of these forms, which are subjected to extreme osmotic stress, that the most elaborate excretory systems are found; these systems are probably associated with water balance. Furthermore, it appears that there may be a very complex interaction between the nervous system (neurosecretions), the circulatory system, and the nephridia to facilitate osmoregulatory mechanisms, but the details remain to be studied. Some members of all nemertean orders except the Palaeonemertea have invaded fresh water and thus must combat water influx from their strongly hypotonic surroundings. Members of some genera (e.g., *Geonemertes*) are terrestrial, although restricted to moist shady habitats where they avoid serious problems of desiccation. In addition, they tend to cover their bodies with a mucous coat that reduces potential water loss. Those forms that inhabit marine subtidal or deep-water environments, or are endosymbiotic (one genus of Heteronemertea, several genera of Hoplonemertea, and the Bdellonemertea), face little or no osmotic stress. But the many species found intertidally do face periods of exposure to air and to lowered salinities. Their soft bodies are largely unprotected, and they are relatively intolerant of fluctuations in environmental conditions. Intertidal nemerteans rely largely on behavioral attributes to survive periods of potential osmotic stress. They tend to remain in moist areas and simply avoid being subjected to adverse conditions. Burrowing in soft, water-soaked substrata, or living among algae or mussel beds, in cracks and crevices, or other areas that retain sea water at low tide are lifestyles illustrating how habitat preference and behavior prevent exposure to stress. In addition, most intertidal nemerteans are somewhat negatively phototactic and restrict their activities to night hours or to foggy or overcast cool mornings and evenings.

Nervous system and sense organs

The basic organization of the nemertean nervous system reflects a relatively active lifestyle. Nemerteans are somewhat more cephalized than flatworms are, especially in the anterior placement of the mouth and feeding structures, and we find related concentrations of sensory and other nervous elements in the head. The ladder-like central nervous system of ribbon worms consists of a complex cerebral ganglion from which arises a pair of ganglionated, lateral (longitudinal) nerve cords (Figure 10A). The cerebral ganglion is formed of four attached lobes that encircle the proboscis apparatus (not the gut, as in many other invertebrates). Each side of the cerebral ganglion includes a dorsal and a ventral lobe; the two sides are attached to one another by dorsal and ventral connectives. Several pairs of sensory nerves provide input directly to the cerebral ganglion from various cephalic sense organs. The main longitudinal nerve cords arise from the ventral lobes of the cerebral ganglion and pass posteriorly; they attach to each other at various points by branched transverse connectives and terminally by an anal commissure. The longitudinal nerves also give rise to peripheral sensory and motor nerves along the length of the body. Elaboration on this basic plan includes additional longitudinal nerve cords, frequently a middorsal one arising from the dorsal commissure of the cerebral ganglion, and a variety of connectives, nerve tracts, and plexi.

As noted in the classification scheme, the positions of the major longitudinal nerve cords vary among the four orders (see Figure 3). These differences are summarized in Table 1. These changes in the position of the lateral nerve cords from epidermal to mesenchymal correspond to general increases in body complexity and tendencies toward specialization. Most workers agree that these differences reflect a plesiomorphic (epidermal) to apomorphic (subepidermal) trend among these taxa.

Ribbon worms possess a variety of sensory receptors, many of which are concentrated at the anterior end and associated with an active, typically hunting lifestyle and with other aspects of their natural history. Nemerteans are very sensitive to touch. This tactile sensitivity plays a role in food handling, avoidance responses, locomotion over irregular surfaces, and mating behavior. Several types of modified ciliated epidermal cells are scattered over the body surface (especially abundant at the anterior and posterior ends) and are presumed to have a tactile function. The cells occur either singly or in clusters; some of the latter types are located in small depressions and can be thrust out from the body surface.

The eyes of ribbon worms are located anteriorly and number from two to several hundred; they can be arranged in various patterns (Figure 10B). Most of these ocelli are of the inverted pigment-cup type, similar to those seen in flatworms, although a few

A

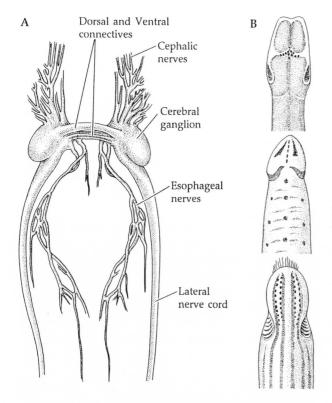

Dorsal and Ventral connectives

Cephalic nerves

Cerebral ganglion

Esophageal nerves

Lateral nerve cord

B

C

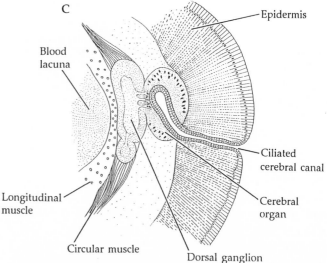

Epidermis

Blood lacuna

Ciliated cerebral canal

Cerebral organ

Longitudinal muscle

Circular muscle

Dorsal ganglion

Figure 10

Nervous system and sense organs of nemerteans. A, Anterior portion of the nervous system of *Tubulanus*; see text for explanation of variations. B, The cephalic slits and grooves and eye spots are visible on the heads of three benthic nemerteans. C, The cerebral organ of *Tubulanus* (cross-section). Note the association of the organ with the cerebral canal, the nervous system, and the blood system. D, Clusters of frontal glands occur in the anterior end of a hoplonemertean (longitudinal section). (A adapted from several sources; B,C from Bayer and Owre 1968, D after Hyman 1951.)

D

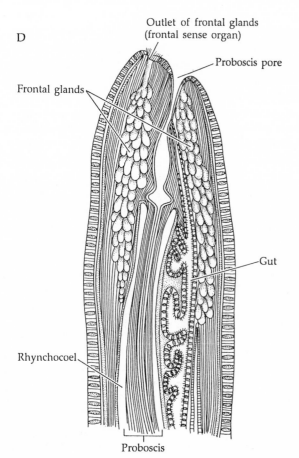

Outlet of frontal glands (frontal sense organ)

Proboscis pore

Frontal glands

Gut

Rhynchocoel

Proboscis

species possess lensed eyes. As discussed in Chapter 3, these types of eyes typically are sensitive to light intensity and light direction. They help the nemerteans avoid bright light and potential exposure to predators or environmental stresses.

Much of the sensory input important to nemerteans is chemosensory. These worms are very sensitive to dissolved chemicals in their environment and employ this sensitivity in food location, probably mate location, substratum testing, and general water analysis. Evidently all nemerteans respond to contact with chemical stimuli, and many are capable of distance chemoreception of materials in solution. At least three different nemertean structures have been implicated (some through speculation) in the initiation of chemotactic responses; these are **cephalic slits** or **grooves**, **cerebral organs**, and **frontal glands** (= **cephalic glands**) (Figures 10B,C,D). Cephalic slits are deep or shallow furrows that occur laterally on the heads of many ribbon worms (see also Figure 1). These furrows are lined with a ciliated sensory epithelium supplied with nerves from the cerebral ganglion. Water is circulated through the cephalic slits and over this presumably chemosensory epithelial lining.

Table One

Location of the lateral longitudinal nerve cords in nemertean orders

TAXON	EPIDERMAL	DERMAL	WITHIN BODY WALL MUSCLES	INTERNAL TO BODY WALL MUSCLES
Palaeonemertea	√	√	√	
Heteronemertea			√	
Hoplonemertea				√
Bdellonemertea				√

Most nemerteans possess a pair of remarkably complex and unique structures called cerebral organs (Figure 10C). The core of each cerebral organ is a ciliated epidermal invagination (the **cerebral canal**), which is expanded at its inner end. These canals lead laterally to pores within the cephalic slits (when present) or else directly to the outside via separate pores on the head. The inner ends of the canals are surrounded by nervous tissue of the cerebral ganglion, and by glandular tissue, and they are often intimately associated with lacunar blood spaces. Cilia in the cerebral canal circulate water through the open portion of the organ; this activity intensifies in the presence of food. Nemerteans presumably use this mechanism when hunting and tracking prey or in other chemotactic responses. The association of the cerebral canals with glandular, nervous, and circulatory structures has led some workers to suggest an endocrine/neurosecretory function for the cerebral organs. Other suggestions have included auditory, gas exchange, excretory, and tactile activities. Cerebral organs are absent in several genera, including the symbiotic *Carcinonemertes* and *Malacobdella* and the pelagic hoplonemerteans.

In the region anterior to the cerebral ganglion large frontal glands open to the outside through a pitlike **frontal sense organ** (Figure 10D). These structures receive nerves from the cerebral ganglion and appear to be chemosensory, but solid evidence for this suggestion is lacking. Statocysts have been found in some nemerteans, including pelagic forms where geotaxis is an obvious advantage.

Reproduction and development

Asexual processes. Many nemerteans show remarkable powers of regeneration, and nearly all species can regenerate at least posterior portions of the body. Those with the greatest regenerative abilities are certain species of *Lineus*, which engage in asexual reproduction on a regular basis by undergoing multiple transverse fission into numerous fragments (Fig-

ure 11). The fragments are often extremely small and the process is sometimes referred to simply as fragmentation. The small pieces often form mucous cysts within which the new worm regenerates; larger pieces grow into new animals without the protection of a cyst. In some nemerteans only anterior fragments can regenerate into new worms.

Sexual reproduction. Nemerteans constitute a relatively small phylum, but they show remarkable variation in reproductive and developmental strategies. Most ribbon worms are gonochoristic, although protandric and even simultaneous hermaphrodites are known. Unlike the complex reproductive systems of flatworms, the reproductive system of nemerteans has gonads that are simply specialized patches of mesenchymal tissue arranged serially along each side of the intestine and alternating with the midgut diverticula (Figure 12). In *Malacobdella* and a few others, the gonads are more-or-less packed within the mesenchyme (Figure 1E). In most nemerteans the development of gonads occurs along nearly the entire

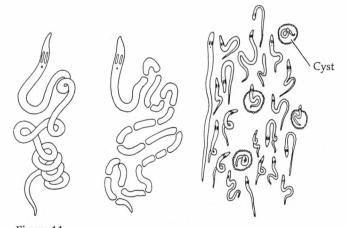

Figure 11

Fragmentation and regeneration in a nemertean, *Lineus vegetus*. Each fragment regenerates into a complete worm. Small fragments may form cysts. (After Coe 1934.)

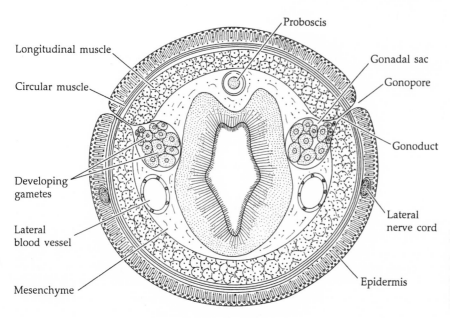

Figure 12

Arrangement of gonads in the nemertean *Carinina* (cross-section). Note the position of a pair of gonads in the mesenchyme. See also Figures 1E,F and 7. (After Hyman 1951.)

length of the body, but in a few species they are restricted to certain regions, usually toward the anterior end. The gonads begin to enlarge and hollow just prior to the onset of breeding activities. Specialized cells in the walls of the rudimentary ovaries and testes proliferate eggs and sperm into the lumina of the enlarging gonadal sacs. In females additional special cells are responsible for yolk production. There is evidence that maturation is under neurosecretory hormonal control, at least in some species. The secretions are probably from the cerebral organ complex (see various publications by Bierne).

With the proliferation of gametes, the gonadal sacs expand to almost fill the area between the gut and the body wall. When the animals are nearly ready to spawn, mating behavior is initiated and the worms become increasingly active. As mentioned earlier, mate location probably depends on chemotactic responses. The same is apparently true of spawning itself, at least for some species, because the presence of a ripe conspecific stimulates the release of gametes from other mature individuals. Experimental evidence indicates that physical contact is not necessary for such a spawning response; thus, some sort of pheromone is probably involved. In nature, however, spawning usually occurs in concert with actual physical contact; tactile responses evidently follow chemotactic mate location. During such mating activities, veritable knots of scores of worms may writhe in a mucus-covered mating mass. The coordinated release of ripe gametes under such conditions ensures successful fertilization. The gametes

are extruded through temporary pores or through ruptures in the body wall. Rupture occurs by contraction of the body wall muscles or of special mesenchymal muscles surrounding the gonads.

Fertilization is often external, either free in the sea water or in a gelatinous mass of mucus produced by the mating worms. In the latter situation, actual egg cases are frequently formed, and part or all of the embryonic development occurs within them (Figure 13). Internal fertilization occurs in certain nemerteans. In some cases the sperm are released into the mucus surrounding the mating worms and then move into the ovaries of the female; once fertilized, the eggs are usually deposited in egg capsules, where they develop. Some terrestrial species are ovoviviporous; the embryos are retained within the body of the female and development is fully direct—an obvious advantage for surviving on land. Ovoviviparity is also known in a few other nemerteans, including deep-sea pelagic forms. Since the population densities of these pelagic worms are extremely low, they must capitalize on the relatively infrequent encounters of males and females and ensure successful fertilization. In a few cases the males are equipped with suckers, which are used to clasp the female, or, rarely, with a protrusible penis, which is used to transfer sperm.

Regardless of the method of fertilization, development through the gastrula is similar among most of the nemerteans studied to date. Cleavage is holoblastic and spiral, producing either three (*Tubulanus*) or, more typically, four quartets of micromeres.

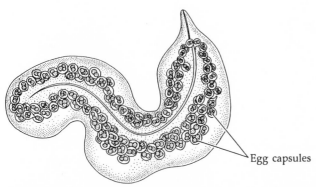

Figure 13

Egg capsules in egg case of *Lineus ruber*. (After Hyman 1951.)

A coeloblastula forms, and this often shows the rudiments of an apical ciliary tuft associated with a slight thickening of the blastula wall at or near the animal pole. Gastrulation is usually by invagination of the macromeres and the fourth micromere quartet to produce a coelogastrula. In at least one genus (*Prostoma*, a hermaphroditic freshwater form), gastrulation is by unipolar ingression of the vegetal macromeres; this movement yields a stereogastrula, which later hollows. Mesoderm may originate in several ways, and in some cases the processes are poorly understood. In a few instances it appears that at least some of the middle layer arises from micromere ectodermal precursors, hence producing ectomesoderm. In *Tubulanus*, which produces only three quartets of micromeres, the mesoderm apparently derives from the 3D macromere; in others it is probably from the usual 4d mesentoblast.

There are quite a few developmental strategies among the nemerteans. Members of the orders Palaeonemertea, Hoplonemertea, and Bdellonemertea undergo direct development within egg cases.* The embryos in these three orders pass through stages much like the larval stages of certain flatworms, but they develop gradually to juvenile worms without any abrupt metamorphosis. These embryos are nourished by yolk until they hatch, whereupon they commence feeding. The hatching forms of some of these nemerteans resemble macrostomids (Figure 14). This is especially true of the palaeonemerteans, in which the proboscis apparatus is not fully formed at hatching (the proboscis of heteronemerteans and bdellonemerteans is functional at the time of hatching).

Hubrechtella dubia is classified as a palaeonemertean but possesses a typical larval stage. It is not known whether this is a developmental oddity or an indication that a taxonomic reassessment is warranted for this species.

The heteronemerteans undergo a bizarre and fascinating pattern of indirect development. Most species of this order produce a free-swimming, planktotrophic larva called the **pilidium** (Figure 15). At this stage the gut is incomplete, consisting of a mouth located between a pair of flaplike ciliated lobes, a stomodeal foregut, and a blind intestine; the anus forms later as a proctodeal invagination. Interestingly, the intestinal diverticula of nemerteans do not form as evaginations of the gut wall but are produced by medial encroachments of mesenchyme, which press in the gut wall, leaving the diverticula. As the pilidium swims and feeds, a series of invaginations in the larval ectoderm (Figure 15B) eventually pinch off internally to produce the presumptive adult ectoderm. Thus, the metamorphosed juvenile develops *within* a larval skin while the animal is still planktonic (Figure 15C). In this way the animal prepares for benthic life before it faces the rigors of settlement. When development is completed, the larval skin is shed and the juvenile assumes its life on the bottom.

Another free-swimming larval form, called the **Iwata larva**, is known among certain heteronemerteans. This larval stage was named after Fumio Iwata of Hokkaido University, who has contributed much to our knowledge of nemertean development. The Iwata larva derives from a yolky egg and undergoes lecithotrophic development. It passes through a stage of ectodermal invaginations and eventually sheds its larval skin, much like the processes described for the pilidium. In many species the emerging juveniles eat their larval ectoderm.

Some heteronemerteans undergo direct development via the production of a larval form (the **Desor larva**) that passes its entire developmental life within a protective egg case. The Desor larva was named after E. Desor, who saw and described this pattern of development in *Lineus* in 1848. Prior to hatching from its egg capsule, the Desor larva undergoes ectodermal inpocketing like that in other heteronemerteans. For this reason, many authorities categorize Desor development as indirect (since a metamorpho-

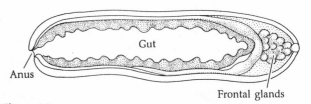

Figure 14

Hatching form produced by direct development in *Prosorhochmus*, a hoplonemertean. (After Hyman 1951.)

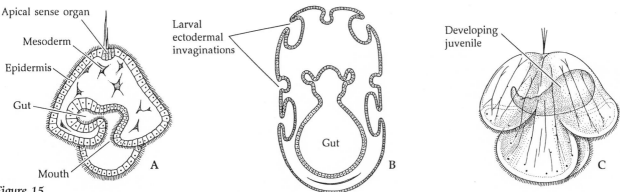

Figure 15

Development of a pilidium larva. A, Pilidium larva (section). B, A pilidium larva (transverse section) during invagination of larval ectoderm to form adult skin. C, Late pilidium larva with juvenile formed within. (A,C redrawn from Gibson 1972, after various sources. B after Gontcharoff 1961.)

sis does occur), exemplifying one of the semantic problems of defining life cycle patterns.

It is worth mentioning here that some mystery surrounds the embryonic origin of the lining of the rhynchocoel. The space forms by a hollowing of tissue, as in schizocoely. If that tissue is true mesoderm, then, as interpreted by Turbeville and Ruppert (1985), the rhynchocoel is a true coelom.

Nemertean phylogeny

The fossil record is of no use in establishing the origin of nemerteans in geological time. They diverged sometime after the evolution of the spiralian bilateral condition. Their origin is puzzling, and one must take into account controversies about the position of the flatworms, the origin of the acoelomate condition (Chapter 10), and the possible coelomic nature of the nemertean rhynchocoel and blood vessels. One view is that nemerteans arose from early archoöphoran turbellarian stock, perhaps sharing common ancestry with the macrostomid flatworms. The nemerteans and turbellarians display a number of similarities beyond their basic acoelomate *Baupläne*. The shared features include such things as protonephridia, types of ocelli, certain histological characteristics (especially of the epidermis), and the general organization of the nervous system. Furthermore, various ciliated slits and depressions among turbellarians may be homologous to the cephalic slits and similar structures of the nemerteans. Some flatworms possess frontal (cephalic) glands thought to be homologous to those of ribbon worms. Some of the

strongest evidence for a flatworm-nemertean relationship comes from examination of their embryogenies. The spiral cleavage of archoöphoran turbellarians (especially of macrostomids and some polyclads) is certainly paralleled in the early development of nemerteans. This commonality is enhanced by a consideration of the similarities among various later developmental and hatching stages. But, if nemerteans are descendants of flatworms, or if they are sister groups, there is no doubt that they have "done more" (evolutionarily) with their acoelomate *Bauplan* than have their platyhelminth cousins. Furthermore, the similarities shared by platyhelminths and nemerteans are not unique, because they occur in other taxa. In other words, their similarities are symplesiomorphies. Their union to a common ancestor (or deriving one from the other) is usually supported by treating them both as acoelomates. However, we have seen in this chapter that the nemerteans probably have coeloms (certain blood vessels and the rhynchocoel), and in Chapter 10 we explored the possibility that the acoelomate condition may even be secondarily derived from some spiralian, schizocoelomate ancestor.

We are left with various ways to interpret the evidence about nemertean and flatworm relationships and ancestry. The alternative cladograms shown in Figures 16A and B illustrate these possibilities. First, the nemerteans and flatworms may (Figure 16A) or may not (Figure 16B) share a common ancestor and represent a monophyletic clade. Second, the acoelomate condition, in one or both groups, may be primitive within triploblastic, spiralian taxa, or it may be secondarily derived from some protostomous, schizocoelomate ancestor. These views encompass ideas about when the coelom arose relative to the origins of these groups and are explained more fully in the legend for Figure 16. The recent discovery (Berg 1985) of a seemingly "segmented" ribbon worm, *Annulonemertes*, further com-

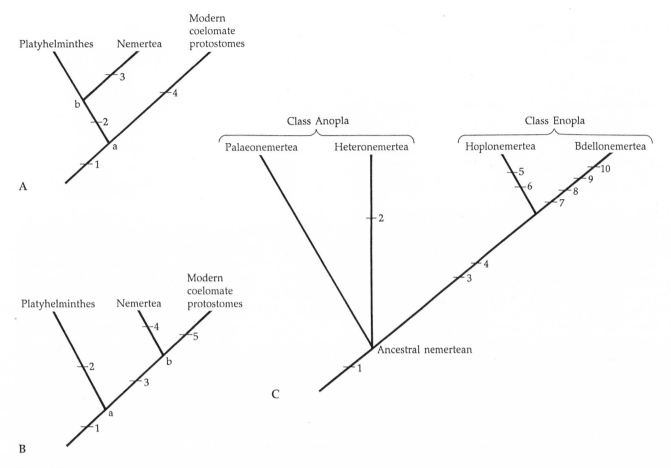

Figure 16

Some ideas about nemertean phylogeny. (See text for additional explanation.) A, A cladogram based on the hypothesis that the platyhelminths and nemerteans constitute a true monophyletic clade, and that both share the acoelomate condition. However, we may interpret this cladogram in two very different ways. First, we may view the events represented by (1) as the origin of the triploblastic, spiralian condition to produce an acoelomate ancestor (a). This condition, viewed here as plesiomorphic within the spiralian line, was retained in ancestor (b) and in the Platyhelminthes, whereas the nemerteans acquired a set of synapomorphies (3), including the proboscis apparatus, a circulatory system, and an anus. In this scenario, the coelom arose (schizocoely) along the line (4) leading to the modern coelomate protostomes. Second, we may view the assumed acoelomate condition of the flatworms and nemerteans as having arisen secondarily from a coelomate, protostome ancestor. In this case, we add schizocoely to the list of synapomorphies defining (a) and occurring at (1). The common ancestor (b) to the platyhelminths and nemerteans, then, becomes defined by a loss or reduction of the coelom (2) to achieve the acoelomate condition.

B, A cladogram based on the hypothesis that the platyhelminths and nemerteans do not constitute a monophyletic clade. Still, this arrangement can be interpreted in more than one way. Assuming (1) to be the origin of the triploblastic, spiralian condition leading to ancestor (a), then the flatworms retain this ancestral condition and are not distinguished by any unique synapomorphies. The remaining spiralian phyla are distinguished by the origin of an anus and a circulatory system (3), and ancestor (b) is an acoelomate creature. The characteristics of ancestor (b) are retained by the nemerteans that acquire the nemertean synapomorphies (4), and the coelom arises at (5). The other interpre-

tation would view the coelom arising at either (1) or (3). In the first case, the flatworms must have lost the coelom (2) to become secondarily acoelomate. In the second case, the flatworms are viewed as primitively acoelomate. In either case, we may view the nemerteans as either losing the coelom (if they are acoelomate) or retaining it in the form of the rhynchocoel (and possibly some blood vessels) (4) from their coelomate ancestor (b), which also gave rise to the modern coelomate protostomes.

C, This cladogram to the orders of nemerteans is applicable regardless of which hypothesis of nemertean origin is accepted, so long as an epidermal nerve cord is considered the primitive condition within the Nemertea. The phylum can be defined by at least the presence of the unique proboscis apparatus (1), and by other things, depending on how one views the taxon's origin (see above). The ancestral nemertean was probably much like a modern-day member of the order Palaeonemertea, and that order has retained many of the ancestral traits. In fact, the lack of any unique synapomorphies distinguishing the Palaeonemertea suggests that the ancestral nemertean was itself what we would recognize as a palaeonemertean. The heteronemerteans are distinguished anatomically by their unique arrangement of body wall muscles in outer longitudinal, middle circular, and inner longitudinal layers (2). The origin of the Enopla is defined by movement of the mouth anterior to the cerebral ganglion (3), and movement of the longitudinal nerve cords to a mesenchymal position (4). The hoplonemerteans are defined by the appearance of proboscis armature (5) and the division of the proboscis into distinct regions (6). The bdellonemerteans are characterized by a suite of synapomorphies associated, in large part, with their endocommensalistic lifestyles. These include a reduction in sensory organs (7), the appearance of a posterior sucker (8), the elongation of the gut (9), and the loss of intestinal diverticula (10).

plicates the situation. (These matters are also addressed in Chapter 24.)

The phylogenetic relationships among the various taxa of the Nemertea are also difficult to assess. We propose one hypothesis, illustrated and explained in Figure 16C. One of the principal structural trends among the nemerteans is the internalization of the major longitudinal nerve cords as depicted in Table 1. If we assume that the earliest ribbon worms possessed epidermal nerve cords, as do some modern palaeonemerteans, then it is likely that the phylum arose prior to the evolution of the subepidermal nerve condition. The Palaeonemertea and Heteronemertea probably appeared much earlier than the Hoplonemertea and Bdellonemertea. The palaeonemerteans and heteronemerteans retain the presumed plesiomorphic feature of the placement of the mouth posterior to the cerebral ganglion, a feature shared with most turbellarians. The relatively simple and unarmed probisces and the placement of the nerve cords external to the mesenchyme suggest further that these orders are relatively primitive among the ribbon worms. The heteronemerteans diverged from this clade with their adoption of indirect development, the unique formation of the double larval and adult ectoderm during metamorphosis, and the evolution of their unique arrangement of body wall muscles. The encapsulation, and thus functionally direct development, of those heteronemerteans with a Desor larva is undoubtedly a secondary abandonment of free larval life. The divergence of the heteronemerteans apparently occurred after the evolution of an intramuscular position of the nerve cords among the palaeonemertean ancestors.

The hoplonemerteans show some distinct changes from the members of orders mentioned above. Most notable are the regional specialization and armature of the proboscis, the movement of the nerve cords to a mesenchymal position, and the movement of the mouth more anteriorly. The bdellonemerteans are a specialized offshoot that displays significant modification for an endosymbiotic lifestyle (simplification of the proboscis, coiling and increased relative length of the gut probably associated with their herbivorous habits, and decreased body length).

Selected References

Berg, G. 1985. *Annulonemertes* gen. nov., a new segmented hoplonemertean. *In* C. Morris et al. (eds.), *The Origins and Relationships of Lower Invertebrates*. Syst. Assoc. Spec. Vol. No. 28, Oxford, pp. 200–209.

Bianchi, S. 1969a. On the neurosecretory system of *Cerebratulus marginatus* (Heteronemertini). Gen. Comp. Endocr. 12:541–548.

Bianchi, S. 1969b. The histochemistry of the neurosecretory system in *Cerebratulus marginatus* (Heteronemertini). Gen. Comp. Endocr. 13:206–210.

Bierne, J. 1966. Localisation dans les ganglions cérébroïdes du centre regulateur de la maturation sexuelle chez la femelle de *Lineus ruber* Muller (Hétéronémertes). C. r. Hebd. Séanc. Acad. Sci. Paris 262:1572–1575.

Bierne, J. 1967. Sur le contrôle endocrinien de la différenciation du sexe chez la Némerte *Lineus ruber* Muller. La masculinisation des ovaries des chimères hétérosexuées. C. r. Hebd. Séanc. Acad. Sci. Paris 265:447–450.

Bierne, J. 1970. Influence des facteurs hormonaux gonado-inhibiteur et androgène sur la différenciation sexuelle des parabiontes hétérosexués chez un némertien. Ann. Biol. 9:395–400.

Borlase, W. 1758. *The Natural History of Cornwall*. Jackson, Oxford. [Includes the first "description" of a ribbon worm.]

Cantell, C.-E. 1969. Morphology, development, and biology of the pilidium larvae from the Swedish west coast. Zool. Bidrag. Fran Uppsala 38:61–111.

Clark, R. B. and J. B. Cowey. 1958. Factors controlling the change of shape of certain nemerteans and turbellarian worms. J. Exp. Biol. 35:731–748.

Coe, W. R. 1931. A new species of nemertean (*Lineus vegetus*) with asexual reproduction. Zool. Anz. 94:54–60.

Coe, W. R. 1934. Regeneration in nemerteans. IV. Cellular changes involved in restitution and reorganisation. J. Exp. Zool. 67:283–314.

Coe, W. R. 1940. Revision of the nemertean fauna of the Pacific coasts of North, Central, and northern South America. Allan Hancock Pac. Exped. 2(13):247–323.

Coe, W. R. 1943. Biology of the nemerteans of the Atlantic coast of North America. Trans. Conn. Acad. Sci. 35:129–328. [This is one of Coe's classic works; see also his dozens of other publications from 1895 through the late 1950s.]

Desor, E. 1848. On the embryology of *Nemertes*, with an appendix on the embryonic development of *Polynoe*, and remarks upon the embryology of marine worms in general. Boston J. Nat. Hist. 6:1–18.

Eakin, R. M. and J. A. Westfall. 1968. Fine structure of nemertean ocelli. Am. Zool. 8:803 (abstract).

Fleming, L. C. and R. Gibson. 1981. A new genus and species of monostiliferous hoplonemerteans, ectohabitant on lobsters. J. Exp. Mar. Biol. Ecol. 52:79–93.

Gibson, R. 1968. Studies on the biology of the entocommensal rhynchocoelan *Malacobdella grossa*. J. Mar. Biol. Assoc. U.K. 48:637–656.

Gibson, R. 1972. *Nemerteans*. Hutchinson University Library, London. [An excellent summary of what was known and what was not known in the early 70s.]

Gibson, R. 1982a. *A Synopsis of British Nemerteans*. Synopses of the British Fauna No. 24. Published for the Linnean Society of London and the Estuarine and Brackishwater Sciences Association, Cambridge University Press, London.

Gibson, R. 1982b. Nemertea. *In* S. Parker (ed.), *Synopsis and Classification of Living Organisms*, Vol. 1. McGraw-Hill, New York, pp. 823–846.

Gibson, R. 1985. The need for a standard approach to taxonomic descriptions of nemerteans. Am. Zool. 25:5–14.

Gibson, R. and J. Jennings. 1969. Observations on the diet, feeding mechanism, digestion and food reserves of the ectocommensal rhynchocoelan *Malacobdella grossa*. J. Mar. Biol. Assoc. U.K. 49:17–32.

Gibson, R. and J. Moore. 1976. Freshwater nemertines. Zool. J. Linn. Soc. 58:177–218.

Gontcharoff, M. 1961. Nemertiens. *In* P. Grassé (ed.), *Traité de Zoologie*, Vol. 4, Pt. 1. Masson et Cie, Paris, pp. 783–886.

Humes, A. G. 1942. The morphology, taxonomy and bionomics of the nemertean genus *Carcinonemertes*. Illinois Biol. Monogr. 18:1–105.

Hylbom, R. 1957. Studies on palaeonemerteans of the Gullmar Fiord area (west coast of Sweden). Ark. Zool. 10:539–582.

Hyman, L. H. 1951. *The Invertebrates*, Vol. 2, Playthelminthes and Rhynchocoela: The Acoelomate Bilateria. McGraw-Hill, New York.

Iwata, F. 1958. On the development of the nemertean *Micrura akkenshiensis*. Embryologia 4:103–131.

Iwata, F. 1960. Studies on the comparative embryology of nemerteans with special reference to their interrelationships. Publ. Akkeshi Mar. Biol. Stn. 10:1–51.

Iwata, F. 1985. Foregut formation of the nemerteans and its role in nemertean systematics. Am. Zool. 25:23–36.

Jennings, J. B. 1960. Observations in the nutrition of the rhynchocoelan *Lineus ruber*. Biol. Bull. 119(2):189–196.

Jennings, J. B. and R. Gibson. 1969. Observations on the nutrition of seven species of rhynchocoelan worms. Biol. Bull. Mar. Biol. Lab. Woods Hole 136:405–433.

Jensen, D. D. 1960. Hoplonemertines, myxinoids and deuterostome origins. Nature 188:649–650.

Jensen, D. D. 1963. Hoplonemertines, myxinoids, and vertebrate origins. *In* E. C. Dougherty (ed.), *The Lower Metazoa: Comparative Biology and Phylogeny*. University of California Press, Berkeley, pp. 113–126.

Karling, T. G. 1965.. *Haplopharynx rostratus* Meixner mit den Nemertinen vergleichen. Z. Zool. Syst. Evol. 3:1–18.

McDermott, J. J. 1976a. Predation of the razor clam *Ensis directus* by the nemertean worm *Cerebratulus lacteus*. Chesapeake Sci. 17(4):299–301.

McDermott, J. J. 1976b. Observations on the food and feeding behavior of estuarine nemertean worms belonging to the order Hoplonemertea. Biol. Bull. 150:57–68.

McDermott, J. J. and P. Roe. 1985. Food, feeding behavior, and feeding ecology of nemerteans. Am. Zool. 25:113–125.

Moore, J. and R. Gibson. 1981. The *Geonemertes* problem (Nemertea). J. Zool. Lond. 194:175–201. [A revision of the world's terrestrial nemerteans.]

Riser, H. W. 1974. Nemertinea. *In* A. Giese and J. Pearse (eds.), *Reproduction of Marine Invertebrates*, Vol. 1. Academic Press, New York, pp. 359–389.

Riser, H. W. 1985. Epilogue: Nemertinea, a successful phylum. Am. Zool. 25:145–151.

Roe, P. 1970. The nutrition of *Paranemertes peregrina*. I. Studies on food and feeding behavior. Biol. Bull. 139:80–91.

Roe, P. 1976. Life history and predator–prey interactions of the nemertean *Paranemertes peregrina* Coe. Biol. Bull. 150–80–106.

Roe, P. and J. L. Norenburg. 1985. Introduction to the symposium: Comparative biology of nemertines. Am. Zool. 25:3. [Introductory comments to this symposium on nemertines; several of the contributed papers are also included in this reference list.]

Schultze, M. S. 1851. *Beiträge zur Naturgeschichte der Turbellarien*. C.A. Koch, Greifswald.

Stricker, S. A. 1982. The morphology of *Paranemertes sanjuanensis* sp. n. (Nemertea, Monostylifera) from Washington, U.S.A. Zool. Scr. 11(2):107–115.

Stricker, S. A. 1983. S.E.M. and polarization microscopy of nemertean stylets. J. Morph. 175:153–169.

Stricker, S. A. 1985. The stylet apparatus of monostyliferous hoplonemerteans. Am. Zool. 25:87–97.

Stricker, S. A. and R. Cloney. 1982. Stylet formation in nemerteans. Biol. Bull. 162:387–403.

Stricker, S. A. and R. A. Cloney. 1983. The ultrastructure of venom-producing cells in *Paranemertes peregrina* (Nemertea, Hoplonemertea). J. Morphol. 177:89–107.

Turbeville, J. M. and E. E. Ruppert. 1985. Comparative ultrastructure and the evolution of nemertines. Am. Zool. 25:53–71.

Wickham, D. E. 1979. Predation by the nemertean *Carcinonemertes errans* on eggs of the Dungeness crab, *Cancer magister*. Mar. Biol. 55:45–53.

Wickham, D. E. 1980. Aspects of the life history of *Carcinonemertes errans* (Nemertea: Carcinonemertidae), an egg predator of the crab *Cancer magister*. Biol. Bull. 159:247–257.

Wourms, J. P. 1976. Structure, composition, and unicellular origin of nemertean stylets. Am. Zool. 16:213 (abstract).

Chapter Twelve

"Pseudocoelomates," Priapulans, Gnathostomulids, and Loriciferans

In the face of this bewildering array of conflicting opinions about the interrelationships of the aschelminth phyla, it is impossible to form a coherent picture of the evolution of the animals.

R. B. Clark
Dynamics in Metazoan Evolution, 1964

In this chapter we examine ten phyla of metazoa. They are placed together here both for convenience and because they provide some interesting examples of comparative biology, the basic subject of this book. We will explore several of the concepts explained in the introductory sections, including systematic problems, clades and grades, convergent evolution, and the exploitation of limiting *Baupläne*. The phyla discussed here include that odd assemblage of seven phyla usually regarded as the pseudocoelomates (Rotifera, Gastrotricha, Kinorhyncha, Nematoda, Nematomorpha, Acanthocephala, and Entoprocta), the controversial Priapula (or Priapulida), the Gnathostomulida, and the Loricifera. Space does not permit us to cover these groups in great detail, but you should be aware that the literature on some of them (e.g., rotifers and nematodes) is vast. While the members of each phylum represent a success story in their own right, some are far more abundant and diverse than others. Some of these phyla comprise only a few known species, whereas others (e.g., Nematoda) include thousands of described species.

Taxonomic history

Two of the phyla dealt with in this chapter were established quite recently and thus do not figure in much of the complex history of the others. The group Gnathostomulida was formally described as an order

of Turbellaria by Peter Ax (1956) and subsequently raised to phylum status by Riedl (1969). The phylum Loricifera was established in 1983 by Reinhardt Kristensen.

Many of the sorts of creatures now described as pseudocoelomate were discovered centuries ago, but for our purposes we may consider their taxonomic history beginning in the late nineteenth century. By that time much of the infrastructure of the classification of most of these groups had been established, and their evolutionary affinities were being examined. Much of this work was brought into focus by Grobben (1908, 1910), who assumed the pseudocoelomate nature of the rotifers, kinorhynchs, gastrotrichs, nematodes, nematomorphs, and acanthocephalans; he used this condition to justify placing these groups as classes within the phylum Aschelminthes (previously Nemathelminthes). Grobben proposed a dozen or so characteristics of these taxa (including the pseudocoelom) that he considered *unique-in-combination*. The general concept of a phylum Aschelminthes was retained by Hyman (1951), but she dropped the Acanthocephala and added the Priapula, a group about which there is still some debate.

Although some authors continue to recognize the phylum Aschelminthes, most do not. It is now generally accepted that the pseudocoelomate condition has probably been derived several times in the metazoa. That is to say, the "Aschelminthes" is almost certainly a polyphyletic taxon.

The discovery of entoprocts dates back to the late eighteenth century. By about 1840 their superficial resemblance to the Ectoprocta (see Chapter 21) was recognized, and the two groups were placed together under the phylum Bryozoa. Hatschek (1888) eventually undertook a detailed embryological study of certain entoprocts, noting fundamental organizational differences between them and the ectoprocts.

On the basis of his observations, he raised the entoprocts to the rank of a separate phylum. In 1921, A. H. Clark also proposed separate phylum status for the entoprocts (under the name Calyssozoa). Other names have been proposed. Some recent texts reflect the attempts of certain workers (e.g., Nielsen 1971, 1977) to reestablish an affinity between the ectoprocts and entoprocts.

Given the number of uncertainties concerning possible relationships among many of these groups and the high probability that they do not constitute a monophyletic group, we consider it most reasonable to treat them here as separate phyla. Be aware, however, that ongoing ultrastructure studies are bringing us closer to an understanding of these creatures, and some revisions should appear in the near future.

The pseudocoelomate condition

Before discussing each phylum separately, it is essential to note some aspects of the pseudocoelomate condition itself. The pseudocoelom is a blastocoel that persists into adulthood, regardless of when it first appears during development. It is, therefore, a retained embryonic feature and may be justly considered as a paedomorphic characteristic. Of course, the appearance of an embryonic blastocoel (hence, a transient or incipient pseudocoelom) is not unique to the groups under consideration here. That is, most metazoa have a blastocoel at some stage in their ontogeny; it is a more-or-less universal trait of animals at the metazoan grade of complexity. Most workers today view the pseudocoelom as a paedomorphic condition secondarily derived from coelomate ancestors rather than as a primitive, precoelomic stage in metazoan evolution.

The advantages of body cavities were discussed in Chapter 3. The benefits related to internal transport, space for organ development, hydrostatic support systems, and pressure-operated protrusion of parts should come to mind in this regard. However, in the absence of an effective circulatory system, these cavity-conferred benefits are generally only realized when body size is kept relatively small or when body shape maintains small diffusion distances. Exceptions to these principles involve either structural modifications of the body cavity or the establishment of a lifestyle (such as endoparasitism) that reduces the problems of the low surface-to-volume ratio.

As we shall see, not all "pseudocoelomates" retain a spacious fluid-filled body cavity. In some cases (e.g., gastrotrichs, entoprocts) the space has been to various degrees invaded by mesenchyme, which sometimes obliterates it completely to produce a solid body construction, functionally similar to that in acoelomates. The trade-off in such instances may be the acquisition of supportive and storage facilities at the expense of transport and hydrostatic qualities.

The rotifers

The phylum Rotifera (Latin *rota*, "wheel"; *fera*, "to bear") includes more than 1,800 described species. These tiny metazoa, first seen by early microscopists such as Antony van Leeuwenhoek in the late seventeenth century, were at first lumped with the protozoa as little *animalcules*. A few species reach lengths of 2–3 mm, but most are less than 1 mm long. Despite their small size, rotifers are quite complex and display a variety of body forms (Figure 1). They are most common in fresh water, but marine forms are known, and some live in damp soil or in the water film on mosses.

The body is typically divided into head, trunk, and foot (Figure 1). The anterior end bears a ciliary organ called the **corona**. When active, the coronal cilia often give the impression of a pair of rotating wheels, hence the derivation of the phylum name; the rotifers were historically called "wheel animalicules." The members of this phylum are further characterized by being pseudocoelomate; having a complete gut (usually); protonephridia; a tendency to eutely; and, in many species, having syncytial tissues or organs (Box 1). The pharynx is modified as a **mastax** comprising sets of internal jaws, or **trophi**.

PHYLUM ROTIFERA

CLASS SEISONIDEA: (Figure 1A). Epizoic on the marine leptostracan crustacean *Nebalia*; corona reduced to bristles; trophi fulcrate; males fully developed; females produce only mictic ova. Monogeneric: *Seison*.

CLASS BDELLOIDEA: (Figure 1B). Fresh water and terrestrial; corona typically well developed; trophi ramate; males unknown; reproduction by parthenogenesis; paired germovitellaria. (e.g., *Adineta, Ceratotrocha, Embata, Habrotrocha, Philodina, Rotaria*)

CLASS MONOGONONTA: (Figures 1C–E). Predominately fresh water; swimmers, creepers, or sessile; corona and trophi variable; males typically reduced in size, complexity, and of short longevity; mictic and amictic ova produced in many species; single germovitellarium. (e.g., *Asplanchna, Collotheca, Euchlanis, Floscularia, Monostyla, Stephanoceros, Testudinella*)

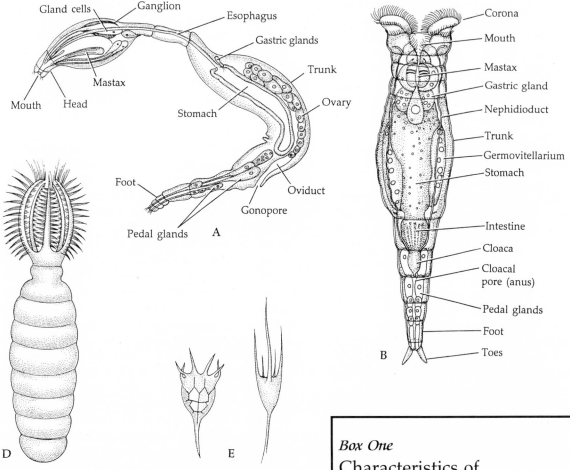

Figure 1

Representative rotifers. A, *Seison annulatus* (class Seisonidea), a marine rotifer from the gills of *Nebalia*. B, *Philodina roseola* (class Bdelloidea). C–F, Members of the class Monogononta. C, *Floscularia*, a sessile rotifer. D, *Stephanoceros*, one of the strange collothecacean rotifers with the corona modified as a trap. E, The cuticular loricae from two loricate rotifers. (A–D after Nogrady 1982; E after various authors.)

General external anatomy and details of the corona

The body of rotifers is surrounded by an intraepidermal cuticle, which may be thin or thick and is often annulated to allow flexibility. The cuticle often bears spines, tubercles, or other sculpturing, and may be developed as a thickened casing, or **lorica** (Figure 1E). Many rotifers bear single dorsal and paired lateral sensory antennae arising from various regions of the body. In most species the foot is elongate and provided with cuticular annuli that permit a telescoping action. The distal portion of the foot often bears spines, or a pair of "toes" through which the ducts from **pedal glands** pass. The secretion from

---

Box One
Characteristics of the Phylum Rotifera

1. Triploblastic, bilateral, unsegmented pseudocoelomates
2. Gut complete and regionally specialized
3. Pharynx modified as a mastax, containing jawlike elements called trophi
4. Anterior end often bears variable ciliated fields as a corona
5. Posterior end often bears toes and adhesive glands
6. Cuticle well developed and secreted within the epidermis
7. With protonephridia, but no special circulatory or gas exchange structures
8. With unique retrocerebral organ
9. Males generally reduced or absent; parthenogenesis common
10. Inhabit marine, freshwater, or semiterrestrial environments; sessile or free-swimming

the pedal glands enables the rotifer to attach temporarily to the substratum. The foot is absent from some swimming forms (e.g., *Asplanchna*) and is modified for permanent attachment in sessile types (e.g., *Floscularia*).

The corona is the most characteristic external feature of rotifers. Its morphology varies greatly and has been used extensively in taxonomic and phylogenetic investigations. The presumed primitive condition is shown in Figure 2A. A well developed patch of cilia surrounds the anteroventral mouth. This patch is the **buccal field**, or **circumoral field**, and extends dorsally around the head as a ciliary ring called the **circumapical field**. The extreme anterior part of the head bordered by this ciliary ring is the **apical field**. The corona has evolved to a variety of modified forms in different rotifer taxa. The most familiar coronal form is that seen in the bdelloids. Here the buccal field is quite reduced, and the circumapical field is separated into two ciliary rings, one slightly anterior to the other (Figure 2B). The anterior-most ring is called the **trochus**, the other the **cingulum**. In many rotifers the trochus is a pair of well defined anterolateral rings of cilia called **trochal discs** (Figure 2C), which may be

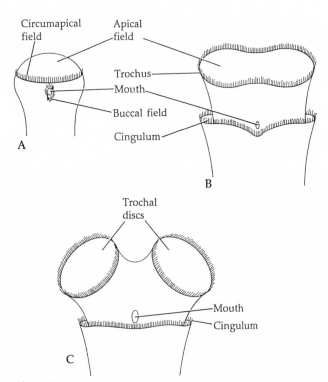

Figure 2

Modifications of the corona among selected rotiferan types. A, Presumed primitive condition has buccal and circumapical fields. B, The circumapical field is separated into trochus and cingulum. C, The trochus is separated into two trochal discs.

retracted or extended for locomotion and feeding. The metachronal ciliary waves along these trochal discs impart the impression of rotating wheels.

Body wall, body cavity, support, and locomotion

The cuticle of rotifers is produced by and lies within the syncytial epidermis. The epidermis exhibits eutely in that the epidermal nuclear number is consistent for each species (usually from about 900 to 1,000 nuclei). Beneath the epidermis are various circular and longitudinal muscle bands (Figure 3); there are no sheets or layers of body wall muscles. The internal organs lie within a typically spacious, fluid-filled pseudocoelom.

In the absence of a thick muscular body wall, body support and shape are maintained by the cuticle and the hydrostatic skeleton provided by the pseudocoelom. In most cases the cuticle is only flexible enough to allow slight changes in shape, so increases in hydrostatic pressure within the body cavity can be used to protrude body parts (e.g., foot, trochal discs). These parts are retracted by various longitudinal retractor muscles (Figure 3).

Although a few rotifers are sessile, most are motile and quite active, moving about by swimming or inchworm-like creeping. Some are exclusively either swimmers or crawlers, but many are capable of both methods of locomotion. Swimming is accomplished by beating the coronal cilia, forcing water posteriorly along the body and driving the animal forward, sometimes in a spiral path. When creeping, a rotifer attaches its foot with secretions of the pedal glands and then elongates its body and extends forward. It attaches the extended anterior end to the substratum, releases its foot, and draws its body forward by muscular contraction.

Feeding and digestion

Rotifers display a variety of feeding methods, depending upon the structure of the corona (Figure 2) and the mastax trophi (Figure 4). Ciliary suspension feeders have well developed coronal ciliature and a grinding mastax. These forms include the bdelloids, which have trochal discs and a **ramate mastax**, and a number of monogonontan rotifers, which have a separate trochus and cingulum and a **malleate mastax**. These forms typically feed on minute organic detritus or small organisms. The feeding current is produced by the action of the cilia of the trochus (or trochal discs), which beat in a direction opposite to that of the cilia of the cingulum. Particles are drawn into a ciliated food groove that lies between these

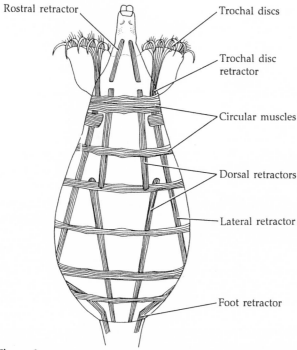

Figure 3

Major muscle bands of the bdelloid, *Rotaria* (dorsal view). (After Hyman 1951.)

balia. These rotifers (*Seison*) crawl around the base of the legs and gills of their host, feeding on detritus and on the host's brooded eggs. Some bdelloids (e.g., *Embata*) also live on the gills of crustaceans (particularly amphipods and decapods). There are isolated examples of endoparasitic rotifers inhabiting hosts such as *Volvox*, some freshwater algae, snail egg cases, and the body cavities of certain annelids and terrestrial slugs. Little is known about nutrition in most of these forms.

The digestive tract of most rotifers is complete and more-or-less straight (Figure 5A). (The anus has been secondarily lost in a few types, and some have a moderately coiled gut.) The mouth leads inward to the pharynx (mastax) either directly or via a short, ciliated, **buccal tube**. Depending on the feeding method and food sources, swallowing is accomplished by various means, including ciliary action of the buccal field and buccal tube or a piston-like pumping action of certain elements of the mastax opposing ciliary bands and are carried to the buccal field and mouth.

Raptorial feeding is common in many species of Monogononta. Coronal ciliation in these rotifers is often reduced or used exclusively for locomotion. Raptorial feeders obtain food by grasping it with protrusible mastax jaws; most possess either a **forcipate mastax** or an **incudate mastax** (Figures 4C and D). Raptorial rotifers feed mainly on small animals but are known to ingest plant material as well. They may ingest their prey whole and subsequently grind it to smaller particles within the mastax, or they may pierce the body of the plant or animal with the tips of the mastax jaws and suck fluid from the prey.

Some monogonontan rotifers have adopted a trapping method of predation. In such cases the corona usually bears spines or setae arranged as a funnel-shaped trap (Figure 1D). The mouth in these trappers is located more-or-less in the middle of the ring of spines (rather than in the more typical anteroventral position); thus, captured prey are drawn to it by contraction of the trap elements. The mastax in trapping forms is often reduced.

A few rotifers have adopted symbiotic lifestyles. As noted in the classification scheme, seisonids live on marine leptostracan crustaceans of the genus *Ne-*

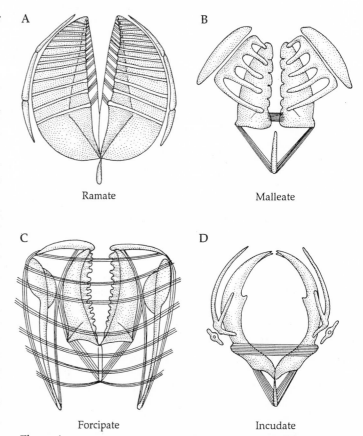

Figure 4

Representative mastax types of rotifers. A–B, Crushing/grinding forms. C–D, Grasping, predatory forms. See text for additional details. (A,B,D after Nogrady 1982; C after Hyman 1951.)

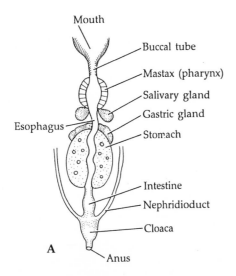

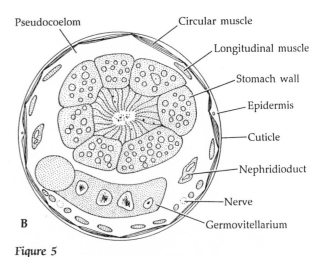

Figure 5

A, Digestive system of a rotifer. B, Cross section through the trunk. (B redrawn from Sherman and Sherman 1976, after Grassé 1948.)

apparatus. The mastax is lined with cuticle and is ectodermal in origin. Opening into the gut lumen just posterior to the mastax are ducts of the **salivary glands**. There are usually two to seven such glands, and they are presumed to secrete digestive enzymes and perhaps lubricants aiding the movement of the mastax trophi.

A short esophagus connects the mastax and stomach. A pair of **gastric glands** opens into the posterior end of the esophagus; these glands apparently secrete digestive enzymes. The walls of the esophagus and gastric glands are often syncytial. The stomach is generally thick-walled and may be cellular or syncytial, comprising a specific number of cells or nuclei in each species (Figure 5B). The intestine is short and leads to the anus, which is located dorsally near the posterior end of the trunk. Except for *Asplanchna*, which lacks a hindgut, an expanded **cloaca** connects the intestine and anus. The oviduct and usually the nephridioducts empty into this cloaca.

Digestion probably begins in the lumen of the mastax and is completed extracellularly in the stomach, where absorption occurs. Although much remains to be learned about the digestive physiology of rotifers, some experimental work indicates that diet has multiple and important effects on various aspects of their biology, including the size and shape of individuals as well as some life cycle activities (Gilbert 1980a).

Circulation, gas exchange, excretion, and osmoregulation

There are no special organs for internal transport or for the exchange of gases between tissues and the environment. The pseudocoelomic fluid provides a medium for circulation within the body, which is aided by general movement and muscular activities. Small body size reduces diffusion distances and facilitates the transport and exchange of gases, nutrients, and wastes. These activities are further enhanced by the absence of linings and partitions within the body cavity, so the exchanges occur directly between the organ tissues and the body fluid. Gas exchange probably occurs over the general body surface wherever the cuticle is sufficiently thin.

Most rotifers possess one pair of flame bulb protonephridia, located far forward in the body. A nephridioduct leads from each flame bulb to a collecting bladder, which in turn empties into the cloaca via a ventral pore. In some forms, especially the bdelloids, the ducts open directly into the cloaca, which is enlarged to act as a bladder (Figure 5A).

The protonephridial system of rotifers is primarily osmoregulatory in function, and is most active in freshwater forms. Excess water from the pseudocoelom and probably from food is also pumped from the body via the anus by muscular contractions of the bladder. This "urine" is significantly hypotonic relative to the body fluids.

Some rotifers (especially the semiterrestrial species) are able to withstand extreme environmental stresses by entering a state of metabolic dormancy. They have been experimentally desiccated and kept in a dormant condition for as long as four years— reviving upon the addition of water. Some have survived freezing in liquid helium at −272°C, and other such severe stresses.

It is likely that the protonephridia also remove

nitrogenous excretory products from the body. This form of waste removal is probably supplemented by simple diffusion of wastes across permeable body wall surfaces.

Nervous system and sense organs

The cerebral ganglion of rotifers is located dorsal to the mastax, in the neck region of the body. Several nerve tracts arise from the cerebral ganglion, some of which bear additional small ganglionic swellings (Figure 6A). There are usually two major longitudinal nerves positioned either both ventrolaterally or one dorsally and one ventrally.

The coronal area generally bears a variety of touch-sensitive bristles or spines and often a pair of ciliated pits presumed to be chemoreceptors (Figure 6B). The dorsal and lateral antennae are probably tactile. Some rotifers bear sensory **flosculi**, which are arranged as a cluster of micropapillae encircling a pore. These flosculi may be tactile or chemosensory. Most of the errant rotifers possess at least one simple ocellus embedded in the cerebral ganglion. In some, this cerebral ocellus is accompanied by one or two pairs of lateral ocelli on the coronal surface, and sometimes by a pair of apical ocelli in the apical field. The lateral and apical ocelli are multicellular epidermal patches of photosensitive cells.

Associated with the cerebral ganglion is the so-called **retrocerebral organ**. This glandular structure gives rise to ducts that lead to the body surface in the apical field (Figure 6B). The function of this organ is unknown, but some workers suggest that it may be sensory.

Reproduction and development

Except for parthenogenesis, asexual reproduction is unknown among rotifers, and most show only very weak powers of regeneration.

Rotifers are gonochoristic; but, except for the genus *Seison*, the males are either reduced in abundance, size, and complexity (Monogononta) or are completely unknown (Bdelloidea). If you find a rotifer, the chances are good that it is a female.

The male reproductive system (Figure 7) includes a single testis (paired in members of the class Seisonidea), a sperm duct, and a posterior gonopore, whose wall is usually folded to produce a copulatory organ. Prostatic glands are sometimes present in the wall of the sperm duct. The males are short lived and they possess a reduced gut unconnected to the reproductive tract.

The female system includes paired (Bdelloidea) or single (Monogononta) syncytial **germovitellaria** (Figure 7); in the Seisonidea, there are no yolk glands. Eggs are produced in the ovary and receive yolk directly from the vitellarium before passing along the oviduct to the cloaca; in those forms that have lost the intestinal portion of the gut (e.g., *Asplanchna*), the oviduct passes directly to the outside via a gonopore.

In rotifers with a male form, copulation occurs either by insertion of the male copulatory organ into the cloacal area of the female or by hypodermic impregnation. In the latter case, males attach to females at various points on the body, and they apparently inject sperm directly into the pseudocoelom. The

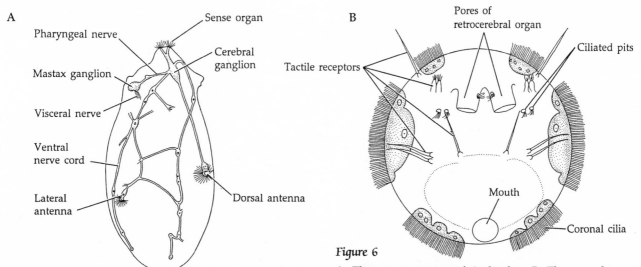

Figure 6

A, The nervous system of *Asplanchna*. B, The coronal area of *Euchlanis* (apical view). Note the various sense organs. (After Hyman 1951.)

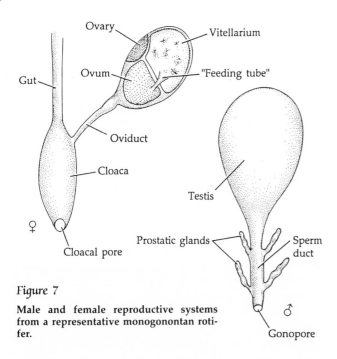

Figure 7

Male and female reproductive systems from a representative monogonontan rotifer.

sperm somehow find their way to the female reproductive tract, where fertilization takes place. The number of eggs produced by an individual female is determined by the original, fixed number of ovarian nuclei, usually 20 or fewer, depending on the species. Once fertilized, the ova produce a series of encapsulating membranes and are then either attached to the substratum or carried externally or internally by the female.

Parthenogenesis is the rule among the bdelloids, but it is also a common and usually seasonal occurrence in the monogonontans, where it tends to alternate with sexual reproduction. This cycle (Figure 8) is an adaptation to freshwater habitats that are subject to severe seasonal changes. During favorable conditions, females reproduce parthenogenetically through the production of mitotically derived diploid ova (**amictic ova**). These eggs develop without fertilization into more females. However, when these ova from amictic females are subjected to particular environmental conditions (so-called **mixis stimuli**) they develop into **mictic females**, which produce haploid ova, or **mictic ova**, by meiosis. The exact stimulus apparently varies among different species and may include such factors as changes in day length, temperature, food sources, or increases in population density. Mictic ova may develop by parthenogenesis to haploid males, which produce sperm by mitosis. These sperm fertilize other mictic ova, producing diploid, thick-walled, resting zygotes. The resting zygotic form is extremely resistant to low temperatures,

desiccation, and other adverse environmental conditions. Upon the return of favorable conditions, the zygotes develop and hatch as amictic females, completing the cycle.

Only a few studies have been conducted on the embryogeny of rotifers (see references, especially Pray 1965). In spite of the paucity of data, and some conflicting interpretations in the literature, it is generally accepted that rotifers have modified spiral cleavage. The isolecithal ova undergo unequal holoblastic early cleavage to produce a stereoblastula. Gastrulation is by epiboly of the presumptive ectoderm and involution of the entoderm and mesoderm; the gastrula gradually hollows to produce the blastocoel, which persists as the pseudocoelom. The mouth forms in the area of the blastopore. Definitive nuclear numbers are reached early in development for those organs and tissues displaying eutely (e.g., epidermis, stomach, ovary).

Errant rotifers undergo direct development, hatching as mature or nearly mature individuals. Sessile forms pass through a short dispersal phase, sometimes called a larva; this form is similar to typical

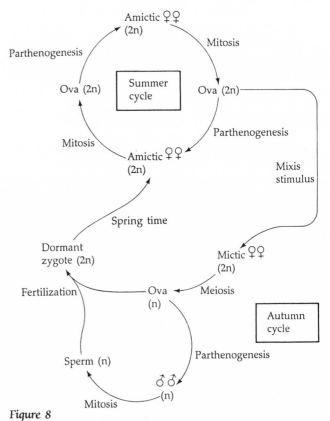

Figure 8

Mictic/amictic alternation in the life cycle of a monogonontan rotifer. See text for explanation.

swimming types. The larva eventually settles and attaches to the substratum.

Many rotifers exhibit **developmental polymorphism**. This phenomenon is also seen in some protozoa, insects, and primitive crustaceans. It is the expression of alternative morphotypes under different ecological conditions, by organisms of a given genetic constitution; the differentiation of certain castes in social insects is one of the most remarkable examples of developmental polymorphism. In all such animals studied to date, the alternative adult morphotypes appear to be products of flexible developmental pathways, which are triggered by various environmental cues and often mediated by internal mechanisms (e.g., hormonal activities). In one well-studied genus of rotifers (*Asplanchna*), the environmental stimulus regulating which of the several adult morphologies is produced is the presence of a specific molecular form of vitamin E—α-tocopherol. *Asplanchna* obtains tocopherol in its diet of algae or other plant material, or when it preys on other herbivores (animals do not synthesize tocopherol). This chemical acts directly on the developing tissues of the rotifer, where it stimulates cytoplasmic growth, mitotic division, and DNA synthesis.

The gastrotrichs

The phylum Gastrotricha (Greek *gasteros*, "stomach"; *trichos*, "hair") comprises about 450 species of small marine, brackish, and freshwater metazoa. Most species are less than 1 mm long, although a few reach 3 mm in length. Many gastrotrichs bear a superficial resemblance to rotifers or even large ciliate protozoa (for which they are often mistaken). Figure 9 illustrates some of the body forms within this phylum and some features of their external anatomy, and Box 2 lists the distinguishing features of this group. Many gastrotrichs live in the interstitial spaces of loose sediments. Others are found in surface detritus or among the filaments of aquatic plants; a few are planktonic.

The gastrotrich body is typically divisible into head and trunk. A few possess an elongate "tail" (e.g., *Urodasys*; Figure 9). Externally, these animals bear various arrangements of spines, bristles, and scales or plates derived from the well developed cuticle. The body bears two or many **adhesive tubes** equipped with glands that secrete attachment and releaser substances used for temporary adherence to objects in the environment. The few gastrotrichs that lack adhesive tubes are planktonic swimmers. Gas-

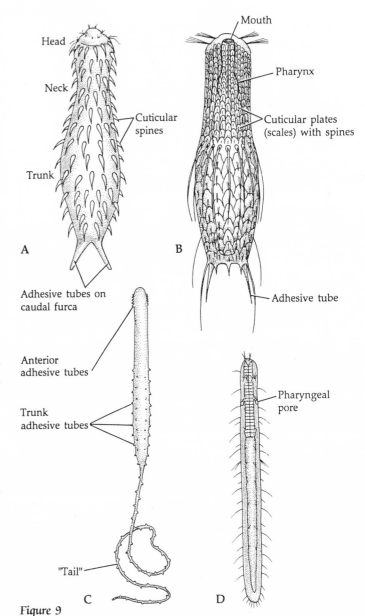

Figure 9

Representative gastrotrichs. A–B, Members of the order Chaetonotida. A, *Chaetonotus*. B, *Aspidophorus*. C–D Members of the order Macrodasyida. C, *Urodasys*. D, *Pleurodasys*. (A redrawn from Sherman and Sherman 1976, after Grassé 1948; B,C after Hyman 1951; D after Hummon 1982.)

trotrichs have a complete digestive tract and protonephridia. There is, however, no body cavity surrounding the internal organs. The "space" that is traditionally considered to be a pseudocoelom is apparently an artifact caused by certain common methods of fixation. For this reason, adult gastrotrichs are more appropriately viewed as functionally acoelomate.

PHYLUM GASTROTRICHA

ORDER CHAETONOTIDA: (Figures 9A and B). Primarily fresh water, but also includes marine, estuarine, and semiterrestrial forms; with a variable number of adhesive tubes; pharyngeal pores absent; protonephridia usually present; most are parthenogenetic (males unknown).

SUBORDER MULTITUBULATINA: Marine; with posterior, anterior and numerous lateral adhesive tubes; hermaphroditic. Monogeneric: *Neodasys*, with two known species.

SUBORDER PAUCITUBULATINA: Mostly fresh water; usually with two adhesive tubes at the ends of posterior caudal furca, but adhesive tubes are lacking in some; hermaphroditic or (usually) parthenogenetic females only. (e.g., *Aspidophorus, Chaetonotus, Dasydytes, Lepidodermella*)

ORDER MACRODASYIDA: (Figures 9C and D). Marine and estuarine gastrotrichs usually bearing numerous adhesive tubes along the head and trunk; pharyngeal pores present; protonephridia absent; hermaphroditic. (e.g., *Dactylopodola, Macrodasys, Platydasys, Pleurodasys, Urodasys*)

Body wall, support, and locomotion

The body is covered by a cuticle of varying thickness and complexity. The outer part of the cuticle comprises from one to many layers, the inner part is fibrous and produces the spines, scales, plates, and other covering structures (Figure 9). The gastrotrich cuticle is unique in that its outer portion is made of several layers of unit membrane-like structures. The epidermis is partly cellular and partly syncytial and is ciliated ventrally (thus the derivation of the phylum name). Some of the epidermal cells are characteristically monociliated, a feature shared with the gnathostomulids. Oddly, the outer cuticle extends over these external cilia. Internal to the epidermis are bands of circular and longitudinal muscles. There is no body cavity in gastrotrichs, and the internal organs are rather tightly packed together—an apparently acoelomate condition.

Support is provided by the cuticle and the compact body construction. Most move by ciliary gliding, although some of the more flexible forms "inch" along by alternately attaching the posterior and anterior adhesive tubes.

Feeding and digestion

Most gastrotrichs feed by pumping small food items into the gut by action of the muscular pharynx, or by ciliary currents in the foregut. In the macrodasyids **pharyngeal tubes** connect the pharynx lumen to the outside (Figure 9D) and allow the release of excess water taken in with the food. Gastrotrichs feed on nearly any organic material, live or dead, of appropriately small size (detritus, protozoa, unicellular algae, bacteria).

The mouth is terminal and often surrounded by a ring of oral spines that aid in food capture. A short buccal cavity connects the mouth to an elongate muscular pharynx lined with cuticle (a stomodeum) (Figure 10). The entodermally derived portion of the gut is a straight tube, which may be differentiated into a stomach and an intestine. The rectum is a proctodeum and leads to the anus on the dorsal surface. Digestion and absorption are quite rapid and occur in the midgut.

Circulation, gas exchange, excretion, and osmoregulation

There are no special circulatory or gas exchange structures in gastrotrichs. These functions are accomplished by simple diffusion, as they are in so many other tiny metazoa. At least some species are capable of anaerobic respiration.

Paired protonephridia are present in the freshwater gastrotrichs but are absent from most marine forms, suggesting that they are primarily osmoregulatory in function. The protonephridia lie on each

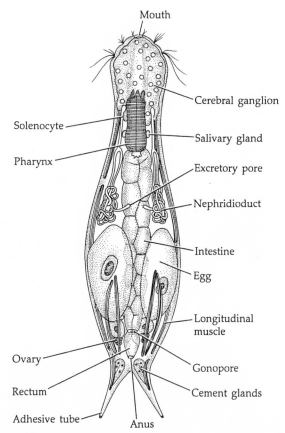

Figure 10

Internal anatomy of a chaetonotid gastrotrich (*Chaetonotus*). (Redrawn from Sherman and Sherman 1976, after Grassé 1948.)

side of the gut as single cells or clusters of cells that lead to long, coiled nephridioducts (Figure 10). These ducts lead to a pair of nephridiopores located near the ventral midline of the body. Each protonephridial cell is a **solenocyte** (sometimes called a **cyrtocyte**), which is characterized in part by bearing only one or two flagella rather than the many flagella found in flame bulbs.

Nervous system and sense organs

The cerebral ganglion is a relatively large bilobed mass connected dorsally over the anterior portion of the pharynx (Figure 10). A ganglionated, lateral longitudinal nerve cord arises from each lobe of the cerebral ganglion and extends to the posterior end of the body.

Sensory receptors are predominantly tactile. They occur as spines and bristles over much of the body but are concentrated on the head. Some species bear ciliated depressions on the sides of the head, which may be chemosensory pits. A few species contain pigmented ocelli in the cerebral ganglion.

Reproduction and development

Because the most primitive gastrotrichs are all hermaphroditic, it has been hypothesized that hermaphroditism is the primitive state in this group. The condition wherein only parthenogenetic females are known is therefore thought to be a derived condition having arisen by evolutionary reduction of the male reproductive structures. This hermaphroditic situation is most prevalent and has been best studied in members of the order Macrodasyida (Figure 11). The male system comprises one or two testes with associated sperm ducts that lead to a single gonopore on the ventral midline. In a few instances paired male pores are present; rarely, the male system connects with the hindgut. In a very few forms (e.g., *Macrodasys*), a copulatory structure, called a **caudal organ**, is present near the male gonopore.

The female system includes one or two ovaries, lying just posterior to the testes in the hermaphroditic forms. These ovaries are rather diffuse organs not bounded by a typical capsule or tissue layer. As eggs are produced they are released into an ill-defined space, called the uterus, adjacent to the ovaries and associated with a yolk-producing tissue or vitellar-

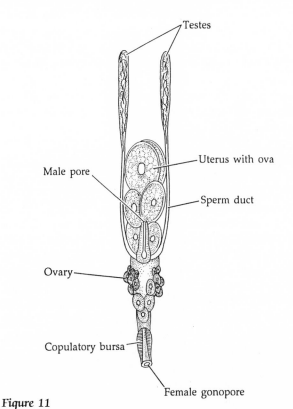

Figure 11

Reproductive system of a hermaphroditic gastrotrich (*Macrodasys*). (After Hyman 1951.)

ium. The eggs are eventually carried to an oviduct, or more commonly to a saclike receiving area called the **X-organ**, which connects with the female gonopore.

Following mutual cross-fertilization, the zygotes are released singly or a few at a time by rupture of the female's body wall. Both thick-shelled dormant eggs and thin-shelled, rapidly developing eggs are produced by gastrotrichs. Parthenogenesis predominates in the freshwater forms, where the appearance of hermaphroditic individuals (and the production of sperm) seems to take place only infrequently. The appearance of hermaphrodites is probably associated with certain environmental conditions, or perhaps with an internal genetic clock.

Few embryological studies have been conducted on gastrotrichs. Cleavage is holoblastic and essentially radial; the early blastomeres arise in such a fashion as to produce a bilaterally symmetrical embryo. A coeloblastula forms, and then gastrulation occurs by the movement of two cells into the blastocoel from the presumptive ventral surface. These "ingressed" cells presumably contribute to the formation of entoderm and, ultimately, of the midgut. Stomodeal and proctodeal invaginations form and connect with the developing gut. Two additional cells drop from the surface to the interior as the presumptive germ cells from which gametes eventually arise.

Development in gastrotrichs is direct, and juveniles hatch from the egg capsules. Although developmental time to hatching varies with the type of egg produced, maturation is rapid, and usually the animals are sexually mature within a few days after hatching.

The kinorhynchs

Among the most intriguing of the "little animalcules" are the members of the phylum Kinorhyncha (Greek *kineo*, "movable"; *rhynchos*, "snout"), sometimes called the Echinodera (Greek *echinos*, "spiney"; *dere*, "neck") or Echinoderida. Since their discovery in the mid-1800s, about 150 species of kinorhynchs have been described, nearly all of which are less than 1 mm in length. Most live in marine sands or muds, from the intertidal zone to a depth of several thousand meters. Some are known from algal mats, holdfasts, and sandy beaches; others live on hydroids, ectoprocts, or sponges.

Externally, most kinorhynchs are similar in appearance (Figure 12), and most of the specific differences are in the details of spination and the arrange-

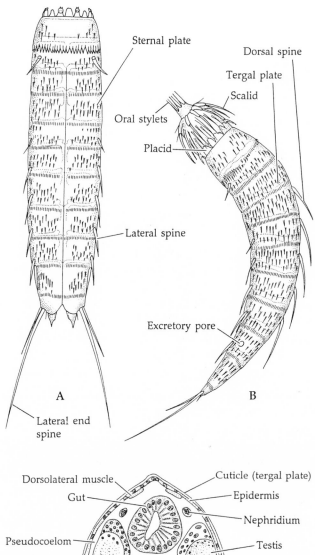

Figure 12

A–B, External anatomy of the kinorhynch *Echinoderes*. A, Ventral view with head retracted. B, Lateral view with head extended. C, A trunk zonite (cross section) of a kinorhynch. Note the arrangement of the body wall structures and the organs within the pseudocoelom. (A,B after Higgins 1982; D after Hyman 1951.)

ment and structure of their thick cuticular plates. The body comprises a distinct head, which is retractable into the anterior portion of the neck, and a trunk. The body is divided into thirteen clearly defined **zonites**, which some workers view as true metameric segments. The head is formed by segment 1 and bears a retractable **oral cone**. The mouth is borne on

adhesive tubes present; trunk segments usually with dense covering of cuticular hairs or denticles. The habitat of cyclorhagids varies. (e.g., *Cateria*, *Centroderes*, *Echinoderes*)

ORDER HOMALORHAGIDA: With 4–8 placids, but ventral plate(s) of first trunk segment assist in closing; trunk triangular in cross section; trunk spines and cuticular hairs few in number or rudimentary. Most homalorhagids live in subtidal muds. (e.g., *Kinorhynchus*, *Neocentrophyes*, *Pycnophyes*)

Body wall

Beneath the well developed cuticle is a nonciliated epidermis (Figure 12C), which contains elements of the nervous system. Largely internal to the epidermis but still attached to the cuticle are bands of dorsolateral and ventrolateral intersegmental muscles (Figure 13). Some of the anterior longitudinal muscle bands serve as head retractors. A series of metamerically arranged dorsoventral muscles create the increased hydrostatic pressure that protracts the head and oral cone when the retractor muscles relax.

The internal organs lie within a pseudocoelom that contains some amebocytes.

Support and locomotion

Body shape in kinorhynchs is more-or-less fixed by the rigid plates of the supportive cuticular exoskeleton, but the animals are able to flex and even twist at the points of articulation between adjacent zonites. In the absence of external cilia, burrowing

---

the oral cone and is surrounded by a ring of anteriorly directed spines called **oral stylets**. Behind the oral cone, the head bears up to seven rings of posteriorly directed spines, called **scalids**. Segment 2 forms the "neck," which consists of a series of plates called **placids** that fold over the head when it is retracted. The remaining eleven segments form the trunk. Most trunk segments are made up of a dorsal (**tergal**) plate and a pair of ventral (**sternal**) plates (Figure 12C). The anus is located on the last segment and is usually flanked by strong **lateral end spines**.

The kinorhynchs are clearly pseudocoelomate, and they possess several characteristics shared by other such groups (Box 3). However, their metameric condition has led to many controversies about their affinities with other taxa. Similarities with the arthropods are striking and include metamerism, nature of the cuticle, and ecdysis.

PHYLUM KINORHYNCHA

ORDER CYCLORHAGIDA: With 14–16 placids as a closing apparatus, although in one group a clamshell-like first trunk segment may assist; trunk oval, round, or slightly triangular in cross section; numerous cuticular trunk spines;

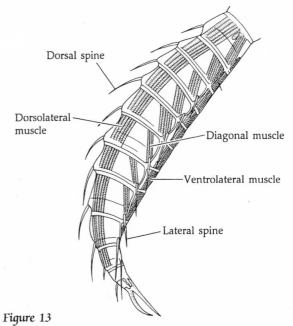

Figure 13

The trunk muscles of a kinorhynch (lateral view). (After Hyman 1951.)

movement is accomplished by sticking the head forward into the substratum, anchoring the anterior spines, and then pulling the rest of the body forward.

Feeding and digestion

Kinorhynchs are probably direct deposit feeders, ingesting the substratum and digesting the organic material or eating unicellular algae contained therein. However, the details of feeding and the exact nature of the food are not known.

The mouth leads into a buccal cavity located within the oral cone and thence to a muscular pharynx (Figure 14A). The buccal cavity, pharynx, and esophagus are lined with cuticle and represent a stomodeum. Various paired glands are often associated with the esophagus, but their functions are as yet uncertain. The esophagus connects with an elongate, straight midgut, which leads to a short, cuticle-lined proctodeal hindgut (rectum) and the anus on the last segment. So-called digestive glands often arise from the midgut. Nothing is known about the digestive physiology in kinorhynchs.

Circulation, gas exchange, excretion, and osmoregulation

Circulation within the body is by diffusion through the pseudocoelom; gas exchange is by diffusion across the body wall. Body movements aid diffusion in accomplishing internal transport. Kinorhynchs possess one pair of solenocyte protonephridia in the region of the tenth and eleventh segments, each with a short nephridioduct to the outside (Figure 14B). Excretory and osmoregulatory physiology remains largely unstudied, although some species can tolerate low salinities for short periods, and some live in salinities as low as 6 parts per thousand in the Gulf of Finland.

Nervous system and sense organs

The central nervous system of kinorhynchs is relatively simple and is intimately associated with the epidermis. A series of ten ganglia is arranged in a ring around the pharynx. These may be viewed collectively as a circumenteric, multilobed, cerebral ganglion. Each ganglion may give rise to a single longitudinal nerve cord. The two midventral nerve cords are most prominent, and they bear ganglia in each segment.

Sensory receptors include tactile bristles, spines, and flosculi on the body. Photoreceptors are present on the pharyngeal nerve ring in at least some species.

Reproduction and development

Kinorhynchs are gonochoristic and possess relatively simple reproductive systems (Figure 14C). Externally, males and females are usually indistinguishable from one another. In both sexes, paired saclike gonads lead to short gonoducts that open separately on the thirteenth segment. In the males, the gonopore is associated with two or three cuticular, penial spines (**spicules**) that presumably aid in copulation. The female gonads comprise both germ cells and nutritive (yolk-producing) cells. Each oviduct bears a diverticulum that forms a seminal receptacle prior to ending at a gonopore.

Mating has never been seen, and egg laying and early development have not been adequately studied. Fertilized ova are deposited in the environment in egg cases within which direct development takes place. The juveniles emerge from the egg case with 11 of the 13 segments already formed (Kozloff 1972), so development is direct. Apparently they do not

Figure 14

Internal anatomy of kinorhynchs. A, The digestive tract of *Pycnophyes*. B, The nephridial arrangement in *Pycnophyes*. C, The simple female reproductive system of a kinorhynch. (After Hyman 1951.)

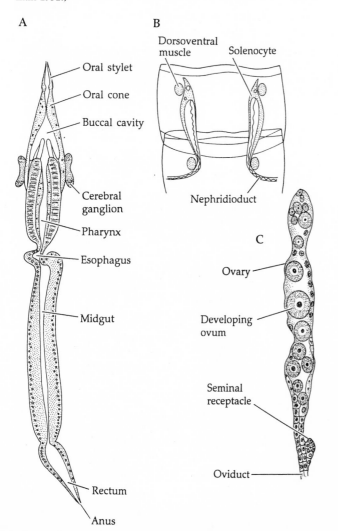

hatch as an unsegmented "larva" that adds all its segments by a sequence of molts (as reported initially by Nyholm 1947). Juveniles molt periodically, passing through six juvenile stages to adulthood; adults do not molt.

The nematodes

An enormous amount of literature exists on the nematodes (roundworms or threadworms), much of it dealing with the parasitic species of economic or medical importance. Many of the parasitic forms have been known since ancient times, but the small, free-living types were not discovered until after the invention of the microscope. Many authorities now prefer the phylum name Nemata (Greek *nema*, "thread"), although Nematoda is still commonly used. With over 12,000 described species (and probably several times that many undescribed species), they are one of the most abundant groups of metazoa on Earth; one study revealed about 90,000 nematodes in a single rotting apple, and another study turned up 236 species in 6.7 cc of coastal mud. Whereas most of the free-living species are very small, many of the parasitic forms are much larger; and members of one species reach lengths of up to 8 m. Nematodes are known from virtually every habitat in the seas, fresh water, and on land. Some are generalists, but many have very specific habitats. One species is known only from felt coasters under beer mugs in a few towns in eastern Europe. Nematodes parasitize nearly all groups of animals and plants, and some cause serious damage to crops and livestock, and some are pathogenic in humans. Marine nematodes are among the most common and widespread groups of animals, occurring from the shore to the abyss. Wherever they are found, they are often the most numerous metazoa, in both numbers of species and individuals. Some environments yield as many as 3 million threadworms per square meter. Despite their abundance, the free-living nematodes are poorly known, and thus their importance in marine benthic systems is little appreciated.

Nematodes are vermiform pseudocoelomates with thin, unsegmented bodies that are usually distinctly round in cross section (Box 4). To the untrained and unaided eye, many nematodes look very much alike, but there are variations in external body form (Figure 15).

Classification

It is beyond the scope of this text to present very much of the exhaustive classification scheme of the

---

Box Four

Characteristics of the Phylum Nematoda

1. Triploblastic, bilateral, vermiform, unsegmented, pseudocoelomates
2. Body round in cross section and covered by a layered cuticle; growth in juveniles usually accompanied by molting
3. With unique cephalic sense organs called amphids; some have caudal sense organs called phasmids
4. Gut complete; mouth surrounded by six lips bearing sense organs (often reduced to three lips, or to a simple ring)
5. Most with unique excretory system, comprised of one or two renette cells or a set of collecting tubules
6. Without special circulatory or gas exchange structures
7. Body wall has only longitudinal muscles
8. Longitudinal muscle cells connected to longitudinal nerve cords by unique muscle arms
9. Epidermis produced into longitudinal cords housing nerve cords
10. Gonochoristic
11. Inhabit marine, freshwater, and terrestrial environments; some are free-living and some parasitic

---

roundworms. The two classes, Adenophorea and Secernentea, are usually subdivided into two and three subclasses respectively, containing several orders and many families. The list below is only of the classes and subclasses; for further information, we refer the reader to Maggenti (1982).

PHYLUM NEMATODA (= NEMATA)

CLASS ADENOPHOREA (= APHASMIDA): (Figures 15A, C, and E). With cephalic chemoreceptors called amphids but lacking caudal phasmids; excretory system comparatively simple, not cuticularized, and without collecting tubules; most are free living (marine, fresh water, terrestrial), although a few parasitic species are known. This class includes the subclasses Enoplia (7 orders) and Chromadoria (5 orders).

CLASS SECERNENTEA (= PHASMIDA): (Figures 15B, D, and F). With cephalic amphids and caudal phasmids; excretory

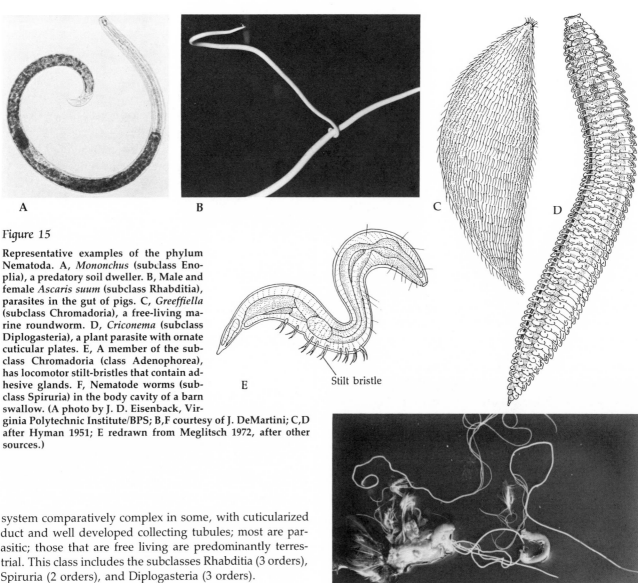

Figure 15

Representative examples of the phylum Nematoda. A, *Mononchus* (subclass Enoplia), a predatory soil dweller. B, Male and female *Ascaris suum* (subclass Rhabditia), parasites in the gut of pigs. C, *Greeffiella* (subclass Chromadoria), a free-living marine roundworm. D, *Criconema* (subclass Diplogasteria), a plant parasite with ornate cuticular plates. E, A member of the subclass Chromadoria (class Adenophorea), has locomotor stilt-bristles that contain adhesive glands. F, Nematode worms (subclass Spiruria) in the body cavity of a barn swallow. (A photo by J. D. Eisenback, Virginia Polytechnic Institute/BPS; B,F courtesy of J. DeMartini; C,D after Hyman 1951; E redrawn from Meglitsch 1972, after other sources.)

system comparatively complex in some, with cuticularized duct and well developed collecting tubules; most are parasitic; those that are free living are predominantly terrestrial. This class includes the subclasses Rhabditia (3 orders), Spiruria (2 orders), and Diplogasteria (3 orders).

Body wall, support, and locomotion

The nematode body is covered by a well developed and complexly layered cuticle secreted by the epidermis (Figure 16D). The cuticle is responsible in part for allowing the invasion of hostile environments, such as dry terrestrial soils and the digestive tracts of hosts, for it drastically reduces the permeability of the body wall. Predominantly terrestrial or parasitic nematodes (class Secernentea) usually have a dense, fibrous, inner layer of the cuticle, whereas most of the free-living marine and freshwater forms (class Adenophorea) lack this inner layer. The texture of the cuticle is highly variable among nematodes. It may be relatively smooth or covered with various sorts of sensory setae and wartlike bumps. The cuticle in many roundworms is ringed or marked with longitudinal ridges and grooves. In many marine forms the cuticle contains radially arranged rods or other inclusions of various shapes.

The epidermis varies from cellular to syncytial among different taxa and often protrudes into the pseudocoelom as dorsal, ventral, and lateral cords (Figures 16A and C). The dorsal and ventral thickenings house longitudinal nerve cords, and the lateral thickenings contain excretory canals (when present, as they are in some secernenteans) and neurons. Internal to the epidermis is a relatively thick layer of obliquely striated longitudinal muscle that is connected to the dorsal and ventral nerve cords by so-called **muscle arms** (Figures 16A and B). This ar-

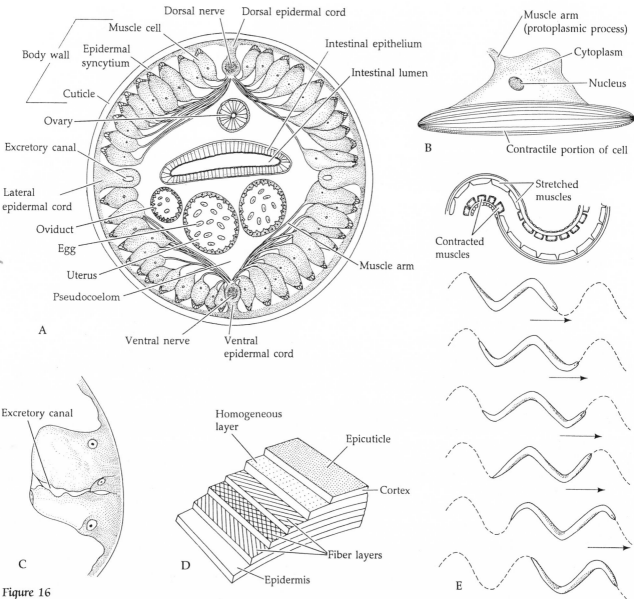

Figure 16

A, Stylized cross section through a female nematode such as *Ascaris* (subclass Rhabditia). B, A single longitudinal muscle cell, illustrating the origin of the muscle arm. C, The lateral epidermal cord of *Cucullanus* (Rhabditia). D, The layers of the cuticle. E, Undulatory locomotion in a free-living nematode results from the action of the longitudinal muscle fibers. The concave areas along the body represent positions of muscle contraction; the convex areas are regions of muscle stretching. Leverage is gained against surrounding objects or the substratum in the environment. (A,B after various sources; C after Hyman 1951; D after Sherman and Sherman 1976; E adapted from Lee and Atkinson 1977.)

rangement is different from the usual neuromuscular junctions in animals; in nematodes the connections are made by extensions of the muscle cells rather than of the neurons. There is no circular muscle layer.

The fluid-filled pseudocoelom is not spacious. The apparently large body cavity shown in many studies is an artifact caused by shrinkage of tissues in alcohol. Modern microscopy techniques reveal that the organs of most nematodes occupy nearly all of the internal space. The cuticle and the hydrostatic qualities of body tissues provide body support in nematodes. In the absence of circular body wall muscles, some types of locomotion, such as peristaltic burrowing, are impossible. The typical pattern of locomotion involves contraction of longitudinal muscles, producing a somewhat undulatory motion (Figure 16E). Among the free-living nematodes this movement pattern relies on contact with environmental substrata, against which the body pushes.

The muscles act against the hydrostatic skeleton and the cuticle, which serve as antagonistic forces to the muscle contractions. The crossed collagenous fibers of the cuticle are nonelastic, but their arrangement allows shape changes as the body undulates. When placed in a fluid environment and deprived of contact with solid objects, benthic nematodes thrash about rather inefficiently. Some actually do swim (but not very well), and some crawl or "inch" along using various cuticular spines, grooves, ridges, and glands to gain purchase on the substratum (Figure 15E).

Feeding and digestion

Nematodes display a great diversity in habits and habitats, and have evolved a number of different feeding strategies that are often reflected in anatomical features of the mouth area. Various labial flaps, papillae, and other armature often are arranged in radially symmetrical patterns (Figure 17).

Many infaunal nematodes are direct deposit feeders. Others are detritivores or microscavengers, living in or on dead organisms or fecal material. Many of these species apparently do not feed directly on the carcasses they inhabit, but on fungi and bacteria growing in the decomposing organic matter. Some of the free-living nematodes are predatory carnivores, feeding on a variety of other small animals. Others

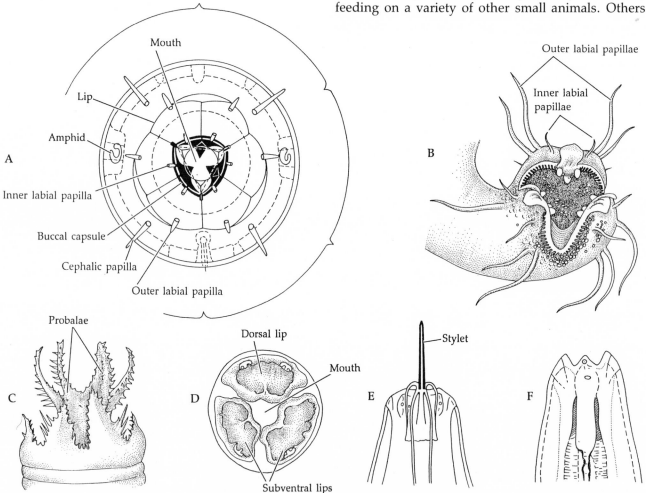

Figure 17

Modifications of the anterior end among selected nematodes. A, The front end of a generalized nematode (anterior view). Note the basic radial symmetry of the parts. B, The anterior end of a free-living marine nematode. C, The anterior end of *Acrobeles* (subclass Rhabditia), a soil nematode bearing modified labia (probalae) apparently used in burrowing or food sorting. D, The tripartite lip of *Ascaris* (subclass Rhabditia) is used to attach the parasite to the host's intestinal wall. E, The anterior end of *Nygolaimus* (subclass Enoplia) has a protruded stylet, used to puncture prey. The stylet is then retracted and the prey's body fluids sucked out. F, The simple anterior end of the free-living nematode *Panagrolaimus* (subclass Rhabditia). (A,C,E after Hyman 1951; B after various sources; D after Noble and Noble 1982; F after Pennak 1953.)

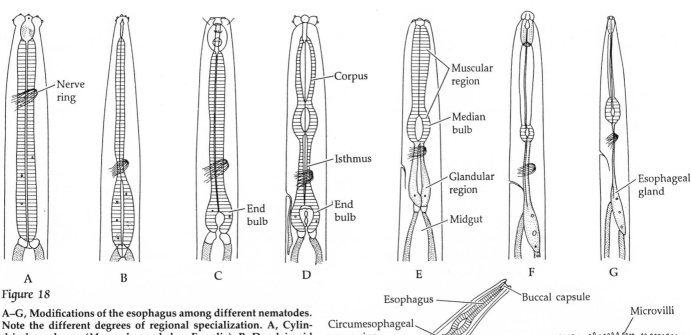

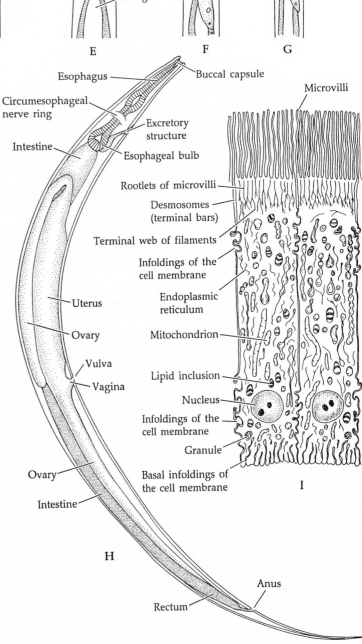

Figure 18

A–G, Modifications of the esophagus among different nematodes. Note the different degrees of regional specialization. A, Cylindrical esophagus (*Mononchus*: subclass Enoplia). B, Dorylaimoid esophagus (*Dorylaimus*: Enoplia). C, Bulboid esophagus (*Ethmolaimus*: subclass Chromadoria). D, Rhabditoid esophagus (*Rhabditis*: subclass Rhabditia). E, Diplogasteroid esophagus (*Diplogaster*: subclass Diplogasteria). F, Tylenchoid esophagus (*Helicotylenchus*: Diplogasteria). G, Aphelencoid esophagus (*Aphelenchus*: Diplogasteria); H, digestive tract and reproductive system of a female *Rhabditis*; I, intestinal epithelium of *Ascaris* (Rhabditia). (A–G modified after several authors, based on Sassar and Jenkins, 1960; H after Sherman and Sherman 1976; I after Meglitsch 1972.)

are herbivorous, feeding on diatoms, algae, and bacteria. Plant parasitic nematodes use an oral stylet to pierce individual root cells and then suck out the contents.

The myriad parasitic nematodes are known from nearly every group of plants and animals. In invertebrates and vertebrates (including people) nematodes parasitize a variety of fluids and body organs, where they may cause extreme tissue damage. Such damage may be caused by feeding on the host's tissues or by various burrowing or encysting activities of the parasite.

The nematode digestive tract varies in complexity and regional specialization from one species to another (Figure 18). The anteriorly located mouth leads to a short buccal cavity connected to a stomodeal esophagus (often called a pharynx). The esophagus is elongate and may be subdivided into distinct muscular and glandular regions, the details of which are of considerable taxonomic importance (Figures 18A–G). Behind the esophagus is a long, straight midgut or intestine, which leads to a short rectum and a subterminal anus on the ventral surface of the body. In males the rectum may be properly termed a cloaca because it receives products of the reproductive system as well as of the gut. The esophageal glands and perhaps the midgut lining secrete digestive enzymes into the gut lumen. Initial digestion is extracellular; final intracellular digestion occurs in the midgut following absorption across the surfaces of the microvilli of the midgut cells (Figure 18I).

Circulation, gas exchange, excretion, and osmoregulation

There are no special circulatory or gas exchange structures in nematodes. As in many of the other groups considered in this chapter, these functions are accomplished by diffusion and movement of the pseudocoelomic fluids. Some parasitic nematodes possess a form of hemoglobin in the body fluids that presumably transports and stores oxygen. Both aerobic and anaerobic metabolic pathways occur in various groups of nematodes, and many of these worms are able to shift from one mechanism to the other according to environmental oxygen concentrations. Facultative anaerobiosis is surely significant in parasitic nematodes and those that live in other anoxic environments.

The excretory structures of nematodes are unique and apparently not homologous to any of the protonephridial types found in other metazoa. There exists a rather clear evolutionary sequence of different excretory structures among nematodes (Figure 19). The presumed plesiomorphic condition occurs in certain free-living taxa and has been modified to various degrees among other groups, especially within specialized parasitic forms. In many free-living threadworms the system comprises one or two glandular **renette cells** that connect directly to a midventral excretory pore, and sometimes a third cell forming an **ampulla** at the opening (Figure 19A). Modifications to this system often include the formation of various arrangements of intracellular collecting ducts within the cytoplasm of extensions of the renette cells (Figure 19B). In many advanced parasitic species the renette cell bodies are lost completely, leaving only the system of tubules in an H or inverted Y pattern (Figures 19C and D). Many members of the subclass Enoplia lack renette cells altogether. Instead, they have numerous unicellular units distributed along the entire length of the body. Each cell opens to the outside via a duct and a pore. If these cells are excretory in function, they may represent nonciliated protonephridia and may be the primitive condition within the phylum Nematoda.

Most nematodes are ammonotelic, but some excrete increased amounts of urea when in a hypertonic environment. Apparently much of the loss of nitrogenous wastes is across the wall of the midgut, and the renette cells are primarily osmoregulatory. Water

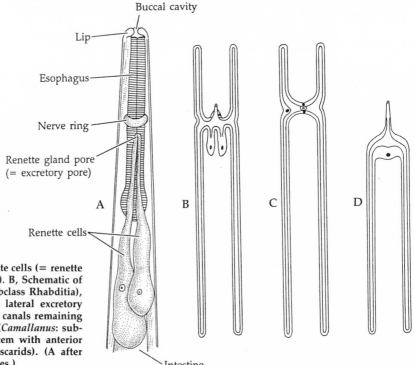

Figure 19

Nematode excretory systems. A, A pair of renette cells (= renette gland) leading to the excretory pore (*Rhabditis*). B, Schematic of the excretory system of *Oesophagostomum* (subclass Rhabditia), wherein the renette cells are associated with lateral excretory canals. C, The so-called H-system of collecting canals remaining after loss of the glandular renette cell bodies (*Camallanus*: subclass Spiruria); D, Modification of the H-system with anterior excretory pore and lateral canals (in many ascarids). (A after Hyman 1951; B–D modified from various sources.)

Buccal cavity

Lip

Esophagus

Nerve ring

Renette gland pore (= excretory pore)

Renette cells

Intestine

A   B   C   D

balance is also aided by the activities of other tissues, organs, and structures. The cuticle of at least some threadworms is differentially permeable to water in that it allows water to enter but not to leave the body. This condition is advantageous under conditions of potential desiccation, but it presents problems in hypotonic environments where excess water must be eliminated. This elimination is apparently accomplished by the renette cells (when present), by the gut lining, and by the epidermis, which mediates the active removal of salts across the body wall. Marine species do not osmoregulate well, and they desiccate rapidly when exposed to air.

Nervous system and sense organs

With some variation, the structure of the central nervous system of nematodes is similar throughout the phylum (Figures 20A and B). The cerebral ganglion comprises a circumesophageal nerve ring and various associated ganglia that contain the majority of the nerve cell bodies. A wreath of sensory and motor nerves extends anteriorly from the nerve ring and serves the cephalic sense organs and mouth structures. Via a series of associated ganglia, longitudinal nerves extend posteriorly through the epidermal cords (Figure 16A). The major nerve trunk is ventral and includes both motor and sensory fibers. It is formed from the union of paired nerve tracts that arise ventrally on the nerve ring and fuse posteriorly, where the main trunk bears ganglia. The dorsal nerve cord is motor, and the less well developed lateral nerve tracts are predominantly sensory. Lateral commissures connecting some or all of the longitudinal nerves occur in many nematodes.

The most abundant sense organs of nematodes are various papillae and setae that serve as tactile

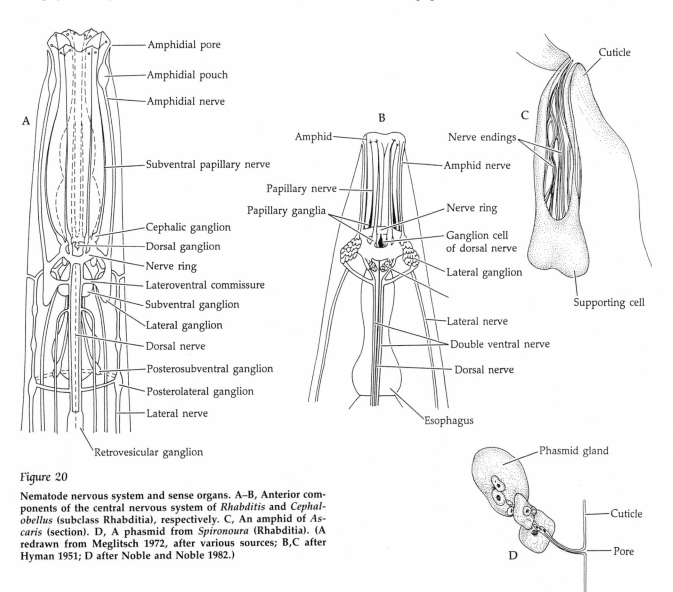

Figure 20

Nematode nervous system and sense organs. A–B, Anterior components of the central nervous system of *Rhabditis* and *Cephalobellus* (subclass Rhabditia), respectively. C, An amphid of *Ascaris* (section). D, A phasmid from *Spironoura* (Rhabditia). (A redrawn from Meglitsch 1972, after various sources; B,C after Hyman 1951; D after Noble and Noble 1982.)

receptors in these worms' highly touch-oriented world (e.g., interstitial, parasitic, and soil habitats). **Amphids** are paired organs located laterally on the head. They consist of an external pore leading inward to a short duct and **amphidial pouch**. The pouch is associated with a unicellular gland and an **amphidial nerve** from the cerebral nerve ring (Figures 20A and C), although there is some variation in structural details among different species. The receptor sites of amphids are actually derived from modified cilia (until this discovery, nematodes were considered to be completely devoid of cilia). Most specialists presume that the amphids are chemosensory in function. Most members of the class Secernentea (the parasitic forms) possess a posteriorly located pair of glandular structures called **phasmids** (Figure 20D). These structures also are considered to be chemoreceptors. Some freshwater and marine free-living nematodes possess a pair of pigment-cup ocelli on the sides of the pharynx. At least some nematodes contain proprioceptors in the lateral epidermal cords. These sensory cells contain a cilium and apparently monitor bending of the body during locomotion (Hope and Gardiner 1982).

Reproduction and development

Most nematodes are gonochoristic and show some degree of sexual dimorphism (Figures 15B and 21). The males tend to be smaller than females and are often sharply curved posteriorly. The male system typically includes one or two threadlike tubular testes, each of which is regionally differentiated into a distal **germinal zone**, a middle **growth zone**, and a proximal **maturation zone** near the junction with the sperm duct. The sperm duct extends posteriorly, where it enlarges as a seminal vesicle leading to a muscular ejaculatory duct that joins the hindgut near the anus. Some have prostatic glands that secrete seminal fluid into the ejaculatory duct. Most male nematodes possess a copulatory apparatus, including one or two cuticular spines, or **spicula**, that can be inserted into the female gonopore to guide the transmission of sperm. This apparatus and other components of the male system are shown in Figures 21B and C.

The female reproductive system (Figure 21A) usually consists of paired elongate ovaries that gradually hollow as oviducts, then enlarge as uteri. The

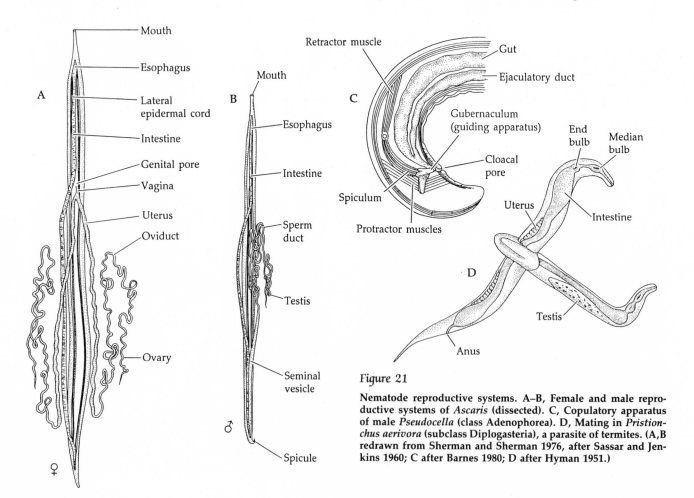

Figure 21

Nematode reproductive systems. A–B, Female and male reproductive systems of *Ascaris* (dissected). C, Copulatory apparatus of male *Pseudocella* (class Adenophorea). D, Mating in *Pristionchus aerivora* (subclass Diplogasteria), a parasite of termites. (A,B redrawn from Sherman and Sherman 1976, after Sassar and Jenkins 1960; C after Barnes 1980; D after Hyman 1951.)

uteri converge to form a short vagina connected to the single gonopore. The female gonopore is completely separate from the anus, opening on the ventral surface near the middle of the body.

Prior to copulation, the males produce sperm and store them in the seminal vesicle, and the females produce eggs, which are moved into the hollow uteri. Potential mates make contact (male-attracting pheromones are produced by females of some species), and the male usually wraps his curved posterior end around the body of the female near her gonopore (Figures 15B and 21D). Thus positioned, the copulatory spines are inserted into the vagina; sperm are transferred by contractions of the ejaculatory duct. Fertilization usually occurs within the uteri. A relatively thick double-layered shell forms around each zygote; the inner layer is derived from the fertilization membrane and the outer layer is produced by the uterine wall. The zygotes are usually deposited in the environment, where development takes place.

In addition to the general description given above, two relatively uncommon reproductive processes occur in nematodes. In the few known hermaphroditic nematodes, sperm and egg production take place within the same gonad (an **ovitestis**). Sperm formation precedes egg production, so the animals are technically protandric; but they do not engage in cross-fertilization as occurs in many sequential hermaphrodites. Rather, the sperm are stored until ova are produced, and self-fertilization occurs. Parthenogenesis occurs in a few species of gonochoristic nematodes. Sperm and eggs are produced by the separate males and females that then engage in typical copulation. However, the sperm do not fuse with the egg nuclei, but apparently serve only to stimulate cleavage.

Development among free-living nematodes typically is direct, although the term *larva* is often used for juvenile stages. Cleavage is holoblastic and subequal. The orientation of blastomeres during early cleavage is fairly consistent among those nematodes that have been studied, but it cannot be readily assigned to a clearly radial or spiral pattern. Figure 22 illustrates this cleavage pattern and explains some details of cell fates. A stereoblastula or slightly hollow coeloblastula forms and undergoes gastrulation by epiboly of the presumptive ectoderm combined with an inward movement of presumptive entoderm and mesoderm. After a specific point in development, few nuclear divisions occur, and most subsequent growth, even after hatching, is by the enlargement of existing cells. Four sequential cuticular molts during juvenile life usually accompany growth.

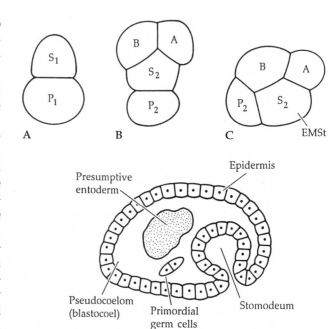

Figure 22

Early embryogeny of *Parascaris equorum* (subclass Rhabditia). A, Two-cell stage, coded S_1 and P_1 cells. B–C, Four-cell stage. B, The S_1 cell divides to produce cells A and B; the P_1 cell divides to produce cells S_2 and P_2. C, The P cell migrates to the presumptive posterior end of the embryo. D, A later stage following gastrulation, during the formation of stomodeum. The S_2 (EMSt, entodermal/mesodermal/stomodeal) cell has divided to produce an E and an MSt cell; the latter has divided again to form an M and an St cell. The E cell forms the entoderm upon gastrulation, the M cell becomes the mesoderm, and the St cell forms the region of the stomodeum; the A and B cells produce most of the ectoderm, the P_2 becomes some of the posterior ectoderm and the primordial germ cells, and some mesoderm. (Modified from Boveri, 1899.)

Life cycles of some parasitic nematodes

The study of parasitic nematode life cycles is a field unto itself, and we have space here to present only a few of these interesting life histories (see Figures 23 and 24). Some parasitic nematodes undergo very simple life cycles. An example is the whipworm *Trichuris trichiura* (class Adenophorea) (Figure 23A). These relatively large (3–5 cm long) nematodes reside and mate in the human gut, and the fertilized eggs are passed with the host's feces. Reinfection occurs when another host ingests these embryos. A more complicated life cycle is that of another adenophoran nematode called the trichina worm (*Trichinella spiralis*), which is acquired by the ingestion of poorly cooked meat containing encysted larvae (Figure 23B).

Most of the parasitic nematodes belong to the class Secernentea, which includes such notables as

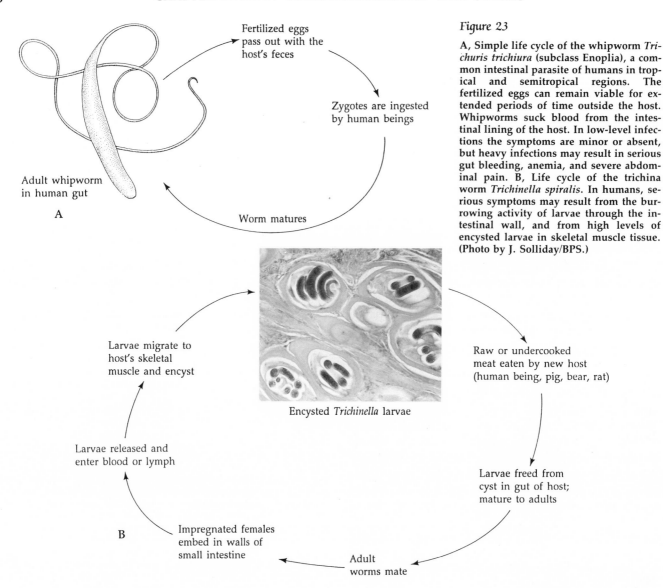

Fertilized eggs pass out with the host's feces

Zygotes are ingested by human beings

Adult whipworm in human gut

A

Worm matures

Figure 23

A, Simple life cycle of the whipworm *Trichuris trichiura* (subclass Enoplia), a common intestinal parasite of humans in tropical and semitropical regions. The fertilized eggs can remain viable for extended periods of time outside the host. Whipworms suck blood from the intestinal lining of the host. In low-level infections the symptoms are minor or absent, but heavy infections may result in serious gut bleeding, anemia, and severe abdominal pain. B, Life cycle of the trichina worm *Trichinella spiralis*. In humans, serious symptoms may result from the burrowing activity of larvae through the intestinal wall, and from high levels of encysted larvae in skeletal muscle tissue. (Photo by J. Solliday/BPS.)

Larvae migrate to host's skeletal muscle and encyst

Encysted *Trichinella* larvae

Raw or undercooked meat eaten by new host (human being, pig, bear, rat)

Larvae released and enter blood or lymph

Larvae freed from cyst in gut of host; mature to adults

B   Impregnated females embed in walls of small intestine

Adult worms mate

hookworms, pinworms, ascarids, and filarial worms, two examples of which are included in Figure 24.

One of the most dramatic nematode infections in humans is **filariasis**, caused by any of a number of nematodes called the **filarids**. These parasites require an intermediate host, typically some blood-sucking insect (e.g., fleas, biting flies, mosquitoes). One such filarid is *Wuchereria bancrofti*, whose vector is a mosquito. When present in high numbers in people, masses of adult *Wuchereria* block lymphatic vessels and cause fluid accumulation (edema) and severe swelling, a condition resulting in grotesque enlargement of body parts and known as **elephantiasis**. Such infections often affect the legs and arms, or the scrotum in males and the breasts in females.

In many coastal areas, humans occasionally acquire infections of nematodes that normally parasitize only marine animals. Such infections result from consuming raw or undercooked fish, such as sashimi or ceviche, that serve as the intermediate host of the parasite.

The nematomorphs

There are about 230 described species in the phylum Nematomorpha (Greek, *nema*, "thread"; *morphe*, "shape"), commonly called the horsehair or gordian worms. The phylum name, and the common name of horsehair worms, derive from the threadlike or

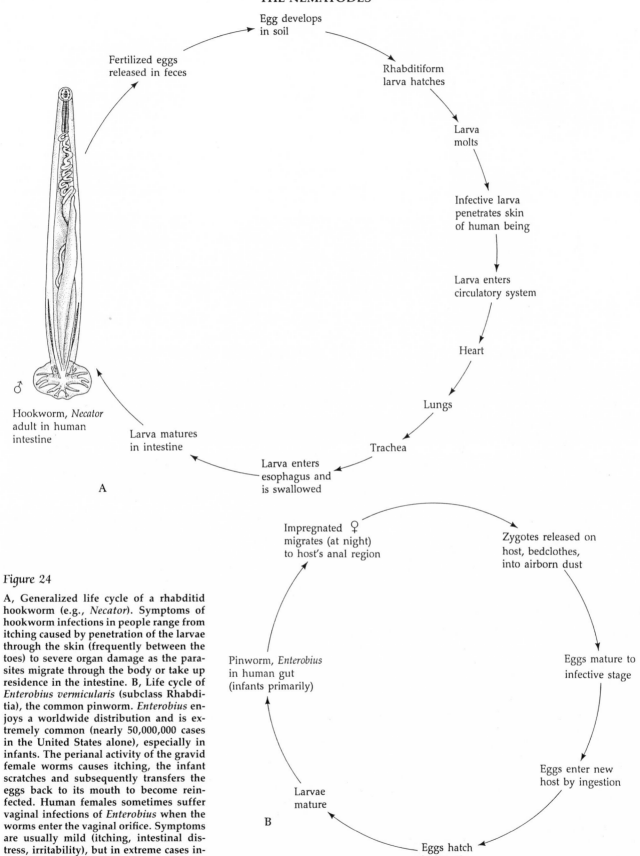

Egg develops
in soil

Fertilized eggs
released in feces

Rhabditiform
larva hatches

Larva
molts

Infective larva
penetrates skin
of human being

Larva enters
circulatory system

Heart

Lungs

Trachea

♂

Hookworm, *Necator*
adult in human
intestine

Larva matures
in intestine

Larva enters
esophagus and
is swallowed

A

Impregnated ♀
migrates (at night)
to host's anal region

Zygotes released on
host, bedclothes,
into airborn dust

Eggs mature to
infective stage

Pinworm, *Enterobius*
in human gut
(infants primarily)

Eggs enter new
host by ingestion

Larvae
mature

B

Eggs hatch
in intestine

Figure 24

A, Generalized life cycle of a rhabditid hookworm (e.g., *Necator*). Symptoms of hookworm infections in people range from itching caused by penetration of the larvae through the skin (frequently between the toes) to severe organ damage as the parasites migrate through the body or take up residence in the intestine. B, Life cycle of *Enterobius vermicularis* (subclass Rhabditia), the common pinworm. *Enterobius* enjoys a worldwide distribution and is extremely common (nearly 50,000,000 cases in the United States alone), especially in infants. The perianal activity of the gravid female worms causes itching, the infant scratches and subsequently transfers the eggs back to its mouth to become reinfected. Human females sometimes suffer vaginal infections of *Enterobius* when the worms enter the vaginal orifice. Symptoms are usually mild (itching, intestinal distress, irritability), but in extreme cases intestinal lesions may form.

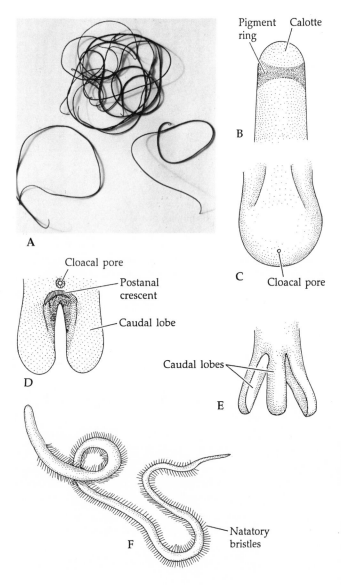

Figure 25

Representative nematomorphs. A, Gordioid nematomorphs. B, Anterior end of *Gordius* (Gordioida). C–D, Posterior end of a female and a male, respectively (*Gordius*). E, Posterior end of female *Paragordius* (Gordioida). F, *Nectonema*. (A courtesy J. DeMartini; B–F adapted from Hyman 1951, after various authors.)

The nematomorph pseudocoelom may be spacious and fluid-filled or nearly obliterated by the invasion of mesenchyme. The larvae are parasitic in arthropods, and the digestive tract appears to be nonfunctional even in adults. In contrast to many other pseudocoelomates, the nematomorphs appear to lack any structural excretory mechanisms.

Most adult nematomorphs live in fresh water among the litter and algal mats near the edges of ponds and streams. A few species are semiterrestrial in damp soil, such as in moist gardens and greenhouses. Members of the genus *Nectonema* are pelagic in coastal marine environments.

PHYLUM NEMATOMORPHA

CLASS NECTONEMATOIDA: (Figure 25F). Marine, planktonic; with a double row of natatory setae along each side of body; with dorsal and ventral longitudinal epidermal cords; pseudocoelom spacious and fluid filled; gonads single; larvae parasitize decapod crustaceans. Monogeneric: *Nectonema*.

CLASS GORDIOIDA: (Figure 25A). Fresh water and semi-terrestrial; lack lateral rows of setae; with a single, ventral epidermal cord; pseudocoelom filled with mesenchyme in young animals but becomes spacious in older individuals; gonads paired; larvae parasitize aquatic and terrestrial insects, such as grasshoppers and crickets. (e.g., *Chordotes, Gordius, Paragordius*)

Body wall, support, and locomotion

The general organization of the body wall of nematomorphs is in many aspects similar to that of the nematodes (Figures 26A, B, and C). The cuticle is very thick (especially in the gordioids) and comprises an outer homogeneous layer and an inner, lamellate, fibrous layer. The homogeneous layer often forms bumps, warts, or papillae (collectively called **areoles**, Figure 26D), some of which bear apical spines or pores. The function of the areoles is unknown, but spined ones may be touch-sensitive and pore-bearing ones may produce a lubricant. Some species have two or three **caudal lobes** at the posterior end.

The epidermis covers the entire body and also forms either a ventral epidermal cord or dorsal and

hairlike shape of these animals (Figure 25A), and from the belief held for some time after their discovery in the fourteenth century that they actually arose from the hairs of horses' tails. They are generally from 1–3 mm in diameter and up to a meter in length. Many of the very elongate forms tend to twist and turn upon themselves in such a way as to give the appearance of complicated knots, and thus the name gordian worms.* The characteristics of this phylum are listed in Box 5. Overall, nematomorphs are somewhat similar to nematodes.

*King Gordius of Phrygia tied a formidable knot (the Gordian knot) and declared that whoever might undo it would be the ruler of all Asia. No one could until Alexander the Great cut it through with his broadsword, settling the issue in a style consistent with the rest of his adventures.

ventral epidermal cords; the ventral cord contains a longitudinal nerve tract. Beneath the epidermis is a single layer of longitudinal muscle cells; as in nematodes, there is no layer of circular muscle in the body wall.

The pseudocoelom varies among nematomorphs from spacious to mesenchyme-filled (Figures 26A and B) and surrounds the internal organs. The hydrostatic or structural qualities of the body cavity, along with the well developed cuticle, provide body support.

Locomotion in the planktonic *Nectonema* is by undulatory swimming using the body wall muscles and the natatory bristles (Figure 25F) or by passive flotation in nearshore currents, aided by the bristles, which provide resistance to sinking. Freshwater and semiterrestrial nematomorphs (class Gordioida) use their longitudinal muscles to move by undulations or by coiling and uncoiling movements.

Feeding and digestion

The general impression is that nematomorphs "feed" only during the parasitic larval stage, when they probably absorb nutrients across their body wall from the host's tissue and fluids. Current work suggests that the adults rely entirely on nutrients stored during their larval and juvenile parasitic life. However, after emergence from their hosts, minute juveniles grow to adults of impressive size. It seems more likely that both juveniles and adults do in fact "feed" in some as yet undiscovered manner.

The digestive tract of adult nematomorphs is more-or-less degenerate and considered nonfunctional (Figure 26E). The basic plan of the gut is a simple elongate tube running the length of the body. However, there is usually no mouth opening, and the pharyngeal region is typically solid. The intestine or midgut is a thin-walled tube and may serve an excretory function rather than a digestive one. The hindgut remains functional as a cloaca and receives the reproductive ducts. In *Nectonema* a tiny mouth and pharynx lead to a midgut that deteriorates posteriorly and is not connected to the cloaca. This anatomy suggests that nematomorphs may be in an early stage of adapting to parasitic lifestyles, and that the gut persists as a vestigial remnant of its ancestral condition.

Circulation, gas exchange, excretion, and osmoregulation

Very little is known about the physiology of nematomorphs. Internal transport is undoubtedly by diffusion through the pseudocoelom and mesenchyme, and is probably aided by body movement. The free-living gordian worms are presumably obligate aerobes as adults and are restricted to moist environments with ample available oxygen. The threadlike body results in short diffusion distances between the environment and the body organs and tissues.

Excretory and osmoregulatory functions probably operate on a strictly cellular level, there being no protonephridia or other known special structures for these functions. Some workers, however, have speculated that the cells of the midgut may function in the excretion of metabolic wastes, and that they may have a structure similar to the Malpighian tubules of insects (see Chapter 17).

Nervous system and sense organs

Like the nervous systems of nematodes and some other pseudocoelomates, the nervous system of nematomorphs is closely associated with the epidermis. The cerebral ganglion is a mass of nervous tissue located in a region of the head called the **calotte** (Figures 25B and 26F). A single midventral nerve cord arises from the cerebral ganglion and extends the

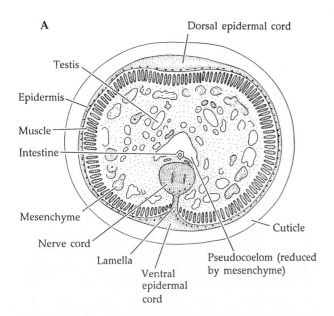

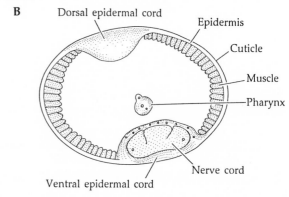

Figure 26
Internal anatomy of nematomorphs. A, A gordioid nematomorph (cross section). B, *Nectonema* (cross section). Note the invasion of mesenchyme in A and the spacious pseudocoelom in B. C, Body wall of *Paragordius*. D, The ornate cuticular areoles of *Chordotes*, a gordioid nematomorph. E, The anterior end of *Paragordius* (sagittal section). F, The anterior end of *Paragordius* (sagittal section). Note the elements of the nervous system and the suspected photoreceptor units. G, The female reproductive system of *Paragordius*. H, A nematomorph larva. (A–G redrawn from Hyman 1951, after various authors; H redrawn from Meglitsch 1972.)

length of the body. In *Nectonema* the ventral nerve cord is entirely within the epidermis, but in gordioids it lies farther inward while remaining attached to the epidermis by a tissue connection called the **epidermal lamella** (Figure 26A and B).

Nematomorphs are touch-sensitive, and some are apparently chemosensitive; adult males are able to detect and track mature females from a distance. However, the structures associated with these sensory functions are only a matter of speculation. Presumably, some of the cuticular areoles are tactile, and perhaps others are chemoreceptive. Members of the genus *Paragordius* possess modified epidermal cells in the calotte that contain pigment and may function as a photoreceptor (Figure 26F).

Reproduction and development

Nematomorphs are gonochoristic and display some sexual dimorphism (Figure 25). The male reproductive system includes one (*Nectonema*) or two (gordioids) testes lying within the pseudocoelom (Figure 26A). Each testis opens to the cloaca via a short sperm duct, which is sometimes swollen as a seminal vesicle. Females of the class Gordioida possess a pair of elongate ovaries. These open to the cloaca through a seminal receptacle (Figure 26G). *Nectonema* contains no discrete ovary; instead, the germinal cells occur as scattered oocytes within the pseudocoelom.

Mating has been studied in some members of the class Gordioida. The females remain relatively inactive, but the males become highly motile during the breeding season and respond to the presence of potential mates in their environment. Once a male lo-

cates a receptive female, he wraps his body around her and deposits a drop of sperm near her cloacal pore. The sperm move into the seminal receptacle and are stored while the ova mature. The eggs are apparently fertilized as they pass from the ovaries to the cloaca, and they are then laid in gelatinous strings.

Early development has been studied in only a few species of nematomorphs. Cleavage is holoblastic, equal, and appears to be modified from a spiral pattern. A coeloblastula forms and then gastrulates by invagination of the presumptive entoderm. Mesodermal cells proliferate into the blastocoel from the area around the blastopore. The anus and cloacal chamber also form from the area of the blastopore. A **nematomorph larva** develops (Figure 26H) and emerges from the egg case. The larva penetrates the body wall of nearly any animal it happens to encounter, but it will develop normally only in an appropriate arthropod host. Within the host's hemocoel, the larva grows to a juvenile nematomorph, which in turn leaves the host to mature. Cuticular molting has been reported in some species.

The priapulans

About fifteen extant species (and six fossil species) are assigned to the phylum Priapula (from *Priapos*, the Greek god of reproduction, symbolized by

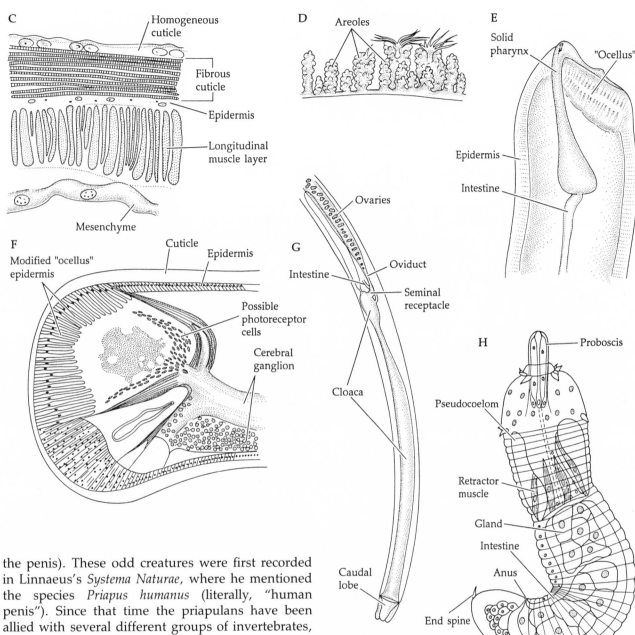

the penis). These odd creatures were first recorded in Linnaeus's *Systema Naturae*, where he mentioned the species *Priapus humanus* (literally, "human penis"). Since that time the priapulans have been allied with several different groups of invertebrates, especially with various other pseudocoelomates. About the time the aschelminth concept was losing favor, William Shapeero (1961) published findings suggesting that the priapulans were truly coelomate. Recently, however, the nature of the body cavity has been questioned again, and most specialists are suggesting anew an affinity of priapulans with various pseudocoelomates. In the absence of complete embryological data, these questions remain unresolved, and the evolutionary relationships of priapulans remain enigmatic.

The major features of these animals are given in Box 6. Priapulans are cylindrical vermiform creatures, from 0.55 mm to 20 cm in length. The body comprises an **introvert** (= proboscis), necklike **collar**, trunk or abdomen, and sometimes a "tail" or **caudal appendage** (Figure 27). When present, the caudal appendage varies in form and function among different species. Larger priapulans are infaunal burrowers in relatively fine marine sediments and appear to be restricted to boreal or cold temperate seas. The meiofaunal species may burrow or live interstitially among sediment particles.

Box Six

Characteristics of the Phylum Priapula

1. Triploblastic, bilateral, unsegmented, or superficially annulated, and vermiform
2. Body cavity lined, but lining probably not peritoneal (coelomate or pseudocoelomate)
3. Nervous system radially arranged and largely intraepidermal
4. Gut complete
5. With many protonephridia associated with the gonads as a urogenital system
6. Many with unique caudal appendage that may serve for gas exchange
7. No circulatory system
8. Thin cuticle periodically molted
9. With unique loricate larva
10. Gonochoristic
11. Marine and benthic; most are burrowers

PHYLUM PRIAPULA

Classes and orders are usually not recognized within the Priapula; the phylum is divided into three extant families (and five extinct families).

FAMILY PRIAPULIDAE: (Figures 27A and B). Relatively large (4–20 cm); abdomen with superficial annulations; caudal appendage (absent from *Halicryptus spinulosus*) either a grapelike cluster of fluid-filled sacs (called vesiculae) or a muscular extension with cuticular hooks. (*Acanthopriapulus*, *Halicryptus*, *Priapulopsis* and *Priapulus*)

FAMILY TUBILUCHIDAE: (Figure 27C). Small (less than 2 mm long); abdomen not annulated; caudal appendage vermiform and muscular. Tubiluchids live in sediments of shallow tropical waters: four species, in two genera: *Tubiluchus* and *Meiopriapulus*.

FAMILY MACCABEIDAE (= CHAETOSTEPHANIDAE): (Figure 27D). Small (less than 3 mm long); meiofaunal; abdomen with rings of tubercles and posterior longitudinal ridges with hooks; no caudal appendage; posterior end of abdomen extensible and mobile, used for burrowing (posterior end first). Maccabeids are found in the Mediterranean and in the Indian Ocean; two described species: *Maccabeus tentaculatus* and *M. cirratus*.

Body wall, support, and locomotion

The priapulan body is covered by a thin, flexible cuticle, which forms a variety of spines, warts, and tubercles (Figure 27). Large hooked spines are often present around the mouth. The cuticle may contain some chitin and is periodically molted as the animal grows. Beneath the cuticle lies an epidermis of thin elongate cells with large fluid-filled intercellular

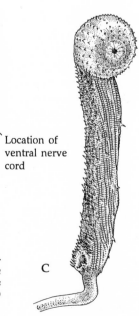

Figure 27

Representative priapulans. **A**, *Priapulus bicaudatus* (family Priapulidae). **B**, *Halicryptus* (Priapulidae). **C**, *Tubiluchus corallicola* (Tubiluchidae), pharynx retracted. **D**, *Maccabeus tentaculatus* (Maccabeidae). (A after Hyman, 1951; B after Meglitsch 1972; C,D redrawn from photographs in Calloway 1982.)

spaces. Beneath the epidermis are well developed layers of circular and longitudinal muscles. There are also complex muscle layers and bands associated with the pharynx, and a set of introvert (proboscis) retractor muscles (Figure 28A).

There is a lining to the body cavity, but its origin and exact nature are unknown. Some authors (e.g., Shapeero 1961) hold that this lining is a cellular peritoneum and that the body cavity is a true coelom, although more recent work suggests that the lining is a simple noncellular membrane secreted by surface cells on the retractor muscles and that the body cavity is a pseudocoelom. In any case, this lining covers not only the inner surface of the body wall, but the internal organs as well, and does form mesentery-like extensions. The fluid of the body cavity contains motile amebocytes and free erythrocytes with hemerythrin.

Maintenance of body form and support are provided by the hydrostatic skeleton of the body cavity. The contraction of circular muscles around this cavity also facilitates protrusion of the introvert by increasing the internal pressure. Those priapulans that move about through the substratum do so largely by peristaltic burrowing, probably using the various hooks and other cuticular extensions to hold one part of the body in place while the rest is pushed or pulled along. *Maccabeus* is thought to use its ring of posterior cuticular spines for anchorage within its burrow (Figure 27D).

Feeding and digestion

The majority of priapulans (i.e., members of the family Priapulidae) live in soft sediments and prey on various soft-bodied invertebrates such as polychaete worms. During feeding, a portion of the toothed, cuticle-lined pharynx is everted through the mouth at the end of the extended introvert. As the prey is grasped, the pharynx is inverted; the introvert retracts, and the prey is drawn into the gut.

Tubiluchus corallicola (Figure 27C) lives in coral sediments and feeds on organic detritus. The pharynx is lined with pectinate teeth, which the animal uses to sort food material from the coarse sediment particles. *Maccabeus tentaculatus*, a tube dweller, is a trapping carnivore. Surrounding the mouth are eight short tentacles, presumed to be touch-sensitive. These are in turn ringed by 25 highly branched spines (Figure 27D). It is suspected that when a potential prey touches the sensory tentacles, the outer spines quickly close as a trap.

The digestive system is complete and either straight or slightly coiled (Figure 28A). The portion of the gut that lies roughly within the bounds of the introvert comprises the buccal tube, the pharynx, and the esophagus, all of which are lined with cuticle and together constitute a stomodeum. In members of the genus *Tubiluchus*, the stomodeum also includes a region behind the esophagus called a **polythridium**. This part of the gut bears a circlet of two rows of plates and may operate to grind food. The midgut or intestine is the only entodermally derived section of the digestive tract and is followed by a short proctodeal rectum. The anus is located at the posterior end of the abdomen, either centrally or slightly to one side. Nothing is known about the digestive physiology of priapulans, although it is likely that digestion and absorption occur in the midgut.

Circulation, gas exchange, excretion, and osmoregulation

Internal transport takes place through diffusion and movement of the fluid in the body cavity. The presence of the respiratory pigment hemerythrin in the body fluid cells suggests an oxygen transport or storage function; many priapulans are known to live in marginally anoxic muds. In those species with a vesiculate caudal appendage, the lumen of that structure is continuous with the main body cavity. Such caudal appendages may function as gas exchange surfaces.

Clusters of solenocyte protonephridia lie in the posterior portion of the body cavity, and are associated with the gonads as a urogenital system or complex (Figure 28A). Various priapulans inhabit both hyper- and hyposaline environments, so the protonephrida may function in both osmoregulation and excretion.

Nervous system and sense organs

The priapulan nervous system is intraepidermal and is constructed for the most part on a radial plan within the cylindrical body (Figure 28B). Although a typical cerebral ganglion is wanting, there is a circumenteric nerve ring within the buccal tube epithelium. The main ventral nerve cord arises from this ring and gives off a series of ring nerves and peripheral nerves along the body. In addition, longitudinal nerves extend from the main nerve ring along the inner pharyngeal lining and are connected by the ring commissures.

Little is known of sense organs in priapulans. The caudal appendage contains tactile receptors, and so may many of the bumps and spines on the body surface. Flosculi are present in the members of the family Tubiluchidae.

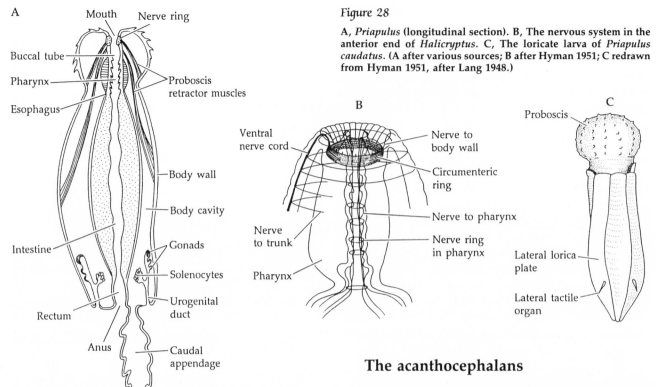

Figure 28

A, *Priapulus* (longitudinal section). B, The nervous system in the anterior end of *Halicryptus*. C, The loricate larva of *Priapulus caudatus*. (A after various sources; B after Hyman 1951; C redrawn from Hyman 1951, after Lang 1948.)

Reproduction and development

Priapulans are gonochoristic, although males are unknown in *Maccabeus tentaculatus*. The reproductive organs are similarly placed and connected in both sexes. The paired gonads are drained by genital ducts, which are joined by collecting tubules from the protonephridia to form a pair of urogenital ducts exiting posteriorly through urogenital pores (Figure 28A).

Priapulans free-spawn, first the males and then the females, and fertilization is external. Cleavage is holoblastic and radial, and results in a coeloblastula (some accounts differ) that undergoes invagination. The origin of the body cavity and many other aspects of morphogenesis remain unknown. A larval form (sometimes called a **loricate larva**) eventually develops (Figure 28C). This larva bears a striking resemblance to loriciferans. The abdomen of the larva is encased within a thick cuticular lorica into which the head can be withdrawn. The larvae live in benthic muds and are probably detritivores. The lorica is periodically molted as the larva grows, and it is finally lost as the animal metamorphoses to a juvenile priapulan. At that time the caudal appendage forms in those species that have this structure.

The acanthocephalans

As adults, the 700 described species of acanthocephalans are obligate intestinal parasites in vertebrates, particularly in freshwater teleosts. Larval development takes place in intermediate arthropod hosts. The name Acanthocephala (Greek *acanthias*, "prickly"; *cephalo*, "head") derives from the presence of recurved hooks located on an eversible proboscis at the anterior end (Figures 29B and D). The rest of the body forms a cylindrical or flattened trunk, often bearing rings of small spines. Most acanthocephalans are less than 20 cm long, but a few species are nearly a meter in length. The digestive tract has been completely lost. And, except for the reproductive organs, there is significant structural and functional reduction of most other systems, related to the parasitic lifestyles of these worms (Box 7). The persisting organs lie within an open pseudocoelom, which is partially partitioned by mesentery-like **ligaments**.

Classification

The phylum Acanthocephala is divided into three classes on the basis of the arrangement of proboscis hooks, the nature of the epidermal nuclei, spination patterns on the trunk, and the nature of the reproductive organs—Palaeacanthocephala, Archiacanthocephala, and Eoacanthecephala. It is beyond our purposes here to detail these differences, but representative members of these classes are shown in Figure 29.

Body wall, support, attachment, and nutrition

Adult acanthocephalans attach to their host's intestinal wall by their proboscis hooks (Figure 29A). In nearly all species the proboscis is retractible into a deep **proboscis receptacle**, an action that enables the body to be pulled close to the host's intestinal mucosa. Nutrients are absorbed through the unique body wall, which comprises a thin outer cuticle underlaid by a syncytial epidermis. The epidermis

Figure 29

Representative acanthocephalans. **A**, *Macracanthorhynchus hirudinaceus*, an archiacanthocephalan, attached to the intestinal wall of a pig. **B**, *Corynosoma*, a palaeacanthocephalan found in aquatic birds and seals. **C**, Longitudinal section through the anterior end of *Acanthocephalus* (class Palaeacanthocephala). **D**, An adult male eoacanthocephalan (*Pallisentis fractus*). **E**, The isolated female reproductive system of *Bolbosoma*. (A after various sources; B,C after Hyman 1951; D after Noble and Noble 1982, from Cable and Dill; E redrawn from Meglitsch 1972, after Yamaguti 1963.)

Box Seven
Characteristics of the Phylum Acanthocephala

1. Triploblastic, bilateral, unsegmented, vermiform pseudocoelomates
2. Anterior end with hook-bearing proboscis
3. Epidermis contains a unique system of channels called the lacunar system
4. Gut absent
5. Protonephridia absent except in a few species
6. With unique system of ligaments and ligament sacs partially partitioning the body cavity
7. Gonochoristic
8. All are obligate parasites in vertebrates; many have complex life cycles

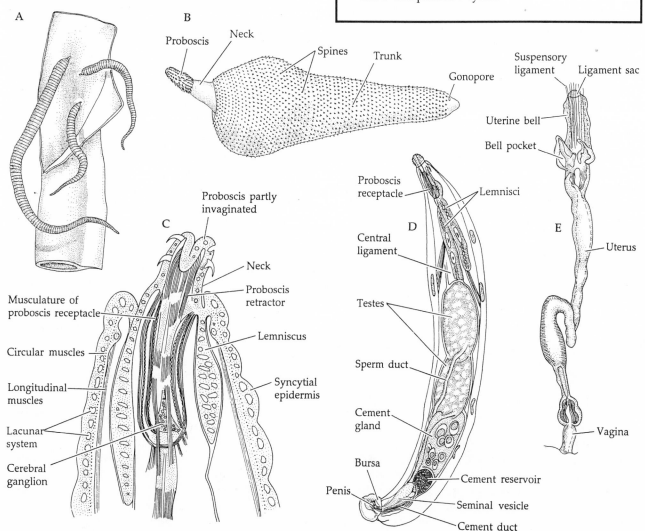

houses a complex system of circulatory channels called the **lacunar system** (Figure 29C). At the junction of the proboscis and the trunk, the epidermis extends inward as a pair of hydraulic sacs (**lemnisci**) that facilitate extension of the proboscis; the proboscis is withdrawn by retractor muscles. Thin layers of circular and longitudinal muscles lie beneath the epidermis of the body wall.

One or two large sacs lined with connective tissue arise from the rear wall of the proboscis receptacle and extend posteriorly in the body. These structures support the reproductive organs and divide the body into dorsal and ventral **ligament sacs** in the archiacanthocephalans and eoacanthocephalans, or produce a single ligament sac down the center of the pseudocoelom in the palaeacanthocephalans (Figure 29D and E). Within the walls of these sacs are strands of fibrous tissue, the ligaments, that may represent remnants of the gut.

The body is supported by the cuticle and the hydrostatic qualities of the pseudocoelom and lacunar system. The muscles and ligament sacs add some structural integrity to this support system.

Circulation, gas exchange, and excretion

Exchanges of nutrients, gases, and waste products occur by diffusion across the body wall (a few species of Archiacanthocephala possess one pair of protonephridia). Internal transport is by diffusion within the body cavity and probably by the epidermal lacunar system.

Nervous system

As in most obligate endoparasites, the nervous system and the sense organs of acanthocephalans are greatly reduced. A cerebral ganglion lies within the proboscis receptacle (Figure 29C) and gives rise to nerves to the body wall muscles, the proboscis, and the genital regions. Males possess a pair of genital ganglia.

The proboscis bears several simple sensory receptors, which are presumed to be tactile. Males have sense organs in the genital area, especially on the penis.

Reproduction and development

Acanthocephalans are gonochoristic and females are generally somewhat larger than males. In both sexes the reproductive systems are associated with the ligament sacs (Figure 29D). In males, paired testes lie within a ligament sac and are drained by sperm ducts to a common seminal vesicle. Entering the seminal vesicle or the sperm ducts are six or eight cement

Figure 30

Life cycle of *Macracanthorhynchus hirudinaceus*, an intestinal parasite in pigs. The adults reside in the intestine of the definitive host and embryos are released with the feces. The encapsulated embryos are ingested by the secondary host, in this case, beetle larvae. Within the secondary host, the embryo passes through the acanthor and acanthella stages, eventually becoming a cystacanth, while the beetle grows. When the beetle is ingested by a pig, the juvenile matures into an adult, thereby completing the cycle. (After various sources.)

glands, whose secretions serve to plug the female genital pore following copulation. The seminal vesicle leads to an eversible penis, which lies within a genital bursa connected to the gonopore. This gonopore is often called a cloacal pore, because the bursa appears to be a remnant of the hindgut.

In females, a single mass of ovarian tissue forms within a ligament sac. Clumps of immature ova are released from this transient ovary and enter the pseudocoelom, where they mature and are eventually fertilized. The female reproductive system comprises a gonopore, a vagina, and an elongate uterus that terminates internally in a complex open funnel called the **uterine bell** (Figure 29E). During mating the male everts the copulatory bursa and attaches it to the female gonopore. The penis is inserted into the vagina, sperm are transferred, and the vagina is capped with cement. Sperm then travel up the female system, enter the pseudocoelom through the uterine bell, and fertilize the eggs.

Much of the early development of acanthocephalans takes place within the body cavity of the female. Cleavage is holoblastic, unequal, and likened to a highly modified spiral pattern. A stereoblastula is produced, at which time the cell membranes break down to yield a syncytial condition. Eventually, a shelled **acanthor larva** is formed (Figure 30). At this or an earlier stage the embryo leaves the mother's body. Remarkably, the uterine bell "sorts" through the developing embryos by manipulating them with its muscular funnel; it accepts only the appropriate embryos into the uterus. Embryos in earlier stages are rejected and pushed back into the body cavity, where they continue development. The selected embryos pass through the uterus and out the genital pore and are eventually released with the host's feces.

Once outside the definitive host, the developing acanthocephalan must then be ingested by an arthropod intermediate host—usually an insect or a crustacean—to continue the life cycle. The acanthor larva penetrates the gut wall of the intermediate host

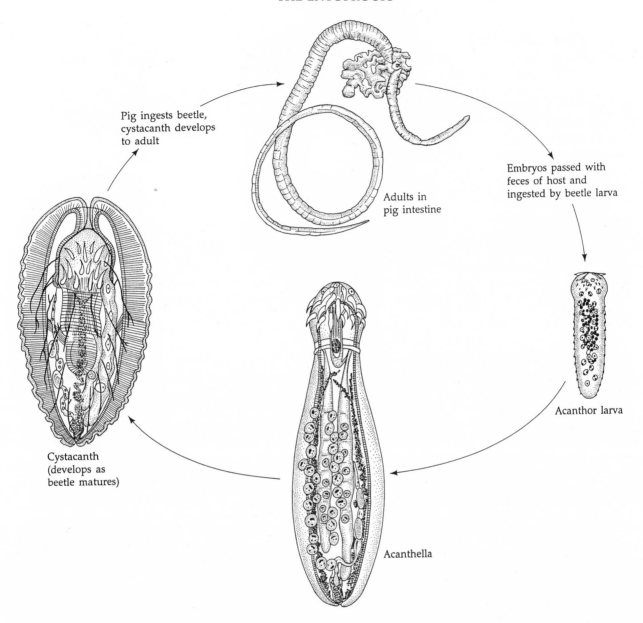

Pig ingests beetle, cystacanth develops to adult

Adults in pig intestine

Embryos passed with feces of host and ingested by beetle larva

Cystacanth (develops as beetle matures)

Acanthella

Acanthor larva

and enters the body cavity, where it develops into an **acanthella** and then into an encapsulated form called a **cystacanth** (Figure 30). When the intermediate host is eaten by an appropriate definitive host, the cystacanth attaches to the intestinal wall of the definitive host and matures into an adult.

The entoprocts

The phylum Entoprocta (Greek *entos*, "inside"; *proktos*, "anus") comprises about 150 species of small, sessile, solitary or colonial creatures that superficially resemble hydroids and ectoprocts (Figure 31). Almost all entoprocts are marine and live attached to various substrata, including algae, shells, and rock surfaces. They are common intertidally, but some forms are known from depths as great as 500 meters. Although technically bilateral, the individual zooids have in many respects assumed a functionally radial form. The body consists of a cuplike **calyx** from which arises a whorl of ciliated tentacles (Figure 31). Both the mouth and anus are located on the surface or **vestibule** of the calyx and are surrounded by the **tentacular crown**. The anus is elevated on a distinct papilla, called the **anal cone**. Each calyx arises from a stalk, which in solitary forms attaches directly to the substratum and in colonial forms attaches to larger branches or to horizontal stolons. Box 8 lists the distinguishing features of entoprocts.

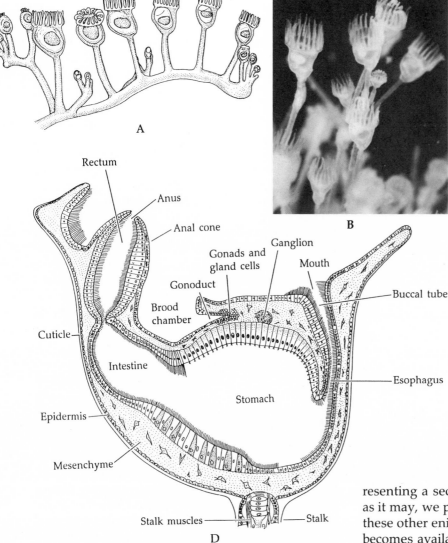

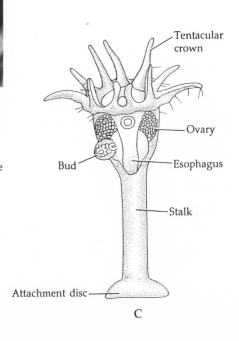

Figure 31

Representative entoprocts. A, A portion of a *Pedicellina* colony. B, A portion of a *Barentsia* colony. C, *Loxosoma*; a solitary individual with a bud. D, A zooid of *Pedicellina* (sagittal section). (A after Nielsen 1964; B courtesy of G. McDonald; C redrawn from Barnes 1980, modified from Hyman 1951; D after Hyman 1951.)

resenting a secondarily filled pseudocoelom. Be that as it may, we present the entoprocts here, along with these other enigmatic groups, until more information becomes available.

PHYLUM ENTOPROCTA

The phylum is divided into four families (Figure 31). Some specialists recognize two orders on the basis of the presence or absence of a septum between the stalk and calyx. See Nielsen (1982) for details.

FAMILY LOXOSOMATIDAE: Without a septum between stalk and calyx; solitary; often commensal on other invertebrates; some are capable of limited movement on suckered base. (e.g., *Loxosoma, Loxosomella*)

FAMILY LOXOKALYPODIDAE: Without a septum between stalk and calyx; colonial; ectocommensal on the polychaete *Glycera nana* in the northeastern Pacific. Monotypic: *Loxokalypus socialis*.

FAMILY PEDICELLINIDAE: With a complete stalk-calyx septum; muscles extend length of stalk; colonial. (e.g., *Myosoma, Pedicellina*)

The phylogenetic relationships of entoprocts with other taxa are controversial. During most of the eighteenth and nineteenth centuries, entoprocts were included with ectoprocts in the phylum Bryozoa. Since the discovery of their noncoelomate nature (Clark 1921), a separate phylum status for the Entoprocta has generally been accepted. Until quite recently the entoprocts were considered to be unquestionably pseudocoelomates, albeit with the body cavity fully invested by mesenchyme. However, some workers (Nielsen 1971, 1977) have raised again the possibility that the entoprocts and ectoprocts (see Chapter 21) may be closely related—in fact, that the former may represent the ancestral condition of the latter. To confuse the issue even more, the possibility has been raised that the entoprocts' solid body structure may be primitively acoelomate rather than rep-

Box Eight

Characteristics of the Phylum Entoprocta

1. Triploblastic, bilateral, unsegmented
2. Sessile and solitary or colonial with zooids borne on stalks
3. Visceral mass housed within a cup-shaped calyx, the ventral surface of which is directed away from the substratum
4. Zooids bear a ring of tentacles that enclose both the mouth and the anus
5. Gut complete and U-shaped
6. Area between gut and body wall filled with gelatinous mesenchyme; thus, functionally acoelomate
7. With one pair protonephridia, but lacking special circulatory or gas exchange structures
8. Most are hermaphroditic
9. All are marine except *Urnatella* (fresh water)

FAMILY BARENTSIIDAE: With an incomplete septum between the stalk and calyx; stalk muscles short and discontinuous; colonial; some live in fresh water. (e.g., *Barentsia*, *Urnatella*)

Body wall, support, and movement

The calyx and stalk are covered by a thin cuticle that does not extend over the ciliated portion of the tentacles or the vestibule. The epidermis is cellular, and the epidermal cells are cuboidal to somewhat flattened. Various subepidermal muscle bands serve to retract the tentacles, compress the body to extend the tentacles, and contract the calyx. Other muscles are located within the stalk and provide some ability to bend. As mentioned above, there is no persistent body cavity, and the area between the gut and body wall is filled with mesenchyme containing sessile and errant amebocytes within its matrix (Figure 31D). The mesenchyme and cuticle provide body support.

Feeding and digestion

Entoprocts are suspension feeders. They extract food particles, mostly phytoplankton, from currents produced by the lateral cilia on the tentacles (Figure 32B). These animals are oriented with their ventral side away from the substratum; the dorsal surface is attached to the stalk. Water currents pass from dorsal to ventral, flowing between the tentacles (Figure 32A). Food is trapped by the lateral cilia and moved in a sheet of mucus to the frontal cilia; once there, it is directed towards the vestibule. The frontal cilia move the mucus and food to ciliated vestibular food grooves at the base of the tentacular ring. Here additional ciliary tracts carry the food to the mouth (Figure 32C).

Food and mucus are moved into the gut by cilia lining the buccal tube and by muscular contractions of the esophagus (Figure 31D). The esophagus leads to a spacious stomach, from which a short intestine extends to the rectum located within the anal cone. Food is moved through the gut by cilia. The stomach lining secretes digestive enzymes and mucus, which are mixed with the food by a tumbling action caused by the ciliary currents. Digestion and absorption probably occur within the stomach and intestine, where food is held for a time by an intestinal-rectal sphincter muscle.

Circulation, gas exchange, and excretion

The gut also apparently serves as an excretory passage. Cells in the ventral stomach wall accumulate precipitations of uric acid and guanine and release them into the stomach lumen, from whence they are discharged through the anus. Entoprocts also possess a pair of flame bulb protonephridia located between the stomach and the vestibule epithelium. The protonephridia drain to a short common nephridioduct that leads to a pore on the surface of the vestibule. The duct is lined by ameboid cells called **athrocytes**; these cells are thought to phagocytize wastes from the mesenchyme and release them to the excretory duct.

Internal transport is largely through the expansive gut; diffusion distances are small between its lumen and the body wall. Other transport occurs through the mesenchyme by diffusion and by means of motile amebocytes. Gas exchange probably occurs over much of the body surface, particularly at the cuticle-free tentacles and vestibule.

Nervous system

As is so often the case in small sessile invertebrates, the nervous system is greatly reduced. A single ganglionic mass lies between the stomach and vestibular surface (Figure 31D), and is called the subenteric ganglion. This ganglion gives rise to several pairs of nerves to the tentacles, calyx wall, and stalk. Unicellular tactile receptors are concentrated on the

Figure 32

Entoproct feeding. A, General ciliary currents in *Loxosoma*. B, A tentacle of *Pedicellina* (cross section). Note the lateral and frontal cilia. C, *Loxosoma* (top view), illustrating details of feeding currents. See text for explanation. (All after Hyman 1951.)

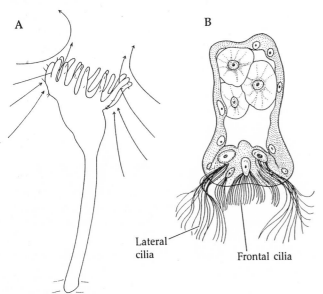

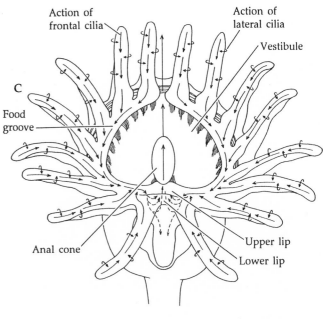

tentacles and scattered over much of the body surface.

Reproduction and development

Colony growth occurs by budding from the stolon or various branches of the stalk (Figure 31A). Solitary forms produce buds that separate from the parent (Figure 31C).

Most, perhaps all, entoprocts are hermaphroditic; some are protandric. Those that are thought to have separate sexes may also be protandric, but with a long temporal separation of the male and female phases. One or two pairs of gonads occur in protandric and simultaneously hermaphroditic species, respectively. The gonads lie just beneath the surface of the vestibule. Short gonoducts lead from the gonads to a common pore opening to a brood chamber (Figure 31D).

Sperm apparently are released into the water and then enter the female reproductive tract, and fertilization occurs in the ovaries or oviducts. As the zygotes move along the oviducts, cement glands secrete stalks by which the embryos are attached to the wall of the brood chamber. A kind of viviparity occurs in some species where the embryos are retained within the female tract. In these and in a few external brooders, special cells of the adult provide nutrition to the developing embryos in a pseudoplacental arrangement.

Cleavage in entoprocts is holoblastic and spiral. Nonsynchronous divisions produce five "quartets" of micromeres at about the 56-cell stage. The cell fates are similar to those in typical protostome development, including the derivation of mesoderm from the 4d mesentoblast. A coeloblastula forms and gastrulates by invagination. A larva develops (Figure 33A) that, according to some authors, is similar to a trochophore, the basic larval type among protostomes (see especially Chapter 13). Most entoproct larvae are free swimming and planktotrophic, but a few produce lecithotrophic or benthic crawling larvae. Upon settling, the larvae of most species attach by the coronal ciliary band and undergo a remarkable rotation of the body mass to direct the ventral, vestibular surface away from the substratum (Figure 33B). In some, however, the larva adheres to the substratum and, without rotating, then produces an appropriately oriented asexual bud that becomes the adult (Figure 33C).

The gnathostomulids

The phylum Gnathostomulida (Greek, *gnathos*, "jaw"; *stoma*, "mouth") comprises 80 or so species of minute vermiform animals (Figure 34). These meiofaunal creatures were first described as turbellarians (Ax 1956), but were later given phylum status by

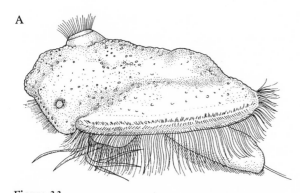

Figure 33

Settlement and metamorphosis in *Loxosoma*. A, A larva of *L. harmeri*. B, Metamorphosis of *L. harmeri*: the larval body forms the first zooid; C, Metamorphosis of *L. leptoclini*: the larva attaches and produces a bud that becomes the first zooid. (After Nielsen 1971.)

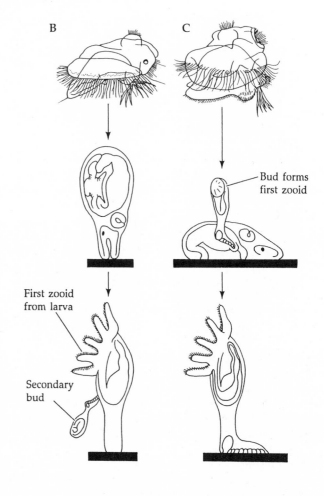

Riedl (1969). Gnathostomulids are found interstitially in marine sands, often occurring in high densities in anoxic sulfide-rich conditions, from the intertidal zone to depths of hundreds of meters. They have been found worldwide. The elongate body (less than 1 mm long) is usually divisible into head, trunk, and narrow tail regions. Distinguishing features of this phylum include a unique jawed pharyngeal apparatus and monociliated epidermal cells (Box 9). Technically the gnathostomulids are acoelomate bilateria, but their relationship to other groups is highly controversial (see Sterrer et al. 1985).

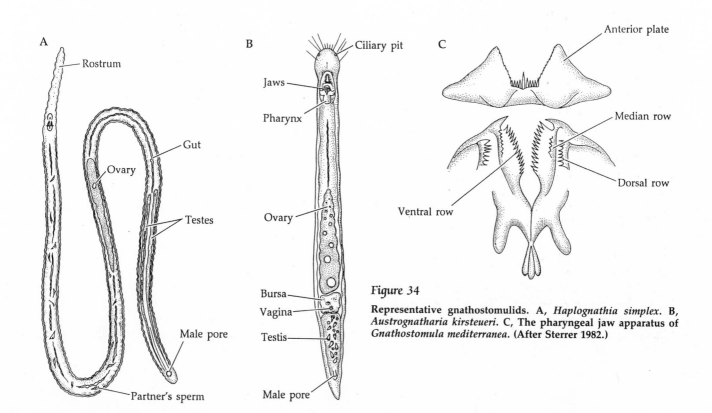

Figure 34

Representative gnathostomulids. A, *Haplognathia simplex*. B, *Austrognatharia kirsteueri*. C, The pharyngeal jaw apparatus of *Gnathostomula mediterranea*. (After Sterrer 1982.)

PHYLUM GNATHOSTOMULIDA

ORDER FILOSPERMOIDEA: Body usually very elongate, with slender rostrum; male system lacks penis; sperm filiform; female system lacks a vagina. (e.g., *Haplognathia, Pterognathia*)

ORDER BURSOVAGINOIDEA: Body usually not extremely elongate relative to width; without slender rostrum; male system with penis; sperm not filiform; female system with bursa and usually a vagina. (e.g., *Austrognatharia, Gnathostomula*)

Body wall, support, and locomotion

Each outer epithelial cell bears a single cilium by which the animals move in a gliding motion. This motion is aided by body contortions produced by the contraction of thin strands of subepidermal (cross-striated) muscle fibers. These actions, plus reversible ciliary beating, facilitate twisting, turning, and crawling among sand grains, and limited swimming in some species. The body is supported by its more-or-less solid construction, with a loose mesenchyme filling the area between the internal organs.

Nutrition, circulation, excretion, and gas exchange

The mouth is located on the ventral surface at the "head–trunk" junction and leads inward to a muscular pharynx armed with an anterior, often comblike, plate and a pair of movable jaws (Figure 34C). The pharynx connects with an elongate, simple, saclike gut. An anus is present in some genera. Gnathostomulids ingest bacteria and fungi by scrap-

ing them into the mouth with the comb plate of the pharynx and swallowing them by the action of the jaws.

These animals probably depend largely on diffusion for circulation, excretion, and gas exchange. Isolated solenocytes have been reported in some species, but they resemble epidermal cells and join to "epidermal canal cells" that eventually open on the body surface.

Nervous system

The nervous system is intimately associated with the epidermis and as yet is incompletely described. A host of sensory ciliary pits and sensory cilia occur on the head. Gnathostomulid specialists have attached a formidable array of names to these structures, which are of major taxonomic significance.

Reproduction and development

Gnathostomulids are protandric or simultaneous hermaphrodites. The male system includes one or two testes generally located in the posterior part of the trunk and tail; the female system consists of a single large ovary (Figure 34). Members of the order Bursovaginoidea possess a vaginal orifice and a sperm-storage bursa, both associated with the female gonopore, and a penis in the male system; members of the order Filospermoidea lack these structures.

Mating has been only superficially studied in gnathostomulids, and the method of sperm transfer is unknown. The penis is glandular and adheres to the partner's body. Although the method of sperm penetration is not certain, suggestions include sperm boring through the body wall and mutual hypodermic impregnation. In any case, these animals appear to be gregarious, to rely on internal fertilization, and to deposit zygotes singly in their habitat. Cleavage is spiral, but the details of later development are lacking; development is direct.

The loriciferans

It may be apparent to you by now that interstitial habitats (the meiobenthic realm) are home for a host of bizarre and specialized creatures. Studies on meiofauna continue to reveal new animals, previously undescribed taxa, and myriad examples of convergent evolution associated with success in this environment. A short note in *Science* magazine (Lewin 1983) announced the discovery of a new phylum and the publication of its description by Reinhardt Kristensen (1983), a Danish zoologist who has studied meiofauna for a number of years. This new phylum has been named the Loricifera (Latin *lorica*, "corset"; *ferre*, "to bear") in reference to the well developed cuticular **lorica** encasing most of the body (Figure 35; Box 10). The description of the phylum was initially based upon a single species, *Nanaloricus mysticus*, but eight other species have since been collected and described by Higgins and Kristensen (1986), and about two dozen more are in the works.

The loriciferan body is small (225–383 μm long) and divided into a head (introvert), neck and thorax, and loricate abdomen. The head, neck and most of the thorax are retractable within the lorica. The mouth is located at the end of an **oral cone** that projects from the head and contains protrusible oral stylets in some species. Rings of spinelike **scalids** of various sorts protrude from the spherical head. The first ring always consists of eight, anteriorly directed **clavoscalids**, the remaining eight rings of **spinoscalids** are directed posteriorly. The lorica comprises six plates, which bear anteriorly directed spines around the base of the neck. Beneath the cuticle and the body wall are various muscle bands, including those responsible for retraction of the anterior parts.

The digestive tract is complete. A long, tubular, buccal canal leads from the mouth to a muscular pharynx bulb (Figure 35F). The lumen of these anterior gut structures is lined with cuticle. Behind the pharynx is a short esophagus and long midgut leading to a short rectum and an anus located on an anal cone. One pair of **salivary glands** is associated with the buccal tube.

The central nervous system includes a large, dorsal, cerebral ganglion and a number of smaller ganglia associated with the various body regions and parts.

Loriciferans are gonochoristic; males and females differ externally in the form of certain scalids. The male reproductive system comprises two dorsal testes in the abdominal body cavity, probably a pseudocoelom. The female system includes a pair of ovaries and probably a seminal receptacle. Fertilization is suspected to be internal.

Nothing is yet known about early development in loriciferans. A feeding larval form develops, which has been named a **Higgins-larva** in honor of Robert Higgins, who discovered a single adult specimen (now *Pliciloricus enigmaticus*) in 1974 but believed it to be a larval priapulan. The larva (Figures 35D and E) is built along the same general body plan as the adult, but in most species it possesses a pair of "toes" at the posterior end that are used for locomotion. These toes are thought to have adhesive glands at their

---

Box Ten

Characteristics of the Phylum Loricifera

1. Bilateral, unsegmented, and probably pseudocoelomate
2. Body divided into a head, neck, and thorax, all retractable into an abdomen
3. Abdomen housed in cuticular lorica
4. Mouth on oral cone beset with stylets
5. Gut complete
6. No apparent circulatory or gas exchange systems
7. One pair of protonephridia
8. Gonochoristic; development includes unique Higgins-larva
9. All inhabit marine, interstitial environments

---

bases. The larva grows by molting, and early stages apparently contain yolk reserves in the cells of the body cavity.

Loriciferans were first found in clean shell-gravel sediments but are now known from mud to depths of several thousand meters. The larvae have been observed alive and are free living. The adults are probably free living, but the biology of this newest phylum is poorly known.

Some phylogenetic considerations

We have alluded to the complex phylogenetic relationships of the ten phyla covered in this chapter with one another and with members of other taxa. In spite of the unresolved issues and difficulties involved, there is a large amount of literature on these matters and many more ideas and speculations than we can detail here.

As we stressed in the introductory sections of this book, one of the major difficulties in sorting out phylogenetic hypotheses is evaluating the nature of shared characteristics among groups. Particularly difficult is the separation of homologous traits from convergently similar ones. The exploitation of similar habitats by animals with different phylogenetic histories is frequently accompanied by the appearance of convergent similarities, characteristics that serve

A

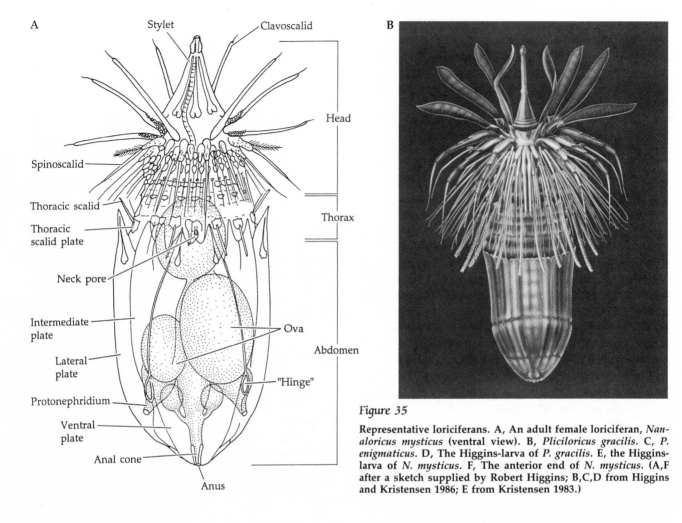

Stylet

Clavoscalid

Head

Spinoscalid

Thoracic scalid

Thoracic
scalid plate

Neck pore

Intermediate
plate

Lateral
plate

Protonephridium

Ventral
plate

Anal cone

Anus

Thorax

Abdomen

Ova

"Hinge"

B

Figure 35

Representative loriciferans. A, An adult female loriciferan, *Nan-aloricus mysticus* (ventral view). B, *Pliciloricus gracilis*. C, *P. enigmaticus*. D, The Higgins-larva of *P. gracilis*. E, the Higgins-larva of *N. mysticus*. F, The anterior end of *N. mysticus*. (A,F after a sketch supplied by Robert Higgins; B,C,D from Higgins and Kristensen 1986; E from Kristensen 1983.)

those different animals in their shared environment. We have seen this principle in the case of parasitism, and it is no less evident in other special environments, such as those encountered by interstitial animals. Furthermore, the constraints of a particular *Bauplan*, such as one associated with small size, can in themselves produce certain convergent similarities among taxa.

It may very well be that the ten phyla discussed in this chapter do in fact represent fewer than ten separate monophyletic clades. But it is almost certainly true also that most of these ten phyla must be regarded as having achieved a grade of body organization that reflects considerable convergent evolution. The problem, of course, is which of these taxa are closely related to one another, and which are not, and which traits are reliable in making such judgments. Along these lines, we offer the following sampling of ideas.

We agree with most zoologists that the presence

of a pseudocoelom is insufficient grounds by itself for hypothesizing that the "pseudocoelomates" constitute a monophyletic group, or clade. The retention of a persistent blastocoel may be nothing more than a paedomorphic characteristic derived convergently in numerous groups. The same may be said of certain other features that tend to accompany the pseudocoelomate *Bauplan*, such as small size, eutely, and external ciliation. Some of the phyla discussed here contain species with a spacious pseudocoelom, as well as other species in which the cavity is invaded by mesenchyme (e.g., Nematomorpha), suggesting extreme plasticity in this feature. We consider the pseudocoelom *sensu lato* not to be a particularly conservative or useful phylogenetic feature, but more likely the repeated product of flexible developmental programs.

Most of the taxa discussed in this chapter appear to show some basic affinity to the protostome line, especially in the tendency toward spiral-like cleav-

C

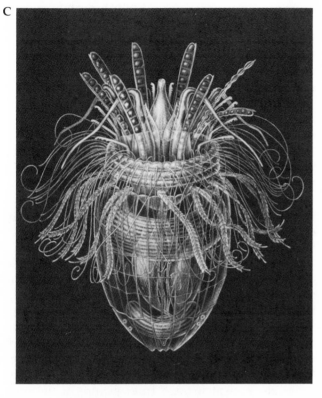

D

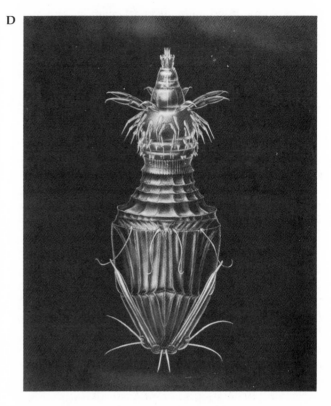

E

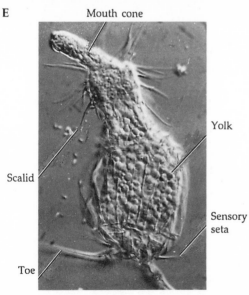

Mouth cone

Yolk

Scalid

Sensory
seta

Toe

F

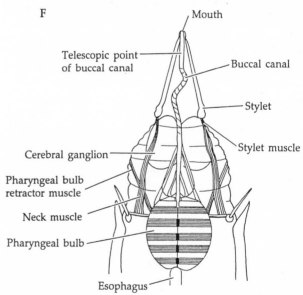

Mouth

Telescopic point
of buccal canal

Buccal canal

Stylet

Stylet muscle

Cerebral ganglion

Pharyngeal bulb
retractor muscle

Neck muscle

Pharyngeal bulb

Esophagus

age. This feature, coupled with small size, blastocoel retention, external ciliation patterns, and the presence of protonephridia in adults, have led many workers to speculate that the "aschelminth" groups may have arisen from protostome coelomate ancestors, perhaps by neoteny (e.g., Lorenzen 1985). Some recent hypotheses (see Kristensen 1983) imply that lumping some of these taxa may be appropriate. For

example, on the basis of certain developmental and adult morphological characteristics, some zoologists propose uniting the loriciferans, priapulans, kinorhynchs, nematomorphs, and perhaps nematodes into a single phylum.

The gnathostomulids have been viewed as being related to annelids, rotifers, turbellarians, or gastrotrichs. The latter two groups have a more-or-less mesenchyme-filled body, external ciliation, and other common features. Possible ties to annelids and rotifers have been based on similarities in the pharynx and jaw apparatus, but these are probably convergent features rather than true homologues. A more traditional hypothesis has allied the gnathostomulids with the turbellarians, but these groups differ in some very basic ways. The gnathostomulids contain rather poorly developed mesenchyme, have monociliated epidermal cells, and have a distinct pharynx apparatus, all of which constitute evidence against a relationship with the flatworms. Ax (1985), however, makes a case for treating the Gnathostomulida and Platyhelminthes as sister taxa.

The evolutionary relationships of the Acanthocephala seem to be a complete mystery. Since no free-living species are known, it is difficult to trace their phylogeny to any free-living ancestors. The phylum may have had its origin in a group that had already established itself in a parasitic lifestyle. Cestodes have been suggested as a possible ancestral group to the acanthocephalans. However, these two taxa differ in fundamental aspects of their anatomy and ontogeny, so there is little to corroborate such a hypothesis. Lorenzen (1985) suggests cautiously that acanthocephalans may be highly derived rotifers. The simplest

known acanthocephalans live in fishes, and the most specialized forms in birds and mammals. From this it has been suggested that the acanthocephalans were parasites of the earliest fishes and subsequently evolved along with their vertebrate hosts. If so, these parasites could be very old (Silurian or older). The absence of an acanthocephalan fossil record and the reduction of their body parts also make phylogenetic studies on this phylum exceedingly difficult.

The entoproct/ectoproct problem remains controversial. The spiral cleavage in entoprocts suggests a protostome affinity, whereas ectoprocts are much more deuterostome-like in their early development. We do not think that the two groups are related to each other, and we expand on this point of view in Chapter 21.

These and other speculations have done little over the years to bring order out of the phylogenetic chaos of these groups. Some hope lies in future studies on ultrastructure and development of those animals where information is still incomplete. With the limited data available, and so many questions still unanswered, we feel it is appropriate to consider these ten groups as separate phyla for the time being, although we watch for a truer picture in the near future.

Selected References

General

Clark, A. H. 1921. A new classification of animals. Inst. Oceanogr. Monaco Bull. 400.

Dougherty, E. C. (ed.) 1963. *The Lower Metazoa: Comparative Biology and Phylogeny.* University of California Press, Berkeley. [See Sections IG and IIIA (in part) for papers by various authors dealing with pseudocoelomate phylogeny.]

Florkin, M. and B. T. Scheer. 1969. *Chemical Zoology* Vol. 3, Sect. 2, Nematoda and Acanthocephala. Academic Press, New York.

Giese, A. C. and J. S. Pearse. 1974. *Reproduction of Marine Invertebrates*, Vol. 1, Acoelomate and Pseudocoelomate Metazoans. Academic Press, New York.

Grassé, P. (ed.) 1965. Némathelminthes, Rotifères, Gastrotriches, et Kinorhynques. *In* P. Grassé, *Traité de Zoologie*, Vol. 4, Pts. 2–3. Masson et Cie, Paris.

Grobben, K. 1908. Die systematische Einteilung des Tierreiches. Verh. K. Zool. Bot. Ges. Wien. 58.

Hulings, N. C. and J. S. Gray (eds.) 1971. A manual for the study of meiofauna. Smithson. Contrib. Zool. 78:1–83.

Hyman, L. H. 1951. *The Invertebrates*, Vol. 3, Acanthocephala, Aschelminthes, and Entoprocta, McGraw-Hill, New York.

Lorenzen, S. 1985. Phylogenetic aspects of pseudocoelomate evolution. *In* S. C. Morris et al. (eds.), *The Origins and Relationships of Lower Invertebrates*. Syst. Assoc. Spec. Vol. No. 28, Oxford, pp. 210–223.

Malakhov, V. V. 1980. Cephalorhyncha, a new type of animal kingdom uniting Priapulida, Kinorhyncha, Gordiacea, and a system of aschelminthes worms. Zool. Zh. 54(4):481–499. [In Russian.]

Marquardt, W. C. and R. S. Demaree Jr. 1985. *Parasitology*. Macmillan, New York.

Morris, C. S., J. D. George, R. Gibson and H. M. Platt (eds.). 1985. *The Origins and Relationships of Lower Invertebrates*. Systematics Association Special Volume No. 28, Oxford.

Wilson, R. A. and L. A. Webster. 1974. Protonephridia. Biol. Rev. 49:127–160.

Rotifera

Aloia, R. and R. Moretti. 1973. Mating behavior and ultrastructural aspects of copulation in the rotifer *Asplanchna brightwelli*. Trans. Am. Microsc. Soc. 92(3):371–380.

Birky, W. 1971. Parthenogenesis in rotifers: The control of sexual and asexual reproduction. Am. Zool. 11:245–266.

Clement, P. 1985. The relationships of rotifers. *In* S. C. Morris et al. (eds.), *The Origins and Relationships of Lower Invertebrates*. Syst. Assoc. Spec. Vol. No. 28, Oxford, pp. 224–247.

Donner, J. 1966. *Rotifers*. Frederick Warne & Co., New York.

Gilbert, J. J. 1974. Dormancy in rotifers. Trans. Am. Microsc. Soc. 93:490–515.

Gilbert, J. J. 1980a. Some effects of diet on the biology of the rotifers *Asplanchna* and *Brachionus*. *In* D. C. Smith and Y. Tiffon (eds.), *Nutrition in the Lower Metazoa*. Pergamon Press, New York, pp. 57–72.

Gilbert, J. J. 1980b. Developmental polymorphism in the rotifer *Asplanchna sieboldi*. Am. Sci. 68:636–646.

Koehler, J. K. 1966. Some comparative fine structure relationships of the rotifer integument. J. Exp. Zool. 162:231–244.

Nogrady, T. 1982. Rotifera. *In* S. Parker (ed.), *Synopsis and Classification of Living Organisms*, Vol. 1. McGraw-Hill, New York, pp. 866–872.

Pray, F. A. 1965. Studies on the early development of the rotifer *Monostyla cornuta* Müller. Trans. Am. Microsc. Soc. 84(2): 210–216.

Tannreuther, G. 1919. The development of *Asplanchna ebbesbornii* (Rotifer). J. Morphol. 33(2): 389–421.

Voigt, M. and W. Koste. 1978. *Rotatoria die Radertiere Mitteleuropas*, 2nd Ed. Borntraeger, Berlin.

Gastrotricha

Boaden, P. J. S. 1985. Why is a gastrotrich? *In* S. C. Morris et al. (eds.), *The Origins and Relationships of Lower Invertebrates*. Syst. Assoc. Spec. Vol. No. 28, Oxford, pp. 248–260.

D'Hondt, J. L. 1971. Gastrotricha. Oceanogr. Mar. Biol. Annu. Rev. 9: 141–192.

Hummon, W. 1971. Biogeography of sand beach Gastrotricha from the northeastern United States. Biol. Bull. 141(2): 390.

Hummon, W. 1982. Gastrotricha. *In* S. Parker (ed.), *Synopsis and Classification of Living Organisms*, Vol. 1. McGraw-Hill, New York, pp. 857–863.

Rieger, G. E. and R. M. Rieger. 1977. Comparative fine structure study of the gastrotrich cuticle and aspects of cuticle evolution within the Aschelminthes. Z. Zool. Syst. Evolutionsforsch. 15: 81–124.

Rieger, R. M. 1976. Monociliated epidermal cells in Gastrotricha: Significance for concepts of early metazoan evolution. Z. Zool. System. Evolutionsforsch. 14(3): 198–226.

Ruppert, E. E. 1977. Zoogeography and speciation in marine Gastrotricha. Mikrofauna Meeresboden 61: 231–251.

Ruppert, E. E. 1978. The reproductive system of gastrotrichs. II. Insemination in *Macrodasys*: A unique mode of sperm transfer in metazoa. Zoomorphologie 89: 207–228.

Ruppert, E. E. 1978. The reproductive system of gastrotrichs. III. Genital organs of Thaumastodermatinae subfam. n. and Diplodasyniae subfam. n. with discussion of reproduction in Macrodasyida. Zool. Scr. 7: 93–114.

Ruppert, E. E. 1982. Comparative ultrastructure of the gastrotrich pharynx and the evolution of myoepithelial foreguts in aschelminthes. Zoomorphologie 99: 181–220.

Teuchert, G. 1973. Die Feinstruktur des Protonephridial systems von *Turbanella cornuta* Remane, einem marinen Gastrotrich der Ordung Macrodasyoidea. Z. Zellforsch. 136: 277–289.

Teuchert, G. 1977. The ultrastructure of the marine gastrotrich *Turbanella cornuta* Remane (Macrodasyoidea). Zoomorphologie 88: 189–246.

Weiss, M. J. and D. P. Levy. 1979. Sperm in "parthenogenetic" freshwater gastrotrichs. Science 205: 302–393.

Kinorhyncha

Higgins, R. P. 1961. Morphological, larval, and systematic studies of Kinorhyncha. Ph.D. dissertation, Duke University.

Higgins, R. P. 1971. A historical overview of kinorhynch research. *In* N. C. Hulings (ed.), *Proceedings of the First International Conference on Meiofauna*. Smithson. Contrib. Zool. 76: 25–31.

Higgins, R. P. 1982. Kinorhyncha. *In* S. Parker (ed.), *Synopsis and Classification of Living Organisms*, Vol. 1. McGraw-Hill, New York, pp. 874–877.

Kozloff, E. 1972. Some aspects of development in *Echinoderes* (Kinorhyncha). Trans. Am. Microsc. Soc. 91(2): 119–130.

Nyholm, K. G. 1947. Contributions to the knowledge of the post-embryonic development in Echinoderida Cyclorhagae. Zool. Didr. Uppsala 25: 423–428.

Zelinka, C. 1928. *Monographie der Echinodera*. Wilhelm Engleman, Leipzig.

Nematoda

Andrassy, I. 1976. *Evolution as a Basis for the Systematization of Nematodes*. Pitman, London.

Arvy, L. 1963. Données sur le parasitisme protélien de *Nectonema* (Nématomorphe), chez les Crustacés. Ann. Parasitol. Paris 38(6): 887–892.

Bird, A. F. 1971. *The Structure of Nematodes*. Academic Press, New York.

Boveri, T. 1899. Die Entwicklung von *Ascaris* mit besonderer Rucksicht auf die Kernverhaltnisse. Festschr. 70.

Chitwood, B. G. and M. B. Chitwood. 1974. *Introduction to Nematology*. University Park Press, Baltimore.

Crofton, H. D. 1966. *Nematodes*. Hutchinson University Library, London.

Croll, N. A. 1970. *The Behavior of Nematodes*. Edward Arnold, London.

Croll, N. A. and B. E. Matthews. 1977. *Biology of Nematodes*. Wiley, New York.

Dailey, M. D., L. A. Jensen and B. W. Hill. 1981. Larval anisakine roundworms of marine fishes from southern and central California, with comments on public health significance. Calif. Fish & Game 67(4): 240–245.

Deutsch, A. 1978. Gut ultrastructure and digestive physiology of two marine nematodes, *Chromadorina germanica* (Butschli, 1874) and *Diplolaimella* sp. Biol. Bull. 155: 317–355.

Heip, C., M. Vinex and G. Vranken. 1985. The ecology of marine nematodes. Oceanogr. Mar. Biol. Annu. Rev. 23: 399–489.

Hope, W. D. and S. L. Gardiner. 1982. Fine structure of a proprioceptor in the body wall of the marine nematode *Donostoma californicum* Steiner and Albin, 1933 (Enoplida: Leptostomatidae). Cell Tissue Res. 225: 1–10.

Hope, W. D. and D. G. Murphy. 1972. A taxonomic hierarchy and checklist of the genera and higher taxa of marine nematodes. Smithson. Contrib. Zool. 137. [Includes an outstanding bibliography of papers dealing with marine nematodes.]

Lee, D. L. and H. J. Atkinson. 1977. *Physiology of Nematodes*, 2nd Ed. Columbia University Press, New York.

Levine, N. D. 1968. *Nematode Parasites of Domestic Animals and Man*. Burgess, Minneapolis.

Maggenti, A. R. 1982. Nemata. *In* S. Parker (ed.), *Synopsis and Classification of Living Organisms*. McGraw-Hill, New York, pp. 880–929.

Nicholas, W. L. 1975. *The Biology of Free-Living Nematodes*. Clarendon Press, Oxford.

Nickle, W. R. 1984. *Plant and Insect Nematodes*. Marcel Dekker, New York.

Poinar, G. O., Jr. 1983. *The Natural History of Nematodes*. Prentice-Hall, Englewood Cliffs, New Jersey.

Roggen, D. R., D. J. Raski and N. O. Jones. 1966. Cilia in nematode sensory organs. Science 152: 515–516.

Sassar, J. N. and W. R. Jenkins. 1960. *Nematology*. University of North Carolina Press, Chapel Hill.

Schaefer, C. 1971. Nematode radiation. Syst. Zool. 20: 77–78.

Somers, J. A., H. H. Shorey and L. K. Gastor. 1977. Sex pheromone communication in the nematode, *Rhabditis pellio*. J. Chem. Ecol. 3(4): 467–474.

Thorne, G. 1961. *Principles of Nematology*. McGraw-Hill, New York.

Wallace, H. R. 1970. The movement of nematodes. *In* A. M. Fallis (ed.), *Ecology and Physiology of Parasites*. University of Toronto Press, Toronto, pp. 201–212.

Yeats, G. W. 1971. Feeding types and feeding groups in plant and soil nematodes. Pedobiologia 11: 173–179.

Zuckerman, B. M., W. F. Mai and R. A. Rhode (eds.) 1971. *Plant Parasitic Nematodes*. Vol. 1, Morphology, Anatomy, Taxonomy, and Ecology. Vol. 2, Cytogenetics, Host-Parasite Interactions, and Physiology. Academic Press, New York.

Nematomorpha

Carvalho, J. C. M. 1942. Studies on some Gordiacea of North and South America. J. Parasitol. 28: 213–222.

Swanson, A. R. 1982. Nematomorpha. *In* S. Parker (ed.), *Synopsis and Classification of Living Organisms*, Vol. 1. McGraw-Hill, New York, pp. 931–932.

Priapulida

Calloway, C. B. 1975. Morphology of the introvert and associated structures of the priapulid *Tubiluchus corallicola* from Bermuda. Mar. Biol. 31:161–174.

Calloway, C. B. 1982. Priapulida. *In* S. Parker (ed.), *Synopsis and Classification of Living Organisms*, Vol. 1. McGraw-Hill, New York, pp. 941–944.

Dawydoff, C. 1959. Classes der Echiuriens et Priapuliens. *In* P. Grassé (ed.), *Traité de Zoologie*, Vol. 5, Pt. 1. Masson et Cie, Paris, pp. 855–926.

Hammon, R. A. 1970. The burrowing of *Priapulus caudatus*. J. Zool. 162:469–480.

Land, J. van der. 1968. A new aschelminth, probably related to the Priapulida. Zool. Meded. Rijksmus. Nat. Hist. Leiden 42(22):237–250.

Land, J. van der. 1970. Systematics, zoogeography, and ecology of the Priapulida. Zool. Verh. Rijksmus. Nat. Hist. Leiden 112:1–118.

Land, J. van der. 1982. A new species of *Tubiluchus* (Priapulida) from the Red Sea. Neth. J. Zool. 32:324–335.

Land, J. van der. 1985. Affinities and interphyletic relationships of the Priapulida. *In* S. C. Morris et al. (eds.), *The Origins and Relationships of Lower Invertebrates*. Syst. Assoc. Spec. Vol. No. 28, Oxford, pp. 261–273.

Lang, K. 1948. On the morphology of the larva of *Priapulus*. Arkiv. Zool. 41A, art. nos. 5 and 9.

Lang, K. 1953. Die Entwicklung des Eies von *Priapulus caudatus* Lam. und die systematische Stellung der Priapuliden. Ark. Zool. Ser. 2 5(5):321–348.

Morris, S. C. 1977. Fossil priapulid worms. Palaeontol. Assoc. Lond. Spec. Pap. Palaeontol. 20:1–95.

Por, F. D. and H. J. Bromley. 1974. Morphology and anatomy of *Maccabeus tentaculatus* (Priapulida: Seticoronaria). J. Zool. 173:173–197.

Shapeero, W. L. 1961. Phylogeny of Priapulida. Science 133(3455):879–880.

Shapeero, W. L. 1962. The epidermis and cuticle of *Priapulus caudatus* Lamarck. Trans. Am. Microsc. Soc. 81(4):352–355.

Acanthocephala

Amin, O. M. 1982. Acanthocephala. *In* S. Parker (ed.), *Synopsis and Classification of Living Organisms*. McGraw-Hill, New York, pp. 933–940.

Baer, J. C. 1961. Acanthocéphales. *In* P. Grassé (ed.), *Traité de Zoologie*, Vol. 4, Pt. 1. Masson et Cie, Paris, pp. 733–782.

Crompton, D. W. T. 1970. *An Ecological Approach to Acanthocephalan Physiology*. Cambridge University Press, New York.

Crompton, D. W. T. and B. B. Nickol. 1985. *Biology of the Acanthocephala*. Cambridge University Press, New York.

Nicholas, W. L. 1973. The biology of Acanthocephala. Adv. Parasitol. 11:671–706.

Whitfield, P. J. 1971. Phylogenetic affinities of Acanthocephala: An assessment of ultrastructural evidence. Parasitology 63:49–58.

Yamaguti, S. 1963. *Systema Helminthum*, Vol. 5, Acanthocephala. Interscience, New York.

Entoprocta

Nielsen, C. 1964. Studies on Danish Entoprocta. Ophelia 1(1):1–76.

Nielsen, C. 1971. Entoproct life cycles and the entoproct/ectoproct relationship. Ophelia 9(2):209–341.

Nielsen, C. 1977. The relationships of Entoprocta, Ectoprocta, and Phoronida. Am. Zool. 17:149–150.

Nielsen, C. 1982. Entoprocta. *In* S. Parker (ed.), *Synopsis and Classification of Living Organisms*, Vol. 1. McGraw-Hill, New York, pp. 771–772.

Nielsen, C. 1989. *Entoprocts. Keys and Notes for the Identification of the Species*. Synopses of the British Fauna. Academic Press, New York.

Gnathostomulida

Ax, P. 1956. Die Gnathostomulida, eine rätselhafte Wurmgruppe aus dem Meeressand. Abh. Akad. Wiss. Lit. Mainz Math. Naturwiss. Kl. 8:1–32.

Ax. P. 1965. Zur Morphologie und Systematik der Gnathostomulida. Untersuchungen an *Gnathostomula paradoxa* Ax. Z. Zool. Syst. Evolutionsforsch. 3:259–296.

Ax, P. 1985. The position of the Gnathostomulida and Platyhelminthes in the phylogenetic system of the Bilateria. *In* S. C. Morris et al. (eds.), *The Origins and Relationships of Lower Invertebrates*. Syst. Assoc. Spec. Vol. No. 28, Oxford, pp. 168–180.

Durden, C., J. Rodgers, E. Yochelson and R. Riedl. 1969. Gnathostomulida: Is there a fossil record? Science 164:855–856.

Kirsteuer, E. 1969. Gnathostomulida—a new component of the benthic invertebrate fauna of lagoons. Mem. Symp. Int. Lagunas Costeras UNAM UNESCO, pp. 537–544.

Knauss, E. B. 1979a. Indication of an anal pore in Gnathostomulida. Zool. Scr. 8:181–186.

Knauss, E. B. 1979b. Fine structure of the male reproductive system in two species of *Haplognathia* Sterrer (Gnathostomulida, Filospermoidea). Zoomorphologie 94:33–48.

Kristensen, R. M. and A. Nørrevang. 1977. On the fine structure of *Rastrognathia macrostoma* gen. et sp. n. placed in Rastro-

gnathiidae fam. n. (Gnathostomulida). Zool. Scr. 6:27–41.

Kristensen, R. M. and A. Nørrevang. 1978. On the fine structure of *Valvognathia pogonostoma* gen. et sp. n. (Gnathostomulida, Onychognathiidae) with special reference to the jaw apparatus. Zool. Scr. 7:179–186.

Mainitz, M. 1979. The fine structure of gnathostomulid reproductive organs I. New characters in the male copulatory organ of *Scleroperalia*. Zoomorphologie 92:241–272.

Riedl, R. J. 1969. Gnathostomulida from America. Science 163:445–452.

Riedl, R. J. 1971. On the genus *Gnathostomula* (Gnathostomulida). Int. Rev. Ges. Hydrobiol. 56:385–496.

Riedl, R. J. and R. Rieger. 1972. New characters observed on isolated jaws and basal plates of the family Gnathostomulidae (Gnathostomulida). Zool. Morphol. Tiere 72:131–172.

Rieger, R. M. and M. Mainitz. 1977. Comparative fine structure of the body wall in Gnathostomulida and their phylogenetic position between Platyhelminthes and Aschelminthes. Z. Zool. Syst. Evolutionsforsch. 15:9–35.

Sterrer, W. 1968. Beiträge zur Kenntnis der Gnathostomulida. I. Anatomie und Morphologie des Genus *Pterognathia* Sterrer. Ark. Zool. 22:1–125.

Sterrer, W. 1971. On the biology of Gnathostomulida. Vie Milieu Suppl. 22:493–508.

Sterrer, W. 1972. Systematics and evolution within the Gnathostomulida. Syst. Zool. 21(2):151–173.

Sterrer, W. 1982. Gnathostomulida. *In* S. Parker (ed.), *Synopsis and Classification of Living Organisms*. McGraw-Hill, New York, pp. 847–851.

Sterrer, W., M. Mainitz and R. M. Rieger. 1985. Gnathostomulida: Enigmatic as ever. *In* S. C. Morris et al. (eds.), *The Origins and Relationships of Lower Invertebrates*. Syst. Assoc. Spec. Vol. No. 28, Oxford, pp. 181–199.

Loricifera

Franzen, Å. and R. M. Kristensen. 1984. Loricifera, en nyupptäckt stam inom djurriket. Fauna Flora 79:56–60.

Higgins, R. P. and R. M. Kristensen. 1986. New Loricifera from southeastern United States coastal waters. Smithson. Contrib. Zool. 438:1–70.

Kristensen, R. M. 1983. Loricifera, a new phylum with Aschelminthes characters from the meiobenthos. Z. Zool. Syst. Evolutionsforsch. 21:163–180.

Lewin, R. 1983. New phylum discovered —named. Science 222:149.

Chapter Thirteen

Phylum Annelida: The Segmented Worms

The study of polychaetes used to be a leisurely occupation, practised calmly and slowly.

Kristian Fauchald

The Polychaete Worms, 1977

Introduction to the protostomes

Chapters 13 to 20 cover a vast assemblage of invertebrates comprising three very large phyla (Annelida, Arthropoda, Mollusca) and several small phyla. These groups are considered by many zoologists to constitute a clade whose members are known as the protostomes, or Protostomia. The monophyly of such a clade has been challenged on various grounds (e.g., Løvtrup 1975). However, the available evidence suggests to us that the challenge is not a strong one (see our analysis in Chapter 24), and we continue to recognize the basic protostome/deuterostome division among the coelomate metazoa, as discussed in Chapter 4.

It is appropriate at this point to remind you of some of the characteristics upon which the protostome lineage is based. While some secondarily derived modifications occur, protostomes generally display spiral cleavage with predictable cell fates (including the origin of mesoderm from a single mesentoblast—the 4d cell), the mouth arises from the blastopore, and they are schizocoelous. As we have seen, some of these traits are shared with other taxa, suggesting a relationship. For example, spiral cleavage, cell fates, and mouth origin in platyhelminths and nemerteans are fundamentally the same as in coelomate protostomes. Even in cnidarians, the mouth arises from the blastopore. Either these shared features are convergences or they represent symplesiomorphies that predate the origin of the coelomate

protostome clade—unless, as we have discussed in earlier chapters, the flatworms and nemerteans arose from true protostomes by loss or reduction of the coelom. Hence, these characters shed little light on the possible monophyly of Protostomia. Of particular importance, then, is the appearance of a particular unique synapomorphy in the protostomes: schizocoely. It was the evolution of this type of coelom formation that most clearly marks and defines the origin of the lineage that we recognize today as the protostomes. The great diversity within this lineage is reflected in various modifications on basic body plans, or *Baupläne*. In this and the next several chapters, we explore this diversity and establish a set of phylogenetic hypotheses that illustrate possible evolutionary pathways within the protostomes.

The annelids

The first phylum in our coverage of the protostomes is the Annelida (Greek, *annulatus*, "annulated," or "ringed"), which comprises the 15,000 or so species of segmented worms. This group includes the familiar earthworms and leeches as well as various marine "sand worms," "tube worms," and an array of other forms. Some are tiny animals of the meiofauna; others, such as certain southern hemisphere earthworms and some marine species, exceed 3 m in length. Three examples of annelids are depicted in Figure 1.

The annelids have successfully invaded virtually all habitats where sufficient water is available. They are particularly abundant in the sea but also abound in fresh water, and many live on land. There are also parasitic, mutualistic, and commensal species. Their success is in part a result of their highly adaptive and evolutionarily flexible *Bauplan* and of their exploita-

Figure 1

Representative annelids. A, *Nereis virens*, an errant polychaete. B, A pair of copulating earthworms (class Oligochaeta). C, A pond leech (class Hirudinida). (A photo by H. W. Pratt/BPS; B photo by R. K. Burnard/BPS; C photo by J. Soliday/BPS.)

A   B   C

tion of several distinct life history strategies. The basic annelid condition is characterized by a segmented body in which many parts are repeated in each of many segments, a situation referred to as **serial homology**. These triploblastic coelomate worms possess a complete gut, a closed circulatory system, a well developed nervous system, and excretory structures in the form of protonephridia or, more commonly, metanephridia (Box 1). Many marine forms produce a characteristic **trochophore larva** (Figure 34), a feature shared with several other protostome taxa (e.g., sipunculans, echiurans, and many molluscs). The story of annelid diversity and success is one of variation on this basic theme.

Taxonomic history and classification

As mentioned in earlier chapters, the roots of modern animal classification can be traced to Linnaeus (1758), who placed all invertebrates except insects in the taxon Vermes. In 1809, Lamarck established the taxon Annelida; he had a reasonably good idea of their unity and of their differences from other groups of worms. He and many other workers recognized especially the affinity between polychaetes and oligochaetes, but the hirudinidans (leeches) were often allied with the trematode platyhelminths. Cuvier (1816) united the annelids and the arthropods under the taxon Articulata, a scheme that remained popular into the twentieth century. The relationship of the leeches to the other annelids was not solidly

established until 1851, by Vogt. The phylum Annelida, as we know it today, has been recognized for about 80 years.

The phylum is usually divided into three or four classes. The class Polychaeta is the largest and most diverse group, with over 10,000 described species. The Myzostomida are sometimes viewed as aberrant, symbiotic polychaetes, although many authorities now view these animals as meriting separate class rank. The classes Oligochaeta and Hirudinida are sometimes united as a single class, the Clitellata, a concept with considerable merit (see phylogeny section).

PHYLUM ANNELIDA

CLASS POLYCHAETA: Sand worms, tube worms, clam worms, and others (Figure 2). With numerous setae (= chaetae) on the trunk segments; most with well developed parapodia; prostomium and peristomium usually bear sensory organs (palps, tentacles, cirri) or extensive feeding and gas exchange tentacular structures; foregut often modified as eversible stomodeal pharynx (proboscis), sometimes armed with chitinous jaws; reproductive structures simple, often transient; without a clitellum; most are gonochoristic; development often indirect, with a free-swimming trochophore larva; mostly marine; burrowers, errant, tube-dwelling, interstitial, or planktonic; some live in brackish water, a few inhabit fresh water or are parasitic. The class is divided into 25 orders and 87 families, some of which are listed below to illustrate the diversity within the class. The annotations are not diagnostic, merely descriptive synopses (see Fauchald 1977 and Pettibone 1982 for complete listings and diagnoses of all orders and families).

ORDER PHYLLODOCIDA

FAMILY PHYLLODOCIDAE: With thin, elongate bodies of up to 700 homonomous segments; most common as active epibenthic predators on solid substrata, a few burrow in mud. (e.g., *Eteone, Eulalia, Notophyllum, Phyllodoce*)

FAMILY ALCIOPIDAE: Body homonomous, but form varies from short and broad to long and slender; body transparent except for pigment spots in some genera; with pair of huge complex eyes on prostomium; planktonic, predaceous. (e.g., *Alciopa, Alciopina, Torrea, Vanadis*)

FAMILY TOMOPTERIDAE: Body flattened, with finlike parapodia; transparent; planktonic, swimming predators. Monogeneric: *Tomopteris.*

FAMILY GLYCERIDAE: Long, cylindrical, tapered, homonomous body; enormous pharynx armed with four hooked jaws used in prey capture; large pharyngeal proboscis also used in burrowing; most are infaunal burrowers in soft substrata. (e.g., *Glycera, Glycerella, Hemipodus*)

FAMILY SYLLIDAE: Mostly small, homonomous worms found on various substrata; active predators on small invertebrates, some eat diatoms; pharynx armed with a single tooth or a ring of small teeth for grasping prey; a few are interstitial. (e.g., *Autolytus, Brania, Odontosyllis, Syllis, Trypanosyllis*)

FAMILY NEREIDAE: Moderate to large polychaetes tending to homonomy; mostly errant predators with well developed parapodia; one pair of large, curved pharyngeal jaws; some burrow, but most are epibenthic in protected habitats: among mussel communities, in holdfasts of algae, crevices, under rocks, etc. (e.g., *Cheilonereis, Dendronereis, Neanthes, Nereis, Platynereis*)

FAMILY NEPHTYIDAE: Often large, with well developed parapodia; burrowers in marine sands and muds; eversible, jawed pharynx used in prey capture and burrowing. (e.g., *Aglaophamus, Micronephtyes, Nephtys*)

FAMILY APHRODITIDAE: Body broad, oval or oblong, with less than 60 segments; with flattened, solelike ventral surface, and rounded dorsum covered with scales (elytra) overlaid by a thick felt- or hairlike layer, giving some the common name of "sea mouse;" slow moving; epibenthic or burrowers; most are omnivorous. (e.g., *Aphrodita, Pontogeneia*)

FAMILY POLYNOIDAE: Most relatively short and somewhat flattened dorsoventrally; one Antarctic species, *Eulagisca gigantea*, reaches a length of nearly 20 cm and a width of about 10 cm; polynoids tend to have relatively few segments of a more-or-less fixed number; the dorsum is covered by scales (elytra), hence the common name "scale worms;" pharynx with one pair of jaws; well developed parapodia; errant but usually

cryptic (under stones, etc.) predators; several forms are commensalistic. (e.g., *Arctonoe, Halosydna, Harmothoe*)

ORDER EUNICIDA

FAMILY EUNICIDAE: Elongate, homonomous, generally large polychaetes, some exceeding 3 m in length; pharynx with complex set of jaw plates; some are sedentary in mucous or parchment-like tubes, many are gregarious in cracks and crevices in hard substrata; some leave their tube areas to feed, most are predatory carnivores, but many omnivorous species are known. (e.g., *Eunice, Marphysa, Palola*)

FAMILY LUMBRINERIDAE: Thin, elongate polychaetes without head appendages and with reduced parapodia; pharynx with complex jaw apparatus of several elements; crawl in algal mats, holdfasts, hydroids, and small cracks in hard substrata; some burrow in sand or mud; lumbrinerids include a variety of feeding types including carnivores, scavengers, detritivores, and deposit feeders. (e.g., *Lumbrinerides, Ninoe*)

FAMILY ARABELLIDAE: Body form similar to that of lumbrinerids; pharynx with complex jaw apparatus; most

---

Box One

Characteristics of the Phylum Annelida

1. Schizocoelous, bilaterally symmetrical, segmented worms
2. Development typically protostomous; segments arise by teloblastic growth
3. Digestive tract complete, usually with regional specialization
4. With a closed circulatory system
5. Nervous system well developed, with a dorsal cerebral ganglion, circumenteric connectives, and ventral ganglionated nerve cord(s)
6. Most possess metanephridia or, less commonly, protonephridia
7. With paired, segmentally arranged, epidermal setae bundles
8. Head composed of presegmental prostomium and peristomium
9. Gonochoristic or hermaphroditic; many have a characteristic trochophore larva (secondarily lost in some groups)
10. Marine, terrestrial, and freshwater species exist

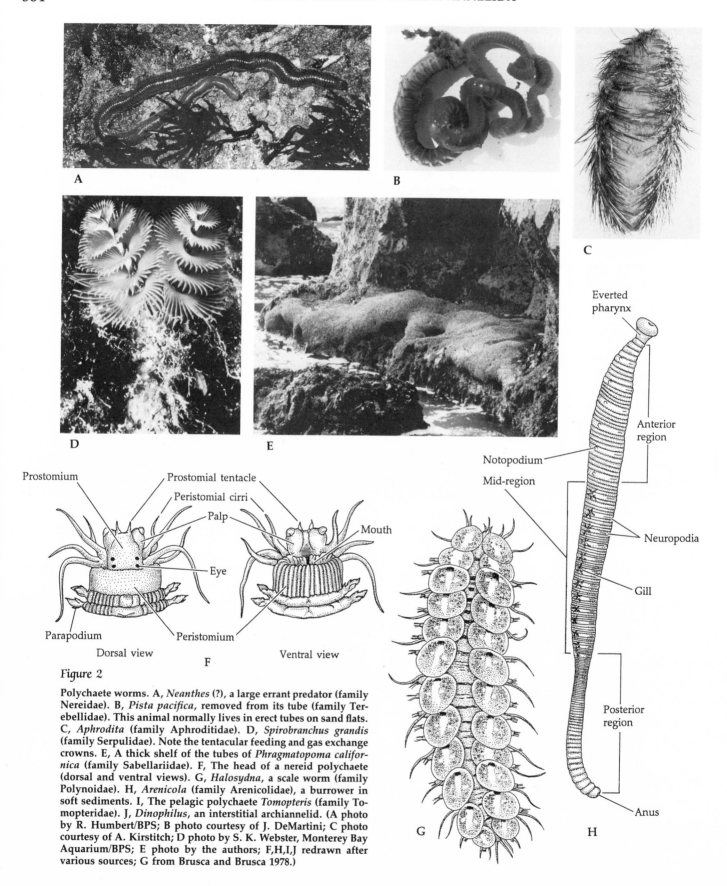

Figure 2

Polychaete worms. A, *Neanthes* (?), a large errant predator (family Nereidae). B, *Pista pacifica*, removed from its tube (family Terebellidae). This animal normally lives in erect tubes on sand flats. C, *Aphrodita* (family Aphroditidae). D, *Spirobranchus grandis* (family Serpulidae). Note the tentacular feeding and gas exchange crowns. E, A thick shelf of the tubes of *Phragmatopoma californica* (family Sabellariidae). F, The head of a nereid polychaete (dorsal and ventral views). G, *Halosydna*, a scale worm (family Polynoidae). H, *Arenicola* (family Arenicolidae), a burrower in soft sediments. I, The pelagic polychaete *Tomopteris* (family Tomopteridae). J, *Dinophilus*, an interstitial archiannelid. (A photo by R. Humbert/BPS; B photo courtesy of J. DeMartini; C photo courtesy of A. Kirstitch; D photo by S. K. Webster, Monterey Bay Aquarium/BPS; E photo by the authors; F,H,I,J redrawn after various sources; G from Brusca and Brusca 1978.)

burrow in soft substrata aided by secretion of copious amounts of mucus; predatory carnivores; some are endoparasitic in various worms, including other polychaetes. (e.g., *Arabella*, *Drilonereis*)

ORDER SPIONIDA

FAMILY SPIONIDAE: Body thin, elongate, homonomous; peristomial palps long and coiled; pharynx unarmed; most burrow, or form delicate sand or mud tubes; a few bore into calcareous substrata, including mollusc shells; most use the grooved peristomial palps to selectively extract food from the sediment surface. (e.g., *Polydora*, *Scolelepis*, *Spio*, *Spiophanes*)

ORDER CHAETOPTERIDA

FAMILY CHAETOPTERIDAE: Body fleshy, relatively large and distinctly heteronomous, divided into two or three functional regions with modified parapodia; chaetopterids live in more-or-less permanent U-shaped burrows lined with secretions from the worm; most are mucous-net filter feeders, eating plankton and detritus passed through the tube by water currents. (e.g., *Chaetopterus*, *Mesochaetopterus*, *Phyllochaetopterus*)

ORDER CIRRATULIDA

FAMILY CIRRATULIDAE: Elongate, relatively homonomous polychaetes with up to 350 segments, each with a pair of threadlike branchial filaments; pharynx unarmed and noneversible; cirratulids are mostly shallow-water burrowers lying just beneath the surface of the sediment, from where they extend their branchiae into the overlying water; most are selective deposit feeders, extracting organic detritus from the surface sediments. (e.g., *Cirratulus*, *Cirriformia*, *Dodecaceria*)

ORDER OPHELIIDA

FAMILY OPHELIIDAE: Homonomous polychaetes with up to 60 segments; general body shape varies among genera from rather short and thick to elongate and somewhat tapered; most opheliids burrow in soft substrata, but many swim by undulatory body movements; pharynx unarmed; most are direct deposit feeders. (e.g., *Armandia*, *Euzonus*, *Ophelia*, *Polyophthalmus*)

ORDER CAPITELLIDA

FAMILY MALDANIDAE: Body elongate and homonomous except that some mid-trunk segments are elongate, hence the common name of "bamboo worms"; burrow head downward and secrete a mucous sheath to which sand particles adhere, thereby forming a tube; proboscis unarmed but eversible and used in burrowing and selective deposit feeding. (e.g., *Clymenella*, *Maldane*, *Praxillella*)

FAMILY ARENICOLIDAE: The so-called lugworms have a rather thick, fleshy, heteronomous body divided into two or three distinguishable regions; pharynx unarmed but eversible and aids burrowing and feeding; arenicolids live in intertidal and subtidal sands and muds in J-shaped burrows; direct deposit feeders. (e.g., *Arbarenicola*, *Arenicola*)

ORDER TEREBELLIDA

FAMILY PECTINARIIDAE: Body short and conical, with only about 20 segments; live in conical sandy tubes open at both ends (the "ice-cream-cone worms") feed on detritus extracted from sediment. (e.g., *Amphictene*, *Pectinaria*, *Petta*)

FAMILY SABELLARIIDAE: Heteronomous tube dwellers; anterior setae modified as operculum; tubes of some may form extensive shelves or "reefs." (e.g., *Phragmatopoma*, *Sabellaria*)

FAMILY TEREBELLIDAE: Moderate-sized tube-dwelling polychaetes with fragile, fleshy bodies; heteronomous; body of two distinct regions; most lack eversible pharynx; most live in various types of permanent tubes (e.g., mud, sand, shell fragments); head bears numerous elongate feeding tentacles; most with 1–3 pairs of well developed branchiae on anterior trunk segments; feed on surface detritus. (e.g., *Amphitrite*, *Pista*, *Polycirrus*, *Terebella*)

ORDER SABELLIDA

FAMILY SABELLIDAE: Tube-dwelling polychaetes commonly called "fan worms" or "feather-duster worms"; body heteronomous, divided into two regions similar to those of terebellids; pharynx unarmed and noneversible; peristomium bears a classy crown of branched, feathery tentacles that projects from the tube and functions in gas exchange and ciliary suspension feeding. (e.g., *Eudistylia*, *Myxicola*, *Sabella*, *Schizobranchia*)

FAMILY SERPULIDAE: Heteronomous body divided into two regions; tube dwellers; anterior end bears a tentacular crown as in sabellids, plus a funnel-shaped

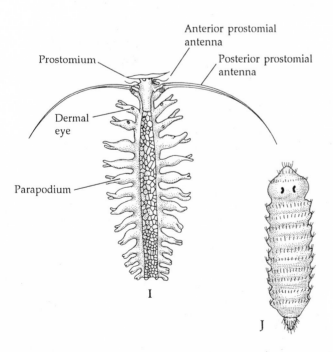

Prostomium

Dermal eye

Parapodium

Anterior prostomial antenna

Posterior prostomial antenna

I

J

operculum that can be pulled into the end of the calcareous tube when the worm withdraws; ciliary suspension feeders. (e.g., *Hydroides, Serpula, Spirobranchus*)

FAMILY SPIRORBIDAE: Small heteronomous polychaetes living in coiled calcareous tubes attached to hard substrata; tubes coil to the right or left, usually depending on species; anterior end with tentacular crown and operculum similar to those of serpulids. (e.g., *Circeis, Paralaeospira, Spirorbis*)

THE "ARCHIANNELIDS": Once considered a single valid taxon, the archiannelids are now divided into several orders (e.g., Nerillida, Dinophilida, Polygordiida, Protodrilida); these odd worms appear to be a polyphyletic assemblage of paedomorphic polychaetes; most are minute, often interstitial, and retain various larval features, including retention of external ciliary bands in some species. (e.g., *Dinophilus, Nerilla, Polygordius, Protodrilus, Saccocirrus*)

CLASS (?) MYZOSTOMIDA: (Figure 6) includes several families of flattened, oval, aberrant annelids, probably derived from polychaetes. The body is drastically modified for symbiotic life; with suckers and hooks; ecto- and endosymbionts of echinoderms, mainly crinoids. (e.g., *Asteriomyzostomum, Cystimyzostomum, Myzostoma, Myzostomum*)

CLASS OLIGOCHAETA: Earthworms and many freshwater worms (Figure 3). With few setae and no parapodia; cephalic sensory structures reduced; body externally homonomous except for clitellum; often with complex reproductive systems; hermaphroditic; often placed as a subclass of the class Clitellata; fresh water, terrestrial, some marine. This class comprises three orders based in part on details of the male reproductive system, the first two contain a single family each. There is some controversy about the taxonomic arrangement of oligochaetes (see comments in Brinkhurst 1982b).

ORDER LUMBRICULIDA (FAMILY LUMBRICULIDAE): Moderate-size, freshwater oligochaetes, many of which are known only from Lake Baikal in the Soviet Union. (e.g., *Lamprodilus, Stylodrilus, Styloscolex, Trichodrilus*)

ORDER MONILIGASTRIDA (FAMILY MONILIGASTRIDAE): Pre-

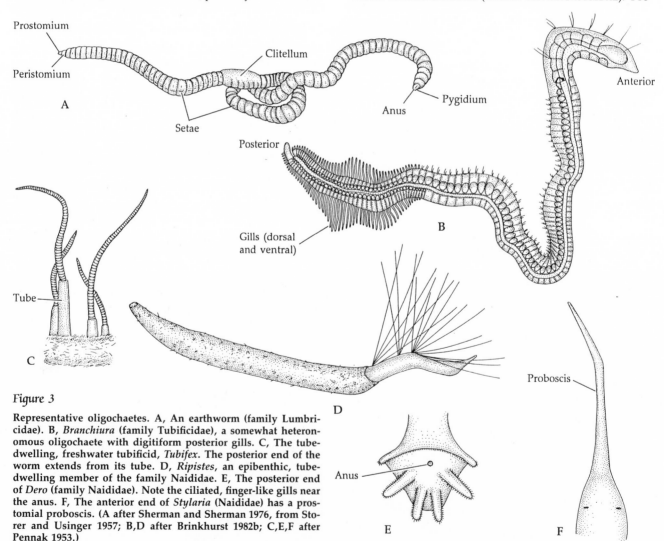

Figure 3

Representative oligochaetes. A, An earthworm (family Lumbricidae). B, *Branchiura* (family Tubificidae), a somewhat heteronomous oligochaete with digitiform posterior gills. C, The tube-dwelling, freshwater tubificid, *Tubifex*. The posterior end of the worm extends from its tube. D, *Ripistes*, an epibenthic, tube-dwelling member of the family Naididae. E, The posterior end of *Dero* (family Naididae). Note the ciliated, finger-like gills near the anus. F, The anterior end of *Stylaria* (Naididae) has a prostomial proboscis. (A after Sherman and Sherman 1976, from Storer and Usinger 1957; B,D after Brinkhurst 1982b; C,E,F after Pennak 1953.)

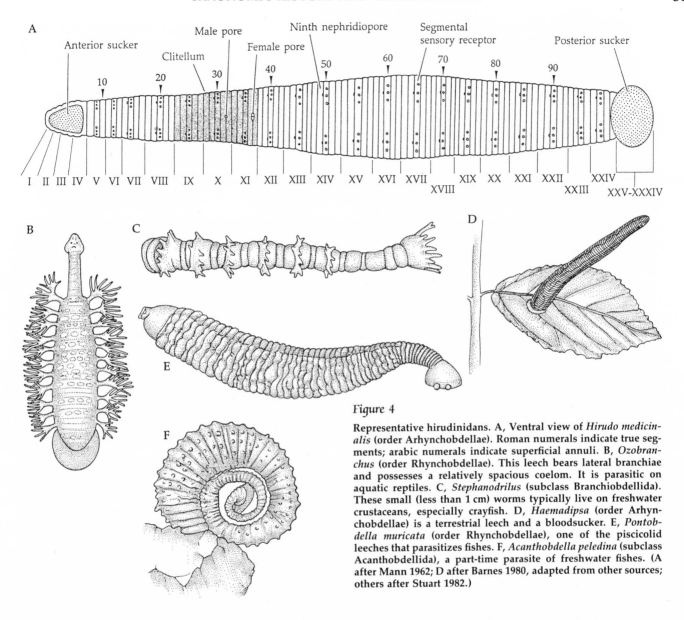

Figure 4

Representative hirudinidans. A, Ventral view of *Hirudo medicinalis* (order Arhynchobdellae). Roman numerals indicate true segments; arabic numerals indicate superficial annuli. B, *Ozobranchus* (order Rhynchobdellae). This leech bears lateral branchiae and possesses a relatively spacious coelom. It is parasitic on aquatic reptiles. C, *Stephanodrilus* (subclass Branchiobdellida). These small (less than 1 cm) worms typically live on freshwater crustaceans, especially crayfish. D, *Haemadipsa* (order Arhynchobdellae) is a terrestrial leech and a bloodsucker. E, *Pontobdella muricata* (order Rhynchobdellae), one of the piscicolid leeches that parasitizes fishes. F, *Acanthobdella peledina* (subclass Acanthobdellida), a part-time parasite of freshwater fishes. (A after Mann 1962; D after Barnes 1980, adapted from other sources; others after Stuart 1982.)

sumed primitive terrestrial oligochaetes (some workers consider these as a suborder of the order Haplotaxida; see Jamieson 1978); most known from damp soil, Asia; a few quite large, exceeding 1 m in length. (e.g., *Desmogaster, Moniligaster*)

ORDER HAPLOTAXIDA: With over 25 families, includes the vast majority of oligochaete species; all habitats; diverse body forms; a few families are listed here as examples.

FAMILY TUBIFICIDAE: So-called sludge worms or tubifex worms; up to 2 cm long; fresh water and marine; some are very common in areas of high pollution. (e.g., *Clitellio, Limnodrilus, Tubifex*)

FAMILY NAIDIDAE: Many freshwater species; some live in marine or brackish water; some parasitic forms; some build tubes; several species bear an elongate prostomial proboscis; a few possess gills; almost all reproduce asexually; but most possess gonads at some stage of development; these fully aquatic oligochaetes are known worldwide. (e.g., *Branchiodrilus, Dero, Ripistes, Slavina, Stylaria*)

FAMILY LUMBRICIDAE: Includes the various terrestrial earthworms; often relatively large; with well developed and complex reproductive systems; most are direct deposit feeders. (e.g., *Allolobophora, Diporodrilus, Eisenia, Lumbricus*)

CLASS HIRUDINIDA: Leeches (Figure 4). Body with fixed number of segments, each with superficial annuli; heteronomous, with clitellum and a posterior and usually anterior sucker; often placed as a subclass of the class Clitellata; complex reproductive systems, hermaphroditic; most are fresh water or marine, a few are semiterrestrial; ecto-parasitic, predaceous or scavenging. The class is presently recognized as comprising three subclasses described briefly below.

SUBCLASS ACANTHOBDELLIDA: With a single family and species (Acanthobdellidae, *Acanthobdella peledina*). To 3 cm long; found in cold, freshwater lakes; part of the animal's life is spent as an ectoparasite on freshwater fishes (notably trout, char, and grayling), and presum-ably the rest of the time is spent in vegetation; body with 30 segments, posterior sucker only; with paired setae on anterior segments; coelom partially reduced, but obvious and with intersegmental septa; considered to represent something of a pre-leech condition.

SUBCLASS BRANCHIOBDELLIDA: With a single family (Branchiobdellidae). Usually less than 1 cm long; ecto-commensal or ectoparasitic on freshwater crayfishes; body with 15 segments; with anterior and posterior suck-ers; setae absent; coelom partially reduced, but spacious throughout most of the body; ultrastructure studies sug-gest these animals may be intermediate between oligo-chaetes and "true" leeches (hirudineans); sometimes placed with the oligochaetes. (e.g., *Cambarincola*, *Ste-phanodrilus*)

SUBCLASS HIRUDINEA: The "true" leeches; with two or-ders and about 12 families. Most are marine or fresh water, a few are semiterrestrial or amphibious; many are ectoparasitic bloodsuckers, others are free-living preda-tors or scavengers; some parasitic forms serve as vectors for pathogenic protozoa, nematodes, and cestodes; body of 34 segments; with anterior and posterior suckers; no setae; coelom reduced to a complex series of channels (lacunae); two principal orders are the Rhynchobdellae and the Arhynchobdellae. (e.g., *Glossiphonia*, *Hirudo*, *Oz-obranchus*, *Pontobdella*)

The annelid *Bauplan*

The annelid body plan represents the classic ex-ample of the metameric, triploblastic, coelomate bi-lateria, and provides a good model for comparison with other protostomes. The elongate body is usually cylindrical, but it has become markedly flattened in some groups, notably the leeches. The head is com-posed of a **prostomium** and a **peristomium**, the latter bearing the mouth. These two regions are primitively presegmental and without setae, but some groups have secondarily incorporated true segments (in which case lateral setae may be present). A terminal, postsegmental part—the **pygidium**—bears the anus (Figures 2 and 3). The gut is separated from the body wall by the coelom, except in those species in which the body cavity is lost. The trunk segmentation is visible externally as rings, or **annuli**, and is reflected

internally by the serial arrangement of coelomic com-partments separated from one another by intersegmental septa (Figure 7). This basic arrangement has been modified to various degrees among the anne-lids, particularly by reduction in the size of the coe-lom or by loss of septa; the latter modification has led to fewer but larger internal compartments. The bodies of some annelids are **homonomous**, bearing segments that are very much alike. Many others, however, have groups of segments specialized for different functions and are thus **heteronomous** (Fig-ures 2, 3, and 4). Heteronomy is facilitated by the presence of serially repeated segments and has con-tributed greatly to the morphological diversification among annelids. The hydraulic properties of different coelomic arrangements have allowed corresponding modifications in patterns of locomotion, which are responsible in part for annelid success in a variety of habitats.

Segmentation is further reflected internally by the metameric arrangement of organs and system components (i.e., serial homology). In the primitive condition, each segment contains a portion of the circulatory, nervous, and excretory systems, in ad-dition to the coelomic compartments. This arrange-ment in the annelids reflects the principle of system compatibility, which we have stressed in various ways as being critical to the success of any *Bauplan*. That is to say, given the relative isolation of the body segments from one another by the septa, it is nec-essary to provide each individual segment with sys-tem components to adequately serve their structural and physiological needs. Thus, the origin of the seg-mented coelomic condition must have involved the coevolution of this serially homologous arrangement of other parts. In general we may view much of annelid success in terms of their evolutionary escape from the limitations of a nonsegmented coelomate *Bauplan*.

The combination of characteristics listed in Box 1, especially the coelom, segmentation, closed cir-culatory system, regionally specialized gut, and na-ture of the excretory structures, act in concert to al-leviate many of the constraints imposed by some other body plans. The annelids are not bound to the small size or the extremely flattened shapes dictated by the necessity for small diffusion distances. Actions of the body wall musculature do not interfere directly with the internal organs as they do in "solid" body constructions, and more active lifestyles are served by the metameric and more efficient systems sup-porting metabolic functions.

Class synopses and body forms

Polychaeta. The class Polychaeta presumably includes the most primitive members of the phylum, at least according to most authorities. Nearly all of these animals are marine, living in habitats ranging from the intertidal zone to extreme depths. But quite a few inhabit brackish or fresh water; at least two species live on land; and there are a number of symbiotic forms. They range in length from less than 1 mm for some interstitial species to over 3 m for some errant forms. Associated with their success is a great diversity of structural types, all evolved from a basic metameric coelomate *Bauplan*. The examples in Figure 2 serve to illustrate some of the variety among members of this class.

The body form of polychaetes often reflects their habits and habitat.* Active (errant) hunting predators (e.g., members of the families Phyllodocidae, Glyceridae, Syllidae, Nereidae) and some burrowing deposit feeders (e.g., members of the Lumbrineridae) are characterized by a more-or-less homonomous body construction. Less active (sedentary) polychaetes include various suspension-feeding tube dwellers (e.g., members of the Sabellidae, Serpulidae, Spirorbidae), those that inhabit permanent burrows (e.g., members of the Chaetopteridae), and certain groups of direct and indirect deposit feeders and detritus feeders (e.g., members of the Arenicolidae, Terebellidae). These sedentary polychaetes usually show some degree of heteronomy, with different body regions specialized for particular functions.

The myriad variations in body form among polychaetes can best be described relative to the basic annelid plan of head, segmented trunk, and pygidium. The head comprises the prostomium and peristomium. It often bears appendages in the form of prostomial **palps** and **antennae** (tentacles) and fleshy peristomial **cirri**, or it may be naked, like that of some infaunal burrowers. The nature of these head appendages varies greatly and often reveals clues as to the worms' habits. The trunk may be homonomous or variably heteronomous as noted above, and each segment typically bears a pair of unjointed appendages called **parapodia**, and bundles of **setae** (Figure 5). The parapodia are primitively biramous, with a dorsal **notopodium** and a ventral **neuropodium**. However, they have evolved a diversity of forms and serve a variety of functions (locomotion, gas exchange, protection, anchorage, creation of water currents). In heteronomous polychaetes the morphology of the parapodia varies in different body regions (Figure 2). Often, for example, some parapodia are modified as gills, others as locomotor structures, and others to assist in food gathering.

Polychaetes are gonochoristic and have simple reproductive structures, generally proliferating gametes from the peritoneum. Many produce free-swimming larvae, but there is a great variety of life history strategies in this class.

Myzostomida. All members of the aberrant group Myzostomida are adapted for symbiotic life (Figure 6). Most live on the arms of crinoids (sea lilies) or on other echinoderms and ingest suspended material picked up by their hosts. Some are truly parasitic and reside within the gut or body cavity of their hosts, apparently feeding on body fluids or gut contents.

The myzostomids are drastically modified from the basic annelid plan, with small flattened bodies and setal hooks on their parapodia for attachment to their hosts. They had long been treated as specialized polychaetes, but recently some annelid specialists have regarded them as a separate class.

Oligochaeta. The class Oligochaeta comprises well over 3,500 species of worms, the majority of which live in freshwater and terrestrial environments. Some 200 or so species of oligochaetes have successfully invaded brackish and marine waters, including both littoral and deep-sea benthic habitats. However, many new species are being described, suggesting that the number of kinds of marine oligochaetes is actually much greater than presently known. Most oligochaetes are burrowers, although some inhabit layers of epibenthic detritus or live among algal filaments; a few are tube dwellers, and some are parasitic. Oligochaetes range in length from less than 1 mm for some aquatic forms to over 3 m for certain Australian earthworms. They differ from polychaetes in a number of respects, many of which are related to their exploitation of land and freshwater environments and their burrowing habits. They lack parapodia, and generally bear fewer or less elaborate head appendages and fewer setae. Furthermore, they are hermaphroditic, and most possess relatively complicated reproductive systems. Oligochaetes typically copulate and produce brooded or encapsulated embryos that develop directly to juveniles. They are far less diverse than the polychaetes

---

*Older classification schemes recognized a division into two subclasses, the motile Errantia and the nonmotile Sedentaria. It is now agreed that the superficial similarities used to unite the members of these "subclasses" are the results of convergence; hence the terms no longer have taxonomic validity.

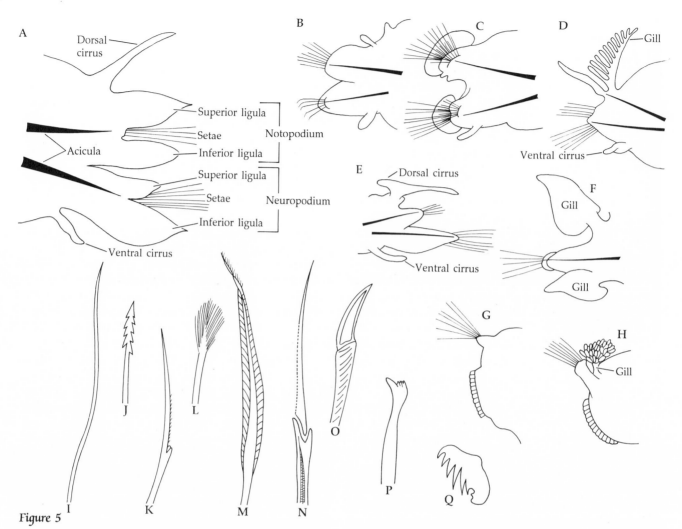

Figure 5

Parapodial and setal types among polychaetes. A, A stylized parapodium. B, The parapodium of a glycerid has reduced lobes. C, The parapodium of a nephtyid. D, The parapodium of a eunicid has a reduced notopodium. Note the presence of a dorsal filamentous gill. E, The parapodium of a polynoid has the dorsal cirrus modified as a scale, or elytron. F, The parapodium of a phyllodocid. Note that the noto- and neuropodia are modified as gill blades. G, The very reduced parapodium of a tube-dwelling sabellid. H, The parapodium of an arenicolid. I–Q, Setae from various polychaetes. The classification of setal types rivals that of sponge spicules in complexity and terminology. A few general types are distinguished here as simple setae (I–M), compound setae (N–O), hooks (P), and uncini (Q). (A–H redrawn from Meglitsch 1972; I–Q redrawn from Smith and Carlton 1975.)

in terms of variation on their basic body plan.

The typical oligochaete body is clearly built on the metameric annelid *Bauplan*. Externally, the body may be homonomous or may include some regional specialization, such as localized gills or variation in setal length (Figure 3). A few segments are modified as the **clitellum**, a secretory region that functions in reproduction. Setae are relatively few in number, oc-

curring in segmentally paired bundles ranging from 1 to as many as 25 setae per bundle. The setae of various species vary in length, shape, and thickness, some being short and quite stout, others long and thin. Most oligochaete setae are movable, a capability often employed in burrowing, as we describe later.

In spite of the apparent differences between oligochaetes and polychaetes, they are both clearly derived from a common stock and show numerous homologies of major body parts arranged according to the metameric plan. Albeit small, the oligochaete head is formed from a prostomium and peristomium, and often it incorporates some anterior body segments. The prostomium is usually very small; but in a few species it is elongated as a tentacle or proboscis (Figure 3F). Posterior to the metameric trunk is the pygidium, which bears the anus.

Hirudinida. The class Hirudinida includes the familiar leeches (subclass Hirudinea) and some less familiar but presumably related animals (subclasses

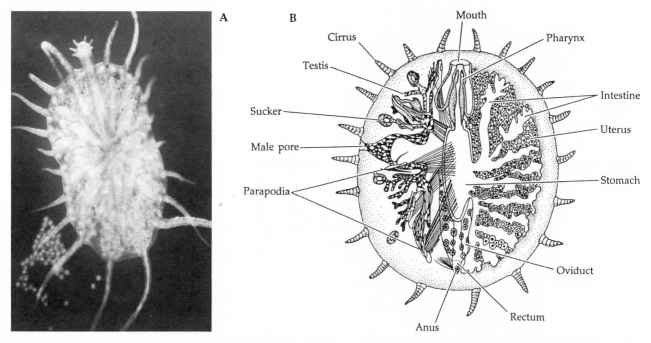

Figure 6

A, A myzostomid from a crinoid. Note the egg-filled body and the everted proboscis. B, *Myzostoma*. The parapodia are reduced to tiny suckers, often armed with setal hooks for attachment to their host. Their reproductive systems, however, are much more complex than those of polychaetes, a feature common to many parasitic animals. (A courtesy of D. Zmarzly; B redrawn from several sources.)

Branchiobdellida and Acanthobdellida). As mentioned in the classification summary, members of the last two subclasses possess characteristics that actually suggest an intermediate position between hirudineans and oligochaetes. Most of the 500 or so species that make up this class are hirudineans (true leeches), and it is to them that we devote most of our attention. The branchiobdellids and acanthobdellids contain but a single family and a single species, respectively, and their special features are mentioned briefly where appropriate.

Like the oligochaetes, the hirudinidans are clitellate annelids. They possess a fixed number of segments, which are traditionally numbered with Roman numerals. We caution you that there is some confusion about the number of true segments in these animals, partly because the embryogeny of the head region is not fully understood. The peristomium is obscure, if it is present at all, and the prostomium is typically fused with some anterior segments. By convention, when numbering "segments," no distinction is made between the presegmental and "truly" segmental portion of the body. Some attempts

have been made to resolve this problem by examining internal, serially repeated structures. For example, in most hirudineans there are 34 ganglia on the ventral nerve cord, but the first one apparently innervates the prostomium, suggesting that there are only 33 real segments. Counting the minute prostomium, the bodies of branchiobdellids are composed of 15 segments, of acanthobdellids 30 segments, and of hirudineans 34 segments. Furthermore, these segments are generally obscured by superficial annulations, giving the impression of many more segments (Figure 4). Externally, the leeches are characterized by anterior and posterior suckers and a clitellum; they lack setae or parapodia. Internally, the coelom is typically reduced to a series of interconnected channels and spaces, without serially arranged septa. Paired lateral setae are lacking on all but the Acanthobdellida.

We generally think of leeches as the large bloodsuckers popularized in horror stories and films. Many, however, are free-living predators, and some are scavengers. Hirudinidans range from less than 0.5 to about 25 cm in length. They occur in both fresh and salt water, and a few even live in moist terrestrial environments. Those that are full- or part-time parasites feed on the body fluids of a variety of vertebrate and invertebrate hosts. Some of these leeches serve as the intermediate hosts and vectors for certain protozoan, nematode, and cestode parasites.

Most leeches are flattened dorsoventrally. In the "true" leeches (Hirudinea), the body is divisible into

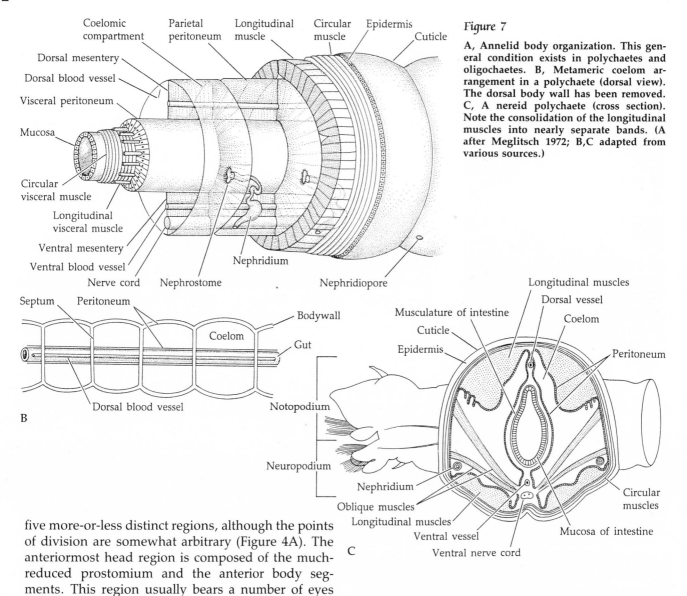

Figure 7

A, Annelid body organization. This general condition exists in polychaetes and oligochaetes. B, Metameric coelom arrangement in a polychaete (dorsal view). The dorsal body wall has been removed. C, A nereid polychaete (cross section). Note the consolidation of the longitudinal muscles into nearly separate bands. (A after Meglitsch 1972; B,C adapted from various sources.)

five more-or-less distinct regions, although the points of division are somewhat arbitrary (Figure 4A). The anteriormost head region is composed of the much-reduced prostomium and the anterior body segments. This region usually bears a number of eyes and a ventral mouth surrounded by the oral or anterior sucker. Segments V–VIII form the **preclitellar region**, followed by the **clitellar region** (segments IX–XI). The clitellum is only apparent during periods of reproductive activity. The **postclitellar** or midbody region comprises segments XII–XXVII. The **posterior region** of the body includes the ventrally directed posterior sucker formed from seven fused segments (XXVIII–XXXIV).

Other than the suckers, gonopores, and nephridiopores, leeches have few distinctive external features. In a few forms the body surface bears tubercles, and members of the family Ozobranchidae possess fingerlike or filamentous gills (Figure 4B). Setae are present on the first few segments of the acanthobdellid *Acanthobdella peledina*, but they do not occur in any other member of the class.

Reproductively, hirudinidans are similar to oligochaetes; they are hermaphroditic and possess complex reproductive systems. They generally copulate, and their embryos undergo direct development.

Body wall and coelomic arrangement

The polychaete body is covered by a thin cuticle of scleroprotein or mucopolysaccharide fibers deposited by epidermal microvilli. The epidermis is a columnar epithelium that is ciliated on certain parts of the body. Beneath the epidermis lies a layer of connective tissue, circular muscles, and thick longitudinal muscles, the latter often arranged as four bands (Figures 7A and C). The circular muscles do not form a fully continuous sheath, but are interrupted at least at the positions of the parapodia. As in all eucoelomates, the inner lining of the body wall is the peritoneum, which surrounds the coelomic spaces and lines the surfaces of internal organs.

The polychaete coelom is primitively arranged as laterally paired (i.e., right and left) spaces, serially arranged within the trunk (Figure 7). Dorsal and ventral mesenteries separate the members of each pair of coeloms, and muscular intersegmental septa isolate each pair from the next along the length of the body. In many polychaetes, the intersegmental septa have been secondarily lost or are perforated, so in these animals the coelomic fluid is continuous among segments. In many small polychaetes the coelomic lining is entirely lost. Such conditions radically alter the hydraulic qualities of the body; the significance of some of these differences is discussed in the next section.

In addition to the main body wall and septal muscles, a variety of other muscles function to retract protrusible and eversible body parts (e.g., branchiae, pharynx), and to operate the parapodia (Figure 7C). Each parapodium is an evagination of the body wall and contains a variety of muscles. Movable parapodia are operated primarily by sets of diagonal (oblique) muscles, which have their origin near the ventral body midline. These muscles branch and insert at various points inside the parapodium. Large parapodia typically contain a pair of chitinous and scleroprotein supporting rods called **acicula** (Figure 5A), on which some muscles insert and operate. The setae are also served by parapodial muscles and can usually be retracted and extended.

The body wall of oligochaetes (Figure 8) is constructed on the same plan as that described for polychaetes. A thin cuticle covers the epidermis, which is usually a columnar epithelium containing various mucus-secreting gland cells. Cilia occur on certain parts of the epidermis, for example, around the prostomium and gills of some small freshwater forms. Beneath the epidermis are the usual circular and longitudinal muscle layers, the latter being bounded internally by the peritoneum. Both muscle layers are usually quite thick and complex, especially in the

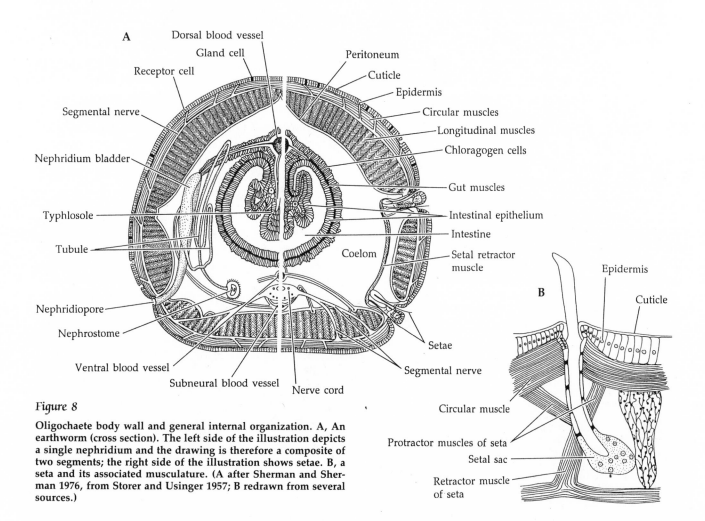

Figure 8

Oligochaete body wall and general internal organization. A, An earthworm (cross section). The left side of the illustration depicts a single nephridium and the drawing is therefore a composite of two segments; the right side of the illustration shows setae. B, a seta and its associated musculature. (A after Sherman and Sherman 1976, from Storer and Usinger 1957; B redrawn from several sources.)

larger terrestrial oligochaetes. The intersegmental septa are generally well developed and muscular. They are for the most part functionally complete, except at the anterior and posterior ends of the worm, and perforations in the septa are regulated by sphincter muscles. This effective isolation of coelomic compartments from one another plays a major role in oligochaete locomotion.

Many terrestrial oligochaetes bear pores connecting individual coelomic spaces to the outside. These pores are guarded by sphincter muscles and allow the escape of coelomic fluids onto the body surface. This controlled loss of body fluids is presumed to function in maintaining a moist film over the body to facilitate gas exchange and to prevent desiccation.

Parapodia are absent in oligochaetes, but setae do occur in nearly all forms. The setae are movable by various muscles and play a role in locomotion.

A cross-sectional view of the body wall of a leech is dramatically different from that of other annelids, in large part because of the presence of a thick dermal connective tissue layer beneath the epidermis and the reduction of the coelom (Figure 9). A thin cuticle covers a single layer of simple epidermal cells. The epidermis contains a number of mucous gland cells, some of which are quite large and extend well below the surface. The usual circular and longitudinal muscles are present, but they are more loosely organized than in polychaetes and oligochaetes. Distinct bands of dorsoventral muscles are also present, as well as diagonal (oblique) muscles between the circular and longitudinal layers. The dense dermis fills the areas between the muscle bands.

Reduction of the coelom is associated with proliferation of connective tissue deep beneath the body surface. Septate coelomic compartments are present only in *Acanthobdella* (in the first five segments) and in the midbody region of branchiobdellids. In all other members of the Hirudinida, the coelomic spaces are represented by channels and spaces that augment the circulatory system in the rhynchobdellid leeches and that completely replace the circulatory system in the arhynchobdellids (see section on circulation, below). In members of the arhynchobdellid family Hirudinidae, the coelomic channels are further reduced by the production of mesenchyme-like **botryoidal tissue** derived from the coelomic lining and proliferated outward around the channels.

Support and locomotion

Polychaeta. Polychaetes provide us with a classic example of the employment of coelomic spaces as a hydrostatic skeleton for body support. Coupled with the well developed musculature, the metameric body plan, and the parapodia, this hydrostatic quality pro-

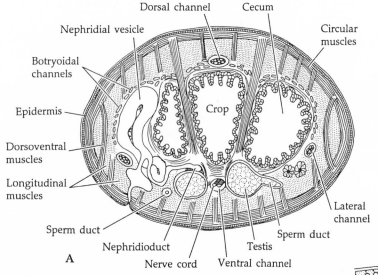

Figure 9

A, A leech, *Hirudo* (cross section), in which the original circulatory system has been lost and replaced by coelomic channels (see also Figure 23). B, The body wall of *Hirudo*. Note in both of these illustrations the effectively "acoelomate" body structure resulting from the reduction of the coelom. (A after Kaestner 1967; B after Mann 1962.)

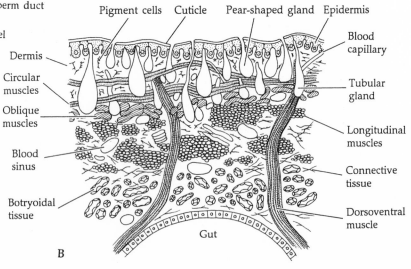

vides the basis for understanding locomotion in these worms. We begin a survey of locomotor patterns by examining the presumed primitive condition, like that in *Nereis*, which is an errant, homonomous, usually epibenthic polychaete (Figure 1A). Keep in mind that in such polychaetes the intersegmental septa are functionally complete, and thus the coelomic spaces in each segment can be effectively isolated hydraulically from each other. Modifications on this fundamental arrangement are discussed later.

In addition to burrowing, *Nereis* can engage in three basic epibenthic locomotor patterns: slow crawling, rapid crawling, and rather inefficient swimming (Figures 10B–D). All of these methods of movement depend primarily on the bands of longitudinal muscles, especially the larger dorsolateral bands, and on the parapodial muscles. The circular muscles are relatively thin and serve primarily to maintain adequate hydrostatic pressure within the coelomic compartments. Each method of locomotion in *Nereis* (and similar forms) involves the antagonistic action of the longitudinal muscles on opposite sides of the body in each segment. During movement, the longitudinal muscles on one side of any given segment alternately contract and relax (and are stretched) in opposing synchrony with the action of the muscles on the other side of the segment (Figure 10A). Thus, the body is thrown into undulations that move in metachronal waves from posterior to anterior. Variations in the length and amplitude of these metachronal waves combine with parapodial movements to produce the different patterns of locomotion. The parapodia and their setae are extended maximally in a power stroke as they pass along the crest of each metachronal wave. Conversely, the parapodia contract maximally in the wave troughs during their recovery stroke. Thus, the parapodia on opposite sides of any given segment are exactly out of phase with one another.

When *Nereis* is crawling slowly, the body is thrown into a high number of metachronal undulations of short wavelength and low amplitude (Figure 10B). The extended parapodial setae on the wave crests are pushed against the substratum and serve as pivot points as the parapodium engages in its power stroke. As the parapodium moves past the crest, it is retracted and lifted from the substratum during its recovery stroke. The main pushing force in this sort of movement is provided by the parapodial muscles.

During rapid crawling, much of the driving force is provided by the longitudinal body wall muscles in association with the longer wavelength and greater amplitude of the body undulations (Figure 10C),

which accentuates the power strokes of the parapodia.

Nereis can leave the substratum and engage in a rather inefficient swimming action. In swimming, the metachronal wavelength and amplitude are even greater than they are in rapid crawling (Figure 10D). When watching a nereid swim, however, one gets the impression that the "harder they try" the less progress is made, and there is some truth to this observation. The problem is that even though the parapodia act as paddles, pushing the animal forward on their power strokes, the large metachronal waves continue to move from posterior to anterior and actually create a water current in that same direction; this current tends to push the animal into reverse. The result is that *Nereis* is able to lift itself off the substratum, but then largely thrashes about in the water. This behavior is used primarily as a short-term mechanism to escape benthic predators rather than as a means to get from one place to another.

With these basic patterns and mechanisms in mind, consider a few more examples of locomotion methods in other polychaetes. *Nephtys* (family Nephtyidae) superficially resembles *Nereis*, but its methods of movement are significantly different. Although *Nephtys* is less efficient than *Nereis* at slow walking, it is a much better swimmer; it is also capable of effective burrowing in soft substrata. The large, fleshy parapodia serve as paddles, and, when swimming, *Nephtys* does not produce long, deep metachronal waves. Rather, the faster it swims, the shorter and shallower the waves become, thus eliminating much of the counterproductive force described for *Nereis*. When initiating burrowing, *Nephtys* swims head first into the substratum, anchors the body by extending the setae laterally from the buried segments, and then extends the proboscis deeper into the sand. A swimming motion is then employed to burrow deeper into the substratum.

In contrast to the above descriptions, various scale worms (family Polynoidae; Figure 2G) have capitalized on the use of their muscular parapodia as efficient walking devices. The body undulates little if at all, and there is a corresponding reduction in the size of the longitudinal muscle bands and their importance in locomotion. In fact, these worms depend almost entirely on the action of the parapodia for walking. Polynoids cannot swim.

Many of the really efficient burrowers have secondarily lost most of the intersegmental septa, or have septa that are perforated (e.g., *Arenicola*, *Polyphysia*). The loss of complete septa means that seg-

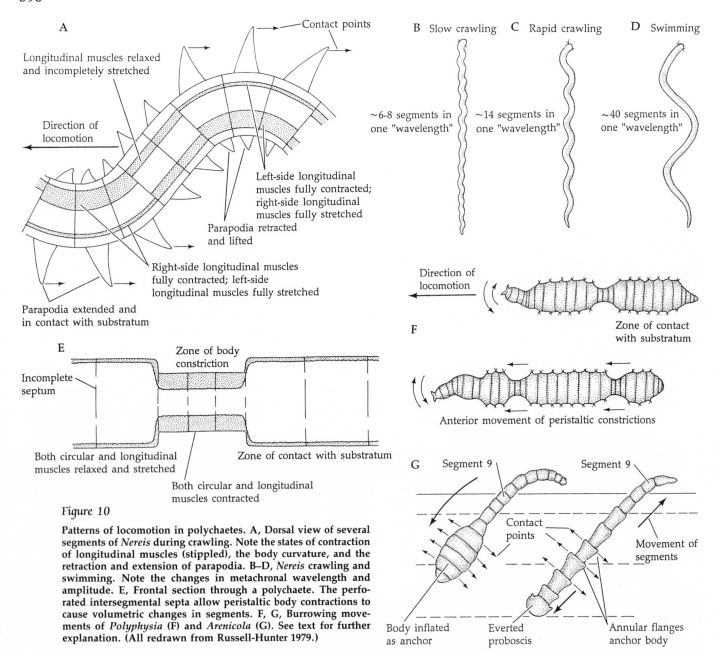

Figure 10

Patterns of locomotion in polychaetes. **A,** Dorsal view of several segments of *Nereis* during crawling. Note the states of contraction of longitudinal muscles (stippled), the body curvature, and the retraction and extension of parapodia. **B–D,** *Nereis* crawling and swimming. Note the changes in metachronal wavelength and amplitude. **E,** Frontal section through a polychaete. The perforated intersegmental septa allow peristaltic body contractions to cause volumetric changes in segments. **F, G,** Burrowing movements of *Polyphysia* (F) and *Arenicola* (G). See text for further explanation. (All redrawn from Russell-Hunter 1979.)

ments are not of constant volume; in other words, a loss of coelomic fluid from one body region causes a corresponding gain in another (Figure 10E). These polychaetes have reduced parapodia. The setae, or simply the surface of the expanded portions of the body, serve as anchor points, while the burrow wall provides an antagonistic force resisting the hydraulic pressure. In *Polyphysia*, peristaltic waves move constricted body regions forward while the anchored parts provide leverage (Figure 10F). The constricted areas are reduced both in diameter and in length by simultaneous contraction of both the circular and the longitudinal muscles.

Arenicola burrows by first embedding and anchoring the anterior body region in the substratum. The anchoring is accomplished by contracting the circular muscles of the posterior portion of the body, thus forcing coelomic fluid anteriorly and causing the first few segments to swell (Figure 10G). Then the posterior longitudinal muscles contract, thereby pulling the back of the worm forward. To continue the burrowing, a second phase of activity is undertaken. As the anterior circular muscles contract and the longitudinal bands relax, the posterior edges of each involved segment are protruded as anchor points to prevent backward movement; the proboscis is thrust forward, deepening the burrow. Then the proboscis is retracted, the front end of the body engorged with fluid, and the entire process repeated.

Other burrowing mechanisms are known among some polychaetes. For example, *Glycera* (family Glyceridae), a long, sleek worm, burrows rapidly by using its large muscular proboscis almost exclusively (Figure 14B). The proboscis is thrust into the substratum and swelled; then the body is drawn in by contraction of the proboscis muscles.

Most tube-dwelling polychaetes (Figure 11) are heteronomous and have rather soft bodies and relatively weak muscles. The parapodia are reduced, so the setae are used to position and anchor the animal within its tube. Movement within the tube is usually accomplished by slow peristaltic action of the body or by setal movements. When the anterior end is extended for feeding, it may be quickly withdrawn by special retractor muscles while the unexposed portion of the body is anchored in the tube.

Polychaete tubes provide protection as well as support for these soft-bodied worms, and also serve to keep the animal oriented properly in relation to the substratum. Some polychaetes build tubes composed entirely of their own secretions. Most notable among these tube builders are the serpulids and spirorbids, which construct their tubes of calcium carbonate secreted by a pair of large glands near a fold of the peristomium called the **collar** (Figure 11B). The crystals of calcium carbonate are added to an organic

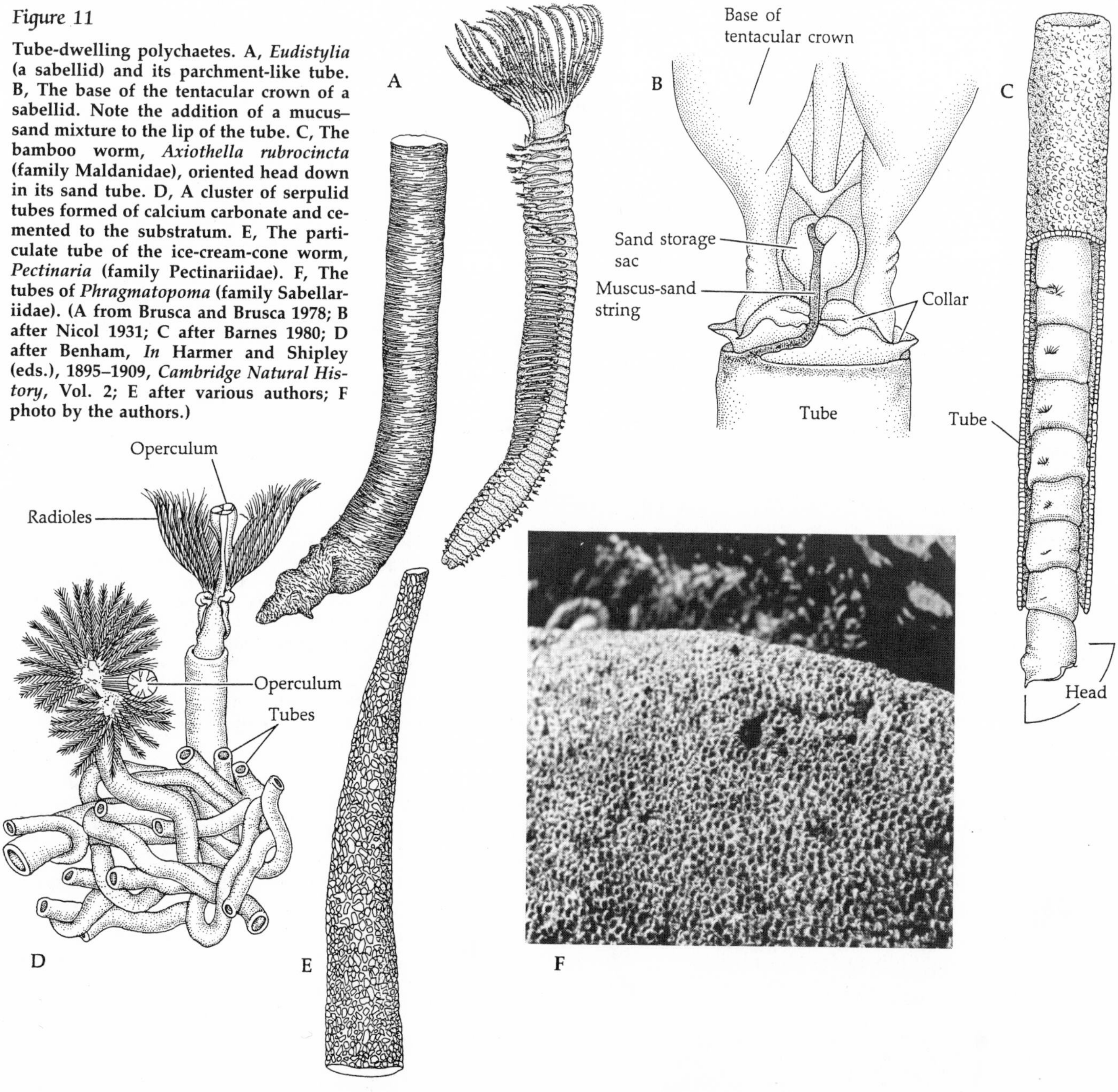

Figure 11

Tube-dwelling polychaetes. A, *Eudistylia* (a sabellid) and its parchment-like tube. B, The base of the tentacular crown of a sabellid. Note the addition of a mucus–sand mixture to the lip of the tube. C, The bamboo worm, *Axiothella rubrocincta* (family Maldanidae), oriented head down in its sand tube. D, A cluster of serpulid tubes formed of calcium carbonate and cemented to the substratum. E, The particulate tube of the ice-cream-cone worm, *Pectinaria* (family Pectinariidae). F, The tubes of *Phragmatopoma* (family Sabellariidae). (A from Brusca and Brusca 1978; B after Nicol 1931; C after Barnes 1980; D after Benham, *In* Harmer and Shipley (eds.), 1895–1909, *Cambridge Natural History*, Vol. 2; E after various authors; F photo by the authors.)

matrix; the mixture is molded to the top of the tube by the collar fold and held in place until it hardens.

Some sabellids produce parchment-like or membranous tubes of organic secretions molded by the collar. Others, such as *Sabella*, mix mucous secretions with size-selected particles extracted from feeding currents, and then lay down the tube with this material. Numerous other polychaete groups form similar tubes of sediment particles collected in various ways and cemented together with mucous secretions (Figure 11).

A few polychaetes are able to excavate burrows by boring into calcareous substrata, such as coral skeletons or mollusc shells (e.g., certain members of the families Eunicidae, Spionidae, Sabellidae). In extreme situations, the activity of the polychaetes may have deleterious effects on the "host." For example, species of *Polydora* (Spionidae) can cause serious damage to commercially raised oysters.

Many sedentary polychaetes use modifications of the basic locomotor actions described above to provide means of moving water through their tubes or burrows. Some of these modifications are discussed in the following section on feeding.

Oligochaeta. Oligochaetes rely heavily on their well developed hydrostatic skeleton both for support and locomotion. The action of the body wall muscles on the coelomic fluids provides the hydraulic changes associated with the typical pattern of oligochaete locomotion. In the absence of parapodial "paddles," oligochaetes depend on peristalsis and setal manipulation for burrowing, for moving through bottom debris, or for crawling over surfaces. There are notable differences between the mechanics of oligochaete and polychaete locomotion, even when comparing relatively similar movement patterns such as burrowing. The most important reason for these differences is that most burrowing polychaetes have lost intersegmental septa (or at least have evolved perforations in them), whereas most of the oligochaetes retain functionally complete septa. In oligochaetes each segment functions more-or-less independently of the others, and constricting one area of the body does not result in the flow of coelomic fluid to another area. Thus each segment of an oligochaete is of fixed volume, and a decrease in the diameter of a particular segment must be accompanied by an increase in its length, and vice versa (Figure 12).

Oligochaete burrowing involves the alternately contracting circular and longitudinal muscles within each segment. The shape of each segment changes from long and thin to short and thick with the re-

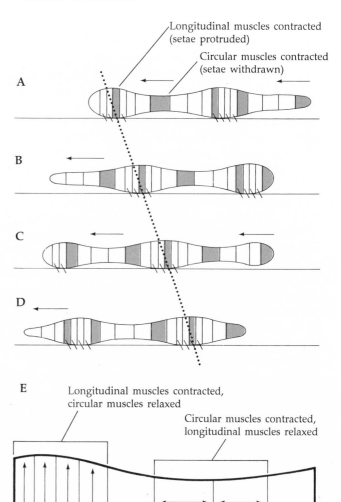

Figure 12

Earthworm locomotion. A–D, An earthworm moving to the left. Every fourth segment is darkened for reference. The dotted line passes through a posteriorly moving point of contact with the substratum. E, Several segments of an earthworm (sagittal section). Since each segment is a functionally isolated compartment, shortening and elongation accompany the contraction of longitudinal and circular muscles, respectively, while each segment essentially maintains a constant volume. See text for further explanation. (A–D after Russell-Hunter 1979, adapted from Gray and Lissmann 1938; E after Russell-Hunter 1979.)

spective muscle actions. These shape changes move along the body in a peristaltic wave generated by a sequence of impulses from the ventral nerve cord and associated motor neurons. So, at any moment during locomotion, the body of the worm appears as alternating thick and thin regions (Figure 12). Without some method of anchoring the body surface, this action would not produce any forward motion. The setae provide this anchorage as they are protruded like so many barbs from the thick portions of the body. When the longitudinal muscles relax and the circular muscles contract, the body diameter decreases and the setae are turned to point posteriorly and lie close to the body. As shown in Figure 12, as the anterior end of the body is extended by circular-muscle contraction, the setae prevent backsliding, the head is pressed into the substratum, and the worm advances. The anterior end then swells by contraction of the longitudinal muscles, and the rest of the body is pulled along.

There are some minor variations on this general scheme. For instance, when moving across relatively smooth surfaces, earthworms may employ the mouth as a sort of sucker. The mouth is pressed against the substratum and provides a temporary attachment point against which the muscles can operate in place of the usual setal anchorage. Also, giant neurons in many oligochaetes may be stimulated and cause the rapid contraction of longitudinal muscles in many segments, thereby eliciting rapid escape or withdrawal responses.

Hirudinida. Body support in leeches is provided by the more-or-less solid body construction, the fibrous connective tissue and included muscle bands, and the hydrostatic qualities of the coelomic channels. The absence of isolated, spacious, and segmentally arranged coelomic compartments precludes certain kinds of locomotion seen in many polychaetes and oligochaetes. Instead, what spaces remain in leeches are continuous, and as such these animals cannot move like a truly segmented worm. We may view the circular and longitudinal muscles as acting antagonistically against a functionally single internal space whose volume remains constant (as does the volume of the whole body).

Leeches do not burrow, but are mostly surface dwellers; thus they move over substrata rather than through them. Without setae or parapodia, the suckers serve as the points of contact with the substratum against which the muscle action can operate. Beginning with the posterior sucker attached, the circular muscles are contracted. Given the mechanics of the creature, the only possible result of this action is for the entire body to elongate as its diameter is reduced. Thus, the body is extended forward, and the anterior sucker is attached. Now the posterior sucker is released and the longitudinal muscles contract, shortening the body (and increasing its diameter) and drawing the posterior end forward. The whole busi-

ness is an "inchworm" movement, as depicted in Figure 13. Some leeches are also capable of swimming by dorsoventral body undulations; this behavior is an important mechanism for locating and contacting nonbenthic hosts.

Feeding

Polychaeta. The great diversity of form and function among polychaetes has allowed them to exploit nearly all marine food resources in one way or another. For convenience we have categorized polychaetes as raptorial, deposit, and suspension feeders. However, there are several feeding methods and dietary preferences within each of these basic designations. Following a discussion of selected examples of these feeding types, we mention a few of the symbiotic polychaetes.

The most familiar raptorial polychaetes are hunting predators (e.g., many phyllodocids, syllids, and nereids). These animals tend toward homonomy and are capable of rapid movement across the substratum. For the most part they feed on various small invertebrates. When prey is located by chemical or mechanical means, the worm everts its pharynx (proboscis) by quick contractions of the body wall muscles in the anterior segments, an action increasing the hydrostatic pressure in the coelomic spaces and causing eversion of the pharynx. As a result of the design of the pharynx, the jaws gape at the anteriormost

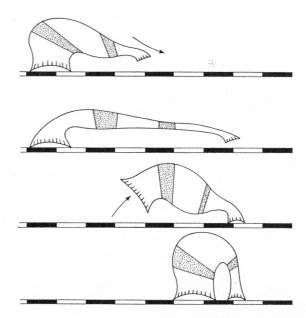

Figure 13

Locomotion in a leech, moving left to right, using the anterior and posterior suckers to progress in "inch worm" fashion. (After Russell-Hunter 1979, adapted from Gray and Lissmann 1938.)

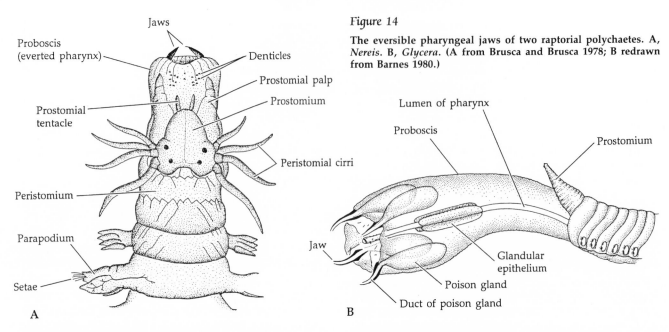

Figure 14

The eversible pharyngeal jaws of two raptorial polychaetes. A, *Nereis*. B, *Glycera*. (A from Brusca and Brusca 1978; B redrawn from Barnes 1980.)

end when the pharynx is everted (Figure 14A). Once the prey is grasped with the jaws, the coelomic pressure is released and the proboscis and captured prey are pulled into the body by large retractor muscles. Many of these raptorial feeders can also ingest plant material and detritus. Some scavenge, feeding on almost any dead organic matter.

Some predatory polychaetes do not actively hunt. Many scale worms (family Polynoidae) sit and wait for passing prey, then ambush it by sucking it into their mouth or grasping it with pharyngeal jaws.

Not all raptorial polychaetes are surface dwellers. Some live in tubes (*Diopatra*) or in complex branched burrows (*Glycera*). Such polychaetes detect the presence of potential prey outside their tubes or burrows by chemosensory or vibration-sensory means and then extend their everted proboscis to capture the prey. Some leave their residence to hunt for short periods of time. Certain forms (e.g., *Glycera*) have poison glands associated with the jaws (Figure 14B).

A number of polychaetes are direct deposit feeders, actually ingesting the substratum and digesting the organic matter contained therein (e.g., members of the families Arenicolidae, Opheliidae, and Maldanidae). The lugworms, such as *Arenicola*, excavate an L-shaped burrow, which they irrigate with water drawn into the open end by peristaltic movements of the worm's body (Figure 15A). The water percolates upward through the overlying sediment and tends to "liquefy" the sand at the blind end of the L, near the worm's mouth. This sand is ingested by the muscular action of a bulbous proboscis. The water brought into the burrow also adds suspended organic

material to the sand at the feeding site. The worm periodically moves to the open end of its tunnel and defecates excess ingested sand outside the burrow in characteristic castings.

Some maldanids live in straight vertical burrows, head down, and ingest the sand at the bottom. They periodically move upward (backward) to defecate. Some maldanids, and perhaps many other polychaetes, take in dissolved organic compounds, especially amino acids, as a significant part of their nutrient supply.

A number of other direct deposit feeders (some opheliids, for example) do not live in constructed burrows but move about through the substratum, ingesting the sediments. In high concentrations, some populations of these polychaetes pass thousands of tons of sediments through their guts each year, having a significant impact on the nature of the deposits in which they live.

Selective deposit feeders are defined by their ability to effectively sort the organic material from the sediment prior to ingestion (e.g., many members of the families Terebellidae, Spionidae, and Pectinariidae, among others). However, the methods used by polychaetes of these families to sort food differ significantly. Most terebellids (e.g., *Amphitrite*, *Pista*, *Terebella*) establish themselves vertically in the sediment, posterior end down, either in shallow burrows or permanent tubes (Figure 15B). The feeding tentacles are modified prostomial appendages that are extended over the substratum. These hollow tentacles are extended by ciliary crawling, and can be retracted by intrinsic muscles. Once extended, the tentacular

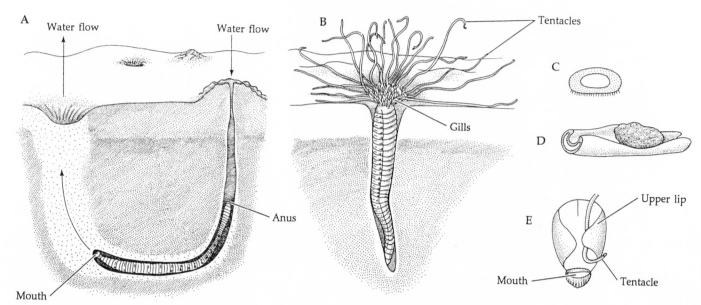

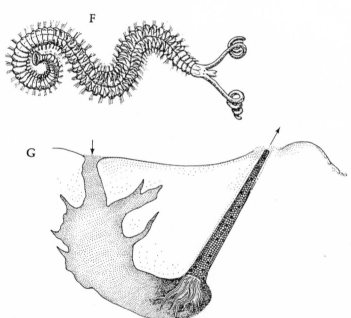

Figure 15

epithelium secretes a mucous coat to which organic material, sorted from the sediment, adheres. The tentacle rolls to form a longitudinal groove along which food and mucus are carried by cilia to the mouth (Figures 15C–E). Tube-dwelling spionids engage in a similar method of feeding. In these animals, the feeding structures are more muscular and are derived from the prostomial palps (Figure 15F). They are swept through the water or brushed through the surface sediments, extracting food and moving it to the mouth.

Pectinaria, the "ice-cream-cone-worm," lives in a tube constructed of sand grains and open at both ends. The animal orients itself head down, with the posterior end of the tube projecting to the sediment surface (Figure 15G). Head appendages partially sort the sediment, and a relatively high percentage of organic matter is ingested. A number of other polychaetes employ these and other methods of selective deposit feeding.

Various forms of suspension feeding are accomplished by many tube-dwelling polychaetes (e.g., members of the families Serpulidae and Sabellidae), and by some that live in relatively permanent burrows (e.g., Chaetopteridae) (Figures 2D; 11A,B,D; and 16). The feeding structures of *Sabella* and many related types are a crown of bipinnately branched peristomial tentacles called **radioles**. Some of these worms generate their own feeding currents, whereas others "fish" their tentacles in moving water. In any event, as food-laden water passes over the tentacles, the water is driven by cilia upward between the **pinnules** (branches) of the radioles (Figures 16A and B).

Deposit feeding polychaetes. A, *Arenicola*, a direct deposit feeder, in its burrow. Arrows indicate direction of the water flow; the substratum around the head is loosened and ingested by the worm (see text for additional explanation). B, A terebellid polychaete in its feeding posture within the substratum. The prostomial tentacles "creep" over the surface of the substratum by ciliary action and accumulate food, which is then passed to the mouth. C, A terebellid tentacle (cross section) has cilia on the underside. D, A section of the tentacle rolls to form a food groove. E, A tentacle is wiped across the oral area, where food is passed to the mouth. Such terebellids are indirect (selective) deposit feeders. F, A spionid, *Polydora*, another selective deposit feeder that uses its tentacle-like prostomial palps for feeding. G, The ice-cream-cone worm, *Pectinaria*, in feeding position. A water current is created (arrows), liquefying the sand around the tentacled head; organic material is removed and ingested. (A,B after Barnes 1980; C–E after Dales 1955; F after Smith and Carlton 1975; G after Kaestner 1967.)

Figure 16

Two strategies of suspension feeding in polychaetes. A–B, Suspension feeding by a sabellid polychaete. A, Tentacular crown extended from tube and water currents (arrows) passing between tentacles. B, A portion of a tentacle (radiole) in section. Various ciliary tracts remove particulate matter and direct it to the longitudinal groove on the radiole axis. Here, sorting by size occurs. The largest particles are mostly rejected, the smallest ones ingested; the medium-size particles are used in tube building. C, *Chaetopterus* in its U-shaped burrow. The ventral view shows details of the anterior end. A water current (arrows) is produced through the burrow by fan-shaped parapodia. Food is removed as the water passes through a secreted mucous bag. The bag is eventually passed to the mouth and ingested, food and all. See text for additional details. (A,B after Newell 1970; C after L. A. Borradaile et al. 1958, *The Invertebrata*, 3rd Ed., Cambridge University Press.)

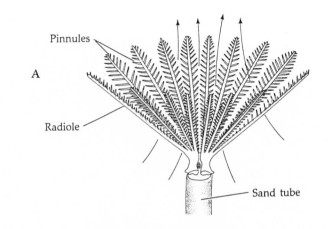

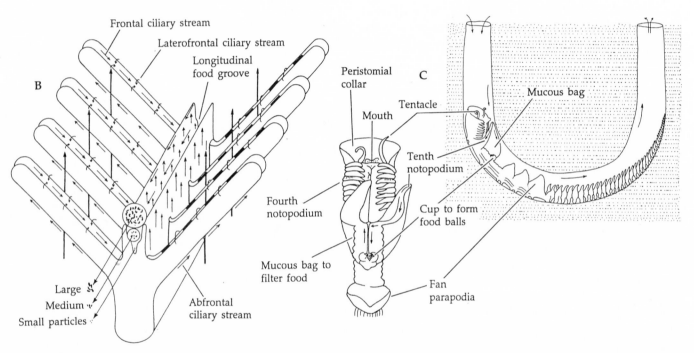

Eddies form on the medial side (inside) of the tentacular crown and between the pinnules, slowing the flow of water, decreasing its carrying capacity, and thus facilitating extraction of suspended particles. The particles are carried, with mucus, along a series of small ciliary tracts on the pinnules to a groove along the main axis of each radiole. This groove is widest at its opening and decreases in width in a stepwise fashion to a narrow slot deep in the groove. By this means, particles are mechanically sorted into three size categories as they are carried into the groove. Typically, the smallest particles are carried to the mouth and ingested, the largest particles are rejected, and the medium-sized ones are stored for use in tube building.

Polychaetes living in blind tubes must be able to discharge fecal material from the tube. Many sabellids bear an external ciliated groove that carries fecal pellets from the posterior anus to the anterior end of the worm where the waste is ejected over the lip of the tube.

Members of the family Chaetopteridae are among the most heteronomous of all polychaetes; the body is distinctly regionally specialized (Figure 16C). They are one of the few animals that truly filter water for food. These animals reside in U-shaped burrows through which they move water, extracting suspended materials. Each body region plays a particular role in this feeding process. For example, in *Chaetopterus*, segments 14–16 bear greatly enlarged notopodial fans that serve as paddles to create the water current through the burrow. These and a few other segments also bear suckers modified from the neuropodia, which help anchor the worm in position within the burrow. A mucous bag, produced by secretions from segment 12, is held as shown in Figure 16C, so water flows into the open end of the bag and through the mucous walls. Particles as small as 1 μm in diameter are captured by this structure, and there is some evidence that even protein molecules are held in the mucous net, probably by ionic charge attraction rather than mechanical filtering. During active feeding, the bag is rolled into a ball, passed to the mouth by a ciliary tract every 15–30 minutes or so, and ingested; then a new bag is produced.

Symbiotic relationships with other animals are the exception, not the rule among polychaetes. Nonetheless, there are some interesting cases that reflect, again, the adaptive diversity of these worms. Many symbiotic polychaetes are hardly modified from their free-living counterparts and do not usually show the drastic adaptive characteristics often associated with this sort of life. For many, the relationship with their host is a loose one, the polychaete often utilizing the host merely as a protective refuge. We have already mentioned certain boring polychaetes that form burrows in the shells of other invertebrates and are quite similar to their nonsymbiotic relatives. Among the most common commensalistic polychaetes are certain polynoid scale worms, especially members of the genera *Halosydna* and *Arctonoe* (Figure 17A). A polynoid has even been discovered living as a commensal in the mantle cavity of giant deep-sea mussels residing near thermal vents on the East Pacific Rise. Various other species reside in the arm grooves of sea stars, beneath the shell edge of large limpets, and in the gill chambers of certain chitons. One scale worm, *Hesperonoe adventor*, inhabits the burrows of the innkeeper worm, *Urechis caupo* (phylum Echiura, Chapter 14). There are many examples of these rather informal associations: certain syllids that live and feed on hydroids, a nereid (*Nereis fucata*) that resides in the shells of hermit crabs, and so on. Most of these animals do not feed upon their hosts, but prey upon tiny organisms that happen into their immediate environment. Others consume detritus or scraps from their host's meals.

A number of other odd associations are known among the polychaetes. Several species of arabellids live in the bodies of echiurans and of other polychaetes. Again, these endosymbionts show little structural modification associated with their lifestyles, other than a tendency for small body size and reduction in the pharyngeal jaws. A clear example of a parasitic polychaete is *Ichthyotomus sanguinarius*.

These small (1 cm long) worms attach to eels by a pair of stylets or jaws. The stylets are arranged so that when their associated muscles contract, the stylets fit together like the closed blades of a scissors. The stylets are thrust into the host, and when the muscles relax they open and anchor the parasite to the fish (Figure 17B). Species of *Polydora* often excavate galleries in various calcareous substrata (e.g., shells) and have been responsible for killing oysters in commercially harvested areas of Europe, Australia, and North America. The Pacific hydrocoral, *Allopora californica*, typically harbors colonies of *Polydora alloporis*, whose paired burrow openings are often mistaken for the hydrozoan's polyp cups.

A most unusual symbiotic relationship exists between the strange "Pompeii worm" (*Alvinella pompejana*) and a variety of marine chemoautotrophic sulfur bacteria. This polychaete is a member of the recently discovered deep-sea hydrothermal vent communities of the East Pacific Rise. It lives closer to the hot water extrusions than any other member of the vent community, building honeycomb-like structures called "snowballs" around the thermal plumes. Temperatures inside the "snowballs" reach an astonishing 250°C. The bodies of the Pompeii worms are covered with various unique vent bacteria. Evidence suggests that the worms continually transport these symbiotic bacteria to the mouth, by ciliary mucous tracts, for consumption.

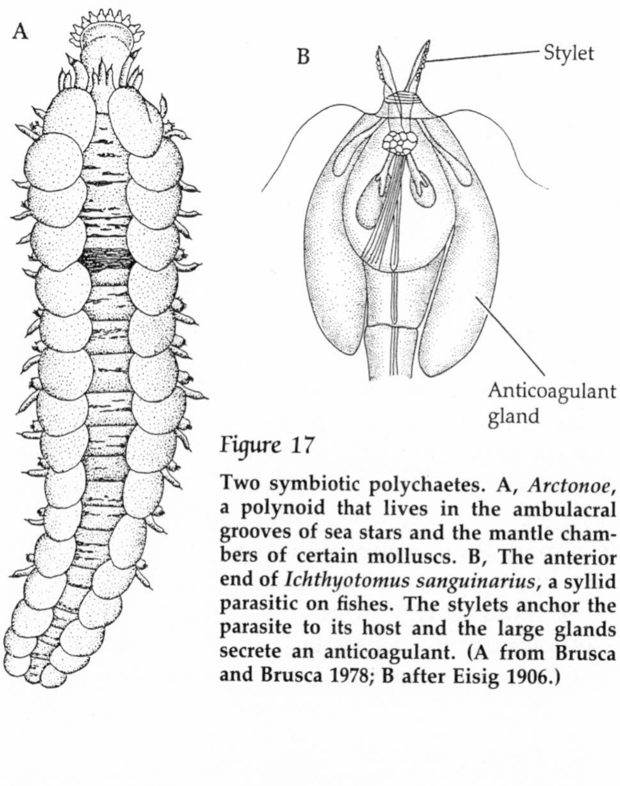

Figure 17

Two symbiotic polychaetes. A, *Arctonoe*, a polynoid that lives in the ambulacral grooves of sea stars and the mantle chambers of certain molluscs. B, The anterior end of *Ichthyotomus sanguinarius*, a syllid parasitic on fishes. The stylets anchor the parasite to its host and the large glands secrete an anticoagulant. (A from Brusca and Brusca 1978; B after Eisig 1906.)

Oligochaeta. Feeding strategies among the oligochaetes are less diverse than among the polychaetes. This difference is not surprising given the general absence of external body elaborations among oligochaetes. The lack of head appendages precludes various forms of tentacular feeding, and the absence of parapodia eliminates the ability to sit in one place and generate water currents from which food can be extracted. Rather, the oligochaetes show specialized modes of feeding that evolved in association with their particular habitats, and they do in fact exploit a variety of food resources. Most can be classified as predators, detritivores, or direct deposit feeders. Predation occurs in some of the freshwater oligochaetes, most of which capture prey by a sucking action of their muscular pharynx. They generally feed upon small invertebrates, such as other worms and tiny crustaceans. Many are able to evert the dorsal portion of the pharynx (as a proboscis), on which are located mucus-secreting glands; prey are stuck to the everted structure by the mucus and withdrawn into the gut with the retraction of the pharynx.

A variety of methods are employed by detritivorous oligochaetes. Many live in the surface layer of organic debris on the bottoms of ponds and streams, where they draw small particles of food into the gut by muscular or ciliary action of the foregut. Most such "detritivores" also ingest live microorganisms along with detrital material. One interesting feeding method is seen in the tube-dwelling members of the genus *Ripistes* (Figure 3D), a freshwater naidid. The long setae located on some of the anterior segments are waved about in the water and small detrital particles adhere to them. Food material is then ingested by wiping the setae across the mouth.

Most terrestrial and many aquatic oligochaetes are at least in part direct deposit feeders. Earthworms burrow through the soil, ingesting the substratum as they move. As the soil is passed along the digestive tract, the organic material is digested and absorbed from the gut. The inorganic, undigestible material passes out the anus. Earthworms are said to "work" the soil in this manner, loosening and aerating it. Many of these terrestrial burrowers, including the common earthworm *Lumbricus*, also retrieve organic material from the surface. These worms can burrow to the surface of the soil and there use their sucker-like mouth to obtain relatively large pieces of food (e.g., partially decomposed leaves), which they carry back underground for ingestion.

Several species of gutless marine oligochaetes have recently been described in shallow coral-sand habitats and in anaerobic, sulfide-rich subsurface sediments. These worms typically harbor subcuticular symbiotic bacteria, whose precise role in the host's nutritional regimen is not yet fully understood. The endosymbiotic bacteria may be very important to the worms; they are passed to the fertilized eggs during oviposition from storage areas next to the female's gonopore.

Hirudinida. Well over half the known species of the Hirudinida are ectoparasites that feed by sucking the blood or other body fluids from their hosts. Most of the remaining members of the class are predators on small invertebrates, and there are a few scavengers that feed on dead animal matter. Some families contain members adapted to a particular feeding mode, but more often feeding methods cut across taxonomic lines. Food-getting involves the structures of the foregut, which generally include either a protrusible pharyngeal proboscis or cutting structures in the form of slicing jaws or stylets. Unfortunately, little work has been done on the details of feeding in most hirudinidans.

The branchiobdellids are tiny worms that live on freshwater crustaceans, especially crayfish. The anterior end of the pharynx bears a pair of toothed jaws. These animals eat other epizoites living on the host, but they also feed on the host's eggs and body fluids. *Acanthobdella peledina*, the only known species of acanthobdellid, lives on the skin of freshwater fishes in cold, high-elevation lakes, particularly in northern Europe and Alaska. It apparently spends only about four months each year attached to its host; the rest of the time it is presumably free living.

The two orders of the subclass Hirudinea are distinguished from one another in part on the basis of the structure of the feeding apparatus. Members of the order Rhynchobdellae possess a pharyngeal proboscis but lack jaws, whereas members of the order Arhynchobdellae lack the proboscis, and all but a few possess jaws (Figure 20). Still, predators and parasites are known among both groups. The predatory forms either grasp their prey with the jaws or pierce them with the stiff proboscis. In either case, the prey is typically swallowed whole, but a few rhynchobdellids attack relatively large prey by piercing them with the proboscis and then sucking out the fluids and soft tissues by a pumping action of the pharynx.

Because of their medical importance, much work has been done on the parasitic leeches, especially those that affect livestock, game animals, or humans. Blood-sucking leeches are not especially host specific, and most do not remain attached to a single host for long periods of time; many of these leeches may feed by other means when not attached to a suitable host.

A few species of leeches feed exclusively on invertebrate hosts, including annelids (even other leeches), gastropods, and crustaceans, but the majority of them parasitize vertebrates. Some leeches are parasitic on members of particular higher taxa of vertebrates. For example, most of the Piscicolidae (order Rhynchobdellae) feed on the blood of fishes (including some deep-sea and hydrothermal vent fishes), whereas the Ozobranchidae (another family of Rhynchobdellae) seem to prefer aquatic reptiles such as turtles and crocodilians.

Of all the leeches, none has been more intensively studied than *Hirudo medicinalis*—the medicinal leech. The common name is derived from the rather gruesome practice of using these leeches to draw blood from humans afflicted with particular diseases for which such "bleeding" was once thought to be an effective treatment. Leeches produce a number of chemicals that are actively being studied for possible use in human medicine, including anticoagulants, anesthetics, and vasodilators. In some areas of the world, including the United States, leeches are still used medicinally, by being placed directly on surface hematomas to draw blood and reduce swelling. (Hopefully your personal physician no longer engages in this practice.) A powerful anticoagulant allows blood to continue draining even after the leech has been removed.

Hirudo and many other members of the family Hirudinidae are relatively common in tropical and temperate freshwater habitats. Most of them favor warm-blooded hosts and take their preferred blood meals from wading mammals (including humans when available; late-night TV buffs should take more serious note of *The African Queen* next time it is aired). These leeches will, however, feed on other vertebrates. When feeding, the leech anchors to the host by the suckers and presses the mouth against the surface of the host's body. Most of these leeches possess three bladelike jaws, each shaped somewhat like a half circle (Figure 20B). The jaws are set at roughly 120° angles to one another so that the cutting edges form a Y-shaped incision. Muscles rock the jaws to and fro, making slices in the host's skin. The leech produces an anesthetic and an anticoagulant (called **hirudin**) that are introduced into the wound. Blood is sucked from the host by the muscular pharynx. Whereas the predatory leeches feed frequently, the bloodsuckers probably take meals at widely spaced, very irregular intervals, depending on the availability of hosts. These long periods of fasting

apparently present no problems to these animals; some can survive well over a year without feeding. When they do feed, they gorge themselves with several times their own weight in blood. The digestive process is very slow.

Digestive system

Polychaeta. The gut of polychaetes is constructed on a basic annelid plan of foregut, midgut, and hindgut; some examples are shown in Figure 18. The foregut is a stomodeum and includes the buccal capsule or tube, the pharynx, and at least the anterior portion of the esophagus. It is lined with cuticle, and the teeth or jaws, when present, are derived from scleroprotein produced along this lining. The jaws are often hardened with calcium carbonate or metal compounds. When present, the eversible portion of this foregut is derived from the buccal tube or the pharynx. Various glands are often associated with the foregut, including poison glands (glycerids), esophageal glands (nereids and others), and mucus-producing glands in several groups.

The entodermally derived midgut generally comprises the posterior portion of the esophagus and a long straight intestine, the anterior end of which may be modified as a storage area, or stomach. The midgut may be relatively smooth, or its surface area may be increased by folds, coils, or many large evaginations (or ceca). The midgut is often histologically differentiated along its length. Typically, the anterior midgut (stomach or anterior intestine) contains secretory cells that produce digestive enzymes. The secretory midgut grades, either abruptly or gradually, to a more posterior absorptive region. Toward the posterior end of the gut, there are frequently additional secretory cells that produce mucus, which is added to the undigested material during the formation of fecal pellets. Food is moved along the midgut by cilia and by peristaltic action of intrinsic gut muscles, usually comprising both circular and longitudinal layers. A short proctodeal rectum connects the midgut to the anus, located on the pygidium.

There has been surprisingly little work done on the digestive physiology of polychaetes, but a variety of digestive enzymes are known from different species. Predators tend to produce proteases, herbivores largely carbohydrases. Some omnivorous forms (e.g., *Nereis virens*) produce a mixture of proteases, carbohydrases, lipases, and even cellulase. Digestion is predominantly extracellular in the midgut lumen, although intracellular digestion is known in some groups (e.g., *Arenicola*).

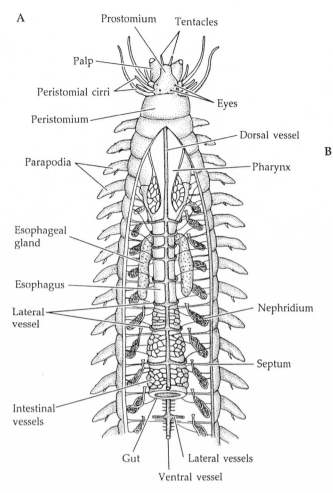

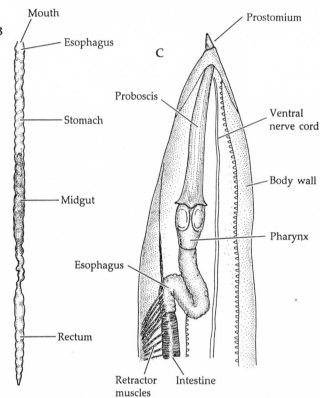

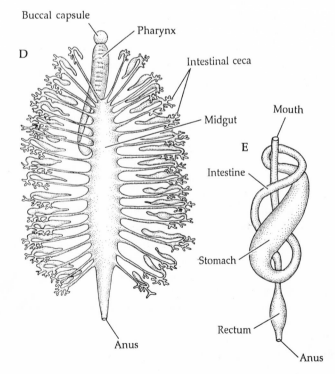

Figure 18

Polychaete digestive systems. A, A dissected nereid (dorsal view). Note the regional specialization of the anterior gut and other internal structures. B, The simple tubular gut of *Owenia*. C, A dissected *Glycera* (dorsal view). D, The multicecate gut of *Aphrodita*. E, The coiled digestive tract of *Petta*. (A after various sources; B redrawn from Barnes 1980, after Dales 1967; C redrawn from Barnes 1980; D,E redrawn from Meglitsch 1972.)

Oligochaeta. The oligochaete digestive system is basically a straight tube with various degrees of regional specialization, particularly toward the anterior end (Figure 19). In an earthworm, for example, the mouth leads inward to a stomodeal foregut comprised of a short buccal tube (or buccal cavity), muscular pharynx, and esophagus. The posterior esophagus often bears enlarged regions forming a **crop**, wherein food is stored, and one or more muscular **gizzards** lined with cuticle and used to mechanically grind ingested material. The esophagus of many oligochaetes also bears thickened portions of the wall in which are located lamellar evaginations lined with glandular tissue (Figure 19D). These **calciferous glands** remove calcium from ingested material. The excess calcium is precipitated by the glands as calcite and then released back into the gut lumen. Calcite is not absorbed by the intestinal wall and so passes out of the body via the anus. In addition, the calciferous

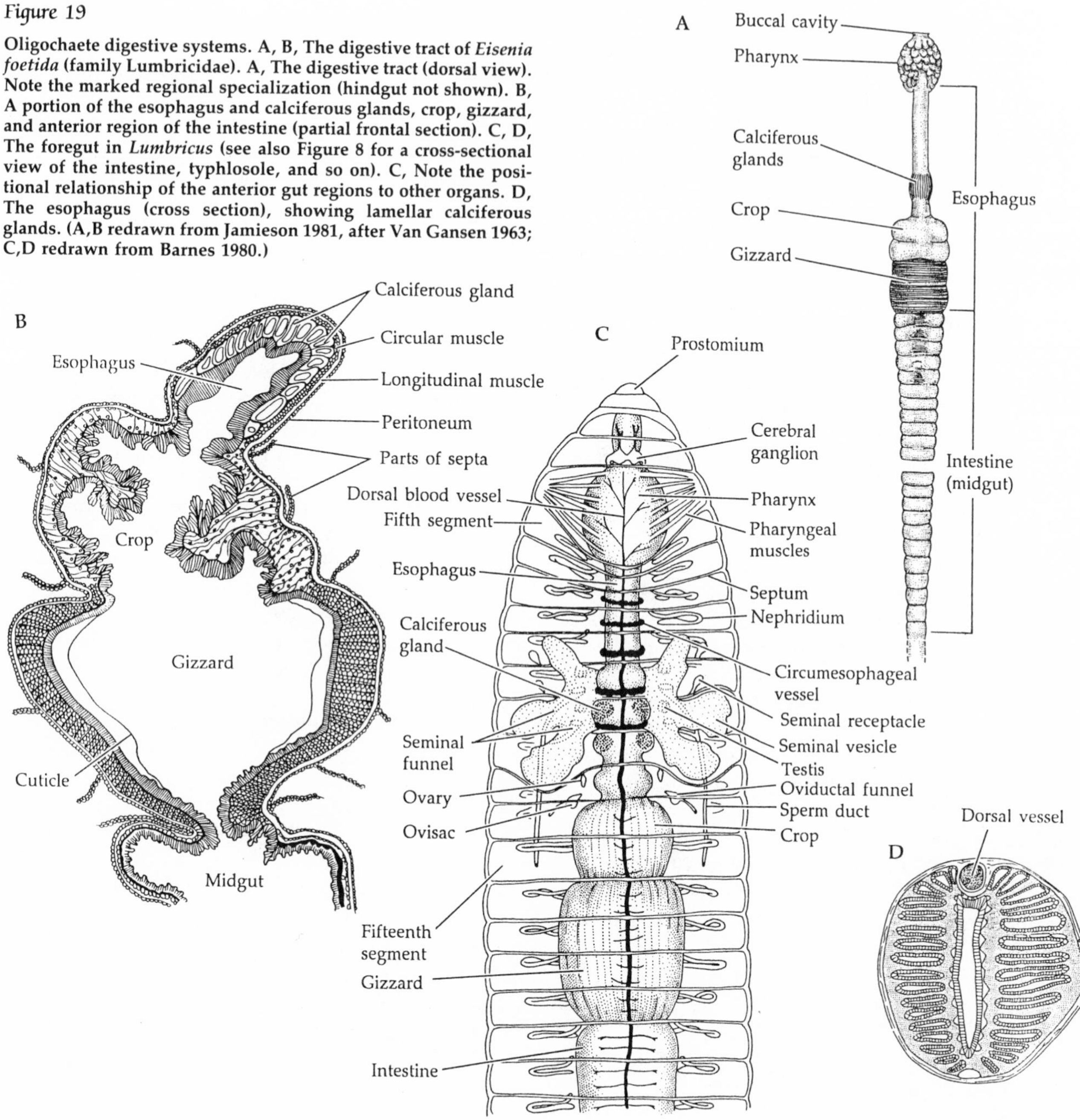

Figure 19

Oligochaete digestive systems. A, B, The digestive tract of *Eisenia foetida* (family Lumbricidae). A, The digestive tract (dorsal view). Note the marked regional specialization (hindgut not shown). B, A portion of the esophagus and calciferous glands, crop, gizzard, and anterior region of the intestine (partial frontal section). C, D, The foregut in *Lumbricus* (see also Figure 8 for a cross-sectional view of the intestine, typhlosole, and so on). C, Note the positional relationship of the anterior gut regions to other organs. D, The esophagus (cross section), showing lamellar calciferous glands. (A,B redrawn from Jamieson 1981, after Van Gansen 1963; C,D redrawn from Barnes 1980.)

glands apparently regulate the level of calcium ions and carbonate ions in the blood and coelomic fluids, thereby buffering the pH of those fluids.

The primary functions of the foregut are ingestion, transport, storage, and mechanical digestion of food. These processes are aided by lubricating mucus produced by glands of the pharynx. In a few oligochaetes, the pharyngeal glands are also thought to produce an amylase and a protease that initiate chemical breakdown.

The remainder of the digestive tube is dominated by a straight, entodermally derived midgut or intestine leading to a short proctodeal hindgut and anus located on the pygidium. The anterior midgut is predominantly secretory and produces a variety of digestive enzymes that are released into the gut lumen. Various authors have reported that carbohydrases, proteases, cellulase, and chitinase are produced from the midgut epithelium. Digestion is mostly extracellular. Much of the absorption of digested food occurs across the posterior half of the intestinal wall into the blood. Undigested materials are passed from the anus, often as characteristic castings or fecal pellets.

In terrestrial species the surface area of the intestine is relatively large, owing to a middorsal groove called the **typhlosole** (Figures 8A and 22B); in addition, the intestine of some oligochaetes bears segmentally arranged lateral diverticula. Food is moved through the digestive tube by peristaltic action of the muscular gut wall and by general body movements associated with locomotion.

Associated with the midgut of many oligochaetes, and some other annelids as well, are masses of pigmented cells called **chloragogen cells**. These modified peritoneal cells contain greenish, yellowish, or brownish globules that impart the characteristic coloration to the **chloragogenous tissue**. This tissue lies within the coelom, but is pressed tightly against the visceral peritoneum of the intestinal wall and typhlosole. This tissue serves as a site of intermediary metabolism (e.g., synthesis and storage of glycogen and lipids, deamination of proteins). It also plays a role in excretion, as we discuss below.

Hirudinida. The leech digestive tract is clearly derived from the basic annelid plan of stomodeal foregut, entodermal midgut, and short proctodeal hindgut (Figure 20). The foregut, as mentioned earlier, variously includes a mouth, jaws, buccal cavity, proboscis, pharynx, and esophagus. This region is lined with cuticle that provides stiffness to the proboscis and forms the jaws. The stomodeum also contains masses of unicellular **salivary glands** that se-

crete the anticoagulant hirudin in the jawed bloodsuckers and may produce enzymes to aid penetration of the proboscis in those parasitic forms that lack jaws.

Posterior to the esophagus is the enlarged midgut, usually called the stomach or crop. This region bears large ceca in most leeches, providing a large storage capacity as well as a high surface area (Figure 20C). In some kinds of leeches, the posterior midgut is structurally differentiated from the anterior portion. A short proctodeal rectum connects the midgut to the anus, located dorsally near the junction of the body and the posterior sucker.

Little is known about digestion in hirudinidans, except for some fragmentary information on bloodsucking leeches. Apparently, midgut enzymes are limited to exopeptidases. It is reasoned that this condition accounts for the extremely slow rate of digestion in these animals (a medicinal leech may take several months to digest the contents of a full blood meal!). Most leeches, including predatory and parasitic species, harbor a rich bacterial gut flora. These bacteria probably aid in the digestive events and may also provide metabolic products, such as vitamins, that are useful to their host.

Circulation and gas exchange

Polychaeta. You will recall the relationship between the presence of a complete, regionally spe-

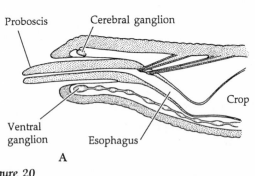

Proboscis Cerebral ganglion

Ventral ganglion Esophagus Crop

A

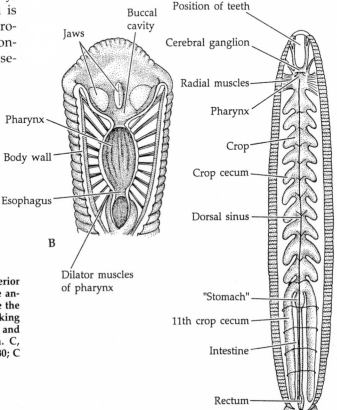

Jaws Buccal cavity Position of teeth

Cerebral ganglion

Pharynx Radial muscles

Body wall Pharynx

Esophagus Crop

Crop cecum

Dorsal sinus

B

Dilator muscles of pharynx

"Stomach"

11th crop cecum

Intestine

Rectum

C

Figure 20

Feeding structures and digestive tract of leeches. A, The anterior end (longitudinal section) of a rhynchobdellid leech. B, The anterior end of an arhynchobdellid leech (cutaway view). Note the arrangement of the jaws and the musculature of the sucking pharynx. The oral aperture is pressed against the host's skin, and the three jaws are rocked to and fro, slicing into the skin. C, Basic gut structure of *Hirudo*. (A,B redrawn from Barnes 1980; C after Mann 1962.)

cialized digestive tract combined with a circulatory system, as discussed in our coverage of the nemerteans (Chapter 11). The same principle applies to polychaetes, and to annelids in general, but takes on additional significance when viewed in association with the coelomate, segmented *Bauplan*. Given the relatively large size of many polychaetes, the compartmentalization of their coelomic chambers, and the fact that only certain portions of their gut serve to absorb digested food products, it is essential that a circulatory mechanism be present for internal transport and distribution. Furthermore, many polychaetes have their gas exchange structures limited to particular body regions; thus they depend on the circulatory system for internal transport of gases.

It is perhaps easiest to understand the circulatory system of polychaetes by considering it in concert with their gas exchange structures. In many polychaetes that lack appendages, the entire body surface functions in gas exchange (e.g., lumbrinerids, arabellids). Some of the active epibenthic forms utilize highly vascularized portions of the parapodia as "gills." Special gas exchange structures, or **branchiae**, are found in the form of trunk filaments (cirratulids), anterior gills (terebellids), and tentacular (branchial) crowns on the head (sabellids, serpulids, and spirorbids). Since the blood generally carries respiratory pigments, the anatomy of the circulatory system has evolved along with the form and location of these gas exchange structures.

We again begin our examination with a homonomous polychaete, such as *Nereis*, in which the parapodia are serially arranged and the notopodia function as gills. The major blood vessels of such an animal include a middorsal longitudinal vessel, which carries blood anteriorly, and a midventral vessel, which carries blood posteriorly. Exchange of blood between these vessels occurs through posterior and anterior vascular networks and serially arranged segmental vessels in the septa and around the gut (Figure 21A). Anterior vessel networks are especially well developed around the muscular pharynx and the region of the cerebral ganglion.

The movement of blood in *Nereis* depends on the action of the body wall muscles and on intrinsic muscles in the walls of the blood vessels, especially the major dorsal vessel. There are no special "hearts" or pumping organs. The blood passing through the various segmental vessels supplies the body wall muscles, gut, nephridia, and parapodia as illustrated in Figure 21A. Note that the oxygenated blood is being returned to the dorsal vessel, thus maintaining a primary supply of oxygen to the anterior end of the animal, including the feeding apparatus and cerebral ganglion.

There are many variations on this basic circulatory scheme, and we mention only a few to illustrate the variety within the polychaetes. Drastic differences are even present among polychaetes of generally similar body forms. Among the homonomous forms, for example, the circulatory system may be reduced or lost. In some cases (e.g., members of the family Syllidae), this reduction is probably associated with small size. This hypothesis, however, cannot be applied to the glycerids, many of which are large and quite active. In these worms, and some others, the circulatory system is greatly reduced and has become fused with remnants of the coelom. Glycerids contain red blood cells (with hemoglobin) in the coelomic fluid. Since glycerids have incomplete septa, the coelomic fluid can pass among segments, moved by body activities and ciliary tracts on the peritoneum. In their burrowing lifestyle, enlarged parapodial gills or delicate anterior gills would be disadvantageous; thus the general body surface has probably taken over the function of gas exchange and the coelom the function of circulation. Reduction or loss of the circulatory system has also occurred in a few sedentary polychaetes and in a few nonsedentary types as well (e.g., the terebellid *Polycirrus* and the archianellids).

Compared with *Nereis*, many polychaetes display additional blood vessels, modification of vessels, differences in blood flow patterns, and the formation of large sinuses. As might be predicted, some striking differences are seen among certain heteronomous polychaetes with reduced parapodia and anteriorly located branchiae (e.g., terebellids, sabellids, and serpulids). In many of these worms, in the region of the stomach and anterior intestine, the dorsal vessel is replaced by a voluminous blood space called the **gut sinus** (Figures 21C and D). Usually, the dorsal vessel continues anteriorly from this sinus and often forms a ring connecting with the main ventral vessel. In the sabellids and serpulids, a single, blind-ended vessel extends into each branchial tentacle. Blood flows in and out of these branchial vessels, which in some forms (e.g., serpulids) are equipped with valves that prevent backflow into the dorsal vessel. This two-way flow of blood within single vessels is quite different from the capillary exchange system in most closed vascular systems.

Specialized pumping structures have evolved in a number of polychaetes. They are usually referred to as "hearts," but are best viewed as independently derived modifications of vessels. They are best developed in certain tube-dwelling forms and compen-

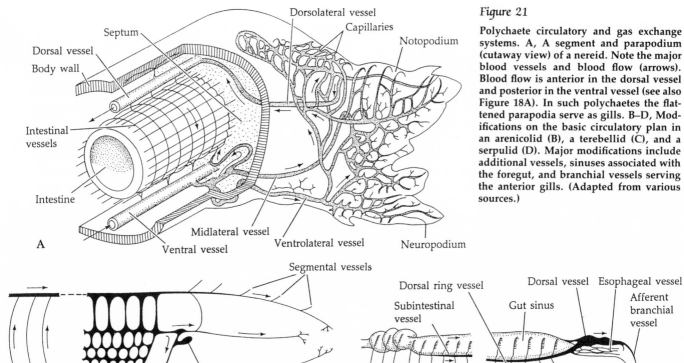

Figure 21

Polychaete circulatory and gas exchange systems. A, A segment and parapodium (cutaway view) of a nereid. Note the major blood vessels and blood flow (arrows). Blood flow is anterior in the dorsal vessel and posterior in the ventral vessel (see also Figure 18A). In such polychaetes the flattened parapodia serve as gills. B–D, Modifications on the basic circulatory plan in an arenicolid (B), a terebellid (C), and a serpulid (D). Major modifications include additional vessels, sinuses associated with the foregut, and branchial vessels serving the anterior gills. (Adapted from various sources.)

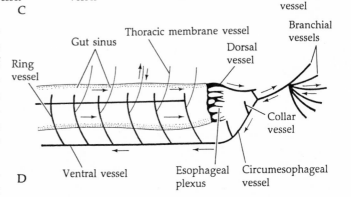

sate for the reduced effect of general body movements on circulation. These "hearts" are often little more than an enlarged and muscularized portion of one of the usual vessels; the dorsal heart of chaetopterids is such a structure. Terebellids possess a "pumping station" at the base of the gills that functions to maintain blood pressure and flow within the branchial vessels (Figure 21C). A variety of other similar structures are known.

Most polychaetes contain some respiratory pigment within their circulatory fluid. Those without any such pigment include various syllids, phyllodocids, polynoids, aphroditids, *Chaetopterus*, and a few others. When a pigment is present, it is usually some type of hemoglobin, although chlorocruorin is common in some families (e.g., certain sabellids and serpulids), and hemerythrin occurs in magelonids. Some polychaetes have more than a single type of pigment; for example, members of the genus *Serpula* contain both hemoglobin and chlorocruorin in the blood.

Polychaete respiratory pigments may occur in the blood itself, the coelomic fluid, or both. With a few

exceptions, blood pigments occur in solution and coelomic pigments are contained within corpuscles. The latter situation is generally associated with a degeneration or loss of the circulatory system (as in glycerids). The incorporation of coelomic pigments, usually hemoglobin, into cells is probably a mechanism to prevent the serious osmotic effects that would result from large numbers of free dissolved molecules in the body fluid. Corpuscular coelomic hemoglobins tend to be of much smaller molecular sizes than those dissolved in the blood plasma. The significance of this difference is not clear, but some interesting ideas on this and related matters are discussed by C. P. Mangum (1976) in relation to the oxygen problems of arenicolids.

The types of respiratory pigments and their disposition within the body are related at least in part to the lifestyles of polychaetes. As we discussed in Chapter 3, different pigments, even different forms of the same basic pigment, have different oxygen loading and unloading characteristics. The nature of the pigments in a particular worm often reflects an ability to store and release oxygen during periods of environmental oxygen depletion. A number of intertidal burrowing polychaetes take up and store oxygen during high tides and dissociate the stored oxygen during low tides. This sort of physiological cycle ameliorates the potential stress of oxygen depletion in the body during periods of low tide. Some have more than one pigment type; for example, one form of hemoglobin for "normal" conditions and another form that stores oxygen and releases it during periods of stress. Some polychaetes (e.g., *Euzonus*) can actually convert to anaerobic metabolic pathways during extended periods of anoxic conditions.

Oligochaeta. The oligochaete circulatory system is structurally similar to that of polychaetes but shows some modifications, especially in the pattern of blood flow. The differences described below are largely adaptations to living in terrestrial and freshwater environments—areas that generally subject the inhabitants to more stress than marine habitats.

In *Lumbricus* and many others, three main longitudinal blood vessels extend most of the body length and are connected to one another in each segment by additional segmentally arranged vessels (Figure 22). The largest longitudinal blood vessel is the **dorsal vessel**; the wall of this vessel is quite thick and muscular, and provides much of the pumping force for blood movement. Suspended in the mesentery beneath the gut is the longitudinal **ventral vessel**. The third longitudinal vessel lies ventral to the nerve cord and is called the **subneural vessel**. Additional, smaller, longitudinal vessels occur in some oligochaetes, especially some of the terrestrial forms. These additional vessels are usually branches from one of the three major vessels to particular parts of the gut, mostly in the anterior region of the worm.

Exchanges between the longitudinal vessels occur in or between each segment through various routes supplying the body wall, gut, and nephridia (Figure 22B). Most of the exchanges between the blood and the tissues take place through capillary beds supplied by afferent and efferent vessels. Blood flows posteriorly in ventral and subneural vessels and anteriorly in the dorsal vessel. Generally, exchange in each segment is from ventral to dorsal through the body wall and the gut, and from dorsal to ventral through the afferent and efferent nephridial vessels. In addition, most oligochaetes possess from two to five pairs of "hearts" that carry blood from the dorsal to the ventral vessel in the esophageal

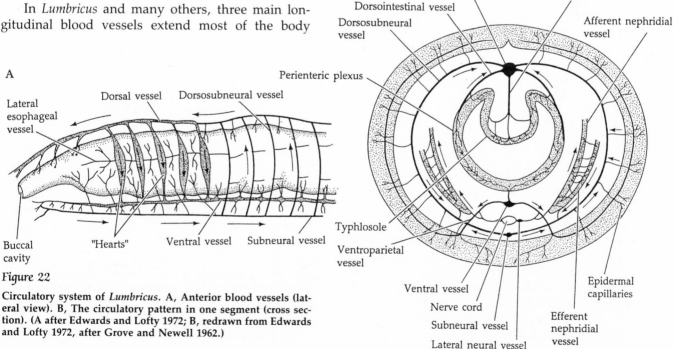

Figure 22

Circulatory system of *Lumbricus*. A, Anterior blood vessels (lateral view). B, The circulatory pattern in one segment (cross section). (A after Edwards and Lofty 1972; B, redrawn from Edwards and Lofty 1972, after Grove and Newell 1962.)

region. These muscular vessels aid in propelling the blood and maintaining blood pressure. The "hearts" and the dorsal blood vessel are often equipped with flap valves to ensure a general one-way blood flow.

Most oligochaetes have hemoglobin dissolved in the plasma; members of some families (e.g., Naididae) lack blood pigments. Various phagocytic amebocytes are also present in the circulatory fluids of most of these worms.

Gas exchange is cutaneous in oligochaetes. A few worms possess extensions of the body wall that increase the surface area and function as simple gills (e.g., *Branchiura, Dero*; see Figure 3), but most depend upon exchange across the general body surface. The body surface is kept moist either by the environment, by mucus, or by coelomic fluid released through pores, as described earlier. Most oligochaetes, especially the relatively large ones, have an extensive intraepidermal capillary network derived from the blood vessels within the body wall (Figure 22). These capillaries provide a constant blood supply from the ventral vessel to the body wall and a high surface area for exchange of gases between the blood and the environment.

Many terrestrial oligochaetes are capable of sufficient gas exchange only when exposed to air; they will drown if submerged. (Remember, air contains far more oxygen than does water.) We have all seen earthworms crawling about the surface following a heavy rain. One particular species of earthworm (*Alma emini*) has evolved a remarkable adaptation that allows it to survive the rainy season in its East African habitat. When rains cause the burrows to flood, the worm moves to the surface of the soil and forms a temporary opening. The worm then projects its posterior end out through the opening and rolls the sides of the body wall into a pair of folds, forming an open chamber that serves as a kind of "lung." The highly vascularized posterior epithelium enhances the exchange of gases. A number of aquatic oligochaetes can tolerate periods of low available oxygen and even anoxic conditions for short periods of time.

Evidence indicates that the hemoglobin in the blood of oligochaetes serves as an oxygen transport mechanism. No doubt some oxygen is also carried in solution, but apparently as much as 40 percent of the worms' oxygen-carrying capacity is accounted for by hemoglobin.

Hirudinida. Certain features of the hirudinidan *Bauplan* demand some sort of circulatory system. Many are relatively large and quite active. The drastic reduction of the coelom and invasion by "solid" tissue results in an analogue of an acoelomate construction, and the gut is regionally specialized. As we have seen before, these sorts of characteristics typically evolve in concert with some mechanism of internal transport or adaptive modification in shape as solutions to the surface-to-volume dilemma.

Evolutionarily, the leeches have approached this problem in several ways, including flattening of the body and the formation of extensive gut ceca or diverticula, both of which reduce internal diffusion distances. However, the most important adaptations for internal transport are structural circulatory vessels and channels. In most of the rhynchobdellids, this system is a combination of the ancestral annelid circulatory system and the reduced coelomic spaces; in the arhynchobdellids the original circulatory system is completely replaced by one derived entirely from the reduced coelom (Figure 23). In both of these arrangements the circulatory fluid is moved through the system by the action of particular contractile vessels and by general body movements.

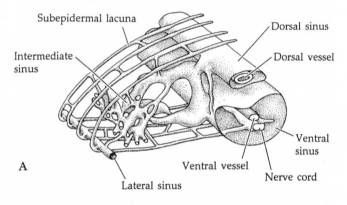

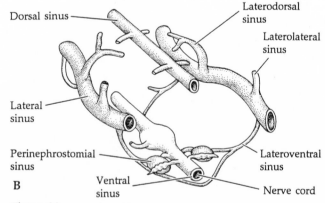

Figure 23

Coelomic and circulatory systems of two leeches. **A,** A portion of the circulatory and coelomic systems of *Placobdella*, a rhynchobdellid leech in which the circulatory system persists and is associated with the coelomic channels. **B,** A portion of the coelomic system in *Hirudo*, an arhynchobdellid leech. Here the circulatory system has been completely replaced by the coelomic channels. (A redrawn from Barnes 1980; B after Mann 1962.)

Gas exchange is accomplished by diffusion across the body wall; gills are present only in the ozobranchids (Figure 4B). Some leeches possess hemoglobin in solution in the circulatory fluid, accounting for approximately 50 percent of their oxygen-carrying capacity.

Excretion and osmoregulation

In Chapter 3 we discussed the structural types of nephridial organs in invertebrates. Thus far, we have seen various types of protonephridia, especially among the acoelomate and certain "pseudocoelomate" metazoa. Most annelids possess some sort of nephromixia, often serially arranged as one pair per segment, with the pore in the segment posterior to the nephridium. However, variations on this theme are great, and we may view the different conditions as having been derived from a basic primitive plan that arose with the evolution of the coelomate metameric *Bauplan*.

We remind you again that the success of an animal with segmentally arranged coelomic compartments depends upon the physical and physiological maintenance of those separate segments. The removal of metabolic wastes (predominantly ammonia) and the regulation of osmotic and ionic balance must occur in each functionally isolated coelomic chamber. The following account is based in part on the analysis by Goodrich (1946). (See also the discussion and figures pertaining to excretion and osmoregulation in Chapter 3.) We must alert you to some recent work in which the phylogenetic relationships among nephridial types are reevaluated (see Smith et al. 1987; Ruppert and Smith 1988; Smith and Ruppert 1988). Our conservative approach should be viewed as tentative.

Polychaeta. The presumed plesiomorphic condition in polychaetes is a serially homonomous body with a pair of complete coelomic spaces in each segment. Primitively, each coelom is served by two pairs of ducts that lead to the exterior; each pair includes a coelomoduct and a nephridioduct. The inner end of each coelomoduct bears an open ciliated funnel through which gametes escape, and the inner end of each nephridioduct bears protonephridia, which function in excretion and osmoregulation. This primitive condition has been lost in all but a very few extant polychaetes, but it persists in *Vanadis* (family Alciopidae). In the other few hundred species of polychaetes that possess protonephridia, the coelomoduct and nephridioduct are united to form a protonephromixium (e.g., various phyllodocids; Figure 24A).

The vast majority of polychaetes, however, possess metanephridia that open to the coelom by a ciliated nephrostome. In some (e.g., the capitellids), these metanephridia are entirely separate from the coelomoducts; but in most either the coelomic and nephridial ducts are united to form a metanephromixium, or there is a single interior opening that leads to a single duct as a mixonephrium (Figure 24B). This last case, generally called simply a metanephridium, may represent the complete incorporation of coelomoduct and nephridium into a single organ. However, much of the implied phylogenetic sequence suspected here has not been clearly retained in the ontogeny of living polychaetes. In all of these cases, the functional significance of different arrangements remains much the same in terms of serving a metameric body. That is, each coelomic compartment is equipped with a mechanism for elimination of wastes, for osmoregulation, and for the discharge of gametes.

In certain groups of polychaetes that possess incomplete septa, or that have lost septa between the coelomic spaces, the number of nephridia is reduced. When the coelomic fluid is continuous throughout the body, it is unnecessary to provide each segment with a set of these organs. In some sedentary polychaetes (e.g., serpulids and sabellids) without complete intersegmental septa, there is only a single pair of nephridia. These are located at the anterior end of the worm and lead to a single nephridioduct and common pore on the head; hence wastes are discharged outside the tube (Figure 24C).

As described in Chapter 3, the open nephrostomes of metanephridia nonselectively pick up coelomic fluids. This action is followed by resorption of materials from the nephridium back into the body, either directly into the surrounding coelomic fluid, or into the blood in cases where extensive nephridial blood vessels are present (e.g., in some nereids and in aphroditids). In either case the composition of the urine is quite different from that of the body fluids; the difference indicates a significant amount of selectivity.

Osmoregulation presents little problem for subtidal polychaetes living in relatively constant osmotic conditions. Intertidal and estuarine forms, however, must be able to withstand periods of stress associated with fluctuations in environmental salinities. There are also polychaetes that inhabit fresh and brackish water, and a few tropical forms that burrow in damp soil and leaf litter. These animals are subjected to

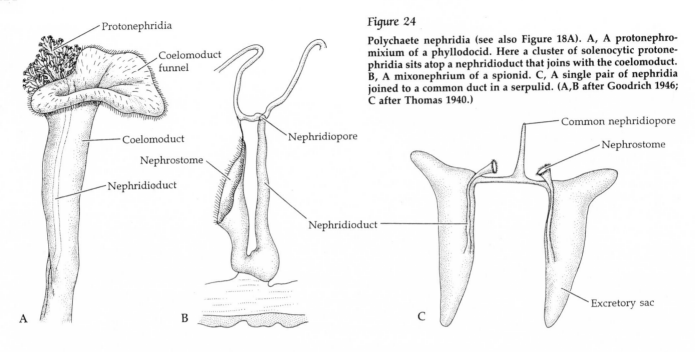

Figure 24

Polychaete nephridia (see also Figure 18A). A, A protonephro-mixium of a phyllodocid. Here a cluster of solenocytic protonephridia sits atop a nephridioduct that joins with the coelomoduct. B, A mixonephrium of a spionid. C, A single pair of nephridia joined to a common duct in a serpulid. (A,B after Goodrich 1946; C after Thomas 1940.)

periodic or constant osmotic stress and solve such problems by means of tolerance or regulation, or both. Many forms are osmoconformers (e.g., *Arenicola*), allowing the tonicity of their body fluids to fluctuate with changes in the environmental salinity. Most polychaete osmoconformers have relatively simple metanephridia, with comparatively short nephridioducts and correspondingly weaker resorptive and regulatory capacities. Some also have relatively thin body wall musculature, and the body swells when in a hypotonic medium. It is likely that burrowers and tube-dwellers face less osmotic stress than epibenthic forms, because the water in their tubes may be less subject to ionic variation than the overlying water.

Osmoregulators, such as a number of estuarine nereids, often have thicker body walls that tend to resist changes in shape and volume. When water enters the body from a hypotonic surrounding, the increased hydrostatic pressure generated within the coelom works against that osmotic gradient. In addition, regulators are able to maintain (within limits) a more-or-less constant internal fluid tonicity because of the greater selective capabilities of their more complex nephridia.

Oligochaeta. Oligochaetes possess paired, segmentally arranged metanephridia, usually in all but the extreme anterior and posterior segments. These nephridia are similar to the mixonephria of polychaetes, but many show various secondary modifications or elaborations.

A typical oligochaete nephridium is composed of a preseptal nephrostome (either open to the coelom or secondarily closed as a bulb), a short canal that penetrates the septum, and a postsegmental nephridioduct that is variably coiled and sometimes dilated as a bladder (Figure 25). The nephridiopores are usually located ventrolaterally on each segment.

Aquatic oligochaetes are ammonotelic, and most terrestrial forms are at least partially ureotelic. These wastes are transported to the nephridia via the circulatory system and by diffusion through the coelomic fluid. Uptake of materials into the nephridial lumen is partly nonselective (in those worms with open nephrostomes) from the coelom, and partly selective across the walls of the nephridioduct from the afferent nephridial blood vessels. A significant amount of selective resorption occurs back into the efferent blood flow along the distal portion of the nephridioduct, facilitating efficient excretion as well as ionic and osmoregulation.

The precise role of the chloragogen cells in excretion is not fully understood. While it is known that protein deamination and nitrogenous waste formation occur within these cells, the method of elimination of this waste is unknown. Individual chloragogen cells break free into the coelom and are probably engulfed by phagocytic amebocytes that apparently accumulate wastes in a precipitated form.

How, or if, these waste-filled cells are ever actually eliminated from the body remains a question.

Ionic and osmoregulation are of utmost importance to freshwater and terrestrial soft-bodied invertebrates such as oligochaetes. The moist, permeable surface necessary for gas exchange and the severe osmotic gradients across the body wall present potentially serious problems of water loss to terrestrial forms, and of water gain to freshwater forms; both face the loss of precious diffusible salts. Passive diffusion of water and salts also occurs across the gut wall.

The major organs of water and salt balance in freshwater oligochaetes are the nephridia. Excess water is excreted and salts are retained by selective and active resorption along the nephridioduct. The problem in terrestrial forms is more serious, and the solution is less well understood than in freshwater species. Surprisingly, earthworms are not absolute osmoregulators; rather they lose and gain water according to the amount of water in their environment. Most species can tolerate a loss of up to nearly 20 percent of their body water, and some can lose up to 75 percent of their body water and still recover. Under normal conditions, water conservation by earthworms is probably accomplished in several ways. The production of urea allows the excretion of a relatively hypertonic urine compared with that of a strictly ammonotelic animal. There may also be active uptake of water and salts from food across the gut wall. Certainly there are behavioral adaptations for remaining in relatively moist environments, in addition to the physiological adaptations that allow these animals to tolerate temporary partial dehydration of their bodies.

Hirudinida. The excretory structures of hirudinidans are structurally different from those of oligochaetes and polychaetes, but they are presumably derived from metanephridia. Leech nephridia are paired and segmentally arranged but are usually absent from several anterior and posterior segments. The nephrostomes are ciliated funnels associated with coelomic circulatory vessels, an arrangement that probably evolved with the reduction in the main body coelom and the loss of septa. Some hirudini-

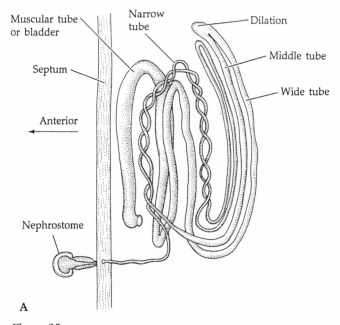

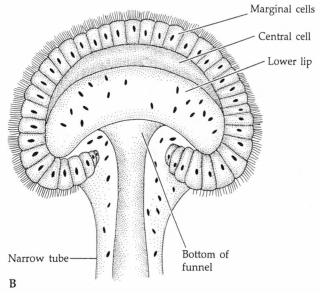

Figure 25

Lumbricus nephridia. A, A single nephridium and its relationship to a septum. B, Details of the nephrostome. Evidence suggests (see Laverack 1963) that earthworm nephridia are highly selective excretory and osmoregulatory units. The nephridioduct is regionally and functionally specialized along its length from the nephrostome to the nephridiopore. The narrow tube receives body fluids and various solutes, first from the coelom through the nephrostome and then from the blood across the walls of capillaries that lie adjacent to the tube. In addition to various forms of nitrogenous wastes (ammonia, urea, uric acid), various coelomic proteins, water, and ions (Na$^+$, K$^+$, Cl$^-$) are also picked up. Apparently, the wide tube serves as a site of selective resorption (probably into the blood) of proteins, ions, and water, leaving the urine rich in nitrogenous compounds as waste. (A,B after Edwards and Lofty 1972, after various sources.)

dans possess clusters of nephrostomes called **ciliated organs**. Each nephridium leads ultimately to a ventrolateral nephridiopore. It is, however, the microscopic structure of each unit between the nephrostome and external pore that is so remarkably different from other metanephridia (Figure 26).

The nephrostome leads not into an open duct but into a blind chamber called the **nephridial capsule**. The capsule is connected to a "nephridioduct" composed of a single row of cells through which runs an intracellular canal. This canal appears to be somewhat transitory, especially near the capsule, in that it forms from the coalescence of tiny intracellular tubules and vacuole-like chambers. Exactly how this arrangement works is unclear, but its structure suggests a good deal of selectivity during urine formation. Selective filtration and resorption would be expected in an animal whose excretory units directly drain the circulatory fluid. The intracellular nephridial canal connects with a short chamber derived from an ectodermal invagination at the nephridiopore. In some true leeches a relatively large bladder is formed near the pore (Figure 26).

Ammonia is the main nitrogenous waste product eliminated via the nephridia. Apparently, particulate waste materials are engulfed by phagocytes, both in the coelomic fluid and in the "mesenchyme," but the eventual disposition of this material is not known.

The nephridia of freshwater leeches are also osmoregulatory. The urine is very dilute, a fact suggesting the excretion of excess water and the retention of various salts. In certain terrestrial leeches the urine from the anterior and posterior nephridiopores is released onto the surfaces of the suckers, thereby providing a moist surface for effective suction.

Nervous system and sense organs

The central nervous system of annelids probably arose from a ladder-like system. The major trends in annelids have been a reduction in the number of longitudinal cords as the nervous system has become more centralized and concentrated, and the development of segmentally arranged ganglia along the longitudinal cord(s) associated with the metameric *Bauplan*. The fundamental plan of the central nervous system in annelids (as in protostomes in general) comprises a dorsal cerebral ganglion, paired circumenteric connectives, and one or more ventral longitudinal nerve cords along the body (Figure 27).

Polychaeta. The cerebral ganglion of polychaetes is usually bilobed and lies within the prostomium.

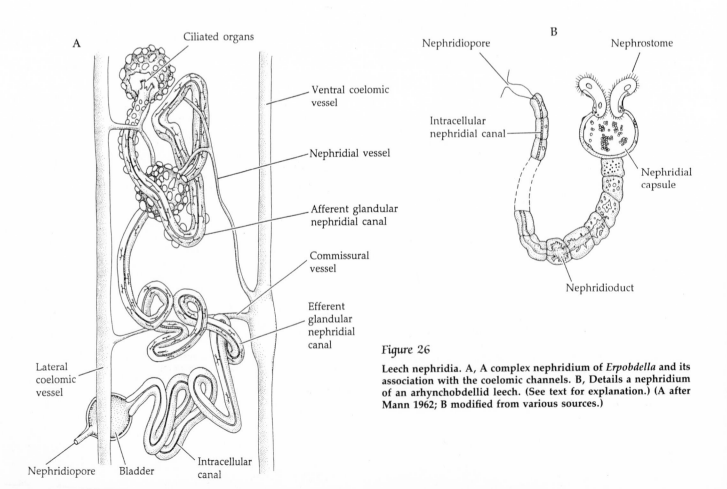

Figure 26

Leech nephridia. A, A complex nephridium of *Erpobdella* and its association with the coelomic channels. B, Details a nephridium of an arhynchobdellid leech. (See text for explanation.) (A after Mann 1962; B modified from various sources.)

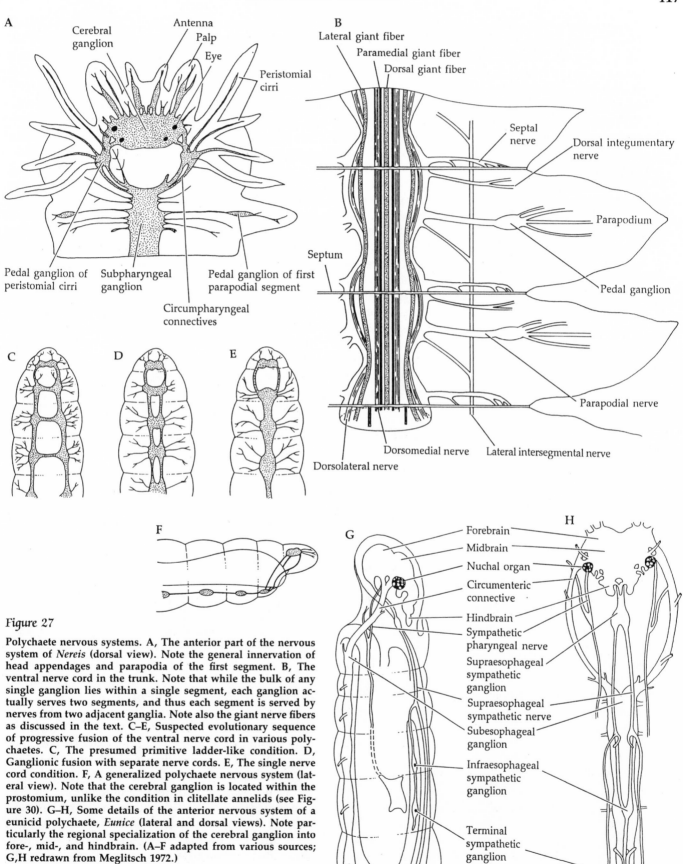

A, Cerebral ganglion, Antenna, Palp, Eye, Peristomial cirri, Pedal ganglion of peristomial cirri, Subpharyngeal ganglion, Pedal ganglion of first parapodial segment, Circumpharyngeal connectives

B, Lateral giant fiber, Paramedial giant fiber, Dorsal giant fiber, Septal nerve, Dorsal integumentary nerve, Parapodium, Pedal ganglion, Septum, Parapodial nerve, Dorsolateral nerve, Dorsomedial nerve, Lateral intersegmental nerve

C D E F

G H, Forebrain, Midbrain, Nuchal organ, Circumenteric connective, Hindbrain, Sympathetic pharyngeal nerve, Supraesophageal sympathetic ganglion, Supraesophageal sympathetic nerve, Subesophageal ganglion, Infraesophageal sympathetic ganglion, Terminal sympathetic ganglion, Ventral nerve cord

Figure 27

Polychaete nervous systems. A, The anterior part of the nervous system of *Nereis* (dorsal view). Note the general innervation of head appendages and parapodia of the first segment. B, The ventral nerve cord in the trunk. Note that while the bulk of any single ganglion lies within a single segment, each ganglion actually serves two segments, and thus each segment is served by nerves from two adjacent ganglia. Note also the giant nerve fibers as discussed in the text. C–E, Suspected evolutionary sequence of progressive fusion of the ventral nerve cord in various polychaetes. C, The presumed primitive ladder-like condition. D, Ganglionic fusion with separate nerve cords. E, The single nerve cord condition. F, A generalized polychaete nervous system (lateral view). Note that the cerebral ganglion is located within the prostomium, unlike the condition in clitellate annelids (see Figure 30). G–H, Some details of the anterior nervous system of a eunicid polychaete, *Eunice* (lateral and dorsal views). Note particularly the regional specialization of the cerebral ganglion into fore-, mid-, and hindbrain. (A–F adapted from various sources; G,H redrawn from Meglitsch 1972.)

One or two pairs of circumenteric connectives extend from the cerebral ganglion around the foregut and unite ventrally in the subenteric ganglion. Primitively, a pair of longitudinal nerve cords arises from the subenteric ganglion and extends the length of the body (Figure 27C). Ganglia are arranged along these nerve cords, one pair in each segment, and are connected by transverse commissures. Lateral nerves extend from each ganglion to the body wall and each bears a so-called **pedal ganglion**. This double nerve cord arrangement is common in certain groups of polychaetes, including sabellids and serpulids. Interestingly, in the amphinomids there are four longitudinal nerve cords, a medial pair and a lateral pair, the latter connecting the pedal ganglia. Some workers consider the amphinomid condition primitive, but others contend that the lateral longitudinal cords have been secondarily derived within this group. Similar but perhaps nonhomologous lateral longitudinal cords appear in some other polychaete taxa that are considered to be relatively advanced.

The general evolutionary trend among most polychaetes has been the fusion of the medial nerve cords to form a single midventral longitudinal cord (Figure 27). The degree of fusion varies among taxa; some retain separate nerve tracts within the single cord. In addition, the position of the ventral nerve cord varies. Primitively the cord is subepidermal, but in more advanced forms it lies internal to the body wall muscle layers.

The cerebral ganglion is often regionally specialized into three regions, typically called the **forebrain**, **midbrain**, and **hindbrain**. Generally, the forebrain innervates the prostomial palps, the midbrain the eyes and prostomial antennae or tentacles, and the hindbrain the chemosensory **nuchal organs** (Figures 27A, G, and H). The circumenteric connectives arise from the fore- and midbrains. The midbrain also gives rise to a complex of motor **stomatogastric nerves** associated with the foregut, especially with the operation of the proboscis or pharynx. The circumenteric connectives often bear ganglia from which nerves extend to the peristomial cirri, or else these appendages are innervated by nerves from the subenteric ganglion. The subenteric ganglion appears to exhibit excitatory control over the ventral nerve cord(s) and segmental ganglia.

The nerves that arise from the segmental ganglia innervate the body wall musculature and parapodia (via the pedal ganglia), and the digestive tract. The ventral nerve cord and sometimes the lateral nerves of most annelids contain some extremely long neurons, or **giant fibers**, of large diameter; these neurons

facilitate rapid, "straight through" impulse conduction, bypassing the ganglia (Figure 27B). Giant fibers are apparently lacking in some polychaetes (e.g., syllids). Giant fibers are well developed in tube-dwellers, such as sabellids and serpulids, permitting rapid contraction of the body and retraction into the tube.

It is now generally accepted that polychaetes possess organs or tissues of neurosecretory or endocrine functions. Most of the secretions appear to be associated with the regulation of reproductive activities.

Polychaetes possess an impressive array of sensory receptors (Figures 28 and 29). As expected, the kinds of sense organs present and the degree of their development vary among polychaetes with different lifestyles. Certainly, the requirements for particular sorts of sensory information are not the same for a tube-dwelling sabellid as they are for an errant predatory nereid or a burrowing arenicolid.

In general, polychaetes are highly touch sensitive. Crawlers, tube dwellers, and burrowers alike depend upon tactile reception for interaction with their immediate surroundings (locomotion, anchorage within their tube, and so on). Touch receptors are distributed over much of the body surface but are concentrated in such areas as the head appendages and parts of the parapodia. The setae are also typically associated with sensory neurons and serve as touch receptors. Some burrowers and tube dwellers have such a strong positive response to contact with the walls of their burrow or tube that the response dominates all other receptor input. Some of these polychaetes will remain in their burrow or tube regardless of other stimuli that would normally produce a negative response.

Most polychaetes possess photoreceptors, although these structures are lacking in many infaunal burrowers. The best developed polychaete eyes occur in pairs on the dorsal surface of the prostomium. In some there is a single pair of eyes (e.g., most phyllodocids); in many there are two or more pairs (e.g., nereids, polynoids, hesionids, many syllids). These prostomial eyes vary in complexity, but are largely constructed as direct pigment cups (as opposed to the inverted pigment cups of animals such as flatworms and ribbon worms). These photoreceptors may be simple depressions in the body surface lined with retinular cells, or they may be quite complex, with a distinct refractive body or lens (Figures 29A–D). In nearly all cases, the eye units are covered by a modified section of the cuticle that functions as a cornea. The eyes of most polychaetes are capable of transmitting information on light direction and intensity, but in certain pelagic forms (e.g., alciopids) the

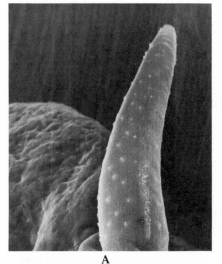

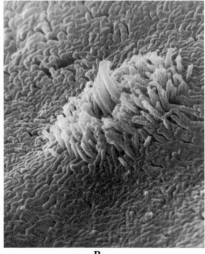

A   B

Figure 28

Epithelial sense organs (probably chemo-sensory) on the polychaete *Nereis*. A, A dorsal cirrus of a parapodium showing distribution of sense organs (scanning electron micrograph). B, A single sense organ (scanning electron micrograph). (From P. J. Mill, 1976, *Structure and Function of Proprioceptors in the Invertebrates*, Halsted Press [Wiley], New York. Photograph by and courtesy of D. A. Dorsett, with the permission of Methuen and Co., Ltd.)

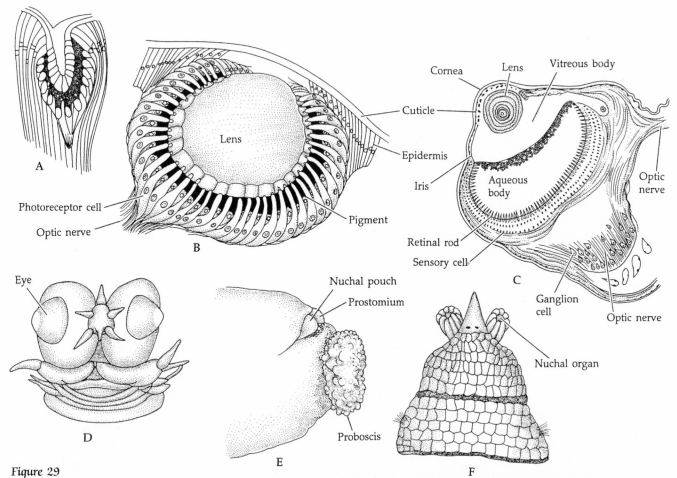

Figure 29

Polychaete photoreceptors and nuchal organs. A, Simple pit-like eye of a chaetopterid. B, Lensed pigment-cup eye of a nereid. C, A complex eye (section) of an alciopid (*Vanadis*). D, The head of *Vanadis* (ventral view). Note the large eye lobes. E, Nuchal organ of *Arenicola*. F, Nuchal organs of *Notomastus*. (A,B,E redrawn from various sources after Fauvel et al. 1959; C,D after Hermans and Eakin 1974; F redrawn from Barnes 1980.)

eyes are huge and possess true lenses capable of accommodation and perhaps image perception (Figures 29C and D).

In addition to, or instead of, the prostomial eyes, some polychaetes bear photoreceptors on other parts of the body. A few bear simple eyespots along the length of the body (e.g., the opheliid *Polyophthalmus*). Pygidial eyespots are known among newly settled sabellariid larvae, and some adult sabellids (small ones such as *Fabricia*). Interestingly, in these cases the animals crawl backward. Many fanworms (sabellids and serpulids) possess complex eyes or simple ocelli on the branchial crown tentacles and react to decreases in light intensity by retracting into their tubes. This "shadow response" is certainly advantageous to such sedentary animals in avoiding predators and can easily be demonstrated by passing one's hand to cast a shadow over a live worm.

Nearly all polychaetes are sensitive to dissolved chemicals in their environment. Most of the chemoreceptors are specialized cells that bear a receptor process extending through the cuticle. Sensory nerve fibers extend from the base of each receptor cell. Such simple chemoreceptors are often scattered over much of the worm's body, but they tend to be concentrated on the head and its appendages. Some polychaetes also possess ciliated pits or slits called nuchal organs, which are presumed to be chemosensory (Figures 29E and F). These structures are typically paired and lie posteriorly on the dorsal surface of the prostomium. In some forms (e.g., certain nereids) the nuchal organs are simple depressions, whereas in others (e.g., archiannelids, opheliids) they are rather complex eversible structures equipped with special retractor muscles. In members of the family Amphinomidae the nuchal organs are elaborate outgrowths of an extension of the prostomium called the **caruncle**.

Statocysts are common in some burrowing and tube-dwelling polychaetes (e.g., certain terebellids, arenicolids, and sabellids). A few forms possess several pairs of statocysts, but most have just a single pair, located near the head. These statocysts may be closed or open to the exterior, and the statolith may be a secreted structure or formed of extrinsic material, such as sand grains. It has been demonstrated experimentally that the statocysts of some polychaetes do serve as georeceptors and help in the maintenance of proper orientation when the bearer is burrowing or tube building.

A number of other structures of presumed sensory function occur in some polychaetes. These structures are often in the form of ciliated ridges or grooves occurring on various parts of the body and associated with sensory neurons. A variety of names have been applied to these structures, but in most cases their function(s) remains unclear.

Oligochaeta. The central nervous system of oligochaetes consists of the usual annelid components: a supraenteric cerebral ganglion joined to a ganglionated ventral nerve cord by circumenteric connectives and a subenteric ganglion (Figure 30). With the reduction in head size, especially of the prostomium, the cerebral ganglion occupies a more posterior position than in the polychaetes, often lying as far back as the third body segment. The pair of ventral nerve cords, one cord arising from each of the circumenteric connectives, is almost always fused as a single tract in oligochaetes. The ventral nerve cord usually contains some giant fibers, similar to those of many polychaetes.

The cerebral ganglion gives rise to several anteriorly-directed prostomial nerves, most of which are sensory. The circumenteric connectives and segmental ganglia give rise to sensory and motor nerves to the body wall and various organs in each segment. As in the polychaetes, it is the subenteric ganglion that appears to be the center for motor control of body movements; the cerebral ganglion acts to mediate these activities by inhibitory influences. The independent but coordinated action of each segment during locomotion depends on a series of stimulus–response reactions involving the segmental ganglia, but these reactions are initiated by the subenteric ganglion. If that ganglion is removed, all movement ceases; but if the cerebral ganglion is removed, normal movement continues, although responses to external stimuli are absent. Oligochaetes also possess special cells within the segmental muscles that serve as stretch receptors. These sensory units supply feedback to the ventral nerve cord about the state of the muscles in each segment, and thus are important to the coordinated contraction and relaxation of the muscles during locomotion.

The sense organs of oligochaetes are clearly associated with their habits. The general name of **epithelial sense organs** has been given to a variety of receptor units distributed over most of the body and innervated by sensory nerves from the central nervous system. These receptors can be free nerve endings within the epidermis or clusters of special receptor cells associated with various bumps and tubercles. Many of these structures are undoubtedly tactile in function, providing an important source of information to burrowers and crawlers. Others are suspected to be chemoreceptors that supply important infor-

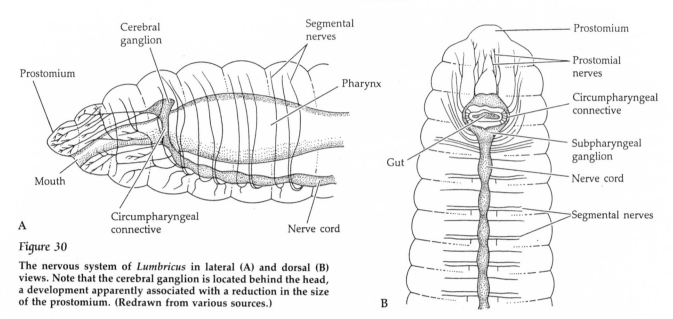

Figure 30

The nervous system of *Lumbricus* in lateral (A) and dorsal (B) views. Note that the cerebral ganglion is located behind the head, a development apparently associated with a reduction in the size of the prostomium. (Redrawn from various sources.)

mation about the relatively unstable freshwater or terrestrial environment of oligochaetes. Many are known to be sensitive to changes in pH and to the secretions of other worms. Chemoreception also probably plays a role in food location and selection.

Some freshwater oligochaetes possess paired pigment-cup ocelli at the anterior end; nearly all others bear simple photoreceptors distributed over the entire body surface. These worms are generally negatively phototactic to high light intensities.

Hirudinida. The nervous system of leeches has received a great deal of attention from zoologists. Even large leeches have nervous systems composed of very few neurons, and these individual cells are sufficiently large that their circuitry has been traced in great detail. For much more information on this matter and for an example of a classic study on one system of one animal, we refer you to Nicholls and Van Essen (1974).

With relatively minor modifications, the hirudinidan nervous system is clearly derived from the basic annelid plan. The cerebral ganglion is usually set back from the anterior end of the body at about the level of the pharynx (Figure 31). The cerebral ganglion, circumenteric connectives, and subenteric ganglion together form a rather thick nerve ring around the foregut. Two longitudinal nerve cords arise from the ventral portion of this ring and extend posteriorly through the body. The nerve cords are separate in some areas, but the segmental ganglia are fused. Peripheral nerves include abundant sensory neurons

from the cerebral ganglion and segmentally arranged motor and sensory neurons from the ventral nerve cord ganglia.

As do other annelids, hirudinidans employ neurosecretions to control certain activities. Some leeches are capable of rapid color changes that are apparently regulated by such means. Reproductive activities may also be under neurosecretory control.

Some hirudinidans, especially blood suckers, are extremely sensitive to certain types of environmental stimuli, although their sensory receptors are relatively simple in structure. These animals possess an array of epidermal sense organs similar to those found in oligochaetes. In addition, leeches have from two to ten dorsal eyes of varying complexity, and special sensory papillae that bear bristles extending from the body surface. The presence of the bristles suggests touch sensitivity, but the exact function of the papillae is unknown. In fact, except for the eyes, the functions of leech sense organs are not well understood at all, and most of the information is based on simple behavioral responses to various stimuli.

Leeches are for the most part negatively phototactic. However, some of the blood-sucking species react positively to light when preparing to feed. This behavioral change may be adaptive by causing the leech to move into areas where a host encounter is more likely. Most leeches can also detect movement in their surroundings, as evidenced by their responses to shadows passing over them. This reaction has been noted particularly in leeches that attack

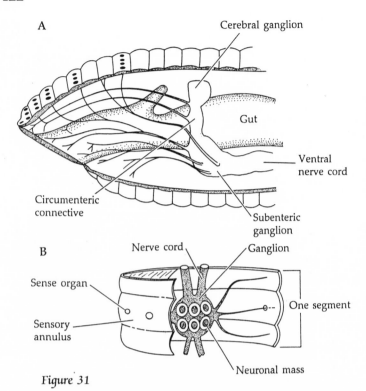

Figure 31

Leech nervous system. A, The anterior nervous system (lateral view). B, A generalized leech segment comprising three annuli, cut away to show segmental ganglion and innervation of epithelial sense organs. (After Mann 1962.)

fishes. Again, the adaptive significance of this behavior may be in facilitating encounters with hosts by responding with increased movement when a fish passes "overhead."

Leeches also respond to mechanical stimulation in the forms of direct touch and vibrations in their environments. They are also chemosensitive and known to be attracted to various secretions of potential hosts. Some aquatic and even terrestrial leeches that prefer "warm-blooded" mammalian hosts are apparently attracted to points of relatively high temperatures in their surroundings, thus aiding in food location.

Regeneration and asexual reproduction

Polychaeta. Different polychaetes show various degrees of regenerative capabilities. Nearly all of them are capable of regenerating lost appendages such as palps, tentacles, cirri, and parapodia. Most of these worms can regenerate posterior body segments if the trunk is severed; this sort of regeneration is a complicated process and is not fully understood.

There are numerous exceptional cases of the re-

generative powers of polychaetes. While regeneration of the posterior end is common, many polychaetes, such as nereids, cannot regenerate lost heads. However, sabellids, syllids, and some others can regrow the anterior end. The most dramatic regenerative powers among the polychaetes occur, oddly, in a few forms with highly specialized and heteronomous bodies. In *Chaetopterus*, for example, the anterior end will regenerate a normal posterior end as long as the regenerating part (the anterior end) includes not more than fourteen segments; if the animal is cut behind the fourteenth segment, regeneration does not occur. Furthermore, any single segment from among the first fourteen can regenerate anteriorly and posteriorly to produce a complete worm (Figure 32A). An even more dramatic example of regenerative power is known among certain species of *Dodecaceria* (Cirratulidae), which are capable of fragmenting their bodies into individual segments, each of which can then regenerate a complete individual!

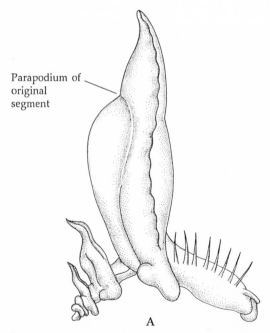

Parapodium of original segment

A

Figure 32

Regeneration, asexual reproduction, and epitoky in polychaetes. A, The remarkable regeneration of a chaetopterid from a single excised segment. B, Asexual reproduction by transverse partitioning in a syllid (see text for explanation). C, A portion of another syllid in which reproductive individuals are budded from the parent's parapodia (*Syllis*). D, The posterior end of *Typhlosyllis*, bearing a cluster of asexually produced epitokes. E, The epitokous palolo worm, *Palola viridis*. F, An epitokous nereid, *Nereis irrorata*. Note the dimorphic condition of the anterior and posterior parapodia. (A,C redrawn from Meglitsch 1972, after various sources; B after Russell-Hunter 1979; D,E,F redrawn from Barnes 1980, after Fauvel et al. 1959.)

Regeneration appears to be controlled by neuroendocrine secretions released by the central nervous system at sites of regrowth and to be initiated by severing the elements of the nervous system. Initiation has been demonstrated experimentally by cutting the ventral nerve cord while leaving the body intact; the result is the formation of an "extra" part at the site of cutting. The actual mechanism of regeneration has been studied in a variety of polychaetes, and although the results are not entirely consistent, a general scenario can be outlined. Normal growth and addition of segments take place immediately anterior to the pygidium, in a region known as the **growth zone**. However, this growth zone is obviously not involved in regeneration, unless of course only the pygidium is lost. Rather, when the trunk is severed, the cut region heals over and then a patch of generative tissue, or **blastema**, forms. The blastema comprises an inner mass of cells originating from nearby tissues that were derived originally from

mesoderm and an outer covering of cells from ectodermally derived tissues such as the epidermis. These two constituent cell masses act somewhat as a growth zone analogue, proliferating new body parts according to their tissue origins. This process is coupled with the growth of the gut, which contributes parts of entodermal origin. In addition, some workers have demonstrated that relatively undifferentiated cells from mesenchyme-like layers of the body migrate to injured areas and contribute to various (and uncertain) degrees to the regenerative process. These so-called **neoblast cells** are ectomesodermal in origin, because they arise embryonically from presumptive ectoderm. During regeneration, they apparently contribute to tissues and structures normally associated with true mesoderm, and perhaps even other germ layers as well. The implication here is that the germ layer of the precursor of a regenerated part may not correspond to the "normal" origin of that part. For example, regenerated coelomic spaces may be lined

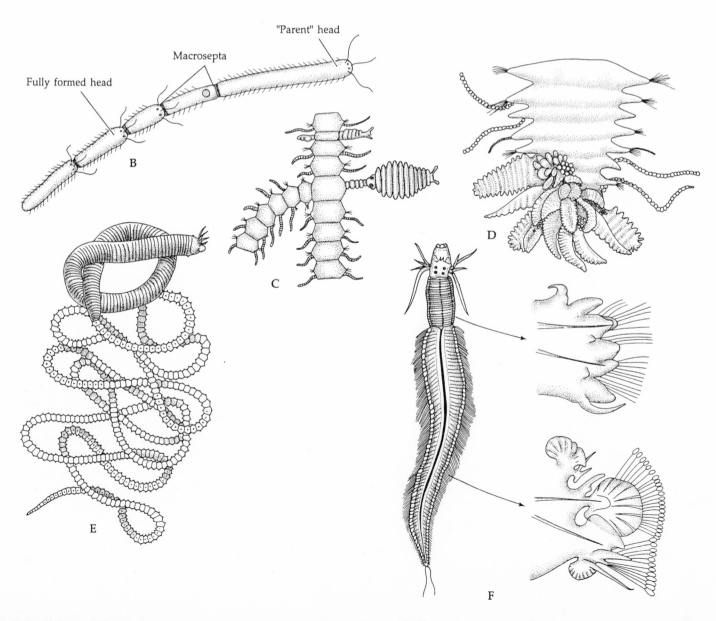

with tissue derived originally from ectoderm rather than from mesoderm.

A number of polychaetes use their regenerative powers for asexual reproduction. A few reproduce asexually by multiple fragmentation. We described above the ability of *Dodecaceria* to regenerate complete individuals from isolated segments; this phenomenon occurs spontaneously and naturally in these animals as an efficient reproductive strategy. Spontaneous transverse fragmentation of the body into two or several groups of segments also occurs in certain syllids, chaetopterids, cirratulids, and sabellids (Figure 32B). The point or points at which the body is fragmented are typically specific for particular species, and they may be anticipated by ingrowths of the epidermis that produce partitions across the body; each partition is called a **macroseptum** and marks the plane of subsequent fragmentation. Asexual reproduction results in a variety of regeneration patterns, including chains of individuals, budlike outgrowths, or direct growth to new individuals from isolated fragments. Asexual reproduction in polychaetes may be under the same sort of neurosecretory control as that postulated for nonreproductive regeneration.

Oligochaeta. The ability of oligochaetes to regenerate parts of the body varies greatly among different species. Many can regenerate almost any excised body part and can regrow both front and back ends, whereas others, such as *Lumbricus terrestris*, have very weak regenerative powers. The mechanisms of regeneration are presumably similar to those of polychaetes. A polarity exists from anterior to posterior, rear segments generally being more easily regrown than front ones.

In contrast to polychaetes, oligochaetes tend to have a more fixed number of body segments, a condition that has implications in the regenerative process. Data from regeneration experiments on certain earthworms indicate that most oligochaetes never regenerate more segments than were possessed by the original worm. In fact, the regenerated worms tend to possess the same number of segments as the original. The mechanism controlling this regeneration of a predictable number of segments is unknown, but Moment (1953) has offered an interesting theory. At least some oligochaetes (e.g., earthworms) have a measurable voltage difference along the length of their bodies, each segment having a slightly different electrical potential from the next. Moment suggests that regeneration ceases when the original, overall voltage potential is regained by the regrowth of the

proper number of segments. Presumably, the normal electrical gradient thus produced imparts an inhibitory effect on the regeneration process.

Most freshwater oligochaetes are capable of asexual reproduction. In some members of the family Naididae, sexual reproduction is very rare, and in a few it may not occur at all. Usually, however, asexual reproduction is a seasonal event, alternating with sexual activity. We have described this sort of cyclic phenomenon in other invertebrates, such as many rotifers, that live in freshwater environments subject to drastic seasonal fluctuations. Rapid asexual reproduction usually occurs in early to mid summer. Thus the worms take advantage of mild conditions and abundant food supplies. The offspring of this activity mature and reproduce sexually in the late summer and early fall and produce overwintering stages that hatch in the spring.

Oligochaetes reproduce asexually by one or more forms of transverse fission. In some it is by fragmentation at one or several points along the body followed by regeneration of each fragment to a new worm. In others "buds" are produced on the parent's body as "offspring precursors." Once the new individuals are partially formed, fission occurs and the offspring break free.

Hirudinida. Asexual reproduction is unknown among the hirudinoidans.

Sexual reproduction and development

Polychaeta. The great majority of polychaetes are gonochoristic; hermaphroditism is known in serpulids, certain freshwater nereids, and isolated cases in other families (see Bergland 1986 and Franke 1986). Some syllids are protogynic, and some eunicids protandric. Polychaetes lack permanent gonads or other complex reproductive organs. Rather, the gametes arise by proliferation of cells from the peritoneum; these cells are released into the coelom as gametogonia or primary gametocytes. Formation of gametes may occur throughout the body or only in particular regions of the trunk. Within a reproductive segment, the production of gametes may occur all over the coelomic lining or only on specific areas.

The gametes generally mature within the coelom and then are released to the outside by various mechanisms (e.g., via coelomoducts, nephridia, or rupture of the parent body wall). Many polychaetes release eggs and sperm into the water, where external fertilization is followed by fully indirect development with a planktotrophic larval stage. Others display mixed life history patterns. In these forms, fertilization is internal, followed by brooding or by the production of floating or attached egg capsules. In most instances, the embryos are released as free-swimming larvae. Some species brood their embryos on the body surface.

Many of the free-spawning polychaetes have evolved methods that ensure relatively high rates of fertilization. One of these methods is the interesting phenomenon of **epitoky**, characteristic of many benthic syllids, nereids, and eunicids. This phenomenon involves the production of a sexually reproductive worm called an **epitokous individual**. Epitokous forms may arise from nonreproductive animals (**atokous forms**) by a transformation of an individual worm, as in most nereids and eunicids, or by the asexual production of new individuals that are epitokous, as in most syllids. In nereids, the posterior body segments (or **epitoke**) become swollen and filled with gametes, and their associated parapodia become enlarged and natatory (Figure 32F). In cases where the epitokous worm is asexually produced, the reproductive individual is often without a head and lacks the atokous anterior end. Such epitokes are formed as single or clusters of outgrowths from particular body regions (Figure 32D).

In any event, the epitokes serve as gamete-carrying bodies capable of swimming from the bottom upward into the water column, where the gametes are released. Epitoky is controlled by neurosecretory activity, and the upward migration of the epitokes is precisely timed to synchronize spawning within a population. Although not completely understood, the reproductive swarming of epitokes appears to be linked with lunar periodicity. This activity not only ensures successful fertilization but establishes the developing embryos in a planktonic habitat suitable for the larvae. Certain South Sea island people predict the swarming and collect the ripe epitokes under a full moon and feast on them.

Early polychaete development exemplifies a classic, protostomous, spiralian pattern. The eggs are telolecithal with small to moderate quantities of yolk, the amount varying with particular developmental strategies. Those with a period of encapsulation or brooding prior to larval release generally contain more yolk than those that free spawn and develop quickly to planktotrophic larvae. In any case, cleavage is holoblastic and clearly spiral. A coeloblastula or, in the cases of more yolky eggs, a stereoblastula (Figure 33) develops and undergoes gastrulation by invagination, epiboly, or a combination of these two events. Gastrulation results in the internalization of the presumptive entoderm (the 4A, 4B, 4C, 4D, and

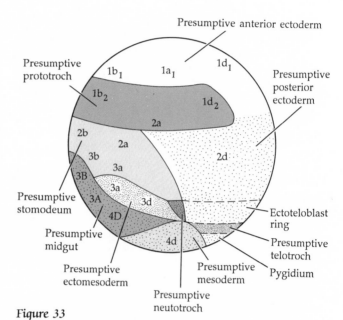

Figure 33

Fate map of a polychaete (*Scoloplos*) blastula (viewed from the left side). (After Anderson 1973.)

the 4a, 4b, 4c cells) and presumptive mesoderm (the 4d mesentoblast). The derivatives of the first three micromere quartets give rise to ectoderm and ectomesoderm, the latter producing various larval muscles between the body wall and the developing gut. As the entoderm hollows to become the archenteron, a stomodeal invagination forms at the site of the blastopore and a proctodeal invagination produces the hindgut.

In many polychaetes, these early ontogenetic events result in a **trochophore larva**, a form characteristic of many protostomous groups (e.g., molluscs, sipunculans, and echiurans). The early, presegmental trochophore (Figure 34A) is characterized by the presence of a locomotor ciliary band just anterior to the region of the mouth. This ciliary band, called the **prototroch**, arises from special cells, called **trochoblasts**, of the first and second quartets of micromeres. Most trochophore larvae also bear an **apical ciliary tuft** associated with an **apical sense organ** derived from a plate of thickened ectoderm at the presumptive anterior end. In addition, there is often a perianal ciliary band called the **telotroch** and a midventral band called the **neurotroch**. By this stage the mesentoblast has divided to form a pair of cells called **teloblasts**, which in turn proliferate a pair of mesodermal bands, one on each side of the archenteron in the region of the hindgut, an area known as the **growth zone** (Figure 35A). Many trochophores bear various larval sense organs such as ocelli, as well as

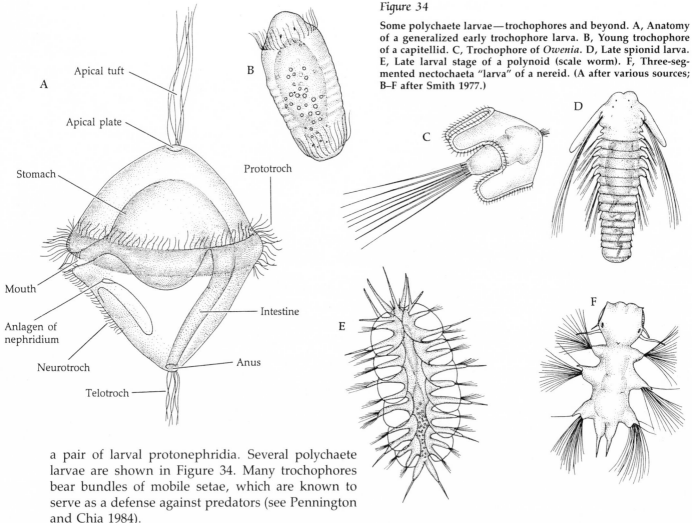

Figure 34

Some polychaete larvae—trochophores and beyond. A, Anatomy of a generalized early trochophore larva. B, Young trochophore of a capitellid. C, Trochophore of *Owenia*. D, Late spionid larva. E, Late larval stage of a polynoid (scale worm). F, Three-segmented nectochaeta "larva" of a nereid. (A after various sources; B–F after Smith 1977.)

a pair of larval protonephridia. Several polychaete larvae are shown in Figure 34. Many trochophores bear bundles of mobile setae, which are known to serve as a defense against predators (see Pennington and Chia 1984).

The larva grows and elongates by proliferation of tissue in the growth zone (Figure 35). Simultaneously, segments are produced by the anterior proliferation of mesoderm from the teloblast derivatives on either side of the gut. These packets of mesoderm hollow (schizocoely) and expand as paired coelomic spaces, which eventually obliterate the blastocoel. Thus, the production of serially arranged coelomic compartments and the formation of segments are one and the same; the anterior and posterior walls of each coelemic compartment form the intersegmental septa. Proliferation of segments by this process is called **teloblastic growth**. Externally, additional ciliary bands are added at each segment. These **metatrochal bands** aid in locomotion as the animal increases in size. Such segmented larvae are sometimes called **polytroch larvae**.

The fates of the various larval regions are now apparent (Figure 35C). The region anterior to the prototrochal ring becomes the prostomium, while the prototrochal area forms the peristomium. Note that these two parts are not involved in the proliferation of segments and are thus presegmental. However, in some polychaetes one or more of the anterior trunk segments may be incorporated into the peristomium during growth. The segmental, metatrochal portion of the larva forms the trunk, and the growth zone and postsegmental pygidium remain as the corresponding adult body parts. The apical sense organ becomes the cerebral ganglion, which is eventually joined with the developing ventral nerve cord by the formation of circumenteric connectives. The body continues to elongate as more segments form, and the juvenile worm finally drops from the plankton and assumes the lifestyle of a young polychaete. This whole affair was beautifully described in verse by the late Walter Garstang (1951), where he explains the development of *Phyllodoce* in the first part of his classic poem, *The Trochophores*:

The trochophores are larval tops the Polychaetes set
  spinning
With just a ciliated ring—at least in the beginning—
They feed, and feel an urgent need to grow more
  like their mothers,
So sprout some segments on behind, first one, and
  then the others.
And since more weight demands more power, each
  segment has to bring
Its contribution in an extra locomotive ring:
With these the larva swims with ease, and, adding
  segments more,
Becomes a *Polytrochula* instead of *Trochophore*.
Then setose bundles sprout and grow, the sequel
  can't be hid:
The larva fails to pull its weight, and sinks—
  an Annelid.

Oligochaeta. Nowhere is the divergence of the
oligochaetes from the polychaetes more apparent
than in the major differences in reproductive and life
history strategies. The evolution of new reproductive
styles unquestionably contributed greatly to the successful invasion of fresh water and land by the oligochaetes.

Oligochaetes are hermaphroditic and usually
possess distinct and complex reproductive systems,
including permanent gonads. Furthermore, various
parts of the reproductive apparatus are restricted to
particular segments, usually in the anterior portion
of the worm (Figure 36). The design of the reproductive systems facilitates mutual cross fertilization followed by encapsulation and deposition of the zygotes, as explained below.

The male system includes one or two pairs of
testes located in one or two specific body segments.
Sperm are released from the testes into the coelomic
spaces, where they mature or are picked up by storage sacs (seminal vesicles) derived from pouches of
the septal peritoneum (Figure 36B). There may be a
single seminal vesicle or as many as three pairs in
some earthworms. When mature, the sperm are released from the seminal vesicles, picked up by ciliated seminal (sperm) funnels, and carried by sperm
ducts to paired gonopores.

The female reproductive system consists of a single pair of ovaries located posterior to the male system (Figures 19C and 36B). Again, the ova are released into the adjacent coelomic space and
sometimes stored until mature in shallow pouches in
the septal wall called the **ovisacs**. Next to each ovisac
is a ciliated funnel that carries the mature ova to an
oviduct and eventually to the female gonopore. Most
oligochaetes also possess one or two or more pairs
of blind sacs called **spermathecae** (seminal receptacles) that open to the outside via separate pores
(Figure 36).

Of major importance to the overall reproductive
strategy of oligochaetes is a region of glandular tissue
called the **clitellum** (Latin for "saddle") (Figure 36A).
This structure is a principal feature unifying members

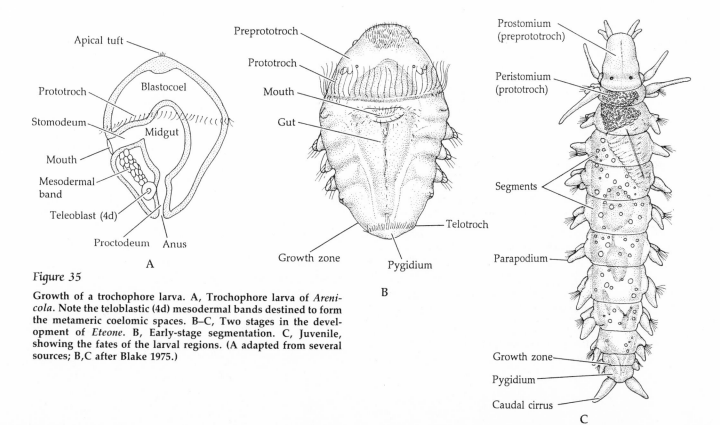

Figure 35

Growth of a trochophore larva. A, Trochophore larva of *Arenicola.* Note the teloblastic (4d) mesodermal bands destined to form
the metameric coelomic spaces. B–C, Two stages in the development of *Eteone.* B, Early-stage segmentation. C, Juvenile,
showing the fates of the larval regions. (A adapted from several
sources; B,C after Blake 1975.)

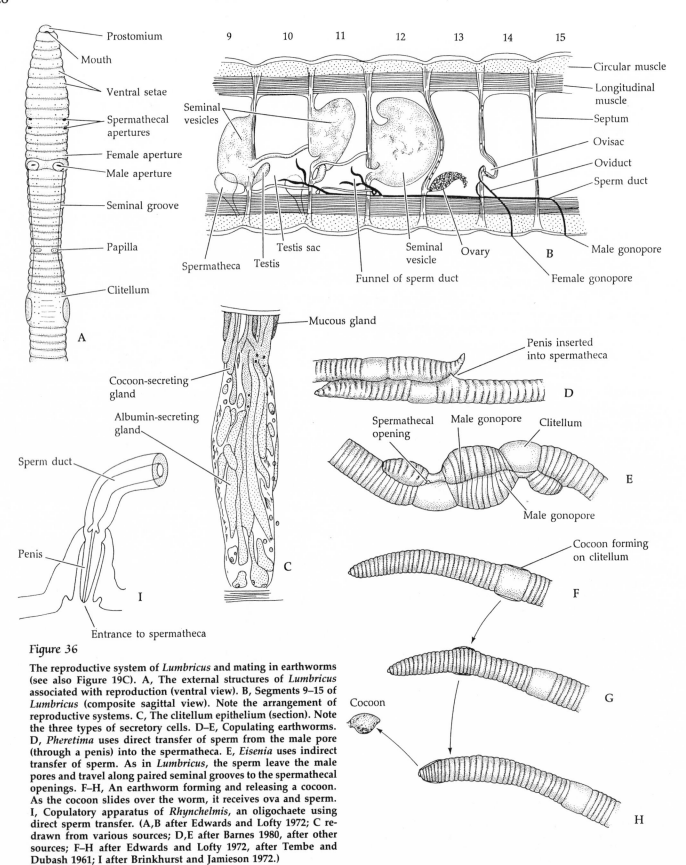

Figure 36

The reproductive system of *Lumbricus* and mating in earthworms (see also Figure 19C). A, The external structures of *Lumbricus* associated with reproduction (ventral view). B, Segments 9–15 of *Lumbricus* (composite sagittal view). Note the arrangement of reproductive systems. C, The clitellum epithelium (section). Note the three types of secretory cells. D–E, Copulating earthworms. D, *Pheretima* uses direct transfer of sperm from the male pore (through a penis) into the spermatheca. E, *Eisenia* uses indirect transfer of sperm. As in *Lumbricus*, the sperm leave the male pores and travel along paired seminal grooves to the spermathecal openings. F–H, An earthworm forming and releasing a cocoon. As the cocoon slides over the worm, it receives ova and sperm. I, Copulatory apparatus of *Rhynchelmis*, an oligochaete using direct sperm transfer. (A,B after Edwards and Lofty 1972; C redrawn from various sources; D,E after Barnes 1980, after other sources; F–H after Edwards and Lofty 1972, after Tembe and Dubash 1961; I after Brinkhurst and Jamieson 1972.)

of the classes Oligochaeta and Hirudinida as the clitellate annelids; and even though we have presented a conventional classification scheme here, there is strong support for uniting these groups as a single class Clitellata (see Phylogeny section). The clitellum has the appearance of a thick sleeve that partially or completely encircles the worm's body. It is formed of secretory cells within the epidermis of particular segments. The exact position of the clitellum and the number of segments involved are consistent within any one species. In freshwater forms the clitellum is located around the position of the gonopores, but in most earthworms it is posterior to the gonopores.

There are three types of gland cells within the clitellum, each secreting a different substance important to reproduction: mucus that aids in copulation, the material forming the outer casing of the egg capsule (or **cocoon**), and albumin deposited with the zygotes inside the cocoon. In most aquatic oligochaetes these cell types all lie in a single layer, but in the terrestrial forms distinct cell layers are present (Figure 36C).

Free spawning and indirect larval development simply would not do in the environments of many oligochaetes, and the success of these animals has depended in large part on contact mating, exchange of sperm, and direct development. The high survival rate of zygotes produced by such methods balances the relatively high parental investment. And, as we have seen, hermaphroditism is one way for slow and sluggish animals, who might encounter sexual partners only infrequently, to increase their reproductive success.

During copulation, the mating worms align themselves side-by-side but facing in opposite directions (Figures 36D and E). Mucous secretions from the clitellum hold the worms in this copulatory posture. Most oligochaetes position themselves so that the male gonopores of one are aligned with the spermathecal openings of the other. In such cases, special copulatory setae near the male pores or eversible penis-like structures aid in anchoring the mates together (Figure 36I). The lumbricid earthworms are not so accurate with their mating, and their copulatory position does not bring the male pores against the spermathecal openings. Instead of sperm being transferred directly into the spermathecae, lumbricids bear external sperm grooves along which the male gametes must travel prior to entering the spermathecal pores (Figures 36A and E). These grooves are actually formed temporarily by the contraction of certain underlying muscles and are covered by a sheet of mucus. Once formed, the grooves undulate

by muscle action and the sperm are transported along the body to their destination. Following the mutual exchange of sperm to the seminal receptacles of each mate, the worms separate, each then functioning as an inseminated female.

From several hours to a few days following copulation, a sheet of mucus is produced around the clitellum and all of the anterior segments. Then the clitellum produces the cocoon itself in the form of a leathery, proteinaceous sleeve. The cocoons of terrestrial species are especially tough and resistant to adverse conditions. Albumin is secreted between the cocoon and the clitellar surface. The amount of albumin deposited with the cocoon is much greater in terrestrial species than in aquatic forms.

Thus formed, the cocoon and underlying albumin sheath are moved toward the anterior end of the worm by muscular waves and backward motion of the body. As it moves along the body, the cocoon first receives eggs from the female gonopores and then sperm (originally from the mate) from their storage place in the spermathecae. Fertilization occurs within the albumin matrix inside the cocoon. The open ends of the cocoon contract and seal as they pass off the anterior end of the body (Figures 36F–H). The closed cocoons are deposited in benthic debris by aquatic oligochaetes. Terrestrial forms deposit their cocoons in the soil at various depths, the particular depth depending on the moisture content of the substratum. The shape and size of the cocoon are often species-specific.

Oligochaetes produce telolecithal ova, but the amount of yolk varies greatly and inversely with the amount of albumin secreted into the cocoon. The eggs of freshwater oligochaetes contain relatively large amounts of yolk but are encased with only a small quantity of albumin. Conversely, the eggs of terrestrial species tend to be weakly yolked but are supplied with large quantities of albumin on which the developing embryos depend for a source of nutrition. In any case, cleavage is holoblastic and unequal. And, although highly modified, evidences of the ancestral spiralian pattern are still apparent in cell placement and fates (e.g., an identifiable 4d mesentoblast homologue gives rise to the presumptive mesoderm).

Development is direct, with no trace of a larval stage. However, the teloblastic production of coelomic spaces and segments is an obvious retained characteristic of the basic annelid plan.

Development time varies from about one week to several months, depending on the species involved and environmental conditions. In climates where relatively severe conditions follow cocoon deposition, development time is generally sufficient to ensure that the juveniles hatch in the spring. Under more stable conditions, development time is shorter and reproduction is less seasonal.

The number of zygotes within each cocoon varies from 1 to about 20, again depending on the species. However, when several zygotes are included, only one or a few actually reach the hatching stage.

Hirudinida. Leeches are clitellate, hermaphroditic annelids, with complex reproductive organs (Figures 37A and B). As in the oligochaetes, their environments and lifestyles are best suited to embryo encapsulation and direct development.

The male reproductive system includes a variable number of paired testes, usually from five to ten pairs in leeches, arranged serially beginning in about segment XI or XII and extending posteriorly (Figure 37B). The testes are drained by a pair of sperm ducts, one on each side of the body, that lead to a copulatory apparatus and a single gonopore located midventrally on segment X.

The copulatory apparatus of leeches is often complex and varies in structure among different species. Each sperm duct is coiled distally and enlarges as an ejaculatory duct. The two ducts join at a common glandular, muscular atrium. In arhynchobdellids the atrium is modified as an eversible penis. The rhynchobdellids lack a penis and the atrium functions as a chamber in which spermatophores are produced.

There is a single pair of ovaries in leeches. The ovaries may be spherical or elongate, extending through several segments (Figures 37A and B). Oviducts extend anteriorly from the ovaries and unite as a common vagina, which leads to the female gonopore on the midventral surface of segment XI, just behind the male pore. In some leeches an **oviducal gland** surrounds a portion of the oviduct and vagina, and apparently functions in egg-laying activity.

Copulation and sperm transfer differ markedly between the rhynchobdellids and arhynchobdellids,

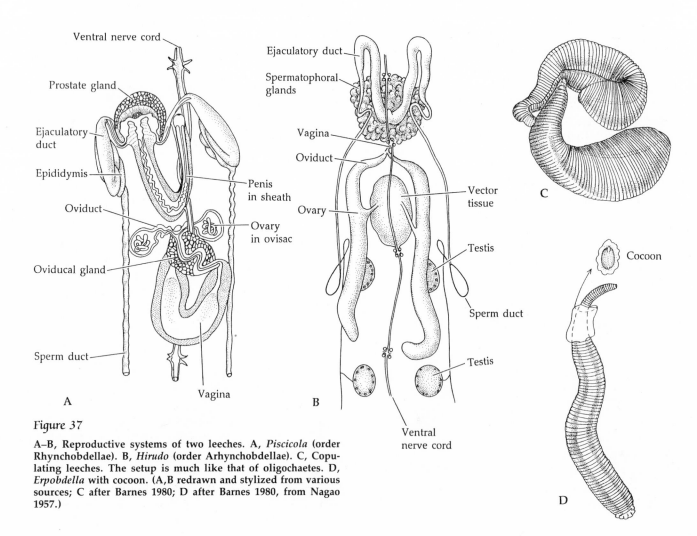

Figure 37

A–B, Reproductive systems of two leeches. A, *Piscicola* (order Rhynchobdellae). B, *Hirudo* (order Arhynchobdellae). C, Copulating leeches. The setup is much like that of oligochaetes. D, *Erpobdella* with cocoon. (A,B redrawn and stylized from various sources; C after Barnes 1980; D after Barnes 1980, from Nagao 1957.)

owing in part to the differences in the male copulatory structures. In those with a penis (Arhynchobdellae), mating worms align themselves so that the male pore of each rests over the female pore of the other. The penis is everted and inserted into the mate's vagina; sperm are then introduced directly into the vagina, which serves as a seminal receptacle. Fertilization takes place within the female reproductive system.

The rhynchobdellid leeches engage in an unusual form of hypodermic impregnation. These animals grasp one another by their anterior suckers and bring their male pores in alignment with a particular region of the mate's body (Figure 37C). In most rhynchobdellids, the spermatophores are released onto the clitellar region of the mate. In some of the piscicolids, however, there is a special "target" area, beneath which is a mass of tissue called **vector tissue** that connects with the ovaries via short ducts. In either case, after the spermatophores are placed on the body of the mate they penetrate the body surface of the recipient and the sperm emerge beneath the epidermis. In those species without vector tissue, the sperm migrate to the ovaries by way of the coelomic channels and sinuses. In the other species, the sperm move through the vector tissue and their attending ducts to the ovaries, where fertilization takes place.

Cocoon formation in leeches is similar to that in oligochaetes, with the clitellum producing a cocoon wall and albumin (Figure 37D). However, as the cocoon slides anteriorly past the female gonopore, it receives the zygotes or young embryos rather than separate eggs and sperm. The cocoons are deposited in damp soil by terrestrial species and even by a few aquatic forms that migrate to land for this process (e.g., *Hirudo*). Most of the aquatic forms deposit their cocoons by attaching them to the bottom or to algae; a few attach them to their hosts (e.g., some piscicolids). A few freshwater leeches display some degree of parental care for their cocoons. Some of these bury their cocoons and then remain over them, generating ventilatory currents. Others attach the cocoons to their own bodies and brood them externally.

The embryogeny of leeches is basically the same as that described for oligochaetes. Except for a few species, the amount of yolk is relatively small and development time quite short.

Annelid phylogeny

A discussion of annelid phylogeny involves, by necessity, some consideration of the origin of the coelomic condition and segmentation. We cannot cover all the myriad details and arguments on these topics; instead we present a set of general ideas about the origin of and radiation within the Annelida and establish a foundation for placing the rest of the protostome phyla in a reasonable and understandable perspective. The scenario described below is expressed in the cladogram and evolutionary tree in Figure 38.

The protostome coelom may have evolved as a response to peristaltic burrowing (see Chapter 4). Within the protostomes, this first schizocoelomate creature may have arisen from a triploblastic, acoelomate ancestor or from a diploblastic precursor. In the latter case, the coelom arose coincidentally with 4d mesoderm and other spiralian features. These ideas were explored in Chapter 10. We may view the protoannelids as homonomous metameric burrowers derived from a segmented, coelomate, ancestral vermiform creature. The nature of the protoannelid remains controversial. Mettam (1985) proposed the possibility that the first annelids were tiny interstitial creatures, but Westheide (1985) suggests that they were moderate to large forms. In any event, the annelids almost certainly share a common ancestor with the arthropods (Chapters 15–19), and we view the two groups as sister taxa (see Chapter 24).

Paleontological evidence obtained over the past two decades indicates that coelomate burrowers existed well over 600 million years ago, in Precambrian times. Some of the Precambrian and early Cambrian fossils include trace fossils and burrows thought to be of annelid origin (see Glaessner 1976 and Brasier 1979). Fossils that are clearly polychaetes appear by the middle Cambrian. These fossils are also the earliest certain records of members of the Annelida, thus the polychaetes are typically considered to be the earliest derived group within the phylum as we know it today. The evolution of parapodia in association with the coelomate metameric *Bauplan* presumably allowed early annelids to "emerge from the muds" and begin to exploit surface locomotion by crawling, as seen in many modern day polychaetes, and eventually by certain forms of undulatory swimming. Tube-dwelling lifestyles, heteronomy, the loss of complete septa, and most of the other variations within members of the Polychaeta are secondarily derived conditions within various annelid lines.

The cladogram in Figure 38A depicts the annelids as comprising two great radiations; these lineages, or sister-groups, are the Polychaeta and the clitellates, or Clitellata (= Oligochaeta + Hirudinida). Although the precise nature of the ancestral group (the "pro-

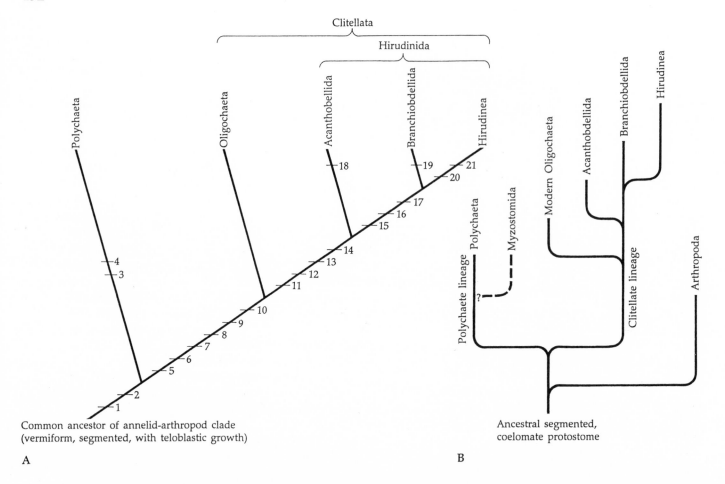

Figure 38

Proposed cladogram (A) and evolutionary tree (B) depicting the evolution of the annelids. The ancestor to the annelid–arthropod line is viewed as a segmented coelomate protostome (see Chapter 24). The origin of the annelids was marked by the appearance of two synapomorphies: (1) the annelid head, and (2) epidermal, paired setae (or setal bundles). Once these features had arisen, the ancestral worm would unquestionably be classified as an annelid, and this protoannelid was the ancestor from which all modern taxa were ultimately derived. Early radiation from this ancestral annelid led to the polychaete lineage, defined by the evolution of parapodia (3), and elaboration of the complex head (4). Some workers have suggested that nuchal organs may also be a synapomorphy of polychaetes (to have been subsequently lost in some modern polychaete taxa).

A second clade led to the clitellate annelids, which lack parapodia and generally retain the basic ancestral homonomous condition. This clade is defined by the appearance of six synapomorphic characters: (5) evolution of the hermaphroditic condition (simultaneous hermaphroditism); (6) development of the clitellum and its complex suite of features; (7) loss of the free larval stage and reliance on direct development; (8) posterior displacement of the cerebral ganglion into the anterior trunk segments; (9) evolution of a distinct, fixed, and complex reproductive system, with specialized organs (e.g., permanent gonads); (10) the tendency toward a fixed number of body segments. This is the evolutionary level at which the oligochaetes exist today, representing the most primitive of the Clitellata.

The hirudinidan line (Acanthobdellida–Branchiobdellida–

Hirudinea) is defined by the appearance of four synapomorphies: (11) the tendency toward reduction of the septal walls and fusion of the segmental coelomic compartments; (12) appearance of a posterior body sucker; (13) subdivision of the body segments by superficial annuli; (14) a major reduction in body setae, initially lost on all but the anterior segments. This is the evolutionary level at which the acanthobdellids have remained, and this group could thus be viewed as a primitive "relict" taxon representative of the transition from an oligochaete (or oligochaete-like ancestor) to hirudineans. Acanthobdellids have the body segment number fixed at 30 segments (18). Next to appear in the evolution of this lineage were (15) an anterior body sucker, (16) loss of the remaining body setae, and (17) the evolution of "mesenchymal" tissue and reduction of the coelom. This is the evolutionary level represented today by the branchiobdellids, which also may be viewed as representative of the transition from oligochaetes to hirudineans. Branchiobdellids have the body segment number fixed at 15 segments (19). Finally, the "true leeches" (subclass Hirudinea) are distinguished by (20) further reduction of the coelom to a series of channels or lacunae, and by (21) having the body segment number fixed at 34 segments.

The cladogram illustrates the genealogy of these five groups by depicting the sequence in which their defining synapomorphies appeared. The evolutionary tree presents these taxa in a more traditional fashion. It should be noted that some of the characters we have used are quite complex and may be viewed as multicharacter suites, themselves subject to more detailed analyses.

toannelids") from which these two lines arose is lost in time, we can reasonably hypothesize that it was a relatively simple, homonomous, metameric burrower with a compartmentalized segmented coelom, epidermal setae, and a head composed of a presegmental prostomium and peristomium. In one of these lines parapodia and head appendages evolved, giving rise to the polychaetes. In terms of life history strategies, the first true polychaetes were probably similar to many of their modern descendants. It is likely that they were gonochoristic broadcast spawners with indirect development. Their ancestors had already established the basic protostome plan of development, complete with rapid development of a trochophore larval stage.

The other line, the clitellate annelids, arose by way of evolution of a complex hermaphroditic life history, permanent gonads, direct development, the remarkable structure known as the clitellum, and other features (see Figure 38A). Within this clade are two major groups: the Oligochaeta and the Hirudinida (all three of the hirudinidan subclasses are depicted in the cladogram). However, note that the line leading to the oligochaetes on the cladogram is not defined by any synapomorphies. We are not aware of any features unique to the oligochaetes as they are currently understood. The oligochaetes are defined entirely by their retention of primitive features (symplesiomorphies) and by the *lack* of the features distinguishing the hirudinidan lineage. Thus, the ancestor of the entire clitellate lineage may be viewed as being an oligochaete (i.e., as we define oligochaetes today). As we noted in Chapter 2, groups such as the Oligochaeta that are defined solely by retention of plesiomorphic features (rather than by synapomorphies) are typically paraphyletic. This is almost certainly the case with the Oligochaeta. Recognition of this fact adds credence to the proposal that the oligochaetes and hirudinidans be combined into a single class, the Clitellata. In this view, oligochaetes are simply primitive clitellates, and hirudinidans are more derived clitellates, evolved from some isolated "oligochaete" ancestor. Although this is a view we

favor, it has not yet been accepted by all annelid biologists.

The date of origin of the clitellate annelids is a somewhat controversial matter. Some workers suggest that the oligochaetes arose quite late in geological time, in association with the evolution of flowering land plants in the Jurassic and Cretaceous periods. However, fossils of oligochaete-like creatures (e.g., *Palaeoscolex*) are known from Cambrian deposits, suggesting that clitellates arose closer to the origin of Polychaetes, in line with the sister group relationships depicted in the cladogram. Of course, it is possible that the clitellate line arose from a preexisting polychaete stock through such evolutionary steps as loss of the parapodia, reduction of the complex head region, and so on, but this is a less parsimonious hypothesis than to view them arising prior to the origin of what we recognize today as polychaetes. We thus envision the oligochaetes arising from some pre-polychaete stock in shallow estuarine or freshwater muds. The evolution of complex hermaphroditic reproductive systems, yolky eggs, and the clitellum to facilitate encapsulation of the developing embryos freed the early oligochaetes from the marine environment and the dependency on a larval phase. These events were certainly of major importance in their successful exploitation of freshwater and land habitats.

The acanthobdellids and branchiobdellids show a mixture of characteristics between the oligochaetes and the hirudineans (the "true leeches"), thus forming an evolutionary series. The evolution of suckers may have initially provided a means of clinging to objects as some oligochaetes left the muds and assumed an epibenthic lifestyle. This ability would have been especially useful in streams and rivers. The use of suckers for temporary attachment to substrata may have preadapted these creatures for later exploitation of ectoparasitism. Some workers have suggested that the odd Acanthobdellida may be no more than degenerate leeches that have lost the anterior sucker. However, the presence of anterior setae in this enigmatic group argues for its position as a primitive hirudinidan. ·

Selected References

General

Anderson, D. T. 1973. *Embryology and Phylogeny in Annelids and Arthropods.* Pergamon Press, Oxford.

Berrill, N. J. 1952. Regeneration and budding in worms. Biol. Rev. 27:401.

Brasier, M. D. 1979. The Cambrian radiation event. *In* M. R. House (ed.), *The Origin of Major Invertebrate Groups.* Academic Press, London, pp. 103–160.

Brinkhurst, R. O. and S. R. Gelder. 1989.

Do the lumbriculids provide the ancestors of the brachiobdellidans, acanthodbellidans and leeches? Hydrobiol. 180: 7–15.

Cather, J. 1971. Cellular interactions in the regulation of development in annelids and molluscs. Adv. Morphog. 9:67–124.

Clark, R. B. 1969. Systematics and phylogeny: Annelida, Echiura, Sipuncula. *In* M. Florkin and B. T. Scheer (eds.), *Chemical Zoology*, Vol. 4. Academic Press, New York, pp. 1–68.

Clark, R. B. 1979. Radiation of the metazoa. *In* M. R. House (ed.), *The Origin of Major Invertebrate Groups*. Academic Press, London, pp. 55–102.

Dales, R. P. 1967. *Annelids*, 2nd Ed. Hutchinson University Library, London.

Fauvel, P., M. Avel, H. Harant, P. Grassé and C. Dawydoff. 1959. Embranchement des Annélides. *In* P. Grassé (ed.), *Traité de Zoologie*, Vol. 5, Pt. 1. Masson et Cie, Paris, pp. 3–686.

Goodrich, E. S. 1946. The study of nephridia and genital ducts since 1895. Q. J. Microsc. Sci. 86:113–392.

Herlant-Meewis, H. 1965. Regeneration in annelids. *In* M. Abercrombie and J. H. Brachet (eds.), *Advances in Morphogenesis*, Vol. 4. Academic Press, New York, pp. 155–215.

Iwanoff, P. P. 1928. Die Entwicklung der Larvalsegmente bei den Anneliden. Z. Morphol. Ökol. Tiere 10:62–161.

Løvtrup, S. 1975. Validity of the protostomia–deuterostomia theory. Syst. Zool. 24:96–108.

Mangum, C. 1970. Respiratory physiology in annelids. Am. Sci. 58(6):641–647.

Mangum, C. 1976. Primitive respiratory adaptations. *In* R. C. Newell (ed.), *Adaptation to Environment*. Butterworth Group Publishing, Boston, pp. 191–278.

Mettam, C. 1985. Functional constraints in the evolution of the Annelida. *In* S. C. Morris et al. (eds.), *The Origins and Relationships of Lower Invertebrates*. Syst. Assoc. Spec. Vol. No. 28, Oxford, pp. 297–309.

Mill, P. J. (ed.). 1978. *Physiology of Annelids*. Academic Press, London.

Moment, G. B. 1953. A theory of growth limitation. Am. Nat. 88(834):139–153.

Newell, R. C. 1970. *Biology of Intertidal Animals*. American Elsevier, New York.

Nicol, J. A. C. 1948. The giant axons of annelids. Q. Rev. Biol. 23:291–323.

Pettibone, M. H. 1982. Annelida. *In* S. Parker, *Synopsis and Classification of Living Organisms*, Vol. 2. McGraw-Hill, New York, pp. 1–43. [Pettibone's contribution includes an introductory section on the phylum and coverage of the Polychaeta.]

Ruppert, E. E. and P. R. Smith. 1988. The functional organization of filtration nephridia. Biol. Rev. 63:231–258.

Wald, G. and S. Rayport. 1977. Vision in annelid worms. Science 196:1434–1439.

Wells, G. P. 1959. Worm autobiographies. Sci. Am. 200(6):132–141.

Polychaeta/Myzostomida

Anderson, D. T. 1966. The comparative embryology of the Polychaeta. Acta. Zool. Stockholm 47:1–41.

Barnes, R. D. 1965. Tube-building and feeding in chaetopterid polychaetes. Biol. Bull. 129:217–233.

Baskin, D. G. 1976. Neurosecretion and the endocrinology of nereid polychaetes. Am. Zool. 16:107–124.

Berglund, A. 1986. Sex change by a polychaete: Effects of social and reproductive costs. Ecology 67(4):837–845.

Blake, J. A. 1975. The larval development of Polychaeta from the northern California coast. III. Eighteen species of Errantia. Ophelia 14:23–84.

Bobin, G. 1944. Morphogénèse des sois chez les annélides polychètes. Ann. Inst. Océanogr. Monaco 22:1–106.

Boilly, B. 1969. Sur l'origine des cellules régénératrices chez les annélides polychètes. Arch. Zool Exp. Gen. 110:127–143.

Brown, S. C. 1975. Biomechanics of water-pumping by *Chaetopterus veriapedatus* Renier: Skeletomusculature and kinematics. Biol. Bull. 149:136–150.

Brown, S. C. 1977. Biomechanics of water-pumping by *Chaetopterus variapedatus* Renier: Kinetics and hydrodynamics. Biol. Bull. 153:121–132.

Caspers, H. 1984. Spawning periodicity and habit of the palolo worm *Eunice viridis* in the Samoan Islands. Mar. Biol. 79:229–236.

Clark, R. B. 1956a. The blood vascular system of *Nephtys* (Annelida, Polychaeta). Q. J. Microsc. Sci. 97:235–249.

Clark, R. B. 1956b. *Capitella capitata* as a commensal, with a bibliography of parasitism and commensalism in polychaetes. Ann. Mag. Nat. Hist. 9(102):433–448.

Clark, R. B. 1961. The origin and formation of the heteronereis. Biol. Rev. 36:199–236.

Clark, R. B. 1965. Endocrinology and the reproductive biology of polychaetes. Oceanogr. Mar. Biol. Annu. Rev. 3:211–255.

Clark, R. B. and W. N. Hess. 1940. Swarming of the Atlantic palolo worm, *Leodice fucato*. Tortugas Lab. Papers 33(2):21–70.

Clark, R. B. and P. J. W. Olive. 1973. Recent advances in polychaete endocrinology and reproductive biology. Ocean. Mar. Bio. Annu. Rev. 11:175–222.

Clark, R. B. and D. J. Tritton. 1970. Swimming mechanisms of nereidiform polychaetes. J. Zool. Lond. 161:257–271.

Dales, R. P. 1955. Feeding and digestion in terebellid polychaetes. J. Mar. Biol. Assoc. U.K. 34:55–79.

Dales, R. P. 1957. Preliminary observations on the role of the coelomic cells in food storage and transport in certain polychaetes. J. Mar. Biol. Assoc. U.K. 36:91–110.

Dales, R. P. 1962. The polychaete stomodeum and the interrelationships of the families of Polychaeta. Proc. Zool. Soc. Lond. 139:389–428.

Dales, R. P. and G. Peter. 1972. A synopsis of the pelagic Polychaeta. J. Nat. Hist. 6(1):55–92.

Desbruyeres, D., F. Gaill, L. Laubier, D. Prieur, and G. Rau. 1983. Unusual nutrition of the "pompeii worm" *Alvinella pompejana* (polychaetous annelid) from a hydrothermal vent environment: SEM, TEM, ^{13}C, and ^{15}N evidence. Mar. Biol. 75:201–205.

Durchon, M. 1965. Sur l'évolution phylogénétique et ontogénétique de l'épitoquie chez les néréidiens. Zool. Jahrb. Syst. 92:1–12.

Eakin, R. M., G. G. Martin and C. T. Reed. 1977. Evolutionary significance of fine structure of archiannelid eyes. Zoomorphologie 88:1–18.

Eisig, H. 1906. *Ichthyostomous sanguinarius*, eine auf Aalen schmarotzend Annelide. Fauna Flora Neapel 28.

Evans, S. M. 1973. A study of fighting reactions in some nereid polychaetes. Anim. Behav. 21:138–146.

Fauchald, K. 1975. Polychaete phylogeny: A problem in protostome evolution. Syst. Zool. 23:493–506.

Fauchald, K. 1977. The polychaete worms: Definitions and keys to the orders, families and genera. Nat. Hist. Mus. Los Angeles Co. Ser. 28:1–190. [A revised version is in preparation.]

Fauchald, K. and P. A. Jumars. 1979. The diet of worms: A study of polychaete feeding guilds. Oceanogr. Mar. Biol. Annu. Rev. 17:193–284. [This and the above paper by Fauchald (1977) are key works in modern polychaetology.]

Fischer, A. and H.-D. Pfannenstiel (eds.). 1984. *Polychaete Reproduction. Progress in Comparative Reproductive Biology*. Gustav Fischer Verlag, Stuttgart and New York. [A symposium volume.]

Fitzharris, T. P. 1976. Regeneration in sabellid annelids. Am. Zool. 16:593–616.

Franke, H. D. 1986. Sex ratio and sex change in wild and laboratory populations of *Typosyllis prolifera* (Polychaeta). Mar. Biol. 90:197–208.

Glaessner, M. F. 1976. Early Phanerozoic annelid worms and their geological and biological significance. J. Geol. Sci. 132:259–275.

Gray, J. 1939. Studies in animal locomotion. VIII. The kinetics of locomotion of *Nereis diversicolor*. J. Exp. Biol. 16:9–17.

Hartman, O. 1959 and 1965. Catalogue of the polychaetous annelids of the world. Occ. Papers Allan Hancock Fnd. Vol. 23. Supplement and Index (1965).

Hartman, O. 1966. Polychaeta, Myzostomidae, and Sedentaria of Antarctica. Contrib. #288, Allan Hancock Foundation, Vol. 7, Antarctic Res. Ser., pp. 1–158.

Hermans, C. O. 1969. The systematic position of the Archiannelida. Syst. Zool. 18: 85–102.

Hermans, C. and R. M. Eakin. 1974. Fine structure of the eyes of an alciopid polychaete *Vanadis tagensis*. Z. Morphol. Tiere 79: 245–267.

Hill, S. D. 1970. Origin of the regeneration blastema in polychaete annelids. Am. Zool. 10: 101–112.

Jacobsen, V. H. 1967. The feeding of the lugworm, *Arenicola marina*. Ophelia 4: 91–109.

Jägersten, G. 1940. Zur Kenntnis der Morphologie, Entwicklung und Taxonomie der Myzostomida. Nova Acta Regiae Soc. Sci. Ups. 4, 11(8): 1–84.

Kay, D. G. 1974. The distribution of the digestive enzymes in the gut of the polychaete *Neanthes virens*. Comp. Biochem. Physiol. 47A: 573–582.

Kristensen, R. M. and T. Niilonen. 1982. Structural studies on *Diurodrilus westheidei* sp. n. from the Arctic interstitial meiobenthos, W. Greenland. Zool. Scr. 11(1): 1–12.

Kristensen, R. M. and A. Nørrevang. 1982. Description of *Psammodrilus aedificator* sp. n. (Polychaeta), with notes on the Arctic interstitial fauna of Disko Island, W. Greenland. Zool. Scr. 11(4)265–279.

Kudenov, J. D. 1977. Brooding behavior and protandry in *Hipponoe gaudichaudi* (Polychaeta: Amphinomidae). Bull. So. Calif. Acad. Sci. 76(2): 85–90.

Kudenov, J. D. 1977. The functional morphology of feeding in three species of maldanid polychaetes. Zool. J. Linn. Soc. 60: 95–109.

Levin, L. A. 1982. Interference interactions among tube-dwelling polychaetes in a dense infaunal assemblage. J. Exp. Mar. Biol. Ecol. 65: 107–119.

MacGinitie, G. E. 1939. The method of feeding of *Chaetopterus*. Biol. Bull. 77: 115–118.

Mangum, C. 1976. The oxygenation of hemoglobin in lugworms. Physiol. Zool. 49(1): 85–99.

Mettam, C. 1967. Segmental musculature and parapodial movement of *Nereis diversicolor* and *Nephtys hombergi* (Annelida: Polychaeta). J. Zool. Lond. 153: 245–275.

Nicol, E. A. T. 1931. The feeding mechanism, formation of the tube, and physiology of digestion in *Sabella pavonia*. Trans. R. Soc. Edinburgh 56(3): 537–598.

Oglesby, L. C. 1965. Water and chloride fluxes in estuarine nereid polychaetes. Comp. Biochem. Physiol. 16: 437–455.

Orrhage, L. 1980. Structure and homologies of the anterior end of the polychaete families, Sabellidae and Serpulidae. Zoomorphologie 96: 113–168.

Pennington, T. J. and F. S. Chia. 1984. Morphological and behavioral defenses of trochophore larvae of *Sabellaria cementarium* (Polychaeta) against four planktonic predators. Biol. Bull. 167: 168–175.

Reish, D. J. 1970. The effects of varying concentrations of nutrients, chlorinity, and dissolved oxygen on polychaetous annelids. Water Res. 4: 721–735.

Reish, D. J. and K. Fauchald (eds.). 1977. *Essays on Polychaetous Annelids in Memory of Dr. Olga Hartman*. Allan Hancock Foundation, Los Angeles.

Robbins, D. E. 1965. The biology and morphology of the pelagic annelid *Poeobius meseres* Heath. J. Zool. 146: 197–212.

Ruby, E. G. and D. L. Fox. 1976. Anerobic respiration in the polychaete *Euzonus* (*Thoracophelin*) *mucronata*. Mar. Biol. 35(2): 149–153.

Schroeder, P. C. and C. O. Hermans. 1975. Annelida: Polychaeta. *In* A. C. Giese and J. S. Pearse (eds.), *Reproduction of Marine Invertebrates*, Vol. 3. Academic Press, New York, pp. 1–205.

Smetzer, B. 1969. Night of the palolo. Nat. Hist. 78: 64–71.

Smith, P. R. and E. E. Ruppert. 1988. Nephridia. *In* W. Westheide and C. O. Hermans (eds.), *Microfauna Marina* 4. Gustave Fischer Verlag, Stuttgart and New York, pp. 231–262.

Smith, P. R., E. E. Ruppert and S. L. Gardiner. 1987. A deuterostome-like nephridium in the mitraria larva of *Owenia fusiformis* (Polychaeta, Annelida). Biol. Bull. 172: 315–323.

Smith, R. I. 1958. On reproductive pattern as a specific characteristic among nereid polychaetes. Syst. Zool. 7: 60–73.

Smith, R. I. 1963. A comparison of salt loss rate in three species of brackish-water nereid polychaetes. Biol. Bull. 125: 332–343.

Spies, R. B. 1975. Structure and function of the head in flabelligerid polychaetes. J. Morphol. 147: 187–208.

Swedmark, B. 1955. Recherches sur la morphologie le développement et la biologie de *Psammodrilus balanoglossoides* polychète sédentaire de la microfaune des Sables. Arch. Zool. Exp. Gen. 92: 141–220.

Wells, G. P. 1950. Spontaneous activity cycles in polychaete worms. Symp. Soc. Exp. Biol. 4: 127–142.

Westheide, W. 1985. The systematic position of the Dinophilidae and the archiannelid problem. *In* S. C. Morris et al. (eds.), *The Origins and Relationships of Lower Invertebrates*. Syst. Assoc. Spec. Vol. No. 28, Oxford, pp. 310–326.

Whitlatch, R. B. 1974. Food-resource partitioning in the deposit feeding polychaete *Pectinaria gouldii*. Biol. Bull. 147(1): 227–235.

Wilson, E. B. 1892. The cell-lineage of *Nereis*. J. Morphol. 6: 361–480.

Oligochaeta/Hirudinida

Anderson, D. T. 1966. The comparative early embryology of the Oligochaeta, Hirudinea and Onychophora. Proc. Linn. Soc. N.S.W. 91: 10–43.

Brinkhurst, R. O. 1982a. *British and Other Marine and Estuarine Oligochaetes*. (Synopsis of British Fauna No. 21). Academic Press, New York.

Brinkhurst, R. O. 1982b. Oligochaeta. *In* S. Parker, *Synopsis and Classification of Living Organisms*, Vol. 2. McGraw-Hill, New York, pp. 50–61. [Includes coverage of some taxa by R. W. Sims.]

Brinkhurst, R. O. and D. G. Cook (eds.). 1980. *Aquatic Oligochaete Biology*. Plenum, New York.

Brinkhurst, R. O. and B. G. Jamieson. 1972. *Aquatic Oligochaeta of the World*. Toronto University Press, Toronto.

Cook, D. G. and R. O. Brinkhurst. 1973. *Marine Flora and Fauna of the Northeastern United States. Annelida: Oligochaeta*. NOAA Tech. Rpt. NMFS CIRC–374.

Edwards, C. A. and J. R. Lofty. 1972. *Biology of Earthworms*. Chapman and Hall, London.

Giere, O. and C. Langheld. 1987. Structural organization, transfer and biological fate of endosymbiotic bacteria in gutless oligochaetes. Mar. Biol. 93: 641–650.

Gray, J. and H. W. Lissmann. 1938. Studies in animal locomotion, VII. Locomotory reflexes in the earthworm. J. Exp. Biol. 15: 506–517.

Gray, J., H. W. Lissmann and R. J. Pumphrey. 1938. The mechanism of locomotion in the leech. J. Exp. Biol. 15: 408–430.

Holt, T. C. 1965. The systematic position of the Branchiobdellida. Syst. Zool. 14: 25–32.

Holt, T. C. 1968. The Branchiobdellida: Epizootic annelids. Biologist 1(3–4): 79–94.

Jamieson, B. G. 1981. *The Ultrastructure of the Oligochaeta*. Academic Press, London.

Jamieson, B. G. 1988. On the phylogeny and higher classification of the Oligochaeta. Cladistics 4: 367–410.

Klemm, D. J. 1982. *Leeches (Annelida: Hirudinea) of North America*. U.S. Environ. Protection Agency, EPA–600/3–82–025.

Lasserre, P. 1975. Clitellata. *In* A. C. Giese and J. S. Pearse (eds.), *Reproduction of Marine Invertebrates*, Vol. 3. Academic Press, New York, pp. 215–275.

Laverack, M. S. 1963. *The Physiology of Earthworms*. Macmillan, New York.

Mann, K. H. 1962. *Leeches (Hirudinea), Their Structure, Ecology, and Embryology*. Pergamon Press, New York.

Nicholls, J. G. and D. Van Essen. 1974. The nervous system of the leech. Sci. Am. 230(1):38–48.

Satchell, J. E. (ed.). 1983. *Earthworm ecology: From Darwin to Vermiculture*. Methuen, New York. [A symposium volume.]

Seymour, M. K. 1969. Locomotion and coelomic pressure in *Lumbricus*. J. Exp. Biol. 51:47.

Stephensen, J. 1930. *The Oligochaeta*. Oxford University Press, New York.

Stuart, J. 1982. Hirudinoidea. *In* S. Parker, *Synopsis and Classification of Living Organisms*, Vol. 2. McGraw-Hill, New York, pp. 43–50.

Tembe, V. B. and P. J. Dubash. 1961. The earthworms: A review. J. Bombay Nat. Hist. Soc. 51(1):171–201.

Van Gansen, P. 1963. Structures et fonctions du tube digestif du lombricien *Eisenia foetida* Savigny. Ann. Soc. R. Zool. Belgique 93:1–121.

Wallwork, J. A. 1983. *Earthworm Biology*. Edward Arnold and University Park Press, Baltimore.

Chapter Fourteen

Four Phyla of Coelomate Worms

The reader will find the echiurids and sipunculids between the annelids and the arthropods, simply because three-dimensional printing is not practical.

Joel W. Hedgpeth
Introduction to Seashore Life, 1962

The coelomate phyla Sipuncula, Echiura, Pogonophora, and Vestimentifera are often dismissed in short fashion as "minor" or "lesser" groups. These four phyla comprise fewer than 600 species that are never as abundant or important ecologically as some other worms, especially the polychaetes and nematodes. Nonetheless, they display *Baupläne* different from any discussed thus far and provide important lessons in functional morphology. We include them in a single chapter for comparison with one another and with other coelomate groups.

The sipunculans and echiurans are clearly protostomes, although their phylogenetic positions within that assemblage are not entirely clear. The pogonophorans and vestimentiferans, on the other hand, have been said to show an apparent mixture of protostome and deuterostome characteristics and have been the subject of serious debate. Recent work, however, has demonstrated that the members of these two groups are almost certainly protostomes, as we explain in this chapter.

The sipunculans

The phylum Sipuncula (Greek *siphunculus*, "little tube") includes about 250 species in 17 genera, most of which are commonly called "peanut worms." In many respects sipunculans are built along an annelidan plan, but they show no evidence of segmentation (Box 1). The body is sausage-shaped and divisible into a retractible **introvert** and a thicker trunk (Figure 1). It is when the introvert is retracted and the body is turgid that some species resemble a peanut—hence the vernacular name. The anterior end of the introvert bears the mouth and feeding tentacles. The tentacles are derived from the regions around the mouth (**peripheral tentacles**) and around the nuchal organ (**nuchal tentacles**); differences in the arrangement are of taxonomic importance. The gut is characteristically U-shaped and highly coiled, and the anus is located dorsally on the body near the introvert–trunk junction. The body surface is usually beset with minute bumps, warts, tubercules, or spines. Sipunculans range in length from less than 1 cm to about 50 cm, but most are 5–10 cm long.

The coelom is well developed and unsegmented, forming a spacious body cavity. Metanephridia are present, with nephridiopores on the ventral body surface. There is no circulatory system, but the coelomic fluid includes cells containing a respiratory pigment. Most sipunculans are gonochoristic and reproduce by epidemic spawning. Development is usually indirect, is typically protostomous, and includes a free-swimming larva.

Sipunculans are benthic and exclusively marine. They are usually reclusive, either burrowing into sediments or living beneath stones or in algal holdfasts. In tropical waters sipunculans are common inhabitants of coral and beachrock communities, where they often burrow into hard calcareous substrata. Some inhabit abandoned gastropod shells, polychaete tubes, and other such structures. They are found from the intertidal zone to depths of over 5,000 meters. In the Far East (e.g., Vietnam) large sand-burrowing species are occasionally consumed as human food. Some representative species are illustrated in Figure 1.

Box One
Characteristics of the Phylum Sipuncula

1. Bilateral, unsegmented, schizocoelomate worms
2. Development protostomous (spiral cleavage, 4d mesoderm, schizocoely)
3. With a complete, U-shaped gut and dorsally placed anus
4. Mouth area with tentacles derived from region around mouth (peripheral tentacles) or nuchal organ (nuchal tentacles)
5. One pair of metanephridia
6. Nervous system constructed on an annelid-like plan, but simple and with no evidence of segmentation
7. With a unique retractable anterior body region called the introvert
8. Coelomic fluid with specialized multicellular structures (urns) for waste collection
9. Development is usually indirect, with a trochophore larva and sometimes a second larval stage, the pelagosphera
10. Entirely marine, benthic

A

Retracted introvert

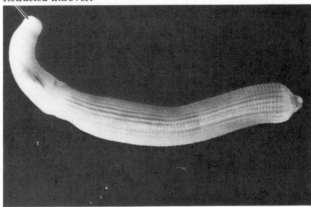

B

The sipunculan *Bauplan* is founded on the qualities of the spacious body coelom. Uninterrupted by septa, the coelomic fluid serves as an ample circulatory medium for these sedentary worms. The coelom and associated musculature also function as a hydrostatic skeleton and as a hydraulic system for locomotion and extension of the introvert.

Taxonomic history and classification

The first published illustrations of sipunculans were produced from woodcuts made in the mid-sixteenth century. Linnaeus included these animals in the twelfth edition of his *Systema Naturae* (1767) and placed them in the Vermes, along with so many other odds and ends. In the nineteenth century, Lamarck and Cuvier considered the sipunculans to be relatives of the holothurian echinoderms (sea cucumbers). No separate taxon was established for these worms until 1828, when DeBlainville introduced the name Sipun-

C   D

Figure 1

Representative sipunculans. A, *Phascolosoma,* **with the tip of the introvert turned inward. B,** *Sipunculus nudus.* **C,** *Themiste pyroides* **has a short introvert and stalked tentacles. D,** *Phascolion* **in a gastropod shell. (A photo by the authors; B photo by A. Kirstitsch; C,D after Hyman 1959.)**

culida and allied the group with certain parasitic helminths.

In 1847 Quatrefages invented the group Gephyrea to include sipunculans, echiurans, and priapulans. The Greek root *gephyra* means "bridge," as Quatrefages regarded these animals intermediate between annelids and echinoderms. The gephyrean concept was founded on superficial characteristics, but it persisted well into the twentieth century even though many authors attempted to raise the constituent groups to individual phylum status. Finally, Hyman (1959), recognizing the polyphyletic nature of the Gephyrea, elevated the sipunculans to separate phylum rank; her view was quickly accepted. At that time, however, no classes, orders, or families were recognized, and the phylum was divided into only genera and species. The Herculean effort by Stephen and Edmonds (1972) and subsequent modifications by other workers (e.g., Rice 1982) led to a classification comprising 4 families and 16 genera. Most recently, Cutler and Gibbs (1985) applied modern phylogenetic methods to the sipunculans and produced a classification scheme of 2 classes, 4 orders, 6 families, and 17 genera. The classification below is based largely upon that work.

PHYLUM SIPUNCULA

CLASS PHASCOLOSOMIDA: (Figure 1A). Without peripheral (oral) tentacles around mouth; nuchal tentacles present in an arc around nuchal organ; introvert hooks usually organized into distinct rings; with two metanephridia.

> ORDER ASPIDOSIPHONIFORMES: Trunk bears anteriorly (and occasionally posteriorly) located cuticular or calcareous shield; one family (Aspidosiphonidae) and three genera (*Aspidosiphon, Cloeosiphon, Lithacrosiphon*).

> ORDER PHASCOLOSOMIFORMES: Without trunk shields; one family (Phascolosomatidae) and three genera (*Antillesoma, Apionsoma, Phascolosoma*).

CLASS SIPUNCULIDA: (Figures 1B–D). Peripheral oral tentacles present (sometimes reduced to a flaplike veil) partially or wholly surrounding the mouth; nuchal tentacles present or absent; introvert hooks not organized into distinct rings.

> ORDER GOLFINGIAFORMES: Longitudinal muscles of body wall not in distinct bands.

>> FAMILY THEMISTIDAE: Peripheral tentacles borne in clusters on stalks produced from oral disc; dorsal retractor muscles absent; with two metanephridia. Monogeneric (*Themiste*).

>> FAMILY PHASCOLIONIDAE: Peripheral tentacles not borne in stalked clusters; dorsal retractor muscles present; with a single metanephridium; spindle muscle absent; most species asymmetrically coiled; Two genera (*Onchnesoma, Phascolion*).

>> FAMILY GOLFINGIIDAE: Peripheral tentacles not borne in stalked clusters; dorsal retractor muscles present; with two metanephridia; spindle muscle present. Three genera (*Golfingia, Nephasoma, Thysanocardia*).

> ORDER SIPUNCULIFORMES: Longitudinal muscles of the body wall in distinct bands; dermis with system of coelomic channels; with two metanephridia. One family (Sipunculidae) and five genera (*Phascolopsis, Siphonomecus, Siphonosoma, Sipunculus, Xenosiphon*).

Body wall, coelom, circulation, and gas exchange

The sipunculan body surface is everywhere covered by a well developed cuticle that varies from thin on the tentacles to quite thick and layered over much of the trunk (Figure 2). The cuticle often bears papillae, warts, or spines of various shapes. Members of the family Aspidosiphonidae are characterized by anterior and sometimes posterior **shields** derived from thickened cuticle or calcareous deposits.

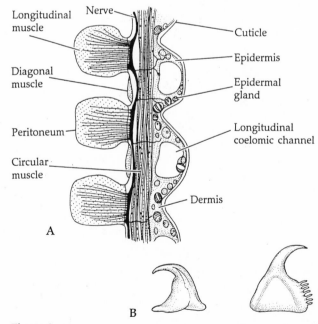

Figure 2

A, The body wall of *Sipunculus nudus* (cross section). **B,** Two types of cuticular spines from sipunculans. (Redrawn from Hyman 1959, B after Fischer 1952.)

Beneath the cuticle lies the epidermis, the cells of which are cuboidal over most of the body but grade to columnar and ciliated on the tentacles. The epidermis contains a variety of unicellular and multicellular glands, some of which project into the cuticle and produce some of the surface papillae. Some of these glands are associated with sensory nerve endings, others are responsible for producing the cuticle or for mucus secretion.

Underlying the epidermis, especially where it is raised, is a connective tissue dermis of fibers and loose cells. In the Sipunculidae, which are primarily large, warm-water species, the dermis also houses a system of coelomic extensions or channels (Figure 2).

These **coelomic channels** may extend entirely through the muscle layers and into the epidermal layer, and may have a gas exchange function. The muscles of the body wall include outer circular and inner longitudinal layers and sometimes a thin, middle layer of diagonal fibers. The longitudinal muscles form a continuous sheet in many sipunculans, but in some genera these muscles form distinct bundles or bands (e.g., *Phascolosoma* and all members of the family Sipunculidae). This arrangement gives the internal surface of the body wall a ribbed appearance that is usually visible through the animal's cuticle. From one to four large **introvert retractor muscles** extend from the body wall into the introvert where they insert on the gut just behind the mouth (Figures 3A and B).

A peritoneum lines the body wall and the internal organs. The coelom is a continuous space, but peritoneal mesenteries form incomplete partitions supporting the organs. In addition to the main body coelom and the coelomic channels in the body wall, a separate fluid-filled "coelom" called the **compensation system** is associated with the tentacles. The

Figure 3

Sipunculan anatomy. A, *Sipunculus nudus*. B, *Golfingia vulgaris*. C–D, Free coelomic urns from *Sipunculus* (C) and *Phascolosoma* (D). E, The gut of *Golfingia* (partial section). Note the ciliated intestinal groove. F, A nephridium of *Phascolosoma* (section). (Redrawn from Hyman 1959, E after Stehle 1953.)

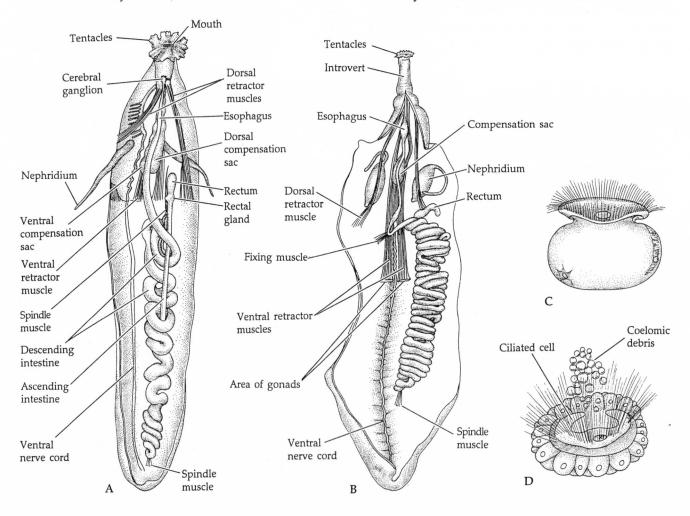

hollow tentacles contain lined spaces that are continuous with one or two sacs (the **compensation sacs**) that lie next to the esophagus (Figure 3A). Upon eversion of the introvert, circular body muscles apply pressure on these sacs and force the contained fluid into the tentacles, causing their erection.

The fluid of the body cavity contains a variety of cells and other inclusions. There are both granular and agranular amebocytes of uncertain function, and red blood cells containing hemerythrin. Also contained in the coelomic fluid are multicellular structures called **urns**, some of which are fixed to the peritoneum and some of which swim free in the fluid (Figures 3C and D). The urns accumulate waste materials and dead cells by trapping them with cilia and mucus.

Support and locomotion

Sipunculans are sedentary creatures. The general body shape is maintained by the muscles of the body wall and the hydrostatic skeleton established by the large coelom. The body is essentially a fluid-filled

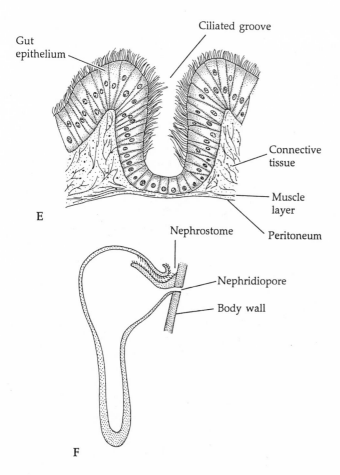

bag of constant volume, so any constriction at one point must be accompanied by an expansion at another. Burrowing in soft substrata is accomplished by peristalsis, driven by the circular and longitudinal muscles of the body wall and by the action of the introvert. Movement through holdfasts and rubble occurs in a similar manner. Some species that possess an anterior cuticular shield burrow into hard substrata and use the shield as an operculum to close the burrow entrance. Burrowing into hard substrata is probably accomplished by both mechanical and chemical means, the latter facilitated by secretions of epidermal glands, the former by using cuticular structures, such as spines and the posterior shield, as rasps.

The introvert is extended when the coelomic pressure is increased by contracting the circular muscles of the body wall. Withdrawal is accomplished by the retractor muscles, which pull from the mouth end, turning the introvert inward on itself. When the introvert is fully extended, the tentacles are erected by increasing the pressure in the compensation sacs.

Sipunculans are well suited for cryptic lifestyles. They require contact with their surroundings and are highly tactile. Placed alone in a glass dish, they are rather inactive except for the rolling in and out of the introvert. However, if several are placed together or with small stones or shell fragments, they soon respond by assuming contact with each other or surrounding objects.

Feeding and digestion

There is surprisingly little information on the details of sipunculan feeding mechanisms. Indirect evidence from anatomy, gut contents, and general behavior suggests that different feeding methods are used in different habitats. Most of the sipunculans that can place their tentacles at a substratum–water interface are selective or nonselective detritivores (e.g., shallow burrowers, algal holdfast dwellers); they use the mucus and cilia on the tentacles to obtain food. Deeper burrowers in sand are direct deposit feeders. Some appear to be ciliary-mucus suspension feeders, using the tentacles to extract organic material from the water. Sipunculans that burrow in calcareous substrata use spines or hooks on their introverts to retrieve organic detritus within reach and ingest the material by retracting the introvert. Limited data suggest that at least some sipunculans take up dissolved organic compounds directly across the body wall. Some workers speculate that up to 10% of these animals' nutritional requirements may be met in this fashion.

Gut
epithelium

Ciliated groove

Connective
tissue

Muscle
layer

E

Nephrostome

Peritoneum

Nephridiopore

Body wall

F

Because the anus is located anteriorly on the dorsal side of the body, the digestive tract is basically U-shaped, although highly coiled (Figures 3A and B). The mouth is at the end of the introvert and is wholly or partially surrounded by the **peripheral tentacles** (Sipunculida) or lies near the **nuchal tentacles** (Phascolosomida). The mouth leads inwards to a short, muscular, stomodeal pharynx, which is followed by an esophagus that extends through the introvert and into the trunk. The midgut consists of a long intestine composed of descending and ascending portions coiled together. It is usually supported by a threadlike **spindle muscle** that extends from the body wall near the anus through the coils to the end of the trunk and by several **fixing muscles** connecting the gut to the body wall. The ascending intestine leads to a short proctodeal rectum terminating in the anus.

The intestine is ciliated and bears a distinct groove along its length (Figure 3E). This ciliated groove leads ultimately to a small pouch or diverticulum (or rectal gland) off the rectum. The function of this groove and diverticulum is unknown. The lumen epithelium of the descending intestine contains a variety of gland cells that are presumably the sources of digestive enzymes.

Circulation and gas exchange

Gas exchange probably occurs across most of the body surface, especially at the tentacles where the cuticle is thin. The coelomic fluid, in the body cavity and in body wall channels, provides a circulatory medium, aided by diffusion and body movements. Hemerythrin in the red blood cells serves to store and transport oxygen.

Excretion and osmoregulation

Most sipunculans possess one pair of metanephridia (nephromixia) (Figure 3F). Two genera (*Onchnesoma* and *Phascolion*) have but a single nephridium. Species in these genera tend to be asymmetrically coiled. The nephridiopores are located ventrally on the anterior region of the trunk. The nephrostome lies close to the body wall, near the pore, and leads to a large nephridial sac that extends posteriorly in the trunk.

Sipunculans are ammonotelic. Nitrogenous wastes accumulate in the coelomic fluid and are excreted through the nephridia. The urns (Figures 3C and D) also play a major excretory role by picking up particulate waste material in the coelom. Urns originate as fixed epithelial cell complexes in the peritoneum, where they function to trap and remove particulate debris. They are also known to secrete mucus

in response to pathogens in the coelomic fluid. Fixed urns regularly detach and become free-swimming in the coelomic fluid. Not only do urns effectively cleanse the coelomic fluid, but they also participate in a clotting process when a sipunculan is injured. Free urns can be seen by preparing a wet slide of fresh coelomic fluid. They are usually obvious, moving about trailing strands of mucus and bits of particulate matter. The fate of wastes accumulated by urns is unknown, but at least some are probably transported to the nephridia.

Sipunculans are basically osmoconformers. Under normal conditions, the coelomic fluid is nearly isotonic to the surrounding sea water. However, when placed in hypotonic or hypertonic environments, the body volume increases or decreases respectively. Interestingly, the rates of volume change differ when the animal is exposed to these two opposing environments, suggesting that sipunculans are better at preventing water loss than at preventing water gain. This situation may be due to a differential permeability of the cuticle, or perhaps some active mechanism of the nephridia. In any case, sipunculans rarely face severe osmotic problems in their usual environments, and even in laboratory experiments they are able to recover nicely from most conditions of osmotic stress.

Nervous system and sense organs

The general structure of the sipunculan nervous system is similar in many respects to that in annelids. A bilobed cerebral ganglion lies dorsally in the introvert, just behind the mouth. Circumenteric connectives extend from the cerebral ganglion to a ventral nerve cord running along the body wall through the introvert and trunk (Figures 3A, 3B, and 4A). The ventral nerve cord is single, and there is no evidence of segmental ganglia. In *Phascolosoma agassizii*, a double ventral nerve cord forms initially, but later fuses as a single cord. This ontogenetic example suggests that the ancestral sipunculan condition may have been a double ventral nerve cord like that in some primitive annelids, although perhaps without segmental ganglia. Lateral nerves arise from the nerve cord and extend to the body wall muscles and sensory receptors in the epidermis.

Sensory receptors are widespread in sipunculans, but many are poorly understood. Tactile receptor cells are scattered over the body within the epidermis and are especially abundant on and around the tentacles. Chemosensory nuchal organs are located on the dorsal side of the introvert in many forms (Figure 4B). Many species possess a pair of

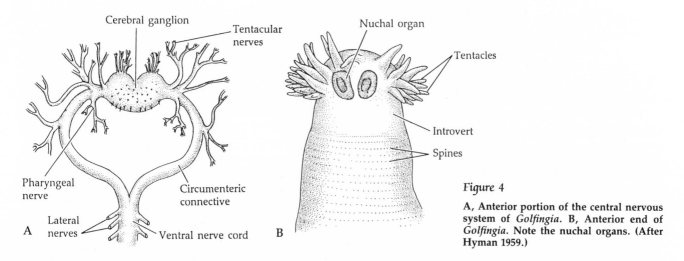

Figure 4

A, Anterior portion of the central nervous system of *Golfingia*. B, Anterior end of *Golfingia*. Note the nuchal organs. (After Hyman 1959.)

pigment-cup ocelli on the dorsal surface of the cerebral ganglion, and some possess a so-called **cerebral organ**, which consists of a ciliated pit projecting inward to the cerebral ganglion. The cerebral organ may be involved in chemoreception or perhaps neurosecretion.

Reproduction and development

Sipunculans possess reasonable powers of regeneration. Most species are able to regrow lost parts of the tentacles and even the introvert, and some can regenerate portions of the trunk and the digestive tract.

It was long believed that sipunculans could not reproduce asexually. However, some studies report the occurrence of asexual reproduction in at least two species (Rice 1970a). The process takes place by transverse fission of the body, whereby the worm divides into a small posterior fragment and a large anterior portion. Both portions then regrow the missing parts. Regeneration from the small posterior part is especially remarkable, since most of the trunk, anterior gut, retractor muscles, nephridia, introvert, and so on must be regrown.

Except for *Golfingia minuta*, sipunculans are gonochoristic. (Facultative parthenogenesis has been reported in one species, *Themiste lageniformis*.) The gametes arise from the coelomic lining, often near the origins of the retractor muscles. Gametes are released into the coelom, where they mature. Ripe eggs and sperm are picked up selectively by the nephridia and stored in the sacs until released. The eggs are encased in a layered, porous covering. Males spawn first, probably in response to some environmental cue, and then the presence of sperm in the water stimulates females to spawn.

Following external fertilization, the zygotes pass through typical protostomous development. Cleavage is spiral and holoblastic, but the relative sizes of micromeres and macromeres differ among species, depending on the amount of yolk in the egg. The cell fates are the same as those in many other protostomes. The first three quartets of micromeres become ectoderm and ectomesoderm; the 4d cell produces entomesoderm; and 4a, 4b, 4c, and 4Q form the entoderm. The mouth opens at the site of the blastopore, and the surrounding ectodermal cells grow inward as a stomodeum. The anus breaks through secondarily on the dorsal surface (Figure 5). The 4d mesoderm proliferates as two bands, as it does in annelids, but yields the major trunk coelom without segmentation.

Four different developmental sequences have been recognized among various sipunculans (see papers by Rice). A few species are known to undergo direct development (e.g., *Golfingia minuta*, *Phascolion crypta*, and *Themiste pyroides*). The eggs of these sipunculans are covered by an adhesive jelly and attach to the substratum after fertilization. The embryo develops directly to a vermiform individual that hatches from the egg coat as a juvenile sipunculan.

The other three developmental patterns are indirect, involving various combinations of larval stages. In some species (e.g., *Phascolion strombi*), a free-living lecithotrophic trochophore larva develops and metamorphoses into a juvenile worm. The other two developmental patterns involve a second larval stage, the **pelagosphera larva**, that forms after a metamorphosis of the trochophore (Figure 5C). In some species both the trochophore and the pelagosphera forms are lecithotrophic and relatively short lived (e.g., various species of *Golfingia* and *Themiste*),

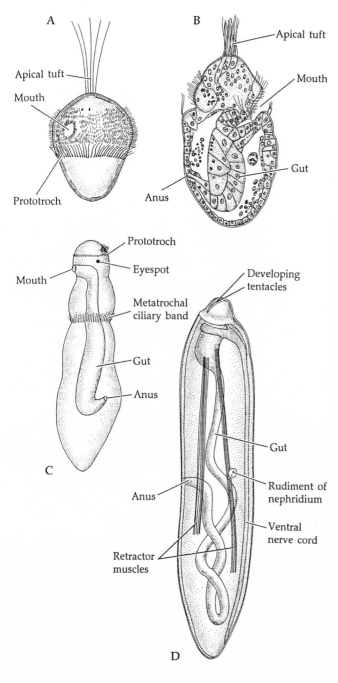

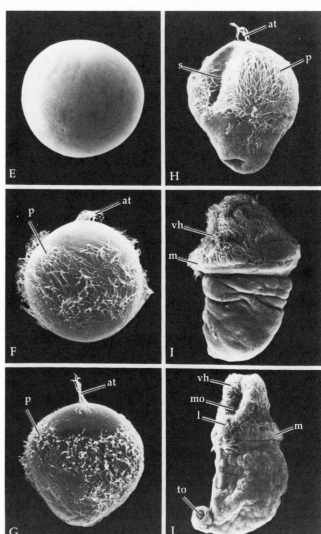

Figure 5

Sipunculan development. A, Young trochophore of *Golfingia*. B, A later larva of *Golfingia* (section), showing gut shape and placement of the anus. C, The pelagosphera larva of *Phascolosoma* has an enlarged metatroch. D, A juvenile sipunculan. E–J, Series of scanning electron micrographs showing growth to the trochophore and metamorphosis to the pelagosphera in *Siphonosoma*. at, apical tuft; m, metatroch; mo, mouth region; l, lower ciliated lip; to, terminal attachment organ; vh, ventral ciliated head; p, prototroch; s, stomodeum. (A–D after Hyman 1959; E–J courtesy of M. Rice.)

while in others the pelagosphera larva is planktotrophic and may live for extended periods of time in the plankton (e.g., *Aspidosiphon parvulus*, *Sipunculus nudus*, and members of the genus *Phascolosoma*).

The transformation of the trochophore to the pelagosphera larva involves a reduction or loss of the prototrochal ciliary band and the formation of a single metatrochal band for locomotion. The pelagosphera eventually elongates, settles, and becomes a juvenile sipunculan (Figure 5D).

The echiurans

Members of the phylum Echiura (Greek *echis*, "serpent-like") are also unsegmented, coelomate worms. They resemble sipunculans in many regards and are often found in similar habitats, such as soft benthic substrata. Biologists who specialize in one group often specialize in both. There are about 135 known species. Like the sipunculans, the echiurans share certain features with the annelid *Bauplan* but lack metamerism (Box 2). The vermiform body is divided into an anterior, preoral **proboscis** and an enlarged trunk (Figure 6). The mouth is located at the anterior end of the trunk at the base of a **proboscis groove**, or **gutter**. The body surface may be smooth or somewhat warty, and sometimes it bears setae (e.g., *Urechis*). Many echiurans are quite large. The trunk may be from a few to as many as 40 cm long, but the proboscis may reach lengths of 1–2 meters (e.g., in *Bonellia* and *Ikeda*). The Pacific *Listriolobus pelodes* may be sexually mature when only 7 mm long, whereas the Japanese *Ikeda taenioides* may reach lengths in excess of 2 meters. Some forms, such as *Bonellia*, show drastic sexual dimorphism, wherein "dwarf" males are less than 1 cm long.

The spacious body cavity contains numerous mesenteries but is not completely partitioned by septa, and there is no evidence of segmentation. From one to many metanephridia lie in the anterior trunk region, and each usually leads to its own ventral pore. Most echiurans possess a simple closed circulatory system. These worms are gonochoristic, and their development reflects a clear protostome affinity, including a trochophore larval stage.

Echiurans are benthic and live exclusively in marine or brackish-water habitats. Most burrow in sand or mud, or live in surface detritus or rubble. Some

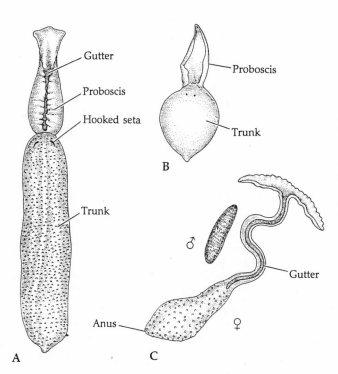

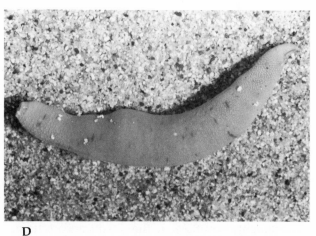

D

Figure 6

Representative echiurans. **A**, *Echiurus*. **B**, *Listriolobus*, with the proboscis contracted. **C**, *Bonellia viridis*. Note the extreme sexual dimorphism between the large female and the tiny male. (See text for details). **D**, *Urechis caupo*, the "fat innkeeper." (A after Barnes 1980; B after Fischer 1946; C after MacGinitie and Mac-Ginitie 1968; D photo by K. Lucas/BPS.)

species typically inhabit rock galleries excavated by boring clams or other invertebrates. They are known from intertidal regions to a depth of 10,000 meters. A few examples of echiurans are shown in Figure 6.

Taxonomic history and classification

Echiurans were first reported in the literature during the eighteenth century. The first few species were described (by Pallas) as annelids. Eventually they were placed in the now-abandoned taxon Gephyrea along with the sipunculans and priapulans. In 1896 Sedgwick suggested raising the sipunculans and priapulans to separate phylum status, but he considered the echiurans to be a class of annelid worms. It was not until 1940—when Newby, and subsequently others, conducted studies on echiuran development—that these worms were established as a separate phylum. The name of the phylum has varied over the years (e.g., Echiurida, Echiuroidea), but Echiura is the current preferred spelling. The phylum comprises three orders.

PHYLUM ECHIURA

ORDER ECHIUROINEA: Body wall muscle layers are, from outer to inner, circular, longitudinal, and oblique; with one to several pairs of metanephridia (some species with a single nephridium); circulatory system present; hindgut not modified as a gas exchange organ; with about 130 species in two families, Bonelliidae (e.g., *Bonellia, Bruunella, Achaetobonellia*), and Echiuridae (e.g., *Echiurus, Listriolobus, Thalassema*).

ORDER XENOPNUESTA: Body wall muscle layers as in Echiuroinea; 2–3 pairs of nephridia; circulatory system absent; hindgut elongate and thin walled, modified for gas exchange; proboscis very short; one family (Urechidae), monogeneric (*Urechis*), with four species.

ORDER HETEROMYOTA: Body wall muscle layers are, from outer to inner, longitudinal, circular, oblique; 200–400 nephridia; circulatory system present and bearing heartlike enlargement; hindgut not enlarged; one family (Ikedaidae), monotypic (*Ikeda taenioides*).

Body wall and coelom

The body wall of echiurans is roughly similar to that of sipunculans, except for variations in muscle arrangement, as outlined in the classification scheme. A thin cuticle covers the epidermis, which is composed of a cuboidal epithelium and contains a variety of gland cells. A fibrous dermis lies beneath the epidermis. Layers of circular, longitudinal, and oblique muscles form the bulk of the body wall, which is lined internally by the peritoneum. The epidermis is ciliated along the proboscis groove, or gutter.

The coelomic cavity is spacious and occupies most of the trunk. It is interrupted only by partial mesenteries between the gut and the body wall. The coelomic fluid contains red blood cells, with hemoglobin in some species, and various types of amebocytes. Cells likened to chlorogogen cells have been reported in a few species.

Support and locomotion

The large trunk coelom provides a hydrostatic skeleton against which the body wall muscles operate. The nonseptate coelom allows peristaltic movements as the animal burrows or moves through shell rubble or gravel. The preoral proboscis is capable of shortening and lengthening, but it does not roll in and out as does the introvert of sipunculans. This characteristic, and the more conventional posterior position of the anus, allow one to distinguish quickly between members of these two similar phyla.

Feeding and digestion

Most echiurans feed on epibenthic detritus. Typically, the animal lies with the trunk more-or-less buried in the substratum, the anterior end directed upward, and the proboscis extended over the sediment (Figures 7A, C, and D). Densely packed gland cells of the proboscis epithelium secrete mucus, to which organic detrital particles adhere. The mucous coating and the food are moved along the ventral proboscis gutter by ciliary action into the mouth.

An interesting exception to the above feeding method occurs in *Urechis*. Members of this genus excavate and reside within U-shaped burrows in soft substrata (Figure 7B). *Urechis* possesses a very short proboscis, unlike those of more typical echiurans, and engages in mucous-net filter feeding. A ring of glands located near the proboscis–trunk junction produces a funnel-shaped mucous net, which is attached to the burrow wall by the proboscis. Water is drawn through the burrow and the sheet of mucus by peristaltic movements of the body, and suspended food particles as small as 1 μm are caught in the finemeshed net. Periodically, the animal grasps the food-laden net with its proboscis and ingests it.

The digestive tract is generally very long and coiled, leading from the mouth at the base of the proboscis to the posterior anus (Figure 8). The foregut is a stomodeum and may be regionally specialized as pharynx, esophagus, gizzard, and stomach, or it may be more-or-less uniform along its length. The midgut usually bears a longitudinal ciliated groove, or **siphon**, which probably aids the movement of materials through the gut. It may also shunt excess water

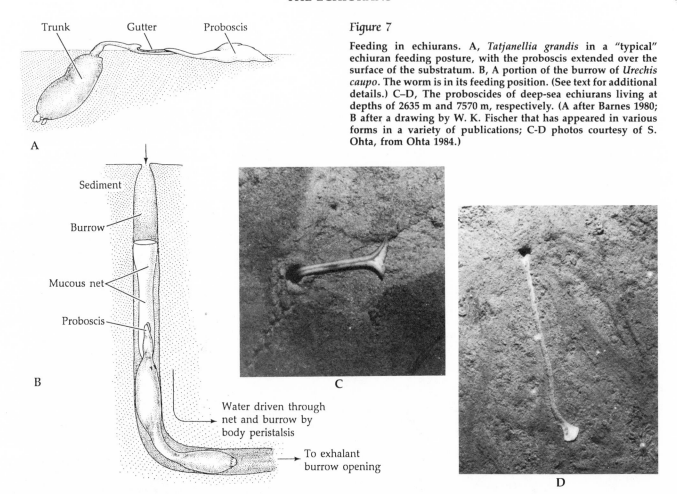

Trunk   Gutter   Proboscis

A

Sediment

Burrow

Mucous net

Proboscis

B

Water driven through
net and burrow by
body peristalsis

To exhalant
burrow opening

C

D

Figure 7
Feeding in echiurans. A, *Tatjanellia grandis* in a "typical" echiuran feeding posture, with the proboscis extended over the surface of the substratum. B, A portion of the burrow of *Urechis caupo*. The worm is in its feeding position. (See text for additional details.) C–D, The proboscides of deep-sea echiurans living at depths of 2635 m and 7570 m, respectively. (A after Barnes 1980; B after a drawing by W. K. Fischer that has appeared in various forms in a variety of publications; C-D photos courtesy of S. Ohta, from Ohta 1984.)

from the main midgut lumen, thereby concentrating food and facilitating digestion. The hindgut, or cloaca, is a proctodeum and varies in structure among different species. In most echiurans the cloaca bears a pair of large excretory diverticula called **anal vesicles** (see below). In some species of *Urechis* the cloaca is enlarged and very thin walled (Figure 8C). In such cases water is pumped in and out of the hindgut for gas exchange.

Not much is known about digestive physiology in echiurans. The epithelium of the midgut is rich in gland cells that presumably produce and secrete digestive enzymes. Digestion and absorption occur mainly in the midgut.

Circulation and gas exchange

Most echiurans possess a simple closed circulatory system, although it is entirely absent from some forms (e.g., *Urechis*). The circulatory system generally includes dorsal and ventral longitudinal vessels in the trunk, and median and lateral vessels in the proboscis (Figures 8A and B). There is no major pumping

organ (except in *Ikeda*); the blood is transported by pressures generated from body movements and by the weak musculature of the vessel walls. The blood is usually colorless. Apparently the main function of the circulatory system is the transport of nutrients and perhaps other metabolites through the body.

The main site of gas exchange in at least some echiurans is probably the hindgut, which is usually provided with oxygenated water by **cloacal irrigation**—water being pumped in and out of the anus by muscle action. It is likely that some gas exchange also occurs across the general body surface, particularly on the proboscis. The coelomic fluid contains red blood cells with hemoglobin.

Excretion and osmoregulation

The excretory structures of echiurans include paired metanephridia and anal vesicles. The number of nephridia varies: one pair in *Bonellia*; two pairs in *Echiurus*; three pairs in *Urechis*; to hundreds of pairs in *Ikeda*. When only one or a few pairs are present, the nephridia are located in the anterior region of the

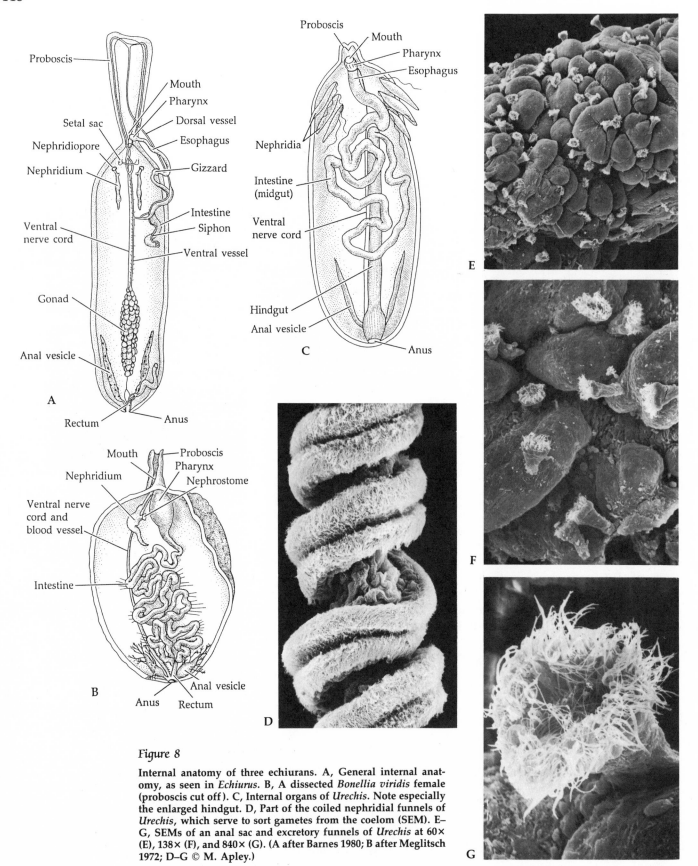

Figure 8

Internal anatomy of three echiurans. A, General internal anatomy, as seen in *Echiurus*. B, A dissected *Bonellia viridis* female (proboscis cut off). C, Internal organs of *Urechis*. Note especially the enlarged hindgut. D, Part of the coiled nephridial funnels of *Urechis*, which serve to sort gametes from the coelom (SEM). E–G, SEMs of an anal sac and excretory funnels of *Urechis* at 60× (E), 138× (F), and 840× (G). (A after Barnes 1980; B after Meglitsch 1972; D–G © M. Apley.)

trunk and lead to nephridiopores on either side of the ventral midline (Figure 8). The degree to which these nephridia function in excretion is debatable. In some, such as *Urechis*, the nephrostomes are long and coiled (Figure 8D); they seem to function primarily in picking up gametes from the coelom, having relinquished the major excretory responsibility to the **anal vesicles**.

The anal vesicles are hollow sacs arising as evaginations of the cloaca near the anus (Figures 8A–C and 8E–G). Each vesicle bears from about a dozen to as many as 300 ciliated funnels that open to the coelom. Few studies have been conducted on the function of these structures, but they apparently pick up wastes from the coelomic fluid and remove the material to the hindgut and anus. The anal vesicles may also function in gas exchange and perhaps in osmoregulation and the control of intracoelomic pressure. Echiurans are relatively poor osmoregulators, but it is unlikely that they encounter severe osmotic stress in their usual marine habitats.

Nervous system and sense organs

The nervous system of echiurans is simple, although constructed in a fashion generally similar to the annelid plan. An anteriorly located nerve ring extends around the gut and dorsally forward into the proboscis. Ventrally, in the trunk, the nerve ring connects with a single ventral nerve cord extending the length of the body. There are no ganglia in this system and no evidence of segmentation. Lateral nerves arise irregularly from the ventral nerve cord and extend to the body wall muscles (Figures 8A–C). In a few species the ventral nerve cord forms as a double cord but fuses during development.

Associated with the simple nervous system and infaunal sedentary lifestyle of echiurans is the absence of major sensory receptors. These animals are mildly touch sensitive, especially on the proboscis and, when present, the setae, and they may possess chemoreceptors.

Reproduction and development

Asexual reproduction is unknown in echiurans, and little work has been done on the powers of regeneration. At least some display remarkable healing capabilities. For example, the "fat innkeeper," *Urechis caupo*, is often found in bay muds associated with areas subjected to heavy pressure from clam diggers. In some of these tidal flats nearly every specimen of *Urechis* bears scars, some nearly completely across the body, signs that the animal has survived the onslaught of clammers' shovels.

The echiuran sexes are separate, and the gametes are produced in special "gonadal" regions of the peritoneum, often near the base of the ventral blood vessel, and released into the coelom to mature. When ripe, the gametes are accumulated by the nephridia and stored until spawning occurs. The nephridia often swell enormously when packed with eggs or sperm. In most cases epidemic spawning takes place and is followed by external fertilization.

Development is basically similar to other protostomes described thus far (e.g., annelids and sipunculans) (Figure 9). Cleavage is holoblastic and spiral. The cell fates follow the typical protostome pattern. A trochophore larva (Figures 9B and C) develops and may drift in the plankton for up to three months as it gradually elongates to produce a young worm (Figures 9D and E). The prototroch and the preprototroch develop into the proboscis, and the postprototroch forms the enlarged trunk. The 4d mesentoblast proliferates the main trunk coelom. A number of earlier works, and many general texts, include reference to transitory segmentation of the larval coelom and cite such evidence as supporting an annelid ancestry to the echiurans. It is now generally agreed that no such segmentation occurs; the superficial annulations on the larvae are tiers of epithelial cells (Figure 9C). Thus, there is no evidence of metamerism in any stage of echiuran life history.

Mention must be made of a strange case of sexual dimorphism and sex determination in the family Bonelliidae. Female bonelliids are quite large, reaching lengths of up to 2 m, including the proboscis. The males, however, are only a few millimeters long, very reduced in complexity, and often retain remnants of larval ciliation. They live on the female's body or in her nephridia. Evidence suggests that the determination of sex in bonelliids is largely controlled environmentally at the time of larval settlement. If a larva settles on or near the proboscis area of an adult female, it will mature rapidly as a "degenerate" or "dwarf" male. However, if the larva settles away from a female, it will burrow and eventually grow and mature as a female. The induction of maleness is apparently caused by a masculinizing hormone produced by the proboscis of the female worm. However, sometimes two larvae clump together prior to settlement, whereupon one becomes a female and the other a male. Development into a male can also be induced by pH changes that cause a slightly acidic condition of the sea water. Interestingly, a larva that settles on the trunk of a female, rather than on the proboscis, develops an intermediate sexual condition.

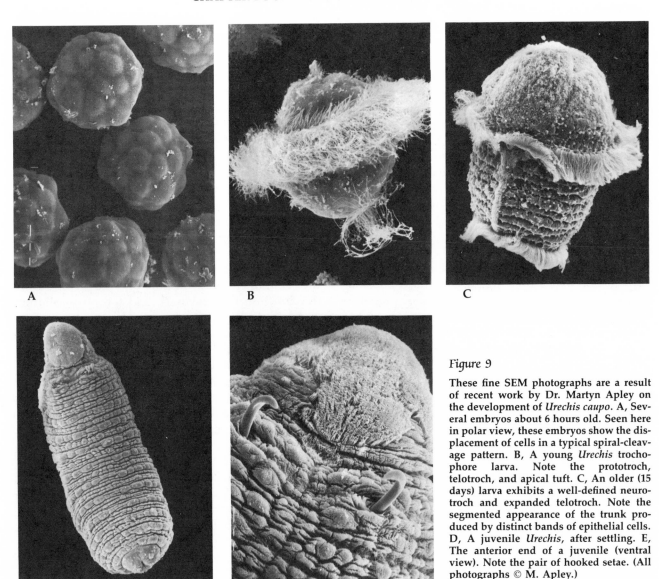

A   B   C

D   E

Figure 9

These fine SEM photographs are a result of recent work by Dr. Martyn Apley on the development of *Urechis caupo*. A, Several embryos about 6 hours old. Seen here in polar view, these embryos show the displacement of cells in a typical spiral-cleavage pattern. B, A young *Urechis* trochophore larva. Note the prototroch, telotroch, and apical tuft. C, An older (15 days) larva exhibits a well-defined neurotroch and expanded telotroch. Note the segmented appearance of the trunk produced by distinct bands of epithelial cells. D, A juvenile *Urechis*, after settling. E, The anterior end of a juvenile (ventral view). Note the pair of hooked setae. (All photographs © M. Apley.)

The pogonophorans and vestimentiferans

The phyla Pogonophora (Greek *pogon*, "beard"; *phor*, "to bear") and Vestimentifera (Latin *vestimentum*, "garment"; *ferre*, "to bear") contain about 145 species of strange vermiform marine creatures called the "beard worms." As described below, these odd animals have been the subject of taxonomic and phylogenetic debate since their discovery around the turn of the century. Pogonophorans live in thin tubes buried in sediments at ocean depths varying from 200 to 10,000 meters (Figure 10). Members of one genus, *Sclerolinum*, reside in wood fragments, pieces of water-logged rope, and other such substrata.

Most species have long thin bodies, usually less than 1 mm in diameter but 10 to 75 cm in length. The tubes are often three to four times as long as the body. One species, *Riftia pachyptila*, that has been assigned to the Vestimentifera and lives in hydrothermal vent communities, reaches the incredible length of 3 meters.

The pogonophoran body is clearly divided into four regions (Figure 10A). These regions usually contain distinct body cavities but the nature of the linings is not fully clear. We view the body cavities as coeloms, but you should be aware of the uncertainty here. The anteriormost body part is called the **prosoma** (or **cephalic lobe**), and it bears from one to over 200 thin tentacles, which bear tiny side branches

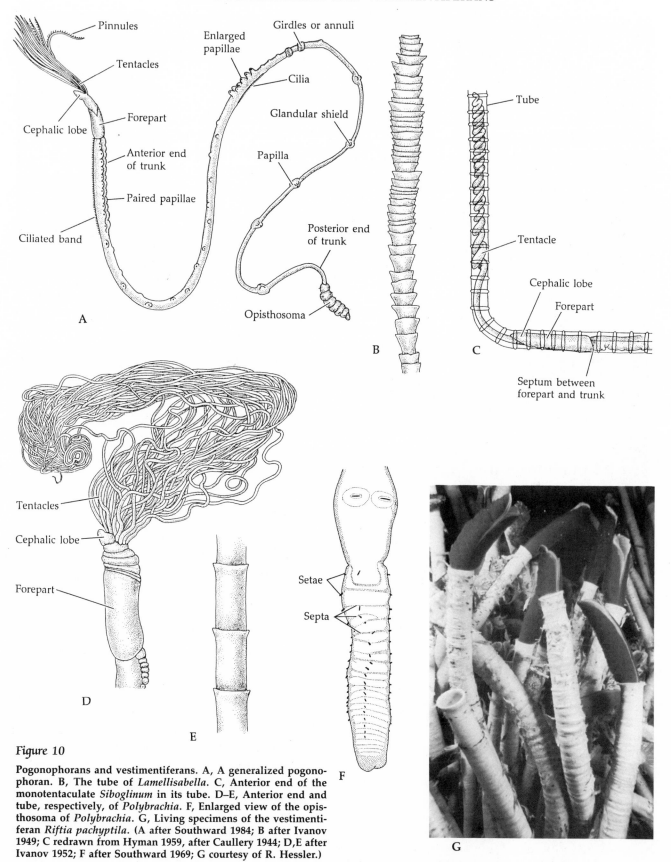

Figure 10

Pogonophorans and vestimentiferans. A, A generalized pogonophoran. B, The tube of *Lamellisabella*. C, Anterior end of the
monotentaculate *Siboglinum* in its tube. D–E, Anterior end and
tube, respectively, of *Polybrachia*. F, Enlarged view of the opisthosoma of *Polybrachia*. G, Living specimens of the vestimentiferan *Riftia pachyptila*. (A after Southward 1984; B after Ivanov
1949; C redrawn from Hyman 1959, after Caullery 1944; D,E after
Ivanov 1952; F after Southward 1969; G courtesy of R. Hessler.)

called **pinnules**. Behind the prosoma is a relatively short **mesosoma** (or **forepart**); the **metasoma** (or **trunk**) is the longest body region and bears various annuli, papillae, and ciliary tracts. These three body regions—the prosoma, mesosoma, and metasoma— give the animal a tripartite, somewhat deuterostome, quality. In fact, until 1964, these were the only parts of pogonophorans known; the posterior portion, the **opisthosoma**, had not yet been collected because the worms fragment easily. However, when entire specimens were finally obtained intact, it was discovered that the opisthosoma appeared to be metameric in structure, with an unpaired coelomic space in each segment, internal serially-arranged septa, and external paired setae. Thus the pogonophorans, at least superficially, possess an enigmatic combination of protostome and deuterostome traits.

The vestimentiferans, including the aforementioned *Riftia*, are viewed by some specialists as being constructed somewhat differently from the pogonophorans. While still a matter of controversy, Jones (1985) interprets the vestimentiferan body as having five regions, as follows. A thick, cylindrical region,

called the **obturaculum**, is the first body part. This is followed by a second region that bears the tentacle-like branchiae. A third body region is defined by the presence of flaplike or winglike projections, the **vestimentum**, from which the phylum name is derived. The main trunk of the body is the fourth region, and the multisegmented opisthosoma the fifth. Furthermore, the first two body regions of vestimentiferans lack very well defined coelomic cavities, and the opisthosoma contains a midsagittal mesentery, which produces a pair of coelomic cavities in each segment—much more polychaete-like than the single cavities in pogonophorans. Based on these and other characteristics, Jones (1985) formally proposed the separation of these worms into two phyla, the Pogonophora and Vestimentifera. While the final word is not yet in, we explain some of the similarities and differences between these two groups in Box 3.

Internally, both pogonophorans and vestimentiferans possess complex closed circulatory systems and well developed nervous systems. The presence of true nephridia is debatable and, surprisingly, there is no digestive tract in the adults. Some of the body

---

Box Three

A Comparison of the Major Characteristics of Pogonophora and Vestimentifera

SHARED TRAITS

1. Bilaterally symmetrical, coelomate worms, with regionally specialized bodies
2. Hind part of body appears metameric (the opisthosoma)
3. Lack a digestive system as adults
4. With a closed circulatory system
5. Usually with coelomoduct-like structures from coeloms to the outside
6. Marine, benthic, tube dwellers

POGONOPHORA	VESTIMENTIFERA
7. Anterior body cavities usually well developed	7. Anterior body cavities not well developed
8. Lack vestimental "wings"	8. With vestimental "wings"
9. Opisthosoma with three nerve cords, probably bearing ganglia	9. Opisthosoma with a single, nonganglionated nerve cord
10. Opisthosomal coeloms serially arranged but unpaired, without medial mesentery	10. Opisthosomal coeloms serially arranged and paired in each segment, with medial mesentery
11. Apparently without a gut at any time	11. With transitory gut in juveniles

cavities connect to the outside via tubes resembling paired coelomoducts.

As if these creatures were not odd enough, there has also been some confusion about the overall body orientation—it was unclear whether the nerve cord is dorsally or ventrally located. The resolution of this question is of extreme evolutionary importance. Until recently we did not know for sure which side is "up" in these animals! Because of this problem, the terms **adneural** (near the nerve cord) and **antineural** (opposite the nerve cord) have been generally preferred over dorsal and ventral. However, recent work by Meredith Jones and Stephen Gardiner (1988) has produced evidence of a transitory gut, at least in vestimentiferans. The arrangement of this structure suggests that the nerve cord is ventral and that the body orientation is very annelid-like.

Taxonomic history and classification

Pogonophorans were first studied in the early 1900s, following the expedition of the Dutch research vessel *Siboga* in Indonesia. The specimens (without opisthosomas) collected on this voyage were given to the French zoologist Maurice Caullery, who studied the material for nearly 50 years and published several papers describing these strange worms. Caullery eventually named the originally collected creatures *Siboglinum weberi*, but he was unable to place them in any known higher taxon. During the middle years of the twentieth century, other zoologists, including C. Dawydoff, suggested that *Siboglinum* was related to the enteropneusts, a group of vermiform hemichordates with a tripartite body architecture (Chapter 23). Other workers studied additional specimens and continued to emphasize a deuterostome relationship, often relating the pogonophorans to the phoronids (Chapter 21). Some species were likened to sabellid polychaetes because of the tentacular crown. The name Pogonophora was first coined (as Pogonofora) to be included as a class of Annelida.

Much of the work on pogonophorans has been done by the Russian specialist A. V. Ivanov (see references), who for years allied them to higher deuterostomes, after studying the incomplete animals. However, in 1964, the collection of whole specimens, with opisthosoma intact, led to a new line of thinking about pogonophoran relationships to other phyla (see especially papers by Southward). It is tempting, of course, to view the pogonophorans as annelids because of the metameric appearance of the opisthosoma and the polychaete-like setae. However, a more conservative approach has been supported by many specialists. Ivanov noted the similarities to members of the phylum Hemichordata, but he also respected the many unique attributes of the pogonophorans and suggested they be placed in their own phylum.

A more recent difficulty has arisen concerning the gigantic pogonophoran-like worms discovered living in certain deep-sea hydrothermal vent environments. Although many authorities consider these huge worms to be true pogonophorans, some view them as aberrant annelids; and, as mentioned, some are convinced they merit separate phylum rank, the Vestimentifera. The first of these hydrothermal vent species was collected from the northeastern Pacific. It was described by Webb (1969) and assigned to a new genus, *Lamellibrachia* (*L. barhami*). At that time, Webb suggested assigning all previously known pogonophorans to the class Frenulata, and the new genus *Lamellibrachia* to the new order Vestimentifera in the class Afrenulata. Since then, however, other genera (e.g., *Riftia*) and species have been discovered off Guyana, from the East Pacific Rise, and elsewhere. Fossil vestimentiferans were discovered in 1984 in Cretaceous deposits of Oman.

Because of the systematic confusion associated with these worms, we present only an abbreviated classification, reflecting the view that two phyla may be warranted. Members of the phyla are illustrated in Figures 10 and 11.

PHYLUM POGONOPHORA

ORDER ATHECANEPHRIA: Prosoma body cavity sacciform, with its paired ducts to the outside placed far apart, near the lateral cephalic blood vessels; exposed end of tube rather flimsy. Includes two families: Oligobrachiidae (*Birsteinia*, *Crassibrachia*, *Nereilinum*, *Oligobrachia*, and *Unibrachium*); Siboglinidae (*Siboglinoides* and *Siboglinum*).

ORDER THECANEPHRIA: Prosoma body cavity U-shaped, with its paired ducts placed close together near the adneural blood vessel; upper end of tube rigid. Includes four families: Polybrachiidae (*Balathealinum*, *Choanophorus*, *Cyclobrachia*, *Diplobrachia*, *Heptabrachia*, *Polybrachia*, and *Zenkevitchiana*); Sclerolinidae (*Sclerolinum*); Lamellisabellidae (*Lamellisabella* and *Siphonobrachia*); Spirobrachiidae (*Spirobrachia*).

PHYLUM VESTIMENTIFERA

For details of the subtaxa see Jones (1985). The phylum (ranked as a class of Pogonophora by some workers) is divided into two classes, the Axonobranchia and Basibranchia, based on the arrangements of the branchiae and branches of the anterior blood vessels. The Axonobranchia contains a single order and family (Riftiida, Riftiidae) and is monogeneric (*Riftia*). The Basibranchia includes two or-

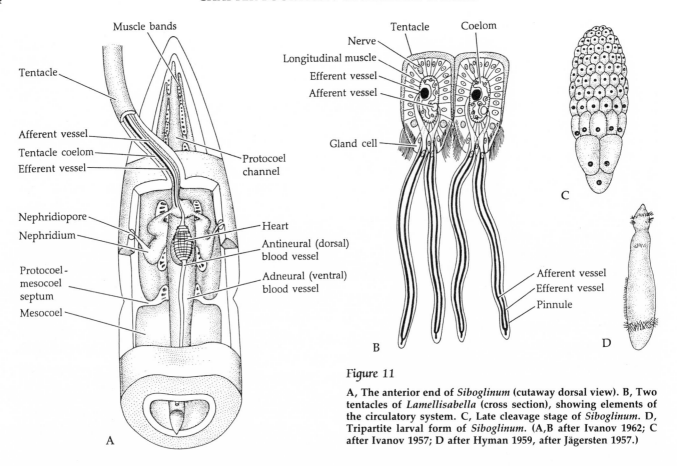

Figure 11

A, The anterior end of *Siboglinum* (cutaway dorsal view). B, Two tentacles of *Lamellisabella* (cross section), showing elements of the circulatory system. C, Late cleavage stage of *Siboglinum*. D, Tripartite larval form of *Siboglinum*. (A,B after Ivanov 1962; C after Ivanov 1957; D after Hyman 1959, after Jägersten 1957.)

ders: the Lamellibrachiida, with the families Lamellibrachiidae (*Lamellibrachia*) and Escarpiidae (*Escarpia*); and the Tevniida, with the families Tevniidae (*Tevnia*, *Oasisia*) and Ridgeiidae (*Ridgeia*).

The tube, body wall, and body cavity

The elongate tubes of pogonophorans and vestimentiferans are composed of chitin and scleroproteins secreted by the epidermis. The tubes are often ringed, flared, or otherwise distinctively shaped and are frequently banded with yellow or brown pigment rings (Figures 10B, C, and E). The upper end of the tube projects above the substratum, so the tentaculate part of the worm can extend into the water. The tubes of pogonophorans are generally open at both ends; those of vestimentiferans are closed basally.

The body surface is covered by a flexible cuticle that is thickened in various patterns, including a collar-like ridge that apparently rests on the rim of the tube when the animal is extended. The epidermis is mostly a cuboidal to columnar epithelium and includes various gland cells, papillae, and some ciliary tracts. Microvilli extend into the cuticle and produce

part of its outer layer. Some large glands extend inward from the epidermis. Some of the trunk epithelium is thought to contain absorptive cells. Beneath the epidermis is a thin layer of circular muscle and a thick layer of longitudinal muscle developed as bands or bundles in some parts of the body. The body cavities are fairly spacious except in the anterior end, where large muscle bands occur. In pogonophorans the prosomal cavity is a small sacciform or U-shaped space with a pair of ducts to the outside. The cavities in the mesosoma and metasoma are paired with sagittally arranged "mesenteries" bisecting the body. A pair of ducts also occurs in the metasoma. The cavities of these first three body parts have been likened to the protocoel, mesocoel, and metacoel of deuterostomes; however, they are also very annelid-like in construction. The opisthosoma comprises several segments, each with cavities in a typical metameric arrangement; these cavities are paired in vestimentiferans and single in pogonophorans. Whether these body cavities are true coeloms (or the remnants of true coeloms) is unclear. Resolution of this important problem awaits detailed developmental studies.

Nutrition

In the absence of a digestive system, the method of nutrition in these relatively large worms is indeed puzzling. Recent experimental work suggests that most pogonophorans are able to absorb dissolved nutrients from the sea water flowing across the tentacles and from the mud surrounding the body surface. Apparently no digestive enzymes are involved in this process; rather the worms take in glucose, amino acids, and fatty acids directly from the environment. Some species may also be capable of epidermal pinocytosis and phagocytosis. The absorption of organic nutrients probably also occurs in the giant vestimentiferans associated with the chemoautotrophic-based ecosystems of hydrothermal vent environments. However, preliminary work (on *Sibo-glinum fiordicum*) indicates that uptake rates of dissolved amino acids are insufficient to account for the animal's metabolic requirements, suggesting augmentation by some other nutritional mechanism. Most recently, evidence has been found in *Riftia* that symbiotic chemoautotrophic bacteria inhabit certain tissues (the **trophosome organ**) of the body of vestimentiferans that live in sulfur-rich environments. The suggestion is that these symbionts generate ATP by carrying out sulfide oxidation and by reducing CO_2 to organic matter, which may in turn be utilized as a nutritional source by the host worm. In at least some vestimentiferans these bacteria are taken in by a transitory digestive tract during the juvenile stages of growth. These bacteria are apparently stored in the young worm's midgut, which persists as the trophosome organ after loss of the rest of the gut (Jones and Gardiner 1988). Recently, chemoautotrophic bacteria have also been discovered in several species of pogonophorans (Southward et al. 1986).

Circulation, gas exchange, excretion, and osmoregulation

As we have seen in members of the phylum Annelida, a body plan that includes internal partitions demands a circulatory mechanism. So it is with the pogonophorans and vestimentiferans, which possess a well developed closed circulatory system (Figure 11A). The major blood vessels are adneural and antineural longitudinal vessels running nearly the entire length of the body. The antineural vessel is swollen as a "heart" in the mesosoma. Blood flows anteriorly in the antineural vessel and posteriorly in the adneural vessel. Within the prosoma the vessels branch, giving rise to lateral cephalic blood vessels and other vessels that supply the anterior end of the

worm. Anteriorly, blood is passed from the antineural to the adneural vessel via afferent and efferent vessels in the tentacles (Figure 11B). Exchange between the main longitudinal vessels also occurs in the posterior body region through a series of connecting blood rings.

If the adneural and antineural positions do in fact correspond to ventral and dorsal, respectively, then this overall circulatory pattern is virtually identical to that of annelids.

The blood contains hemoglobin in solution and a variety of cells and cell-like inclusions. Gas exchange probably takes place across the thin walls of the tentacles. For the vestimentiferans living in deep-sea hydrothermal vent environments, problems of oxygen supply and sulfur toxicity are especially critical. These worms, such as *Riftia pachyptila*, live where hot, anoxic, sulfurous vent water mixes with the cold surrounding oxygenated water. The worms are thus constantly exposed to dramatic fluctuations in oxygen and sulfide availability and ambient temperature. These worms possess very high concentrations of hemoglobin in the fluid of the body cavities as well as in the blood. Vestimentiferan hemoglobin appears to retain its high affinity for oxygen across a wide temperature range. Furthermore, a unique sulfide-binding protein occurs in the blood. This protein serves to concentrate the sulfur and avoid sulfide toxicity, and also to transport it to the chemoautotrophic symbionts. These features are obvious adaptations to the unique environment inhabited by these worms. Interestingly, similar adaptations are known in the burrowing polychaete, *Arenicola*, which typically burrows through anoxic muds in shallow marine environments.

Little is known about excretion and osmoregulation in these animals. Some workers believe that the coelomoduct-like structures associated with the prosomal body cavity may function in these capacities, especially given their association with the circulatory system. Current opinion views these ducts as protonephridia in pogonophorans, but as metanephridia in vestimentiferans. However, uncertainty remains.

Nervous system and sense organs

The nervous system is intraepidermal, a feature considered to be primitive in many groups of invertebrates. A large ganglion lies along the adneural side of the prosoma and gives rise to a prosomal nerve ring. A single, presumably ventral (adneural) nerve cord arises from the posterior edge of the prosomal ganglion and extends through all the body regions.

In many cases there are ganglia (or at least "nerve bulges") at the junctions of the various body regions, and there is an enlargement of the nerve cord in the anterior metasoma. In the pogonophorans, three distinct nerve cords have been identified in the opisthosoma, apparently bearing ganglia. The vestimentiferans have a single, nonganglionated opisthosomal nerve cord.

Sense organs are poorly developed. The tentacles apparently contain tactile receptors, and an adneural (ventral?) ciliary tract on the body surface may be chemosensory.

Reproduction and development

Nothing is known regarding asexual reproduction or regeneration in these groups. They are gonochoristic and possess a pair of gonads in the metasoma. In males, paired sperm ducts extend from the testes to gonopores located anterolaterally on the metasoma. As sperm move along the ducts, they are packed as spermatophores of various shapes—generally pyriform, cylindrical, or leaf-shaped. The female reproductive system includes a pair of ovaries from which arise oviducts leading to gonopores on the sides of the metasoma.

Fertilization has not been observed. In pogonophorans, at least, males apparently release their spermatophores, which then drift to the open tubes of nearby females. Fertilization must take place within the female's body or tube, because developing embryos are located within the tube itself. The eggs are elongate and moderately telolecithal. The zygotes are deposited in the upper end of the tube, where they are all oriented similarly; the animal pole is directed upward in the tube and will become the anterior end of the young worm.

From what little work has been done on pogonophorans, it appears that cleavage is holoblastic, unequal, and somewhat irregular, but cannot be related clearly to a radial or spiral pattern. The blastomeres are larger at the vegetal pole than at the animal pole, and their overall orientation produces a distinctly bilaterally symmetrical, solid blastula (Figure 11C). From the scant evidence available, it appears that the inner cells of the blastula differentiate as entoderm, although apparently a gut never forms in pogonophorans. These inner cells apparently give rise to a single prosomal cavity from which cavities in the mesosoma and metasoma eventually arise. This pattern of body cavity formation is strikingly similar to deuterostome enterocoely. Inner cells located near the posterior end of this early tripartite embryo proliferate spaces within the opisthosoma as it develops,

a development roughly similar to the teloblastic growth and schizocoely seen in polychaetes. Thus, developmental patterns appear reminiscent of both deuterostomes and protostomes! However, this developmental model is based on very few studies and includes a good deal of inference.

Pogonophorans release ciliated motile larvae (Figure 11D) at the tripartite stage. These larvae swim for a very short time before settling and secreting tubes of their own. Development is thus mixed, with a long brooding phase and a short dispersal phase. This developmental strategy serves these deep-sea sedentary worms well. The production of a few yolky eggs followed by protected brooding is clearly adaptive in an environment where long-lived planktotrophic larvae would have little selective value. The dispersal phase, though necessary to these sedentary animals, is of limited duration, reducing the risks associated with free-living larvae in the deep sea.

Some comments on phylogeny

The evolutionary relationships of the four phyla discussed in this chapter—to one another or to other taxa—are difficult to evaluate for several reasons. Being soft-bodied and wormlike, they have left almost no fossil record; their enigmatic nature yields no widely agreed-upon ancestors or sister-groups. As in all such cases, much of the difficulty involves the determination of character state polarity. That is, are the fundamental features that are used to define these groups primitive or derived within the protostome clade? For example, the general embryonic and adult characteristics of members of the phyla Sipuncula and Echiura clearly suggest a relationship to the protostomes. However, some of their traits can be viewed as either primitive or derived when compared with the general annelid *Bauplan*. In the past, some workers have treated the sipunculans and echiurans as offshoots of an early polychaete line. In this scenario, we might view the absence of segmentation as a secondary loss of a partitioned coelom associated with the exploitation of a sedentary, burrowing lifestyle. The reduction in sensory receptors and simplification of the nervous system in general are explainable on this same basis. However, many species of annelids have evolved burrowing lifestyles while retaining their basic metamerism. It is more parsimonious to view the absence of segmentation as a retained primitive state, or symplesiomorphy, and thus derive these nonmetameric worms from a preannelidan, burrowing, unsegmented, coelomate protostome ancestor. Sipunculan/echiuran attributes such as the trochophore larva, paired ventral nerve cords in some juveniles (fused into a single cord in adults), nephromixia, and multiple nephridia in certain species all seem to speak of an ancient, preannelid protostome relationship. This same ancestral line may have given rise to the annelid/arthropod clade, as described in Chapters 13 and 24, and also to the molluscs, as we shall see in Chapter 20.

The Pogonophora and Vestimentifera are also difficult to assess, and a number of issues must be addressed. At face value, the pogonophoran *Bauplan* can be viewed in various ways. We may consider the major body regions as truly tripartite and thus akin to the deuterostome plan, with the opisthosoma being a derived feature convergent with metamerism in annelids. Or we may envision the pogonophorans and vestimentiferans as having arisen from polychaete stock by developing extreme heteronomy associated with a tube-dwelling lifestyle, a feature common to many polychaetes. Depending on which of these ideas one supports, it becomes necessary to view the body orientation in one of two possible ways. If a deuterostome affinity is accepted, then the adneural side of the worm must be dorsal. But, if a protostome relationship is supported, one must flip the worms over so that the nerve cord is ventral and the pattern of blood flow is like that in annelids. We support the latter view as being the most parsimonious and most consistent with the available evidence.

A third possibility is to view the pogonophorans

and vestimentiferans as actually possessing a mixture of real protostome and deuterostome characteristics and accept that this unique combination of traits was established early on, around the time of evolutionary experimentation with cleavage patterns, methods of coelom formation, metamerism, and different *Baupläne*.

Regardless of one's views on this matter, we are left to explain the absence of an adult digestive tract in these worms. This is clearly a secondary loss. In all other phyla with regionally specialized bodies and closed circulatory systems, these features occur in concert with complete and functional guts. It is of course tempting to posit the loss of the gut in these animals as being evolutionarily tied to the exploitation of habitats rich in low molecular weight dissolved nutrients that could be absorbed across the body wall. Given this possibility and the fact that pogonophorans and vestimentiferans are quite long, it is reasonable to assume that if an efficient circulatory system were present, it would be retained as an internal transport mechanism for moving nutrients (as well as gases) through the body even in the absence of a digestive tract.

In our present state of understanding, it seems most reasonable to view the sipunculans and echiurans as early premetameric protostomes, and the pogonophorans and vestimentiferans as having arisen from some early polychaete line. These views are supported by the cladistic analysis presented in Chapter 24.

Selected References

General

Clark, R. B. 1969. Systematics and phylogeny: Annelida, Echiura, Sipuncula. *In* M. Florkin and B. T. Scheer (eds.), *Chemical Zoology*, Vol. 4. Academic Press, New York, pp. 1–68.

Hyman, L. 1959. *The Invertebrates*, Vol. 5, Smaller Coelomate Groups. McGraw-Hill, New York.

Jägersten, G. 1972. *Evolution of the Metazoan Life Cycle*. Academic Press, London.

Kohn, A. J. and M. E. Rice. 1971. Biology of Sipuncula and Echiura. Bioscience 21: 583–584.

Rice, M. E. and M. Todororic (eds.) 1970. *Proceedings of the International Symposium on the Biology of the Sipuncula and Echiura*, Vols. 1 and 2. Nauchno Delo Press, Belgrade, for Institute for Biological Research, Yugoslavia and the Smithsonian Institution, Washington, D.C.

Stephen, A. C. and S. J. Edmonds. 1972. *The Phyla Sipuncula and Echiura*. British Museum (Natural History), London.

Storch, V. 1984. Echiura and Sipuncula. *In* J. Bereiter-Hahn, A. G. Matolsky and K. S. Richards (eds.), *Biology of the Integument*, Springer-Verlag, Berlin, pp. 368–375.

Sipuncula

Abercrombie, R. K. and R. M. Bagby. 1984. An explanation for the folding fibers in proboscis retractor muscles of *Phascolopsis* (=*Golfingia*) *gouldi*. II. Structural evidence from SEM of teased, glycerinated muscles and light microscopy of KOH-isolated fibers. Comp. Biochem. Physiol. 77A(1):31–38.

Åkesson, B. 1958. A study of the nervous system of the Sipunculoidea with some remarks on the development of two species *Phascolion strombi* Montegu and *Golfingia*

minuta Keferstein. C.W.K. Gleerup, Lunds (Unders. over Oresumd, 38).

Bang, B. G. and F. G. Bang. 1980. The urn cell complex of *Sipunculus nudus*: A model for study of mucus-stimulating substances. Biol. Bull. 159:571–581.

Cutler, E. B. 1973. Sipuncula of the Western North Atlantic. Bull. Am. Mus. Nat. Hist. 152:105–204.

Cutler, E. B. 1977. *Sipuncula: Marine Flora and Fauna of the Northeastern U.S.* NOAA Technical Report NMFS Circular 403. U.S. Government Printing Office, Washington, D.C.

Cutler, E. B. and N. J. Cutler. 1988. A revision of the genus *Themiste* (Sipuncula). Proc. Biol. Soc. Wash. 101:741–766. [This paper lists the many other generic revisions published by the Cutlers.]

Cutler, E. B. and P. E. Gibbs. 1985. A phylogenetic analysis of higher taxa in the phylum Sipuncula. Syst. Zool. 34:162–173.

Dybas, L. 1981. Cellular defense reactions of *Phascolosoma agassizii*, a sipunculan worm: Phagocytosis by granulocytes. Biol. Bull. 161:104–114.

Fischer, W. K. 1952. The sipunculid worms of California and Baja California. Proc. U.S. Nat. Mus. 102:371–450.

Gibbs, P. E. 1977. *British Sipunculans*. Synopsis of the British Fauna No. 122. Academic Press, New York.

Hansen, M. D. 1978. Food and feeding behavior of sediment feeders as exemplified by sipunculids and holothurians. Helgol. Wiss. Meeresunters. 31:191–221.

Hermans, C. O. and R. M. Eakin. 1969. Fine structure of the cerebral ocelli of a sipunculid, *Phascolosoma agassizii*. Z. Zellforsch. Mikrosk. Anat. 100:325–399.

Mangum, C. P. and M. Kondon. 1975. The role of coelomic hemerythrin in the sipunculid worm *Phascolopsis gouldi*. Comp. Biochem. Physiol. 50A:777–785.

Nicosia, S. V. 1979. Lectin-induced release in the urn cell complex of the marine invertebrate *Sipunculus nudus* (Linnaeus). Science 206:698–700.

Pilger, J. F. 1982. Ultrastructure of the tentacles of *Themiste lageniformis* (Sipuncula). Zoomorphologie 100:143–156.

Pilger, J. F. 1987. Reproductive biology and development of *Themiste lageniformis*, a parthenogenetic sipunculan. Bull. Mar. Sci. 41:59–67.

Pörter, H.-O., U. Kreutzer, B. Siegmund, N. Heisler and M. K. Grieshaber. 1984. Metabolic adaptation of the intertidal worm *Sipunculus nudus* to functional and environmental hypoxia. Mar. Biol. 79:237–247.

Rice, M. E. 1967. A comparative study of the development of *Phascolosoma agassizii*, *Golfingia pugettensis*, and *Themiste pyroides* with a discussion of developmental patterns in the Sipuncula. Ophelia 4:143–171.

Rice, M. E. 1969. Possible boring structures of sipunculids. Am. Zool. 9:803–812.

Rice, M. E. 1970a. Asexual reproduction in a sipunculan worm. Science 167:1618–1620.

Rice, M. E. 1970b. Observations on the development of six species of Caribbean Sipuncula with a review of development in the phylum. Proc. Int. Symp. Biol. Sipuncula and Echiura. I. Kotor:18–25.

Rice, M. E. 1976. Sipunculans associated with coral communities. Micronesica 12(1):119–132.

Rice, M. E. 1978. Morphological and behavioral changes at metamorphosis in the Sipuncula. *In* F. S. Chia and M. E. Rice (eds.), *Settlement and Metamorphosis of Marine Invertebrate Larvae*. Elsevier, New York, pp. 83–102.

Rice, M. E. 1981. Larvae adrift: Patterns and problems in life histories of sipunculans. Am. Zool. 22:605–619.

Rice, M. E. 1982. Sipuncula. *In* S. P. Parker (ed.), *Synopsis and Classification of Living Organisms*, Vol. 2. McGraw-Hill, New York, pp. 67–69.

Rice, M. E. 1985. Sipuncula: Developmental evidence for phylogenetic inference. *In* S. C. Morris et al. (eds.), *The Origins and Relationships of Lower Invertebrates*. Syst. Assoc. Spec. Vol. No. 28, Oxford, pp. 274–296.

Rice, M. E. 1988. Observations on development and metamorphosis of *Siphonosoma cumanense* with comparative remarks on *Sipunculus nudus* (Sipuncula, Sipunculidae). Bull. Mar. Sci. 42:1–15.

Stehle, G. 1953. Anatomie und Histologie von *Phascolosoma elongatum*. Ann. Univ. Saraviensis, Naturwiss. 2, No. 3.

Stephen, A. C. 1964. A revision of the classification of the phylum Sipuncula. Ann. Mag. Nat. Hist. Ser. 13, 7:457–462.

Storch, V. and U. Welsh. 1979. Zur ultrastruktur der metanephridien des landlebenden sipunculiden *Phascolosoma (Physcosoma) lurco*. Kiel. Meeresforsch. 28(2):227–231.

Thompson, B. E. 1980. A new bathyl sipunculan from southern California, with ecological notes. Deep-Sea Res. 27A:951–957.

Uexkull, J. V. 1903. Die biologische Bauplan von *Sipunculus*. Z. Biol. 44.

Valembois, P. and D. Bioledieu. 1980. Fine structure and functions of haemerythrocytes and leucocytes of *Sipunculus nudus*. J. Morphol. 163:69–77.

Walter, M. D. 1973. Feeding and studies on the gut content in sipunculids. Helgol. Wiss. Meeresunters. 25(4):486–494.

Williams, J. A. and S. V. Margolis. 1974. Sipunculid burrows in coral reefs: Evidence for chemical and mechanical excavation. Pac. Sci. 28(4):357–359.

Zuckerkandl, E. 1950. Coelomic pressures in *Sipunculus nudus*. Biol. Bull. 98:161.

Echiura

Bosch, C. 1981. La musculature, le squelette conjonctif et les mouvements de la trompe de la Bonellie (*Bonellia viridis* Rol., Echiurida). Ann. Sci. Nat. Zool. Paris 3:203–229.

Dawydoff, C. 1959. Classes des Echiuriens et Priapuliens. *In* P. Grassé (ed.), *Traité de Zoologie*, Vol. 5, Pt. 1. Masson et Cie, Paris.

Edmonds, S. J. 1982. Echiura. *In* S. P. Parker (ed.), *Synopsis and Classification of Living Organisms*, Vol. 2. McGraw-Hill, New York, pp. 65–66.

Fischer, W. K. 1946. Echiuroid worms of the North Pacific Ocean. Proc. U.S. Natl. Mus. 96:215–292.

Fischer, W. K. 1949. Additions to the echiuroid fauna of the north Pacific Ocean. Proc. U.S. Natl. Mus. 99(3248):479–497.

Gislen, T. 1940. Investigations on the ecology of *Echiurus*. Lunds Univ. Arsskr. Avd. 36(10):1–39.

Gould-Somero, M. C. 1975. Echiura. *In* A. C. Giese and J. S. Pearse (eds.), *Reproduction of Marine Invertebrates*, Vol. 3. Academic Press, New York, pp. 277–311.

Gould-Somero, M. C. and L. Holland. 1975. Oocyte differentiation in *Urechis caupo* (Echiura): A fine structural study. J. Morphol. 147:475–506.

Jaccarini, V., L. Agius, P. J. Schembri and M. Rizzo. 1983. Sex determination and larval sexual interaction in *Bonellia viridis* (Echiura, Bonelliidae). J. Exp. Mar. Biol. Ecol. 66:25–40.

Jose, K. V. 1964. The morphology of *Acanthobonellia pirotaensis*, n. sp., a bonellid from the Gulf of Kutch, India. J. Morphol. 115:53.

Newby, W. W. 1932. The early embryology of the echiuroid, *Urechis*. Biol. Bull. 63:387–399.

Newby, W. W. 1940. The embryology of the echiuroid worm *Urechis caupo*. Mem. Am. Philos. Soc. 16:1–213.

Ohta, S. 1984. Star-shaped feeding traces produced by echiuran worms on the deep-sea floor of the Bay of Bengal. Deep-Sea Res. 31(12):1415–1432.

Pilger, J. F. 1978. Settlement and metamorphosis in the Echiura: A review. *In* F. S. Chia and M. E. Rice (eds.), *Settlement and Metamorphosis of Marine Invertebrate Larvae*. Elsevier, New York, pp. 103–111.

Pilger, J. F. 1980. The annual cycle of oogenesis, spawning, and larval settlement of the echiuran *Listriolobus pelodes* off southern California. Pac. Sci. 34:129–142.

Pritchard, A. and F. N. White. 1981. Metabolism and oxygen transport in the innkeeper worm *Urechis caupo*. Physiol. Zool. 54:44–54.

Redfield, A. C. and M. Florkin. 1931. The respiratory function of the blood of *Urechis caupo*. Biol. Bull. 11:85–210.

Suer, A. L. 1984. Growth and spawning of *Urechis caupo* (Echiura) in Bodega Harbor, California. Mar. Biol. 78:275–284.

Suer, A. L. and D. W. Phillips. 1983. Rapid, gregarious settlement of the larvae of the marine echiuran *Urechis caupo* Fisher and MacGinitie 1928. J. Exp. Mar. Biol. Ecol. 67:243–259.

Pogonophora/Vestimentifera

Arp, A. J. and J. J. Childress. 1981. Blood function in the hydrothermal vent vestimentiferan tube worm. Science 213:342–344.

Arp, A. J. and J. J. Childress. 1983. Sulfide bonding by the blood of the hydrothermal vent tube worm *Riftia pachyptila*. Science 219:295–297.

Bakke, T. 1977. Development of *Siboglinum fiordicum* Webb (Pogonophora) after metamorphosis. Sarsia 63:65–73.

Bakke, T. 1980. Embryonic and post-embryonic development in the Pogonophora. Zool. Jb. Anat. 103:276–284.

Caullery, M. 1944. *Siboglinum*. Siboga Exped. Monogr. 25 Bis, Livr. 138.

Cavanaugh, C. M., S. L. Gardiner, M. L. Jones, H. W. Janasch and J. B. Waterburg. 1981. Prokaryotic cells in the hydrothermal vent tube worm *Riftia pachyptila* Jones: Possible chemoautotrophic symbionts. Science 213:340–342.

Corliss, J. B. et al. 1979. Submarine thermal springs on the Galapagos Rift. Science 203:1073–1083.

Cutler, E. B. 1974. The phylogeny and systematic position of the Pogonophora. Syst. Zool. 24:512–513.

Cutler, E. B. 1982. Pogonophora. *In* S. P. Parker (ed.), *Synopsis and Classification of Living Organisms*, Vol. 2. McGraw-Hill, New York, pp. 63–64.

Felbeck, H. 1981. Chemoautotrophic potential of the hydrothermal vent tube worm, *Riftia pachyptila* Jones (Vestimentifera). Science 213:366–338.

Felbeck, H., M. A. Powell, S. C. Hand and G. N. Somero. 1985. Metabolic adaptations of vent animals. *In* M. L. Jones (ed.), *The Hydrothermal Vents of the Eastern Pacific: An Overview*. Bull. Biol. Soc. Wash. No. 6.

Fisher, C. R., Jr. and J. J. Childress. 1984. Substrate oxidation by trophosome tissue from *Riftia pachyptila* Jones (phylum Pogonophora). Mar. Biol. Lett. 5:171–183.

Fry, B., H. Gest and J. M. Hayes. 1983. Sulphur isotopic compositions of deep-sea hydrothermal vent animals. Nature 306:51–52.

George, J. D. and E. C. Southward. 1973. A comparative study of the setae of Pogonophora and polychaetous Annelida. J. Mar. Biol. Assoc. U.K. 53(2):403–424.

Grassle, J. F. 1982. The biology of hydrothermal vents: A short summary of recent findings. Mar. Technol. Soc. J. 16(3):33–38.

Grassle, J. F. 1985. Hydrothermal vent animals: Distribution and biology. Science 229:713–717.

Haymon, R. M., R. A. Koski and C. Sinclair. 1984. Fossils of hydrothermal vent worms from Cretaceous sulfide ores of the Sawail Ophiolite, Oman. Science 223:1407–1409.

Ivanov, A. V. 1949. A new representative of the class Pogonophora. Zool. Zhurn. 28.

Ivanov, A. V. 1952. New Pogonophora from the far eastern seas. Zool. Zhurn. 31.

Ivanov, A. V. 1957. Materials on the embryonic development of Pogonophora. Zool. J. Moskau 36.

Ivanov, A. V. 1962. *Pogonophora*. Consultants Bureau, New York.

Ivanov, A. V. 1963. *Pogonophora*. Academic Press, London. [English translation by D. B. Carlisle.]

Jägersten, G. 1957. On the larva of *Siboglinum*. Zool. Bidrag 32.

Jones, M. L. 1980. *Rifita pachyptila*, new genus, new species, the vestimentiferan worm from the Galapagos Rift geothermal vents (Pogonophora). Proc. Biol. Soc. Wash. 93:1295–1313.

Jones, M. L. 1981. *Riftia pachyptila* Jones: Observations on the vestimentiferan worm from the Galapagos Rift. Science 213:333–336.

Jones, M. L. 1984. The giant tube worms. *In* P. R. Ryan (ed.) Deep-sea hot springs and cold seeps. Oceanus 27(3):47–52. [With 13 other papers on various aspects of deep-sea vents and seeps.]

Jones, M. L. 1985. On the Vestimentifera, new phylum: Six new species and other taxa, from hydrothermal vents and elsewhere. Biol. Soc. Wash. Bull., 6:117–158.

Jones, M. L. 1985. Vestimentiferan pogonophorans: Their biology and affinities. *In* S. C. Morris et al. (eds.), *The Origins and Relationships of Lower Invertebrates*. Syst. Assoc. Spec. Vol. No. 28, Oxford, pp. 327–342.

Jones, M. L. (ed.). 1985. Hydrothermal vents of the eastern Pacific: An overview. Bull. Biol. Soc. Wash. 6:1–57.

Jones, M. L. 1987. On the status of the phylum-name, and other names, of the vestimentiferan tube worms. Proc. Biol. Soc. Wash. 10:1049–1050.

Jones, M. L. and S. L. Gardiner. 1988. Evidence for a transient digestive tract in Vestimentifera. Proc. Biol. Soc. Wash. 11:423–433.

Kirkegaard, J. B. 1956. Pogonophora. Galathea Report 2:79–83.

Land, J. van der and A. Nørrevang. 1977. Structure and relationships of *Lamellibrachia* (Annelida, Vestimentifera). Kongel. Dans. Vidensk. Selsk. Biol. Skr. 21(3):1–102.

Little, C. and B. L. Gupta. 1969. Studies on Pogonophora. III. Uptake of nutrients. J. Exp. Biol. 51:759–773.

Nørrevang, A. 1970a. The position of Pogonophora in the phylogenetic system. Z. Zool. Syst. Evolutionsforsch. 8(3):161–172.

Nørrevang, A. 1970b. On the embryology of *Siboglinum* and its implications for the systematic position of the Pogonophora. Sarsia 42:7016.

Nørrevang, A. (ed.) 1975. The phylogeny and systematic position of Pogonophora.

Z. Zool. Syst. Evolutionsforsch. [Includes 10 papers resulting from a symposium on the pogonophorans held at the University of Copenhagen in 1973.]

Powell, M. A. and G. N. Somero. 1983. Blood components prevent sulfide poisoning of respiration of the hydrothermal vent tube worm *Riftia pachyptila*. Science 219:297–299.

Southward, A. J., E. C. Southward, P. R. Dando, R. L. Barrett and R. Ling. 1986. Chemoautotrophic function of bacterial symbionts in small pogonophora. J. Mar. Biol. Assoc. U.K. 66:415–437.

Southward, E. C. 1969. Growth of a pogonophore: A study of *Polybrachia canadensis* with a discussion of the development of taxonomic characters. J. Zool. Lond. 157:449–467.

Southward, E. C. 1971a. Pogonophora of the Northwest Atlantic: Nova Scotia to Florida. Smithson. Contrib. Zool. 88:1–29.

Southward, E. C. 1971b. Recent researches on the Pogonophora. Oceanogr. Mar. Biol. Annu. Rev. 9:193–220.

Southward, E. C. 1972. On some Pogonophora from the Caribbean and the Gulf of Mexico. Bull. Mar. Sci. 22:739–776.

Southward, E. C. 1975. Pogonophora. *In* A. C. Giese and J. S. Pearse (eds.), *Reproduction of Marine Invertebrates*, Vol. 2, Entoprocts and Lesser Coelomates. Academic Press, New York, pp. 129–156.

Southward, E. C. 1978. Description of a new species of *Oligobrachia* (Pogonophora) from the North Atlantic, with a survey of the Oligobrachiidae. J. Mar. Biol. Assoc. U.K. 58:357–365.

Southward, E. C. 1979. Horizontal and vertical distribution of Pogonophora in the Atlantic Ocean. Sarsia 64:51–56.

Southward, E. C. 1980. Regionation and metamerism in Pogonophora. *In* R. Siewing (ed.), Structuranalyse und Evolutionforschung: Das Merkmal. Zool. Jahrb. Abt. Anat. Ontog. Tiere 103:264–2275.

Southward, E. C. 1982. Bacterial symbionts in Pogonophora. J. Mar. Biol. Assoc. U.K. 62:889–906.

Southward, E. C. 1984. Pogonophora. *In* J. Bereiter-Hahn, A. G. Matolsky and K. S. Richard (eds.), *Biology of the Integument*, Vol. 1. Springer-Verlag, Berlin, pp. 376–388.

Southward, E. C. and J. K. Cutler. 1986. Discovery of Pogonophora in warm shallow waters of the Florida shelf. Mar. Ecol. Prog. Ser. 28:287–289.

Webb, M. 1969. *Lamellibrachia barhami*, gen. nov., sp. nov. (Pogonophora) from the northeast Pacific. Bull. Mar. Sci. 19:18–47.

Webb, M. 1980. The Pogonophora: A critical assessment. Ann. Univ. Stellenbosch Ser. A2 Zool. 2:29–33.

Chapter Fifteen

Phylum Arthropoda: The Arthropod *Bauplan* and the Trilobites

[T]here came great swarms of flies into the house of Pharaoh and into his servants' houses, and in all the land of Egypt the land was ruined by reason of the flies.

Exodus 9:24

If we live out our span of life on earth without ever knowing a crab intimately we have missed having a jolly friendship. Life is a little incomplete if we can look back and recall these small people only as supplying the course after soup and with the Chablis.

William Beebe

Nonsuch: Land of Water, 1932

The first arthropods arose in ancient Precambrian seas over 600 million years ago. Since that time they have evolved to become the most abundant and diverse group of metazoa on Earth. They have successfully invaded virtually every habitat and exploited every imaginable lifestyle and developmental strategy. They range in size from tiny mites and crustaceans less than 1 mm long, to the great Japanese spider crabs with leg spans exceeding 4 m. At present there are about a million described species of arthropods. Most specialists believe that there are, at least among the insects, millions more yet to be discovered. Recent studies have estimated that 20–50 million insect species still remain to be described in the tropics alone! No other group of animals or plants approaches this magnitude of species richness. Arthropods are so abundant, so diverse, and play such vital roles in all the Earth's environments, that we devote five chapters to them. Despite this overwhelming diversity, the arthropods share a suite of fundamental unifying similarities. The first part of this chapter is devoted to an exploration of these

basic characters, the features that constitute the arthropod *Bauplan*, and to understanding how this combination of features has led to their preeminent success.

The arthropods and annelids share close evolutionary ties. The former may have evolved from the latter or, more likely, they are closely related groups derived from an ancient, Precambrian, metameric lineage. These evolutionary ideas are only touched upon in this chapter but are discussed in detail in Chapter 19. The unmistakable similarities between annelids and arthropods are reflected most notably in their segmented bodies and the general architecture of their nervous systems. As we will see, the major differences between these two phyla derive largely from the presence of a rigid exoskeleton and evolution of jointed appendages in arthropods. The basic features of the arthropod *Bauplan* are listed in Box 1, and a few common arthropods are shown in Figure 1. Some of these features are unique to the phylum Arthropoda and thus represent defining synapomorphic attributes; others occur in closely related taxa, such as the annelids, the onychophorans and the tardigrades, and hence are symplesiomorphies within the Arthropoda. In Chapter 19 we elaborate on these and other features that may be used to analyze the phylogeny of the arthropods and their allies.

Taxonomic history and classification

Linnaeus (1758) recognized six groups of animals (Vermes, Insecta, Pisces, Amphibia, Aves, and Mammalia), placing all of the invertebrates except the insects into a single group, the Vermes. In the early 1800s such famous zoologists as Lamarck and Cuvier presented substantial reorganizations of Linnaeus's earlier scheme, and it was during this period that the

---

Box One

Characteristics of
the Phylum Arthropoda[a]

1. Bilateral, triploblastic protostomes
2. Body segmented, both internally and externally; segments arise by teloblastic growth
3. Most groups have a strong tendency toward regional body specialization and tagmosis
4. Cuticle forms well developed exoskeleton, generally with thick sclerotized plates (sclerites) consisting of dorsal tergites, lateral pleurites (in most groups), and ventral sternites; the chitinous cuticle is calcified in many groups
5. Each true body segment primitively bears a pair of segmented (jointed) appendages
6. Usually with a pair of lateral faceted (compound) eyes and one to several simple median eyes (although this is presumably the primitive arthropod condition, compound or simple eyes, or both, may be absent in some groups)
7. Coelom reduced to portions of the reproductive and excretory systems; main body cavity is an open hemocoel
8. Circulatory system open; dorsal heart is a muscular pump with lateral ostia for blood return
9. Gut complete, with well developed stomodeum and proctodeum
10. Nervous system annelid-like, with dorsal (supraenteric) brain (= cerebral ganglia), circumenteric (circumesophageal) connectives, and paired, ganglionated, ventral nerve cords, the latter often fused to some extent
11. Functional cilia suppressed, except in sperm of a few groups
12. Growth by process of ecdysis (molting); with cephalic ecdysial glands
13. Muscles striated and arranged in isolated, segmental bands
14. Most are gonochoristic; with direct, indirect, or mixed development

[a]Chapter 19 elaborates on the nature of these characters by discussing ideas concerning the phylogeny of the arthropods and their allies.

---

various arthropod taxa began to emerge. Lamarck recognized four basic arthropod groups: Cirripedia (barnacles), Crustacea, Arachnida, and Insecta. He placed the ostracods with the brachiopods and certain molluscs in the Conchifera, and of course did not realize the crustacean nature of the barnacles. The close relationship between arthropods and annelids was recognized by Cuvier, who included both in his Articulata, and by Lankester, who classified them together with rotifers in his Appendiculata. The great zoologists Hatschek, Haeckel, Beklemishev, Snodgrass, Tiegs, Sharov, and Remane all recognized the Articulata as a discrete phylum, including in it at various times the groups Echiura, Sipuncula, Onychophora, Tardigrada, and Pentastomida. Leuckart, in 1848, separated the arthropods as a distinct phylum and Von Siebold coined the name Arthropoda in the same year, noting as the group's principal distinguishing attribute the jointed legs (Greek *arthro*, "jointed"; *pod*, "foot"). Haeckel published the first evolutionary tree of the arthropods in 1866.

In the past few decades biologists have witnessed an effort on the part of several specialists to support the hypothesis that arthropods are a polyphyletic group. This effort was spearheaded by S. M. Manton and D. T. Anderson. In such a scheme, the four groups we herein consider as subphyla would each be raised to the rank of phylum: Trilobitomorpha, Cheliceriformes, Crustacea, and Uniramia. Our opinion, however, is that the evidence more strongly supports a view of the Arthropoda as a monophyletic clade, and we therefore retain its single-phylum status. Our arguments on this matter are presented in Chapter 19.

A more-or-less agreed upon current classification scheme for the phylum Arthropoda follows. We avoid descriptive detail below the level of subphylum here, but include synopses of the three extant subphyla in the following chapters. It is important to offer a word of caution about terminology among the various groups of arthropods. Because the Arthropoda is such a vast and diverse assemblage, specialists usually concentrate on only one or a few groups. Thus, over time slightly different terminologies have evolved for the different groups. One obvious problem is that students sometimes feel overwhelmed with the array of terms, often used for seemingly similar parts. For example, the hindmost region of the body may be called an **abdomen** or **pleon** (as in many crustaceans), an **opisthosoma** (in cheliceriformes), or a **pygidium** (in trilobites). But there is a more subtle danger. Different terms for similar parts or regions in different taxa do not necessarily imply that those parts or regions are nonhomologous,

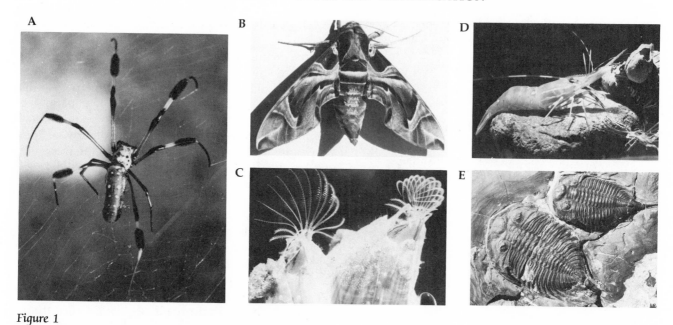

Figure 1

Representative arthropods. A, An orb-weaver spider (subphylum Cheliceriformes). B, A sphinx moth (subphylum Uniramia). C– D, Two very different crustaceans (subphylum Crustacea): an acorn barnacle (C) and a pandalid shrimp (D). E, fossils of two Silurian trilobites (subphylum Trilobitomorpha). (A,B photos by the authors; C,D courtesy of G. McDonald; E, photo by J. N. A. Lott, McMaster Univ./BPS.)

whereas the same term applied to similar parts of different arthropods does not always imply homology. To deal with these problems in the five chapters on arthropods, we have made an effort to achieve consistency in terminology, simplify word use and spelling, and indicate homologies as much as possible. When confusion could arise from these problems of names and homology, we try to point out the nature of each situation.

PHYLUM ARTHROPODA

SUBPHYLUM TRILOBITOMORPHA: Trilobites and their kin (Figure 19). Wholly extinct; body divided into cephalon, thorax, and pygidium (abdomen); segments of cephalon and pygidium fused, those of thorax free; body demarcated by two longitudinal grooves into a median and two lateral lobes; cephalon with one pair preoral antennae; all other appendages postoral and more-or-less similar to one another, with a robust telopod to which is attached at the base a long filamentous branch.

SUBPHYLUM CHELICERIFORMES: Scorpions, spiders, "sea spiders," and so on (see Chapter 16). Body divided into anterior prosoma (cephalothorax) and posterior opisthosoma (abdomen); both regions vary and may be further subdivided; appendages uniramous; legs often without extensor

muscles; first prosomal appendages are postoral chelicerae (often situated preorally in adults); second appendages often modified as pedipalps; remaining prosomal and opisthosomal appendages vary among groups; brain bipartite (without a deutocerebrum).

CLASS CHELICERATA

SUBCLASS MEROSTOMATA: Horseshoe crabs.

SUBCLASS ARACHNIDA: Spiders, scorpions, mites, ticks, and a host of lesser known forms.

CLASS PYCNOGONIDA: "Sea spiders."

SUBPHYLUM UNIRAMIA: Insects and myriapods (see Chapter 17). Body divided into distinct cephalon and a variably regionalized trunk; trunk homonomous and elongate (myriapods), or regionalized as a thorax and abdomen (insects); 4 pairs cephalic appendages—one pair of preoral antennae and 3 pairs of postoral appendages (mandibles, maxillules, maxillae); maxillae absent or partly (or wholly) fused; maxillules free or coalesced; all appendages uniramous; brain tripartite (with a deutocerebrum).

CLASS MYRIAPODA

SUBCLASS CHILOPODA: Centipedes.

SUBCLASS DIPLOPODA: Millipedes.

SUBCLASS SYMPHYLA: Symphylans.

SUBCLASS PAUROPODA: Pauropodans.

CLASS INSECTA: Insects.

SUBPHYLUM CRUSTACEA: Crustaceans (see Chapter 18). Appendages uniramous or biramous; brain tripartite (with a deutocerebrum); body divided into distinct cephalon and a variably regionalized trunk; trunk homonomous and elongate, or regionalized into a thorax and abdomen; 5 pairs of cephalic appendages—the preoral first antennae (antennules), and 4 pairs of postoral appendages: second antennae (which have "migrated" to a preoral position in adults), mandibles, maxillules (first maxillae), and maxillae (second maxillae).

CLASS REMIPEDIA: Remipedes.

CLASS CEPHALOCARIDA: Cephalocarids.

CLASS BRANCHIOPODA: Fairy shrimps, brine shrimps, tadpole shrimps, cladocerans, clam shrimps, etc.

CLASS MAXILLOPODA

    SUBCLASS OSTRACODA: Mussel or seed shrimps.

    SUBCLASS MYSTACOCARIDA: Mystacocarids.

    SUBCLASS COPEPODA: Copepods.

    SUBCLASS BRANCHIURA: Fish lice (argulids).

    SUBCLASS CIRRIPEDIA: Barnacles and their kin.

    SUBCLASS TANTULOCARIDA: Tantulocarids.

CLASS MALACOSTRACA

    SUBCLASS PHYLLOCARIDA: Leptostracans (nebaliaceans).

    SUBCLASS EUMALACOSTRACA: Crabs, shrimps, mantis shrimps, lobsters, crayfish, beach hoppers, pill bugs, euphausiids, etc.

The following three taxa have been variously regarded by different authors as classes, subphyla, or phyla. We discuss their classification and biology in Chapter 19.

    ONYCHOPHORA: Onychophorans.

    TARDIGRADA: Tardigrades ("water bears").

    PENTASTOMIDA: Pentastomids ("tongue worms").

The arthropod *Bauplan*

Remember that the *Bauplan* concept includes not only the theme of body form and function but also the idea that all components of the system must be compatible to produce a whole, functional animal. In addition, one must consider the constraints imposed on form and lifestyle by various combinations of features. If we are to somehow comprehend the "essence of arthropod," then we must first understand the impact of one of the major synapomorphies of this phylum—the hard, jointed exoskeleton—that distinguishes these animals from their living relatives, in particular the annelids. Recall that annelids are characterized by segmented bodies, serially arranged coelomic spaces, and well developed circular and longitudinal muscles in the body wall. Now imagine an annelid-like creature encased in a rigid exoskeleton. What kinds of structural and functional problems would *have* to be solved in order for such an animal to survive? Of a number of possible solutions, the arthropods evolved one particular suite of highly successful adaptations, which is referred to as **arthropodization.**

Given the exoskeleton and the absence of mobile body cilia, several problems are immediately evident if such an animal is going to do more than sit around in a hard shell. The basic problem of locomotion was solved, evolutionarily, by the development of joints and regionalized muscles in the body and in the appendages. Flexibility was provided by thin intersegmental areas ("the joints") in the otherwise rigid covering. Muscles became organized as intersegmental bands associated with the individual body segments and appendage joints, and circular muscles were lost almost entirely (Figure 2).

With the loss of peristaltic capabilities as a result of body rigidity and the absence of circular muscles, the coelom became nearly useless as a hydrostatic skeleton. Loss of the coelom is associated with the formation of an open circulatory system and the use of the body cavity as a **hemocoel** or blood chamber in which the internal organs are bathed directly in body fluids. Large body size still demanded some sort of pumping organ for moving the blood around through the hemocoel, hence retention of the annelid-like dorsal vessel was necessary. However, in the absence of body wall muscles to help move blood about (as it is moved in annelids), the dorsal vessel became a highly muscularized pumping structure—a heart. Instead of the open metanephridia typical of polychaetes (see Chapter 13), excretory organs became closed internally, thereby preventing the blood from being drained from the body. Surface sense organs evolved (the "arthropod setae") with various devices for transmitting sensory impulses to the nervous system in spite of the hard exoskeleton. Gas exchange structures evolved in various ways that overcame the barrier of the cuticle. Growth was no longer a simple process of gradual increase in body size, for these animals were now encased in a rigid, nonliving, outer covering. The complex process of **molting**, or **ecdysis**, evolved in which the skeleton is periodically shed to allow for an increase in real body size.

If we add to this complicated suite of events the notion of arthropods exploiting land and fresh water as well as their ancestral seas, the problems are compounded by exposure to osmotic and ionic stress, aerial gas exchange difficulties, and the need for structural support and appropriate reproductive strategies. Not all of these problems were solved at once, and we discuss some hypothesized sequences of appearance of various arthropod features in Chapter 19.

While the origin of the exoskeleton demanded a host of coincidental changes to overcome new constraints, it clearly endowed these animals with great

Figure 2

"Looking through" a segment of a generalized arthropod. Note the positions of the major organs within the hemocoel and the typical arrangement of body muscles.

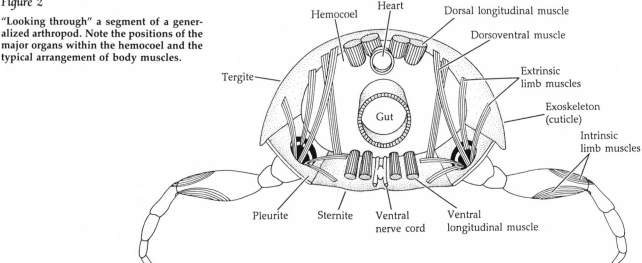

selective advantages, as evinced by their enormous success. One of the key advantages is the protection provided by the exoskeleton. Arthropods are armored not only against physical injury but also against physiological stress. In many cases, the cuticle provides an effective barrier against osmotic and ionic gradients, and as such is a major means of homeostatic control.

The undisputed evolutionary success of arthropods is dramatically reflected in their diversity. If we view a generalized, rather homonomous arthropod prototype with a fairly high number of segments and paired appendages on each of those segments, we can set the stage for just such a dramatic diversification. The diversity seen today has resulted largely from differential specialization of various segments and appendages. We saw a hint of this in our examination of the polychaetes, but nothing of the magnitude evident among the arthropods, in which segment and appendage diversification has reached its zenith. The spectacular radiation within the Arthropoda is also not unlike the situation seen in another highly successful animal group that also relied heavily on appendage modification—the vertebrates.

The arthropod body has itself undergone various forms of regional specialization, or **tagmosis**, to produce segment groups specialized for a host of different functions. These specialized body regions are called **tagmata** (e.g., the head, thorax, and abdomen). Tagmosis may be viewed as an extreme form of heteronomy. Even the head, which primitively bears a presegmental **acron** and four or five fused segments, varies greatly among extant groups. Thus the built-in evolutionary plasticity of regional specialization has, like limb variability, been of paramount impor-

tance in establishing the diversity and dominant position of the arthropods in the animal world. As we discuss the various aspects of the arthropod *Bauplan*, do not lose sight of the "whole animal" and the "essence of arthropod" just described.

A word about appendages

Arthropod appendages are articulated outgrowths of the body wall, equipped with sets of extrinsic (connecting the limb to the body) and intrinsic (wholly within the limb) muscles for moving the various limb segments or pieces (called **articles** or **podites**) (Figure 2). Beyond this fundamental plan, however, the variations in arthropod limbs (and an attending legion of terms) can be overwhelming to students and experts alike. Thus, we take the opportunity offered by this introductory section to outline some generalities and terminology regarding arthropod limbs.

In general, arthropod appendages can be either **uniramous** (with only a single branch, or **ramus**) or **biramous** (Y-shaped, with two rami). The former condition is characteristic of living cheliceriformes and uniramians, the latter of trilobites and crustaceans, although in a number of crustacean taxa the appendages have secondarily become uniramous. Figure 3 illustrates several appendage types and the terms applied to their various parts. The uniramous legs of cheliceriformes and uniramians (and some crustaceans) are typically ambulatory in function. They are generally long and thin, and thus are often called **stenopodia** (Greek *steno*, "narrow"; *podia*, "feet"). Biramous limbs may also be stenopodous; in these limbs the inner branch is long and thin and used for walking, whereas the outer branch is vari-

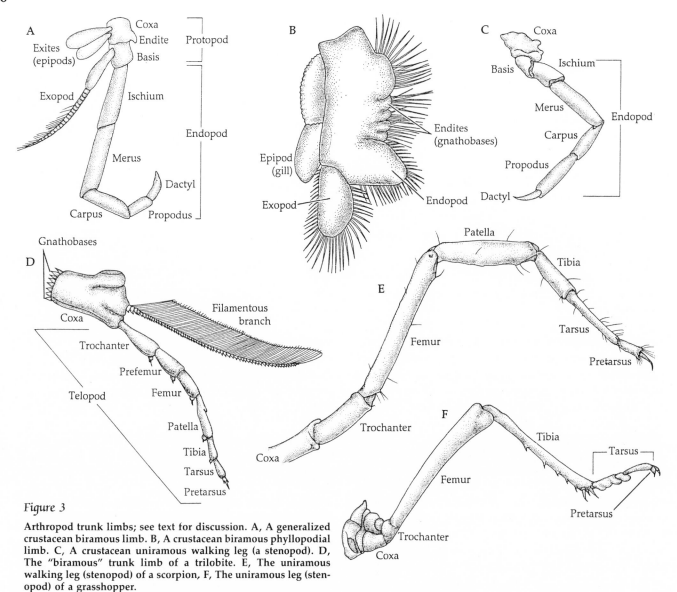

Figure 3

Arthropod trunk limbs; see text for discussion. A, A generalized crustacean biramous limb. B, A crustacean biramous phyllopodial limb. C, A crustacean uniramous walking leg (a stenopod). D, The "biramous" trunk limb of a trilobite. E, The uniramous walking leg (stenopod) of a scorpion, F, The uniramous leg (stenopod) of a grasshopper.

ably reduced. The biramous natatory limbs of some crustaceans (e.g., Cephalocarida, Branchiopoda, Phyllocarida) are greatly expanded and flattened and are called foliacious limbs, or **phyllopodia** (Greek, *phyllo*, "leaf-shaped"; *podia*, "feet").

One approach to understanding the comparative anatomy of arthropod legs is by examination of a generalized biramous limb condition, such as an idealized crustacean limb (Figure 3A). The most proximal limb article (which articulates with the body) is the **coxa**, and together with the next article (the **basis**) these two articles form a functional unit called the **protopod**, or **sympod**. Many specialists believe the primitive arthropod protopod comprised three articles, and that the basalmost article of primitive ar-

thropod appendages was a **precoxa**, which articulated on the flexible pleural membrane between the **tergite** and the **sternite** (Figure 2). At an early evolutionary stage, the precoxa may have been lost, or transformed into small sclerites, or pleural plates, that lie near the limb bases of modern arthropods. A small, third, protopodal article does occur in a few living arthropods, and many specialists regard it as a precoxa. Others view this article as an "extra" podite, perhaps derived through an evolutionary subdivision of the coxa. On the other hand, many of the earliest known Cambrian arthropod and crustacean fossils have protopods of a single article, suggesting that both two- and three-articulate protopods might be derived conditions. In any case, in a biramous limb, two rami arise from the basis of the protopod: a lateral (outer) **exopod** and a medial (inner) **endopod**. Additional structures may arise from the protopod either laterally (collectively called **exites**) or medially (collectively called **endites**). Exites often form gills or gill cleaners (e.g., **epipods**), and coxal endites may form grinding surfaces, or "jaws" (e.g., **gnathobases**), or other special structures.

The exopod may be reduced or lost, and the persisting endopod often forms a stenopodial walking leg (Figure 3C). The endopod and exopod may together form flattened natatory plates, which may also be specialized for simultaneous feeding and swimming (phyllopodia; Figure 3B). Many groups have a number of epipods on the protopod, and sometimes crustaceans have reduced exopods (or endopods) that are no larger than the epipods (Figure 3B). Such phyllopodous limbs have been called "triramous" or "polyramous limbs" by various authors. Some authorities regard phyllopodous limbs as the primitive condition for the Crustacea. Others specialists consider such limbs to be derived within the Crustacea, distinguishing a line that includes cephalocarids, branchiopods, and perhaps phyllocarids. This matter is far from settled.

It is usually inferred that the "biramous" legs of trilobites are quite different from biramous legs in crustaceans. In trilobites, the outer **filamentous branch** (often referred to as the gill ramus) arises from the first article (coxa), not from the second article as it does in crustaceans (Figure 3D). A true trilobite "exopod" would presumably arise from the second limb article, which is called the trochanter in this subphylum and corresponds in position to the crustacean basis. Thus, if we are to regard the trilobite limb as homologous to a crustacean limb, the trilobite gill ramus must be considered an epipod, not an exopod, and the limb would be viewed as uniramous (not biramous). Considerable difference of opinion exists on this issue, and the true phylogenetic affinities of the trilobites may ultimately hinge on the resolution of leg homologies among extant and extinct arthropod groups. A good summary of the problem can be found in Hessler and Newman (1975). Table 1 lists comparable limb articles among selected arthropods. Given the above disclaimers, the reader will realize that we have taken some license with this matter, but in doing so we have tried to establish some clarity and standardization of terms.

The body wall and body cavity

A cross section through a body segment of an arthropod reveals a good deal about their overall architecture (Figure 2). The body cavity is an open hemocoel; the organs are bathed directly in the hemocoelic fluid, called the **hemolymph**, or blood. The

Table One

Comparison of limb segments of members of major arthropod taxa[a]

BASIC BIRAMOUS CRUSTACEAN LIMB	UNIRAMOUS CRUSTACEAN LIMB	TRILOBITE LIMB	CHELICERIFORM LIMB	UNIRAMIAN LIMB
ENDOPOD	ENDOPOD	TELOPOD	MAIN LIMB	MAIN LIMB
coxa	coxa	coxa	coxa	coxa
basis	basis	trochanter	trochanter	trochanter
ischium	ischium	prefemur-femur	femur	femur
merus	merus	patella	patella	(absent)
carpus	carpus	tibia	tibia	tibia
propodus	propodus	tarsus	tarsus	tarsus
dactyl	dactyl	pretarsus	pretarsus	pretarsus
EXOPOD				
present	absent	absent	absent	absent
EPIPOD				
usually present	absent or present (usually as a gill)	present as a gill-bearing branch	absent	absent

[a]See Figure 3. As explained in the text, similarity in position or terminology does not necessarily imply homology, but the listing provides an exercise in comparative anatomy. In this table we view the trilobite limb as uniramous, and its filamentous branch as an epipod.

body wall is composed of the complex, layered **cuticle** secreted by the underlying epidermis (Figure 4). The epidermis, often referred to in arthropods as the **hypodermis**, is typically a simple cuboidal epithelium. In general, each body segment (or **somite**) is "boxed" by four skeletal plates, or **sclerites**: a dorsal **tergite**, two lateral **pleurites**, and a ventral **sternite**.*

*The terms **tergum, sternum**, and **pleuron** are often interchangeably with tergite, sternite, and pleurite. Technically, however, the term *tergum* refers more precisely to the "tergal region," *sternum* to the "sternal region," etc. For this reason, we use the terms tergite (pl., tergites), sternite (pl., sternites). and pleurite (pl., pleurites) when refering to the specific sclerites. In some groups (e.g., the thorax of trilobites; the abdomen of some decapods), the tergites expand ventrolaterally to form side plates that overhang the leg bases. These tergal lobes are often (unfortunately) called "**pleura**" or "**pleural lobes**," but they are *not* pleurites.

However, numerous variations exist, such as fusion of adjacent tergites, fragmentation of the pleurites or sternite to increase the number of sclerites on a given body somite, and loss of pleurites or sternites. Discrete pleurites are apparently absent in the Trilobitomorpha and Cheliceriformes, and the side walls in these groups are membranous and only lightly sclerotized. Muscle bands are attached at points where the inner surface of sclerites projects inward as ridges or tubercles, called **apodemes**.

The exoskeleton varies somewhat among the arthropods, and to some degree it varies on the basis of differences in body size and habitat (e.g., marine versus terrestrial groups). The structure of the cuticle has been most extensively studied in insects and decapod crustaceans (e.g., crabs and lobsters), and the following account is based largely on those groups.

Figure 4 illustrates the cuticles of an insect and a decapod crustacean. The outermost layer is the **epicuticle**, which is itself multilayered. The external sur-

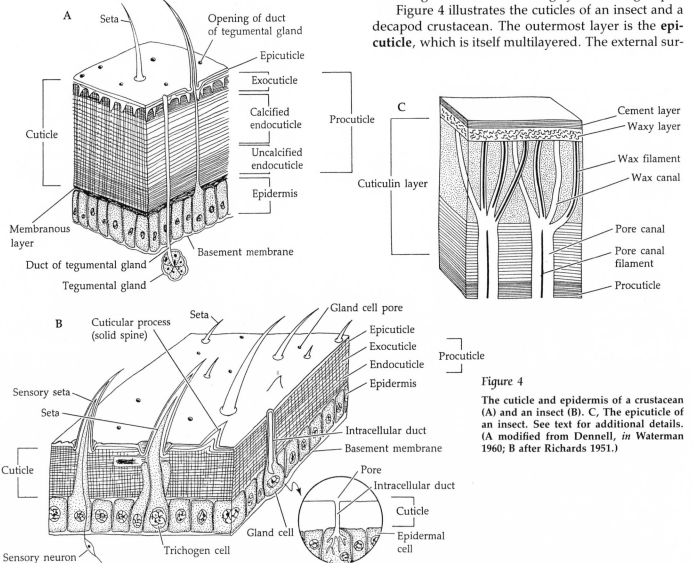

Figure 4

The cuticle and epidermis of a crustacean (A) and an insect (B). C, The epicuticle of an insect. See text for additional details. (A modified from Dennell, *in* Waterman 1960; B after Richards 1951.)

face of the epicuticle is a protective layer of lipoprotein, sometimes called the **cement layer**. Beneath this is a **waxy layer** that is especially well developed in arachnids and insects. The waxes, which are long-chain hydrocarbons and the esters of fatty acids and alcohols, provide an effective barrier to water loss and, coupled with the outer lipoprotein layer, protection against bacterial invasion. These layers of the epicuticle largely isolate the arthropod's internal environment from the external environment. No doubt the development of the epicuticle was critical to the invasion of land and fresh water by various arthropods. The innermost layer of the epicuticle is a **cuticulin layer**, which consists primarily of proteins and is particularly well developed in insects. The cuticulin layer usually has two components: an outer, thin but dense layer and a thicker, somewhat less dense inner layer. The cuticulin layer is involved in hardening of the exoskeleton, as discussed below, and contains canals through which wax reaches the waxy layer near the surface.

Beneath the epicuticle is the relatively thick **procuticle**, which may be subdivided into an outer **exocuticle** and an inner **endocuticle** (Figures 4A and B).* The procuticle consists primarily of layers of protein and chitin. The procuticle is intrinsically tough but flexible. In fact, certain arthropods possess rather soft and pliable exoskeletons (e.g., many insect larvae, parts of spiders, some small crustaceans). However, in most arthropods the cuticle is hard and inflexible except at the joints, a condition brought about by one or both of two processes: **sclerotization** and **mineralization**. Cuticular hardening by sclerotization is a tanning process that probably occurs to various degrees in nearly all arthropods, but most notably in the insects and arachnids. Mineralization of the skeleton in crustaceans is accomplished by the deposition of calcium carbonate in the outer layers of the procuticle.

The layered arrangement of untanned proteins yields a flexible structure. To produce a rigid sclerotized structure, these protein layers are cross-bonded to one another by ortho-quinone linkages. The bonding agent is typically produced from polyphenols and catalyzed by polyphenol oxidases present in the protein layers of the cuticle. Sclerotization generally begins in the protein layer of the epicuticle and progresses into the procuticle to various degrees, where it is associated with a distinct darkening in color. The

relationship between cuticular hardening, joints, and molting is discussed in the next sections, dealing with support, locomotion, and growth.

The epidermis is responsible for the secretion of the cuticle (discussed under molting) and as such contains various unicellular glands, some of which bear ducts to the surface of the cuticle (Figure 4). The epidermis is underlain with a distinct basement membrane, which forms the outer boundary of the body cavity or hemocoel.

The hemocoel contains the circulatory fluid, which bathes the organs. This cavity is not a true coelom, either evolutionarily or ontogenetically, but it may be viewed as a persistent blastocoelic remnant. Thus, one might at first reason that the arthropods technically are "pseudocoelomates." However, the absence of a large body coelom in arthropods is a secondary condition resulting from a loss of the ancestral coelomic body cavity during the evolution of the arthropod *Bauplan* and not a primary condition like that seen in the true "pseudocoelomates" (Chapter 12), at least some of which may never have had a true coelom in their ancestry. A similar secondary loss of the coelom has occurred in the molluscs (Chapter 20).

Support and locomotion

The arthropods, having largely abandoned the hydrostatic skeleton of their coelomate ancestors, lack discrete coelomic spaces and the associated muscle sheets that act upon them. Instead, they rely on the exoskeleton for support and maintenance of body shape. Muscle sheets simply would not work in the presence of the exoskeleton. Hence, the muscles are arranged as short bands that extend from one body segment to the next, or across the joints of appendages and other regions of articulation. An understanding of the nature of the articulation points, areas where the cuticle is notably thin and flexible, is crucial to an understanding of the action of the muscles and hence of locomotion.

In contrast to most of the exoskeleton, joints between body and limb segments are bridged by very thin flexible cuticle in which the procuticular layer is much reduced and unhardened (Figure 5). These thin areas are called **arthrodial** or **articular membranes**. Generally, each articulation is bridged by one or more pairs of antagonistic muscles. One set of muscles, the flexors, acts to bend the body or appendage at the articulation point; the opposing set of muscles, the extensors, serve to straighten the body or appendage.

Joints that operate as described above generally articulate in only a single plane. Such movement is

*Caution: Some authors use the term endocuticle alone to refer to the entire procuticle of crustaceans, and use the two subdivision terms only when referring to insects.

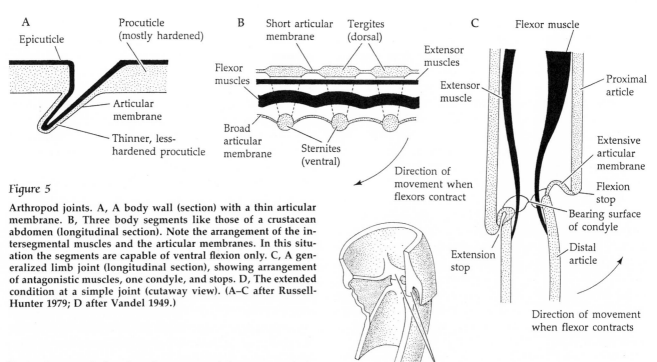

Figure 5

Arthropod joints. A, A body wall (section) with a thin articular membrane. B, Three body segments like those of a crustacean abdomen (longitudinal section). Note the arrangement of the intersegmental muscles and the articular membranes. In this situation the segments are capable of ventral flexion only. C, A generalized limb joint (longitudinal section), showing arrangement of antagonistic muscles, one condyle, and stops. D, The extended condition at a simple joint (cutaway view). (A–C after Russell-Hunter 1979; D after Vandel 1949.)

limited not only by the placement of the antagonistic muscles, but also by the details of the hard parts of the cuticle that border the articular membrane. In such cases the articular membrane may not form a complete ring of flexible material but is interrupted by points of contact between hard cuticle on either side of the joint. These contact points, or bearing surfaces, are called **condyles** and serve as the fulcrum for the lever system formed by the joint. Thus, a **dicondylic joint** allows movement in one plane, but not at angles to that plane. The motion at a joint is also usually limited by hard cuticular processes called **locks** or **stops**, which prevent overextension and overflexion (Figure 5).

Some joints are constructed to allow movement in more than one plane, much like a ball-and-socket joint. For example, in most arthropods the joints between walking legs and the body (the coxal-pleural joints) lack large condyles, and the articular membranes form complete bands around the joints. Also, occasionally two adjacent dicondylic limb joints will articulate at 90° to one another, forming a gimbal-like arrangement that facilitates movement in two planes.

Arthropods have evolved a plethora of locomotor devices for movement in water, on land, and in the air (for a recent review, see the collection of papers in Herreid and Fourtner 1981). Only the vertebrates can boast a similar range of abilities, albeit utilizing a far less diverse set of mechanisms. As in so many aspects of arthropod biology, methods of movement reflect extreme evolutionary plasticity and adaptive qualities associated with the segmented body and appendages. In addition, many arthropods possess the ability to self-amputate appendages (**autotomy**), particularly the thoracic appendages, and they utilize this behavior when escaping from predators. Such appendages typically contain a "fracture plane" at which the limb breaks off, usually near its base.

Movement through water involves various patterns of swimming that include smooth paddling by shrimps, jerky stroking by certain insects and small crustaceans, and startling backward propulsion by tail flexion in lobsters and crayfishes. Aerial locomotion has been mastered by the pterygote (winged) insects but is also practiced by certain spiders that drift on gossamer threads of silk. Many arthropods burrow or bore into various substrata (e.g., ants, termites, burrowing crustaceans). Some terrestrial arthropods that are normally associated with the ground engage in short-term aerial movements that serve as escape responses. Some, like fleas, simply jump, whereas others jump and glide, giving us a possible clue as to the evolutionary origin of flight. Some crustaceans jump as well, such as the familiar beach hoppers, which bound away over the sand when disturbed. Then there are those that move in contact with the surface of the substratum, under water or on land, by various forms of walking, creeping, crawling, or running—such arthropods are referred to as reptant or pedestrian.

All of the common forms of arthropod locomotion except flight depend on the use of typical appendages and thus are based on the principles of joint articulation described above, coupled with specialized architecture of the appendages. Below we discuss aspects of two fundamental types of appendage-dependent locomotion in arthropods—swimming and pedestrian—exploring variations on these methods and others in subsequent chapters.

Good examples of swimming arthropods are found among the crustaceans. Most swimming crustaceans (e.g., anostracans and shrimps) and even those that swim only infrequently (e.g., isopods and amphipods) employ ventral, flaplike setose appendages as paddles. The appendages used for swimming may be restricted to particular body regions (e.g., the abdominal swimmerets of shrimps, stomatopods, and isopods and the metasomal limbs of swimming copepods) or may occur along much of the trunk (e.g., the appendages of anostracans, remipedes, and cephalocarids) (Figure 6). Each appendage engages in a backward power stroke and then a forward recovery stroke. The appendages are constructed in such a way that, as they are brought forward on the recovery stroke, the flaps and marginal setae passively "collapse," to reduce the coef-

ficient of friction. On the power stroke the limbs are held erect, their largest surface facing the direction of limb movement, thus increasing thrust efficiency (by increasing the coefficient of friction and the distance through which the limb travels). Less sophisticated swimming is accomplished in other arthropods by use of various other appendages, including the antennae of many minute crustaceans and larvae and the thoracic stenopods of many aquatic insects. These swimming appendages typically articulate with the body only on a plane parallel to the body axis.

Pedestrian locomotion in arthropods is highly variable, both among different groups and even in individual animals. With the exception of a few strongly homonomous "vermiform" types (e.g., centipedes and millipedes), most arthropods are incapable of lateral body undulations. Thus, they cannot amplify the stride length of their appendages by body waves as many polychaetes do. Walking arthropods depend almost entirely on the mobility of specialized groups of appendages. The structure of these ambulatory legs is quite different from that of paddlelike swimming appendages, and their action is much more complex and variable.

Consider the general movement of an ambulatory leg as it passes through its power (propulsive) and recovery strokes (Figure 7). At the completion of

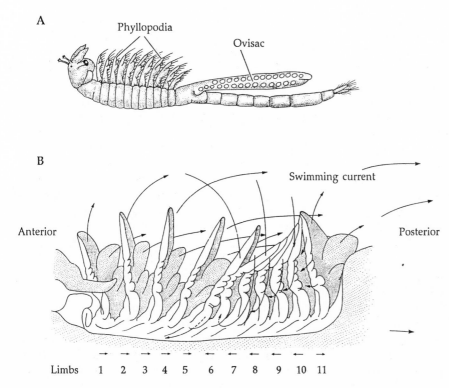

Figure 6

Swimming motions in a primitive crustacean. A, A fairy shrimp (Anostraca) on its back in its normal swimming posture. B, The appendages "in motion," producing a posteriorly directed flow of water that propels the animal forward. Arrows near the bases of the appendages indicate feeding currents (see Chapter 18). The small arrows below the drawing indicate the direction of movement of each numbered appendage at this moment in the anterior progression of the metachronal wave. Water is drawn into the interlimb spaces as adjacent appendages move away from one another, and water is pressed out of the spaces as adjacent limbs move together. The lateral articles of these phyllopodial appendages are hinged in such a way that they extend on the power stroke to present a large surface area and collapse on the recovery stroke, thereby producing less drag. (A modified from Calman 1909; B redrawn from Barrington 1967.)

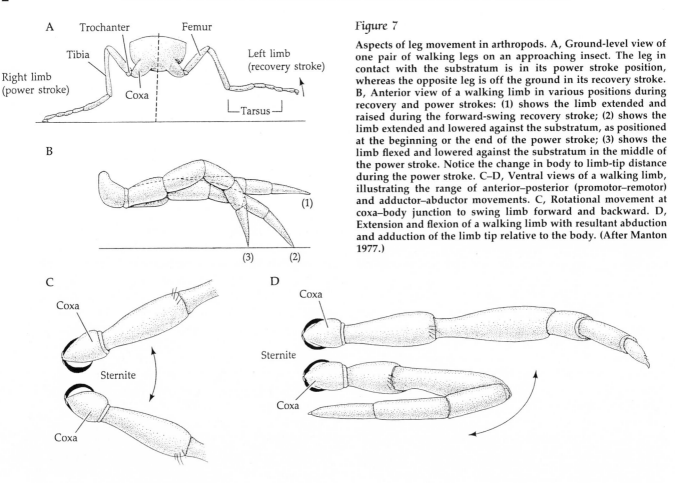

Figure 7

Aspects of leg movement in arthropods. A, Ground-level view of one pair of walking legs on an approaching insect. The leg in contact with the substratum is in its power stroke position, whereas the opposite leg is off the ground in its recovery stroke. B, Anterior view of a walking limb in various positions during recovery and power strokes: (1) shows the limb extended and raised during the forward-swing recovery stroke; (2) shows the limb extended and lowered against the substratum, as positioned at the beginning or the end of the power stroke; (3) shows the limb flexed and lowered against the substratum in the middle of the power stroke. Notice the change in body to limb-tip distance during the power stroke. C–D, Ventral views of a walking limb, illustrating the range of anterior–posterior (promotor–remotor) and adductor–abductor movements. C, Rotational movement at coxa–body junction to swing limb forward and backward. D, Extension and flexion of a walking limb with resultant abduction and adduction of the limb tip relative to the body. (After Manton 1977.)

the power stroke, the appendage is extended posteriorly and its tip is in contact with the substratum. The recovery stroke involves lifting the limb, swinging it forward, and placing it back down on the substratum; by then the limb is extended anterolaterally. The power stroke is accomplished by first flexing and then extending the leg while the tip is held in place against the substratum. Thus the body is first pulled and then pushed forward by each limb.

The ability to perform these complicated movements obviously would not be possible if all of the limb joints and limb/body joints were dicondylic articulations in the same plane, parallel to the body axis. The leg must be able to move up and down as well as forward and backward, and the action at each joint must be coordinated with the actions of all the others. In general, the distal limb joints are dicondylic, with articulation planes parallel to the limb axis. They allow the appendage to flex and extend, that is, to move the tip closer to (adduction) or farther from (abduction) the point of limb origin. The actions typically involve the usual sets of antagonistic flexor and extensor muscles described earlier. In some arachnids and a few crustaceans, however, certain limb joints lack extensor muscles, and the limbs are extended by an increase in blood pressure. Raising and lowering of the limb are also accomplished by extensor and flexor muscles, which thus serve as levators and depressors, respectively; the muscles in the proximal leg joints usually serve these purposes.

Anterior–posterior limb movements are accomplished in two basic ways. First, the ball-and-socket type of joint at the point of limb–body articulation typically carries out these actions in most crustaceans and uniramians. Promotor and remotor muscles that are associated with these ball-and-socket joints rotate the limb forward and backward, respectively. Second, many arachnids accomplish multidirectional limb movements by using only uniplanar dicondylic joints. In these arthropods, one or more of the proximal joints articulate perpendicular to the limb axis, and thus to the rest of the limb joints, providing forward and backward movement.

Understanding how a single limb moves does not describe locomotion of the whole animal. The various patterns of pedestrian locomotion in arthropods, called **gaits**, are the results of many factors (e.g., leg number, leg movement sequences, stride lengths, speed). The number of patterns is great, but it is limited by certain biological and physical constraints. Speed is limited by rates of muscle contraction and the necessity to coordinate leg movements to avoid tangling. Furthermore, the animal must at all times maintain an appropriate distribution of legs in various phases of power and recovery strokes so that its weight is fully supported.

The gaits of uniramians, particularly insects, have been more extensively studied than those of crustaceans and cheliceriformes. Studies on uniramians led to an attempt to establish principles under which all pedestrian arthropod locomotion could be unified. The most frequently used descriptions of arthropod walking, crawling, and running are based on Wilson's (1966) "metachronal" or "insect model." The basic idea of Wilson's model is that the legs on each side of the body move in metachronal waves from back to front and that the waves overlap to various degrees, depending on the speed of movement. This model does work for some arthropods, some of the time, but things are not so simple and attempts to overgeneralize have been misleading. A good deal of the work on crustaceans and arachnids (and even insects) indicates that leg movement sequences, stepping patterns, stride lengths, and other characteristics are extremely variable, even within individuals, and depend on a host of factors other than speed. The actions of the joints are coordinated by information supplied to the central nervous system by proprioceptors in the joints themselves. Detailed analyses are beyond the scope of this text, and we refer you again to Herreid and Fourtner (1981) for more information. Additional information also is presented in Chapters 16–19.

Growth

The imposition of a rigid exoskeleton precludes growth by means of a gradual increase in external body size. Rather, overall increase in body size takes place in staggered increments associated with the periodic loss of the old exoskeleton and the deposition of a new, larger one (Figure 8). This process of shedding the exoskeleton is called molting or **ecdysis**, and it is a phenomenon characteristic of arthropods and a few other invertebrates with thick cuticles (e.g., onychophorans, tardigrades, kinorhynchs, nematodes, nematomorphs, loriciferans, and priapulans). It is unlikely, however, that the complex physiological basis of molting in all these groups is homologous. The molting process varies in detail even among the

arthropods. It has been best studied in certain insects and crustaceans, and the description below is based primarily on those two groups. We first outline the basic steps in the molt cycle and then briefly discuss the hormonal control of these events.

The **intermolt stages** between molts are called **instars**. It is during the intermolt phase that real tissue growth occurs, albeit with no increase in external size. When such tissue growth reaches the point where the body "fills" its exoskeletal case, the animal usually enters a physiological state known as **premolt** or **proecdysis**. During this stage there is active preparation for the molt, including accelerated growth of any regenerating parts. During this premolt period, certain epidermal glands secrete enzymes that begin digesting the old endocuticle, thus separating the exoskeleton from the epidermis. In many crustaceans some of the calcium is removed from the cuticle during this period and stored within the body for later redeposition. As the old cuticle is loosened and thinned, the epidermis begins secreting a soft new cuticle. Figure 9 depicts some of these events.

Once the old cuticle has been substantially loosened and a new cuticle formed, actual ecdysis occurs. The old cuticle splits in such a way that the animal can wriggle free and pull itself out. The lines along which the cuticle splits vary among different arthropods but are consistent within particular groups. It is important to remember that all cuticular linings are lost during ecdysis, including the linings of the fore- and hindgut, the eye surfaces, and the cuticle that lines every pit, groove, and spine on the body surface. When you see a cast-off intact exoskeleton, or

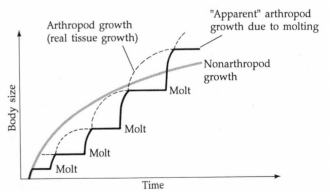

Figure 8

Arthropod versus nonarthropod growth. The heavy solid line indicates incremental ("stair-step") growth of an arthropod as measured by changes in external body size associated with molts. The dotted line depicts real tissue growth in the same arthropod. The gray line depicts the typical growth of a nonarthropod.

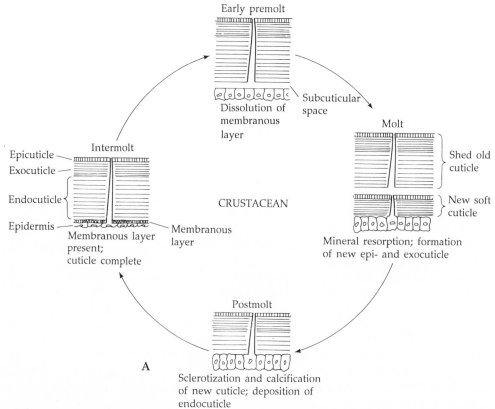

Early premolt

Dissolution of membranous layer

Subcuticular space

Intermolt

Epicuticle
Exocuticle
Endocuticle
Epidermis
Membranous layer present; cuticle complete
Membranous layer

CRUSTACEAN

Molt

Shed old cuticle

New soft cuticle

Mineral resorption; formation of new epi- and exocuticle

Postmolt

Sclerotization and calcification of new cuticle; deposition of endocuticle

A

Figure 9

Arthropod molting. Schematic representations of some events in the molting of a crustacean (A) and an insect (B). The separation of the old cuticle from the body is generally accomplished by dissolution of the membranous layer in large crustaceans and by digestion of the inner boundary of the cuticle in insects. See text for additional details. C, A swimming crab in the final stage of extracting itself from the old, molted exoskeleton; only the chelipeds remain to be pulled out of the exuviae. D, The cast-off exoskeleton of a tarantula. (A after Roer and Dillaman 1984; B after Wigglesworth 1954; C, photo by the authors; D, from Foelix 1982, reprinted with the permission of Harvard University Press.)

exuviae,* of a large crab or lobster, you are bound to be impressed with its wonderfully perfect detail. It is, in fact, at first difficult to imagine how the animal could extricate itself from each and every tiny part of the old cuticle (Figures 9C and D). The ability to do so depends, of course, on the great flexibility of the body within its new and unhardened skeleton.

As soon as the arthropod emerges from its old cuticle, and while the new cuticle is still soft and pliable, the body swells rapidly by uptake of air or water. Once the new casing is thus enlarged, the animal enters a **postmolt period (postecdysis)** during which the new cuticle is hardened by sclerotization and/or the redeposition of calcium salts. The excess water (or air) is then actively pumped from the body as real tissue growth occurs during the subsequent intermolt period. During the sclerotization process, the cuticle becomes drier, stiffer, and resistant to chemical and physical degradation through the molecular cross-linking process in the protein-chitin matrix.

We have stressed the adaptive significance of the arthropod exoskeleton in terms of its protective and supportive qualities. However, during the postmolt period, before the new skeleton is hardened, the

animal is quite vulnerable to injury, predation, and osmotic stress. Many arthropods become reclusive, hiding in protective nooks and crannies and not even feeding when in this "soft-shelled" condition. The time required for hardening of the new skeleton varies greatly among arthropods, generally being longer in larger animals. The well known and delectable "soft-shell crabs" of the eastern United States are simply blue crabs (*Callinectes*) caught during their soft postmolt period.

The events of the molt cycle outlined above are controlled by a complex hormonal system. Several models have been proposed to explain the hormonal pathways involved in molting in insects and crustaceans, and recent work (e.g., Chang 1985) suggests

* The word "exuviae" is always spelled in the plural form; there is no such word as "exuvium."

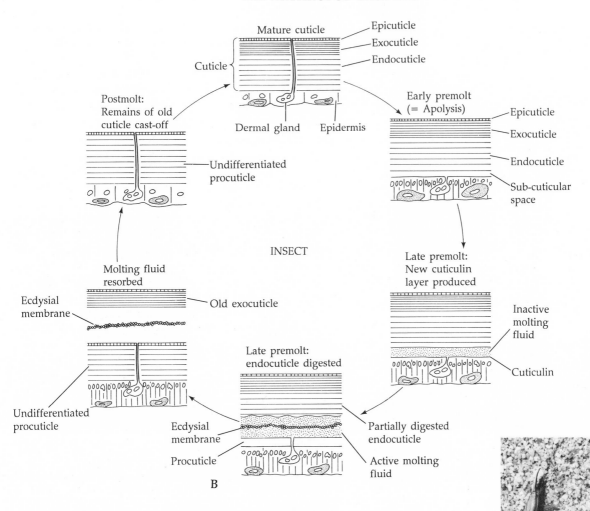

Mature cuticle
Epicuticle
Exocuticle
Endocuticle
Cuticle

Postmolt:
Remains of old
cuticle cast-off

Dermal gland   Epidermis

Undifferentiated
procuticle

Early premolt
(= Apolysis)
Epicuticle
Exocuticle
Endocuticle
Sub-cuticular
space

INSECT

Molting fluid
resorbed

Late premolt:
New cuticulin
layer produced

Ecdysial
membrane
Old exocuticle

Inactive
molting
fluid

Cuticulin

Undifferentiated
procuticle

Ecdysial
membrane

Procuticle

Late premolt:
endocuticle digested

Partially digested
endocuticle

Active molting
fluid

B

that the picture is still incomplete. The hormonal activities of the crustacean ecdysial cycle have been most extensively studied in decapods. In some (e.g., lobsters and crayfish) molting occurs periodically throughout the animal's life, but in many (e.g., copepods and some crabs) molting, and therefore growth, ceases at some point and a maximum size is attained. Animals that have engaged in their final molt are said to have entered a state of **anecdysis**, or permanent intermolt. Among insects, molting is largely associated with metamorphosis from one developmental stage to the next (e.g., pupa to adult).

In both crustaceans and insects (and probably all arthropods), the initiation of molting, beginning with the events of proecdysis, is brought about by the action of the **molting hormone**, called **ecdysone**. Apparently, however, the pathways controlling the secretion of ecdysone are different in insects and crustaceans, as diagrammed in Figure 10. In crustaceans, ecdysone is secreted by an endocrine gland called the **Y-organ** located at the base of the antennae or near

C

D

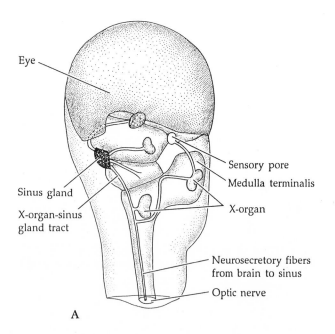

Eye

Sinus gland

X-organ–sinus gland tract

Sensory pore

Medulla terminalis

X-organ

Neurosecretory fibers from brain to sinus

Optic nerve

A

the mouth parts. The action of the Y-organ is controlled by a complex neurosecretory apparatus located in the eyestalks. During the intermolt period, a **molt-inhibiting hormone** (**MIH**) is produced by neurosecretory cells of the **X-organ**, located in a region of the eyestalk nerve (or ganglion) called the **medulla terminalis** (Figure 10A). MIH is carried by axonal transport to a storage area called the **sinus gland**, which appears to control MIH release into the blood. As long as sufficient levels of MIH are present in the blood, the production of ecdysone by the Y-organ is inhibited.

The active premolt and subsequent molt phases are initiated by sensory input into the central nervous system. The stimulus is external for some crustaceans (e.g., day length for certain crayfish), and internal for others (e.g., growth of soft tissues in certain crabs). External stimuli are transmitted via the central nervous system to the medulla terminalis and X-organ (Figure 10B). Appropriate stimuli inhibit the se-

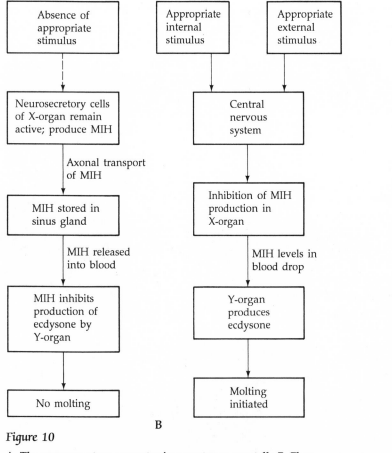

Absence of appropriate stimulus

Appropriate internal stimulus

Appropriate external stimulus

Neurosecretory cells of X-organ remain active; produce MIH

Central nervous system

Axonal transport of MIH

MIH stored in sinus gland

Inhibition of MIH production in X-organ

MIH released into blood

MIH levels in blood drop

MIH inhibits production of ecdysone by Y-organ

Y-organ produces ecdysone

No molting

Molting initiated

B

Figure 10

A, The neurosecretory apparatus in a crustacean eyestalk. B, Flow diagram of events inhibiting and initiating molting in crustaceans. C, Flow diagram of events initiating molting in an insect. See text for further details.

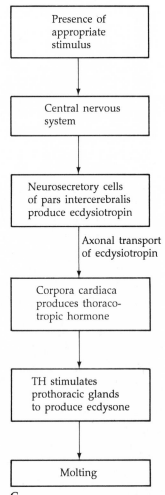

Presence of appropriate stimulus

Central nervous system

Neurosecretory cells of pars intercerebralis produce ecdysiotropin

Axonal transport of ecdysiotropin

Corpora cardiaca produces thoracotropic hormone

TH stimulates prothoracic glands to produce ecdysone

Molting

C

cretion of MIH, ultimately resulting in the production of ecdysone and initiation of a new molt cycle.

The sequence of events in insects is somewhat different from that in crustaceans in that a molt inhibitor is apparently not involved. When an appropriate stimulus is introduced to the central nervous system, certain neurosecretory cells in the cerebral ganglion are activated. These cells, which are located in the **pars intercerebralis**, secrete **ecdysiotropin**. This hormone is carried by axonal transport to the **corpora cardiaca**, a neural mass associated with the brain. Here, **thoracotropic hormone** is produced and carried to the **prothoracic glands**, stimulating them to produce and release ecdysone (Figure 10C). Some additional details of neuroendocrine action in insect development are discussed in Chapter 17.

The digestive system

It will come as no surprise that the great diversity among arthropods is reflected in their display of nearly every feeding method imaginable. As a group, their only real constraint in this regard is the absence of external, functional cilia. Evolutionarily, many arthropods have overcome even this limitation and suspension feed by other means. So varied are the feeding strategies that we postpone discussion of them to the sections and chapters on particular taxa and here attempt only to generalize about the basic structure and function of arthropod digestive systems.

The digestive tract of arthropods is complete and generally straight, extending from a ventral mouth on the head to a posterior anus. Regional specializations occur in various taxa, being most complex in the uniramians and crustaceans. In almost all cases there is a well developed, cuticle-lined, stomodeal **foregut** and proctodeal **hindgut**, connected by an entodermally derived **midgut** (Figure 11). In general, the foregut serves for ingestion, transport, storage, and mechanical digestion of food; the midgut for enzyme production, chemical digestion, and absorption; and the hindgut for water absorption and the preparation of fecal material. The midgut typically bears one or more evaginations in the form of **digestive ceca** (often referred to as the "**digestive gland**," "**liver**," or "**hepatopancreas**"). The number of ceca and the arrangement of the other gut regions varies among the different taxa.

The various terrestrial arthropods have convergently evolved many similar features as adaptations to life on land. Many of these convergent structures are associated with (although not necessarily derived from) the gut. For example, certain excretory structures called **Malpighian tubules** that develop from the mid- or hindguts of insects, arachnids, and tardigrades appear to be convergences (i.e., nonhomologous structures). The excretory structures of onychophorans also used to be called Malpighian tubules, but they have recently been shown to be complex metanephridia with secondarily derived, closed end-sacs. Many unrelated terrestrial taxa have special **repugnatorial glands**, which may or may not be associated with the gut and which produce noxious substances used to deter predators. Many different groups of terrestrial arthropods also have evolved the ability to produce silks or silklike substances for use outside their bodies. Among these are the larvae of commercial silkworm moths (*Bombyx* and *Anaphe*), which spin silk into cocoon fibers. Many other silklike fibers are produced by nonhomologous structures among arthropods, and although the silks vary greatly in chemical composition, all share a common molecular feature that gives them strength and elasticity: they are composed of regular assemblies of long-chain macromolecules (most being fibrous proteins), and many incorporate collagens. Modified salivary glands are common silk-producing organs, but silks are also sometimes secreted by the digestive tract, Malpighian tubules, accessory reproductive glands, and assorted dermal glands. Silk production occurs in chelicerates (false scorpions, spiders, and mites) and uniramians (many insect orders, chilopods, diplopods, and pauropodans). Arthropod silks are used in the production of cocoons, egg cases, webs, larval "houses," flotation rafts, prey entrapment threads, draglines, spermatophore receptacles, intraspecific recognition devices, and other sundry items. The truly spectacular array of silk uses by spiders is discussed in the next chapter.

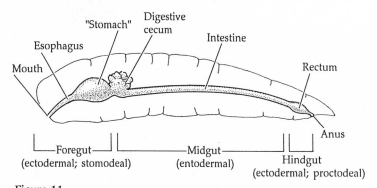

Figure 11

The major gut regions of arthropods. The myriad variations on this theme are discussed in subsequent chapters on particular taxa.

Circulation and gas exchange

A major aspect of the arthropod *Bauplan* is reflected in the nature of the circulatory system. The open hemocoelic system is in part an indirect result of the imposition of the exoskeleton and the loss of an internally segmented and fluid-filled coelom. We have seen that the presence of isolated coelomic spaces (like those in annelids) essentially requires a closed circulatory system, but that this requirement is eliminated in the arthropods. Furthermore, without a muscular, flexible body wall to augment blood movement, a pumping mechanism becomes necessary in the arthropods, thus the elaboration of a muscular heart. The result is a system wherein the blood is driven from the heart chamber through short vessels and into the hemocoel, where it bathes the internal organs. The blood returns to the heart via a noncoelomic **pericardial sinus** and perforations in the heart wall called **ostia** (Figure 12). The flow of blood back to the heart is along a decreasing pressure gradient resulting from lowered pressure within the pericardial sinus as the heart contracts. The complexity of the circulatory system varies greatly among arthropods, the differences being dependent in large part on body size and shape. These differences include variations in the size and shape of the heart (Figure 12), number of ostia, length and number of vessels, arrangement of hemocoelic sinuses, and circulatory structures associated with gas exchange.

Arthropod blood or hemolymph serves to transport nutrients, wastes, and usually gases. It includes a variety of types of amebocytes and, in some groups, clotting agents. The blood of many kinds of small arthropods may be colorless, wherein gases are simply carried in solution. Most larger forms, however, contain hemocyanin and a few contain hemoglobin. Both pigments are always dissolved in the fluid rather than contained within cells. In most groups of arthropods, the circulatory route takes at least some of the blood past the gas exchange surfaces before returning to the heart.

One of the major evolutionary problems arising from the acquisition of a relatively impermeable exoskeleton concerns gas exchange, particularly for terrestrial arthropods, where any increase in cuticular

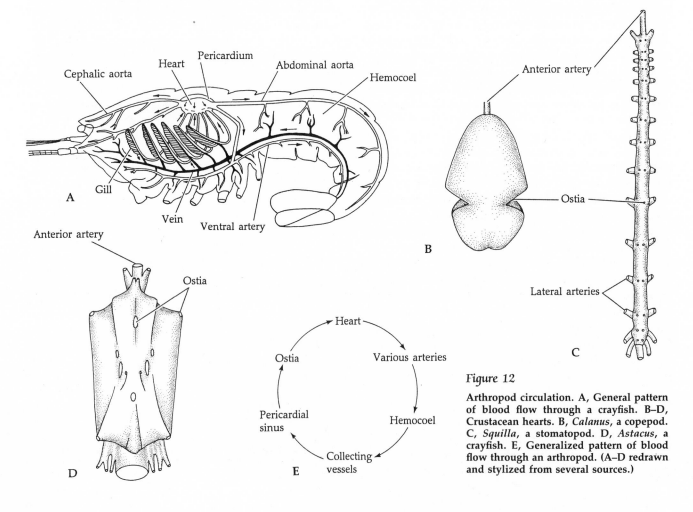

Figure 12

Arthropod circulation. A, General pattern of blood flow through a crayfish. B–D, Crustacean hearts. B, *Calanus*, a copepod. C, *Squilla*, a stomatopod. D, *Astacus*, a crayfish. E, Generalized pattern of blood flow through an arthropod. (A–D redrawn and stylized from several sources.)

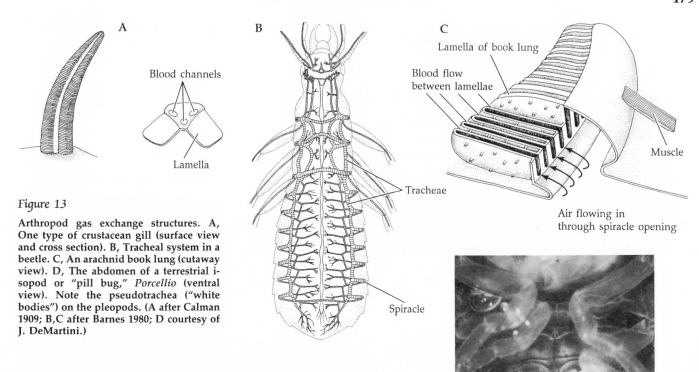

Figure 13

Arthropod gas exchange structures. A, One type of crustacean gill (surface view and cross section). B, Tracheal system in a beetle. C, An arachnid book lung (cutaway view). D, The abdomen of a terrestrial isopod or "pill bug," *Porcellio* (ventral view). Note the pseudotrachea ("white bodies") on the pleopods. (A after Calman 1909; B,C after Barnes 1980; D courtesy of J. DeMartini.)

permeability to facilitate gas exchange also increases the threat of water loss. Remember that gas exchange surfaces not only must be permeable but also must be kept moist (see Chapter 3). Evolutionarily, the dilemma for the arthropods becomes one of disrupting the integrity of the exoskeleton in such a way as to allow gas exchange to occur without seriously jeopardizing the survival of the animal by abandoning the principal benefits of that skeleton.

The design of arthropod gas exchange structures has taken one form in aquatic groups and quite another in terrestrial taxa (Figure 13). The former is best exemplified by the crustaceans, and the latter by the insects and terrestrial chelicerates. Some very tiny crustaceans (e.g., copepods) with a low surface-to-volume ratio exchange gases cutaneously across the general body surface or at the usual thin areas such as articular membranes. Most larger crustaceans have evolved various types of gills in the form of thin-walled, blood-filled cuticular evaginations. Gills are usually branched or folded, providing large surface areas (Figure 13A). The gills of some crustaceans are exposed, unprotected, to the surrounding medium (e.g., euphausiids), whereas in others the gills are carried beneath protective shields of the exoskeleton (e.g., crabs and lobsters).

The most successful terrestrial arthropods—insects and arachnids—have evolved gas exchange structures in the form of invaginations of the cuticle, rather than the evaginations seen in aquatic crusta-

ceans. Obviously, external gills would be unacceptable in dry conditions; but placed internally, these kinds of gas exchange structures are kept moist by acting as humidity chambers, allowing oxygen to enter solution for uptake. Many arachnids possess invaginations called **book lungs**, which are sacciform pockets with thin, highly folded walls called **lamellae**—sort of "dry, internal gills" (Figure 13C). Insects possess inwardly directed branching tubules called **tracheae**, which open externally through pores called **spiracles** (Figure 13B). Tracheal systems differ from the other structures mentioned so far in that the inner ends of the tubules open directly to the hemocoel or to specific tissues, allowing direct gas exchange between the air and the blood or internal organs. Some arachnids also have tracheae, and some terrestrial crustaceans (isopods) have "pseudotracheae" on the abdominal appendages; these structures are short, branching tubes that bring air close to the blood-filled spaces in these appendages (Figure 13D).

Excretion and osmoregulation

With the evolution of a hemocoelic circulatory system, nephridia with open nephrostomes (such as those occuring in most annelids) become functionally untenable. It simply would not do to drain the blood directly from an open hemocoel to the outside. Arthropods have evolved a variety of highly efficient excretory structures that share a common adaptive feature in that they are internally closed. In addition to this major difference between the nephridia of arthropods and those of segmented and other coelomate protostomes (e.g., annelids, sipunculans, and echiurans), there has been a reduction in the overall number of excretory units.

In many arthropods, portions of the excretory units are coelomic remnants and are formed in various segments during development. In most adult crustaceans only a single pair of nephridia persist, and these are usually associated with particular segments of the head (i.e., **antennal glands** and **maxillary glands**) (Figure 14A). In arachnids there may be as many as four pairs of nephridia (and in onychophorans, many more), opening at the bases of the walking legs (i.e., **coxal glands**).

A second type of arthropod excretory structure occurs in many terrestrial forms (e.g., arachnids and insects). These structures arise as blind tubules extending into the hemocoel from the gut wall and are known as **Malpighian tubules** (Figure 14C).

The various arthropod nephridial types of coelomic origin are functionally much more complex and efficient than open metanephridia. The uptake of materials from the hemocoel by the inner ends of these nephridia apparently involves filtration pressure as well as active transport. The fluid entering the nephridium is generally similar to the blood itself, but as it passes along the plumbing system of the nephridium, a good deal of selective resorption occurs, particularly of salts and nutrients such as glucose. Thus the urine exiting the nephridial pores is markedly different from the blood and represents a concentration of nitrogenous waste products (Figure 14B).

Malpighian tubules accomplish the same process, but they must rely on assistance from the gut. Malpighian tubule uptake from the hemocoel is relatively nonselective, and this "primary urine" (with nutrients, water, salts, and so on) is emptied directly into the gut. The ability of the gut to resorb water plays a critical role in terrestrial arthropods. Very little resorption of nonwaste material occurs along the length of the tubule itself. The hindgut is mostly responsible for concentrating the urine by reabsorbing the nonwaste fractions.

As in most aquatic invertebrates, marine crustaceans excrete most (about 70–90 percent) of their nitrogenous waste as ammonia, the remainder being in the forms of urea, uric acid, amino acids, and some other compounds. Terrestrial arachnids and insects excrete predominantly uric acid (via the hindgut and anus). In Chapter 3 we reviewed some of the rela-

A

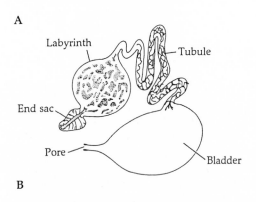

B

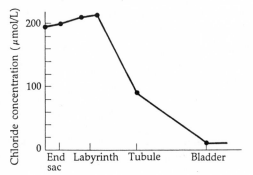

Figure 14

Arthropod excretory structures. A, The antennal gland of a decapod crustacean (section). B, Changes in chloride content of the excretory fluid in different regions of a decapod antennal gland. Note the active resorptive capabilities of the structure. C, An insect gut. Note the attachments of Malpighian tubules. (A after various sources; B after Parry, *in* Waterman 1960; C after Snodgrass 1952.)

C

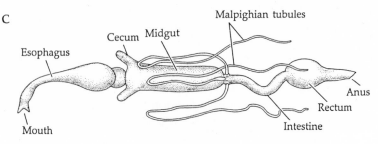

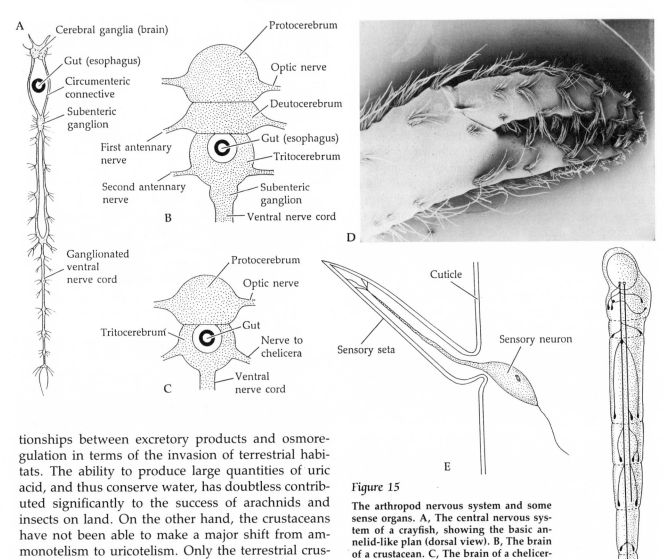

Figure 15

The arthropod nervous system and some sense organs. A, The central nervous system of a crayfish, showing the basic annelid-like plan (dorsal view). B, The brain of a crustacean. C, The brain of a chelicerate. D, Sensory setae on the walking leg of a lobster (*Homarus*). E, A typical arthropod tactile seta. F, Distribution of proprioceptors in a spider leg. (A after various sources; B,C after Vandel 1949; D from Derby 1982; F after Foelix 1982.)

tionships between excretory products and osmoregulation in terms of the invasion of terrestrial habitats. The ability to produce large quantities of uric acid, and thus conserve water, has doubtless contributed significantly to the success of arachnids and insects on land. On the other hand, the crustaceans have not been able to make a major shift from ammonotelism to uricotelism. Only the terrestrial crustaceans (i.e., isopods—wood lice and pill bugs) show but a slight increase in uric acid excretion over that of their marine counterparts.

Nervous system and sense organs

The general plan of the arthropod nervous system is strikingly similar to that of annelids, and many general homologies exist (Figure 15A). The arthropod **brain** actually comprises two or three regions, each of which is a separate ganglion.* The anterior ganglia form the dorsal (supraesophageal) **protocerebrum** and, when present, the **deutocerebrum**. The poste-

*The concentration of nervous tissue in the arthropod head has been called a "brain," "cerebrum," "cerebral ganglion," and "cerebral ganglionic mass." These terms may seem somewhat misleading because they are either anthropomorphic or suggest the presence of a single head ganglion. In fact, the head has either two or three separate but closely associated ganglia (concentrations of nervous tissue composed primarily of nerve cell bodies).

rior-most ganglion, the **tritocerebrum**, usually forms circumenteric connectives around the esophagus to a ventral subesophageal (subenteric) ganglion. The latter is formed by the coalescence of several other head ganglia, usually those associated with the mandibles and maxillae. A double or single, ganglionated ventral nerve cord extends through some or all of the body segments. Crustaceans and uniramians possess a tripartite brain composed of the two anterior ganglia (protocerebrum and deutocerebrum), and the posterior ganglion (tritocerebrum). In cheliceriformes the deutocerebrum is absent. As shown in Figures

15B and C, each of these regions gives rise to a major pair of nerves leading to particular head appendages. There is a good deal of argument about the homology of these regions and their associated head segments in the major arthropod subphyla.

The segmental ganglia of the ventral nerve cord show various degrees of linear fusion with one another in different groups of arthropods. Hence, just as tagmosis is reflected in the joining of body segments externally, so it is apparent in the union of groups of ganglia along the ventral nerve cord. These modifications of the central nervous system are examined more closely in the following chapters on arthropod subphyla.

Although the presence of an exoskeleton has had little evolutionary effect on the structure of the brain and nerve cord, it has had a major impact on the nature of sensory receptors. Unmodified, the exoskeleton would impose an effective barrier between the environment and the epidermal sensory nerve endings. Hence, most of the external mechanoreceptors and chemoreceptors are actually cuticular processes (setae, hairs, bristles), pores, or slits, collectively called **sensilla**. It seems probable that remnants of cilia form the nonmotile nerve basis for many of these sensory structures.

Most arthropod tactile receptors (mechanoreceptors) are movable bristles or setae, the inner ends of which are associated with sensory neurons (Figures 15D and E). When the cuticular projections are touched, that movement is translated into a deformation of the nerve ending, thereby initiating an impulse. Sensitivity to environmental vibrations is similar to tactile reception. Sensilla in the form of fine "hairs" or setae are mechanically moved by external vibrations and impart that movement to underlying sensory neurons. Some terrestrial arthropods bear thin membranous cuticular windows overlying depressions lined with sensory nerves. When struck by airborne vibrations (e.g., sound), these windows vibrate in turn and impart the stimulus to the chamber and thence to the nerves below.

We have seen in most soft-bodied invertebrates that chemoreception is associated with ciliated epithelial structures (e.g., nuchal organs and ciliated pits), across which dissolved chemicals diffuse to nerve endings. In arthropods, in the presence of a relatively impermeable cuticle and in the absence of free cilia, it is obvious that such arrangements are not possible. Still, the basic conditions for chemoreception must be met, namely, that dissolved chemicals must be brought into contact with sensory neurons. Thus, many arthropods possess special thin setae with permeable cuticular coverings, often associated with the head appendages. Others possess minute slits or pores through the cuticle, often on special hollow setae, that bring the environment into contact with chemoreceptor neurons.

Proprioception is of particular importance to animals with jointed appendages, such as arthropods and vertebrates. The way in which these stretch receptors span joints enables them to convey information to the central nervous system about the relative positions of appendage articles or body segments. (Figure 15F). Through this system, an arthropod (or vertebrate) knows where its appendages are, even without seeing them.

Arthropods possess three basic kinds of photoreceptors, including **simple ocelli**, **complex lensed ocelli**, and faceted or **compound eyes**. Although compound eyes occur in all four arthropod subphyla, they have been lost or modified in various groups throughout the phylum. Because of their unique structure and function, compound eyes are described here in some detail.

As their name indicates, compound eyes comprise from a few to many distinct photoreceptive units, called **ommatidia**. Each ommatidium is supplied with its own nerve tracts leading to the major optic nerve, and each has its own field of vision through square or hexagonal cuticular **facets** on the eye surface. Figure 16 illustrates this and other aspects of the structure and function of compound eyes, and should be consulted throughout the following discussion. The visual fields of neighboring ommatidia overlap to various degrees, with the result that a shift in position of an object within the total visual field causes changes in the impulses from several ommatidia. In part on the basis of this phenomenon, compound eyes are especially suitable for detecting movement. In general, compound eyes with many small facets probably produce higher-resolution images than eyes with fewer, larger facets. It has recently been demonstrated that the compound eyes of some trilobites may have been fundamentally different from those of other arthropods. They may have been much more efficient at detecting depth of field, object shape, and even color than previously thought (see Fordyce and Cronin 1989). The description below is based on a generalized crustacean or insect compound eye.

Each ommatidium is covered by a modified portion of the exoskeleton called the **cornea**; the special epidermal cells that produce the corneal elements are called **corneagen cells**. When viewed externally, facets on the surface of each cornea produce the characteristic mosaic pattern so frequently photographed by microscopists. The core of the ommatidium comprises a **crystalline cone**, a group of **crystalline cone cells** (usually four), sometimes a **crystalline cone stalk**, and a basal **retinula**, or **retinular element**. The retinula is a complex structure formed from several **retinular cells**, which are the photosensitive units and give rise to the sensory nerve tracts. These retinular cells are arranged as a cylinder along the long axis of the ommatidium. The core of this cylinder is the **rhabdome**, which is made up of rhodopsin-containing microtubular folds (microvilli) of the cell membranes of the retinular cells. Each retinular cell's contribution to the rhabdome is called a **rhabdomere**. The initiation of an impulse depends on light striking the rhabdome portion of the retinular element. Light that enters through the facet of a particular ommatidium is directed to the rhabdome by the lenslike qualities of the cornea and the crystalline cone. These

lenses have a fixed focal length, so accommodation to objects at different distances is not possible. Light is shared among all rhabdomeres of a given rhabdome, although not necessarily equally.

The versatility of compound eyes is to a large degree a result of distal and proximal **screening pigments** located in cells that wholly or partially surround the core of the ommatidium (Figure 16). **Distal screening pigments** are located in the **iris cells**, and **proximal pigments** often in the retinular cells. In many cases, these screening pigments are capable of migrating in response to varying light conditions and thus changing their positions somewhat along the length of the ommatidium. In bright light the screening pigments may disperse so that nearly all of the light that strikes a particular rhabdomere must have entered through the facet of that ommatidium. In other words, the screening pigment prevents light that strikes at an angle from passing through one ommatidium and into another. Many crustacean eyes are fixed in this condition. Such **appositional eyes** (= light-adapted eyes) are thought to maximize resolution, in that the image resulting from the visual field of each ommatidium is maintained as a discrete unit. Conversely, under conditions of dim light,

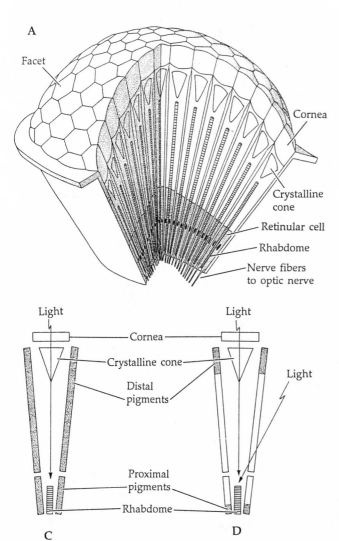

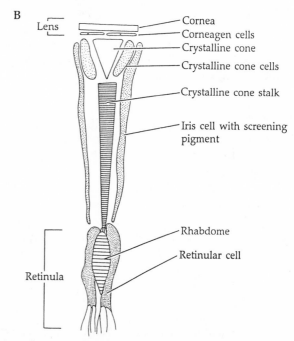

Figure 16

Compound eyes; see text for details. A, A compound eye (cutaway view). B, A single ommatidium. C–D, Major ommatidial elements in an appositional, or light-adapted, eye (C) and a superpositional, or dark-adapted, eye (D). (A after Pearse et al. 1987; B–D after various sources.)

screening pigments may concentrate, usually distally, thereby allowing light entering at angles to the corneal surface to pass through more than one ommatidium before striking rhabdomeres. The result is that the image formed by each ommatidium is superimposed on the images formed by neighboring ommatidia. This design has the advantage of producing enhanced irradiances on the retina, but at the cost of reduced resolution. Visual acuity is presumably affected by the degrees of overlap among the fields of vision of neighboring ommatidia—the greater the overlap, the poorer the visual acuity. Many crustacean eyes are fixed in this condition also. Such **superpositional eyes** (= dark-adapted eyes) function as efficient light-gathering structures while apparently sacrificing visual acuity or image-formation capabilities.

Both insects and crustaceans have compound eyes occurring in appositional and superpositional conditions. Some arthropod groups possess compound eyes that are always either appositional or superpositional; thus they lack the ability to switch back and forth in varying light conditions. For example, maxillopodans and branchiopods apparently all possess appositional eyes. However, within the two principal malacostracan clades, Eucarida and Peracarida, both types occur (e.g., isopods and amphipods have appositional eyes, but mysids have superpositional eyes). Furthermore, crustacean larvae that possess compound eyes almost always have the appositional type, which metamorphose into superpositional eyes in those groups that possess them in adulthood.

Reproduction and development

The great diversity of adult form and habit among arthropods is also reflected in their reproductive and developmental strategies. The extreme evolutionary and ontogenetic plasticity of arthropods has clearly contributed greatly to their success in a wide range of habitats. It has also led to a great deal of convergence and parallel evolution among the ar-

thropods, as different groups have developed similar structures under similar selective conditions or pressures.

Nearly all arthropods are gonochoristic, and most engage in some sort of formal mating. Fertilization is usually internal and often followed by brooding or some other form of parental care, at least during early development. Development is frequently mixed, with brooding and encapsulation followed by larval stages, although direct development occurs in many groups.*

Arthropod eggs are centrolecithal, but the amount of yolk varies greatly and results in different patterns of early cleavage. Cleavage is holoblastic in the relatively weakly-yolked eggs of xiphosurans, some scorpions, and various crustaceans (e.g., copepods and barnacles), and is meroblastic in strongly-yolked eggs of most insects and many other crustaceans. A number of arthropods exhibit a unique form of meroblastic cleavage that begins with nuclear divisions within the yolky mass (Figure 17). These intralecithal nuclear divisions are followed by a migration of the daughter nuclei to the periphery of the cell and subsequent partitioning of the nuclei by cell membranes. These processes typically result in a periblastula that consists of a single layer of cells around an inner yolky mass.

Holoblastic cleavage may be more-or-less radial or may show traces of a spiral pattern (e.g., in barnacles). However, the latter case appears to be some-

*Direct development in arthropods is often called **amorphic development**, and indirect development is often called **anamorphic development**. These terms seem unnecessarily cumbersome and even a bit misleading, and we avoid their use in this text.

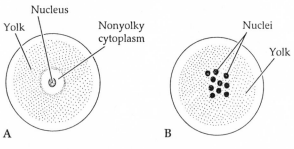

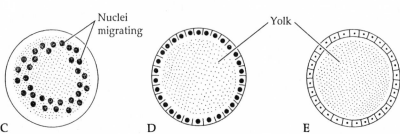

Figure 17

Superficial cleavage of a centrolecithal egg, and the formation of a periblastula in arthropods. See text for other cleavage patterns. **A,** Centrolecithal egg. **B,** Intralecithal nuclear divisions following fertilization. **C,** Migration of nuclei to the periphery of the cell. **D–E,** The periblastula is produced by a partitioning of nuclei as cell membranes form.

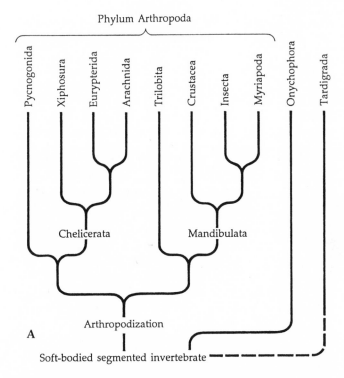

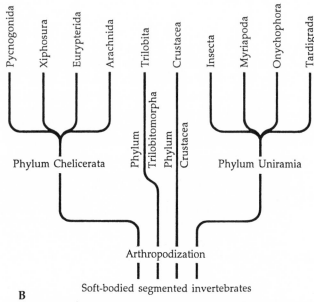

Figure 18

Two generalized views of arthropod phylogeny. A, A monophyletic scheme wherein the phenomenon of arthropodization occurred only once. B, A polyphyletic scheme wherein arthropodization occurred four separate times. For a detailed discussion of arthropod phylogney see Chapter 19. (Modified from Herreid, in Herreid and Fourtner 1981.)

what different from the typical spiralian pattern seen in sipunculans, echiurans, polychaetes, and most molluscs (see Chapter 18).

A comment on arthropod phylogeny

Since the works of Darwin appeared, biologists have debated ideas concerning the origin and radiation of arthropods. While we present somewhat detailed aspects of this topic in Chapter 19, it is worth noting here one major area of disagreement concerning arthropod evolution. The questions are whether the arthropods constitute a polyphyletic or monophyletic group, and, if they are monophyletic, how the major groups are related to one another. The traditional view has been to treat them as a monophyletic taxon at the phylum rank. However, nearly 100 years after the establishment of the phylum, some workers began to question arthropod monophyly. The idea of arthropod polyphyly was initiated by O. W. Tiegs and Sidnie M. Manton in the 1940s and 1950s, and eventually it resulted in major supporting works on comparative anatomy (Manton 1977) and comparative embryology (Anderson 1973).

In spite of the complexities and details addressed in the arguments for and against polyphyly, the fundamental question at issue is: Did the combined suite of synapomorphies that define "Arthropoda" evolve only once, or more than once? If polyphyly is correct,

then arthropodization occurred two, three, or perhaps even four separate times, as opposed to only once (Figure 18). We view monophyly as both the most parsimonious explanation of the origin of the arthropods and the most biologically reasonable hypothesis based on current knowledge. Our argument in support of monophyly is based, in part, on the premise that the "polyphylists" have relied too heavily on differences among groups, rather than seeking to identify the groups' unifying similarities. We present our arguments for monophyly in Chapter 19.

The trilobites

Of all fossil invertebrates, trilobites are perhaps the most symbolic as ancient and exotic faunas in the minds of most people. The subphylum Trilobitomorpha (Latin, *trilobito*, "three-lobed"; Greek *morph*, "form") includes nearly 4,000 species of arthropods known only from the fossil record (Figure 19). They were very abundant in early Paleozoic seas (Cambrian and Ordovician, 500–600 million years ago) and persisted in smaller numbers into the Permian (280

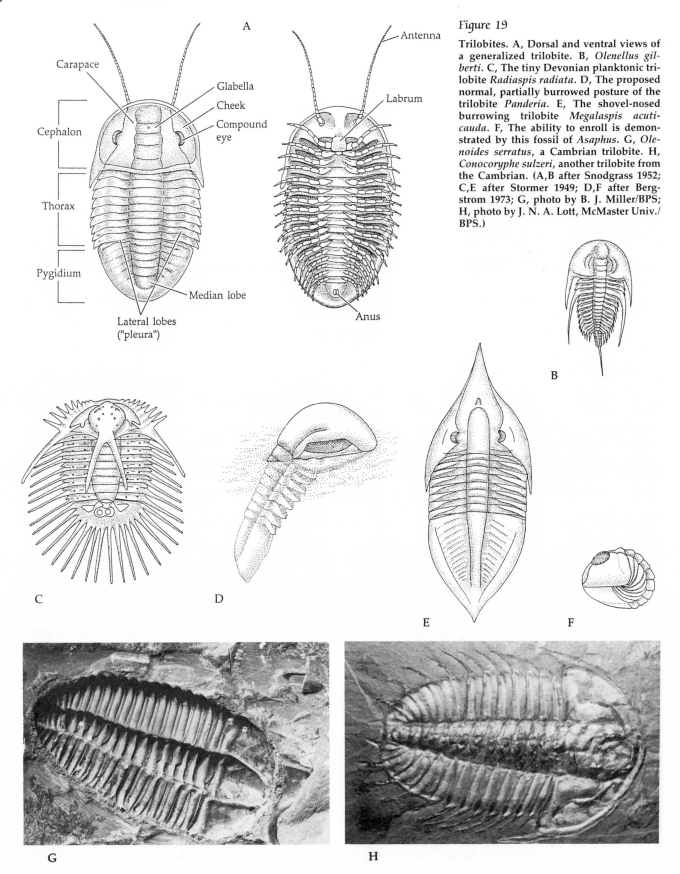

Figure 19

Trilobites. A, Dorsal and ventral views of a generalized trilobite. B, *Olenellus gilberti*. C, The tiny Devonian planktonic trilobite *Radiaspis radiata*. D, The proposed normal, partially burrowed posture of the trilobite *Panderia*. E, The shovel-nosed burrowing trilobite *Megalaspis acuticauda*. F, The ability to enroll is demonstrated by this fossil of *Asaphus*. G, *Olenoides serratus*, a Cambrian trilobite. H, *Conocoryphe sulzeri*, another trilobite from the Cambrian. (A,B after Snodgrass 1952; C,E after Stormer 1949; D,F after Bergstrom 1973; G, photo by B. J. Miller/BPS; H, photo by J. N. A. Lott, McMaster Univ./ BPS.)

Carapace

Glabella

Cheek

Compound eye

Cephalon

Thorax

Pygidium

Median lobe

Lateral lobes ("pleura")

Antenna

Labrum

Anus

A

B

C

D

E

F

G

H

million years ago). Because of their hard exoskeletons, great abundance, and wide distribution, the trilobites* have left a rich fossil record, and more is known about them than about many other extinct taxa.

Although trilobites were exclusively marine, they exploited a variety of habitats and lifestyles. Most were benthic, either crawling about over the bottom or plowing through the top layer of sediment. Most of these benthic trilobites were probably scavengers or direct deposit feeders, although some species may have been predators that lay partially burrowed in soft sediments and grabbed passing prey. Some workers speculate that trilobites may have suspension fed by using the filamentous parts of their appendages. Most of the benthic forms were a few centimeters long, although some giants reached lengths of 60–70 centimeters. A few trilobites appear to have been planktonic—they were mostly small forms, less than 1 cm long and equipped with spines that aided in flotation.

General body form

The trilobite body was broadly oval and somewhat flattened dorsoventrally, and some were capable of rolling into a ball—a behavior called enrolling or conglobation (Figure 19). Two longitudinal furrows on the dorsum divided the body into a median and two lateral lobes—thus the name trilobite. The lateral lobes, sometimes referred to as "pleural lobes," were produced by outgrowths of the dorsal body wall (the tergites) that extended over the appendages. The exoskeleton of the dorsal surface was much thicker than that of the ventral surface. Trilobites were divided into three tagmata: **cephalon**, **thorax**, and **pygidium**, with each region bearing a number of appendages. The pygidium presumably comprised a number of fused or partially budded body segments plus a postsegmental part called the **telson**.

The cephalon was commonly composed of five or six fused segments covered dorsally by a solid **cephalic shield** (or **carapace**), which in most species bore well developed compound eyes on the lateral lobes. The longitudinal furrows divided the cephalon into a median **glabella** and lateral **cheeks**. The glabella was usually further subdivided into three to five lobes by a series of transverse furrows (Figure 19A). The cephalic shield was distinctly rolled under along its anterior and lateral edges, thereby producing a concavity on the ventral surface of the cephalon. Ventrally the cephalon commonly bore four or five pairs of appendages, plus a median, unpaired flap called the **labrum** projecting posteriorly over the mouth. Only the first pair of appendages, the multisegmented antennae, were preoral in adults. However, evidence indicates that even these appendages may have arisen postorally during embryogenesis, migrating to a preoral position later in development. Behind the antennae there were typically three or four pairs of "biramous" postoral appendages, similar in structure to the limbs on the rest of the body (Figures 3D, 19A, and 20). Recall from our earlier discussion that the biramous condition of these limbs may not be homologous to that in crustaceans. In some cases the number of cephalic somites (and hence appendages) varied from this typical arrangement. For example, in the highly unusual genus *Marrella* there was but one postoral somite, and in *Rhenops* and *Emeraldella* there were five. These variations suggest that the fusion of somites into a cephalic tagma was never firmly fixed in the developmental program of these primitive arthropods. With the exception of the antennae and the tagmosis of the body, the trilobites were relatively homonomous.

The thorax comprised a variable number of segments, each with a pair of "biramous" appendages and supported ventrally by a transverse **tendinous bar**. The pygidium included from a few to as many as two dozen appendage-bearing segments (Figure 19). Only the terminal piece of the trilobite pygidium (the telson) is considered homologous to the postsegmental pygidium of annelids or the telson of other arthropods (see subsequent chapters).

As we have mentioned, much controversy surrounds the nature of trilobite limbs and their homologous relationships to the appendages of other arthropods. The trilobite leg bore an ambulatory branch called the **telopod**, and a laterally directed, presumably gill-bearing **filamentous branch**. Both rami arose from the basal article, or coxa, although some earlier workers suggested that another article, the precoxa, may have existed between the coxa and the body wall, and originally gave rise to the filamentous branch. The coxae also bore well-developed endites that served as masticatory or grasping gnathobases (Figures 3D and 20). A fundamental question of homology that remains unanswered is whether or not the filamentous branch of the leg is simply an elaborate coxal epipod or is actually homologous to the crustacean exopod (i.e., is the trilobite limb uniramous or biramous?). Because the true crustacean ex-

---

*Although a variety of different taxa are grouped within the phylum Trilobitomorpha, we use the diminutive "trilobites" to collectively refer to them all.

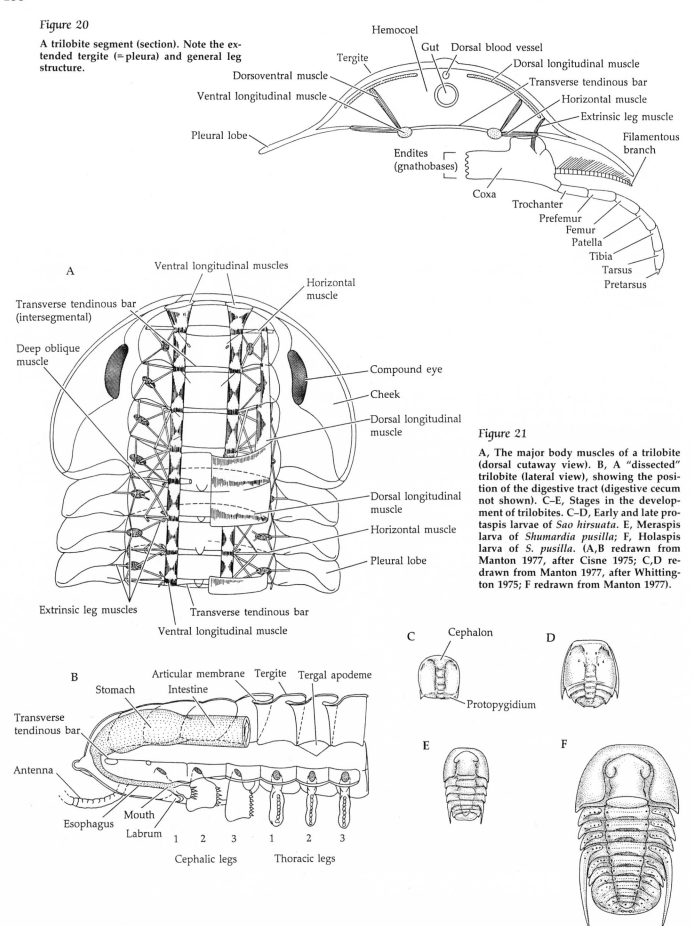

Figure 20

A trilobite segment (section). Note the extended tergite (=pleura) and general leg structure.

Figure 21

A, The major body muscles of a trilobite (dorsal cutaway view). B, A "dissected" trilobite (lateral view), showing the position of the digestive tract (digestive cecum not shown). C–E, Stages in the development of trilobites. C–D, Early and late protaspis larvae of *Sao hirsuata*. E, Meraspis larva of *Shumardia pusilla*; F, Holaspis larva of *S. pusilla*. (A,B redrawn from Manton 1977, after Cisne 1975; C,D redrawn from Manton 1977, after Whittington 1975; F redrawn from Manton 1977).

opod arises from the second limb segment (the basis), most workers do not accept a homology with the crustacean exopod. However, some specialists have argued that the trilobite protopod is composed of a fused coxa and basis (or perhaps a coxa–basis distinction had not yet evolved from a primitive, single articulate, arthropod protopod), in which case the filamentous branch could indeed be homologous with the crustacean exopod (even though the two kinds of rami are not at all similar in appearance).

Internal anatomy

Over the past few years X-ray techniques have provided information about the structure of the digestive tract and arrangement of the muscles in trilobites. As in other arthropods, the muscles occurred in bands associated with body and limb articulations (Figure 21A). The trilobite body was incapable of lateral bending, but the animals could easily flex ventrally. The digestive tract was relatively simple (Figure 21B), extending from the mouth on the ventral side of the cephalon to the anus on the pygidium. From the mouth, the esophagus extended forward in the cephalon and then dorsally to meet an enlarged area variably called the stomach or crop. A digestive cecum apparently arose from the anterior part of the stomach. The intestine arose from the posterior end of the stomach and extended straight to the anus.

Development

Trilobites had a mixed life history pattern. After they hatched, they passed through at least three larval stages of several instars each (Figures 21C–F). The larval stages are well represented in the fossil record and appear to be consistent throughout the subphylum. The hatching form, or **protaspis larva**, was less than 1 mm long and consisted of most or all the cephalic segments fused to the "protopygidium" and covered by a dorsal shield. Segments were added through several molts until the cephalon was complete, and the trunk and pygidium were distinct (**meraspis larva**). Through subsequent molts the animal eventually took on the form of a miniature trilobite (**holaspis larva**). A further series of molts brought the holaspis to the juvenile form by the addition of segments and an increase in size.

A comment on phylogeny

The characteristics of the trilobites, and their antiquity, make them prime candidates for the role of ancestor to at least some (if not all) of the major groups of arthropods. Trilobites are often considered to have been closely allied with early cheliceriform or crustacean stock, but the true nature of their affinities remains uncertain. Much of this uncertainty hinges on the nature of the head and trunk appendages, and their possible homologies to the appendages of the other major arthropod subphyla. These matters are discussed in Chapter 19.

Selected References

Below we provide a small sampling of the vast literature on arthropods. We have tried to select works of importance on the entire phylum and on trilobites in particular, and certain papers that deal with general features and processes discussed in this chapter. Several of the listings are collections of papers, and we have generally not cited the contents separately (e. g., Abele 1982; Gupta 1979, 1983; Bernhard 1966; Herreid and Fourtner 1981). More extensive reference lists for each subphylum are included in Chapters 16, 17, and 18. Chapter 19 includes a list of references on arthropod phylogeny.

General References

American Zoologist. 1984. Cellular mechanisms of ion regulation in arthropods. Vol. 24, No. 1. [13 papers presented at a symposium in Louisville, Kentucky in 1982.]

Abele, L. G. (ed.). 1982. *The Biology of Crustacea*, Vol. 1. Systematics, the Fossil Record, and Biogeography. Academic Press, New York.

Anderson, D. T. 1973. *Embryology and Phylogeny in Annelids and Arthropods*. Pergamon Press, Oxford.

Anderson, D. T. 1982. Embryology. *In* L. G. Abele (ed.), *The Biology of Crustacea*, Vol. 2, Embryology, Morphology, and Genetics. Academic Press, New York, pp. 1–41. [We single this paper out because it is the most recent summary statement of Anderson's views on crustacean embryology and its relationship to that of other arthropods; see paper by Weygoldt 1979 for opposing views.]

Bernhard, G. C. (ed.). 1966. *The Functional Organization of the Compound Eye*, Pergamon Press, New York.

Bliss, D. E. (gen. ed.). 1982– . *The Biology of Crustacea*. Academic Press, New York. [A continuing series; nine volumes had appeared at the time of this writing.]

Briggs, D. E. G. and R. A. Fortey. 1989. The early radiation and relationships of the major arthropod groups. Science 246:241–243.

Bowerman, R. F. 1977. The control of arthropod walking. Comp. Biochem. Physiol. 56A:231–247.

Cameron, J. N. 1985. Molting in the blue crab. Sci. Am. 252(5):102–109.

Calman, W. T. 1909. Crustacea. *In* R. Lankester (ed.), *A Treatise on Zoology*. Adam and Charles Black, London. [A classic synopsis of the Crustacea; reprinted by A. Asher & Co., Amsterdam, 1964.]

Carthy, J. D. 1965. *The Behaviour of Arthropods*. W. H. Freeman, San Francisco.

Chang, E. S. 1985. Hormonal control of molting in decapod Crustacea. Am. Zool. 25:179–185.

Clarke, K. U. 1973. *The Biology of Arthropoda*. American Elsevier, New York.

Cloudsley-Thompson, J. L. 1958. *Spiders, Scorpions, Centipedes and Mites: The Ecology and Natural History of Woodlice, Myriapods and Arachnids*. Pergamon Press, New York.

Derby, C. 1982. Structure and function of cuticular sensilla of the lobster *Homarus americanus*. J. Crust. Biol. 2:1–21.

Edney, E. B. 1957. *The Water Relations of Terrestrial Arthropods*. Cambridge University Press, London.

Eguchi, E. and T. H. Waterman. 1966. Fine structure patterns in crustacean rhabdomes. *In* Proc. Internat. Symp. on Functional Organization of the Compound Eye, Stockholm, Oct. 25–27, 1965. Pergamon Press, New York, pp. 105–124.

Fairbridge, R. W. and D. Jablonski. 1979. *The Encyclopedia of Paleontology*. Dowden, Hutchinson and Ross, Stroudsburg, Pennsylvania.

Florkin, M. and B. T. Scheer (eds.). 1970, 1971. *Chemical Zoology*, Vols. 5 and 6. Academic Press, New York.

Foelix, R. 1982. *Biology of Spiders*. Harvard University Press, Cambridge, Massachusetts.

Grenacher, H. 1879. *Untersuchungen über das Sehorgan der Arthropoden inbesondere der Spinnen, Insekten, und Crustaceen*. Vanderhoek and Ruprecht, Gottingen.

Gupta, A. P. (ed.) 1979. *Arthropod Phylogeny*. Van Nostrand Reinhold Co., New York. [Views of many authors on the origin and evolution of the arthropods, including summary papers by Anderson and Manton espousing polyphyly and by others supporting monophyly.]

Gupta, A. P. (ed.) 1983. *Neurohemal Organs of Arthropods: Their Development, Evolution, Structures, and Functions*. Charles C. Thomas, Springfield, Illinois.

Hassell, M. P. 1978. *The Dynamics of Arthropod Predator-Prey Systems*. Princeton University Press, Princeton, New Jersey.

Herreid, C. F. and C. R. Fourtner (eds.) 1981. *Locomotion and Energetics in Arthropods*. Plenum Press, New York.

Herrnkind, W. F. 1972. Orientation in shore-living arthropods, especially the sand fiddler crab. *In* H. E. Winn and B. L. Olla (eds.), *Behavior of Marine Animals*, Vol. 1, Invertebrates. Plenum Press, New York.

Hessler, R. R. 1964. The Cephalocarida: Comparative skeletomusculature. Mem. Conn. Acad. Arts Sci. 16:1–97.

Hessler, R. R. 1981. Evolution of arthropod locomotion: A crustacean model. *In* C. F. Herreid and C. R. Fourtner (eds.), *Locomotion and Energetics in Arthropods*. Plenum, New York, pp. 9–30.

Hessler, R. R. 1982. The structural morphology of walking mechanisms in eumalacostracan crustaceans. Philos. Trans. R. Soc. London Ser. B 296:245–298.

Hessler, R. R. 1985. Swimming in Crustacea. Trans. R. Soc. Edinburgh 76:115–122.

Hessler, R. R. and W. A. Newman. 1975. A trilobitomorph origin for the Crustacea. Fossils and Strata 4:437–459.

Horridge, G. A. 1978. The separation of visual axes in apposition compound eyes. Phil. Trans. Roy. Soc. London Ser. B 285:1–59.

Kaestner, A. 1968, 1970. *Invertebrate Zoology*, Vols. 2 and 3. Wiley, New York. [Volumes 2 and 3 of this three-volume work do an excellent job of describing the arthropods; adroitly translated from the German by H. W. and L. R. Levi.]

Kenchington, W. 1984. Biological and chemical aspects of silks and silk-like materials produced by arthropods. South Pac. J. Nat. Sci. 5:10–45.

Krishnakumaran, A. and H. A. Schneiderman. 1968. Chemical control of molting in arthropods. Nature 220:601–603.

Kunze, P. 1979. Apposition and superposition eyes. *In* H. Autrum (ed.), Comparative physiology and evolution of vision in invertebrates, A: Invertebrate photoreceptors. *Handbook of Sensory Physiology* VII/6A: 441–502.

Linzen, B. and 16 others. 1985. The structure of arthropod hemocyanins. Science 229:519–524. [A review of these large copper-based, oxygen-carrying proteins in arthropods.]

Lockwood, A. P. M. 1967. *Aspects of the Physiology of Crustacea*. W. H. Freeman, San Francisco.

Manton, S. M. 1950–1973. The evolution of arthropod locomotory mechanisms. Parts 1–11. All published in J. Linn. Soc. London (Zool.).

Manton, S. M. 1973. Arthropod phylogeny—a modern synthesis. J. Zool 171:111–130.

Manton, S. M. 1977. *The Arthropoda: Habits, Functional Morphology, and Evolution*. Clarendon Press, Oxford.

Mikulic, D. G., D. E. G. Briggs and J. Kluessendorf. 1985. A Silurian soft-bodied biota. Science 228:715–717.

McIver, S. B. 1975. Structure of cuticular mechanoreceptors of arthropods. Annu. Rev. Entomol. 20:381–397.

Moore, R. C. (ed.) 1959. *Treatise on Invertebrate Paleontology*, Part O, Arthropoda 1; Part R, Arthropoda 4. Geological Society of America and the University of Kansas Press, Lawrence.

Müller, K. J. and D. Walossek. 1987. Morphology, ontogeny, and life habit of *Agnostus pisiformis* from the Upper Cambrian of Sweden. Fossils and Strata 19:1–124.

Müller, K. J. and D. Walossek. 1988. External morphology and larval development of the Upper Cambrian maxillopod *Bredocaris admirabilis*. Fossils and Strata 23:1–70.

Parker, S. P. (ed.) 1982. *Synopsis and Classification of Living Organisms*, Vol. 2. Mc-Graw-Hill, New York. [Pages 71–739 are devoted to the Arthropoda and present the views of many experts on matters of classification.]

Richards, A. G. 1951. *The Integument of Arthropods*. University of Minnesota Press, Minneapolis.

Rivindranth, M. H. 1980. Haemocytes in haemolymph coagulation of arthropods. Biol. Rev. 55:139–170.

Roer, R. and R. Dillaman 1984. The structure and calcification of the crustacean cuticle. Am. Zool. 24:893–909.

Sanders, H. L. 1955. The Cephalocarida, a new subclass of Crustacea from Long Island Sound. Proc. Natl. Acad. Sci. U.S.A. 41(1):61–66.

Sanders, H. L. 1963. The Cephalocarida: Functional morphology, larval development, comparative external anatomy. Mem. Conn. Acad. Arts Sci. 15:1–80.

Sargent, W. 1987. *The Year of the Crab: Marine Animals in Modern Medicine*. W. W. Norton, New York. [Emphasis is on *Limulus* research.]

Schram, F. R. (gen. ed.) 1983–. *Crustacean Issues*. A. A. Balkema, Rotterdam. [A continuing series; five issues had appeared at the time of this writing.]

Schram, F. R. 1986. *Crustacea*. Oxford University Press, New York.

Sharov, A. G. 1966. *Basic Arthropodan Stock*. Pergamon Press, Oxford.

Shaw, S. R. 1969. Optics of arthropod compound eyes. Science 165:88–90.

Skinner, D. M. 1985. Interacting systems in the control of the crustacean molt cycle. Am. Zool. 25:275–284.

Snodgrass, R. E. 1938. The evolution of Annelida, Onycyhophora and Arthropoda. Smithson. Misc. Coll. 97:1–59.

Snodgrass, R. E. 1952. *A Textbook of Arthropod Anatomy*. Comstock Publishing Association, Ithaca, New York. [Still one of the best!]

Snyder, A. W. 1979. Physics of vision in compound eyes. *In* H. Autrum (ed.), Comparative physiology and evolution of vision in invertebrates, A: Invertebrate photoreceptors. *Handbook of Sensory Physiology*, VII/6A:225–314.

Snyder, A. W., S. B. Laughlin and D. G. Stavenga 1977. Information capacity in compound eyes. Vision Res. 17:1163–1175.

Snyder, A. W., D. G. Stavenga and S. B. Laughlin. 1977. Spatial information capacity of compound eyes. J. Comp. Physiol. 116:183–207.

Størmer, L. 1977. Arthropod invasion of land during late Silurian and Devonian times. Science 197:1362–1364.

Tickle, C. 1981. Limb regeneration. Am. Sci. 69:639–646.

Tiegs, O. W. and S. M. Manton 1958. The evolution of the Arthropoda. Biol. Rev. 33:255–337.

Vandel, A. 1949. Généralités sur les Arthropodes. *In* P. Grassé (ed.), *Traité de Zoologie*, Vol. 6. Masson et Cie, Paris, pp. 159–216.

Waterman, T. H. 1960, 1961. *The Physiology of Crustacea*, Vols. 1 and 2. Academic Press, New York.

Waterman, T. H. and A. S. Pooley. 1980. Crustacean eye fine structure seen with scanning electron microscopy. Science 209:235–240.

Wehner, R. 1981. Spatial vision in arthropods. *In* H. Autrum (ed.), Comparative physiology and evolution of vision in invertebrates, C: Invertebrate visual centers and behavior II. *Handbook of Sensory Physiology*, VII/6C:287–616.

Wilson, D. M. 1966. Insect walking. Annu. Rev. Entomol. 11:103–122.

Weygoldt, P. 1979. Significance of later embryonic stages and head development in arthropod phylogeny. *In* A. P. Gupta (ed.), *Arthropod Phylogeny*, Van Nostrand-Reinhold, New York, pp. 107–135.

Trilobites

Bergstrom, J. B. 1969. Remarks on the appendages of trilobites. Lethaia 2: 395–414.

Bergstrom, J. B. 1973. Organization, life, and systematics of trilobites. Fossils and Strata No. 2, Universitetsforlaget, Oslo.

Cisne, J. L. 1973. Life history of an Ordovician trilobite, *Triarthrus eatoni*. Ecology 54:135–142.

Cisne, J. L. 1974. Trilobites and the origin of arthropods. Science 186: 13–18.

Cisne, J. L. 1975. Anatomy of *Triarthrus* and the relationships of the Trilobita. Fossils and Strata 4:45–63.

Cisne, J. L. 1979. The visual system of trilobites. Palaeontology 22:1–22.

Eldredge, N. 1971. Patterns of cephalic musculature in the Phacopina (Trilobita) and their phylogenetic significance. J. Paleontol. 45:52–67.

Eldredge, N. 1973. Systematics of lower and lower middle Devonian species of the trilobite *Phacops* Emmrich in North America. Bull. Amer. Mus. Nat. Hist. 151:285–338.

Eldredge, N. 1977. Trilobites and evolutionary patterns. *In* A. Hallam (ed.), *Patterns of Evolution*. Elsevier, Amsterdam, pp. 305–332.

Fordyce, D. and T. W. Cronin. 1989. Comparison of fossilized schizochroal compound eyes of phacopid trilobites with eyes of modern marine crustaceans and other arthropods. J. Crust. Biol. 9:554–569.

Fortey, R. A. 1975. Early Ordovician trilobite communities. Fossils and Strata 4:339–360.

Harrington, H. J. et al. 1959. Part O, Arthropoda 1. *In* R. C. Moore (ed.), *Treatise on Invertebrate Paleontolgoy*. Geological Society of America and the University of Kansas Press, Lawrence, Kansas.

Levi-Setti, R. 1975. *Trilobites: A Photographic Atlas*. University of Chicago Press, Chicago. [Coffee-table paleontology at its best.]

Martinsson, A. (ed.) 1975. *Evolution and Morphology of the Trilobita, Trilobitoida and Merostomata*. Universitentsforlaget, Oslo.

McNamara, K. J. 1986. The role of heterochrony in the evolution of Cambrian trilobites. Biol. Rev. 61:121–156.

Størmer, L. 1939, 1941, 1951. Studies on trilobite morphology, I, II, and III. Nor. Geol. Tidsskr. 19:143–273, 21:49–163, and 29:108–157.

Stubblefield, C. J. 1926. Notes on the development of a trilobite, *Shumardia pusilla* (Sars). J. Linn. Soc. London Zool. 36:345–472.

Sturmer, W. and J. Bergstrom. 1973. New discoveries on trilobites by X-rays. Paleontol. Z. 47 1/2:104–141.

Towe, K. M. 1973. Trilobite eyes: Calcified lenses in vivo. Science 179:1007–1009.

Whittington, H. B. 1957. The ontogeny of trilobites. Biol. Rev. 32:421–469.

Whittington, H. B. 1971. Redescription of *Marrella splendens* (Trilobitoidea) from the Burgess Shale, Middle Cambrian, British Columbia. Bull. Geol. Surv. Can. 209:1–24.

Whittington, H. B. 1975. Trilobites with appendages from the Burgess Shale, Middle Cambrian, British Columbia. Fossils and Strata 4:97–136.

Chapter Sixteen

Phylum Arthropoda: The Cheliceriformes

The skin of it is so soft, smooth, polished and neat, that she precedes the softest skin'd Mayds, and the daintiest and most beautiful Strumpets . . . she hath fingers that the most gallant Virgins desire to have theirs like them, long, round, of exact feeling, that there is no man, nor any creature, that can compare with her.

The Reverend E. Topsell
Circa 1607, describing a house spider

The arthropod subphylum Cheliceriformes includes the classes Chelicerata (horseshoe "crabs," spiders, scorpions, mites, ticks, and many less familiar groups) and Pycnogonida (sea "spiders"). The 65,000 or so described species also include some impressive fossil forms, such as the Paleozoic eurypterids, or "water scorpions," some of which were over 2 meters long. The rich fossil record left by marine cheliceriformes indicates that this group had its origin in ancient Cambrian or Precambrian seas. Today, however, most live on land, where they are second in number only to the insects. A few kinds, such as some mites, have secondarily invaded various aquatic habitats. On land, they have adapted to virtually every imaginable situation and lifestyle.

In addition to the basic characteristics common to all arthropods, members of the subphylum Cheliceriformes are distinguished by several unique features. The body is typically divided into two main regions, called **prosoma** and **opisthosoma** (or **cephalothorax** and **abdomen**, respectively), the latter usually bearing a postsegmental **telson** (Figures 1, 2, and 3). In contrast to most other arthropods, it is not possible to delineate a discrete "head" in cheliceriformes. There are no antennae, but generally all six segments of the prosoma bear appendages. The first

pair of appendages are the **chelicerae**, followed by the **pedipalps** and four pairs of walking legs. The chelicerae and pedipalps are specialized for an enormous variety of roles in the various cheliceriform groups, including sensation, feeding, defense, locomotion, and copulation. A detailed discussion of anatomy is provided in the *Bauplan* section (see Box 1).

The subphylum Cheliceriformes is a large and highly diverse group. We first provide a classification, with characteristics and lists of some common genera, and then a more detailed synopsis of the major taxa of the class Chelicerata. Because of their many unique and distinct features, the pycnogonids are treated separately from the other Cheliceriformes, at the end of this chapter. We caution readers that, although most workers consider pycnogonids to be cheliceriformes, their precise taxonomic position within the Arthropoda is still a bit controversial.

CLASS CHELICERATA
SUBCLASS MEROSTOMATA

Prosoma covered by a large, carapace-like shield; compound eyes usually reduced; pedipalps similar to walking legs; opisthosoma divided—as **mesosoma** and **metasoma** —or undivided, with flaplike appendages as book gills; telson long and spiked.

ORDER EURYPTERIDA: Extinct "water scorpions" (Figure 1); opisthosoma segmented and divided, with scalelike appendages on mesosoma; metasoma narrow. (e.g., *Eurypterus, Pterygotus*)

ORDER XIPHOSURA: Certain extinct forms and the living horseshoe "crabs" (Figure 2). Opisthosoma unsegmented and undivided, but with six pairs of flaplike appendages, the first pair fused medially as a genital operculum over the gonopores; the last five pairs are book gills (or gill books); pedipalps and walking legs chelate; last (fourth) pair of legs modified distally for support on soft substrata. (e.g., *Carcinoscorpius, Limulus, Tachypleus*)

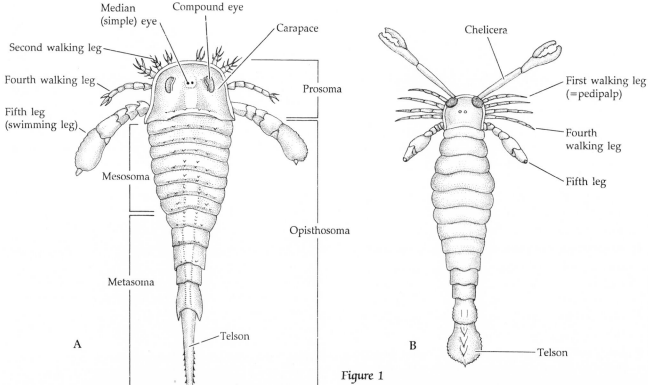

Figure 1

"Water scorpions" (subclass Merostomata, order Eurypterida), an extinct group of chelicerates. A, *Eurypterus* (dorsal view). B, *Pterygotus buffaloensis*, which reached lengths of nearly 3 meters. C, *Eurypterus remipes*, a Silurian species. Eurypterids flourished in Paleozoic seas; some may have invaded fresh water and perhaps even land. (A,B after Snodgrass 1952; C photo by B. J. Miller/BPS.)

SUBCLASS ARACHNIDA

Prosoma wholly or partly covered by carapace-like shield; without compound eyes; opisthosoma segmented or unsegmented, divided or undivided; opisthosomal appendages absent or modified as spinnerets (spiders) or pectines (scorpions); gas exchange by tracheae, book lungs, or both; nearly all terrestrial; over 60,000 species.

ORDER SCORPIONES: True scorpions (Figure 3). Prosoma fused and covered by carapace-like shield; opisthosoma elongate, segmented, and divided into mesosoma and metasoma of 7 and 5 segments, respectively; telson spinelike, with poison gland; chelicerae of 3 articles; pedipalps large, chelate, of 6 articles; with pair of median eyes and sometimes additional pairs of lateral eyes; first mesosomal segment bears a gonopore covered by genital operculum; second mesosomal segment bears a pair of unique appendages called pectines; third through sixth mesosomal segments each with a pair of book lungs. (e.g., *Androctonus, Bothriurus, Buthus, Centruroides, Chactus, Chaerilus, Diplocentrus, Hadrurus, Hemiscorpion, Nebo, Parabuthus, Paruroctonus, Tityus, Vaejovis*)

Box One
Characteristics of the Subphylum Cheliceriformes

1. Body composed of two tagmata: the prosoma and the opisthosoma (or "cephalothorax" and "abdomen"); prosoma composed of a presegmental acron and six somites, often covered by a carapace-like dorsal shield; opisthosoma composed of up to 12 somites and a postsegmental telson, subdivided into two parts in some groups
2. Appendages of prosoma are chelicerae, pedipalps, and four pairs of walking legs; antennae absent
3. Appendages multiarticulate and uniramous
4. Gas exchange by book gills, book lungs, or tracheae
5. Excretion by coxal glands and/or Malpighian tubules (Note: Cheliceriform Malpighian tubules are not homologous to those of insects or tardigrades.)
6. With simple medial eyes and sometimes lateral compound eyes
7. Gut with 2–6 pairs of digestive ceca
8. Most are gonochoristic

ORDER UROPYGI: Whip scorpions and vinegaroons (Figure 4A). Prosoma elongate and covered by a carapace-like shield; opisthosoma segmented and divided; mesosoma broad with two pairs of book lungs; metasoma short with long, whiplike telson; first walking legs elongate and multiarticulate distally; with one pair of median and four to five pairs of lateral eyes. (e.g., *Albaliella, Chajnus, Mastigoproctus*)

ORDER SCHIZOMIDA: Schizomids (Figures 4B and C). Prosoma divided; first four segments (the **proterosoma**) covered by short carapace-like shield called the **propeltidium**, followed by two free segments, the **meso-** and **metapeltidium**, respectively; opisthosoma segmented and divided; mesosoma with one pair of book lungs; metasoma with short, thin telson; eyes present or absent. (e.g., *Agastoschizomus, Megaschizomus, Nyctalops, Protoschizomus, Schizomus*)

ORDER AMBLYPYGI: Tailless whip scorpions and whip spiders (Figures 4D and E). Prosoma undivided, covered by carapace-like shield and connected to opisthosoma by narrow pedicel; opisthosoma segmented but undivided; with two pairs of book lungs; telson absent; chelicerae with spider-like fangs; pedipalps raptorial; first pair of legs greatly elongate, sensory appendages; most with 8 eyes. (e.g., *Acanthophrynus, Damon, Heterophrynus, Stegophrynus, Tarantula*)

ORDER PALPIGRADI: Palpigrades (Figure 4F). Prosoma divided into proterosoma, covered by the propeltidium, followed by two free segments, and joined to opisthosoma by narrow pedicel; opisthosoma segmented and divided into broad mesosoma and short narrow metasoma, the latter bearing a long multiarticulate telson; eyes absent. (e.g., *Allokoenenia, Eukoenenia, Koenenia, Leptokoenenia, Prokoenenia*)

ORDER ARANEAE: Spiders (Figure 5). Prosoma undivided, covered by carapace-like shield and attached to opisthosoma by narrow pedicel; opisthosoma undivided and unsegmented except in liphistiids and some mygalomorph families, which have discrete tergites (i.e., "segmented" abdomens); chelicerae modified as fangs, usually with venom glands; opisthosoma bears book lungs and/or trachea, and silk-producing glands and spinnerets; most with 8 eyes.

SUBORDER MESOTHELAE: "Segmented spiders;" one family, Liphistiidae (*Heptathela* and *Liphistius*).

SUBORDER OPISTHOTHELAE: "Modern" spiders.

SUPERFAMILY MYGALOMORPHAE: Tarantula-like spiders; about 15 families including the following:

FAMILY CTENIZIDAE: Trap-door spiders. (e.g., *Cyclocosmia, Ummidia*)

FAMILY ATYPIDAE: Purse-web spiders. (e.g., *Atypus*)

FAMILY THERAPHOSIDAE: Tarantulas and bird spiders. (e.g., *Acanthoscurria, Aphonopelma*)

FAMILY DIPLURIDAE: Funnel-web spiders. (e.g., *Diplura*)

SUPERFAMILY ARANEOMORPHAE: "True" or "typical" spiders; about 75 families including the following:

FAMILY LOXOSCELIDAE: Brown spiders. (e.g., *Loxosceles*)

FAMILY THERIDIIDAE: Cobweb and widow spiders. (e.g., *Argyrodes, Episinus, Latrodectus, Ulesanis*)

FAMILY ULOBORIDAE: Orb-weavers. (e.g., *Hyptiotes, Uloborus*)

FAMILY ARANEIDAE: Orb-weavers. (e.g., *Araneus, Argiope, Cyrtophora, Mastophora, Nephila, Pasilobus, Zygiella*)

FAMILY TETRAGNATHIDAE: Orb-weavers. (e.g., *Dolichognatha, Eucta, Leucauge, Meta, Pachygnatha*)

FAMILY CLUBIONIDAE: Sack-spiders. (e.g., *Clubiona*)

FAMILY LINYPHIIDAE: Sheet-web spinners. (e.g., *Erigone, Dicymbium, Linyphia*)

FAMILY AGELENIDAE: Funnel-weavers. (e.g., *Agelena, Coelotes*)

FAMILY ARGYRONETIDAE: Water spiders; monotypic: *Argyroneta aquatica*.

FAMILY LYCOSIDAE: Wolf spiders. (e.g., *Lycosa, Pardosa, Pirata*)

FAMILY PISAURIDAE: Nursery-web spiders. (e.g., *Dolomedes, Pisaura*)

FAMILY OXYOPIDAE: Lynx spiders. (e.g., *Oxyopes*)

FAMILY THOMISIDAE: Crab spiders. (e.g., *Thomisus, Xysticus*)

FAMILY HETEROPODIDAE: Crab spiders. (e.g., *Heteropoda*)

FAMILY SALTICIDAE: Jumping spiders. (e.g., *Portia, Salticus*)

FAMILY DINOPIDAE: Ogre-faced spiders. (e.g., *Dinopis*)

FAMILY SCYTODIDAE: Spitting spiders. (e.g., *Scytodes*)

ORDER RICINULEI: Ricinuleids (Figure 4K). Prosoma fully covered by carapace-like shield and broadly joined with opisthosoma; opisthosoma unsegmented, with paired tracheae; chelicerae pincer-like, covered with flaplike cucullus; pedipalps small with coxae fused medially; eyes reduced; third legs of males modified for sperm transfer; cuticle very thick. Three genera: *Cryptocellus, Pseudocellus,* and *Ricinoides.*

ORDER PSEUDOSCORPIONIDA: False scorpions (Figure 4G). Prosoma covered by dorsal carapace-like shield, but clearly segmented ventrally; opisthosoma undivided but with 11–12 segments and broadly joined to prosoma; chelicerae chelate and bearing spinnerets; pedipalps large and scorpion-like; eyes present or absent. (e.g., *Chelifer, Chitrella, Chthonius, Dinocheirus, Garypus, Menthus, Pseudogarypus*)

ORDER SOLPUGIDA: Solpugids (Figure 4H). Prosoma divided into a proterosoma covered by a carapace-like shield and two free segments; opisthosoma undivided, but with 11 segments, bearing three pairs of tracheae; chelicerae huge and held forward; pedipalps long and leglike; propeltidium with one pair of eyes. (e.g., *Biton, Branchia, Dinorhax, Galeodes, Solpuga*)

ORDER OPILIONES: Daddy long-legs, or harvestmen (Figures 4I and J). Prosoma divided into a proterosoma and two free segments broadly joined to a segmented opisthosoma; chelicerae small, of three articles, pincer-like; pedipalps long and leglike; one pair eyes; one pair tracheae on opisthosoma; male with penis. (e.g., *Caddo, Leiobunum, Trogulus*)

ORDER ACARI: Mites, ticks and chiggers (Figure 6). Prosoma undivided, covered by carapace-like shield and broadly joined with opisthosoma; junction of prosoma and opisthosoma sometimes indicated by a disjugal furrow, but more often segments indistinguishably fused to one another; sometimes with a secondary sejugal furrow across body between second and third walking legs; chelicerae pincer-like or styliform; pedipalpal coxae uniquely fused with head elements; opisthosoma unsegmented except in a few primitive forms; gas exchange cutaneous or by tracheae; eyes present or absent; males of some species with a penis.

SUBORDER OPILIOACARIFORMES: Primitive mites. (e.g., family Opiliocaridae)

SUBORDER PARASITIFORMES: Mites and ticks. (e.g., *Aponomma, Argas, Boophilus, Dermacentor, Ornithodorus, Zeroseius*)

SUBORDER ACARIFORMES: Mites and chiggers. (e.g., *Demodex, Halotydeus, Penthaleus, Scirus, Tydeus*)

Synopses of chelicerate taxa

Subclass Merostomata

The merostomates probably had their origin in the Late Precambrian, diversifying during the great Cambrian invertebrate radiation. Fossil eurypterids and xiphosurans are known from the Ordovician, and these creatures flourished in the Silurian and Devonian. Alas, only five species of merostomates have survived to modern times, all xiphosurans—the "horseshoe crabs."

Order Eurypterida. The eurypterids represented a zenith in arthropod body size (Figure 1), some being nearly 3 m long (e.g., *Pterygotus*). These giant chelicerates, called "water scorpions," roamed ancient seas and freshwater environments well into the Permian, and were very abundant during their heyday. There is evidence that some species even became amphibious, or semiterrestrial.

Eurypterids were probably capable of swimming as well as crawling. The last pair of prosomal limbs were greatly enlarged and flattened distally, and may have been used as paddles. Another hypothesis suggests that these animals swam ventral side up, using the flaplike opisthosomal appendages as paddles and the enlarged prosomal legs for balance.

The chelicerae were extremely reduced in some species, but they were well developed and chelate in others, evidence supporting the idea that the eurypterids exploited a variety of food resources and feeding strategies.

Order Xiphosura. The extant members of the order Xiphosura (the horseshoe crabs) are regarded as "living fossils" and are, of course, far better known than their extinct relatives. Especially well studied is *Limulus polyphemus* (Figure 2), the common horseshoe crab of the Atlantic and Gulf coasts of North America and a favorite laboratory animal of physiologists. The five living species of Xiphosura belong to three genera in the family Limulidae, and all inhabit shallow marine waters, generally on clean sandy bottoms where they crawl about or burrow just beneath the surface, preying on other animals or scavenging.

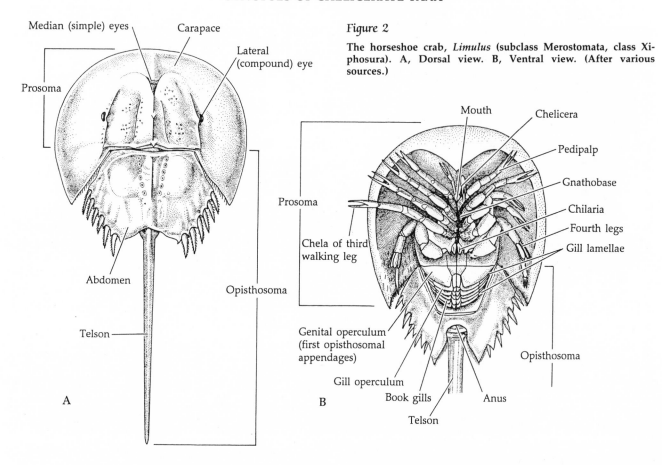

Figure 2

The horseshoe crab, *Limulus* (subclass Merostomata, class Xiphosura). A, Dorsal view. B, Ventral view. (After various sources.)

The chelicerae of horseshoe crabs are smaller than the other appendages, being composed of only three or four articles. Each walking leg is formed of seven articles (coxa, trochanter, femur, patella, tibia, tarsus, pretarsus), the last two of which form the chela (Figure 2). The coxal endites of the pedipalps and first three walking legs are modified as **gnathobases**. Arising from the coxae of the fourth walking legs are tiny appendage-like processes called **flabella**, which function as gill cleaners. In addition, just posterior and medial to the last walking legs is a pair of reduced appendages called **chilaria**. Their function is unknown, and there is some controversy about their evolutionary significance—some specialists say that they may reflect an additional opisthosomal segment.

The three living genera of limulids have distinct geographical ranges and are restricted to eastern coasts of the New and Old worlds. *Limulus* (*L. polyphemus*) is restricted to eastern North America, ranging from Nova Scotia to the Yucatan region of Mexico; *Tachypleus* (*T. tridendatus*) occurs in Southeast Asia; *Carcinoscorpius* has been collected only from Malaysia, Siam, and the Philippine islands.

Subclass Arachnida

Order Scorpiones. The scorpions (Figure 3) are considered to be among the most ancient terrestrial arthropods and the most primitive arachnids. They evolved from aquatic ancestors, perhaps even the eurypterids or a common ancestor, and invaded land during the Carboniferous period. All of the roughly 1,200 known living species are terrestrial predators. They inhabit a variety of environments, particularly deserts and tropical rain forests where some arboreal species occur. They are notably absent from colder regions of the world. Scorpions include the largest living arachnids, with some reaching lengths of about 18 centimeters.

The scorpion body is clearly divided into three regions: prosoma, mesosoma, and metasoma. The chelicerae are short, chelate structures, bearing gnathobases for grinding food. The pedipalps are large and composed of six segments, the last two of which form the grasping chelae. The walking legs each comprise eight articles (coxa, trochanter, femur, patella, tibia, metatarsus, tarsus, pretarsus). The ventral surface of the mesosoma bears the genital pore, a pair of sensory **pectines**, and four pairs of spiracular

A

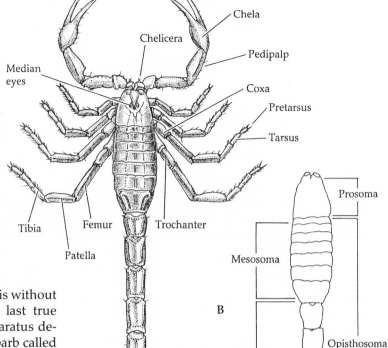

B

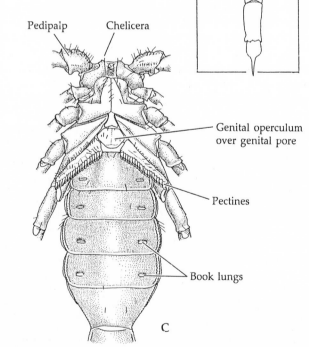

C

openings to the book lungs. The metasoma is without appendages. The anus is located on the last true segment and is followed by a stinging apparatus derived from the telson and bearing a sharp barb called the **aculeus**.

Order Uropygi. The whip scorpions (Figure 4A) are moderately large arachnids reaching lengths of 1 to 8 cm. There are fewer than 100 known species of living uropygids, most of which occur in Southeast Asia; a few are known from the southern United States and parts of South America; and some apparently introduced species occur in Africa. With the exception of a few desert species, whip scorpions live under rocks, in leaf litter, or in burrows in relatively humid tropical and subtropical habitats. The telson is sensitive to light, and most are negatively phototactic, active only at night.

The elongate first walking legs are held forward as "feelers," aiding the animals in their nocturnal hunting excursions. They feed on various small invertebrates by grasping prey with the pedipalps and grinding it with the chelicerae.

Whip scorpions possess a pair of **repugnatorial glands** that open near the anus. When a uropygid is threatened by a potential predator, it raises the opisthosoma and sprays on the would-be attacker an acidic liquid secreted by these glands. Some forms (e.g., *Mastigoproctus*) produce a secretion high in acetic acid, earning them the common name "vinegaroons."

Order Schizomida. There are about 80 species of small (less than 1 cm) arachnids in this order (Figures

Figure 3

Scorpions (subclass Arachnida, order Scorpiones). A, *Androctonus australis*. B–C, *Buthus martensi* (dorsal and ventral views). (A courtesy R. F. Foelix; B,C after Keegan 1980.)

4B and C). Although some authors classify them as a suborder of Uropygi, they are distinguished from true whip scorpions by the divisions of the prosoma and the shorter telson. Schizomids live in leaf litter, under stones, and in burrows and are most common in tropical and subtropical areas of Asia, Africa, and the Americas; a few temperate species are known.

The first walking legs are sensory and resemble those of uropygids. Like the uropygids, the schizomids are predators on small invertebrates, and they also possess opisthosomal repugnatorial glands.

Order Amblypygi. The 70 or so species of amblypygids are commonly known as whip spiders or tailless whip scorpions, reflecting similarities to both spiders and uropygids (Figures 4D and E). Externally they resemble whip scorpions, but internally they are very much like spiders except that they lack spinnerets and poison glands. Amblypygids are widely distributed in warm humid areas, where they are found under bark or in leaf litter, and in similar protected habitats; several species are cave dwellers.

Most species are less than 5 cm in length, but the first pair of legs may be as long as 25 cm each! These limbs are used as touch (and chemo-) receptors (like antennae in other, nonchelicerate arthropods) and are important in prey location because these animals hunt at night. Amblypygids walk sideways with the long "feelers" extended, sensing for potential prey. Once located, prey is grasped by the pedipalps and torn open by the chelicerae. The body fluids are then sucked from the victim and ingested.

Order Palpigradi. There are about 60 species of palpigrades, all of which are less than 3 mm long (Figure 4F). These arachnids have undergone a great deal of evolutionary reduction in association with their small size and cryptic habits. They are colorless, have very thin cuticles, and have lost the circulatory and gas exchange organs. Most are found under rocks or in caves, and one species lives on sandy beaches. They have been recorded from widely separated parts of the world. These tiny creatures are quite rare, and their biology remains poorly known.

Order Araneae. Spiders comprise about 35,000 known species, in two suborders (Figure 5). The suborder Mesothelae includes the primitive, segmented spiders, which are characterized by persisting segmentation of the opisthosoma (Figure 5E). All of the segmented spiders are placed in a single family, the Liphistiidae. Most are 1–3 cm long and construct burrows with a trapdoor over the entrance. All are predators, feeding on animals that pass within striking distance of the burrow opening.

The other suborder, Opisthothelae, comprises two large groups: the mygalomorph (tarantula-like) spiders and the araneomorph ("typical") spiders. Most members of both groups have unsegmented opisthosomas, and are differentiated by the nature of the chelicerae (Figures 5H and I). Mygalomorph spiders possess chelicerae that articulate in a manner that enables movement of the appendages parallel to the body axis (i.e., **orthognathous** movement), whereas araneomorph spiders have chelicerae that move at right angles to the body axis (i.e., **labidognathous** jaws). Many authors treat the opisthothelan spiders as two suborders (Orthognatha and Labidognatha) on the basis of this difference. Members of the suborder Mesothelae have orthognathous chelicerae.

Spiders are the most familiar of all chelicerates, of course, and they are one of the most abundant groups of land animals. They have successfully exploited nearly every terrestrial habitat, and many freshwater and intertidal areas as well, and display a truly staggering array of lifestyles—all utilizing a fairly uniform *Bauplan*.

The body is divided into a fused prosoma and a fleshy opisthosoma (Figure 5). The first opisthosomal segment forms the narrow **pedicel**, which joins the two body regions. The chelicerae each have two segments; the basal article is short and conical and the distal article is a hard, curved fang, usually bearing a pore from the duct of the poison gland (Figure 16A). The **labium** is a medial flap projecting ventrally over the mouth. The pedipalps are composed of six articles each. In most spiders the proximal articles of the pedipalps are enlarged as lobes (or endites) called **maxillae** that bear gnathobases used to handle and grind food.* These gnathobases flank the mouth laterally. Beyond the maxillae, the pedipalps extend forward as tactile organs in females and juveniles, but they are highly modified as sperm transfer organs in mature males. In a few species of spiders (e.g., tarantulas) the pedipalps function as walking "legs." The ventral surface of the prosoma bears a cuticular plate, or **sternum**, around which arise the four pairs of walking legs (Figure 5); each of the legs is formed of eight articles.

The opisthosoma bears openings for gas exchange, the reproductive system, the spinnerets, and the anus (Figure 5). On the ventral surface posterior

---

*Not to be confused with the "true" maxillae of crustaceans, a (probably) nonhomologous kind of appendage of the same name.

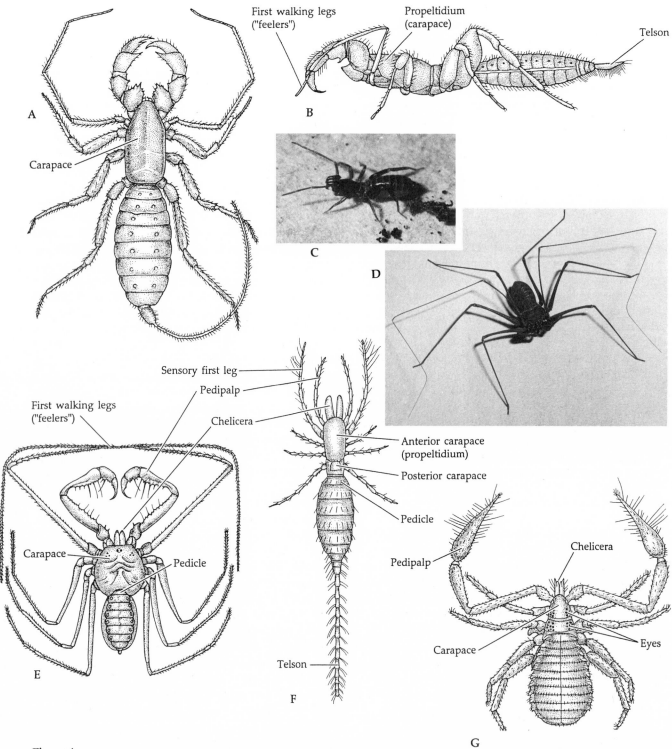

Figure 4

Some examples of arachnids other than spiders and scorpions. A, A whip scorpion, *Mastigoproctus* (order Uropygi). B, *Nyctalops crassicaudatus* (order Schizomida) (side view). C, *Schizomus* (about 4 mm long). D, *Heterophrynus longicornis* (order Amblypygi). E, *Stegophrynus dammermani* (order Amblypygi). F, *Koenenia* (order Palpigradi). G, The false scorpion *Chelifer cancroides* (order Pseudoscorpionida). H, *Galeodes arabs* (order Solpugida). I, An unnamed opilionid. J, A female daddy longlegs (order Opiliones) depositing eggs. K, *Ricinoides crassipalpe* (order Ricinulei) (dorsal view). (A,F redrawn from Barnes 1980, after Millot et al. 1949 and other sources; B,E,G,H,K after Savory 1977; C,D, courtesy of R. F. Foelix; I after Barth and Broshears 1982, from Krantz 1978; J after Barth and Broshears 1982, from Kaestner 1968.)

to the pedicel is a transverse **epigastric furrow** in which the gonopores are located. In most females, a slightly elevated plate, the **epigynum**, lies in front of the furrow and bears the openings to the seminal receptacles. Lateral to the epigastric furrow are openings (spiracles) to the **book lungs** or anterior trachea. The hind part of the opisthosoma bears the anus and the **spinnerets**. Many also possess one or two spiracles leading to a posterior tracheal system. The ability of spiders to produce silk and fashion it into a great variety of functional devices (e.g., webs) has been a major factor in their evolutionary success (see discussion below).

Order Ricinulei. There are only about 35 species of ricinuleids (Figure 4K). They are less than 1 cm long and live in caves and tropical forest leaf litter in West Africa and tropical America. Ricinuleids are slow-moving predators on tiny invertebrates.

Order Pseudoscorpionida. There are about 2,000 described species of false scorpions, the largest of which reach lengths of only 7 mm (Figure 4G). They are cosmopolitan and found in a great variety of habitats—under stones, in litter, in soil, under bark, and in animal nests. One genus, *Garypus*, is found on sandy marine beaches, and one species, *Chelifer cancroides*, typically cohabits with humans.

These strange little creatures resemble scorpions in general appearance, but they lack the elongation of the opisthosoma and telson and do not have a stinging apparatus. However, they do possess poison glands in the pedipalps with which they immobilize their prey, usually other tiny arthropods (e.g., mites). Once captured, the victim is torn open by the chelicerae and the body fluids sucked out.

A number of false scorpions use larger arthropods as temporary "hosts" for purposes of dispersal. This phenomenon of "hitchhiking," known as **phoresy**, typically involves females that grab onto the larger animal with their pedipalps. The host is often a flying insect. *Chelifer cancroides*, for example, is frequently encountered as a phoretic "guest" on house flies.

Order Solpugida. Most of the 900 or so species of solpugids (or solifugids) live in tropical and subtropical desert environments of America, Asia, and Africa. In contrast to many arachnids, they are often daytime hunters, hence the common name "sun spiders." They are also known as "wind spiders," because the males run at very high speeds, "like the wind." Although some are only a few millimeters long, many reach lengths of up to 7 cm. The feeding habits of many solpugids are unknown. Among those that have been studied, most are omnivorous, but they frequently show a preference for termites or other arthropods. Lacking poison, they rip their prey apart with strong chelicerae (Figure 4H).

Order Opiliones. The order Opiliones is a large and diverse group of about 5,000 described species.

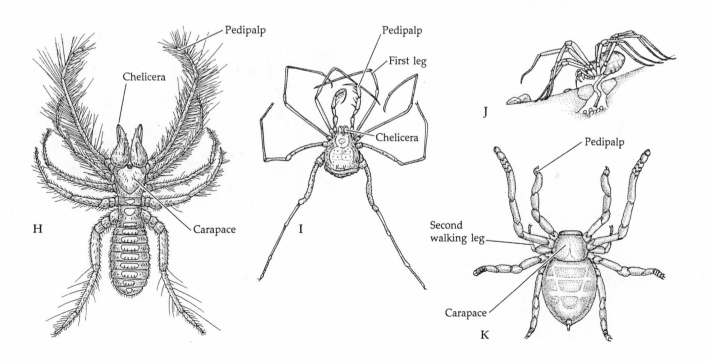

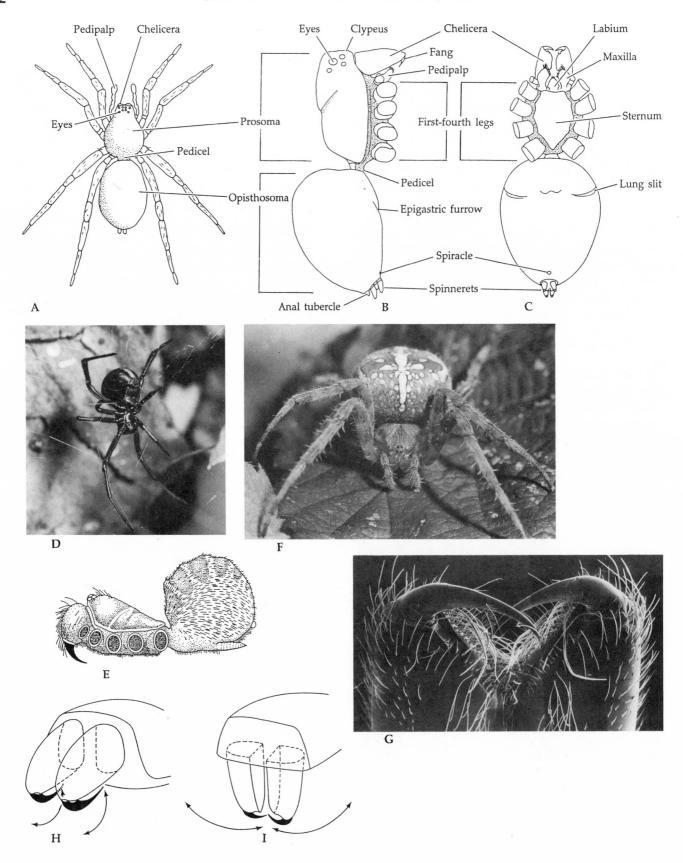

Figure 5

Spiders. A–C, A generalized spider. A, Dorsal view. In B (lateral view) and C (ventral view) the legs have been omitted except the coxae. D, The notorious black widow spider, *Latrodectus mactans* (family Theridiidae). E, *Liphistius* (suborder Mesothelae) (side view with legs removed). Note the evidence of persistent opisthosomal segments. F, An orb-weaving, or "garden," spider, *Araneus diadematus* (family Araneidae). G, The chelicerae of *Tegenaria* (superfamily Araneomorphae). H–I, The orientation and plane of motion of the chelicerae of an orthognath (H) and a labidognath (I) spider. (A–C adapted from Gertsch 1979 and Foelix 1982; D,G courtesy of R. F. Foelix; E after Millot et al. 1949; F from Foelix 1982, reprinted with permission of Harvard University Press; H–I redrawn from Foelix 1982, after Kaestner 1969.)

They are generally considered to be closely related to mites (order Acari). These arachnids are commonly called harvestmen or daddy longlegs (Figures 4I and J). They are known from nearly all climatic regions of the world, including subarctic areas, but are most abundant in tropical South America and Southeast Asia. Most species have small bodies, less than 2 cm in length, but usually they have very long legs (up to 10 cm).

Opilionids prefer damp, shaded areas and are commonly found in leaf litter, on trees and logs in dense forests, and in caves. They feed on a variety of small invertebrates and also scavenge on dead animal and plant matter. Food is grasped by the pedipalps and passed to the chelicerae for chewing. These animals are among the few arachnids capable of ingesting small solid particles as well as the usual liquefied food. Opilionids possess a pair of defensive repugnatorial glands, which produce noxious secretions containing quinones and phenols.

As discussed later, most arachnids mate by indirectly passing sperm from the male to the female. A unique attribute of the opilionids is the presence on the males of a penis used for direct copulation.

Order Acari. The mites and ticks comprise the largest group of arachnids (Figure 6). There are approximately 30,000 known species, and some experts suggest that there may be as many as a million or more yet to be described—we will not discuss all of them! The order Acari is a difficult taxon to characterize, and it may represent a polyphyletic assemblage. Because of the size and diversity of this group, we give a somewhat extended account. Even so, it is not possible to do justice here to the vast range of forms and lifestyles represented by its members.

Members of this order occur worldwide. Most are terrestrial, many are parasitic, and some have invaded aquatic environments. The tremendous evolutionary success of the members of Acari, particularly the mites, is reflected in their species diversity and tremendously varied lifestyles. This success is probably due, at least in part, to their compact body and reduced size. By coupling these attributes with the inherent arthropod quality of segment and appendage specialization, mites have exploited myriad microhabitats unavailable to many other animals. Thus the Acari have evolutionarily turned the general arachnid *Bauplan* toward small size and extreme specialization—with great success. If the group is in fact polyphyletic, then we may envision several ancestral "arachnid lineages" evolving convergently to capitalize on miniaturization and the acarine *Bauplan*.

Recent classification schemes divide the members of Acari into three groups. The most primitive mites are omnivorous and predatory forms in the suborder Opilioacariformes. They are characterized by the retention of opisthosomal segmentation (at least ventrally) and the presence of a transverse groove (the **disjugal furrow**) separating the prosoma from the opisthosoma. These mites are found on tropical forest floors and in arid temperate habitats.

The remainder of the thousands of kinds of mites and ticks are placed in the suborders Parasitiformes and Acariformes, the latter housing most of the species. The body of parasitiform mites is undivided dorsally, there being no clear transverse groove (Figure 6A). The suborder Parasitiformes includes free-living and symbiotic forms in all parts of the world. The free-living types inhabit various terrestrial habitats, including leaf litter, decaying wood and other transient organic material, mosses, insect and small mammal nests, and soil. Most of these mites are predators on small invertebrates. Many species are full- or part-time symbionts on other animals, either as immatures or as adults. The hosts are frequently other arthropods, such as centipedes, millipedes, ants, and especially beetles. In some cases the relationship is truly parasitic, in others phoretic; in many instances the nature of the association is unknown.

The most familiar members of the Parasitiformes are the ticks (families Argasidae and Ixodidae; Figure 6F). Ticks are blood-sucking ectoparasites on vertebrates (one species, *Aponomma ecinctum*, lives on beetles). They are the largest members of the entire order, and some swell to 2–3 cm in length during a blood meal. The chelicerae are smooth and adapted for slicing skin. The ixodids are called the hard ticks because of a sclerotized shield covering the entire dorsum. They parasitize reptiles, birds, and mammals and generally remain attached to their hosts for

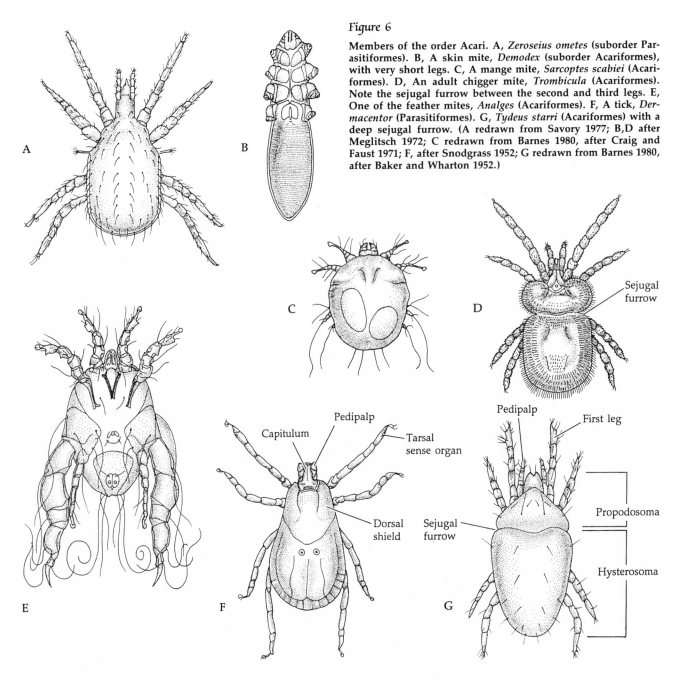

Figure 6

Members of the order Acari. A, *Zeroseius ometes* (suborder Parasitiformes). B, A skin mite, *Demodex* (suborder Acariformes), with very short legs. C, A mange mite, *Sarcoptes scabiei* (Acariformes). D, An adult chigger mite, *Trombicula* (Acariformes). Note the sejugal furrow between the second and third legs. E, One of the feather mites, *Analges* (Acariformes). F, A tick, *Dermacentor* (Parasitiformes). G, *Tydeus starri* (Acariformes) with a deep sejugal furrow. (A redrawn from Savory 1977; B,D after Meglitsch 1972; C redrawn from Barnes 1980, after Craig and Faust 1971; F, after Snodgrass 1952; G redrawn from Barnes 1980, after Baker and Wharton 1952.)

days or even weeks, feeding on blood. Some are vectors of important diseases, for example, *Dermacentor andersoni* (vector of Rocky Mountain spotted fever) and *Boophilus annulatus* (vector of Texas cattle fever).

The soft ticks (Argasidae) lack the heavily sclerotized dorsal shield. They are usually rather transient on avian and mammalian hosts (particularly bats), typically feeding for less than an hour at a time. When not attached to a host, these ticks remain hidden in cracks and crevices or buried in soil. Disease-carrying soft ticks include *Argus persicus* (vector of fowl spirochetosis) and *Ornithodoros moubata* (vector of African relapsing fever, or "tick fever").

The vast and diverse suborder Acariformes is itself generally considered to be polyphyletic, and the unifying features thus convergent. These mites usually have their bodies divided into two regions, but not as the normal prosoma and opisthosoma. Rather, the disjugal furrow is lost and a secondarily evolved **sejugal furrow** partly or wholly traverses the dorsum between the origins of the second and third pairs of walking legs (Figures 6D and G). The front part of the body is called the **propodosoma**, the hind part the **hysterosoma**. In some acariform mites this division is secondarily lost.

Myriad free-living forms inhabit terrestrial, freshwater, and marine environments and include herbivorous (some fungivorous) and predacious species. Feeding methods are diverse. Many ingest solid as well as liquid food, and a few aquatic forms are known to suspension feed. Free-living acariform mites are found in virtually every conceivable situation: in soil, leaf litter, decaying organic matter, mosses, lichens, and fungi; under bark; on freshwater algae; in sands; on seaweeds; at all altitudes and most ocean depths. Certain groups are serious pests and destroy stored grain crops and other food products. On the other hand, some predatory acariform mites have been used for biological pest control, to eat other arthropods—even other mites!

Most of the symbiotic acariform mites are parasites on various vertebrate and invertebrate hosts, although in most cases details of the relationships are unknown. Various species parasitize marine and freshwater crustaceans, freshwater insects, marine molluscs, the pulmonary chambers of terrestrial snails and slugs, terrestrial arthropods, the outer surfaces of all groups of terrestrial vertebrates, and the nasal passages of amphibians, birds, and mammals. In addition to direct parasitism, many mites are phoretic, using their hosts for dispersal. There are also many plant-eating mites that are considered parasitic.

A great many acariform mites cause economic or medical problems, either by direct predation and parasitism, by acting as disease vectors, or by feeding on stored food products. The family Penthaleidae includes the red-legged earth mite (*Halotydeus destructor*) and the winter grain mite (*Penthaleus major*), both of which are serious pests of many important crops. Members of the superfamily Eriophyoidea are vermiform mites adapted to feeding on various plants. This group includes the gall and leaf-curl mites, and a number of others that serve as vectors for certain disease-causing viruses (e.g., wheat and rye mosaic viruses). Another family of mites (Demodicidae) includes parasites of hair follicles and sebaceous glands of mammals. Two species, *Demodex folliculorum* and *D. brevis*, occur specifically in the sebaceous glands and follicles, respectively, of the human forehead. A third species, *D. canis*, causes mange in dogs. A brief sampling of other problems caused by acariform mites includes subcutaneous tumor-like growths in humans, various sorts of skin irritations, mange in many domestic animals, reduced wool production in sheep, and feather loss in birds.

With a mixture of reluctance and relief, we leave the mites; interested readers should consult the literature list at the end of this chapter for some leading references.

The chelicerate *Bauplan*

By now you should have a good idea of the general nature of arthropods and, specifically, some idea of the diversity within the class Chelicerata. This section covers the general biology and structure of members of this class, with an emphasis on the xiphosurans, spiders, and scorpions; other taxa are mentioned incidentally. This treatment, following the systematic synopses, should convey an impression not only of the diversity but also of the unity within this group, and it should also reinforce the concept of the evolutionary plasticity of the arthropod *Bauplan* in general.

The chelicerate body is typically divided into two main regions, the prosoma and the opisthosoma. A discrete head is not recognizable (Figures 1–6). The prosoma includes a presegmental **acron** and six segments; the opisthosoma includes up to twelve segments and a postsegmental, postanal telson. As in other arthropod groups, these basic body regions have undergone various degrees of specialization and tagmosis. In most chelicerates the entire prosoma is fused and covered by a carapace-like shield. However, in certain groups (e.g., palpigrades and solpugids) the prosoma is divided into three parts: a **proterosoma**, comprising the acron and the first four segments, all fused and covered by a carapace; and two free segments. The opisthosoma may be undivided, as it is in spiders; or it may be divided into an anterior mesosoma and posterior metasoma, as it is in the eurypterids and scorpions.

The appendages further distinguish the chelicerates from other arthropods. There are no antennae, but all six segments of the prosoma usually bear appendages. The first pair of appendages are embryologically postoral, often pincer-like, chelicerae. The chelicerae migrate during embryogeny to reach a position lateral to the mouth, or even preoral, in adults of most groups; here they serve as fangs or grasping structures during feeding. The chelicerae are followed by a pair of postoral pedipalps, which are usually elongate or more rarely in the form of pincers. The pedipalps are usually sensory in function, but in some groups (e.g., scorpions) they aid in food-getting and defense. The remaining four segments of the prosoma typically bear the walking legs.

The numbers of segments and appendages on the opisthosoma vary. In general, appendages are absent or very reduced, although in the horseshoe crabs they persist as large platelike limbs, called **book gills**, that function in locomotion and gas exchange. In most other chelicerates the opisthosomal limbs persist only as specialized structures, such as the silk-producing spinnerets in spiders or the pectines of scorpions.

In summary, then, we may define the Chelicerata as cheliceriform arthropods in which the body is divided into two regions (or two tagmata), a prosoma and an opisthosoma, and in which the first two pairs of appendages are chelicerae and pedipalps and the remaining four pairs are walking legs. Evolutionarily, this *Bauplan* has been a highly successful one.

As mentioned earlier, the great success of spiders seems to have been due, in large part, to the evolution of complex behaviors associated with silk and web production. Because we are paying special attention to members of the order Araneae in this chapter and because silk production is so important to nearly all facets of their lives, we present a special section on spider silk and its uses.

Spinnerets, spider silk, and spider "webs"

Spider silk is a complex fibrous protein composed mostly of the amino acids glycine, alanine, and serine. The silk is produced in a liquid, water-soluble form that transforms into an insoluble thread after it leaves the body. This transformation involves as much as a tenfold increase in molecular weight of the silk protein, brought about at least in part by the formation of intermolecular bonds. How this transformation is accomplished is not completely understood, but evidence suggests that physical stretching of the threads is involved. Spider silk is nearly as strong as nylon, and about twice as elastic.

A spider's silk-producing apparatus is located in the opisthosoma and comprises various sets of glands. The liquid silk produced by these glands is secreted into ducts through which it passes to the spinnerets (Figure 7). Each spinneret bears some combination of small and large tubes called **spools** and **spigots**, respectively, which open to the outside. The apparatus spins the silk into threads of different thicknesses. The spinnerets are actually highly modified opisthosomal appendages, and they retain some musculature that allows for their movement during spinning (Figure 7D).

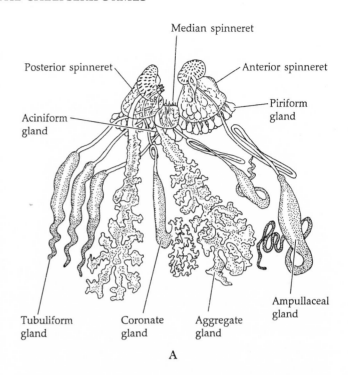

A

Recent studies indicate that there are about six different kinds of silk glands, each producing a different kind of silk. By spinning the various types of silks, at different diameters, spiders produce a variety of threads of different qualities for different functions. The numbers and kinds of glands and spinnerets vary among groups in the order. Evolutionarily, spiders probably originally had four pairs of spinnerets, a condition seen today on some of the primitive "segmented spiders" (suborder Mesothelae). Most other spiders have three pairs, although some have only one or two.

Another spinning organ, called the **cribellum** (Figure 7E) occurs in several families of araneomorph spiders. Some authors divide the araneomorph spiders into two groups based on the presence or absence of the cribellum, but it is now generally agreed that loss of the cribellum has occurred several times and that such a division is artificial. The cribellum is a platelike structure, anterior to the usual spinnerets, that bears numerous small spigots (up to 40,000 in some species). The cribellate silk is emitted in many extremely fine threads and then combed into a delicate mesh by a row of bristles (the **calamistrum**) on the fourth walking legs (Figure 7F). The resulting mesh, known as a "hackle band," is used as a prey-capture net by some spiders (e.g., the Uloboridae).

The varied uses of silk are intimately associated with nearly every aspect of the lifestyles and habits of spiders, as explained more fully in following sec-

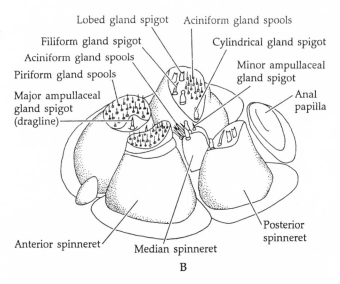

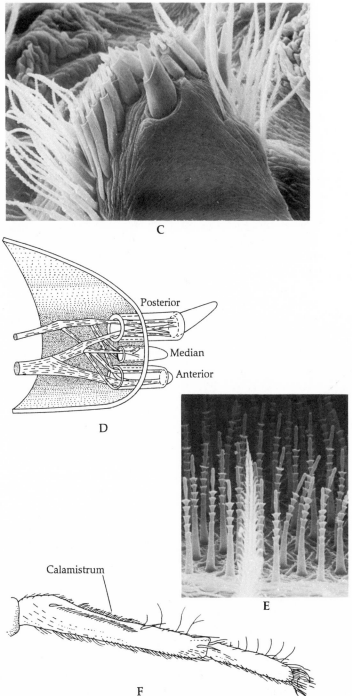

Figure 7

Silk glands and spinnerets in spiders. See text for additional details. A, The silk glands and spinnerets of *Nephila*, the golden silk spider. Only one member of each pair of glands is shown. **B,** The spinnerets of the orb-weaver *Araneus* (external view). **C,** Median spinneret of *Uloborus*. Note the different sizes of spigots (382×). **D,** Cutaway view of the posterior end of the opisthosoma (*Tegenaria*). Note the spinneret muscles. **E,** The cribellar spigots of *Hypochilus* (scanning electron micrograph, 1600×). **F,** The comblike calamistrum on the fourth walking legs of *Amaurobius*, used to brush the silk threads as they emerge from the cribellum. (A redrawn from Foelix 1982, after Peters 1955; B after Barnes 1980, from Wilson 1969; C,E, from Foelix 1982, reprinted with permission of Harvard University Press; D redrawn from Foelix 1982, after Peters 1967; F after Meglitsch 1972.)

tions. Various kinds of silk are used as safety lines and climbing lines; for the construction of nests, cocoons, and traps; to wrap prey for brief storage; as egg sacs and sperm platforms; and to line burrows. Many newly hatched spiderlings spin long, thin, gossamer threads to ride the wind for aerial dispersal. Perhaps the most interesting and familiar use of silk by spiders is prey capture. Various sorts of snares and nets are spun, most of which can be loosely referred to as webs. Prey-capture web construction and use is discussed in more detail in the section on feeding, but in general these webs serve both to trap the prey and, through vibrations, to signal the spider of its presence. Indeed, most spiders live in a world dominated by vibrations, a world where a potential meal, or predator, or mate reveals itself with characteristic resonance patterns.

Nentwig and St. Heimer (1983) present a general theory of the evolution of web construction in which primitive spiders are viewed as using silk primarily in burrow construction and to protect their devel-

oping young. Subsequent advances in web building were, first, to extend threads to the burrow entrance; next, to extend them upward, with attachments to foliage; and finally, to move the entire structure off the ground and into the aerial environment. This last example, of presumably advanced web types, is well expressed by the familiar orb-weaver spiders. Thus,

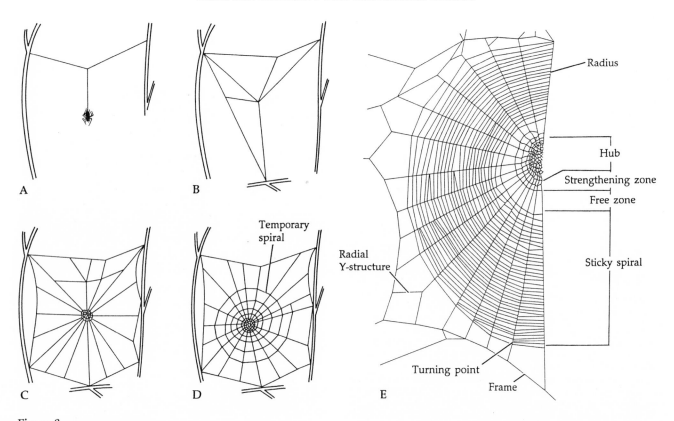

Figure 8

Construction of an orb web. A, Formation of the Y-shaped frame. B–C, Addition of radial threads. D, Addition of the temporary spiral or working platform. E, A portion of the completed orb web, with the temporary spiral replaced by the sticky spiral. (Adapted from Foelix 1982, Gertsch 1979, and others.)

we present here a brief description of the construction of an orb web, as commonly produced by members of the families Araneidae and Tetragnathidae.

The spinning of an orb web takes place in three phases that are apparently genetically programmed (Figure 8). The first phase is the construction of a supportive Y-shaped framework and a series of radiating threads. The upper branches of the Y are initially laid down as a horizontal thread between two objects in the spider's environment. The spider sits in one spot and secretes the thread into the air; the loose end is then carried by air movements and flutters about until it contacts and sticks to an object. The spider then tacks down its end of the thread, moves to the center of the horizontal line, and drops itself on a vertical thread. The vertical thread is pulled taut and attached, thereby producing the Y-shaped frame. The intersection of the three branches of the Y becomes the hub of the final web, and it is from this point that the radii are extended. Radial threads are attached to the frame threads; once this initial

phase is complete, the spider quickly lays down a temporary spiral thread, starting from the center, as the second stage of construction. This spiral thread, along with the framework, serves as a working platform during the final stage of web building—the production of the sticky spiral or prey trap. This final thread is always coated with a glue that automatically assumes a beaded distribution after it is deposited (Figure 9A). As the sticky spiral is laid down, the temporary scaffold-spiral is removed or eaten.

Some orb-weavers (e.g., *Argiope*) produce a dense mesh of silk, called a **stabilimentum**, across the hub of their webs (Figure 9B). Often considered to function as a structural stabilizer, as a patch of camouflage for the spider, or as a device to capture drinking water, recent work suggests that the stabilimentum may serve as a visual signal to warn away large flying animals, such as birds, that could easily damage the web.

Many orb-weavers (e.g., *Araneus*) can produce an entire web in less than 30 minutes, and most build a new web every night. They do not "waste" the silk of the old web, but eat it before or during the production of a new one. Radioactive labeling experiments show that the silk proteins from eaten webs appear at the spigots in new threads soon after ingestion, often within a few minutes!

The whole business of orb web production is remarkably well programmed; in fact, angles between radii are constant. Web building appears to depend entirely on tactile input, for normal webs are produced at night (and by experimentally blinded spiders). Even gravity does not seem to be necessary, as illustrated by two famous orb-weaver spiders sent into space aboard Skylab.

It has recently been discovered that some tropical spider species cooperate in web building, prey capture, and even spiderling rearing. Such social activity has been reported for some 20 species, in at least six families. Sociality ranges from loose overwintering aggregations to more permanent aggregations of certain orb-weavers (e.g., *Cyrtophora moluccensis*) in which the webs of several individuals share common support lines. In *Anelosimus eximius*, and some other species, eggs are positioned in the web and tended by many different females; hatched spiderlings also are fed by regurgitation by various females.

Locomotion

As in all arthropods, chelicerate locomotion follows the principles of joint articulation and leg movements discussed in Chapter 15. Except for the xiphosurans and a few aquatic arachnids, the legs must also be strong enough to support the body on land.

The xiphosurans are slow benthic crawlers and shallow burrowers, utilizing their stout prosomal limbs to push their heavy bodies over and through the sand. The legs are close together (Figure 2), and thus coordination of movement sequences is essential. Horseshoe crabs are also able to swim, upside down, by beating the opisthosomal appendages.

Walking by terrestrial chelicerates demands that the body be supported off the substratum and that the four pairs of legs move in sequences that maintain the animal's balance. More information on the problems of terrestrial arthropod locomotion is discussed in Chapter 17 (see Figure 13 in Chapter 17). The details of scorpion walking patterns have been examined by Root and Bowerman (1978). During simple forward walking, each of the eight limbs moves through the usual power and recovery strokes. In scorpions the joints between the coxae and the body are virtually immovable and do not contribute to the overall motion of the limbs. Not all of the walking legs move in the same pattern. The tips of the anterior legs are brought quite high off the ground during their recovery strokes; they may be used to feel ahead as the animal moves. Also, the tips of each pair of legs extend different distances from the body, allowing for stride overlap without contact between legs

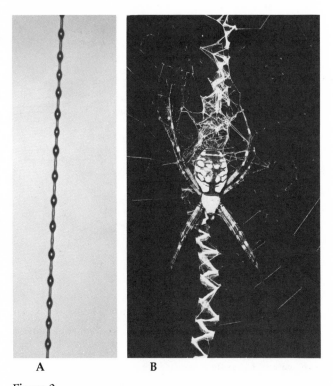

Figure 9

A, A thread of the sticky spiral of an orb web with evenly distributed adhesive droplets. B, *Argiope* on its orb web with a stabilimentum. (A from Foelix 1982, reprinted with permission of Harvard University Press; B courtesy R. F. Foelix.)

(Figure 10). Leg movements in scorpions do not follow the usual metachronal model. Rather, the typical sequence of movement along one side of the body is leg 4, then 2, then 3, and finally 1, with the legs on the opposite side being generally, but not precisely, out of phase. The somewhat staggered overlap of the movement sequences produces a smooth gait during forward walking, unlike the jerky motion of an insect moving by the alternating tripod pattern (Figure 14 in Chapter 17). Like many arthropods, scorpions are also able to change speeds, turn abruptly, walk backward, and burrow in loose sand.

Spiders have evolved a number of methods of locomotion, all involving the usual leg motions inherent in jointed limbs, except that limb extension in spiders is assisted in part by hydrostatic pressure (involving the femur–patella and tarsus–metatarsus joints). Normal spider walking is similar to the alternating tripod gait of insects described in Chapter 17. The eight ambulatory limbs move in a so-called **diagonal rhythm** sequence (Figure 11). That is, legs 2 and 3 on one side of the body are moving simultaneously with legs 1 and 4 on the opposite side. This gait maintains a four-point stance while distributing

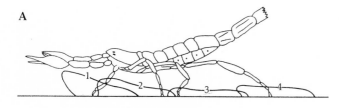

A

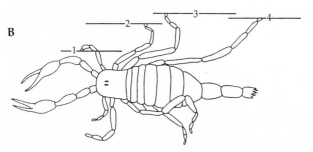

B

Figure 10

Walking in scorpions. Numbers denote legs 1–4. A, The paths of the leg tips during their recovery strokes. Notice that the anterior legs are lifted higher off the ground than are the posterior legs. B, Distance of each leg tip from the body. See text for additional explanation. (After Root and Bowerman 1978.)

the body weight more-or-less equally among those appendages in contact with the substratum. During very slow walking, however, the posterior to anterior metachronal waves of limb motion are detectable. The stride lengths of the legs overlap somewhat and vary with speed and direction of movement. The placement of the legs on the body, and the arcs through which they swing, prevent actual limb contact.

A number of spiders (e.g., members of the family Salticidae) are also capable of jumping (Figures 12A and B). Propulsion is achieved primarily by a rapid extension of the fourth pair of legs. Once the spider is airborne, the front legs are extended forward and used in landing. Salticids jump during normal locomotion, and also when capturing prey and escaping from predators.

Silk plays an important role in various methods of spider locomotion. When walking or jumping, most spiders continually produce a strong thread behind them called a **dragline** (Figure 12C). The dragline is periodically tacked to the substratum, providing a safety rope for the wandering spider. Thus, a spider brushed from a surface does not fall to the ground; instead it pays out dragline silk and dangles like a mountaineer who has lost his footing. Silk is also used to provide a substratum on which spiders move, including the aerial dispersal of young spiders on very thin threads. Web spinners crawl over their webs with limb movements more-or-less like those used in normal walking, except that stride lengths must correspond to the distances between threads in the web. Many spiders are able to move about on a single thread (Figure 12D). This may involve dropping vertically as a thread is paid out from the spinnerets, climbing up a vertical thread, or moving while hanging upside down on a horizontal thread (they do not walk atop single threads, tightrope fashion). Most spiders that are capable of these sorts of activ-

A

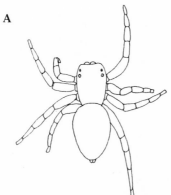

B

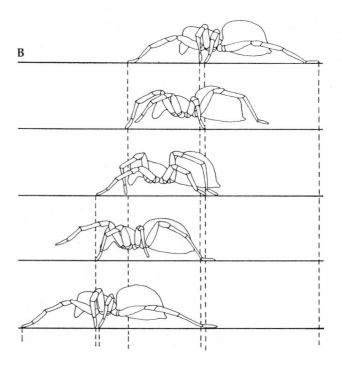

Figure 11

Walking in spiders. A, A salticid with legs in walking positions (dorsal view). B, A lycosid during slow walking (lateral view). Vertical dotted lines show leg tip contact position relative to forward progress of the body. Note the large degree of overlap in leg movements, particularly of the first two legs. Tangling is prevented in part by keeping the tips of adjacent legs at different distances from the body. (After Foelix 1982.)

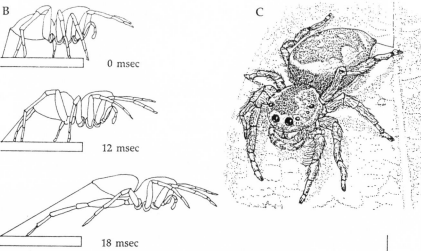

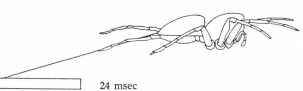

0 msec

12 msec

18 msec

24 msec

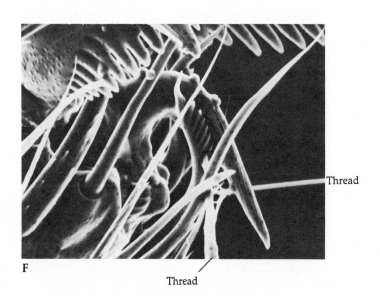

Figure 12

More aspects of spider locomotion. A, A salticid preparing to jump. B, A salticid jumping. Note the safety line. C, A salticid paying out its dragline while walking. D, The orb-weaver *Zygiella* climbing up a single thread. E–F, The tarsal tip of *Araneus* grasping a single thread. (A from Foelix 1982, reprinted with permission of Harvard University Press; B redrawn from Foelix 1982, after slow-motion pictures by Parry and Brown 1959; C after Levi and Levi 1968; D redrawn from Foelix 1982, after Frank 1957; E redrawn from Foelix 1982, after Foelix 1970; F courtesy of R. F. Foelix.)

Thread

Thread

ities possess intricately fashioned "thread clamps" on some or all of the leg tips (Figures 12E and F).

Many spiders, especially small ones, also excavate burrows, and a few lycosids and pisaurids can walk on water (Figure 13). One species, *Argyroneta aquatica* (family Argyronetidae), actually lives under water, where it walks on submerged plants or swims. There are also several intertidal species that can tolerate submergence at high tides.

Feeding and digestion

Feeding. The basic chelicerate feeding strategy is one of prey capture followed by extensive external digestion and then ingestion of liquefied food or, more rarely, small particulate material. There are exceptions, of course, most of which involve drastic modifications of the mouth parts. For example, we have already mentioned the varied feeding habits among members of the order Acari (mites, ticks, and chiggers), many of which are herbivores or parasites

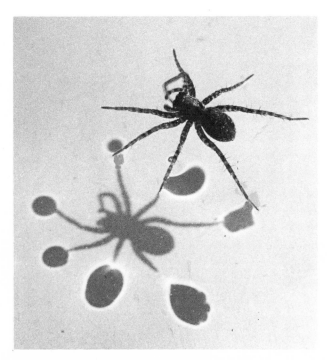

Figure 13

The wolf spider, *Pardosa amentata*, on the water surface. The right front leg is raised from the water surface for cleaning. Notice the unequal distribution of body weight on the rest of the legs, indicated by the shadows of the "dimples" in the water surface; the weight is normally almost equally distributed when the spider rests on all eight legs. (From Foelix 1982, reprinted with permission of Harvard University Press.)

and have piercing mouth parts. Whereas many chelicerates are highly specialized in terms of food-getting behavior, others are generalists. The horseshoe crabs, for example, commonly feed on a variety of invertebrates, including worms, molluscs, and other infaunal and epibenthic creatures; but they will also scavenge on almost any organic material they find. Food is gathered by any of the chelate appendages, passed to the gnathobases along the ventral midline, where it is ground to small bits, and then moved forward to the mouth.

Scorpions feed mostly on insects, although some large species occasionally eat snakes and lizards. Most are nocturnal and detect prey mainly by highly sensitive mechanoreceptors. A fascinating method of insect-prey location by the Mojave Desert sand scorpion, *Paruroctonus mesaensis*, has been described by Phillip Brownell. Brownell noticed that *Paruroctonus* ignores both airborne vibrations (e.g., wingbeats) and visual input but responds immediately to nearby prey that are in contact with the sand. This scorpion is even able to detect buried prey, which it immedi-

ately uncovers and attacks. Apparently the scorpion senses subtle mechanical waves set up in the loose sand by movements of the prey. Special mechanoreceptors in the walking legs are stimulated as the waves pass beneath the scorpion's limbs. The information is processed to determine the direction and approximate distance to the prey source.

Once a scorpion locates a victim, the prey is grasped by the chelate pedipalps. The opisthosoma is arched over the prosoma, bringing the telson and stinging apparatus into a position to inject the venom (Figure 14). Muscular contractions force the venom through a pore in the aculeus and into the prey. The venom is a neurotoxin that can quickly paralyze and kill most prey; in fact some scorpion venoms can even kill large animals, including humans. The most familiar of these dangerous scorpions is probably *Centruroides sculpturatus*, a common species in Mexico and the southwestern United States. A North African species, *Androctonus australis*, produces a venom considered to be as potent as that of cobras. Other genera of potentially dangerous scorpions include *Buthus* and *Parabuthus* (Africa and the Middle East) and *Tityus* (South America). A well written and superbly illustrated account of medically important scorpions has been compiled by H. L. Keegan (1980).

The sequence of events in the ingestion of food by scorpions is typical of many arachnids. After being captured and stung, the prey is passed to the chelicerae, which tear it into small bits. The gnathobases grind and mash the food as digestive juices are released through the mouth; this process reduces the food to a semiliquid form. As this organic "soup" is ingested, the hard parts are discarded and more bits of food are moved between the gnathobases for processing.

Virtually all spiders are predatory carnivores,* and with the exception of a few families (e.g., Salticidae, Oxyopidae, Thomisidae, and Lycosidae), all hunt or feed mainly at night. We can separate most spiders into two broad categories based on prey-capture strategies. First are the more sedentary spiders that use some sort of silken web, trap, or net to catch prey. Second are the "wandering" spiders that actively hunt or ambush prey without the direct use of silk (although many wrap their victims after capture).

We have already briefly described the construction of the familiar orb web. When a potential prey item, such as a flying insect, strikes and adheres to a web, its movements send out vibrations that alert

---

*There is growing evidence that young spiderlings in certain families may consume only pollen that is caught on web silk.

the spider to the presence of food. The spider then moves rapidly to the victim and bites it. Aranaeologists have likened the speed and grace of this attack in many species to that of mammalian predators such as cheetahs and leopards. Most spiders are solitary, but a few live and feed communally (Figure 15A).

Another type of silken trap is the horizontal **sheet web** produced by members of the families Linyphiidae and Agelenidae. The linyphiids spin a dome-shaped sheet web suspended by a network of supporting threads (Figure 15B). Insects become entangled in the supporting threads, and the spider (resting on the underside of the web) shakes the whole structure until the prey drops to the sheet. The spider bites through the web from below, pulls the prey through, and repairs the hole in the web. Agelenid sheet webs are flat, with a network of supporting threads. The spider sits in wait in a funnel-shaped home at the side of the web. Insects may hit the net directly or strike the supporting threads and fall to the sheet, whereupon the spider runs directly to the prey and captures it.

Most theridiids build vertical **frame webs** (Figure 15C). Near their attachment to the substratum, the vertical trap threads are beaded with sticky liquid from special silk glands. Walking prey contact these sticky droplets and become trapped. Upon sensing movement in the web, the spider rushes to the prey, wraps it in silk and bites it.

Members of the genus *Hyptiotes* (family Uloboridae) spin abbreviated orb webs of only three sectors (Figure 15D). The spider produces a tension thread from the point of convergence of the radii and a short attachment thread stuck to some solid object; the spider's body acts as a bridge between these two threads. When an insect strikes the web, the spider releases the tension thread and the web, called a "spring trap," collapses around the prey.

The most primitive kinds of "webs" are formed as simple silken tubes with a single opening from which threads radiate out (Figure 15F), a web form used by various lyphistiids. The spider resides in the tube and the threads serve as "fishing lines" or trip lines that allow the spider to detect cursorial prey. An interesting modification of this system is seen in the purse-web of *Atypus* (family Atypidae) (Figure 15G). Here the silken tube is mostly buried beneath the soil, with just a short portion of the blind end lying above the surface. Insects crawling over the exposed tube are detected by the spider, and the orthognathous chelicerae make two parallel slices in the tube wall near the prey. The cheliceral fangs are then extended through the incisions to grab the vic-

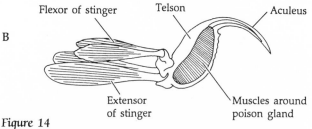

Flexor of stinger   Telson   Aculeus

Extensor of stinger   Muscles around poison gland

Figure 14

A, The scorpion *Androctonus australis,* applying its sting apparatus to a grasshopper while holding the prey with its pedipalps. B, The scorpion's telson and sting apparatus. The sting is normally held flexed; extension drives the spine into the prey. (A adapted from several sources; B after Meglitsch 1972.)

tim and pull it through the tube wall. After the prey is killed, the tear in the tube is repaired and the "trap" is ready again.

Most spiders simply set their "traps" and wait for prey, but others actually manipulate silken structures to catch insects. The Australian ogre-faced spider, *Dinopis* (family Dinopidae), produces a rectangular web of cribellate threads and holds it between its front walking legs. When an insect is visually detected, the spider sweeps the net around the prey. The bolas spider, *Mastophora* (family Araneidae), is among the most bizarre hunting specialists. While hanging from a suspension thread, the bolas spider "throws" a catching thread tipped with sticky liquid to "lasso" its prey (Figure 15H). Bolas spiders hunt at night and specialize in feeding on male moths of the genus *Spodoptera*. *Mastophora* releases an airborne chemical that mimics the sex pheromone of the female *Spodoptera*, thus attracting males within reach of the catch thread and greatly increasing the likelihood of capture success.

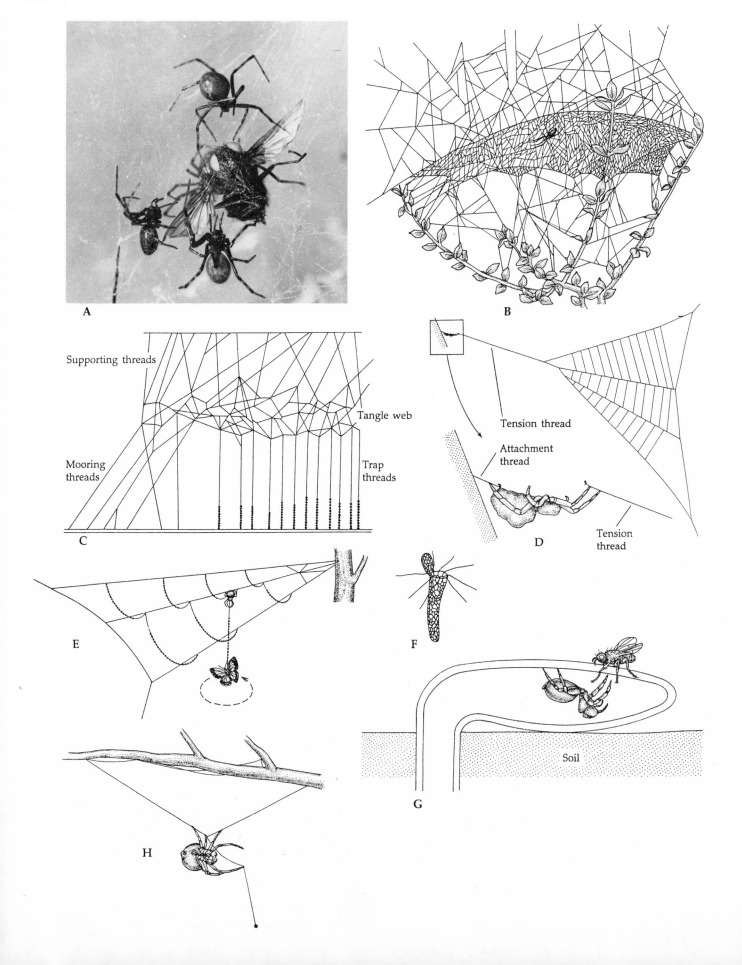

Supporting threads

Tangle web

Mooring threads

Trap threads

C

Tension thread

Attachment thread

Tension thread

D

A

B

E

F

G

Soil

H

Figure 15

Figure 15

The use of spider silk for prey capture. A, Several *Anelosimus* spiders and their prey, an unusual case of communal prey capture. B, The horizontal sheet web of *Linyphia triangularis*. C, The vertical frame web of *Steatoda castanea*. D, The abbreviated orb web of *Hyptiotes*. The spider is stretched between the attachment thread and the tension thread. E, *Pasilobus* "reeling in" an insect caught on a sticky thread. F, The silken tube and trip threads ("fishing lines") of a primitive spider, *Liphistius* (suborder Mesothelae). G, The purse-web spider *Atypus* inside its silken tube. The spider has sensed the presence of an insect on the tube and is about to grab it. H, The bolas spider, *Mastophora*, swinging its capture thread. See text for explanations of these prey-capture devices. (A courtesy of R. F. Foelix; B,C,D,E,G redrawn from Foelix 1982, after various other sources; F after Kaston 1964.)

Whereas prey detection by most sedentary spiders is usually through sensing web vibrations, the "wandering" spiders use a variety of methods. Some wolf and most jumping spiders locate prey visually, while many others sense vibrations (e.g., wing beats or walking movements), or rely simply on accidental contact. Some actually chase prey; others, such as certain trap-door spiders, lie in ambush and wait for victims to come close enough to grab.

Regardless of the method of prey location and capture, once contact is made the spider pulls the victim to the chelicerae and bites it, inserting the fangs and injecting venom from poison glands within the prosoma (Figure 16A). The prey is quickly immobilized or killed. An interesting exception to this grabbing-and-biting pattern is displayed by the spitting spiders (family Scytodidae). The poison glands of *Scytodes* include a glue-producing portion along with the usual venom-secreting cells. A mixture of venom and glue is shot out of pores in the cheliceral fangs by muscular contraction of the glands; thus the prey is captured without direct contact (Figure 16B). Many spiders wrap their prey in silk to some extent prior to feeding, even if silk is not used in actual capture. Many hold the victim and wrap it prior to biting (e.g., theridiids and araneids). Very active insects caught in orb webs are generally wrapped immediately, thereby preventing possible damage to the web or its owner. Potentially dangerous prey, such as large stinging insects, are generally handled in this manner.

Nearly all spiders possess poison glands that produce proteinaceous neurotoxins. The toxicity of spider venom is quite variable, and only about two dozen species are considered dangerous to people. Among these are the American black widow (*Latrodectus mactans*, Figure 5D), a Brazilian wolf spider (*Lycosa erythrognatha*), the brown recluse spider (*Loxosceles reclusa*), an Australian funnel-web spider

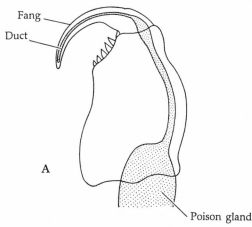

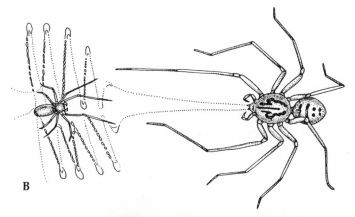

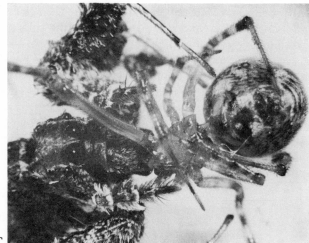

Figure 16

More aspects of prey capture in spiders. A, A spider chelicera. Note the poison gland and duct leading to the tip of the fang. B, The spitting spider, *Scytodes*, captures prey by spraying a combination of poison and adhesive over its victim. C, The salticid *Portia* capturing a cob-web spider. (A redrawn from Foelix 1982, after Millot et al. 1949; B after Foelix 1982; C from Foelix 1982, reprinted with permission of Harvard University Press.)

(*Atrax robustus*), and some species of hunting spiders belonging to the family Ctenidae (e.g., *Phoneutria fera*).

Spiders, like most other chelicerates, ingest their food in a liquid or semiliquid form. The chelicerae of most spiders have dentate gnathobases with which the prey is mechanically pulverized, while at the same time the food is flooded with digestive juices. Except for the hard parts, the prey is thus reduced to a partially digested broth. Bristles bordering the mouth and thousands of overlapping cuticular plates in the pharynx serve as filters so that only very small particles (<1 μm) enter the gut. Members of some families (e.g., Theridiidae and Thomisidae) lack cheliceral teeth. These spiders only puncture their prey and then flush digestive juices in and out of the wound. The liquefied innards of the victim are then sucked out of its body and ingested.

The digestive tract. The digestive system of chelicerates is constructed along the basic arthropod plan of foregut, midgut and hindgut, the first and last parts being lined with cuticle (Figure 17). The foregut is often regionally specialized. In the xiphosurans (e.g., *Limulus*), the foregut loops anteriorly as an esophagus, crop, and gizzard, the last part bearing sclerotized ridges that grind ingested particles (Figure 17A). In many arachnids portions of the foregut are modified as pumping organs for sucking in liquefied food. In scorpions this function is served by a muscular pharynx, and in spiders by an elaborate sucking stomach (Figure 17). The pharynx of spiders may contain chemosensory cells that function as "taste" receptors.

The chelicerate midgut bears paired digestive ceca and is the site of final chemical digestion and absorption (Figure 17C). Xiphosurans have two pairs of ceca arising from the anterior part of the midgut, followed by an intestine, short rectum (the hindgut), and the anus on the posterior margin of the opisthosoma. In *Limulus*, enzymes are produced by the midgut wall and secreted into the lumen. Apparently only preliminary protein digestion occurs extracellularly, and final breakdown takes place in the cells of the digestive ceca after absorption.

Most spiders possess four pairs of digestive ceca in the prosoma and frequently additional branched ceca in the opisthosoma (Figures 17B and C). The midgut expands as a spacious **stercoral pocket** near its junction with a short rectum. Malpighian tubules arise from the midgut wall near the origin of the stercoral pocket. The anus is located on the opisthosoma near the spinnerets.

The midgut of scorpions bears six pairs of digestive ceca (Figure 17D). The first pair, called salivary glands, lies within the prosoma and produces much of the digestive juice used in preliminary external digestion. The remaining five pairs are highly convoluted and lie in the opisthosoma. These produce the enzymes for final digestion and are the site of absorption of the digestive products. Two pairs of Malpighian tubules arise from the posterior region of the midgut, just in front of the short rectum. The anus is on the last opisthosomal segment.

Circulation and gas exchange

As in other arthropods, the chelicerate circulatory system comprises a dorsal ostiate heart situated within a pericardial sinus and giving rise to various open-ended vessels (Figures 17 and 18). The blood leaves these vessels and enters the hemocoel, where it bathes the organs and supplies the structures of gas exchange before returning to the heart. The complexity of the system is primarily a function of body size; some very tiny chelicerates (e.g., palpigrades and some mites) have lost much or all of their circulatory structures—for them, gas exchange is cutaneous. On the other hand, the xiphosurans are large animals, and their *Bauplan* demands a substantial circulatory mechanism to move the blood around inside the rigid body covering. The large tubular xiphosuran heart bears eight pairs of ostia and is attached to the body wall by nine pairs of ligaments that extend through the pericardium (Figure 17A). The organs of these big creatures are supplied with blood through an extensive arterial system arising from the heart and opening into the hemocoel close to the organs themselves. In the opisthosoma a major ventral vessel gives rise to a series of afferent branchial vessels to the book gills. Efferent vessels carry oxygenated blood to a large **branchiopericardial vessel** leading back toward the heart.

The gas exchange organs of xiphosurans are unique among the chelicerates. The presence of gills is, of course, associated with their aquatic lifestyle. The structure of these opisthosomal book gills provides an extremely large surface area, a necessity for adequate gas exchange by these large animals (Figures 2B and 18A). Each gill bears hundreds of thin lamellae, like pages in a book. The blood within the lamellae is separated from the surrounding sea water by only a thin wall. Water is moved over the lamellae by metachronal beating of the gills; these movements also cause blood flow into (on the forward strokes) and out of (on the backward strokes) the gill sinuses.

The heart of a spider lies within the opisthosoma

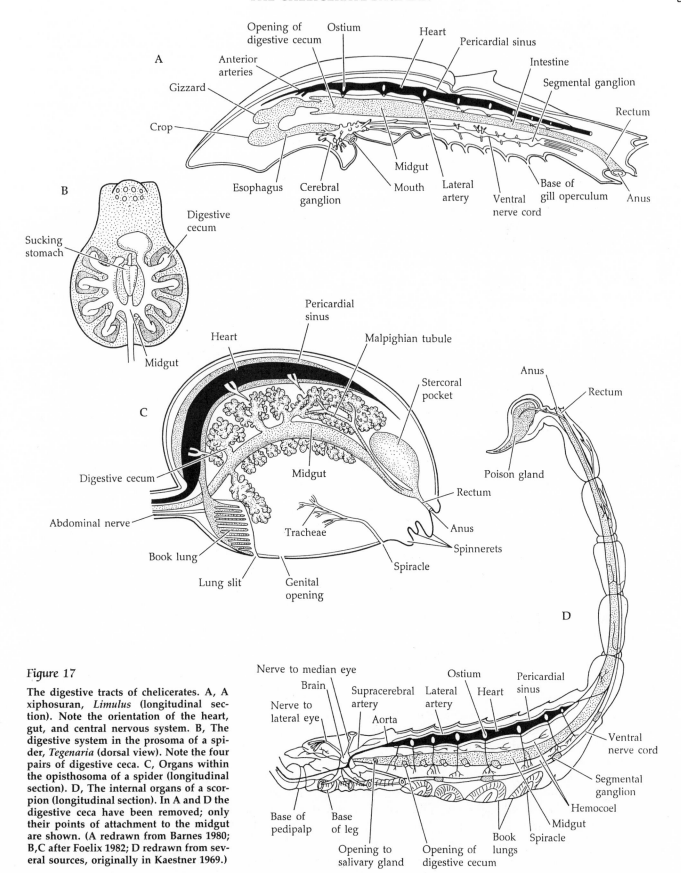

Figure 17

The digestive tracts of chelicerates. A, A xiphosuran, *Limulus* (longitudinal section). Note the orientation of the heart, gut, and central nervous system. B, The digestive system in the prosoma of a spider, *Tegenaria* (dorsal view). Note the four pairs of digestive ceca. C, Organs within the opisthosoma of a spider (longitudinal section). D, The internal organs of a scorpion (longitudinal section). In A and D the digestive ceca have been removed; only their points of attachment to the midgut are shown. (A redrawn from Barnes 1980; B,C after Foelix 1982; D redrawn from several sources, originally in Kaestner 1969.)

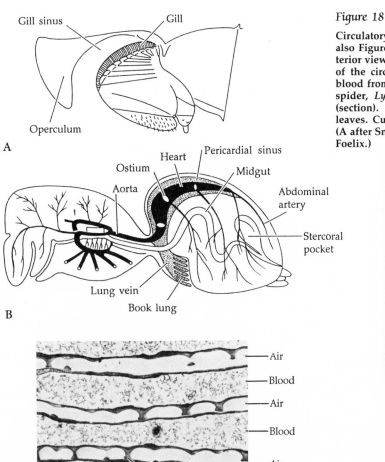

A

B

D

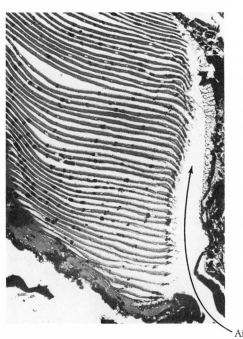

C

Figure 18

Circulatory and gas exchange structures in some chelicerates. See also Figure 17. A, An opisthosomal appendage of *Limulus* (posterior view). Note the gill and opercular parts. B, Major elements of the circulatory system in a spider. Note the direct route of blood from the book lung to the heart. C, The book lung of the spider, *Lycosa* sp. (section). D, The "leaves" of a book lung (section). Note the separation of air spaces and blood-filled leaves. Cuticular pedestals keep the air spaces from collapsing. (A after Snodgrass 1952; B after Foelix 1982; C,D courtesy of R. F. Foelix.)

and bears two to five pairs of ostia. It is suspended within the pericardial sinus by several ligaments attached to the inside of the exoskeleton (Figure 18B). These suspensory ligaments are stretched during systole as blood is pumped from the heart into arteries. The elasticity of the ligaments then effects diastole, expanding the heart and drawing blood into it from the pericardial sinus. The routes of the major arteries ensure that an ample supply of oxygenated blood reaches the major organs, particularly the central nervous system, muscles, and hemocoelic spaces in the limbs where blood pressure aids in leg extension. From the hemocoel, blood is channeled back to the pericardial sinus and the heart.

Gas exchange structures in spiders include book lungs and tracheae, organs adapted to life on land. Generally, the more primitive spiders (suborder Mesothelae and superfamily Mygalomorphae) possess two pairs of book lungs but not tracheae, whereas members of the superfamily Araneomorphae usually

have one pair of book lungs and a system of tracheal tubes. Since the tracheae occur only in advanced groups among spiders, it is likely that they evolved separately from those of uniramians and represent a convergent feature in the two subphyla.

Spider book lungs are located in the second or the second and third segments of the opisthosoma. They open to the outside via spiracles, or **lung slits**, near the epigastric furrow (Figure 5C). Since they are localized and do not extend far into the body, book lungs must receive sufficient circulating blood to ensure adequate distribution of oxygen throughout the body and removal of carbon dioxide from the internal organs. Just inside each lung slit is an expanded chamber, the **atrium**, from which numerous flattened air spaces extend into the hemocoel (Figures 18C and D). These leaflike pockets of air are separated from one another by thin blood-filled extensions of the hemocoel. Although the book lungs are themselves relatively small, this structural arrangement provides a very large surface area between the "pages" of the book lungs and the circulatory fluid. Blood that has passed these surfaces returns directly to the pericardial sinus via the **lung vein** (Figure 18B).

Spider tracheae open to the outside through one or two spiracles located posteriorly on the third opisthosomal segment (Figures 5B, 5C, and 17C). They probably evolved from the book lungs and muscle apodemes on this segment in more primitive spiders. The spiracles open inward to simple or branched tubes. In spiders the open inner ends of the tracheae do not bring the oxygen supply into direct contact with tissues as they do in many insects; rather, a small amount of blood is necessary as a diffusion medium. When the tracheal system is extensive, there is usually a reduction in the structural components of the circulatory system.

The circulatory system of scorpions is directly comparable to that of spiders, except that it is constructed to accommodate an elongate body. The tubular heart bears seven pairs of ostia and extends through most of the mesosoma (Figure 17D). An extensive set of arteries delivers blood to the hemocoel throughout the body and to four pairs of mesosomal book lungs.

Except where reduced or absent, the circulatory system of other chelicerates is built along the same general plan as described above. Gas exchange is cutaneous in the palpigrades and some mites, but in others it occurs through book lungs (Uropygi, Schizomida, and Amblypygi) or tracheae (Ricinulei, Pseudoscorpionida, Solpugida, Opiliones, and Acari).

The blood of chelicerates has been most extensively studied in spiders and in xiphosurans; little is known about the circulatory physiology of scorpions. Because *Limulus* is large and readily available, its blood chemistry has been very well studied, and many a horseshoe crab has made the supreme sacrifice for science at the hands of laboratory physiologists. Hemocyanin, the common respiratory pigment in chelicerates, is dissolved in the blood plasma. At least in spiders, the hemocyanin serves primarily for oxygen storage rather than for immediate transport and delivery of oxygen to the tissues. Hemocyanin has a very high affinity for oxygen, releasing it only when surrounding oxygen levels are low. At least some spiders are able to survive for days after their air supplies have been cut off experimentally by covering the spiracles. Apparently they obtain sufficient oxygen from their hemocyanin-bound stores and some cutaneous exchange.

Chelicerate blood also contains various cellular inclusions, but the functions of most of them are not well understood. The blood of *Limulus* includes amebocytes that may provide clotting agents. Several kinds of blood cells occur in spiders. Interestingly, it seems that all of these originate as undifferentiated cells from the muscular portion of the heart wall; these cells are released into the blood, where they mature and differentiate. Functions that are attributed to the blood cells include clotting, storage, combating infections, and aiding in sclerotization of the cuticle.

Excretion and osmoregulation

The excretory structures of chelicerates are coxal glands and Malpighian tubules, although many groups possess additional supplementary waste-removal mechanisms.

Xiphosurans have two sets of four coxal glands each, arranged along each side of the prosoma near the coxae of the walking legs. The glands on each side of the body converge to a coelomic sac from which a long, convoluted duct arises. The duct leads to a bladder-like enlargement that connects to an excretory pore at the base of the last walking legs. Surprisingly little is known about excretory physiology in xiphosurans. Apparently the coxal glands extract nitrogenous wastes from the surrounding hemocoelic sinuses and carry them to the outside. The coxal glands and their associated tubule system also function in osmoregulation, as evidenced by the production of a dilute urine when the animal is in a hypotonic medium. The digestive ceca probably aid in excretion of excess calcium by removing it from the blood and releasing it into the gut lumen.

The problems of excretion and water balance are obviously much more critical to terrestrial chelicerates than to horseshoe crabs, and land-dwelling arachnids display a variety of structural, physiological, and behavioral adaptations in this regard. Coxal glands persist in many arachnids (spiders, scorpions, palpigrades); in these animals the glands lie within the prosoma and open on the coxae of certain walking legs. The degree to which coxal glands function in excretion and osmoregulation varies among arachnids, but they are considered much less important than the Malpighian tubules.

The Malpighian tubules of arachnids (spiders, scorpions, ricinuleids, and so on) arise from the posterior midgut; thus they are not homologous with the Malpighian tubules of uniramians, which arise from the hindgut and are thus of ectodermal origin. The tubules branch within the hemocoel of the opisthosoma, where they actively accumulate nitrogenous waste products and release them into the gut for elimination along with the feces (Figure 17C). In spiders, the wastes from the tubules and the gut are mixed in the stercoral pocket prior to release from the anus. The excretory action of the Malpighian tubules is often supplemented by other mechanisms, such as the coxal glands. Nitrogenous wastes also are accumulated in the cells of the midgut wall and released into the lumen. In addition, waste products are picked up and stored by special cells called **nephrocytes**, which form distinct clumps in various parts of the prosoma.

The excretory physiology of terrestrial arachnids results in the formation of complex, insoluble, nitrogen-containing compounds of low toxicity. These compounds can be stored and eliminated in semisolid form, thus conserving water. The major excretory product is guanine, although uric acid and other compounds also occur.

Terrestrial arachnids also display various behavioral adaptations to avoid desiccation. Most arachnids are nocturnal, remaining in cooler or more humid, protected places during the daytime. Some spiders actively drink water during dry periods or when they lose blood through injury. Desert scorpions must tolerate not only low humidities but also extreme elevated daytime temperatures. Desert scorpions typically bury themselves in sand during the day. In addition, some species exhibit an adaptive behavior called **stilting**, wherein they raise their bodies off the substratum and allow air to circulate under them. While thought to be mainly a cooling device, the lowered body temperature probably also slows the rate of evaporative desiccation. Some scorpions are also able to withstand large losses of body water—as much as 40 percent of their body weight—with no ill effects.

Nervous system and sense organs

As in all arthropods, the external body form of chelicerates is generally reflected in the structure of the central nervous system. These animals show various degrees of compaction and fusion of the body somites and the associated nervous system while still conforming to the basic segmental annelid–arthropod plan. The cerebral ganglia, or brain, includes the protocerebrum and the tritocerebrum, the deutocerebrum being absent. The tritocerebrum generally contributes to the circumenteric connectives, which unite ventrally with a large ganglionic mass formed in part by fusion of paired anterior ganglia of the ventral nerve cord. In xiphosurans and scorpions this subenteric neuronal mass includes all of the prosomal ganglia, whereas in spiders even the opisthosomal

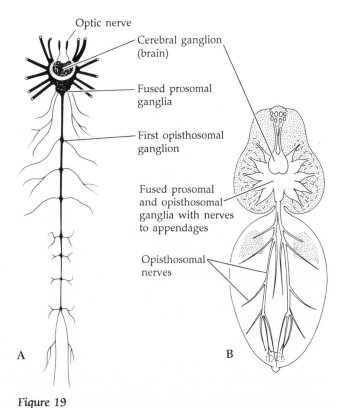

Figure 19

The central nervous system of a scorpion (A) and a spider (B). See also Figure 17. (A after Millot et al. 1949; B after Foelix 1982.)

ganglia fuse anteriorly (Figures 17 and 19). Thus in most spiders the adult nervous system is no longer obviously segmented (except in some members of the suborder Mesothelae), although a chain of ventral ganglia is evident during development. The ventral nerve cord persists in the opisthosoma of xiphosurans and has five segmental ganglia; in scorpions it has seven ganglia.

The protocerebrum and tritocerebrum give rise to nerves to the eyes and chelicerae, respectively. In spiders the cheliceral nerves actually emerge from the protocerebrum, but they can be traced histologically to their origin in the hind part of the cerebral ganglia below the gut. The ventral ganglionic mass, which includes the fused segmental prosomal ganglia, gives rise to nerve tracts to the walking legs and, in spiders, bears a pair of abdominal ganglia from which arise branching nerves to the opisthosoma. Segmental ganglia on the ventral nerve cord in xiphosurans and scorpions serve the opisthosomal appendages, muscles, and sense organs.

In Chapter 15 we discussed some of the qualities of arthropod sense organs in terms of the imposition of the exoskeleton. Xiphosuran sense organs include tactile mechanoreceptors in the form of various spines and bristles; proprioceptors in the joints; chemoreceptors; and photoreceptors. The prosoma bears two simple eyes near the dorsal midline, and two compound eyes placed laterally (Figure 2). The median eyes are pigment cups, but each contains a distinct cuticular lens. The lateral eyes of xiphosurans are compound rhabdomeric units—structures not found in any other members of the Cheliceriformes. These compound eyes differ somewhat from those of crustaceans and insects in that the ommatidia are rather loosely packed together. Each ommatidium includes a dozen or so light-sensitive retinular cells grouped around the usual rhabdome. These cells give rise to the sensory axons forming the optic nerves. The degree of resolving power of xiphosuran eyes has long been debated. Certainly they can detect movement and changes in light intensity and direction. They may also be capable of perceiving clear images, as male horseshoe crabs have been experimentally shown to be attracted to "models" of females.

Arachnids possess well developed sense organs upon which much of their complex behavior depends. Most of the "hairs" on a spider or scorpion body are mechanoreceptors, collectively called **hair sensilla**. **Simple tactile hairs** (or setae) cover much of the body surface and respond to direct physical contact. A second type, called **trichobothria**, are found on the appendages. They are much thinner than simple tactile hairs and are extremely sensitive (Figures 20A and B). Trichobothria are stimulated by airborne vibrations, such as those caused by beating insect wings, natural air currents, and probably some sound frequencies.

Additional mechanoreceptors of arachnids include **slit sense organs** (Figure 20C). These structures occur as single **slit sensilla** or in groups of parallel slits called **lyriform organs**. Slit sensilla are deep grooves in the cuticle associated with sensory neurons. They detect a variety of mechanical stimuli that impose physical deformation on the cuticle around the slit. Depending on their location and orientation, spider slit sensilla serve as proprioceptors (by sensing leg movements and position), as georeceptors (by measuring bending of the pedicel under the weight of the opisthosoma), as direct mechanoreceptors (by sensing direct external pressure on the cuticle), as vibration sensors, and even as phonoreceptors.

Efficient proprioceptors occur in the walking legs and pedipalps of all chelicerates, and are particularly well developed in arachnids. By virtue of their position and number in different joints, they convey information about the direction and velocity of appendage motion as well as the position of the limbs relative to the body and to each other. These "true" proprioceptors appear to work in concert with the lyriform organs.

Chemoreception in arachnids involves sensing both liquid and airborne chemicals that contact the body. This dual ability can be likened to capacities for both taste (contact chemosensitivity) and smell, or olfaction (distance chemosensitivity). This olfactory sense plays major roles in mating for species in which the females release sex pheromones, and in prey location. The most important chemoreceptors are probably the hundreds of erect, hollow hairs with open tips that are present on the pedipalps and other areas around the mouth and are most abundant on the tips of limbs that contact the substratum. Dendrites of sensory neurons extend through the hollow hair shaft to the open tip, where they are directly stimulated by chemicals (Figure 20D).

Scorpions bear a pair of large, unique, comblike structures, the pectines, on the ventral surface of the mesosoma (Figure 3). After detailed studies of innervation, Foelix and Schabronath (1983) suggested that the pectines act as both mechanoreceptors and chemoreceptors. Other work has shown them to be capable of detecting subtle differences in sand grain size. These versatile structures are usually held laterally erect, free to swing back and forth while the scorpion is actively moving about.

The importance of vision varies greatly among different arachnids, although most forms possess some sort of photoreceptor. At least some species in certain groups are blind (e.g., some members of the orders Schizomida, Palpigradi, Ricinulei, Pseudoscorpionida, and Acari). Certain spiders depend on photoreception for prey and mate location, particularly the errant hunters (some lycosids and most salticids). Vision is relatively unimportant to many sedentary species, such as many web builders, which depend more on tactile cues and chemosensitivity. Web builders are not blind, however; many respond behaviorally to variations in light intensity, and some exhibit distinct escape responses when they visually detect potential predators. Nevertheless, these spiders are generally able to build their webs, mate, and capture prey with little or no visual input. Some spiders are capable of perceiving polarized light, presumably as a means of spatial orientation.

Spiders possess single-lensed rhabdomeric eyes, but the sensory units of each are simple ocelli in a cluster. Thus they are quite different from the compound eyes of xiphosurans and other arthropods. Each eye includes a thickened **cuticular lens** over a

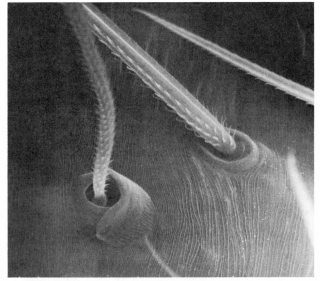

A

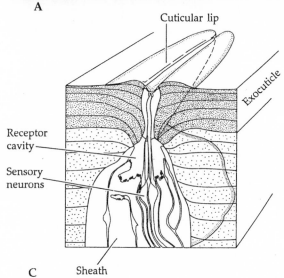

Cuticular lip

Exocuticle

Receptor
cavity

Sensory
neurons

C   Sheath

Figure 20

Mechanoreceptors and chemoreceptors in spiders. A, Two types of mechanoreceptors on the leg of a wolf spider, *Lycosa* sp. (scanning electron micrograph, 1250×). The thicker hair is a direct-contact tactile receptor; the thinner, curved one is a trichobothrium. B, The organization of a trichobothrium in its socket; from *Tegenaria*. C, A slit sense organ from the leg of *Cupiennus* (cutaway view). D, A chemosensitive hair. The mechanoreceptor neurons terminate at the base of the hair (arrow), whereas the chemoreceptor neurons extend through the hollow shaft to the tip. (A from Foelix 1982, reprinted with permission of Harvard University Press; B redrawn from Foelix 1982, after Görner 1965; C redrawn from Foelix 1982, after Barth 1971; D after Foelix and Chu-Wang 1973.)

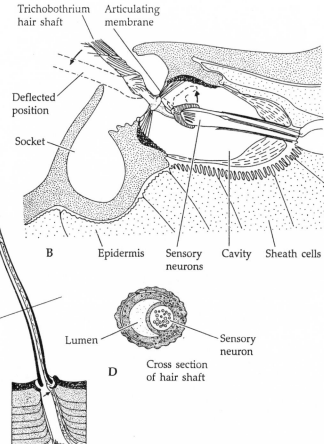

Trichobothrium
hair shaft

Articulating
membrane

Deflected
position

Socket

B   Epidermis   Sensory   Cavity   Sheath cells
                              neurons

Lumen   Sensory
               neuron
          Cross section
D   of hair shaft

vitreous body, which is a layer of cells derived from the epidermis and which covers the **retina** (Figure 21B). The retina is composed of the **sensory (receptor) cells** (ocelli) and the **pigment cells**. The membranes of the sensory cells bear interdigitating microvilli, structures confirming the rhabdomeric nature of the eyes.

Two forms of this basic eye structure occur among spiders. The anterior median eyes—the **main eyes**—have the light-sensitive portions of the sensory cells directed toward the lens, whereas other eyes, **secondary eyes**, are inverted, with the light-receptor elements directed away from the lens (Figure 21). Most of these secondary eyes contain a crystalline reflective layer, called the **tapetum**, which may serve to collect and concentrate light in poorly lit

conditions (e.g., during night hunting). The reflective nature of the tapetum produces the effect in some spiders of eyes that "shine in the dark."

Scorpion eyes are of the direct type and differ from spider eyes in having the retinal layer external to the epidermis. Most work suggests that scorpions depend much more on mechano- and chemoreception than on visual input.

There is no question that spiders, and perhaps other arachnids, are capable of modifying their be-

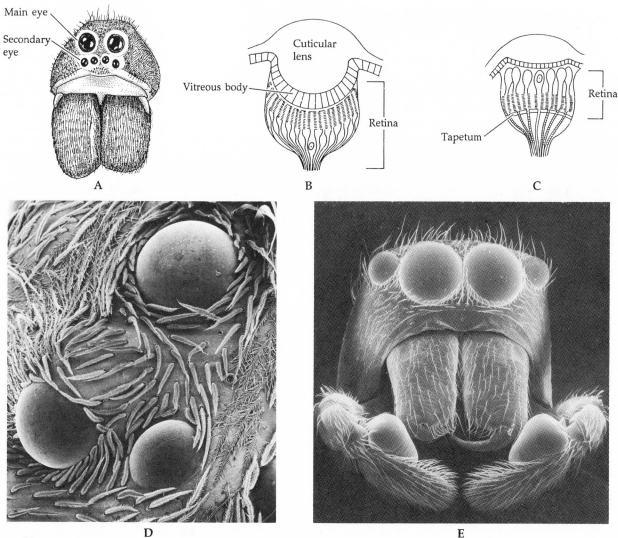

Figure 21

Spider eyes. A, Head-on view of a wolf spider. Note the eye size and position. B, Section through a main eye. C, Section through a secondary eye, showing inverted arrangement of the retinular cells and tapetum. D, Three eyes (and tactile hairs) on the face of the nursery-web spider, *Pisaura*. E, "Looking into the eyes" of *Heliophanus* (family Salticidae). (A redrawn from Barnes 1980, after Kaston, 1978; B,C redrawn from Foelix 1982, after Hoffman 1971; D courtesy of R. F. Foelix; E from Foelix 1982, reprinted with permission of Harvard University Press.)

havior based on experience—that is, they can learn. Memory and association centers in the protocerebrum are responsible for much of this integrative activity. No doubt, at least in spiders, their ability to remember, learn, and make appropriate adaptations in behavior has played an important role in their evolutionary success. The fascinating field of spider behavior is admirably presented in a collection of papers edited by Witt and Rovner ("Spider Communication," 1982).

Reproduction and development

Chelicerates are gonochoristic and generally engage in complex mating behaviors that ensure fertilization. A few are known to be parthenogenetic (e.g., some scorpions and schizomids). Males with a penis occur only in opilionids and some acarids. Free spawning never occurs, and fertilization takes place either internally or as the eggs leave the body of the female. The eggs are generally very yolky; and, except for the xiphosurans, the developmental strategy is direct in spite of the various juvenile stages through which most chelicerates pass prior to maturation. We first present a summary of reproduction and development of *Limulus* and then of arachnids, concentrating again on spiders and scorpions.

The reproductive systems of xiphosurans are similar in males and females. In both sexes the gonad is a single, irregularly branched mass of tissue (Figure 22). Paired gonoducts lead from the gonad to a pair of pores near the ventral midline. The first pair of opisthosomal appendages lies over the gonopores, forming a **genital operculum**.

At the onset of the breeding season, horseshoe crabs migrate into shallow water of protected bays and estuaries. On the east coast of North America, this migration takes place in the spring and summer, and huge numbers of *Limulus* can be seen in suitable areas in preparation for mating. Mating is initiated when the male climbs aboard the back of the female and grasps her with his modified first walking legs. The clasped pair moves to the high water area, usually at a high spring tide, and the female excavates one or more shallow depressions in the sand and deposits her eggs (2,000 to 30,000 eggs per mating). The male releases sperm directly onto the eggs as they are deposited, then the mates separate and the fertilized eggs are covered with sand.

Early development takes place in the sand "nests." Cleavage is holoblastic, producing a stereoblastula with most of the yolk contained in the inner cells. As development continues, the surface cells at the anterior and posterior ends of the embryo divide rapidly, forming two germinal centers. Some of these rapidly proliferating cells migrate inward as the presumptive entoderm and mesoderm. The anterior germinal center gives rise to the first four segments of the prosoma, and the posterior center to the rest of the body. All of the prosomal segments fuse and are eventually covered by the developing carapace-like dorsal shield.

As the yolk reserves become depleted, the embryo emerges from the sand as a **euproöps larva** (or "trilobite larva"), so named because of its resemblance to the fossil Coal Age xiphosuran *Euproöps* (which superficially resembles trilobites; see Figure 22). The larvae swim about and periodically burrow in the sand. Segments are formed and appendages added through a series of molts until the adult form is reached. We view this developmental pattern as a mixed life history. The early developmental stages are supplied with an investment of yolk by the female and are protected by the nest constructed by the parent, but the young emerge as independent feeding larvae prior to maturation.

The reproductive biology of arachnids is directly related to their success on land. They have evolved a variety of sophisticated mating behaviors, clever methods of sperm transfer, and various devices for protecting developing embryos, thereby ensuring successful procreation in terrestrial habitats. A comparison of the functional anatomy of the reproductive systems in spiders and scorpions provides a background for discussing courtship behavior and developmental patterns.

The male reproductive system of spiders consists of a pair of coiled tubular testes in the opisthosoma, which lead to a common sperm duct that opens into the epigastric furrow (Figure 23A). Each developing sperm usually bears a distinct flagellum with an odd

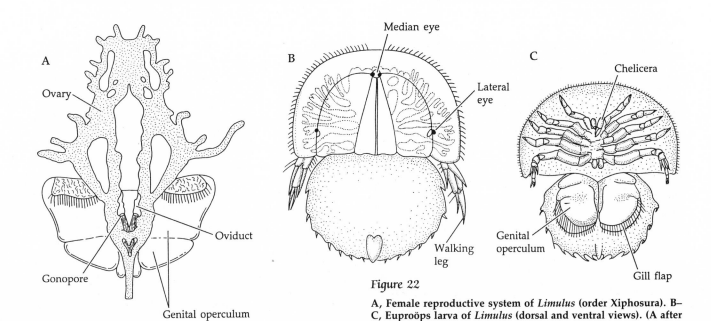

Figure 22

A, Female reproductive system of *Limulus* (order Xiphosura). **B–C**, Euproöps larva of *Limulus* (dorsal and ventral views). (A after Fage 1949; B adapted from several sources; C after Kaestner 1968.)

Figure 23

Spider reproduction. A, Male reproductive system of the tarantula *Grammostola*. The glandular mass near the gonopore is called the ventral spinning field, because it produces the sperm web. B, Sperm of *Oxyopes*. Note the unusual 9 + 3 arrangement of axial filaments in the flagellum. C, Encapsulated form of sperm. D, Simple male pedipalp copulatory structure (palpal organ) (*Segestria*). E, Complex palpal organ (*Araneus*). F, A male *Tetragnatha* on its sperm web, drawing sperm into its palpal organs. G, Female spider reproductive system. H, Seminal receptacle of a lycosid (section), showing sperm inside. (A redrawn from Foelix 1982, after Melchers 1964; B redrawn from Foelix 1982, after Osaki 1969; C,D,F after Foelix 1982; E redrawn from Barnes 1980, after Millot et al. 1949; G after Kaestner 1969; H courtesy of R. F. Foelix.)

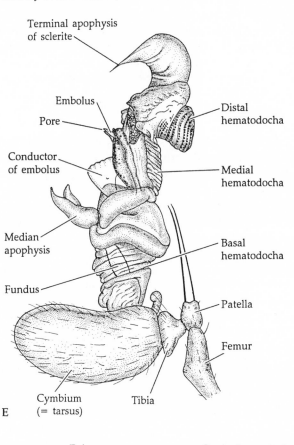

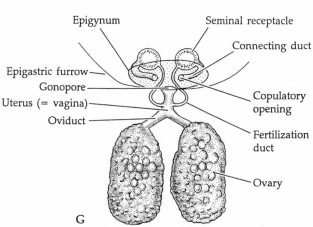

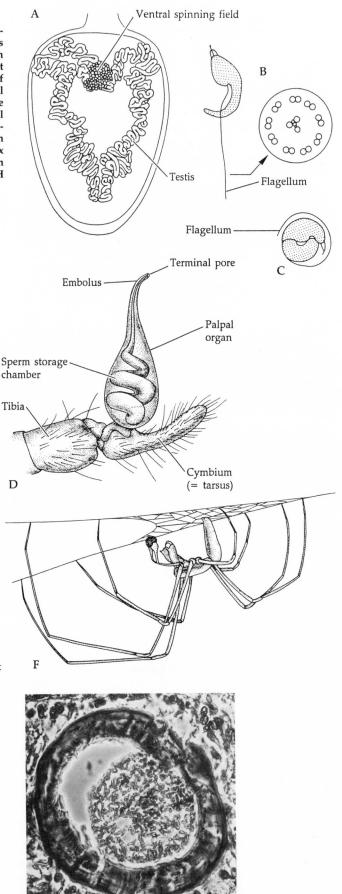

9 + 3 arrangement of axial filaments (Figure 23B). Prior to copulation, the flagellum wraps around the head of the sperm and a protein capsule forms around the gamete (Figure 23C). The sperm remains in this nonmotile state until after copulation.

Although male spiders, like other arachnids, lack a penis, the pedipalps are modified for sperm storage and transfer and serve as copulatory organs. Sperm released from the male gonopore are placed on a specially constructed, silken **sperm web** (Figure 23F). From here the sperm are picked up by the pedipalps, where they are held in special pouches or chambers and eventually transferred to the female. The pedipalps of male spiders vary greatly in form and complexity, generally being simple in certain mygalomorphs and more complex in araneomorphs. In their simplest form each pedipalp bears on its tarsus (called the **cymbium**) a teardrop-shaped process known as the **palpal organ** (Figures 23D and E). A pointed tip, or **embolus**, bears a pore that leads inward to a blind coiled sperm storage chamber. Sperm are drawn into this tube from the sperm web and held until transferred to the female.

The more complex palpal organs are formed of hard and soft parts called **sclerites** and **hematodochae**, respectively; the sclerites bear processes called **apophyses** (Figure 23E). The apophyses are variously shaped in different species of spiders, and some workers believe that each particular architecture has evolved to fit with the mating organs of conspecific females as a mechanism that prevents interspecific mating. Other specialists support the idea that nonproductive encounters are prevented primarily by species-specific courtship behavior (discussed below). Probably both methods of reproductive isolation are utilized. Following the uptake of sperm through the embolus, and insertion of the palpal organ into the female's receiving structure (epigynum), the soft hematodochae are inflated with hemocoelic fluid, thereby causing an erection of the sclerites within the female parts. Once the partners are thus coupled, sperm are injected into the female's copulatory openings.

Female spiders possess a pair of ovaries in the opisthosoma. The lumen of each ovary leads to an oviduct, and the two oviducts unite to form a uterus (also called the vagina), which opens to the outside in the epigastric furrow (Figure 23G). Eggs are produced mainly on the exterior of the ovaries, giving them a bubbled texture; how they move to the internal ovarian lumen is not well understood.

Just inside, or lateral to, the female gonopore a pair of copulatory openings usually occurs, leading through coiled **connecting ducts** to paired seminal receptacles. A second pair of tubes, called the **fertilization ducts**, connects the seminal receptacles to the uterus (Figures 23G and H). Many spiders possess a complexly structured sclerotized plate just in front of the epigastric furrow. This plate, called the **epigynum**, extends over the genital pore and bears the copulatory openings to the seminal receptacles. The form of the epigynum, the position and length of the receiving pores and connecting ducts, and other external features provide a particular topography that matches the palpal organs of conspecific males. These differences in external anatomy, as well as body size and overall courtship behavior, result in a variety of species-specific copulatory positions among spiders. Once sperm are inside the seminal receptacles they are stored there until the female deposits her eggs, which may be months after copulation. At that time, the sperm pass through the fertilization ducts to join the ova during egg-laying.

The reproductive systems of scorpions lie within the mesosoma, and both testes and ovaries are in the form of interconnected tubules (Figure 24). The gonads are drained by lateral sperm ducts or oviducts. The sperm ducts bear various storage chambers (seminal vesicles) and accessory glands, and unite as a **genital chamber** just inside the gonopore on the first segment of the mesosoma. Certain of the accessory glands are responsible for the production of spermatophores. Each of the oviducts is enlarged as a genital chamber, or seminal receptacle, near their point of union with the gonopore.

With few exceptions, sperm transfer in arachnids is always indirect. That is, the sperm leave the body of the male and are then somehow manipulated into the body of the female or are deposited on the eggs outside the female's body. The only exceptions to this rule occur in members of the Opiliones and some Acari, where the male possesses a penis through which sperm pass directly to the reproductive tract of the female. In all other arachnids, the sperm are either placed into the female's body by modified appendages of the male or are placed on the ground in spermatophores and then retrieved by the female. The use of appendages to transfer sperm occurs in the orders Araneae (pedipalps), Uropygi (pedipalps), Ricinulei (third walking legs), and some members of Solpugida (chelicerae) and Acari (chelicerae or third legs). In other arachnids (orders Scorpionida, Schizomida, Amblypygi, Pseudoscorpionida, and many Solpugida and Acari), the males deposit spermatophores on the ground and the females pick them up. We hasten to point out, however, that the reproduc-

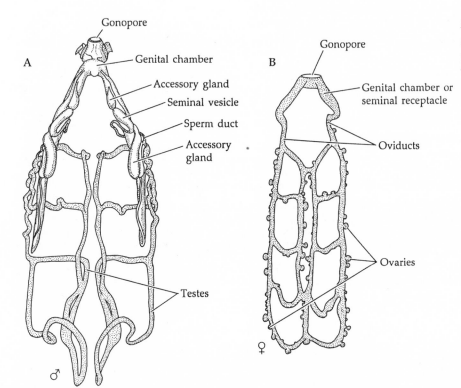

Figure 24

Scorpion reproductive systems. A, Male system in *Buthus*. B, Female system in *Parabuthus*. (After Millot et al. 1949.)

tive biology of many species in these taxa, and in some entire groups (e.g., Palpigradi), has been little studied.

The events leading up to insemination often include species-specific courtship behaviors that must, of course, be compatible with the particular type of sperm transfer (Figures 25 and 26). Again, we can look to spiders and scorpions for examples. Among spiders, these recognition behaviors not only ensure proper conspecific copulation, but also prevent the usually smaller males from being mistaken for prey by the females. Platnick (1971) has classified spider courtship behaviors into three general levels. First-level courtship involves necessary contact between male and female. Among many thomisids and clubionids, mating simply involves the male climbing over the female, positioning her abdomen, and inserting a palpal organ. The males of some thomisids (e.g., *Xysticus*) and at least one genus of araneid (*Nephila*) attach silk threads over the bodies or legs of the females preparatory to copulation. These threads are apparently only part of the recognition ritual, because they are not strong enough to actually restrain the female (Figure 25C). A few other spiders, including certain tarantulas, also use body contact and leg-touching as courtship behavior.

Second-level precopulatory activities involve the release of sex pheromones by the female spider. Some of the most complex behavior patterns occur in male spiders that detect females by olfaction, although other recognition devices may also be involved. Male araneids are apparently led to the female's orb web by her pheromones, and then the web is recognized by contact chemoreception. Once in touch with the edge of the web, the male announces his presence to the female by plucking the threads of the orb itself or a special **mating thread**, which he attaches to the web (Figure 25E). If properly orchestrated, the male's "tune" eventually attracts the female and contact is made.

Males of some species of wolf spiders (Lycosidae) respond to pheromones emitted along with the female's dragline. Upon detection of a female, the male begins a specific set of actions in an effort to win her favor, as the attraction changes from chemical to visual. These male behaviors involve abdomen bobbing and pedipalp waving, coupled with patterns of drumming the pedipalps on the substratum, and stridulation (Figure 25F). If attracted, the female responds to the visual and auditory cues by slowly approaching the male and sending out signals of her own in the form of particular leg movements. Stridulation, using modified pedipalps, also occurs in some uloborids (Figure 25G). Among the most inter-

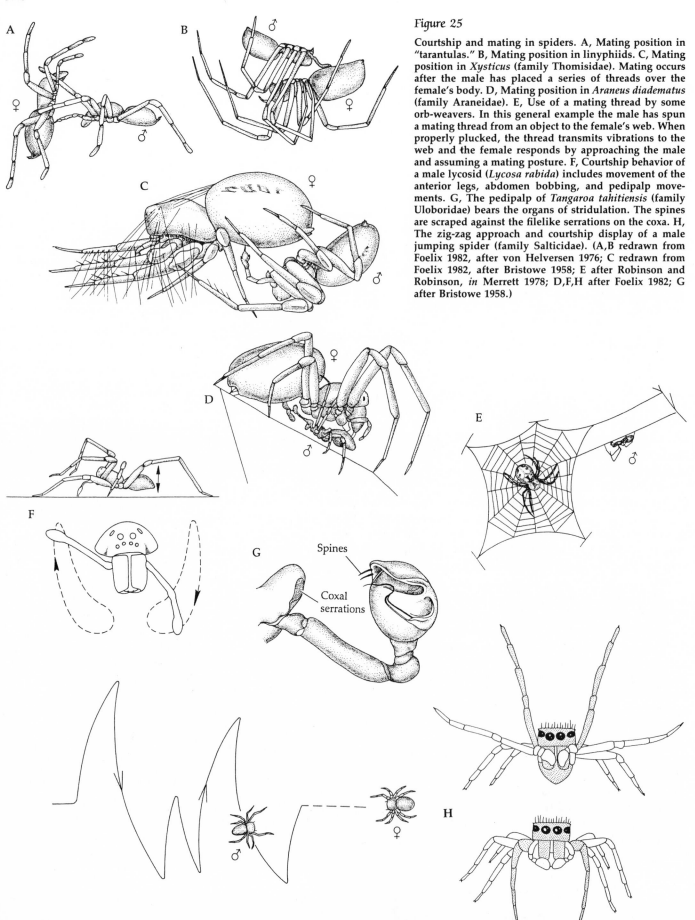

Figure 25

Courtship and mating in spiders. A, Mating position in "tarantulas." B, Mating position in linyphiids. C, Mating position in *Xysticus* (family Thomisidae). Mating occurs after the male has placed a series of threads over the female's body. D, Mating position in *Araneus diadematus* (family Araneidae). E, Use of a mating thread by some orb-weavers. In this general example the male has spun a mating thread from an object to the female's web. When properly plucked, the thread transmits vibrations to the web and the female responds by approaching the male and assuming a mating posture. F, Courtship behavior of a male lycosid (*Lycosa rabida*) includes movement of the anterior legs, abdomen bobbing, and pedipalp movements. G, The pedipalp of *Tangaroa tahitiensis* (family Uloboridae) bears the organs of stridulation. The spines are scraped against the filelike serrations on the coxa. H, The zig-zag approach and courtship display of a male jumping spider (family Salticidae). (A,B redrawn from Foelix 1982, after von Helversen 1976; C redrawn from Foelix 1982, after Bristowe 1958; E after Robinson and Robinson, *in* Merrett 1978; D,F,H after Foelix 1982; G after Bristowe 1958.)

Spines

Coxal serrations

esting second-level courtship behaviors is that of certain nursery-web spiders (*Pisaura*; Pisauridae). After locating a female emitting pheromones, the male captures an insect (usually a fly), spins a silk wrapping around it, and offers it to the female. Acceptance of the gift and of the male are one and the same, for the successful male copulates with the female while she devours the insect. Alas, unsuccessful males are eaten along with the offering.

Another interesting second-level precopulatory behavior has been documented for the Sierra dome spider (*Linyphia litigiosa*) of western North American mountains (Watson 1986). Upon encountering a mature virgin female, a male attacks her pheromone-laden web and packs it into a small tight mass. This behavior hinders the evaporation and dispersal of the male-attracting pheromone, thereby reducing the likelihood of a second male locating the female and competing for her favors.

Third-level courtship behaviors depend primarily on visual recognition between prospective mates and are best known in jumping spiders (Salticidae). A male locates a female and then begins a series of behaviors that identify him as a conspecific individual. Usually the male approaches the female along a zig-zag path and then performs specific movements of the opisthosoma, pedipalps, and front walking legs (Figure 25H). The female signals her approval and receptiveness by sitting still in a visually recog-

nizable position. The male eventually contacts her, caresses her briefly, mounts her, and copulates.

Sex-related behaviors among spiders are not restricted to male–female encounters. Conspecific males frequently exhibit agonistic behavior in competition for a mate. When males encounter one another in the presence of a female, or even in a mating "territory" (such as on the web of a female), they assume various threatening postures and in some cases actually engage in combat. Usually, however, one of the males retreats before any real damage is done, leaving the dominant male free to pursue his sexual interests.

The courtship behavior of scorpions does not involve direct copulation in any form but involves instead deposition of spermatophores onto the ground. Courtship behaviors appear to be relatively similar among those species that have been studied, although subtle species-specific differences allowing recognition undoubtedly occur. In a typical case the male initiates the ritual by grasping the female's pedipalps in his, and in this face-to-face position dances her around in a series of back-and-forth steps (Figure 26A). Eventually the male releases a spermatophore and cements it to the ground. He then continues to move the female around until she is precisely positioned with her genital operculum over the packet of sperm. The spermatophore, a complex, species-specific structure, bears a special process called an **open-**

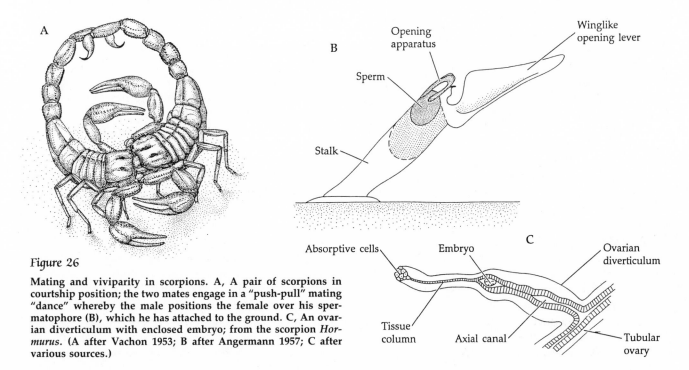

Figure 26

Mating and viviparity in scorpions. A, A pair of scorpions in courtship position; the two mates engage in a "push-pull" mating "dance" whereby the male positions the female over his spermatophore (B), which he has attached to the ground. C, An ovarian diverticulum with enclosed embryo; from the scorpion *Hormurus*. (A after Vachon 1953; B after Angermann 1957; C after various sources.)

ing lever (Figure 26B). The pressure of the female's body on this lever causes the spermatophore to burst, releasing the sperm, which can then enter her gonopore.

Arachnids undergo a variety of developmental patterns, all of which may be considered direct in terms of their life history strategies. Most species produce very yolky eggs, providing the embryos with nourishment through much of their development. By the time of hatching, many resemble miniature adults, or they still contain enough yolk to carry them through subsequent development to the juvenile stage. In some acarids and all ricinuleids, the young hatch as six-legged "larvae"; these immature individuals add the last pair of legs later through molting. In most arachnids the developing embryos are protected by some sort of egg case or cocoon, or they are brooded in or on the body of the female.

Nearly all spiders cement their eggs together in clusters and wrap them in silken cocoons, the sizes and shapes of which vary among different species (Figure 27). The cocoons provide physical protection for the embryos and also insulate them from fluctuations in environmental conditions, particularly changes in humidity. Additional protection results from placing the cocoon underground, in nests, or in other secluded spots. Some species camouflage their cocoons with bits of detritus, others guard their cocoons or actually carry them on their bodies.

Early development of spiders includes intralecithal nuclear divisions followed by migration of the nuclei to the periphery of the embryo. The nuclei are then isolated by cytoplasmic partitioning, a process that produces a periblastula around an inner yolky mass. Gastrulation follows by formation of a germinal center of presumptive entodermal and mesodermal cells that migrate inward. Additional germinal centers produce the precursors of segments and limbs. The immature individuals of different spider species hatch from their egg membranes at different stages, but they always remain inside the cocoon and utilize their yolk reserves until they are able to feed. Most workers recognize three postembryonic, preadult stages in spider development (Figures 27I, J, and K). The majority of spiders hatch from their egg membranes as immobile "prelarvae," characterized by incomplete segmentation and poorly developed appendages. The "prelarva" matures into a "larva" and then into a "nymph," which is the equivalent of a juvenile and physically resembles the adult. In some spiders these early developmental changes take place in a special **molting chamber** inside the cocoon (Figure 27B). (It is important to note that the terms here

Figure 27

Egg cases, hatching, and parental care in some spiders. A–B, formation and general structure of the cocoon of *Agroeca brunnea*. A, The female constructs the cocoon and coats it with bits of soil. B, The cocoon (section). Note the eggs above the molting chamber. C, The subterranean, silk-lined nest of a jumping spider (*Heliophanus cupreus*). D, Female *Theridion* tending a young one and feeding it with regurgitated food. E, A female *Pardosa* carrying her egg case. F, A female wolf spider, *Lycosa*, carrying young on her back. G, Surface of the cocoon of *Ero furcata*. H, A spider hatchling emerging from its egg case. I–K, Prelarval, larval, and nymphal forms of a spider. (A redrawn from Foelix 1982, after Lüters 1966; B,C redrawn from Foelix 1982, after Holm 1940; D redrawn from Meglitsch 1972, after Bristowe 1958; E,G,H from Foelix 1982, reprinted with permission of Harvard University Press; F courtesy of R. F. Foelix; I–K redrawn from Foelix 1982, after Vachon 1957.)

—prelarva, larva, nymph—do not carry the same meanings as they do when used to describe truly indirect development, where the larva is an independent, free-living individual.) The emergence from the cocoon usually occurs at an early "nymphal" stage, when the young are fully formed spiderlings. Many female spiders even engage in postnatal care, by carrying their young on their bodies or by feeding them (Figures 27D and F).

Scorpion development is ovoviviparous or viviparous. Viviparity is perhaps best studied in the Asian scorpion *Hormurus australasiae*. In this species the zygotes lie in tiny diverticula on the walls of the ovarian tubules (Figure 26C). Certain cells of the tubule wall absorb nutrients from the adjacent digestive ceca and supply them as nourishment to the developing embryos. The eggs of *Hormurus* contain very little yolk and undergo holoblastic, equal cleavage. Ovoviviparous scorpions produce yolky eggs that cleave meroblastically. The embryos are brooded in the ovarian tubules but depend on their yolk supplies for nutrients.

In both cases development is clearly direct, with obvious advantages for survival on land. The young eventually emerge from the female's gonopore and crawl onto her back. Here they stay until old enough to make periodic excursions away from their parent and finally assume an independent life. The juveniles molt through several instars until they mature, about one year after birth.

Many other arachnids also brood their embryos, usually externally on the female's body (e.g., Amblypygi, Uropygi, Ricinulei, Schizomida). Members of these groups carry their young in some sort of sac near the female gonopore. False scorpions spin cocoons from their cheliceral silk glands. Solpugids and opilionids are oviparous and deposit their eggs in the

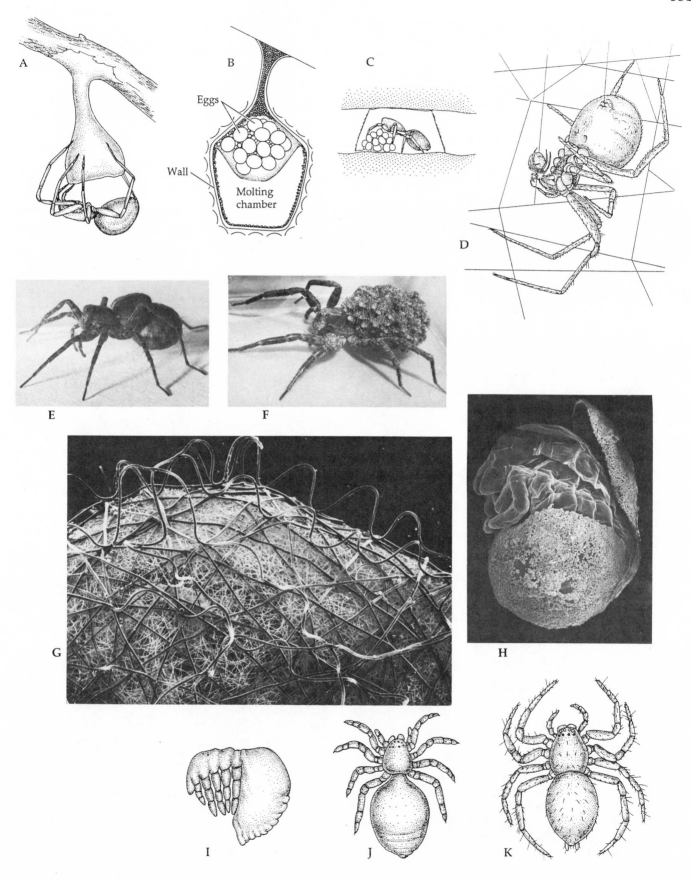

Eggs

Wall

Molting chamber

A   B   C   D

E   F

G   H

I   J   K

soil. In all cases the young hatch and pass through a few or many instars before they mature, and the maturation process may take several years. Again, except for the hexapod "larvae" of mites and ricinuleids, arachnids hatch as more-or-less small adults; and although various names are given to these immature stages, development is strategically direct.

The class Pycnogonida

Pycnogonids (Greek, *pyc*, "thick," "knobby"; *gonida*, "knees") are usually called sea spiders because of their superficial similarity to true, terrestrial spiders (Figures 28 and 29). The class Pycnogonida has been a problematical group in terms of its placement among the other arthropod taxa. Since the turn of the century, this group has been associated at one time or another with virtually every major group of arthropods, as well as with the onychophorans and polychaetes. The principal problem has been uncertainty concerning homologies of the various body regions and appendages. The unique pycnogonid

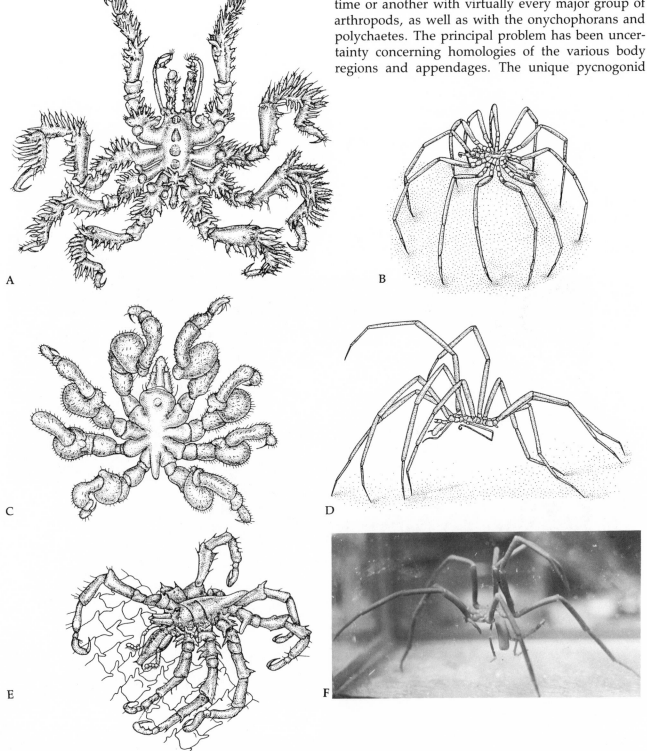

"proboscis," for example, has been "homologized" with everything from the prostomium of polychaete worms (Henry 1953), to the lips of onychophorans (Tiegs and Manton 1958), to various anterior regions of other arthropods.

Recent opinion tends to recognize the closest relationship to be that between pycnogonids and chelicerates, and most specialists have concluded that the pycnogonids probably arose as an early offshoot of the line leading to the chelicerates (e.g., Savory 1977; Bergstrom et al. 1980). Schram (1978) proposed that the two groups should be classified together as sister-groups in the subphylum Cheliceriformes. He postulated that several characters were synapomorphies shared between the chelicerates and pycnogonids: lack of a deutocerebrum; first appendages chelicerae/chelifores and second appendages pedipalps/palps (based on the assumption that these appendage pairs, and their somites, are indeed homologous; nervous innervation tends to support this contention); legs uniramous stenopods (with certain functional similarities); feeding method largely liquid/suctorial; and four postoral segments in the earliest embryonic stages. The pycnogonids also possess several strikingly unique features, synapomorphies not found in any chelicerate, or in any other arthropod group, such as the odd anterior "proboscis," **ovigers** (special appendages for brooding) in males, multiple gonopores (on the second coxal segment of some or all of the walking legs), and the unique body form discussed below.

All 1,000 or so species of pycnogonids are marine. They occur intertidally and to depths of nearly 7,000 m and are distributed worldwide. Most are less than a centimeter in length, but some deep-sea species have bodies that are several centimeters long and leg spans to 70 cm! Many are errant benthic animals, but others live on seaweeds or on other invertebrates, particularly sea anemones, hydroids, ectoprocts, and tunicates. One or two species are thought to live on the bells of pelagic medusae, and another has been seen in the hydrothermal vent communities of the Galapagos rift on the huge vestimentiferan worms.

Having noted that the systematic position of pycnogonids is still a bit uncertain, we treat them here as a class in the subphylum Cheliceriformes. Furthermore, classification within the Pycnogonida is also quite unsettled. Historically, the few known fossil species (which date back to the Devonian) were "dumped" into the single order Palaeopantopoda, and the living forms into the order Pantopoda. A major revision was proposed by Fry (1978), in which the 73 living genera were reassigned to 30 families in 5 orders. The more recently published scheme of Hedgpeth (1982) is based largely on the structure, and particularly the reduction or loss of certain appendages; it is outlined below. There are currently over 1,000 described species in 86 genera. A cladistic phylogenetic analysis of the class Pycnogonida is yet to be published.

CLASS PYCNOGONIDA

FAMILY NYMPHONIDAE: Ovigers 10-segmented and present in both sexes; chelifores 2-segmented; palps 5-segmented; polymerous species exist with 5–6 pairs of legs. (*Nymphon, Pentanymphon,* and *Sexanymphon*)

FAMILY CALLIPALLENIDAE: Ovigers present in both sexes; chelifores present; palps reduced or absent. (e.g., *Callipallene, Pallenopsis*)

FAMILY PHOXICHILIDIIDAE: Ovigers present only in males; chelifores present; palps absent. (e.g., *Anoplodactylus, Halosoma, Phoxichilidium*)

FAMILY AMMOTHEIDAE: Ovigers 9- or 10-segmented and present in both sexes; chelifores 2-segmented or reduced to small bumps; palps present, length and segment number variable. (e.g., *Achelia, Ammothella, Eurycyde, Rhynchothorax*)

FAMILY TANYSTYLIDAE: Similar to and previously incorporated with Ammotheidae; distinguished by reduced chelifores and palps. Monogeneric: *Tanystylum.*

FAMILY ENDEIDAE: Ovigers present in males only; chelifores and palps absent; legs long and slender. Monogeneric: *Endeis.*

FAMILY PYCNOGONIDAE: Ovigers present in males only; chelifores and palps absent; legs short and thick; polymerous species known with 5 pairs of legs. (e.g., *Pentapycnon, Pycnogonum*)

FAMILY COLOSSENDEIDAE: Ovigers 10-segmented and present in both sexes; chelifores present (in polymerous forms) or absent; palps 9- or 10-segmented and elongate; polymerous forms exist with 5 and 6 pairs of legs; includes the largest known pycnogonids (*Dodecolopoda*) and two other genera (*Colossendeis* and *Decolopoda*).

Several other families have been erected over the years to accommodate certain odd pycnogonids that do not fit conveniently elsewhere. Among these is the family Austrodecidae, whose members are found only on ectoproct colonies.

Figure 28

Representative pycnogonids. A, *Nymphopsis spinososssima* (family Ammotheidae). B, A polymerous pycnogonid with six pairs of legs, *Dodecolopoda mawsoni* (family Colossendeidae), in a "tip-toe" stance. C, *Tanystylum anthomasti* (family Tanystylidae). D, The ten-legged *Decolopoda australis* (family Colossendeidae) (side view of animal walking). E, *Achelia echinata* (family Ammotheidae) feeding on an ectoproct colony, one zooid at a time. F, A pycnogonid walking. (A,C after Hedgpeth, in Parker 1982; B,D after Schram and Hedgpeth 1978; E after Wyer and King 1974; F courtesy of J. W. Hedgpeth.)

The pycnogonid *Bauplan*

External anatomy

The bodies of pycnogonids are not clearly divided into recognizable tagmata as are those of other arthropods (Figures 28 and 29). The first body "region" bears an anteriorly directed **proboscis**, which varies in size and shape among species. The proboscis contains a chamber and bears an opening at its distal end. The actual mouth is probably the connection between the proboscis chamber and the esophagus (Figure 30A), although there is some confusion on this matter. This anteriormost "region" also bears paired appendages in the form of **chelifores**, **palps**, **first walking legs**, and, when present, **ovigers**. The chelifores may be chelate or achelate, or absent altogether. The ovigers are modified legs that carry the developing embryos and, in some groups (e.g, *Nymphon* and *Colossendeis*), are used to clean other appendages (Figure 29C). Also located on the first segment of most species is a tubercle with four simple eyes. The following body segments form the "trunk" and may be variably fused, but each bears a pair of lateral processes on which are borne the walking legs. Because of the orientation of these lateral processes, the walking legs are arranged somewhat radially around the body. The posteriormost body segment carries a dorsally inserted **tubercle**, which may be a vestigial abdomen, or opisthosoma, and which bears the anus. Perhaps the most distinctive feature of the pycnogonids, compared with other arthropods, is the presence of multiple gonopores, found on some or all of the walking legs.

One of the most unusual aspects of pycnogonid morphology is the existence of **polymerous species**, which possess more than four pairs of walking legs. This phenomenon is unique among arthropods and occurs in the pycnogonid genera *Pentanymphon*, *Pentapycnon*, and *Decolopoda* (with five pairs of legs), and *Sexanymphon* and *Dodecolopoda* (with six pairs of legs).

Variability among different pycnogonids also occurs in appendage shape and length, spination, proboscis structure, reduction or loss of chelifores and palps, and many other external features. Several examples are shown in Figures 28 and 29 to illustrate this diversity. In all, however, the body is remarkably

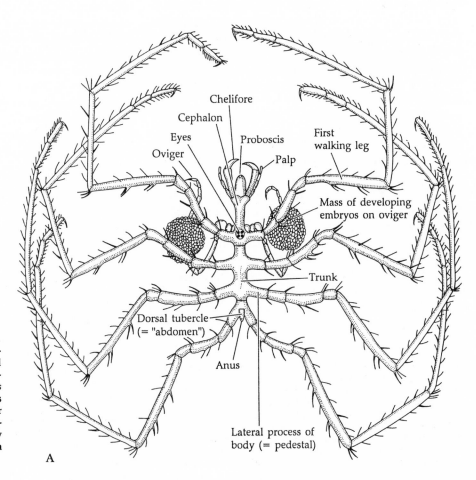

Figure 29

External features of pycnogonids. A, A male *Nymphon rubrum* carrying embryos on its ovigers (dorsal view). B, Walking leg of *Colossendeis australis*. C, *Colossendeis scotti*, using its oviger to clean its limbs. (A after Fage 1949 and other sources; B after Schram and Hedgpeth 1978; C original drawing by J. W. Hedgpeth, from a photo of a live specimen in an aquarium.)

reduced and narrow, a feature compensated for by extensions of the gut ceca and gonads into the legs!

Locomotion

The walking legs of pycnogonids are typically nine-segmented (Figure 29) and are borne on large, fixed lateral processes (or **pedestals**) along the sides of the body. The junction of the first coxal segment and the pedestal is a more-or-less immovable joint and does not contribute to the action of the leg. The joint between the first and second coxae is hinged to provide promotion and remotion, and the rest of the joints provide the usual flexion and extension. However, the coxal joints also allow a certain amount of "twisting" and thus accentuate the anterior–posterior swing of the appendage tips during the power and recovery strokes. Some joints lack extensor muscles, and limb extension may be aided by hydrostatic pressure, as in many arachnids. Note that pycnogonid specialists have given names to the leg articles that do not parallel those used for any other arthropod group (Figure 29B).

Many of the commonly encountered intertidal pycnogonids are quite sedentary and move very slowly. Most of these small forms have short thick legs that are somewhat prehensile and serve more for clinging to other invertebrates or algae than for rapid locomotion.

Deep-water benthic pyncogonids tend, in general, to be more active than their intertidal counterparts. However, some of the very large deep-sea forms (e.g., *Colossendeis*) may depend more on slow, deep-ocean currents to roll them around on the bottom than on their own locomotor powers. Under-

water photographs often reveal these pycnogonids in stationary postures on the tips of some of their legs (Figures 28B and 29C). Schram and Hedgpeth (1978) noted two ambulatory patterns among Antarctic benthic pycnogonids. In the "normal" forms with four pairs of legs, walking involves a variety of irregular leg movement patterns and does not conform clearly to any metachronal gait. However, the polymerous species tend to show a definite pattern of leg movement that involves posterior to anterior metachronal waves. The need for greater coordination to avoid leg tangling is a likely explanation for the more regular pattern in polymerous pycnogonids. These more errant pycnogonids have longer and thinner legs than the sedentary forms and tend to walk on the tips of the legs.

Many pycnogonids are also known to swim periodically by employing leg motions similar to those used in walking. Some species are known to "hang" from the water's surface, utilizing their light body weight and the water's surface tension. Several studies also report a characteristic "sinking behavior" in a number of species. When dropping to the bottom, these pycnogonids elevate all of their appendages over the dorsal surface of their bodies in what has been called a "basket configuration." This behavior eliminates much of the frictional resistance to sinking and allows the animal to drop quickly through the water column.

Feeding and digestion

As far as is known, the feeding habits of pycnogonids are largely dictated by the form of their proboscis, and food is limited to material that can be

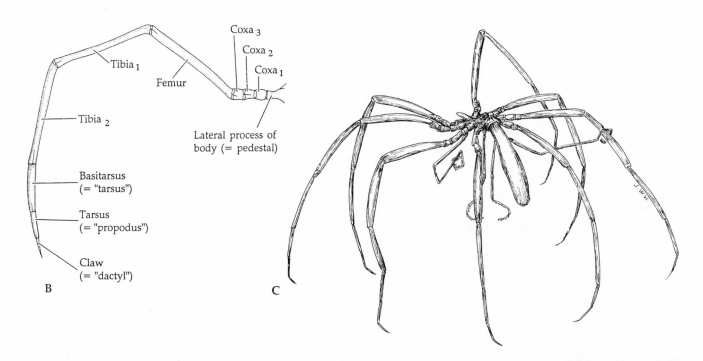

sucked into the gut. Even with this basic structural constraint, pycnogonids feed on a variety of organisms.

A few pycnogonids feed on algae, but most are carnivorous. Species that consume other animals usually use three cuticular teeth at the tip of the proboscis to pierce the body of their hosts; then they suck out body fluids and tissue fragments. Pycnogonids that live on hydroids use their chelifores to pick off pieces of their host and pass them to the proboscis opening. One species (*Achelia echinata*) feeds on ectoprocts by inserting the proboscis into the chamber housing an individual and sucking out the zooid (Figure 28E).

Very little is known about the feeding habits of the deep-sea benthic forms. Undersea photographs and direct observations of aquarium specimens of the giant *Colossendeis colossea* indicate that it may walk slowly along the bottom, sweeping its palps across the substratum to sense prey that might be sucked from the mud.

The digestive tract extends from the mouth at the base of the proboscis to the anus, which opens on a posterodorsal tubercle on the last "trunk" segment (Figure 30A). A chamber within the proboscis bears a region of dense bristles that serve to strain and mechanically mix food. The muscles of the foregut supply the suction for ingestion. A short esophagus connects the mouth to the elongate midgut or intestine from which digestive ceca extend into each leg, providing a high surface area for digestion and absorption. A short proctodeal rectum leads to the anus.

Digestion is predominantly, if not exclusively, intracellular. The cells of the midgut and cecal walls include phagocytes that engulf ingested food materials. Some of these cells may actually break free from the gut lining and phagocytize food particles while drifting in the gut lumen. Apparently these loose cells reattach to the gut wall after they have "fed." It has been suggested that upon reattachment these errant cells first pass their digested food contents to the fixed cells of the gut wall and then assume an excretory function by picking up metabolic wastes, detaching again and being eliminated via the anus.

Circulation, gas exchange, and excretion

Pycnogonids lack special organs for gas exchange or excretion. The digestive ceca and the overall *Bauplan* together present a very high surface area:volume ratio, and exchange of gases and the elimination of wastes probably occur largely by diffusion across the body and gut wall. The special wandering cells of the midgut may help in excretion.

The circulatory system includes an elongate heart with incurrent ostia, but no blood vessels. As in other arthropods, the heart is located dorsally, within a pericardial chamber separated from the ventral hemocoel by a perforated membrane. Blood leaves the heart anteriorly and flows through the hemocoelic spaces of the body and appendages. Contraction of the heart causes a lowered pressure within the dorsal pericardial body chamber, and blood is thus drawn through the perforations in the membrane and toward the heart. Upon relaxation, the blood flows through the ostia into the heart lumen.

Nervous system and sense organs

The central nervous system of pycnogonids includes a dorsal brain above the esophagus; circumenteric connectives; a subenteric ganglion; and a ganglionated ventral nerve cord (Figures 30A and B). The nerve cord bears a ganglion for each pair of walking legs, and additional ganglia are present in the polymerous species.

The cerebral ganglia include a proto- and tritocerebrum; pycnogonids lack a deutocerebrum (as do chelicerates). The protocerebrum innervates the eyes and the tritocerebrum innervates the chelifores, an arrangement similar to that in chelicerates. The cerebral ganglia also give rise to a well-developed ganglionated proboscis nerve.

Little work has been done on pycnogonid sense organs. Tactile reception is provided by touch-sensitive hairs and probably by the palps. On the body surface, just dorsal to the cerebral ganglia, is a tubercle, with four simple eyes (some deep-sea species lack eyes). The eyes, when present, are set in such positions as to provide 360° vision.

Reproduction and development

Pycnogonids are gonochoristic. Mating is typically followed by a period of brooding, where the embryos are held by the male ovigers, followed by the release of unique **protonymphon larvae** (Figure 30D). The protonymphon is a curious creature with three pairs of appendages. It usually lives in a symbiotic relationship with cnidarians, molluscs, or echinoderms. These relationships are poorly understood, but in some cases they appear to be parasitic or commensalistic. This mixed developmental strategy has been replaced by direct patterns in some species wherein the larval stage is passed within an egg case.

Sexual dimorphism is common among most species of pycnogonids. The males bear the unique ovi-

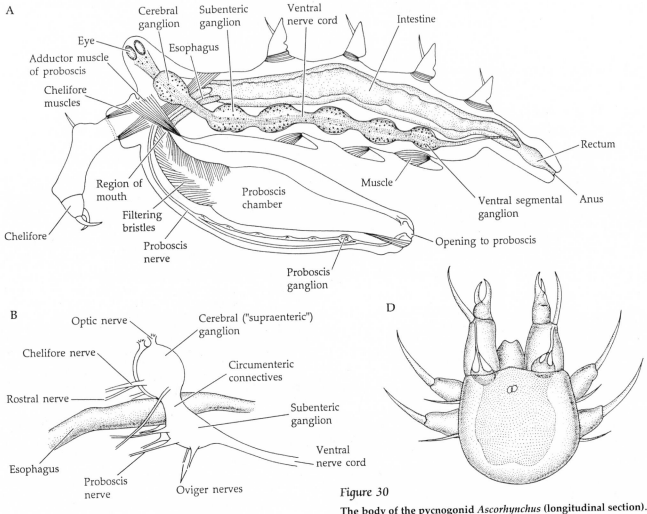

A

Eye

Adductor muscle
of proboscis

Chelifore muscles

Chelifore

Region of mouth

Filtering bristles

Proboscis nerve

Cerebral ganglion

Esophagus

Proboscis ganglion

Subenteric ganglion

Ventral nerve cord

Intestine

Rectum

Muscle

Ventral segmental ganglion

Anus

Opening to proboscis

Proboscis chamber

B

Optic nerve

Chelifore nerve

Rostral nerve

Esophagus

Proboscis nerve

Oviger nerves

Cerebral ("supraenteric") ganglion

Circumenteric connectives

Subenteric ganglion

Ventral nerve cord

D

Figure 30

The body of the pycnogonid *Ascorhynchus* (longitudinal section). B, Anterior portion of the nervous system of *Nymphon*. C, A female pycnogonid with developing ova stored in the femoral portions of the gonad diverticula. D, Protonymphon larva. (A,C redrawn from several sources after Fage 1949; B after Schram and Hedgpeth 1978; D after Hedgpeth, in Parker 1982.)

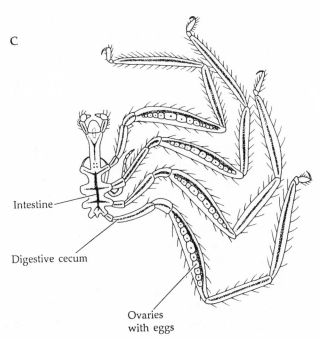

C

Intestine

Digestive cecum

Ovaries with eggs

gers associated with the first body segment; these appendages are absent in females of some families (e.g., Phoxicilidiidae, Endeidae, and Pycnogonidae), and often reduced in the females of other families. Female pycnogonids usually have enlarged femora.

Internally, the reproductive systems of males and females are similar and relatively simple. In both, the gonad is single and U-shaped, with extensions into the legs where gametes are produced and stored. The expanded femora in females provide space for storing unfertilized eggs (Figure 30C). The multiple gonopores are usually located on the ventral surface of the second coxae of two or all pairs of legs and are thus close to the regions of gamete storage. During mating, the male typically hangs beneath the female

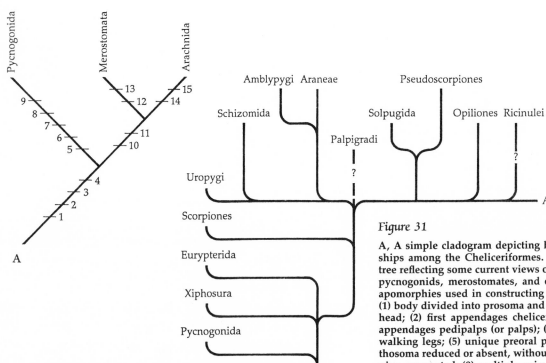

Figure 31

A, A simple cladogram depicting hypotheses about the relationships among the Cheliceriformes. B, A traditional evolutionary tree reflecting some current views on the relationships among the pycnogonids, merostomates, and orders of arachnids. The synapomorphies used in constructing the cladogram are as follows: (1) body divided into prosoma and opisthosoma, without distinct head; (2) first appendages chelicerae (or chelifores); (3) second appendages pedipalps (or palps); (4) typically with four pairs of walking legs; (5) unique preoral proboscis; (6) ovigers; (7) opisthosoma reduced or absent, without appendages; (8) walking legs nine-segmented; (9) multiple pairs of gonopores (borne on some or all legs); (10) prosomal carapace-like shield; (11) first or second opisthosomal segment modified as genital somite; (12) opisthosomal appendages modified as book gills; (13) telson long and spiked; (14) opisthosomal appendages reduced, lost, or modified as spinnerets or pectines; (15) with tracheae, book lungs, or both.

or assumes a stance over her back. As the female releases her eggs, the male fertilizes them. Using a sticky secretion from special femoral glands, the male cements the eggs into a spherical mass and carries them on his ovigers (Figure 29A).

Knowledge of pycnogonid development is based on relatively few studies. There is some variation among members of different genera, but cleavage is usually holoblastic and leads to the formation of a stereoblastula. The inward movement of the presumptive entodermal and mesodermal cells is frequently accompanied by a disappearance of some cell membranes, a process leading to the formation of some syncytial tissues in the gastrula. Germinal centers become apparent as appendage buds take form, and the protonymphon stage follows. Through a series of molts, segments and appendages are added to produce the juvenile. Nothing is known about reproduction in most of the deep-sea species. Their eggs, however, are very small, an observation suggesting that the young stages may also be parasitic.

Cheliceriform phylogeny

Although over 150 years have passed since the discovery of pycnogonids, arguments still ensue as

to their phylogenetic relationships. Most evidence, however, seems to point to a distant cheliceriform ancestry. Their chelifores and palps are probably homologous to the chelicerae and pedipalps of the chelicerates. In the most recent phylogenetic analysis, Schram (1978) placed the pycnogonids as a sister-group to the Chelicerata and erected the name "Cheliceriformes" for the entire group. We have generally followed that scheme here.

Figure 31A is a simple cladogram depicting one view of relationships among the three major cheliceriform taxa. The subclass Arachnida is probably a monophyletic group, although some workers have suggested that it might be diphyletic, having arisen from two separate invasions of land (one leading to the scorpions, the second to all other orders). Such a hypothesis is not incompatible with the tree depicted in Figure 31A.

Like the uniramians, the arachnids probably invaded land early in the Silurian, at the same time the land plants were becoming established. The earliest Paleozoic scorpions were aquatic, respiring by book gills which were probably the predecessors (homologues) to the book lungs of modern terrestrial arachnids. Silurian and Devonian scorpions shared shallow marine or estuarine environments with their close relatives, the eurypterids.

Figure 31B is a traditional evolutionary tree depicting the relationships shown in the cladogram and also current views on the position of lower taxa. The relationship of the Cheliceriformes to other arthropod subphyla is discussed in Chapter 19.

Selected References

General References

Grassé, P. 1949. *Traité de Zoologie*, Vol 6, Onychophores–Tardigrades–Arthropodes–Trilobitomorphes–Chelicerataes. Masson et Cie, Paris.

Kraus, O. 1976. On the phylogenetic position and evolution of the Chelicerata. Entomol. Ger. 3:1–12.

Levi, H. W. et al. 1982. Chelicerata. *In* S. P. Parker (ed.), *Synopsis and Classification of Living Organisms*, Vol. 2. McGraw-Hill, New York, pp. 71–173.

Sherman, R. G. 1981. Chelicerates. *In* N. A. Ratcliffe and A. F. Rowley (eds.), *Invertebrate Blood Cells*, Vol. 2. Academic Press, New York.

Snodgrass, R. E. 1952. *A Textbook of Arthropod Anatomy.* Cornell University Press, Ithaca, New York.

Størmer, L., A. Petrukenvitch and J. W. Hedgpeth. 1955. Chelicerata, with sections on Pycnogonida and *Palaeoisopus*. *In* R. C. Moore (ed.), *Treatise on Invertebrate Paleontology*, Part P, Arthropoda 2. Geological Society of America and the University of Kansas Press, Lawrence.

Van der Hammen, L. 1977. The evolution of the coxa in mites and other groups of Chelicerata. Acarologia 19(1):12–19.

Merostomata

Bonaventura, J., C. Bonaventura and S. Tesh (eds.). 1982. *Physiology and Biology of Horseshoe Crabs.* A. R. Liss, New York.

Cohen, J. A. and H. J. Brockmann. 1983. Breeding activity and mate selection in the horseshoe crab, *Limulus polyphemus.* Bull. Mar. Sci. 33:274–281.

Fage, L. 1949. Classe des Merostomaces. *In* P. P. Grassé (ed.), *Traité de Zoologie*, Vol. 6. Masson et Cie, Paris, pp. 219–262.

Fahrenbach, W. H. 1979. The brain of the horseshoe crab (*Limulus polyphemus*). III. Cellular and synaptic organization of the corpora pedunculata. Tissue Cell 11:163–200.

Fisher, D. C. 1984. The Xiphosuridae: Archetypes of bradytely? *In* N. Eldredge and S. Stanley (eds.), *Living Fossils.* Springer-Verlag, New York, pp. 186–213.

Gregg, R. A. and D. A. Eagles. 1984. The relationship between sensory activity and muscle tension in the anterior flexor muscle of horseshoe crab walking legs. Mar. Behav. Physiol. 10:283–301.

Griffin, A. J. and W. H. Fahrenbach. 1977. Gill receptor arrays in the horseshoe crab (*Limulus polyphemus*). Tissue Cell 9:745–750.

Riska, B. 1981. Morphological variation in the horseshoe crab (*Limulus polyphemus*), a "phylogenetic relic." Evolution 35:647–658.

Rudloe, A. 1981. Aspects of the biology of juvenile horseshoe crabs, *Limulus polyphemus.* Bull. Mar. Sci. 31:125–133.

Sekiguchi, K. and H. Sugita. 1980. Systematics and hybridization in the four living species of horseshoe crabs. Evolution 34:712–718.

Shuster, C. N. 1950. Natural history of *Limulus.* Contrib. Woods Hole Oceanogr. Inst. 564:18–23.

Arachnida

Abushama, F. T. 1964. On the behavior and sensory physiology of the scorpion *Leiurus quinquestriatus.* Anim. Behav. 12:140–153.

Anderson, J. F. 1970. The metabolic rate of spiders. Comp. Biochem. Physiol. 33:51–72.

Anderson, J. F. 1978. Energy content of spider eggs. Oecologia 37:41–58.

Arthur, D. R. 1962. *Ticks and Disease.* Pergamon Press, London.

Baerg, W. 1958. *The Tarantula.* University of Kansas Press, Lawrence.

Baker, E. W. and G. W. Wharton. 1952. *An Introduction to Acarology.* Macmillan, New York.

Balogh, J. 1972. *The Oribatid Genera of the World.* Akademiai Kiado, Budapest, Hungary.

Barth, F. G. 1971. Der sensorische Apparat der Spaltsinnesorgane (*Cupiennius salei* Keys., Araneae). Z. Zellforsch. 112:212.

Barth, F. G. and P. Pickelmann. 1975. Lyriform slit sense organs—modeling an arthropod mechanoreceptor. J. Comp. Physiol. 103:39–54.

Barth, F. G. and J. Stagl. 1976. The slit sense organs of arachnids. Zoomorphologie 86:1–23.

Barth, F. G. (ed.). 1985. *Neurobiology of Arachnids.* Springer-Verlag, New York.

Beck, L., R. Foelix, E. Gödeke and R. Kaiser. 1977. Morphologie, Larvalentwicklung und Haarsensillen der Geisselspinne *Heterophrynus longicornis* Butler (Arach., Amblypygi). Zoomorphologie 88:259–276.

Binns, E. S. 1983. Phoresy as migration—some functional aspects of phoresy in mites. Biol. Rev. 57:571–620.

Bosenberg, H. 1905. Beitrage zur kenntuis der spermatogenese beiden Arachnoiden. Zool. Jahrb. Anat. Zool. 21:505.

Bowerman, R. F. and M. Burrows. 1981. The morphology and physiology of some walking leg motor neurons in a scorpion. J. Comp. Physiol. 140:31–42.

Braunitzer, G. and D. Wolff. 1955. Vergleichende chemische Untersuchungen über die Fibroine von *Bombyx mori* und *Nephila madagascariensis.* Z. Naturforsch. 106:404.

Bristowe, W. S. 1958. *The World of Spiders.* New Naturalist Series, Collins, London.

Brownell, P. H. 1977. Compressional and surface waves in sand: Used by desert scorpions to locate prey. Science 197:479–482.

Brownell, P. H. 1984. Prey detection by the sand scorpion. Sci. Am. 251(6):86–97.

Brownell, P. H. and R. D. Farley. 1979. Detection of vibrations in sand by the tarsal organs of the nocturnal scorpion *Paruroctonus mesaensis.* J. Comp. Physiol. 131:23–30.

Brownell, P. H. and R. D. Farley. 1979. Orientation to vibrations in sand by the nocturnal scorpion *Paruroctonus mesaensis*: mechanism of target localisation. J. Comp. Physiol. 131:31–38.

Bub, K. and R. F. Bowerman. 1979. Prey capture by the scorpion *Hadrurus arizonensis* Ewing (Scorpionea; Vaejovidae). J. Arachnol. 7:243–253.

Buchli, H. H. R. 1969. Hunting behavior in the Ctenizidae. Am. Zool. 9:175–193.

Burgess, J. W. 1976. Social spiders. Sci. Am. 234(3):101–106.

Carthy, J. D. 1968. The pectines of scorpions. *In* J. D. Carthy and G. E. Newell (eds.). *Invertebrate Receptors.* Academic Press, New York, pp. 251–261.

Cloudsley-Thompson, J. L. 1961. Some aspects of physiology and behavior of *Galeodes arabs.* Entomol. Exp. Appl. 4:257–263.

Cloudsley-Thompson, J. L. 1968. *Spiders, Scorpions, Centipedes and Mites*. Pergamon Press, New York.

Cooke, J. A. L. 1967. The biology of Ricinulei. Zoologica 151:31–42.

Dill, L. M. 1975. Predatory behavior of the zebra spider, *Salticus scenicus*. Can. J. Zool. 53:1284–1289.

Dumpert, K. 1978. Spider odor receptor: Electrophysiological proof. Experimentia 34:754–756.

Edgar, A. L. 1971. Studies on the biology and ecology of Michigan Phalangida (Opiliones). Misc. Publ. Mus. Zool., University of Michigan 144:1–64.

Edgar, W. D. 1970. Prey and feeding behavior of adult females of the wolf spider *Pardosa amentata*. Neth. J. Zool. 20(4):487–491.

Eisner, T. 1962. Survival by acid defense. Nat. Hist. 71:10–19.

Eisner, T. and S. Nowicki. 1983. Spider web protection through visual advertisement: Role of the stabilamentum. Science 219:185–187.

Ellis, C. H. 1944. The mechanism of extension in the legs of spiders. Biol. Bull. 86:41–50.

Evans, G. O. and E. Browning. 1955. *Some British Mites of Economic Importance*. British Museum of Natural History, London.

Evans, G. O., J. G. Sheals and D. MacFarlane. 1961. *The Terrestrial Acari of the British Isles. An Introduction to Their Morphology, Biology and Classification*, Vol. 1, Introduction and Biology. British Museum of Natural History, London.

Foelix, R. F. 1970. Chemosensitive hairs in spiders. J. Morphol. 132:313–334.

Foelix, R. F. 1970. Structure and function of tarsal sensilla in the spider *Araneus diadematus*. J. Exp. Zool. 175:99–124.

Foelix, R. F. 1982. *Biology of Spiders*. Harvard University Press, Cambridge, Massachusetts. [Highly recommended.]

Foelix, R. F. and A. Choms. 1979. Fine structure of a spider joint receptor and associated synapses. Eur. J. Cell Biol. 19:149–159.

Foelix, R. F. and I.-W. Chu-Wang. 1973. The morphology of spider sensilla. II. Chemoreceptors. Tissue Cell 5(3):461–478.

Foelix, R. F. and J. Schabronath. 1983. The fine structure of scorpion sensory organs. I. Tarsal sensilla. II. Pecten sensilla. Bull. Br. Arachnol. Soc. 6(2):53–74.

Foelix, R. F. and D. Troyer. 1980. Giant neurons and associated synapses in the peripheral nervous system of whip spiders. J. Neurocytol. 9:517–535.

Foil, L. D., L. B. Coons, and B. R. Norment. 1979. Ultrastructure of the venom gland of the brown recluse spider, *Loxosceles reclusa*. Int. J. Insect Morphol. Embryol. 8:325–334.

Forster, L. 1982. Vision and prey–catching strategies in jumping spiders. Am. Sci. 70:165–175.

Forster, R. R. 1980. Evolution of the tarsal organ, the respiratory system and the female genitalia in spiders. Int. Congr. Arachnol. 8:269–285.

Francke, O. F. 1979. Spermatophores of some North American scorpions. J. Arachnol. 7:19–32.

Frank, H. 1957. Untersuchungen zur funtionellen Anatomie der lokomotorischen Extremitäten von *Zygiella x-notata*, einer Radnetzspinne. Zool. Jahrb. Abt. Anat. Ontog. Tiere 76:423.

Gardner, B. T. 1965. Observations on three species of jumping spiders. Psyche 72:133–147.

Gertsch, W. J. 1979. *American Spiders*, 2nd Ed. Van Nostrand Reinhold, New York.

Görner, P. 1965. Mehrfach innervierte Mechanorezeptoren bei Spinnen. Naturwissenschaften 52:437.

Griffiths, D. A. and C. E. Bowman (eds.). 1983. *Acarology VI*, Vols. 1–2. Wiley, New York.

Hadley, N. F. 1974. Adaptational biology of desert scorpions. J. Arachnol. 2:11–23.

Herreid, C. F. and R. J. Full. 1980. Energetics of running tarantulas. Physiologist 23:40.

Hoffman, H. 1971. Die Augen der Araneae. Z. Morphol. Tiere 69:201.

Holm, A. 1940. Studien über die Entwicklung und Entwicklungsbiologie der Spinnen. Zool. Bijdr. 19:1.

Kaestner, A. 1929. Bau und Funktion der Fächertracheen einiger Spinnen. Z. Morphol. Ökol. Tiere 13:463–558.

Kaston, B. J. 1964. The evolution of spider webs. Am. Zool. 4:191–207.

Kaston, B. J. 1970. The comparative biology of American black widow spiders. Trans. San Diego Soc. Nat. Hist. 16:33–82.

Kaston, B. J. 1978. *How to Know the Spiders*, 3rd Ed. W. C. Brown, Dubuque, Iowa.

Keegan, H. L. 1980. *Scorpions of Medical Importance*. University Press of Mississippi.

Kovoor, J. 1977. Silk and the silk glands of Arachnida. Annee Biol. 16:97–172.

Krantz, G. W. 1978. *A Manual of Acarology*, 2nd Ed. Oregon State University, Corvallis.

Kullmann, E. J. 1972. Evolution of social behavior in spiders. Am. Zool. 12(3):419–426.

Land, M. F. 1972. Stepping movements made by jumping spiders during turns mediated by the lateral eyes. J. Exp. Biol. 57:15–40.

Levi, H. W. 1948. Notes on the life history of the pseudoscorpion *Chelifer cancroides*. Trans. Am. Microsc. Soc. 67:290–299.

Levi, H. W. 1967. Adaptations of respiratory systems of spiders. Evolution 21:571–583.

Levi, H. W. 1978. Orb-weaving spiders and their webs. Am. Sci. 66:734–742.

Levi, H. W. and L. R. Levi. 1968. *A Guide to Spiders and Their Kin*. A Golden Nature Guide, Golden Press, New York.

Linzen, B. and R. Loewe. 1971. Hemocyanin in a spider. Naturwissenschaffen 58:269.

Lüters, H. 1966. Der Lebenszyklus von *Agroeca brunnea* Blackwall (Araneae, Clubionidae) unter besonderer Berucksichtigung des kokonbau und des Häutungsverhaltens. Ph.D. Diss., Göttingen.

McCrone, J. D. 1969. Spider venoms: Biochemical aspects. Am. Zool. 9:153–156.

Melchers, M. 1964. Zur Biologie der Vogelspinnen (Fam. Aviculariidae). Z. Morphol. Ökol. Tiere 58:517.

Merrett, P. (ed.). 1978. *Arachnology*. Symposia of the Zoological Society of London, Number 42. Academic Press. [A fine collection of work by many authors.]

Millot, J. et al. 1949. Classe de Arachnides. *In* P. Grassé (ed.), *Traiteé de Zoologie*, Vol. 6. Masson et Cie, Paris, pp. 263–905.

Moffett, S. and G. S. Doell. 1980. Alteration of locomotor behavior in wolf spiders carrying normal and weighted egg cocoons. J. Exp. Zool. 213:219–226.

Nentwig, W. and St. Heimer. 1983. Orb webs and single–line webs: An economic consequence of space web reduction in spiders. Z. Zool. Syst. Evolutionsforsch. 21:26–37.

Osaki, H. 1969. Electron microscope study on the spermatozoan of the Liphistiid spider *Heptathela*. Acta Arachnol. 22:1.

Parry, D. A. and R. H. J. Brown. 1959. The jumping mechanism of salticid spiders. J. Exp. Biol. 36:654.

Paulus, H. F. 1979. Eye structure and the monophyly of Arthropoda. *In* A. P. Gupta, *Arthropod Phylogeny*. Van Nostrand Reinhold, New York, pp. 299–377.

Petrunkevitch, A. 1955. Arachnida. *In* R. C. Moore (ed.), *Treatise on Invertebrate Paleontology*, Vol. 2. Geological Society of America, New York, pp. 42–162.

Platnick, N. I. 1971. The evolution of courtship behavior in spiders. Bull. Br. Arachnol. Soc. 2:40–47.

Platnick, N. I. 1976. Drifting spiders or continents? Vicariance biogeography of the spider subfamily Laroniinae. Syst. Zool. 25:101–109.

Platnick, N. I. and W. J. Gertsch. 1976. The suborders of spiders: A cladistic analysis. Am. Mus. Novit. 2607:1–15.

Polis, G. A. and R. D. Farley. 1980. Population biology of a desert scorpion (*Paruroctonus mesanensis*): Survivorship, microhabitat, and the evolution of life history strategy. Ecology 61:620–629.

Pollock, J. 1966. Life of the ricinulid. Animals 8:402–405.

Robinson, M. H. and B. Robinson. 1980. Comparative studies of the courtship and mating behavior of tropical araneid spiders. Pac. Insects Monogr. 36:1–218.

Rodriguez, J. G. (ed.). 1979. *Recent Advances in Acarology*, Vols. 1–2. Academic Press. [The two volumes total well over 1,000 pages and include contributions on both applied and basic mite research.]

Root, G. and R. F. Bowerman. 1978. Intra-appendage movements during walking in the scorpion *Hadrurus arizonensis*. Comp. Biochem. Physiol. 59A:57–63.

Rovner, J. S. 1971. Mechanisms controlling copulatory behavior in wolf spiders. Psyche 78(1):150–165.

Rovner, J. S. 1975. Sound production by nearctic wolf spiders: A substratum-coupled stridulatory mechanism. Science 190:1309–1310.

Rovner, J. S., G. A. Higashi and R. F. Foelix. 1973. Maternal behavior in wolf spiders: The role of abdominal hairs. Science 182:1153–1155.

Sabu, L. S. 1965. Anatomy of the central nervous system of arachnids. Zool. Jahrb. Abt. Anat. Ontog. Tiere 82:1–154.

Sauer, J. R. and J. A. Hair (eds.). 1986. *Morphology, Physiology, and Behavioral Biology of Ticks*. Eillis Horwood, Chichester.

Savory, T. H. 1962. "Daddy Longlegs." Sci. Am. 207:119.

Savory, T. H. 1977. *Arachnida*, 2nd Ed. Academic Press, New York.

Schaefer, M. 1977. Winter ecology of spiders. Z. Angew. Entomol. 83:113–134.

Seyfarth, E. A. and F. G. Barth. 1972. Compound slit sense organs on the spider leg: Mechanoreceptors involved in kinesthetic orientation. J. Comp. Physiol. 78:176–191.

Shear, W. A. (ed.). 1981. *Spiders: Webs, Behavior and Evolution*. Stanford University Press, Stanford, California.

Snodgrass, R. E. 1948. The feeding organs of Arachnida, including mites and ticks. Smithson. Misc. Coll. 110:1–93.

Snow, K. R. 1970. *The Arachnids: An Introduction*. Columbia University Press, New York.

Stewart, D. M. and A. W. Martin. 1974. Blood pressure in the tarantula, *Dugesiella hentzi*. J. Comp. Physiol. 88:141–172.

Tolbert, W. W. 1975. Predatory avoidance behavior and web defensive structures in the orb weavers *Argiope aurantia* and *Argiope trifasciata*. Psyche 82:29–52.

Turnbull, A. L. 1973. Ecology of true spiders. Annu. Rev. Entomol. 18:305–348.

Vachon, M. 1953. The biology of scorpions. Endeavor 12:80–89.

Von Helversen, O. 1976. Gedanken zur Evolution der Paarungsstellungen bei den Spinnen (Arachnida, Araneae). Ent. Germ. 3:13.

Walcott, C., and W. G. van der Kloot. 1959. The physiology of the spider vibration receptor. J. Exp. Zool. 141:191–244.

Watson, P. J. 1986. Transmission of a female sex pheromone thwarted by males in the spider *Linyphia litigiosa* (Linyphiidae). Science 233:219–221.

Weygoldt, P. 1969. *The Biology of Pseudoscorpions*. Harvard University Press, Cambridge, Massachusetts.

Weygoldt, P. 1972. Geisselskorpione und Geisselspinnen (Uropygi and Amblypygi). Z. des Kölner Zoo 15(3):95–107.

Weygoldt, P. 1974. Indirect sperm transfer in arachnids. Verh. Dtsch. Zool. Ges. 67:308–313.

Weygoldt, P. 1979. Significance of later embryonic stages and head development in arthropod phylogeny. *In* A. P. Gupta (ed.), *Arthropod Phylogeny*. Van Nostrand Reinhold, New York, pp. 107–136.

Wilson, D. M. 1967. Stepping patterns in tarantula spiders. J. Exp. Biol. 47:133–151.

Wilson, R. S. 1969. Control of drag-line spinning in certain spiders. Am. Zool. 9:103–111.

Witt, P. N. 1975. The web as a means of communication. Biosci. Commun. 1:7–23.

Witt, P. N., C. F. Reed and D. B. Peakall. 1968. *A Spider's Web*. Springer-Verlag, New York.

Witt, P. N. and J. S. Rovner (eds.). 1982. *Spider Communication: Mechanisms and Ecological Significance*. Princeton University Press, New Jersey.

Witt, P. N., M. B. Scarboro, D. B. Peakall and R. Gause. 1977. Spider web-building in outer space: Evaluation of records from the Skylab spider experiment. Am. J. Arachnol. 4:115.

Pycnogonida

Arita, K. 1937. Beiträge zur Biologie der Pantopoden. J. Dep. Agric. Kyushu Imp. Univ. 5:271–288.

Bergström, J., W. Stürmer and G. Winter. 1980. *Palaeoisopus, Palaeopantopus* and *Palaeothea*, pycnogonid arthropods from the Lower Devonian Hunsrück Slate, West Germany. Palaeont. Zh. 54:7–54.

Child, C. A. 1979. Shallow–water Pycnogonida from the Isthmus of Panama and the coasts of middle America. Smithson. Contrib. Zool. 293:1–86.

Cole, L. J. 1905. Ten-legged pycnogonids, with remarks on the classification of the Pycnogonida. Ann. Mag. Nat. Hist. 15:405–415.

Cole, L. J. 1910. Peculiar habit of a pycnogonid new to North America with observations on the heart and circulation. Biol. Bull. 18:193–203.

Fage, L. 1949. Classe des Pycnogonides. *In* P. Grassé (ed.), *Traité de Zoologie*, Vol. 6. Masson et Cie, Paris, pp. 906–914.

Fry, W. G. 1965. The feeding mechanisms and preferred foods of three species of Pycnogonida. Bull. Br. Mus. Nat. Hist. Zool. 12:195–223.

Fry, W. G. 1978. A classification within the Pycnogonida. Zool. J. Linn. Soc. 63:35–78.

Fry, W. G. (ed.). 1978. Sea Spiders (Pycnogonida). Zool. J. Linn. Soc. 63:1–238.

Fry, W. G. and J. H. Stock. 1978. A pycnogonid bibliography. Zool. J. Linn. Soc. 64:197–238.

Hedgpeth, J. W. 1941. A key to the Pycnogonida of the Pacific coast of North America. Trans. San Diego Soc. Nat. Hist. 9:253–264.

Hedgpeth, J. W. 1948. The Pycnogonida of the Western North Atlantic and the Caribbean. Proc. U.S. Natl. Mus. 97:157–342.

Hedgpeth, J. W. 1954. On the phylogeny of the Pycnogonida. Acta Zool. 35:193–213.

Hedgpeth, J. W. 1978. A reappraisal of the Palaeopantopoda with description of a species from the Jurassic. Zool. J. Linn. Soc. 63:23–34.

Hedgpeth, J. W. 1982. Pycnogonida. *In* S. P. Parker, *Synopsis and Classification of Living Organisms*, Vol. 2. McGraw-Hill, New York, pp. 169–173.

Hedgpeth, J. W. and W. G. Fry. 1964. Another dodecapodous pycnogonid. Ann. Mag. Nat. Hist. 13:161–169.

Henry, L. M. 1953. The nervous system of the Pycnogonids. Microentomology 18:16–36.

Hilton, W. A. 1942. Pycnogonids from Allan Hancock expeditions. Allan Hancock Pac. Expeds. 5:277–339.

King, P. E. 1973. *Pycnogonids*. Hutchinson University Library, London.

McCloskey, L. R. 1973. Marine flora and fauna of northeastern United States. Pycnogonida. NOAA Tech. Rept., Natl. Mar. Fish. Serv. Circular 386.

Morgan, E. 1971. The swimming of *Nymphon gracile* (Pycnogonida): The mechanics of the leg-beat cycle. J. Exp. Biol. 55:273–287.

Morgan, E. 1972. The swimming of *Nymphon gracile* (Pycnogonida): The swimming gait. J. Exp. Biol. 56:421–432.

Nakamura, K. 1981. Postembryonic development of a pycnogonid *Propallene longiceps*. J. Nat. Hist. 15:49–62.

Schram, F. R. and J. W. Hedgpeth. 1978. Locomotory mechanisms in Antarctic pycnogonids. Zool. J. Linn. Soc. 63:145–169.

Wirèn, E. 1918. Zur Morphologie und Phylogenie der Pantopoden. Zool. Bidr. Uppsala 6:41–181.

Wyer, D. W. and P. E. King. 1974. Relationships between some British littoral and sublittoral bryozoans and pycnogonids. Estuarine Coastal Mar. Sci. 2:177–184.

Chapter Seventeen

Phylum Arthropoda: The Insects and Myriapods

Even in matters about which man is wont to especially pride himself, such as those touching social organisation, he might with advantage go to the ant to learn wisdom, since many of the problems of modern civilisation involved in the questions concerned in the regulation of increase of population, the proper division of labour, and the support of useless individuals, have been satisfactorily solved by many of the species of insects that live habitually in communities.

Richard Lydekker

The Royal Natural History, Volume 6 (1896)

The arthropod subphylum Uniramia comprises two distinct classes of closely related animals, the Insecta and the Myriapoda. The latter group includes the centipedes, millipedes, symphylans, and pauropodans. Modern uniramians are fundamentally terrestrial arthropods, and living forms that inhabit aquatic environments have secondarily invaded these habitats through modifications of their air-breathing gas exchange systems. However, as we discuss in Chapter 19, the fossil record suggests that the earliest uniramians (many-segmented, myriapod-like animals) were actually benthic marine creatures. Figures 1 and 2 illustrate a variety of uniramian types.

The most spectacular evolutionary radiation among the uniramians has, of course, been within the insects, which inhabit nearly every conceivable terrestrial and freshwater habitat and, less commonly, even the sea's surface and the littoral region. Insects are also found in many unlikely places such as oil swamps and seeps, sulfur springs, glacial streams, and brine ponds. They often live where few other metazoa can exist. Close to a million species of insects have been described, and authorities estimate 20–50 million remain undescribed. The beetle family Curculionidae (the weevils) contains about 65,000 described species and is itself larger than any other animal phylum.

Insects pollinate most of the flowering plants and food crops, and provide us with luxuries such as honey, beeswax, and silk. Honeybees alone pollinate $20 billion worth of crops in the United States and produce $200 million worth of honey annually. Insects are key items in the diets of many terrestrial animals and play a major role as reducer-level organisms. However, they also consume about one-third of our potential annual harvest and transmit major diseases such as malaria and yellow fever. Every year billions of dollars are spent on insect control of various sorts. Needless to say, the subject of insect biology, or **entomology**, is a discipline in its own right, and a multitude of books and college courses exist on the subject. If we apportioned pages to animal groups on the basis of numbers of species, overall abundance, or economic importance, insect chapters could easily fill 90 percent of this text. References at the end of this chapter provide entry into some of the current literature on insects.

Because the subphylum Uniramia comprises such a large and diverse assemblage of arthropods, we first present a brief outline of their classification, followed by a more detailed synopsis of selected higher taxa. The synopsis provides a brief diagnosis and comments on the biology of each taxon. These two sections serve as a preface to the *Bauplan* discussion that follows and as a reference that the reader can turn to when needed.

Uniramian classification

As noted in Chapter 15, and examined in more depth in Chapter 19, the uniramians have traditionally been classified alongside members of the subphylum Crustacea in a group called the Mandibulata. The suitability of a taxon Mandibulata, which implies that the uniramians and crustaceans represent a monophyletic clade (and are sister-groups), has recently been questioned by several authorities. For this reason, we treat the uniramians and crustaceans as two distinct subphyla, in two separate chapters, and discuss their evolutionary relationship in Chapter 19.

The classification scheme given below is largely accepted by contemporary specialists. Some disagreement exists regarding the categorical ranking of some taxa, but as long as the hierarchy of the arrangement remains relatively stable, this is of little consequence. A few recent workers, for example, regard the Uniramia as a phylum, in which case the subclasses indicated below might be elevated to the rank of class, and so on. This particular issue is also discussed in Chapter 19. The name Uniramia alludes to the uniramous (unbranched) nature of the appendages in members of this subphylum (Latin *uni*, "one"; *rami*, "branch"). An older name, Atelocerata, alluded to the "loss" of a second pair of antennae in this group, in contrast with members of the subphylum Crustacea, which possess two pairs of antennae.

With nearly 1 million named species in the taxon Uniramia, we have opted not to include representative genera in the classification scheme below. Even introductory entomology courses typically require students to recognize uniramian arthropods only to the level of order or family.

SUBPHYLUM UNIRAMIA

CLASS MYRIAPODA

SUBCLASS DIPLOPODA: Millipedes

SUBCLASS PAUROPODA: Pauropodans

SUBCLASS CHILOPODA: Centipedes

SUBCLASS SYMPHYLA: Symphylans

CLASS INSECTA (HEXAPODA)*

SUBCLASS MYRIENTOMATA (ORDER PROTURA): Proturans

SUBCLASS OLIGOENTOMATA (ORDER COLLEMBOLA): Springtails

---

*The late Sidnie Manton suggested that the class Insecta, as here used, may not be monophyletic. She preferred the term "Hexapoda" to include all five insect subclasses, reserving the term "Insecta" for the Pterygota (winged insects) only.

SUBCLASS DIPLURATA (ORDER DIPLURA): Diplurans

SUBCLASS ZYGOENTOMATA

    ORDER THYSANURA: Silverfish and bristletails

    ORDER ARCHAEOGNATHA: Rockhoppers

SUBCLASS PTERYGOTA: Winged insects

  INFRACLASS PALEOPTERA: Ancient winged insects

    ORDER EPHEMEROPTERA: Mayflies

    ORDER ODONATA: Dragonflies and damselflies

  INFRACLASS NEOPTERA: Modern, wing-folding insects

  SUPERORDER ORTHOPTERODEA

    ORDER BLATTODEA: Cockroaches

    ORDER MANTODEA: Mantids and praying mantids

    ORDER ISOPTERA: Termites

    ORDER PLECOPTERA: Stone flies

    ORDER ORTHOPTERA: Locusts, katydids, crickets, grasshoppers

    ORDER DERMAPTERA: Earwigs

    ORDER GRYLLOBLATTODEA: Grylloblattids

    ORDER EMBIOPTERA: Embiopterans

    ORDER PHASMIDA (= PHASMATOPTERA): Stick and leaf insects

  SUPERORDER HEMIPTERODEA

    ORDER ZORAPTERA: Zorapterans

    ORDER PSOCOPTERA: Book and bark lice

    ORDER PHTHIRAPTERA: Phthirapterans

    ORDER HEMIPTERA: True bugs

    ORDER THYSANOPTERA: Thrips

    ORDER ANOPLURA: Sucking lice

    ODER MALLOPHAGA: Biting lice, bird lice

    ORDER HOMOPTERA: Plant bugs, including white flies, aphids, scale insects, cicadas

  SUPERORDER HOLOMETABOLA

    ORDER STREPSIPTERA: Strepsipterans, twisted-wing insects

    ORDER COLEOPTERA: Beetles

    ORDER NEUROPTERA: Lacewings, ant lions, snake flies, dobsonflies, fish flies, alder flies, owl flies

    ORDER RAPHIDIOPTERA: Raphidiopterans

    ORDER MEGALOPTERA: Megalopterans

    ORDER HYMENOPTERA: Ants, bees, wasps

    ORDER MECOPTERA: Scorpion flies, snow flies

    ORDER SIPHONAPTERA: Fleas

    ORDER DIPTERA: True flies, mosquitoes, gnats

    ORDER TRICHOPTERA: Caddisflies

    ORDER LEPIDOPTERA: Butterflies, moths

Synopses of major uniramian taxa

Class Myriapoda

Postcephalic segments usually numerous, not differentiated into thorax and abdomen, typically each with one pair of walking legs; mandible with articulating endite; maxillules (first maxillae) and maxillae without palps; living species without compound eyes (except perhaps members of the order Scutigeromorpha, subclass Chilopoda); coxae of legs with single articulation with sternum; with lateral repugnatorial glands on trunk segments; without entodermally derived digestive ceca; with organs of Tömösvary; copulation indirect; development direct. (See Figure 1.)

Subclass Diplopoda. Trunk segments fused into pairs called diplosegments; each diplosegment with two pairs of legs, ganglia, and heart ostia; anteriormost segments often with legs suppressed except for some internal musculature; first trunk segment legless; antennae simple, seven-jointed; first maxillae (maxillules) fused into a gnathochilarium; second

maxillae absent; gonopores opening on or near coxae of second pair of legs (third trunk segment); cuticle normally calcified but without waxy epicuticle; most capable of rolling into a tight coil. Millipede classification is unstable, with many families of uncertain relationship and validity. About 10,000 species have been described.

Subclass Pauropoda. Small (0.5–1.5 mm); eyeless; 9–11 leg-bearing trunk segments (with some partly fused as diplosegments), plus free telson; without limbs on first trunk segment; mouthparts poorly developed; first maxillae (maxillules) fused as a gnathochilarium; second maxillae absent; antennae branched; most without tracheal or circulatory systems; blind; gonopores on third trunk segment; terga present, often large and extended over two segments; cuticle soft, uncalcified. Although found in all parts of the world, pauropodans occur mainly in moist soils and woodland litter; they are not especially common, numbering only about 500 species in 5 families.

Subclass Chilopoda. Numerous unfused trunk segments, all but the last two with one pair of legs, the first pair modified as large poison fangs; antennae simple, of varying segmentation; both pairs of maxillae medially coalesced; cuticle uncalcified; gonopores on last true body segment. Chilopods are generally divided into about 20 families; about 2,500 species have been described.

Figure 1

Representative myriapods. A, A California centipede (Chilopoda). B, *Scutigera coleoptrata*, the common house centipede (Chilopoda). C, A millipede (Diplopoda) from East Africa. D, *Pauropus silvaticus* (Pauropoda). E, *Scutigerella* (Symphyla). (A courtesy of G. McDonald; B photo by R. A. Wyttenbach/BPS; C photo by R. K. Burnard/BPS; D,E after Snodgrass 1952.)

Subclass Symphyla. Small (0.5–8.0 mm); eyeless; trunk with 14 segments, last fused to telson; first 12 trunk segments each with a pair of legs; penultimate segment with cerci and pair of long sensory hairs; dorsal surface with 15–22 tergal plates; soft uncalcified cuticle; antennae long, simple, threadlike; first maxillae medially coalesced; second maxillae completely fused as complex labium; one pair of spiracles (on head); tracheae supply first 3 trunk segments; gonopores open on third trunk segment. Symphylans are generally uncommon arthropods that occur in soil and rotting vegetation. About 120 species have been described in 2 families.

Class Insecta

Postcephalic segments differentiated into thorax (of 3 segments) and abdomen (of 11 or fewer segments); cephalon with one pair lateral compound eyes and (usually) a triad or pair of medial ocelli; with one pair of multiarticulate antennae, mandibles, and first maxillae; second maxillae fused to form complex labium; maxillae and labium usually with palp; each thoracic segment with one pair of legs; wings often present on second and third thoracic segments; gonopores open terminally or subterminally on abdomen (on seventh, eighth, or ninth abdominal segment); paired cerci often present; males often with intromittent and clasping structures; development direct, involving relatively slight changes in body form (ametabolous or hemimetabolous), or indirect, with striking changes (holometabolous). (See Figure 2.)

About 32 orders of insects are currently recognized; we include only some of these in the following synopses.

Subclass Myrientomata. Minute (smaller than 2 mm); whitish; wingless; without eyes, abdominal spiracles, hypopharynx, or cerci; Malpighian tubules vestigial, reduced to small papillae; sucking mouthparts; stylet-like mandibles with single articulation; vestigial antennae; first pair of legs carried in elevated position and used as surrogate "antennae"; abdomen 11-segmented, with well developed telson; first three abdominal segments with small appendages; without external genitalia, but male gonopores on protrusible phallic complex; gonopores terminal; with or without tracheae; simple development. Rare insects, occurring in leaf litter, moist soils, and rotting vegetation. One order, Protura, with about 100 species.

Subclass Oligoentomata. Small (most less than 6 mm); wingless; without Malpighian tubules; often without tracheae; biting–chewing mouthparts; mandibles with single articulation; abdomen of 6 or fewer segments, without spiracles or cerci; with terminal gonopores; antennae 4-articulate, first 3 articles with muscles; first abdominal segment with ventral tube (collophore) of unknown function; third abdominal segment with small, partly fused appendages; appendages of fourth or fifth abdominal segment form springlike structure (furcula) operated by hemocoelic fluid pressure; with or without small compound eyes; ocelli vestigial; simple development. One order (Collembola, the springtails), 11 families, and about 2,000 species.

Subclass Diplurata. Small (less than 4 mm); whitish; wingless; eyeless; without external genitalia or Malpighian tubules; chewing mouthparts; mandibles with single articulation; abdomen 11-segmented when embryonic, tenth and eleventh segments fuse before hatching; gonopores on ninth segment; 7 pairs of lateral abdominal styli; 1 pair caudal cerci; with tracheae and up to 7 pairs of abdominal spiracles; antennae multiarticulate, each article with musculature; simple development. One order (Diplura) composed of 7 families and about 100 species.

Subclass Zygoentomata. Small; wingless; without ocelli; compound eyes reduced; mandibles biting–chewing, with two articulations; antennae multiarticulate but only basal article with musculature; abdomen 11-segmented, with 3–8 pairs of lateral styli and 2–3 caudal filaments; female gonopores on eighth segment, male gonopores on tenth; without copulatory organs; with tracheae; simple development. Thysanurans occur in leaf litter or under bark or stones; some occur in buildings where they may become pests. The two orders, Thysanura (= Zygoentoma; silverfish) and Archaeognatha (rockhoppers), comprise about 700 described species.

Subclass Pterygota. With paired wings on the second and third thoracic segments (may be secondarily lost in one or both sexes, or modified for functions other than flight); mandibles with two articulations; only basal articles of antennae with musculature; without abdominal styli except on genital segments; female gonopores on eighth abdominal segment, on tenth in male; female often with ovipositor; molting ceases at maturity.

About 27 orders of pterygotes are recognized (depending on the authority). We diagnose below only the more important and/or common of these.

Infraclass Paleoptera. Wings cannot be folded back over body; when at rest, wings either held straight out to the side or vertically above the abdomen, with dorsal surfaces pressed together; wings always membranous, with many longitudinal and cross-veins; wings also tend to be fluted, or accordion-like; antennae highly reduced or vestigial in adults; hemimetabolous development. Two extant orders; many extinct groups.

Order Ephemeroptera. Adults with vestigial mouthparts, minute antennae, and soft bodies; wings held vertically over body when at rest; forewings present; hindwings present or absent; long, articulated, lateral cerci and usually a medial caudal filament; male with first pair of legs elongated for clasping female in flight; second and third legs of male, and all legs of female, may be vestigial or absent; abdomen 10-segmented; larvae aquatic; young (nymphs) with paired articulated lateral gills, caudal filaments, and well developed mouthparts; adults preceded by winged subimago stage.

Mayflies are primitive winged insects in which the aquatic nymphal stage has come to dominate the life cycle. Larvae hatch in fresh water and are long-lived, passing through many instars. Mayfly nymphs are important food for many fishes. Adults (imagos) live but a few hours or days, do not feed, and undertake aerial copulation, sometimes in large nuptial swarms. About 2,100 species are known.

Order Odonata. Adults with small filiform antennae, large compound eyes, and chewing mouthparts, including massive powerful mandibles; labium modified into prehensile organ; two pairs of large wings, at rest held outstretched (dragonflies) or straight up over body (damselflies); abdomen slender and elongate, 10-segmented; male with accessory genitalia on second and third abdominal sternites; eggs and larvae aquatic, with caudal or rectal gills.

Dragonflies and damselflies are primitive insects; the adults are highly active predators on other flying insects. Many species are 7–8 cm long, and some extinct forms were nearly a meter in length. The aquatic nymphs are also predaceous. About 5,500 species have been described.

Infraclass Neoptera. Wings at rest can be folded backward to cover the body; accordion-like wing construction seen in paleopterans has been lost in all but the most primitive neopteran orders.

Superorder Orthopterodea. Primitive neopterans

with biting–chewing mouthparts, two pairs of wings, and hemimetabolous development.

Order Blattodea. Body flattened dorsoventrally; pronotum large, with expanded margins and extending over head; forewings, when present, leathery; hindwings expansive and fanlike; ovipositor reduced; cerci multiarticulate; legs adapted for running; eggs laid in cases (ootheca).

Of the 3,700 described species of cockroaches, fewer than 40 are domestic (household inhabitants). Some species are omnivores, while others are restricted in diet. Some live in and feed on wood, and have intestinal flora that aid in cellulose digestion. Most species are tropical, but some live in temperate habitats. They are sometimes found in caves, deserts, and ant and bird nests. Cockroaches are often classified together with the mantids (below).

Order Mantodea. First pair of legs large and raptorial; prothorax elongate and often markedly elaborated; head highly mobile, with very large compound eyes, not covered by pronotum; forewings thickened, hindwings membranous; abdomen 11-segmented; with reduced ovipositor, but complex male genitalia; one pair multiarticulate cerci.

Mantids prey on insects and spiders. The digestive tract is short and straight, but includes a large crop, ribbed or toothed proventriculus, and 8 midgut ceca. Malpighian tubules are numerous, over 100 in some species. The order is primarily tropical, with only a few of the 1,800 species occurring in temperate regions.

Order Isoptera. Small; soft-bodied; wings equal-sized, elongate, membranous, dehiscent (shed by breaking at basal line of weakness); antennae short, filamentous, 11- to 33-segmented; cerci small to minute; ovipositor reduced or absent; many with rudimentary or no external genitalia; marked polymorphism.

Termites ("white ants") are strictly social insects, usually with three distinct types of individuals, or castes: reproductives, soldiers, and workers. Workers are generally sterile, blind, individuals with normal mandibles, responsible for foraging, nest construction, and caring for members of the other castes. Soldiers are blind, usually sterile, wingless forms with powerful enlarged mandibles used to defend the nest. Reproductives have wings and fully formed compound eyes and are produced in large numbers at certain times of the year, whereupon they emerge from the colony in swarms. Mating occurs at this

Figure 2

Representative insects and their orders. A, The common silverfish, *Lepisma saccharina* (Thysanura). B, A mayfly (Ephemeroptera). C, A damselfly (Odonata). D, A praying mantis, *Stagmomantis californica* (Mantodea). E, The California giant stone fly, *Pteronareys californica* (Plecoptera). F, The differential grasshopper, *Melanoplus differentialis* (Orthoptera). G, The European earwig, *Forficula auricularia* (Dermaptera). H, A webspinner (Embioptera). I, A walking stick, *Diapheromera* (Phasmida). J, The western box-elder bug, *Leptocoris rubrolineatus* (Hemiptera). K, A woodland cicada, *Platypedia* (Homoptera). L, The western pine-borer beetle, *Chalcophora angulicornis* (Coleoptera). M, An ant lion, *Brachynemurus* (Neuroptera). N, A flea, *Dolichopsyllus* (Siphonaptera). O, A robber fly (Diptera). P, The silver-striped caddis fly, *Hesperophylax* (Trichoptera). Q, A congregation of monarch butterflies (Lepidoptera). (A,B,D,F,G,H,J,K,L,M,O,P photos by P. J. Bryant/BPS; C photo by R. A. Wyttenbach/BPS; E photo by R. Humbert/BPS; I photo by B. J. Miller/BPS; N photo by J. DeMartini; Q photo by G. McDonald.)

time, and individual pairs start new colonies. Wings are shed after copulation. Colonies form nests (**termitaria**) in wood that is in or on the ground.

Workers in the presumed primitive termite families harbor a variety of symbiotic cellulose-digesting flagellate protozoa in special chambers in the hindgut (termites are thought to have evolved from a form of wood-eating cockroach). Other families contain symbiotic bacteria that serve the same purpose.

Termites often occur in enormous numbers; one spectacular estimate suggests that there are about three-quarters of a ton of termites for every person on earth! Their numbers are probably growing rapidly as continual deforestation produces ever increasing food and habitat supplies. About 2,000 termite species have been described.

Order Plecoptera. Adults with reduced mouthparts, elongate antennae, (usually) long articulated cerci, soft bodies, and a 10-segmented abdomen; without ovipositor; wings membranous, pleated, folding over and around abdomen when at rest; wings with primitive venation; nymphs aquatic, with gills.

Stone flies, like mayflies, have experienced considerable evolutionary radiation of the larval (nymphal) stage. Stone fly nymphs are important consumers in freshwater systems and also serve as major prey items for various fishes and invertebrates. Adults of most species feed, but are short-lived and die soon after mating. About 1,600 species have been described.

Order Orthoptera. Pronotum unusually large, extending posteriorly over mesonotum; forewings thickened and leathery; hindwings fanlike, occasionally modified for stridulation or camouflage; hindlegs often large, adapted for jumping and, rarely, for stridulation; auditory tympana present on forelegs and abdomen; ovipositor large; male genitalia complex; cerci distinct, short, and jointed.

Grasshoppers, locusts, crickets, and their kin are common and abundant insects at all but the coldest latitudes. This order includes some of the largest living insects, to 12 cm in length and twice that in wingspan. About 20,000 species have been described. Most are herbivores, but many are omnivorous and some are predatory. Stridulation is common among males, usually by rubbing specially modified forewings (tegmina) together, or rubbing a ridge on the inside of the hind femur against a special vein of the tegmen. None stridulate by rubbing the hindlegs together, as is commonly thought.

Order Dermaptera. Cerci form heavily sclerotized posterior forceps; forewings (when present) are short and leathery, the tegmina lack veins; forewings serve as elytra covering the semicircular, membranous hindwings (when present); ovipositor reduced or absent.

M

N

O

P

Q

Earwigs are common but poorly understood insects. Most appear to be nocturnal scavenging omnivores. The forceps are used in predation, for defense, to hold the mate during courtship, for grooming the body, and for folding the hindwings under the elytra. Some species eject a foul-smelling liquid from abdominal glands when disturbed. Most species are tropical, although many also inhabit temperate regions. About 1,100 species have been described.

Order Phasmida. Body cylindrical or markedly flattened dorsoventrally, usually elongate; prothorax short; meso- and metathorax elongate; forewings absent, or forming small and leathery tegmina; hindwings fanlike; biting–chewing mouth parts; short unsegmented cerci; ovipositor weak; caudal cerci minute.

Walking sticks, stick and leaf insects, specters and other phasmids are some of the oddest of all the insects. Although resembling orthopterans in basic form, they are clearly a distinct radiation of their own. Certain walking sticks reach a body length of nearly 35 cm, though most are less than 4 cm. Sexual dimorphism is so striking that males and females have often been given different names. Over 2,500 species have been described.

Superorder Hemipterodea. Antennae usually short, with reduced number of articles; leg tarsi of three or fewer articles; no cerci; true male gonopods wanting; ocelli absent in immature stages, but present in winged adults of some taxa; when present, wings have reduced venation; four or fewer Malpighian tubules; most species small, although a few hemipterans attain lengths of several centimeters.

The overall trend in this superorder has been toward a reduction of anatomical features. Some orders have specialized piercing–sucking mouthparts for fluid feeding. Development is hemimetabolous and relatively gradual, although metamorphosis in several groups includes one or two inactive pupa-like stages.

Order Zoraptera. Wingless or with two pairs of wings; termite-like in appearance; wings eventually shed; colonial; antennae moniliform, with 9 articles; abdomen short, oval, 10-segmented; chewing mouthparts; simple development; about 25 species of minute (to 3 mm), uncommon insects, usually found in colonies in dead wood; they feed chiefly on mites and other small arthropods.

Order Psocoptera. Small (1–10 mm long); antennae long, filiform, of many articles; prothorax short; meso- and metathorax often fused; chewing mouthparts; abdomen of nine segments; cerci absent.

Psocids—the book and bark lice—generally feed on microflora such as algae and fungi, and occur in moist areas suitable to such plant growth (e.g., under bark, in leaf litter, under stones, and in human habitations where humid climates prevail). They are often pests that get into various stored food products or consume insect and plant collections, and some species live in books and eat the bindings. About 2,600 species have been described.

Order Hemiptera. Piercing–sucking mouthparts form an articulated beak, generally arising from the anterior part of the head; mandibles and first maxillae stylet-like, lying in dorsally grooved labium; dorsum somewhat flattened, with both pairs of wings lying flat on abdomen at rest; hindwings membranous; forewings hardened basally, membranous distally, and smaller than hindwings; pronotum large.

Most true bugs are herbivorous, although many prey on other arthropods or vertebrates, and some are specialized vertebrate ectoparasites. Sound production is widespread through a variety of different mechanisms. Many exhibit cryptic coloration or mimic other insects; ant mimicry is especially common. Hemipterans are of considerable economic importance, many being serious crop pests or carriers of human disease organisms (e.g., Chagas disease). They occur worldwide and in virtually all habitats. About 35,000 species have been described. Common families of Hemiptera include Nepidae (water scorpions), Belostomatidae (giant water bugs or "toe biters"), Corixidae (water boatmen), Notonectidae (backswimmers), Gerridae (water striders), Saldidae (shore bugs), Cimicidae (bedbugs), and Reduviidae (assassin and kissing bugs).

Order Thysanoptera. Slender, minute (0.5–1.5 mm) insects with long narrow wings (when present) bearing long marginal setal fringes; mouthparts form a conical, asymmetrical, sucking beak; left mandible a stylet, right mandible vestigial; with compound eyes; antennae of 4–10 articles; abdomen 10-segmented; without cerci.

About 4,000 species of thrips have been described. Most are herbivores or predators, and many pollinate flowers and are known to transmit plant viruses and fungal spores; hence the group is of considerable economic significance.

Order Anoplura. Wingless, blood-sucking ectoparasites of mammals; less than 4 mm in length; thorax completely fused; cuticle largely membranous and expandable to permit engorgement; compound eyes absent or of a single ommatidium; ocelli absent; piercing–sucking mouthparts retractable into a buccal trophic pouch; antennae short (3–5 articles); with one pair dorsal thoracic spiracles; females lack ovipositor; without cerci.

Commonly called "sucking lice," these insects spend their entire life on one host. Eggs (nits) are usually attached to the hair of the host, although the human body louse may attach eggs to clothing. Posthatching development comprises three nymphal instars. About 500 species are known.

Order Mallophaga. Wingless, obligate ectoparasites on birds and mammals; mostly less than 5 mm long; antennae of 5 or fewer articles, exposed or concealed in grooves beneath the head; chewing mouthparts; cuticular sclerites more rigid and distinct than in anoplurans; meso- and metathorax often partly fused; one pair ventral thoracic spiracles and 6 or fewer abdominal spiracles; compound eyes with only two ommatidia; no ocelli; female without ovipositor.

The biting or chewing lice have thick cuticles and their color usually matches that of the fur or feathers of their hosts. Like sucking lice, biting lice glue their large eggs on the fur or feathers of the host, and hatching is followed by 3 nymphal instars. Some species that infest domestic birds and mammals are of economic importance. None infests humans and no human diseases are known to be transmitted by biting lice. About 2,800 species have been described.

Order Homoptera. With piercing–sucking mouthparts that form a beak, generally arising from posterior part of the head; mandibles and first maxillae stylet-like, lying in labium; both pairs of wings membranous, usually held tentlike over abdomen when at rest (some families wingless); pronotum large; hindlegs often adapted for jumping; body often protected by waxy secretions; most produce saccharine anal secretions (honeydew).

Homopterans are another large and abundant group, with over 33,000 species known from most of the world's environments. They closely resemble hemipterans and are sometimes classified with them in a single order (see Figure 49). All are plant feeders, hence the common name "plant bugs." The homopterans are of considerable economic importance. Heavy infestations on plants may cause wilting, stunting or even death, and some are vectors of important plant diseases. Common families include the Cicadidae (cicadas), Cicadellidae (leafhoppers), Fulgoridae (plant hoppers), Membracidae (tree hoppers), Cercopidae (spittlebugs and froghoppers), Aleyrodidae (whiteflies), Aphidae (aphids), as well as coccoids, scale insects, mealybugs, and many others.

Superorder Holometabola. Development holometabolous (complete), with distinct egg, larval, pupal, and adult stages; larvae with internal wing buds; mouth parts usually biting–chewing; Malpighian tubules numerous in adult.

The holometabolans are the dominant living insects, occurring worldwide in great numbers. Larvae and adults generally inhabit quite different habitats.

Order Coleoptera. Body usually heavily sclerotized; forewings sclerotized, leathery or horny, and modified as rigid covers (elytra) over hindwings; hindwings membranous, often reduced or absent; biting mouthparts; antennae usually with 8–11 articles; prothorax large and mobile; mesothorax reduced; abdomen typically of 5 (or up to 8) segments; without ovipositor; male genitalia retractable.

Beetles comprise the largest order of insects, with over 300,000 described species—almost a third of all known animals! They range from minute to large, and occur in all the world's environments (except the open sea). The order Coleoptera includes several common families such as the Cicindelidae (tiger beetles), Carabidae (ground beetles), Dytiscidae (predaceous diving beetles), Gyrinidae (whirligig beetles), Hydrophilidae (water scavenger beetles), Staphylinidae (rove beetles), Cantharidae (soldier beetles), Lampyridae (fireflies and lightning bugs), Phengodidae (glowworms), Elateridae (click beetles), Buprestidae (metallic wood-boring beetles), Coccinellidae (ladybird beetles), Meloidae (blister beetles), Tenebrionidae (darkling beetles), Scarabaeidae (scarab beetles, dung beetles, and june "bugs"), Cerambycidae (long-horned beetles), Chrysomelidae (leaf beetles), Curculionidae (weevils), and Brentidae (primitive weevils).

Order Neuroptera. Soft-bodied; two pairs of similar, highly veined wings held tentlike over the abdomen when at rest; larvae with well developed legs; adults with biting–chewing mouthparts; Malpighian tubules secrete silk at pupation; abdomen 10-segmented; without cerci; with a group of glandular sensilla, called trichobothria, on the abdomen.

Lacewings, ant lions, snake flies, and dobsonflies form a complex group of over 4,700 described species. Adults of many species are important predators of pest insects (e.g., aphids). Larvae of some (hellgrammites) are commonly used as fish bait. In many, larvae have piercing–sucking mouthparts, and in others, the larvae are predaceous and have biting mouthparts. Pupae are often unusual in possessing free appendages and functional mandibles; they may actively walk about prior to the adult molt, but do not feed.

Order Hymenoptera. Mouthparts often elongate and modified for ingesting floral nectar, although the mandibles usually remain functional; labium often (bees) distally expanded as a glossa; usually with two pairs of membranous wings; hindwings small, coupled to forewings by hooks; wing venation highly reduced; antennae well developed, of various forms and with 3–70 articles; reduced metathorax usually fused to first abdominal segment; males with complex genitalia; females with ovipositor (in most), modified for sawing, piercing or stinging.

Ants, bees, wasps, sawflies, and their relatives are all active insects with a tendency to form polymorphic social communities. Two suborders are generally recognized. Included in the suborder Symphyta are the primitive, wasplike, "thick-waisted" hymenopterans (sawflies, horntails, and their kin). They rarely show conspicuous sexual dimorphism and are always fully winged. The first and second abdominal segments are broadly joined. Larvae are mostly caterpillar-like, with a well developed head capsule, true legs, and often abdominal prolegs. About 4,700 species have been described.

The suborder Apocrita comprises the "narrow-waisted" hymenopterans (true wasps, bees, and ants), in which the first and second abdominal segments are joined by a distinct and often elongate constriction. Adults tend to be strongly social and display marked polymorphism. Social communities often include distinct castes of haploid males, parthenogenetic females, queens, and indivuduals with other sex-related aberrations, as well as worker and soldier forms. Larvae are legless, usually soft, white, and grublike. They feed within or upon the body of a host arthropod or its egg, in a plant gall, or fruit or seed. Some live in nests constructed by the adults and, as in bees, are fed by the adults. About 125,000 species have been described.

Order Mecoptera. Two pairs of similar, narrow, membranous wings, carried horizontally from sides of body when at rest; antennae long, slender, and of many articles (about half of the body length); head with ventral rostrum and reduced biting mouthparts; long slender legs; meso-, metathorax, and first abdominal tergum fused; abdomen 11-segmented; female with two cerci; male genitalia prominent and complex, at apex of attenuate abdomen and often resembling a scorpion's stinger.

Scorpion flies and their allies are usually found in moist places, often in forests, where most are diurnal flyers. Some feed on nectar, others prey on insects or are scavengers. Just under 500 species have been described worldwide; they are best represented in the Holarctic region.

Order Siphonaptera. Small (less than 5 mm long); wingless; body laterally compressed and heavily sclerotized; short antennae lie in deep grooves on sides of head; mouthparts piercing–sucking; compound eyes often absent; legs modified for clinging and (especially hindlegs) jumping; abdomen 11-segmented; abdominal segment 10 with distinct dorsal pincushion-like sensillum, containing a number of sensory organs; without ovipositor; pupal stage passed in cocoon.

Adult fleas are ectoparasites on mammals and birds, from which they take blood meals. They occur wherever suitable hosts are found, including the Arctic and Antarctic. Larvae usually feed on organic debris in the nest or dwelling place of the host. Host specificity is often weak, particularly among the mammal parasites, and fleas regularly commute from one host species to another. Fleas act as intermediate hosts and vectors for such diseases as plague, dog and cat tapeworms, and various nematodes. About 1,750 species have been described.

Order Diptera. Adults with one pair membranous mesothoracic forewings and a metathoracic pair of clublike halteres (organs of balance); head large and mobile; compound eyes large; antennae primitively filiform, with 7–16 articles, and often secondarily annulated (reduced to only a few articles in some groups); mouthparts adapted for sponging, sucking, or lapping; mandibles of blood-sucking females developed as piercing stylets; hypopharynx, laciniae, galeae, and mandibles variously modified as stylets in parasitic and predatory groups; labium forms a proboscis, consisting of distinct basal and distal portions, the latter in higher families forming a spongelike pad (labellum) with absorptive canals; mesothorax greatly enlarged; abdomen primitively 11-segmented, but reduced or fused in many higher forms; male genitalia complex; females without true ovipositor, but many with secondary ovipositor composed of telescoping posterior abdominal segments; larvae lack true legs, although ambulatory structures (prolegs and pseudopods) occur in many.

The true flies (including mosquitoes and gnats) are a large and diverse group. They are noted for their excellent vision and aeronautic capabilities. The mouthparts and digestive system are modified for a fluid diet, and several groups feed on blood or plant juice. Dipterans are important carriers of diseases, such as sleeping sickness, yellow fever, African river blindness, and various enteric diseases. Myiasis—the infestation of living tissue by dipteran larvae—is often a problem with livestock and occasionally humans. Many dipterans are also beneficial (to people) as parasites/predators of other insects and as pollinators of flowering plants. Dipterans occur worldwide and in virtually every major environment (except the open sea). Some breed in extreme environments, such as hot springs, saline desert lakes, oil seeps, tundra pools, and even shallow benthic marine habitats. About 150,000 species have been described.

Some common dipteran families include Tipulidae (crane flies), Culicidae (mosquitoes), Simuliidae (black flies and buffalo gnats), Chironomidae (midges), Tabanidae (horse flies, deerflies, and clegs), Bombyliidae (bee flies), Asilidae (robber flies), Syrphidae (hover flies and flower flies), Tephritidae (fruit flies), Otitidae (picture-winged flies), Coelopidae (kelp flies), Drosophilidae (pomace or vinegar flies; often also called "fruit flies"), Ephydridae (shore flies and brine flies), Scatophagidae (dung flies), Muscidae (house flies, stable flies, etc.), Calliphoridae (blowflies, bluebottles, greenbottles, etc.), Sarcophagidae (flesh flies), Tachinidae (tachinid flies), and Glossinidae (tsetse flies).

Order Trichoptera. Adults resemble small moths, but with the body and wings covered with short hairs (rather than scales); two pairs of wings, tented in oblique vertical plane (rooflike) over abdomen when at rest; compound eyes present; mandibles minute or absent; antennae usually as long or longer than body, setaceous; legs long and slender; larvae and pupae mainly in fresh water, adults terrestrial; larvae with abdominal prolegs on terminal segment.

The freshwater larvae of caddis flies construct fixed or portable "houses" (cases) made of sand grains, wood fragments, or other material bound together by silk emitted through the labium. Some also use silk to produce food-filtering devices. Most larvae inhabit benthic habitats of temperate streams, ponds, and lakes. Larvae are primarily herbivorous scavengers. Adults are strictly terrestrial and have liquid diets. About 7,000 species have been described.

Order Lepidoptera. Minute to large; sucking mouthparts; mandibles usually vestigial; first maxillae coupled, forming a tubular sucking proboscis, coiled between labial palps when not in use; head, body, wings, and legs usually densely scaled; compound eyes well developed; usually with two pairs of membranous and colorfully scaled wings, coupled to one another by various mechanisms; male genitalia complex; females with ovipositors.

Butterflies and moths are among the best known and most colorful of all the insects. Adults are primarily nectar feeders, and many are important pollinators. Larvae are wormlike caterpillars that feed on green plants. Caterpillars have three pairs of thoracic legs and a pair of soft prolegs on each of abdominal segments 3–6. The anal segment bears a pair of prolegs or claspers. About 120,000 species have been described.

The uniramian *Bauplan*

In Chapter 15 we briefly discussed the various advantages and constraints imposed by the phenomenon of arthropodization, including those associated with the establishment of a terrestrial lifestyle. Departure from the ancestral aquatic environment necessitated the evolution of stronger and more efficient support and walking appendages, of special adaptations to withstand osmotic and ionic stress, and of aerial gas exchange structures. Reproductive behavior became increasingly complex, and in many cases highly evolved social systems developed. Within the class Insecta, many taxa underwent intimate coevolutionary radiation with land plants, particularly angiosperms. The basic arthropod *Bauplan* included many preadaptations to life in a "dry" world. As we have seen, the arthropod exoskeleton inherently provides physical support and protection from predators; and by incorporating waxes into the epicuticle, the insects, like the arachnids, acquired an effective barrier to water loss. Similarly, the highly adaptable, serially arranged arthropod limbs evolved into a variety of specialized locomotor and food-capturing appendages in the uniramians (Figures 3, 4, and 5).

The key features that distinguish the members of the Uniramia are listed in Box 1. The appendages of the head are, from anterior to posterior: **antennae,**

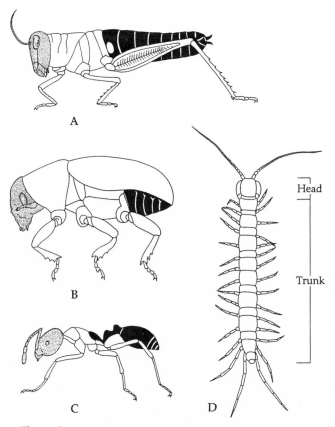

Figure 3

Principal body regions of insects and myriapods. A–C, Three kinds of insects (class Insecta). A, Grasshopper. B, Beetle. C, Ant. D, A centipede (class Myriapoda). The stippled region is the head; the white region is the thorax; the black region is the abdomen.

> **Box One**
> # Characteristics of the Subphylum Uniramia
>
> 1. Body of three (Insecta) or two (Myriapoda) tagmata; trunk (or abdomen) of many segments
> 2. All appendages are multiarticulate and uniramous; head appendages often reduced to a single article[a]
> 3. Head appendages, from anterior to posterior, are antennae, mandibles, maxillules (first maxillae), and maxillae (second maxillae); the second maxillae may be fused into a single flaplike structure called a labium (not homologous to the crustacean "labium"), or may be wanting
> 4. Mandibles are whole-limb jaws (i.e., of a single article, with a biting surface at the tip)[b]
> 5. With an aerial gas exchange system composed of tracheae and spiracles
> 6. With ectodermally derived Malpighian tubules (proctodeal evaginations)
> 7. Most with simple ocelli, at least in some stage of the life cycle
> 8. Gut without digestive ceca in Myriapoda; with ceca in Insecta
> 9. Without a carapace
> 10. Gonochoristic

[a]Pauropodan antennae are branched, but whether this represents a vestige of a primitive biramous conditon (as in crustaceans) or is secondarily derived is not certain.
[b]An articulating mandibular endite occurs in some myriapods, but never in insects.

mandibles, **first maxillae*** (= maxillules), and **second maxillae**. The second maxillae may be either free, or fused into a flaplike **labium†** that covers the buccal field, or absent altogether. In contrast to all other arthropod groups, a carapace never develops, and there is no evidence that one ever existed. In adult insects, the head is followed by a clearly differentiated thorax and limbless abdomen. Myriapods, on the other hand, have but two body tagmata—the head and a long, homonomous trunk.

The three segments of the insect thorax are termed **prothorax**, **mesothorax**, and **metathorax** (Figure 4). As is typical for arthropods, each thoracic

segment is divided radially into four regions, each region composed of one or more sclerites: the dorsal **notum** (pl., **nota****); the ventral **sternum**, typically comprising a single **sternite**; and the lateral **pleurites**. Thoracic sclerites are also referred to by the particular segment on which they occur. Thus, the notum of the prothorax is the **pronotum**, the pleurites of the mesothorax the **mesopleurites**, and so on. The pronotum of many insects is a large and conspicuous sclerite extending from the head to the base of the

*Entomologists (and entomology texts) generally refer to the first maxillae (maxillules) of insects simply as *maxillae*, and the fused second maxillae as the *labium*.

†The "labium" of uniramians is not homologous to the "labium" of crustaceans; the former is derived from true appendages (the second maxillae); the latter is not (see Chapter 18).

**In insects, the thoracic tergites are called nota.

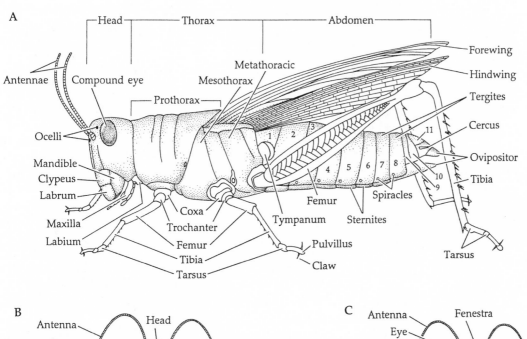

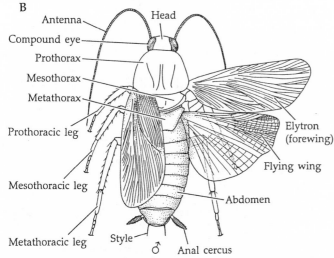

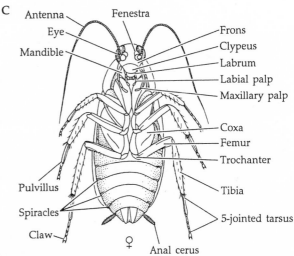

Figure 4

General body anatomy of insects. **A,** Grasshopper (order Orthoptera). **B–C,** Dorsal and ventral views of a cockroach (order Blattodea). (**A** after Snodgrass 1942; **B,C** after Sherman and Sherman 1976, from other sources.)

wings. The three somites of the thorax each bears a pair of walking legs that articulate with the pleural regions. Wings, when present, are borne on the meso- and metathorax and articulate with processes on the notum and pleural regions of these two somites. The pronotum is usually a single sclerite. The meso- and metanota are often divided into three sclerites each. In winged orders the pleural regions are separated into anterior and posterior sclerites (pleurites) by an arthrodial membrane, called a **pleural suture**, extending from the base of the leg to the base of the wing. Each sternite is also often divided into two or three sclerites.

The insect abdomen typically consists of 11 segments, but the last segment is often vestigial, and as a result of fusion many insects have fewer than 11

functional abdominal segments. Each abdominal segment generally includes only two sclerites, a tergite and a sternite, the pleurites being minute or absent. Although the abdominal segments of adult insects lack appendages, transitory limb buds or rudiments appear fleetingly in the embryos and developmental forms of some species, harking back to a more homonomous, myriapod-like ancestral condition. A pair of sensory **cerci** (sing., **cercus**) often project from the last abdominal segment.

Insects have a waxy layer in the epicuticle that has allowed them to fully exploit even the driest of

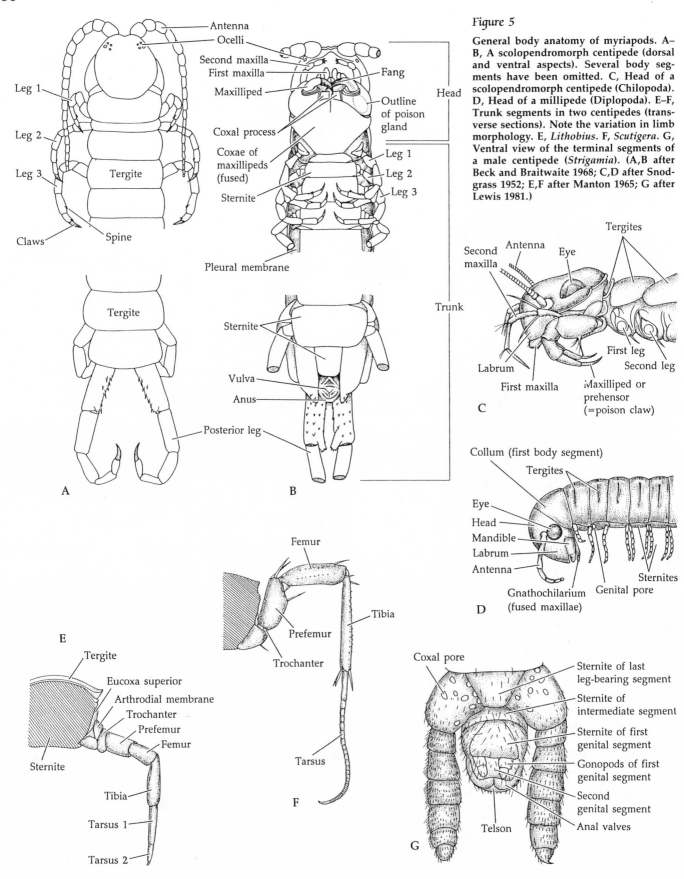

Figure 5

General body anatomy of myriapods. A–B, A scolopendromorph centipede (dorsal and ventral aspects). Several body segments have been omitted. C, Head of a scolopendromorph centipede (Chilopoda). D, Head of a millipede (Diplopoda). E–F, Trunk segments in two centipedes (transverse sections). Note the variation in limb morphology. E, *Lithobius*. F, *Scutigera*. G, Ventral view of the terminal segments of a male centipede (*Strigamia*). (A,B after Beck and Braitwaite 1968; C,D after Snodgrass 1952; E,F after Manton 1965; G after Lewis 1981.)

terrestrial environments. Myriapods never acquired a cuticular wax layer and are largely relegated to more humid environments, although a few kinds of millipedes live in semiarid conditions. Most insects are small, between 0.5 and 3.0 cm in length. The smallest are the thrips, feather-winged beetles, and certain parasitic wasps, which are all nearly microscopic. The largest are certain beetles, orthopterans, and walking sticks, the latter attaining lengths greater than 30 cm in some species.

The myriapods differ from insects primarily in their retention of a more homonomous condition, e.g., lack of a differentiated thorax, possession of paired appendages on all or most postcephalic segments, and absence of wings (Figure 5). The head appendages and legs are similar to those of the insects. Metamerism is strong in myriapods and evident internally in structures such as the segmental heart ostia, tracheae, and ganglia.

The chilopods, or centipedes, bear one pair of walking legs per segment, except on the last two segments, which are limbless. There may be up to 173 trunk segments. Most species are 1–2 cm in length, although some tropical giants attain lengths of 20–25 cm.

The diplopods, or millipedes, display an interesting modification of basic uniramian metamerism. The head is followed by an expanded limbless segment called the **collum**, which forms a conspicuous, heavily sclerotized collar between head and trunk (Figure 5). Each of the next three body segments bears a single pair of legs. The rest of the trunk is composed of **diplosegments**, each formed by fusion of two somites and bearing a double complement of metameric organs and structures, including two pairs of legs. Each diplosegment has one tergite, two pleurites, and two or three sternites. Like centipedes, millipedes range in length from 1 to 30 cm. The cuticle of millipedes is particularly robust, being well sclerotized and often strengthened by the addition of calcium salts.

Pauropodans are minute, soft-bodied, eyeless, soil-inhabiting myriapods that resemble millipedes (Figure 1D). Most are less than 2 mm in length and have at most 11 trunk segments, nine of which have legs. As in millipedes, segmental pairing occurs and usually only six tergites are visible dorsally.

Symphylans (Figure 1E) are also minute, eyeless myriapods, never exceeding 1 cm in length. The

Figure 6

The mouth appendages of a typical biting–chewing insect, a grasshopper.

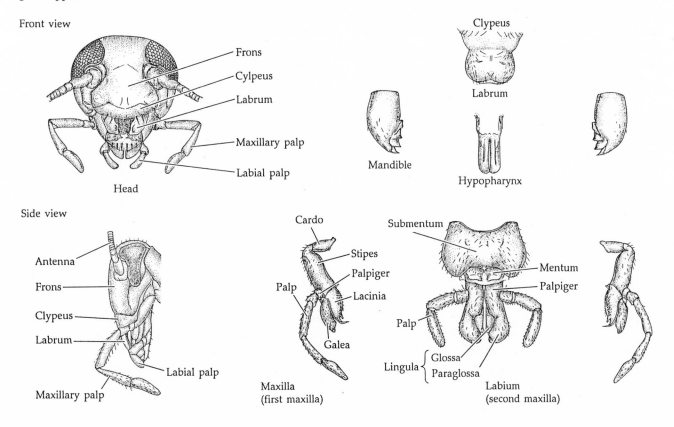

trunk has 12 pairs of leg-bearing segments. Some tergites are divided, and 15–22 tergites are usually visible dorsally. The thirteenth body segment bears a pair of long sensory hairs (cerci) that serve as spinnerets, and a tiny postsegmental telson. Like pauropodans, symphylans inhabit loose soil and humus.

The uniramian head and mouth appendages

As you have no doubt realized by now, much discussion and evolutionary speculation regarding the Uniramia (and the arthropods in general) focus on the nature of the head and its appendages (Figures 5, 6, and 7). The basic arthropod head of a presegmental acron plus a number of fused somites (see Table 2 in Chapter 19) is retained in the uniramians, as follows. The acron bears the eyes and contains the protocerebrum. The first postacronal cephalic somite (the **antennary segment**) bears the antennae (Figure 8) and houses the deuterocerebrum. The second true cephalic segment is called the **premandibular segment**; it lacks appendages but houses the tritocerebrum. The third true segment (**mandibular segment**) carries the mandibles, and the fourth (**first maxillary segment**) bears the first pair of maxillae (= maxillules). The fifth cephalic segment (**second maxillary segment**) bears the second pair of maxillae (= maxillae). The second maxillae are fused as a labium in insects and symphylans, but they have been lost altogether in millipedes and pauropodans. In insects the first maxillae and labium usually bear small palps. In diplopodans and pauropodans the first maxillae fuse to form a flaplike **gnathochilarium**. The ganglia of the mandibular and maxillary segments are generally fused to form the subesophageal ganglion.

Some specialists argue that the first postacronal somite of uniramians is a limbless preantennary somite, thus making the antennae the appendages of the second (rather than the first) postacronal cephalic somite. At the heart of this disagreement is the question of whether the appendages of the first or second segment (i.e., first or second antennae) were lost in the evolutionary history of this group. Innervation from the brain and the presence of unique sensory "peg organs" on uniramian antennae (which may be homologous to the aesthetascs of crustacean first antennae) argue in favor of uniramian antennae being appendages of the first postacronal segment.

Insects and myriapods possess two additional head structures (Figures 5 and 6). Projecting off the front of the head, anterior to the mouth field, is a platelike process called the **labrum**, or **upper lip**. The labrum arises from a frontal head sclerite called the

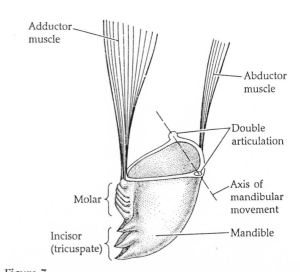

Figure 7

The musculature of an insect mandible. (After Snodgrass 1952.)

clypeus and hangs down over the mandibles, much as our upper lip covers our teeth. In insects the clypeus attaches to another sclerite, the **frons** (= **epistome**), which often bears an ocellus. Posterior to the buccal field, arising behind the first maxillae and near the base of the labium, is another flaplike or tonguelike structure called the **hypopharynx**. The **salivary glands** open through the hypopharynx. Neither the labrum nor the hypopharynx are true appendages; both are independent outgrowths of the body wall. The clypeus, labrum, and hypopharynx also occur in crustaceans; these structures are probably true homologues in the two groups, although this matter also is being debated.

In insects the orientation of the head and mouthparts with respect to the rest of the body varies (Figure 9). The situation in which the mouth and mouthparts occupy the rear portion of the head, more-or-less in a continuous series with the legs, is considered the most primitive condition, called the **hypognathous condition**. The hypognathous condition occurs mostly in herbivorous groups and bloodsuckers. In the **prognathous condition**, the mouth and mouthparts are shifted forward; this condition is typical of carnivorous species that actively pursue their prey and in larvae, particularly larvae of members of the order Coleoptera, which use their mandibles for burrowing. In insects of the orders Hemiptera and Homoptera, the mouthparts are **opisthorhynchous**, forming an elongate proboscis that slopes backward between the front legs. More will be said on insect mouthpart specialization in the section on feeding.

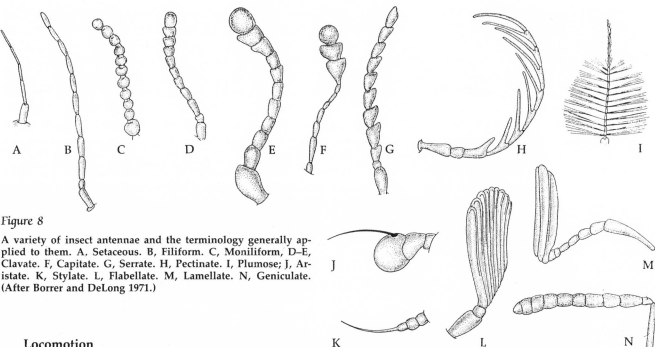

Figure 8

A variety of insect antennae and the terminology generally applied to them. A. Setaceous. B, Filiform. C, Moniliform, D–E, Clavate. F, Capitate. G, Serrate. H, Pectinate. I, Plumose; J, Aristate. K, Stylate. L, Flabellate. M, Lamellate. N, Geniculate. (After Borrer and DeLong 1971.)

Locomotion

Pedestrian locomotion. Uniramians rely heavily on their well-sclerotized exoskeleton for both body and limb support on land, and the limbs provide the physical support to lift the body clear of the ground during locomotion. In order to accomplish this, the limbs must be long enough to hold the body up high, but not so high as to endanger stability. Most uniramians maintain stability by keeping the legs in positions that suspend the body in a slinglike fashion and by keeping the overall center of gravity low (see Figures 10–14).

The basic design of arthropod limbs was described in Chapter 15. Recall that in most arachnids

the coxae are immovably fixed to the body, and limb movement occurs at more distal joints. However, in uniramians (as in most crustaceans) antero-posterior limb movements take place between the coxae and body proper. Like the power controlled by the range of gears in an automobile, the power exerted by a limb is greatest at low speeds and less at higher speeds. At lower speeds, the legs are in contact with the ground for longer periods of time, and *more* legs are contacting the ground at any given moment. Thus, in burrowing forms, such as most millipedes, the legs are short and the gait is slow and powerful as they literally bulldoze their way through soil or rotting wood. At high speeds, less than half the limbs

Figure 9

Different positions of head and mouthparts relative to the rest of the body. A, Hypognathous condition (grasshopper). B, Prognathous condition (beetle larva). C, Opisthorhynchous condition (aphid). (After Chapman 1982.)

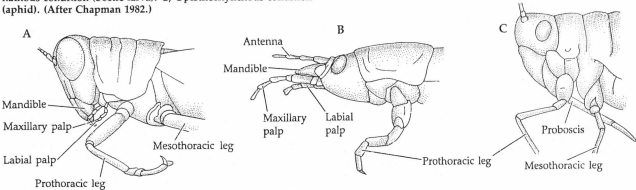

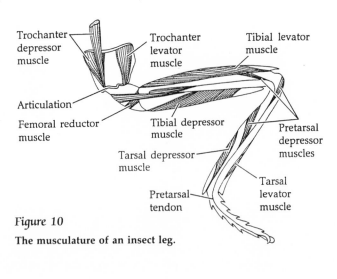

Figure 10

The musculature of an insect leg.

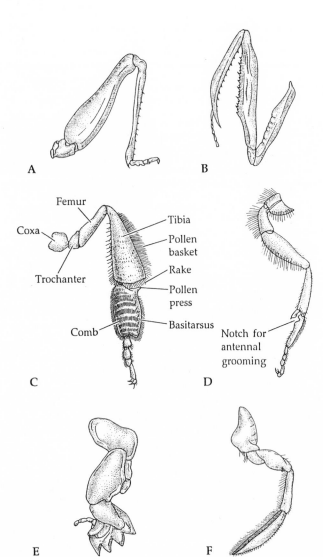

Figure 11

Leg modifications in some insects. A, The hindleg of a grasshopper (order Orthoptera). This leg is adapted for jumping. B, The raptorial foreleg of a praying mantis (order Mantodea). This leg is adapted for prey capture. C, The hindleg of a honeybee (order Hymenoptera). This leg is adapted for collecting and holding pollen. D, The foreleg of a worker honeybee has a notch for antennal grooming. E, The foreleg of a mole cricket (order Orthoptera) is modified for digging. F, The hindleg of a backswimmer (order Coleoptera) is modified for swimming. (A,B,D,E,F after Barth and Broshears 1982; C after Chapman 1982.)

may touch the ground at any given moment, and for shorter periods of time (Figure 12). Longer limbs increase the speed of a running gait, as do limbs capable of swinging through a greater angle. Thus, limbs long in length and stride are typical features of the fastest running uniramians, such as many insects, centipedes, and symphylans. Among the pauropodans, a range of movement speeds is seen—from species that are "slow plowers" to those that are fast runners.

Locomotor repertoires typically evolve in concert with overall habits of animals, particularly feeding behaviors. For example, many centipedes are largely surface-dwelling predators and predictably tend to run quickly across the surface of the substratum. *Scutigera* is a small centipede that qualifies as "world class runner," reaching speeds up to 42 cm/sec when in pursuit of its favorite prey—flies. On the other hand, most millipedes are herbivores or scavengers that burrow through soil, leaf litter, or rotten timber in search of food. Thus, millipedes tend to have shorter legs and slower, more powerful gaits.

Power may also be increased by increasing the number of legs, and thus the number of body segments. In this regard, physical limits are reached when the body achieves a length-to-width ratio that could result in buckling of the trunk. This constraint is partly overcome in millipedes and pauropodans in two principal ways: by increasing the cross-sectional body diameter (becoming "fatter") and by uniting segments into pairs (diplosegments). In some species the body wall is further strengthened by having certain trunk segments fused into a solid ring. Also, when millipedes and pauropodans burrow, they tuck the head ventrally so that the collum is thrust forward like the broad blade of a bulldozer (Figure 5D).

One odd group of centipedes lacks the high-speed modifications of most chilopods. The geophilomorphs burrow, aided by dilation of the body and use of the trunk musculature in a fashion similar to that of earthworms. This peristalsis-like motion is rare in arthropods because of the rigid exoskeleton.

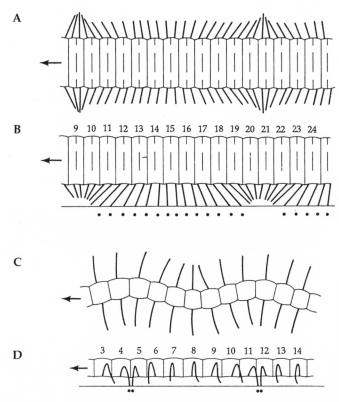

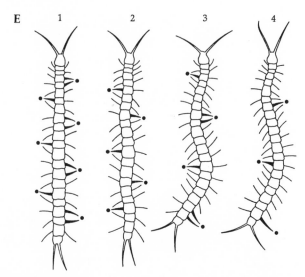

Figure 12

Locomotion in millipedes and centipedes. A–B, A typical millipede (such as *Spirostreptus* or *Gymnostreptus*) in motion. Note the 16 diplosegments (with 32 pairs of legs). The upper figure (dorsal view) shows left and right leg sets exactly in phase with each other. The lower figure (lateral view of the same animal) shows that the majority of the limb tips are on the ground at once, an arrangement yielding a slow but powerful gait. C–D, A typical centipede (such as *Scolopendra* or *Cryptops*) in motion. Note the 12 segments, each with one pair of legs. The upper figure (dorsal view) shows the limb pairs in opposite phase and the undulations of the body that accentuate stride length. The lower figure (lateral view of the same animal) illustrates that fewer than one-third of the limb tips are on the ground at any one time, an arrangement yielding the short, swift strokes typical of a rapid but weak gait. Arrows indicate the animal's direction of travel; dots are points of leg-tip contact with substratum. E, Locomotion in a scolopendrid centipede at various speeds. Centipedes are one of the few groups of arthropods capable of accentuating stride length by lateral undulations. 1–4 show the body waves and leg actions at increasing speeds. Limbs shown with heavy lines are in their power strokes, with the tips against the substratum (•); limbs depicted by thin lines are in various stages of their recovery strokes. Notice that at maximum speed the animal is still supported by a tripod stance. (A–D after Russell-Hunter 1969; E after Manton 1965.)

However, geophilomorphs have enlarged areas of flexible cuticle on the sides of the body between the tergites and sternites; these enlarged and flexible pleural areas allow them to significantly alter their body diameter. Other centipedes have smaller, flexible pleural areas that allow some degree of lateral undulatory motion. Centipede legs attach to these flexible regions, an arrangement that increases the range of limb motion in these fast-running surface dwellers. This condition is in marked contrast to that of millipedes, in which the legs are short and arise from the ventral sternites, limiting their range of movement (Figure 12).

One of the principal problems associated with increased limb length is that the field of movement of one limb may overlap that of adjacent limbs. Potential leg interference is prevented by having limbs of different lengths, so that the tips of adjacent legs move at different distances from the body (Figure 13). Fast-running insects usually have limbs of slightly different lengths; and in many centipedes the limbs of each succeeding segment are slightly longer than the immediately anterior pair. Insects usually move their legs in an **alternating tripod sequence** (Figure 14). Balance is maintained by always having three legs in contact with the ground. The legs of most myriapods move in clear metachronal waves from posterior to anterior (Figure 12), augmented in the centipedes by lateral body undulations. Unlike most arthropods, millipedes move the two pairs of legs on each diplosegment synchronously. Stability is not a problem in these elongate "thousand-leggers."

The evolution of wings in insects

Among the many remarkable advances of insects, flight is the most impressive—no other invertebrates truly fly. The wingless insects belong either to groups that have secondarily lost the wings (e.g., fleas, lice, certain scale insects) or to primitive taxa

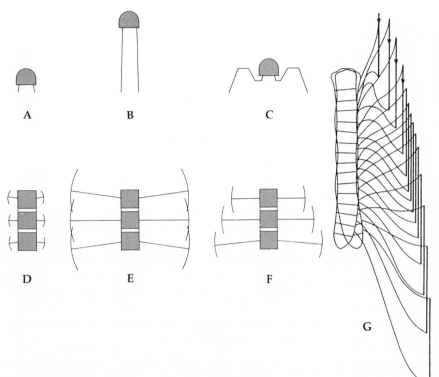

Figure 13

Problems associated with increased limb length in terrestrial arthropods. A–C, Instability: A, Short legs, stable configuration. B, Long limbs, high center of gravity, unstable configuration. C, Long limbs bent in stabilizing configuration. D–F, Overstepping: D, short limbs, no overstepping; E, Long limbs, marked overstepping when the full swing of the limb is used. F, Long limbs differing in length, no overstepping. G, The field of leg movement in a running centipede, *Scutigera*. The heavy vertical lines trace the movement of the tips of each leg during the propulsive backstroke. Note the gradual increase in limb length posteriorly, an arrangement that prevents overstepping of limb tips. (A–F after Clarke 1973; G after Barth and Broshears 1982.)

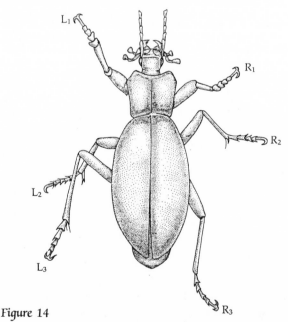

Figure 14

A beetle walking. The alternating tripod gait consists of alternate stepping with two sets of three legs; thus the body is always supported by a triad of legs. Three legs (L_1, R_2, and L_3) are moving forward while the other three (R_1, L_2, and R_3) are moving backward. (After Barth and Broshears 1982.)

that probably arose prior to the evolution of wings (e.g., collembolans and thysanurans). Wings are carried on the second (mesothoracic) and third (metathoracic) segments of the thorax. In two orders the wings are effectively reduced to a single pair, the other pair having become modified as protective dorsal covers (**elytra**) in beetles or as organs of balance (**halteres**) in dipterans. Halteres beat with the same frequency as the forewings, functioning as gyroscopes to assist in flight performance and stability—flies fly very well.

The origin of insect flight, like the origin of wings and flight in vertebrates, is a wonderful mystery yet to be completely solved. The basic problem is one of explaining the intermediate stages—animals with winglike appendages that are not yet fully functional for flight. Certainly, the primitive wingless groups (the "apterygota") evolved prior to the origin of the winged pterygote line that dominates the Insecta today. The most primitive winged insects (e.g., mayflies, dragonflies, orthopterans) have fluted, fan-shaped membranous wings with many trusslike supporting veins (Figure 15). Studies of fossil insects indicate that these netlike wings are reminiscent of the most primitive wing architecture. The earliest known insect fossils (early Carboniferous or Devon-

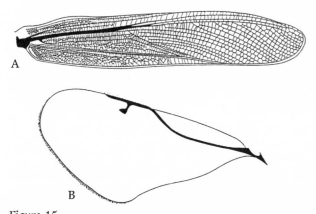

Figure 15

Primitive and advanced wings. A, The many-veined forewing of a locust (order Orthoptera). B, The forewing of a hymenopteran. Note the extreme reduction of venation.

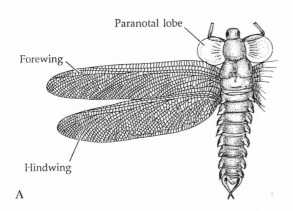

Paranotal lobe

Forewing

Hindwing

A

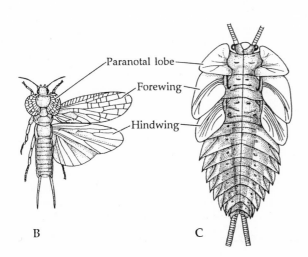

Paranotal lobe

Forewing

Hindwing

B　　C

ian) include both winged and wingless types—but no intermediates or partly winged fossil forms have yet been found. Thus, theories on the origin of wings are matters of considerable speculation and controversy. The most widely accepted theory (the **paranotal theory**) suggests that wings originated as lateral aerodynamic flaps of the thoracic nota that enabled insects to alight right-side-up when jumping or when blown about by winds. According to this theory, these stabilizing paranotal lobes gradually enlarged, first functioning in gliding, and later evolving hinged structures and muscles at their bases. The occurrence of fixed paranotal lobes in certain ancient fossil insects (Palaeodictyoptera) has been cited in support of the paranotal theory (Figure 16). Some recent studies suggest that these primitive paranotal lobes might have been used for other purposes: to cover the spiracular openings or gills in amphibious insects; to protect or conceal the insects from predators; for sexual (courtship) display; or for thermoregulation by absorption of solar heat.

A second theory of wing origin—the **branchial theory**—suggests that the pterygote insects are derived from ancient aquatic forms. In this scenario, the forerunners of wings were thoracic, highly tracheate gills that gradually sclerotized and functioned as stabilizers during swimming. An intermediate stage might have involved the acquisition of a hinge line and musculature, transforming the gills into swimming paddles or "agitators" to increase water flow over them, thus preadapting them for use as wings when these archaic insects invaded land. Recent studies on nervous system fine structure have revealed the presence of interneurons, nerves that are important components of the flight motor, in the first three abdominal ganglia of locusts. This discovery has been cited as evidence that wings originated from serially distributed segmental structures, perhaps pleural "appendages" of some sort, as suggested by the branchial theory.

Flight. Wings of modern insects develop as evaginations or outpocketed folds of the integument, and thin cuticular membranes form the upper and

Figure 16

Fossil insects with paranotal lobes on the prothorax. A, *Stenodictya lobata.* B, *Lemmatophora typa.* C, Nymphal stage of *Rochdalia parkeri*, a Paleozoic terrestrial palaeodictyopteran. In this species, all three thoracic segments appear to have had "articulated" thoracic lobes. (A after B. Rodendorf (ed.), 1962, Arthropoda-Tracheata and Chelicerata, in *Textbook of Paleontology*, Academy of Sciences, U.S.S.R.; B after Snodgrass 1952; C after R. J. Wooten, 1972, Paleontology 15:662.)

Figure 17

The nomenclature of basic wing venation in insects. Although the "cells" formed within the veins also have names, only the names of the veins are given here. Longitudinal veins are coded with capital letters, cross-veins with lowercase letters. Longitudinal veins: costa (C); subcosta (SC); radius (R); radial sector (RS); media (M); cubitus (CU); anal veins (A). Cross veins: humeral (h); radial (r); sectorial (s); radiomedial (rm); medial (m); mediocubital (m-cu); cubitoanal (cu-a).

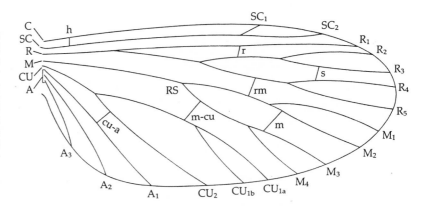

lower surfaces of each wing. Wing veins anastomose and eventually open into the body and contain circulating hemolymph. Veins are thought to originally have been trachea that now serve as thickened, tubular, skeletal support rods. The arrangement of veins in insect wings is an important diagnostic character at all taxonomic levels. The principal veins, their branches, and the internal cells are systematically named and numbered, allowing entomologists to communicate about such matters (Figure 17).

Like birds and bats, insects fly by flapping their wings and creating air currents from which they gain lift. The precise mechanism of insect flight is still not fully understood, but it is almost certainly not the same as that of fixed-wing flight (e.g., soaring animal flight or airplane flight). Each wing articulates with the edge of the notum (thoracic tergite), but its proximal end rests on a dorsolateral pleural process, which acts as a fulcrum (Figure 18). The wing hinge itself is composed in large part of **resilin**, a highly

elastic protein that allows for rapid, sustained movement. By complex actions of wing orientation, insects can hover, fly forward and backward, and negotiate highly sophisticated aerial maneuvers. The complex wing movements are made possible by the flexibility of the wing itself and by the action of a number of different muscle sets that run from the wing base to the inside walls of the thoracic segment on which it is borne. These **direct flight muscles** serve to raise and lower the wings and to tilt their plane at different angles (somewhat like altering the blade angles on a helicopter) (Figure 19). However, except for dragonflies, the direct flight muscles are not the main source of power for insect wing movements. Most of the force comes from two sets of **indirect flight muscles**, which neither originate nor insert on the wings themselves (Figures 19 and 20). This remarkable adaptation is described below.

Dorsal longitudinal muscles run between apodemes at the anterior and posterior ends of each winged segment. When these muscles contract, the segment is shortened, a change resulting in a dorsal arching of the segment roof and a downstroke of the wings. **Dorsoventral muscles** extend from the notum to the sternum (or to basal leg joints) in each wing-bearing segment and are antagonistic to the longitudinal muscles. Contraction of the dorsoventral muscles lowers the roof of the segment and in doing so raises the wings, and incidentally pulls the legs up during flight. Thus, wing flapping is primarily generated by rapid changes in the walls and overall

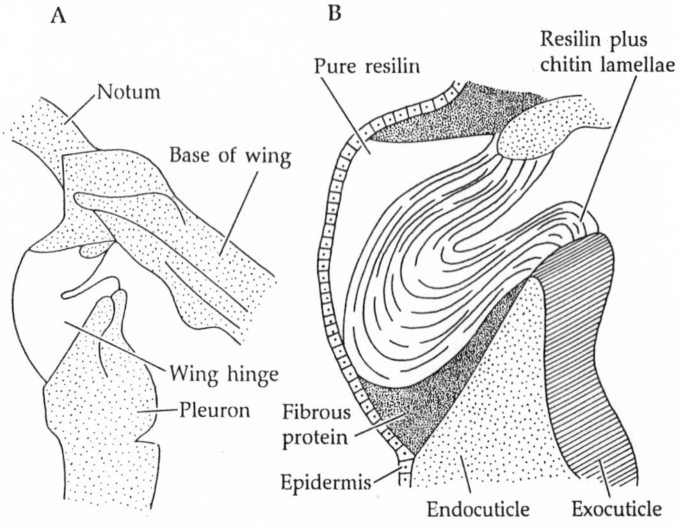

Figure 18

A typical insect wing hinge arrangement. This transverse section through the thoracic wall of a grasshopper shows the base of the wing and the wing hinge. A, Entire hinge area. B, Enlargement of hinge section. (After Anderson and Weis-Fogh 1964.)

Figure 19

Wing movements in an insect such as a dragonfly, in which direct wing muscles cause depression of the wings. A, The dorsoventral muscles contract to depress the notum as the basalar muscles relax, a combination forcing wings into an upstroke. B, The dorsoventral muscles relax as the basalar muscles contract, a combination pulling the wings into a downstroke and relaxing (and raising) the notum.

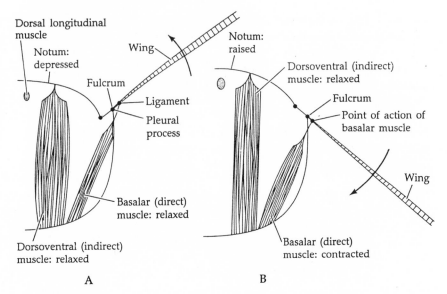

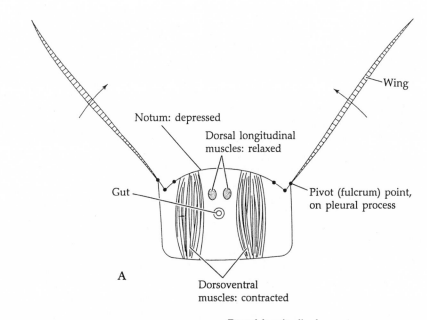

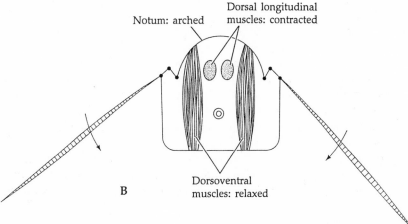

Figure 20

Wing movements in an insect such as a fly or hemipteran, in which both up and down movements of the wings are produced by indirect flight muscles. In these transverse sections of a thoracic segment, dots represent pivot points and arrows indicate direction of wing movement. Only two sets of muscles are shown. A, The dorsoventral muscles contract, depressing the thoracic notum and forcing the wings into upstroke. B, The dorsoventral muscles relax as the dorsal longitudinal muscles contract to "pop up" the notum, elevating it and forcing the wings into a downstroke.

shape of the mesothorax and metathorax. Other, smaller thoracic muscle sets are present and serve to make minor adjustments to this basic operation.

Insects with low wing-flapping rates (e.g., dragonflies, orthopterans, mayflies, and lepidopterans) are limited by the rate at which nerves can repeatedly fire and muscles can execute contractions. However, in insects with high wing-flapping rates (e.g., dipterans, hymenopterans, and some coleopterans), an entirely different regulatory mechanism has evolved. Once flight has been initiated and a high wing-flapping rate attained (up to 100 beats/second), myogenic control takes over. This mechanism exploits the elastic–mechanical properties of the exoskeleton. When one set of indirect muscles contracts, the thorax is deformed, which stretches the second set of indirect muscles and thus directly stimulates their contraction. This establishes a second deformation, which in turn stretches and stimulates the first muscle group. The elastic rebound after each deformation of the exoskeleton thus contributes to the alternating stimulation of the antagonistic sets of muscles. Once initiated, this mechanism is nearly self-perpetuating, the nonsynchronous firing of nerves serving only to keep it in action.

Not all insects utilize wings to fly. Many small and immature insects are effectively dispersed by wind power alone. Some first instar lepidopterans use silk threads for dispersal (as do ballooning spiders and mites). Tiny scale insects are commonly taken in "aerial plankton" nets. In fact, studies have revealed the existence of a large "aerial plankton" of insects and other minute arthropods, extending to altitudes as high as 14,000 feet. Most are tiny winged forms, but wingless species are also common.

Feeding and digestion

Feeding. Every conceivable kind of diet is exploited within the Uniramia, including herbivory, carnivory, scavenging, and various kinds of symbiosis. The myriapods are somewhat limited in their nutritional habits, whereas insects display an enormous range of feeding strategies, commensurate with their marvelous diversity.

Most centipedes are active, aggressive predators on smaller invertebrates, particularly worms, snails, and other arthropods. The first trunk appendages (actually **maxillipeds**) are greatly enlarged, clawlike structures called **prehensors**, or **forcipules**. These raptorial limbs are directed forward alongside the mouth field (Figure 5), and are used to stab prey and inject poison. The poison is produced in large **venom glands** located in the basal articles of the prehensors. The poison is so effective that large centipedes such as the tropical *Scolopendra* can subdue small lizards and frogs. The bite of even the most dangerous centipede is normally not fatal to people, but the poison can cause a reaction similar to that accompanying a serious wasp sting. A few deaths have been reported from *Scolopendra gigantea*, but in these cases circumstances were probably complicated by other health problems or extreme sensitivity. Some centipedes actually rear up on their hind legs and capture flying insects! The prehensors and second maxillae hold the prey, while the mandibles and first maxillae bite and chew. Geophilomorph centipedes possess defensive weapons called **repugnatorial glands** on the ventral side of the segments. Some lithobiomorphs bear large numbers of unicellular repugnatorial glands on the last four pairs of legs, which they kick in the direction of any enemy. This motion throws out sticky droplets of the noxious secretions.

Millipedes are herbivores with a preference for dead and decaying plant material, and they play an important role in the recycling of leaf litter in many parts of the world. Most bite off large pieces of vegetation with their powerful mandibles, mix it with saliva as they chew it, and then swallow it. Some, such as the tropical siphonophorids, are specialized to feed on plant juices. In these groups the labrum, gnathochilarium, and reduced mandibles are modified into a suctorial piercing beak. In addition, a few odd groups of millipedes are predators and feed as centipedes do. Most trunk diplosegments in millipedes have lateral repugnatorial glands that secrete volatile toxic liquids used for defense.

Pauropodan feeding biology is not well understood, but most appear to be scavengers. These minute, blind creatures crawl through soil and humus, feeding on fungi and decaying plant and animal matter. Symphylans are primarily herbivores, although a few have adopted a carnivorous, scavenging lifestyle. Many consume live plants. One species, *Scutigerella immaculata*, is a serious pest in plant nurseries and flower gardens, where it has been reported in densities greater than 90 million per acre.

One of the key factors in the phenomenal radiation within the insects has been their evolution of a multitude of nutritional strategies. A comprehensive survey of insect feeding biology alone could easily fill a book this size. In a very general sense, insects can be classified as (1) biters–chewers, (2) suckers, or (3) spongers (Figure 21). Biters–chewers, such as the grasshoppers, have the least modified mouth parts, so we describe them first (see Figure 6). In these

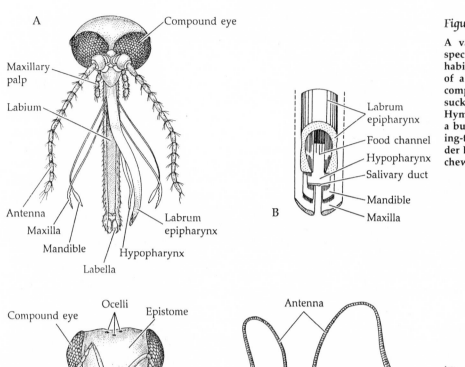

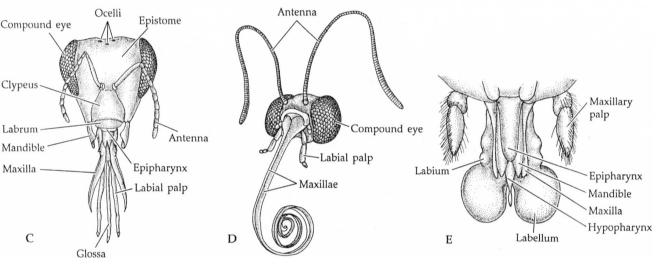

Figure 21

A variety of insect mouth appendages, specialized for different types of feeding habits. A–B, piercing-sucking mouthparts of a mosquito (order Diptera). Note the complex stylet structure in (B). C, Piercing-sucking mouthparts of a honeybee (order Hymenoptera). D, Sucking mouthparts of a butterfly (order Lepidoptera). E, Sponging-type mouthparts of a false blackfly (order Diptera). For an illustration of biting–chewing mouthparts, see Figure 6.

insects the maxillae and labium have well developed leglike palps that help hold the food in place, while powerful mandibles cut off and chew bite-sized pieces. The mandibles lack palps (in all insects) and typically bear small, sharp teeth that work in opposition as the appendages slide against one another in the transverse (side to side) fashion characteristic of most arthropod jaws (Figure 7). Biting–chewing insects may be carnivores, herbivores, or scavengers. In many plant eaters, the labrum bears a notch or cleft into which a stem or leaf edge may be affixed while being eaten. Some of the best examples of this feeding strategy are seen among the Orthoptera (locusts, grasshoppers, and crickets); most people have witnessed the efficient fashion in which these insects consume green plants.

In sucking insects the mouthparts are markedly modified for the consumption of liquid foods, generally plant saps or nectars, or animal blood or cell fluids (Figures 21 and 22). Sucking mouthparts and liquid diets have clearly evolved many times in different insect lines—further testimony to the commonness of evolutionary convergence in arthropods and the adaptability of their appendages. In some sucking insects, such as mosquitoes, feeding is initiated by piercing the victim's epidermal tissue; this mode of feeding is thus referred to as "piercing–sucking." Other insects, such as butterflies and moths that feed on flower nectar, do not pierce anything and are merely "suckers." In all sucking insects the mouth itself is very small and well hidden. The mouthparts, instead of being adapted for handling and chewing solid pieces of food, are elongated into a needle-like beak. Different combinations of mouth appendages constitute the beak in different taxa. Plant bugs (Homoptera) and true bugs (Hemiptera) are piercer–suckers and have a beak composed of five elements: an outer trough-like element (the **labium**) and, lying in the trough, four very sharp **stylets** (the two mandibles and two maxillae). The stylets are often barbed to tear the prey's tissues and enlarge the wound. The labrum is in the form of a small flap covering the base of the grooved labium. When piercer–suckers feed, the labium remains stationary and the stylets do the work of puncturing the plant (or animal) and drawing out the liquid meal.

Different variations of piercing–sucking mouthparts are found in other insect taxa. For example, in mosquitoes, midges, and certain "biting" flies (e.g., horse flies) there are six stylets, the labrum-epipharynx and the hypopharynx being long and slender like the mandibles and first maxillae. Other "biting flies," such as the stable fly, have mosquito-like mouthparts, but lack mandibles and maxillae altogether. Fleas (Siphonaptera) have three stylets: the labrum-epipharynx and the two mandibles. Thrips have unusual mouthparts; in these insects the right mandible is greatly reduced, making the head somewhat asymmetrical; the left mandible, the first maxillae, and the hypopharynx make up the stylet. Blood-sucking lice have two piercing stylets but, because of the extreme head modifications of lice, it is uncertain which mouth appendages these actually are!

Lepidopterans are nonpiercing sucking insects in which the paired first maxillae are enormously elongated, coiled, and fused to form a tube through which flower nectar is sucked; the mandibles are vestigial or absent (Figure 21D). Bees are similar; they have the first maxillae and labium modified together to form a nectar-sucking tube, but the mandibles are retained and used for wax manipulation during hive construction (Figure 21C). The nectar is converted into honey, which is stored in the hive as a food reserve. Bees in an average hive consume about 500 pounds of honey per year—we get the leftovers.

Associated with sucking mouthparts are various mechanisms for drawing liquid food into the mouth. Most piercer–suckers rely largely on capillary action. Others have developed feeding "pumps." Often the feeding pump is developed by elongation of the preoral cavity, or **cibarium**, which by extension of the cuticle around the mouth becomes a semiclosed chamber connecting with the alimentary canal (Figure 22). In these cases **cibarial muscles** from the

clypeus are enlarged to make a powerful pump. In lepidopterans, dipterans, and hymenopterans the cibarial pump is combined with a **pharyngeal pump**, which operates by means of muscles arising on the front of the head. Specialized salivary glands are also often associated with sucking mouthparts. In homopterans a salivary pump forces saliva through the feeding tube and into the prey, softening tissues and predigesting the liquid food. In mosquitoes the saliva carries blood thinners and anticoagulants (and often parasites such as *Plasmodium*, which causes malaria).

In spongers, such as most flies (order Diptera), the labium is expanded distally as a fleshy porous mass, called a **labellum** (Figure 21E). Fluid nutrients are transported by capillary action along minute surface channels from the labella to the mouth. In many spongers, such as house flies, saliva is exuded onto the food to partly liquefy it. In strict spongers, the mandibles are absent. In "biting" spongers, such as horse flies, the mandibles serve to slice open a wound in the flesh or other food substance, thus exposing the blood and cellular fluids to be sponged up by the labella.

Many insects are scatophagous, feeding on animal feces. Most of these groups have biting mouth-

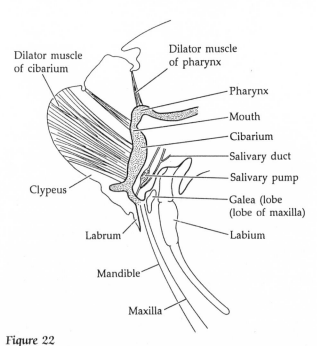

Figure 22

The head (vertical section) of a piercing–sucking insect, a cicada (order Homoptera). Note the enlargement of the cibarial dilator muscles, which activate a cibarial sucking pump. (After Snodgrass 1944.)

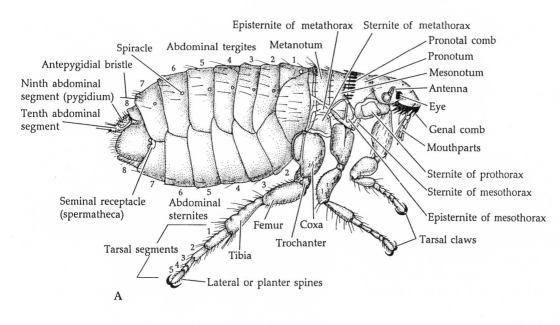

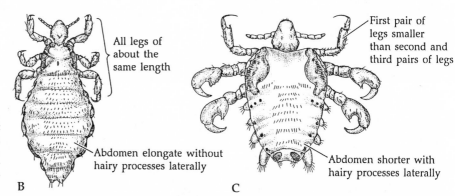

Figure 23

Three common household insect parasites. A, The cat flea (*Ctenocehalides felis*), adult female. B, *Pediculus humanus*, the human head and body louse. C, *Phthirus pubis*, the human pubic louse. (After a Department of Health and Human Services Bulletin, Public Health Service, Atlanta.)

All legs of about the same length

Abdomen elongate without hairy processes laterally

First pair of legs smaller than second and third pairs of legs

Abdomen shorter with hairy processes laterally

B

C

parts, but some (such as certain flies) have sucking mouthparts. Perhaps the most famous of the scatophagous insects are the dung beetles, or tumblebugs (certain beetles in the families Scarabaeidae and Histeridae). These remarkable insects "harvest" animal dung by biting off pieces or by slicing off chunks with specialized head or leg structures, and working them to a ball. They then roll the dung ball a considerable distance, and eventually bury it in the soil, whereupon females deposit eggs within it. Larvae are thus assured of a ready food supply. Dung balls are often maneuvered by a pair of dung beetles pushing and pulling in a cooperative effort.

There are many symbiotic insects and three orders are composed entirely of wingless parasites, most of which spend their entire lives on their host: Mallophaga (biting lice), Anoplura (sucking lice), and Siphonaptera (fleas). Other orders contain primarily free-living insects but include various families of parasitic or micropredatory forms, or groups in which the larval stage is parasitic but the adults free living. Members of the order Mallophaga are particularly common on birds, thus they are often referred to as the "bird lice." However, many also occur on dogs, cats, horses, cattle, and other mammals. These biting lice have broad heads and biting mouthparts used to chew the epithelial cells and other structures on the host's skin. Sucking lice (order Anoplura) have narrow heads and piercing–sucking mouthparts used to suck blood and tissue fluids from the host, always a mammal. Unlike most arthropod parasites, lice (both types) spend their entire lives on the bodies of their hosts, and transmission to new hosts is generally by direct contact. For this reason most lice show a high degree of host specificity. Lice eggs, or **nits**, are attached by the female to the feathers or hairs of the host, where they develop without a marked metamorphosis. Many lice, particularly those whose diet is chiefly keratin, possess symbiotic intracellular bacteria that appear to aid in the digestion of their food. These bacteria are passed to the offspring by way of the insects' eggs. Similar bacteria occur in ticks, mites, bedbugs, and some blood-sucking dipterans.

None of the biting lice are known to infest people or to transmit human disease microorganisms, although one species acts as an intermediate host for certain dog tapeworms. The sucking lice, on the other hand, include two genera that commonly infest humans (*Pediculus* and *Phthirus*) (Figures 23B and C). The latter genus includes the notorious *P. pubis*, the human pubic "crab" louse (which often occurs on other parts of the body as well). A number of sucking lice are vectors for human disease organisms, including those causing tularemia, trench fever, leishmaniasis, and epidemic typhus. The most common reaction to lice infestation—a condition known as pediculosis, or **lousy**—is simple irritation and itching caused by the anticoagulant injected by the parasite during feeding. Chronic lice infestation among certain footloose travelers is manifested by leathery, darkened skin—a condition known as "vagabond's disease."

Fleas (order Siphonaptera) are perhaps the best known of all insect parasites (Figure 23A). Nearly 1,800 species have been described from birds and mammals. Unlike lice, fleas are holometabolous, passing through egg, larva, pupa, and adult stages. Some species of fleas live their entire lives on their host, although eggs are generally deposited in the host's environment and larvae feed on local organic debris. Larvae of human fleas feed on virtually any organic crumbs they find in the household furniture or carpet. Upon metamorphosis to the adult stage, fleas may undergo a quiescent period until an appropriate host appears. A number of serious disease organisms are transmitted by fleas. At least 8 of the 60 or so species of fleas associated with household rodents are capable of acting as vectors for bubonic plague bacteria. Other flea-transmitted diseases include murine (nonepidemic) typhus, tularemia, salmonellosis, and dog, cat, and certain human tapeworms.

Most "parasites" in many other insect orders do not live continuously upon a host and have feeding behaviors that fall in that "gray zone" between true obligate parasitism and predation. These insects are often classed as **parasitoids**, intermittent parasites, or micropredators. For example, bedbugs (order Hemiptera, family Cimicidae) are minute flattened insects that feed upon birds and mammals. However, most live in the nest or sleeping area of their host, emerging only periodically to feed. The common human bedbug (*Cimex*) hides in bedding, in cracks, or under rugs by day and feeds on its host's blood at night. They are piercer–suckers, much like the anoplurans. Bedbugs are not known to transmit any human diseases. Mosquitoes (family Culicidae), on the other hand, are vectors for a large number of disease-causing microorganisms, including those responsible for malaria (see Figure 36 in Chapter 5), yellow fever, viral encephalitis, dengue, and filariasis (with its gross symptom, elephantiasis).

Kissing bugs (order Hemiptera, family Reduviidae) also have a casual host relationship. They live in all kinds of environments, but often inhabit the burrows or nests of mammals. Most feed on the blood of mammals, and the mouthparts are modified as efficient piercing–sucking devices. Host specificity is low. Several species are vectors of mammalian trypanosomiasis (*Trypanosoma cruzi*, the causative agent of Chagas disease).

Many insect plant parasites cause abnormal growth of plant tissues caused **galls**. Many fungi and nematodes also produce plant galls, but most are caused by mites and insects. Parasitic adults may bore into the host plant or, more commonly, deposit their eggs in plant tissues, where they undergo larval development. The presence of an insect or its larva stimulates the plant tissues to grow rapidly, forming the gall.

Digestive system. Like the guts of all arthropods, the long, usually straight uniramian gut is divisible into a stomodeal foregut, entodermal midgut, and proctodeal hindgut (Figure 24). **Salivary glands** are associated with one or several of the mouth appendages (Figure 25). The salivary secretions soften and lubricate solid food, and in some species contain enzymes that initiate chemical digestion. In moths (caterpillars), and in larval bees and wasps, the salivary glands secrete silk used to make pupal cells.

All uniramians, as well as many crustaceans and cheliceriformes that consume solid foods, produce a **peritrophic membrane** in the midgut (Figure 24B). This is a sheet of thin chitinous material that may line the midgut or pull free to envelop and coat the food particles as they pass through the gut. The peritrophic membrane serves to protect the delicate midgut epithelium from abrasion. The membrane is permeated by microscopic pores that allow passage of

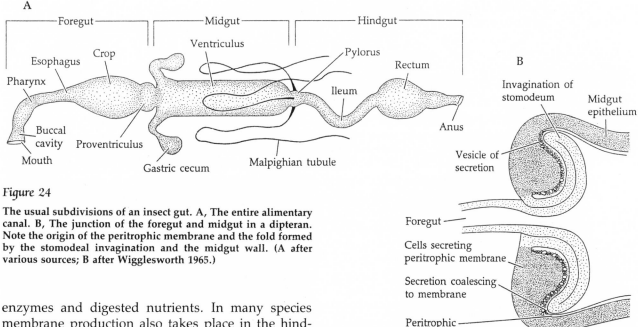

Figure 24

The usual subdivisions of an insect gut. A, The entire alimentary canal. B, The junction of the foregut and midgut in a dipteran. Note the origin of the peritrophic membrane and the fold formed by the stomodeal invagination and the midgut wall. (A after various sources; B after Wigglesworth 1965.)

enzymes and digested nutrients. In many species membrane production also takes place in the hindgut, where it encapsulates the feces as discrete pellets.

Along with their vast range of feeding habits, insects have evolved a number of specialized digestive structures. The foregut is typically divided into a well-defined **pharynx, esophagus, crop,** and **proventriculus** (Figures 24 and 25). The pharynx is muscular, particularly in the sucking insects where it forms a pharyngeal pump. The crop is a storage center whose walls are highly extensible in species that consume large but infrequent meals. The proventriculus regulates food passage into the midgut, either as a simple valve that strains the semifluid foods of suckers or as a grinding organ, called a **gizzard** or **gastric mill**, that masticates the chunks ingested by biters. Well developed gastric mills have strong cuticular teeth and grinding surfaces that are gnashed together by powerful proventricular muscles.

The midgut (= stomach) of most insects bears **gastric ceca** that lie near the midgut–foregut junction and resemble those of crustaceans. These evaginations serve to increase the surface area available for digestion and absorption. In some cases the ceca also house mutualistic microorganisms (bacteria and protozoa). The insect hindgut serves primarily to regulate the composition of the feces and perhaps to absorb some nutrients. Digestion of cellulase in termites and certain wood-eating roaches is made possible by enzymes produced by protozoa and bacteria that inhabit the hindgut.

Fat bodies occur in the hemocoel of many insects and appear to function in much the same way as

chlorogogen tissue in annelids, storing certain food reserves, particularly glycogen. Many insects do not feed during their adult life; instead, they rely on stored nutrients accumulated in the larval or juvenile stages.

The guts of myriapods are long and straight, and often show regional specializations similar to those of insects (Figure 26). Salivary glands (= mandibular glands) open near the mouth in most centipedes and millipedes. Their secretions serve to soften and sometimes to partially digest food before ingestion. The mouth leads inward to a long esophagus, which is sometimes expanded posteriorly as a storage area, or **crop** and **gizzard** (as in most centipedes). The gizzard often contains cuticular spines that help strain large particles from the food entering the midgut, where absorption occurs (Figure 26C). The midgut connects to a short, proctodeal hindgut that terminates at the anus.

Circulation and gas exchange

The uniramian circulatory system includes a dorsal tubular heart that pumps blood toward the head. The heart narrows anteriorly to a vessel-like aorta, from which blood flows posteriorly through large hemocoelic chambers before returning to the **pericardial sinus** and then back to the heart via paired lateral **ostia** (Figure 27). In diplopods, the heart bears two

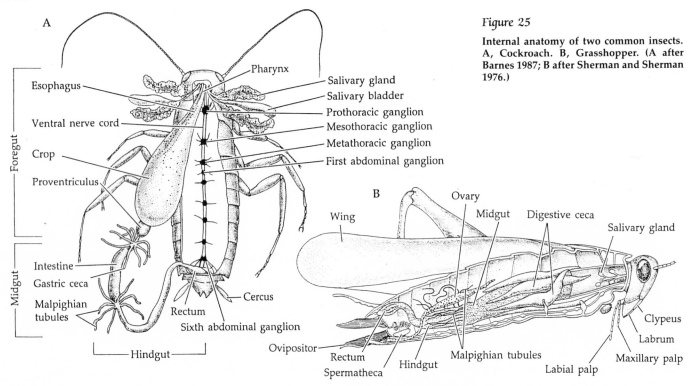

Internal anatomy of two common insects. A, Cockroach. B, Grasshopper. (A after Barnes 1987; B after Sherman and Sherman 1976.)

A

Pharynx
Salivary gland
Salivary bladder
Prothoracic ganglion
Mesothoracic ganglion
Metathoracic ganglion
First abdominal ganglion

Esophagus
Ventral nerve cord
Crop
Proventriculus

Foregut
Midgut

Intestine
Gastric ceca
Malpighian tubules
Rectum
Cercus
Sixth abdominal ganglion
Hindgut

B

Ovary
Midgut
Digestive ceca
Salivary gland

Wing

Clypeus
Labrum
Maxillary palp
Labial palp
Malpighian tubules
Hindgut
Rectum
Spermatheca
Ovipositor

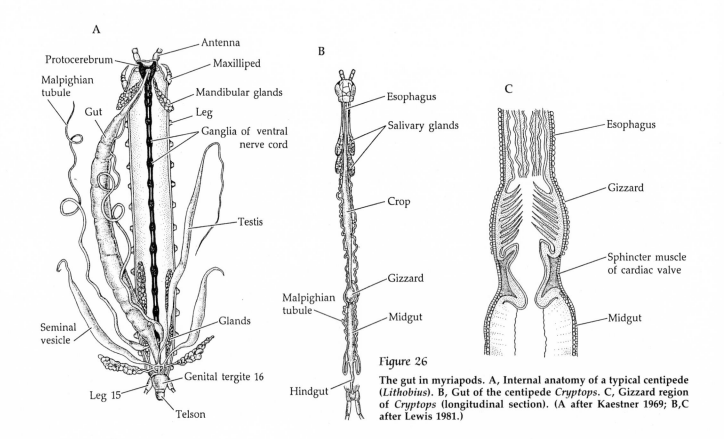

A

Antenna
Maxilliped
Mandibular glands
Leg
Ganglia of ventral nerve cord

Protocerebrum
Malpighian tubule
Gut

Testis

Seminal vesicle

Glands
Genital tergite 16
Leg 15
Telson

B

Esophagus
Salivary glands

Crop

Gizzard

Malpighian tubule
Midgut

Hindgut

C

Esophagus

Gizzard

Sphincter muscle of cardiac valve

Midgut

Figure 26

The gut in myriapods. A, Internal anatomy of a typical centipede (*Lithobius*). B, Gut of the centipede *Cryptops*. C, Gizzard region of *Cryptops* (longitudinal section). (A after Kaestner 1969; B,C after Lewis 1981.)

pairs of ostia in each diplosegment; one pair of ostia occurs in each segment in the chilopods.

In most insects the heart extends through the first nine abdominal segments; the number of ostia is variable. Accessory pumping organs, or **pulsatile organs**, often occur at the bases of wings and of especially long appendages, such as the hind legs of grasshoppers, to assist in circulation and maintenance of blood pressure. The uniramian heart is a rather weak pumping organ, and blood is moved primarily by routine muscular activity of the body and appendages. Hence, circulation is slow and system pressure is relatively low.

Like many arachnids, some uniramians use hydraulic pressure of the hemocoelic system in lieu of extensor muscles. In this way, for example, butterflies and moths unroll their maxillary feeding tubes.

Many types of hemocytes have been reported from the blood of insects. None function in oxygen storage or transport, but several are apparently important in wound healing and clotting. Nutrients, wastes, and hormones can be efficiently carried by this system, but respiratory gases cannot. The active lifestyles of these animals require special structures to carry out the tasks of respiratory gas exchange and of excretion. These are the tracheal system and Malpighian tubules described below.

Desiccation is one of the principal dangers faced by terrestrial invertebrates, and even though the general body surface of the uniramians may be largely waterproof, the gas exchange surfaces cannot be. Such situations always involve some degree of compromise between water loss and gas exchange with the atmosphere.

In some minute insects, gas exchange occurs by direct diffusion across the body surface. However, the vast majority of uniramians rely on a **tracheal system** (Figures 28 and 29). As explained in Chapter 15, tracheae are extensive tubular invaginations of the body wall opening through the cuticle by pores called **spiracles**. Since tracheae are epidermal in origin, their linings are shed with each molt. The cuticular wall of each trachea is sclerotized and usually strengthened by rings or spiral thickenings called **taenidia**, which keep the tube from collapsing but allow changes in length that may accompany body

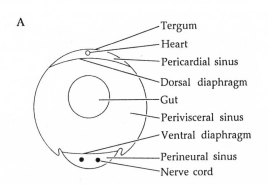

A
- Tergum
- Heart
- Pericardial sinus
- Dorsal diaphragm
- Gut
- Perivisceral sinus
- Ventral diaphragm
- Perineural sinus
- Nerve cord

Figure 27

The circulatory system of insects. A, An insect abdomen (cross section). Note the division of the hemocoel into three chambers (a dorsal pericardial sinus, a ventral perineural sinus, and a central perivisceral sinus). These chambers are separated by diaphragms lying on frontal planes. B, Blood circulation in an insect with a fully developed circulatory system (longitudinal section). Arrows indicate the circulatory course. C, A cockroach (ventral dissection). Note the dorsal and segmental vessels. The dorsal diaphragm and aliform muscles are continuous over the ventral wall of the heart and vessels, but they are omitted from the diagram for clarity. (B after Wigglesworth 1965; C after Chapman 1971.)

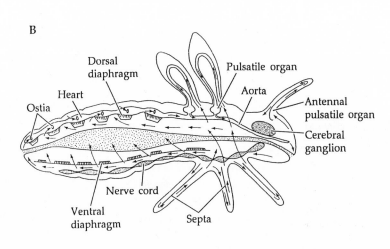

B
- Dorsal diaphragm
- Pulsatile organ
- Heart
- Aorta
- Ostia
- Antennal pulsatile organ
- Cerebral ganglion
- Nerve cord
- Ventral diaphragm
- Septa

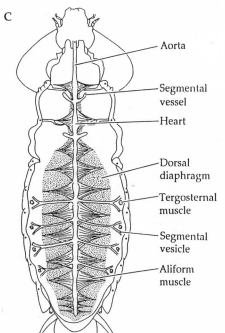

C
- Aorta
- Segmental vessel
- Heart
- Dorsal diaphragm
- Tergosternal muscle
- Segmental vesicle
- Aliform muscle

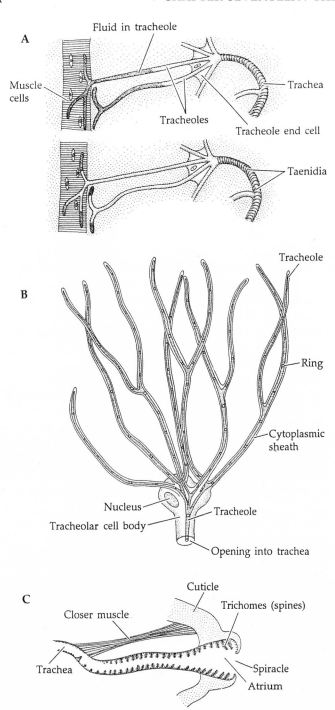

Figure 28

The tracheal system of insects. A, Tracheoles and flight muscle cells. Note the region where the tracheoles become functionally intracellular within the muscle fibers. The upper figure depicts a situation in which the muscle cells are well oxygenated, oxygen demand is low, and fluid accumulates in the tracheoles. The lower figure depicts muscle cells that are oxygen-deficient. Decreased fluid volumes allow tissues increased access to oxygen. B, A tracheolar cell. The taenidia are rings that serve to keep the lumen of the tracheoles open. C, A generalized insect spiracle (longitudinal section). Note the dust-catching spines (trichomes) within the atrium. (A after Snodgrass 1935; B after Clarke 1973.)

movements. The tracheae originating at one spiracle often anastomose with others to form branching networks penetrating most of the body. In some insects it appears that air is taken into the body through the thoracic spiracles and released through the abdominal spiracles, thus creating a "flow-through system."

Spiracles are placed segmentally, although not necessarily on all segments. In chilopods the spiracles are located on the membranous pleural side walls just above and behind the base of each, or most, legs. Diplopods usually bear two pairs of spiracles per diplosegment, just in front of each leg base. In the symphylans a single pair of spiracles opens on the sides of the head, and the tracheae supply only the first three trunk segments. Except for a few primitive species, most pauropodans lack a tracheal system. In insects the number and placement of spiracles varies, but most have a pair located above the second and third pairs of legs (or only above the third pair), and a pair laterally on each of the first seven or eight abdominal segments.

Each spiracle is usually recessed in an **atrium**, the walls of which are lined with setae or spines (**trichomes**) that prevent dust, debris, and parasites from entering the tracheal tubes. In myriapods the spiracles are often surrounded by a sclerotized rim or lip, the **peritrema**, which also aids in excluding foreign particles. A muscular valve or other closing device is often present and is under control of internal partial pressures of O_2 and CO_2. In resting insects most of the spiracles are generally closed.

Ventilation of the tracheal system is accomplished by simple diffusion gradients, as well as by pressure changes induced by the animal itself. Almost any movement of the body or gut causes air to move in and out of some tracheae. Telescopic elongation of the abdomen is used by some insects to move air in and out of the tracheal tubes. Some insects have expanded tracheal regions called **tracheal pouches**, which function as sacs for air storage. Muscular activity of certain body regions contracts the air sacs (via hemocoelic pressure), thus initiating bursts of rhythmic ventilatory waves that may be sustained in part by the elasticity of the tracheal walls.

Unlike the situation in most animals, the blood of uniramians appears to play no significant role in oxygen transport. Instead, the air-filled tracheae extend directly to each organ, where the ends lie close to or within the tissues. In the case of flight muscles, where oxygen demand is high, the tracheal tubes actually invade the muscle fibers themselves.

The innermost parts of the tracheal system are the **tracheoles**, which are thin-walled, fluid-filled

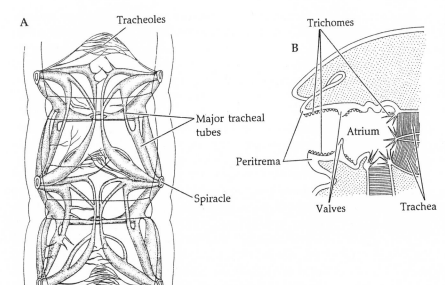

Figure 29

The tracheal system of centipedes and millipedes. A, Tracheal system of three body segments of a typical centipede, *Scolopendra cingulata*. B, A spiracle of *S. cingulata* (transverse section). (After Lewis 1981.)

channels that end as a single cell, the **tracheole end cell** (= **tracheolar cell**) (Figure 28). Unlike tracheae, tracheoles are not shed during ecdysis. The tracheoles are so minute (0.2–1.0 μm) that ventilation is impossible and gas transport here relies on aqueous diffusion. This ultimate constraint on the rate of gas exchange may be the primary reason terrestrial arthropods never achieved extremely large sizes.

In aquatic insects the spiracles are usually non-functional, and gases simply diffuse across the body wall directly to the tracheae. A few species retain functional spiracles; they hold an air bubble over each opening, through which oxygen from the surrounding water diffuses. The air bubbles are held by secreted waxes and by patches of hydrophobic hairs in densities that may exceed 2 million per square millimeter. A few geophilomorph centipedes live in intertidal habitats, under rocks or in algal mats. Presumably, air retained within the tracheal system (and in the spiracular atrium) is sufficient to last during submergence at high tides. Most aquatic insects, particularly larval stages, have external projections of the body wall that are covered by thin, unsclerotized cuticle and contain blood, tracheae, or air bubbles (Figure 30). These **gills** contain channels that lead to the main tracheal system. In some aquatic insects, such as dragonfly nymphs, the rectum bears tiny branched tubules called **rectal accessory gills**. By pumping water in and out of the anus, these insects exchange gases across the increased surface area of the thin gut wall. There are analogous examples of hindgut respiratory irrigation in other, unrelated groups (e.g., echiurans and holothurians).

Excretion and osmoregulation

The major problem of water conservation and the nature of the circulatory and gas exchange systems in uniramians necessitated the evolution of entirely new structures to remove metabolic wastes. In many terrestrial arthropods the solution is Malpighian tub-

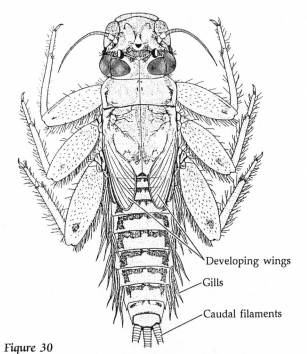

Figure 30

Aquatic nymph of a mayfly, *Thraulodes salinus* (order Ephemoptera), with lateral abdominal gills. (From Allen and Brusca, 1978, Canad. Entomol. 110: 413–433.)

ules. These unbranched outgrowths of the gut usually arise near the junction of the midgut and hindgut (Figures 24, 25, 26, and 31). The blind distal ends extend into the hemocoel and lie among various organs and tissues. From two to several hundred Malpighian tubules may be present. Like the gas exchange surfaces, the excretory system is a site of potential water loss, because nitrogenous wastes initially occur in a dissolved state. These problems are compounded in small terrestrial organisms, such as most uniramians, by their large body surface-to-volume ratios. And water loss problems are even more severe in flying insects, because flight is probably the most metabolically demanding of all locomotor activities.

The uniramian cuticle is sclerotized/tanned to various degrees, adding a small measure of waterproofing. This sclerotization is the only protection that exists in the myriapods, and these creatures must rely to a considerable extent on behavioral strategies to avoid desiccation. Many live in humid or wet environments, or are active only during cool periods. Others stay hidden in cool or moist habitats, such as under rocks, during hot hours or dry periods. Insects have within the epicuticle a waxy layer, which greatly increases their resistance to desiccation and frees them to more fully exploit dry environments. In many terrestrial arthropods (including primitive insects), an eversible **coxal sac** projects from the body wall near the base of each leg (not to be confused with the coxal glands of arachnids). It is thought that the coxal sacs assist in maintaining body hydration by taking up water from the environment (e.g., dew drops).

In the absence of sufficient blood pressure for typical excretory filtration, uniramians use osmotic pressure to achieve the same result. Various ions, especially potassium, are actively transported across the Malpighian tubule epithelium from the blood into the tubule lumen (Figure 31). The osmotic gradient maintained by this ion transport mechanism enables water and solutes to move from the body cavity into the tubules, and thence to the gut. Water and other metabolically valuable materials are selectively reabsorbed into the blood across the wall of the rectum, while the Malpighian filtrate left behind is mixed with the other gut contents. Reabsorption of water, amino acids, salts, and other nutrients may be enhanced by the action of special cells in thickened regions called **rectal glands**. The soluble potassium urate of the Malpighian tubules has, at this point in the gut, been precipitated out as solid uric acid as a result of the low pH of the rectum (pH 4–5). Uric acid crystals cannot be reabsorbed into the blood, hence they pass out the gut with the feces. Insects also possess special cells called **nephrocytes** or **pericardial cells** that move about in certain areas of the hemocoel, engulfing and digesting particulate or complex waste products.

Despite their length, centipedes and millipedes usually possess only one and two pairs of Malpighian tubules, respectively. However, because they are largely confined to moist habitats and nocturnal activity patterns, a significant portion of the excretory wastes may be ammonia rather than uric acid.

Nervous system and sense organs

The uniramian nervous system conforms with the basic arthropod plan described in Chapter 15 (Figures 32 and 33). Very little secondary fusion of ganglia occurs in the myriapods, but considerable concentration is seen in the more heteronomous insects. In the myriapods the ventral nerve cord retains much of its primitive double nature, with a pair of fused ganglia in each segment. Millipedes possess two pairs of fused ganglia in each diplosegment. In insects the two strands of the ventral nerve cord as well as the segmental ganglia are often largely fused. In the dipterans, for example, even the three thoracic ganglia are fused into a single mass. The largest number of free ganglia occurs in the primitive wingless

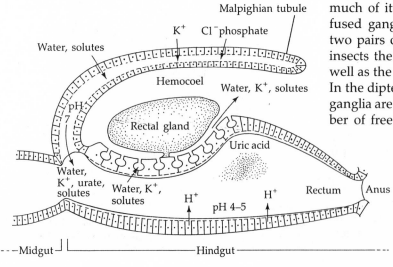

Figure 31

Excretion and osmoregulation in an insect. A single Malpighian tubule opening into the hindgut at its junction with the midgut. Arrows indicate the flow of materials. See text for further explanation. (Modified from Fretter and Graham 1976.)

A

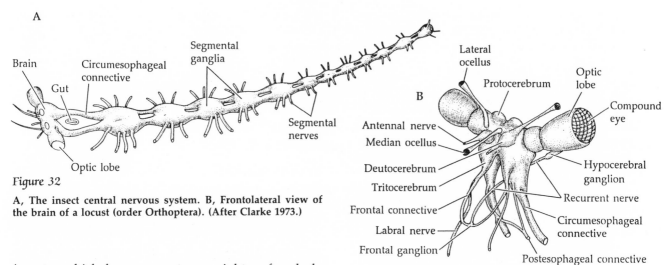

Figure 32

A, The insect central nervous system. B, Frontolateral view of the brain of a locust (order Orthoptera). (After Clarke 1973.)

insects, which have as many as eight unfused abdominal ganglia. Giant fibers have also been reported from several insect orders.

Like the brains of other arthropods, the cerebral ganglia (the "brain") comprise three distinct regions: the protocerebrum (associated with the eyes), the deutocerebrum (associated with the antennae), and the tritocerebrum. The subesophageal ganglion is composed of fused ganglia of the third, fourth, and perhaps the fifth head segments and controls the mouthparts, salivary glands, and some other local musculature.

Insects possess a **hypocerebral ganglion** between the cerebral ganglion and the foregut. Associated with this ganglion are two pairs of glandular bodies called the **corpora cardiaca** and the **corpora allata** (Figure 34). These two organs work in concert with the **prothoracic glands** and certain neurosecretory cells in the protocerebrum. The whole complex is a major endocrine center that regulates growth, metamorphosis, and other functions (see Chapter 15).

Uniramians typically possess simple ocelli at some stage in their life cycle (Figure 35). In living myriapods these are the only eyes that ever develop. Centipedes possess a few to many ocelli. In the scutigeromorph centipedes, up to 200 ocelli may cluster to form a sort of pseudo-compound eye. However, despite many convergent similarities to the true, image-forming compound eyes of insects and crustaceans, the eyes of living centipedes appear to function only in the detection of light and dark. Many parasitic and troglobitic uniramians have lost their eyes completely. Symphylans and pauropodans also lack eyes. Many burrowing diplopodans may also be eyeless, whereas others have from 2 to 80 ocelli arranged variously on the head. Some diplopods also possess integumental photoreceptors, and many eye-

less species exhibit negative phototaxis. In insects, simple ocelli may occur in the larval, juvenile, and adult stages. When present in adults, they usually form a triad or a pair on the anterodorsal surface of the head.

Whether or not the compound eyes of insects, certain fossil myriapods, crustaceans, trilobites, and xiphosurans are homologous structures is a hotly debated question. We think they are, and we address this subject in Chapter 19. Most adult insects have a

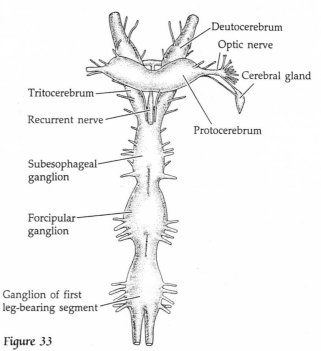

Figure 33

The brain and anterior ganglia of a centipede, *Lithobius forficatus* (dorsal view). (After Lewis 1981.)

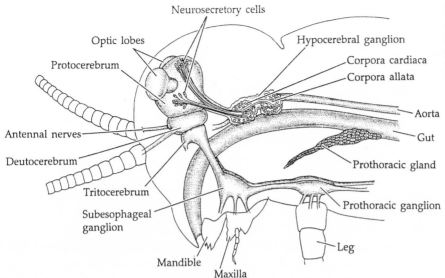

Figure 34

Endocrine organs and central nervous system in the head and thorax of a generalized insect. These cells all play some part in the control of molting and metamorphosis. (After Wells 1968.)

pair of compound eyes, which bulge out to some extent, giving these animals a wide field of vision in all directions (Figure 36A). Compound eyes are greatly reduced or absent in parasitic groups and in many cave-dwelling forms. The general anatomy of the arthropod compound eye was described in Chapter 15, but several distinct structural trends occur in insect eyes, as we describe below.

The number of ommatidia apparently determines the overall visual acuity of compound eyes; hence, larger eyes are typically found on active, predatory insects such as dragonflies and damselflies (order

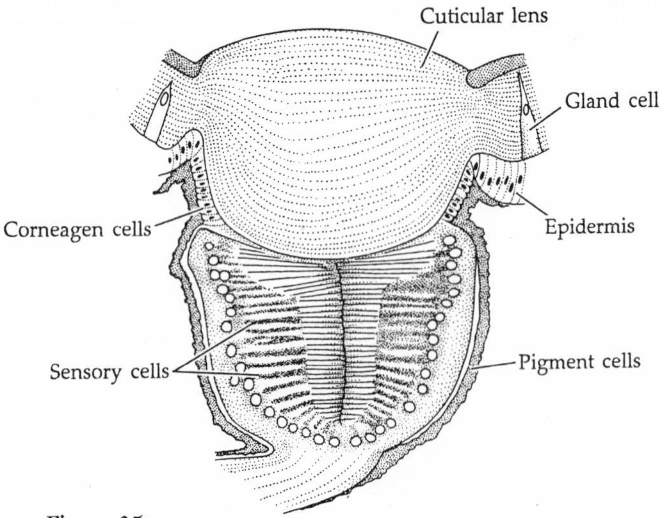

Figure 35

The ocellus of a centipede (section). (After Grenacher 1880, Arch. Mikrosk. Anta. Entwmech. 18:415–467.)

Odonata), which may have over 10,000 ommatidia in each eye. On the other hand, workers of some ant species have but a single ommatidium per compound eye! Similarly, larger facets capture more light and are typical of nocturnal insects. In all cases, a single ommatidium consists of two functional elements: an outer light-gathering part comprising a lens and crystalline cone, and an inner sensory part composed of a rhabdome and sensory cells (Figure 36B).

In some insects the outer surface of the lens is covered with minute conical nipples about 0.2 μm high and arranged in a hexagonal pattern. It is thought that these projections might decrease reflection from the surface of the lens, thus increasing the proportion of light transmitted through the facet. The **cornea** (lens), like the rest of the cuticle, is secreted by epidermal cells, each lens being the product of two **corneagen cells**, which later may become withdrawn to the sides of the ommatidium to form the two **primary pigment cells**. Beneath the cornea are four crystalline cone cells called **Semper cells**, which in most insects produce the **crystalline cone**. The crystalline cone is a hard, clear, intracellular structure bordered laterally by the primary pigment cells. Insect eyes in which the crystalline cone is present are called **eucone eyes** (Figure 36B). Immediately behind the crystalline cone (in eucone eyes) are the sensory elements. These are elongate sensory neurons called **retinular cells**. Primitively, each ommatidium probably contained eight retinular cells arising from three successive divisions of a single cell. This number is found in some insects today, but in most it is reduced to six or seven, the other one or two persisting as

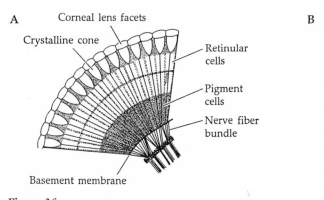

A

Corneal lens facets
Crystalline cone
Retinular cells
Pigment cells
Nerve fiber bundle
Basement membrane

Figure 36

Compound eyes of insects. A, A generalized insect compound eye (cross section). B, A single ommatidium from a compound eye (a eucone ommatidium).

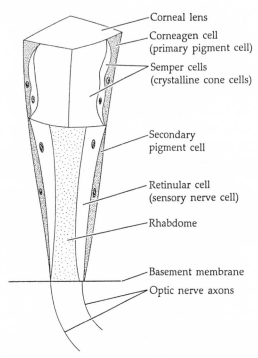

B

Corneal lens
Corneagen cell (primary pigment cell)
Semper cells (crystalline cone cells)
Secondary pigment cell
Retinular cell (sensory nerve cell)
Rhabdome
Basement membrane
Optic nerve axons

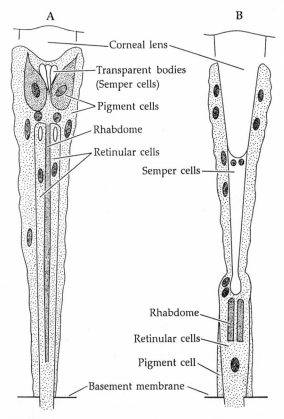

A   **B**

Corneal lens
Transparent bodies (Semper cells)
Pigment cells
Rhabdome
Retinular cells
Semper cells
Rhabdome
Retinular cells
Pigment cell
Basement membrane

Figure 37

Two other types of ommatidia that occur in the compound eyes of different insect groups. See text for details. A, Acone ommatidium of a beetle (order Coleoptera). B, Exocone ommatidium of a beetle (Coleoptera). (After Chapman 1971.)

short basal cells in the proximal region of each ommatidium. Arising from each retinular cell is a neuronal axon that passes out through the basement membrane at the back of the eye into the optic lobe. There is no true optic nerve in insects; the eye connects directly with the optic lobe of the brain.

The **rhabdomeres** consist of tightly packed microvilli that are about 50 nm in diameter, have a hexagonal cross section, and extend toward the central axis of the ommatidium at right angles to the long axis of the retinular cell. Collectively, the rhabdomeres of each ommatidium form the **rhabdome**. The retinular cells are surrounded by 12 to 18 **secondary pigment cells**, which isolate each ommatidium from its neighbors.

In some hemipterans, coleopterans, and dipterans, the Semper cells do not form a crystalline cone but become transparent and undergo only a little modification (Figure 37A). Ommatidia of this type are called **acone ommatidia**. The acone condition may be primitive, because the eyes of many apterygotes are of this type. In most dipterans and some odonatans, the Semper cells produce cones that are liquid-filled or gelatinous rather than crystalline. Ommatidia of this type are called **pseudocone ommatidia**. In some beetles the lens is formed from an inward extension of the cornea, and the Semper cells form a refractile structure between the cuticle and the retinular cells. This type is an **exocone ommatidium** (Figure 37B).

One of the features that distinguishes the myriapods from the insects is the **organ of Tömösvary** (although not all myriapods possess this organ). Located at the base of each antenna, each of this pair of organs consists of a disc with a central pore where the ends of sensory neurons converge (Figure 38). The exact role of this organ has yet to be clearly established, but it may be chemosensory, pressure sensory, humidity sensory, or auditory. The last idea is probably the most popular today. Whether the organ of Tömösvary might detect aerial vibrations (auditory impulses), or only ground vibrations, is also a debated question.

The general body surface of uniramians, like that of other arthropods, bears a great variety of microscopic sensory hairs and setae. The incredible diversity of these cuticular surface structures has only begun to be explored with the advent of scanning electron microscopy. These **sensilla** are most heavily concentrated on the antennae, mouthparts, and legs. Most appear to be tactile or chemosensory. Club- or peg-shaped chemosensory setae, usually called **peg organs** and resembling the aesthetascs of crustaceans, are particularly common on the antennae of uniramians (Figure 39). Centipedes and millipedes are noted for their highly sensitive antennae, which are richly supplied with tactile and chemosensory setae.

Insects have internal proprioreceptors called **chordotonal organs**. These structures stretch across joints and monitor the movement and position of various body parts. **Phonoreceptors** occur in most insect orders. These may be simple modified body or

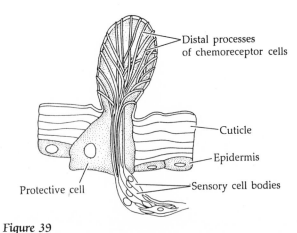

Figure 39

A chemosensory peg organ from the antenna of a grasshopper. (After Slifer et al. 1959, J. Morphol. 105:145–191.)

appendage setae, or antennae, or complex structures called **tympanic organs** (Figure 40). Tympanic organs generally develop from the fusion of parts of a tracheal dilation and the body wall, which forms a thin **tympanic membrane** (= **tympanum**). The receptor nerves in an underlying air sac, or attached directly to the tympanic membrane, respond to vibrations in much the same fashion as they do in the cochlea of the human inner ear. Some insects can discriminate among different sound frequencies, but others are tone deaf. Tympanic organs occur on either the abdomen, the thorax, or the forelegs.

Sound communication in insects, like light communication in fireflies, is a species-specific means of mate communication. Several insect groups (e.g., orthopterans, dipterans, and homopterans) possess sound-producing structures. Male flies of the genus *Drosophila* create species-specific mating songs by rapidly vibrating their wings or abdomen. These "love songs" attract conspecific females for copulation. It has been demonstrated that the rhythm of the male's song is encoded in genes inherited from his mother, on the X chromosome, whereas the song's "pulse interval" is controlled by genes on autosomal chromosomes.

Cicadas may possess the most complex sound-producing organs in the animal kingdom (Figure 41). The ventral metathoracic region of male cicadas bears two large plates, or **opercula**, that cover a complex system of vibratory membranes and resonating chambers. One membrane, the **tymbal**, is set vibrating by special muscles, and other membranes in the resonating chambers amplify the vibrations. The sound leaves the cicada's body through the metathoracic spiracle.

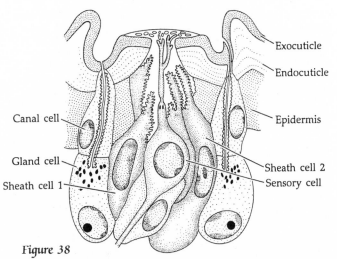

Figure 38

The organ of Tömösvary from the centipede *Lithobius forficatus*. (After Lewis 1981, based on Tichy 1973, Zool. Jahrb. Anat. 91: 93–139.

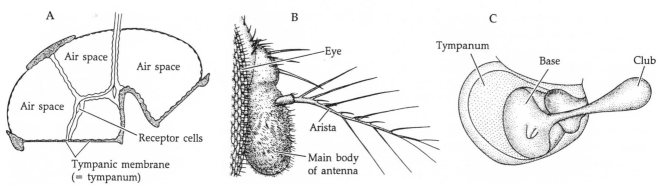

Figure 40

Insect "ears." Insect auditory organs (phonoreceptors) differ widely in their anatomy and location. A, The "ear" of noctuid moths (order Lepidoptera), is a pressure receiver used to detect the ultrasonic cries of hunting bats. It is similar to most other insect "ears" in comprising a tympanic membrane backed by a tracheal air space. Two receptor cells attach to the tympanum. B, In fruit flies (order Diptera), a feathery seta called an arista arises on the third antennal segment. The arista detects air movements, thus responding to sound through interaction with vibrating air particles. It is used to detect the calling song of the species. C, In the "ear" of a water boatman (order Hemiptera), the tympanum is covered by the base of a club-shaped cuticular body that protrudes outside the animal. The club performs rocking movements that allow some frequency analysis of the songs of other water boatmen. (After Michelsen 1979.)

A somewhat similar method of communication occurs in insects that live on the surface of water. Numerous families of beetles and bugs utilize water surfaces as a substratum both for locomotion and communication by waves or ripples. Such insects produce a signal by vertical oscillations of one or more pairs of legs, and sometimes also with distinct vertical body motions. Wave patterns produced are species-specific. Potential prey trapped on a surface film may also be recognized in this fashion, like spider prey-recognition by web vibrations. Limited data suggest that the receptor organs for ripple communication are either specialized sensilla on the legs or special proprioreceptors between joints of the legs or antennae, perhaps similar to the tarsal organs of scorpions (see Chapter 16).

A number of insects are bioluminescent, the most familiar being beetles of the family Lampyridae, known as lightning bugs or fireflies. In the tropics, where they are especially abundant, fireflies are sometimes kept in containers and used as natural flashlights, or women may wrap them in gauze bags to be worn as glowing hair ornaments. The light of fireflies and other luminescent insects ranges from green through red and orange, depending on the species and the precise chemical nature of the luciferin–luciferase system involved. Light-producing or-

gans are typically composed of clusters of light-producing cells, or **photocytes**, backed by a layer of reflecting cells and covered with a thin, transparent epidermis. Photocytes are richly supplied with tracheae, oxygen being necessary for the chemical reaction, and are provided with direct nervous innervation. Each species of firefly and most other glowing insects have distinct flash patterns, or codes, to facilitate mate recognition and communication.

One of the most sophisticated communication behaviors among insects has been said to be the famous "honeybee dance." Each day, forager bees leave the colony to locate new food sources (e.g.,

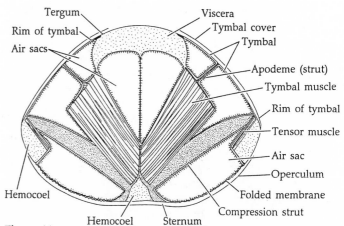

Figure 41

Sound production in cicadas (order Homoptera) from the first abdominal segment (section). Sound is produced by buckling of the tymbal, a thin disc of cuticle. The tymbal muscle is connected to the tymbal by a strut. Contraction of this muscle causes the tymbal to buckle inward, thereby producing a click that is amplified by resonance in the underlying air sacs. On relaxation, the elasticity of the muscle causes the tymbal to buckle out again. On the underside of this abdominal segment, a folded membrane can be stretched to tune the air sacs to the resonant frequency of the tymbal. (After Blaney 1976.)

fresh flower blooms). They fly meandering search forays until a good source is located. Then they return to the hive along a straight flight path (a "bee line"); while doing so, they may imprint a navigational "map" of the direction from the colony to the food. Most behaviorists believe that this information is communicated to hivemates in a complex "tail wagging" dance that allows other bees from the hive to fly directly to the new feeding ground.

A large body of research on bee navigation has accumulated since the pioneering "dancing bees" studies of Karl von Frisch early in this century. We now know that honeybees have both outstanding vision and magnetic detectors in their bodies. Much of the bee's daily activity, including navigation and flower recognition, relies strongly on ultraviolet vision. Honeybees appear to utilize a hierarchical series of flight orientation mechanisms. When the primary mechanism is blocked, a bee can switch to a secondary system. The primary navigation system utilizes the pattern of polarized ultraviolet sunlight in the sky. The pattern depends on the location of the sun as determined by two coordinates, the azimuth and the elevation. Bees and many other animals that orient with the sun have a built-in ability to compensate for changes in the sun's position with time, both hourly changes (elevation) and seasonal changes (azimuth). On cloudy days, when the sun's light is largely depolarized, bees cannot rely on their ultraviolet celestial navigation mechanism and thus may switch to the second order navigational system—navigation by landmarks (foliage, rocks, and so on) that were also imprinted during the most recent flight to the food source. Limited evidence suggests that some form of tertiary backup system may also exist.

Thus, if the honeybee dance hypothesis is correct, honeybees must simultaneously process information concerning time, the direction of flight relative to the sun's azimuth, the movement of the sun, the distance flown, and local landmarks (not to mention complications from other factors such as crosswinds), and in doing so reconstruct a straight-line heading to inform hivemates. If recent evidence is correct, bees (like homing pigeons) may also detect Earth's magnetic fields with iron compounds (magnetite) located in their abdomens. Bands of cells in each abdominal segment of the honeybee contain iron-rich granules, and nerve branches from each segmental ganglion appear to innervate these tissues.

However, the honeybee dance hypothesis has not been validated to every scientist's satisfaction, and some workers doubt its existence altogether (see references by Frisch, Gould, Rosin, Wells, and Wen-

ner). The history of the honeybee dance controversy has been one of the more interesting debates in biology.

In some insects the ocelli are the principal navigation receptors. Some locusts and dragonflies and at least one walking ant species utilize their ocelli to read compass information from the blue sky. As in bees, the pattern of polarized light in the sky seems to be the main compass cue. In some species, both ocelli and compound eyes may function in this fashion.

Reproduction and development

Uniramians are gonochoristic, and most are oviparous; a few insects are ovoviviparous. Parthenogenesis is known in one group of millipedes, some centipedes and symphylans, and many insects. The myriapods all have direct development; the young hatch as "miniature adults," although usually with fewer body segments. The most primitive insects also have direct development. However, advanced insects undergo complex mixed development that typically includes radically different larval, pupal, and adult forms. Mixed development in insects is surely a secondarily derived phenomenon, not derived from larval strategies of marine forms, whose planktonic stages would never work on land. Myriapods, like many arachnids, rely on indirect copulation and insemination. Their spermatophores are deposited in the environment, or are held by the male and later picked up by the female. Most insects utilize direct copulation, injecting the sperm into the female's reproductive system. Beyond these generalizations, uniramian reproductive behaviors are so varied that we discuss each major taxon separately, followed by a short section on early uniramian embryogeny.

Chilopoda. Female centipedes possess a single elongate ovary located above the gut, whereas males have from 1 to 26 testes similarly placed (Figure 42). The oviduct joins with the openings from several **accessory glands** and a pair of **seminal receptacles** just internal to the gonopore situated on the **genital segment** (the legless segment in front of the pygidium). The pore is usually flanked by a pair of small grasping appendages, or **gonopods**. In males the testes join the ducts of several accessory glands and a pair of **seminal vesicles** near the gonopore, which opens on the ventral surface of the genital segment. The male gonopore also lies between a pair of small gonopods. Sperm are packed into spermatophores. Like the other myriapods, chilopods rely on indirect sperm transfer. The genital glands of all female cen-

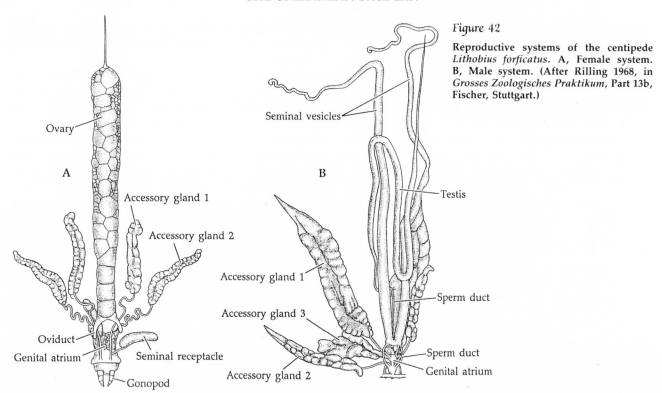

Figure 42

Reproductive systems of the centipede *Lithobius forficatus.* **A,** Female system. **B,** Male system. (After Rilling 1968, in *Grosses Zoologisches Praktikum,* Part 13b, Fischer, Stuttgart.)

tipedes except the scutigeromorphs produce silky threads that are spun into a weblike nuptial net, onto which the male deposits the spermatophores (Figure 43). Mating pairs of centipedes typically go through some courtship behavior, stroking each other with their antennae and moving the large (to several millimeters) spermatophore about the nuptial net. Eventually the female picks up the spermatophore with her gonopods and inserts it into her gonopore. The female coats the eggs with moisture and fungicides before depositing them in the ground or in rotting vegetation. Some degree of parental guarding is common. Some species hatch as juveniles with all body segments; others add new segments with the posthatch molts until the adult body form is attained.

Diplopoda. Millipedes possess a single pair of elongate gonads in both sexes. Unlike chilopods, millipede gonads lie between the gut and the ventral nerve cord. The gonopores open on the third (genital) trunk segment. In females each oviduct opens separately into a **genital atrium** (= **vulva**) near the limb coxae. A groove in the vulva leads into one or more seminal receptacles. Males possess one or a pair of penes, also on the third trunk segment. The mandibles are used to transfer sperm in some millipedes (the "pill millipedes"), but more commonly one or

both pairs of legs of the seventh trunk segment are modified as gonopods and serve this purpose. Males bend the anterior body segments back under the trunk until the penes and gonopods make contact, whereupon the spermatophore is picked up in preparation for copulation. Copulating millipedes embrace with their legs and mandibles, and may coil their trunks together. Sometimes a pair remains in copula for up to two days. Antennal tapping, head drumming, and stridulation may be included in the precopulatory repertoire, and pheromones are known to play an important role in many species. During mating, an enzyme dissolves the spermatophore casing and the sperm move into the seminal receptacles of the female millipede.

Eggs are fertilized as they are laid in the ground. Some species fashion a nest of soil and humus reinforced with parental feces, or the entire nest may be composed of fecal material. One to 300 eggs are laid at a time. Hatchling millipedes typically possess seven or fewer body segments and only three pairs of legs. Additional segments and legs are added at the rear of the last somite with succeeding molts.

Pauropoda. In pauropodans a single ovary lies beneath the gut, but the testes are located above the gut. As in millipedes, the gonopores are on the third

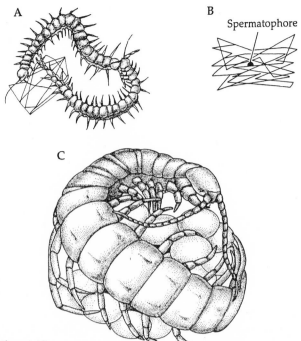

Figure 43

Mating in the common tropical centipede *Scolopendra cingulata*. A, A male and female in mating position over the sperm web. B, A sperm web with a spermatophore. C, A female and her brood. (After Lewis 1981.)

(genital) trunk segment. Females take up a spermatophore, often after the male has suspended it from one or a few silk threads spanning two stones or leaves. Fertilization is internal. The yolky eggs are laid in decaying wood. Hatchlings lack many adult segments and usually have only three pairs of legs.

Symphyla. Paired gonads discharge through gonopores on the third trunk segment. At least some symphylans construct stalklike structures, topped with spermatophores. When a female encounters one of these elevated sperm packages, she bends the stalk over with her antennae, bites off the spermatophores, and stores the sperm in her preoral cavity. When her own eggs are ripe, the female removes them from her gonopore with her mouthparts and cements them to moss or some other substratum. Fertilization occurs during the process, for she coats each egg with the stored sperm. Young symphylans hatch with only about half the adult number of trunk segments and appendages.

Insects. Like other uniramians, insects are gonochoristic, but unlike the myriapods, most insects rely on direct copulation and insemination. Repro-

ductively mature insects are termed **imagos**. Females have one pair of ovaries formed of clusters of tubular **ovarioles** (Figure 44A). The oviducts unite as a common duct before entering a **genital chamber**. Seminal receptacles (**spermathecae**) and **accessory glands** also empty into the genital chamber. The genital chamber opens, via a short **copulatory bursa** (= **vagina**), on the sternum of the eighth or occasionally the seventh or ninth abdominal segment. The male system is similar, with a pair of testes, each comprising a number of **sperm tubes** (Figure 44B). Paired sperm ducts dilate as seminal vesicles (where sperm are stored) and then unite as a single ejaculatory duct. Near this duct, accessory glands discharge seminal fluids into the reproductive tract. The lower end of the ejaculatory duct is housed within a penis, which extends posteroventrally from the ninth abdominal sternite.

Courtship behaviors in insects are extremely diverse and often quite elaborate, and each species has its own recognition methods. The courtship can be simple chemical or visual attraction, but more typically it involves pheromone release, followed by a variety of displays, tactile stimulation, songs, flashing lights, or other rituals that may last for hours. The subject of insect courtship is a large and fascinating study of its own. Although the field of pheromone biology is still in its infancy, sexual attractant or aggregation pheromones have been identified from about 450 different insect species (about 250 of which are synthesized and sold commercially for pest control purposes).

Most insects transfer sperm directly as the male inserts either his penis (Figure 44D) or a gonopod into the genital chamber of the female. Special abdominal claspers, or other articulated cuticular structures on males, often augment his copulatory grip. Such morphological modifications are species-specific and thus valuable recognition characters, both for insect mates and insect taxonomists.

Copulation often takes place mid-flight. In some of the primitive wingless insects and in the odonatans, sperm transfer is indirect, as it is in the myriapods. In these cases, a male may deposit his sperm on specialized regions of his body, to be picked up later by the female, or he may simply leave the sperm on the ground, where they are found and taken up by females. In bedbugs (order Hemiptera, family Cimicidae) males use their swollen penis to pierce a special region of the female's body wall. Sperm are then deposited directly into an internal organ (the **organ of Berlese**). From there sperm migrate to the ovaries, where fertilization takes place as eggs are released.

Figure 44

Insect reproductive systems. A, Female system. B, Male system. C, The posterior end of the abdomen of a mature female insect. D, The posterior end of the abdomen of a mature male insect. (After Snodgrass 1935.)

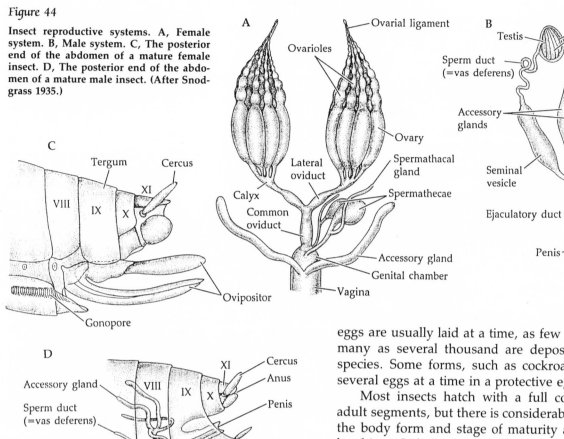

Sperm may be suspended in an accessory gland secretion, or, more commonly, these secretions harden about the sperm to produce spermatophores. Females of many insect species store large quantities of sperm within the seminal receptacles. In some cases sperm from a single mating is sufficient to fertilize a female's eggs for her entire reproductive lifetime, which may be a few days to several years.

Insect eggs are protected by a thick membrane (the **chorion**) produced within the ovary. Fertilization occurs as the eggs pass through the oviduct to be deposited. Accessory glands contribute adhesives or secretions that harden over the zygotes. In many species, cuticular extensions around the female gonopore form an **ovipositor** (Figure 44C), which places the eggs in a brooding site that will afford suitable conditions for the young once they hatch (such as in a shallow underground chamber, in a plant stem, or within the body of a host insect). Although 50–100 eggs are usually laid at a time, as few as one and as many as several thousand are deposited by some species. Some forms, such as cockroaches, enclose several eggs at a time in a protective egg case.

Most insects hatch with a full complement of adult segments, but there is considerable variation in the body form and stage of maturity at the time of hatching. Only in the primitive wingless insects do the young hatch out as juveniles closely resembling the adult, or imago condition. In insects, such direct development is called **simple development**. In the pterygotes, development is mixed, and hatching stages undergo a series of morphological changes (metamorphoses) before the adult condition is achieved. Such growth may occur by gradual changes known as **hemimetabolous development** (= **incomplete metamorphosis**) (Figure 45), or by a dramatic series of metamorphoses called **holometabolous development** (= **complete metamorphosis**) (Figure 46).

Hemimetabolous insects have young that resemble adults in that they possess compound eyes, antennae, and similar feeding and walking appendages. Functional wings and sexual structures, however, are always lacking. The immature forms are often called **nymphs**; they usually have wing rudiments called **wing pads** that expand and form functional wings with the preadult molt. Many hemimetabolous insects have aquatic gilled nymphs, called **naiads** (e.g., mayflies, dragonflies, damselflies). The principal changes during growth are in body size and proportions and in the development of wings and sexual structures. Nymphs and adults often live in the same general habitat.

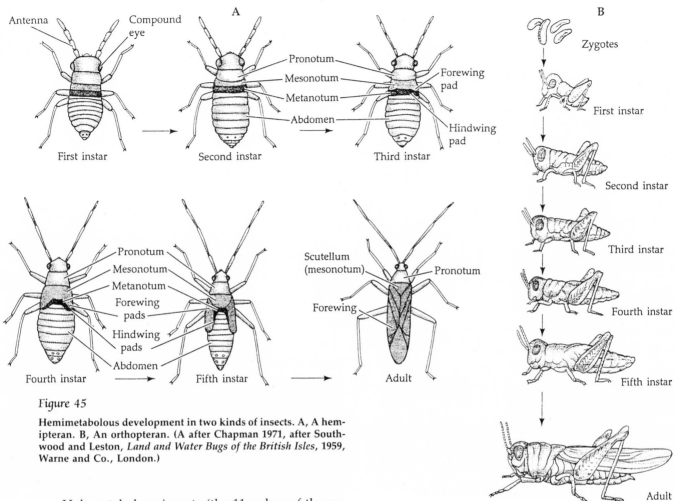

Figure 45

Hemimetabolous development in two kinds of insects. A, A hemipteran. B, An orthopteran. (A after Chapman 1971, after Southwood and Leston, *Land and Water Bugs of the British Isles*, 1959, Warne and Co., London.)

Holometabolous insects (the 11 orders of the superorder Holometabola) hatch as vermiform larvae that bear no resemblance whatsoever to adult forms. So different are they from the adults that larvae are often given separate vernacular names. For example, butterfly larvae are called caterpillars, fly larvae maggots, and beetle larvae grubs. Larvae lack compound eyes and antennae, and their natural history differs markedly from that of the adults. The mouthparts may be wholly unlike those of the adult, and external wing pads are never present. Often the greater part of an insect's lifetime is spent in a series of larval instars. Larvae typically consume vast quantities of food and attain a larger size than the adult. Termination of the larval stage is accompanied by **pupation**, during which (in a single molt) the **pupa** stage is entered. Pupae do not feed or move about very much. They often reside in protective niches in the ground, within plant tissues, or housed in a cocoon. Energy reserves stored during the long larval life are utilized by the pupa to undergo wholesale transformation of the body. Many structures are broken down and reorganized to attain the adult form; external wings and sexual organs are formed. The remarkable transformation from larval stage to adult stage in holometabolous insects is one of the most impressive achievements of animal evolution.

The success of the holometabolous lifestyle is demonstrated by the fact that such insects outnumber hemimetabolous species 10 to 1. There is a popular theory among evolutionary biologists that claims mixed and indirect development, including holometabolous development in insects, is selectively advantageous because it results in the ecological segregation of adults from young, thus avoiding intraspecific competition and allowing each stage to develop its own suite of specific survival strategies. Satisfactory confirmation of this theory has been difficult to come by. We have seen that such developmental strategies are common in marine and some freshwater invertebrates, but only the insects have managed to exploit this habit so successfully on land.

The role of ecdysone in initiating molting is described in Chapter 15. This hormone works in conjunction with a second endocrine product in controlling the sequence of events in insect metamorphosis. The second product, **juvenile hormone**, is manufactured and released by the corpora allata, a pair of glandlike structures associated with the brain (Figure 34). When ecdysone initiates a molt in an early larval instar, the accompanying concentration of juvenile hormone is high. Such a high concentration ensures a larva-to-larva molt. After the last larval instar is reached, the corpora allata ceases to secrete juvenile hormone. Low concentrations of juvenile hormone result in a larva-to-pupa molt. Finally, when the pupa is ready to molt, juvenile hormone is absent altogether from the hemolymph; this deficiency leads to a pupa-to-adult molt.

A number of insect groups, particularly those living in unstable environments, undergo seasonal parthenogenesis. In some groups (e.g., honeybees and aphids), fertilized eggs develop into females, and unfertilized (parthenogenetic) eggs develop into males. **Polyembryony** also occurs in a number of insect taxa, particularly parasitic Hymenoptera; in this form of development the early embryo splits to give rise to more than one developing embryo. Thus, from two to thousands of larvae may result from a single fertilized egg, which is often deposited in the body of another (host) insect.

Uniramian embryogeny

As discussed in Chapter 15, the large centrolecithal eggs of arthropods are often very yolky, a condition resulting in dramatic modifications of the presumed ancestral spiralian cleavage pattern. Although vestiges of holoblastic spiral cleavage are still discernible in some crustaceans, the uniramians show almost no trace of spiral cleavage at all; they have shifted almost entirely to meroblastic cleavage. Most

Figure 46

Holometabolous development in two kinds of insects. A, A lepidopteran. B, A coleopteran. (A after Chapman 1971, after Urquhart 1960; B after Ross 1965.)

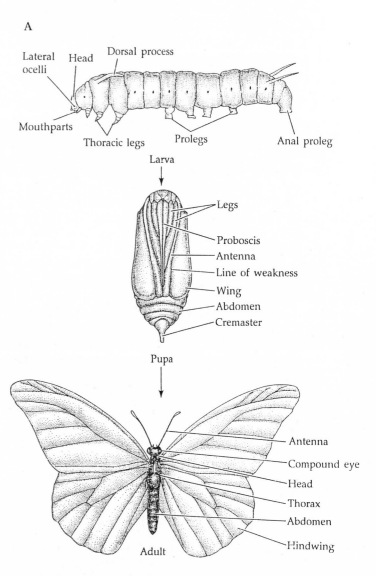

A

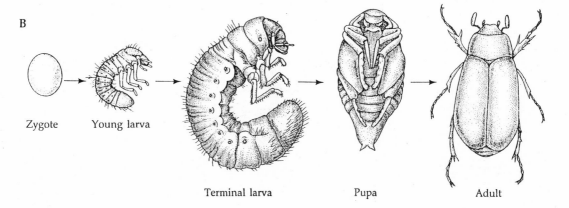

B

uniramians undergo early cleavage by intralecithal nuclear divisions, followed by a migration of the daughter nuclei to the peripheral cytoplasm (=**periplasm**). Here they continue to divide until the periplasm is dense with nuclei, whereupon cell membranes begin to form, partitioning uninucleate cells from one another. At this point the embryo is a periblastula, comprising a yolky sphere that has a few scattered nuclei and is covered by a thin cellular layer, or **blastoderm** (see Figure 17 in Chapter 15).

Along one side of the blastula a patch of columnar cells forms a **germinal disc**, sharply marked off from the thin cuboidal cells of the remaining blastoderm (Figure 47A). From specific regions of this disc, presumptive entodermal and mesodermal cells begin to proliferate as germinal centers. These cells migrate inward during gastrulation to lie beneath their parental cells, which now form the ectoderm. The mesoderm proliferates inward as a longitudinal **gastral groove** (Figure 47B); the cells of the developing gut usually surround and gradually begin absorbing the central yolky mass of the embryo, and paired coelomic spaces appear in the mesoderm.

As segments begin to demarcate and proliferate, each receives one pair of mesodermal pouches and eventually develops **appendage buds**. As the mesoderm contributes to various organs and tissues, the "coeloms" merge with the small blastocoel to produce the hemocoelic space. The mouth and anus arise by ingrowths of the ectoderm that form the fore- and hindguts, which eventually establish contact with the developing entodermal midgut.

Variations on this basic scheme occur in groups with secondary yolk reduction, such as pauropodans, symphylans, and collembolans. In these groups a largely holoblastic cleavage takes place, and a coeloblastula forms. In the chilopods and a few insects, a type of cleavage occurs that is somewhat intermediate between total and superficial cleavage. After a few intralecithal nuclear divisions, the yolk breaks up into blocks called **yolk pyramids** (Figure 47C). Gradually the yolk pyramids disappear, and development shifts to the superficial type.

Within this very general developmental framework, different uniramian groups show considerable variation, especially with regards to embryonic cell fates—much more variation than is seen among the crustaceans. For details on these and related matters, see Anderson (1973).

Uniramian phylogeny

We favor an evolutionary model that views uniramians as the sister-group of the crustaceans, an association implying that the two taxa share an immediate common ancestor. Not all biologists agree with this view. Those that do often classify these two groups together in the taxon Mandibulata. We also view insects as the sister-group of the myriapods. Few specialists would dispute this latter hypothesis. The relative popularity of any given theory, of course, does not mean that it is necessarily any "more correct" than any other, simply that it happens to be the one currently in vogue. Usually, but not always, current popularity derives from the substantiating weight of the most recently discovered evidence. We discuss the evolution of arthropods in Chapter 19, including the arguments surrounding the concept of the "Mandibulata." Figure 48 is a cladogram depict-

Figure 47

Early stages of uniramian development. A, The blastoderm (blastula) of a generalized insect (cross section). Note the thickened germinal disc. B, An early gastrula of a honeybee (cross section). Note the gastral groove and the proliferation of mesoderm. Entoderm is derived from the front and back ends of the gastral groove. C, A yolk pyramid embryo of the centipede *Scolopendra*. (Modified after Anderson 1973.)

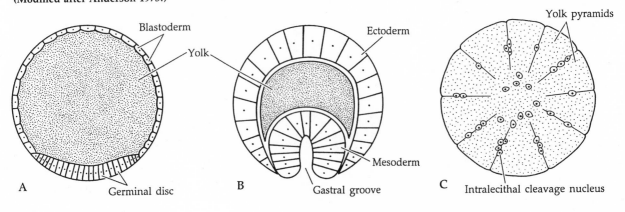

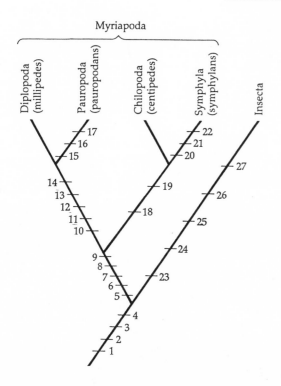

Myriapoda

Diplopoda (millipedes)
Pauropoda (pauropodans)
Chilopoda (centipedes)
Symphyla (symphylans)
Insecta

Figure 48

A cladogram depicting the phylogeny of the subphylum Uniramia. In this cladogram, the insects and myriapods are hypothesized to be monophyletic sister-groups. Defining synapomorphies are indicated on the cladogram and listed below.

The subphylum Uniramia is characterized by the possession of at least four unique synapomorphies: (1) mandibles are whole-limb jaws; (2) gas exchange system is tracheal; (3) Malpighian tubules are ectodermally derived; and (4) one pair of antennae (probably the second antennae) is lost. A fifth possible synapomorphy of uniramians is loss of the carapace; Uniramia is the only arthropod subphylum in which a carapace or distinct head shield does not occur. The question of uniramous limbs is addressed in Chapter 19.

The class Myriapoda is characterized by the following synapomorphies: (5) presence of organs of Tömösvary; (6) presence of myriapod repugnatorial glands; (7) loss of the entodermally derived digestive ceca; (8) loss of the compound eyes (at least in extant species); and (9) loss of palps on the first and second maxillae. Within the Myriapoda, the subclasses Diplopoda and Pauropoda are hypothesized to form a sister-group to the subclasses Chilopoda and Symphyla. The Diplopoda–Pauropoda line is distinguished by (10) fusion of the first maxillae to form a gnathochilarium; (11) loss of the second maxillae; (12) suppression of the anterior pair of trunk appendages; (13) tendency to form diplosegments; and (14) modification of the first trunk segment into a collum. Pauropodans are distinguished by having branched antennae (15); loss of the eyes (16); and loss of the circulatory system (17).

The Chilopoda–Symphyla line is distinguished by (18) medial coalescence of both pairs of maxillae and (19) modification of the first pair of trunk appendages into raptorial poison fangs (maxillipeds). Symphylans are characterized by (20) fusion of the second maxillae to form a labium; (21) loss of the eyes; and (22) the presence of spinnerets on the thirteenth trunk segment.

The class Insecta is distinguished by the following synapomorphies: (23) fusion of the second maxillae as a labium (a condition also seen in the Symphyla, hypothesized by this cladogram to be an evolutionary convergence); (24) loss of the articulating endite on the mandible; (25) formation in the postcephalic region of a distinct three-segmented thorax and an abdomen; (26) fixation of the number of abdominal somites at 11 (or fewer by reduction); and (27) loss of all abdominal appendages.

No doubt other characters exist that can be used to test these cladistic hypotheses. Other such characters might corroborate the relationships depicted in this tree, or they might suggest changes in the tree that would yield more parsimonious hypotheses.

ing phylogenetic relationships within Uniramia; it includes subtaxa of the Myriapoda. Figure 49 provides a cladogram of the relationships among the Insecta.

We have already commented on the spectacular evolutionary radiation of the insects. In addition to the living orders of insects, at least 10 other orders arose and radiated in late Paleozoic and early Mesozoic times, and then went extinct. In this chapter we have seen examples of insects that live in every conceivable terrestrial and freshwater habitat, and even in some marine habitats. You have been introduced to insects that are scavengers, parasites, predators, and herbivores, and to species that have evolved highly specialized diets and others that are extreme generalists. This enormous diversity has led to a great many different hypotheses concerning insect phylogeny. Nevertheless, a very general theory of insect evolution has gradually emerged over the past few decades (see Kristensen 1975 and Hennig 1981 for good reviews of this history). One of the best concise descriptions of insect phylogeny appeared in a phylogenetic analysis constructed by Daly et al. (1978) in their introductory entomology text. Their cladogram is shown in Figure 49. Although we have not included the synapomorphies on the figure—because many involve detailed characters beyond the scope of this text—a few of the principal ones are noted below.

The genealogical relationships depicted in the in-

sect cladogram roughly parallel the generally accepted classification scheme provided at the beginning of this chapter, although a precise match would require some reorganization of the primitive wingless orders at the class/subclass level. Proturans, collembolans, and diplurans are generally agreed to be the most primitive living insects, and most specialists view them as a monophyletic sister-group to all other living insects. These three orders share the highly primitive character trait of having the mandibles and first maxillae deeply retracted into pouches in the head capsule, with only the tips protruding for feeding. This condition is termed **entognathy** (or **endognathy**). Entognathous mandibles are monocondylic, articulating at a single point on the head, whereas all

Figure 49

A cladogram depicting the hypothetical relationships of the extant insect orders. General relationships of the major groups are discussed in the text. This cladogram is adapted from Daly et al. (1978), to which the reader is referred for a detailed discussion of the synapomorphies used to construct the tree. The cladogram depicts the three primitive entognathous orders (Protura, Collembola, and Diplura) as the sister-group to all other orders (the more advanced, ectognathous insects). Among the ectognathous insects, the orders Archaeognatha and Thysanura are hypothesized to have evolved prior to the "invention" of flight. The remaining insects, in the subclass Pterygota, all have wings and are divisible into four or five monophyletic groups on the basis of their developmental patterns and other features: Paleoptera, Orthopterodea, Hemipterodea, and Holometabola. Characters that distinguish each of these groups are listed in the synopses appearing early in this chapter. Note that the wingless insects (the "apterygotes") do not form a natural (monophyletic) group.

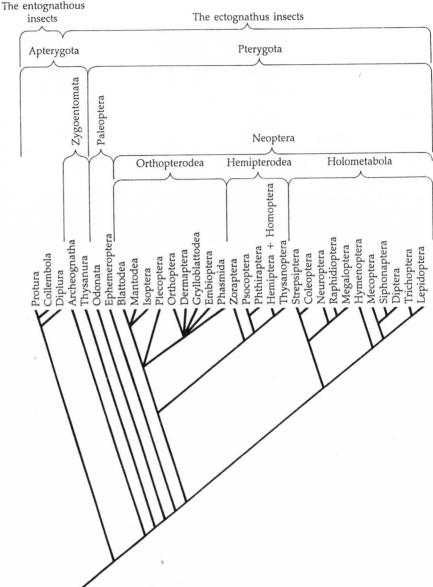

other insects (**ectognathous** insects) primitively possess dicondylic mandibles, articulating with the head at two points (Figure 7). In addition, all antennal articles of entognathous insects are muscular, whereas in ectognathous insects only the basalmost one or two articles retain musculature. Like the myriapods, entognathous insects primarily inhabit forest litter and upper soil layers, requiring high relative humidity or sources of standing water. Furthermore, some retain apparent vestiges of abdominal appendages, and sperm transfer is indirect, again like that seen in myriapods. Many other similarities exist between myriapods and the primitive entognathous insects.

The more advanced, ectognathous insects comprise two large groups: the wingless ectognaths (subclass Zygoentomata: orders Thysanura and Arachaeognatha) and the winged insects (subclass Pterygota). Note that the cladogram implies that the Zygoentomata is a paraphyletic group and the Pterygota a monophyletic taxon. Thysanurans appear to be living representatives of an "intermediate stage" in insect evolution—between the development of ectognathous mouthparts and the development of wings. Like other ectognaths, thysanurans have lost the antennal musculature in all but the two basal articles. Like the entognaths, however, they have rudimentary abdominal appendages, employ indirect insemination, and molt throughout their lives. In some thysanurans the elongate monocondylic mandible has the cutting (incisor) lobe separated by a considerable distance from the grinding (molar) lobe, a condition similar to that seen in most crustaceans.

One very important hypothesis to note in this cladogram is that the wingless insects (the apterygotes) do not constitute a natural, or monophyletic group. They are merely five primitive orders that apparently evolved prior to the invention of wings in insects. Although the taxon Apterygota was commonly used in older entomology texts, it has largely been abandoned in the last decade, because of its assumed paraphyletic nature.

The winged insects (subclass Pterygota), broadly speaking, comprise four groups, or evolutionary lineages. The most primitive are the dragonflies and damselflies (order Odonata), which are usually combined with the mayflies (order Ephemeroptera) in the infraclass Paleoptera. Characteristic of members of the Paleoptera is the inability to flex the wings over the back when at rest. Another primitive attribute of the odonatans, mentioned earlier in this chapter, is their direct flight musculature. Extant paleopterans have aquatic larval stages (naiads). This has led some workers to suggest that the pterygotes may have arisen from aquatic ancestors whose gills evolved into wings, a theory of flight origin also discussed earlier in this chapter. However, this argument is considerably weakened by the fact that many extinct paleopterans are thought to have had terrestrial nymphs.

The "modern" insects (infraclass Neoptera) comprise three large groups, usually ranked as superorders: Orthopterodea, Hemipterodea, and Holometabola. Good discussions of evolution within each of these major insect lineages may be found in the references cited above.

Selected References

The amount of published information on insects and myriapods is overwhelming. We list primarily the most recent literature below, emphasizing selected identification guides, texts, and research papers that represent major publications or reviews in their respective fields. The most recently published books listed below will provide access to the primary literature.

Insects

American Zoologist. 1981. Insect Systems. Vol. 21, No. 3, pp. 623–791. [Eighteen papers on insect physiology, emphasizing the endocrine system.]

Anderson, S. O. and T. Weis-Fogh. 1964. Resilin: A rubberlike protein in arthropod cuticle. Adv. Insect Physiol. 2:1–66.

Arnett, R. H. 1985. *American Insects: A Handbook of the Insects of North America.* Van Nostrand Reinhold, New York.

Arnett, R. H., N. M. Downie and H. E. Jaques. 1980. *How to Know the Beetles.* Wm. C. Brown, Dubuque, Iowa.

Barth, F. G. 1985. *Insects and Flowers. The Biology of a Partnership.* Translated by M. A. Bierderman-Thorson. Princeton University Press, Princeton, New Jersey.

Bell, W. J. and R. T. Cardé (eds.). 1984. *Chemical Ecology of Insects.* Sinauer, Sunderland, Massachusetts.

Bennet-Clark, H. C. and E. C. A. Lucey. 1967. The jump of the flea: A study of the energetics and a model of the mechanisms. J. Exp. Biol. 47:59–76.

Bland, R. G. 1978. *How to Know the Insects.* Wm. C. Brown, Dubuque, Iowa.

Blaney, W. 1976. *How Insects Live.* Elsevier-Phaidon, Phaidon Press, Littlegate House, Oxford. [A general primer of insect ecology, in coffee-table style.]

Blum, M. S. (ed.). 1985. *Fundamentals of Insect Physiology.* Wiley-Interscience, New York.

Blum, M. S. and N. A. Blum. 1979. *Sexual Selection and Reproductive Competition in Insects.* Academic Press, New York.

Borrer, D. J. and D. M. DeLong. 1971. *An Introduction to the Study of Insects,* 3rd Ed. Holt, Rinehart & Winston, New York. [A standard reference and identification guide for many generations of entomology students.]

Borrer, D. J. and R. E. White. 1970. *A Field Guide to the Insects of America North of Mexico.* Houghton Mifflin, Boston.

Bromenshenk, J. J., S. R. Carlson, J. C. Simpson and J. M. Thomas. 1985. Pollution monitoring of Puget Sound with honey bees. Science 227:632–634.

Brown, W. L., Jr. et al. 1982. Insecta. *In* S. P. Parker (ed.), *Synopsis and Classification of Living Organisms.* McGraw-Hill, New York, pp. 326–680.

Brues, C. T., A. L. Melander and F. M. Carpenter. 1954. Classification of insects. Bull. Mus. Comp. Zool. Harv. Univ. 108: 1–913. [An old but still useful compendium of keys and diagnoses of insect orders and families.]

Butler, C. G. 1967. Insect pheromones. Biol. Rev. 42:42–87.

Chapman, R. F. 1982. *The Insects,* 3rd Ed. Harvard University Press, Cambridge, Massachusetts. [One of the best references on general insect anatomy.]

Cheng, L. 1976. *Marine Insects.* North-Holland, Amsterdam/American Elsevier, New York.

Christiansen, K. and P. Bellinger. 1988. Marine littoral Collembola of North and Central America. Bull. Mar. Sci. 42:215–245.

Chu, H. F. 1949. *How to Know the Immature Insects.* Wm. C. Brown, Dubuque, Iowa.

Clarke, K. U. 1973. *The Biology of the Arthropoda.* American Elsevier, New York.

Crosland, M. W. J. and R. Crozier. 1986. *Myrmecia pilosula,* an ant with only one pair of chromosomes. Science 231:1278–1284.

Daly, H. V., J. T. Doyen and P. R. Ehrlich. 1978. *Introduction to Insect Biology and Diversity.* McGraw-Hill, New York. [One of the best.]

Davey, K. G. 1965. *Reproduction in the Insects.* Oliver and Boyd, Edinburgh.

Denno, R. F. and H. Dingle (eds.). 1979. *Insect Life History Patterns: Habits and Geographic Variation.* Springer-Verlag, New York.

Dethier, V. G. 1963. *The Physiology of Insect Senses.* Methuen, London.

Dyer, F. C. and J. L. Gould. 1983. Honey bee navigation. Am. Sci. 71:587–597.

Edney, E. B. 1966. The impact of the atmospheric environment on the integument of insects. Proceedings of the Fourth International Biometerology Congress, 1966. pp. 71–81.

Edson, K. M., S. B. Vinson, D. B. Stoltz, and M. D. Summers. 1981. Virus in a parasitoid wasp: Suppression of the cellular immune response in the parasitoid's host. Science 211:582–583.

Ellington, C. P. 1984. The aerodynamics of flapping animal flight. Am. Zool. 24:95–105.

Evans, H. E. 1984. *Insect Biology: A Textbook of Entomology*. Addison-Wesley, Reading, Massachusetts.

Evans, P. D. and V. B. Wigglesworth (ed.). 1986. *Advances in Insect Physiology*, Vol. 19. University of Cambridge, Cambridge, U.K. [The latest in a landmark series of volumes.]

Fent, K. and R. Wehner. 1985. Ocelli: A celestial compass in the desert ant *Cataglyphis*. Science 228:192–194.

Fletcher, D. J. C. and M. Blum. 1983. Regulation of queen number by workers in colonies of social insects. Science 219:312–314.

Frisch, K. von. 1967. *The Dance Language and Orientation of Bees*. Cambridge University Press, Cambridge.

Garth, J. S. and J. W. Tilden. 1986. *California Butterflies*. University of California Press, Berkeley.

Gillott, C. 1980. *Entomology*. Plenum, New York.

Gilmour, D. 1965. *The Metabolism of Insects*. Oliver and Boyd, Edinburgh.

Gould, J. L. 1976. The dance-language controversy. Q. Rev. Biol. 51:211–243.

Gould, J. L. 1985. How bees remember flower shapes. Science 227:1492–1494.

Gould, J. L. 1986. The locale map of honey bees: Do insects have cognitive maps? Science 232:861–863.

Hermann, H. R. (ed.). 1984. *Defense Mechanisms in Social Insects*. Praeger, New York.

Herreid, C. F., II and C. R. Fourtner. 1981. *Locomotion and Energetics in Arthropods*. Plenum, New York.

Hinton, H. E. 1981. *Biology of Insects Eggs*, Vols. 1–3. Pergamon Press, Elmsford, New York. [A benchmark study of insect egg morphology and biology; no embryology.]

Holt, V. M. 1973. *Why Not Eat Insects?*. Reprinted from the original (1885) by E. W. Classey Ltd., 353 Hanworth Rd., Hampton, Middlesex, England. [Ninety-nine pages of fun and recipes.]

House, H. L. 1965. Insect nutrition. *In* M. Rockstein (ed.), *The Physiology of Insecta*, Vol. 2. Academic Press, New York.

Hoyle, G. (ed.). 1977. *Identified Neurons and Behavior of Arthropods*. Plenum, New York.

Huffaker, C. B. and R. L. Rabb (eds.). 1984. *Ecological Entomology*. Wiley-Interscience, New York.

Jaques, H. E. 1951. *How to Know the Beetles*. Wm. C. Brown, Dubuque, Iowa.

Jenkin, P. M. 1966. Apolysis and hormones in the moulting cycles of Arthropoda. Ann. Endocrinol. 27:331–341.

Jones, J. C. 1964. The circulatory system of insects. *In* M. Rockstein (ed.), *The Physiology of Insecta*, Vol. 3. Academic Press, New York.

Kerkut, G. A. and L. I. Gilbert. 1984. *Comprehensive Insect Physiology, Biochemistry and Pharmacology*. Pergamon Press, Elmsford, New York. [In 13 volumes.]

Kettlewell, H. B. D. 1961. The phenomenon of industrial melanism in Lepidoptera. Annu. Rev. Entomol. 6:245–262.

Klots, A. B. and E. B. Klots. 1975. *Living Insects of the World*. Doubleday, Garden City, New York. [An excellent picture-oriented guide to the major insect groups; limited but accurate text.]

Lehmkuhl, D. M. 1979. *How to Know the Aquatic Insects*. Wm. C. Brown, Dubuque, Iowa.

Lewis, T. (ed.). 1984. *Insect Communication*. Academic Press, Orlando, Florida. [A definitive review.]

Locke, M. and D. S. Smith (eds.). 1980. *Insect Biology in the Future*. Academic Press, Orlando, Florida.

Matsuda, R. 1965. Morphology and evolution of the insect head. Mem. Am. Entomol. Inst. Ann Arbor 4:1–334.

Merritt, R. W. and K. W. Cummins (eds.). 1978. *An Introduction to the Aquatic Insects of North America*. Kendall/Hunt, Dubuque, Iowa. [Excellent keys to most North American groups.]

Michelsen, A. 1979. Insect ears as mechanical systems. Am. Sci. 67:696–706.

Michener, C. D. 1974. *The Social Behavior of the Bees: A Comparative Study*. Belknap Press/Harvard University Press, Cambridge, Massachusetts.

Novak, V. J. A. 1966. *Insect Hormones*. Methuen, London.

Phelan, P. L. and T. C. Baker. 1987. Evolution of male pheromones in moths: Reproductive isolation through sexual selection? Science 235:205–207.

Prestwick, G. D. 1987. Chemistry of pheromone and hormone metabolism in insects. Science 238:999–1006.

Pringle, J. W. S. 1965. Locomotion: Flight. *In* M. Rockstein (ed.), *The Physiology of Insecta*, Vol. 2. Academic Press, New York.

Resh, V. and D. Rosenberg (eds.). 1984. *The Ecology of Aquatic Insects*. Praeger, New York.

Robinson, G. E. 1985. The dance language of the honey bee: The controversy and its resolution. Am. Bee J. 126:184–189.

Roeder, K. D. 1963. *Nerve Cells and Insect Behaviour*. Harvard University Press, Cambridge, Massachusetts.

Rosin, R. 1984. Further analysis of the honey bee "Dance Language" controversy. I. Presumed proofs for the "Dance Language" hypothesis by Soviet scientists. J. Theor. Biol. 107:417–442.

Rosin, R. 1988. Do honey bees still have a "dance language"? Am. Bee J. 128:267–268.

Saunders, D. S. 1982. *Insect Clocks*, 2nd Ed. Pergamon Press, Elmsford, New York. [Good introduction to photoperiodism.]

Schmitt, J. B. 1962. The comparative anatomy of the insect nervous system. Annu. Rev. Entomol. 7:137–156.

Scott, J. A. 1986. *The Butterflies of North America: A Natural History and Field Guide*. Stanford University Press, Stanford, California.

Shorrocks, B. 1980. *Drosophila*. Pergamon Press, Elmsford, New York. [Everything you ever wanted to know about *Drosophila*, and more.]

Snodgrass, R. E. 1935. *Principles of Insect Morphology*. McGraw-Hill, New York. [An early classic; still useful.]

Snodgrass, R. E. 1944. The feeding apparatus of biting and sucking insects affecting men and animals. Smithson. Misc. Coll. 104(7):1–113.

Snodgrass, R. E. 1952. *A Textbook of Arthropod Anatomy*. Cornell University Press, Ithaca, New York. [Generations of subsequent books and reports have relied heavily on the information and figures contained in this, and Snodgrass's 1935 text.]

Snodgrass, R. E. 1960. Facts and theories concerning the insect head. Smithson. Misc. Coll. 142:1–61.

Somps, C. and M. Luttges. 1985. Dragonfly flight: Novel uses of unsteady separated flows. Science 228:1326–1329.

Stobbart, R. H. and J. Shaw. 1964. Salt and water balance: Excretion. *In* M. Rockstein (ed.), *The Physiology of Insecta*, Vol. 3. Academic Press, New York.

Tauber, M. J., C. A. Tauber and S. Masaki. 1986. *Seasonal Adaptations of Insects*. Oxford University Press, New York. [A comprehensive treatment of insect life cycles.]

Treherne, J. E. and J. W. L. Beament (eds.). 1965. *The Physiology of the Insect Central Nervous System*. Academic Press, New York.

Treherne, J. E., M. J. Berridge and V. B. Wigglesworth. 1963–1985. *Advances in Insect Physiology*, Vols. 1–18. Academic Press, New York.

Tu, A. T. (ed.). 1984. *Handbook of Natural Toxins*, Vol. 2, Insect Poisons, Allergens, and Other Invertebrate Venoms. Marcel Dekker, New York. [Summarizes information on insects, chilopods, and arachnids.]

Unarov, B. P. 1966. *Grasshoppers and Locusts*. Cambridge University Press, Cambridge.

Urquhart, F. 1960. *The Monarch Butterfly*. University of Toronto Press, Toronto.

Usinger, R. L. (ed.). 1968. *Aquatic Insects of California, with Keys to North American Genera and California Species*. University of California Press, Berkeley.

Vane-Wright, R. I. and P. R. Ackery. 1989. *The Biology of Butterflies*. University of Chicago Press, Chicago.

Weis-Fogh, T. and M. Jensen. 1956. Biology and physics of locust flight. I. Basic principles in insect flight. A critical review. Philos. Trans. R. Soc. London Ser. B 239:415–458.

Wells, P. H. and A. M. Wenner. 1973. Do bees have a language? Nature 241:171–174.

Wenner, A. M. and P. H. Wells. 1987. The honey bee dance language controversy: The search for "truth" vs. the search for useful information. Am. Bee J. 127:130–131.

Wenner, A. M., P. H. Wells and D. L. Johnson. 1969. Honeybee recruitment to food sources: Olfaction or language? Science 164:84–86.

Wigglesworth, V. B. 1954. *The Physiology of Insect Metamorphosis*. Cambridge University Press, Cambridge. [A long-standing classic; dated but still useful.]

Wigglesworth, V. B. 1984. *Insect Physiology*, 8th Ed. Chapman and Hall, London. [The seventh edition is longer and also useful, but not as current.]

Williams, C. B. 1958. *Insect Migration*. Collins, London.

Wilson, E. O. 1963. The social biology of ants. Annu. Rev. Entomol. 8:345–368.

Wilson, E. O. 1963. Pheromones. Sci. Am. 208:2–11.

Wilson, E. O. 1971. *The Insect Societies*. Harvard University Press, Cambridge, Massachusetts.

Myriapods

Albert, A. M. 1983. Life cycle of Lithobiidae, with a discussion of the *r*- and *K*-selection theory. Oecologia 56:272–279.

Blower, G. 1951. A comparative study of the chilopod and diplopod cuticle. Q. J. Microsc. Sci. 92:141–161.

Blower, J. G. (ed.). 1974. *Myriapoda*. Academic Press, London.

Camatini, M. (ed.). 1980. *Myriapod Biology*. Academic Press, New York.

Cloudsley-Thompson, J. L. 1958. *Spiders, Scorpions, Centipedes and Mites*. Pergamon Press, London.

Hoffman, R. L. et al. 1982. Chilopoda–Symphyla–Diplopoda–Pauropoda. *In* S. P. Parker (ed.), *Synopsis and Classification of Living Organisms*. McGraw-Hill, New York, pp. 681–726.

Lewis, J. G. E. 1961. The life history and ecology of the littoral centipede *Strigamia* (= *Scolioplanes*) *maritima* (Leach). Proc. Zool. Soc. London 137:221–247.

Lewis, J. G. E. 1965. The food and reproductive cycles of the centipedes *Lithobius variegatus* and *Lithobius forficatus* in a Yorkshire woodland. Proc. Zool. Soc. London 144:269–283.

Lewis, J. G. E. 1981. *The Biology of Centipedes*. Cambridge University Press, New York.

Rajulu, G. S. 1970. A study on the nature and formation of the spermatophore in a centipede *Ethmostigmus spinosus*. Bull. Mus. Hist. Nat. Paris 41 (Suppl. 2):116–121.

Rosenberg, J. and G. Seifert. 1977. The coxal glands of Geophilomorpha (Chilopoda): Organs of osmoregulation. Cell Tiss. Res. 182:247–251.

Uniramian Evolution and Phylogeny

Alexander, R. D. 1964. The evolution of mating behaviour in arthropods. Symp. R. Entomol. Soc. London 2:78–94.

Anderson, D. T. 1973. *Embryology and Phylogeny in Annelids and Arthropods*. Pergamon Press, New York.

Andersson, M. 1984. The evolution of eusociality. Annu. Rev. Ecol. Syst. 15:165–189.

Bennett, D. and S. M. Manton. 1963. Arthropod segmental organs and Malpighian tubules with particular reference to their function in Chilopoda. Ann. Mag. Nat. Hist. (13)5:545–556.

Douglas, M. M. 1980. Thermoregulatory significance of thoracic lobes in the evolution of insect wings. Science 211:84–86.

Futuyma, D. J. and M. Slatkin (eds.). 1983. *Coevolution*. Sinauer, Sunderland, Massachusetts.

Goodchild, A. J. P. 1966. Evolution of the alimentary canal in the Hemiptera. Biol. Rev. 41:97–140.

Gupta, A. P. (ed.). 1979. *Arthropod Phylogeny*. Van Nostrand Reinhold, New York. [Contributed chapters, mostly on insects.]

Hennig, W. 1981. *Insect Phylogeny*. Wiley, New York. [Perhaps the best modern reference on the subject.]

Hoy, R. R., A. Hoikkala and K. Kaneshiro. 1988. Hawaiian courtship songs: Evolutionary innovation in communication signals of *Drosophila*. Science 240:217–220.

Kristensen, N. P. 1975. The phylogeny of hexapod "orders." A critical review of recent accounts. Z. Zool. Syst. Evolutionsforsch. 13:1–44.

Kukalova-Peck, J. 1983. Origin of the insect wing and wing articulation from the insect leg. Can. J. Zool. 61:1618–1669.

Magnum, C. P. et al. 1985. Centipedal hemocyanin: Its structure and its implications for arthropod phylogeny. Proc. Natl. Acad. Sci. U.S.A. 82:3721–3725.

Manton, S. M. 1964. Mandibular mechanisms and the evolution of arthropods. Philos. Trans. R. Soc. London Ser. B 247:1–183.

Manton, S. M. 1965. The evolution of arthropod locomotory mechanisms. Part 8. Functional requirements and body design in Chilopoda, . . . and its bearing upon the evolution of the arthropodan haemocoel. J. Linn. Soc. London Zool. 46:251–483.

Manton, S. M. 1977. *The Arthropoda: Habits, Functional Morphology, and Evolution*. Clarendon Press, Oxford.

Manton, S. M. and N. Heatley. 1937. The feeding, digestion, excretion and food storage of *Peripatopsis*. Philos. Trans. R. Soc. London Ser. B 227:411–464.

Robertson, R. M., K. G. Pearson and H. Reichert. Flight interneurons in the locust and the origin of insect wings. Science 217:177–179.

Shinohara, K. 1970. On the phylogeny of Chilopoda. Proc. Japan. Soc. Syst. Zool. 6:35–42.

Thornhill, R. and J. Alcock. 1983. *The Evolution of Insect Mating Systems*. Harvard University Press, Cambridge, Massachusetts.

Wilson, E. O. 1985. Invasion and extinction in the West Indian ant fauna: Evidence from the Dominican amber. Science 229:265–267.

Wilson, E. O. 1985. The sociogenesis of insect colonies. Science 228:1489–1425.

Chapter Eighteen

Phylum Arthropoda: The Crustaceans

Creatures of such unsuspected importance and numbers stir our imagination and invite us to find out more about them. But let us beware of lightly following our curiosity in this matter. The attempt to obtain a clearcut definition of the class Crustacea has left many a student bewildered and dissatisfied.

Waldo L. Schmitt

Crustaceans, 1965

Crustaceans are, in our opinion, the most exciting of all invertebrates. They are certainly one of the most popular invertebrate groups, even among nonbiologists, for they include some of the world's most delectable gourmet fare, such as lobster, crab, and shrimp. There are more than 30,000 described living species, and probably four times that number waiting to be discovered and named. They exhibit an incredible diversity of form and habit, ranging in size from tiny interstitial and planktonic forms less than a millimeter in length to giant crabs with leg spans of 4 m and lobsters attaining weights up to 17 pounds (*Homarus americanus*). Crustaceans are found at all depths in every marine, brackish, and freshwater environment known; a few have become successful on land, the most notable being sow bugs, or pill bugs. Crustaceans are the most widespread and diverse group of invertebrates inhabiting the world's oceans. In addition to the well known groups noted above, the subphylum Crustacea includes beach hoppers, barnacles, fish lice, fairy shrimps, krill, and many other lesser known taxa. The range of morphological diversity among crustaceans far exceeds that of even the insects. In fact, because of their diversity in form and number, it is often said that crustaceans are the "insects of the sea." We prefer to think of insects as the "crustaceans of the land."

Despite the enormous morphological diversity seen among crustaceans (Figure 1), they do display a suite of fundamental unifying features (see Table 1 and Box 1). In an effort to introduce both the diversity and unity of this large group, we first present a classificatory scheme and a short synopsis of the major taxa and then discuss the biology of the group as a whole, with examples drawn from the various members. As in the two preceding chapters, we ask that you keep in mind the general account of the arthropods presented in Chapter 15.

Crustacean classification

Crustaceans have been known since ancient times and have served humans as sources of both food and legend. It is somewhat comforting to carcinologists (those who study crustaceans) to note that *Cancer*, one of the two invertebrates represented in the zodiac, is a crab (the other of course is *Scorpio*— another arthropod). A modern understanding of the subphylum Crustacea as a taxon can be traced to Lamarck's scheme in the early nineteenth century. He recognized most crustaceans as such, but placed the barnacles (subclass Cirripedia) and a few others in separate groups. For many years barnacles were classified with molluscs, because of their calcareous outer shell. Crustacean classification as we know it today was more-or-less established during the second half of the last century, although internal revisions continue. The basic classification scheme presented here is adapted from several recent sources, in particular Bowman and Abele (1982), although parts are modified in accordance with the work of other specialists. The higher level relationships and taxonomy of crustaceans are currently in a state of reassessment (see phylogeny section at the end of this chapter). Most workers recognize five classes.

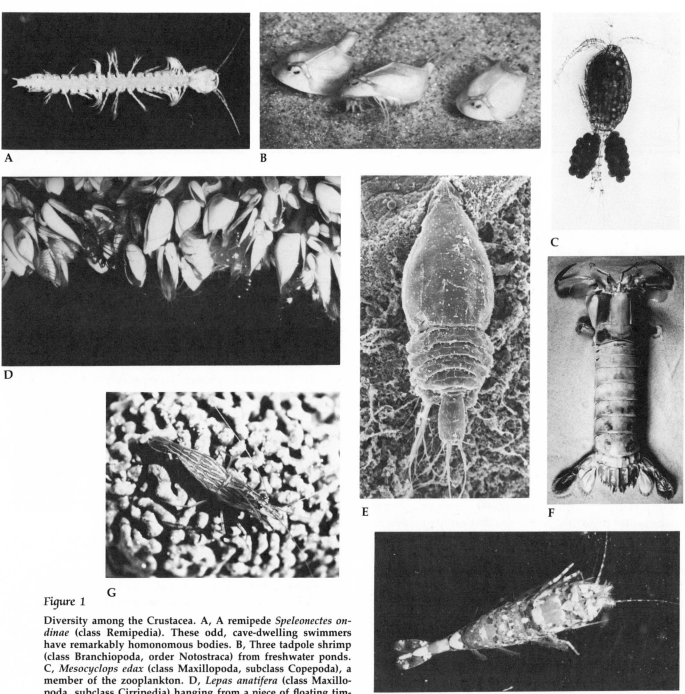

Figure 1

Diversity among the Crustacea. A, A remipede *Speleonectes ondinae* (class Remipedia). These odd, cave-dwelling swimmers have remarkably homonomous bodies. B, Three tadpole shrimp (class Branchiopoda, order Notostraca) from freshwater ponds. C, *Mesocyclops edax* (class Maxillopoda, subclass Copepoda), a member of the zooplankton. D, *Lepas anatifera* (class Maxillopoda, subclass Cirripedia) hanging from a piece of floating timber. E, *Deoterthron* (class Maxillopoda, subclass Tantulocarida), a parasite on other crustaceans. F, A mantis shrimp (class Malacostraca, order Stomatopoda). G–H, Two caridean decapod shrimps (class Malacostraca, superorder Eucarida): *Lysmata californica*, resting on a tunicate (G), and *Crangon* (H). I–J, Two brachyuran decapod crabs, *Pachygrapsus crassipes* and *Chionoecetes tanneri* (class Malacostraca, superorder Eucarida). K–M, Three different anomuran decapods (class Malacostraca, superorder Eucarida). K, A hermit crab, *Paguristes bakeri*. L, The coconut crab, *Birgus latro*, climbing a tree. M, *Pleuroncodes planipes*, a pelagic galatheid crab. N–O, Two members of the order Isopoda (class Malacostraca, superorder Peracarida). N, *Ligia occidentalis*, the rock "louse," an inhabitant of the high spray zone on rocky

shores. O, *Idotea resecata*, often found on eel grass and surf grass. P–Q, Two rather strange amphipods (class Malacostraca, superorder Peracarida): P, *Cystisoma*, a huge (some exceed 10 cm), transparent, pelagic hyperiid amphipod. Q, *Cyamus scammoni*, a parasitic caprellid amphipod that lives on whales. Most amphipods belong to the suborder Gammaridea (see Figure 19). (A photo by D. Williams, courtesy of J. Yager; C courtesy of G. Wyngaard; D courtesy A. Kirstitch; E from Boxshall and Lincoln 1987, courtesy of G. Boxshall; G,N photos by the authors; H,I,J,K,M,O,P,Q courtesy of G. McDonald; L courtesy of P. Fankboner)

CLASS CEPHALOCARIDA: Cephalocarids (Figures 3A and B). (e.g., *Chiltoniella, Hutchinsoniella, Lightiella, Sandersiella*)

CLASS BRANCHIOPODA: Branchiopods (Figure 4).

    ORDER ANOSTRACA: Fairy shrimps. (e.g., *Artemia, Branchinecta, Branchinella*)

    ORDER NOTOSTRACA: Tadpole shrimps. (*Lepidurus* and *Triops*)

    ORDER CLADOCERA: Water "fleas." (e.g., *Anchistropus, Daphnia, Leptodora, Moina, Polyphemus*)

    ORDER CONCHOSTRACA: "Clam" shrimps. (e.g., *Caenestheria, Cyzicus, Eulimnadia, Lynceus*)

CLASS MAXILLOPODA (Figures 5–8)

  SUBCLASS OSTRACODA: Ostracods.

    SUPERORDER MYODOCOPA: (e.g., *Cypridina, Gigantocypris*)

    SUPERORDER PODOCOPA: (e.g., *Candona, Cypris, Darwinula*)

    SUPERORDER PALAEOCOPA: Only shells known; may be extinct.

SUBPHYLUM CRUSTACEA

CLASS REMIPEDIA: Remipedes (Figures 1A and 3C–E).

    ORDER NECTIOPODA: Living remipedes. (e.g., *Cryptocorynectes, Godzillius, Lasionectes, Speleonectes*)

    ORDER ENANTIOPODA: Extinct remipedes. Monotypic: *Tesnusocaris.*

Table One

Comparison of distinguishing features of the five crustacean classes

TAXON	CARAPACE OR CEPHALIC SHIELD	BODY TAGMATA AND NUMBER OF SEGMENTS IN EACH (EXCLUDING TELSON)	THORACOPODS	MAXILLIPEDS
Remipedia	Cephalic shield	Cephalon (5) Trunk (up to 32 segments)	Not phyllopodous; protopod of one article	1 pair
Cephalocarida	Cephalic shield	Cephalon (5) Thorax (8) Abdomen (11)	Phyllopodous; protopod of one article	None
Branchiopoda				
Anostraca	None	Cephalon (5) Thorax (usually 11) Abdomen (usually 8)	Phyllopodous; protopod of one article	None
Notostraca	Carapace	Cephalon (5) Thorax (11) Abdomen (many segments)	Phyllopodous; protopod of one article	None
Cladocera	Folded carapace	Cephalon (5) Trunk (segmentation obscured)	Phyllopodous; protopod of one article	None
Conchostraca	Bivalved carapace	Cephalon (5) Trunk (10–32)	Phyllopodous; protopod of one article	None
Maxillopoda	Carapace, or cephalic shield	Cephalon (5) Thorax (6) Abdomen (4)	Not phyllopodous; often reduced; protopod multiarticulate or of one article	0–1 pair
Malacostraca				
Phyllocarida	Large "folded" carapace covers thorax	Cephalon (5) Thorax (8) Abdomen (7)	Phyllopodous; protopod of 2 articles	None
Eumalacostraca	Carapace large or secondarily reduced or lost	Cephalon (5) Thorax (8) Abdomen (6)	Not phyllopodous; uniramous in many; protopod of 1, 2, or 3 articles	0–3 pairs

SUBCLASS MYSTACOCARIDA: Mystacocarids. (*Ctenocheilocaris* and *Derocheilocaris*)

SUBCLASS COPEPODA: Copepods

  ORDER CALANOIDA: (e.g., *Calanus, Diaptomus, Eucalanus, Euchaeta, Platycopia*)

  ORDER HARPACTICOIDA: (e.g., *Cancricola, Harpacticus, Psammus, Sunaristes, Tisbe*)

  ORDER PLATYCOPIOIDA: (e.g., *Antrisocopia, Platycopia*)

ORDER CYCLOPOIDA: (e.g., *Cyclopina, Cyclops, Eucyclops, Lernaea, Myticola, Notodelphys*)

ORDER POECILOSTOMATOIDA: (e.g., *Chondracanthus, Ergasilus, Pseudanthessius*)

ORDER SIPHONOSTOMATOIDA: (e.g., *Clavella, Nemesis, Penella, Pontoeciella, Trebius*)

ORDER MONSTRILLOIDA: (e.g., *Monstrilla, Stilloma, Thespesiopsyllus*)

  ORDER GELYELLOIDA: (e.g., *Gelyella*)

ANTENNULES	ANTENNAE	COMPOUND EYES	ABDOMINAL APPENDAGES	GONOPORE LOCATION
Biramous	Biramous	Absent	All trunk appendages similar	♂: protopods of thoracopod 14; ♀: on thoracopod 7
Uniramous	Biramous	Absent	None	Common pores on protopods of thoracopod 6
Uniramous	Uniramous	Present	None	"Genital region" of anterior abdomen (usually thoracomere 12/13)
Uniramous	Vestigial	Present	Present (posteriorly reduced)	Thoracomere 11
Uniramous	Biramous	Present	Posterior trunk limbless	Variable; usually on apodous posterior region
Uniramous	Biramous	Present	All trunk appendages similar	Trunk segment 11
Uniramous	Uniramous or biramous	Present or absent	None	Variable; on thorax or abdomen; usually sternal
Biramous	Uniramous	Present	Pleopods (posteriorly reduced)	♂: coxae of thoracopod 8; ♀: coxae of thoracopod 6
Uniramous, biramous, or triramous	Uniramous or biramous	Present	Usually 5 pairs pleopods, 1 pair uropods	♂: coxae of thoracopod 8, or sternum of thoracomere 8; ♀: coxae of thoracopod 6, or sternum of thoracomere 6

ORDER MISOPHRIOIDA: (e.g., *Boxshallia, Misophria*)

ORDER MORMONILLOIDA: Monogeneric: *Mormonilla.*

SUBCLASS BRANCHIURA: Branchiurans or argulids. (*Argulus, Chonopeltis, Dipteropeltis,* and *Dolops*)

SUBCLASS CIRRIPEDIA: Barnacles and their kin.

ORDER THORACICA: Goose and acorn barnacles. (e.g., *Balanus, Chthamalus, Conchoderma, Coronula, Koleolepas, Lepas, Pollicipes, Tetraclita, Verruca*)

ORDER ASCOTHORACICA: Parasitic "barnacles." (e.g., *Ascothorax, Dendrogaster, Laura, Synagoga*)

ORDER ACROTHORACICA: Boring barnacles. (e.g., *Cyptophialus, Kochlorine, Trypetesa*)

ORDER RHIZOCEPHALA: Parasitic "barnacles." (e.g., *Heterosaccus, Mycetomorpha, Peltogaster, Sacculina, Sylon, Thompsonia*)

SUBCLASS TANTULOCARIDA: Tantulocarids. (e.g., *Basipodella, Deoterthron, Microdajus, Stygotantulus*)

Box One

Characteristics of
the Subphylum Crustacea

1. Body composed of a 5-segmented head (plus acron), with a long postcephalic trunk; trunk usually divided into more-or-less distinct thorax and abdomen
2. Cephalon (head) composed of (anterior to posterior): presegmental acron, antennular somite, antennal somite, mandibular somite, maxillulary somite, and maxillary somite; one or more anterior thoracomeres may fuse with the head in members of the classes Remipedia, Maxillopoda, and Malacostraca, their appendages forming maxillipeds
3. Cephalic shield or carapace generally present
4. Appendages are multiarticulate, biramous (or secondarily uniramous) limbs (the nature of the primitive crustacean antennule is uncertain)
5. Mandibles are multiarticulate limbs that usually function as gnathobasic jaws
6. Gas exchange typically by aqueous diffusion across branchial ("gill") surfaces
7. Excretion by true nephridial structures (e.g., antennal glands, maxillary glands)
8. Both simple ocelli and compound eyes occur in most taxa, at least at some stage of the life cycle; compound eyes often elevated on stalks
9. Gut with digestive ceca
10. With nauplius larva; development usually mixed or direct

CLASS MALACOSTRACA (Figures 9–19)

SUBCLASS PHYLLOCARIDA

ORDER LEPTOSTRACA: Nebaliids. (e.g., *Dahlella, Epinebalia, Nebalia, Nebaliella, Nebaliopsis, Paranebalia*)

SUBCLASS EUMALACOSTRACA

SUPERORDER HOPLOCARIDA

ORDER STOMATOPODA: Mantis shrimps. (e.g., *Gonodactylus, Hemisquilla, Lysiosquilla, Squilla*)

SUPERORDER SYNCARIDA: Syncarids.

ORDER BATHYNELLACEA: (e.g., *Bathynella*)

ORDER ANASPIDACEA: (e.g., *Anaspides, Psammaspides*)

SUPERORDER EUCARIDA

ORDER EUPHAUSIACEA: Euphausiaceans, "krill." (e.g., *Bentheuphausia, Euphausia, Meganyctiphanes, Nyctiphanes*)

ORDER AMPHIONIDACEA: Amphionidaceans. Monotypic: *Amphionides reynaudii.*

ORDER DECAPODA: Crabs, shrimps, lobsters, etc.

SUBORDER DENDROBRANCHIATA: Penaeid and sergestid shrimps (e.g., *Lucifer, Penaeus, Sergestes, Sicyonia*)

SUBORDER PLEOCYEMATA

INFRAORDER CARIDEA: Caridean and procaridean shrimps. (e.g., *Alpheus, Crangon, Hippolyte, Lysmata, Macrobrachium, Palaemon, Pandalus, Pasiphaea, Procaris*).

INFRAORDER STENOPODIDEA: Stenopodid shrimps. (e.g., *Spongicola, Stenopus*)

INFRAORDER THALASSINIDEA: Mud and ghost shrimps . (e.g., *Axius, Callianassa, Thalassina, Upogebia*)

INFRAORDER ASTACIDEA: Crayfish and chelate lobsters. (e.g., *Astacus, Cambarus, Homarus*)

INFRAORDER PALINURA: Palinurid, spiny, and Spanish (slipper) lobsters. (e.g., *Evibacus, Panulirus, Scyllarus, Stereomastis, Willemoesia*)

INFRAORDER ANOMURA: Hermit crabs, galatheid crabs, sand crabs, porcelain crabs, etc. (e.g., *Birgus, Coenobita, Emerita, Galathea, Hippa, Lithodes, Lomis, Paguristes, Pagurus, Petrolisthes, Pleuroncodes*).

INFRAORDER BRACHYURA: "True" crabs. (e.g., *Callinectes, Cancer, Cardisoma, Corystes, Grapsus, Hemigrapsus, Maja, Macrocheira, Ocypode, Pachygrapsus, Pinnotheres, Portunus, Xantho*).

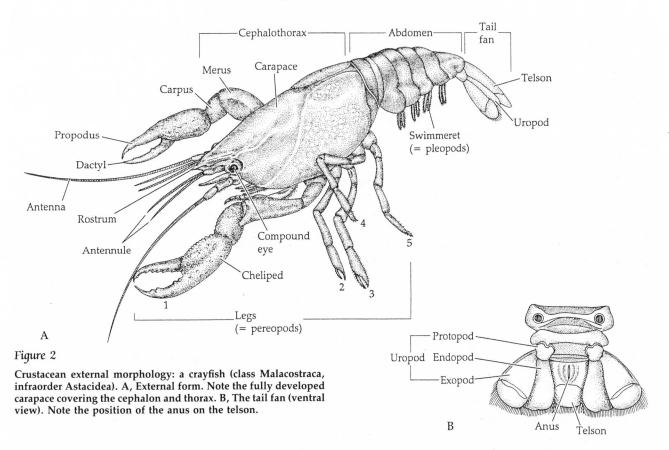

Figure 2

Crustacean external morphology: a crayfish (class Malacostraca, infraorder Astacidea). A, External form. Note the fully developed carapace covering the cephalon and thorax. B, The tail fan (ventral view). Note the position of the anus on the telson.

SUPERORDER PERACARIDA

ORDER MYSIDA: Mysids or opossum shrimps (e.g., *Acanthomysis, Hemimysis, Mysis, Neomysis*)

ORDER LOPHOGASTRIDA: Lophogastrids. (e.g., *Gnathophausia, Lophogaster*)

ORDER CUMACEA: Cumaceans. (e.g., *Campylaspis, Cumopsis, Diastylis, Diastylopsis*)

ORDER TANAIDACEA: Tanaids. (e.g., *Apseudes, Heterotanais, Paratanais, Tanais*)

ORDER MICTACEA: Mictaceans (3 known species: *Hirsutia bathyalis, H. sandersetalia*, and *Mictocaris halope*)

ORDER SPELAEOGRIPHACEA: Spelaeogriphaceans. (2 known species: *Potiicoara brazilienses* and *Spelaeogriphus lepidops*)

ORDER THERMOSBAENACEA: Thermosbaenaceans. (e.g., *Halosbaena, Limnosbaena, Monodella, Theosbaena, Thermosbaena, Tulumella*).

ORDER ISOPODA: Isopods (sea slaters, rock lice, pill bugs, etc.)

SUBORDER GNATHIIDEA: (e.g., *Gnathia, Paragnathia*)

SUBORDER ANTHURIDEA: (e.g., *Anthura, Bathura, Colanthura, Cyathura, Mesanthura, Neoanthura*)

SUBORDER FLABELLIFERA: (e.g., *Aega, Bathynomus, Cirolana, Limnoria, Lironeca, Sphaeroma, Exosphaeroma*)

SUBORDER ONISCIDEA: (e.g., *Armadillidium, Ligia, Oniscus, Porcellio, Trichoniscus, Tylos*)

SUBORDER VALVIFERA: (e.g., *Arcturus, Edotea, Erichsonella, Idotea, Saduria, Synidotea*)

SUBORDER PHREATOICIDEA: (e.g., *Mesamphisopus, Phreatoicopis, Phreatoicus*)

SUBORDER ASELLOTA: (e.g., *Asellus, Eurycope, Ianiropsis, Jaera, Janira, Microcerberus, Munna*)

SUBORDER EPICARIDEA: (e.g., *Aporobopyrus, Bopyrus, Dajus, Ione, Phryxus*)

SUBORDER CALABOZOIDEA: (*Calabozoa*)

ORDER AMPHIPODA: Amphipods (beach hoppers, sand fleas, scuds, skeleton shrimps, etc.).

SUBORDER GAMMARIDEA: (e.g., *Ampithoe, Anisogammarus, Corophium, Eurythenes, Gammarus, Niphargus, Orchestia, Phoxocephalus, Rhachotropis, Talitrus*)

SUBORDER HYPERIIDEA: (e.g., *Cystisoma, Hyperia, Leptocotis, Phronima, Primno, Rhabdosoma, Scina, Streetsia, Vibilia*).

SUBORDER CAPRELLIDEA: (e.g., *Caprella, Cyamus, Metacaprella*)

SUBORDER INGOLFIELLIDEA: (*Ingolfiella*)

Synopses of crustacean taxa

The following descriptions of higher crustacean taxa will give you some idea of the range of diversity within the group, and the variety of ways in which these successful animals have exploited the basic crustacean *Bauplan*. A specific diagnosis of each taxon is followed by some general comments.*

Class Remipedia, Order Nectiopoda

Body divided into cephalon and elongate, homonomous trunk of up to 32 segments; cephalon with paired, preantennular frontal processes; antennules biramous; antennae paddle-like; trunk limbs laterally directed, biramous, paddle-like, but without large epipods; rami of trunk limbs (exopod and endopod) of three or more articles; anal segment (telson) with caudal rami; without a carapace, but cephalic shield covers head; living species without eyes; midgut with serially arranged digestive ceca; first trunk segment fused with head and bearing one pair prehensile maxillipeds; labrum very large, forming a chamber (atrium oris) in which reside "internalized" mandibles; maxillules may function as hypodermic fangs; last trunk segment partly fused dorsally with telson; segmental double ventral nerve cord; male gonopore on trunk limb 14, female on 7; to about 30 mm in length (Figures 3C, D, and E).

The above diagnosis is for the living remipedes; the fossil order (Enantiopoda) is currently based on a single poorly preserved specimen. The discovery of living remipedes, strange vermiform crustaceans first collected from a cavern on Grand Bahama Island, has given the carcinological world a turn (see Yager 1981 and Schram et al. 1986). The combination of features distinguishing these creatures is puzzling, for they possess characteristics that are certainly very primitive (e.g., long, homonomous trunk; double ventral nerve cord; segmental digestive ceca; paddle-like antennae) as well as attributes traditionally recognized as advanced (e.g., maxillipeds; nonphyllopodous, biramous limbs). They swim about on their backs as a result of metachronal beating of the trunk appendages, a style of locomotion similar to that of anostracans. The remipedes are reminiscent of both branchiopods and cephalocarids. The laterally directed limbs are unlike those of any other crustacean, and the "internalized" mandibles and hypodermic maxillules also appear unique. The nature of the preantennular "processes" is also puzzling.

The few species of living remipedes discovered thus far are found in caves with connections to the sea. The water in these caves is often distinctly stratified, with a layer of fresh water overlying the salt water in which the remipedes live. There is still a great deal to learn about the biology of these animals.

Class Cephalocarida

Head followed by an 8-segmented thorax, an 11-segmented abdomen, and a telson with caudal rami; common gonopore on protopods of sixth thoracopods; carapace absent but head covered by cephalic shield; without eyes; thoracopods 1–7 biramous and phyllopodous, with large, flattened exopods and epipods (exites) and stenopodous endopods; thoracopod 8 reduced or absent; maxillae resemble thoracopods; no maxillipeds (Figures 3A and B).

The cephalocarids are tiny, elongate crustaceans ranging in length from 2.0–3.7 mm. There are only nine described species, distributed among four genera. All are benthic marine detritus feeders. Most are associated with sediments covered by a layer of flocculent organic detritus; others have been found in clean sands, and still others in mud rich in organic material. They have been collected from the intertidal zone to depths of over 1500 m. Most specialists view the cephalocarids as very primitive crustaceans, largely because of their relatively homonomous body form, undifferentiated maxillae, and "generalized" appendage structure.

Class Branchiopoda

Number of segments and appendages on thorax and abdomen vary, the latter usually lacking appendages; carapace present or absent; telson usually with caudal rami; body appendages generally phyllopodous; maxillules and maxillae reduced or absent; no maxillipeds; typically with paired compound and single simple eyes.

The branchiopods are difficult to describe in a general way. Most are small, freshwater forms with minimal body tagmosis and leaflike legs. The four orders of Branchiopoda are considered separately below (Figure 4).*

---

*In conformity with current usage we use the terms **thoracomeres** and **thoracopods** when referring to the segments and appendages, respectively, of the thorax, regardless of whether or not any of these segments are fused to the head. The term **pereon** refers to that portion of the thorax *not* fused with the head (when any fusion occurs), and the terms **pereonites** (= **pereomeres**) and **pereopods** for the segments and appendages, respectively, of the pereon. Hence, on a eumalacostracan with the first thoracic segment (thoracomere 1) fused to the head, thoracomere 2 may be referred to as pereonite 1, and the first pair of pereopods represents the second pair of thoracopods, and so on. Be assured that we are trying to simplify, not confuse, this issue.

*A recent study by Fryer (1987) divides the Branchiopoda into 10 orders.

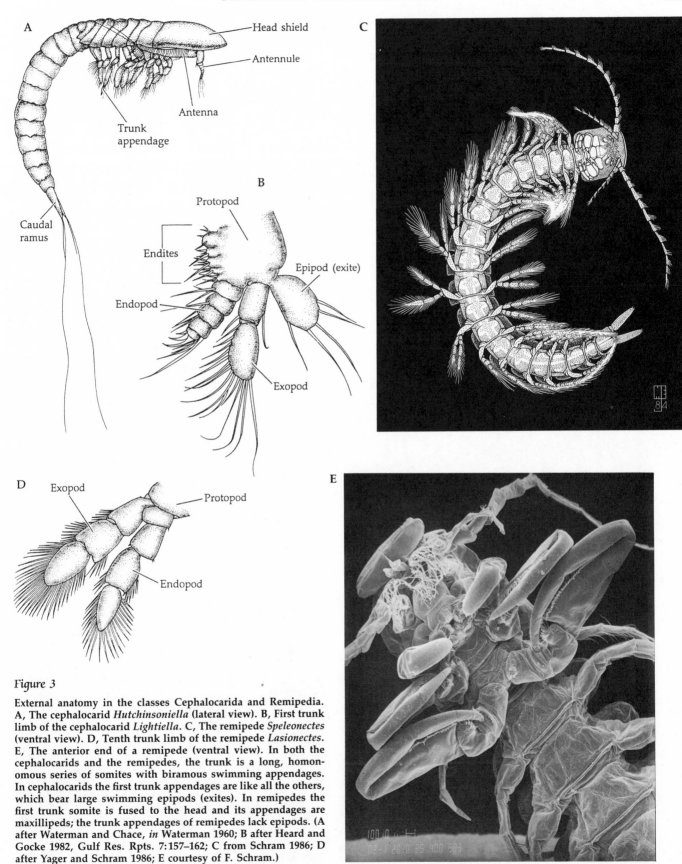

Figure 3

External anatomy in the classes Cephalocarida and Remipedia. A, The cephalocarid *Hutchinsoniella* (lateral view). B, First trunk limb of the cephalocarid *Lightiella*. C, The remipede *Speleonectes* (ventral view). D, Tenth trunk limb of the remipede *Lasionectes*. E, The anterior end of a remipede (ventral view). In both the cephalocarids and the remipedes, the trunk is a long, homonomous series of somites with biramous swimming appendages. In cephalocarids the first trunk appendages are like all the others, which bear large swimming epipods (exites). In remipedes the first trunk somite is fused to the head and its appendages are maxillipeds; the trunk appendages of remipedes lack epipods. (A after Waterman and Chace, *in* Waterman 1960; B after Heard and Gocke 1982, Gulf Res. Rpts. 7:157–162; C from Schram 1986; D after Yager and Schram 1986; E courtesy of F. Schram.)

Figure 4

External anatomy and diversity in the class Branchiopoda. A, An anostracan (*Branchinecta*) in swimming posture. B, Trunk limb of an anostracan (*Linderiella*). C–D, The notostracan *Triops* (dorsal and ventral views). E–F, Two cladocerans: *Daphnia* (E) and *Leptodora* (F). G, The shed carapace, or ephippium, of *Daphnia*. Embryos are enclosed. H, Two extreme stages in the seasonal change in head form of *Daphnia* (cyclomorphosis). I–K, The concostracan *Lynceus*. I, The valves are partially open (ventral view). J, The head (ventral view). K, Whole animal. One valve has been removed. (A after various authors; B after Thiery and Champeau 1988, J. Crust. Biol. 8:70–78; C,D,H after Pennak 1978; E,F after Belk, *in* Parker 1982; I–K from Martin et al. 1986, Zool. Scripta 15:221–232.)

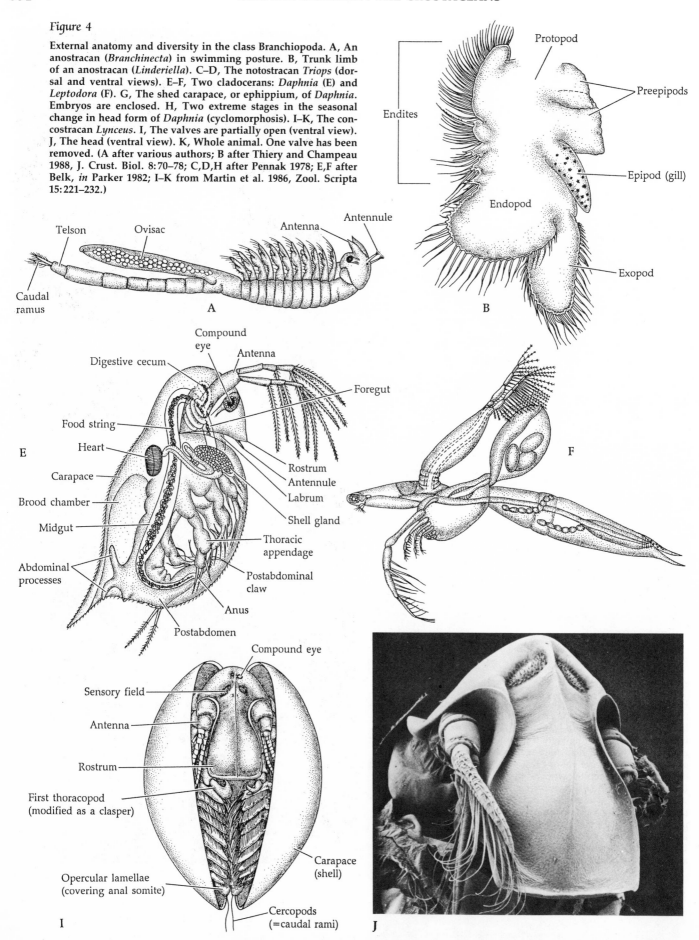

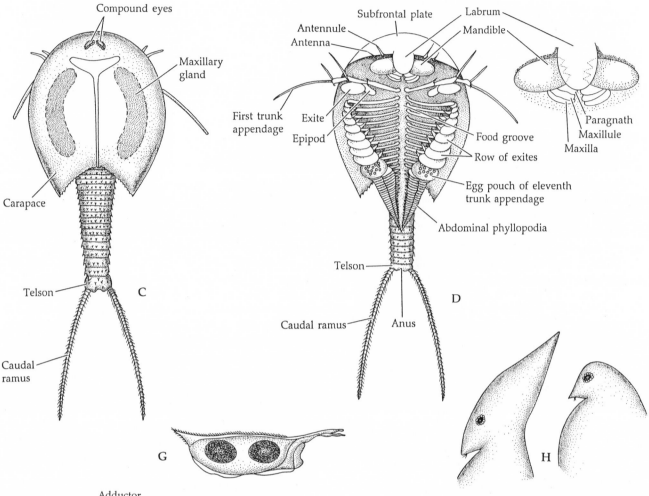

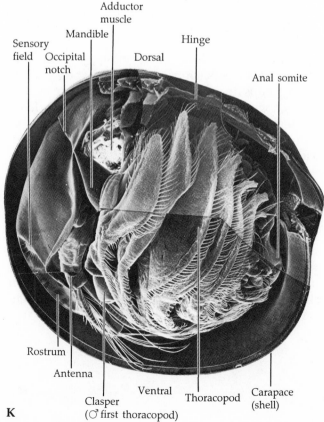

Order Anostraca. Postcephalic trunk divisible into appendage-bearing thorax of 11 segments (17 or 19 in members of the family Polyartemiidae), and abdomen of 8 segments plus telson with caudal rami; gonopores on genital region of abdomen; trunk limbs biramous and phyllopodous; carapace absent; paired, large, stalked compound eyes and single median simple eye (Figure 4A).

The anostracans are commonly called fairy shrimps (including *Artemia*, the brine shrimp). They differ from other branchiopods in lacking a carapace or well developed head shield. There are about 185 species in the order, most of which are small, to about 1 cm in length, although some attain lengths of 10 cm. These cosmopolitan animals inhabit ephemeral ponds, hypersaline lakes, and marine lagoons. In many areas they are an important food resource for water birds. Anostracans are sometimes united with the extinct order Lipostraca in the subclass Sarsostraca.

Anostracans swim ventral side up by metachronal beating of the trunk appendages. Many use the locomotor limb movements to suspension feed. Some other species scrape organic material from submerged surfaces, and one species (*Branchinecta gigas*) is specialized as a predator on other fairy shrimps.

Order Notostraca. Thorax of 11 segments, each with a pair of phyllopodous appendages; abdomen of "rings," each formed by fusion of several segments; each anterior ring with several pairs of appendages, posterior rings lack appendages; telson with long caudal rami; gonopores on last thoracomere; broad, shieldlike carapace fused only with head, but extending to cover thorax and part of abdomen; paired, sessile compound eyes and a single simple eye lie close together near anterior midline on carapace (Figures 4C and D).

Notostracans are commonly called "tadpole shrimps." There are nine living species placed in a single family, Triopsidae, most of which are 2–10 cm long. The common name derives from the general body shape wherein the broad carapace and narrow "trunk" give the animals a superficial tadpole-like appearance. Notostracans are sometimes placed with two other, extinct, orders in the subclass Calmanostraca.

Notostracans inhabit inland waters of all salinities, but none occur in the ocean. Of the two known genera, *Triops* lives only in temporary waters, and their eggs are capable of surviving extended dry periods. Most species of *Lepidurus* live in temporary ponds, but at least one species (*L. arcticus*) inhabits permanent ponds and lakes. *Triops* is of some economic importance in that large populations often occur in rice paddies and destroy the crop by burrowing into the mud and dislodging young plants.

Tadpole shrimps mostly crawl, but they are also capable of swimming for short periods of time by beating the thoracic limbs. They feed on organic material stirred up from the sediments, although some scavenge or prey on other animals, including molluscs, other crustaceans, frog eggs, and even frog tadpoles and small fishes.

Some species of notostracans are exclusively gonochoristic, but others may include hermaphroditic populations (often those populations living at high latitudes). Some European populations appear to reproduce solely by parthenogenesis.

Order Cladocera. Body segmentation generally reduced; thorax and abdomen fused as a "trunk" bearing 4–6 pairs of appendages anteriorly and terminating in a flexed "postabdomen" with clawlike caudal rami; trunk appendages usually phyllopodous; gonopores on sides of trunk; carapace single, but folded along dorsal midline giving bivalved appearance; carapace usually encloses entire trunk but not cephalon (reduced to brood chamber in some); single median compound eye (Figures 4E and F).

The cladocerans or "water fleas" include about 400 species of predominantly freshwater crustaceans. Most cladocerans are 0.5–3 mm long, but *Leptodora kindtii* reaches 18 mm in length. Except for the cephalon and large natatory antennae, the body is enclosed by a folded carapace, which is fused with at least some of the trunk region. The carapace is greatly reduced in members of the families Polyphemidae and Leptodoridae and forms a brood chamber.

Members of this order are distributed worldwide in nearly all inland waters, and several genera are marine. Most are benthic crawlers or burrowers, others are planktonic and swim by means of their large antennae. One genus, *Scapholeberis*, is typically found in the surface film of ponds, and another, *Anchistropus*, is ectoparasitic on *Hydra*. Most of the benthic forms feed by scraping organic material from sediment particles or other objects, and the planktonic species are suspension feeders.

In sexual reproduction, fertilization generally occurs in a brood chamber between the dorsal surface of the trunk and the inside of the carapace. Members of most families release the fertilized eggs into the water when the parent molts, but in the family Daphnidae they are retained in the shed carapace, which is modified as an egg case called an **ephippium** (Figure 4G).

Cladoceran life histories are often compared with those of animals such as rotifers and aphids. In all three groups dwarf males occur in many species, and parthenogenesis is common. Members of two families that undergo parthenogenesis (Moinidae and Polyphemidae) produce eggs with very little yolk. In these groups the floor of the brood chamber is lined with glandular tissue that secretes a fluid rich in nutrients that are absorbed by the developing embryos. Periods of overcrowding, adverse temperature, or food scarcity can induce parthenogenetic females to produce male offspring. Occasional periods of sexual reproduction are known to occur in most parthenogenetic species. Many cladocerans are also known to undergo seasonal changes of body form through succeeding generations of parthenogenetically produced individuals, a phenomenon known as cyclomorphosis (Figure 4H).

Order Conchostraca. Body divided into cephalon and trunk, the latter with 10–32 segments with no apparent separation into thorax and abdomen; trunk limbs phyllopodous, decreasing in size posteriorly; males with trunk limbs 1, or 1–2, modified for grasping females during mating; trunk typically terminates in spinous anal somite or telson, usually with robust caudal rami (**cercopods**); gonopores on eleventh trunk segment; bivalved carapace completely encloses body; valves folded (hinged in *Lynceus*) dorsally and attached by adductor muscles and a suspensory membrane arising from the maxillary somite; usually with a pair of sessile compound eyes and a single, median, simple (naupliar) eye (Figures 4I, J, and K).

There are about 200 described species of conchostracans, or "clam shrimps," in five extant families. The common name derives from the clamlike appearance of the valves, which usually bear concentric growth lines reminiscent of bivalved molluscs. The conchostracans are easily distinguished from the similar cladocerans by the fact that the carapace covers the head as well as the trunk. The presence of appendages the entire length of the trunk further characterizes clam shrimp. Conchostracans are often classified together with cladocerans in the subclass Diplostraca.

Conchostracans live in permanent and ephemeral freshwater habitats worldwide, except in Antarctica. Most species are benthic, but many conchostracans swim during reproductive periods. Some are direct suspension feeders, whereas others stir up detritus from the substratum and feed on suspended particles.

Class Maxillopoda

With 5 cephalic, 6 thoracic, and usually 4 abdominal somites and a telson; thoracomeres variously fused with cephalon; usually with caudal rami; thoracic segments with biramous (sometimes uniramous) limbs, lacking epipods (except in many ostracods); abdominal segments lack typical appendages; (reductions of this basic 5-6-4 body plan are common); carapace present or reduced; eyes variable, but naupliar eyes (when present) unique, with three cups, each with tapetal cells (= maxillopodan eye) (Figures 5–8).

Although the class Maxillopoda is accepted by most specialists as a valid monophyletic taxon, there is still some controversy about its diagnosis and component groups (see for examples Grygier 1983; Newman 1983; Schram 1983, 1986; and others). We include here the taxa Mystacocarida, Ostracoda,

Copepoda, Branchiura, Tantulocarida, and Cirripedia. Pentastomids, certain fossil forms, and some larvae for which adults are unknown are also placed here by some workers. Without belaboring the issue, we warn you that different specialists sometimes interpret the nature of maxillopodan tagmata in different ways, leading to some confusion. Inclusion of members of the subclass Tantulocarida is controversial because of the six- or seven-segmented abdomen in the juveniles of some species.

Maxillopodans are mostly small crustaceans, barnacles being a notable exception. They are recognizable by their shortened bodies, especially the reduced abdomen, and absence of a full complement of legs. The reductions in body size and leg number, emphasis on the naupliar eye, minimal appendage specialization, and certain other features have led biologists to hypothesize that paedogenesis may have played a large role in maxillopodan origin. That is, in many ways they resemble early postlarval forms that evolved sexual maturity before attaining all the adult features.

Subclass Ostracoda. Body segmentation reduced; trunk with 1–3 pairs of limbs of variable structure; caudal rami present; gonopores on lobe anterior to caudal rami; carapace bivalved, hinged dorsally, enclosing body and head; most with one median simple (maxillopodan) eye, and sometimes sessile or weakly stalked compound eyes (Figure 5).

Ostracods comprise about 8,000 living species of bivalved crustaceans ranging in length from 0.1 to 32 mm. They superficially resemble conchostracans in having the entire body enclosed within the valves of the carapace. However, ostracod valves lack the concentric growth rings of conchostracans, and there are major differences in the body segmentation and appendages. A good deal of confusion exists about the nature of ostracod limbs, and this confusion is reflected in the variety of names applied by different authors. We have adopted terms here that allow the easiest comparison with other taxa.

Ostracods possess fewer limbs than found in any other crustacean subclass. The 5 head appendages are followed by 1–3 trunk limbs. Superficially, the (second) maxillae often appear to be absent; however, the highly modified fifth limbs are in fact these appendages. The trunk seldom shows external evidence of segmentation, although all 11 postcephalic somites are discernable in some taxa. The trunk limbs vary in structure among taxa. The third pair bear the gonopores and form the "copulatory organs."

Ostracods are one of the most successful groups

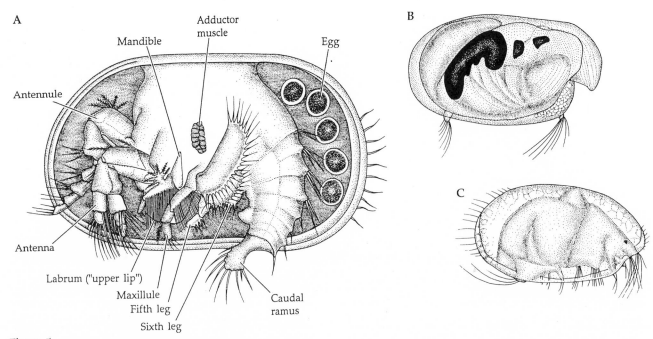

Figure 5

External anatomy in the class Maxillopoda, subclass Ostracoda. A, The ostracod *Cytherella*, with the left valve removed. B—C, two myodocopid ostracods. (A after Moore, 1969, *Treatise on Invertebrate Paleontology*; B—C after Smith 1977.)

of crustaceans. Most are benthic crawlers or burrowers, but many have adopted a planktonic life, and a few are terrestrial in moist habitats. They are abundant worldwide in all aquatic environments and are known to depths of 7,000 m in the sea. Some are commensal on echinoderms or other crustaceans, and a few podocopans have invaded supralittoral sandy regions (members of the family Terrestricytheridae) and terrestrial mosses and humus (some members of the families Cyprididae and Cypridopsidae). There are three principal taxa (ranked as superorders here) recognized within the subclass Ostracoda (the superorders Myodocopa, Podocopa, and Palaeocopa), and although we do not fully describe each of these, some of the differences between them are mentioned in the following discussion.

The myodocopans are all marine. Most are benthic, but the group also includes all of the marine planktonic ostracods, and some of these are demersal. The largest of all ostracods is a member of this group, the planktonic *Gigantocypris*, which exceeds 3 cm in length. Myodocopans include scavengers, detritus feeders, suspension feeders, and some predators.

Members of the superorder Podocopa include predominantly benthic forms; and although some are capable of temporary swimming, none are fully planktonic. Feeding methods include suspension feeding, herbivory, and detritus feeding. The superorder Palaeocopa includes many extinct taxa but only a single living family, Puniciidae, known largely from empty valves dredged in the South Pacific. To date, these shells have been assigned to two genera and are similar in form to those of extinct members of the family Eurychilinidae, a group known from Ordovician to Devonian deposits.

Subclass Mystacocarida. Body divided into cephalon and 10-segmented trunk; telson with claw-like caudal rami; cephalon characteristically cleft; all cephalic appendages nearly identical, but antennae and mandibles biramous, antennules, maxillules, and maxillae uniramous; first trunk segment bears maxillipeds but is not fused with cephalon; no carapace; gonopores on fourth trunk segment; trunk segments 2–5 with short, single-segment appendages (Figure 6A).

There are only ten described species of mystacocarids, eight in the genus *Derocheilocaris* and two in *Ctenocheilocaris*. The body is generally of uniform thickness along its length. Most are less than 0.5 mm long, although *D. ingens* reaches 1 mm. The head is marked by a transverse "cephalic constriction" between the origins of the first and second antennae, a feature suggestive of a retention of primitive head segmentation. In addition, the lack of fusion of the cephalon and maxillipedal trunk segment, simplicity of mouth appendages, absence of trunk regionalization, and other features have led some workers to propose that the mystacocarids are among the most primitive crustaceans. These attributes may, however, simply be related to a paedomorphic origin and specialization for interstitial habitats.

Mystacocarids are marine, interstitial crustaceans that live in littoral and sublittoral sands. They are known from western and southern Africa, the eastern coast of the United States, the Gulf of Mexico, Chile, the Mediterranean Sea, and southern Europe.

The rather vermiform body and small size of mystacocarids are clearly adaptations to interstitial life and movement among sand grains. As far as is known, mystacocarids feed by scraping organic material from the surfaces of sand grains with their setose mouthparts.

Subclass Copepoda. Without a carapace, but cephalon with a well developed cephalic shield; single, median, simple maxillopodan eye (sometimes lacking); one or more thoracomeres fused to head to form cephalosome; thorax of 6 segments, first fused to head and bearing maxillipeds; abdomen of 5 segments, including anal somite (= telson); well developed caudal rami; abdomen without appendages, except an occasional reduced pair on the first segment, associated with the gonopores; point of main body flexure varies among major groups; antennules uniramous, antennae uni- or biramous; 4–5 pairs natatory thoracopods; posterior thoracopods always biramous (Figures 6B–K).

There are about 9,000 described species of copepods. Most are small, 0.5–10 mm long, but some free-living forms exceed 1.5 cm in length, and certain highly modified parasites may reach 25 cm. The bodies of most copepods are distinctly divided into three tagmata, the names of which vary in the literature. The first region includes the five fused head segments and one or two additional fused thoracic somites; it is called a "cephalosome" and bears the usual head appendages and a pair of maxillipeds. All of the other limbs arise on the remaining thoracic segments, which together constitute the "metasome." The abdomen, or "urosome," bears no limbs. The appendage-bearing regions of the body (cephalosome and metasome) are frequently collectively called the "prosome." In our discussion of copepods, we generally restrict ourselves to the terms cephalon, thorax, and abdomen, a practice allowing easier comparisons with other crustaceans and among the various copepod taxa.

The classification of copepods has been subjected to several revisions over the past decade or so, but a general acceptance of the ten orders listed earlier seems to be emerging. We hasten to point out that you will certainly encounter other schemes, because detailed phylogenetic analyses are just beginning to be accomplished and will probably be accompanied by further refinements.

The majority of the free-living copepods, and those most frequently encountered, belong to the orders Calanoida, Harpacticoida, and Cyclopoida, although even some of these are parasitic (Figure 6I). We focus here on these three groups and then briefly discuss some of the other, smaller orders and their modifications for parasitism.

The calanoids are characterized by having the point of major body flexure between the thorax and the abdomen, marked by a distinct narrowing of the body (Figure 6C). They possess one pair of maxillipeds and greatly elongate antennules. Most of the calanoids are planktonic, and as a group they are extremely important as primary consumers in freshwater and marine food webs. The point of body flexure in members of the orders Harpacticoida and Cyclopoida is between the last two (fifth and sixth) thoracic segments (Figures 6D and E). (It should be noted that some authors define the urosome as that region of the body posterior to this point of flexure; thus in these two groups urosome does not necessarily equate with abdomen.) Harpacticoids are generally rather vermiform, the posterior segments not much narrower than the anterior ones, whereas the cyclopoids generally narrow abruptly at the major body flexure. Both the antennules and the antennae are quite short in harpacticoids, but the latter are moderately long in cyclopoids (although never as long as the antennules of calanoids). Furthermore, the antennae are uniramous in cyclopoids but biramous in the other two groups. Most harpacticoids are benthic, and those that have adapted to a planktonic life show modified body shapes. Harpacticoids occur in all aquatic environments; encystment is known to occur in at least a few freshwater and marine species. Cyclopoids are known from fresh and salt water, and most are planktonic.

The nonparasitic copepods move by crawling or swimming, using some or all of the thoracic limbs. Many of the planktonic forms have highly setose appendages (Figure 6F), offering a high resistance to sinking. Calanoids are predominantly plankton feeders. Benthic harpacticoids are often reported as detritus feeders, but many feed predominantly on microorganisms living on the surface of detritus or sediment particles (e.g., diatoms, bacteria, and protozoa).

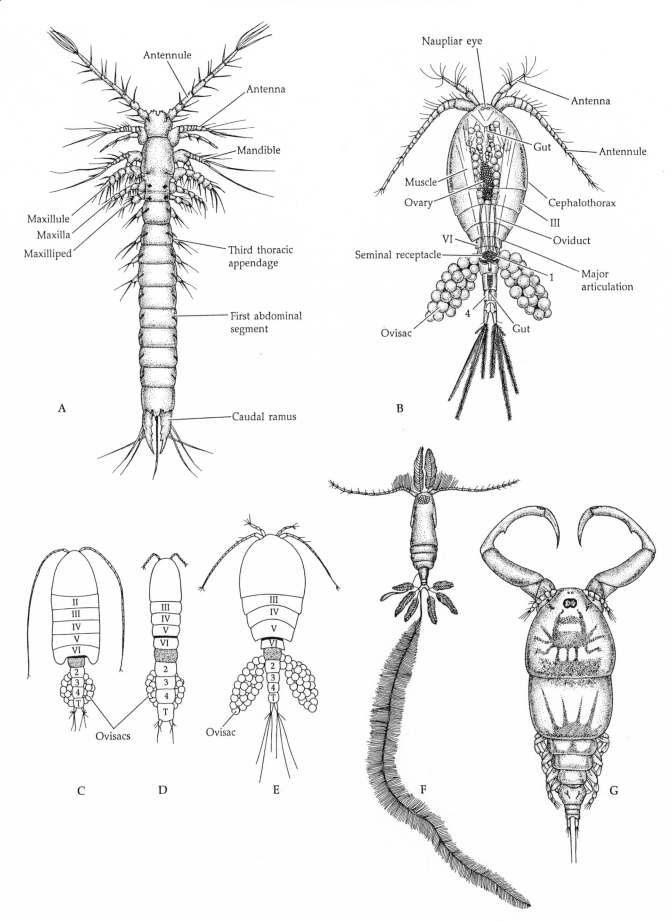

Antennule

Antenna

Mandible

Maxillule
Maxilla
Maxilliped

Third thoracic
appendage

First abdominal
segment

Caudal ramus

A

Naupliar eye

Antenna

Antennule

Gut

Muscle

Ovary

Cephalothorax

III

VI

Oviduct

Seminal receptacle

1

Major
articulation

Ovisac

4

Gut

B

II
III
IV
V
VI

2
3
4
T

Ovisacs

C

III
IV
V
VI

2
3
4
T

Ovisacs

D

III
IV
V
VI

2
3
4
T

Ovisac

E

F

G

Of the remaining orders, the Misophrioida and Mormonilloida are planktonic; members of the Monstrilloida are planktonic as adults, but the larval stages are endoparasites of certain gastropods, polychaetes, and occasionally echinoderms. Members of the orders Siphonostomatoida and Poecilostomatoida are exclusively parasitic, and often show highly modified body forms (Figures 6G, H, and J). Siphonostomatids are endo- or ectoparasites of various invertebrates as well as marine and freshwater fishes; they are often very tiny and show a reduction or loss of body segmentation. Poecilostomatids parasitize invertebrates and marine fishes, and may also show a reduced number of body segments.

Subclass Branchiura. Body oval, head and most of trunk covered by broad carapace; antennules and antennae reduced, the latter sometimes absent; mouthparts modified for parasitism; no maxillipeds; thorax reduced to four segments, with paired biramous appendages; abdomen unsegmented, bilobed, limbless, but with minute caudal rami; female gonopores at bases of fourth thoracic legs, male with single gonopore on midventral surface of last thoracic somite; paired, sessile compound eyes, and 1–3 median simple eyes (Figure 6L).

The Branchiura includes about 130 species of ectoparasites on marine, freshwater, and brackish water fishes. Their bodies are compact, broadly oval,

Figure 6

External anatomy of the class Maxillopoda, subclasses Mystacocarida, Copepoda, and Branchiura. A, The mystacocarid *Derocheilocaris*. B, General anatomy of a cyclopoid copepod. C–E, General body forms of a calanoid (C), harpacticoid (D), and cyclopoid (E) copepod. Note the points of body articulation (dark band) and the position of the genital segment (shaded segment). Roman numerals are thoracic segments; arabic numerals are abdominal segments; T, telson. F, An elaborately setose calanoid copepod adapted for flotation. G, *Ergasilus pitalicus*, a recently described poecilostomatid copepod ectoparasitic on cichlid fishes. H, A

female *Trebius heterodonti* with egg sacs. This siphonostomatid copepod parasitizes horn sharks in California. I, *Notodelphys*, a cyclopoid copepod adapted for endoparasitism in tunicates. J, *Clavella adunca*, a siphonostomatid copepod, shows extreme body reduction. It attaches to the gills of fishes by its elongate maxillae. K, Two siphonostomatid copepods attached to the dorsal fin of a shark. L, *Argulus foliaceus*, a branchiuran that parasitizes fishes. (A after Delamare-Deboutteville and Chappuis 1954; G after Thatcher 1984; H after M. Dojiri; I after McLaughlin 1980; J redrawn from Marcotte 1982, after Kabata 1979; K courtesy of G. McDonald; L adapted from Kaestner 1980.)

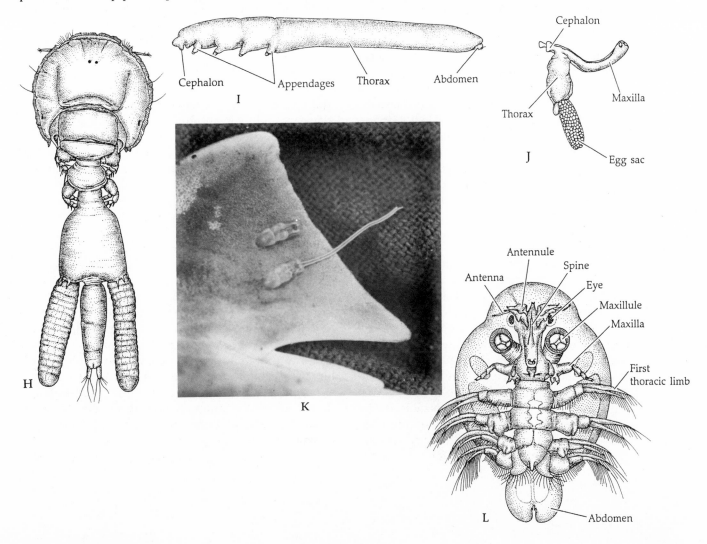

and highly flattened. The mouthparts are modified for parasitic life. The antennae are reduced, sometimes absent, and the antennules generally bear hooks or spines for attachment. The mandibles are reduced in size and complexity, bear cutting edges, and are housed within a styliform "proboscis" apparatus. The maxillules are clawed in *Dolops*, but they are modified as stalked suckers in the other genera (*Argulus*, *Chonopeltis*, and *Dipteropeltis*). The uniramous maxillae usually bear attachment hooks. The thoracopods are biramous and used for swimming when the animal is not attached to a host. Branchiurans feed by piercing the skin of their hosts and sucking blood or tissue fluids. Once they locate a host, they crawl toward the fish's head and anchor in a spot where water flow turbulence is low (e.g., behind a fin or gill operculum).

Members of the genus *Argulus* occur worldwide, but members of the other three genera have restricted distributions. *Chonopeltis* is found only in Africa, *Dipteropeltis* in South America, and *Dolops* in South America, Africa, and Tasmania.

Subclass Cirripedia. Primitively with tagmata as in the class, but in most the adult body is greatly modified for attached or parasitic life; thorax of six segments with paired biramous appendages; abdomen without limbs; telson absent in most, although caudal rami persist on abdomen in some; carapace "bivalved" (folded) or forming fleshy mantle; first thoracomere often fused with cephalon and bearing maxillipeds, or "oral appendages"; female gonopores near bases of first thoracic limbs, male gonopore on median penis on last thoracic or first abdominal segment; compound eyes lost in adults; median eyes reduced or absent in adults (Figure 7).

The 1,000 or so described species are mostly free-living barnacles, but this group also includes some strange parasitic cirripedes rarely seen except by specialists. The common acorn and goose barnacles (Figures 7A–E) are members of the order Thoracica. Members of the order Acrothoracica are minute animals that burrow into calcareous substrata, including corals, and mollusc shells (Figure 7G). Members of the Ascothoracica and Rhizocephala are parasites, the first of anthozoans and echinoderms, the second of crustaceans, especially decapods (Figures 7F and H).

The maxillopodan body plan has been so extensively modified and exploited by the cirripedes that the basic features are nearly unrecognizable in the sessile and parasitic adults. Only some ascothoracicans retain the full complement of thoracic and abdominal segments and it is largely for this reason that they are regarded as the most primitive living Cirripedia. In all others, at least the abdomen is greatly reduced in adults (and most cyprids). In cyprid larvae and ascothoracicans, the carapace is present and "bivalved," but in all others it is lost (order Rhizocephala) or modified as a membranous, saclike mantle (barnacles). In the barnacles (order Thoracica), it is this mantle that produces the familiar calcareous plates that enclose the body.

Locomotion in barnacles is generally confined to the larval stages. A few adult thoracicans are specifically adapted to live attached to floating objects (e.g., seaweeds, pumice, and logs) or nektonic marine animals (e.g., whales and sea turtles); others are often found on the shells and exoskeletons of various errant invertebrates (e.g., crabs and gastropods), which inadvertently provide a means of transportation from one place to another. The parasitic forms also enjoy free rides on their hosts. In at least one species of Ascothoracica, *Synagoga mira*, males retain the ability to swim throughout their lives, attaching only temporarily while feeding on corals.

Thoracican and acrothoracican barnacles use their feathery thoracopods ("cirri") to suspension feed. The thoracican family Coronulidae is composed of suspension-feeding barnacles that attach to whales and turtles (e.g., *Chelonibia*, *Platylepas*, *Stomatolepas*, *Coronula*, and *Xenobalanus*). Ascothoracicans generally have mouth parts modified for piercing and sucking body fluids, but some live inside other animals and absorb the host's tissue fluids. Most rhizocephalans are endoparasitic and are the most highly modified of all cirripedes (Figure 7, 24, and 25). They mainly inhabit decapod crustaceans, but a few are known from isopods, cumaceans, and even thoracican barnacles. The body consists of an external reproductive part (the externa) and an internal, ramifying, nutrient-absorbing part (the interna).

Most species of barnacles are hermaphrodites, although separate sexes are the rule in acrothoracicans, ascothoracicans, and rhizocephalans.

Subclass Tantulocarida. Juveniles with cephalon, six-segmented thorax, and abdomen of up to seven segments; cephalon lacking appendages but with internal median stylet; thoracopods 1–5 biramous, 6 uniramous; abdomen without appendages but with caudal rami; adults highly modified, with "unsegmented" sacciform thorax and a reduced abdomen bearing a uniramous penis on the first segment; female gonopores on fifth thoracic segment; parasitic on other crustaceans (Figure 8).

Until recently the members of this group had

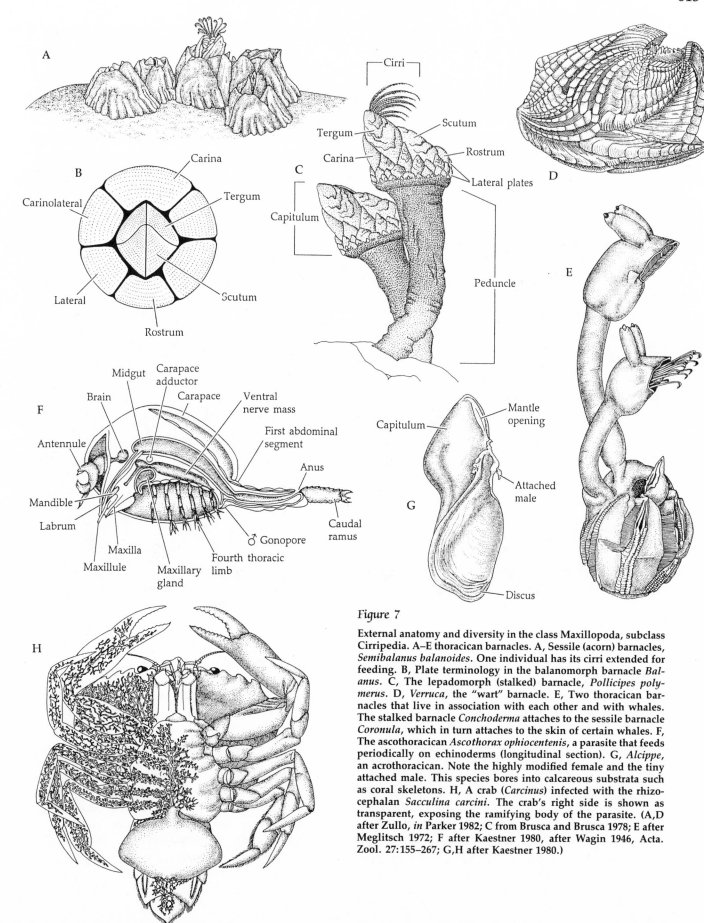

Figure 7

External anatomy and diversity in the class Maxillopoda, subclass Cirripedia. A–E thoracican barnacles. A, Sessile (acorn) barnacles, *Semibalanus balanoides*. One individual has its cirri extended for feeding. B, Plate terminology in the balanomorph barnacle *Balanus*. C, The lepadomorph (stalked) barnacle, *Pollicipes polymerus*. D, *Verruca*, the "wart" barnacle. E, Two thoracican barnacles that live in association with each other and with whales. The stalked barnacle *Conchoderma* attaches to the sessile barnacle *Coronula*, which in turn attaches to the skin of certain whales. F, The ascothoracican *Ascothorax ophiocentenis*, a parasite that feeds periodically on echinoderms (longitudinal section). G, *Alcippe*, an acrothoracican. Note the highly modified female and the tiny attached male. This species bores into calcareous substrata such as coral skeletons. H, A crab (*Carcinus*) infected with the rhizocephalan *Sacculina carcini*. The crab's right side is shown as transparent, exposing the ramifying body of the parasite. (A,D after Zullo, *in* Parker 1982; C from Brusca and Brusca 1978; E after Meglitsch 1972; F after Kaestner 1980, after Wagin 1946, Acta. Zool. 27:155–267; G,H after Kaestner 1980.)

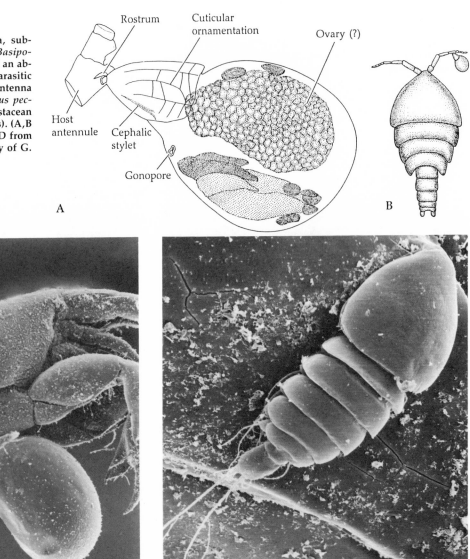

Figure 8

Anatomy in the class Maxillopoda, subclass Tantulocarida. A, An adult *Basipodella atlantica*. Note the absence of an abdomen and the modifications for parasitic life. B, *Basipodella* attached to the antenna of a copepod host. C–D, *Microdajus pectinatus* ("larva") on a tanaid crustacean host (scanning electron micrographs). (A,B after Boxshall and Lincoln 1983; C,D from Boxshall and Lincoln 1987, courtesy of G. Boxshall.)

been assigned to various parasitic groups of the subclasses Copepoda and Cirripedia. In 1983 and 1987 Geoffrey Boxshall and Roger Lincoln published analyses of the known genera and proposed the new taxon Tantulocarida. Subsequent work supports a view of these animals as maxillopodans, although the presence of six or seven abdominal segments in juveniles of some species is inconsistent with this view. The absence of cephalic appendages and other striking modifications for parasitic life obscure affinities with other maxillopodans. In a recent cladistic analysis Boxshall and Huys (1989) support a relationship between the tantulocarids and an ostracod–branchiuran–cirripede line.

Tantulocarids are less than 0.5 mm long. They attach to their hosts, typically deep-water crustaceans, by penetrating the body with a protruding cephalic stylet. The young bear natatory thoracopods. About a dozen species have been described.

Class Malacostraca

Body of 19–20 segments, including five-segmented cephalon, eight-segmented thorax, and six-segmented pleon (seven-segmented in leptostracans), plus telson; with or without caudal rami; carapace covering part or all of thorax, or reduced, or absent; 0–3 pairs of maxillipeds; thoracopods primitively biramous, uniramous in some groups, phyllo-

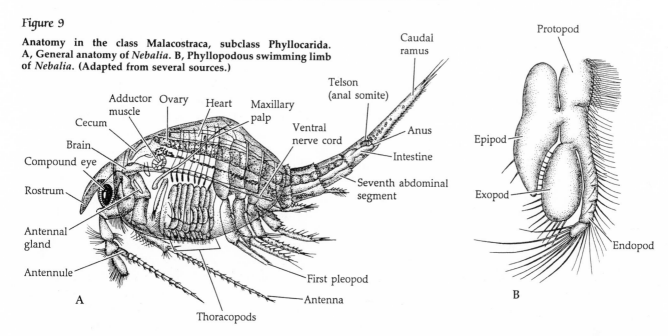

Figure 9

Anatomy in the class Malacostraca, subclass Phyllocarida. A, General anatomy of *Nebalia*. B, Phyllopodous swimming limb of *Nebalia*. (Adapted from several sources.)

podous only in members of the subclass Phyllocarida; antennules and antennae usually biramous; abdomen (pleon) usually bears 5 pairs biramous pleopods and 1 pair biramous uropods; eyes usually present, compound, stalked or sessile; mostly gonochoristic; female gonopores on sixth, and male pores on eighth thoracic segments.

Most recent classification schemes divide the more than 20,000 species of malacostracans into two subclasses, Phyllocarida and Eumalacostraca. The phyllocarids are typically viewed as representing the primitive malacostracan condition (5-8-7, plus telson; Figure 9). The basic eumalacostracan *Bauplan*, characterized by the 5-8-6 (plus telson) arrangement of body segments, was recognized in the early 1900s by W. T. Calman, who termed the defining features of the Eumalacostraca "caridoid facies" (see Figure 10).

Much work has been done since Calman's day, and the group has been reorganized many times. Still, the basic elements of Calman's "caridoid facies" are present in all members of the subclass Eumalacostraca. The hoplocarids (stomatopods) have several unique synapomorphies that have suggested to some workers that they may have diverged independently from an early malacostracan phyllocarid-like ancestor.

Subclass Phyllocarida

Order Leptostraca. With the typical malacostracan characteristics, except notable for presence of seven free pleomeres (plus telson), phyllopodous thoracopods (all similar to one another), and no maxillipeds; with a bivalved carapace (without a hinge) covering the thorax; cephalon with a movable, artic-

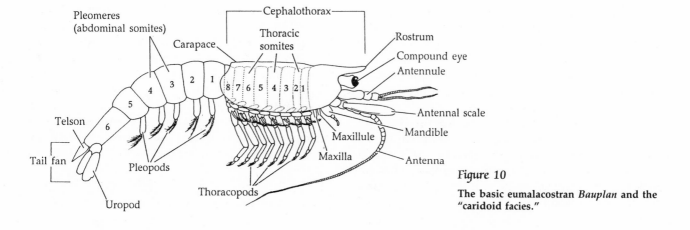

Figure 10

The basic eumalacostran *Bauplan* and the "caridoid facies."

ulated rostrum; pleopods 1–4 similar and biramous, 5–6 uniramous (no uropods); paired stalked compound eyes; antennules biramous; antennae uniramous (Figure 9).

The subclass Phyllocarida includes fewer than 20 species in six genera. Most are 5–15 mm long, but *Nebaliopsis typica* is a giant at nearly 4 cm in length. The body form is distinctive, with a loose bivalved carapace covering the thorax, a protruding head, and an elongate abdomen (Figure 9). All leptostracans are marine, and most are epibenthic from the intertidal zone to a depth of 400 m; *Nebalia typhlops* is bathypelagic. Most species seem to occur in low-oxygen environments. One recently described species, *Dahlella caldariensis*, is associated with the hydrothermal vents of the Galapagos and the East Pacific Rise.

Most of the benthic leptostracans suspension feed by stirring up bottom sediments. They are also capable of grasping relatively large bits of food directly with the mandibles.

Subclass Eumalacostraca

Head, thorax, and abdomen of 5-8-6 somites respectively (plus telson); with 0, 1, 2, or 3 thoracomeres fused with head, their respective appendages usually modified as maxillipeds; antennules and antennae primitively biramous; antennae often with scalelike exopod; most with well developed carapace, secondarily reduced in syncarids and some peracarids; gills primitively as thoracic epipods; tail fan composed of telson plus paired uropods; abdomen typically long and muscular. Four superorders: Hoplocarida, Syncarida, Eucarida, and Peracarida.

Superorder Hoplocarida, Order Stomatopoda. Carapace covering portion of head and fused with thoracomeres 1–4; head with movable, articulated rostrum; thoracopods 1–5 uniramous and subchelate, second pair massive and raptorial (all five sometimes called "maxillipeds" or gnathopods, because they are involved in feeding); thoracopods 6–8 biramous, ambulatory; pleopods biramous, with dendrobranchiate-like gills on exopods; antennules triramous; antennae biramous; with large, paired, stalked compound eyes (Figures 11A and B). Some authorities place the Hoplocarida as a subclass, coequal with the Phyllocarida and Eumalacostraca.

All 350 or so living hoplocarids are placed within the order Stomatopoda, known as "mantis shrimps." They are relatively large crustaceans, ranging in length from 2 to 30 cm. Compared with most malacostracans, the muscle-filled abdomen is thick and robust relative to the cephalon and thorax. Some of the anterior thoracopods are especially reduced in size, but a distinctive feature is the pair of huge raptorial subchelate second thoracopods.

Most stomatopods are found in shallow tropical or subtropical marine environments. Nearly all of them live in burrows excavated in soft sediments, or in cracks and crevices, among rubble, or in other protected spots. They are raptorial carnivores, preying on fishes, molluscs, cnidarians, and other crustaceans. The large subchelae of the second thoracopods act either as crushers or spears (Figure 11B).

Stomatopods crawl about using the posterior thoracopods and the flaplike pleopods. They also can swim by metachronal beating of the pleopods (the swimmerets).

For these relatively large animals, living in narrow burrows requires a high degree of maneuverability. The short carapace and flexible, muscular abdomen allow these animals to twist double and turn around within their tunnels or in other cramped quarters. This ability facilitates an escape reaction whereby a mantis shrimp darts into its burrow rapidly, head first, and then turns around to face the entrance.

Stomatopods are among the few malacostracans that possess pleopodal gills. Only the isopods share the trait of abdominal gills, but they are quite different in the two groups. The tubular, thin, highly branched gills of stomatopods provide a large surface area for gas exchange in these active animals.

Superorder Syncarida. With 0 (Bathynellacea) or one pair (Anaspidacea) of maxillipeds; no carapace; pleon bears telson with or without furcal lobes; at least some thoracopods biramous, eighth often reduced; pleopods variable; compound eyes present (stalked or sessile), or absent (Figures 11C and D).

There are about 150 described species of syncarids in two orders, Anaspidacea and Bathynellacea.* To many workers, the syncarids represent a key group in malacostracan evolution. Through studies of the fossil record and extant members of the family Anaspididae (e.g., *Anaspides*), it has been suggested that syncarids may encompass the most primitive living eumalacostracan *Bauplan*. Among the living syncarids, members of the family Anaspididae are endemic to Tasmania, where they inhabit various freshwater environments, such as open lake surfaces, streams, ponds, and crayfish burrows. All of the other syncarids are interstitial or live in subsurface

---

*Until recently a third order, the Stygocaridacea, was recognized. Most workers now agree that the stygocarids should be reduced to the rank of family within the order Anaspidacea.

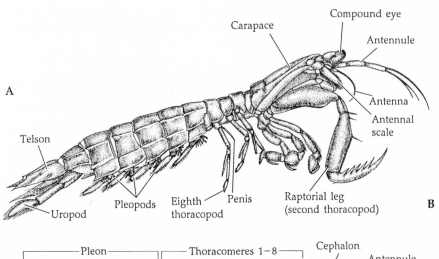

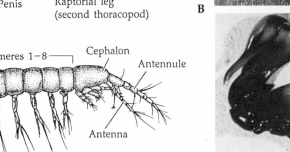

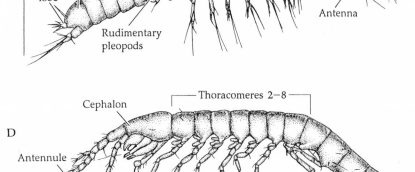

Figure 11

External anatomy in the class Malacostraca, superorders Hoplocarida and Syncarida. A, External anatomy of the hoplocarid (stomatopod) *Squilla*. B, "Spearing" claw and "clubbing" claw (second thoracopod) of stomatopods. C, A bathynellacean, *Bathynella*. D, An anaspidacean, *Stygocarella*. (A adapted from various sources; B courtesy of R. Caldwell; C,D after Schminke, *in* Parker 1982.)

ground water. None are marine. These reclusive eumalacostracans show various degrees of what some have regarded as paedomorphism, including small size (Anaspididae includes members to 5 cm; most others are less than 1 cm long), eyelessness, and reduction or loss of pleopods and some posterior pereopods. The bathynellaceans possess a pleotelson formed by the fusion of the telson to the last pleonite (Figure 11C).

Syncarids either crawl or swim. Little is known about the biology of most species, although some are considered omnivorous. Unlike most other crustaceans, which carry their eggs and develop early embryos, syncarids lay their eggs or shed them into the water following copulation.

Superorder Eucarida. Telson without caudal rami; 0, 1, or 3 pairs maxillipeds; carapace present, covering and fused dorsally with head and entire thorax; usually with stalked compound eyes; gills thoracic (Figures 12–15).

Although members of this group are highly diverse, they are united by the presence of a complete carapace that is fused with all thoracic segments, forming a characteristic cephalothorax. Most of the thousands of species belong to the order Decapoda. The other two orders are the Euphausiacea (krill) with about 90 species and the monotypic Amphionidacea (*Amphionides reynaudii*).

Order Euphausiacea. The euphausiaceans are distinguished among the eucarids by the absence of maxillipeds, the exposure of the thoracic gills external

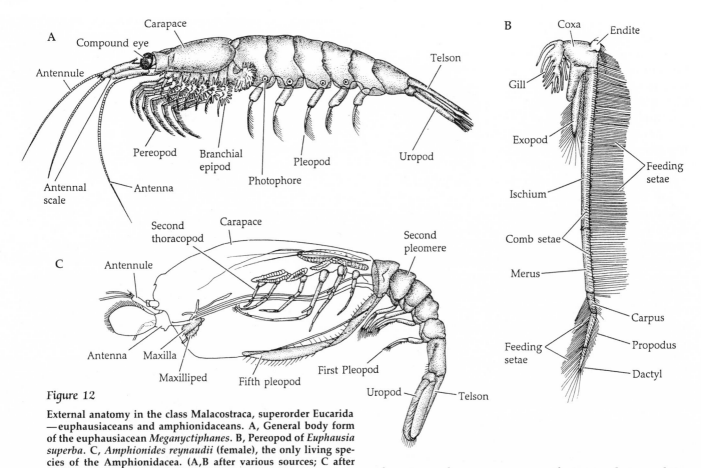

Figure 12

External anatomy in the class Malacostraca, superorder Eucarida —euphausiaceans and amphionidaceans. **A,** General body form of the euphausiacean *Meganyctiphanes*. **B,** Pereopod of *Euphausia superba.* **C,** *Amphionides reynaudii* (female), the only living species of the Amphionidacea. (A,B after various sources; C after Abele and Felgenhauer, *in* Parker 1982.)

to the carapace, and the possession of biramous thoracopods (the last one or two pairs sometimes being reduced). They are shrimplike in appearance. Most of them have photophores on the eyestalks, the bases of the second and seventh thoracopods, and between the first four pairs of abdominal limbs (Figures 12A and B).

All euphausiaceans are pelagic; they range in length from about 4 to 15 cm. The pleopods function as "swimmerets." They are known from all oceanic environments to depths of 5,000 m. Most species are distinctly gregarious, and where they occur in huge schools (krill) they provide a major source of food for larger nektonic animals (baleen whales, squids, fishes) and even some marine birds. Krill densities often exceed 1,000 animals/m^3 (614 g wet weight/m^3). Euphasiaceans are generally suspension feeders, although predation and detritivory also occur.

Order Amphionidacea. The single species of the order Amphionidacea, *Amphionides reynaudii*, possesses an enlarged cephalothorax covered by a thin,

almost membranous carapace that extends to enclose the thoracopods (Figure 12C). The thoracopods are biramous with short exopods. The first pair is modified as maxillipeds, and the last pair is absent from females. Some of the mouthparts are highly reduced in females. The pleopods are biramous and natatory, except that the first pair in females is uniramous and greatly enlarged, perhaps functioning to form a brood pouch extending under the thorax. Females have a reduced gut and apparently do not feed. *Amphionides* is a worldwide member of marine oceanic plankton and occurs to a depth of 1,700 m.

Order Decapoda. The decapods are among the most familiar eumalacostracans. They possess a well developed carapace enclosing a branchial chamber, but they differ from other eucarid orders in generally possessing three pairs of maxillipeds, usually leaving five pairs of functional uniramous or weakly biramous pereopods (hence the name, Decapoda). Rearrangement of the subtaxa within this order is a popular carcinological pastime, and the classification is far from being stable (see references section for an entry into the vast literature on decapod classification). In vernacular terms, nearly every decapod may be recognized as some sort of "shrimp," "crab," "lobster," or "crayfish."

In addition to the general terminology already introduced here and in Chapter 15, the reader should understand a few details about the structure and placement of decapod gills, because these features are important taxonomically. All decapod gills arise as thoracic coxal exites (epipods), but their final placement varies. Those that remain attached to the coxae are **podobranchs** (= "foot gills"), but others eventually become associated with the articular membrane between the coxae and body and are thus called **arthrobranchs** (= "joint gills"). Some actually end up on the lateral body wall, or surface of the thoracic pleurites, as **pleurobranchs** (= "side gills"). The sequence by which some of these gills arise ontogenetically varies. For example, in members of the suborder Dendrobranchiata and the infraorder Stenopodidae, the arthrobranchs appear before the pleurobranchs, whereas in members of the infraorder Caridea the reverse is true. In most of the other decapods the arthrobranchs and pleurobranchs tend to appear simultaneously. These developmental differences may be minor heterochronic dissimilarities and of less phylogenetic importance than actual gill anatomy.

Among the decapods the gills can be one of three basic structural types, described as **dendrobranchiate, trichobranchiate,** and **phyllobranchiate** (Figures 27B, C, and D). All three of these gill types include a main axis carrying afferent and efferent blood vessels, but they differ markedly in the nature of the side filaments or branches. Dendrobranchiate gills bear two principal branches off the main axis, each of which is divided into multiple secondary branches. Trichobranchiate gills bear a series of radiating unbranched tubular filaments, and phyllobranchiate gills are characterized by a double series of platelike or leaflike branches arising from the axis. The occurrences of these gill types among various taxa are noted below.

Close inspection of the proximal parts of the pereopods usually reveals another decapod feature. In most forms the basis and ischium are fused (as a basi-ischium), the point of fusion indicated by a suture line.

The decapods are a highly diverse group. They occur in all aquatic environments at all depths, and a few spend most of their lives on land. Many are pelagic, but others have adopted benthic sedentary, errant, or burrowing lifestyles. Feeding strategies include nearly every habit imaginable: suspension feeding, predation, herbivory, scavenging, and more.

Suborder Dendrobranchiata. This group includes about 450 species of decapods, most of which are penaeid and sergestid shrimps (Figure 13A). As the name indicates, these decapods possess dendrobranchiate gills, a unique synapomorphy of the taxon. One genus, *Lucifer*, has secondarily lost the gills completely. The dendrobranchiate shrimps are further characterized by chelae on the first three pereopods, copulatory organs modified from the first pair of pleopods in males, and ventrolateral expansions of the abdominal tergites (called "pleural lobes"). Generally, none of the chelipeds are greatly enlarged. In addition, females of this group do not brood their eggs; fertilization is external and the embryos hatch as nauplius larvae (see section on development). Many of these animals are quite large, over 30 cm long. The sergestids are pelagic and all marine, whereas penaeids are pelagic or benthic, and some occur in brackish water. Some (e.g., *Penaeus*, *Sergestes*, and *Acetes*) are of major commercial importance in the world's shrimp fisheries.

Suborder Pleocyemata. All of the remaining decapods belong to the suborder Pleocyemata. Members of this taxon never possess dendrobranchiate gills. The embryos are brooded on the female's pleopods and hatch at some stage later than the nauplius larva. Included in this suborder are several kinds of shrimps, plus the crabs, crayfish, lobsters, and a host of less familiar forms (Figures 13B–E, 14, and 15). Most current workers recognize seven infraorders within the Pleocyemata, as we have done below, but a number of other schemes have been proposed and persist in the literature. One older approach was to divide these (and sometimes all) decapods into two large groups, called the Natantia and Reptantia. These terms refer to swimming decapods and walking decapods, respectively. Although these terms have largely been abandoned as formal taxa, they still serve a useful descriptive purpose (much as the adjectives *errant* and *sedentary* do for polychaete worms), and you will often see references in the literature to natant decapods and reptant decapods.

INFRAORDER CARIDEA. The nearly 2,000 living species in this infraorder are generally referred to as the caridean shrimps (Figures 13B, C, and D). These natant decapods have phyllobranchiate gills. The first one or two pairs of pereopods are chelate and variably enlarged, except in the unique genus *Procaris*, which lacks chelation of any limbs. The first pleopods are generally somewhat reduced, but not much modified in the males. The ventrolateral expansions

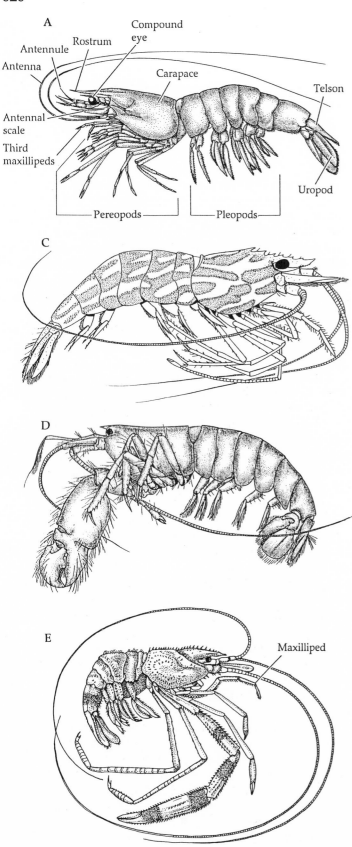

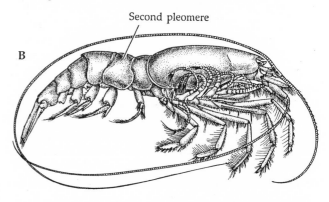

External anatomy and diversity in shrimps: class Malacostraca, superorder Eucarida, order Decapoda. A, A penaeid shrimp (suborder Dendrobranchiata), *Penaeus setiferus*. B, A procarid shrimp (suborder Pleocyemata, infraorder Caridea), *Procaris ascensionis*. C, A hippolytid shrimp (infraorder Caridea, family Hippolytidae), *Lysmata californica*. D, An alpheid shrimp (infraorder Caridea, family Alpheidae), *Alpheus*. E, A stenopodid shrimp (infraorder Stenopodidea), *Stenopus*. (A,B,E after Abele and Felgenhauer, *in* Parker 1982; C,D from Brusca 1980.)

("pleural lobes") of the second abdominal tergite are distinctly enlarged to overlap both the first and third "pleura."

INFRAORDER STENOPODIDEA. All 20 or so species in this infraorder belong to a single family, the Stenopodidae (Figure 13E). Stenopodid shrimps bear trichobranchiate gills. The first three pairs of pereopods are chelate, and the third pair is significantly larger than the others. The first pleopods are uniramous in males and females but are not strikingly modified. The second abdominal "pleura" are not expanded as they are in carideans.

These shrimps are usually only a few centimeters long (2–7 cm). Most species are tropical and associated with benthic environments, especially with coral reefs. Many are commensal, and the group includes the cleaner shrimps (e.g., *Stenopus*) of tropical reefs, which are known to remove parasites from local fishes. Stenopodids often occur in male–female couples. Perhaps the most noted example of this bonding is the glass sponge shrimp, *Spongicola venusta*: a young male and female shrimp enter the atrium of a host sponge, usually the glass sponge *Euplectella*, and eventually grow too large to escape, and thus spend the rest of their days together.

INFRAORDER THALASSINIDEA. The mud and ghost shrimps are particularly difficult to place within the decapods, and they may not represent a monophyletic group. Sometimes they are included with the

Figure 14

External anatomy and diversity in some reptant decapods (class Malacostraca, superorder Eucarida). A, A mud shrimp, *Callianassa* (infraorder Thalassinidea). B, A spiny lobster, *Panulirus* (infraorder Palinura). C, A hermit crab, *Paguristes*, in its shell (infraorder Anomura). D, A hermit crab, *Pagurus*, removed from its shell to expose the soft abdomen. E, A porcelain crab, *Petrolisthes* (Anomura), with the reduced posterior pereopods extended. F, A sand or mole crab, *Emerita* (Anomura). G, The umbrella crab *Cryptolithodes* (Anomura). (A–C from Brusca 1980; D–G from Brusca and Brusca 1978.)

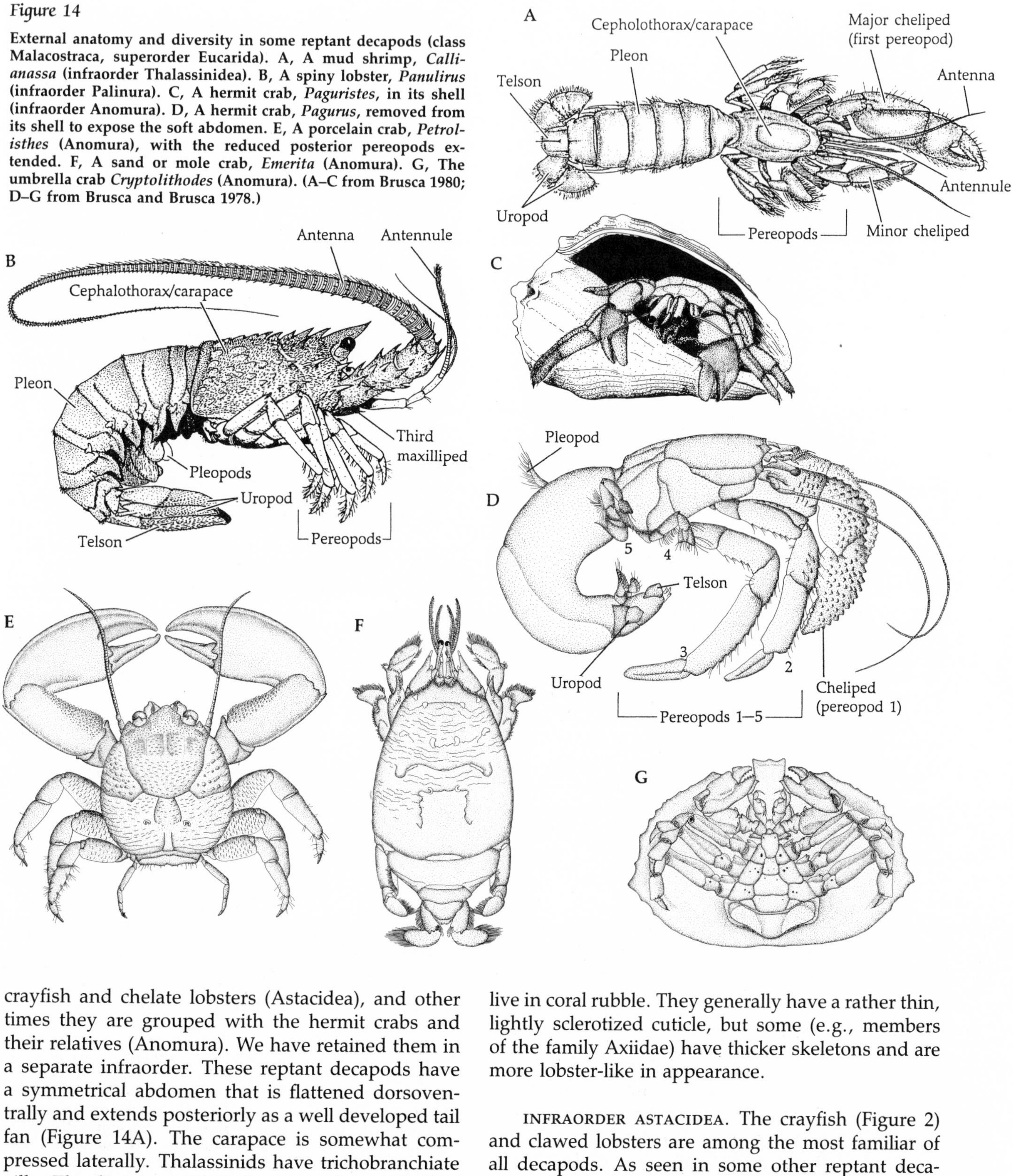

crayfish and chelate lobsters (Astacidea), and other times they are grouped with the hermit crabs and their relatives (Anomura). We have retained them in a separate infraorder. These reptant decapods have a symmetrical abdomen that is flattened dorsoventrally and extends posteriorly as a well developed tail fan (Figure 14A). The carapace is somewhat compressed laterally. Thalassinids have trichobranchiate gills. The first two pairs of pereopods are chelate, and the first pair is generally much enlarged. Most of these animals are marine sediment burrowers or

live in coral rubble. They generally have a rather thin, lightly sclerotized cuticle, but some (e.g., members of the family Axiidae) have thicker skeletons and are more lobster-like in appearance.

INFRAORDER ASTACIDEA. The crayfish (Figure 2) and clawed lobsters are among the most familiar of all decapods. As seen in some other reptant decapods, the dorsoventrally flattened abdomen terminates in a strong tail fan. The gills are trichobranchiate. The first three pairs of pereopods are always chelate, and the first pair is greatly enlarged. Most crayfish live in fresh water, but a few species live in damp soil, where they may excavate extensive and complex burrow systems. *Homarus*, the Maine lobster, is strictly marine.

INFRAORDER PALINURA. This group includes the spiny lobsters and slipper lobsters (Figure 14B). The flattened abdomen bears a tail fan; the carapace may be cylindrical or somewhat flattened dorsoventrally. The gills are trichobranchiate. The chelation of the pereopods varies: the first four pairs may be chelate; only the fifth pair is chelate; or none are chelate. These animals are all marine and are found in a variety of habitats.

INFRAORDER ANOMURA. This group includes hermit crabs, porcelain crabs, mole and sand crabs, and a few other kinds of crablike decapods (Figures 14C–G). The abdomen of anomurans is either soft and asymmetrically twisted (as in hermit crabs), or it is symmetrical, short, and flexed beneath the thorax (as in porcelain crabs and others). Those with twisted abdomens typically inhabit gastropod shells or other empty "houses" not of their own making. The carapace shape and gill structure vary. The first pereopods are chelate; the third pereopods are never chelate. The second, fourth, and fifth pairs are usually simple, but occasionally they are chelate or subchelate. The fifth pereopods (and sometimes the fourth) are generally much reduced and do not function as walking limbs; the fifth pereopods function as gill cleaners and often are not visible externally. The pleopods are reduced or absent. The eyes are medial to the antennae. Most anomurans are marine, but a few freshwater and semiterrestrial species are known.

INFRAORDER BRACHYURA Included here are the so-called true crabs (Figure 15). The abdomen is symmetrical but reduced and flexed beneath the thorax, and uropods are usually absent. The cephalothorax is distinctly flattened dorsoventrally and usually expanded laterally. The gills are phyllobranchiate. The first pereopods are chelate and usually enlarged. The second and third pereopods are simple; the fourth and fifth pairs are rarely subchelate. Pereopod 5 is well developed and not modifed as a gill cleaner, and the eyes are lateral to the antennae. The males lack the third, fourth, and fifth pairs of pleopods. The "true" crabs are mostly marine, but freshwater and semiterrestrial species occur in the tropics. The land crabs (certain species of the families Gecarcinidae,

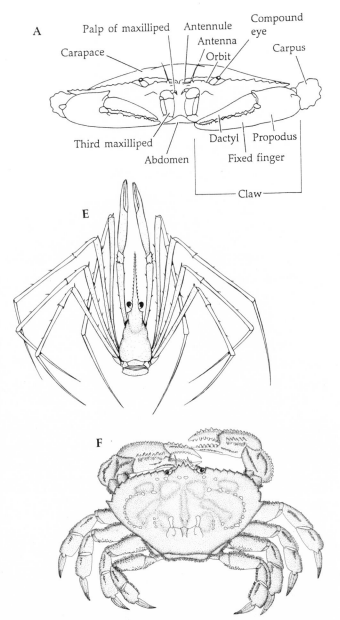

Figure 15

External anatomy and diversity of the "true," or brachyuran, crabs (class Malacostraca, superorder Eucarida, order Decapoda). A–B, General crab anatomy: frontal and ventral views of a swimming crab (family Portunidae). C, A spider crab (family Majidae), *Loxorhynchus*. D, A kelp crab (Majidae), *Pugettia*. E, An arrow crab (Majidae), *Stenorhynchus*. F, A cancer crab (family Cancridae), *Cancer*. G, A grapsid crab (family Grapsidae), *Pachygrapsus*. H, A pinnotherid or pea crab (family Pinnotheridae), *Parapinnixa*. I, A xanthid crab (family Xanthidae), *Trapezia*. Members of this species are obligate commensals in scleractinian corals. J, A fiddler crab (family Ocypodidae), *Uca*. K, A ghost crab (Ocypodidae), *Ocypode*. L, A calappid crab (family Calappidae), *Hepatus* (anterior view). M, A dromiid crab (family Dromiidae), *Hypoconcha* (anterior view). Members of this genus carry bivalve mollusc shells on their backs. N, Female (top) and male (bottom) brachyuran crabs. Note the difference in the abdomen width in these two specimens. (C,D,F,G from Brusca and Brusca 1978; E, H–M from Brusca 1980; N courtesy of J. Demartini.)

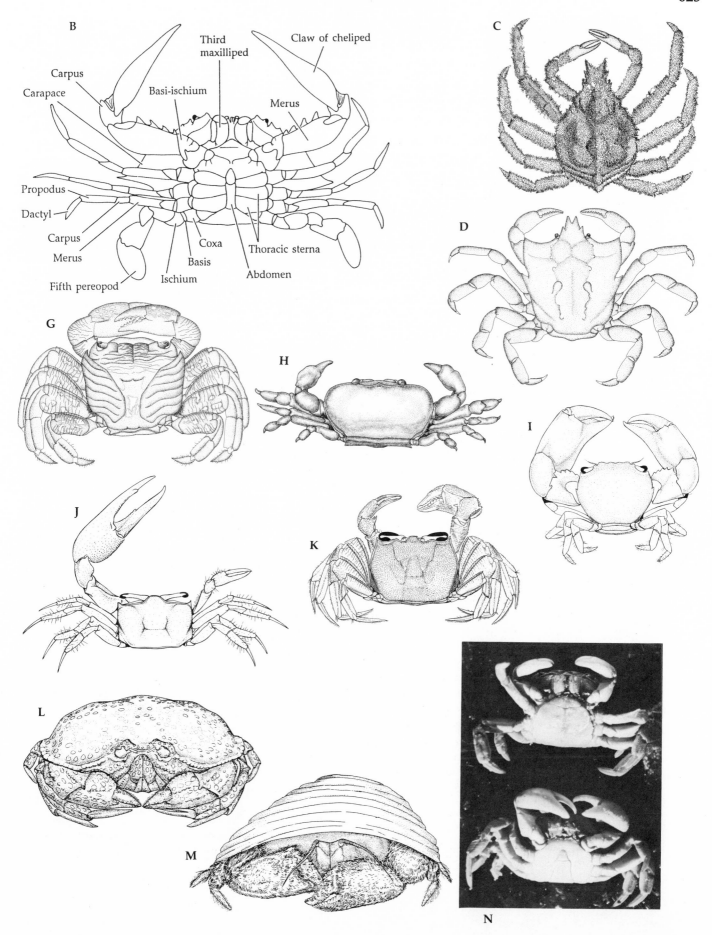

Ocypodidae, and Grapsidae) are still dependent on the ocean for breeding and larval development.

Superorder Peracarida. Telson without caudal rami; 1 (rarely 2–3) pair maxillipeds; maxilliped basis typically produced into an anteriorly directed, blade-like endite; mandibles with articulated accessory processes in adults, between molar and incisor processes, the **lacinia mobilis**; carapace, when present, not fused with posterior pereonites and usually reduced in size; gills thoracic or abdominal; with unique thoracic coxal endites, called oostegites, that form a ventral brood pouch or marsupium in females of all species except members of the order Thermosbaenacea, the latter using the carapace to brood embryos; in most groups, young hatch as mancas, a prejuvenile stage lacking the last pair of thoracopods (no true larval forms) (Figures 16–19).

A few specialists have argued against the monophyletic nature of the superorder Peracarida. However, the presence of numerous unique features (e.g., oostegites, marsupial brooding, bladelike endite of mandible, direct development, manca stage, adult lacinia mobilis), coupled with the clear trend towards reduction of the carapace, eyestalks, and other features seem to indicate that the group is a valid monophyletic clade. The roughly 11,000 species are presently divided among nine orders, including the recently established Mictacea. The orders Mysida and Lophogastrida were formerly combined (as the Mysidacea), but many authorities now treat them separately.

The peracarids are an extremely successful group of malacostracan crustaceans and are known from many habitats. Although most are marine, many occur on land and in fresh water, and several species live in hot springs at temperatures of 30–40°C! Aquatic forms include planktonic as well as benthic species at all depths. The group includes the most successful terrestrial crustaceans—the pill bugs and sow bugs of the order Isopoda, and a few amphipods that have invaded land and live in damp forest leaf litter or gardens. Peracarids range in size from tiny interstitial forms only a few millimeters long to planktonic amphipods over 12 cm long (*Cystisoma*), a lophogastrid 35 cm long (*Gnathophausia*), and benthic isopods growing to 44 cm in length (*Bathynomus*). These animals exhibit all sorts of feeding strategies; a number of them, especially isopods and amphipods, are symbionts.

Order Mysida. Carapace well developed, covering most of thorax, but never fused with more than four anterior thoracic segments; maxillipeds (1–2 pairs) not intimately associated with cephalic appendages; thoracomere 1 separated from head by internal skeletal bar; abdomen with well developed tail fan; pereopods biramous, except last pair, which is sometimes reduced; pleopods reduced or, in males, modified; compound eyes stalked, sometimes reduced; discrete gills absent; usually with a statocyst in each uropodal endopod (Figure 16A).

There are nearly 700 species of mysids, ranging in length from about 2 mm to about 8 cm. Most swim by action of the thoracic exopods. Mysids are shrimp-like crustaceans that are often confused with the superficially similar euphausiaceans (which lack things such as oostegites and uropodal statocysts). Mysids are generally pelagic or demersal but are known from all ocean depths. Some species are intertidal and burrow in the sand during low tides. Most are omnivorous suspension feeders, eating algae, zooplankton, and suspended detritus.

Order Lophogastrida. Similar to mysids, except maxillipeds (1 pair) associated with cephalic appendages; thoracomere 1 not separated from head by internal skeletal bar; pleopods well developed; gills present; without statocysts; all 7 pairs pereopods well developed and similar (except among members of the family Eucopiidae, in which their structure varies) (Figures 16B and C).

There are about 40 known species of lophogastrids, most of which are 1–8 cm long, although *Gnathophausia ingens* is a giant, reaching 35 cm (Figure 16B). All are pelagic swimmers, and the group has a cosmopolitan oceanic distribution. Lophogastrids are largely predators on zooplankton.

Order Cumacea. Carapace present, covering and fused to first three thoracic segments; 3 pairs maxillipeds, the first with modified branchial apparatus associated with branchial cavity formed by carapace; pereopods 1–5 ambulatory, simple, 1–4 may be biramous; pleopods usually absent from females and present in males; telson sometimes fused with sixth pleonite, forming pleotelson; uropods styliform; compound eyes absent, or sessile and usually fused (Figure 16D).

Cumaceans are odd-appearing crustaceans, with a large bulbous anterior end and a long slender posterior—they are shaped rather like horizontal commas! The famous carcinologist Waldo L. Schmitt liked to refer to cumaceans as "little wonders and queer blunders." They are distributed worldwide and include about 850 small species (0.5–2 cm long). Most

Figure 16

External anatomy and diversity in some peracarid crustaceans (class Malacostraca, superorder Peracarida)—mysids, lophogastrids, cumaceans, and tanaids. A, Anatomy of a generalized mysid (order Mysida). B, A lophogastrid, *Gnathophausia*. C, Second pereopod of *Gnathophausia*. D, A cumacean, *Diastylis*, in its typical partially buried position. The arrows indicate the feeding and ventilation current. E, A tanaid, *Apseudes*. (Adapted from various sources.)

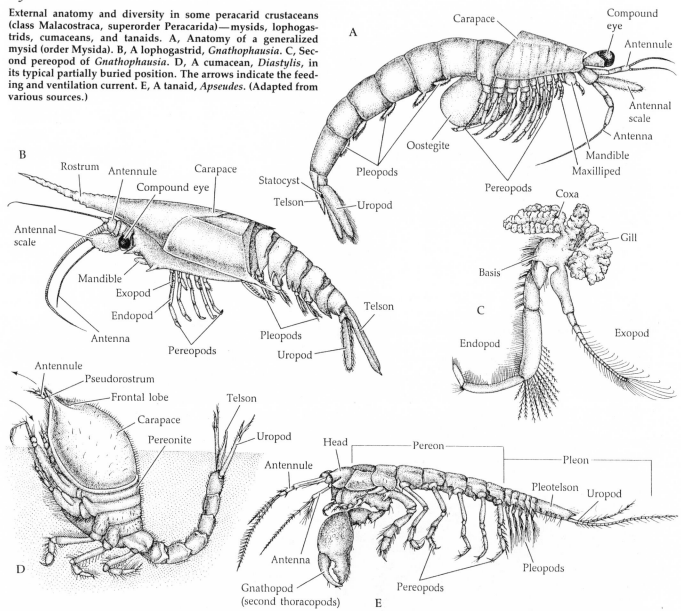

are marine, although a few freshwater and brackish-water species are known. They live in association with bottom sediments, but are capable of swimming and probably leave the bottom to breed and to molt. Some are suspension feeders, and others eat the organic film on sand grains.

Order Tanaidacea. Carapace present and fused with first two thoracic segments; thoracopods 1–2 are maxillipeds, the second being chelate; thoracopods 3–7 simple, ambulatory pereopods; pleopods present or absent; uropods biramous or uniramous; telson and last 1 or 2 pleonites fused as pleotelson; com-

pound eyes absent, or present and on "cephalic lobes" (Figure 16E).

Members of this order are known worldwide from benthic marine habitats; a few live in brackish or nearly fresh water. Most of the 850 or so species are small, ranging from 0.5 to 2 cm in length. They often live in burrows or tubes and are known from all ocean depths. Many are suspension feeders, others are detritivores, and still others are predators.

Order Spelaeogriphacea. Carapace short, fused with first thoracomere; one pair maxillipeds; pereopods 1–7 simple, biramous, with shortened exopods;

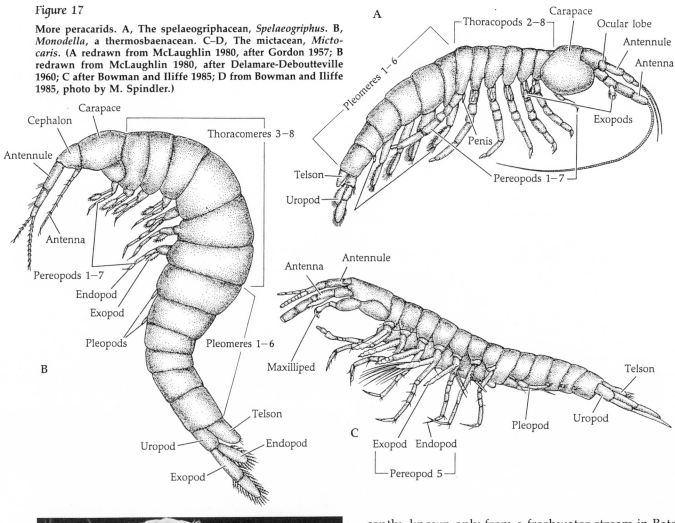

Figure 17
More peracarids. A, The spelaeogriphacean, *Spelaeogriphus*. B, *Monodella*, a thermosbaenacean. C–D, The mictacean, *Mictocaris*. (A redrawn from McLaughlin 1980, after Gordon 1957; B redrawn from McLaughlin 1980, after Delamare-Deboutteville 1960; C after Bowman and Iliffe 1985; D from Bowman and Iliffe 1985, photo by M. Spindler.)

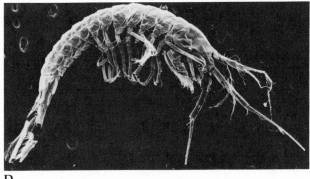

exopods on legs 1–3 modified for producing currents, on legs 4–7 as gills; pleopods 1–4 biramous, natatory; pleopod 5 reduced; tail fan well developed; compound eyes nonfunctional or absent, but eyestalks often persist (Figure 17A).

The order Spelaeogriphacea is currently known from only two troglobitic species. These small (less than 1 cm), transparent peracarids were, until re-

cently, known only from a freshwater stream in Bats Cave on Table Mountain, South Africa; a second species has been recently reported from Brazil. Little is known about the biology of these animals, but they are suspected to be detritus feeders.

Order Thermosbaenacea. Carapace present, fused with first thoracomere and extending back over 2–3 additional segments; one pair maxillipeds; pereopods biramous, simple, lacking epipods and oostegites; carapace forms dorsal brood pouch; two pairs uniramous pleopods; uropods biramous; telson free or forming pleotelson with last pleonite; eyes set on platelike eyestalks but lacking pigment (Figure 17B).

About eleven species of thermosbaenaceans are recognized, in six genera. *Thermosbaena mirabilis* is known from freshwater hot springs in North Africa, where it lives at temperatures in excess of 40°C. Several species in other genera occur in much cooler fresh waters, typically ground waters or caves. Other species are marine, or inhabit anchialine pools (land-locked saltwater habitats). From a few studies, it appears that most thermosbaenaceans feed on plant detritus. Some workers place thermosbaenaceans in their own superorder, Pancarida.

Order Mictacea. Without a carapace, but with a well developed head shield fused with first thoracomere and produced laterally over bases of mouthparts; one pair maxillipeds; pereopods simple, 1–5 or 2–6 biramous, exopods natatory; gills lacking; pleopods reduced, uniramous; uropods biramous, with 2–5 segmented rami; telson not fused with pleonites; eyestalks present (*Mictocaris*) but lacking any evidence of visual elements, or absent (*Hirsutia*) (Figures 17C and D).

The Mictacea is the most recently established peracarid order (see Bowman et al. 1985; Sanders et al. 1985; Bowman and Iliffe 1985; Just and Poore 1988). The order was erected to accommodate two species of unusual crustaceans: *Mictocaris halope* (discovered in marine caves in Bermuda) and *Hirsutia bathyalis* (from a benthic sample, 1,000 m deep in the Guyana Basin off northeastern South America). A third species, *Hirsutia sandersetalia*, was described in 1988 from a benthic sample (1,500 m) off Australia. Mictaceans are small, 2-3.5 mm in length. More is known about *Mictocaris* because many specimens have been recovered, and some have been studied alive, whereas *Hirsutia* is thus far represented by a single damaged female dredged from the deep sea. *Mictocaris* is pelagic in cave waters and swims by using its pereopodal exopods.

Order Isopoda. Carapace absent; first thoracomere fused with head; one pair maxillipeds; seven pairs uniramous pereopods, first of which is sometimes subchelate, others usually simple (gnathiids have only five pereopods, as thoracopod 2 is a maxilliped (the "pylopod") and thoracopod 8 is missing); pereopods variable, modified as ambulatory, prehensile, or natatory limbs; pereopodal coxae often expanded as lateral side plates (coxal plates); pleopods biramous and well developed, natatory and for gas exchange (as gills in aquatic taxa, and with pseudotracheae in the terrestrial Oniscidea); telson fused with 1–6 pleonites, forming pleotelson; eyes usually sessile and compound, absent from some, and on "cephalic lobes" in the family Gnathiidae (Figure 18).

The isopods include over 4,000 marine, freshwater, and terrestrial species, ranging in length from 0.5 to 440 mm, the largest being the huge benthic *Bathynomus*. They are common inhabitants of nearly all environments, and some groups are exclusively (Epicaridea) or partly (Flabellifera) parasitic. Members of the suborder Oniscidea include a number of species that have wholly invaded land—they are the most successful terrestrial crustaceans. Their direct development, flattened shape, osmoregulatory capabilities, thickened cuticle, behavioral adaptations, and aerial gas exchange organs (pseudotracheae) allow most oniscideans to live completely divorced from aquatic environments.

Isopod feeding habits are extremely diverse, depending upon their environment and mouthpart structure. Many are herbivorous or omnivorous scavengers, but direct plant feeders, detritivores, and predators are also common. Some are parasites that feed on host tissue fluids (e.g., on fishes or on other crustaceans).

Order Amphipoda. Carapace absent; first thoracomere fused to head; one pair maxillipeds; seven pairs uniramous pereopods, with first, second, and sometimes others frequently modified as chelae or subchelae; pereopodal coxae expanded as lateral side plates (coxal plates); gills thoracic (medial pereopodal epipods); abdomen "divided" into two regions of three segments each, an anterior pleon and posterior urosome, with anterior appendages typical pleopods and urosomal appendages modified as uropods; telson free or fused with last urosomite; other urosomites sometimes fused; compound eyes sessile, absent in some, huge in many (but not all) members of the suborder Hyperiidea (Figure 19). The classification of the Amphipoda is in need of major reanalysis; it is likely that none of the existing suborders are monophyletic.

Isopods and amphipods share many unique attributes and are probably closely related. Earlier workers long ago recognized these fundamental similarities (e.g., sessile compound eyes, complete loss of carapace, and presence of coxal plates) and classified them together as the "Edriopthalma" or "Acarida." The roughly 6,000 species of amphipods range in length from tiny 1-mm forms to giant deep sea benthic species reaching 25 cm, and one group of planktonic forms exceeding 10 cm (Figure 1P). They have invaded most marine and freshwater habitats and often comprise a large portion of the biomass in an area. A few gammarideans are semiterrestrial in moist forest leaf litter or on supralittoral sandy beaches (e.g., beach hoppers); a few others live in moist gardens and greenhouses (e.g., *Talitrus sylvaticus* and *T. pacificus*). Most of the gammaridean amphipods are benthic, but a few have adopted a pelagic

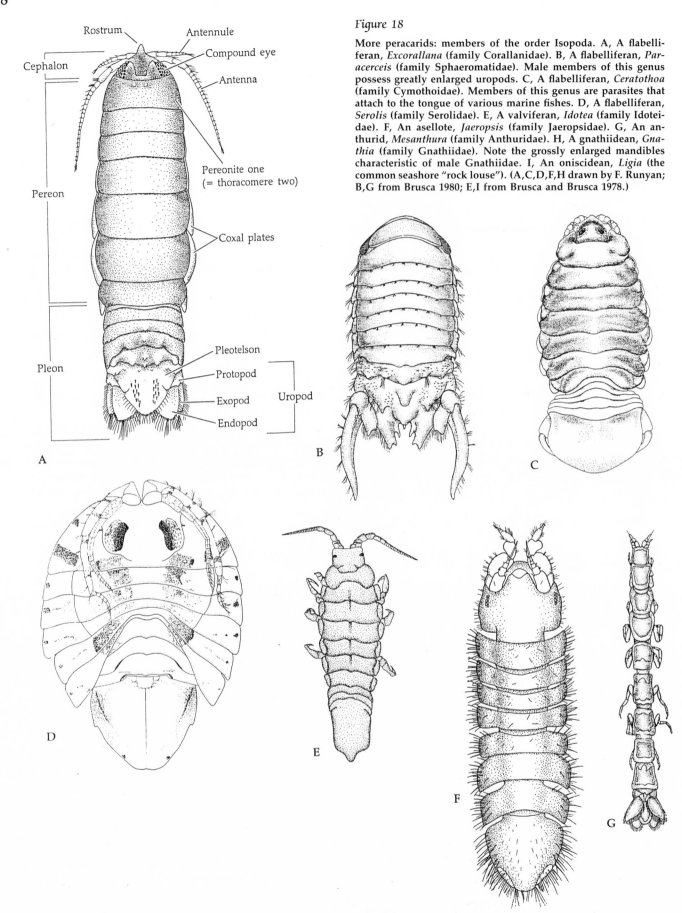

Figure 18

More peracarids: members of the order Isopoda. A, A flabelliferan, *Excorallana* (family Corallanidae). B, A flabelliferan, *Paracerceis* (family Sphaeromatidae). Male members of this genus possess greatly enlarged uropods. C, A flabelliferan, *Ceratothoa* (family Cymothoidae). Members of this genus are parasites that attach to the tongue of various marine fishes. D, A flabelliferan, *Serolis* (family Serolidae). E, A valviferan, *Idotea* (family Idoteidae). F, An asellote, *Jaeropsis* (family Jaeropsidae). G, An anthurid, *Mesanthura* (family Anthuridae). H, A gnathiidean, *Gnathia* (family Gnathiidae). Note the grossly enlarged mandibles characteristic of male Gnathiidae. I, An oniscidean, *Ligia* (the common seashore "rock louse"). (A,C,D,F,H drawn by F. Runyan; B,G from Brusca 1980; E,I from Brusca and Brusca 1978.)

lifestyle, usually in deep oceanic waters. There are many intertidal species, and a great many of these live in association with other invertebrates or seaweeds.

The suborder Hyperiidea includes exclusively pelagic amphipods that have apparently escaped the confines of benthic life by becoming associated with other plankters, particularly gelatinous zooplankton such as medusae, ctenophores, and salps (Figure 19G). The hyperiideans are usually characterized by huge eyes (and a few other inconsistent features), but several groups bear eyes no larger than those of most gammarideans. The precise nature of the relationships between hyperiideans and their zooplankton hosts remains controversial. Although some actually do eat host tissue, or kill the host to fashion a floating "home," many others may utilize the host merely for transport or as a nursery for newly hatched young. This fascinating problem was reviewed by Laval (1980).

There are two other small amphipod suborders: Ingolfiellidea and Caprellidea. The first contains only about 30 species, most of which live in freshwater and brackish-water caves and other subterranean waters; a few are marine and interstitial. Little is known about their biology. The caprellid amphipods are highly modified for clinging to other organisms, including filamentous algae and hydroids, and some

(family Cyamidae) are parasitic on whales and dolphins (Figures 19H and I).

In addition to parasitism, amphipods exhibit a vast array of feeding strategies, including scavenging, herbivory, carnivory, and suspension feeding.

The crustacean *Bauplan*

We realize that the above synopses are rather extensive, but the diversity of the crustaceans demands emphasis before attempting to generalize about their biology. While reading the following sections, you should keep in mind that there are many exceptions to the examples given. As in other arthropods, crustacean success has been closely tied to modifications of the jointed exoskeleton and appendages. They have exploited an evolutionary flexibility that has allowed an extensive range of modification of these body parts for a great variety of functions.

The most basic crustacean body plan is that of a head (**cephalon**) followed by a long body (a **trunk**) with many similar appendages, as seen in the presumed primitive class Remipedia (Figures 1A and 3C). In the other crustacean classes, however, various degrees of tagmosis occur, and the cephalon is typically followed by a trunk that is divided into two distinct regions, a **thorax** and an **abdomen**. All crustaceans possess, at least primitively, a **cephalic shield** (**head shield**) or a **carapace**. The cephalic shield results from the fusion of the dorsal head tergites to form a solid cuticular plate, often with ventrolateral folds ("pleural folds") on the sides. Head shields are characteristic of the classes Remipedia and Cephalocarida, and they also occur in some branchiopods, maxillopodans and malacostracans. The carapace is a more expansive structure, comprising the head shield and a large fold of the exoskeleton that arises (primitively) from the maxillary somite. The carapace may extend over the body dorsally and laterally, as well as posteriorly, and it often fuses to one or more thoracic segments, thereby producing a **cephalothorax** (Figure 2). In addition, the carapace may grow forward beyond the head as a **rostrum**.

Most of the differences among the major groups of crustaceans, and the basis for much of their classification, arise from variations in the number of somites in the thorax and abdomen, the form of their appendages, and the size and shape of the carapace. A brief skimming of the synopses and corresponding figures will give you some idea of the range of variability in these characteristics.

Uniformity within the subphylum Crustacea is

H   I

Figure 19

And still more peracarids: amphipod diversity. A, General anatomy of a hyperiidean amphipod. B, General anatomy of a gammaridean amphipod. C–D, Two gammarideans: *Hyale*, a beach hopper (C), and *Heterophlias*, an unusual, dorsoventrally flattened amphipod (D). E–G, Three hyperiideans: *Primno* (E); *Leptocotis* (F); a hyperiid on its host medusa (G). H, A caprellid, *Caprella*. I, A caprellid, *Cyamus* (family Cyamidae), that is parasitic on whales. J, An ingolfiellid, *Ingolfiella*. (B,F after Bowman and Gruner 1973; C,D after Barnard 1969; E after Bowman 1978; I after Laval 1972; J after Delamare-Deboutteville 1960.)

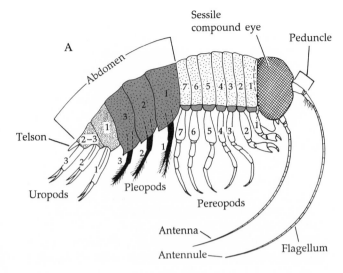

evidenced particularly by the elements of the cephalon and by the primitively biramous appendages, the latter explained in Chapter 15. Except for a few cases of secondary reduction, the head of any crustacean always has five pairs of appendages. From anterior to posterior, these are **antennules** (first antennae); **antennae** (second antennae); multiarticulate gnathobasic **mandibles**; **maxillules** (first maxillae); and **maxillae** (second maxillae). The presence of two pairs of antennae is, among arthropods, unique to the crustaceans, as is the nauplius larva and, in many, stalked compound eyes. A few crustaceans are blind, but most have eyes. Although the eyes of some are simple, most possess a pair of well developed compound eyes, either set directly on the head (**sessile eyes**) or borne on distinct movable stalks.

In many crustaceans (e.g., members of the classes Remipedia, Maxillopoda, and Malacostraca) from one to three anterior thoracic segments (**thoracomeres**) are fused with the cephalon. The appendages of these fused segments are typically incorporated into the head as additional mouthparts called **maxillipeds**. In the class Malacostraca, the remaining free thoracomeres are together termed the **pereon**, each of its segments being a **pereomere** or **pereonite**, and their appendages being **pereopods**. The pereopods may be specialized for walking, swimming, gas exchange, feeding, and defense. Crustacean thoracic (and pleonal) appendages are probably primitively biramous, although a "reduction" to a uniramous condition is seen in a variety of taxa. As described in Chapter 15, the basic crustacean limb is composed of a basal **protopod** (= **sympod**), from which may arise medial endites (e.g., gnathobases), lateral exites (e.g., epipods), and two rami, the **endopod** and **exopod**. Members of the classes Remipedia, Cephalocarida, Branchiopoda, and some ostracods possess appendages with uniarticulate (single-segment) protopods; the remaining classes (most Maxillopoda and Malacostraca) usually have appendages with multiarticulate protopods (see Table 1). The term **peduncle** is a general name occasionally applied to the basal

portion of certain appendages; it is occasionally (but not always) used as a synonym of protopod.

The abdomen, called a **pleon** in malacostracans, is composed of several segments (**pleomeres** or **pleonites**), followed by a postsegmental plate or lobe, the **anal somite** or **telson**, bearing the anus (Figure 2). In primitive crustaceans the anal somite bears a pair of appendage-like or spinelike processes conventionally called **caudal rami** (see, for example, Figure 3A). In the Eumalacostraca the anal somite is a flattened telson, which may bear terminal lobes but never caudal rami.

In general, distinctive abdominal appendages (**pleopods**) occur only in the malacostracans. These appendages are almost always biramous, and often they are flaplike, setose, and used for swimming (Figure 2). The posteriorly directed last pair(s) of abdominal appendages are usually different from the pleopods; these are called **uropods**. Together with the telson, the uropods form a distinct **tail fan** in many malacostracans (Figure 2).

Crustaceans produce a characteristic larval stage called the **nauplius larva** (Figure 32D), which bears a median simple (naupliar) eye and three pairs of appendages (destined to become the antennules, antennae, and mandibles). In many groups, however, the free-living nauplius stage is lost or suppressed. In such cases, development is either fully direct or mixed, with larval hatching taking place at some postnaupliar stage (Table 2). Often other larval stages follow the nauplius or other hatching stage as the individual passes through a series of molts during which segments and appendages are gradually added.

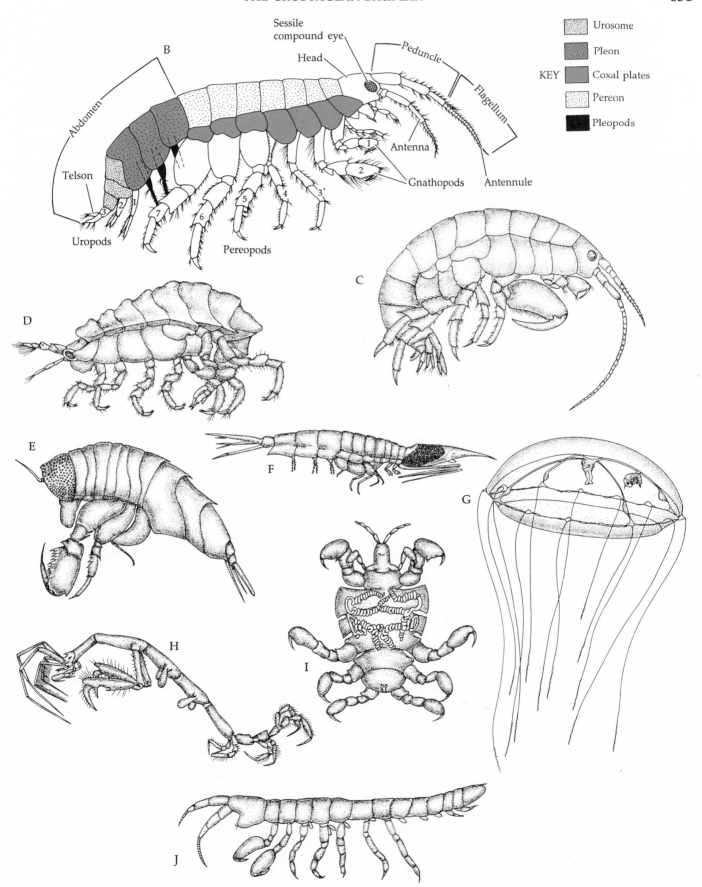

Locomotion

Crustaceans move about primarily by use of their limbs; lateral body undulations are unknown. They crawl or swim, or more rarely burrow, "hitch-hike," or jump. Many of the ectoparasitic forms (e.g., certain isopods and copepods) are sedentary on their hosts, and most cirripedes are fully sessile.

Swimming is usually accomplished by a rowing action of the limbs. Archetypical swimming is exemplified by crustaceans with relatively undifferentiated trunks and high numbers of similar biramous appendages (e.g., remipedes, anostracans, and notostracans). In general, these animals swim by posterior to anterior metachronal beating of the trunk limbs (see Figure 6 in Chapter 15, and Figures 21A, B, and F in this chapter). The appendages in such crustaceans are often broad and flattened, and usually bear fringes of setae that increase the effectiveness of the power stroke. On the recovery stroke the limbs are flexed and the setae may collapse, effectively reducing resistance. In members of some groups (e.g., Cephalocarida, Branchiopoda, and Leptostraca), large exites or epipods arise from the base of the leg, producing broad, "leafy" limbs called **phyllopodia**. These flaplike structures aid in locomotion and may also serve as gas exchange surfaces (Figures 20A, B, and C). Although such epipods increase the surface area on the power stroke, they also are hinged so that they collapse on the recovery stroke, reducing resistance. Metachronal limb movements are retained in many of the "higher" swimming crustaceans, but they tend to be restricted to selected appendages (e.g., the pleopods of shrimps, stomatopods, amphipods, and isopods; the pereopods of euphausiaceans and mysids). In swimming euphausiaceans and mysids, the thoracopods beat in a metachronal rowing fashion, with the exopod and setal fan extended on the power stroke and flexed on the recovery stroke. The movements and nervous-muscular coordination of crustacean limbs is deceivingly complex. In the common mysid *Gnathophausia ingens*, for example, 12 separate muscles power the thoracic exopod alone (three that are extrinsic to the exopod, five in the limb peduncle, and four in the exopodal flagellum).

Recall from our discussions in Chapters 3 and 15 that at the low Reynolds numbers small crustaceans (such as copepods or larvae) swim about in, the net-like setal appendages act not as a filtering net but as a paddle, pushing water in front of them and dragging the surrounding water along with them due to the thick boundary layer adhering to the limb. Only in larger organisms, with Reynolds numbers ap-

Figure 20

Generalized thoracic appendages of various crustaceans. A, Cephalocarida. B, Branchiopoda. Dashed lines indicate fold or "hinge" lines. C, Leptostraca (Phyllocarida). D, Remipedia. E, Euphausiacea. F, Caridea (Decapoda). G, Lophogastrida (Peracarida). H, Spelaeogriphacea (Peracarida). I, Isopoda (Peracarida). A–C are biramous, phyllopodous thoracopods; D is a biramous, flattened, but nonphyllopodous thoracopod; E–I are stenopodous thoracopods.

Because of the presence of large epipods on the legs of the cephalocarids, branchiopods, and phyllocarids, some authors refer to them as "triramous" appendages. However, smaller epipods also occur on many typical "biramous" legs, so this distinction seems unwarranted (and confusing). Note that in the four presumably primitive groups of crustaceans (cephalocarids, branchiopods, phyllocarids, and remipedes) the protopod is composed of a single article, whereas in all other crustaceans it consists of two or three articles. And in branchiopods and leptostracans, the articles of the endopod are not clearly separated from one another. In the higher crustaceans (classes Maxillipoda and Malacostraca), the protopod comprises 2–3 separate articles, although in most maxillopodans these may be reduced and not easily observed. In the lophogastrid (G), the large marsupial oostegite characteristic of female peracarids is shown arising from the coxa. In two groups (amphipods and isopods) all traces of the exopods have disappeared, and only the endopod remains as a long, powerful, uniramous walking leg. (Adapted from various sources.)

proaching 1, do setose appendages begin to act like filters, or rakes, as the surrounding water is less viscous and the boundary layer thinner. Of course, the closer together the setae and setules are placed, the more likely it is that their individual boundary layers will overlap; thus densely setose appendages are more likely to act as paddles.

Not all swimming crustaceans move by typical metachronal waves of limb action. For example, certain planktonic copepods move haltingly and otherwise depend on their long antennules and dense setation for flotation (Figure 6F). Watch living calanoid copepods and you will notice that they may move slowly by use of the antennae and other appendages, or in short jerky increments, often sinking slightly between these movements. The jerky forward motion results from an extremely rapid and condensed metachronal wave of power strokes along the trunk limbs. Although the long antennae may appear to be acting as paddles, they actually collapse against the body an instant prior to the beating of the limbs, thus reducing resistance to forward motion. Some other planktonic copepods create swimming currents by rapid vibrations of cephalic appendages, and the body moves smoothly through the water. Exceptions to "rowing" occur in the swimming crabs (family Portunidae) and some deep-sea asellote isopods (e.g., family Eurycopidae), which use paddle-shaped posterior thoracopods to scull about.

Most eumalacostracans with well developed ab-

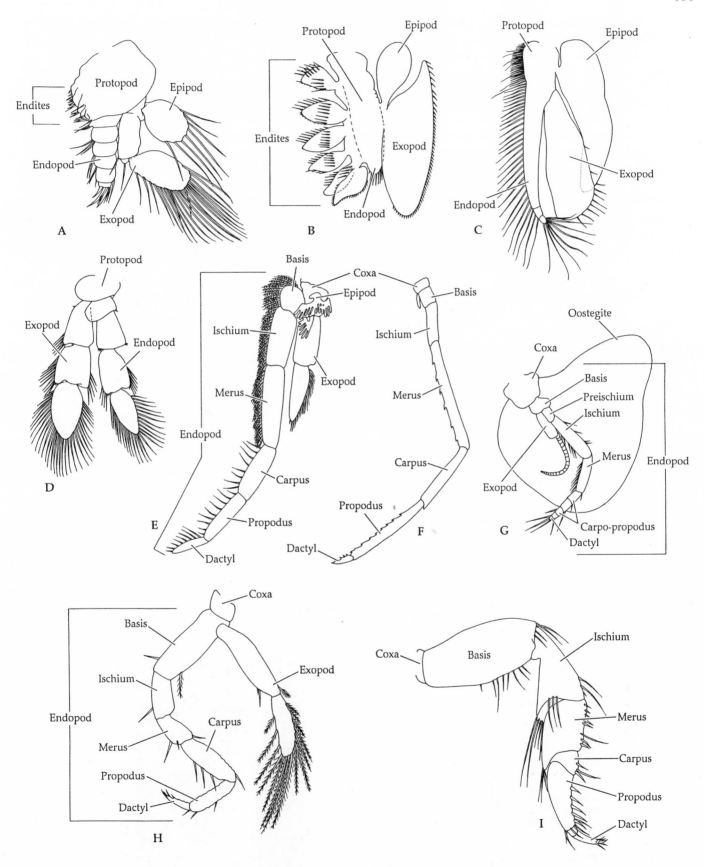

domens exhibit a form of temporary, or "burst," swimming that serves as an escape reaction (e.g., mysids, syncarids, euphausiaceans, shrimps, lobsters, and crayfish). By rapidly contracting the ventral abdominal (flexor) muscles, such animals shoot quickly backward, the spread tail fan providing a large propulsive surface (Figures 21C, D, and E). This behavior is called the "caridoid escape reaction."

Surface crawling by crustaceans is accomplished by the same general sorts of leg movements described in the preceding chapters for insects and other arthropods—by flexion and extension of the limbs to pull or push the animal forward. Walking limbs are typically composed of relatively stout, more-or-less cylindrical articles (i.e., **stenopodous** limbs) as opposed to the broader, often phyllopodous limbs of swimmers (see Figure 20 for a comparison of crustacean limb types). Walking limbs are lifted from the substratum and moved forward during their recovery strokes; then they are placed against the substratum, which provides purchase as they move posteriorly through their power strokes—pulling and then pushing the animal forward. As in many other arthropods, crustaceans generally lack lateral flexibility

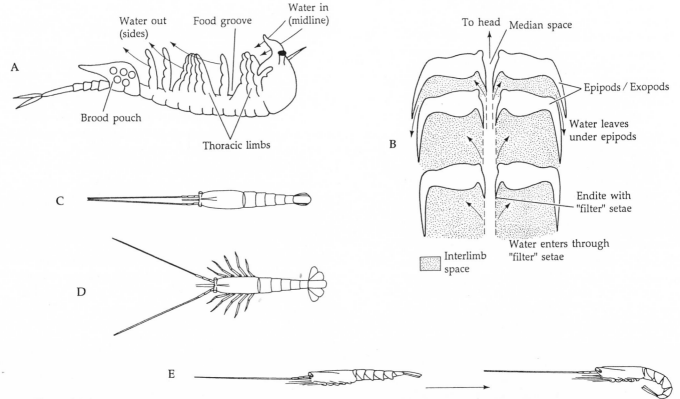

Figure 21

Aspects of locomotion (and feeding) in three crustaceans (see also Chapter 15). A–B, Generation of swimming and feeding currents in an anostracan. A, An anostracan swimming on its back by metachronal beating of the trunk limbs. The limbs are hinged to fold on the recovery stroke, thereby reducing resistance. Water is drawn into the interlimb spaces (B), and food particles are trapped on the medial sides of the endites; excess water is pressed out laterally and the trapped food is moved anteriorly to the mouth. C–E, Locomotion in the postlarva of the spiny lobster, *Panulirus argus*. C, Normal swimming posture when moving forward slowly. D, Sinking posture with appendages flared to reduce sinking rate. E, A quick retreat by rapid tail flexure (the "caridoid escape reaction"), a method commonly employed by crustaceans with well developed abdomens and tail fans. F, A swimming remipede, *Lasionectes*. Note the metachronal waves of appendage movement. (A,B modified from Russell-Hunter 1979, after various sources; C–E after Calinski and Lyons 1983; F photo by D. Williams, courtesy of J. Yager.)

at the body joints, so turning is accomplished by reducing the stride length or movement frequency on one side of the body, toward which the animal turns (like slowing one tread on a tractor or tank).

Most walking crustaceans can also reverse the direction of leg action and move backward, and most brachyuran crabs can walk sideways. Brachyuran crabs are perhaps the most agile of all crustaceans. The extreme reduction of the abdomen in this group allows for very rapid movement because adjacent limbs can move in directions that avoid interference with one another. Brachyuran crab legs are hinged in such a way that most of their motion involves lateral extension (abduction) and medial flexion (adduction) rather than anterior-posterior rotation. As a crab moves, the limbs move in various sequences, as in normal crawling, but those on the leading side exert their force by flexing and pulling the body toward the limb tips, while the opposite, trailing, legs exert propulsive force as they extend and push the body away from the tips. Still, this is simply a mechanical variation on the common arthropod walking behavior.

In addition to these basic locomotor methods ("typical" walking, and swimming by metachronal beating of limbs), many crustaceans move by other specialized means. Ostracods, cladocerans, and conchostracans, which are almost entirely enclosed by their carapaces (Figures 4E, 4I, and 5), swim by rowing with the antennae; mystacocarids crawl in interstitial water using various head appendages; certain semiterrestrial amphipods known as "beach hoppers" (e.g., *Orchestia* and *Orchestoidea*) execute dramatic jumps by rapidly extending the urosome and its appendages (uropods); most caprellid amphipods (Figure 19H) move about in inchworm fashion, using their subchelate appendages for clinging. There are also a number of crustacean burrowers and even some that build their own tubes or "homes" from material in their surroundings. Many benthic amphipods, for example, spin silk-lined mud burrows in which they reside. One species, *Pseudamphithoides incurvaria* (suborder Gammaridea) constructs and lives in an unusual "bivalved pod" cut from the thin blades of the same alga on which it feeds (Figure 22A). Another gammaridean amphipod, *Photis conchicola*, actually uses empty gastropod shells in a fashion similar to that of hermit crabs (Figure 22B). "Hitchhiking" (phoresis) occurs in various ectosymbiotic crustaceans, including isopods that parasitize fishes or shrimps, and hyperiidean amphipods that ride on gelatinous drifting plankters.

All in all, the evolutionary ability to specialize body regions and appendages has provided crustaceans with a repertoire of locomotor capabilities more diverse than that of any other group of arthropods; and this ability has no doubt enhanced their tremendous success. In addition to simply getting from one place to another in their usual day-to-day activities, many crustaceans exhibit various migratory behaviors, employing their locomotor skills to avoid stressful situations or to remain in areas of optimal conditions. A number of planktonic crustaceans undergo daily vertical migrations, typically moving upward at night and to greater depths during the day. Such vertical migrators include various copepods, ostracods, and hyperiid amphipods (the last migrating perhaps by riding on their hosts). Such movements place the animals in the near-surface feeding grounds during the dark hours, when there is probably less danger of them being detected by visual predators. In daytime, they move to deeper, perhaps "safer," water. Many intertidal errant crustaceans routinely use their abilities to move to change their behaviors during high and low tides. Some anomuran and brachyuran crabs simply move in and out with the tide, or they seek shelter beneath rocks when the tide is out, thus avoiding problems of exposure.

One of the most interesting locomotor behaviors among crustaceans is the mass migration of the spiny lobster, *Panulirus argus*, in the Gulf of Mexico and north Caribbean. Each autumn lobsters queue up in single file to march in long lines for several days. They move from shallow areas to the edge of deeper oceanic channels. This behavior is apparently triggered by Arctic storm fronts moving into the area, and it may be an innate behavior for avoiding rough water conditions in the shallows.

Feeding

With the exception of ciliary mechanisms, crustaceans have exploited virtually every feeding strategy imaginable. Even without cilia, many crustaceans generate water currents and engage in various types of suspension feeding. We have selected a few examples to demonstrate the range of mechanisms that occur in this group.

In some crustaceans the locomotor action of the thoracic limbs simultaneously creates the suspension feeding currents (Figures 21A and B). As the metachronal wave of appendage motion passes along the body adjacent limb pairs are alternately moved apart and then pressed together, thus changing the size of each interlimb space (see Figure 13E in Chapter 3 and Figure 6 in Chapter 15). Surrounding water is drawn into an interlimb space as the adjacent appendages move away from one another, and water-borne particles are trapped by setae on the endites. From here,

A

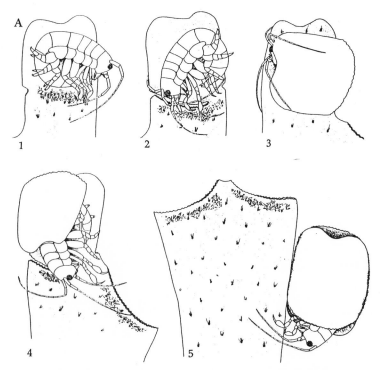

1   2   3

4   5

Figure 22

Amphipod "houses." A, The complex sequence of steps in construction of a "bivalve pod" from the brown alga *Dictyota*, by the Caribbean amphipod *Pseudampithoides incurvaria*: (1) initiation of cut and notch; the upper flap of the alga forms the first "valve"; (2) continuation of the cut across the algal thallus; (3) measuring and clearing algal hairs off the second branch tip; (4) cutting of the second valve; (5) completed "pod" with valves attached along margins by threadlike secretions. (6) *P. incurvaria* in its algal "pod" (scanning electron micrograph). The antennules and antennae are protruding from the pod entrance on the right. B, *Photis conchicola*, a temperate, eastern Pacific amphipod that spins its silken tube inside minute snail shells, which are then carried about in the style of hermit crabs. (A from Lewis and Kensley 1982; B from Carter 1982.)

Figure 22

Amphipod "houses." A, The complex sequence of steps in construction of a "bivalve pod" from the brown alga *Dictyota*, by the Caribbean amphipod *Pseudampithoides incurvaria*: (1) initiation of cut and notch; the upper flap of the alga forms the first "valve"; (2) continuation of the cut across the algal thallus; (3) measuring and clearing algal hairs off the second branch tip; (4) cutting of the second valve; (5) completed "pod" with valves attached along margins by threadlike secretions. (6) *P. incurvaria* in its algal "pod" (scanning electron micrograph). The antennules and antennae are protruding from the pod entrance on the right. B, *Photis conchicola*, a temperate, eastern Pacific amphipod that spins its silken tube inside minute snail shells, which are then carried about in the style of hermit crabs. (A from Lewis and Kensley 1982; B from Carter 1982.)

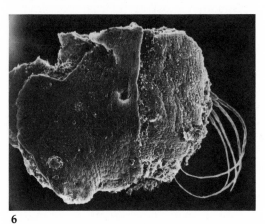

6

the trapped particles are moved to a midventral food groove and then anteriorly, toward the head. This mechanism of forming a boxlike "filter" press with setose phyllopodous limbs is the typical suspension feeding strategy of cephalocarids and phyllocarid malacostracans.

Planktonic copepods were long thought to "filter" feed by generating lateral feeding gyres or currents by locomotor movements of the antennae and mouth appendages. It was believed that these gyres swept in small particles that were directly filtered by the maxillae. This classic idea of maxillary filtration, built on work by H. G. Cannon in the 1920s, has subsequently been questioned by several workers (e.g., Fryer, Gurney, Koehl, Strickler); but the model persists and is still commonly presented in general books. As mentioned in Chapter 3, we now know that copepods and other small planktonic crustaceans live in a world of low Reynolds numbers, a world dominated by viscosity rather than by inertia, which is the force that dominates the world of larger animals. Movements of the antennae and mouth appendages establish patterns of water flow past the body of a copepod, and the second maxillae capture food particles without actually "touching" them. The setae on the limbs behave more like paddles than sieves, with a water layer near the limb adhering to it and forming part of the "paddle." As the maxillae move apart, parcels of water containing food are drawn

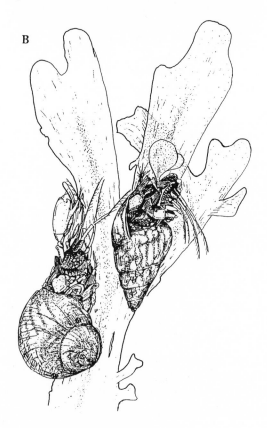

B

into the interlimb space. As the maxillae press together, the "package" is moved forward to the endites of the maxillules, which push it into the mouth. Thus, food particles are not actually filtered from the water but are captured in small packets of water (e.g., Koehl and Strickler 1981). High-speed cinematography indicates that copepods may capture individual algal cells, one at a time, by this "vacuum" method.

Sessile thoracican barnacles feed by using their long, feathery, biramous thoracopods, called **cirri**, to remove suspended material from the surrounding water (Figures 7A, C, and E). Studies indicate that barnacles are capable of trapping food particles ranging from 2 μm to 1 mm, including detritus, bacteria, algae, and various zooplankters. Many barnacles are also capable of preying on larger planktonic animals by coiling a single cirrus around the prey, in tentacle fashion. In moderately moving or very quiet water, most barnacles actively feed by extending the last three pairs of cirri, fanlike, and sweeping them rhythmically through the water. The setae on adjacent limbs and limb rami overlap to form an effective filtering net. The first three pairs of cirri serve to remove trapped food from the posterior cirri and pass it to the mouthparts. In areas of high-water movements, such as wave-swept rocky shores, barnacles often extend their cirri into the backwash of waves, allowing the moving water to simply run through the "filter," rather than moving the cirri through the water. In such areas you will often see clusters of barnacles in which all the individuals are oriented similarly, taking advantage of this labor-saving device.

Most krill (euphausiaceans) feed while swimming. The thoracopods form a "feeding basket" that expands as the legs move outward, sucking food-laden water in from the front. Once inside the "basket," particles are retained on the setae of the legs as the water is squeezed out laterally. Other setae comb the food particles out of the "trap" setae, while yet another set brushes them forward to the mouth region.

Sand crabs of the genus *Emerita* (Anomura) use their long setose antennae in a fashion similar to that of barnacle cirri that "passively" strain wave backwash (see Figure 13B in Chapter 3 and Figure 23A in this chapter). *Emerita* is adapted to living on wave-swept sand beaches. Their compact oval shape and strong appendages facilitate burrowing in the unstable substratum. They burrow posterior-end first in the area of shallow wave wash, with their anterior ends facing upward. Following a breaking wave, as the water rushes seaward, *Emerita* unfurls its antennae into the moving water along the surface of the sand. The fine setose mesh traps bacteria, protozoa, and phytoplankton from the water, and then the antennae brush the collected food onto the mouthparts. Many porcelain and hermit crabs also engage in suspension feeding. By twirling their antennae in various patterns (Figures 23B and C) these anomurans create spiraling currents that bring food-laden water toward the mouth. Food particles become entangled on the setae near the base of the antennae and then are brushed into the mouth by the endopods of the third maxillipeds. Many of these animals also feed on detritus by simply picking up particles with their chelipeds.

Mud and ghost shrimps, such as *Callianassa* and *Upogebia* (Thalassinidea), suspension feed within their burrows. Water is driven through the burrow by beating the pleopods, and the first two pairs of pereopods remove food with medially directed setal brushes. The maxillipeds then comb the captured particles forward to the mouth.

Most of the other crustacean feeding mechanisms are less complicated than suspension feeding and usually involve direct manipulation of food by the mouthparts and sometimes the pereopods, especially chelate or subchelate anterior legs.

Many small crustaceans may be classified as microphagous selective deposit feeders, removing food from the sediments in which they live. Mystacocarids, many harpacticoid copepods, and some cumaceans and gammaridean amphipods are referred to as "sand grazers" or "sand lickers." By various methods these animals remove detritus, diatoms, and other microorganisms from the surfaces of sediment particles. For example, interstitial mystacocarids simply brush sand grains with their setose mouthparts. On the other hand, some cumaceans pick up an individual sand grain with their first pereopods and pass it to the maxillipeds, which in turn rotate and tumble the particle against the margins of the maxillules and mandibles. The maxillules brush and the mandibles scrape, removing organic material. Some sand-dwelling isopods may utilize a similar feeding behavior.

Predatory crustaceans include stomatopods, probably remipedes, and most lophogastrids, as well as many species of anostracans, cladocerans, copepods, ostracods, cirripedes, anaspidaceans, euphausiaceans, decapods, tanaids, isopods, and amphipods. Predation typically involves grasping the prey with chelate or subchelate pereopods (or sometimes directly with the mouth appendages) followed by tearing, grinding, or shearing with various mouth-

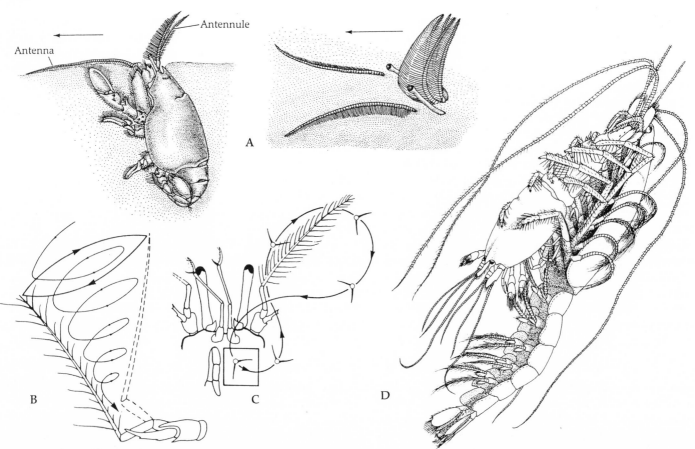

Figure 23

Some crustacean feeding mechanisms. See also Figure 21. A, Suspension feeding in the sand crab *Emerita*. The arrows point seaward and indicate the direction of water movement as waves recede. The antennules are held erect and direct water through branchial chambers. The antennae serve in feeding by removing food particles from the water and then brushing them onto the mouthparts. B–C, The suspension-feeding hermit crabs, *Australeremus cooki* (B) and *Paguristes pilosus* (C) twirl their antennae, either in a circle or a figure eight, and create water currents that pull food particles to the mouth region. D, The predatory shrimp *Procaris ascensionis* is shown here munching on another shrimp (*Typhlatya*) as it holds the prey in a "cage" formed by the pereopodal endopods. (A after Barnes 1980; B,C from Schembri 1982; D from Abele and Felgenhauer 1985.)

parts, particularly the mandibles. Perhaps the most clearly adapted predatory specialists are the stomatopods (Figures 11A and B; also see Figure 20D in Chapter 3). These hunters or ambushers possess greatly enlarged, raptorial subchelate limbs used to stab or to club and smash prey. Some species search out prey, but many sit in ambush at their burrow entrance. The actual attack generally follows visual detection of potential prey items, including other crustaceans, molluscs, and even small fishes. Once captured and stunned or killed by the raptorial claws, the prey is held against the mouthparts and shredded to ingestible-sized pieces.

Although the cave-dwelling caridean shrimp *Procaris* is omnivorous, its predatory behavior is particularly interesting. Its prey includes other crustaceans, in particular amphipods and shrimps. After *Procaris* locates a potential victim (probably by chemoreception), it moves quickly to the prey and grasps it within a "cage" formed by the pereopodal endopods (Figure 23D). Once captured, the prey is eaten while the shrimp swims about. Apparently the third maxillipeds press the prey against the mandibles, which bite off chunks and pass them to the mouth.

The remipedes may capture prey with their raptorial mouth appendages (Figure 3E), and then immobilize the victim with an injection from the hypodermic maxillules. It is suspected that tissues are then sucked out of the prey through a mandibular mill by the remipede's muscular foregut.

Another fascinating adaptation for predation is seen in the snapping shrimp *Alpheus* (Figure 13D). Generally, one of the chelipeds is much larger than the other, and the movable finger is hinged in such a way that it can be snapped closed; this forceful closing produces a loud popping sound and a pressure or "shock" wave in the surrounding water. At least some species (e.g., *A. californiensis*) use this mechanism in ambushing prey. The shrimp sits at its burrow entrance with its antennae extended. When a potential prey approaches (usually a small fish), the shrimp "pops" its cheliped and the resulting pressure wave stuns the victim, which is then quickly pulled into the burrow, killed, and consumed. *Alpheus* usually lives in pairs within the burrow, and prey captured by one individual is shared with the cohabitant.

Macrophagous herbivorous and scavenging crustaceans generally feed by simply hanging on to the food source and biting off bits with their mandibles (similar to the feeding technique of grasshoppers). Notostracans, some ostracods, and many decapods, isopods and amphipods are scavengers and herbivores. A number of crustaceans are full or part-time detritivores, including some or most cephalocarids, conchostracans, some ostracods, bathynellaceans, thermosbaenaceans; many peracarids; and some decapods. Some scavenge directly on detritus, but others (e.g., cephalocarids) stir up the sediment to remove organic particles by suspension feeding.

Finally, several groups of crustaceans have adopted various degrees of parasitism. These rela-

tionships vary from the highly modified and fully parasitic rhizocephalan cirripedes, whose bodies ramify throughout the host tissue and absorb nutrients directly (Figures 7H, 24, and 25), to various ectoparasites with mouthparts modified for piercing and sucking body fluids (e.g., many copepods, branchiurans, tantulocarids, and epidaridean isopods).

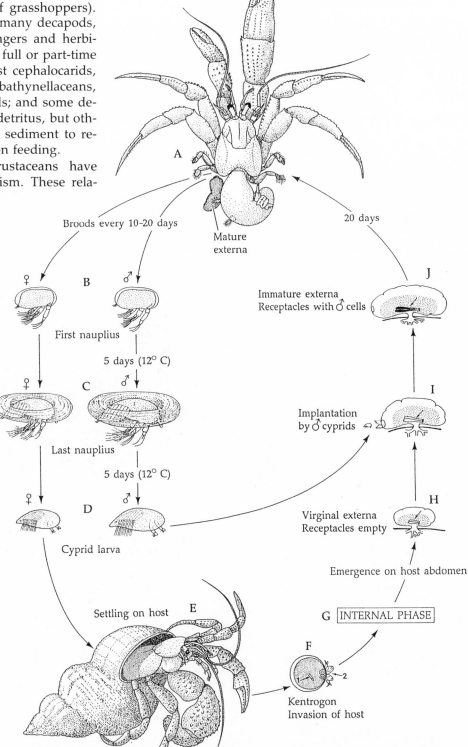

Figure 24

The remarkable life cycle of the rhizocephalan cirripede *Peltogaster paguri*, a parasite of hermit crabs. A, Mature reproductive portion of parasite (externa) produces numerous broods of male and female larvae, which are released as nauplii (B–C) and eventually metamorphose into cyprid larvae (D). Female cyprids settle on the thorax and limbs of host crabs (E) and undergo a major internal metamorphosis into the kentrogon form (F), which is provided with a pair of antennules (2) and an injection stylet (1). Part of the kentrogon's viscera (the invasion cell) is transferred to the host through the hollow stylet, where it becomes a small internal form (the interna) (G). Eventually the female parasite emerges on the abdomen of the host as a virginal externa (H). When the externa acquires a mantle pore, or aperture, it becomes attractive to male cyprids (I). Male cyprids settle within the aperture, transform into a trichogon form and implant part of their body contents into the female's receptacles (J). The deposit proceeds to differentiate into spermatozoa, which fertilizes the eggs of the female. (From Høeg and Lutzen 1985.)

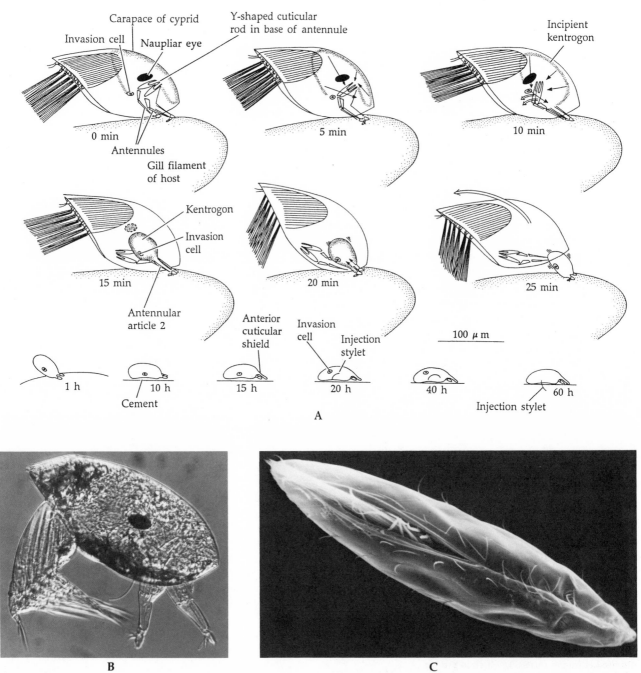

Figure 25

Various stages in the life cycle of the parasitic rhizocephalan barnacle *Lernaeodiscus porcellanae*. This parasite infects the anomuran crab *Petrolisthes cabrilloi* in southern California. See Figure 24 for a full description of a rhizocephalan life cycle. A, Settlement of the female cyprid larva on the host's gill filament and subsequent kentrogon formation and host invasion. The time after initial settlement is indicated. At 0 minutes the cyprid is shown walking (by means of its antennules) on the host's gill cuticle. After 5 minutes the cyprid is permanently attached. After 10 minutes the cyprid's epidermis (stippled outline) begins contracting inward to form the "incipient kentrogon." After 15 minutes formation of the kentrogon is complete and the old cyprid tissues begin to disintegrate. After 25 minutes the remnant of the old cyprid is cast off. After about 15 hours the kentrogon is firmly cemented to the host's gill, and after about 20 hours formation of the injection stylet commences. After 40 hours the stylet is fully formed, and after 60 hours the stylet has penetrated and injected the invasion cell into the host. B, A live cyprid. C, A cyprid (ventral view; scanning electron micrograph). D–E, Diagrams of cyprid larvae before and after settlement (right side of carapace removed; naupliar eye omitted); the dotted line in the second antennular article indicates the primordial kentrogon cuticle, and the placement of muscle fibers in the cyprid are indicated by hatched arrows with numbers; the muscles are hypothesized to effect formation of the kentrogon and separation of the old cyprid from the kentrogon. In E, kentrogon formation is complete. F, A whole gill (scanning electron micrograph) of the host crab *P. cabriolli* with several attached kentrogons (arrows). G, A 2-hour-old kentrogon (sagittal section). (From Høeg 1985.)

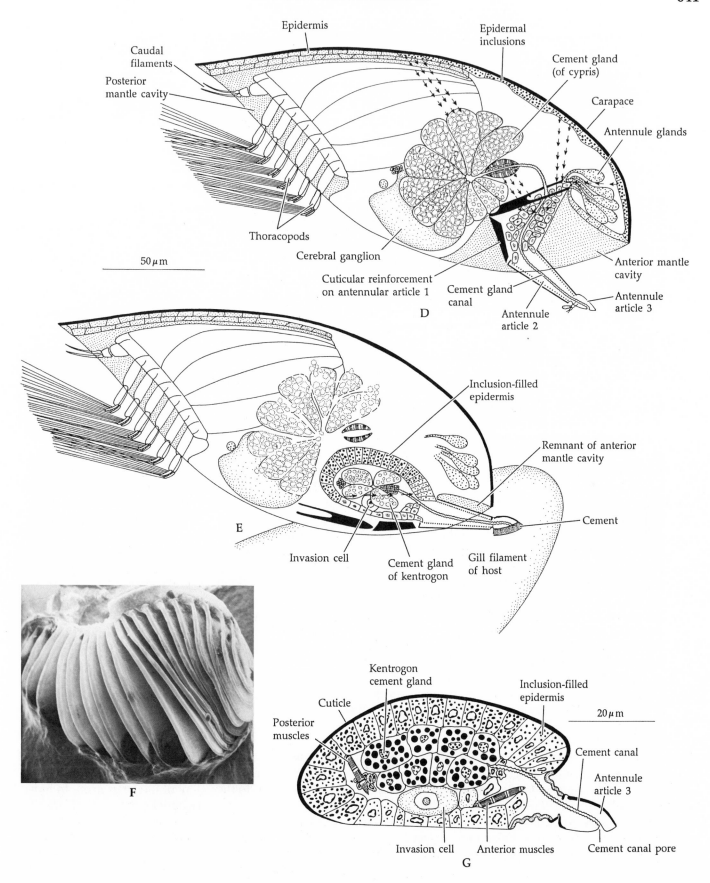

Caudal filaments

Posterior mantle cavity

Epidermis

Epidermal inclusions

Cement gland (of cypris)

Carapace

Antennule glands

Thoracopods

Cerebral ganglion

Cuticular reinforcement on antennular article 1

Cement gland canal

Antennule article 2

Anterior mantle cavity

Antennule article 3

50 μm

D

Inclusion-filled epidermis

Remnant of anterior mantle cavity

Cement

E

Invasion cell

Cement gland of kentrogon

Gill filament of host

F

Kentrogon cement gland

Inclusion-filled epidermis

20 μm

Cuticle

Posterior muscles

Cement canal

Antennule article 3

Invasion cell

Anterior muscles

Cement canal pore

G

Digestive system

The digestive system of crustaceans includes the usual arthropod foregut, midgut, and hindgut. The foregut and hindgut are lined with cuticle that is continuous with the exoskeleton and molted with it. The stomodeal foregut is modified in different groups, but usually includes a relatively short pharynx–esophagus region followed by a stomach. The stomach often has chambers or specialized regions for storage, grinding, and sorting; these structures are best developed in the malacostracans (Figure 26H). The midgut forms a short or long intestine—the length depending mainly on overall body shape and size—and bears variably placed digestive ceca. The ceca are serially arranged only in the remipedes. In some malacostracans, such as crabs, the ceca fuse to form a solid glandular mass (= digestive gland) within which are many branched, blind tubules. The hindgut is usually short, and the anus is generally borne on the anal somite or telson, or on the last segment of the abdomen (when the anal somite or telson is reduced or lost).

Examples of crustacean digestive tracts are shown in Figure 26. After ingestion, the food material is usually handled mechanically by the foregut. This may involve simply transporting the food to the midgut or, more commonly, processing the food in various ways prior to chemical digestion. For example, the complex foregut of decapods (Figure 26H) is divided into an anterior **cardiac stomach** and a posterior **pyloric stomach**. Food is stored in the enlarged portion of the cardiac stomach and then moved a bit at a time to a region containing a **gastric mill**, which usually bears heavily sclerotized teeth. Special muscles associated with the stomach wall move the teeth, grinding the food to smaller particles. The macerated material then moves into the back part of the pyloric stomach, where sets of filtering setae prevent large particles from entering the midgut. This type of foregut arrangement is most well developed in macrophagous decapods (scavengers, predators, and some herbivores). Thus the food can be taken in quickly, in big bites, and mechanically processed afterward.

Circulation and gas exchange

The basic crustacean circulatory system usually comprises a dorsal ostiate heart within a pericardial cavity, and variously developed vessels emptying to an open hemocoel (Figure 26). The heart is lacking in most ostracods, many copepods, and many cirripedes. In some groups the heart is replaced or supplemented by accessory pumping structures derived from muscular vessels.

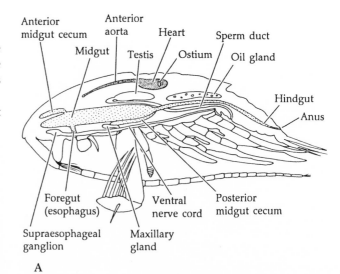

Figure 26

Internal anatomy of representative crustaceans. **A,** A calanoid copepod. **B,** A lepadomorph barnacle. **C,** A balanomorph barnacle. **D,** A stomatopod. **E,** A crayfish. **F,** An isopod. **G,** An amphipod. **H,** The stomach of a crayfish. **I,** A brachyuran crab. (A, D–G, I after McLaughlin 1980; B,C adapted from several sources; H after Kaestner 1980.)

The primitive heart structure in crustaceans is a long tube with segmental ostia, a condition retained in part in cephalocarids, and in some branchiopods, leptostracans, and stomatopods. However, the general shape of the heart and the number of ostia are also closely related to body form and location of gas exchange structures. The heart may be relatively long and tubular, and extend through much of the post-cephalic region of the body as it does in the remipedes, anostracans, and leptostracans; or it may tend toward a globular or box shape and be restricted to the thorax, in association with the thoracic gills (as in cladocerans, maxillopodans, and decapods). The intimate coevolution of the circulatory system in relation to body form and gill placement is best exemplified when comparing closely related groups. For instance, although isopods and amphipods are probably closely related to one another, the hearts are located largely in the pleon and pereon, respectively, corresponding to the pleopodal and pereopodal gill locations.

The number and length of blood vessels and the presence of accessory pumping organs are related to body size and to the extent of the heart itself. In most nonmalacostracans, for example, there are no arterial vessels at all. The heart pumps blood directly into the hemocoel from both ends. These animals tend to have short bodies, or long hearts, or both, an arrangement facilitating circulation of the blood to all

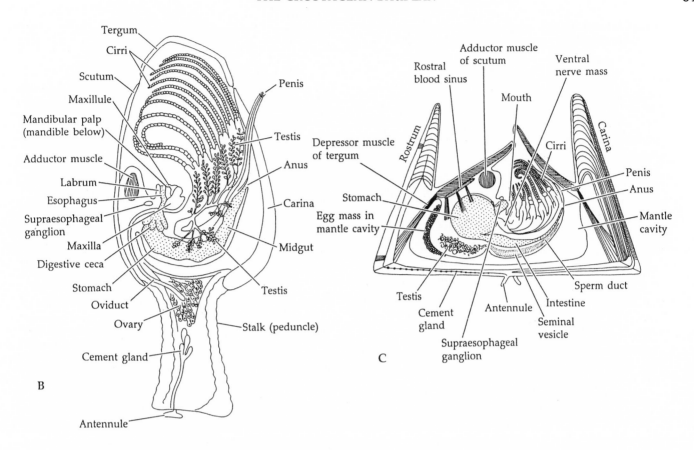

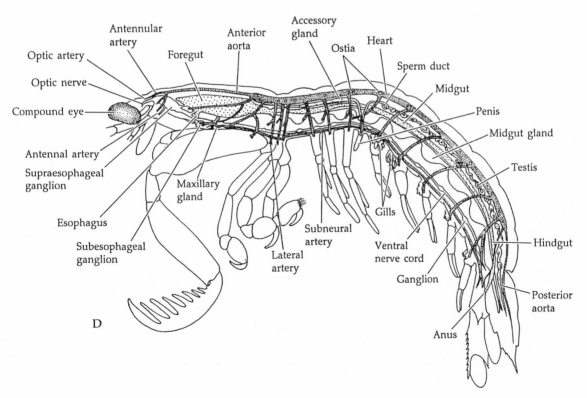

(Figure 26 continued on next page)

(Figure 26 continued from preceding page)

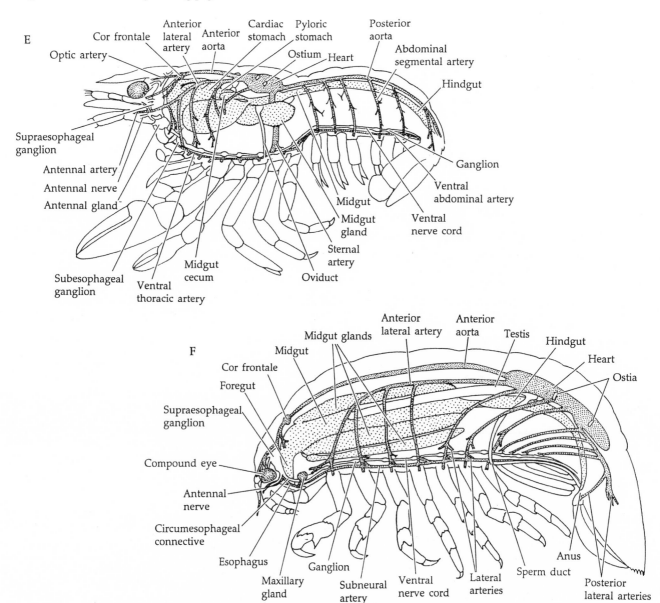

body parts. Sessile forms, such as most cirripedes, have lost the heart altogether, although it is replaced by a vessel "pump" in the thoracicans. Large malacostracans tend to have well developed vessel systems, thus ensuring that blood flows throughout the body and hemocoel and to the gas exchange structures (Figures 26D, E, and G). Large or active crustaceans may also possess an anterior accessory pump called the **cor frontale** (Figures 26E and F), which helps maintain blood pressure, and often a venous system for returning blood to the pericardial chamber.

Crustacean blood contains a variety of cell types, including phagocytic and granular amebocytes and special wandering **explosive cells** that release a clotting agent at sites of injury or autotomy. In nonmalacostracans, oxygen is either carried in solution or attached to dissolved hemoglobin; most malacostracans possess hemocyanin in solution (although some contain hemoglobin within tissues).

We have mentioned the form and position of gas exchange organs (gills) for some groups of crustaceans in the taxonomic synopses. Some small forms (e.g., copepods) lack distinct gills and rely on cuta-

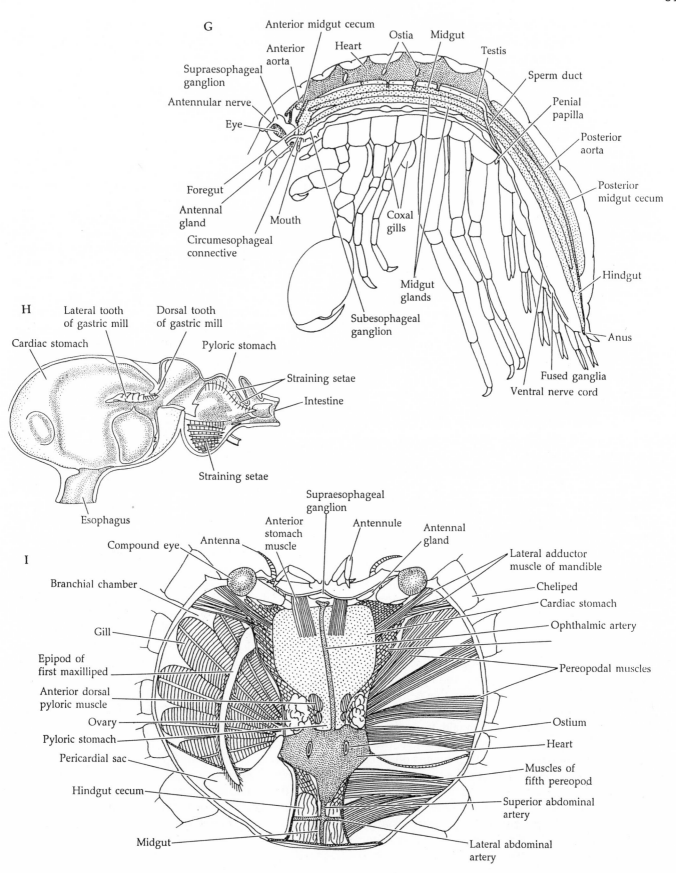

G

Anterior midgut cecum

Anterior aorta

Heart

Ostia

Midgut

Testis

Sperm duct

Supraesophageal ganglion

Antennular nerve

Eye

Penial papilla

Posterior aorta

Posterior midgut cecum

Foregut

Antennal gland

Circumesophageal connective

Mouth

Coxal gills

Midgut glands

Subesophageal ganglion

Hindgut

Anus

Fused ganglia

Ventral nerve cord

H

Lateral tooth of gastric mill

Dorsal tooth of gastric mill

Cardiac stomach

Pyloric stomach

Straining setae

Intestine

Straining setae

Esophagus

Supraesophageal ganglion

Anterior stomach muscle

Antennule

Antennal gland

Compound eye

Antenna

Lateral adductor muscle of mandible

I

Branchial chamber

Cheliped

Cardiac stomach

Ophthalmic artery

Gill

Epipod of first maxilliped

Anterior dorsal pyloric muscle

Pereopodal muscles

Ovary

Pyloric stomach

Ostium

Heart

Pericardial sac

Muscles of fifth pereopod

Hindgut cecum

Superior abdominal artery

Midgut

Lateral abdominal artery

neous exchange facilitated by their relatively thin cuticles and high surface area-to-volume ratio. In the small forms of other groups a thin, membranous inner lining of the carapace serves this purpose (e.g., Conchostraca, Cladocera, Cirripedia, Leptostraca, Cumacea, Mysida, and even some members of the Decapoda).

Most crustaceans, however, possess distinct gills of some sort (Figure 27). These structures are typically derived from thoracic epipods that have been modified in various ways to provide a large surface area to the environment. The inner hollow chambers or channels of these appendages are confluent with the hemocoel or supply vessels. Although their structure varies considerably (recall the various decapod gills described earlier), they all operate on the basic principles of gas exchange organs addressed in Chapter 3 and throughout this text. That is, the circulatory fluid is brought close to the oxygen source in an organ with a relatively high surface area. The gills provide a thin, moist, permeable surface between the internal and external environments. The gills of stomatopods (Figure 26D) and isopods (Figure 26F) are pleopodal. In the first case they are branched processes off the base of the appendages; but in the isopods the flattened pleopods themselves are vascularized and provide the necessary surface area for exchange. Stomatopods also have epipodal gills on the thoracopods, but these are highly reduced.

For gills to be efficient, a flow of water must be maintained across them. In stomatopods and aquatic isopods the current is generated by the beating of the pleopods. Similarly, the pereopodal gills of euphausiaceans are constantly flushed by water as the animal swims. However, in many crustaceans the gills are concealed to various degrees, and require special mechanisms in order to produce the ventilating currents. In many decapods, for example, the gills are contained in **branchial chambers** formed between the carapace and the sides of the body (Figure 27). While such an arrangement provides protection from damage to the fragile gill filaments, the openings to the chambers are generally small, and this restricts much passive flow of water. Not surprisingly, the solution to this dilemma comes once again from the evolutionary plasticity of crustacean appendages. Most decapods have elongate exopods on the maxillae, called **gill bailers** or **scaphagnathites**, that vibrate to create ventilating currents through the branchial chambers (Figure 27A). These currents typically enter from the sides and rear, through small openings around the coxae of the pereopods, and exit anteriorly from under the carapace in the vicinity of the mouth field (and antennal glands). This current can be easily seen by observing a crab or lobster in quiet water. The flow rate of this current can be altered, depending on environmental factors, and can also be reversed, thus allowing certain decapods to burrow in sand or mud with only their front ends exposed to the water.

Having the gills in branchial chambers protects them from desiccation and enables many crustaceans to live in intertidal regions, where they are frequently exposed to air. By avoiding direct exposure to very dry conditions, the branchial chambers always remain moist, so diffusion of respiratory gases continues even during low tides. Other decapods have invaded land, especially certain crayfish and the anomuran and brachyuran crabs known as "land crabs" (e.g., the hermit crab *Coenobita* and the coconut crab *Birgus*; Figure 1L). Again, the gills of these crustaceans are kept moist in their semienclosed chambers. Another striking decapod adaptation to life in air is displayed by the "sand-bubbler" crabs of the Indo-Pacific region (family Ocypodidae: *Scopimera*, *Dotilla*). These crabs possess on their legs or sternites membranous discs that were once thought to be auditory organs (tympana) but are now thought to function as gas exchange surfaces (Maitland 1986).

The most successful crustaceans on land are not the decapods, but the familiar sow bugs and pill bugs. The success of these oniscidean isopods (e.g., *Porcellio*) is due in large part to the presence of aerial gas exchange organs called **pseudotracheae** (Figures 27H and I). These organs are inwardly directed, moderately branched, thin-walled, blind sacs in some of the pleopodal exopods. These tubes connect to the outside via small pores. Air circulates through these sacs, and gases are exchanged with the blood in the pleopods. Thus, in these animals the original aquatic pleopodal gills have been refashioned for air breathing by moving the exchange surfaces inside, where they remain moist. The superficially similar tracheal systems of isopods, insects, and arachnids evolved independently, by convergence, in association with other adaptations to life on land.

Excretion and osmoregulation

Like other fundamentally aquatic invertebrates, crustaceans are ammonotelic, whether in fresh water or sea water, or on land. Ammonia is released both through nephridia and by way of the gills. As discussed in Chapter 15, most crustaceans possess nephridial excretory organs in the form of either **antennal glands*** or **maxillary glands** (Figures 4C, 7F, 9A,

---

*Antennal glands are sometimes called "green glands."

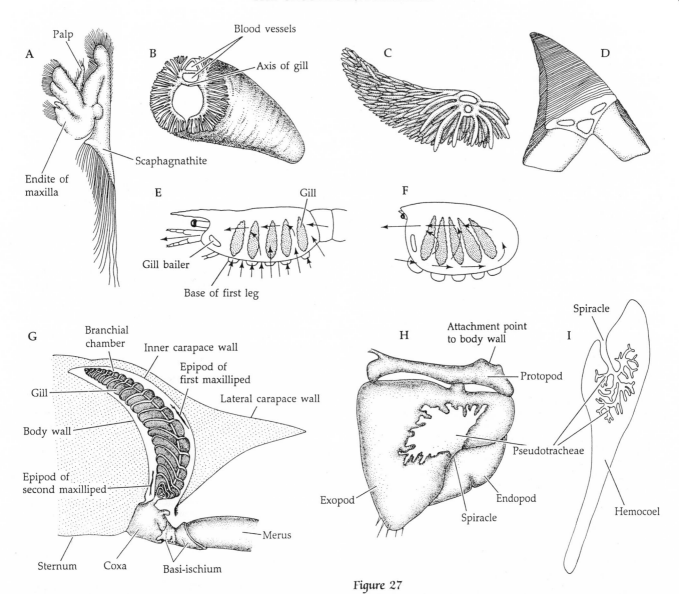

Figure 27

Gas exchange structures. A, Maxilla of the shrimp *Pandalus*. Note the setose scaphagnathite used to generate the ventilating current. B–D, Cross sections of types of decapod gills. B, Dendrobranchiate. C, Trichobranchiate. D, Phyllobranchiate. E–F, Paths of ventilating currents through the left branchial chambers of a shrimp (E) and a brachyuran crab (F). G, The branchial chamber (cross section) of a brachyuran crab showing the position of a single phyllobranchiate podobranch. H–I, A pleopod of the terrestrial isopod *Porcellio* (surface view and section). Note the pseudotracheae. (A–D after Meglitsch 1972; E–I after Kaestner 1980.)

and 26). These are serially homologous structures, constructed similarly but differing in the position of their associated pores (at the base of the second antennae or second maxillae, respectively). The inner blind end is a coelomic remnant of the nephridium called the **sacculus**, which leads through a variably coiled duct to the pore. The duct may bear an enlarged bladder near the opening (Figure 14A in Chapter 15).

Most crustaceans have only one pair of these nephridia, but lophogastrids and mysids have both antennal and maxillary glands, and a few others (cephalocarids and a few tanaids and isopods) have well developed maxillary and rudimentary antennal glands. Except for myodocopan ostracods (which

possess antennal glands), most nonmalacostracans have maxillary glands, as do stomatopods, cumaceans, and most tanaids. All of the other malacostracans have antennal glands.

Blood-filled channels of the hemocoel intermingle with branched extensions of the sacculus epithelium, creating a large surface area across which filtration occurs. The cells of the sacculus wall also actively take up and secrete material from the blood into the lumen of the excretory organ. These processes of filtration and secretion are to some degree selective, but most of the regulation of urine composition is accomplished by active exchange between the blood and the excretory tubule. These activities not only regulate the loss of metabolic wastes but are also extremely important in water and ion balance, particularly in freshwater and terrestrial crustaceans.

The excretion and osmoregulation carried out by antennal and maxillary gland activity are supplemented by other mechanisms. The cuticle itself acts as a barrier to exchange between the internal and external environments and, as we have mentioned, is especially important in preventing water loss on land or excessive uptake of water in fresh water. Moreover, thin areas of the cuticle, especially the gill

surfaces, serve as sites of waste loss and ionic exchange. Phagocytic blood cells and certain regions of the midgut are also thought to accumulate wastes. In some terrestrial isopods, ammonia actually diffuses from the body in gaseous form.

Nervous system and sense organs

The central nervous system of crustaceans is constructed in concert with the segmented *Bauplan*, along the same lines as we have seen in annelids and other arthropods (Figure 28). In the more primitive condition it is ladder-like, the segmental ganglia being largely separate and linked by transverse **commissures** and longitudinal **connectives** (Figure 28A). As described in Chapter 15, the crustacean brain is composed of three fused ganglia, the two anterior being the dorsal (supraesophageal) **protocerebrum** and **deutocerebrum**, which are thought to be preoral in origin. From the protocerebrum, **optic nerves** innervate the eyes. From the deutocerebrum, **antennulary nerves** run to the antennules, while smaller nerves innervate the eyestalk musculature. The third ganglion of the brain is the posterior **tritocerebrum**, which presumably represents the first postoral somite ganglion. The tritocerebrum forms a pair of circumenteric connectives that extend around the esophagus to a **subesophageal** or **subenteric ganglion** and link the brain with the ventral nerve cord

Figure 28

Central nervous systems of four crustaceans. A, The ladder-like system of an anostracan. Note the absence of well developed ganglia in the posterior, apodous, portion of the trunk. B, Elongate metameric system of a crayfish. C, Highly compacted system of a brachyuran crab, wherein all thoracic ganglia have fused and the abdominal ganglia are reduced. D, Nervous system of a hyperiid amphipod. Note the loss of the urosomal ganglia typical of amphipods. (A after Pennak 1978; B,C after Kaestner 1980; D after Brusca 1981.)

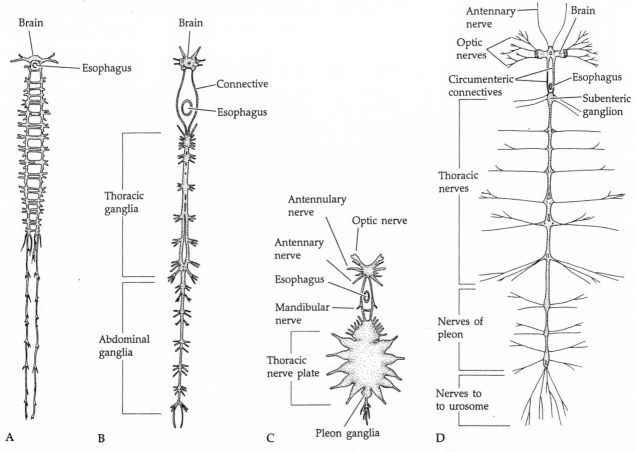

A   B   C   D

bearing the segmental body ganglia. From the trito-cerebrum also arise the **antennary nerves** as well as certain sensory nerves from the anterior region of the head.

The nature of the ventral nerve cord often clearly reflects the influence of body tagmosis. In primitive crustaceans with relatively homonomous bodies (e.g., remipedes, cephalocarids, and many branchiopods), the ganglia associated with each postantennary cephalic segment remain separate along the ventral nerve cord. In advanced forms, however, a single large, subenteric ganglionic mass is formed by the fusion of ganglia associated with the postoral cephalic segments (e.g., those of the mandibles, maxillules, maxillae, and, when present, maxillipeds). The ganglia of the thorax and abdomen may also be variably fused, depending on segment fusion and body compaction. For example, in most long-bodied decapods (lobsters and crayfish), the thoracic and abdominal ganglia are mostly fused across the body midline but remain separate from one another longitudinally (Figure 28B). However, in short-bodied decapods (e.g., crabs), all of the thoracic segmental ganglia are fused to form a large ventral nerve plate and the abdominal ganglia are much reduced (Figure 28C). Even in the spiny lobster (*Panulirus*), the ganglia are largely fused and the whole system is concentrated in the anterior region of the body.

Most crustaceans have a variety of sensory receptors that transmit information to the central nervous system in spite of the imposition of the exoskeleton (as we have explained for arthropods in general) (Figure 29). Among the most obvious of these sensory structures are the many innervated setae or sensilla that cover various regions of the body and appendages (Figure 30). Recent studies using electron mi-

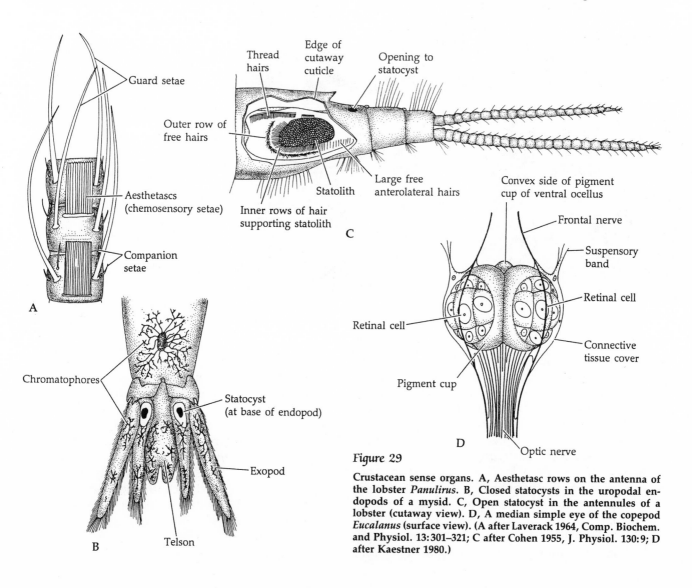

Figure 29

Crustacean sense organs. A, Aesthetasc rows on the antenna of the lobster *Panulirus*. B, Closed statocysts in the uropodal endopods of a mysid. C, Open statocyst in the antennules of a lobster (cutaway view). D, A median simple eye of the copepod *Eucalanus* (surface view). (A after Laverack 1964, Comp. Biochem. and Physiol. 13:301–321; C after Cohen 1955, J. Physiol. 130:9; D after Kaestner 1980.)

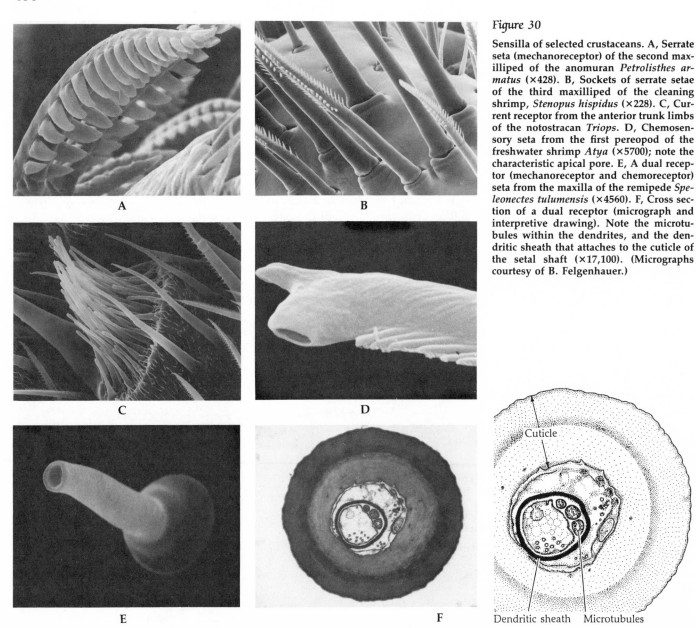

Figure 30

Sensilla of selected crustaceans. A, Serrate seta (mechanoreceptor) of the second maxilliped of the anomuran *Petrolisthes armatus* (×428). B, Sockets of serrate setae of the third maxilliped of the cleaning shrimp, *Stenopus hispidus* (×228). C, Current receptor from the anterior trunk limbs of the notostracan *Triops*. D, Chemosensory seta from the first pereopod of the freshwater shrimp *Atya* (×5700); note the characteristic apical pore. E, A dual receptor (mechanoreceptor and chemoreceptor) seta from the maxilla of the remipede *Speleonectes tulumensis* (×4560). F, Cross section of a dual receptor (micrograph and interpretive drawing). Note the microtubules within the dendrites, and the dendritic sheath that attaches to the cuticle of the setal shaft (×17,100). (Micrographs courtesy of B. Felgenhauer.)

croscopes have resulted in categorization systems based on function (see Bush and Laverack, in Atwood and Sandman 1982; Derby 1982). Many of these sensilla are mechanoreceptors (sensing touch and currents); others are chemoreceptors. Most crustaceans also possess special chemoreceptors in the form of clumps or rows of spinelike cuticular processes called **aesthetascs** (Figure 29A), located on the antennae or, more rarely, on the mouthparts. Thermoreceptors may occur in some crustaceans but are not yet documented.

Like all other arthropods, crustaceans contain well developed proprioceptors that provide infor-

mation about body and appendage position and movement during locomotion. A few taxa within the class Malacostraca possess statocysts, which either are fully closed and contain a secreted statolith (mysids, some anthurid isopods) or open to the outside through a small pore and contain a statolith formed of sand grains (many decapods) (Figures 29B and C). In the latter case the statocyst not only serves as a georeceptor but also detects the angular and linear acceleration of the body relative to the surrounding water and the movement of water past the animal (i.e., the statocyst is rheotactic).

There are two types of rhabdomeric photoreceptors among crustaceans, median simple eyes and lateral compound eyes; both are innervated by the protocerebrum. Many species possess both kinds of eyes, either simultaneously or at different stages of development. The compound eyes may be sessile or stalked.

The median eye generally first appears during the nauplius larval stage; thus, it is often called a **naupliar eye**. Like the nauplius larva itself, the median eye is thought to be a primitive feature within the Crustacea; it is secondarily reduced or lost in many taxa where the corresponding larval stage is suppressed. Median eyes are in a sense "compound" in that they are composed of more than one photoreceptor unit (Figure 29D). There are typically three such units in the median eyes of nauplii and up to seven in the eyes of adults in which they persist. Except for their basic rhabdomeric nature, however, the structure of median eye units is unlike the ommatidia of true compound eyes. The former are inverse pigment cups, each with relatively few retinular (photoreceptor) cells. Cuticular lenses are present over the median eyes of most ostracods and some copepods. All members of the class Maxillopoda with a medial eye have a tapetum. In spite of these incidental specializations, simple crustacean eyes probably only function to detect light direction and intensity. Such information is very useful as a means of orientation in planktonic forms without compound eyes, such as nauplius larvae and many copepods.

The structure and function of compound eyes were reviewed in Chapter 15. In terms of visual capacity, much more work has been done on the eyes of insects than on those of crustaceans, and we are left with a good deal of speculation in terms of what crustaceans actually "see." Although they probably lack the visual acuity of many insects, some can discern shapes, patterns, and movement; color vision has been demonstrated in some species. Certain structural differences occur between the compound eyes of insects and crustaceans, probably due in part to adaptations to aerial and aquatic vision requirements. Under water, light has a more restricted angular distribution, a lower intensity, and a narrower range of wavelengths than it does in air. Contrast is also somewhat reduced in water. All of these factors place a premium on enhancing the sensitivity and contrast perception of eyes of aquatic creatures. Mounting the eyes on stalks is one dramatic way in which many crustaceans increase the amount of information available to the eyes, and hence to the central nervous system—by increasing the field of view and binocular range.

Compound eyes are lacking altogether in some crustacean taxa (e.g., Copepoda, Mystacocarida, Cephalocarida, Tantulocarida, and nectiopodan Remipedia). Members of some other groups possess compound eyes only in late larval stages and lose them at metamorphosis (e.g., cirripedes). Reduction or loss of eyes is common in many deep-sea species, burrowers, meiofauna, cave dwellers, and parasites.

There is some disagreement as to whether the compound eye is a primitive or derived feature for crustaceans. This problem is one of many associated with the bigger controversy of monophyly versus polyphyly of the arthropods in general (i.e., the question of homology of the compound eyes of trilobites, cheliceriformes, crustaceans, and insects). This topic is addressed in Chapter 19. We favor the hypotheses that arthropod compound eyes are homologous structures, and that the crustacean compound eye is a primitive feature in the subphylum and the absence of compound eyes in certain crustaceans is a secondary (derived) condition associated with adaptation to special habitats. Whether the unique stalked condition of many crustacean eyes is a primitive or derived condition within the subphylum is still uncertain.

Crustaceans have complex endocrine and neurosecretory systems, but the understanding of these matters is far from complete. In general, the phenomena of molting (see Chapter 15), chromatophore activity, and various aspects of reproduction are under hormonal and neurosecretory control. A recent series of papers has emphasized not only what is known, but more important, the many avenues for future research in this area (*American Zoologist* 25(1):155–284). Interesting recent work (see Laufer et al. 1987) indicates that juvenile hormone-like compounds, long thought to occur only in insects, may also occur in at least some crustaceans. (Juvenile hormones are a family of compounds that regulate adult metamorphosis and gametogenesis in insects.) Bioluminescence also occurs in several crustacean groups. It is common among pelagic decapods and it has also been reported in certain ostracods, hyperiid amphipods, and copepod larvae.

Reproduction and development

We have repeatedly mentioned the relationship of an animal's reproductive/developmental pattern to its lifestyle and overall survival strategy. With the exception of purely vegetative processes like asexual budding, the crustaceans have managed to exploit virtually every life history scheme imaginable. The sexes are usually separate, although hermaphroditism is the rule in remipedes, cephalocarids, most cirripedes, and a few decapods. In addition, parthenogenesis is common among many branchiopods and certain ostracods.

The reproductive systems of crustaceans are generally quite simple (Figure 26). The gonads are derived from coelomic remnants and lie as paired, elongate structures in various regions of the trunk. In many cirripedes, however, the gonads lie in the cephalic region. In some cases the paired gonads are partially or wholly fused into a single mass. A pair of gonoducts extends from the gonads to genital pores located either on a trunk sternite, on the arthrodial membrane between the sternite and the protopod, or on the protopod itself. In many crustaceans the paired penes are fused into a single median penis (e.g., in tantulocarids, cirripedes, and some isopods). The female system sometimes includes seminal receptacles. The position of the gonopores varies among the five classes (see Table 1).

Most crustaceans copulate, and many have evolved courtship behaviors, the most elaborate and well known of which occur among the decapods (see the chapter by Salmon in Rebach and Dunham 1983 for a review). Although many crustaceans are gregarious, (e.g., certain planktonic species, barnacles, many isopods and amphipods), most decapods live singly except during the mating season. More-or-less permanent, or at least seasonal, pairing is known among many crustaceans [e.g., stenopodid shrimp, pinnotherid ("pea") crabs that often live as pairs in the mantle cavities of bivalve molluscs or in the burrows of thalassinid shrimps, and certain parasitic and commensal isopods].

Mating in nonpaired crustaceans must involve mechanisms that facilitate location and recognition of the partners. Among decapods, and perhaps many other crustaceans, scattered individuals apparently find one another either by distance chemoreception (pheromones) or by synchronized migrations associated with lunar periodicity, tidal movements, or some other environmental cue. Males of some marine ostracods produce complex, species-specific, bioluminescent displays (similar to those of fireflies) to attract females (see papers by Cohen and Morin). Once prospective mates are near each other, recognition of conspecifics of the opposite sex may involve several mechanisms. Apparently, most decapods employ chemotactic cues requiring actual contact. Vision is known to be important in the stenopodid shrimps (most of which live in pairs) and certain anomurans (family Porcellanidae) and brachyurans (many grapsids and ocypodids). A good deal of work has been done on fiddler crabs, *Uca* (family Ocypodidae). In *Uca*, males engage in dramatic cheliped waving (of their enlarged chela, or **major claw**) to attract females and repel competitive males (Figures 31A, B, and C). In addition, males produce sounds by stridulation and substratum-thumping that are thought to attract potential mates. Mating generally takes place once the male has enticed the female into his burrow.

Among many crustaceans, the external sexual characteristics are frequently associated with the actual mating process. In some males, particular appendages are modified for grasping the female, such as the antennae of male anostracans, and some ostracods, cladocerans, and copepods. Additionally, males frequently bear special sperm transfer structures in the form of either modified appendages or special penes like those of the thoracican barnacles (Figure 31D), anostracans, and ostracods. Examples of modified appendages include the last trunk limbs of copepods and the anterior pleopods of most male malacostracans (called **gonopods** in most malacostracans, or **petasma** in Dendrobranchiata). Sperm are transferred either loose in seminal fluid or packaged in spermatophores. Motile flagellated sperm occur only in some maxillopodans; in other crustaceans the sperm are nonmotile. (The sperm of some freshwater ostracods are some of the longest in the animal kingdom.) The sperm are deposited directly into the oviduct, or into a seminal receptacle in or near the female system. In some crustaceans, females can store sperm for long periods (e.g., several years in the lobster *Homarus*), thus facilitating multiple broods.

The great majority of crustaceans brood their eggs until hatching occurs. However, the syncarids, almost all dendrobranchiate decapods, and most euphausiaceans shed the zygotes directly into the water. A few others deposit their fertilized eggs in the environment, usually attaching them to some object (e.g., branchiurans, some ostracods, many stomatopods). These deposited embryos may be abandoned or, in the case of stomatopods, carefully tended by the female. Nonetheless, parental protection of the embryos until they hatch as larvae or juveniles is typically the rule in crustaceans. Thus, if we can generalize at all, it can be said that crustaceans usually engage in mixed or direct life histories (see Table 2).

A common theme of embryonic development occurs among crustaceans. It begins with some form of a modified spiral cleavage, passes through a blastula stage with a characteristic arrangement of presumptive areas, and proceeds to development of a nauplius larva followed by gradual completion of the postnaupliar region of the body. The eggs are centro-

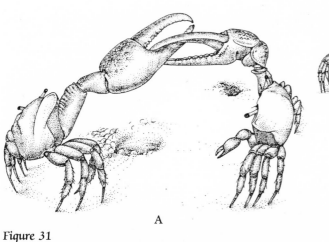

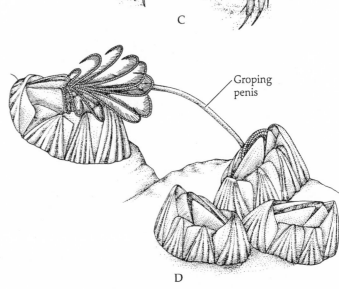

Figure 31

A–C, Mating behaviors by the fiddler crab *Uca*. A, Two males in ritualized combat for the favor of a female (B). C, A single male waving his enlarged cheliped to attract a female. D, A balanomorph barnacle, with cirri and groping penis extended, impregnating a neighbor. The advantage of a long penis in sessile animals is made obvious by this illustration. (After Barnes 1980.)

lecithal, with various amounts of yolk. The amount of yolk greatly influences the type of early cleavage and is often related to the time of hatching (see Chapter 4).

As far as is known, the zygotes of most nonmalacostracans undergo some form of holoblastic cleavage, as do those of syncarids, euphausiaceans, penaeids, amphipods, and parasitic isopods. However, the cleavage pattern is extremely variable, ranging from equal to unequal and from "semi-radial" to spiral. The occurrence of modified spiral cleavage (Figure 32A) in many crustaceans is generally viewed as evidence of close ties between the crustaceans and other spiralian groups such as the annelids. However, in his Herculean monograph, Anderson (1973) argued convincingly that the cell lineages and germ layer origins in crustaceans are significantly different from those of the usual (protostome) spirally cleaving embryo. For example, whereas the typical case involves a 4d origin of mesoderm, in barnacles this germ layer arises from the 3A, 3B, and 3C cells, and the 4d contributes to ectoderm (Figure 32B). Other differences between various crustacean and other arthropod taxa involve the relative positions of the presumptive germ layers to one another, especially the entoderm and mesoderm. Although Anderson and others have given great weight to this point in their support of a polyphyletic origin of the arthropods, we again emphasize that embryological variations in such a diverse and ancient taxon as the Arthropoda do not negate the fundamental similarities that unite the group.

Meroblastic cleavage is the rule among many malacostracans. Here again, the exact pattern varies, but it largely involves intralecithal nuclear divisions followed by nuclear migration to the periphery of the embryo and subsequent partitioning of the nuclei into a cell layer around a central yolky mass (Figure 32C).

The form of the blastula and the method of gastrulation are dependent primarily on the preceding cleavage pattern and hence ultimately on the amount of yolk. Holoblastic cleavage may lead to a coeloblastula that undergoes invagination (as in syncarids) or ingression (as in many copepods and some cladocerans and anostracans). Others form a stereoblastula followed by epibolic gastrulation (e.g., cirripedes). Most cases of meroblastic cleavage result in a periblastula and the subsequent formation of germinal centers.

In most crustaceans, embryonic development is interrupted when there are three pairs of appendage-bearing somites. The embryo then may hatch as a **nauplius larva** (Figure 32D). The presence of a free-

Table Two

Summary of crustacean reproductive biology

TAXON	DEVELOPMENT TYPE OR LARVAL TYPE AT TIME OF HATCHING	HERMAPHRODITIC	GONOCHORISTIC	PARTHENOGENESIS (IN AT LEAST SOME SPECIES)	COMMENTS
Remipedia	?	+	−	?	Development not yet studied
Cephalocarida	Metanauplius	+	−	−	Two eggs at a time fertilized and carried on genital processes of first pleonites
Branchiopoda					
Anostraca	Nauplius or metanauplius	−	+	+	Embryos usually shed from ovisac early in development; resistant (cryptobiotic) fertilized eggs accommodate unfavorable conditions
Notostraca	Nauplius or metanauplius	+	+	+	Eggs briefly brooded, then deposited on substrata; resistant (cryptobiotic) fertilized eggs accommodate unfavorable conditions
Cladocera	Nauplius, metanauplius; or direct development	−	+	+	All but *Leptodora* undergo direct development; *Leptodora* hatches as nauplii or metanauplii
Conchostraca	Nauplius or metanauplius	−	+	+	Embryos carried on thoracopods, then released
Maxillopoda					
Ostracoda	Direct, or bivalved, nauplius or metanauplius	−	+	+	Embryos usually deposited directly on substrata; many myodocopans and some podocopans brood embryos between valves until hatching as a reduced adult; no metamorphosis; up to 8 preadult instars
Mystacocarida	Nauplius or metanauplius	−	+	−	Little is known concerning this group; eggs apparently laid free; 6 naupliar stages (?)
Copepoda	Nauplius	−	+	−	Usually with 6 naupliar stages leading to a second series or 5 "larval" stages called copepodites
Branchiura	Nauplius or direct development	−	+	−	Embryos deposited; only *Argulus* known to hatch as nauplii; others have direct development and hatch as juveniles
Cirripedia	Nauplius	+	+	−	Six naupliar stages followed by unique larval form called a cyprid larva
Tantulocarida	?	−	+	?	Development entails metamorphosis

swimming nauplius larva is generally viewed as a primitive feature within the Crustacea and occurs in those groups considered primitive on morphological grounds. In a number of these forms, the hatching stage (see Table 2) is planktotrophic, and its release corresponds to the depletion of stored yolk. In more advanced groups of crustaceans, the nauplius either persists (although it sometimes exhibits lecithotrophy, as it does in euphausiaceans and dendrobranchiate shrimps), or is passed during a brood period,

or is lost altogether. The suppression or total loss of the early free larval stages is associated either with longer periods of brooding and other forms of parental care or with a heterochronic "shortening" of development.

Crustacean development is either direct, with the embryos hatching as juveniles that resemble miniature adults, or mixed, with embryos brooded for a brief or prolonged period and then hatching as a distinct larval form that may pass through several

Table Two
(continued)

TAXON	DEVELOPMENT TYPE OR LARVAL TYPE AT TIME OF HATCHING	HERMAPHRODITIC	GONOCHORISTIC	PARTHENOGENESIS (IN AT LEAST SOME SPECIES)	COMMENTS
Malacostraca					
Phyllocarida	Direct development	−	+	−	All undergo direct development in female brood pouch, hatching as postlarval "manca" (juvenile)
Eumalacostraca					
Hoplocarida	Antizoea or pseudozoea	−	+	−	Eggs brooded or deposited in burrow; hatch late as a clawed pseudozoea larva, or earlier as an unclawed antizoea larva; both go through several molts before settling on bottom as juveniles
Syncarida	Direct development	−	+	−	Free larval stages lost; eggs deposited on substratum
Eucarida					
Euphausiacea	Nauplius	−	+	−	Embryos shed or briefly brooded; typically undergo nauplius → zoea → megalopa → juvenile → adult transition series
Amphionidacea	Nauplius	−	+	−	Apparently brooded under thorax, but held by anterior pleopods; typically undergo nauplius → zoea → megalopa → juvenile → adult transition series
Decapoda	Protozoea or zoea (nauplius in Dendrobranchiata)	+	+	−	Dendrobranchiata shed embryos to hatch in water as nauplii or protozoea; all others brood embryos (on pleopods), which do not hatch until at least the zoea stage
Peracarida	Direct development	−	+	−	Embryos brooded in marsupium, typically formed from ventral coxal plates called oostegites; usually released as mancas (subjuveniles with incompletely developed eighth thoracopods and certain differences in body proportions and pigmentation). Brood pouch (marsupium) in Thermosbaenacea formed by dorsal carapace chamber

subsequent stages before the adult condition is achieved. Direct development occurs in some ostracods, cladocerans, and branchiurans and in all phyllocarids, syncarids, and peracarids. All other crustaceans have some form of mixed development.* The larval stages that have been recognized in crustacean

*Crustacean development is occasionally described as being either epimorphic, metamorphic, or anamorphic. **Epimorphic development** is direct; in crustaceans it is thought to result from a delay in the hatching of the embryo, thereby causing the nauplius (and any other possible larval stages) to be suppressed or absent. **Metamorphic development** is the type of extreme mixed development seen among the Eucarida; it includes dramatic transitions in body form from one life history stage to another (this pattern is similar to holometabolous development in insects—for example, the transformation of a caterpillar into a butterfly). **Anamorphic development** is a less extreme type of mixed development in which the embryo hatches as a nauplius larva, but the adult form is achieved through a series of gradual changes in body morphology as new segments and appendages are added. In other words, the postnaupliar stages gradually take on the adult form with succeeding molts; the classic example of anamorphic development is often said to be the Anostraca.

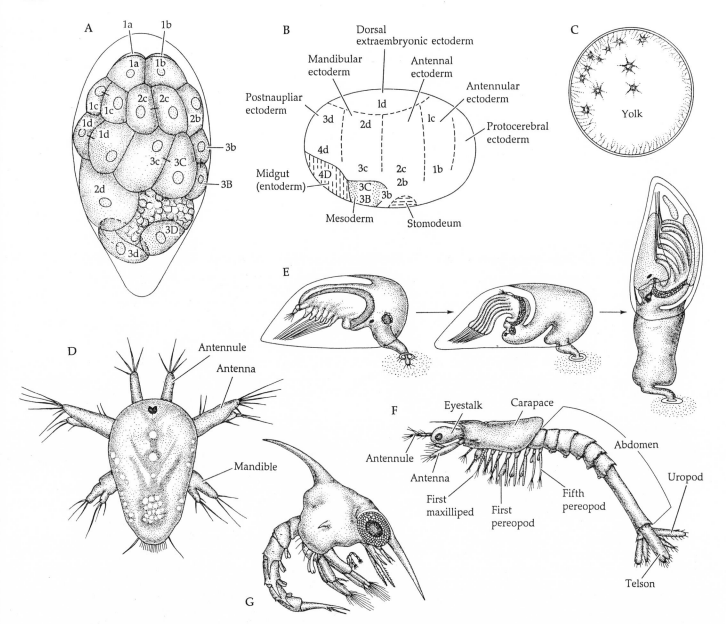

Figure 32

**Cleavage and posthatching stages of representative crustaceans.
A, Modified holoblastic spiral cleavage has produced a 28-cell
embryo of the cirripede *Tetraclita*. The cells are labeled as in
Wilson's coding system. B, A fate map of a cirripede blastula
(right-side view). Compare this map with the annelid fate map
in Figure 33 of Chapter 13. C, Intralecithal nuclear divisions in
the early cleavage of a mysid. D, Newly hatched copepod nauplius larva. E, Settling and metamorphosis of a lepadomorph
barnacle. F, The zoea (mysis) stage larva of the dendrobranchiate
shrimp *Penaeus*. G–H, Zoea larva and megalopa forms of the
brachyuran crab *Callinectes sapidus*. I, Zoea larva of a porcelain
crab. J, The characteristic antizoea larva of a stomatopod. K, The
translucent, paper-thin phyllosoma larva of the lobster *Jassa*. L,
Cryptoniscus stage of the epicaridean isopod *Probopyrus bithynis*
(ventral view). (A–C,F after Anderson 1973; D,K after Green 1961;
G,H after Cameron 1985; J after Brooks 1878; L after Dale and
Anderson 1982.)**

groups that undergo mixed development have been
assigned a plethora of names, and the homologies
among these forms are not well understood. The
more commonly encountered developmental forms
are summarized below (and in Table 2 and Figure
32), but we do not attempt to describe them all.

In the classes Cephalocarida and Branchiopoda,
and in most Maxillopoda, the nauplius larva grows
by a series of molts that adds new segments and
appendages gradually as the adult morphology appears. In many groups, hatching may be somewhat
delayed, and the emergent nauplius larva is termed
a **metanauplius**. The basic nauplius possesses only
three body somites, while the metanauplius has a

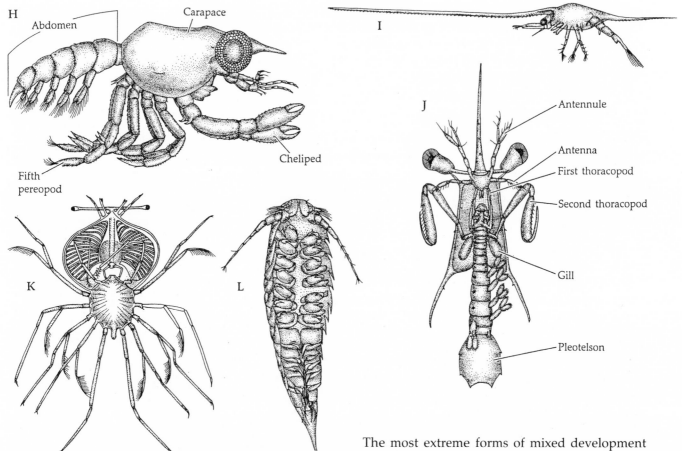

Abdomen
Carapace
H
Cheliped
Fifth
pereopod

K

L

I

J
Antennule
Antenna
First thoracopod
Second thoracopod
Gill
Pleotelson

few more; however, both possess only three pairs of similar-appearing appendages (which become the adult antennules, antennae, and mandibles). The end of the naupliar/metanaupliar stage is defined by the appearance of the fourth pair of functional limbs, the maxillules. In the cirripedes a distinct bivalved **cypris larva** follows the nauplius, and in copepods a post-naupliar stage called a **copepodite** (simply a small juvenile) is often recognized.

Some of the most dramatic morphological changes to occur during crustacean larval development affect the mouth appendages of the parasitic Branchiura. In the case of the maxillules, these appendages are narrow, barbed, stylet-like structures in the newly hatched larvae, but later transform into the broad sucking disks characteristic of adults (Figure 6L). The barbed larval maxillules are used to pierce the host's flesh for initial attachment; the adult suckers, which flank the mouth opening, facilitate attachment but also allow for mobility on the host by means of shuffling movements.

The most extreme forms of mixed development occur in the malacostracan superorder Eucarida. The most complex developmental sequences are seen among the dendrobranchiate shrimps, which hatch as a typical nauplius larva that eventually undergoes a metamorphic molt to be come a **protozoea larva**, with sessile compound eyes and a full complement of head appendages. The protozoea, after several molts, becomes a **zoea larva**, with stalked eyes and three pairs of thoracopods (as maxillipeds). The zoea eventually yields a juvenile stage (the **postlarva**) that resembles a miniature adult but is not sexually mature. In some groups (Amphionidacea, Caridea, and Brachyura) the postlarva is called a **megalopa**, and in the Anomura it is often called a **glaucothöe**. In other eucarids some (or all) of these stages are absent.

Various other terms have been coined for different (or similar) developmental stages. For example, the modified zoeal stages of stomatopods are called **antizoea** and **pseudozoea**, and the advanced zoeal stage of many other malacostracans is often called a **mysis larva**. In euphausiaceans, the nauplius is followed by two stages, the **calyptopis** and the **furcilia**, which roughly correspond to protozoea and zoea stages, before attaining the juvenile morphology.

From this wealth of terms and diversity of developmental sequences, we can draw two important generalizations concerning the biology and evolution of the crustaceans. First, developmental strategies reflect adaptations to different lifestyles. In spite of many exceptions, we can cite the early release of dispersal larvae by groups with limited adult mobility, such as thoracican barnacles, or by those whose resources may not permit production of huge quantities of yolk, such as the copepods. At the other end of this adaptive spectrum is the direct development of peracarids—a major factor allowing the invasion of land by certain isopods. And, between these extremes, we see all degrees of mixed life histories, with larvae being released at various stages following brooding and care. Second, because developmental stages also evolve, an analysis of developmental sequences can sometimes provide us with information about the radiation of the principal crustacean lineages. For example, the evolution of oostegites and direct development combine as a unique synapomorphy of the Peracarida. Similarly, the addition of a unique larval form, such as the **cyprid larva** of the cirripedes that follows the nauplius, can be viewed as a unique specialization (synapomorphy) that demarcates that group.

Lastly, it should be noted that among the branchiopods and some freshwater ostracods there have evolved specialized ways of coping with the harsh conditions of many freshwater environments. These adaptations include production of special over-wintering forms, usually eggs or zygotes, that can survive extreme cold, lack of water, or anoxic conditions. Perhaps most remarkable in this respect is the brine shrimp, *Artemia*, whose encysted embryos are capable of an extreme state of anaerobic quiescence. During these resistant stages, the metabolic rate of the embryos may drop to less than 10 percent of their normal rate.

Crustacean phylogeny

Zoologists have been debating about crustacean phylogeny for many generations. Much of the current debate involves two competing views on the nature of the primitive, or ancestral, crustacean body plan (Figure 33). One view holds that the first crustaceans had leaflike (phyllopodous) thoracic legs that were used both for swimming and for suspension feeding, as seen in the living cephalocarids, many of the branchiopods, and the phyllocarid malacostracans. The other view holds that the first crustaceans had nonphyllopodous, simple, paddle-like legs that

Figure 33

Theories of evolution within the Crustacea. See text for discussion. A, A traditional view of crustacean relationships, represented in an evolutionary tree. B, A recently published cladogram (Schram 1986) depicting the phylogeny of the crustaceans. This phylogenetic hypothesis views the ancestral crustaceans as remipede-like creatures, with cephalic feeding and simple (nonphyllopodous), paddle-like thoracic legs. C, An alternative cladogram depicting the phylogeny of the Crustacea. This phylogenetic hypothesis views the first crustacean as a thoracic suspension-feeding animal, with phyllopodous thoracic legs.

The synapomorphies used for the two cladograms are listed below. On Cladogram C, in addition to the synapomorphies that define each of the five crustacean classes, we have noted (in parentheses) some additional characters that may not be synapomorphies at this level but that originated *within* the indicated classes subsequent to the origins of the group themselves. These characters are parallel (or convergent) features that arose independently in several different crustacean classes.

Synapomorphies in Cladogram B: 1, mouthparts in posteriorly directed atrium; 2, raptorial mouthparts; 3, subrectangular head shield; 4, postcephalic trunk tagmosis; 5, thorax of 8 or fewer segments; 6, mandibular palp generally lacking; 7, antennules uniramous; 8, without abdominal appendages; 9, foliacious limbs; 10, polyramous limbs; 11, maxillopodan naupliar eye; 12, bulbous heart; 13, thorax of 6 or fewer segments; 14, trunk of 11 or fewer segments; 15, carapace does not cover limbs; 16, with uropods; 17, thoracopod endopods stenopodous; 18, polyramous limbs; 19, malacostracan naupliar eye.

Synapomorphies in Cladogram C: 1, biramous antennae (second antennae); 2, nauplius larva; 3, gnathobasic mandibles; 4, raptorial mouth appendages; 5, mouth appendages situated in posteriorly directed atrium; 6, first postcephalic appendages maxillipeds; 7, trunk appendages oriented laterally; 8, postcephalic trunk regionalized as thorax and abdomen; 9, loss of internal organ homonomy (e.g., segmental gut ceca); 10, reduction in number of body segments; 11, loss of abdominal appendages; 12, reduction of abdomen to 11 segments; 13, fully developed carapace; 14, reduction of abdomen to fewer than 9 segments; 15, maxillae reduced or absent; 16, thorax shortened to fewer than 11 segments; 17, abdomen shortened to fewer than 8 segments; 18,

were used for swimming but not for feeding; instead, the tasks of feeding were undertaken by the cephalic appendages. This plan is best represented among living crustaceans by the remipedes. Both views agree that the ancestral crustacean probably possessed a long, many-segmented, highly homonomous body.

Crustaceans, like the other arthropod subphyla, are an ancient group. Their fossil record dates back to the early Cambrian (over 550 million years ago), but their origin was certainly earlier than that, perhaps in the Ediacaran Period or even in the Precambrian Era. So much time has passed and so many unknown groups of arthropods have no doubt come and gone that the origin and relationships of the Crustacea may never be understood with certainty. However, phylogenetic hypotheses can still be proposed and tested by various means. The origin of the crustaceans and their relationship to the other ar-

A

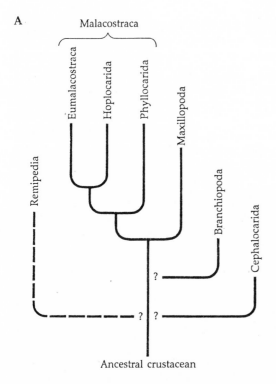

B

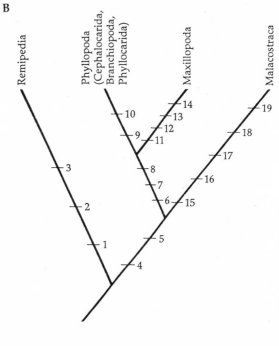

thorax of 6 or fewer segments; 19, abdomen of 4 or fewer segments; 20, carapace reduced; 21, genital appendages on the first abdominal somite (associated with male gonopores); 22, maxillopodan naupliar eye; 23, loss of frontal organs; 24, short, bulbous heart; 25, 8-segmented thorax and 6/7-segmented abdomen, plus telson; 26, male gonopores fixed on thoracomere 8, female gonopores on thoracomere 6; 27, replacement of thoracic suspension feeding and phyllopodous thoracic limbs with cephalic feeding and nonphyllopodous thoracic limbs.

thropods are discussed in Chapter 19; below we briefly discuss relationships within the subphylum Crustacea.

Subsequent to their discovery in 1955, the cephalocarids were regarded by most specialists as the most primitive living crustaceans, with the phyllopodous branchiopods and the phyllocarids (leptostracans) representing the most primitive members of the classes Branchiopoda and Malacostraca, respectively. The evolutionary tree shown in Figure 33A depicts this post-1955 view of crustacean phylogeny. However, with the discovery of the remipedes in 1981, some workers began favoring nonphyllopodous forms as probable ancestral crustaceans. Remipedes possess a suite of features that appear to be extremely primitive, notably a very long body with no postcephalic tagmosis, a double ventral nerve cord, serially arranged digestive ceca, and perhaps the cephalic shield.

C

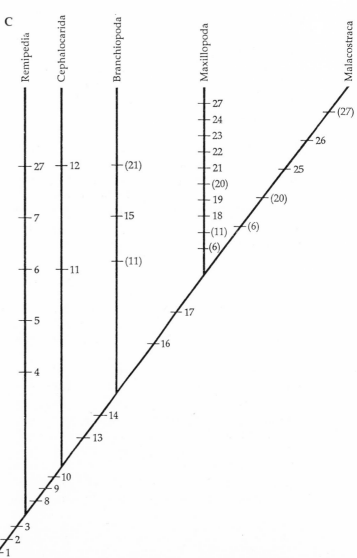

In 1986, F. R. Schram published the first *cladistic* analysis of crustacean classes and orders (Figure 33B). This hypothesis viewed the Remipedia as the most primitive living crustacean class and the phyllopodous crustaceans as an independent monophyletic lineage. Although the idea of remipedes as the most primitive living crustaceans has received some support, Schram's grouping of the thoracic-feeding cephalocarids, branchiopods, and phyllocarids together as a separate and uniquely derived lineage (which he called the Phyllopoda) has been somewhat controversial. Critics claim that removal of the group Phyllocarida from the class Malacostraca ignores many fundamental malacostracan features of phyllocarids. Other specialists argue that the synapomorphies used to define Schram's Phyllopoda–Maxillipoda line are known to be frequently convergent or otherwise variable within the Crustacea. Despite these criticisms, the cladogram in Figure 33B has merit—it is highly parsimonious and deals effectively with the problems raised by the discovery of the remipedes.

Figure 33C presents an alternative to Schram's cladogram. This cladogram is based on the five crustacean classes presented in this text. It was constructed without assumptions regarding which character states might be primitive and which advanced; that is, a phylogenetic computer program was asked to simply construct the shortest possible tree. This tree hypothesizes that phyllopodous feeding limbs (such as those of cephalocarids, many branchiopods, and leptostracans) represent the primitive crustacean limb type. Although the remipedes are viewed as the most primitive living crustaceans, their unique attributes primarily represent adaptations to their specialized habitat and direct feeding behavior. The cephalocarids also possess unique adaptations that represent specializations to their interstitial lifestyle. Given the size of the data set and the degree of uncertainty regarding which characters are primitive and which derived, we see no compelling reason to accept one of these trees over the other. In fact, recent evidence suggests that some characteristics long thought to be conservative, homologous features in Crustacea may actually be genetically derived through a variety of pathways (see Ferrari 1988). Furthermore, several other possible cladograms of similar lengths also exist. However, trees in which the remipedes are not placed at the base are markedly longer and less parsimonious than any others, an outcome suggesting that remipedes may indeed be the most primitive living crustaceans.

It is clear that we are still a long way from understanding the relationships among the crustacean classes. Like the arthropods in general, crustaceans exhibit high levels of evolutionary parallelism and convergence and many apparent reversals of character states. This genetic flexibility is no doubt due, in part, to the nature of the segmented body and serially homologous appendages, which, as we have stressed, provide enormous opportunity for evolutionary experimentation. Any conceivable cladogram of crustacean phylogeny will require the acceptance of such variability and homoplasy. For example, if the ancestor of all crustaceans is assumed to have had phyllopodous limbs (as implied in Figure 33C), then loss of this leg type must be hypothesized in the lines leading to the taxa Remipedia, Maxillopoda, and higher Malacostraca. However, if the ancestral crustacean is assumed to have had nonphyllopodous limbs, evolution of phyllopodous limbs must be hypothesized to have occurred in the lines leading to the taxa Cephalocarida, Branchiopoda, and primitive Malacostraca (Phyllocarida). Similar complexities arise regarding the evolution of the carapace from a cephalic shield, and the subsequent loss of the carapace in various distantly related groups, such as anostracans, isopods, and amphipods. (For example, is the absence of a carapace in cephalocarids a primitive or derived feature?) As you are no doubt beginning to realize, pursuit of these matters quickly becomes labyrinthine, and there are too many possible combinations of tree topologies and character changes (millions actually) to analyze without the aid of computers.

In the 1950s, Russian biologist W. N. Beklemischev and Swedish carcinologist E. Dahl independently proposed that copepods and several related classes represent a monophyletic clade. Dahl proposed the class Maxillopoda for these taxa. In 1983, American zoologists W. Newman and M. Grygier reviewed the concept of Maxillopoda, and Grygier provided the first cladistic hypothesis for the relationships within this class (see also Boxshall and Huys 1989). The shortening of the thorax to six or fewer segments and of the abdomen to four or fewer segments, reduction of the carapace (or in the case of ostracods and cirripedes, extreme modification of the carapace), loss of abdominal appendages, and other associated changes in the maxillopodans are now thought to be tied to paedomorphic events during the larval stage, or at an early postlarval stage, as this group began to radiate; this idea was proposed in 1942 by R. Gurney. Associated with reduction of the trunk and trunk limbs, the head appendages took on a larger role in both feeding and locomotion (as is the case in crustacean larvae). Other synapomorphies that define the class Maxillopoda are listed on the cladogram.

The monophyletic nature of the class Malacostraca has rarely been questioned. The class is defined by several synapomorphies. Within the Malacostraca are two principal groups: phyllocarids (Leptostraca), which have phyllopodous limbs and seven abdominal somites; and Eumalacostraca, which lack phyllopodous limbs and have but six abdominal segments. The eumalacostracans also have the sixth abdominal appendages modified as uropods (which work in conjunction with the telson as a "tail fan"). Relationships among the four main eumalacostracan lines (hoplocarids, syncarids, peracarids, and eucarids) are far from settled and have provided zoologists with many generations of lively debate (e.g., see papers by Dahl, Hessler, and Schram).

The class Branchiopoda is difficult to define on the basis of unique synapomorphies because it shows such great morphological variability. Apparently some have secondarily lost the carapace, and others have secondarily lost most or all of the abdominal appendages. It is quite possible that the Branchiopoda does not constitute a monophyletic taxon.

Selected References

Abele, L. G. and B. E. Felgenhauer. 1985. Observations on the ecology and feeding behavior of the anchialine shrimp *Procaris ascensionis*. J. Crust. Biol. 5:15–24.

Adiyodi, K. G. and R. G. Adiyodi. 1970. Endocrine control of reproduction in decapod Crustacea. Biol. Rev. 45:121–165.

Allison, E. C. and J. C. Holden. 1971. Recent ostracodes from Clipperton Island, eastern tropical Pacific. Trans. San Diego Soc. Nat. Hist. 16(7):165–214.

American Zoologist. 1985. Advances in Crustacean Endocrinology. Am. Zool. 25(1): 155–284. [12 papers devoted to current work on the crustacean endocrine system.]

Ameyaw-Akumfi, C. E. and B. A. Hazlett. 1975. Sex recognition in the crayfish *Procambarus clarkii*. Science 190:1225–1226.

Barnard, J. L. 1969. The families and genera of marine gammaridean Amphipoda. U. S. Nat. Mus. Bull. 271:1–535.

Barnard, J. L. and C. M. Barnard. 1983. *Freshwater Amphipoda of the World*. Hayfield Association, Mt. Vernon, Virginia.

Bauer, R. T. 1981. Grooming behavior and morphology in the decapod Crustacea. J. Crust. Biol. 1:153–173.

Bauer, R. T. 1987. Stomatopod grooming behavior: Functional morphology and amputation experiments in *Gonodactylus oerstedii*. J. Crust. Biol. 7:414–432.

Belk, D. 1987. Embryonic cuticles of *Artemia* during diapause and hatching: Insights from comparison with other Branchiopoda. J. Crust. Biol. 7:691–696.

Berkes, F. 1975. Some aspects of feeding mechanisms of euphausiids. Crustaceana Leiden 29:266–270.

Bliss, D. E. 1982. *Shrimps, Lobsters and Crabs*. New Century, Piscataway, New Jersey.

Bliss, D. E. (gen. ed.) 1982–1987. *The Biology of Crustacea*, Vols. 1–9. Academic Press, New York. [These volumes represent some of the most recent comprehensive treatments of crustaceans, and are a rich source of information on virtually all aspects of the group, from taxonomy to neurobiology and more. Dorothy Bliss served as editor-in-chief, but each volume is edited by an appropriate specialist and contains contributions from various authors.]

Bousfield, E. L. 1973. *Shallow-water Gammaridean Amphipoda of New England*. Comstock Publishing Association, Ithaca, New York.

Bowman, T. E. 1971. The case of the non-ubiquitous telson and the fraudulent furca. Crustaceana Leiden 21:165–175.

Bowman, T. E. 1978. Revision of the pelagic amphipod genus *Primno* (Hyperiidea: Phrosinidae). Smithson. Contrib. Zool. 275:1–23.

Bowman, T. E. and L. G. Abele. 1982. Classification of the recent Crustacea. In L. G. Abele (ed.), *The Biology of Crustacea*. Vol. 1, Systematics, the Fossil Record, and Biogeography. Academic Press, New York, pp. 1–27.

Bowman, T. E. and H.-E. Gruner. 1973. The families and genera of Hyperiidea (Crustacea: Amphipoda). Smithson. Contrib. Zool. 146:1–64.

Bowman, T. E. and T. M. Iliffe. 1985. *Mictocaris halope*, a new unusual peracaridan crustacean from marine caves on Bermuda. J. Crust. Biol. 5:58–73.

Bowman, T. E. and T. M. Iliffe. 1988. *Tulumella unidens*, a new genus and species of thermosbaenacean crustacean from the Yucatan Peninsula, Mexico. Proc. Biol. Soc. Wash. 101:221–226.

Bowman, T. E., S. P. Garner, R. R. Hessler, T. M. Iliffe, and H. L. Sanders. 1985. Mictacea, a new order of Crustacea Peracarida. J. Crust. Biol. 5:74–78.

Boxshall, G. A. and R. Huys. 1989. New tantulocarid, *Stygotantulus stocki*, parasitic on harpacticoid copepods, with an analysis of the phylogenetic relationships within the Maxillopoda. J. Crust. Biol. 9:126–140.

Boxshall, G. A. and R. J. Lincoln. 1983. Tantulocarida, a new class of Crustacea parasitic on other crustaceans. J. Crust. Biol. 3:1–16.

Boxshall, G. A. and R. J. Lincoln. 1987. The life cycle of the Tantulocarida (Crustacea). Philos. Trans. R. Soc. London Ser. B 315: 267–303.

Bresciani, J. 1986. The fine structure of the integument of free-living and parasitic copepods: A review. Acta Zool. 67:125–145.

Brooks, H. K. 1962. On the fossil Anaspidacea, with a revision of the classification of the Syncarida. Crustaceana Leiden 4: 229–242.

Brooks, H. K. et al. 1969. Arthropoda 4: Crustacea (except Ostracoda), Myriapoda, Hexapoda. In R. C. Moore (ed.), *Treatise on Invertebrate Paleontology* Pt. R. 36, University of Kansas Press, Lawrence.

Brusca, G. J. 1981. Annotated keys to the Hyperiidea (Crustacea: Amphipoda) of North American coastal waters. Allan Hancock Found. Tech. Rep. 5:1–76.

Brusca, G. J. 1981. On the anatomy of *Cystisoma* (Amphipoda: Hyperiidea). J. Crust. Biol. 1:358–375.

Brusca, R. C. 1981. A monograph on the Isopoda Cymothoidea (Crustacea) of the Eastern Pacific. Zool. J. Linn. Soc. 73:117–199.

Brusca, R. C. and M. Gilligan. 1983. Tongue replacement in a marine fish (*Lutjanus guttatus*) by a parasitic isopod (Crustacea: Isopoda). Copeia 3:813–816.

Brusca, R. C. 1984. Phylogeny, evolution and biogeography of the marine isopod subfamily Idoteinae (Crustacea: Isopoda: Idoteidae). Trans. San Diego Soc. Nat. Hist. 20(7):99–134.

Buchholz, F. 1982. Drach's molt staging system adapted for euphausiids. Mar. Biol. 66:301–305.

Burkenroad, M. D. 1947. Reproductive activities of decapod Crustacea. Am. Nat. 81:392–398.

Burkenroad, M. D. 1963. The evolution of the Eucarida (Crustacea: Eumalacostraca) in relation to the fossil record. Tulane Stud. Geol. 2(1):3–16.

Burkenroad, M. D. 1981. The higher taxonomy and evolution of Decapoda (Crustacea). Trans. San Diego Soc. Nat. Hist. 19(17):251–268.

Burnett, B. R. and R. R. Hessler. 1973. Thoracic epipodites in the Stomatopoda (Crustacea): a phylogenetic consideration. J. Zool. 169:381–392.

Burukovskii, R. N. 1985. *Key to Shrimps and Lobsters*. A. A. Balkema, Rotterdam.

Butler, T. H. 1980. *Shrimps of the Pacific Coast of Canada*. Can. Dept. Fish. Oceans Bull. 202:1–280.

Caine, E. A. 1978. Habitat adaptations of North American caprellid Amphipoda. Biol. Bull. 155:288–296.

Caine, E. A. 1983. Community interactions of *Caprella penantis* Leach (Crustacea: Amphipoda) on sea whips. J. Crust. Biol. 3:497–504.

Caine, E. A. 1989. Relationship between wave activity and robustness of caprellid amphipods. J. Crust. Biol. 9:425–431.

Caldwell, R. L. 1987. Assessment strategies in stomatopods. Bull. Mar. Sci. 41:135–150.

Caldwell, R. L. and H. Dingle. 1976. Stomatopods. Sci. Am. 234:81–89.

Caldwell, R. L. and H. Dingle. 1978. Ecology and morphology of feeding and agonistic behavior in mudflat stomatopods. Biol. Bull. 155:134–149.

Calinski, M. D. and W. G. Lyons. 1983. Swimming behavior of the puerulas of the spiny lobster *Panulirus argus* (Latreille, 1804) (Crustacea: Palinuridae). J. Crust. Biol. 3:329–335.

Calman, R. T. 1909. Crustacea. *In* R. Lankester (ed.), *A Treatise on Zoology*, Pt. 7. Adam and Charles Black, London. [A classic treatise; still very useful.]

Cameron, J. N. 1985. Molting in the blue crab. Sci. Am. 252:102–109.

Carlisle, D. B. and F. K. Knowles. 1959. *Endocrine Control in Crustaceans*. Cambridge University Press, Cambridge.

Carter, J. W. 1982. Natural history observations on the gastropod shell-using amphipod *Photis conchicola* Alderman, 1936. J. Crust. Biol. 2:328–341.

Chace, F. A., Jr. 1972. The shrimps of the Smithsonian-Bredin Caribbean expeditions with a summary of the West Indies shallow-water species. Smithson. Contrib. Zool. 98:1–180.

Chace, F. A., Jr. and H. H. Hobbs, Jr. 1969. The freshwater and terrestrial decapod crustaceans of the West Indies with special reference to Dominica. U.S. Nat. Mus. Bull. 292:1–258.

Chang, E. S. 1985. Hormonal control of molting in decapod Crustacea. Am. Zool. 25:179–185.

Chapman, M. A. and M. H. Lewis. 1976. *An Introduction to the Freshwater Crustacea of New Zealand*. Collins, Auckland.

Christy, J. H. 1982. Burrow structure and use in the sand fiddler crab, *Uca pugilator* (Bosc). Anim. Behav. 30:687–694.

Cobb, J. S. and B. F. Phillips (eds.). 1980. *The Biology and Management of Lobsters*. Vols. 1–2. Academic Press, New York.

Cohen, A. C. 1982. Ostracoda. *In* S. P. Parker (ed.), *Synopsis and Classification of Living Organisms*. McGraw-Hill, New York, pp. 181–202.

Cohen, A. C. and J. G. Morin. 1989. Six new luminescent ostracodes of the genus *Vargula* from the San Blas region of Panama. J. Crust. Biol. 9:297–340.

Coull, B. C. 1972. Species diversity and faunal affinities of meiobenthic Copepoda in the deep sea. Mar. Biol. 14:48–51.

Coull, B. C. 1977. Copepoda: Harpacticoida. Marine flora and fauna of the northeastern U. S. NOAA Tech. Rpt., Nat. Mar. Fish. Serv. Circular 399.

Coull, B. C. and J. Grant. 1981. Encystment discovered in a marine copepod. Science 212:342–344.

Crane, J. 1975. *Fiddler crabs of the world (Ocypodidae: Genus Uca)*. Princeton University Press, Princeton, New Jersey.

Cressey, R. F. 1978. Marine flora and fauna of the northeastern United States: Crustacea: Branchiura. NOAA Tech. Rpt., Nat. Mar. Fish. Serv. Circular 413.

Cronin, T. W. 1986. Optical design and evolutionary adaptation in crustacean compound eyes. J. Crust. Biol. 6:1–23.

Dahl, E. 1956. Some crustacean relationships. *In* K. G. Wingstrand (ed.), *Bertil Hanstrom, Zoological Papers in Honor of his 65th Birthday*. Zoological Insitute, Lund, pp. 138–147.

Dahl, E. 1956. On the differentiation of the topography of the crustacean head. Acta Zool. 37:123–192.

Dahl, E. 1976. Structural plans as functional models exemplified by Crustacea Malacostraca. Zool. Scr. 5:163–166.

Dahl, E. 1987. Malacostraca maltreated—the case of the Phyllocarida. J. Crust. Biol. 7:721–726.

Dahl, E. and R. R. Hessler. 1982. The crustacean *Lacinia mobilis*: A reconsideration of its origin, function and phylogenetic implications. Zool. J. Linn. Soc. 74:133–146.

Dale, W. E. and G. Anderson. 1982. Comparison of morphologies of *Probophyrus bithynis, P. floridensis,* and *P. pandalicola* larvae reared in culture (Isopoda: Epicaridea). J. Crust. Biol. 2:392–409.

Darwin, C. 1851–1854. *A Monograph on the Subclass Cirripedia*. Vols. 1–2. Ray Society, London.

Delamare-Deboutteville, C. 1960. *Biologie des eaux souterraines littorales et continentales*. Hermann, Paris.

Delamare-Deboutteville, C. and P. A. Chappuis. 1954. Morphologie des Mystacocarides. Arch. Zool. Exp. Gen. 91:7–24.

Dennell, R. 1937. On the feeding mechanism of *Apseudes talpa*, and the evolution of the peracaridan feeding mechanisms. Trans. Roy. Soc. Edinburgh 59:57–78.

Derby, C. D. 1982. Structure and function of cuticular sensilla of the lobster *Homarus americanus*. J. Crust. Biol. 2:1–21.

Derby, C. D. and J. Atema. 1982. The function of chemo- and mechanoreceptors in lobster (*Homarus americanus*) feeding behaviour. J. Exp. Biol. 98:317–327.

Drach, P. 1939. Mue et cycle d'intermue les Crustacés Décapodes. Ann. Inst. Océan. Monaco 19:103–391.

Drach, P. and C. Tchernigovtzeff. 1967. Sur la méthode de détermination des stades d'intermue et son application générale aux Crustaces. Vie Milieu Ser. A 18:595–610.

Dunham, P. J. 1978. Sex pheromones in Crustacea. Biol. Rev. 53:555–583.

Efford, I. E. 1966. Feeding in the sand crab *Emerita analoga*. Crustaceana Leiden 10:167–182.

Elofsson, R. 1965. The nauplius eye and frontal organs in Malacostraca. Sarsia 19:1–54.

Elofsson, R. 1966. The nauplius eye and frontal organs of the non-Malacostraca. Sarsia 25:1–128.

Felder, D. L. 1973. An annotated key to the crabs and lobsters from the coastal waters of the northwestern Gulf of Mexico. Louisiana State University LSU-SG-73-02.

Felgenhauer, B. E. and L. G. Abele. 1985. Feeding structures of two atyid shrimps with comments on caridean phylogeny. J. Crust. Biol. 5:397–419.

Felgenhauer, B. E., L. G. Abele and W. Kim. 1988. Reproductive biology of the anchialine shrimp *Procaris ascensionis* (Decapoda: Procarididae). J. Crust. Biol. 8:333–339.

Ferrari, F. D. 1988. Developmental patterns in numbers of ramal segments of copepod post-maxillipedal legs. Crustaceana Leiden 54:256–293.

Fincham, A. A. and P. S. Rainbow (eds.). 1988. *Aspects of Decapod Crustacean Biology*. Published for the Zoological Society of London by Clarendon Press, Oxford.

Fitzpatrick, J. F., Jr. 1983. *How to Know the Freshwater Crustacea*. Wm. C. Brown, Dubuque, Iowa.

Fryer, G. 1964. Studies on the functional morphology and feeding mechanism of *Monodella argentarii* Stella (Crustacea: Thermosbaenacea). Trans. R. Soc. Edinburgh 66(4):49–90.

Fryer, G. 1983. Functional ontogenetic changes in *Branchinecta ferox*. Philos. Trans. R. Soc. London Ser. B 303:229–343.

Fryer, G. 1987. A new classification of the branchiopod Crustacea. Zool. J. Linn. Soc. 91:357–383.

Garth, J. S. 1958. Brachyura of the Pacific coast of America. Oxyrhyncha. Allan Hancock Pac. Expeds. 21(1,2):1–854.

Ghiradella, H. T., J. Case, and J. Cronshaw. 1968. Structure of aesthetascs in selected marine and terrestrial decapods: Chemoreceptor morphology and environment. Am. Zool. 8:603–621.

Gilchrist, S. and L. A. Abele. 1984. Effects of sampling parameters on the estimation of population parameters in hermit crabs. J. Crust. Biol. 4:645–654. [For those interested in shell selection by hermit crabs, this paper includes a good literature list.]

Glaessner, M. F. 1960. The fossil decapod Crustacea of New Zealand and the evolution of the order Decapoda. N.Z. Geol. Surv. Paleontol. Bull. 31:1–63.

Glaessner, M. F. 1969. Decapoda. *In* R. C. Moore (ed.), *Treatise on Invertebrate Paleontology*, Part R, Arthropoda 4, Vol. 2. Geological Society of America and University of Kansas Press, pp. R399–R651.

Gordon, I. 1957. On *Spelaeogriphus*, a new cavernicolous crustacean from South Africa. Bull. Br. Mus. Nat. Hist. Zool. 5:31–47.

Gotto, R. V. 1969. *Marine Animals*. American Elsevier, New York.

Govind, C., M. Quigley and K. Mearow. 1986. The closure muscle in the dimorphic claws of male fiddler crabs. Biol. Bull. 170:481–493.

Green, J. 1961. *A Biology of Crustacea*. Quadrangle Books, Chicago.

Greenaway, P. 1985. Calcium balance and moulting in the Crustacea. Biol. Rev. 60(3):425–454.

Grey, D. L., W. Dall, and A. Baker. 1983. *A Guide to the Australian Penaeid Prawns*. North Territory Government Printing Office, Australia.

Grindley, J. R. and R. R. Hessler. 1970. The respiratory mechanism of *Spelaeogriphus* and its phylogenetic significance. Crustaceana Leiden 20:141–144.

Grygier, M. J. 1981. Sperm of the ascothoracican parasite *Dendrogaster*, the most primitive found in Crustacea. Int. J. Invert. Reprod. 3:65–73.

Grygier, M. J. 1982. Sperm morphology in Ascothoracida (Crustacea: Maxillopoda): Confirmation of generalized nature and phylogenetic importance. Int. J. Invert. Reprod. 4:323–332.

Grygier, M. J. 1983. Ascothoracica and the unity of the Maxillopoda. *In* F. R. Schram (ed.), *Crustacean Issues*, Vol. 1, Crustacean Phylogeny. A. A. Balkema, Rotterdam, pp. 73–104.

Grygier, M. J. 1987. Classification of the Ascothoracica (Crustacea). Proc. Biol. Soc. Wash. 100:452–458.

Grygier, M. J. 1987. New records, external and internal anatomy, and systematic position of Hansen's Y-larvae (Crustacea: Maxillopoda: Facetotecta). Sarsia 72:261–278.

Gurney, R. 1942. *Larvae of Decapod Crustacea*. Ray Society, London. [A benchmark survey that desperately needs to be updated.]

Haig, J. 1960. The Porcellanidae (Crustacea: Anomura) of the eastern Pacific. Allan Hancock Pac. Expeds. 24:1–440.

Hallberg, E. and R. Elofsson. 1983. The larval compound eye of barnacles. J. Crust. Biol. 3:17–24.

Hand, S. C. and E. Gnaiger. 1988. Anaerobic dormancy quantified in *Artemia* embryos: A colorimetric test of the control mechanism. Science 239:1425–1427.

Hamner, W. M. 1988. Biomechanics of filter feeding in the Antarctic krill *Euphausia superba*: Review of past work and new observations. J. Crust. Biol. 8:149–163.

Harbison, G. R., D. C. Biggs and L. P. Madin. 1977. The associations of Amphipoda Hyperiidea with gelatinous zooplankton. II. Associations with Cnidaria, Ctenophora and Radiolaria. Deep-Sea Res. 24(5):465–488.

Hart, J. F. L. 1982. *Crabs and Their Relatives of British Columbia*. British Columbia Prov. Mus., Handbook No. 40.

Hartmann, G. 1966, 1967, 1975. *Ostracoda. In Bronns Klassen und Ordnungen des Tierreichs 5*. Arthropoda, Abt. 1 Crustacea, Buch 2, 4 Teil, Lief. 1, 2, 4.

Hartnoll, R. G. 1969. Mating in Brachyura. Crustaceana Leiden 16:161–181.

Herreid, W. F., II and C. R. Fourtner (eds.). 1981. *Locomotion and Energetics in Arthropods*. Plenum, New York.

Herrnkind, W. F. 1985. Evolution and mechanisms of single-file migration in spiny lobster: Synopsis. Contrib. Mar. Sci. 27:197–211.

Hessler, R. R. 1964. The Cephalocarida. Comparative skeletomusculature. Mem. Conn. Acad. Arts Sci. 16:1–97.

Hessler, R. R. 1982. The structural morphology of walking mechanisms in eumalacostracan crustaceans. Philos. Trans. R. Soc. Lond. Ser. B 296:245–298. [An outstanding review.]

Hessler, R. R. 1982. Evolution of arthropod locomotion: A crustacean model. *In* C. F. Herreid and C. R. Fourtner (eds.), *Locomotion and Energetics in Arthropods*. Plenum, New York, pp. 9–30.

Hessler, R. R. 1984. *Dahlella caldariensis*, new genus, new species: A leptostracan (Crustacea, Malacostraca) from deep-sea hydrothermal vents. J. Crust. Biol. 4:655–664.

Hessler, R. R. 1985. Swimming in Crustacea. Trans. R. Soc. Edinburgh 76:115–122.

Hessler, R. R. and W. A. Newman. 1975. A trilobitomorph origin for the Crustacea. Fossils Strata 4:437–459.

Hessler, R. R. et al. 1982. Evolution within the Crustacea. *In* L. G. Abele (ed.), *The Biology of Crustacea*, Vol. 1. Systematics, the Fossil Record, and Biogeography. Academic Press, New York, pp. 149–239.

Ho J. S. 1977. Copepoda: Lernaeopodidae and Sphyriidae. Marine flora and fauna of the northeastern U. S. NOAA Tech. Rpt., Nat. Mar. Fish. Serv. Circular 406.

Ho, J. S. 1978. Copepoda: Cyclopoids parasitic on fishes. Marine flora and fauna of the northeastern U. S. NOAA Tech. Rpts., Nat. Mar. Fish. Serv. Circular 409.

Hobbs, H. H., Jr. 1971. The entocytherid ostracods of Mexico and Cuba. Smithson. Contrib. Zool. 81:1–55.

Hobbs, H. H., Jr. 1972. Crayfishes (Astacidae) of North and Middle America. Biota of Freshwater Ecosystems. Identification Manual No. 9. U. S. Environ. Protection Agency.

Høeg, J. T. 1985. Cypris settlement, kentrogon formation and host invasion in the parasitic barnacle *Lernaeodiscus porcellanae* (Muller) (Crustacea: Cirripedia: Rhizocephala). Acta Zool. 66:1–46.

Høeg, J. T. 1987. Male cypris metamorphosis and a new male larval form, the trichogen, in the parasitic barnacle *Sacculina carcini* (Crustacea: Cirripedia: Rhizocephala). Philos. Trans. R. Soc. Lond. Ser. B 317:47–63.

Høeg, J. T. and J. Lutzen. 1985. *Crustacea Rhizocephala*. Marine Invertebrates of Scandinavia No. 6. Norwegian University Press, Oslo. [Provides an excellent summary of rhizocephalan life history.]

Holdich, D. M. and D. A. Jones. 1983. *Tanaids: Keys and Notes for the Identification of the Species [of England]*. Cambridge University Press, Cambridge.

Holthius, L. B. 1955. The recent genera of caridean and stenopodidean shrimps (class Crustacea, order Decapoda, supersection Natantia) with keys for their determination. Zool. Verh. Rijksmus. Nat. Hist. Leiden 26:1–157.

Holthius, L. B. 1980. FAO Species Catalogue, Vol. 1. Shrimps and prawns of the world, an annotated catalogue of species of interest to fisheries. FAO Fish. Synopses 125:1–261.

Horch, K. W. and M. Salmon. 1969. Production, perception and reception of acoustic stimuli by semiterrestrial crabs. Forma Functio 1:1–25.

Humes, A. G. 1982. A review of Copepoda associated with sea anemones and anemone-like forms (Cnidaria, Anthozoa). Trans. Am. Philos. Soc. 72:1–120.

Humes, A. G. and M. Dojiri. 1983. Copepoda (Xarifidae) parasitic in scleractinian corals from the Indo-Pacific. J. Nat. Hist. 17:257–301.

Humes, A. G. and R. V. Gooding. 1964. A method for studying the external anatomy of copepods. Crustaceana Leiden 6:238–240.

Ingle, R. W. 1980. *British Crabs*. Oxford University Press, Oxford.

Itô, T. 1989. Origin of the basis in copepod limbs, with reference to remipedian and cephalocarid limbs. J. Crust. Biol. 9:85–103.

Ivanov, B. G. 1970. On the biology of the antarctic krill *Euphausia superba*. Mar. Biol. 7:340.

Jones, N. S. 1976. *British Cumaceans*. Academic Press, New York.

Just, J. and G. C. B. Poore. 1988. Second record of Hirsutiidae (Peracarida: Mictacea): *Hirsutia sandersetalia*, new species, from southeastern Australia. J. Crust. Biol. 8:483–488.

Kabata, Z. 1979. *Parasitic Copepoda of British Fishes*. Ray Society, London.

Kaestner, A. 1970. *Invertebrate Zoology*, Vol. 3, Crustacea. [Translated from the 1967 German second edition by H. W. Levi and L. R. Levi.] Wiley, New York.

Kensley, B. and M. Schotte. 1989. *Guide to the Marine Isopod Crustaceans of the Caribbean*. Smithsonian Institution Press, Washington, D.C.

King, M. G. and R. B. Moffitt. 1984. The sexuality of tropical deepwater shrimps (Decapoda: Pandalidae). J. Crust. Biol. 4:547–571.

Koehl, M. A. R. and J. R. Strickler. 1981. Copepod feeding currents: food capture at low Reynolds numbers. Limnol. Oceanogr. 26:1062–1073.

Kornicker, L. S. 1976. Benthic marine Cypridinacea from Hawaii (Ostracoda). Smithson. Contrib. Zool. 231:1–24.

Kornicker, L. S. 1981. Rutidermatidae of southern California (Ostracoda: Myodocopina). Smithson. Contrib. Zool. 334:1–35.

Kornicker, L. S. 1987. Supplementary description of *Cypridina americana* (Müller, 1890), a luminescent mydocopid ostracode

from the East Pacific. Proc. Biol. Soc. Wash. 100:173–181.

Kornicker, L. S. 1988. Myodocopine Ostracoda of the Alaskan continental shelf. Proc. Biol. Soc. Wash. 101:549–567.

Kornicker, L. S. and I. G. Sohn. 1976. Phylogeny, ontogeny, and morphology of livng and fossil Thaumatocypridacea (Myodocopa: Ostracoda). Smithson. Contrib. Zool. 219:1–124.

Land, M. F. 1984. Crustacea. *In* M. A. Ali (ed.), *Photoreception and Vision in Invertebrates*. Plenum, New York, pp. 401–438.

Land, M. F. 1981. Optics of the eyes of *Phronima* and other deep-sea amphipods. J. Comp. Physiol. 145:209–226.

Lang, K. 1948. Monographie der Harpacticoiden. Hakan Ohlssons, Lund., pp. 1–1682. [Whew!]

Laufer, H. et al. 1987. Identification of a juvenile hormone-like compound in a crustacean. Science 235:202–205.

Laval, P. 1972. Comportement, parasitisme et écologie d'*Hyperia schizogeneios* (Amphipode Hypéride) dans le plancton de Villefranch-sur-Mer. Ann. Inst. Océnogr. Paris 48:49–74.

Laval, P. 1980. Hyperiid amphipods as crustacean parasitoids associated with gelatinous zooplankton. Oceanogr. Mar. Biol. Annu. Rev. 18:11–56.

Lewis, S. M. and B. Kensley. 1982. Notes on the ecology and behavior of *Pseudoamphithoides incurvaria* (Just) (Crustacea, Amphipoda, Ampithoidae). J. Nat. Hist. 16:267–274.

Lockwood, A. P. M. 1967. *Physiology of Crustacea*. W. H. Freeman, San Francisco. [Dated, but still useful.]

Lowry, J. K. (ed.) 1983. *Papers from the Conference on the Biology and Evolution of Crustacea*. Aust. Mus. Sydney Mem. 18:1–218. [Eighteen papers on aspects of crustacean evolution.]

Madin, L. P. and G. R. Harbison. 1977. The associations of Amphipoda Hyperiidea with gelatinous zooplankton. I. Associations with Salpidae. Deep-Sea Res. 24:449–463.

Maitland, D. P. 1986. Crabs that breathe air with their legs—*Scopimera* and *Dotilla*. Nature 319:493–495.

Manning, R. B. 1969. *Stomatopod Crustacea of the Western Atlantic*. University of Miami Press, Coral Gables.

Manning, R. B. 1974. Crustacea: Stomatopoda. Marine flora and fauna of the northeastern U. S. NOAA Tech. Rpt., Nat. Mar. Fish. Serv. Circular 386.

Manton, S. M. 1977. *The Arthropoda*. Oxford University Press, London.

Marcotte, B. M. 1982. Copepoda. *In* D. E. Bliss (gen. ed.), *The Biology of Crustacea*,

Vol. 1, Systematics, the Fossil Record, and Biogeography. Academic Press, New York, pp. 185–197.

Markl, J. and B. Kempter. 1981. Subunit heterogeneity in crustacean hemocyanins as deduced by two-dimensional immunoelectrophoresis. J. Comp. Physiol. B141:495–502.

Marshall, S. M. 1973. Respiration and feeding in copepods. Adv. Mar. Biol. 11:57–120.

Martin, J. W. and D. Belk. 1988. Review of the clam shrimp family Lynceidae (Stebbing, 1902) (Branchiopoda: Concostraca) in the Americas. J. Crust. Biol. 8:451–482.

Mauchline, J. 1980. The biology of mysids and euphausiids. Adv. Mar. Biol. 18:1–681.

McCain, J. C. 1968. The Caprellidae (Crustacea: Amphipoda) of the western North Atlantic. U.S. Nat. Mus. Bull. 278:1–147.

McLaughlin, P. A. 1974. The hermit crabs of northwestern North America. Zool. Verh. Rijksmus. Nat. Hist. Leiden 130:1–396.

McLaughlin, P. A. 1983. Hermit crabs—are they really polyphyletic? J. Crust. Biol. 3:608–621.

McLaughlin, P. A. 1980. *Comparative Morphology of Recent Crustacea*. W. H. Freeman, San Francisco.

McLay, C. L. 1988. Crabs of New Zealand. Leigh Lab. Bull. 22:1–463.

Miller, D. C. 1961. The feeding mechanism of fiddler crabs with ecological considerations of feeding adaptations. Zoologica 46:89–100.

Molenock, J. 1975. Evolutionary aspects in the courtship behavior of anomuran crabs (*Petrolisthes*). Behaviour 53(1,2):1–30.

Morgan, S. G. and J. W.Goy. 1987. Reproduction and larval development of the mantis shrimp *Gonodactylus bredini* (Crustacea: Stomatopoda) maintained in the laboratory. J. Crust. Biol. 7:595–618.

Morin, J. G. 1986. "Fireflies" of the sea: Luminescent signaling in marine ostracode crustaceans. Florida Entomol. 69:105–121.

Müller, K. J. 1983. Crustaceans with preserved soft parts from the Upper Cambrian of Sweden. Lethaia 16:93–109.

Nelson, K. and D. Hedgecock. 1980. Enzyme polymorphism and adaptive strategy in the decapod Crustacea. Am. Nat. 116(2):238–280.

Newman, W. A. 1983. Origin of the Maxillopoda: Urmalacostracan ontogeny and progenesis. *In* F. R. Schram (ed.), *Crustacean Issues*, Vol. 1, Crustacean Phylogeny. A. A. Balkema, Rotterdam, pp. 105–120.

Newman, W. A. and A. Ross. 1976. Revision of the balanomorph barnacles; including a catalog of the species. San Diego Soc. Nat. Hist. Mem. 9:1–108.

Newman, W. A. and M. D. Knight. 1984. The carapace and crustacean evolution—a rebuttal. J. Crust. Biol. 4:682–687.

Nolan, B. A. and M. Salmon. 1970. The behavior and ecology of snapping shrimp (Crustacea: *Alpheus heterochelis* and *Alpheus normanni*). Forma Functio 2:289–335.

Omori, M. 1974. The biology of pelagic shrimps in the ocean. Adv. Mar. Biol. 12: 233–324.

Parker, S. P. (ed.) 1982. *Synopsis and Classification of Living Organisms*, Vol. 2. McGraw-Hill, New York, pp. 173–326. [A number of specialists contributed to the crustacean section of this compendium.]

Pennak, R. W. 1978. *Fresh-water Invertebrates of the United States*, 2nd Ed. Wiley, New York.

Pennak, R. W. and D. J. Zinn. 1943. Mystacocarida, a new order of Crustacea from intertidal beaches in Massachusetts and Connecticut. Smithson. Misc. Coll. 103:1–11.

Perry, D. M. and R. C. Brusca. 1989. Effects of the root-boring isopod *Spaeroma peruvianum* on red mangrove forests. Mar. Ecol. Prog. Ser. 57:287–292.

Persoone, G., P. Sorgeloos, O. Roels and E. Jaspers (eds.) 1980. *The Brine Shrimp Artemia*. Universa Press, Wetteren, Belgium.

Provenzano, A. J. 1978. Feeding behavior of the primitive shrimp *Procaris* (Decapoda, Procarididae). Crustaceana Leiden 35:170–176.

Quackenbush, L. S. and W. F. Herrnkind. 1983. Partial characterization of eyestalk hormones controlling molt and gonadal development in the spiny lobster *Panulirus argus*. J. Crust. Biol. 3:34–44.

Rathbun, M. J. 1918. The grapsoid crabs of America. U.S. Nat. Mus. Bull. 97:1–461.

Rathbun, M. J. 1925. The spider crabs of America. U.S. Nat. Mus. Bull. 129:1–613.

Reaka, M. L. 1979. The evolutionary ecology of life history patterns in stomatopod Crustacea. *In* S. Stancyk (ed.), *Reproductive Ecology of Marine Invertebrates*. University of South Carolina Press, Columbia, pp. 239–260.

Reaka, M. L. and R. B. Manning. 1981. The behavior of stomatopod Crustacea, and its relationship to rates of evolution. J. Crust. Biol. 1:309–327.

Rebach, S. and D. W. Dunham (eds.) 1983. *Studies in Adaptation: The Behavior of Higher Crustacea*. Wiley, New York.

Regnault, M. 1987. Nitrogen excretion in marine and freshwater Crustacea. Biol. Rev. 62: 1–24.

Richardson, H. 1905. A monograph of the isopods of North America. U.S. Nat. Mus. Bull. 54:1–727. [Badly out of date, but still a benchmark.]

Roer, R. and R. Dillaman. 1984. The structure and calcification of the crustacean cuticle. Am. Zool. 24:893–909.

Rolfe, W. D. I. 1981. Phyllocarida and the origin of the Malacostraca. Geobios 14:17–27.

Saint Laurent, M. 1979. Vers une nouvelle classification des Crustacés Decapodés Reptantia. Bull. Off. Peces Tunisie 3(1):15–31.

Sanders, H. L. 1955. The Cephalocarida, a new subclass of Crustacea from Long Island Sound. Proc. Natl. Acad. Sci. U.S.A. 41:61–66.

Sanders, H. L. 1957. The Cephalocarida and crustacean phylogeny. Syst. Zool. 6: 112–128.

Sanders, H. L. 1963. The Cephalocarida: Functional morphology, larval development, comparative external anatomy. Mem. Conn. Acad. Arts Sci. 15:1–80.

Sanders, H. L., R. R. Hessler and S. P. Garner. 1985. *Hirsutia bathylis*, a new unusual deep-sea benthic peracaridan crustacean from the tropical Atlantic. J. Crust. Biol. 5:30–57.

Schembri, P. J. Feeding behavior of 15 species of hermit crabs (Crustacea: Decapoda: Anomura) from the Otago region, southeastern New Zealand. J. Nat. Hist. 16:859–878.

Schminke, H. K. 1976. The ubiquitous telson and the deceptive furca. Crustaceana Leiden 30:292–300.

Schmitt, W. L. 1965. *Crustaceans*. University of Michigan Press, Ann Arbor. [Excellent reading.]

Schram, F. R. 1969. Polyphyly in Eumalacostraca? Crustaceana Leiden 16:243–250.

Schram, F. R. (gen. ed.) 1983–1987. *Crustacean Issues*, Vols. 1–5. A. A. Balkema, Rotterdam. [A continuing series of topical symposium volumes, each edited by a specialist in the respective field, e.g., phylogeny, biogeography, growth, barnacle biology.]

Schram, F. R. 1986. *Crustacea*. Oxford University Press, New York. [The most recent comprehensive treatment of the group.]

Schram, F. R., J. Yager and M. J. Emerson. 1986. The Remipedia. Part I. Systematics. San Diego Soc. Nat. Hist. Mem. 15.

Schultz, G. A. 1969. *How to Know the Marine Isopod Crustaceans*. Wm. C. Brown, Dubuque, Iowa. [An error-filled book, but with a relatively complete list of species.]

Siewing, R. 1956. Untersuchungen zur Morphologie der Malacostraca (Crustacea). Zool. Jahrb. Anat. 75(1):39–176.

Skinner, D. M. 1985. Interacting factors in the control of the crustacean molt cycle. Am. Zool. 25:275–284.

Smaldon, G. 1979. *British Coastal Shrimps and Prawns*. Academic Press, New York.

Smirnov, N. N. and B. V. Timms. 1983. A revision of the Australia Cladocera (Crustacea). Rec. Aust. Mus. Suppl. 1:1–132.

Snodgrass, R. E. 1956. Crustacean metamorphosis. Smiths. Misc. Contrib. 131(10):1–78. [Somewhat dated, but still a good introduction to the subject.]

Stebbing, T. R. R. 1893. *A History of Crustacea*. D. Appleton and Co., London. [Obviously dated but still highly informative and very enjoyable reading.]

Stebbins, T. D. 1989. Population dynamics and reproductive biology of the commensal isopod *Colidotea rostrata* (Crustacea: Isopoda: Idoteidae). Mar. Biol. 101:329–337.

Steinsland, A. J. 1982. Heart ultrastructure of *Daphnia pulex* De Geer (Crustacea, Branchiopoda, Cladocera). J. Crust. Biol. 2:54–58.

Stepien, C. A. and R. C. Brusca. 1985. Nocturnal attacks on nearshore fishes in southern California by crustacean zooplankton. Mar. Ecol. Prog. Ser. 25:91–105.

Stevcic, Z. 1971. The main features of brachyuran evolution. Syst. Zool. 20:331–340.

Stock, J. 1976. A new genus and two new species of the crustacean order Thermosbaenacea from the West Indies. Bijdr. Dierkdl. 46:47–70.

Stoner, A. W. 1983. Distributional ecology of amphipods and tanaidaceans associated with three sea grass species. J. Crust. Biol. 3:505–518.

Strickler, R. 1982. Calanoid copepods, feeding currents and the role of gravity. Science 218:158–160.

Stubbings, H. G. 1975. *Balanus balanoides*. Liverpool Mar. Biol. Comm. Mem. 37.

Sutton, S. L. 1972. *Woodlice*. Ginn and Co., London.

Sutton, S. L. and D. M. Holdich (eds.) 1984. *The Biology of Terrestrial Isopods*. Clarendon Press, Oxford.

Thatcher, V. E. 1984. *Ergasilus pitalicus*, new species (Copepoda: Poecilostomatoida: Ergasilidae), a gill parasite of a cichlid fish from the Pacific coast of Colombia. J. Crust. Biol. 4:495–501.

Thiery, A. and A. Champeau. 1988. *Linderiella Massaliensis*, new spcies (Anostraca: Linderiellidae), a fairy shrimp from southeastern France, its ecology and distribution. J. Crust. Biol. 8:70–78.

Thurman, C. L. 1984. Ecological notes on fiddler crabs of South Texas, with special reference to *Uca subcylindrica*. J. Crust. Biol. 4:665–681.

Tomlinson, J. T. 1969. The burrowing barnacles (Cirripedia: Order Acrothoracica). U.S. Nat. Mus. Bull. 259:1–162.

Van Dover, C. L., J. R. Factor and R. H. Gore. 1982. Developmental patterns of scaphagnathites: An aid to the classification of anomuran and brachyuran Crustacea. J. Crust. Biol. 2:48–53.

Van Name, W. G. 1936. The American land and freshwater isopod Crustacea. Bull. Am. Mus. Nat. Hist. 71:1–535.

Van Niekerk, J. P. and D. J. Kok. 1989. *Chonopeltis australis* (Branchiura): Structural, developmental and functional aspects of the trophic appendages. Crustaceana 57:51–56.

Vinogradov, M. E., A. F. Volkov, and T. N. Semenova. 1982. *Amphipody-Giperiidy (Amphipoda, Hyperiidea) Mirovogo Okeana.* Nauka, Leningrad. [In Russian.]

Warner, G. F. 1977. *The Biology of Crabs.* Van Nostrand Reinhold, New York.

Waterman, T. H. (ed.). 1960–1961. *The Physiology of Crustacea,* Vols. 1–2. Academic Press, New York.

Wenner, A. M. (ed.) 1985. *Crustacean Growth: Larval Growth.* A. A. Balkema, Rotterdam.

Wenner, A. M. (ed.) 1985. *Crustacean Growth: Factors in Adult Growth.* A. A. Balkema, Rotterdam.

Wenner, A. M. and S. R. Haley. 1981. On the question of sex reversal in mole crabs (Crustacea, Hippidae). J. Crust. Biol. 1: 506–517.

Whittington, H. B. and W. D. U. Rolfe (eds.) 1963. *Phylogeny and Evolution of Crustacea.* Museum of Comparative Zoology, Cambridge. [Still a benchmark volume in crustacean phylogenetic ideas.]

Williams, A. B. 1984. *Shrimps, Lobsters, and Crabs of the Atlantic Coast of the Eastern United States, Maine to Florida.* Smithsonian Instution Press, Washington, D.C. [An outstanding reference for the region.]

Williams, A. B. and F. A. Chace, Jr. 1982. A new caridean shrimp of the family Bresiliidae from thermal vents of the Galapagos Rift. J. Crust. Biol. 2:136–147.

Williamson, D. I. 1973. *Amphionides reynaudii* (H. Milne Edwards), representative of a proposed new order of eucaridan Malacostraca. Crustaceana Leiden 25(1):35–50.

Wittmann, K. J. 1981. Comparative biology of marsupial development in *Leptomysis* and other Mediterranean Mysidacea (Crustacea). J. Exp. Mar. Biol. and Ecol. 52:243–270.

Yagamuti, S. 1963. *Parasitic Copepoda and Branchiura of Fishes.* Wiley, New York.

Yager, J. 1981. Remipedia, a new class of Crustacea from a marine cave in the Bahamas. J. Crust. Biol. 1:328–333.

Yager, J. and F. R. Schram. 1986. *Lasionectes entrichoma,* new genus, new species (Crustacea: Remipedia) from anchialine caves in the Turks and Caicos, British West Indies. Proc. Biol. Soc. Wash. 99:65–70.

Zinn, D. J., B. W. Found, and M. G. Kraus. 1982. A bibliography of the Mystacocarida. Crustaceana Leiden 42(3):270–274.

Chapter Nineteen

Three Enigmatic Groups and a Review of Arthropod Phylogeny

There are endless elaborations of the cuticle in the form of articulations and other skeletal features which are just as striking as is the form of the vertebrate skeleton.

Sidnie M. Manton
The Arthropoda, 1977

In the first part of this chapter we discuss three invertebrate groups that show close relationships to the arthropods—Onychophora, Pentastomida, and Tardigrada. In the second part of this chapter we present an overview of arthropod phylogeny, one of the most controversial topics in invertebrate zoology today. Hundreds of research papers and many books have been written on the subject of arthropod evolution, many of them highly speculative. We present a general summary of this topic and assess the principal competing views being debated today.

Phylum Onychophora

The first onychophoran (Greek *onycho*, "talon"; *phora*, "bearer") was described by the Reverend Lansdown Guilding in 1826 as a leg-bearing "slug" (a mollusc). Since that initial discovery, 80 or so species have been discovered. Living onychophorans are similar to two controversial middle Cambrian marine fossils: *Aysheaia pedunculata* from the famous Burgess Shale deposits of British Columbia, Canada; and *Aysheaia prolata* from a similar deposit in Utah. Contemporary judgment of these two fossil forms is still mixed. Sharov (1966) and others regard them as polychaetous annelids. Other specialists feel that they may have been true onychophorans, or perhaps animals belonging to a line including the Onychophora

(and perhaps some other primitive "near-arthropods," such as the tardigrades). In any event, the Onychophora may be an old group that has changed very little in its long history. In addition, members of this group have features that are somewhat intermediate between those of annelids and arthropods. For these reasons some workers regard onychophorans as "living fossils," or "missing links" between the annelids and arthropods.

Box 1 lists some features shared by the onychophorans and the annelids and arthropods. Those features that are shared with the annelids are symplesiomorphies—primitive characters that define a larger group containing both the annelids and the onychophorans (and some other protostome taxa). Those features shared with the arthropods are characters of two kinds. They are either onychophoran–arthropod convergences, or they are true homologies and thus synapomorphies that unite the onychophorans with the arthropods in a clade. Both types of characters are represented in the list.

Living onychophorans (Figure 1) are confined to humid habitats, and most, but not all, species occur in the tropics. During unfavorable environmental conditions (extreme wet or dry periods), they retire to protective burrows or other retreats and become inactive. The phylum comprises two families, Peripatidae and Peripatopsidae. The former is circumtropical in distribution, whereas the latter is circumaustral (confined to the temperate Southern Hemisphere).

The onychophoran *Bauplan*

Onychophorans resemble caterpillars or, perhaps more appropriately, soft-bodied, unsegmented centipedes (Figure 1). They range from 1.5 to 15 cm in length. Males are always smaller than females and

667

Box One

Annelid-Like and Arthropod-Like Features of the Onychophorans

ANNELID-LIKE FEATURES

Noncalcified and weakly sclerotized cuticle
  (soft, extensible body wall)
Cuticle not of articulating plates
Body wall musculature
Simple gut
Little cephalization (and strong homonomy)
Sensory organs
Fine structure of eyes
Paired segmental nephridia (with cilia)
Ciliated reproductive tract
Unjointed appendages

ARTHROPOD-LIKE FEATURES

Growth by ecdysis
Elongate dorsal gonads
Highly reduced coelom
Arrangement of hemocoel and
  elongate, dorsal, ostiate heart
"Coxal glands" (without cilia)
Trachea–spiracular system
Large, well-developed brain
Antennae (?)

A

B

Figure 1

Onychophorans. A, A live onychophoran. B, *Peripatopsis sedgwicki* **feeding on a piece of meat. The tips of the jaws are visible within the distended lips. (A photo by D. R. Paulson/BPS; B after Manton 1977.)**

have fewer legs. Little cephalization is present and the body is strongly homonomous. Three paired appendages adorn the head: one pair of fleshy annulated **antennae**, a single pair of **jaws**, and a pair of fleshy **oral papillae** ("slime papillae"), which lie adjacent to the jaws (Figure 2). These are followed by 14–43 pairs of simple **lobopod** walking legs. Although the head appendages and walking legs are superficially annulated, they are not jointed or segmented, nor do they possess intrinsic musculature (all muscles have their origins in the body proper).

The homology of onychophoran head structures with those of annelids or arthropods has long been a matter of debate. The unjointed fleshy nature of the head appendages and lobopod legs may appear more like those of polychaetes than arthropods. However, many specialists feel that the appendages are directly homologous to those of arthropods, particularly those of the uniramians, despite the absence of segmentation and intrinsic musculature that is characteristic of arthropods (see Table 1). Other specialists regard onychophoran appendages simply as primitive, serially arranged, lobopodal appendages of uncertain homology.

The body is covered by a chitinous cuticle that is molted, as it is in the arthropods. However, like that of the annelids, the cuticle of onychophorans is noncalcified, thin, flexible, very permeable, and not divided into articulating plates or sclerites. Beneath the cuticle is a thin epidermis, which overlies a connec-

Table One

Appendage distribution in the four arthropod subphyla[a]

SOMITE (AND ASSOCIATED CEREBRAL GANGLION)	TRILOBITO-MORPHA	CHELICERIFORMES		UNIRAMIA	CRUSTACEA
		CHELICERATA	PYCNOGONIDA		
Acron (protocerebrum)	—	—	—	—	—
1[b] (deutocerebrum)	Antennae	Chelicerae arise postoral embryo-logically but usually migrate preoral in adults	● Chelifores	Antennae	First antennae (= antennules)
2[c] (tritocerebrum)	● First pair legs	Pedipalps (or first legs)	Palps	● Embryonic (with or without transient limb buds)	● Second antennae (migrate to preoral position in adult)
3	Second pair legs	Second pair legs (first pair in arachnids)	Ovigers	Mandibles	Mandibles
4	Third pair legs	Third pair legs (second pair in arachnids)	First pair legs	Maxillules (= first maxillae) (= gnathochilarium of Diplopoda and Pauropoda)	Maxillules (= first maxillae)
5	Fourth pair legs	Fourth pair legs (third pair in arachnids)	Second pair legs	Maxillae (= second maxillae) (= labium of insects and symphylans) (lost in Diplopoda and Pauropoda)	Maxillae (= second maxillae)
6	Fifth pair legs	Fifth pair legs (fourth pair in arachnids)	Third pair legs	Insects, symphylans, pauropodans: First pair legs Chilopods: maxillipeds Diplopods: collum (no legs)	First pair thoracic appendages (walking/swimming legs or maxillipeds)
7	Sixth pair legs	Legs present, reduced, or absent	Fourth pair legs	Insects: second pair legs Myriapods: first pair legs	Second pair thoracic appendages (walking/swimming legs or maxillipeds)
8	Seventh pair legs	Legs present, reduced, or absent	—	Insects: third pair legs Myriapods: second pair legs	Third pair thoracic appendages (walking/swimming legs or maxillipeds)
9	Eighth pair legs	Legs present, reduced, or absent	—	Insects: — Myriapods: third pair legs	Walking/swimming legs
10	Ninth pair legs	Legs present, reduced, or absent	—	Insects: — Myriapods: fourth pair legs	Walking/swimming legs

[a]Appendages of the first 10 (postacronal) somites are listed. Readers are cautioned that the subject of head appendage homology among the arthropods is quite unsettled and highly controversial; different schemes can be found in Manton (1977) and Schram (1978). (For further discussion on these subphyla, see their respective chapters.)

[b]The black circle indicates embryonic placement of the mouth.

[c]The second postacronal (first postoral) somite of the symphylan *Hanseniella* (Uniramia) goes through an embryonic stage in which it bears transitory appendages that later shrink and disappear. This somite is the head segment housing the tritocerebrum. Although all uniramians lack second antennae as adults, it is generally believed that this transitory embryonic appendage is homologous to the second antenna of crustaceans and may be taken as evidence that the primitive uniramians descended from an ancestor whose head appendage arrangement was essentially identical to that of primitive crustaceans. It is this feature that led to the rarely used name Atelocerata (= "incompletely provided with antennae") for the Uniramia. The recent discovery of an ancient (Silurian) marine myriapod further reinforces the concept of a marine origin for the uniramians (Mikulic et al. 1985).

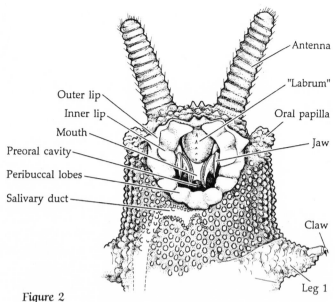

Figure 2

Buccal region (ventral view) of *Peripatus*. (Adapted from several sources.)

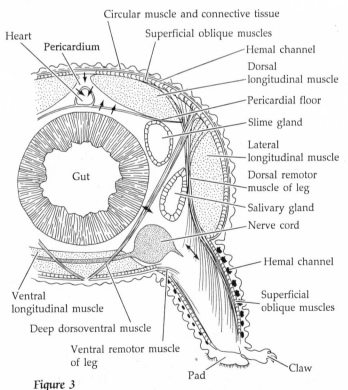

Figure 3

Body segment and leg of *Peripatopsis* (transverse section). The arrows indicate the direction of blood flow. (After Manton 1977.)

tive tissue dermis and layers of circular, diagonal, and longitudinal muscles (Figures 3 and 5C). The body surface of onychophorans is covered with wart-like tubercles, usually arranged in rings or bands around the trunk and appendages. The tubercles are covered with minute scales. Most onychophorans are distinctly colored blue, green, orange, or black, and the papillae and scales give the body surface a velvety sheen.

The coelom, like that of arthropods, is restricted almost entirely to the gonadal cavities. The hemocoel is also arthropod-like, being partitioned into sinuses, including a dorsal **pericardial sinus**.

Locomotion. The paired walking legs are conical, unjointed, ventrolateral lobes with a multispined terminal claw, sometimes called "hooks." When standing or walking, each leg rests on three to six distal transverse pads (Figure 3). Onychophoran legs are considered by some workers to be the only true lobopods in the "animal kingdom," in reference to the saclike nature of these appendages, which are filled with hemocoelomic fluid and contain only extrinsic muscle insertions.

Walking is accomplished by leg mechanics combined with wavelike extension and contraction of the body by hydrostatic forces exerted via the hemocoel. When a segment is elongated, the legs are lifted from the ground and moved forward; when a segment contracts, a pushing force is exerted and the legs are held against the substratum. The overall effect is reminiscent of some types of polychaete locomotion, wherein the parapodia are used mainly for purchase rather than as legs or paddles. However, limb motion in onychophorans is not as coordinated as it is in annelids or arthropods.

The muscles are a combination of smooth and obliquely striated fibers and are arranged similarly to those of annelids. The thin cuticle, soft body, and hydrostatic *Bauplan* allow onychophorans to crawl and force their way through narrow passages in their environs. As we have seen, the efficiency of a hydrostatic skeleton is enhanced by internal longitudinal communication of body fluids. The ancestors of the onychophorans apparently eliminated the intersegmental septa of their annelidan predecessors and expanded the blood vascular system at the expense of the coelom—thus converting from a true coelomic hydrostatic skeleton to a hemocoelic hydrostatic skeleton. As we noted in Chapter 13, a somewhat similar trend toward reduction in the size of the coelom and loss of internal septa occurs in some annelids.

Feeding and digestion. Onychophorans occupy a niche similar to that of centipedes. Most prey on small invertebrates, such as snails, worms, termites, and other insects, which they pursue into cracks and crevices. Special **slime glands** (presumably modified nephridia) open at the ends of the oral papillae (Figure 2); through these openings an adhesive is discharged in two powerful streams, sometimes to a distance of 30 cm. The adhesive hardens quickly, entangling prey (or would-be predators) in a weblike net for subsequent leisurely dining.

The jaws are used to grasp and cut up prey. Paired salivary glands, also thought to be modified nephridia, open into a median dorsal groove on the jaws (Figures 2 and 4). Salivary secretions pass into the body of the prey and partly digest it; the semi-liquid tissues are then sucked into the mouth. The mouth opens into a chitin-lined foregut, composed of a pharynx and esophagus. A large, straight intestine is the principal site of digestion and absorption.

The hindgut (rectum) usually loops forward over the intestine before passing posteriorly to the anus, which is located ventrally or terminally on the last body segment.

Circulation and gas exchange. The circulatory system of onychophorans is arthropod-like and linked to the hemocoelic *Bauplan*. A tubular heart is open at each end and bears a pair of lateral ostia in each segment. The heart lies within a pericardial sinus. Blood leaves the heart anteriorly and then flows posteriorly within the large hemocoel via body sinuses, to eventually reenter the heart by way of the ostia. The blood is colorless, containing no oxygen-binding pigments. Onychophorans possess a unique system of subcutaneous vascular channels, called **hemal channels** (Figure 3). The channels are situated beneath the transverse rings, or ridges, of the cuticle (Figure 5C). A bulge in the layer of circular muscle forms the outer wall of each channel, and the oblique muscle layer forms the inner wall. These channels may to be important in the functioning of the hydrostatic skeleton. The superficial annulations of the onychophoran body (and the body of *Aysheaia*) are external manifestations of these subcutaneous hemal channels.

Gas exchange is by **tracheae** that open to the outside through the many small **spiracles** located between the bands of body tubercles. Each tracheal unit is small and supplies only the immediate tissue near its spiracle (Figure 5B). Anatomical data suggest the tracheal system may not be homologous to that of uniramians or arachnids.

Excretion and osmoregulation. A pair of nephridia lies in each body segment, located in the ventrolateral hemal sinuses (Figure 4). The nephrostome is surrounded by an end sac, which represents a remnant of the coelom (Figure 5A). The tubule enlarges to form a contractile bladder just before opening to the outside via a nephridiopore at the base of each leg, except for those of the fourth and fifth segments; in these segments, the pores sit atop a more distal tubercle. The nature of the excretory wastes is not known. The anterior nephridia are thought to be represented by the salivary glands and slime glands, and the posterior ones by gonoducts in females. Recent work suggests that the nephridia are modified metanephridia.

The legs of some onychophorans, such as *Peripatus*, bear thin-walled eversible sacs or vesicles that open to the exterior near the nephridiopores by way

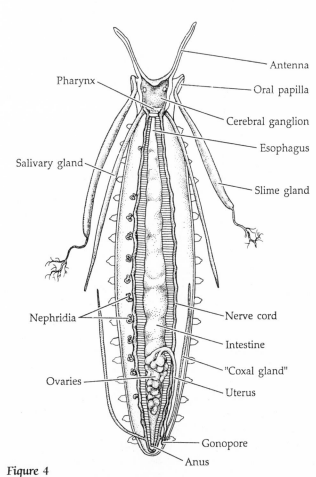

Figure 4
Internal anatomy of a generalized female onychophoran.

Labels: Pharynx, Salivary gland, Nephridia, Ovaries, Antenna, Oral papilla, Cerebral ganglion, Esophagus, Slime gland, Nerve cord, Intestine, "Coxal gland", Uterus, Gonopore, Anus

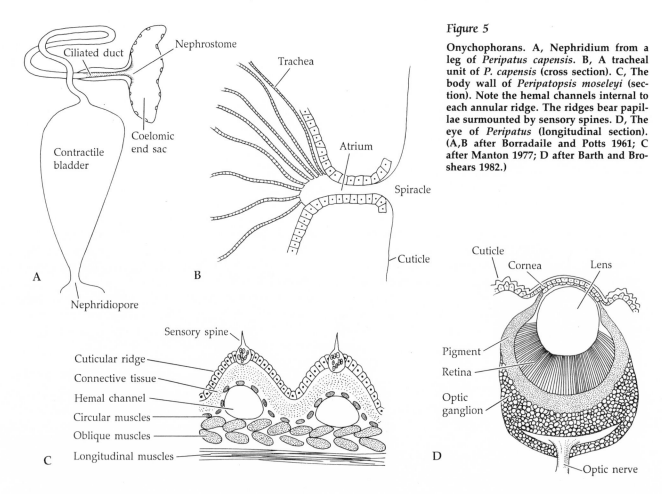

Figure 5

Onychophorans. A, Nephridium from a leg of *Peripatus capensis*. **B,** A tracheal unit of *P. capensis* (cross section). **C,** The body wall of *Peripatopsis moseleyi* (section). Note the hemal channels internal to each annular ridge. The ridges bear papillae surmounted by sensory spines. **D,** The eye of *Peripatus* (longitudinal section). (A,B after Borradaile and Potts 1961; C after Manton 1977; D after Barth and Broshears 1982.)

of minute pores or slits. These vesicles may function in taking up moisture, as do the coxal glands of many myriapods, insects, and arachnids. They are everted by hemocoelic pressure and pulled back into the body by retractor muscles.

Nervous system and sense organs. The nervous system of onychophorans is somewhat ladder-like in structure. A large bilobed cerebral ganglion ("brain") lies dorsal to the pharynx. A pair of ventral nerve cords are connected by transverse segmental commissures. The cerebral ganglion supplies nerves to the antennae, eyes, and oral region. A ganglionic swelling occurs in each segment and gives rise to paired nerves to the appendages and the body wall. The general body surface, especially the larger tubercles, are supplied with sensory hairs.

There is a small dorsolateral eye at the base of each antenna. The eyes are of the direct type, with a large chitinous lens and a relatively well developed retinal layer (Figure 5D). Onychophorans are nocturnal and photonegative.

Reproduction and development. Onychophorans are gonochoristic. Females have a pair of largely fused ovaries in the posterior region of the body (Figure 4). Each ovary connects to a gonoduct; the gonoducts fuse as a uterus. The end of the uterus opens through a gonopore at the back end of the body. Males have a pair of elongate, separate testes. Paired sperm ducts join to form a single tube in which sperm are packaged into spermatophores up to 1 mm in length. The male gonopore is also located poster-oventrally.

Copulation has been observed in only a few onychophorans. In the southern African *Peripatopsis* the male deposits spermatophores seemingly at random on the general body surface of the female. The presence of the spermatophores stimulates special amebocytes in the blood to bring about a localized breakdown of the integument beneath the spermatophore. Sperm then pass from the body surface of the female into her hemocoelic fluid, through which they eventually reach the ovaries where fertilization takes place. In some onychophorans a portion of the uterus is expanded as a seminal receptacle, but sperm transfer in these species is not well understood.

The limited embryological work done on onychophorans suggests that, despite fundamental similarities between the onychophorans and both annelids and arthropods, they also possess some very unusual features. For example, onychophorans may be oviparous, viviparous, or ovoviviparous. Females of oviparous species (e.g., *Oöperipatus*) bear an ovipositor and produce large, oval, yolky eggs with chitinous shells. Evidence suggests that this is the primitive onychophoran condition, even though living oviparous species are rare. The eggs of oviparous onychophorans contain so much yolk that intralecithal superficial early cleavage takes place, with the eventual formation of a germinal disc similar to that seen in many terrestrial arthropods.

Most living onychophorans, however, are viviparous and have evolved a highly specialized mode of development associated with small, spherical, yolkless eggs. Interestingly, most Old World viviparous species develop at the expense of maternal nutrients but lack a placenta, whereas all New World viviparous species have a placental attachment to the oviducal wall (Figure 6). Placental development is viewed as the most advanced condition in the onychophorans.

The yolky eggs of lecithotrophic species have a typical centrolecithal organization. Cleavage is by intralecithal nuclear divisions, similar to that seen in many groups of arthropods. Some of the nuclei migrate to the surface and form a small disc of blastomeres that spreads to eventually cover the embryo as blastoderm, thus producing a periblastula. Simultaneously, the yolk mass divides into a number of "yolk spheres" (Figures 6B, C, and D).

Yolkless and yolk-poor eggs are initially spherical, but once within the oviduct they swell to become ovate. As cleavage ensues, the cytoplasm breaks up into a number of spheres. Nucleate spheres are the

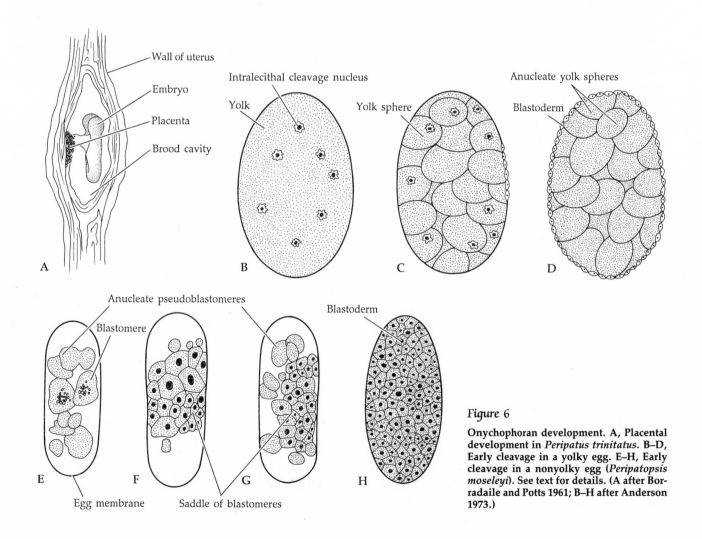

Figure 6

Onychophoran development. A, Placental development in *Peripatus trinitatus*. B–D, Early cleavage in a yolky egg. E–H, Early cleavage in a nonyolky egg (*Peripatopsis moseleyi*). See text for details. (A after Borradaile and Potts 1961; B–H after Anderson 1973.)

blastomeres, and the anucleate ones are called **pseudoblastomeres** (Figures 6E and F). The blastomeres divide and form a saddle of cells on one side of the embryo (Figure 6G). The pseudoblastomeres disintegrate and are absorbed by the dividing blastomeres. The disc expands to cover the embryo with a one-cell-thick blastoderm around a fluid-filled center.

Placental species have even smaller eggs than nonplacental species do, and the egg does not swell after release from the ovary. Further, these eggs are not enclosed in membranes. Cleavage is total and equal, yielding a coeloblastula. The embryo then attaches to the oviducal wall and proliferates as a flat **placental plate**. As development proceeds, the embryo moves progressively down the oviduct and eventually attaches in the uterus. Gestation may be quite long, up to 15 months, and the oviduct/uterus often contains a series of developing embryos of different ages.

Development after formation of the blastula is remarkably similar in the few species of onychophorans that have been studied. Unlike gastrulation in annelids, which involves the movement of presumptive areas of the blastula into their organ-forming positions, gastrulation in onychophorans involves very little actual cell migration. Cells of the presumptive areas undergo immediate organogenesis by direct proliferation. This process involves the proliferation of small cells into the interior of the embryo through and around the yolk mass or the fluid-filled center, and the production by surface cells of the germinal centers of limb buds and other external structures. All onychophorans have direct development. There is no substantial evidence suggesting that they ever had larval stages that might have become suppressed in their evolutionary past. In all species that have been studied, the full complement of segments and adult organ systems are attained before they hatch or are born as juveniles.

Onychophorans are also unusual in that neither a presegmental acron nor a postsegmental pygidium or telson can be clearly differentiated. Recall that in annelids and arthropods a segmental growth zone occurs in front of the anus, and the ectoderm around the anus eventually forms a terminal, appendageless telson or pygidium. In onychophorans, even though growth is teloblastic, the growth zone from which the trunk segments arise appears to be postanal. When the last mesoderm has been formed, the growth-zone ectoderm apparently develops directly into the anal somite with no postsegmental ectoderm remaining. Nevertheless, the last body segment of onychophorans, like that of arthropods, does lack appendages, a fact that leaves the matter a bit unsettled.

Onychophorans probably live for several years, during which time periodic molting takes place, as often as every two weeks in some species.

Phylum Tardigrada

The tardigrades make up a small phylum that appears to be closely tied to the annelid–arthropod line. The first tardigrade was discovered in 1773. Since then over 400 species have been described. Most live in semiaquatic habitats such as the water films on mosses, lichens, liverworts, and certain angiosperms, or soil and forest litter. Others live in various freshwater and marine benthic habitats, both deep and shallow, often interstitially or among shore algae. A few have been reported from hot springs. Some marine species are commensals on the pleopods of isopods or gills of mussels; others are parasites on the epidermis of holothurians or barnacles. They occasionally occur in high densities, to 300,000 per square meter in soil and more than 2,000,000 per square meter in moss. All are small, usually on the order of 0.1–0.5 mm in length, although some 1.7-mm giants have been reported.

Under the microscope, tardigrades resemble miniature eight-legged bears, and even move with an ursine lumbering gait—hence the name Tardigrada (Latin *tardus*, "slow"; *gradus*, "step"). Their locomotion, paunchy body, and clawed legs have earned them the nickname "water bears" (Figure 7). A few fossil specimens from Cretaceous amber have been tentatively identified as tardigrades, but, as in the Onychophora, a clear positive fossil record is lacking. The phylum comprises eight families in three orders: Heterotardigrada, Mesotardigrada, and Eutardigrada. The orders are defined largely on the basis of details of the head appendages, the nature of the leg claws, and the presence or absence of "Malpighian tubules."

Most tardigrade species are widespread and many are cosmopolitan. A major factor in their wide distribution is perhaps the fact that their eggs, cysts, and **tuns** (see below) are light enough and resistant enough to be carried great distances either by the winds or on sand and mud clinging to the feet of insects, birds, and other animals. The minute size and precarious habitats of water bears have resulted in their acquisition of numerous traits also seen in some pseudocoelomate groups that live in similar habitats. These features are discussed below.

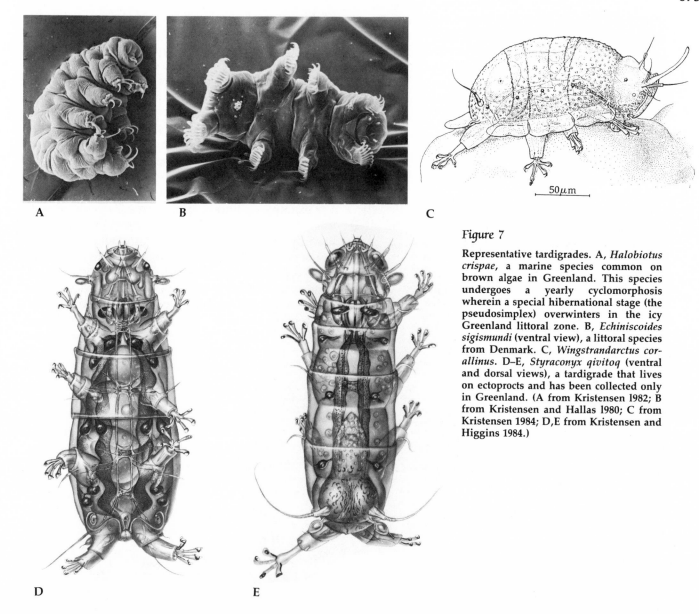

A   B   C

50μm

D   E

Figure 7

Representative tardigrades. A, *Halobiotus crispae*, **a marine species common on brown algae in Greenland. This species undergoes a yearly cyclomorphosis wherein a special hibernational stage (the pseudosimplex) overwinters in the icy Greenland littoral zone. B,** *Echiniscoides sigismundi* **(ventral view), a littoral species from Denmark. C,** *Wingstrandarctus corallinus*. **D–E,** *Styraconyx qivitoq* **(ventral and dorsal views), a tardigrade that lives on ectoprocts and has been collected only in Greenland. (A from Kristensen 1982; B from Kristensen and Hallas 1980; C from Kristensen 1984; D,E from Kristensen and Higgins 1984.)**

Tardigrades are well known for their remarkable powers of **anabiosis** (a state of greatly reduced metabolic activity during unfavorable environmental conditions) and **cryptobiosis** (an extreme state of anabiosis, or dormancy, in which all external signs of metabolic activity are absent). During dry periods, when the vegetation inhabited by terrestrial tardigrades becomes desiccated, these little creatures can encyst by pulling in their legs, losing body water, and secreting a double-walled cuticular envelope around the shriveled body. Such cysts maintain a very low basal metabolism. Further reorganization of the body (or "deorganization") can result in a single-walled tun stage, in which body metabolism is undetectable (a cryptobiotic state).

The resistant qualities of the tardigrade tun have been demonstrated by experiments in which individuals have recovered after immersion in extremely toxic compounds such as as brine, ether, absolute alcohol, and even liquid helium. They have survived temperatures ranging from +149°C to −272°C, on the brink of absolute zero. They have also survived high vacuums, intense ionizing radiation, and long periods with no environmental oxygen whatsoever. Some Soviet zoologists claim that tardigrades have survived experiments in outer space.

Following desiccation, when water is again available, the animals swell and become active within a few hours. Many rotifers, nematodes, mites, and a few insects are also known for their anabiotic powers, and these groups often occur together in the surface water of plants such as mosses and lichens. One marine tardigrade (*Echiniscoides*) survives quite well with a life cycle that regularly alternates between the active and tun stages, including an experimentally induced cycle forcing it to undergo cryptobiosis every

six hours! Evidence indicates that the tardigrade aging process largely ceases during cryptobiosis, and by alternating active and cryptobiotic periods a tardigrade life span may be extended to several decades. One rather sensational report described a dried museum specimen of moss that yielded living tardigrades when moistened after 120 years on the shelf!

In certain areas of extreme environmental conditions, marine tardigrades may undergo an annual cycle of **cyclomorphosis** (rather than the cryptobiosis typical of terrestrial and freshwater forms). During cyclomorphosis, two distinct morphologies alternate. For example, a littoral species from Greenland, *Halobiotus crispae* (Figure 7A), has a summer morph and a winter morph, the latter being a special hibernational stage that is resistant to freezing temperatures and perhaps low salinities. In contrast to cryptobiotic tuns, winter cyclomorphosis stages are active and motile. Cyclomorphosis is coupled with gonadal development, and in *H. crispae* only the summer morph is sexually mature.

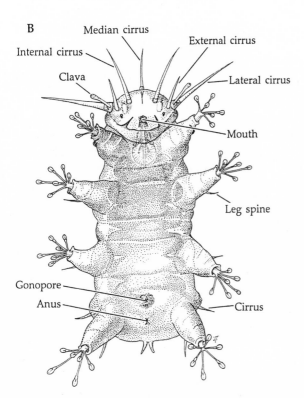

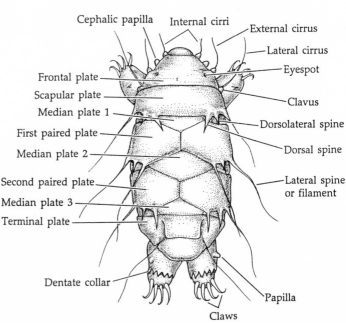

Figure 8

Tardigrade anatomy. A, *Wingstrandarctus corallinus* (ventral view), an inhabitant of shallow, sandy marine habitats in Australia and Florida. B, *Batillipes noerrevangi* (ventral view). C, Generalized *Echiniscus* (dorsal view). (A from Kristensen 1984; B contributed by R. Kristensen; C after Morgan and King 1976.)

The tardigrade *Bauplan*

The body of a tardigrade bears four pairs of ventrolateral legs. The legs are short hollow extensions of the body wall, and each terminates in one or as many as a dozen or so "toes," with adhesive pads or with claws resembling those of onychophorans (Figures 8, 9, and 10). In some, the legs are partially telescopic.

The body is covered by a thin, uncalcified cuticle that is periodically molted. It is often ornamented and occasionally divided into symmetrically arranged dorsal and lateral (rarely ventral) plates. The plates may be homologous with the sclerites of arthropods, but this is not certain. The cuticle shares some features with both annelids and arthropods, but it is also unique in certain ways. It comprises up to seven distinguishable layers, contains various tanned proteins and chitin, occasionally has a wax layer, and lines the foregut and rectum. The cuticle is secreted by an underlying epidermis composed of a constant cell number in many (but not all) species. Eutely is common in minute metazoa, and we discussed several examples among the pseudocoelomate phyla, notably rotifers. Growth proceeds by molts, as in onychophorans and arthropods, with sexual maturity being attained after three to six instars.

Although the body is quite short, it is nevertheless homonomous and rather weakly cephalized. Nonmarine tardigrades are often colorful animals, exhibiting shades of pink, purple, green, red, yellow, gray, and black. Color is determined by cuticle pigments, the color of the food in the gut, and the presence of granular bodies suspended in the hemocoel.

Like the coelom of arthropods and onychophorans, the coelom of tardigrades is greatly reduced, and in adults it is confined largely to the gonadal cavities. The main body cavity is thus a hemocoel, and the body fluid directly bathes the internal organs and body musculature. The musculature of tardigrades is very different from the annelid-like arrangement seen in onychophorans, in which the body wall muscles are in sheetlike layers—in tardigrades there is no circular muscle layer in the body wall, and the muscles occur in separate bands extending between subcuticular attachment points, as they do in arthropods (Figure 9).

It was long thought that tardigrades possessed only smooth muscle, in contrast to the striated muscles of arthropods, and in the past this feature was used as an argument against a close relationship between these two phyla. However, recent work by the Danish zoologist R. M. Kristensen has shown that both smooth and striated muscles occur in tardigrades, the latter predominantly in the most primitive species. The striated muscles are of the arthropod type, being cross-striated, rather than obliquely striated like those in the onychophorans. Numerous fine-structural details of the muscle attachment regions are also shared between tardigrades and arthropods. Kristensen has suggested that a partial shift from arthropod-like striated muscle to smooth muscle in some tardigrades might have accompanied a transition from the marine to the terrestrial environment, perhaps being functionally tied to the phenomenon of cryptobiosis. Furthermore, both slow and fast nerve fibers may occur in tardigrades, the former predominating in the somatic musculature, the latter in the leg musculature. However, the leg musculature appears to be entirely extrinsic, like that of onychophorans, with one attachment near the tip of the leg, and the other within the body proper. Most of the muscle bands in tardigrades consist of only a single or a few large muscle cells each.

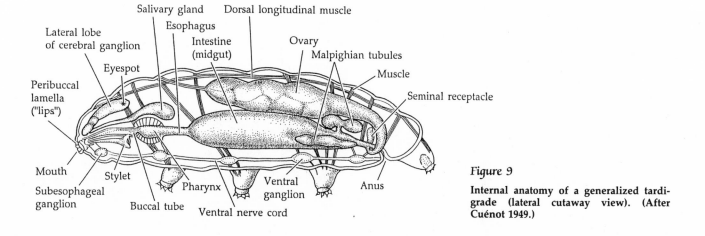

Salivary gland Dorsal longitudinal muscle
Esophagus
Lateral lobe Intestine Ovary
of cerebral ganglion (midgut) Malpighian tubules
Eyespot Muscle
Peribuccal Seminal receptacle
lamella
("lips")
Mouth
Subesophageal Stylet Pharynx Ventral Anus
ganglion ganglion
Buccal tube Ventral nerve cord

Figure 9

Internal anatomy of a generalized tardigrade (lateral cutaway view). (After Cuénot 1949.)

Locomotion. The concentration of muscles as discrete units and the thickening of the cuticle resulted in a major shift in locomotor strategies away from the primarily hydrostatic system used in annelids and onychophorans. Instead, tardigrades use a step-by-step gait controlled by independent antagonistic sets of muscles or by flexor muscles that work against hemocoelic pressure. The claws, pads, or discs at the ends of the legs are used for purchase and for clinging to objects, such as strands of vegetation or sediment particles (Figure 10). At least one marine species is capable of limited jellyfish-like swimming, by use of a bell-shaped expansion of the cuticle margin to keep it suspended just above the sediment.

Feeding and digestion. Water bears usually feed on the fluids inside plant or animal cells, by piercing the cell walls with a pair of **oral stylets**. Soil-dwelling species feed on bacteria, algae, and decaying plant matter or are carnivorous on small invertebrates. Predatory and omnivorous tardigrades have a terminal mouth; herbivorous and detritivorous ones have a ventral mouth.

The mouth opens into a short stomodeal buccal tube, which leads to a bulbous, muscular pharynx (Figure 9). A pair of large salivary glands flanks the esophagus and produces digestive secretions that empty into the mouth cavity; these glands also are responsible for production of a new pair of oral stylets with each molt (hence they are often referred to as **"stylet glands"**). The muscular pharynx produces a suction that attaches the mouth tightly upon the prey item during feeding and pumps the cell fluids out of the plant or animal prey and into the gut. In many species there is a characteristic arrangement of chitinous **rods**, or **placoids**, within an expanded region of the pharynx. These rods provide "skeletal" support for the musculature of that region and possibly contribute to the masticating action. The pharynx empties into an esophagus, which in turn opens into a large intestine (midgut), where digestion and absorption take place. The short hindgut (the cloaca or rectum) leads to a terminal anus. In some species defecation accompanies molting, the feces and cuticle being abandoned together.

At the intestine–hindgut junction in freshwater species are three or four large glandular structures that are called Malpighian tubules, each consisting of only about three to nine cells. The precise nature of these organs is not well understood, but they are probably not homologous to Malpighian tubules of arthropods. In at least one tardigrade genus (*Halobiotus*), the Malpighian tubules are greatly enlarged and have an osmoregulatory function. It is probable that some excretory products may be absorbed through the gut wall and eliminated with the feces, and other products may be deposited in the old cuticle prior to molting.

Circulation and gas exchange. Perhaps because of their small size and moist habitats, tardigrades have no traces of discrete blood vessels or gas exchange structures; consequently, they rely on diffusion through the body wall and extensive body cavity. The body fluid contains numerous cells credited with a storage function.

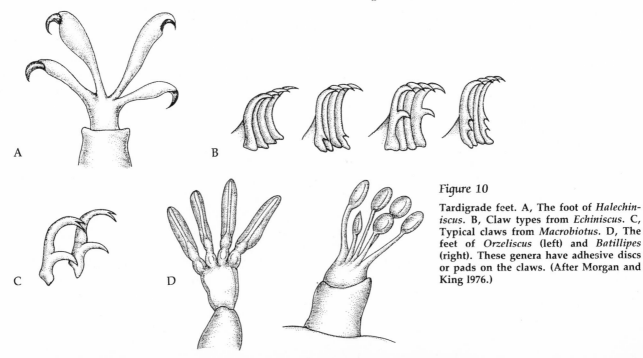

Figure 10

Tardigrade feet. A, The foot of *Halechiniscus*. B, Claw types from *Echiniscus*. C, Typical claws from *Macrobiotus*. D, The feet of *Orzeliscus* (left) and *Batillipes* (right). These genera have adhesive discs or pads on the claws. (After Morgan and King 1976.)

Nervous system and sense organs. The nervous system is built on the annelid–arthropod plan and is distinctly metamerous. A large, lobed, dorsal cerebral ganglion is connected to a subesophageal ganglion by a pair of commissures surrounding the buccal tube (Figure 9). From the subesophageal ganglion, a pair of ventral nerve cords extends posteriorly, connecting a chain of four ganglia that serve the four pairs of legs. Sensory bristles or spines occur on the body, particularly in the anterior and ventral region and on the legs. The structure of these bristles is probably homologous to that of arthropod setae (Figure 11). A pair of sensory eyespots is often present, each consisting of five cells, one of which is a pigmented light-sensitive cell. The anterior end of many tardigardes bears long sensory cirri, and most species also have a pair of hollow anterior cirri called **clava** that are probably chemosensory in nature. The clava appear structurally similar to the olfactory setae of many arthropods.

Reproduction and development. Tardigrades are typically gonochoristic, with both sexes possessing a single saclike gonad lying above the gut. In males, the gonad terminates as two sperm ducts that extend to a single gonopore, opening just in front of the anus or into the rectum. In females the single oviduct (right or left) either opens through a gonopore dorsal to the anus or opens into the rectum (in this case, a **cloaca**) (Figure 9). There are either two complex seminal receptacles that open separately or a single, small seminal receptacle that opens into the rectum near the cloaca.

Males are unknown in some genera, but most tardigrades that have been studied copulate and lay eggs. Dwarf males have been recently discovered in several genera. In some tardigrades, sperm are deposited directly into the female's seminal receptacle (or cloaca) or into the body cavity by cuticular penetration. In the latter case fertilization takes place in the ovary. In others, a wonderfully curious form of indirect fertilization takes place; the male deposits sperm beneath the cuticle of the female prior to her molt and fertilization occurs when she later deposits eggs into the shed cuticular cast. Several studies have shown tardigrade sperm to be flagellated. In at least a few species, a very primitive courtship behavior exists, wherein the male strokes the female with his cirri. Thus stimulated, the female deposits her eggs on a sand grain, upon which the male then spreads his sperm.

Females lay from 1 to 30 eggs at a time, depending on the species. In strictly aquatic species the fer-

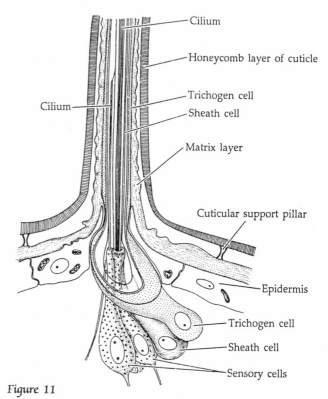

Figure 11

Base of an external "bristle" of *Batillipes noerrevangi* (longitudinal section), showing the relationship between the various cells and the cuticle. (After Kristensen 1981.)

tilized eggs are either left in the shed cuticle or glued to a submerged object. The eggs of terrestrial species bear thick sculptured shells that resist dry periods (Figure 12). Some species alternate between thin- and thick-walled eggs, depending on environmental conditions. Parthenogenesis may be common in some species, notably those in which males are unknown. Hermaphroditism has also been reported in a few genera.

The only reasonably complete studies of tardigrade embryology were published by E. Marcus in the 1920s. According to Marcus, whose work is now questioned by some specialists, development is direct and rapid. A blastula develops, with a small blastocoel, and eventually it proliferates an inner mass of entoderm, which later hollows to form the archenteron. Cleavage is described as holoblastic. Stomodeal and proctodeal invaginations develop, completing the digestive tube. Subsequent to gut formation, five pairs of archenteric coelomic pouches appear off the gut, reminiscent of the enterocoelous development of many deuterostomes. The first pair arises from the stomodeum (ectoderm) and the last four pairs from the midgut (entoderm). The two posterior

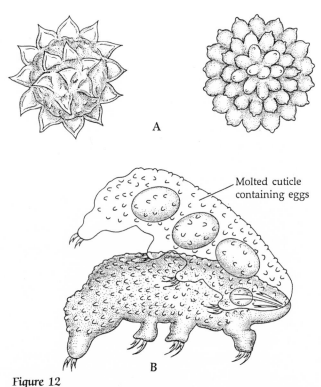

Figure 12

A, Sculptured eggs of tardigrades. B, A female *Hypsibius annulatus* in the process of molting an egg-containing cuticle. (After Morgan 1982.)

pouches fuse to form the gonad; the others disappear as their cells disperse to form the body musculature. Development is typically completed in 14 days or less, whereupon the young use their stylets to break out of the shell.

Juveniles lack adult coloration, have fewer lateral and dorsal spines and cirri, and can have reduced numbers of claws. At birth, the number of cells in the entire body is relatively fixed, and growth is primarily by increase in cell size rather than in cell number. In nature, water bears may live only a few months or may survive for a great many years.

The Pentastomida

The phylum Pentastomida comprises about 95 species of odd parasitic metazoa known as "tongue worms." These animals are all assigned to a single class (Pentastomata), divided into seven families in two orders, Cephalobaenida and Porocephalida (see Self 1982 for details). All are inhabitants of the lungs or nasal passageways of vertebrates, primarily rep-

tiles but also some mammals and birds.* The body is highly modified and reduced to a wormlike structure 2–13 cm in length (Figure 13). Adult appendages are reduced to two pairs of small anteroventral "legs," which are lobelike and bear chitinous claws, somewhat reminiscent of those in onychophorans and tardigrades. The "legs," often reduced to merely the claws, allow the parasite to cling to the host tissue.

The cuticle is nonchitinous, highly porous, and molted periodically, at least until the full adult size is attained. The body musculature shows characteristics of both annelids and arthropods. The muscles are arranged in circular and longitudinal sheets within the body wall, some running from segment to segment in a fashion reminiscent of annelids, but the cells are clearly cross-striated as they are in arthropods. The coelom is highly reduced and the principal body cavity is a hemocoel.

The mouth is without jaws and in some species sits on the end of a snoutlike anterior projection. Coupled with the two pairs of legs, these animals appear to have five "snouts" and this led to the original designation Pentastomida (Greek *penta*, five; *stomida*, mouths). The gut is a simple straight tube. The anterior end is muscular and serves as a pumping organ used to suck blood from the hosts. **Frontal glands**, present in some species, are thought to produce secretions that break down host tissue and serve as an anticoagulant. The gut and reproductive organs are suspended within the hemocoel.

The nervous system is constructed on the typical annelid–arthropod plan, with up to five identifiable ganglia concentrated in the head region. A few sensory papillae may occur in the head and caudal regions. The gas exchange, circulatory and excretory organs have been entirely lost, a feature characteristic of many metazoan endoparasites.

Pentastomids are gonochoristic, and females are larger than males. Fertilization is internal and takes place in the definitive host. The sperm are transferred to the female's vagina by way of the male cirrus during copulation. Early embryos have three pairs of legs and clearly defined coelomic pouches. The early embryo metamorphoses into a **primary larva** with two pairs of double-clawed legs and one or more piercing stylets (Figure 13E). There is a striking resemblance between pentastomid larvae and crustacean nauplii, suggesting a possible homology between the two.

The primary larvae are autoinfective to the defin-

*There are some recent reports of pentastomids from the human eye.

itive host, but they can also migrate to the host's gut and pass out with the feces. Many pentastomids require an intermediate host, which may be almost any kind of vertebrate. Larvae bore through the gut wall of the intermediate host, where they undergo further development to the infective stage. Once the intermediate host is consumed by a definitive host (usually a predator), the parasite makes its way from the new host's stomach up the esophagus or bores through the intestinal wall, eventually settling in the lungs, nasal passages, or bronchi. Here they feed on the host's blood.

Convincing cases that pentastomids are actually highly modified crustacean parasites, probably branchiurans (or at least maxillopodans), have been presented by Wingstrand (1972), Riley et al. (1978), Grygier (1983), Abele et al. (1989), and others. In addition to larval similarities, pentastomids have branchiuran-like sperm morphology and embryogenesis. Cuticular fine structure and arrangement of the nervous system is also crustacean-like. The somewhat anne-lid-like musculature may be an adaptation associated with a vermiform body and parasitic lifestyle.

Phylogeny and evolution of the arthropods and their allies

From annelids to arthropods

As indicated in the preceding four chapters, we favor a monophyletic view of the taxon Arthropoda, and we give the group phylum status. Although the view that arthropods are a natural group is both the traditional and probably the most popular view, it is by no means a universally held opinion. However, we support monophyly not out of tradition but out of evaluation of the evidence. The rest of this chapter is devoted to this problem and related issues, and examines the hypothesis that "arthropodization" occurred only once. We discuss why this hypothesis is the most parsimonious and biologically reasonable one among several alternatives. In discussing arthro-

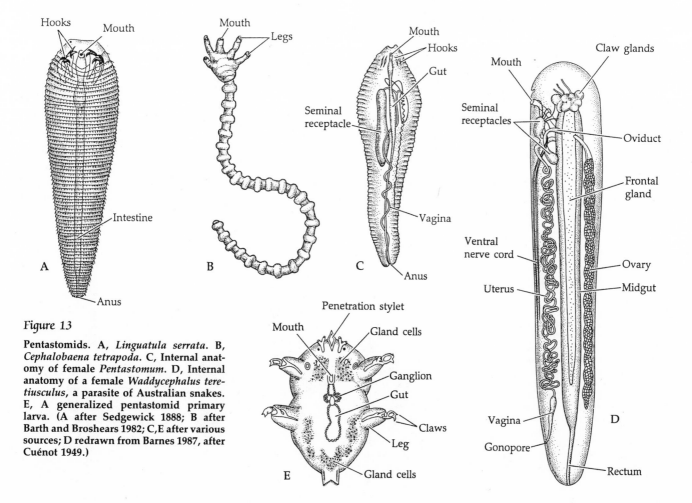

Figure 13

Pentastomids. **A**, *Linguatula serrata*. **B**, *Cephalobaena tetrapoda*. **C**, Internal anatomy of female *Pentastomum*. **D**, Internal anatomy of a female *Waddycephalus teretiusculus*, a parasite of Australian snakes. **E**, A generalized pentastomid primary larva. (A after Sedgewick 1888; B after Barth and Broshears 1982; C,E after various sources; D redrawn from Barnes 1987, after Cuénot 1949.)

pod evolution we tread somewhat on thin ice, for the matter is fraught with controversy. The older literature (and some recent work) largely comprises narrative scenarios about how one group might have evolved from some other group, often based on preconceived notions of hypothetical ancestors or prototypes. Some workers have confused issues by their method of emphasizing *differences* among groups, rather than searching for shared similarities among groups (see Chapter 2)* This is not to say that the problem will become entirely tractable by changes in analytical methodology, but it will certainly become less ambiguous, more reasonable, and, most important of all, more testable.

Problems also exist within the data base itself. For example, there is a shortage of good fossil material for onychophorans, tardigrades, pentastomids, and many primitive arthropod groups. On the other hand, many arthropod-like fossils are so bizarre as to defy reasonable connection with any existing taxa (e.g., Figure 14). There is also a paucity of comparative embryological data for many extant groups. Conflicting theories of arthropod evolution have been based largely on comparative anatomy and, to a lesser degree, development of living groups, and the situation is not likely to change until other kinds of data become available.

Recently a few workers have begun to establish testable hypotheses of arthropod evolution by way of strict genealogy—that is, by uniting groups on the basis of their unique, shared derived characteristics (i.e., synapomorphies). We have constructed a cla-

*Reliance upon differences among taxa in the construction of phylogenetic theories typically demands the invocation of numerous ad hoc hypotheses of convergence, and results in the creation of paraphyletic taxa. Reliance on similarities among taxa in the construction of phylogenetic theories typically reduces the need for such ad hoc hypotheses, and results in the creation of monophyletic taxa.

dogram of the annelid–arthropod line in this fashion (Figure 15) on the basis of information presented in the preceding discussions of arthropods and their allies; and we discuss below the evolutionary implications of the phylogeny it depicts.

There is little argument that the annelids and arthropods are closely related, although ancient, phyla (both groups are probably Precambrian in origin). The body plans of these two phyla are more similar to one another than to any other major protostome group. The shared features of the segmented body, with its paired segmental external and internal structures, are powerful evidence of close ancestral ties. Further homologies are seen in the nature of the regionalized or lobed brain and its circumenteric connectives, meeting with a subenteric ganglion from which the paired ventral nerve cords arise. Strong similarities are also seen in the patterns of development and growth, including the addition of body segments by teloblastic growth just in front of the postsegmental pygidium (in annelids) or telson (in most arthropods).

There is also little argument with the hypothesis that the onychophorans and tardigrades are related to this annelid–arthropod line in some fashion. There

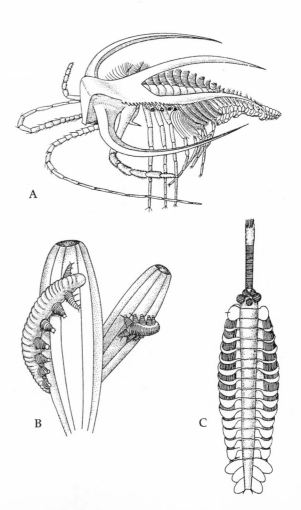

Figure 14

Three unusual extinct arthropods from the fossil record. A, *Marrella splendens* (reconstruction in oblique view), standing on the anterior three walking legs (legs of right side omitted; gill branches 1–4 and 10–26 removed). B, *Aysheaia pedunculata*. This reconstruction suggests that the animals preyed on marine sponges. C, *Opabinia regalis* (dorsal view). Note the astonishing apparent presence of multiple compound eyes. It is unclear whether the mouth sits at the end of the long cephalic proboscis or whether this structure merely grabbed food and transported it to a mouth located on the main head region. (A after Whittington 1971, Geol. Surv. Can. Bull. 209:1–24; B after Whittington 1978, Philos. Trans. R. Soc. London Ser. B 284:165–197; C after Whittington 1975, Philos. Trans. R. Soc. London B Ser. 271:1–43.)

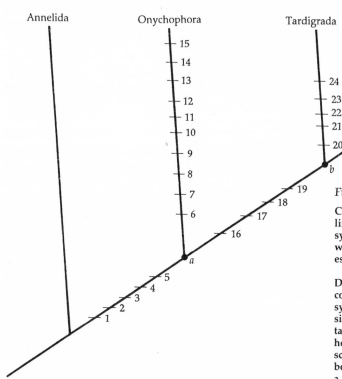

Annelida Onychophora Tardigrada First Arthropoda (protoarthropod)

Figure 15

Cladogram depicting the relationships of the annelid–arthropod line and the monophyletic origin of the phylum Arthropoda. The synapomorphies used to construct the cladogram are listed below, with some annotations. See the text for additional discussion, especially regarding hypothetical ancestors *a* and *b*.

1, Loss of ectodermal cilia; 2, Elongate dorsal gonads; 3, Development of ventrolateral appendages; 4, Reduction of the coelom (and development of the hemocoel and open circulatory system with dorsal, tubular, ostiate heart within a pericardial sinus); 5, Growth by ecdysis; 6, Suppression of external segmentation; 7, Oblique muscle layer in body wall; 8, Subcutaneous hemal channels; 9, Unique oral papillae; 10, Body papillae and scales; 11, Slime glands;* 12, Nonmigratory gastrulation; 13, Lobopods, with pads and claws; 14, Coxal glands (viewed here as a convergent feature to the "true" coxal glands in certain arthropods); 15, Tracheal system (probably convergent to the tracheal system of arthropods); 16, Origin of arthropod setae; 17, Loss of serially arranged nephridia; 18, Shift from lobopodal locomotion to true leg-gait movement; 19, Loss of sheets of annelid-like musculature and evolution of discrete, segmental, cross-striated muscle bands attached to cuticular apodemes; 20, Tardigrade "Malpighian tubules"†; 21, Tardigrade leg claws; 22, Buccal stylets; 23, Suite of character states associated with miniaturization (e.g., reduction in somite number, loss of heart and gas exchange organs, and eutely); 24, Cryptobiosis; 25, Lateral compound eyes; 26, Calcification of the cuticle; 27, Fully segmental sclerites; 28, Articulating, jointed appendages with intrinsic musculature; 29, Appendages more ventrally positioned; 30, Increased regional specialization of the gut; 31, Loss of all motile cilia and flagella, except in sperm of some groups; 32, Cephalic ecdysial glands.

are, however, two principal areas where controversies rage. First is the exact phylogenetic position of the onychophorans and tardigrades. That is, are these two groups separate phyla, and if so, where do they belong in an evolutionary tree depicting the origin of the arthropods? Or, should they actually be considered as members of the phylum Arthropoda itself? The second controversy centers on whether the arthropods are themselves a monophyletic, polyphyletic, or paraphyletic group. Just about every conceivable combination of possibilities has been proposed at one time or another in answer to these questions.

Occasionally one still sees reference to a possible "aschelminth" (i.e., pseudocoelomate) relationship for the tardigrades. Let us dispense quickly with this idea by pointing out that such attributes as cellular eutely of the epidermis, absence of circulatory and gas exchange organs, similar pharyngeal structures, reduction of the coelom, and cryptobiosis almost certainly represent features acquired by way of convergent evolution, probably largely as a result of the lifestyle and extreme miniaturization of these animals. It is unlikely that possession of these features has anything to do with direct evolutionary ties to the rotifers, nematodes, or any other pseudocoelomate group. Tardigrades are coelomate animals, and

they possess characteristics that would appear to link them inextricably to the annelid–arthropod line, as discussed below. It is also worth noting, if only to point out that phylogenetic theories are indeed tested and falsified, that throughout most of the nineteenth century the notion persisted that the vertebrates arose from the arthropods. The last serious treatment of this now curious theory was presented in a 1912

---

*There is no evidence that slime glands occurred in the marine, fossil, onychophoran-like genus *Aysheaia*, and slime-squirting would probably have been rather ineffective in the sea. Hence, these organs may have evolved subsequent to the origin of what we now recognize as terrestrial onychophorans, thus defining a distinct terrestrial clade within the taxon Onychophora.

†"Malpighian tubules" may not represent a synapomorphy for the tardigrades as a whole, but rather for a freshwater/terrestrial clade within the taxon Tardigrada.

book by paleontologist William Patten, *The Evolution of the Vertebrates and Their Kin.*

The key to understanding arthropod evolution lies in establishing the relationships among the four subphyla covered in Chapters 15 through 18 and the two small groups Onychophora and Tardigrada. Thus, our analysis focuses on evidence of evolutionary change in *Baupläne* from the annelid to the arthropod condition.

In reviewing the *Baupläne* of the onychophorans and tardigrades, it is clear that they share certain features with the annelids and other protostome phyla. These features include an uncalcified cuticle not broken up into distinct articulating segmental sclerites, certain aspects of body musculature, a relatively simple gut, minimal cephalization, unjointed appendages that are simple saclike outgrowths of the body wall, paired segmental nephridia (in onychophorans), and a ciliated reproductive tract. Because these features are shared with other, more distantly related protostome phyla, they must be regarded as primitive characters within the onychophorans and tardigrades; that is, features of basic body organization that have persisted from an earlier lineage predating the appearance of these two taxa. As such, these plesiomorphic features hint only at *distant* evolutionary relationships. To examine the *immediate* evolutionary relationships of the onychophorans and tardigrades, we must search for attributes shared uniquely by them or with the arthropods. Figure 15 is a cladogram depicting the appearance of some of these features—synapomorphies that define sister-groups within the annelid–arthropod line.

By viewing these taxa in this fashion, one quickly sees that, indeed, an entire body plan is uniquely shared among the onychophorans, tardigrades, and arthropods. This basic plan is that of an annelid-like body in which the true coelom has become reduced to nothing more than chambers housing the gonads and part of their ducts, and often a portion of the nephridia. In place of the coelom, a large hemocoel remains to fill the body and bathe the internal organs and musculature in hemocoelic fluid. This radical shift in body organization was associated with an increased sclerotization of the cuticle and concomitant loss of external cilia, establishment of growth by ecdysis (molting), loss of a closed circulatory system (with retention of, but alterations in, the dorsal tubular blood vessel), and appearance of ventrolateral body appendages. Of course, each of these newly derived features represents a complex suite of many interrelated characters. For example, growth by ecdysis involves a highly integrated system of glands,

neurotransmitters, and programmed cellular responses throughout the body. The possibility that all of these complex features evolved independently, by way of convergent evolution, seems highly unlikely.

As we mentioned in Chapter 15, these complex and fundamental transformations set the stage for the evolution of a series of new features that is referred to as **arthropodization**. These and other characteristics that may be used to demarcate the sequence of appearances of these taxa are portrayed in Figure 15. Although other synapomorphies exist, we have restricted the cladogram primarily to those features discussed in the previous four chapters. Figure 16A illustrates a more traditional expression of the relationships depicted in the cladogram, by way of an evolutionary tree set against a geological time scale. Figures 16B–G show several other, different evolutionary trees that have been proposed for these taxa in the past.

The body plan summarized above—a largely homonomous, annelid-like, hemocoelic creature with growth punctuated by periodic molting—describes the ancestral onychophoran–tardigrade–arthropod *Bauplan*. The hypothetical ancestral creature predicted by the cladistic analysis is indicated by "*a*" on Figure 15. Whether this ancestral form was itself derived from a polychaete-like ancestor, an oligochaete-like ancestor, or something that was not even an annelid (as we define annelids) is uncertain (see Chapter 24). In any case, this ancestral creature lacked lateral body appendages (such as the parapodia of polychaetes), and instead probably had ventrolateral lobopodal legs. It is unlikely that ventrolateral lobopods evolved directly from parapodia, as the two kinds of appendages do not appear to be homologous.

From this ancestral animal arose the onychophorans. Most textbooks place the onychophorans in a phylum by themselves largely to avoid the confusing issues of their relationship. Manton and others (Tiegs, Hessler, Newman) have suggested they be included in the taxon Uniramia as the sister-group of the insects and myriapods. Still others have suggested that the onychophorans should be allied with the tardigrades, and perhaps even with the pentastomids, in a separate clade. This confusion has existed partly because previous workers have tended to treat all the characters shared among the onychophorans and other taxa more-or-less equally. That is, primitive features (symplesiomorphies) as well as derived features have been used in attempts to assess evolutionary relationships. By restricting the phylogenetic analysis to uniquely derived attributes (synapomorphies), a clearer picture emerges. The position of the onychophorans in this cladogram (Figure 15) depicts their origin *prior* to the appearance of the tardigrades or the arthropods.

Figure 15 denotes the synapomorphies that uniquely define the Onychophora. The tracheal system and perhaps the coxal glands appear to be convergent to similar structures in some other terrestrial arthropods; clearly, they are adaptations to life on land. Onychophorans retain the serial, segmental, paired body nephridia, hydrostatic locomotion, extrinsic appendage musculature, and basic body musculature of the metameric ancestor at the base of the entire annelid–arthropod clade.

Subsequent to the appearance of the onychophorans, an ancestral line arose in which the sheet-like body musculature was replaced by discrete muscle bands attached to nodes (apodemes) on the inner cuticular surface. Accompanying this transition was a loss of the serially-repeated segmental body nephridia, and evolution of what may be called a true leg-gait type of locomotion. These changes demarcated a major step in the direction of arthropodization. However, this creature was not yet a true arthropod. It was from this ancestral form (indicated by "*b*" on the cladogram in Figure 15) that both the tardigrades and the arthropods might have arisen.

The most primitive living tardigrades are marine species, a fact suggesting that this group also arose from an early marine stock. The evolution of the tardigrades was largely a process of miniaturization, and many of the unique attributes recognized in tardigrades appear to be linked to the dramatic reduction in body size and the shift from a marine ancestor to a freshwater–terrestrial lifestyle. Because of their small size and semiaquatic habits, tardigrades apparently experienced no great evolutionary pressure favoring aerial gas exchange structures (e.g., tracheae) or specific environmental water absorption devices (e.g., coxal glands) as in some other terrestrial lineages arising from the annelid–arthropod line.

Thus, it was probably not until after the appearance of both the onychophorans and tardigrades that the important final steps of arthropodization actually began. Figure 15 (see legend) notes the major synapomorphies that define the arthropods; most of these are the traditional characters used to diagnose the phylum. Many of these characters are actually complex mixes of anatomical and physiological features, and each represents a major evolutionary commitment toward the development of the arthropod *Bauplan*. Three of the most notable advances were the appearance of a hardened cuticle arranged in distinct articulating sclerites, the segmentation of the appendages, and the appearance of intrinsic appendage musculature. The acquisition of these features resulted in primitive Precambrian creatures that may be considered the first true arthropods—the *protoarthropods*.

The arthropods: Monophyletic or polyphyletic?

Some zoologists have proposed that the taxon Arthropoda is a polyphyletic assemblage, that is, a grade rather than a clade. O. W. Tiegs proposed a **diphyletic theory** of origin for the arthropods in 1947, suggesting the onychophoran–myriapod–insect line belonged to one phylum, and the trilobitomorph–crustacean–cheliceriform line to another. Certain similarities between trilobites and cephalocarid crustaceans prompted Hessler and Newman (1975) to elaborate on the derivation of the Crustacea from the Trilobitomorpha (Figure 16E). The diphyletic theory views the independent origins of these two arthropod lines from two *different* annelid ancestors, the result being a fundamental split between early terrestrial (onychophoran–uniramian) and marine (trilobitomorph–crustacean–cheliceriform) lineages.

The late Sidnie M. Manton, one of this century's great comparative anatomists, and the embryologist D. T. Anderson expanded the diphyletic theory still further, envisioning the arthropods not as two but as four separate phyla. Manton and Anderson established their **polyphyletic theory** in the form of narrative expositions and defended it (against monophyly) primarily on the basis of the absence of data and on differences among the major arthropod groups rather than on the presence of unifying similarities among them. The essence of the polyphyletic theory is that the four major arthropod groups—cheliceriformes, trilobites, uniramians, and crustaceans—do *not* share a common ancestor that *was itself an arthropod*. Rather, the theory holds that all four groups sprang independently from different annelid or annelid-like ancestors.

Manton was convinced that the mandibles of uniramians and crustaceans were so different that the two groups could not be closely related. She pointed out that, not only is the muscle arrangement somewhat different, but crustacean mandibles are multiarticulate appendages in which the biting and chewing surfaces are highly developed coxal endites, or gnathobases. Uniramian mandibles, Manton argued, are of a single piece, not multiarticulate, and biting and chewing takes place with the distal portion of the "entire limb." The crustacean type of mandible

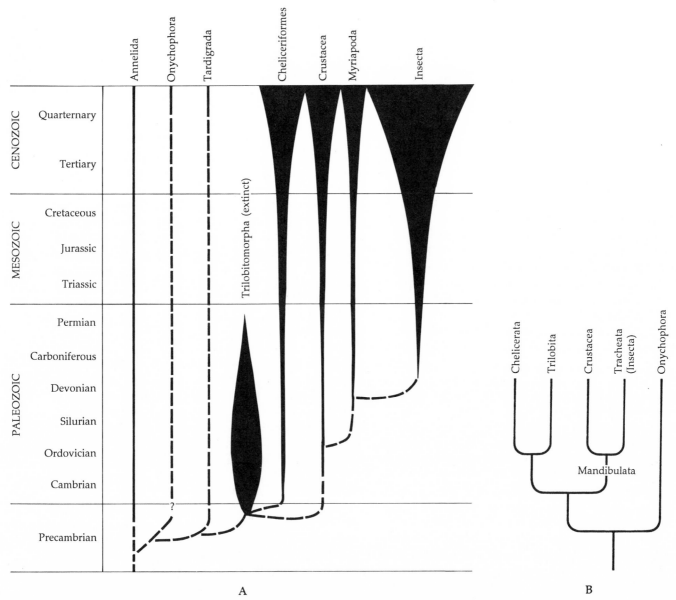

Figure 16

Six proposed evolutionary trees of the annelid–arthropod line. A is based on the cladogram in Figure 15 and discussions in this chapter. The unbroken lines represent the extent of the known fossil record for each group; the dashed lines indicate hypothesized ages of origin. Shapes of areas for the five principal arthropod lines roughly reflect numbers of species at any given time, although these are not entirely proportional (e.g., trilobites flourished in the early Paleozoic but were extinct by the end of that period; the other four lines have increased gradually in diversity from the Paleozoic to recent times, the insects being the most speciose group of all).

B is after Weygoldt (in Gupta 1979), who undertook a survey of embryonic stages and head development in the arthropods and concluded that the arthropods were monophyletic and the sister-group of the onychophorans. Tree C is after Sharov (1966), who

regarded the arthropods as a monophyletic clade but considered the onychophorans to be an independent offshoot from the annelids. However, Sharov also felt that the annelids, onychophorans, tardigrades, and arthropods taken together should also be viewed as a monophyletic group, which he suggested should be ranked as a phylum (Articulata). D is after Paulus (in Gupta 1979), who concluded on the basis of an analysis of eye structure that the arthropods were a monophyletic group. E is after Hessler and Newman (1975). This tree is said to represent a phylogeny of the arthropods that combines the diphyletic hypothesis of Tiegs and Manton (1958) with evolutionary ideas concerning the Trilobitomorpha, Chelicerata, and Crustacea as suggested in their study. F is after Manton (1977 and earlier works). This tree depicts the independent origins of the four arthropod groups.

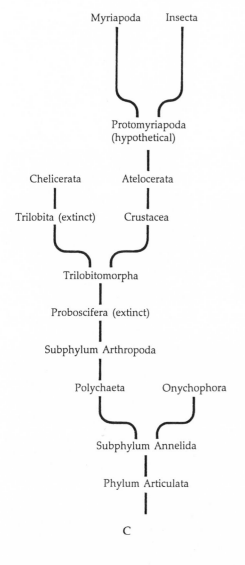

C

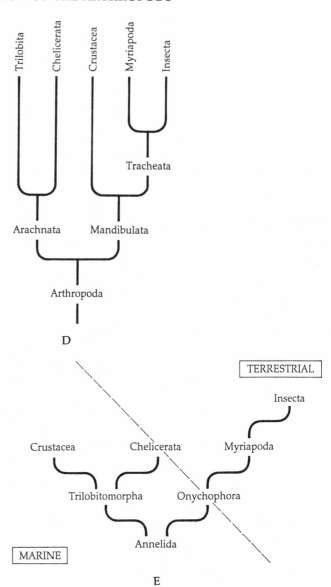

D

E

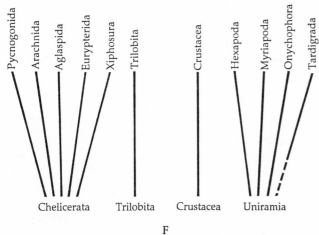

F

she called a **gnathobasic "jaw"**; the uniramian type of mandible she called a **whole-limb "jaw."**

Other authors have noted that the mandibles of crustaceans and uniramians can be viewed as fundamentally similar (homologous), in that they represent the appendages of the second postoral somite that simply possess modifications upon a basic structural type. In this view, the jaws represent not *convergences*, as Manton claimed, but normal evolutionary *divergence* resulting from radiation and specialization within the two groups. More precisely, most crustacean mandibles retain vestiges of the primitive multiarticulate limb (in the form of a palp), whereas uniramians long ago lost all vestiges of the multiarticulate form, and over millennia the precise nature of the jaw musculature and movement changed subtly in association with their terrestrial lifestyle and direct feeding behavior. Both mandibular types, however, could easily have originated as a simple jointed appendage of the third postacronal head somite, rather than separately from fleshy unjointed annelid-like palps, as suggested by Manton. That subtle changes in "jaw" anatomy might have occurred over the 550-million-year history of the mandibulate taxa is hardly surprising. In fact, various genera in many different crustacean subclasses and orders have lost the mandibular palp, and in many of these cases the jaws are strikingly similar to those of the uniramians.* As we have seen, the evolutionary plasticity of arthropod limbs is manifested in a host of ways throughout the phylum.

Figure 17 depicts Manton's concept of a polyphyletic origin for the arthropods in the form of a cladogram. As can be seen, the theory proposes an unresolved polytomous tree of several (perhaps as many as seven) branches. Hence, the tree says almost nothing about the phylogenetic relationships of the taxa, other than that they are *somehow* related to one another and to the annelids. Three kinds of evidence have been used to support the polyphyletic theory: (l) lack of fossil organisms representing the "missing links" or intermediate forms between these groups; (2) the resistance of Manton and Anderson to proposing such "workable" intermediate forms; and, (3) perceived mutually exclusive differences in structure and function, particularly of the mouthparts and developmental processes, among the various groups. Based largely on *differences* among these groups, Manton and Anderson reckoned that they cannot be very closely related and thus should not be ranked in a single phylum. The principal prediction of this polyphyletic theory is that all the characters we recognize as defining an "arthropod" are evolutionary convergences in as many as four (or more) separate lines of organisms descended from several different annelid-like ancestors.

Manton and Anderson were well aware of the numerous unique and complex features collectively shared by the arthropods, including the synapomor-

*The idea that geological time sets some specifiable or predictable constraint on the amount of evolution that can take place in any given structure is an erroneous notion. The morphological changes in jaw structure necessitated by a monophyletic theory seem trivial in comparison to other documented morphological radiations. For example, the entire skull of many vertebrate families has diversified to a much greater extent in far less time.

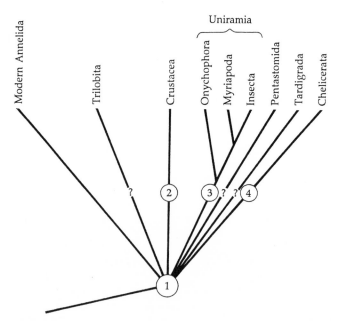

Figure 17

The relationships of the annelid–arthropod assemblage by Manton's polyphyletic theory, depicted in a cladogram. Ancestors numbered 1 through 4 are all different ancient annelid or annelid-like creatures, without features typically recognized as arthropodous. The question marks leading to three of the taxa indicate a lack of reasonable hypotheses of relationship for these groups.

phies noted on the cladogram in Figure 15. Despite this evidence, however, they argued that virtually all of these features arose independently in each of the three living arthropod lines (and probably also in the fourth line, the trilobites); that is, "arthropodization" occurred several times, independently. The polyphyletic theory thus views the Arthropoda as the product of a massive convergent evolutionary phenomenon, at an incredibly detailed and complex level. In our opinion, a monophyletic origin of the arthropods is far more strongly supported by the evidence than is a polyphyletic theory. To properly refute the monophyletic hypothesis, one must show that one or more of the four major arthropod lines is more closely related to (i.e., shares more unique characters with) some group of nonarthropod organisms than it is to the other arthropods. No such evidence has ever been presented. Major evolutionary works of both Manton and Anderson are cited at the end of this chapter for those students wishing to pursue the polyphyletic theory in greater depth. Weygoldt (1979) provides one of the best, most succinct criticisms of Manton's and Anderson's ideas.

The monophyletic theory of arthropod evolution depicted by Figure 15 is based soley on suites of shared derived features (synapomorphies). The cladogram reveals only the *sequence* of appearances of the groups shown, and predicts the probable character states of the ancestral stock from which each arose. It does not specify what the actual ancestors were, and owing to the great age of these taxa and the nature of the fossil record, we may never know this. Having established a monophyletic hypothesis of arthropod origin, we can now briefly examine the evolution of the four major arthropod subphyla—Trilobitomorpha, Cheliceriformes, Crustacea, and Uniramia.

Evolution of the arthropod subphyla

The fossil record of arthropods, meager as it is, dates back to earliest Paleozoic. By that time, all four subphyla were in existence and had already undergone substantial radiation. In other words, the principal clades within the Arthropoda had already begun to diverge before the first arthropods were fossilized. The earliest trilobite, crustacean, and cheliceriform fossils date from the Cambrian; myriapods had appeared by at least the Silurian, apterygote insects by the Devonian, and winged insects by the Carboniferous. The ancestral stocks from which these groups evolved have long since disappeared in the abyss of Precambrian and early Paleozoic history. Unless significant and much older fossil material is found, paleontology will tell us little about the origins of the four groups. However, by careful comparative anatomical and phylogenetic analyses of the living groups one can construct reasonable and testable evolutionary hypotheses for the main arthropod lines.

Several workers have proposed cladograms of the arthropods and their allies based on various data sets. For example, trees have been constructed on the basis of antennal structure, eye structure, the internal anatomy of various organ systems, sperm morphology, hemocoelic cellular structures, embryology of the arthropod head and, of course, gross external morphology. Most of these proposals have supported a monophyletic origin of the phylum Arthropoda. Beyond this, however, things remain unsettled. In considering the four arthropod subphyla (Trilobitomorpha, Cheliceriformes, Uniramia, and Crustacea), there are 15 possible resolved cladograms and literally hundreds of possible evolutionary trees that can be proposed. However, only four general cladograms are strongly supported by the data, and these tend to recur in the literature (Figure 18). It would take many pages to discuss the merits of each of these cladograms, or even to go over the characters that apply to each. Much of the debate about these competing theories rests on two thorny and unre-

solved issues: (1) whether or not the biramous limbs of trilobites and crustaceans are homologous, and (2) whether the mandibles of uniramians and crustaceans are homologous.

Most earlier studies proposed that uniramians and crustaceans represent a monophyletic clade, with the cheliceriformes standing apart from that line at a considerable distance (Figures 18A, B, and C). The uniramian–crustacean clade is called the "Mandibulata," a name coined by the great comparative morphologist R. E. Snodgrass. (Mandibulata is often given subphylum ranking, moving the Uniramia and Crustacea each to the rank of class.) Largely as a result of the efforts of Manton and Anderson, many workers now avoid use of the taxon Mandibulata. The position of the taxon Trilobitomorpha has also been a hotly debated subject. As mentioned above, Hessler and Newman (1975) presented a case for a trilobite origin of the Crustacea (Figures 16E and 18D). The four alternative hypotheses shown in Figure 18 posit the trilobites as (A) the primitive sister-group to all other arthropods; (B) the sister-group of the Cheliceriformes; (C) the sister-group of the mandibulates; and (D) the sister-group of a crustacean–cheliceriform line.

The hypotheses of arthropod phylogeny depicted in Figures 16A–D and 18A–C suggest that the arthropods diverged early (probably Precambrian) into two lines, one in which the head lacked true "jaws," the other in which "jaws" (mandibles) were present (i.e., the Mandibulata). The former line, which includes the groups Cheliceriformes and Trilobitomorpha, is occasionally called the "Amandibulata," "Arachnata," or "Arachnomorpha." Unlike the mandibulates, there is less compelling evidence that the cheliceriformes and trilobites represent a monophyletic clade (as posited, for example, in Figure 18B); hence, the "taxon" Amandibulata has rarely been used in formal classifications. It also may have been in the Precambrian that the mandibulates split into what we recognize today as the crustaceans and uniramians.

Within each of the four major arthropod clades, which we recognize in this text as subphyla, numerous parallelisms and convergences have appeared. Given the great age and sheer size of the group, a high level of convergence and parallelism is not surprising. For example, air-breathing tracheae probably evolved independently at least four times—in the Onychophora, the Arachnida, the terrestrial isopod Crustacea, and the Uniramia. Also, the Malpighian tubules of tardigrades, insects, and arachnids each appear to be derived in a different fashion (i.e., to

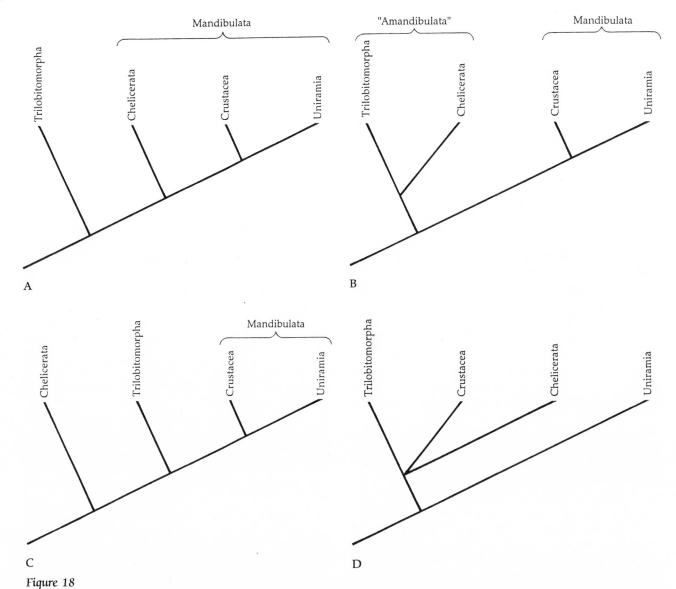

Figure 18

The four major competing hypotheses of relationship among the principal arthropod taxa. A–C view the Crustacea and Uniramia as sister-groups (the Mandibulata). D views the trilobites, crustaceans, and chelicerates as sister-groups.

be nonhomologous, convergent structures). Also, within each line at least some groups have undergone parallel loss of the lateral compound eyes. Similarly, within each line the primitive compound eye underwent distinct modifications unique to each individual subphylum. Nevertheless, the fundamental photoreceptive plan of two lateral faceted eyes and a cluster of simple median eyes occurs in fossil arthropod groups as well as in all three extant subphyla. In fact, the basic ommatidial structure of the mandibulates is

one of the strongest unifying features of a Uniramia–Crustacea clade.*

Much of the evolutionary radiation among the arthropods has centered on appendage modification and body tagmosis (Table 1). Indeed, the four arthropod subphyla are defined largely on those bases. In the uniramians and crustaceans, the head primitively comprises five somites, each with a pair of appen-

*The basic mandibulate ommatidium consists of a cornea made by two corneagen cells, a tetrapartite eucone crystalline cone, and a retinula of eight cells (see Chapter 15). Within each of the two mandibulate subphyla, the eyes have undergone specific modifications. However, primitive taxa of both groups retain the basic ommatidial plan. Other synapomorphies uniting the Crustacea and Uniramia in a monophyletic clade (the Mandibulata) are noted in Figure 19.

dages, all but the first antennae being modified specifically as feeding structures at some stage in their life histories. Primitive uniramians (e.g., myriapods) are similar to primitive crustaceans (e.g., remipedes) in having a long, undivided trunk of many undifferentiated somites. Characters defining the taxa Uniramia and Crustacea are indicated in Figure 19, and were discussed in Chapters 15, 17, and 18.

Given the "mandibulate hypothesis," the transition from an aquatic crustacean-like ancestor to a primitive aquatic uniramian might have taken place in late Precambrian or early Paleozoic seas. The oldest known uniramian fossils are marine myriapods from the Silurian, and the oldest insect fossils are Devonian collembolans. Terrestrial myriapods also first appear in the fossil record in the Devonian and represent some of the earliest known land animals. A reasonable theory of uniramian origin would view the myriapods arising from a shallow-water or littoral, benthic Ordovician or Silurian marine ancestor. The ancestor might have been a carapaceless, highly homonomous, perhaps remipede-like creature. Recall that there is no evidence that uniramians ever possessed a carapace. Myriapods developed an aerial tracheal system and various water conservation structures (e.g., Malpighian tubules) that helped to avoid desiccation. Loss of the compound eyes in myriapods probably took place after the origin of the first insects, since Devonian myriapods still possessed faceted eyes.

A 19 + 2 body plan (19 true somites, plus the pre- and postsegmental acron and telson) occurs in what many people feel are the most primitive myriapods, as well as in some remipedes and in the malacostracan crustaceans. A few specialists have actually suggested that perhaps the 19 + 2 body plan might be ancestral to all living arthropod taxa. Some evidence suggests that primitive cheliceriformes also may have possessed the 19 + 2 body plan. This hypothesis implies that the highly multisegmented arthropods (e.g., cephalocarids, chilopods, and diplopods) independently evolved more elongate bodies with many trunk somites.

One striking example of the ontogenetic and evolutionary flexibility of arthropods is seen in the nature of their appendages. Uniramous legs characterize the taxa Cheliceriformes and Uniramia, two groups now fundamentally terrestrial in habit, and also in certain crustaceans (e.g., amphipods and isopods). These uniramous conditions almost certainly arose independently in at least some of these groups. Isopods, of course, are notable as the only crustacean

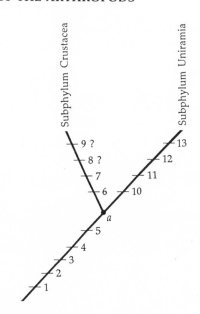

Figure 19

A cladogram depicting the relationships of the two mandibulate subphyla, Crustacea and Uniramia. This cladogram views the Crustacea and Uniramia as sister-groups, united as a monophyletic clade by the presence of numerous unique, shared, derived features (synapomorphies) that define their common ancestor ("*a*"): (1) a fundamental "mandibulate" ommatidial structure; (2) a tripartite brain; (3) mandibles present as appendages of the third postacronal head somite; (4) two pairs maxillae present on postacronal head somites four and five; and, (5) all head appendages except antennules used for feeding at some stage in life history. Crustaceans are distinguished by a nauplius larva (6) and "gnathobasic" mandibles (7). The presence of stalked compound eyes (eyes elevated off the head on a stalk) (8), and biramous second antennae (9), may also be unique synapomorphies of crustaceans, although the primitive versus derived nature of these features is uncertain. The uniramians lost the second pair of antennae (10), developed "whole-limb" mandibles (11), and possess ectodermally derived Malpighian tubules (12), and a unique tracheal respiratory system (13).

Numerous modifications of the primitive (ancestral) *Baupläne* that have occurred define subgroups within both subphyla, particularly with regards to body tagmosis and the appendages of the head and body. The origin of biramous appendages may be viewed as either a unique synapomorphy of the crustacean lineage or as an additional feature arising prior to ancestor *a*. In the latter case, uniramous limbs arose secondarily on the uniramian line. Characters defining the various uniramian and crustacean subgroups are discussed in Chapters 17 and 18.

taxon to have successfully invaded land, and uniramy may be a prerequisite for arthropod success outside the aquatic world. It is also of interest to note that the fossil record indicates the coincidental origins of terrestrial cheliceriformes and uniramians and terrestrial vegetation approximately 400 million years ago. Indeed, this Silurian–Devonian transition period witnessed massive invasions of the terrestrial world by higher plants and animals of all kinds.

Selected References

Tardigrades, Onychophorans, and Pentastomids

Abele, L. G., W. Kim and B. E. Felgenhauer. 1989. Molecular evidence for inclusion of the phylum Pentastomida in the Crustacea. Mol. Biol. Evol. 6:685–691.

Banaja, A. A., J. L. James and J. Riley. 1975. An experimental investigation of a direct life-cycle in *Reighardia sternae* (Diesing 1864), a pentastomid parasite of the herring gull. Parasitolology 71:493–503.

Bussers, J. C. and C. Jeuniaux. 1973. Structure et composition de la cuticle de *Macrobiotus* et de *Milnesium tardigradum*. Ann. Soc. R. Zoo. Belg. 103:271–279.

Carvalho, A. L. 1942. Sobre *"Peripatus heloisae,"* do Brasil Central. Zool. Bol. Mus. Nac. Rio de Janeiro (n. s.) 2:57–73.

Claude-Joseph, F. 1928. Observations sur un Péripate de Chile. Ann. Sci. Nat. Zool. (Ser. 10) ll:285–298.

Cuénot, L. 1952. Phylogénèse de règne animal. *In* P. Grassé (ed.), *Traité de Zoologie*, Vol. 1. Masson et Cie, Paris, pp. 1–33.

Doucet, J. 1965. Contribution à l'étude anatomique, histologique et histochimique des Pentastomes. Mem. l'Office Recherche Sci. Tech. Outre Mer 14:1–150.

Grassé, P. (ed.). 1949. *Traité de Zoologie*, Vol. 6, Onychophores, Tardigrades, Arthropodes, Trilobitomorphes, Chelicerates. Masson et Cie, Paris.

Heymons, R. 1935. Pentastomida. *In* H. G. Bronn (ed.), *Klassen und Ordnungen des Tierreichs*, Bd. 5, Abt. 4. Akad. Verlagsgesellschaft, Frankfurt.

Hoyle, G. and M. Williams. 1980. The musculature of *Peripatus dominicae* and its innervation. Philos. Trans. R. Soc. London Ser. B 288:481–510.

Janvier, H. 1975. Une péripate du Chile chassezur de termites. Entomologiste 31:63.

King, C. E. 1962. The occurrence of *Batillipes mirus* Richters in the Gulf of Mexico. Bull. Mar. Sci. 12:201–203.

Kristensen, R. M. l978. On the structure of *Batillipes nøerrevangi* Kristensen, 1978. 2. The muscle attachments and the true cross-striated muscles. Zool. Anz. Jena 200(3/4):173–184.

Kristensen, R. M. 1982. New aberrant eutardigrades from homothermic springs on Disko Island, west Greenland. *In* D. R. Nelson (ed.), *Proceedings of the Third International Symposium on Tardigrada*. East Tennessee State University Press, Johnson City, Tennessee, pp. 203–220.

Kristensen, R. M. 1981. Sense organs of two marine arthrotardigrades (Heterotardigrada, Tardigrada). Acta Zool. 62:27–41.

Kristensen, R. M. 1982. The first record of cyclomorphosis in Tardigrada, based on a new genus and species from Arctic meiobenthos. Z. Zool. Syst. Evolutionsforsch. 20:249–270.

Kristensen, R. M. 1983. A new marine interstitial eutardigrade from east Greenland, with comments on habitat and biology. Vidensk. Medd. Dan. Naturhist. Foren. Khobenhavn 144:99–144.

Kristensen, R. M. 1984. On the biology of *Wingstrandarctus corallinus* nov. gen. et spec., with notes on the symbiontic bacteria in the subfamily Florarctinae (Arthrotardigrada). Vidensk. Medd. Dan. Naturhist. Foren. Khobenhavn 145:201–218.

Kristensen, R. M. and T. E. Hallas. 1980. A tidal genus *Echiniscoides* and its variability, with erection of Echiniscoididae fam. n. (Tardigrada). Zool. Scr. 9:113–127.

Kristensen, R. M. and R. P. Higgins. 1984. A new family of Arthrotardigrada (Tardigrada: Heterotardigrada) from the Atlantic Coast of Florida, U.S.A. Trans. Am. Microsc. Soc. 103:295–311.

Kristensen, R. M. and R. P. Higgins. 1984. Revision of *Styraconyx* (Tardigrada: Halechiniscidae), with descriptions of two new species from Disko Bay, west Greenland. Smithson. Contrib. Zool. 391:1–40.

Kristensen, R. M. and T. Nilonen. 1982. Structural studies on *Diurodrilus* Remane (Diurodrilidae fam. n.), with descriptions of *Diurodrilus westheidei* sp. n. from the Arctic interstitial meiobenthos, W. Greenland. Zool. Scr. 11:1–12.

Kristensen, R. M. and J. Renaud-Mornant. 1983. Existence d'arthrotardigrades semibenthiques de genres nouveaux de la sousfamille des Styraconyxinae subfam. nov. Can. Biol. Mar. 24:337–353.

Marcus, E. 1929. Tardigrada. *In* H. G. Bronn (ed.), *Klassen und Ordnungen des Tierreichs*, Bd. 5, Abt. 4. Akad. Verlagsgesellschaft, Frankfurt.

Marcus, E. 1937. Sobre os Onychophoros. Inst. Biol. Sec. Agric. São Paulo. Arch. 8:255–266.

Marcus, E. 1957. Tardigrada. *In* W. T. Edmondson, H. B. Ward and G. C. Whiple (eds.), *Freshwater Biology*, 2nd Ed. Wiley, New York, pp. 508–521.

Morgan, C. I. 1982. Tardigrada. *In* S. P. Parker (ed.), *Synopsis and Classification of Living Organisms*, Vol. 2. McGraw-Hill, New York, pp. 731–739.

Morgan, C. I. and P. E. King. 1976. *British Tardigrades: Keys and Notes for the Identification of the Species*. Synopsis of British Fauna No. 9. Academic Press, New York.

Nelson, D. R. (ed.) 1982. *Proceedings of the Third International Symposium on Tardigrada*. East Tennessee State University Press, Johnson City, Tennessee.

Nicoli, R. M. 1963. Phylogenese et systematique le phylum des Pentastomida. Ann. Parasitol. Hum. Comp. 38:483–516.

Nørrevang, A. 1972. Oogenesis in Pentastomida. Acta Zool. 53:57–72.

Osche, G. 1963. Die systematische Stellung und Phylogenie der Pentastomida. Z. Morphol. Oekol. Tiere 52:487–596.

Peck, S. B. 1975. A review of the New World Onychophora, with the description of a new cavernicolous genus and species from Jamaica. Psyche 82:341–358.

Pollock, L. W. 1976. Tardigrada. Marine flora and fauna of the northeastern U.S. NOAA Tech. Rpt., Nat. Mar. Fish. Serv. Circular 394:1–25.

Renaud-Mornant, J. and M.-N. Anselme-Moizan. 1969. Stades larvaires du tardigrade marin *Stygarctus bradypus* Schulz et position systematique des Stygarctidae. Bull. Mus. Nat. Hist. Nat. Zool. 41:883–893.

Riley, J. 1983. Recent advances in our understanding of pentastomid reproductive biology. Parasitology 86:59–83.

Self, J. T. 1969. Biological relations of the Pentastomida: A bibliography on the Pentastomida. Exp. Parasitol. 24:63–119.

Self, J. T. 1982. Pentastomida. *In* S. P. Parker (ed.), *Synopsis and Classification of Living Organisms*, Vol. 2. McGraw-Hill, New York, pp. 726–728.

Young, A. M. 1980. On the patchy distribution of onychophorans in two cacao plantations in northeastern Costa Rica. Brenesia 17:143–148.

Arthropod Phylogeny

Selected references on phylogeny within each of the five arthropod subphyla may be found below, as well as in their respective chapters (Chapters 16–18).

Almond, J. E. 1985. The Silurian–Devonian fossil record of the Myriapoda. Philos. Trans. R. Soc. London Ser. B 309:227–237.

Anderson, D. T. 1973. *Embryology and Phylogeny in Annelids and Arthropods*. Pergamon Press, New York. [A review of Anderson's comparative embryological studies, including his argument for arthropod polyphyly.]

Bergstrom, J. 1972. Appendage morphology of the trilobite *Cryptolithus* and its implications. Lethaia 5:85–94.

Bourdreaux, H. B. 1979. *Arthropod Phylogeny with Special Reference to Insects*. Wiley, New York.

Briggs, D. E. G. and R. A. Fortey. 1989. The early radiation and relationships of the major arthropod groups. Science 246:241–243.

Burnett, B. R. and R. R. Hessler. 1973. Thoracic epipodites in the Stomatopoda (Crustacea): A phylogenetic consideration. J. Zool. 169:381–392.

Cisne, J. L. 1974. Trilobites and the origin of arthropods. Science 186:13–18.

Cisne, J. L. 1975. Anatomy of *Triarthrus* and the relationships of the Trilobita. Fossils Strata 4:45–63.

Eldredge, N. and S. M. Stanley (eds). 1984. *Living Fossils*. Springer-Verlag, New York.

Grygier, M. J. 1981. Sperm of the ascothoracican parasite *Dendrogaster*, the most primitive found in Crustacea. Int. J. Invert. Reprod. 3:65–73.

Grygier, M. J. 1983. Ascothoracida and the unity of Maxillopoda. *In* F. R. Schram (ed.), *Crustacean Phylogeny*. A. A. Balkema, Rotterdam, pp. 73–104.

Gupta, A. P. (ed.) 1979. *Arthropod Phylogeny*. Van Nostrand Reinhold, New York. [Thirteen contributors review aspects of arthropod evolution; eight favor a monophyletic origin, three are uncommitted, and two—Manton and Anderson—favor polyphyly.]

Hessler, R. R. and W. A. Newman. 1975. A trilobitomorph origin for the Crustacea. Fossils and Strata 4:437–459.

House, M. R. (ed.). 1979. *The Origin of Major Invertebrate Groups*. Academic Press, New York. [Includes review papers by Whittington, Manton, Anderson, and others.]

Labandeira, C. C., B. S. Beall and F.M. Hueber. 1988. Early insect diversification: Evidence from a Lower Devonian bristletail from Québec. Science 242:913–916.

Lauterbach, K. E. 1973. Schlusselereignisse in der Evolution der Stammgruppe der Euarthropoda. Zool. Beitr. 19:251–259.

Lauterbach, K. E. 1986. Zum Grundplan der Crustacea. Verh. Naturwiss. Ver. Hamburg 28:27–63.

Manton, S. M. 1949. Studies on the Onychophora VII. The early embryonic stages of *Peripatopsis*, and some general considerations concerning the morphology and phylogeny of Arthropoda. Philos. Trans. R. Soc. London Ser. B 223:483–580.

Manton, S. M. 1958. Habits of life and evolution of body design in Arthropoda. J. Linn. Soc. (Zool.) 44:58–72.

Manton, S. M. 1960. Concerning head development in the arthropods. Biol. Rev. 35:265–282.

Manton, S. M. 1964. Mandibular mechanisms and the evolution of arthropods. Philos. Trans. R. Soc. London Ser. B 247: 1–183.

Manton, S. M. 1973. Arthropod phylogeny—a modern synthesis. J. Zool. 171:11–130.

Manton, S. M. 1977. *The Arthropoda: Habits, Functional Morphology, and Evolution*. Clarendon Press, Oxford. [A detailed summary of the evidence that Manton claims supports the polyphyletic theory of arthropod origin.]

Martinsson, A. (ed.). 1975. Evolution and Morphology of the Trilobita, Trilobitoidea and Merostomata: Proc. NATO Adv. Study Inst., Oslo, July 1973. Fossils and Strata 4: 1–468.

McKenzie, K. G. 1983. On the origin of Crustacea. Aust. Mus. Mem. 18:21–43.

McNamara, K. J. 1986. The role of heterochrony in the evolution of Cambrian trilobites. Biol. Rev. 61:121–156.

Mikulic, D. G., D. E. G. Briggs and J. Kluessendorf. 1985. A Silurian soft-bodied biota. Science 228:715–717.

Mikulic, D. G., D. E. G. Briggs and J. Kluessendorf. 1985. A new exceptionally preserved biota from the lower Silurian of Wisconsin, U.S.A. Philos. Trans. R. Soc. London Ser. B 311:75–85.

Müller K. J. and D. Walossek. 1986. Arthropod larvae from the Upper Cambrian of Sweden. Trans. R. Soc. Edinburgh Earth Sci. 77:157–179.

Newman, W. A. and M. D. Knight. 1984. The carapace and crustacean evolution—a rebuttal. J. Crust. Biol. 4:682–687.

Platnick, N. I. 1978. [Review of] *The Arthropoda: Habits, Functional Morphology, and Evolution* by S. M. Manton, 1977, Oxford University Press. Syst. Zool. 27(2):252–255.

Retallack, G. J. and C. R. Feakes. 1987. Trace fossil evidence for Late Ordovician animals on land. Science 235:61–63.

Riley, J., A. A. Banaja and J. L. James. 1978. The phylogenetic relationships of the Pentastomida: The case for their inclusion within the Crustacea. Int. J. Parasitol. 8: 245–254.

Robison, R. A. 1985. Affinities of *Aysheaia* (Onychophora), with description of a new Cambrian species. J. Paleontol. 59:226–235.

Rolfe, W. D. I. 1981. Phyllocarida and the origin of the Malacostraca. Geobios 14:17–27.

Rolfe, W. D. I. 1985. Early terrestrial arthropods: A fragmentary record. Philos. Trans. R. Soc. London Ser. B 309:207–218.

Sanders, H. L. 1957. The Cephalocarida and crustacean phylogeny. Syst. Zool. 6: 112–128.

Schram, F. R. 1974. Paleozoic Pericarida of North America. Fieldiana Geol. 33:95–124.

Schram, F. R. 1978. Arthropods: A convergent phenomenon. Fieldiana Geol. 39:61–108.

Schram, F. R. (ed.) 1983. *Crustacean Phylogeny*. A. A. Balkema, Rotterdam.

Schram, F. R. and J. W. Hedgpeth. 1978. Locomotory mechanisms in Antarctic pycnogonids. Zool. J. Linn. Soc. 63:145–169.

Schram, F. R. and W. D. I. Rolfe. 1982. New euthycarcinoid arthropods from the Upper Pennsylvanian of France and Illinois. J. Paleontol. 56:1434–1450.

Sharov, A. G. 1966. *Basic Arthropodan Stock*. Pergamon Press, New York.

Shear, W. A. et al. 1984. Early land animals in North America: Evidence from Devonian age arthropods from Gilboa, New York. Science 224:492–494.

Siewing, R. 1956. Untersuchungen zur Morphologie der Malacostraca (Crustacea). Zool. Jahrb. Anat. 75:39–176.

Snodgrass, R. E. 1952. *A Textbook of Arthropod Anatomy*. Cornell University Press, Ithaca, New York. [Dated, but still one of the best references for general arthropod anatomy and morphology.]

Snodgrass, R. E. 1960. Facts and theories concerning the insect head. Smithson. Misc. Coll. 142:1–61.

Störmer, L. 1977. Arthropod invasion of land during Late Silurian and Devonian times. Science 197:1362–1364.

Tiegs, O. W. 1940. The embryology and affinities of the Symphyla, based on a study of *Hanseniella agilis*. Q. J. Microsc. Sci. 82: 1–225.

Tiegs, O. W. 1945. The post-embryonic development of *Hanseniella agilis* (Symphyla). Q. J. Microsc. Sci. 85:191–328.

Tiegs, O. W. 1947. The development and affinities of the Pauropoda, based on a study of *Pauropus sylvaticus*. Q. J. Microsc. Sci. 88:165–336.

Tiegs, O. W. and S. M. Manton. 1958. The evolution of the Arthropoda. Biol. Rev. 33:255–337. [The polyphyletic theory of arthropod evolution.]

Waterson, C. D. (ed.). 1985. Fossil arthropods as living animals. Trans. R. Soc. Edinburgh Earth Sci. 76:103–399. [An important synopsis of current opinion; resulting from an international conference held in 1984.]

Weygoldt, P. 1979. Significance of later embryonic stages and head development in arthropod phylogeny. *In* A. P. Gupta (ed.), *Arthropod Phylogeny*. Van Nostrand Reinhold, New York, pp. 107–135. [We cite this paper separately from the Gupta volume because it presents one of the most lucid criticisms of arthropod polyphyly we have read.]

Weygoldt, P. 1986. Arthropod interrelationships—the phylogenetic–systematic approach. Z. Zool. Syst. Evolutionsforsch. 24:19–35.

Weygoldt, P. and H. F. Paulus. 1979. Untersuchungen zur Morphologie, Taxonomie und Phylogenie der Chelicerata. II. Cladogramme und die entfaltung der Chelicerata. Z. Zool. Syst. Evolutionsforsch. 17:177–200.

Whittington, H. B. 1980 The significance of the fauna of the Burgess Shale, Middle Cambrian, British Columbia. Proc. Geol. Assoc. 91:127–148.

Whittington, H. B. and W. D. I. Rolfe (eds.). 1963. *Phylogeny and Evolution of Crustacea*. Proceedings of a conference held at Cambridge, Massachusetts, March 6–8, 1962. Spec. Publ. Mus. Comp. Zool. Harv. Univ. [A benchmark symposium publication; still an important reference although many new ideas on the subject will be found in the post-1970 literature, especially Schram (ed.) 1983.]

Wilson, M. A. 1985. Disturbance and ecologic succession in an Upper Ordovician cobble-dwelling hardground fauna. Science 228:575–578.

Wingstrand, K. G. 1972. Comparative spermatology of a pentastomid *Raillietiella hemidactyli* and a branchiuran crustacean *Argulus foliaceus* with a discussion of pentastomid relationships. Biol. Skr. 19:1–72.

Chapter Twenty

Phylum Mollusca

In essentials molluscs are one of the most compact groups of animals; but there can be few phyla that show such wide diversity imposed on such a uniform plan.

J. E. Morton
Molluscs: An Introduction to Their Form and Function, 1960

Orange and speckled and fluted nudibranchs slide gracefully over the rocks, their skirts waving like the dresses of Spanish dancers.

John Steinbeck
Cannery Row, 1945

Molluscs include some of the best known invertebrates; almost everyone is familiar with snails, clams, slugs, squids, and octopuses. Molluscan shells have been popular since ancient times, and some cultures still use them as tools, containers, musical devices, money, fetishes, and decorations. Many aboriginal groups have for centuries relied on molluscs for a substantial portion of their diet. Evidence of historical use and knowledge of molluscs is seen in ancient texts and hieroglyphics, on coins, in tribal customs, and in "kitchen middens" or "shell mounds" (piles of discarded material, especially shells and pottery, from former aboriginal habitation sites). Today, many molluscs are harvested commercially. The annual world squid and octopus fishery exceeds 2 million metric tons per year.

There are over 50,000 described, living mollusc species and about 60,000 known fossil molluscs. However, many species still await names and descriptions, especially those from poorly studied regions such as the deep sea, polar areas, and terrestrial and freshwater habitats in the tropics. It has been estimated that only about half of the living molluscs have so far been described. In addition to the three familiar molluscan classes comprising the clams (Bivalvia), snails and slugs (Gastropoda), and squids and octopuses (Cephalopoda), five other classes exist: chitons (Polyplacophora), tusk shells (Scaphopoda), *Neopilina* and its kin (Monoplacophora), solenogasters (Aplacophora), and the aberrant group Caudofoveata. Although members of these eight classes differ enormously in superficial appearance, they are remarkably similar in their fundamental *Bauplan* (see Box 1).

In order to describe both the diversity and the unity of the members of the Mollusca, we present a summary classification of the phylum, including short synopses of key characters that define the major taxa. Following this, we discuss in detail molluscan biology and the molluscan *Bauplan*—the basic unifying attributes that distinguish the phylum.

Taxonomic history and classification

Molluscs* carry the burden of a very long and convoluted taxonomic history, in which hundreds of names for various taxa have come and gone. Aristotle was probably the first scientist to formally recognize

---

*The name of the phylum derives from the Latin *molluscus*, meaning "soft," in allusion to the similarity of some molluscs to the mollusca, a kind of Old World soft nut with a thin but hard shell. The vernacular for Mollusca is often spelled *mollusks* in the United States, whereas in the rest of the world it is generally spelled *molluscs*. In biology, a vernacular or diminutive name is generally derived from the proper Latin name; thus, the American custom of altering the spelling of Mollusca by changing the *c* to *k* seems to be an aberration (although it may have its historic roots in the German language, which does not have the free-standing *c*; e.g., Molluskenkunde). We prefer the spelling *molluscs*, which seems to be the proper vernacularization and is in line with other accepted terms, such as *molluscan*, *molluscoid*, and *molluscivore*.

Box One
Characteristics of the Phylum Mollusca

1. Bilaterally symmetrical (or secondarily asymmetrical), coelomate protostomes
2. Coelom generally reduced to vestiges around nephridia, heart, gonads, and part of intestine
3. Principal body cavity is a hemocoel (open circulatory system)
4. Viscera often concentrated as a "visceral mass"
5. Body covered by thick epidermal–cuticular sheet of skin, the mantle, which forms a cavity (the mantle cavity) in which are housed the ctenidia, osphradia, nephridiopores, gonopores, and anus
6. Mantle with shell glands that secrete calcareous epidermal spicules, shell plates, or shells
7. Heart lies in pericardial chamber and is composed of distinct ventricle and atria
8. With large, well defined muscular foot, often with a flattened creeping sole
9. Buccal region provided with a radula
10. Complete gut, with marked regional specialization, including very large digestive ceca
11. With large, complex metanephridia ("kidneys")
12. Embryogeny typically protostomous
13. With trochophore larva, and usually a veliger larva

ied animals, including not only cephalopods, slugs, and pteropods but also tunicates, anemones, medusae, echinoderms, and polychaetes. Under Testacea, Linnaeus included chitons, bivalves, univalves, nautiloids, barnacles, and the serpulid polychaetes (which secrete calcareous tubes). In 1795 Cuvier published a revised classification of the Mollusca that was the first to approximate modern views. De Blainville (1825) altered the name Mollusca to Malacozoa, which won little favor but survives in the terms *malacology, malacologist*, etc.

Much of the nineteenth century passed before the phylum was purged of all extraneous groups. In the 1830s J. Thompson and C. Brumeister identified the larval stages of barnacles and revealed them to be crustaceans, and in 1866 A. Kowalevsky removed the tunicates from the Mollusca. Separation of the brachiopods from the molluscs was a long and controversial ordeal that was not resolved until near the end of the nineteenth century.

Aplacophorans were discovered in 1841 by the Swedish naturalist Lovén. He classified them with holothurian echinoderms because of their vermiform bodies and the presence of calcareous spicules in the body walls of both groups. Graff (1875) recognized aplacophorans as molluscs, and shortly thereafter it became fashionable to classify the chitons and aplacophorans together in the class Amphineura. This scheme persisted until the 1950s when the two groups were again separated; however, a few recent workers have once more suggested that these two groups might represent a monophyletic clade unto themselves. In any event, the name Aplacophora has precedence for the solenogasters, and the name Polyplacophora has precedence for chitons.

The history of classification of species in the class Gastropoda has been volatile, undergoing constant change since Cuvier's time. Most modern malacologists adhere more-or-less to the basic schemes of Milne-Edwards (1848) and Spengel (1881). The former, basing his classification on the respiratory organs, recognized the groups Pulmonata, Opisthobranchia, and Prosobranchia. Spengel based his scheme on the nervous system and divided the gastropods into the Streptoneura and Euthyneura. Streptoneura is equivalent to Prosobranchia; Euthyneura embraces Opishthobranchia and Pulmonata. The bivalves have been called Bivalvia, Pelecypoda, and Lamellibranchiata.

Molluscan classification at the generic and species levels is incredibly jumbled. Many species of gastropods and bivalves are burdened with several names (synonyms), and some species actually bear

molluscs, dividing them into two groups: Malachia (the cephalopods) and Ostrachodermata (the shelled forms), the latter being divided into univalves and bivalves. Jonston (or Jonstonus) created the name Mollusca in 1650 for the cephalopods and barnacles, but this name was not accepted until it was resurrected and redefined by Linnaeus. As we have discussed elsewhere, Linnaeus regarded all invertebrates except insects as Vermes, a group divided into Intestina, Mollusca, Testacea, Lithophyta, and Zoophyta. His Mollusca was a potpourri of soft-bod-

hundreds of different synonyms. This tangle is partly the result of a long history of amateur participation in the field of molluscan taxonomy, especially with gastropods, which comprise about three-quarters of the molluscs. It is unfortunately still common practice in the molluscan literature to describe new species of molluscs on the basis of the shell alone, ignoring the body of the animal altogether, and the major mollusc collections of the world house largely shell-only specimens. This situation creates enormous problems for those specialists interested in phylogenetic studies of molluscs. To make matters worse, when new species of molluscs are first described it is not uncommon for some (or most) of the specimens of the type series to be deposited with private shell collectors rather than strictly into institutional research collections (such as those of museums and universities). In such instances, valuable type material may never be available to the professional scientific community because these specimens are subsequently traded, sold, or lost.

Only taxa with extant members are included in the classification synopsis.* Some important fossil groups are discussed later in the chapter. Examples of the major molluscan taxa appear in Figure 1. The eight classes of extant molluscs are listed below, followed by a detailed classification and synopses.

Class Caudofoveata
Class Aplacophora
Class Monoplacophora
Class Polyplacophora
Class Gastropoda
    Subclass Prosobranchia
    Subclass Opisthobranchia
    Subclass Pulmonata
Class Bivalvia
    Subclass Protobranchia
    Subclass Lamellibranchia
    Subclass Anomalodesmata
Class Scaphopoda
Class Cephalopoda
    Subclass Nautiloidea
    Subclass Coleoidea

*A multitude of extinct molluscs have been described. Perhaps the most well known are some of the groups of cephalopods that had hard external shells, similar to those of living *Nautilus*. One of these groups was the ammonites (a term reserved for Jurassic and Cretaceous members of the order Ammonoidea). They differed from nautiloids in having shell septa that were highly fluted on the periphery, forming complex mazelike septal sutures. Ammonoids also had the siphuncle lying against the outer wall of the shell, as opposed to the condition seen in nautiloids where the siphuncle runs through the center of the shell whorls.

PHYLUM MOLLUSCA

CLASS CAUDOFOVEATA (= CHAETODERMATOMORPHA) Caudofoveatans (Figure 2I). Vermiform, cylindrical, shell-less; body wall bears a chitinous(?) cuticle and imbricating scale-like calcareous spicules; without eyes, tentacles, statocysts, crystalline style, or foot; with a pair of bipectinate ctenidia.

Small, wormlike, infaunal molluscs that live upside-down in vertical burrows on the deep-sea floor. Little is known of their ecology. There are about 70 known species. (e.g., *Chaetoderma, Falcidens, Limifossor, Psilodens, Scutopus*)

CLASS APLACOPHORA (= SOLENOGASTRES): Aplacophorans or solenogasters. Vermiform, cylindrical or compressed, shell-less; mantle cavity rudimentary; body wall imbued with calcareous spicules; without eyes, tentacles, statocysts, crystalline style, or nephridia; without ctenidia or with 1 pair folded (but not bipectinate) ctenidia; with or without radula; without a flattened foot, but usually with ventral furrow or "pedal groove" (thought to be homologous to the foot of other molluscs, being hydrostatic and possessing large mucus-secreting pedal glands) (Figures 2A–H).

Odd marine molluscs living primarily at depths greater than 200 m. They are often abundant components of deep-sea epifauna or infauna; many reside and feed upon cnidarians. About 250 species have been described. (e.g., *Chevroderma, Dondersia, Lepidomenia, Lophomenia, Neomenia, Prochaetoderma, Proneomenia, Rhopalomenia*)

CLASS MONOPLACOPHORA: Monoplacophorans. With a single, caplike shell; foot forms weak ventral muscular disc, with 8 pairs retractor muscles; shallow mantle cavity around foot encloses 5–6 pairs ctenidia; 2 pairs gonads; 6–7 pairs metanephridia; 2 pairs heart atria; with radula and distinct but small head; without eyes; tentacles present only around mouth; with a crystalline style and posterior anus; mostly extinct (Figures 1A and 3).

Until the first living species (*Neopilina galatheae*) was discovered by the Danish "Galathea Expedition" in 1952, monoplacophorans were known only from lower Paleozoic fossils. Since then, its unusual anatomy has been a source of much evolutionary speculation. Monoplacophorans are limpet-like in appearance, living species are less than 3 cm in length, and most live at considerable depths. About 11 described species in 3 genera (*Monoplacophorus, Neopilina, Vema*).

CLASS POLYPLACOPHORA: Chitons. Flattened, elongated molluscs with a broad ventral foot and eight dorsal shell plates; shells with unique articulamentum layer; mantle forms thick girdle that borders and may partly or entirely cover shell plates; epidermis of girdle usually with calcareous or chitinous spines, scales, or bristles; mantle cavity encircles foot and bears from 6 to more than 80 pairs of ctenidia; 1 pair nephridia; without eyes, tentacles, or crystalline style; nervous system lacking discrete ganglia, except in buccal region; radula present (Figures 1B and 4).

Marine, primarily rocky intertidal grazing herbivores; a few deep-sea species are known. They are unique in their

A

B

E

F

I

J

K

M

N

Figure 1

Diversity among the molluscs. A, The cap-shaped shell of *Neopilina* (class Monoplacophora). B, *Tonicella*, the lined chiton (class Polyplacophora, order Ischnochitonida). C, *Perotrochus*, a primitive gastropod, or "slit shell" (class Gastropoda, order Archaeogastropoda, family Pleurotomaridae). D, The red abalone *Haliotis rufescens* (order Archaeogastropoda). Note the exhalant holes in the shell. E, *Epitonium scalare*, the precious wentletrap (order Mesogastropoda). F, *Conus* (order Neogastropoda). Note the siphon extending from the front of the shell aperture. G, *Chromodoris*, a nudibranch (subclass Opisthobranchia). H, *Monadenia infumata*, a terrestrial snail from California (subclass Pulmonata). I, A small intertidal *Octopus* (class Cephalopoda, subclass Coleoidea, order Octopoda). J, *Histioteuthis*, a pelagic squid (order Teuthoida). K, *Fustiaria*, a tusk shell (class Scaphopoda). L, Scallops (class Bivalvia, superorder Filibranchia, family Pectinidae). M, The giant clam *Tridacna* (superorder Eulamellibranchia, family Tridacnidae). N, *Clinocardium*, a cockle (superorder Eulamellibranchia, family Cardiidae). O, *Chaceia ovoidea*, the wart-necked piddock (superorder Eulamellibranchia, family Pholadidae). Some individuals have bored into sandstone; one is exposed (A courtesy of W. Jorgensen; B,D,F,I,J courtesy of G. McDonald; C,E,H,O photos by K. Lucas/BPS; G courtesy of A. Kerstitsch; K,L,M,N courtesy of P. Fankboner.)

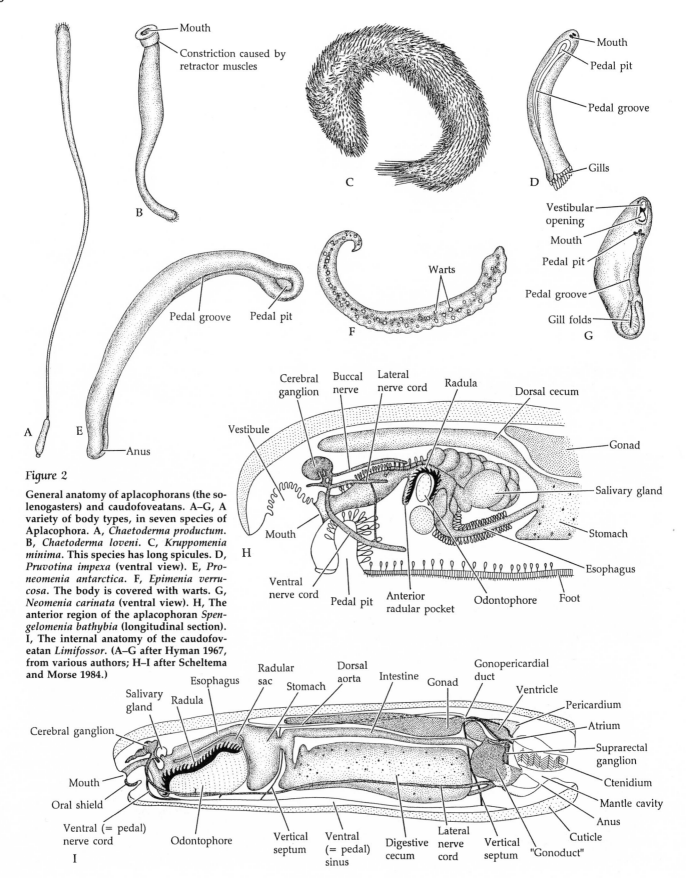

Figure 2

General anatomy of aplacophorans (the so-lenogasters) and caudofoveatans. A–G, A variety of body types, in seven species of Aplacophora. A, *Chaetoderma productum*. B, *Chaetoderma loveni*. C, *Kruppomenia minima*. This species has long spicules. D, *Pruvotina impexa* (ventral view). E, *Proneomenia antarctica*. F, *Epimenia verrucosa*. The body is covered with warts. G, *Neomenia carinata* (ventral view). H, The anterior region of the aplacophoran *Spengelomenia bathybia* (longitudinal section). I, The internal anatomy of the caudofoveatan *Limifossor*. (A–G after Hyman 1967, from various authors; H–I after Scheltema and Morse 1984.)

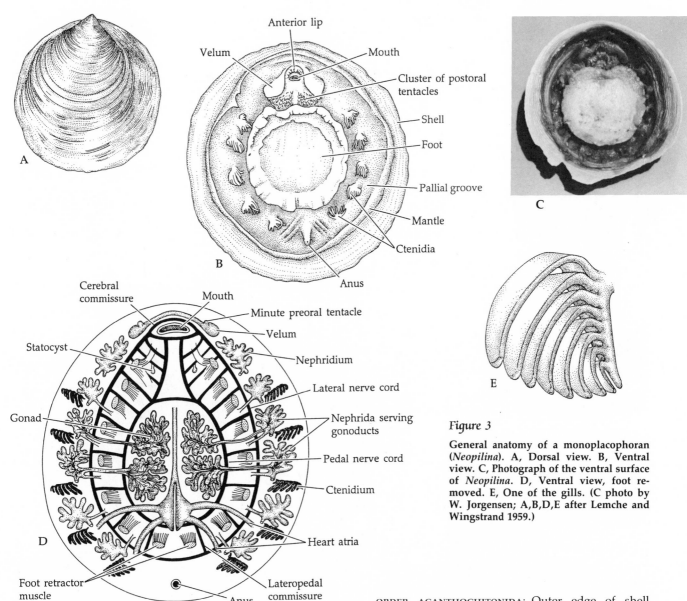

Figure 3

General anatomy of a monoplacophoran (*Neopilina*). A, Dorsal view. B, Ventral view. C, Photograph of the ventral surface of *Neopilina*. D, Ventral view, foot removed. E, One of the gills. (C photo by W. Jorgensen; A,B,D,E after Lemche and Wingstrand 1959.)

possession of 8 (sometimes 7) separate shell plates, called valves, and a thick marginal girdle; about 600 species have been described.

ORDER LEPIDOPLEURIDA: Primitive chitons with outer edge of shell plates lacking attachment teeth; girdle not extending over plates; ctenidia limited to a few posterior pairs. (e.g., *Choriplax, Lepidopleurus, Lepidochiton, Oldroydia*)

ORDER ISCHNOCHITONIDA: Outer edges of shell plates with attachment teeth; girdle not extending over plates, or extending partly over plates; ctenidia occupying most of mantle groove, except near anus. (e.g., *Callistochiton, Chaetopleura, Ischnochiton, Katharina, Lepidozona, Mopalia, Nuttallina, Placiphorella, Schizoplax, Tonicella*)

ORDER ACANTHOCHITONIDA: Outer edge of shell plates with well developed attachment teeth; shell valves partially or completely covered by girdle; ctenidia do not extend full length of foot. (e.g., *Acanthochitona, Cryptochiton, Cryptoplax*)

CLASS GASTROPODA: Snails and slugs. Asymmetrical molluscs with single, usually spirally coiled shell into which body can be withdrawn; shell lost or reduced in many groups; during development, visceral mass and mantle rotate 90–180° on foot (torsion), so mantle cavity lies anterior, and gut and nervous system are twisted; some taxa have partly or totally reversed the rotation (detorsion); with muscular creeping foot (modified in swimming and burrowing taxa); head with statocyst and eyes (often reduced or lost), and 1–2 pairs tentacles; most with complex radula and crystalline style, the latter being lost in most predatory groups; 1–2 nephridia; mantle (= pallium) usually forms cavity housing ctenidia, osphradia, and hypobranchial

Figure 4

General anatomy of chitons (class Polyplacophora). A–B, A typical chiton (dorsal and ventral views). C, The underside of the gumboot chiton, *Cryptochiton stelleri*. D, Dorsal view, shell plates (valves) removed. E, Dorsal view, dorsal musculature removed to reveal internal organs. F, Dorsal view showing extensive kidneys. G, The arrangement of internal organs (lateral view). (C photo by authors; others adapted from various sources.)

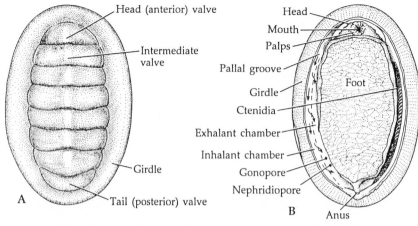

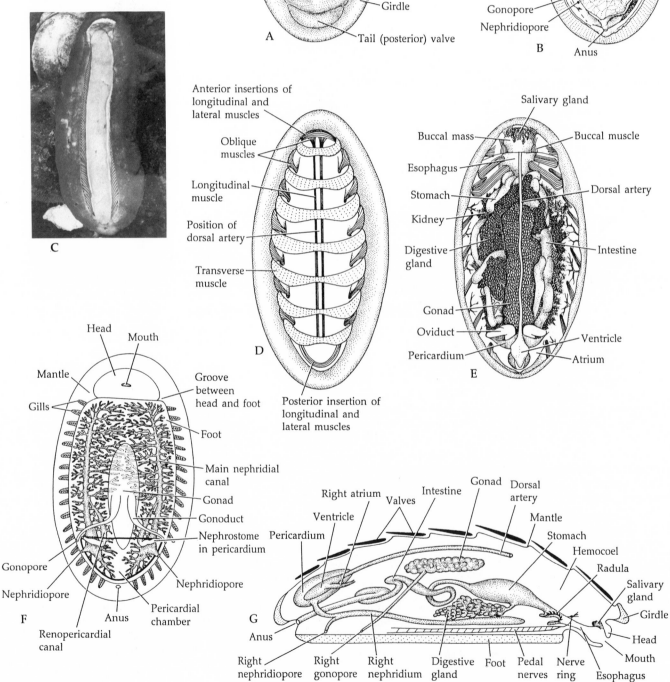

glands; ctenidia sometimes lost and replaced with secondary gas exchange structures (Figures 1C–H, 5, 6, and 7).

Gastropods comprise about 40,000 living species of marine, terrestrial, and freshwater snails and slugs. The class is usually divided into three subclasses: prosobranchs (largely shelled marine snails); opisthobranchs (marine slugs); and pulmonates (terrestrial snails and slugs). However, this arrangement is viewed by some authorities as artificial, and numerous revisionary schemes have appeared (e.g., Haszprunar 1985), although none has met with broad acceptance.

SUBCLASS PROSOBRANCHIA: Usually with a spirally coiled shell, sometimes with cap-shaped or tubular shell; mantle cavity usually anteriorly directed, near head, containing osphradia, ctenidia, hypobranchial glands, anus, and nephridiopores; head generally with tentacles bearing basal eyes; foot with creeping planar sole and typically with corneous or calcareous operculum to close shell aperture upon retraction of head and foot; radula variable or absent; nervous system streptoneurous.

ORDER ARCHAEOGASTROPODA: Primitive prosobranchs; shell primitive with nacreous layer; radula modified for herbivory, often with numerous teeth in transverse rows, usually rhipidoglossate or docoglossate; 1–2 bipectinate ctenidia; mantle cavity without siphon; primitively with 2 hypobranchial glands, 2 osphradia, 2 atria, and 2 metanephridia; sexes usually separate; male generally without penis; nervous system weakly concentrated.

Primarily marine, although a few freshwater and terrestrial species are known; virtually all are herbivores. About 3,000 species have been described in 26 families, including Pleurotomaridae and Scissurellidae (slit-shelled molluscs, the most primitive living prosobranchs: e.g., *Perotrochus, Pleurotomaria, Scissurella*); Haliotidae (abalones, *Haliotis*); Fissurellidae (keyhole limpets: e.g., *Diodora, Fissurella, Lucapinella, Megathura, Puncturella*); Acmaeidae, Patellidae, Lottiidae, and two other families (the "true" limpets: e.g., *Acmaea, Lepeta, Lottia, Patella*); Trochidae (trochids: e.g., *Calliostoma, Margarites, Tegula, Trochus*); Turbinidae (turbans: e.g., *Astraea*); Neritidae (nerites: e.g., *Nerita, Theodoxus*); Helicinidae (helicinids: e.g., *Alcadia, Helicinia*).

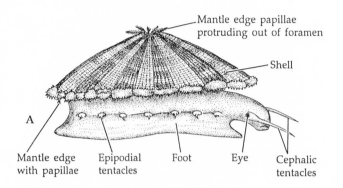

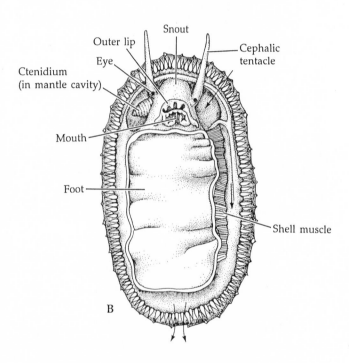

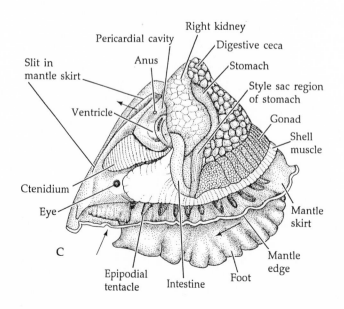

Figure 5

General anatomy of limpet-like archaeogastropods. A, The volcano limpet *Fissurella* (family Fissurellidae) (lateral view). B, A true limpet (family Lottiidae) (ventral view). Arrows indicate the direction of water currents. C, The keyhole limpet *Puncturella* (Fissurellidae), removed from shell and seen from the left. Arrows indicate water currents. Certain structures are visualized through the mantle skirt: ctenidium, eye, anus, and epipodial tentacles. (A after Hyman 1967; B,C after Fretter and Graham 1962.)

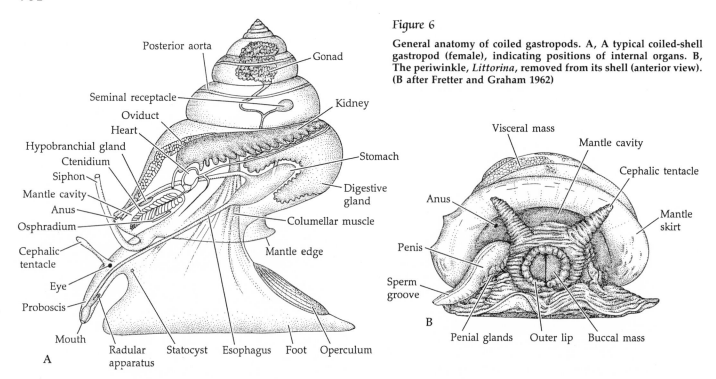

Figure 6

General anatomy of coiled gastropods. **A**, A typical coiled-shell gastropod (female), indicating positions of internal organs. **B**, The periwinkle, *Littorina*, removed from its shell (anterior view). (B after Fretter and Graham 1962)

ORDER MESOGASTROPODA: Shell mainly porcelaneous and nonnacreous; operculum usually present and corneous, rarely calcified; head with pair of cephalic tentacles, usually with basal eyes; mantle cavity asymmetrical, with incurrent opening on anterior left, often elaborated into an inhalant siphon; right ctenidium lost; left ctenidium usually monopectinate; hypobranchial glands often lost on left; right nephridium often lost; radula generally taenioglossate, occasionally lost; most are gonochoristic; higher forms with concentrated ganglia.

Includes marine, freshwater, and terrestrial forms; there are about 10,000 species in nearly 100 families, including Hydrobiidae (e.g., *Hydrobia*); Viviparidae (e.g., *Viviparus*); Littorinidae (periwinkles: e.g., *Littorina*); Turritellidae (tower or turret shells: e.g., *Turritella*); Caecidae (caecids: e.g., *Caecum*); Vermetidae (vermetids or "worm" gastropods: e.g., *Serpulorbis, Tripsycha, Vermetus, Vermicularia,*); Cerithiidae (ceriths: e.g., *Cerithium, Liocerithium*); Potamididae (potamids or horn shells: e.g., *Cerithidea*); Strombidae (conchs or strombids: e.g., *Strombus*); Epitoniidae (wentletraps or epitoniids: e.g., *Epitonium*); Janthinidae (janthinids: e.g., *Janthina*); Hipponicidae (horse hoof limpits: e.g., *Hipponix*); Capulidae (cap limpets: e.g., *Capulus*); Calyptraeidae (cup and saucer shells and slipper shells: e.g., *Calyptraea, Crepidula, Crucibulum*); Carinariidae (one of several families of pelagic molluscs collectively called heteropods: e.g., *Carinaria*); Naticidae (moon snails: e.g., *Natica, Polinices*); Eratoidae (coffee bean shells: e.g.,

Erato, Trivia); Cypraeidae (cowries: e.g., *Cypraea*); Ovulidae (ovulas or egg shells: e.g., *Jenneria, Ovula, Simnia*); Tonnidae (tun shells: e.g., *Malea*); Cassididae (helmet shells: e.g., *Cassis*); Ficidae (fig shells: e.g., *Ficus*).

ORDER NEOGASTROPODA: Shell without nacreous layer; radula with 1–3 teeth in each row; 1 (left) monopectinate ctenidium; 1 osphradium; radula rachiglossate or toxoglossate; mantle forms siphon, carried within siphonal canal or notch of shell; sexes separate, male with penis; nervous system concentrated; operculum, if present, chitinous; heart with left atrium only; right nephridium lost.

About two dozen families of marine snails, including Buccinidae (whelks: e.g., *Buccinum, Cantharus, Macron, Metula*); Columbellidae (dove shells: e.g., *Anachis, Columbella, Mitrella, Nassarina, Pyrene, Strombina*); Coralliophilidae (e.g., *Coralliophila, Latiaxis*); Fasciolariidae (tulip shells and spindle shells: e.g., *Fasciolaria, Fusinus, Leucozonia, Troschelia*); Harpidae (harp shells: e.g., *Harpa*); Marginellidae (marginellas, e.g., *Granula*); Melongenidae (whelks, false trumpets: e.g., *Melongena*); Mitridae (miter shells: e.g., *Mitra, Subcancilla*); Muricidae (rock shells: e.g., *Ceratostoma, Hexaplex, Murex, Phyllonotus, Pteropurpura, Pterynotus, Acanthina, Morula, Neorapana, Nucella, Purpura, Thais*); Nassariidae (dog whelks and basket shells: e.g., *Nassarius*); Olividae (olive shells: e.g., *Agaronia, Oliva, Olivella*); Volutidae (volutes: e.g., *Cymbium, Lyria, Voluta*); Cancellariidae (e.g., *Cancellaria, Admete*); Conidae (cone shells: e.g.,

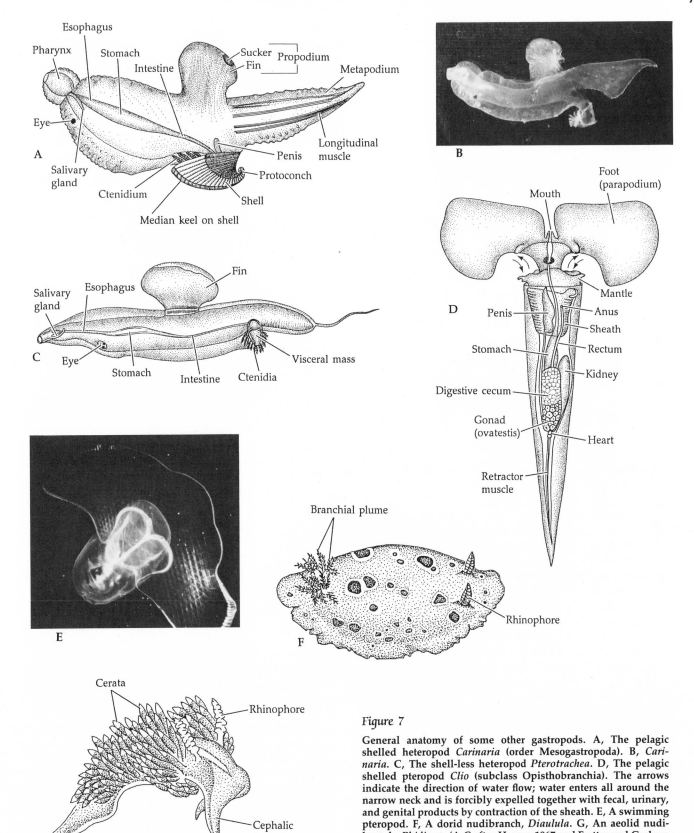

Figure 7

General anatomy of some other gastropods. A, The pelagic shelled heteropod *Carinaria* (order Mesogastropoda). B, *Carinaria*. C, The shell-less heteropod *Pterotrachea*. D, The pelagic shelled pteropod *Clio* (subclass Opisthobranchia). The arrows indicate the direction of water flow; water enters all around the narrow neck and is forcibly expelled together with fecal, urinary, and genital products by contraction of the sheath. E, A swimming pteropod. F, A dorid nudibranch, *Diaulula*. G, An aeolid nudibranch, *Phidiana*. (A,C after Hyman 1967 and Fretter and Graham 1962; B courtesy of G. McDonald; E courtesy of J. King; F,G from Brusca and Brusca 1978.)

Conus), Turridae (tower shells: e.g., *Crassispira, Lora*), Terebridae (auger shells: e.g., *Terebra*).

SUBCLASS OPISTHOBRANCHIA: Sea slugs and their kin. Body variously detorted; shell external, internal, reduced, or lost; ctenidia and mantle cavity usually reduced or lost; usually without operculum; head with 1–2 pairs of rhinophores or tentacles; hermaphroditic; euthyneurous with various degrees of nervous system concentration. Primarily marine, benthic; a few freshwater species.

Traditional (conservative) classifications include nine orders (and over 100 families) of opisthobranchs. However, some authorities feel that the Opisthobranchia may not be a monophyletic taxon. In any case, shell loss almost certainly occurred several times within this subclass. Several alternative classifications of this group have been suggested but a consensus has not yet been achieved. The nine orders and some common genera are Acochlidioidea (e.g., *Acochlidium, Unela*); Cephalaspidea (e.g., *Acteon, Aglaja, Bulla, Chelidonura, Haminoea, Navanax, Retusa, Rictaxis, Scaphander*); Runcinoidea (e.g., *Ilbia, Ildica, Runcina*); Sacoglossa (e.g., *Berthelinia, Elysia, Oxynoe, Tridachia*); Anaspidea (the sea hares: e.g., *Aplysia, Dolabella, Stylocheilus*); Thecosomata (the shelled pteropods: e.g., *Clio, Limacina*); Gymnosomata (the naked pteropods: e.g., *Clione*); Notaspidea (e.g., *Berthellina, Gymnotoplax, Pleurobranchus, Tylodina, Umbraculum*); Nudibranchia (the "true" nudibranchs: e.g., *Acanthodoris, Aegires, Aeolidia, Aldisa, Armina, Bornella, Calma, Chromodoris, Corambe, Coryphella, Dendrodoris, Dendronotus, Diaulula, Doris, Embletonia, Fiona, Glaucus, Hexabranchus, Hopkinsia, Janolus, Phidiana, Phyllidia, Platydoris, Polycera, Rostanga, Scyllaea, Tambja, Trinchesia*).

SUBCLASS PULMONATA: Land snails and slugs. Shell usually present and coiled, lost in some groups; mantle cavity forms lung with contractile opening; without ctenidia (except perhaps in *Siphonaria*); body detorted to various degrees; highly concentrated nervous system (euthyneurous); hermaphroditic; without larvae; mainly terrestrial and freshwater forms, a few marine species.

ORDER ARCHAEOPULMONATA: Primitive pulmonates with spirally coiled shell, but no operculum; mainly littoral. (e.g., *Cassidula, Ellobium, Otina*)

ORDER BASOMMATOPHORA: Shell variable, minute or moderate-sized, generally spirally coiled (or planospiral) or patelliform; usually without an operculum; eyes at bases of sensory stalks; freshwater and intertidal; includes freshwater limpets. (e.g., *Bulinus, Carychium, Chilina, Lanx, Physa, Planorbis, Siphonaria, Trimusculus*)

ORDER STYLOMMATOPHORA: Shell absent or present; if present usually spirally coiled and often partly or completely enveloped by dorsal mantle; eyes on tips of sensory stalks; terrestrial; an enormous group with over 15,000 described species. (e.g., *Achatina, Arion, Bulimulus, Cepaea, Haplotrema, Helix, Liguus, Limax, Megaspira, Oreohelix, Pupilla, Rachis, Succinea, Vertigo*)

Figure 8

General anatomy of bivalves. **A,** *Tresus*, a deep-burrowing eulamellibranch, with a strong foot and long, fused siphons. **B,** A typical eulamellibranch (cross section). **C,** The eulamellibranch *Mercenaria*, with the left valve and mantle removed. **D,** Internal anatomy of *Mercenaria*. The visceral mass is opened up, the foot is dissected, and most of the gills are cut away. **E,** The common mussel, *Mytilus*, seen from the right side after removal of the right shell and the mantle. **F,** *Mytilus*, with the visceral mass opened up, the foot dissected, and most of the gills cut away. (**A** from Brusca and Brusca 1978; others adapted from various sources.)

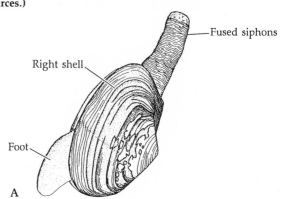

ORDER SYSTELLOMMATOPHORA: Sluglike, without internal or external shell; dorsal mantle integument forms a keeled or rounded notum; head usually with 2 pairs tentacles, upper ones forming contractile stalks bearing eyes. (e.g., *Onchidium, Onchidella, Rhodope*)

CLASS BIVALVIA (= PELECYPODA; ALSO = LAMELLIBRANCHIATA): Clams, oysters, etc. Laterally compressed; shell typically of two valves hinged together dorsally by elastic ligament and shell-teeth; shells closed by adductor muscles; head rudimentary, without eyes or radula, but eyes and statocysts may occur elsewhere on body; foot typically laterally compressed, usually without a sole; 1 pair large bipectinate ctenidia, used in combination with labial palps in ciliary feeding; large mantle cavity; posterior edges of mantle often fused to form inhalant and exhalant siphons; 1 pair nephridia (Figures 1L–O and 8).

Bivalves are marine or freshwater molluscs, primarily microphagous or suspension feeders. The class is large, with over 8,000 living species represented at all depths and in all marine environments. Bivalve classification has been in a state of turmoil over the past 50 years. Hardly any two authors today utilize exactly the same classification or nomenclature. Some workers delimit the higher taxa on the basis of shell characters alone (e.g., hinge anatomy, position of muscle scars), others rely solely on internal organ anatomy (e.g., ctenidia, stomach), and still others use ecological characters (e.g., feeding methods, adaptations to various habitats). For some alternatives to the classification below, see Purchon (1977), Morton (1979), and Moore (1960).

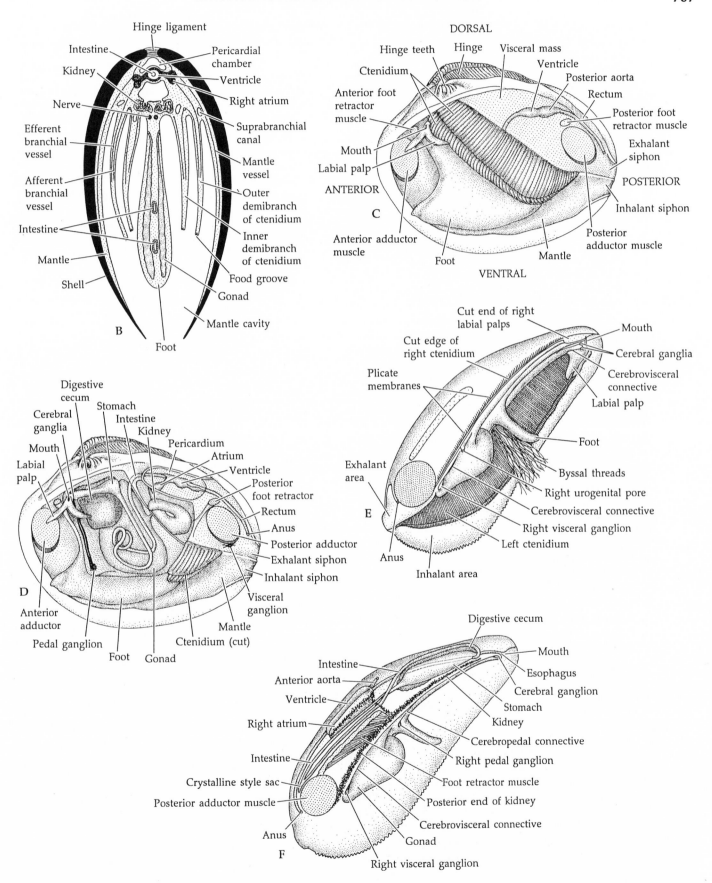

SUBCLASS PROTOBRANCHIA: Ctenidia are simple, unfolded, bipectinate, platelike leaflets suspended in the mantle cavity.

ORDER NUCULOIDA (= PALAEOTAXODONTA): Shell aragonitic, interior nacreous or porcelaneous; periostracum smooth; shell valves equal and taxodont (i.e., the valves have a row of short teeth along hinge margin); adductor muscles equal in size; with large palp proboscides used for food collection; ctenidia small, strictly for gas exchange; foot longitudinally grooved and with a plantar sole, adults without byssus threads; nervous system primitive, often with incomplete union of cerebral and pleural ganglia; marine, mainly infaunal detritivores. (e.g., *Malletia, Nucula, Yoldia*)

ORDER SOLEMYOIDA (= CRYPTODONTA): Shell valves thin, elongate, and equal in size; uncalcified along outer edges, without hinge teeth; anterior adductor muscle larger than posterior one; ctenidia large, used both for gas exchange and feeding; palps small. (e.g., *Solemya*)

SUBCLASS LAMELLIBRANCHIA: Ctenidia with very long filaments that fold back on themselves so that each row of filaments forms two lamellae; adjacent filaments usually attached to one another by ciliary tufts (filibranch condition), or by tissue bridges (eulamellibranch condition); labial palps small.

SUPERORDER FILIBRANCHIA (= PTERIOMORPHIA): Ctenidia simple, outer fold not connected dorsally to visceral mass, with free filaments or with adjacent filaments attached by ciliary tufts; shell aragonitic or calcitic, sometimes nacreous; mantle margin unfused, with weakly differentiated incurrent and excurrent apertures or siphons; foot well developed or extremely reduced; usually attached by byssal threads or cemented to substratum (or secondarily free).

Primitive bivalves, including mussels (Mytilidae: e.g., *Adula, Brachidontes, Lithophaga, Modiolus, Mytilus*) and other clams, such as the ark shells (Arcidae: e.g., *Anadara, Arca, Barbatia*), glycymerids (Glycymerididae: e.g., *Glycymeris*), true oysters (Ostreidae: e.g., *Crassostrea, Ostrea*), pearl oysters (Pteriidae: e.g., *Pinctada, Pteria*), hammer oysters (Malleidae: e.g., *Malleus*), pen shells (Pinnidae: e.g., *Atrina, Pinna*), file shells (Limidae: e.g., *Lima*), scallops (Pectinidae: e.g., *Chlamys, Lyropecten, Pecten*), thorny oysters (Spondylidae: e.g., *Spondylus*), and jingle shells (Anomiidae: e.g., *Anomia, Pododesmus*).

SUPERORDER EULAMELLIBRANCHIA (= HETERODONTA): Complex ctenidia, outer fold completely connected dorsally to roof of mantle cavity, with adjacent filaments attached by tissue bridges; shell generally aragonitic, without nacreous layer; shell valves equal to subequal, with a few large cardinal teeth separated from the elongated lateral teeth by a toothless space; mantle more-or-less fused posteroventrally and form-

ing incurrent and excurrent apertures that are frequently drawn out into siphons; foot usually lacks byssal threads in adult. Advanced bivalves, mainly marine, including three main groups (treated here as orders).

ORDER PALEOHETERODONTA: Shell aragonitic, pearly internally; periostracum usually well developed; valves usually equal, with few hinge teeth; elongate lateral teeth (when present) are not separated from the large cardinal teeth; usually dimyarian; mantle opens broadly ventrally, mostly unfused posteriorly but with excurrent and incurrent apertures. About 1,200 species of marine and freshwater clams. Includes the nearly extinct family Trigoniidae (with fewer than six living species, in the Australasian region), and the family Unionoidea (freshwater bivalves; e.g., *Anodonta*).

ORDER VENEROIDA: Usually thick-valved, equivalved, and isomyarian. Includes the following families: cockles (Cardiidae: e.g., *Clinocardium, Laevicardium, Trachycardium*), little heart shells (Carditidae: e.g., *Cardita*), giant clams (Tridacnidae: e.g., *Tridacna*), surf clams (Mactridae: e.g., *Mactra*), solens (Solenidae: e.g., *Ensis, Solen*), tellinids (Tellinidae: e.g., *Florimetis, Macoma, Tellina*), semelids (Semelidae: e.g., *Leptomya, Semele*), wedge shells (Donaci-

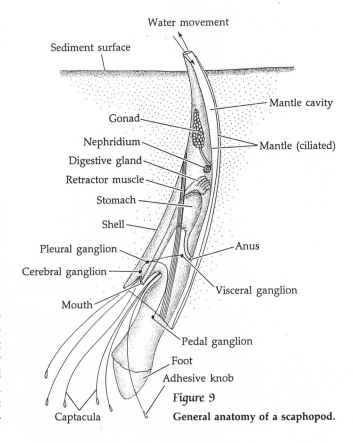

Figure 9

General anatomy of a scaphopod.

dae: e.g., *Donax*), venus clams (Veneridae: e.g., *Chione, Dosinia, Tivela, Pitar, Protothaca*), and the freshwater families Sphaeriidae (e.g., *Sphaerium*) and Corbiculidae (e.g., *Corbicula*).

ORDER MYOIDA: Thin-shelled burrowing forms with well develped siphons; shell with 0–1 cardinal teeth. Includes the soft-shell clams, shipworms, and others: families Pholadidae (piddocks: e.g., *Barnea, Chaceia, Martesia, Pholas*), Teredinidae (shipworms: e.g., *Bankia, Teredo*), Corbulidae (e.g., *Corbula, Mya*).

SUBCLASS ANOMALODESMATA: Includes septibranchs. Shells equivalved, aragonitic, of 2–3 layers, innermost consisting of sheet nacre; periostracum often incorporates granulations; with 0–1 hinge teeth; generally isomyarian, rarely amyarian; posterior siphons usually well developed; mantle usually fused ventrally, with anteroventral pedal gape, and posteriorly with ventral incurrent and dorsal excurrent apertures or siphons; ctenidia eulamellibranchiate or septibranchiate (modified as a horizontal septum). Marine bivalves; one order (Pholadomyoida), and about 12 families, including Cuspidariidae (e.g., *Cuspidaria*), Poromyidae (e.g., *Poromya*), and Pandoridae (e.g., *Pandora*).

CLASS SCAPHOPODA: Tusk shells (Figures 1K and 9). Shell of one piece, tubular, usually tapering, open at both ends; head rudimentary, projecting from larger aperture; mantle cavity large, extending along entire ventral surface; without ctenidia or eyes; with radula, proboscis, crystalline style; with paired clusters of clubbed contractile tentacles (captacula) that serve to capture and manipulate prey; heart absent; foot somewhat cylindrical. About 350 species of marine, benthic molluscs in eight families, including Dentaliidae (e.g., *Dentalium, Fustiaria*), Laevidentaliidae (e.g., *Laevidentalium*), Pulsellidae (e.g., *Pulsellum, Annulipulsellum*), and Gadilidae (e.g., *Cadulus, Gadila*).

CLASS CEPHALOPODA (= SIPHONOPODA) : Nautilus, squids, cuttlefish, and octopuses (Figures 1I–J, 10, 11, 12, and 22). With linearly chambered shell, usually reduced or lost; if external shell present, animal inhabits last (youngest) chamber, with a filament of living tissue (the siphuncle) extending through older chambers; body cavity large; circulatory system largely closed; head with large, complex eyes and circle of prehensile arms or tentacles around mouth; with radula and beak; 1–2 pairs ctenidia, and 1–2 pairs complex nephridia; mantle forms large ventral pallial cavity containing ctenidia; with muscular funnel (the siphon) through which water is forced, providing jet propulsion; some tentacles of male modified for copulation; benthic or pelagic, marine; about 650 living species.

SUBCLASS NAUTILOIDEA (= TETRABRANCHIATA): The pearly nautilus. Shell external, many chambered, coiled in one plane, exterior porcelaneous, interior nacreous (pearly); head with many (80–90) suckerless tentacles (4 modified as a spadix in male for copulation) and protected by a fleshy, lobelike hood; 13-element radula; beak of chitin and calcium carbonate; funnel of 2 separate

Figure 10

The anatomy of *Nautilus* (saggital section).

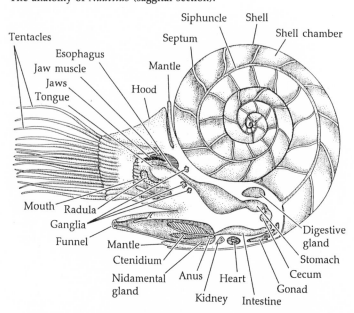

folds; 2 pairs ctenidia ("tetrabranchiate"); 2 pairs nephridia; eyes like a pinhole camera, without cornea or lens; nervous system rather diffuse; with a simple, primitive statocyst; without chromatophores or ink sac. Fossil record rich and diverse, but represented today by a single genus, the chambered or pearly nautilus (*Nautilus*), with five or six Indo-Pacific species.

SUBCLASS COLEOIDEA (= DIBRANCHIATA): Octopuses, squids, and their kin. Shell reduced, internal or absent; head and foot united into a common anterior structure bearing 8–10 prehensile suckered appendages (arms and tentacles), 1 pair modified in male for copulation; 7-element radula; with chitinous beak; funnel a single closed tube; 1 pair ctenidia ("dibranchiate"); 1 pair nephridia; eyes complex, with cornea and lens; nervous system well developed and concentrated; with a complex statocyst; with chromatophores and ink sac.

ORDER SEPIOIDA: Cuttlefish. Body short, dorsoventrally flattened, with lateral fins; shell absent or internal, calcareous, often chambered, straight, or coiled; 8 short arms and 2 long tentacles; tentacles with suckers borne only on spooned tips and retractable into pits; suckers lack hooks. (e.g., *Rossia, Sepia, Spirula*)

ORDER TEUTHOIDA (= DECAPODA): Squids. Body elongate, tubular, with lateral fins; shell internal, reduced to cartilaginous pen; with 8 arms and 2 elongate nonretractile tentacles; suckers often with hooks. Numerous families and genera. (e.g., *Architeuthis, Bathyteuthis, Chiroteuthis, Doryteuthis, Dosidicus, Gonatus, Histioteuthis, Illex, Loligo, Lycoteuthis, Octopoteuthis, Ommastrephes*)

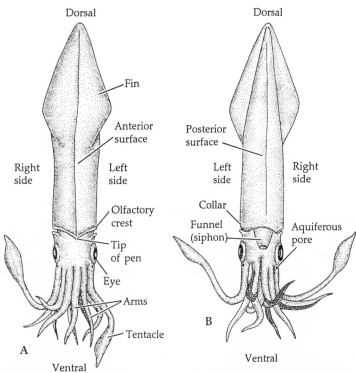

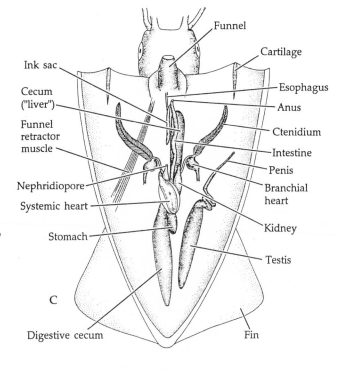

Figure 11

The anatomy of a squid (*Loligo*). A, External morphology (anterior view). B, External anatomy (posterior view) C, Internal anatomy of a male. The mantle is dissected open and pulled aside. D, A large squid stranded on a beach. (A–C after various sources; D courtesy of J. King.)

ORDER OCTOPODA: Octopuses. Body short, round, usually without fins; internal shell vestigial or absent; 8 similar arms joined by web of skin (interbrachial web); most are benthic. About 200 species. (e.g., *Argonauta, Octopus, Opisthoteuthis, Stauroteuthis*)

ORDER VAMPYROMORPHA: Vampire squids. Body plump, with 1 pair fins; shell reduced to thin, leaf-shaped, uncalcified, transparent vestige; 4 pairs equal-sized arms, each with one row of unstalked distal suckers; arms joined by extensive web of skin (the interbrachial membrane); fifth pair of arms represented by 2 tendril-like, retractile filaments; hectocotylus lacking; radula well developed; ink sac degenerate; mostly deep water. (e.g., *Vampyroteuthis*)

The molluscan *Bauplan*

The phylum Mollusca is one of the most morphologically diverse animal groups. Molluscs range in size from microscopic bivalves to giant clams (Tridacnidae) that reach 1 m in length, giant squids (*Architeuthis*) reaching 20 m in overall length, and giant octopuses (*Octopus dofleini*) attaining arm spans of 6 m and weights over 40 kg. Despite their differences, giant squids, cowries, garden slugs, eight-plated chitons, and wormlike aplacophorans are all closely related, and they all share a common and unmistakable body plan (Box 1). In fact, the myriad ways in which evolution has shaped the basic molluscan *Bauplan* provide some of the best lessons in homology and adaptive radiation in the animal kingdom.

Molluscs are bilaterally symmetrical, coelomate protostomes, but the coelom is generally reduced to small vestiges around the heart (the **pericardial chamber**), the gonads, parts of the kidneys, and occasionally part of the intestine (the "**perivisceral coelom**"). The principal body cavity is a hemocoel composed of several large sinuses of the open circulatory system. In general, the body comprises three distinguishable regions: head, foot, and centrally concentrated **visceral mass** (Figure 13). The head may bear

Figure 12

The anatomy of *Octopus*. A, General external anatomy. B, Right-side view of the internal anatomy. C, Arm and sucker (cross section). D, Tip of the hectocotylus arm. (B after Lane 1960; C after Winkler and Ashley 1954.)

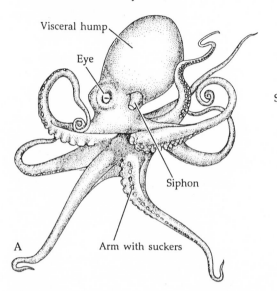

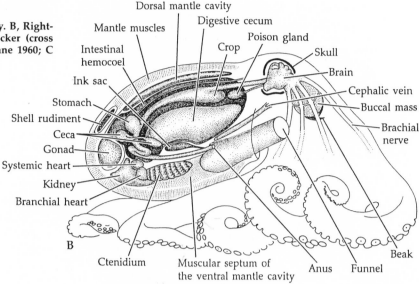

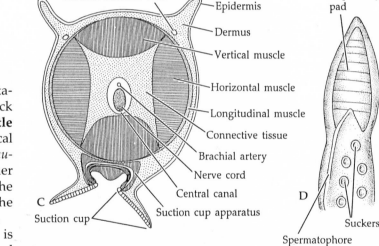

various sensory structures, most notably eyes, statocysts, and tentacles. The body is covered by a thick epidermal–cuticular sheet of skin called the **mantle** (also known as the **pallium**), which plays a critical role in the organization of the general molluscan *Bauplan*. It secretes the hard calcareous skeleton, either as minute sclerites or plates that are embedded in the body wall or as a solid internal or external shell. The body usually bears a large, muscular, ventral **foot**.

Surrounding or posterior to the visceral mass is a cavity—a space between the visceral mass and folds of the mantle itself. This **mantle cavity** (also known as the **pallial cavity**) often houses the gills, or **ctenidia**, the openings of the gut, nephridial, and reproductive systems, and special sense organs called **osphradia**. Water is circulated through this cavity, passing over the ctenidia, excretory pores, and other structures.

The molluscan gut is complete and regionally specialized. The buccal region of the foregut typically bears a uniquely molluscan structure called the **radula**, a toothed, rasping, tonguelike strap that is used in feeding. The open circulatory system usually includes a heart in a pericardial cavity and a few large vessels that empty into or drain hemocoelic spaces. The excretory system consists of one or more pairs of metanephridial kidneys, with nephrostomes usually in the pericardial cavity. The nervous system

typically includes a dorsal cerebral ganglion, circumenteric nerve ring, two pairs of longitudinal ladder-like nerve cords, and several paired ganglia showing various degrees of fusion.

Fertilization may be external or internal. Development is typically protostomous, with spiral cleavage and one or two distinct trochophore larval stages. One of these larval forms is unique to molluscs and is called the **veliger**.

Although this general summary describes the basic *Bauplan* of most molluscs, notable modifications occur and are discussed throughout this chapter. The classes are briefly summarized below.

Some of the most bizarre molluscs are the caudofoveatans and aplacophorans (Figure 2). Members

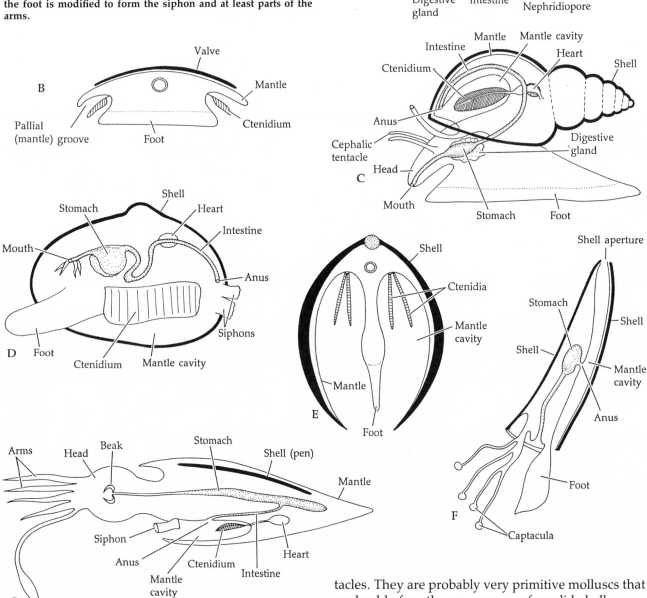

Figure 13

Modifications of the shell, foot, gut, ctenidia, and mantle cavity in five classes of molluscs. A–B, Lateral and cross sections of a chiton (class Polyplacophora). C, Side view of a snail (class Gastropoda). D–E, Cutaway side view and cross section of a clam (class Bivalvia). F, Lateral view of a tusk shell (class Scaphopoda). G, Lateral view of a squid (class Cephalopoda). In cephalopods the foot is modified to form the siphon and at least parts of the arms.

of both groups are small and wormlike, and either burrow in deep-sea sediments or, in the case of most aplacophorans, spend their entire lives living on the branches of various cnidarians, such as gorgonians, upon which they feed. These animals lack a functional foot and have no distinct head, eyes, or ten-

tacles. They are probably very primitive molluscs that evolved before the appearance of a solid shell.

Monoplacophorans are limpet-like molluscs with a single cap-shaped shell ranging from several millimeters to about 4 cm in length (Figures 1A and 3). These benthic, marine animals live at modest to great depths. Their most notable feature is the repetitive arrangement of many organs, a condition that has led some biologists to speculate that they represent a link to an ancient, segmented ancestor of the Mollusca.

Polyplacophorans, or chitons, are oval molluscs that bear not one, but usually eight separate articulating shell plates on their backs (Figures 1B and 4). They range in length from about 7 mm to over 35 cm. These strictly marine animals are common inhabitants of intertidal regions around the world.

Gastropods are probably the best known molluscs (Figures 1C–H, 5, 6, and 7). This class includes the common snails and slugs found in all marine and many freshwater and terrestrial environments. They are the only molluscs that undergo **torsion**, a strange twisting of the body on top of the untwisted foot.

Bivalves include the clams and their kin (Figures 1L–O and 8). The shell is composed of two plates, called **valves**. The smallest bivalves are members of the freshwater family Sphaeriidae and rarely exceed 2 mm; the largest are giant tropical clams (*Tridacna*), one species of which (*T. gigas*) may weigh over 400 kg! Like gastropods, bivalves inhabit all marine environments and many freshwater habitats.

Scaphopods, the tusk shells, live in marine surface sediments at various depths. They bear a distinctive single, tall (2–15 cm) uncoiled shell that is open at both ends (Figures 1K and 9).

The cephalopods are among the most highly modified molluscs and include the pearly nautilus, squids, cuttlefish, octopuses, and a host of extinct forms (Figures 1I–J, 10, 11, 12, and 22). This group includes the largest of all living invertebrates, the giant squid, with body and tentacle lengths exceeding 20 m. Among living cephalopods, only the nautilus has retained an external shell. The cephalopods differ markedly from other molluscs in several ways; for example, they have a spacious body cavity, from which are derived the pericardium, gonadal cavity, nephridiopericardial connections, gonoducts, and various other channels and spaces, all of which form an interconnected system representing a highly modified but true coelom. Also, unlike all other molluscs, cephalopods have a functionally closed circulatory system. Most of these and other modifications are associated with the adoption of an active predatory lifestyle by these remarkable creatures.

The body wall

The body wall of molluscs comprises three recognizable layers: the cuticle, the epidermis, and muscles (Figure 14A). The cuticle is composed largely of various amino acids and sclerotized proteins, but it

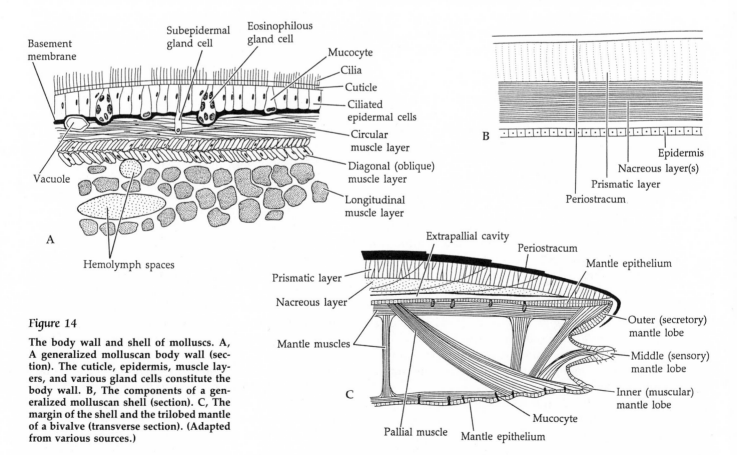

Figure 14

The body wall and shell of molluscs. A, A generalized molluscan body wall (section). The cuticle, epidermis, muscle layers, and various gland cells constitute the body wall. B, The components of a generalized molluscan shell (section). C, The margin of the shell and the trilobed mantle of a bivalve (transverse section). (Adapted from various sources.)

apparently does not contain chitin (except perhaps in the class Caudofoveata). The cuticular protein of molluscs was originally named conchiolin, but is now often simply called **conchin**. The epidermis is usually a single layer of cuboidal to columnar cells; over much of the body these epidermal cells bear cilia. Many of the epidermal cells participate in secretion of the cuticle, while others appear to be different kinds of secretory gland cells. The function of most of these gland cells is not known, but some secrete mucus and are very abundant, especially on the ventral body surface. Other areas of specialized epidermal cells occur on the dorsal body wall, or **mantle**; these cells constitute the important molluscan **shell glands**, which produce the calcareous spicules or shells characteristic of this phylum. Still other epidermal cells form sensory **epidermal papillae** or other receptor structures. The epidermis and outer muscle layer are usually separated by a basement membrane. Rarely is a dermis present between the basement membrane and the muscle layer.

The body wall usually includes three distinct layers of smooth muscle fibers: an outer circular layer, a middle diagonal layer, and an inner longitudinal layer. The diagonal muscles are often in two layers, their fibers running at right angles to each other. The degree of development of each of these muscle layers differs among the classes (for example, in solenogasters the diagonal layers are frequently absent).

The mantle and mantle cavity

We have already hinted at the significance of the mantle cavity and its importance in the success of the molluscan *Bauplan*. Here we offer a brief summary of the nature of the mantle cavity and its disposition in the major groups of molluscs.

The mantle is the dorsal body wall, and in most molluscs it grows during development as one or two folds that contain muscle layers and hemocoelic channels (Figure 14C). This outward growth creates a space lying between the mantle fold and the body proper. This space, the mantle (or pallial) cavity, may be in the form of shallow grooves, or one or two large chambers through which water is passed by ciliary or muscular action. In most cases the mantle cavity houses the gills, or ctenidia, and receives the fecal material from the anus and products of the excretory and reproductive systems. In some instances the incoming water also carries food for suspension feeding.

The mantle cavity of chitons is a pair of pallial grooves lying along the sides of the foot (Figures 13A and B). Water enters these grooves from the front and sides, passing medially over the ctenidia and then posteriorly between the cetnidia and the foot. After passing over the gonopores and nephridiopores, water exits the back end of the grooves and carries away fecal material from the anus.

The single mantle cavity of gastropods originates during development as a posteriorly located chamber. As development proceeds, however, most gastropods undergo a unique twisting (torsion) of the shell and visceral mass to bring the mantle cavity forward, over the head (Figure 13C). Again, water passing through this chamber flows over the ctenidia, anus, gonopores, and nephridiopores. A great many secondary modifications on this plan have evolved in the Gastropoda, including rerouting of current patterns, loss of certain associated structures, and even "detorsion," as discussed in later sections of this chapter.

Bivalves possess a pair of large mantle cavities, one on each side of the foot and visceral mass (Figures 13D and E). The mantle folds line the laterally placed shells and are often produced posteriorly as inhalant and exhalant siphons, through which water enters and leaves the mantle cavity. The water passes over and through the ctenidia, which extract suspended food material and accomplish gas exchange, across the gonopores and nephridiopores, and over the anus as it exits through the exhalant siphon.

Scaphopods bear a tapered, tubular shell, open at both ends (Figure 13F). Water enters and leaves the elongate mantle cavity through the small opening in the shell and flushes over the mantle surface, which, in the absence of ctenidia, is the site of gas exchange. The anus, nephridiopores, and gonopores also empty into the mantle cavity.

In all of the above cases, water is moved through the mantle cavity by the action of cilia. In the cephalopods, however, well developed mantle muscles perform this function. The exposed, fleshy body surface of squids and octopuses is, in fact, the mantle itself (Figure 13G). Unconstrained by an external shell, the mantle of these molluscs expands and contracts to draw water into the mantle cavity and then force it out through a narrow muscular funnel or siphon. This jet of exhalant water provides a means of rapid locomotion for most cephalopods. As usual, in the mantle cavity the water passes over the ctenidia, anus, reproductive pores and excretory openings.

The remarkable adaptive qualities of the molluscan body plan is manifested in these variations in the position and function of the mantle cavity and its associated structures. In fact, even the nature of many other structures is influenced by mantle cavity arrangement, as shown schematically in Figure 15.

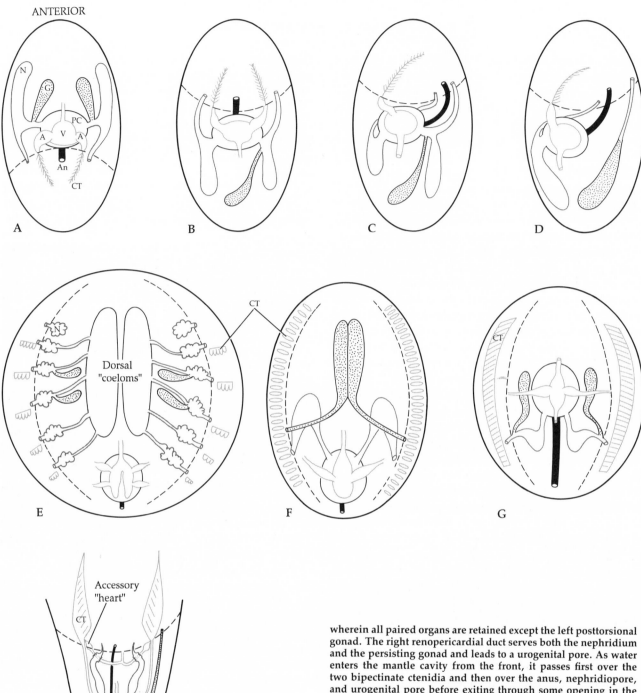

ANTERIOR

KEY

A	Atrium
An	Anus
CT	Ctenidium
G	Gonads
N	Nephridium
PC	Posterior mantle cavity
V	Ventricle

Figure 15

Variations in the mantle cavity, circulatory system, ctenidia, nephridia, reproductive system, and position of the anus in molluscs (dorsal views). Although highly schematic, these drawings give some idea of the evolutionary changes in arrangement of these structures and systems in the phylum Mollusca. A, A hypothetical, untorted, gastropod-like mollusc with a posterior mantle cavity and symmetrically paired atria, ctenidia, nephridia, and gonads. B, A primitive posttorsional archaeogastropod wherein all paired organs are retained except the left posttorsional gonad. The right renopericardial duct serves both the nephridium and the persisting gonad and leads to a urogenital pore. As water enters the mantle cavity from the front, it passes first over the two bipectinate ctenidia and then over the anus, nephridiopore, and urogenital pore before exiting through some opening in the shell (e.g., the holes or slits in primitive snails). C, A more advanced archaeogastropod such as a true limpet. Here the posttorsional right ctenidium and right atrium are lost and the nephridiopore, anus, and urogenital pore are shifted to the right side of the mantle cavity, thus allowing a one-way, left-to-right water flow. D, Most meso- and neogastropods have a single, posttorsional left, monopectinate ctenidium, often suspended from the roof of the mantle cavity. The right renopericardial duct has typically lost its association with the pericardium and is committed entirely to serving the gonad. Such isolation of the gonad and gonoduct from the excretory plumbing has allowed the evolution of elaborate reproductive systems among "higher" gastropods (e.g., neogastropods, opisthobranchs, pulmonates) and was probably a major event in the story of gastropod success. E, The condition in monoplacophorans includes the serial repetition of several organs. F, In polyplacophorans, the gonoducts and nephridioducts open separately into the exhalant regions of the lateral pallial grooves. G, A generalized bivalve condition. The gonads and nephridia may share common pores, as shown here, or else open separately into the lateral mantle chambers. H, The condition in a generalized cephalopod with a single, isolated reproductive system and an effectively closed circulatory system.

The fact that molluscs have been able to successfully exploit a broad range of habitats and lifestyles can be explained in part by these variations, which are central to the story of molluscan evolution. We will have a great deal more to say about these matters throughout this chapter.

The molluscan shell

Although molluscan shells vary greatly in shape and size, they all adhere to the basic construction plan of calcium carbonate laid down in layers and often covered by a thin organic surface coating called a **periostracum** (Figure 14B). The periostracum is composed of a type of conchin (largely quinone-tanned proteins) similar to that found in the epidermal cuticle. The calcium layers are generally of two types, an outer chalky **prismatic** portion and an inner pearly **lamellar** or **nacreous** layer; the latter layer has been lost in many groups. Both layers may incorporate conchin in various ways, often to help bind the calcareous crystals together. Shells of the various higher taxa are often composed of different numbers of calcareous sublayers.

Molluscs are noted for their wonderfully intricate and often flamboyant shell color patterns and sculpturing (Figure 16), but very little is known about the evolutionary origins and functions of these features. Current theory views most molluscan pigments as metabolic by-products, and thus shell colors may primarily represent food residues. Molluscan shell pigments include such compounds as pyrroles and porphyrins. Melanins are common in the integument (cuticle and epidermis), the eyes, and internal organs, but they are rare in shells.

A few shell sculpture patterns are correlated to specific behaviors or habitats. For example, shells with low spires are more stable in areas of heavy wave shock or on vertical rock surfaces. Similarly, the low, cap-shaped shells of limpets (Figures 16H and I) are presumably adapted for withstanding exposure to strong waves. Heavy ribbing, thick or inflated shells, and a narrow gape in bivalves are all possible adaptations to predator protection. In some gastropods, fluted shell ribs help them land right-side-up when they are dislodged from rocks. Several groups of soft-bottom benthic gastropods and bivalves have long spines on the shell that may help stabilize the animals in loose sediments. Many molluscs, particularly clams, have shells covered with living epizootic organisms such as sponges, tube worms, ectoprocts, and hydroids. Some studies suggest that predators have a difficult time recognizing such camouflaged molluscs as potential prey (Figure 16M).

Figure 16

Shell morphology and terminology. A–F, Chiton shells (class Polyplacophora): A, A chiton with eight valves (dorsal view). B, Isolated valves of *Chryptochiton stelleri*, the giant gumboot chiton. C, An anterior valve (ventral view). D–E, An intermediate valve (dorsal and ventral views). F, A posterior valve (ventral view). G, Internal and external features of a spiral gastropod shell. H, A limpet shell (side view). I, The shell of a keyhole limpet (top view). J, The tubular shell of a scaphopod. K, Inside view of the left valve of a clam (class Bivalvia). L, Dorsal view of a clam. M, A rock scallop (*Hinnites*) whose shell is overgrown with sponges, worm tubes, and other things. (A,H,I from Brusca and Brusca 1978; B courtesy of G. McDonald; M photo by authors; others modified from various sources.)

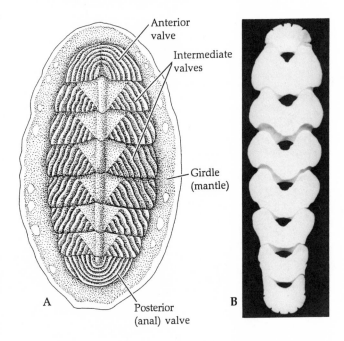

Molluscs may have one shell, two shells, eight shells, or no shell (Figure 16). In the latter case the outer body wall may contain calcareous spicules of various sorts. In aplacophorans, for example, the cuticular spicules ("spines") vary in shape and range in length from microscopic to about 4 mm. These spicules are essentially crystals composed almost entirely of calcium carbonate. Caudofoveatans produce platelike cuticular spicules that give their body surface a scaly texture and appearance. The spicules in both taxa appear to be secreted by a diffuse network of specialized groups of cells; perhaps these clusters of cells constitute primitive shell glands.

In chitons (class Polyplacophora), the shell is produced as eight transverse plates, called **valves** (Figures 16A–F). The valves are encircled by and embedded in a thickened region of the mantle called the **girdle**. The size of the girdle varies from narrow to broad and may cover much of the valves. In the giant Pacific chiton, *Cryptochiton stelleri*, the girdle com-

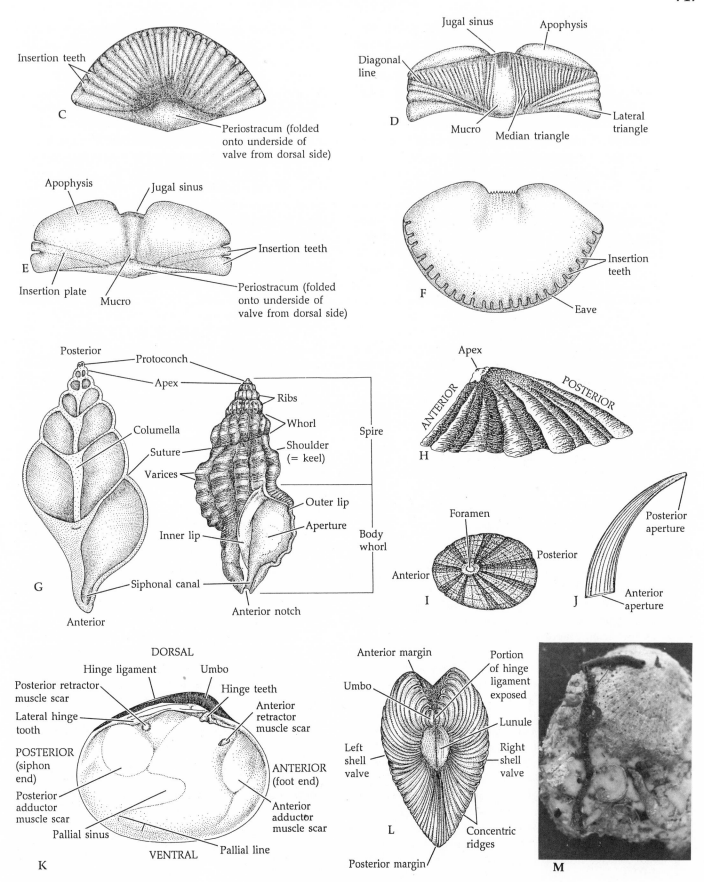

C　Insertion teeth　Periostracum (folded onto underside of valve from dorsal side)

D　Jugal sinus　Apophysis　Diagonal line　Mucro　Median triangle　Lateral triangle

E　Apophysis　Jugal sinus　Insertion teeth　Insertion plate　Mucro　Periostracum (folded onto underside of valve from dorsal side)

F　Insertion teeth　Eave

G　Posterior　Protoconch　Apex　Columella　Suture　Varices　Inner lip　Ribs　Whorl　Shoulder (= keel)　Spire　Outer lip　Aperture　Body whorl　Siphonal canal　Anterior notch　Anterior

H　Apex　ANTERIOR　POSTERIOR

I　Foramen　Anterior　Posterior

J　Posterior aperture　Anterior aperture

K　DORSAL　Hinge ligament　Umbo　Posterior retractor muscle scar　Hinge teeth　Lateral hinge tooth　Anterior retractor muscle scar　POSTERIOR (siphon end)　ANTERIOR (foot end)　Posterior adductor muscle scar　Anterior adductor muscle scar　Pallial sinus　Pallial line　VENTRAL

L　Anterior margin　Portion of hinge ligament exposed　Umbo　Lunule　Left shell valve　Right shell valve　Concentric ridges　Posterior margin

M

pletely covers the valves (Figure 4C). The girdle is thick, heavily cuticularized, and usually beset with calcareous spicules, spines, scales, or bristles secreted by specialized epidermal cells. They are probably homologous with the spicules in the body wall of caudofoveatans and aplacophorans.

The anterior and posterior valves of chitons are referred to as the end valves, or **cephalic** (= anterior) and **anal** (= posterior) **plates**; the six other valves are called the **intermediate valves**. Some details of chiton valves are shown in Figures 16A–F. The shells of chitons are three-layered, with an outer periostracum, a colored **tegmentum**, and an inner calcareous layer, or **articulamentum**. The periostracum is a very thin, delicate organic membrane and is not easily seen. The tegmentum is composed of organic material (probably a form of conchin) and calcium carbonate suffused with various pigments. It is penetrated by vertical canals that lead to minute pores in the surface of the valves. The pores are of two sizes, called **megalopores** and **micropores**, and house special sensory organs called **aesthetes**. The vertical canals arise from a layer of horizontal canals between the tegmentum and articulamentum (Figure 43C). The articulamentum is a thick, calcareous, pearly layer that differs in subtle but fundamental ways from the shell layers of other molluscs.

Monoplacophorans have a single, large, limpet-like shell with the apex situated far forward (Figures 1A and 3). As in chitons, the mantle encircles the body and foot as a circular fold, forming lateral pallial grooves.

The bivalves possess two shells, or valves, that are hinged dorsally and enclose the body and spacious mantle cavity (Figures 1L–O and 16K and L). Shells of bivalves have a thin periostracum, covering two to four calcareous layers that vary in composition and structure. The calcareous layers are often aragonite or an aragonite/calcite mixture, and they typically incorporate a substantial organic framework. The periostracum and organic matrix may account for over 70 percent of the shell's dry weight. Each valve bears a dorsal protuberance called the **umbo**, which is the oldest part of the shell. Concentric growth lines radiate from the umbo. The two valves are attached by an elastic, proteinaceous **hinge ligament**. When the valves are closed by contraction of the **adductor muscles**, the outer part of the hinge ligament is stretched and the inner part is compressed. Thus, when the adductor muscles relax, the resilient ligament causes the valves to open. The hinge apparatus comprises various sockets or tooth-like arrangements (**hinge teeth**) that prevent slipping

of the valves. In most bivalves, the adductor muscles contain both striated and smooth fibers, facilitating both rapid and sustained closure of the valves.

The mantle of bivalves is thin and sheetlike, lining the inner valve surfaces and separating the visceral mass from the shell. The edge of the mantle bears three longitudinal ridges or folds—the inner, middle, and outer folds (Figure 14). The innermost fold is the largest and contains radial and circular muscles, some of which attach the mantle to the shell. The line of mantle attachment appears on the inner surface of each valve as a scar called the **pallial line** (Figure 16K). The middle mantle fold is sensory in function, and the outer fold is responsible for secreting parts of the shell. The cells of the outer fold are specialized: the medial cells lay down the periostracum, and the lateral cells secrete the first calcareous layer. The entire mantle surface is responsible for secreting the remaining innermost calcareous portion of the shell. A thin **extrapallial space** lies between the mantle and the shell, and it is into this space that materials for shell formation are secreted and mixed. Should a foreign object, such as a sand grain, lodge between the mantle and the shell, it may become the nucleus around which are deposited concentric layers of smooth nacreous shell. The result is a pearl, either free in the extrapallial space or partly embedded in the growing shell.

Scaphopod shells are elongated tapering tubes, resembling miniature elephants' tusks, hence the vernacular names "tusk shell" and "tooth shell" (Figures 1K and 16J). The shell is open at both ends, with the smaller opening at the posterior end of the body. Most tusk shells are slightly curved, the concave side being the dorsal surface. As in bivalves, the mantle is large, lining the entire ventral surface of the shell. The posterior aperture serves for both inhalant and exhalant water currents.

Most extant cephalopods have a reduced shell or are shell-less. A completely developed shell is found only in fossil forms and the six or so surviving species of *Nautilus*. In squids and cuttlefish the shell is reduced and internal, and in octopuses it is entirely lacking or present only as a small rudiment.

The shell of *Nautilus* is coiled in a planospiral fashion (whorls lie on a single plane) and lacks a periostracum (Figures 10, 17A, and 22B). *Nautilus* shells (like all cephalopod shells) are divided into internal chambers by transverse septa, but only the last chamber is occupied by the body of the living animal. The animal can withdraw into this chamber and then cover the shell aperture with a fleshy fold of tissue called the **hood**. As the animal grows, it

Figure 17

Two cephalopod shells. A, The chambered shell of *Nautilus*, cut in longitudinal section. B, The egg case "shell" of the paper nautilus, *Argonauta*. (Photos by K. Lucas/BPS.)

periodically moves forward, and the posterior part of the mantle secretes a new septum behind it. Each septum bears a central perforation through which extends a cord of tissue called the **siphuncle**. The siphuncle regulates buoyancy of the animal by varying the amounts of gases and fluids in the shell chambers. The shell is composed of an inner nacreous layer and an outer porcelain layer containing prisms of calcium carbonate and an organic matrix. The outer surface may be pigmented or pearly white. The junctions between septa and the shell wall are called **sutures**, which may be simple and straight, slightly waved (as in *Nautilus*), or highly convoluted (as in the extinct ammonoids). In cuttlefish (order Sepioida), the shell is reduced and internal, with spaces separated by thin septa. Like *Nautilus*, a cuttlefish regulates its buoyancy by altering the relative amounts of fluid and gas in the shell chambers. The small gas-filled shells of the cuttlefish *Spirula* are often found washed up on tropical beaches.

Fossil data suggest that the first cephalopod shells were probably curved cones. From these ances-

tors both straight and coiled shells evolved, although secondary uncoiling probably occurred in several groups. Some straight-shelled cephalopods from the Ordovician Period exceeded 5 m in length, and some Cretaceous coiled species had shell diameters of 3 m.

Gastropod shells are extremely diverse in size and shape (Figure 1C–F). The smallest are microscopic and the largest may exceed 70 cm. The "typical" shape is the familiar conical spiral wound around a central axis or **columella** (Figure 16G). The turns of the **spire** form **whorls**, demarcated by lines called **sutures**. The largest whorl is the **body whorl**, which bears the **aperture** through which the foot and head protrude. The aperture is anterior and the apex of the shell spire is posterior. The first few, very small, whorls at the apex are the remnant of the larval shell, or **protoconch**, which usually differs in sculpturing and color from the rest of the shell. The body whorl and aperture may be drawn out into an anterior **siphonal canal**, to house a **siphon** when present. A smaller posterior canal may also be present on the rear edge of the aperture.

Every imaginable variation on the basic spiraled shell occurs among the gastropods: the shell may be long and slender (e.g., tower shells) or short and plump (e.g., trochids); the spire may be more-or-less incorporated into the body whorl and eventually disappear from view; the shell may be flattened, with all whorls in one plane (e.g., sundials); the last body whorl may completely overgrow the older whorls, reducing the aperture to an elongated slit as the two lips are brought together (Figure 1F) (e.g., cowries, olives, and cones); the shell may coil so loosely as to form a meandering wormlike tube (Figure 19E) (e.g., vermetids); the shell may be reduced and overgrown by the mantle, or it may disappear entirely (e.g., many opisthobranchs and pulmonates). Most gastropods spiral clockwise; that is, they show right-handed or dextral coiling. Some are sinistral (left-handed), and some species can coil in either direction. In limpets the shell is **patelliform**, with a low conical shape with no visible coiling at all (Figures 16H and I). The limpet form was probably derived from coiled ancestors on numerous occasions during gastropod evolution.

Gastropod shells consist of an outer thin organic periostracum and two or three calcareous layers: an outer **prismatic** (or **palisade**) **layer**, a middle **lamellate layer**, and an inner **nacreous layer** (also called the **hypostracum**). The nacreous layer is composed of calcareous lamellae layered with thin films of conchin; it has been lost in many archaeogastropods and almost all meso- and neogastropods. In some gastropods up to six calcareous layers are distinguishable. Gastropods in which the shell is habitually covered by mantle lobes lack a periostracum (e.g., olives and cowries), but in some other groups the periostracum is very thick and "hairy." The prismatic and lamellate layers consist largely of calcium carbonate, either as calcite or aragonite. These two forms of calcium are chemically identical, but they crystallize differently and can be identified by microscopic examination of sections of the shell. Small amounts of other inorganic constituents are incorporated into the calcium carbonate framework, including chemicals such as phosphate, calcium sulfate, magnesium carbonate, and salts of aluminum, iron, copper, strontium, barium, silicon, manganese, iodine, and fluorine. The prismatic layer has the calcium carbonate deposited as vertical crystals, each surrounded by a thin protein matrix. The nacreous layer has the calcium carbonate deposited as thin lamellae, which are always interleaved with conchin.

An intriguing puzzle of gastropod evolution is the appearance of shell-lessness, or the "slug" form. Despite the fact that evolution of the coiled shell led to great success for the gastropods—80 percent of all living molluscs are snails—secondary loss of the shell occurred many times in this class. In forms such as the land and sea slugs, the shell may persist as a small vestige covered by the dorsal mantle (e.g., some sea hares and notaspids), or it may be lost altogether (e.g., the true nudibranchs). In the latter case it is first covered, then resorbed by the mantle during ontogeny. The two primary examples of shell-lessness are the pulmonate land slugs and the marine opisthobranch slugs. Although they are shell-less now, shell loss probably occurred numerous times in both groups. Shells, of course, are energetically expensive to produce and require a large source of calcium in the environment, so it might be advantageous to do away with them if compensatory mechanisms exist (e.g., most, if not all, sea slugs produce chemicals that make them distasteful to predators).

Torsion, or "How the Gastropod Got its Twist"

One of the most remarkable and dramatic steps taken during the course of molluscan evolution was the advent of **torsion**, a unique synapomorphy of modern gastropods. Torsion takes place during development in all gastropods, usually during the late veliger larval stage. It is a rotation of the visceral mass and its overlying mantle and shell as much as 180° with respect to the head and foot (Figures 15A–D, 18, and 52). The twisting is always in a counterclockwise direction (viewing the animal from above) and

is completely different from the phenomenon of coiling. During torsion, the mantle cavity and anus are moved from a posterior to a more anterior position, somewhat above and behind the head. Visceral structures and incipient organs that were on the right side of the larval animal end up on the left side of the adult. The gut is twisted into a U shape, and when the longitudinal nerve cords connecting the pleural to the visceral ganglia develop, they are crossed rather like a figure-eight. Most veligers have nephridia, which reverse sides, but the adult gills and gonads are not fully developed when torsion occurs.

Torsion is usually a two-step process. During larval development, an asymmetrical **velar/foot retractor muscle** develops. It extends from the shell on the right, over the gut dorsally, and attaches on the left side of the head and foot. At a certain stage in the veliger's development, contraction of this muscle causes the shell and enclosed viscera to twist about 90° in a counterclockwise direction. This first 90° twist is usually rapid, taking place in a few minutes to a few hours. The second 90° twist is typically much slower and results from differential tissue growth. By the end of the process, the viscera have been pulled from above toward the left, ultimately leading to the figure-eight arrangement of the adult visceral nerves. But the figure-eight arrangement is not perfect. The left intestinal ganglion usually comes to lie dorsal to the gut and is thus called the **supraintestinal** (= **supraesophageal**) ganglion; however, the right intestinal ganglion lies ventral to the gut, as a **subintestinal** (= **subesophageal**) ganglion (Figures 18 and 40). The torted, figure-eight configuration of the nervous system is referred to as **streptoneury**. The detorted condition, in which the visceral nerves are parallel, is referred to as **euthyneury**.

Gastropods that retain torsion into adulthood are said to be **torted**; those that have secondarily reverted back to a partially or fully untorted state in adulthood are **detorted**. Detorted gastropods, such as most opisthobranchs, undergo a postveliger series of changes through which the original torsion is reversed to various degrees. The process shifts the mantle cavity and at least some of the pallial organs about 90° back to the right, or in some cases all the way back to the rear of the animal.

Evolutionarily speaking, after torsion, the anus lay in front, and the animal could no longer grow in length easily. Subsequent increase in body size thus occurred by the development of loops or bulges in the middle portion of the gut region, producing thereby a characteristic visceral hump. The first signs of torsion and coiling occur at about the same time

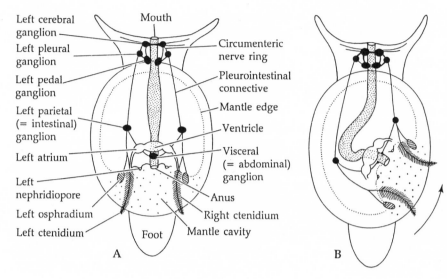

Left cerebral ganglion
Mouth
Left pleural ganglion
Circumenteric nerve ring
Left pedal ganglion
Pleurointestinal connective
Left parietal (= intestinal) ganglion
Mantle edge
Left atrium
Ventricle
Left nephridiopore
Visceral (= abdominal) ganglion
Left osphradium
Anus
Left ctenidium
Right ctenidium
Foot
Mantle cavity

A

B

Figure 18

Pre- and posttorsional adult gastropods. A–D, Dorsal views. A, Hypothetical untorted gastropod. B–C, Stages of torsion as they might appear in an adult snail. D, The fully torted condition. Note that the mantle cavity, gills, anus, and nephridiopores are moved from a posterior to an anterior orientation, just above and behind the head. Furthermore, many structures that were on the right side of the animal in the pretorsional condition (e.g., the right gill, osphradium, heart atrium, and nephridiopore) are located on the left side after torsion has taken place (and the pretorsional left gill, osphradium, atrium, and nephridiopore subsequently occur on the right side). E, A prosobranch veliger larva, before and after torsion (lateral view). Note that after torsion the head can be withdrawn into the anterior mantle cavity. F, Configuration of the principal ganglia and connectives of a hypothetical untorted and a torted adult gastropod. (A–D after Lang, 1900, Lehrbuch der vergleichenden Anatomie der wirbellosen Thiere 3:1–509; E after Morton 1979; F after Barnes 1980.)

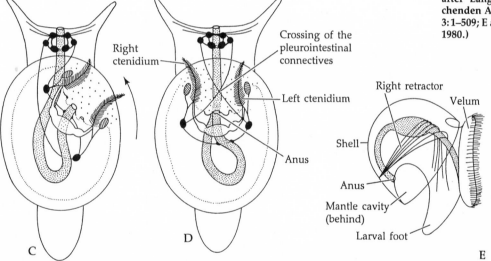

Right ctenidium
Crossing of the pleurointestinal connectives
Left ctenidium
Anus

C

D

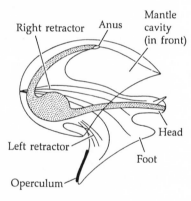

Right retractor
Velum
Shell
Anus
Mantle cavity (behind)
Larval foot

Right retractor
Anus
Mantle cavity (in front)
Left retractor
Head
Foot
Operculum

E

during gastropod development. However, the fossil record suggests that the first coiled gastropod shells were planospiral and that these forms may have predated the appearance of torsion in gastropods. Once both features were established, they coevolved in various ways to produce what we see today in modern living gastropods.

The evolution of asymmetrically coiled shells had the effect of restricting the right side of the mantle cavity, a restriction that led to reduction or loss of the pallial structures on the adult right side (the original left ctenidium, atrium, and osphradium). At the same time, these structures on the adult left side (the original right ctenidium, atrium, and osphradium)

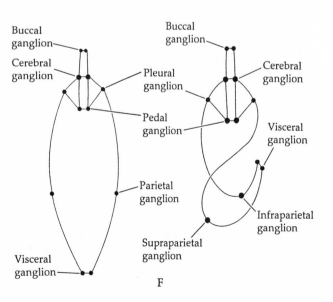

Buccal ganglion
Buccal ganglion
Cerebral ganglion
Pleural ganglion
Cerebral ganglion
Pedal ganglion
Visceral ganglion
Parietal ganglion
Infraparietal ganglion
Supraparietal ganglion
Visceral ganglion

F

tended to enlarge. After torsion and coiling had appeared, the left posttorsional gonad was lost. The single remaining gonad opens on the right side via the posttorsional right nephridial duct and nephridiopore. Most archaeogastropods retain two functional nephridia. In most higher gastropods the posttorsional right nephridium is reduced or lost, but its duct and pore remain associated with the reproductive tract. This isolation of the reproductive system from the excretory system probably allowed the great elaboration of the reproductive organs seen among mesogastropods and neogastropods.

Such profound changes in spatial relations between major body regions, as brought about by torsion and spiral coiling, are rare in the animal kingdom. Several theories on the adaptive significance of torsion have been proposed. The great zoologist Walter Garstang suggested that torsion was an adaptation of the veliger larva that served to protect the soft head and larval velum from predators (see section on development). When disturbed, the immediate reaction of a veliger is to withdraw the head and foot into the larval shell, whereupon the larva begins to sink rapidly. This theory may seem reasonable for evasion of very small planktonic predators, but it seems illogical as a means of escape from larger predators in the sea, which no doubt consume veligers whole. Two zoologists recently tested Garstang's theory by offering torted and untorted abalone veligers to various planktonic predators; they found that, in general, torted veligers were not consumed any less frequently than untorted ones (Pennington and Chia 1985). Garstang first presented his theory in verse, in 1928, as he was often taken to do with his zoological ideas.

The Ballad of the Veliger
or
How the Gastropod Got Its Twist

The Veliger's a lively tar, the liveliest afloat,
A whirling wheel on either side propels his little boat;
But when the danger signal warns his bustling submarine,
He stops the engine, shuts the port, and drops below unseen.

He's witnessed several changes in pelagic motor-craft;
The first he sailed was just a tub, with a tiny cabin aft.
An Archi-mollusk fashioned it, according to his kind,
He'd always stowed his gills and things in a mantle-sac behind.

Young Archi-mollusks went to sea with nothing but a velum—
A sort of autocycling hoop, instead of pram—to wheel 'em;

And, spinning round, they one by one acquired parental features,
A shell above, a foot below—the queerest little creatures.

But when by chance they brushed against their neighbours in the briny,
Coelenterates with stinging threads and Arthropods so spiny,
By one weak spot betrayed, alas, they fell an easy prey—
Their soft preoral lobes in front could not be tucked away!

Their feet, you see, amidships, next the cuddly-hole abaft,
Drew in at once, and left their heads exposed to every shaft.
So Archi-mollusks dwindled, and the race was sinking fast,
When by the merest accident salvation came at last.

A fleet of fry turned out one day, eventful in the sequel,
Whose left and right retractors on the two sides were unequal:
Their starboard halliards fixed astern alone supplied the head,
While those set aport were spread abeam and served the back instead.

Predaceous foes, still drifting by in numbers unabated,
Were baffled now by tactics which their dining plans frustrated.
Their prey upon alarm collapsed, but promptly turned about,
With the tender morsel safe within and the horny foot without!

This manoeuvre (vide Lamarck) speeded up with repetition,
Until the parts affected gained a rhythmical condition,
And torsion, needing now no more a stimulating stab,
Will take its predetermined course in a watchglass in the lab.

In this way, then, the Veliger, triumphantly askew,
Acquired his cabin for'ard, holding all his sailing crew—
A Trochosphere in armour cased, with a foot to work the hatch,
And double screws to drive ahead with smartness and despatch.

But when the first new Veligers came home again to shore,
And settled down as Gastropods with mantle-sac afore,
The Archi-mollusk sought a cleft his shame and grief to hide,
Crunched horribly his horny teeth, gave up the ghost, and died.

Other workers have hypothesized that torsion was an adult adaptation that might have also created more space for retraction of the head into the shell (perhaps also for protection from predators), or for directing the mantle cavity with its gills and water-sensing osphradia anteriorly. Still another theory asserts that torsion evolved in concert with the evolution of a coiled shell—as a mechanism to align the tall spiraling shells from a position in which they stuck out to one side (and were presumably poorly balanced and growth limiting), to a position more in alignment with the longitudinal (head–foot) axis of the body. The latter position would theoretically allow for greater growth and elongation of the shell while reducing the tendency of the animal to topple over sideways.

No matter what the evolutionary forces leading to torsion in the earliest gastropods were, the results were to move the adult anus, nephridiopores, and gonopores to a more anterior position, corresponding to the new position of the mantle cavity. It should be noted, however, that the actual position and arrangement of the mantle cavity and its associated structures show great variation, and in many gastropods these structures, while pointing forward, may actually be positioned near the middle or even the posterior region of the animal's body.

Most of the stories of gastropod evolution focus on changes in the mantle cavity and its associated structures, and many of these changes seem to have been driven by the impact of torsion. Many anatomical modifications of gastropods appear to be adaptations to avoid fouling, for without changing the original flow of water through the mantle cavity, waste from the anus (and perhaps the nephridia) would be dumped on top of the head and pollute the mouth and ctenidial region. Perhaps the first step, subsequent to the evolution of torsion, was to develop slits or holes in the shell, altering water flow so that a one-way current passed first over the ctenidia, then over the anus and nephridiopores, and finally out the slit or shell holes. This arrangement is seen in certain primitive gastropods, such as the slit shells (Pleurotomariacea) and certain archaeogastropods (abalone and keyhole limpets) (Figures 1C and D). Once reduction or loss of the gill and osphradium on the right side had taken place, a different antifouling strategy was achieved—that of a directed water flow through the mantle cavity from left to right, passing across the functional gill and osphradium first, then across the nephridiopores and anus, and on out the right side (Figure 6B). This strategy also had the effect of allowing the left side to enlarge and eventually to develop into structures such as long siphons. While the prosobranchs have retained full or partial torsion, other gastropods (opisthobranchs and pulmonates) have undergone various degrees of detorsion, loss of ctenidia, and a host of other modifications, perhaps in response to the problems originally brought on by torsion.

Locomotion

The caudofoveatans and aplacophorans lack a distinct foot (Figure 2), and locomotion is primarily by slow ciliary gliding movements through or upon the substratum. Neither group moves about very much, caudofoveatans being mostly infaunal burrowers, and aplacophorans being largely symbiotic on various cnidarians. With the exception of these two classes, most molluscs possess a distinct and obvious foot. The foot often forms a flat, ventral, creeping **sole**, like that of snails, slugs, chitons, and monoplacophorans (Figure 19). The sole is ciliated and imbued with numerous gland cells that produce a mucous trail over which the animal glides. In gastropods, a large **pedal gland** supplies substantial amounts of slime, especially in terrestrial species that must glide on relatively dry surfaces. Very small molluscs may move largely by ciliary propulsion. However, most molluscs move primarily by waves of muscular contractions that sweep along the foot.

The gastropod foot possesses sets of **pedal retractor muscles**, which attach to the shell and dorsal mantle at various angles and act in concert to raise and lower the sole or to shorten it in either a longitudinal or a transverse direction. Contraction waves may move from back to front (direct waves), or from front to back (retrograde waves) (Figures 19A and B). Direct waves depend on contraction of longitudinal and dorsoventral muscles beginning at the posterior end of the foot; successive sections of the foot are thus "pushed" forward. Retrograde waves involve contraction of transverse muscles interacting with hemocoelic pressure to extend the anterior part of the foot forward, followed by contraction of longitudinal muscles. The result is that successive areas of the foot are "pulled" forward. In some gastropods the muscles of the foot are separated by a midventral line, so the two sides of the sole operate somewhat independently of each other. The right and left sides of the foot alternate in their forward motion, almost in a stepping fashion, resulting in a sort of bipedal locomotion (Figure 19C).

Modifications of this general benthic locomotory scheme occur in many groups. Some gastropods, such as moon snails (Figure 19D), plow through the

Figure 19

A–B, Locomotion in a benthic gastropod moving to the right by waves of muscle contraction (solid arrow indicates direction of animal movement; dashed arrow indicates direction of muscle wave). In A, the waves of contraction are moving in the same direction as the animal, from back to front (direct waves). Muscles at the rear of the animal contract to lift the foot off the substratum; the foot shortens in the contracted region and then elongates as it is placed back down on the substratum after the wave passes. In this way, successive sections of the foot are "pushed" forward. In B, the animal moves forward as the contraction waves pass in the opposite direction, from front to back (retrograde waves). In this case, the pedal muscles lift the anterior part of the foot off the substratum, the foot elongates, is placed back on the substratum, then contracts to "pull" the animal forward, rather like "stepping." C, *Calliostoma*, a gastropod adapted to crawling on hard substrata. Note the line separating the right and left muscle masses in the rear of the foot; this separation allows a somewhat "bipedal-like" motion as the animal moves. D, The moon snail, *Polinices*, has a huge foot that can be inflated by incorporating water into a network of channels in its tissue, thus allowing the animal to plow through the surface layer of soft sediments. E, *Tenagodus*, a sessile vermetid worm snail. (A–B modified after Miller 1974; C–D courtesy of G. McDonald; E after Hyman 1967.)

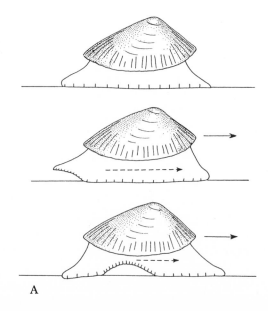

A

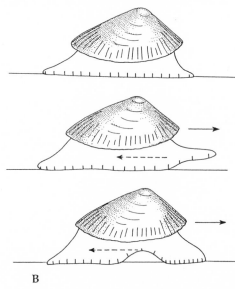

B

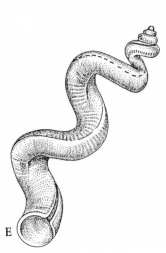

E

C

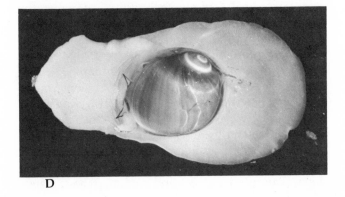

D

sediment by brute force, and they can even burrow beneath the sediment surface. Such gastropods often possess a **propodium**, a thick anterior region of the foot shaped like a blade of a bulldozer, as well as a dorsal flaplike fold of the foot that covers the head as a protective shield. Other burrowers, such as turritellids, dig by jerky side-to-side movements of the projected foot or by thrusting the foot into the substratum, anchoring it by engorgement with hemolymph, and then pulling the body forward by contraction of longitudinal muscles. In the conch *Strombus*, the operculum forms a large "claw" that digs into the substratum and is used as a pivot point as the animal thrusts itself forward like a pole vaulter.

Some molluscs (e.g., chitons and limpets) that inhabit high-energy littoral habitats have a very broad foot that can adhere tightly to hard substrata. Chitons also use their broad girdle for adhesion to the substratum by clamping down tightly and raising the inner margin to create a slight vacuum. Some snails, such as the so-called worm shells (Vermetidae and Siliquariidae), are entirely sessile (Figure 19E). These gastropods have typical larval and juvenile shells, but after they settle and start to grow, the shell whorls become increasingly separated from one another, resulting in a corkscrew or irregular shape. Other gastropods, such as slipper limpets, are sedentary. They tend to remain in one location and, like bivalves, survive by feeding on organic particulates from the surrounding water.

In bivalves the foot is usually bladelike and laterally compressed (the word *pelecypod* means "hatchet foot"), as is the body in general. The pedal retractor muscles in bivalves are somewhat different from those of gastropods, but they still run from the foot to the shell (Figure 8). The foot is directed anteriorly and used primarily in burrowing and anchoring. It operates through a combination of muscle action and hydraulic pressure (Figures 20A–D). Extension of the foot is accomplished by engorgement with hemolymph, coupled with the action of a pair of pedal protractor muscles. With the foot extended, the valves are pulled together by the shell adductor muscles. More hemolymph is forced from the visceral mass hemocoel into the foot hemocoel, causing the foot to expand and anchor in the substratum. Once the foot is anchored, the anterior and posterior pairs of pedal retractor muscles contract and pull the shell downward. Withdrawal of the foot into the shell is accomplished by contraction of the pedal retractors coupled with relaxation of the shell adductor muscles. Many clams burrow upward in this same manner, but others back out by using hydraulic pressure to push against the anchored end of the foot. Most motile bivalves possess well developed anterior and posterior adductor muscles (the **dimyarian** condition).

Most bivalves live in soft benthic habitats, where they burrow to various depths in the substratum (Figures 20E–I). However, several groups have epifaunal lifestyles (e.g., Pectinidae, Limidae) and live free upon the sea floor (Figure 1L). Some are capable of short bursts of "jet-propelled" swimming, which is accomplished by clapping the valves together. Others permanently attach to the substratum either by fusing one valve to a hard surface (e.g., rock oysters, rock scallops) or by using special anchoring lines called **byssal threads** (e.g., mussels [Figures 21A and b], ark shells, and certain oysters such as mangrove oysters and winged oysters). Oysters of the family Ostreidae (including the edible American and European rock oysters) initially anchor as a settling veliger larva, called a **spat**, by secreting a drop of adhesive from the **byssus gland**. Adults, however, have one valve permanently cemented to the substratum.

Byssal threads are secreted as a liquid by the byssus gland in the foot. The liquid flows along a groove in the foot to the substratum, where each thread becomes tightly affixed. The threads are emplaced by the foot; once attached they quickly harden by a tanning process, whereupon the foot is withdrawn. A byssal thread **retractor muscle** may assist the animal in pulling against its anchorage. Mussels have a small, finger-like foot whose principal function is generation and placement of the byssal threads. Giant clams (Tridacnidae) initially attach by byssal threads, but usually lose these as they mature and become heavy enough not to be cast about by currents (Figure 1M). In jingle shells (Anomiidae), the byssal threads run from the upper valve through a hole in the lower valve to attach to the substratum. Byssal threads probably represent a primitive and persisting larval feature in those groups that retain them into adulthood, for many bivalves lacking byssal threads as adults utilize them for initial attachment during settlement.

In many families of attached bivalves, such as mussels and rock scallops, the foot and anterior end are reduced. This often leads to a reduction (**anisomyarian condition**) or loss (**monomyarian condition**) of the anterior adductor muscle. Mantle fusion and siphon development are also greatly reduced in attached bivalves.

Great variation occurs in shell shape and size among attached molluscs. Among bivalves, some of the most remarkable were the Mesozoic rudist clams, in which the lower valve was hornlike and often curved, and the upper valve formed a much smaller hemispherical or curved lid (Figure 21C). Young rudist clams attached to the substratum or to other individuals, but adults of many species apparently lived free on the sea floor, perhaps rolling about with the currents. Rudists were large heavy creatures that

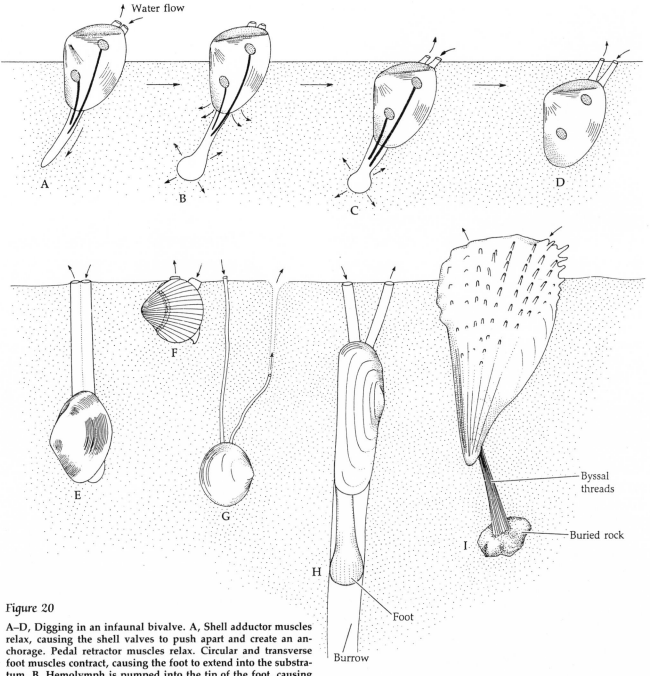

Figure 20

A–D, Digging in an infaunal bivalve. A, Shell adductor muscles relax, causing the shell valves to push apart and create an anchorage. Pedal retractor muscles relax. Circular and transverse foot muscles contract, causing the foot to extend into the substratum. B, Hemolymph is pumped into the tip of the foot, causing it to expand and form an anchorage. Siphons close and withdraw as the shell adductor muscles contract, closing the shell and forcing water out between the valves and around the foot. C, Anterior and posterior pedal retractor muscles contract, pulling the clam deeper into substratum. D, The shell adductor muscles relax to allow shell valves to push apart and create an anchorage in the new position. The foot is withdrawn. E–I, Five bivalves in soft sediments. E, A deep burrower with long, fused siphons (e.g., *Tresus*). F, A shallow "nestler" with very short siphons (e.g., *Clinocardium*). G, A deep burrower with long, separate siphons (e.g., *Scrobicularia*). H, The razor clam (*Tagelus*) lives in unstable sands and maintains a burrow into which it can rapidly escape. I, The pen shell, *Atrina*, attaches its byssal threads to solid objects buried in soft sediments. (A–D after Trueman 1966.)

often formed massive reeflike aggregations, either by somehow attaching to the substratum or by simply accumulating in large numbers on the seabed, in "log jams."

The habit of boring into hard substrata has evolved in many different bivalve lines. In all cases, excavation begins quickly after larval settlement. As the animal bores deeper, it grows in size and soon becomes permanently trapped, with only the siphons

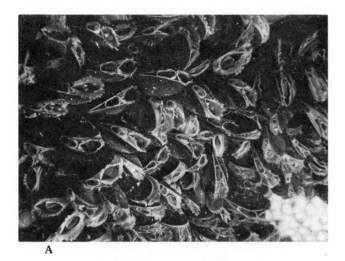

A

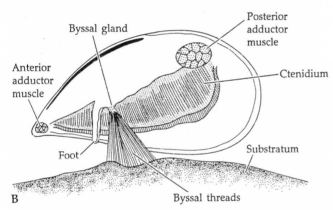

B

Figure 21

A, A "bed" of bay mussels (*Mytilus edulis*) attached by byssal threads. B, A mussel (lateral view, with left valve removed). C, Shell shape in a Mesozoic rudist clam. D, Anatomy of the wood-boring bivalve *Teredo*. The pallets (only one is shown) are plates that close over the siphons when they are retracted. (A photo by H. W. Pratt/BPS; B–D after various sources.)

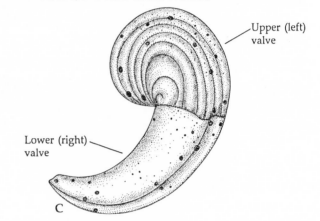

C

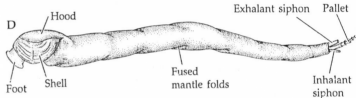

D

protruding out of the original small opening (Figure 1O). Boring is usually by a mechanical process; the animal uses serrations on the anterior region of the shells to abrade or scrape away the substratum. Some species also secrete an acidic mucus that partially dissolves or weakens hard substrata. Numerous species in the family Pholadidae bore into wood (e.g., *Martesia, Xylophaga*), soft stone (e.g., *Pholas*), or a variety of substrata (e.g., *Barnea*). Species in the family Teredinidae (e.g., *Bankia, Teredo*) are known as shipworms because of their preference for wood, including the wooden hulls of ships. Only small remnants of the shells remain and serve as drilling structures in shipworms, with the vermiform body trailing behind (Figure 21D). Some species in the family Mytilidae also are borers, such as *Lithophaga*, which bores into hard calcareous rocks, shells of various other molluscs (including chitons) and corals, and the genus *Adula*, which bores into soft rocks.

Scaphopods are adapted to infaunal habitats, burrowing vertically by the same basic mechanism used by many bivalves (Figures 1K and 9). The elongate foot is projected downward into soft substrata, whereupon the tip is expanded to serve as an anchoring device; contraction of the pedal retractor muscles pulls the animal downward.

Perhaps the most remarkable locomotor adaptation of molluscs is swimming, which has evolved in several different taxa. In some of these groups, the foot is modified as the swimming structure. In the unique gastropod group known as heteropods, the body is laterally compressed, the shell is highly reduced, the foot forms a ventral fin, and the animal swims upside down (Figures 7A–C). In another unusual group of gastropods, the pteropods (sea butterflies), the foot forms two long lateral fins called **parapodia** that are used like oars (Figure 7D and E). Some opisthobranchs also swim by graceful undulations of flaplike folds (also called parapodia) along the body margin. Violet snails (*Janthina*) float about the ocean's surface on a raft of bubbles secreted by the foot, and some planktonic opisthobranchs (e.g., *Glaucus, Glaucilla*) stay afloat by use of an air bubble held in the stomach!

The champion swimmers are, of course, the cephalopods (Figure 22). These animals have abandoned the generally sedentary habits of other molluscs and have become effective high-speed predators. They have turned virtually all aspects of their biology to exploiting this lifestyle. Most cephalopods swim by rapidly expelling water from the mantle cavity. The mantle has both radial and circular muscle layers. Contraction of the radial muscles and relaxation of the circular muscles draws water into the mantle cavity. Reversal of this muscular action forces water out of the mantle cavity. The mantle edge is clamped tightly around the head to channel the escaping water through a ventral tubular **funnel**, or **siphon** (Figure 11). The funnel is highly mobile and can be manipulated to point in nearly any direction, thus allowing the animal to turn and steer. Squids attain the greatest swimming speeds of any aquatic invertebrates, and several species are known to leap many feet into the air. Most octopuses are benthic and lack the fins and streamlined bodies characteristic of squids. Although they still use water-powered jet propulsion, they more commonly rely on the long suckered arms for crawling about the sea floor. Cuttlefish are slower than squids, and not only use their fins for stabilization, but also undulate them to assist in steering and propulsion. Nautiloids and sepioids actively regulate their buoyancy by secretion and reabsorption of shell chamber gases (chiefly nitrogen) by the cells of the siphuncle. Most species move up and down in the water on a diurnal cycle, often traveling hundreds of meters in each direction.

Figure 22

Swimming cephalopods. A, *Sepia*, the cuttlefish. B, *Nautilus*. C, *Vampyroteuthis*, a "vampire" squid. (A photo by authors; B photo by D. J. Wrobel, Monterey Bay Aquarium/BPS; C courtesy of G. McDonald.)

The unoccupied chambers of the shell are filled partly with gas and partly with a liquid called the **cameral fluid**. The septa act as braces, giving the shells enough strength to withstand pressures at great depths. As discussed earlier, each septum in nautiloid shells is perforated in the center by a small hole, through which runs a cord of living body tissue, called the siphuncle, which originates in the viscera and is enclosed in a porous calcareous tube. Various ions dissolved in the cameral fluid can be pumped through the porous outer layers into the cells of the siphuncular epithelium. When the cellular concentration of ions is high enough, the diffusion gradient thus created draws fluid from the shell chambers into the cells of the siphuncle while the fluid is replaced by gas. The result is an increase in buoyancy. By regulating this process, nautiloids are able to change depth or remain neutrally buoyant wherever they are.

Feeding

Two basic and fundamentally different types of feeding occur among the species of Mollusca: herbivory or predation (macrophagy), and suspension feeding (suspension microphagy). In Chapter 3 we reviewed the basic mechanics of these two feeding modes. Here we briefly summarize the ways in which these feeding behaviors are employed by molluscs. In this section we also discuss a uniquely molluscan anatomical feature, the **radula**, which is used in both herbivory and predation, and has become modified in a variety of unusual and interesting ways.

The molluscan radula and macrophagy. The radula is usually a ribbon of recurved chitinous teeth stretched over a supportive base (Figures 23–26). The

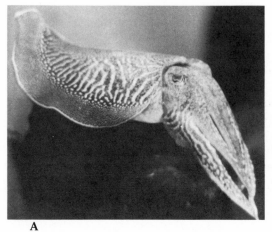

A

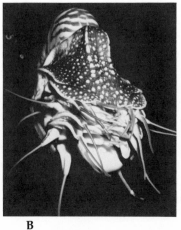

B

C

Figure 23

A generalized molluscan radula and associated buccal structures (longitudinal section).

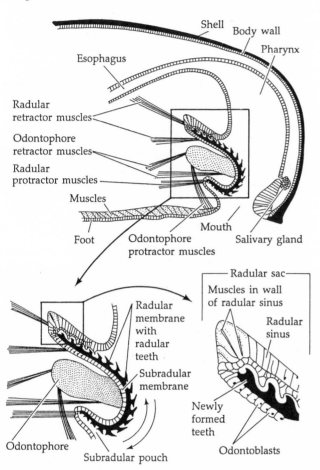

(Figures 23 and 24A). The ribbon, called a **radular membrane**, is moved back and forth over a cartilaginous or hemocoelic **odontophore** by sets of **radular protractor** and **retractor muscles**. The radula is usually housed in a **radular sac**, in which the radular membrane and new teeth are continually being produced by special cells called **odontoblasts**, to replace

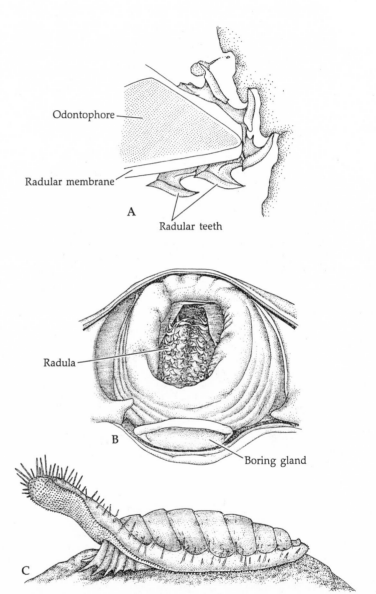

Figure 24

Feeding in macrophagous molluscs. A, Cutting and scraping action of a gastropod radula. **B,** A boring gastropod, the moon snail *Natica*, with radula visible in the mouth and the boring gland exposed (oral view). **C,** The Pacific chiton *Placiphorella velata* in feeding position, with head flap raised. (A after Solem 1974; B after Hyman 1967; C after McLean 1962, Proc. Malac. Soc. London 35: 23–26.)

teeth may be simple, serrate, pectinate, or otherwise modified. The radula often functions as a scraper to remove food particles for ingestion, although in many groups it has become adapted for other actions. A radula is present in at least some of the most primitive living molluscs and is therefore assumed to have originated in the earliest stages of molluscan evolution. In solenogasters (class Aplacophora) and probably the caudofoveatans, the teeth, when present, may not be borne on a ribbon *per se* but on a basal expansion of the foregut epithelium—perhaps the evolutionary forerunner of the ribbonlike radula. In some solenogasters, the teeth form simple plates embedded in either side of the lateral foregut wall, while in others they form a transverse row, or up to 50 rows, with as many as 24 teeth per row.

In gastropods and most other molluscs the radula projects from the pharynx or buccal cavity floor as a complex tooth-bearing ribbon and associated muscles

material lost by erosion during feeding. Measurements of radular growth indicate that up to five rows of new teeth may be added daily in some species. The odontophore itself is moved in and out of the buccal cavity by sets of **odontophore protractor** and **retractor muscles**, which also assist in applying the radula firmly against the substratum being attacked. The number of teeth ranges from a few to thousands and serves as an important taxonomic character in many groups.

Like mammalian dentition, radular teeth show adaptations to the type of food eaten. In primitive archaeogastropods (e.g., keyhole limpets, abalones, top shells), the radulae bear large numbers of fine marginal teeth in each row (Figures 25A and 26A).

As the radula is pulled over the bending plane of the odontophore, these teeth act like brushes, sweeping small particles to the midline where they are caught on the recurved parts of the central teeth, which draw the particles into the buccal cavity. These snails are mostly intertidal foragers that live on diatoms and other algae growing on the substratum. Such radulae are called **rhipidoglossate**. Some archaeogastropods (e.g., acmaeid and patellid limpets) possess **docoglossate** radulae, which bear relatively few teeth in each transverse row. In acmaeid radulae, for example, there are no central or marginal teeth, and only two pairs of lateral teeth per row (Figure 26B). The mucous trails left by some limpets (e.g., homing species such as the Pacific *Lottia gigantea* and *Collisella*

Figure 25

Various arrangements of radular teeth. A, The rhipidoglossate condition of an abalone, *Haliotis*. The marginals on the right side are not shown. B, The taenioglossate condition of the mesogastropod *Viviparus*. C, The taenioglossate condition of the mesogastropod *Littorina*. D, The highly modified taenioglossate condition of the heteropod *Pterotrachea*. Only one transverse row of teeth is shown. E, The rachiglossate condition of the neogastropod *Buccinum*. F, The toxoglossate condition of neogastropod *Conus* (a single tooth). (A,C,E,F after Fretter and Graham 1962; B, D after Hyman 1967.)

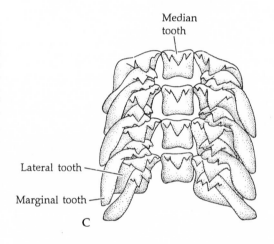

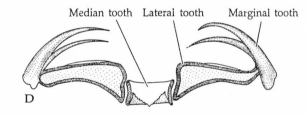

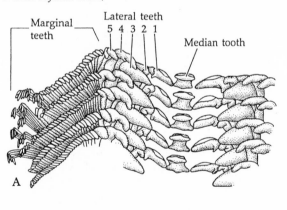

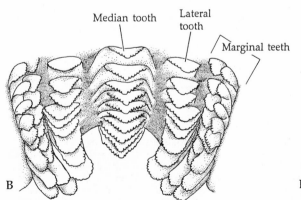

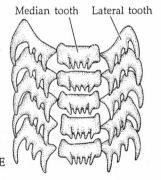

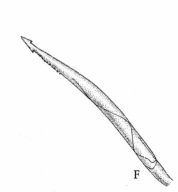

scabra) actually serve as adhesive traps for the microalgae that are their primary food resource.

The radulae of mesogastropods are **taenioglossate**, that is, the number of marginal teeth is reduced (Figures 25B–D). In conjunction with the elaboration of muscular jaws, taenioglossate radulae are capable of powerful rasping action, and snails such as some littorines feed by directly scraping off the surface cell layers of algae.

The most advanced prosobranch gastropods (Neogastropoda) usually have **rachiglossate** radulae, which lack marginal teeth altogether (Figures 25E and 26C and D). They use the remaining (medial) teeth for rasping, tearing, or pulling. These snails are usually carnivores or carrion feeders. Neogastropods of the families Muricidae and Naticidae feed upon other molluscs by boring through the prey's calcareous shell to obtain the underlying flesh. The boring is mainly mechanical; the predator bores with its radula while holding the prey with the foot. The mechanical activity may be complemented by chemical action whereby the predator secretes an acidic chemical from a **boring gland** (also called the "accessory boring organ") and periodically applies the chemical to the drill hole to weaken the calcareous matrix (Figure 24B). Muricids such as the American drill (*Urosalpinx*)

and the Japanese drill (*Rapana*) cause a loss of millions of dollars annually for oyster farms.

Some carnivorous gastropods (e.g., *Janthina*) do not gnaw or rasp their prey, but swallow it whole. In these gastropods a **ptenoglossate** radula forms a covering of strongly curved spines over the buccal mass. The prey is seized by the quickly extruded buccal mass and simply pulled whole into the gut. Pyramidellids have lost the radula altogether and feed by sucking blood or other fluids from their prey by use of a hypodermic stylet on the tip of an elongate proboscis. One of the most unusual feeding strategies occurs in the neogastropod *Cancellaria cooperi*, which attaches to Pacific electric rays and makes small cuts in the skin through which the proboscis is inserted to feed on the ray's blood and cellular fluids.

In terms of feeding, the most specialized gastropods may be the cone snails (*Conus*), which have vastly modified radulae that comprise only a few isolated poison-injecting teeth (**toxoglossate radulae**). The harpoon-like teeth (Figure 25F) are discharged from the end of a long proboscis that can be thrown out rapidly to capture prey, usually a fish, a worm, or another gastropod, which is then pulled into the gut (Figure 27). The venom is rapidly injected through the hollow, curved radular teeth by contrac-

Figure 26

Gastropod radulae. A, A closeup view of the rhipidoglossate radula of the abalone, *Haliotis rufescens* (order Archaeogastropoda). Note the many hooklike marginal teeth. B, The docoglossate radula of a limpet. C, The serrated central teeth of a rachiglossate radula from *Nucella emarginata*, a prosobranch gastropod that preys on small mussels and barnacles. D, The worn radular teeth of *Nucella*. E, The radula of the opisthobranch *Triopha*, seen here in dorsal view as it rests in the animal. (Photos courtesy of C. DiGiorgio.)

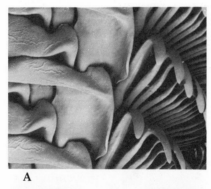

A

B

C

D

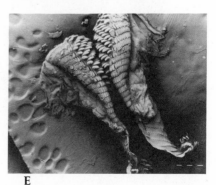

E

Figure 27

Sequence of photographs of the eastern Pacific cone *Conus purpurescens* capturing and swallowing a small fish. The proboscis is extended and swept back and forth above the substratum in search of prey; when a fish is encountered, it is quickly paralyzed and ingested. (Photographs by A. Kerstitsch.)

tion of the venom gland. A few Indo-West Pacific cones produce a potent neuromuscular poison that has caused human deaths.

Certain opisthobranchs and pulmonates also show various radular modifications. Groups that feed on cnidarians, ectoprocts, and sponges and those that scrape algae (e.g., aplyssiids) usually have typical rasping radulae. In sacoglossans, however, the radula is modified as a single row of lancelike teeth that can pierce the cellulose wall of filamentous algae, allowing the gastropod to suck out the cell contents. In most pulmonates and many limpets, the radula is no longer a thin ribbon, but a broad band that has many similar teeth per row and functions much like sandpaper. But the pulmonate slug *Testacella* uses its toothed radula to prey on earthworms.

Aeolid nudibranchs (Figure 7G) have a well deserved reputation for their particular mode of feeding, in which portions of their cnidarian prey are held by the muscular jaws while the radula rasps off pieces for ingestion. Some of the prey's nematocysts are ingested unfired, passed through the nudibranch's gut, and eventually transported to the processes of the digestive gland in the dorsal **cerata** (singular, **ceras**) (Figure 32D and E). How the nematocysts withstand this transport without firing is still a mystery. Popular hypotheses are that mucous secretions by the nudibranch limit the discharge, or that a form of acclimation occurs (like that suspected to occur between anemone fishes and their host anemones), or perhaps that only immature nematocysts survive, to later undergo maturation in the dorsal cerata. It may also be that, once the cnidocytes are digested, the nematocyst's firing threshold is raised, thereby preventing discharge. Once in the cerata, the nematocysts are stored in structures called **cnidosacs**, and they may help the nudibranch to fend off attackers, who depart with a mouthful of discharged nematocysts. Discharge might even be under control of the host nudibranch, perhaps by means of pressure exerted by circular muscle fibers around each cnidosac. A somewhat similar relationship exists between dorid nudibranchs and their prey. Many dorids secrete complex toxic compounds that are incorporated into mucus released from the mantle surface. These noxious chemicals act to deter potential predators. The chemicals may be manufactured by the dorids themselves, but in most cases it appears that they are obtained from the sponges or ectoprocts on which they feed.

In chitons (class Polyplacophora), the radular teeth are also in numerous transverse rows, generally of 17 teeth each (a central tooth flanked by eight on each side). Most chitons are strictly herbivorous grazers. Notable exceptions are certain members of the order Ischnochitonida (family Mopaliidae: e.g., *Mopalia*, *Placiphorella*), which are known to feed both on algae and small invertebrates. *Mopalia* consumes sessile invertebrates, such as barnacles, ectoprocts, and hydroids. *Placiphorella* captures live microinvertebrates (particularly crustaceans) by trapping them beneath its head-flap, a large anterior extension of the girdle (Figure 24C).

In monoplacophorans the radula consists of a ribbon-like membrane bearing a succession of transverse rows of 11 teeth each (a slender median tooth flanked on each side by five broader lateral teeth). Monoplacophorans are probably generalized grazers that feed on minute organisms coating the substratum on which they live.

Cephalopods are predatory carnivores. Squids are some of the most voracious creatures in the sea, successfully competing with fishes for their meals. Octopuses are active generalized carnivores but prey primarily on crabs and clams. Some species of *Octopus* have the radula modified as a drill to bore through the shells of their molluscan prey in a fashion similar to that of gastropod drills. Using their locomotor skills, most cephalopods hunt and catch active prey. Some octopuses, however, hunt "blindly," by "tasting" beneath stones with their chemosensitive suckers. In any event, once a victim is captured and held by the arms, the cephalopod bites it with its horny beak and injects a neurotoxin from modified salivary glands. The ability to quickly immobilize prey helps the soft-bodied cephalopod avoid a potentially dangerous struggle. At least one species of octopus, *Haplochlaena maculosa* (the blue-ringed octopus), injects the potent neurotoxin tetrodotoxin with its bite. The toxin is produced by symbiotic bacteria (*Vibrio*) that inhabit the salivary glands.

Microphagy and suspension feeding. Suspension feeding evolved numerous times in molluscs, but in most cases it involves modifications of the ctenidia that enable the animal to trap particulate matter carried in the mantle cavity current. Many molluscs generate a single current for both gas exchange and feeding. The lamellar nature of molluscan gills preadapted them for extracting suspended food. Increasing the size of the gills and the degree of folding also increases the surface area available for trapping particulates. In suspension feeders, at least some of the gill cilia, which otherwise serve to remove sediment, function to transport particulate matter from the gills to the mouth region. Suspension feeding occurs in some gastropods and most bivalves.

There are three principal groups of suspension-feeding gastropods: pteropods, certain errant prosobranch snails, and vermetids. In the planktonic sea butterflies (pteropods), expanded, ciliated outgrowths ("wings" or "parapodia") of the swimming foot function as food-collecting surfaces or may cooperate with the mantle to produce large mucous sheets that capture small zooplankton (Figures 7D and E). From the foot, ciliary currents carry mucus and food to the mouth. In some, the mucous web may be as much as 2 m across. A different approach is taken by suspension-feeding prosobranch snails. Gas exchange currents carry particulate matter into the mantle cavity. Normally these particulates are then wrapped in mucus and ejected from the mantle cavity as pellets called **pseudofeces**, but several groups have mechanisms to retain the smaller organic particles and carry them to the mouth by way of ciliary-mucus transport. Various versions of this basic plan occur throughout the prosobranchs. For example, in calyptraeid limpets (e.g., *Calyptraea*, *Crepidula*), the ctenidium is greatly enlarged and coated with a thick layer of mucus produced by a special gland called the **endostyle**. Particles are trapped in the ctenidial mucus and carried to the mouth. The radula in all suspension-feeding gastropods is reduced, serving only to pull mucus and food into the mouth. Suspension-feeding gastropods are usually rather sedentary animals and rely on their own cilia-generated water currents to bring them food.

Perhaps the zenith of adaptation to a suspension-feeding lifestyle among gastropods has been achieved by the vermetids, or "worm gastropods." The vermetid shell, coiled in youth, becomes partly or wholly noncoiling in adults and permanently affixed to the substratum (Figure 19E). A special pedal gland produces copious amounts of mucus that is spread outside the shell aperture as a sticky web or net to trap plankton. Periodically the net is hauled in by the foot and pedal tentacles, and a new net is quickly secreted. *Serpulorbis gigas*, a large Mediterranean species, casts out individual threads up to 30 cm long, whereas the gregarious California species *S. squamigerus* forms a communal net shared by many individuals.

Figure 28

Arrangement of ctenidia in some bivalves (transverse sections). A, Protobranch. B, Lamellibranch. C, Septibranch anomalodesmatan. (After Barnes 1980.)

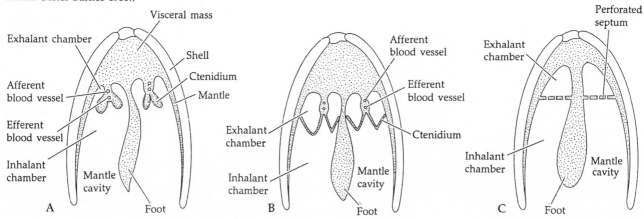

The radula apparently disappeared early in the course of bivalve evolution, and most modern species use their large ctenidia for suspension feeding. However, some primitive species in the subclass Protobranchia are not suspension feeders but engage in a type of deposit-feeding microphagy. Protobranchs live in soft marine sediments and maintain contact with the overlying water either directly (e.g., *Nucula*) or by means of siphons (e.g., *Nuculana, Yoldia*). The two ctenidia are small, conforming to the primitive molluscan **bipectinate** plan of an elongated axis carrying a double row of lamellae (Figure 28A). Protobranchs feed by means of two pairs of palp-like structures flanking the mouth. The two innermost palps are the short **labial palps**, and the two outermost palps are formed into tentacular processes called **proboscides** (each called a **palp proboscis**), which can be extended beyond the shell (Figure 29). During feeding, the proboscides are extended slightly into the bottom sediments. Detrital material adheres to the mucus-covered surface of the proboscides and is then transported by cilia to the labial palps, which function as sorting devices. Low-density particles are carried to the mouth; heavy particles are carried to the palp margins where they are ejected into the mantle cavity.

Two basic kinds of suspension feeding occur in nonprotobranch bivalves. One type occurs in a small, unusual group of the subclass Anomalodesmata known as the septibranchs, which trap small animals. Septibranchs are sessile predators. Their ctenidia are modified as a perforated, but muscular **septum** that divides the mantle cavity into dorsal and ventral

chambers (Figures 28C and 30). The muscles are attached to the shell such that the septum can be raised or lowered within the mantle cavity. Raising the septum causes water to be sucked into the mantle cavity by way of the inhalant siphon; lowering the septum causes water to pass dorsally through the pores into

Figure 29

Feeding in the primitive bivalve *Nucula* (subclass Protobranchia). The clam is seen from the right side, in its natural position in the substratum (right valve and right mantle skirt removed). Arrows show direction of water currents in the mantle cavity. I, inhalant region; E, exhalant region. (After Fretter and Graham 1962.)

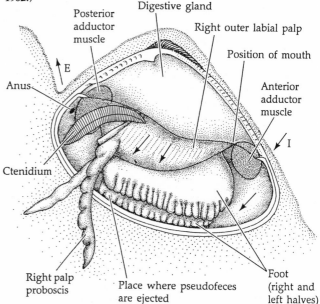

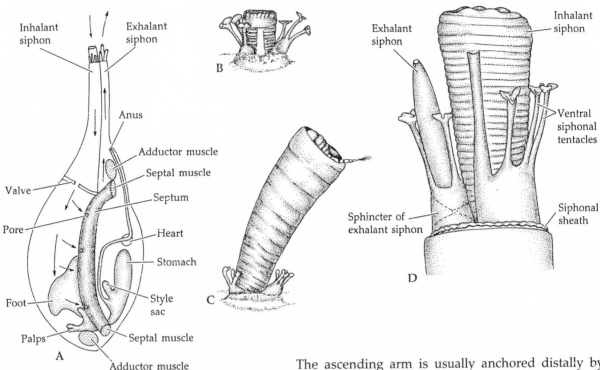

Figure 30

Feeding in the septibranch bivalve *Cuspidaria* **(subclass Anomalodesmata). A, General anatomy of** *Cuspidaria rostrata.* **Arrows indicate water flow. B, Siphonal apparatus protruding from the substratum, but largely contracted. C, Siphonal apparatus extended, capturing a microcrustacean. D, Anatomy of the siphonal apparatus. (A after Yonge and Thompson 1976; B–D after Reid and Reid 1974.)**

the exhalant chamber. These movements also force hemolymph from mantle sinuses into the siphonal sinuses, thereby causing a rapid protrusion of the inhalant siphon, which can be directed toward potential prey. In this fashion, small animals such as microcrustaceans are sucked into the mantle cavity, where they are grasped by muscular labial palps and thrust into the mouth; at the same time, the mantle tissue is serving as the gas exchange surface.

In the second type of suspension feeding, which occurs in members of the large groups Filibranchia and Eulamellibranchia, cilia on the ctenidia generate a water current from which suspended particles are gleaned. Increased efficiency is achieved by various ctenidial modifications. The primary modification has been the conversion of the original, small, triangular plates into V-shaped filaments with extensions on either side (Figures 28B and 31B). The arm of this V-shaped filament that is attached to the central axis of the ctenidium is called the **descending arm**; the arm forming the other half of the V is the **ascending arm**.

The ascending arm is usually anchored distally by ciliary contacts or tissue junctions to the roof of the mantle, or to the visceral mass. Taken together, the two V-shaped filaments, with their double row of leaflets, form a W-shaped structure when seen in cross section. Most filibranchs (e.g., mussels and oysters) have ctenidia wherein adjacent filaments are interlocked to one another by periodic clumps of special cilia, leaving long narrow slits in between (**interfilament spaces**) (Figures 31C and D). The spaces between the arms of the W's are exhalant **suprabranchial chambers**, which communicate with the exhalant area of the mantle edge; the spaces ventral to the W's are inhalant and communicate with the inhalant area of the mantle edge. Most filibranchs are restricted to epibenthic life. Their mantles are not formed into elongate siphons, thus they cannot burrow deeply (see Figure 21A).

Eulamellibranch bivalves have a similar ctenidial design, but neighboring filaments are actually fused to one another by tissue junctions at numerous points along their length, an arrangement resulting in interfilament spaces that are rows of **ostia** rather than the long narrow slits of filibranchs (Figures 31B, E, and F). In addition, the ascending and descending halves of some filaments may be joined by tissue bridges that provide firmness and strength to the gill. Many eulamellibranch bivalves live buried in soft sediments, where long siphons are utilized to maintain contact with the overlying water (Figures 8A and 20).

Figure 31

Ctenidial structure in bivalve molluscs. In all panels, solid arrows indicate the direction of water flow (from inhalant space, between ctenidial filaments, to exhalant space). A, Section through the gill axis in a protobranch, with four alternating filaments (leaflets) on each side. Dashed arrows indicate direction of hemolymph flow in the filament. B, Schematic cutaway view showing five ctenidial filaments on one side of the body of a eulamellibranch. C, Four ctenidial filaments of a filibranch (surface side view). D, Cross section through ascending and descending arms of four filibranch ctenidial filaments. E, Four filaments of a eulamellibranch (surface side view). F, Cross section through ascending and descending arms of four eulamellibranch ctenidial filaments. (Modified after various sources.)

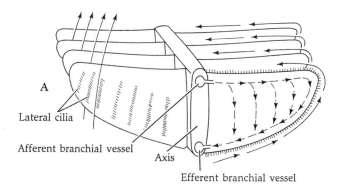

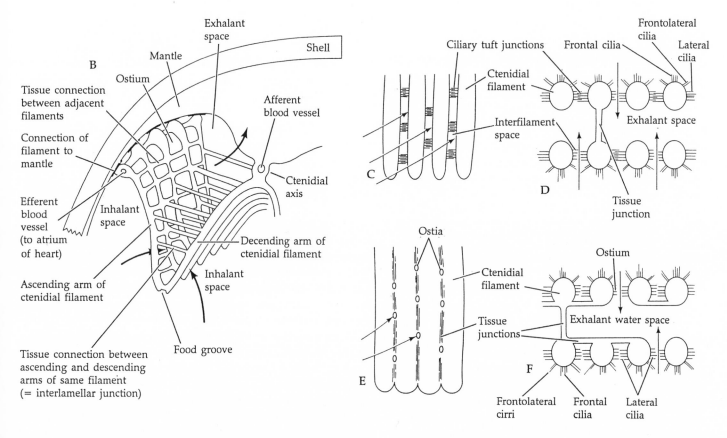

Both filibranch and eulamellibranch bivalves use their ctenidia to capture food. Water is driven from the inhalant to the exhalant parts of the mantle cavity by lateral cilia all along the sides of filaments in filibranchs, or by special lateral ostial cilia in eulamellibranchs (Figures 31C–F). As the water passes through the interfilament spaces it flows through rows of frontolateral cilia, which flick particles from the water onto the surface of the filament facing into the current. These feeding cilia are called **compound cirri**; they have a pinnate structure that probably increases their catching power. Mucus presumably plays some part in trapping the particles and keeping them close to the gill surface, although its precise

role is uncertain. Bivalve ctenidia are not covered with a continuous sheet of mucus, as occurs in many other suspension-feeding invertebrates (e.g., gastropods, tunicates, amphioxus). Once on the filament surface, particles are moved by frontal cilia toward a food groove on the free edges of the ctenidium, and then anteriorly to the labial palps. The palps sort the material by size and perhaps also by quality before passing the food to the mouth. Rejected particles fall off the gill edges into the mantle cavity as pseudofeces. This "filtration" of water by bivalves is quite efficient. The American oyster (*Crassostrea virginica*), for example, can process up to 37 liters of water per hour (at 24°C), and can capture particles as small as 1 μm in size. Studies on the common mussels *Mytilus edulis* and *M. californianus* suggest that these bivalves maintain pumping rates of about 1 liter per hour per gram of (wet) body weight.

Scaphopods are selective deposit feeders that consume minute particulate matter in the surrounding sediment, or occassionally ingest the sediment itself. Two lobes flank the head, each bearing numerous (up to several hundred) long tentacles called **captacula** (Figures 9 and 13F). The captacula are extended out of the mantle cavity and into the substratum by metachronal beating of cilia on the **terminal bulb**. Within the substratum, organic particles and microorganisms (particularly diatoms and forams) adhere to the sticky terminal bulb; small food particles are transported to the mouth by way of ciliary tracts along the tentacles, while larger particles are transported directly to the mouth by muscular contraction of the captacula. The captacula manipulate food inside the scaphopod's shell and move freely through all parts of the mantle cavity. A well developed radula pulls the food into the mouth, perhaps partially macerating it in the process.

Several forms of symbiotic relationships have evolved within the molluscs and are intimately tied to the host's nutritional biology. One of the most interesting of these relationships exists between molluscs and sulfur bacteria. Although not yet fully understood, many molluscs (and some other invertebrates) appear to derive a portion of their nutritional needs from symbiotic, CO_2-fixing sulfur bacteria. The bacteria usually reside on the host mollusc's gill lamellae. This mollusc–bacteria symbiosis has been recently documented from a variety of sulfide-rich anoxic habitats, including deep-sea thermal vents where geothermally produced sulfide is present, and reduced sediments where microbial degradation of organic matter leads to the reduction of sulfate to sulfide (e.g., anoxic marine basins, seagrass bed and mangrove swamp sediments, pulp mill effluent sites, and sewage outfall areas).

One recent study discovered that the gutless clam *Solemya reidi*, which harbors sulfur bacteria on its gills, has the ability to directly oxidize sulfide (Powell and Somero 1986). It does this by means of a special sulfide oxidase enzyme in the mitochondria. This clam inhabits reduced sediments near sewage outfalls and pulp mill effluents, where free sulfides are abundant. The ability to oxidize sulfide not only provides *S. reidi* with a source of energy to drive ATP synthesis but also enables the clam to rid its body of the toxic sulfide molecules. So far, this is the only known metazoan example of direct exploitation of the energy in sulfide for ATP synthesis, but more examples may eventually be found in other invertebrates inhabiting these unusual areas.

The relationship between certain molluscs (and other invertebrates) and dinoflagellates was described in Chapter 3. One of the most notable of these partnerships exists between giant clams (family Tridacnidae) and their symbiotic zooxanthellae (the dinoflagellate *Symbiodinium*). These clams live with their dorsal hinge area affixed to the substratum. They expose their fleshy mantle to sunlight through the large shell gape. The mantle tissues harbor the zooxanthellae. Many species have special lenslike structures that focus light on zooxanthellae living in the deeper tissues. Certain opisthobranchs also maintain a symbiotic relationship with *Symbiodinium*. Several species of *Melibe*, *Pteraeolidia*, and *Berghia* are known to harbor colonies of the dinoflagellate in "carrier" cells associated with their digestive glands. Experiments indicate that when sufficient light is available, host nudibranchs utilize photosynthetically fixed organic molecules produced by the alga to supplement their usual diet of prey. The dinoflagellates are probably not transmitted with the zygotes of the nudibranchs, each new generation thus requiring reinfection from the environment. An even stranger phenomenon occurs in several sacoglossan opisthobranchs (e.g., *Placobranchus*). These sea slugs obtain functional chloroplasts from the algae upon which they feed and incorporate them into their tissues, where they remain active for a period of time and produce photosynthetically fixed carbon molecules utilized by the hosts.

Still another unusual symbiosis was recently discovered between a new form of bacteria and marine shipworms (bivalves of the family Teredinidae) (Figure 21D). Shipworms bore into wood structures and are capable of living on a diet of wood alone. Recently, a cellulose-decomposing, nitrogen-fixing bacterium was discovered in these animals. The bivalve "cultures" the bacterium in pure form in a special organ that is associated with ctenidial blood vessels and is called the **gland of Deshayes**. The bacterium is an aerobic chemoheterotroph that breaks down cellulose and makes its products available to its host. Nitrogen-fixing bacteria occur as part of the gut flora in many animals whose diet is rich in carbon but deficient in nitrogen (e.g., termites). However, shipworms are the only animals known to harbor a nitrogen fixer as a pure culture in a specialized organ (as in the host nodule–*Rhizobium* symbiosis of leguminous plants).

In addition to the above feeding strategies, certain molluscs (notably some bivalves) probably obtain a significant portion of their nutritional needs by direct uptake of dissolved organic material (DOM), such as amino acids. A few bivalves actually lack digestive tracts altogether (e.g., *Solemya*), and their nutritional requirements are probably met to some extent by active absorption of DOM across their ctenidia.

Digestion

Molluscs possess complete guts, several of which are illustrated in Figure 32. The mouth leads inward to a buccal cavity, within which the radula apparatus is located, and sometimes to a muscular pharynx. The esophagus is generally a straight tube connecting the foregut to the stomach. Various glands are often associated with this anterior gut region, including some that produce enzymes and others that secrete

Figure 32

Molluscan guts. A, The digestive system of the prosobranch gastropod *Murex*. B, The digestive system of the land snail *Helix*. C, The intestnal wall of a gastropod (section). D, A nudibranch (*Embletonia*) in which large digestive ceca fill the dorsal cerata. E, A ceras of the nudibranch *Trinchesia* (longitudinal section). The nematocysts (not shown) from this animal's cnidarian prey are stored in the terminal cnidosac. F, The digestive tract and nearby organs of the clam *Anodonta* (longitudinal section). G, The digestive system of the cuttlefish *Eledone*. H, The digestive system of the squid *Loligo*. (A–E after Hyman 1967; F after Bullough 1958; G,H after various sources.)

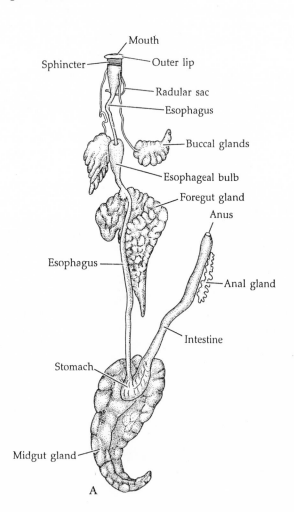

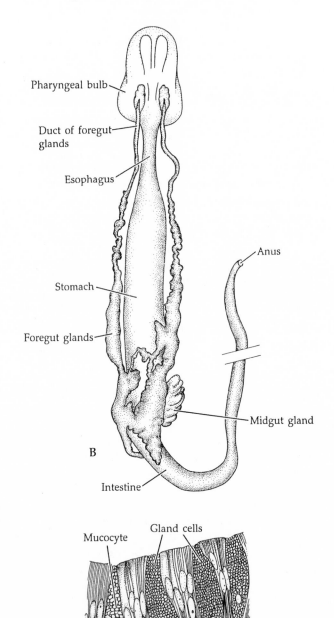

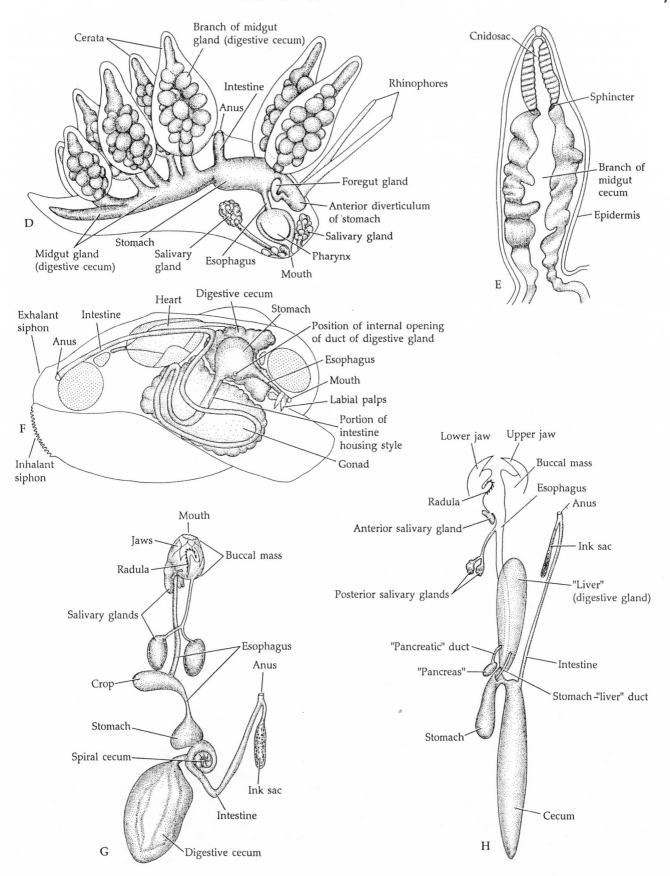

D

Cerata

Branch of midgut gland (digestive cecum)

Intestine

Anus

Rhinophores

Foregut gland

Anterior diverticulum of stomach

Salivary gland

Pharynx

Mouth

Esophagus

Salivary gland

Stomach

Midgut gland (digestive cecum)

E

Cnidosac

Sphincter

Branch of midgut cecum

Epidermis

F

Exhalant siphon

Anus

Intestine

Heart

Digestive cecum

Stomach

Position of internal opening of duct of digestive gland

Esophagus

Mouth

Labial palps

Portion of intestine housing style

Gonad

Inhalant siphon

G

Mouth

Jaws

Radula

Buccal mass

Salivary glands

Esophagus

Anus

Crop

Stomach

Spiral cecum

Intestine

Ink sac

Digestive cecum

H

Lower jaw

Upper jaw

Buccal mass

Esophagus

Anus

Radula

Ink sac

Anterior salivary gland

"Liver" (digestive gland)

Posterior salivary glands

"Pancreatic" duct

Intestine

"Pancreas"

Stomach–"liver" duct

Stomach

Cecum

a lubricant over the radula—often called salivary glands. In many herbivorous species (e.g., certain pulmonates and opisthobranchs), a muscular **gizzard** may be present for grinding up tough vegetable matter. The stomach usually bears one or more ducts that lead to large glandular **digestive ceca**. Several sets of digestive ceca may be present (variously called the digestive diverticula, digestive glands, foregut glands, midgut glands, liver, or other similar terms). The intestine leaves the stomach and terminates as the anus, which is typically located in the mantle cavity near the exhalant water flow.

Once food has entered the buccal cavity of most molluscs, it is carried in mucous strings into the esophagus and then to the stomach. In cephalopods, chunks of food are swallowed by muscular action of the esophagus. The food is stored in the stomach or, in octopuses and *Nautilus*, in an expanded region of the esophagus called the **crop**. In many molluscs the stomach wall bears a chitinous **gastric shield** and a ciliated, ridged sorting area (Figure 33). The posterior stomach region houses a **style sac**, which is lined with cilia and contains the **crystalline style**. This structure is a rodlike matrix of proteins and enzymes (especially amylase), which are slowly released to aid in digestion as the projecting end of the style rotates and grinds against the abrasive gastric shield. The gastric cilia and rotating style wind up the mucus and food into a string called the **protostyle** and draw it along the esophagus to the stomach. The style is produced by special cells of the style sac. In some intertidal bivalves, the style may dissolve completely at low tide when the animal is not feeding, to be reformed when the tide rises. The style of some bivalves is enormous, one-third to one-half the length of the clam itself. Particulate matter is swept against the stomach's anterior sorting region, which sorts mainly by size. Small particles are carried into the digestive ceca, which arise from the stomach wall. Larger particles are passed along ciliated grooves of the stomach to the intestine.

Extracellular digestion takes place in the stomach and digestive ceca, while absorption and intracellular digestion are confined to the cecal and intestinal walls. Extracellular digestion is accomplished by enzymes produced in foregut and stomach glands (e.g., salivary glands, esophageal pouches, pharyngeal glands—often called "sugar glands" because they produce amylase—the stomach, and digestive ceca). In primitive groups, intracellular digestion tends to predominate. In most molluscs, ciliated tracts line the digestive ceca and carry food particles to minute ducts, where they are engulfed by phagocytic cells

of the duct wall. The same cells dump digestive wastes back into the ducts, to be carried by other ciliary tracts back to the stomach, from there to be passed out of the gut via the intestine and anus. In most advanced groups (e.g., cephalopods), extracellular digestion predominates. Enzymes secreted primarily by the ceca and stomach digest the food, and absorption occurs in the stomach, ceca, and intestine. In some cases the stomach has lost some of its primitive features, such as the gastric shield, the sorting area, and the style sac.

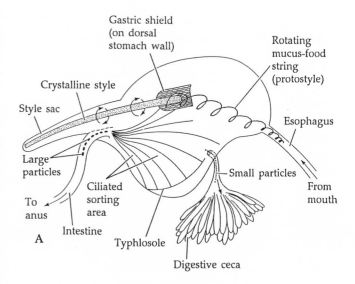

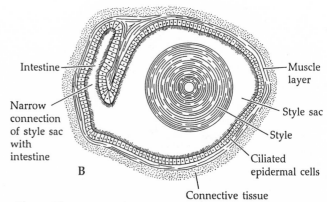

Figure 33

The molluscan stomach and style sac. A, The stomach and style apparatus of a bivalve. The crystalline style rotates to grind against the gastric shield, releasing digestive enzymes and winding up the mucus–food string to assist in pulling it from the esophagus. Food particles are sorted in the ciliated, grooved sorting area: small particles are carried (in part on a ciliated ridge called the typhlosole) to the digestive ceca for digestion; large particles are carried to the intestine for eventual elimination. B, The style sac of a typical prosobranch gastropod (cross section). (A after various sources; B after Hyman 1967.)

Circulation and gas exchange

Although molluscs are coelomate protostomes, the coelom is greatly reduced. The main body cavity is an open circulatory space or hemocoel, which comprises several separate sinuses, and a network of vessels in the gills, where gas exchange takes place. The blood of molluscs contains various cells, including amebocytes, and is referred to as hemolymph. It is responsible for picking up the products of digestion from the sites of absorption and for delivering these nutrients throughout the body. It usually carries in solution the copper-containing respiratory pigment hemocyanin. Many molluscs also use hemoglobin and/or myoglobin to bind oxygen.

The heart lies dorsally, within the **pericardial chamber**, and comprises a pair of **atria** (sometimes called auricles) and a single **ventricle**. In monoplacophorans and in *Nautilus* there are two pairs of atria. The atria receive the **efferent ctenidial** (= branchial) **vessels**, drawing oxygenated hemolymph from each ctenidium and passing it into the muscular ventricle, which pumps it anteriorly through a large **anterior artery** (also called the anterior or cephalic aorta). The anterior artery branches and eventually opens into various sinuses within which the tissues are bathed in oxygenated hemolymph. Return drainage through the sinuses eventually funnels the hemolymph back

into the **afferent ctenidial vessels**. This basic pattern of molluscan circulation is shown diagrammatically in Figure 34; it is modified to various degrees in different classes (Figure 35). In cephalopods, the circulatory system is secondarily closed (Figure 35C).

Most molluscs have true gills, or ctenidia. However, many have lost the ctenidia and rely either on secondarily derived "gills" or on gas exchange across the mantle or general body surface. The presumed primitive gill condition is expressed in several living groups, for example, many of the primitive gastropods (archaeogastropods, such as *Pleurotomaria*) and primitive bivalves (such as protobranchs), and can serve to explain how molluscan gills work. In these cases the gill, or ctenidium, is built around a long, flattened axis projecting from the wall of the mantle cavity (Figure 31A). To each side of the axis are attached triangular or wedge-shaped filaments that alternate in position with filaments on the opposite side of the axis. This arrangement, in which filaments project on both sides of the central axis, is called the **bipectinate**, or **aspidobranch**, **condition**. There is one gill on each side of the mantle cavity, held in position by membranes that divide the mantle cavity into upper and lower chambers (Figures 28A and B). Cilia covering the gill surface draw water into the **inhalant (ventral) chamber**, from which it passes upward be-

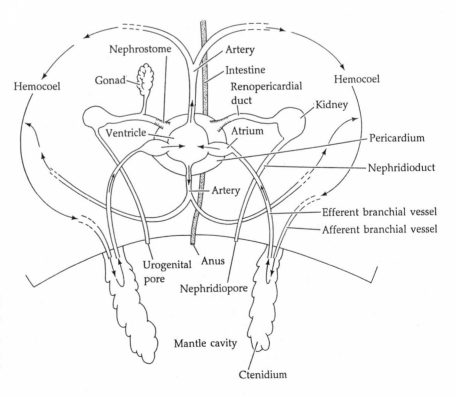

Figure 34

Hemolymph flow in a generalized mollusc. Oxygenated hemolymph is pumped from the ventricle to the hemocoel, where it bathes the organs; eventually it drains into various channels and chambers and then into the afferent branchial vessels, which enter the ctenidia. Oxygen is picked up in the ctenidia and is transported by way of the efferent branchial vessels to the left and right atria, through the ventricle, and then back to the hemocoel. Auxiliary pumping vessels occur in several taxa, particularly in active groups such as cephalopods.

Figure 35

The circulatory systems of three molluscs. A, A typical neogastropod (hemolymph sinuses not shown). B, A typical eulamellibranch bivalve. C, A squid, *Loligo*. (A after Cox, in Moore (ed.) 1960; B after Pearse et al. 1987; C after various sources.)

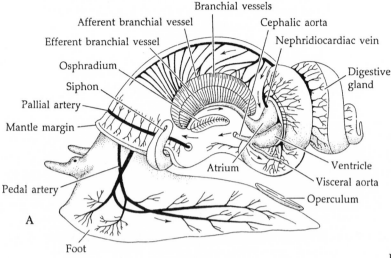

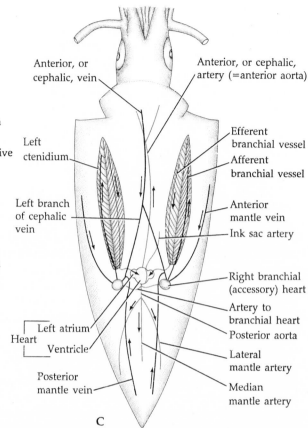

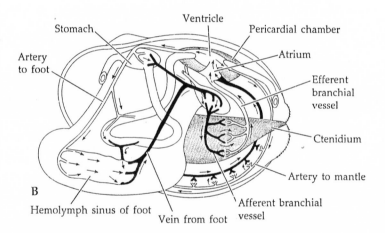

tween the gill filaments to the **exhalant** (dorsal) chamber and then out of the mantle cavity (Figure 31A).

Two vessels run through the gill axis. The **afferent vessel** carries oxygen-depleted hemolymph into the gill, and the **efferent vessel** drains freshly oxygenated hemolymph from the gill to the atria of the heart. Hemolymph flows through the filaments from the afferent to the efferent vessel. Ctenidial cilia carry water over the gill filaments in a direction opposite to that of the flow of the underlying hemolymph in the ctenidial vessels. This countercurrent phenomenon maximizes the diffusion gradients of O_2 and CO_2 between the hemolymph and water, and thus ensures efficient gas exchange (Figure 31A).

Recall from our discussion on torsion that gastropods have evolved novel ways to circulate water over the gills and still avoid fouling from gut or nephridial discharges (see Figure 15). Primitive archaeogastropods with two bipectinate ctenidia accomplish this by circulating water in across the gills, then past the anus and nephridiopore, and away from the body via slits or holes in the shell (Figures 1C, 1D, 6C, and 36). This circulation pattern is used by the slit shells, abalones, and volcano limpets. Many specialists regard the slit shells (family Pleurotomariacea) as "living fossils," animals that reflect an archetypical gastropod condition. Most other gastropods have lost the right ctenidium and with it the right atrium; they circulate water in from the left side of the head and then straight out the right side, where the anus and nephridiopore open (Figure 37A). Others gastropods have lost both ctenidia and utilize secondary respiratory regions, either the mantle surface itself or secondarily derived gills of one kind or another. Limpets of the genus *Patella* have rows of secondary gills in the pallial groove along each side of the body, similar to the condition seen in chitons and monoplacophorans.

Figure 36

The primitive archaeogastropod **Scissurella, one of the slit shells (Pleurotomariacea). (After Fretter and Graham 1962.)**

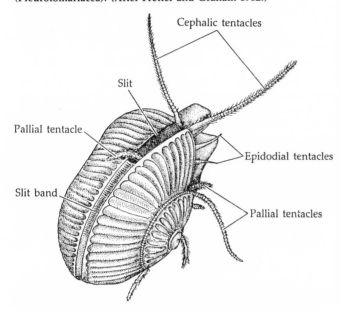

In more advanced gastropods, such as the mesogastropods and neogastropods, one ctenidium is almost always missing, as are the dorsal and ventral suspensory membranes of the remaining gill, which attaches directly to the mantle wall by the gill axis. The gill filaments on the attached side have been lost, while those of the opposite side project freely into the mantle cavity. This advanced arrangement of filaments on only one side of the central axis is referred to as the **monopectinate**, or **pectinobranch**, **condition** (Figure 15D). The dorsal attachment of the monopectinate ctenidium in some species helps prevent fouling in soft sediments. Some advanced mesogastropods and neogastropods have also evolved **inhalant siphons** by extension and rolling of the mantle margin (Figures 1F and 37A). In these cases the margin of the shell may be notched, or drawn out as a canal to house the siphon. The siphon provides access to surface water in burrowing species, and may also function as a mobile, directional sense organ.

Opisthobranch gastropods are largely detorted. In most, the one remaining gill is **plicate** (or folded) rather than filamentous, and in fact may not be homologous with the prosobranch ctenidium. Trends toward detorsion, loss of the shell and ctenidia, and reduction of the mantle cavity occur in many opisthobranchs, and the process has apparently occurred several times within this group. Some nudibranchs have evolved secondary dorsal gas exchange struc-

tures called **cerata**. Many others have a circlet of dorsal gills around the anus that may be homologous to the true ctenidia (Figures 1G and 7F–G).

Wholly terrestrial gastropods generally lack gills, and exchange gases directly across a vascularized region of the mantle, usually within the mantle cavity. The whole arrangement is often referred to as a **lung**. In terrestrial pulmonates, the edges of the mantle cavity have become sealed to the back of the animal except for a small opening on the right side called a **pneumostome** (Figure 37B). There are no

Figure 37

A, A neogastropod with a fleshy siphon, depicted with the shell removed. The animal has lost the posttorsional right ctenidium (and atrium and nephridium). Water flows into the mantle cavity through the siphon from the left, passes over the osphradium and ctenidium and then over the excretory pore and anus before leaving the mantle cavity to the right. B, The terrestrial pulmonate, *Helix*, showing the pneumostome that leads to the "lung." (Photograph by the authors.)

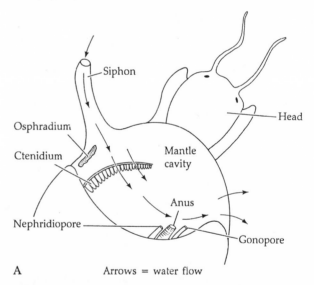

A   Arrows = water flow

B

gills; instead, the roof of the mantle cavity is highly vascularized. By arching and flattening the mantle cavity floor, air is moved in and out of the lung via the pneumostome.

In the polyplacophorans (chitons), the mantle cavity forms a groove (called the **pallial groove**) extending along the body margin encircling the foot (Figure 4). A large number of simple gills lie laterally in the pallial groove. The mantle is held tight against the substratum, largely enclosing this pallial chamber. However, the mantle is raised on either side at the anterior end to form incurrent channels, and is raised in one or two places at the posterior end to form excurrent areas. Water enters and flows along the inhalant region of the pallial chamber lateral to the gills, then passes medially between the gills into the exhalant region along the sides of the foot. Moving posteriorly, the current passes over the gonopores, nephridiopores, and anus before exiting (Figure 4B).

In bivalves the capacious mantle cavity allows the ctenidia to develop a greatly enlarged surface area, serving in most species for both gas exchange and feeding. We discussed earlier many of the morphological modifications of bivalve gills in our coverage of suspension feeding. In addition to the folded, W-shaped ctenidial filaments seen in many bivalves (Figure 28B), some forms (e.g., oysters) bear **plicate ctenidia**. A plicate ctenidium is thrown into vertical ridges or folds, each ridge consisting of several ctenidial filaments. The grooves between these ridges of ordinary filaments bear so-called **principal filaments** whose cilia are important in sorting sediments from the ventilation and feeding currents. All-in-all, the plicate condition gives the ctenidium a corrugated appearance and further increases the surface area for gas exchange.

In spite of these modifications, the basic system of circulation and gas exchange in bivalves is similar to that seen in gastropods (Figure 8 and 35B). In most bivalves, the heart ventricle folds around the gut, so the pericardial cavity encloses not only the heart but also a short section of the digestive tract. The large mantle lines the valves and provides an additional surface area for gas exchange, which in some groups may be as important as the gills in this regard. In septibranchs, which have reduced gills, the mantle surface is the principal area of gas exchange. Most bivalves appear to lack respiratory pigments in the hemolymph, although globins have been found in a few species and recently hemocyanin was identified in protobranchs.

Scaphopods have lost the ctenidia, heart and virtually all vessels. The circulatory system is reduced to a simple system of hemolymph sinuses, and gas exchange takes place mainly across the mantle and body surface (Figure 9).

No doubt associated with their large size and active life style, cephalopods have a circulatory system that is effectively closed, with many discrete vessels, secondary pumping structures, and even capillaries (Figures 10, 11, 12B, and 35C). The result is increased pressure and efficiency of hemolymph flow. In most cephalopods, the vessels leading into the ctenidia are enlarged into powerful accessory **branchial hearts**, which boost the low venous pressure as the hemolymph enters the gills. The gills are folded, increasing their surface area for greater gas exchange associated with a high metabolic rate.

In the enigmatic aplacophorans, gills are usually absent or, if present, form a ciliated, lamellar pouch arising directly off the posterior region of the pericardial chamber. Caudofoveatans have a similar posterior gill. Whether or not these gills are homologous to, or early forerunners of, the ctenidia of other molluscs is uncertain. Monoplacophoran gills have a basic design and histology similar to those of gastropods, but they occur in five or six pairs, aligned bilaterally within the pallial groove, reminiscent of chitons. Well developed lamellae occur only on one side of the monoplacophoran gill axis, an arrangement similar to the monopectinate condition of advanced gastropods.

Excretion and osmoregulation

The basic excretory structures of molluscs are paired tubular metanephridia (often called kidneys) that are primitively similar to those of annelids, echiurans, and sipunculans. Six or seven pairs of metanephridia occur in monoplacophorans, two pairs in the nautiloids, and a single pair in all other molluscs (except where one is lost in advanced gastropods) (see Figure 15). The nephrostome typically opens into the pericardial coelom via a **renopericardial duct**, and the nephridiopore discharges into the mantle cavity, often near the anus (Figure 34). Pericardial fluids pass through the nephrostome and into the nephridium, where selective resorption occurs along the tubule wall until the final urine is ready to pass out the nephridiopore. The pericardial sac and heart wall act as selective barriers between the open nephrostome and the hemolymph in the surrounding hemocoel and heart. Mollusc nephridia are rather large and saclike, and their walls are greatly folded to increase surface area. In many forms, afferent and efferent nephridial vessels carry hemolymph to and from the kidney tissues (Figure 38). Often a short bladder is present just before the nephridiopore.

In at least some species it has been demonstrated that urine formation involves pressure filtration, active secretion, and active resorption. Aquatic molluscs excrete mostly ammonia, and most marine species are osmoconformers. In freshwater species the nephridia are capable of excreting a hyposomotic urine by resorbing salts and by passing large quantities of water. Terrestrial gastropods conserve water by converting ammonia to uric acid, although the degree to which conservation is accomplished varies depending on local environmental conditions. Land snails are capable of surviving a considerable loss of body water, brought on in part by production of the slime trail.

We have already mentioned some variations on the primitive molluscan excretory *Bauplan*. In most gastropods, torsion is accompanied by loss of the adult right nephridium, except for a small remnant that contributes to part of the gonoduct. Some gastropods have lost the direct connection of the nephrostome to the pericardial coelom. In such cases the nephridium is often very glandular and served by afferent and efferent hemolymph vessels, and wastes are removed largely from the circularory fluid. In many gastropods and some other molluscs, the gonoduct fuses with the renopericardial canal, and the nephridiopore functions as a **urogenital pore** and discharges both excretory wastes and gametes. In pulmonates, where the mantle cavity serves as a lung, the excretory duct is elongate and the nephridiopore opens outside the mantle cavity. In monoplacophorans and chitons, the nephridia open into the excurrent regions of the pallial grooves and in scaphopods, the paired nephridia open near the anus.

In bivalves, the two nephridia are located beneath the pericardial cavity and are folded in a long U-shape. One arm of the U is glandular and opens into the pericardial cavity; the other arm forms a bladder and opens through a nephridiopore in the suprabranchial cavity. In protobranchs, the unfolded walls of the tube are glandular throughout.

Cephalopods retain the basic nephridial plan, in which the kidneys drain the pericardial coelom by way of renopericardial canals and empty via nephridiopores into the mantle cavity. However, the metanephridia bear enlarged regions called **renal sacs**. Before reaching the branchial heart, a large vein passes through the renal sac, wherein numerous thin-walled evaginations, called **renal appendages**, project off the vein. As the branchial heart beats, hemolymph is drawn through the renal appendages, and wastes are filtered across their thin walls into the nephridia. The overall result is an increase in excre-

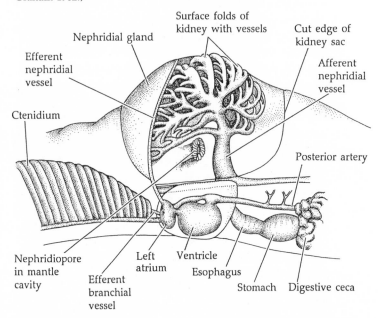

Figure 38

The kidney and nearby organs of *Littorina* (cutaway view). The nephridial sac has been slit open. (Modified from Fretter and Graham 1962.)

tory efficiency over the basic metanephridial plan of other molluscs.

The fluid-filled kidneys of cephalopods are inhabited by a variety of commensals and parasites. The epithelium of the convoluted renal appendages provides an excellent surface for attachment, and the renal pores provide a simple exit to the exterior. Symbionts identified from cephalopod kidneys include viruses, fungi, ciliate protozoa, dicyemids, trematodes, larval cestodes, and juvenile nematodes.

Nervous system

The molluscan nervous system is derived from the basic protostome plan of an anterior circumenteric arrangement of ganglia and paired ventral nerve cords. In molluscs, there are typically two pairs of ventral nerve cords. The more ventral and medial of the two pairs of nerve cords are called the **pedal cords** (or ventral cords); they innervate the muscles of the foot. The more lateral pair of nerves are the **visceral cords** (or lateral cords); they serve the mantle and viscera. Transverse commissures interconnect these longitudinal nerve cord pairs, creating a ladder-like nervous system. This basic plan is most easily seen in primitive molluscs, such as aplacophorans and polyplacophorans (Figure 39). However, the molluscan nervous system is not regarded as segmental, so it differs from the paired ventral nerve cord plan of annelids, which bears segmentally arranged ganglia.

In the simplest molluscs—such as aplacophorans, monoplacophorans, and polyplacophorans—ganglia are poorly developed (Figure 39). A simple nerve ring surrounds the esophagus, often with small cerebral ganglia on either side. Each cerebral ganglion, or the nerve ring itself, issues small nerves to the buccal region and also gives rise to the pedal and the visceral nerve cords. Most other molluscs have more well defined ganglia. Their nervous systems are built around three pairs of large ganglia that interconnect to form a partial or complete nerve ring around the gut (Figures 40 and 41). Two pairs, the **cerebral** and **pleural ganglia**, lie dorsal or lateral to the esophagus, and one pair, the **pedal ganglia**, lies ventral to the gut, in the anterior part of the foot. In cephalopods, bivalves, and advanced gastropods, the cerebral and pleural ganglia are typically fused. From the cerebral ganglia, peripheral nerves innervate the tentacles, eyes, statocysts, and general head surface, as well as **buccal ganglia** with special centers of control for the buccal region, radular apparatus and esophagus. The pleural ganglia give rise to the visceral cords, which extend posteriorly, supplying peripheral nerves to the viscera and mantle. The visceral cords eventually meet with a pair of **parietal** (= **intestinal**, = **pallial**) **ganglia** and from there pass on to terminate in paired **visceral ganglia**. The parietal ganglia innervate the gills and osphradium, and the visceral ganglia serve organs in the visceral mass. The pedal ganglia also give rise to a pair of pedal nerve cords that extend posteriorly and provide nerves to muscles of the foot.

As a result of torsion, the posterior portion of the gastropod nervous system is twisted into a figure-eight, a condition known as streptoneury (Figures 40A and B). In addition to twisting the nervous system, torsion brings the posterior ganglia forward. In many advanced gastropods this anterior concentration of the nervous system is accompanied by a shortening of certain nerve cords and fusion of certain ganglia. In most detorted gastropods the nervous system displays a secondarily derived bilateral symmetry and more-or-less straight, parallel, visceral nerve cords—a condition known as euthyneury (Figure 40C).

In bivalves, the nervous system is clearly bilateral, and fusion has reduced it to three large distinct ganglia. Anterior **cerebropleural ganglia** give rise to two pairs of nerve cords, one extending posterodorsally to the **visceral ganglia**, the other leading ventrally to the **pedal ganglia** (Figure 41). The two cerebropleural ganglia are joined by a dorsal commissure over the esophagus. The cerebropleural ganglia send nerves to the palps, anterior adductor muscle, and mantle. The visceral ganglia issue nerves to the gut, heart, gills, mantle, siphon, and posterior adductor muscle.

The degree of nervous system development within the Cephalopoda is unequaled among invertebrates. Although the paired ganglia seen in other molluscs are also recognizable in cephalopods, extreme cephalization has occurred. Most of the ganglia have shifted forward and are concentrated as lobes of a large brain encircling the anterior gut (Figure

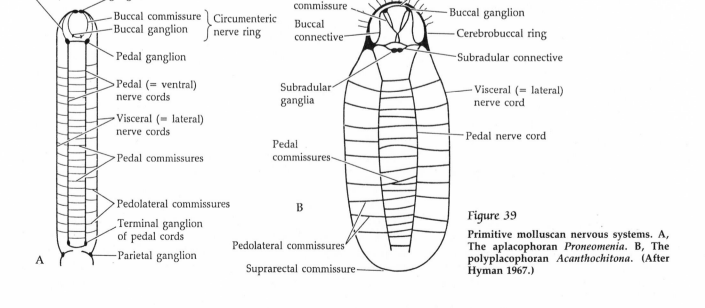

Figure 39

Primitive molluscan nervous systems. A, The aplacophoran *Proneomenia*. B, The polyplacophoran *Acanthochitona*. (After Hyman 1967.)

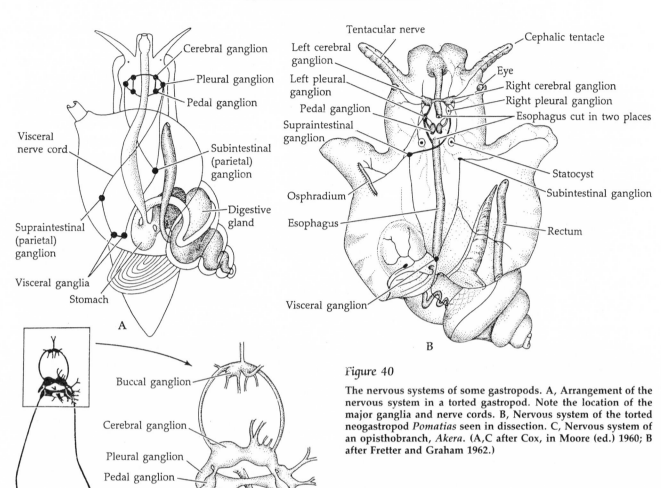

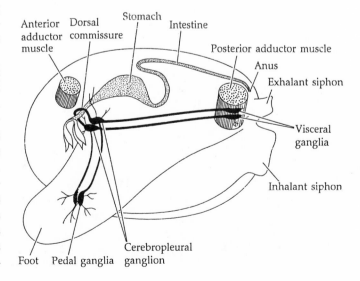

Figure 40

The nervous systems of some gastropods. A, Arrangement of the nervous system in a torted gastropod. Note the location of the major ganglia and nerve cords. B, Nervous system of the torted neogastropod *Pomatias* seen in dissection. C, Nervous system of an opisthobranch, *Akera*. (A,C after Cox, in Moore (ed.) 1960; B after Fretter and Graham 1962.)

Figure 41

The reduced and concentrated nervous system of a typical bivalve.

42A). In addition to the usual head nerves originating from the cerebral ganglion, a large optic nerve extends to each eye. In most cephalopods, much of the brain is enclosed in a cartilaginous case, or **cranium**. The pedal lobes supply nerves to the funnel, and anterior divisions of the pedal ganglia, called **brachial lobes**, send nerves to each of the arms and tentacles, an arrangement suggesting that the funnel and tentacles are derived from the molluscan foot. Octopuses may be the "smartest" invertebrates, for they can be taught some memory-dependent tasks fairly quickly.

Many cephalopods display a rapid escape behavior that depends on a system of giant motor fibers that control powerful and synchronous contractions of the mantle muscles. The command center of this system is a pair of very large first-order giant neurons in the lobe of the fused visceral ganglia. Here, con-

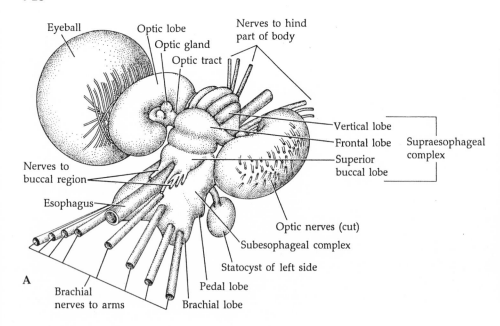

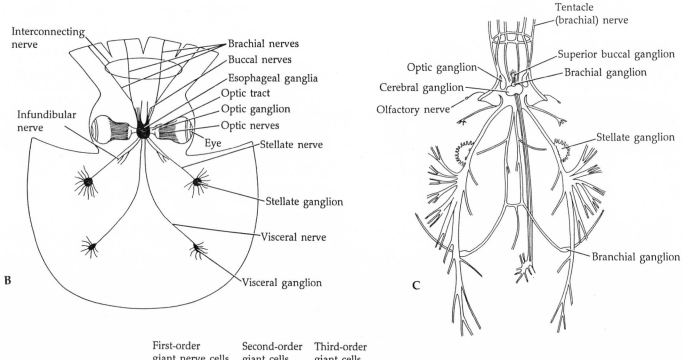

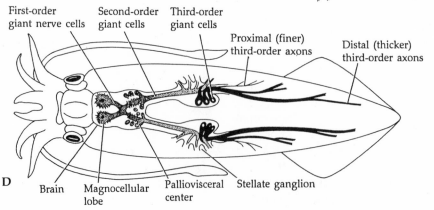

Figure 42

The highly developed nervous system of cephalopods. A, The brain of an octopus. The lobes of the supraesophageal complex correspond to the cerebral and buccal ganglia of other molluscs; the subesophageal complex comprises the fused pedal and pleurovisceral ganglia. About 15 structurally and functionally distinct pairs of lobes have been identified in the brain of octopuses. B, Nervous system of an octopus. C, Nervous system of a squid (*Loligo*). D, Giant fiber system of a squid. Note that the first-order giant neurons possess an unusual cross connection, and that the third-order giant neurons are arranged so that motor impulses can reach all parts of the mantle-wall musculature simultaneously (as a result of the fact that impulses travel faster in thicker axons). (A after Wells 1963 and other sources; B after Winkler and Ashley 1954; D after Russell–Hunter 1979.)

nections are made to second-order giant neurons that extend to a pair of large **stellate ganglia**. At the stellate ganglia, connections are made with third-order giant neurons that innervate the circular muscle fibers of the mantle (Figures 42B–D). Other nerves extend posteriorly from the brain and terminate in various ganglia that innervate the viscera and structures in the mantle cavity.

For several decades neurobiologists have utilized the giant axons of *Loligo* as an experimental system for the study of nerve physiology and mechanics, so much of our fundamental knowledge of how nerve cells work is based on squid neurology. The sea hare *Aplysia*, an opisthobranch, has been used in the same fashion and, although it does not possess giant axons, it possesses exceptionally large neurons and ganglia that can be easily impaled with microelectrodes.

Sense organs

With the exception of the primitive classes Aplacophora and Caudofoveata, molluscs possess various combinations of sensory tentacles, photoreceptors, statocysts, and osphradia. **Osphradia** are patches of sensory epithelium, located on or near the gill, or on the mantle wall (Figure 40B). They function as chemoreceptors and perhaps also as monitors of the amount of sediment in the inhalant current (Figures 43A and B). Little is known about the biology of osphradia, and their anatomy differs markedly throughout the phylum.

In primitive archaeogastropods, an osphradium is present on each gill; in the prosobranchs that possess but one gill, there is only one osphradium, and it lies on the mantle cavity wall anterior and dorsal to the attachment of the gill itself. Osphradia are reduced or absent in gastropods that have lost both gills, that possess a highly reduced mantle cavity, or that have taken up a strictly pelagic existence. Osphradia are best developed in benthic predators and scavengers, such as neogastropods.

Most gastropods have one pair of **cephalic tentacles**, but higher pulmonates and many opisthobranchs possess two pairs. Many archaeogastropods (e.g., limpets, abalone) also have short **epipodial tentacles** on the margin of the foot or mantle. The cephalic tentacles may bear eyes as well as tactile and chemoreceptor cells. Many opisthobranchs have a pair of branching or folded anterior dorsal processes called **rhinophores**, which are highly effective chemoreceptors (Figures 7F and G). Opisthobranchs typically produce a mucopolysaccharide slime trail as they crawl. In many species the trail contains chem-

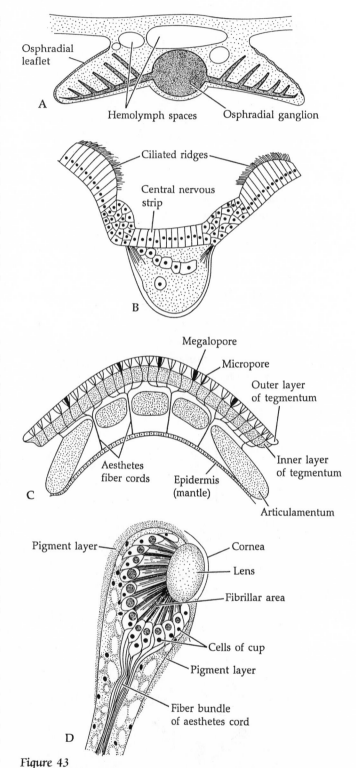

Figure 43

Two sensory organs of molluscs: osphradia and aesthetes. A, Two leaflets (cross section) of a bipectinate osphradium of a gastropod (*Vermetus*). B, The osphradium (cross section) of a heteropod (*Pterotrachea*). C, One valve of a polyplacophoran (*Tonicia*). The aesthetes extend to the shell surface through megalopores and micropores. D, Eye-bearing aesthetes (longitudinal section) in a megalopore of a chiton (*Acanthopleura*). (After Hyman 1967.)

ical messengers that other members of the species "read" by means of their excellent chemoreception. These may be simple trail markers, so one animal can follow or locate another, or they may include alarm substances that serve to warn others of possible danger on the path ahead. For example, when the carnivorous slug *Navanax* (= *Aglaja*) is attacked by a predator, it quickly releases a yellow chemical mixture on its trail that causes other members of the species to abort their trail-following activity.

Recent laboratory experiments have shown that at least one nudibranch (*Tritonia diomedea*) possesses geomagnetic orientation to the earth's magnetic field, the response being modulated by a lunar rhythm (Lohmann and Willows 1987). Motile gastropods usually possess a pair of closed statocysts in the anterior region of the foot.

Most gastropods have a small eye at the base of each cephalic tentacle, but in some, such as the conch *Strombus*, the eyes are enlarged and on long stalks. The higher pulmonates also have eyes placed on the tips of special **optic tentacles**. Primitive gastropods have simple pigment-cup eyes, while some advanced groups have evolved complex eyes with a cornea and lens (Figure 44A, B, and D).

Figure 44

Molluscan eyes. A, The simple pigment-cup arrangement of an archaeogastropod, *Haliotis* (the abalone). B–E, Eyes with lenses. B, The eye of a garden snail (*Helix*). C, The eye of a scallop (*Pecten*). D, The eye of a marine prosobranch (*Littorina*). E, The eye of an octopus (*Octopus*). (After Hyman 1967, Wells 1968, and others.)

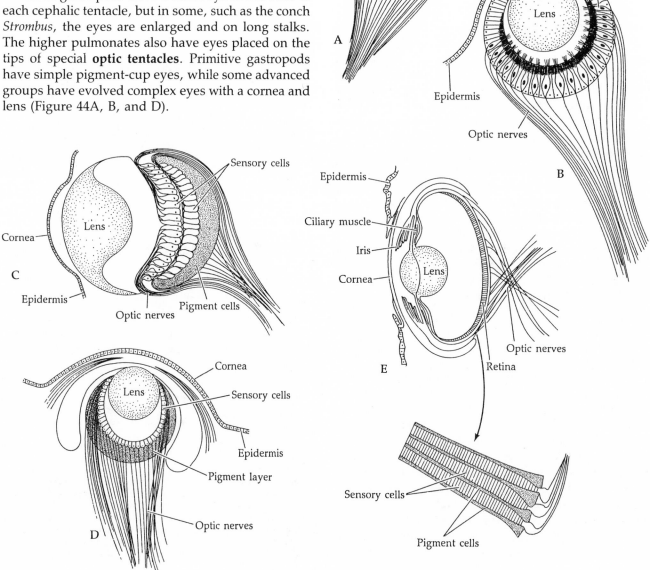

Scaphopods have lost the eyes, tentacles, and osphradia typical of most epibenthic and motile molluscs. The captacula may function as tactile (as well as feeding) structures, but little is known about scaphopod sense organs.

Bivalves carry most of their sensory organs along the middle lobe of the mantle edge (Figure 14C). These receptors include the **pallial tentacles**, which contain both tactile and chemoreceptor cells. The tentacles are commonly restricted to the siphon areas, but in some swimming clams (e.g., *Pecten*, *Lima*) they may line the entire mantle margin. Paired statocysts usually occur in the foot, and are of particular importance in georeception by burrowing bivalves. Ocelli may also be present along the mantle edge. In the spiny oyster *Spondylus* and the swimming clam *Pecten*, the ocelli have a complex cornea–lens arrangement (Figure 44C). The bivalve osphradium lies in the exhalant chamber, beneath the posterior adductor muscle. Because of its positioning, some specialists do not think that the bivalve osphradium is homologous with that of gastropods.

Chitons lack statocysts, cephalic eyes, and tentacles. Instead, they rely largely on two special sensory structures. The **subradular organ** is a modified region of the foregut that serves a chemosensory function. **Aesthetes** are a specialized system of photoreceptors unique to the class Polyplacophora. Aesthetes occur in high numbers across the dorsal surface of the shell plates. They are mantle cells that extend into minute vertical canals that open as megalopores and micropores in the upper tegmentum of the shell (Figures 43C and D). The canals and sensory endings terminate beneath a cap on the shell surface. Little is known about the functioning of aesthetes, but they apparently mediate light-regulated behavior. In at least one family (Chitonidae), some of them are modified as simple eyes. The outer mantle surface of many chitons is liberally supplied with tactile and photoreceptor cells, and the inner mantle cavity usually bears patches of chemosensory epithelium.

Like the rest of their nervous system, the sense organs of cephalopods are very highly developed. The eyes are superficially similar to those of vertebrates (Figure 44E), and the two are often cited as classic examples of convergent evolution. The cephalopod eye sits in a cartilaginous socket associated with the brain case. The cornea, iris, and lens arrangement is much like that of vertebrate eyes. The lens is suspended by ciliary muscles but has a fixed shape and focal length. An iris diaphragm controls the amount of light entering the eye, and the pupil is a horizontal slit. The retina comprises closely packed, long, rodlike photoreceptors whose sensory ends point toward the front of the eye; hence the cephalopod eye is the direct type rather than the indirect type seen in vertebrates. The rods connect to retinal cells that supply fibers to the optic ganglia at the distal ends of the optic nerves. Unlike the eyes of vertebrates, the cephalopod cornea probably contributes little to focusing because there is almost no light refraction at the corneal surface (as there is at an air–cornea interface). The cephalopod eye accommodates to varying light conditions by changes in the size of the pupil and by migration of the retinal pigment. Cephalopod eyes form distinct images, although octopuses are probably quite nearsighted, and experimental work suggests that they may also see colors. In addition, cephalopods can discriminate among objects by size, shape, and vertical versus horizonal orientation. The eyes of *Nautilus* are rather primitive relative to the eyes of other cephalopods. They are carried on short stalks, lack a lens, and are open to the water by way of the pupil.

Nautiloids and coleoids have statocysts that provide information on static body position and on body motion. The arms are liberally supplied with chemosensory and tactile cells, especially on the suckers of benthic hunting octopuses, which have extremely good chemical and textural discrimination capabilities. *Nautilus* is the only cephalopod with osphradia.

Cephalopod coloration and ink

Cephalopods are noted for their striking pigmentation and often dramatic color displays. The integument contains many pigment cells, or **chromatophores**. Most are probably under control of the nervous system and perhaps hormones. Such chromatophores can be individually expanded or contracted by means of tiny muscles attached to the periphery of the cells. Contraction of these muscles pulls the cell and its internal pigment out into a flat plate, thereby displaying the color; relaxation of the muscles causes the cell and pigment to concentrate into a small, inconspicuous dot. Because these chromatophores are displayed or concealed by muscle action, their activity is extremely rapid and cephalopods can change color (and pattern) almost instantaneously. Chromatophore pigments are of several colors—black, yellow, orange, red, and blue. The chromatophore color may be enhanced by deeper layers of iridocytes that both reflect and refract light in a prismatic fashion. Some species, such as the cuttlefish *Sepia* and some octopuses, are capable of closely mimicking their background coloration. Most cephalopods, however, appear to undergo color changes in relation to behavioral rituals, such as courtship and aggression. In octopuses, many color changes are accompanied by modifications in the surface texture of the body, mediated by muscles beneath the skin—sort of like elaborate, controlled "goose-flesh."

In addition to the color patterns formed by chromatophores, some cephalopods are bioluminescent. When present, the photophores are arranged in various patterns on the body, and in some cases even occur on the eyeball. The luminescence is sometimes due to symbiotic bacteria, but in most cases it is intrinsic. The photophores of some species have a complex reflector and focusing-lens arrangement, and some even have an overlying color filter or chromatophore shutter to control the color or flashing pattern. Most luminescent species are deep-sea forms, but little is known about the role of light production in their lives. Some appear to use the photophores to create a countershading effect, so as to appear less visible to predators (and prey) from below and above. Others living below the photic zone probably use their glowing or flashing patterns as a means of communication, the signals serving to keep animals together in schools or to attract prey. The flashing may also play a role in mate attraction. The fire squid, *Lycoteuthis*, can produce several colors of light: white, blue, yellow, and pink. At least one genus of squid, *Heteroteuthis*, secretes a luminescent ink. The light comes from luminescent bacteria cultured in a small gland near the **ink sac**, from which ink and bacteria are ejected simultaneously.

In most nonnautiloid cephalopods, a large ink sac is located near the intestine (Figure 32H). An ink-producing gland lies in the wall of the sac, and a duct runs from the sac to a pore on an ampulla in the rectum. The gland secretes a brown or black fluid that contains a high concentration of melanin pigment and mucus; the fluid is stored in the ink sac. When alarmed, the animal releases the ink through the anus and mantle cavity and out into the surrounding water. The cloud of inky material hangs together in the water, forming a "dummy" image that serves to confuse predators. The alkaloid nature of the ink may also act to deter predators, particularly fishes, and may interfere with their chemoreception.

Like virtually all other aspects of cephalopod biology, the ability to change color and to defend against predators are part and parcel of their active hunting lifestyles. Having lightened their bodies by abandoning the protection of an external shell, non-nautiloid cephalopods have exposed their fleshy parts to predators. The evolution of camouflage and ink production, coupled with high mobility and com-plex behavior, played a major role in the success of these animals in their radical modification of the basic molluscan *Bauplan*.

Reproduction

Primitively, molluscs are gonochoristic, with a pair of gonads that discharge their developing gametes to the outside, either through the nephridia or through separate ducts. In species that free-spawn, fertilization is external and development is indirect. Many molluscs have separate gonoducts to store or transport the gametes, and have various means of internal fertilization. From the primitive case of fully indirect development, various direct and mixed life-history patterns have evolved.

Aplacophorans may be either gonochoristic with single or paired gonads, or hermaphroditic with a pair of gonads—one functioning as an ovary, the other as a testis (Figure 45). In all cases the gonads discharge gametes by way of short **gonopericardial ducts** into the pericardial chamber, where they are picked up by the kidneys and eventually released into the surrounding sea water. Monoplacophorans possess two pairs of gonads, each with a gonoduct connected to one of the pairs of metanephridia (Figure 3D), and fertilization is external.

Most chitons are gonochoristic, although a few hermaphroditic species are known. The two gonads are fused into one, which is situated medially in front of the pericardial cavity (Figure 4F). Gametes are

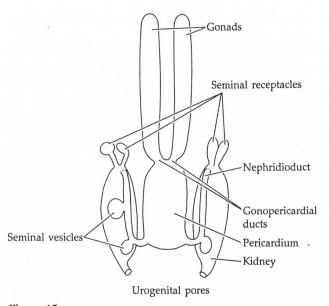

Figure 45

An aplacophoran urogenital system. (After Hadfield, in Giese and Pearse 1979.)

transported directly to the outside by two separate gonoducts, which are not associated with the nephridia. The gonopores are located in the exhalant region of the pallial groove, one in front of each nephridiopore. Fertilization is external but often occurs in the mantle cavity of the female. The eggs are enclosed within a spiny, buoyant membrane and are released into the sea individually or in strings. A few groups brood their embryos in the pallial groove, and in one species (*Callistochiton viviparus*) development takes place entirely within the ovary.

In living gastropods, one of the gonads is always lost and the remaining one is usually coiled within the visceral mass. The gonoduct is always developed in association with the right nephridium (Figures 46 and 47). In cases where the right nephridium is still functional in transporting excretory products, as in primitive archaeogastropods, the genital duct is properly called a **urogenital duct**, because it discharges both gametes and urine. Gastropods may be gonochoristic or hermaphroditic, but even in the lat-

ter case only a single gonad (an **ovotestis**) exists. The commitment of the right nephridial plumbing entirely to serving the reproductive system was a major step in higher gastropod evolution. The isolation of the reproductive tract freed it from the excretory system and allowed its independent evolution. Were it not for this singular event, the great variety of reproductive and developmental patterns in gastropods may never have been realized.

In many gastropods with isolated reproductive tracts, the female system bears a ciliated fold or tube that forms a **vagina** and **oviduct** (or **pallial duct**). The tube develops inwardly from the mantle wall and connects with the genital duct. The oviduct may bear specialized structures for sperm storage or egg case secretion. A **seminal receptacle** often lies near the ovary at the proximal end of the oviduct. Eggs are fertilized at or near this point in the oviduct, prior to entering the long secretory portion of the duct. Many female systems also have a **copulatory bursa** at the distal end of the oviduct, where sperm are received

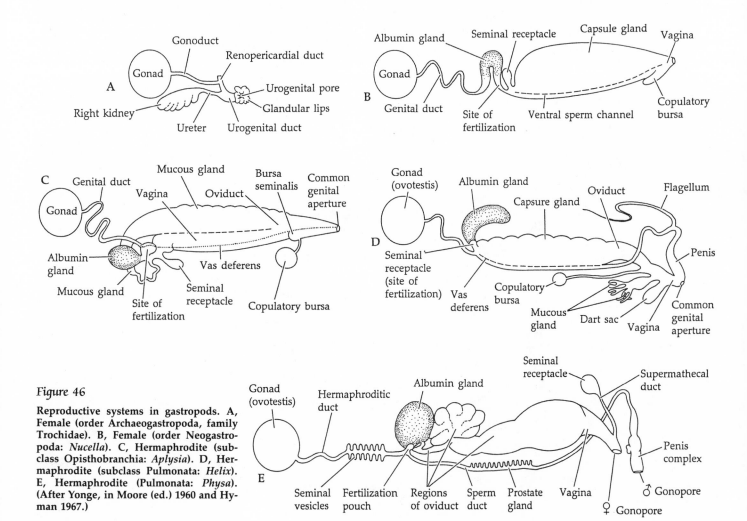

Figure 46

Reproductive systems in gastropods. A, Female (order Archaeogastropoda, family Trochidae). B, Female (order Neogastropoda: *Nucella*). C, Hermaphrodite (subclass Opisthobranchia: *Aplysia*). D, Hermaphrodite (subclass Pulmonata: *Helix*). E, Hermaphrodite (Pulmonata: *Physa*). (After Yonge, in Moore (ed.) 1960 and Hyman 1967.)

during mating. In such cases the sperm are later transported along a ciliated groove in the oviduct to the seminal receptacle, where fertilization takes place. The secretory section of the oviduct may be modified to form an **albumin gland** and a **mucous** or **capsule gland**. Many opisthobranchs lay the fertilized eggs in jelly-like mucopolysaccharide masses or strings produced by these glands. Most terrestrial pulmonates produce a small number of large, individual, yolky eggs, which often are provided with very thin calcareous shells. Other pulmonates brood their embryos internally and give birth to juveniles. Many advanced marine prosobranchs produce **egg capsules** in the form of leathery or hard cases that are attached to objects in the environment, thereby protecting the developing embryos. A ciliated groove is often present to conduct the soft egg capsules from the female gonopore down to a gland in the foot, where they are molded and attached to the substratum.

In gastropods that produce egg capsules or egg cases, the males usually have a penis to facilitate transfer of sperm or spermatophores (Figures 6B and 47), and internal fertilization takes place prior to for-

Figure 47

Reproductive systems (dissections) in some more gastropods. A, The periwinkle *Littorina* (order Mesogastropoda), removed from its shell. B–C, Reproductive systems of the slipper shell *Crepidula* (male and female). D, The land snail *Helix aspera*. (A after Fretter and Graham 1962; B,C after Hyman 1967; D after Cameron and Redfern 1976.)

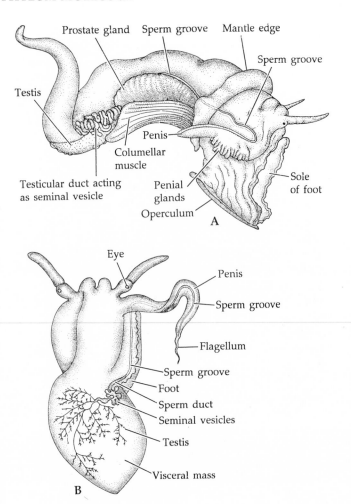

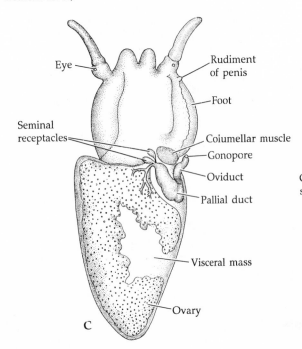

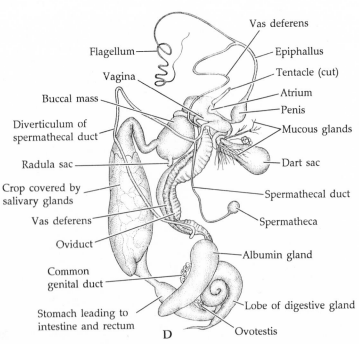

mation of the egg case. The penis is a long extension of the body wall arising behind the right cephalic tentacle and the male genital duct, or **vas deferens**, and may include a **prostate gland** for production of seminal secretions. In many molluscs the proximal region of the vas deferens functions as a sperm storage area, or seminal vesicle.

Both simultaneous and sequential hermaphrodites are common among gastropods. In both cases copulation is the rule, either with one individual acting as the male and the other as the female, or with a mutual exchange of sperm between the two. Sedentary species, such as many limpets and slipper shells, are often protandric hermaphrodites. In slipper shells (*Crepidula*), individuals may stack one atop the other (Figure 48), males generally on top of the stack, females on the bottom. Males use their long penis to inseminate the females below. Males that are in association with females tend to remain male for a relatively long period of time. Eventually, or if isolated from a female, the male develops into a female. Female slipper shells cannot switch back to males, as the masculine reproductive system degenerates during the sex change.

Pulmonates are simultaneous hermaphrodites; opisthobranchs may be either simultaneous or occasionally protandric hermaphrodites. In most simultaneous hermaphrodites a single complex gonad, the ovotestis, coincidentally produces both eggs and sperm (Figure 46C–E and 47D). The genital duct draining an ovotestis is called the **hermaphroditic duct**. There may be separate male and female gonopores, or only a single common gonopore. Such reproductive systems are amazingly complex and variable.

Distinct precopulatory behavior patterns occur in a few groups of gastropods. These primitive courtship routines are best documented in land pulmonates and include behaviors such as oral and tentacular stroking, and intertwining of the bodies. In some pulmonates (e.g., *Helix*) the vagina contains a **dart sac**, which secretes a calcareous dart. When a pair of snails is intertwined, one will drive its dart into the body wall of the other, apparently as a means of sexually arousing its partner.

Most bivalves are gonochoristic and retain the primitive paired gonad plan. The gonads are large, soft organs, closely invested with the viscera and with each other, so an apparently single gonadal mass results (Figure 8D). The gonoducts are simple tubes, and fertilization is external, although some freshwater species brood their embryos for a time. In the protobranchs and other primitive bivalves, the gonoducts join the nephridia, which release the gametes through urogenital pores. In advanced bivalves, the gonoducts open into the mantle cavity separately from the nephridiopores. Hermaphroditism occurs in some bivalve species, including shipworms and some species of cockles, oysters, scallops, and others. Hermaphroditic scallops have ovotestes. Oysters of the genus *Ostrea* are sequential hermaphrodites, most being capable of switching sex in either direction.

Cephalopods are almost all gonochoristic, with a single gonad in the posterior region of the visceral mass (Figures 11, 12, and 49). The testis releases

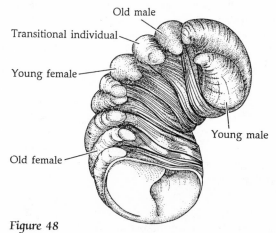

Figure 48

A stack of *Crepidula fornicata*, a slipper shell displaying sequential hermaphroditism.

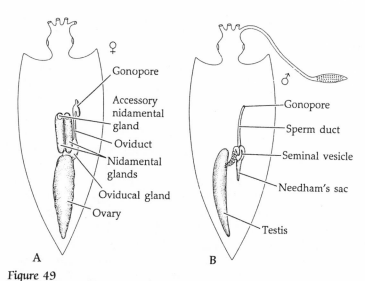

Figure 49

Reproductive systems in squids. Female (A) and male (B) squid (*Loligo*).

sperm to a coiled vas deferens, which leads anteriorly to a seminal vesicle. Here various glands assist in packaging the sperm into elaborate spermatophores, which are stored in a large reservoir called **Needham's sac**. From here the spermatophores are released into the mantle cavity via a sperm duct. In females the oviduct terminates in one **oviducal gland** in squids, and two in octopuses. This gland secretes a protective membrane around each egg.

The highly developed nervous system of cephalopods has facilitated the evolution of some very sophisticated precopulatory behaviors, which culminate in the transfer of spermatophores from the male to the female. Because the oviducal opening of females is deep within the mantle chamber, male cephalopods use one of the arms as an intromittent organ to transfer the spermatophores. The morphological modifications of such arms is called **hectocotyly** (Figure 12D). In squids and cuttlefish the right or left fourth arm is used; in octopuses it is the right third arm. In *Nautilus* four small arms form a conical organ, called the **spandix**, that functions in sperm transfer. Hectocotylous arms have special suckers, spoonlike depressions, or superficial chambers for holding the spermatophores during the transfer, which may be a brief or very lengthy process.

Each spermatophore comprises an elongate sperm mass, a **cement body**, a coiled, "spring-loaded" **ejaculatory organ**, and a **cap**. The cap is pulled off as the spermatophore is removed from the Needham's sac in squids or by uptake of sea water in octopuses. Once the cap is removed, the ejaculatory organ everts, pulling the sperm mass out with it. The sperm mass adheres by means of the cement body to the seminal receptacle or mantle wall of the female, where it begins to disintegrate and liberate sperm for up to two days.

Precopulatory rituals in cephalopods almost always involve striking changes in coloration, as the male tries to attract the female (and discourage other males in the area). Male squids often seize their female partner with the tentacles, and the two swim head-to-head through the water. Eventually the male hectocotylus grabs a spermatophore and inserts it into the mantle chamber of his partner, near or in the oviducal opening. Mating in octopuses can be a savage affair. The exuberance of the copulatory embrace may result in the couple tearing at each other with their sharp beaks, or even strangulation of one partner by the other as the former's arms wrap around the mantle cavity of the latter, cutting off ventilation. In many octopuses (e.g., *Argonauta, Philonexis*) the tip of the hectocotylous arm may break

off and remain in the female's mantle chamber. The detached arm was mistakenly first described as a parasitic worm and given the generic name *Hectocotylus* (hence the origin of the term).

As the eggs pass through the oviduct, they are covered with a capsule-like membrane produced by the oviducal gland. Once in the mantle cavity, various kinds of **nidamental glands** may provide additional layers or coatings on the eggs. In the oceanic squid *Loligo*, which migrates to shallow water to breed, the nidamental glands coat the eggs to form an oblong gelatinous mass, each containing about 100 eggs. The female holds these egg cases in her arms and fertilizes them with sperm ejected from her seminal receptacle. The egg masses harden as they react with sea water and are then deposited on the substratum. The adults die after mating and egg laying. Cuttlefish deposit single eggs and attach them to seaweed or other substrata. Many open-ocean pelagic cephalopods have floating eggs, and the young develop entirely in the plankton. Octopuses usually lay grapelike egg clusters in rocky areas, and many species care for the developing embryos by protecting and cleaning them and by flushing the egg mass with jets of water. Octopuses and squids tend to grow quickly to maturity, reproduce, and then die. The pearly nautilus, however, is long-lived (perhaps to 20 years), slow growing, and able to reproduce for many years after maturity.

One of the most astonishing reproductive behaviors among the invertebrates occurs in the pelagic cephalopod genus *Argonauta*, whose members are known as paper nautiluses. Female argonauts use two specialized arms to secrete a beautiful, coiled, calcareous shell into which the eggs are deposited (Figure 17B). The shell is carried by the female and serves as her temporary home and as a brood chamber for the embryos. The much smaller male often cohabits the shell with the female.

Development

Development in the molluscs is similar in many fundamental ways to that of the other protostomes (Figures 50–53). Most molluscs undergo typical spiral cleavage, with the mouth and stomodeum developing from the blastopore and the anus forming as a new opening on the gastrula wall. Cell fates are typically spiralian, including a 4d mesentoblast.

Development may be direct, mixed, or indirect. During indirect development, the free-swimming trochophore larva that develops is remarkably similar to that seen in annelids (Figure 50). Like the annelid larva, the molluscan trochophore bears an apical sen-

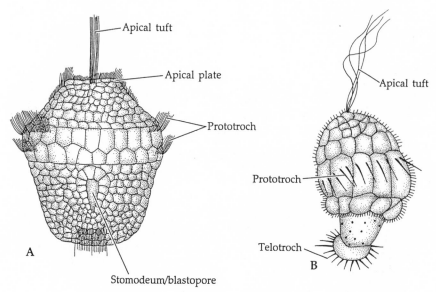

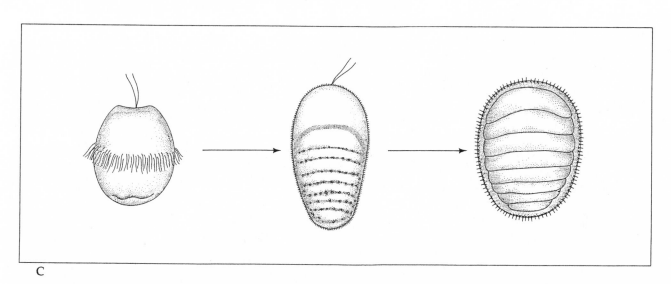

Figure 50

Molluscan trochophores. A, Generalized molluscan trochophore larva. B, Trochophore of an aplacophoran. C, Metamorphosis of a polyplacophoran from trochophore to juvenile. (A after Sherman and Sherman 1976; B after Hadfield, in Giese and Pearse 1979; C after Hyman 1967.)

sory plate with a tuft of cilia and a girdle of ciliated cells—the prototroch—just anterior to the mouth.

In some free-spawning molluscs (e.g., chitons), the trochophore is the only larval stage, and it metamorphoses directly into the juvenile (Figure 50C). In many groups (e.g., gastropods and bivalves), the trochophore is usually followed by a more highly developed, uniquely molluscan larva called a **veliger** (Figure 51). The veliger larva may possess a foot, shell, operculum, and other adult-like structures. The most characteristic feature of the veliger is the swimming and feeding organ, or **velum**, which consists of two large ciliated lobes developed from the prototroch. In some species the velum is subdivided into four, five, or even six separate lobes (Figure 51C).

Eventually eyes and tentacles appear, and the veliger transforms into a juvenile, settles down, and assumes an adult existence. The settling site is critical to adult survival, and larvae of most species will delay final metamorphosis until the preferred specific substratum is detected.

Some bivalves have long-lived planktotrophic veligers, whereas others have short-lived lecithotrophic veligers. Many widely distributed species have very long larval lives that allow dispersal over great distances. A few bivalves have mixed development and brood the developing embryos in the suprabranchial cavity through the trochophore period; then the embryos are released as veliger larvae. Some marine and freshwater clams have direct development. Species

Figure 51

Molluscan veliger larvae. A-B, Side and front views of the veliger larva of *Crepidula* (Gastropoda). C, A prosobranch veliger with four velar lobes. D, A generalized bivalve veliger. E, Glochidium larva of a freshwater bivalve. F, Late veliger of a scaphopod. (A–B redrawn from Hyman 1967, after Werner 1955, Helg. Wissensch. Meeresuntersuchungen, 5; C redrawn from Hyman 1967 after Dawydoff 1940; E after Brusca 1975; F after various sources.)

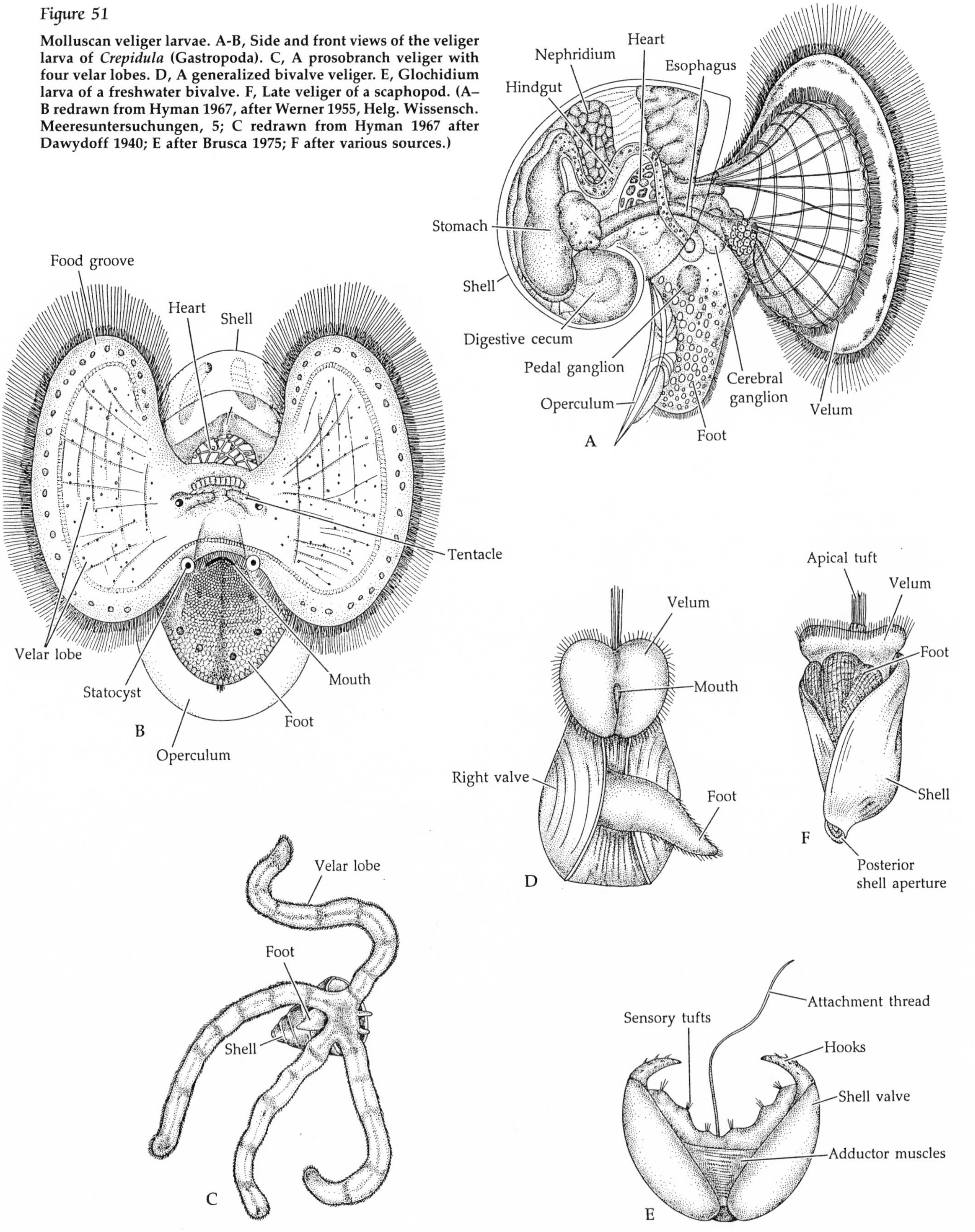

in the freshwater family Sphaeriidae brood the embryos between the gill lamellae and shed juveniles into the water after development is completed. Several unrelated marine groups have independently evolved a similar brooding behavior (e.g., *Arca vivipara* and many members of the family Carditidae).

In the freshwater groups Unionacea and Mutelacea, the embryos are also brooded between the gill lamellae, where they develop to the veliger stage. The veligers of these groups are often highly modified for a parasitic life on fishes, thereby facilitating dispersal. Various names have been given to these specialized parasitic veligers. In the Unionacea they are called **glochidia** (Figure 51E). They attach to the skin or gills of the host fish by a sticky mucus, hooks, or other attachment devices. Most glochidia lack a gut and absorb nutrients from the host by means of special phagocytic mantle cells. The host tissue often forms a cyst around the glochidium. Eventually the larva matures, breaks out of the cyst, drops to the bottom, and assumes its adult life.

Among the gastropods, only the primitive archaeogastropods that still rely on external fertilization have retained a free-swimming trochophore larva. All others suppress the trochophore or pass through it quickly before hatching. In many groups embryos hatch as veligers (e.g., opisthobranchs). Some of these gastropods have planktotrophic veligers that may have brief or extended (to several months) free-swimming lives. Others have lecithotrophic veligers that remain planktonic only for short periods. Planktotrophic veligers feed by use of the **velar cilia**, whose beating drives the animal forward and draws minute planktonic food particles into contact with shorter cilia of a food groove. Once in the food groove, the particles are trapped in mucus and carried along ciliary tracts to the mouth.

Almost all pulmonates and many advanced marine prosobranchs (e.g., neogastropods) have direct development, and the veliger stage is passed in the egg case, or capsule. Upon hatching, tiny snails crawl out of the capsule into their adult habitat. In some neogastropods (e.g., certain species of *Nucella*), the encapsulated embryos are cannibalistic on their siblings, a phenomenon called **adelphophagy**; consequently, only one juvenile eventually emerges from the capsule. It has long been thought that certain species of snails vary their developmental strategies, releasing veligers under some conditions but retaining them within the egg capsule under other conditions, a trick that would be valuable in groups inhabiting a wide variety of habitats. However, this idea has recently been challenged because of possible earlier misidentifications of sibling species.

It is usually during the veliger stage that gastropods undergo torsion; that is, the shell and visceral mass twist 180° relative to the head and foot (Figure 52). As we have seen, this phenomenon is still not fully understood, but it has played a major role in gastropod evolution.

Cephalopods produce large, yolky, telolecithal eggs. Development is always direct, the larval stages being lost entirely or passed within an egg case. Early cleavage is meroblastic and eventually produces a cap of cells (a discoblastula) at the animal pole. The embryo grows in such a way that the mouth opens to the yolk sac, and the yolk is directly "consumed" by the developing animal (Figure 53). As the mantle develops, it eventually overgrows and encloses the funnel, which derives originally from the foot but ends up within the mantle chamber of the adult.

Figure 52

Settled larva of the abalone (*Haliotis*) undergoing torsion. **A**, Left-side view after about 90° of torsion, with mantle cavity on the right side. **B**, Torsion continues as the mantle cavity and its associated structures twist forward over the head. (After Fretter and Graham 1962.)

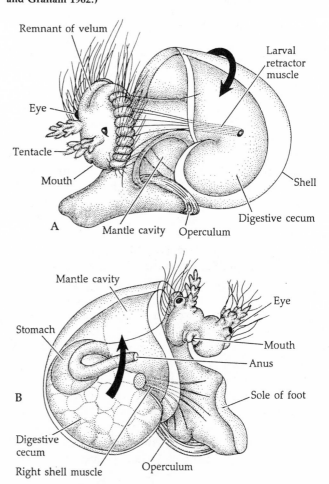

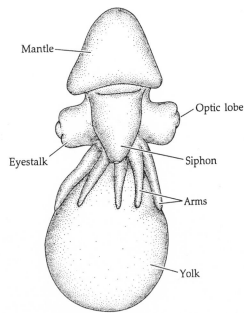

Mantle

Optic lobe

Eyestalk

Siphon

Arms

Yolk

Figure 53
Juvenile cephalopod attached to and consuming its sac of yolk.

Molluscan evolution and phylogeny

The Mollusca is such a diverse phylum, and so many taxa below the class level are apparently artificial (i.e., polyphyletic or paraphyletic), that efforts to trace their evolutionary history have often led to frustration. One popular way of dealing with complex phylogenetic problems is to construct a hypothetical ancestral form and then derive each known lineage from this supposed ancestor. Although sometimes useful as a learning device this procedure typically leads to circular reasoning, as discussed in Chapter 2. By predetermining the nature of the ancestor, the course of subsequent phylogenetic events in a group is largely established a priori. Such hypothetical ancestors are usually constructed by assembling a "paper animal" comprising all the characters suspected to be primitive in the group in question. A phylogenetic tree built by this method is thus rooted by first assembling suites of shared primitive characters, or symplesiomorphies—a method that typically leads to the creation of "unnatural groups" (e.g., paraphyletic taxa).

Until recently, most molluscan specialists entertained the idea of a "hypothetical ancestral mollusc" (affectionately known as HAM), the nature of which derived largely from early work of the eminent British biologist C. M. Yonge. Detailed and sometimes highly imaginative descriptions of this hypothetical ancestral mollusc were proposed by various workers, even including speculations on its ecology and behavior. Although the usefulness of HAM in molluscan phylogenetic analysis is now questioned by most specialists, a brief glimpse at this hypothetical creation can still be instructive. We provide below a description of the essential morphological features of the hypothetical mollusc. Following this brief description, we provide an independent cladistic analysis of the phylum.

The hypothetical ancestral mollusc was viewed as a small, bilaterally symmetrical, benthic marine creature inhabiting Precambrian seas (Figure 54). The ventral surface was a flattened, muscular, creeping sole or foot. The dorsal surface was covered by a single shell that protected the internal visceral mass. A series of pedal retractor muscles, extending from inside the shell to the sides of the foot, could pull the shell down tightly against the substratum. The posterior portion of the shell and mantle overhung the body enough to create a chamber, the mantle cavity. Within this protective chamber were one or more pairs of gills, the anus, and a pair of nephridiopores. The entire epidermis was ciliated, especially the foot. It is presumed that the creature glided over the substratum by use of ventral cilia and mucus. The gills of this hypothetical mollusc were usually described as bipectinate and thus were similar to those of primitive living gastropods and bivalves. From this hypothetical ancestor, all major lineages of molluscs were once thought to have evolved.

The discovery of caudofoveatans and the vexing problem of how to accommodate the aplacophorans eventually made such a hypothetical ancestor for all molluscs increasingly untenable, and HAM is probably now best relegated to the category of examples of the danger of relying on hypothetically reconstructed ancestors in phylogenetic analysis. Figure 55 is a popular "traditional" evolutionary tree of the molluscs, with the groups Caudofoveata and Aplacophora included. The placement of HAM in this scheme is shown at three possible positions. Positions 1 and 2 are ancestral to all shelled molluscs and would thus require the complete loss and reappearance of the shell in the phylum, an unlikely scenario. Position 3 postulates HAM to be simply a primitive monoplacophoran (or prepolyplacophoran).

Most modern workers attempt to avoid the pitfalls of a priori construction of a hypothetical ancestral form. Instead, they have begun to analyze the phylogenetic history of the molluscs by cladistic methods (e.g., Götting 1980; Salvini-Plawen 1980;

Figure 54

The hypothetical ancestral mollusc (HAM), as proposed by C. M. Yonge and espoused by many subsequent workers. Most modern workers no longer recognize HAM as a plausible molluscan ancestral form.

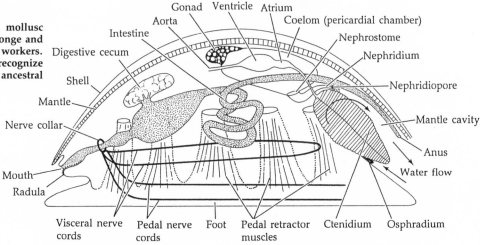

Lauterbach 1983; Wingstrand 1985). Although these studies differ in some details, the phylogenetic sequences resulting from their analyses have been very similar. Our own cladogram (Figure 56A) parallels this current thinking on molluscan evolution. The characters used to construct the cladogram are enumerated in the figure legend and briefly summarized in the following discussion. Because the relationships of the molluscs to other protostomes is uncertain, the characters utilized in our analysis of the molluscan classes were left unpolarized; that is, the cladogram was generated by allowing character states to change in any direction, with no a priori assumptions as to which states were ancestral and which were derived (see Chapter 2 to refresh your memory regarding cladistic analyses). The nodes on the cladogram have been lettered to facilitate the following discussion.

Molluscs share most of their typical protostome features with the sipunculans, echiurans, and annelids, for example spiral cleavage, schizocoely, and the trochophore larva. These are all symplesiomorphic characters within the phylum Mollusca. Perhaps the most fundamental differences between the molluscs and annelids involve segmentation and circulation: annelids are segmented and have a well developed coelom and a closed circulatory system, whereas molluscs are unsegmented and have a reduced coelom

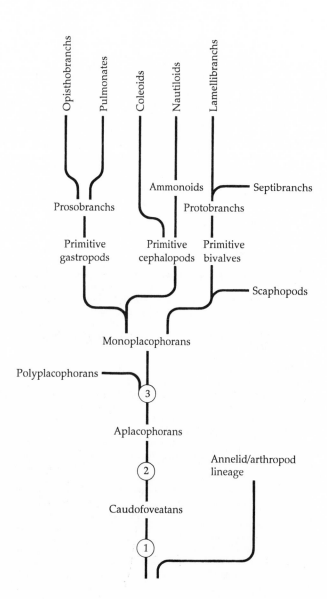

Figure 55

A popular "traditional" view of molluscan evolution, depicted in a general evolutionary tree. The numbers (1), (2), and (3) indicate three possible positions of the hypothetical ancestral mollusc (HAM), as discussed in the text.

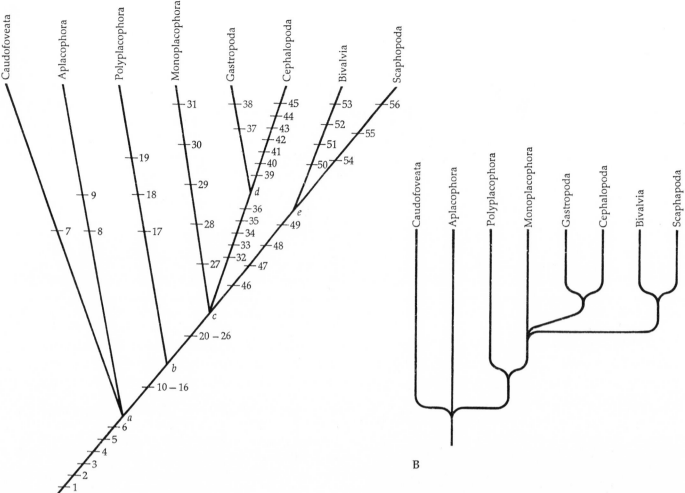

A

B

Figure 56

A cladogram (A) and evolutionary tree (B) depicting one view of the phylogeny of the Mollusca (see text for additional discussion). The numbers on the cladogram indicate suites of synapomorphies defining each hypothesized line or clade. Synapomorphies of the phylum Mollusca: (1) reduction of the coelom and development of an open hemocoelic circulatory system; (2) dorsal body wall forms a mantle; (3) production of calcareous spicules (and ultimately a shell) by mantle shell gland; (4) ventral body wall muscles develop as muscular foot (or foot precursor); (5) radula; (6) chambered heart with separate atria and ventricle.

Synapomorphy of the class Caudofoveata: (7) calcareous spicules of body wall form imbricating scales. Synapomorphies of the class Aplacophora: (8) ventral surface with a pedal groove; (9) posterior end of reproductive system with copulatory spicules.

Synapomorphies of the shelled molluscs defining node b: (10) concentration of the diffuse shell glands into one or a few discrete glands, to produce solid shell(s); (11) development of creeping sole on large, muscular, ventral foot; (12) preoral tentacles; (13) mantle folds form pallial groove around the body; (14) increase in gut complexity, with large mass of digestive ceca; (15) multiple pedal retractor muscles; (16) mobile radular membrane.

Synapomorphies of the class Polyplacophora: (17) unique shell with 7–8 plates (with 7–8 shell gland areas), articulamentum layer, and aesthetes; (18) multiple gills (perhaps not homologous to the ctenidia of other molluscs); (19) expanded and highly cuticularized mantle girdle that "fuses" with shell plates. Synapo-

morphies of the sister-group to the Polyplacophora, defining node c: (20) presence of a single, well defined shell gland and larval shell (protoconch); (21) shell univalve, of a single piece (*note*: the bivalve shell is taken to be derived from the univalve condition); (22) shell of the three-layered design (periostracum, prismatic layer, nacreous layers); (23) without (loss of) calcareous spines in body wall; (24) mantle margin of three parallel folds, each specialized for specific functions; (25) crystalline style; (26) statocysts.

Synapomorphies of the class Monoplacophora: (27) 5–6 pairs ctenidia; (28) 6–7 pairs nephridia; (29) 8 pairs pedal retractor muscles; (30) 2 pair gonads; (31) 2 pair heart atria.

Synapomorphies of the gastropod–cephalopod line, defining node d: (32) high, dorsoventrally elongated body; (33) viscera concentrated dorsally; (34) shell coiling; (35) well developed, clearly demarcated head; (36) mantle cavity restricted to anal region. Synapomorphies of the class Gastropoda: (37) torsion and its associated anatomical conditions; (38) further concentration of internal organs as visceral hump. Synapomorphies of the class Cephalopoda: (39) expansion of the coelom and closure of the circulatory system; (40) septate shell; (41) ink sac; (42) siphuncle; (43) beaklike jaws; (44) foot modified as prehensile arms/tentacles and siphon; (45) extensive fusion of ganglia as brain, housed in cartilaginous cranium.

Synapomorphies of the bivalve–scaphopod line, defining node e: (46) reduction of the head; (47) decentralization of the nervous system; (48) expansion of the mantle cavity to essentially surround the entire body; (49) modification of the foot to a more spatulate form. Synapomorphies of the class Bivalvia: (50) bivalve shell and its associated mantle and ctenidial modifications; (51) loss of radula; (52) byssus; (53) lateral compression of body. Synapomorphies of the class Scaphopoda: (54) tusk-shaped, open-ended shell; (55) loss of ctenidia; (56) captacula.

and an open circulatory system. The most striking synapomorphies distinguishing modern molluscs from annelids and other protostomes are the reduction of the coelom and the concomitant conversion of the closed circulatory system to an open hemocoelic one, the elaboration of the body wall into a mantle capable of secreting calcareous spicules or shell(s), and the unique molluscan radula.

The major steps in the evolution of what we generally think of as a "typical" mollusc—that is, a shelled mollusc—took place after the origin of the two primitive groups Caudofoveata and Aplacophora, perhaps as molluscs adapted to active epibenthic lifestyles. These steps centered largely on the elaboration of the mantle and mantle cavity, the refinement of the ventral surface as a muscular foot, and the evolution of a consolidated dorsal shell gland and solid shell(s) (in place of the independent calcareous spicules seen in members of the taxa Caudofoveata and Aplacophora). Most, although not all, specialists believe that the solid molluscan shell probably evolved by way of consolidation of the diffuse shell gland and subsequent "coalescence" of the free dermal spicules.

The description of a single larval aplacophoran in 1890, in which the dorsal surface may have borne seven transverse bands of parallel spicules (described as "composite plates," reminiscent of chitons), led earlier workers to postulate that aplacophorans and polyplacophorans might be sister-groups and that the multivalve condition was ancestral to the single-shell condition of more derived molluscs. A second aplacophoran was also described as having three such dorsal "plates," in which the spicules actually appeared to have fused. However, neither of these two observations has ever been repeated and no subsequent studies of aplacophoran larval development have verified the presence of "composite spicule plates" in that group. Furthermore, there are fundamental embryological, histological, and constructional differences between the shells of polyplacophorans and those of all other molluscs, an observation suggesting that the chitons stand alone as a unique radiation off the early molluscan line. In fact, some specialists have suggested that the shells of polyplacophorans and all other molluscs are not even homologous structures.

So, three explanations have been offered to explain the "shell problem" in molluscan evolution: (1) The multiplate shell may have been ancestral, the single-shell design having evolved by fusion (or loss) of plates. (2) The single shell may have been ancestral, and the multiplate design arose by subdivision of the single shell. (3) The single-shell and multishell designs arose independently from a shell-less ancestor, but perhaps both arose by means of the mantle shell gland and "consolidation" of calcareous spicules in the body wall. The presence of eight pairs of pedal retractor muscles in both polyplacophorans and monoplacophorans has been taken as evidence in favor of the first explanation. Acceptance of this first hypothesis would postulate that the ancestor at node b in the cladogram in Figure 56A was a multivalved chiton-like creature. Acceptance of the second hypothesis would imply that the ancestor at node b was a univalved, monoplacophoran-like ancestor. The third hypothesis postulates that the ancestor at node b lacked a solid shell altogether.

The primitive, post-caudofoveatan/aplacophoran mantle and foot arrangement was probably similar to what we see today in living polyplacophorans or monoplacophorans. That is, a large flattened sole that was surrounded by a pallial groove and would have contributed to the evolutionary exploitation of epifaunal lifestyles. This primitive pallial arrangement was eventually lost in both the gastropod–cephalopod line and the bivalve–scaphopod line. Secondary modifications on the shape of the foot and other features in bivalves and scaphopods allowed some of these animals to "return" to infaunal life in soft sediments, and many members of both of these taxa are highly adapted to burrowing. (We saw a similar evolutionary transition among annelids.)

Monoplacophorans share the character of a single-plate (univalve) shell with all other molluscs above the level of the polyplacophorans in the cladogram. They also share a similar shell histology and a host of other features. The only unique synapomorphies defining the monoplacophorans seem to be their repetitive organs (multiple gills, nephridia, pedal muscles, gonads, and heart atria). The question of whether this multiplicity arose uniquely in the monoplacophorans or represents a symplesiomorphic retention of ancestral features from some unknown metameric ancestor (at node c in the cladogram), has still not been fully resolved.

The gastropod–cephalopod line is defined by the following features: dorsoventral elongation of the body with dorsal concentration of the viscera; elaboration of a separate and well-defined head; and restriction of the mantle cavity to the posterior region of the body. The bivalve–scaphopod line in the cladogram is defined by reduction of the head region; decentralization of the nervous system and associated reduction or loss of certain sensory structures; and expansion and deepening of the mantle cavity.

The lateral mantle flaps, developing from the back of the scaphopod larva, grow down along the sides to enclose the body and are thought to be homologous with the left and right mantle sheets in bivalves. (Other synapomorphies defining the molluscan classes are listed in the legend to Figure 56.)

Cephalopods are the most specialized molluscs and possess a large number of complex synapomorphies. Primitive shelled cephalopods are represented today by only six species of *Nautilus*, although over 17,000 fossil species of shelled nautiloid cephalopods have been described. This highly successful molluscan class probably arose about 450 million years ago. The nautiloids underwent a series of radiations during the Paleozoic, but were largely replaced by the ammonoids after the Devonian period (325 million years ago). The ammonoids, in turn, became extinct around the Cretaceous–Tertiary boundary (65 million years ago). The origin of the dibranchiate (coleoid) cephalopods (octopuses, squids, and cuttlefish) is obscure, possibly dating back to the Devonian. They diversified mainly in the Mesozoic and became a highly successful group by exploiting, as we have seen, a new lifestyle.

Other interpretations of molluscan phylogeny exist, of course. Some authors view the monoplacophorans and gastropods as sister-groups, primarily on the basis of similarities in shell structure. Other specialists regard the caudofoveatans and aplacophorans as sister-groups, largely on the basis of their vermiform body shape. However, this characteristic may simply reflect the ancestral molluscan condition (i.e., at node a on the cladogram) or be the result of convergent evolution. Some workers feel that the aplacophorans are more closely related to the polyplacophorans (as noted above).

The question of ancestral metamerism in molluscs has been debated since the discovery of the first living monoplacophoran (*Neopilina galatheae*) in 1952. Monoplacophorans have five or six pairs of ctenidia, six or seven pairs of nephridia, two pairs of gonads, two pairs of heart atria, and eight pairs of pedal retractor muscles (even though they have but one shell). However, monoplacophorans are not the only molluscs that have multiple, repeated organs reminiscent of metamerism (or "pseudometamerism," as some prefer to call it). Polyplacophorans have a great many serially repeated gills in the pallial groove, and also possess eight pairs of pedal retractor muscles and eight shell plates. The two pairs of heart atria, nephridia, and ctenidia in *Nautilus* (and two pairs of retractor muscles in some fossil forms) have also been regarded by some malacologists as primitive "metameric" features.

The question is whether organ repetition in these molluscs represents vestiges of a true, or fundamental, metamerism in the phylum. If so, it represents remnants of an ancestral metameric *Bauplan* and may indicate a close relationship to other marine protostomes that display metamerism (i.e., the annelid–arthropod line). On the other hand, organ repetition in certain molluscan groups may be the result of independent convergent evolution and not a fundamental molluscan attribute at all. Absence of metamerism in the primitive groups Caudofoveata and Aplacophora argue against fundamental metamerism as a primitive feature of the molluscs. The genetic/evolutionary potential for serial repetition of organs is a common feature in many animals and occurs in at least four other groups: platyhelminths, nemerteans, the annelid–arthropod line, and chordates. Furthermore, the paired ventral nerve cords of molluscs are not segmental (i.e., do not have segmentally arranged ganglia). In fact, the ganglia of primitive molluscs are rather poorly developed.

We must still address the problem of the phylogenetic origin of the molluscs themselves. The excellent fossil record of molluscs extends back some 500 million years and indicates that the origin of this phylum lies hidden in Precambrian history, and we may never know with much certainty what the first mollusc was like. Numerous ideas on the origin of molluscs have been offered over the past 100 years, but they can be conveniently grouped into four basic theories. (For good reviews of these theories, see Vagvolgyi 1967 and Wingstrand 1985). The **turbellarian theory** views molluscs as having evolved from acoelomate turbellarian platyhelminths, independent of the annelids and other coelomate spiralian groups. This theory is based on a supposed homology and similarity in mode of locomotion between molluscs and flatworms by means of a "ventral mucociliary gliding surface"—a tenuous premise at best. The **modified turbellarian theory** holds, similarly, that the molluscs were derived, together with annelids, from a common stem with its root in the taxon Turbellaria, or in turbellarian ancestry. The **coelomate theory** views the molluscs as a sister-group of the annelids (or perhaps the annelid–arthropod line), the annelid–mollusc common ancestor being equipped with a coelom and some level of, or potential for, primitive metamerism. The **annelid theory** suggests that the molluscs evolved from the annelid line itself, after the development of true metamerism. This last theory requires considerable "de-evolution" of the metameric *Bauplan* and has little current support.

Molluscs are clearly allied with the spiralian protostomes, which are characterized mainly by certain ontogenetic features associated with development (e.g., spiral cleavage, 4d mesentoblast, and trochophore-like larvae). The large pericardial spaces present in most primitive molluscs (e.g., Aplacophora, Monoplacophora, Polyplacophora) also point to a coelomate rather than an acoelomate (turbellarian) ancestry. The question is, Where along the spiralian line did the origin of the Mollusca take place? This problem is still far from being solved, and about the best we can say at this time is that the molluscs probably arose early in the protostome clade, soon after the origin of the coelom but *before* the origin of annelid–arthropod metamerism (see Chapter 24). In spite of the many similarities among protostome groups, there appear to be no synapomorphies clearly linking molluscs as a sister-group to any other single spiralian phylum.

Selected References

The field of malacology is so large and has had such a long history that dealing with the literature is an overwhelming task. Many molluscs are of commercial importance (e.g., *Mytilus*, *Loligo*) and for these groups hundreds of studies appear each year; others are important laboratory/experimental organisms (e.g., *Loligo*, *Octopus*, *Aplysia*) and many papers are also published annually on these groups. New taxonomic monographs on various groups or geographical regions also appear each year, as do countless shell guides and coffee-table picture books. Distilling all of this into a small set of relevant references useful for entry into the original literature is difficult; the list below is our attempt to do so.

General References

Abbott, R. T. 1974. *American Seashells*, 2nd Ed. Van Nostrand Reinhold, New York. [An identification guide.]

Andrews, J. 1971. *Sea Shells of the Texas Coast*. University of Texas Press, Austin. [A good identification guide.]

Boss, K. J. 1971. Critical estimate of the number of recent Mollusca. Occ. Pap. Mollusks Mus. Comp. Zool. Harv. Univ. 3(40):81–135.

Boss, K. J. 1982. Mollusca. In S. P. Parker (ed.), *Synopsis and Classification of Living Organisms*, Vol. 1. McGraw-Hill, New York, pp. 945–1166.

Browne, R. A. and W. D. Russell-Hunter. 1978. Reproductive effort in molluscs. Oecologia 37:23–28.

Cheng, T. C. 1967. Marine mollusks as hosts for symbioses. Adv. Mar. Biol. 5:1–424. [An exhaustive summary.]

Clench, W. J. 1959. Mollusca. In W. T. Edmondson, H. B. Ward and G. C. Whipple (eds.), *Freshwater Biology*, 2nd Ed. Wiley, New York, pp. 1117–1160. [An identification guide.]

Florkin, M. and B. T. Scheer (eds.). 1972. *Chemical Zoology*, Vol. 7, Mollusca. Academic Press, New York. [Includes a good overview of the basic molluscan *Bauplan*.]

Fretter, V. (ed) 1968. *Studies in the Structure, Physiology and Ecology of Mollusks*. Academic Press, London.

Fretter, V. and A. Graham. 1976. *A Functional Anatomy of Invertebrates*. Academic Press, New York. [With an excellent discussion of gastropods and bivalves.]

Galstoff, P. S. 1961. Physiology of reproduction in mollusks. Am. Zool. 1:273–289.

Giese, A. C. and J. S. Pearse (eds.). 1977. *Reproduction of Marine Invertebrates*, Vol. 4. Academic Press, New York. 369 pp. [Covers gastropods and cephalopods.]

Giese, A. C. and J. S. Pearse (eds.). 1979. *Reproduction of Marine Invertebrates*, Vol. 5, Molluscs: Pelecypods and Lesser Classes. Academic Press, New York.

Götting, K.-J. 1974. *Malakozoologie*. G. Fischer, Stuttgart.

Graham, A. 1955. Molluscan diets. Proc. Malacol. Soc. London 31:144–159.

Grassé, P. (ed.). 1968. *Traité de Zoologie*, Vol. 5 (pts. 2 and 3). Masson et Cie, Paris.

Hochachka, P. W. (ed). 1983. *The Mollusca*, Vol. 1, Metabolic Biochemistry and Molecular Biomechanics; Vol. 2, Environmental Biochemistry and Physiology. Academic Press, New York.

Hyman, L. H. 1967. *The Invertebrates*, Vol. 6, Mollusca I. Aplacophora, Polyplacophora, Monoplacophora, Gastropoda. The Coelomate Bilateria. McGraw-Hill, New York. [Still one of the best general surveys of molluscan anatomy in English.]

Jones, A. M. and J. M. Baxter. 1987. *Molluscs: Caudofoveata, Solengastres, Polyplacophora and Scaphopoda*. Synopses of the British Fauna (new ser.). Linnean Society, London.

Jørgensen, C. B. 1966. *Biology of Suspension Feeding*. Pergamon Press, New York. [Much of this classic treatise deals with molluscs.]

Keen, A. M. 1971. *Sea Shells of Tropical West America. Marine Mollusks from Baja California to Peru*, 2nd Ed. Stanford University Press, Stanford, California. [Identificaiton guide for tropical eastern Pacific molluscs.]

Keen, A. M. and E. Coan. 1974. *Marine Mollucan Genera of Western North America: An Illustrated Key*. 2nd Ed. Stanford University Press, Stanford, California. [Technical keys.]

Kniprath, E. 1981. Ontogeny of the molluscan shell field. Zool. Scr. 10:61–79.

Moore, R. C (ed.). 1957–71. *Treatise on Invertebrate Paleontology. Mollusca*, Parts 1–6 (Vols. I–N). University of Kansas Press and Geological Society of America, Lawrence, Kansas.

Morris, P. A. 1966. *A Field Guide to Shells of the Pacific Coast and Hawaii*. Houghton Mifflin Co., Boston.

Morton, J. E. 1979. *Molluscs*, 5th Ed. Hutchinson University Press, London. [Somewhat dated but still one of the best introductions to general molluscan biology.]

Morton, J. E. and C. M. Yonge. 1964. Classification and structure of the Mollusca. In K. M. Wilbur and C. M. Yonge (eds.), *Physiology of Mollusca*, Vol. 1. Academic Press, New York, pp. 1–58.

Pennak, R. W. 1978. *Freshwater Invertebrates of the United States*. Wiley, New York. [Good identification guide.]

Potts, W. T. W. 1967. Excretion in molluscs. Biol. Rev. 42:1–41.

Purchon, R. D. 1977. *The Biology of the Mollusca*, 2nd ed. Pergamon Press, New York.

Raven, C. P. 1958. *Morphogenesis: The Analysis of Molluscan Development*. Pergamon Press, New York. [Review of molluscan embryology.]

Jones, D. S. 1983. Sclerochronology: Reading the record of the molluscan shell. Am. Sci. 71:384–391.

Salvini-Plawan, L. von. 1980. Proposed classification of Mollusca. Malacologia 19:249–278.

Solem, A. *The Shell Makers: Introducing Mollusks*. Wiley, New York. [Emphasizes gastropods.]

Vaught, K. C. 1989. *A Classification of the Living Mollusca*. Edited by R. T. Abbott and K. J. Boss. American Malacologists, Melbourne, Florida.

Wilbur, K. M. (gen. ed.). 1983–present. *The Mollusca*. Vols. 1–12. Academic Press, New York. [A continuing series of volumes edited and authored by luminaries in the field of malacology; volumes 1–10 cover biochemistry, molecular biology, biomechanics, metabolism, physiology, ecology, development, and evolution.]

Wilbur, K. M. and C. M. Yonge (eds). 1964, 1966. *Physiology of Mollusca*, Vols. 1 and 2. Academic Press, New York.

Yonge, C. M. 1932. The crystalline style of the Mollusca. Sci. Prog. 26: 643–653.

Yonge, C. M. and T. E. Thompson. 1976. *Living Marine Molluscs*. Collins, London. [An excellent modern review.]

Aplacophora and Caudofoveata

Baba, K. 1951. General sketch of the development in a solenogastre, *Epimenia verrucosa* (Nierstr.). Misc. Rep. Res. Inst. Nat. Resour. Tokyo 19/21:38–46.

Salvini-Plawen, L. von. 1968. Uber Lebendbeobachtungen an Caudofoveata (Mollusca, Aculifera), nebst Bermerkungen zum System der Klasse. Sarsia 31: 105–126.

Salvini-Plawen, L. von. 1978. Antarktische und subantarktische Solenogastres. Zoologica 128:1–315.

Scheltema, A. H. 1978 Position of the class Aplacophora in the phylum Mollusca. Malacologia 17:99–109.

Scheltema, A. H. 1981. Comparative morphology of the radulae and the alimentary tracts in the Aplacophora. Malacologia 20: 361–383.

Monoplacophora

Clarke, A. H. and R. J. Menzies. 1959. *Neopilina (Vema) ewingi*, a second living species of the Paleozoic class Monoplacophora. Science 129:1026–1027.

Lemche, H. 1957. A new living deep-sea mollusc of the Cambro-Devonian class Monoplacophora. Nature 179:413–416. [Report of the first discovery of living monoplacophorans.]

Lemche, H. and K. G. Wingstrand. 1959. The anatomy of *Neopilina galatheae* Lemche, 1957. Galathea Rpt. 3:9–71.

Menzies, R. J. and W. Layton. 1962. A new species of monoplacophoran mollusc, *Neopilina (Neopilina) veleronis* from the slope of the Cedros Trench, Mexico. Ann. Mag. Nat. Hist. (13)5:401–406.

Warén, A. 1988. *Neopilina goesi*, a new Caribbean monoplacophoran mollusk dredged in 1869. Proc. Biol. Soc. Wash. 101:676–681.

Wingstrand, K. G. 1985. On the anatomy and relationships of recent Monoplacophora. Galathea Rpt. 16:7–94.

Polyplacophora

Burghardt, G. and L. Burghardt. 1969. *A Collector's Guide to West Coast Chitons*. Spec. Publ. No. 4, San Francisco Aquarium Society.

Boyle, P. R. 1977. The physiology and behavior of chitons. Annu. Rev. Oceanogr. Mar. Biol. 15:461–509.

Eernisse, D. J. 1986. The genus *Lepidochitona* Gray, 1821 (Mollusca: Polyplacophora) in the northeastern Pacific Ocean (Oregonian and Californian Provinces). Zool. Verh. Rijksmus. Nat. Hist. Leiden 228:1–52.

Ferreira, A. J. 1986. A revision of the genus *Acanthopleura* Guilding, 1829 (Mollusca: Polyplacophora). Veliger 28:221–279.

Fisher, v.-F. P. 1978. Photoreceptor cells in chiton esthetes. Spixiana 1: 209–213.

Kass, P. 1972. *Polyplacophora of the Caribbean Region*. Studies of the Fauna of Curacao and other Caribbean Islands, Vol. 41. Martinus Nijhoff, The Hague.

Kass, P. 1985, 1988. *Monograph of Living Chitons*, Vols. 1–3. E. J. Brill, Leiden. [A planned 10–volume treatise.]

Nesson, M. H. and H. A. Lowenstam. 1985. Biomineralization processes of the radula teeth of chitons. *In* J. L. Kirschvink et al. (eds.), *Magnetite Biomineralization and Magnetoreception in Organisms*. Plenum, New York, pp. 333–363.

Smith, A. G. 1966. The larval development of chitons (Amphineura). Proc. Calif. Acad. Sci. 32:433–346.

Gastropoda

Alkon, D. L. 1983. Learning in a marine snail. Sci. Am. 249:70–84.

Barnhart, M. C. 1986. Respiratory gas tensions and gas exchange in active and dormant land snails, *Otala lactea*. Physiol. Zool. 59:733–745.

Behrens, D. W. 1980. *Pacific Coast Nudibranchs. A Guide to the Opisthobranchs of the Northeastern Pacific*. Sea Challengers, Los Osos, California.

Brace, R. C. 1977. Anatomical changes in nervous and vascular systems during the transition from prosobranch to opisthobranch organization. Trans. Zool. Soc. London 34:1–26.

Branch, G. M. 1981. The biology of limpets: Physical factors, energy flow and ecological interactions. Annu. Rev. Oceanogr. Mar. Biol. 19:235–380.

Burch, J. B. 1962. *How to Know the Eastern Land Snails*. Wm. C. Brown Co., Dubuque, Iowa.

Cameron, R. A. D. and M. Redfern. 1976. *British Land Snails*. Academic Press, London.

Carriker, M. R. and D. Van Zandt. 1972. Predatory behavior of a shell-boring muricid gastropod. *In* H. Winn and B. Olla (eds.), *Behavior of Marine Animals*, Vol. 1. Plenum, New York.

Conklin, E. J. and R. N. Mariscal. 1977. Feeding behavior, ceras structure, and nematocyst storage in the aeolid nudibranch, *Spurilla neapolitana*. Bull. Mar. Sci. 27:658–667.

Connor, V. M. 1986. The use of mucous trails by intertidal limpets to enhance food resources. Biol. Bull. 171:548–564.

Cook, S. B. 1971. A study in homing behavior in the limpet *Siphonaria alternata*. Biol. Bull. 141:449–457.

Croll, R. P. 1983. Gastropod chemoreception. Biol. Rev. 58:293–319.

Fretter, V. and M. A. Graham. 1962. *British Prosobranch Molluscs. Their Functional Anatomy and Ecology*. Ray Society, London.

Fretter, V. and J. Peake (eds.). 1975, 1978. *Pulmonates*, Vol. 1, Functional Anatomy and Physiology; Vol. 2A, Systematics, Evolution and Ecology. Academic Press, New York.

Fursich, F. T. and D. Jablonski. 1984. Late Triassic naticid drill holes: Carnivorous gastropods gain a major adaptation but fail to radiate. Science 224:78–80.

Gilmer, R. W. and G. R. Harbison. 1986. Morphology and field behavior of pteropod molluscs: Feeding methods in the families Cavoliniidae, Limacinidae and Peraclididae (Gastropoda: Thecosomata). Mar. Biol. 91: 47–57.

Golikov, A. N. and Y. I. Starobogatov. 1975. Systematics of prosobranch gastropods. Malacologia 15:185–232.

Gosliner, T. 1987. *Nudibranchs of Southern Africa. A Guide to Opisthobranch Molluscs of Southern Africa*. E. J. Brill, Leiden.

Gould, S. J. 1985. The consequences of being different: Sinistral coiling in *Cerion*. Evolution 39:1364–1379.

Gould, S. J. 1987. Systematics and levels of covariation in *Cerion* from the Turks and Caicos. Bull. Mus. Comp. Zool. Harv. Univ. 151:321–363.

Graham, A. 1971. *British Prosobranch and Other Operculate Gastropod Molluscs*. Academic Press, London.

Greenwood, P. G. and R. N. Mariscal. 1984. Immature nematocyst incorporation by the aeolid nudibranch *Spurilla neapolitana*. Mar. Biol. 80: 35–38.

Haszprunar, G. 1985. The Heterobranchia: New concept of the phylogeny of the higher Gastropoda. Z. Zool. Syst. Evolutionsforsch. 23:15–37.

Haszprunar, G. 1985. The fine morphology of the osphradial sense organs of the Mollusca. I. Gastropoda, Prosobranchia. Philos. Trans. Roy. Soc. Lond. B 307:457–496.

Haszprunar, G. 1988. On the origin and evolution of major gastropod groups, with special reference to the Streptoneura. J. Moll. Stud. 54:367–441.

Haszprunar, G. 1989. Die Torsion der Gastropoda—ein biomechanischer Prozess. Z. Zool. Syst. Evolutionsforsch. 27: 1–7.

Hickman, C. S. 1983. Radular patterns, systematics, diversity and ecology of deep-sea limpets. Veliger 26:7–92.

Houston, R. S. 1976. The structure and function of neogastropod reproductive systems, with special reference to *Columbella fuscata* Sowerby, 1932. Veliger 19:27–46.

Kempf, S. C. 1984. Symbiosis between the zooxanthella *Symbiodinium* (= *Gymnodinium*) *microadriaticum* (Freudenthal) and four species of nudibranchs. Biol. Bull. 166:110–126.

Lohmann, K. J. and A. O. D. Willows. 1987. Lunar-modulated geomagnetic orientation by a marine mollusk. Science 235:331–334.

Marcus, E. and E. Marcus. 1967. American Opisthobranch Mollusks. Studies in Tropical Oceanography Series, No. 6. University of Miami Press, Coral Gables, Florida. [Although badly out of date due to the rapid pace at which new opisthobranchs are described, this monograph remains a benchmark work for the American fauna.]

McDonald, G. R. and J. W. Nybakken. 1980. *Guide to the Nudibranchs of California*. American Malacologists, Melbourne, Florida.

Miller, S. L. 1974. Adaptive design of locomotion and foot form in prosobranch gastropods. J. Exp. Mar. Biol. Ecol. 14:99–156.

Miller, S. L. 1974. The classification, taxonomic distribution, and evolution of locomotor types among prosobranch gastropods. Proc. Malacol. Soc. London 41:233–272.

Morton, B. 1978. Feeding and digestion in ship worms. Annu. Rev. Oceanogr. Mar. Biol. 16:107–144.

Norton, S. F. 1988. Role of the gastropod shell and operculum in inhibiting predation by fishes. Science 241:92–94.

Olivera, B. M. et al. 1985. Peptide neurotoxins from fish-hunting cone snails. Science 230:1338–1343.

O'Sullivan, J. B., R. R. McConnaughey and M. E. Huber. 1987. A blood-sucking snail: The cooper's nutmeg, *Cancellaria cooperi* Gabb, parasitizes the California electric ray, *Torpedo californica* Ayres. Biol. Bull. 172: 362–366.

Palmer, A. R. 1977. Function of shell sculpture in marine gastropods: Hydrodynamic destabilization in *Ceratostoma foliatum*. Science 197:1293–1295.

Perry, D. M. 1985. Function of the shell spine in the predaceous rocky intertidal snail *Acanthina spirata* (Prosobranchia:Muricacea). Mar. Biol. 88:51–58.

Ponder, W. F. 1973. The origin and evolution of the Neogastropoda. Malacologia 12:295–338.

Ponder, W. F. (ed.) 1988. Prosobranch phylogeny. Malacol. Rev., Supp. 4. [Contributed papers from a symposium.]

Potts, G. W. 1981. The anatomy of respiratory structures in the dorid nudibranchs, *Onchidoris bilamellata* and *Archidoris pseudoargus*, with details of the epidermal glands. J. Mar. Biol. Assoc. U.K. 61:959–982.

Runham, N. W. and P. J. Hunter. 1970. *Terrestrial Slugs*. Hutchinson University Library, London. [A lovely review of slug biology.]

Salvani-Plawan, L. von and G. Haszprunar. 1986. The Vetigastropoda and the systematics of streptoneurous Gastropoda. J. Zool. 211:747–770.

Seapy, R. and R. E. Young. 1986. Concealment in epipelagic pterotracheid heteropods (Gastropoda) and cranchiid squids (Cephalopoda). J. Zool. 210:137–147.

Sleeper, H. L., V. J. Paul and W. Fenical. 1980. Alarm pheromones from the marine opisthobranch *Navanax inermis*. J. Chem. Ecol. 6:57–70.

Stanley, S. M. 1982. Gastropod torsion: Predation and the opercular imperative. Neues. Jahrb. Geol. Palaeontol. Abh. 164: 95–107. [Succinct review of ideas on why and how torsion evolved.]

Taylor, J. D., N. J. Morris and C. N. Taylor. 1980. Food specialization and the evolution of predatory prosobranch gastropods. Paleontology 23:375–410.

Thiriot-Quievreux, C. 1973. Heteropoda. Annu. Rev. Oceanogr. Mar. Biol. 11: 237–261.

Thompson, T. E. 1976. *Biology of Opisthobranch Molluscs*, Vol. 1. Ray Society, London.

Thompson, T. E. and G. H. Brown. 1984. *Biology of Opisthobranch Molluscs*, Vol. 2. Ray Society, London.

Thompson, T. E. and G. H. Brown. 1976. *British Opisthobranch Molluscs*. Academic Press, London.

Bivalvia

Ansell, A. D. and N. B. Nair. 1969. A comparison of bivalve boring mechanisms by mechanical means. Am. Zool. 9:857–868.

Bayne, B. L. (ed.) 1976. *Marine Mussels: Their Ecology and Physiology*. Cambridge University Press, Cambridge.

Boulding, E. G. 1984. Crab-resistant features of shells of burrowing bivalves: Decreasing vulnerability by increasing handling time. J. Exp. Mar. Biol. Ecol. 76:201–223.

Carpenter, E. J. and J. L. Culliney. 1975. Nitrogen fixation in marine shipworms. Science 187:551–552.

Childress J. J. et al. 1986. A methanotrophic marine molluscan (Bivalvia, Mytilidae) symbiosis: Mussels fueled by gas. Science 233:1306–1308.

Deaton, L. E. 1981. Ion regulation in freshwater and brackish water bivalve mollusks. Physiol. Zool. 54:109–121.

Ellis, A. E. 1978. *British Freshwater Bivalve Mollusks*. Academic Press, London.

Foster-Smith, R. L. 1978. The function of the pallial organs of bivalves in controlling ingestion. J. Moll. Stud. 44:83–99.

Goreau, T. F., N. I. Goreau and C. M. Yonge. 1973. On the utilization of photosynthetic products from zooxanthellae and of a dissolved amino acid in *Tridacna maxima*. J. Zool. 169:417–454.

Jørgensen, C. B. 1974. On gill function in the mussel *Mytilus edulis*. Ophelia 13:187–232.

Judd, W. 1979. The secretions and fine structure of bivalve crystalline style sacs. Ophelia 18:205–234.

Kennedy, W. J., J. D. Taylor and A. Hall. 1969. Environmental and biological controls on bivalve shell mineralogy. Biol. Rev. 44:499–530.

Kristensen, J. H. 1972. Structure and function of crystalline styles in bivalves. Ophelia 10:91–108.

Manahan, D. T., S. H. Wright, G. C. Stephens and M. A. Rice. 1982. Transport of dissolved amino acids by the mussel, *Mytilus edulis*: Demonstration of net uptake from natural seawater. Science 215:1253–1255.

Marincovich, L., Jr. 1975. Morphology and mode of life of the Late Cretaceous rudist, *Coralliochama orcutti* White (Mollusca: Bivalvia). J. Paleontology 49(1):212–223. [Describes the famous Point Banda deposits of Baja California, Mexico]

Morse, M. P., E. Meyhöfer, J. J. Otto and A. M. Kuzirian. 1986. Hemocyanin respiratory pigments in bivalve mollusks. Science 231:1302–1304.

Morton, B. 1978. The diurnal rhythm and the processes of feeding and digestion in *Tridacna crocea*. J. Zool. 185:371–387.

Morton, B. 1978. Feeding and digestion in shipworms. Annu. Rev. Oceanogr. Mar. Biol. 16:107–144.

Owen, G. 1974. Feeding and digestion in the Bivalvia. Adv. Comp. Physiol. Biochem. 5:1–35.

Pojeta, J., Jr. and B. Runnegar. 1974. *Fordilla troyensis* and the early history of pelecypod mollusks. Am. Sci. 62:706–711.

Powell, M. A. and G. N. Somero. 1986. Hydrogen sulfide oxidation is coupled to oxidative phosphorylation in mitochondria of *Solemya reidi*. Science 233: 563–566.

Ray, D. L. (ed) 1959. *Marine Boring and Fouling Organisms*. University of Washington Press, Seattle. [Numerous contributions on shipworms.]

Reid, R. G. B. and F. R. Bernard. 1980. Gutless bivalves. Science 208: 609–610.

Reid, R. G. B. and A. M. Reid. 1974. The carnivorous habit of members of the septibranch genus *Cuspidaria* (Mollusca: Bivalvia). Sarsia 56: 47–56.

Stanley, S. M. 1968. Post-Paleozoic adaptive radiation of infaunal bivalve molluscs —a consequence of mantle fusion and siphon formation. J. Paleontol. 42: 214–229.

Stanley, S. M. 1970. Relation of shell form to life habits of the Bivalvia (Mollusca). Geol. Soc. Am. Mem. 125: 1–296.

Stanley, S. M. 1975. Why clams have the shape they have: An experimental analysis of burrowing. Paleobiology 1: 48.

Taylor, J. D. 1973. The structural evolution of the bivalve shell. Paleontology 16: 519–534.

Trueman, E. R. 1966. Bivalve mollusks: Fluid dynamics of burrowing. Science 152: 523–525.

Vetter, R. D. 1985. Elemental sulfur in the gills of three species of clams containing chemoautotrophic symbiotic bacteria: A possible inorganic energy storage compound. Mar. Biol. 88: 33–42.

Vogel, K. and W. F. Gutmann. 1980. The derivation of pelecypods: Role of biomechanics, physiology, and environment. Lethaia 13: 269–275.

Waterbury, J. B., C. B. Calloway and R. D. Turner. 1983. A cellulolytic nitrogen-fixing bacterium cultured from the Gland of Deshayes in shipworms (Bivalvia: Teredinidae). Science 221: 1401–1403.

Wilkens, L. A. 1986. The visual system of the giant clam *Tridacna*: Behavioral adaptations. Biol. Bull. 170: 393–408.

Yonge, C. M. 1953. The monomyarian condition in the Lamellibranchia. Trans. R. Soc. Edinburgh 62 (p. II): 443–478.

Yonge, C. M. 1973. Giant clams. Sci. Am. 232: 96–105.

Scaphopoda

Bilyard, G. R. 1974. The feeding habits and ecology of *Dentalium stimpsoni*. Veliger 17: 126–138.

Gainey, L. F. 1972. The use of the foot and captacula in the feeding of *Dentalium*. Veliger 15: 29–34.

Trueman, E. R. 1968. The burrowing process of *Dentalium*. J. Zool. 154: 19–27.

Cephalopoda

Barber, V. C. and F. Grazialdei. 1967. The fine structure of cephalopod blood vessels. Z. Zellforsch. Mikrosk. Anat. 77: 162–174. [Also see earlier papers by these authors in the same journal.]

Boycott, B. B. 1965. Learning in the octopus. Sci. Am. 212: 42–50.

Boyle, P. R. (ed.). 1983. *Cephalopod Life Cycles*, Vols. 1–2. Academic Press, New York.

Clarke, M. A. 1966. A review of the systematics and ecology of oceanic squids. Adv. Mar. Biol. 4: 91–300.

Cloney, R. A. and S. L. Brocco. 1983. Chromatophore organs, reflector cells, iridocytes and leucophores in cephalopods. Am. Zool. 23: 581–592.

Denton, E. J. and J. B. Gilpin-Brown. 1973. Flotation mechanisms in modern and fossil cephalopods. Adv. Mar. Biol. 11: 197–264.

Donovan, D. T. 1964. Cephalopod phylogeny and classification. Biol. Rev. 39: 259–287.

Fields, W. G. 1965. The structure, development, food relations, reproduction, and life history of the squid *Loligo opalescens* Berry. Calif. Dept. Fish Game Bull. 131: 1–108.

Hochberg, F. G. 1983. The parasites of cephalopods: A review. Mem. Nat. Mus. Victoria Melbourne 44: 109–145.

House, M. R. and J. R. Senior. 1981. *The Ammonoidea: The Evolution, Classification, Mode of Life and Geological Usefulness of a Major Fossil Group*. Academic Press, New York.

Lane, F. W. 1960. *Kingdom of the Octopus*. Sheridan House, New York. [A somewhat dated, but delightful and informative introduction to cephalopods.]

Lehmann, U. 1981. *The Ammonites: Their Life and Their World*. Cambridge University Press, New York.

Mutvei, H. 1964. On the shells of *Nautilus* and *Spirula* with notes on the shell secretion in non-cephalopod molluscs. Ark. Zool. 16(14): 223–278.

Nixon, M. and J. B. Messenger (eds.). 1977. *The Biology of Cephalopods*. Academic Press, New York.

Packard, A. 1972. Cephalopods and fish: The limits of convergence. Biol. Rev. 47: 241–307.

Roper, C. F. E. and K. J. Boss. 1982. The giant squid. Sci. Am. 246: 96–104.

Roper, C. F., R. E. Young and G. L. Voss. 1969. An illustrated key to the families of the order Teuthoidea (Cephalopoda). Smithson. Contrib. Zool. 13: 1–32.

Saunders, W. B. 1983. Natural rates of growth and longevity of *Nautilus belauensis*. Paleobiology 9: 280–288.

Sheumack, D. D. et al. 1978. Maculotoxin: A neurotoxin from the venom glands of the octopus *Hapalochlaena maculosa* identified as tetrodotoxin. Science 199: 188–189.

Voss, G., L. Opresko and R. Thomas. 1973. The potentially commercial species of octopus and squid of Florida, the Gulf of Mexico and the Caribbean Sea. University of Miami Sea Grant Program Field Guide Ser. 2: 1–33. [Key, descriptions, and photographs of common cephalopods of the region.]

Ward, R., L. Greenwald, and O. E. Greenwald. 1980. The buoyancy of the chambered nautilus. Sci. Am. 243: 190–204.

Wells, M. J. 1963. Heirarchic control of movement in the octopus. Ergeb. Biol. 26: 40–53.

Wells, M. J. 1978. *Octopus: Physiology and Behaviour of an Advanced Invertebrate*. Chapman and Hall.

Winkler, L. R. and L. M. Ashley. 1954. The anatomy of the common octopus of northern Washington. Walla Walla College Publ. Biol. 10: 1–30.

Yarnall, J. L. 1969. Aspects of the behavior of *Octopus cyanea* Gray. Anim. Behav. 17: 747–754.

Young, A. E. and C. F. Roper. 1976. Bioluminescent countershading in midwater animals: Evidence from living squid. Science 191: 1046–1048.

Young, J. Z. 1972. *The Anatomy of the Nervous System of Octopus vulgaris*. Oxford University Press, New York.

Young, R. E. and F. M. Mencher. 1980. Bioluminescence in mesopelagic squid: Diel color change during counterillumination. Science 208: 1286–1288.

Molluscan Evolution and Phylogeny

Batten, R. L., H. B. Rollins and S. J. Gould. 1967. Comments on "The adaptive significance of gastropod torsion." Evolution 21: 405–406.

Beedham, G. and E. Trueman. 1968. The cuticle of the Aplacophora and the evolutionary significance in the Mollusca. Proc. Zool. Soc. London 154: 443–451.

Eldredge, N. and S. M. Stanley (eds.). 1984. *Living Fossils*. Springer-Verlag, New York. [Includes chapters on Monoplacophora, *Pleurotomaria*, and *Nautilus*.]

Garstang, W. 1928. *Origin and Evolution of Larval Forms*. Rpt. British Assoc. Sci, Sec. D.

Garstang, W. [Introduction by Sir A. Hardy] 1951. *Larval Forms and Other Zoological Verses*. Basii Blackwell, Oxford. [Reprinted in 1985 by the University of Chicago Press.]

Ghiselin, M. T. 1966. The adaptive significance of gastropod torsion. Evolution 20: 337–348.

Ghiselin, M. T. 1988. The origin of molluscs in the light of molecular evidence. Oxford Surv. Evol. Biol. 5:66–95.

Götting, K. 1980. Arguments concerning the descendence of Mollusca from metameric ancestors. Zool. Jahrb. Abt. Anat. 103:211–218.

Götting, K. 1980. Origin and relationships of the Mollusca. Z. Zool. Syst. Evolutionsforsch. 18:24–27.

Graham, A. 1979. Gastropoda. *In* M. R. House (ed.), *The Origin of Major Invertebrate Groups*. Academic Press, New York, pp. 359–365.

Gutmann, W. F. 1974. Die Evolution der Mollusken-Konstruction: Ein phylogenetisches Modell. Aufsätze Red. Senckenb. Naturf. Ges. 25:1–24.

Haas, W. 1981. Evolution of calcareous hardparts in primitive molluscs. Malacologia 21:403–418.

Holland, C. H. 1979. Early Cephalopoda. *In* M. R. House (ed.), *The Origin of Major Invertebrate Groups*. Academic Press, New York, pp. 367–379.

Jägersten, G. 1959. Further remarks on the early phylogeny of the Metazoa. Zool. Bidr. Uppsala 30:321–354.

Knight, J. B. and E. L. Yochelson. 1958. A reconstruction of the relationships of the Monoplacophora and the primitive Gastropoda. Proc. Malacol. Soc. London 33:37–48.

Lauterbach, K.-E. von. 1983. Erörterungen zur Stammesgeschichte der Mollusca, insbesondere der Conchifera. Z. Zool. Syst. Evolutionsforsch. 21: 201–216.

Lindbergh, D. R. 1986. Name changes in the "Acmaeidae." Veliger 29(2): 142–148.

Linsley, R. M. 1978. Shell form and the evolution of gastropods. Am. Sci. 66:432–441.

Morton, J. E. 1963. The molluscan pattern: Evolutionary trends in a modern classification. Proc. Linn. Soc. London 174:53–72.

Runnegar, B. and J. Pojeta. 1974. Molluscan phylogeny: The paleontological viewpoint. Science 186:311–317.

Ruppert, E. E. and J. Carle. 1983. Morphology of metazoan circulatory systems. Zoomorphologie 103:193–208.

Salvini-Plawen, L. von. 1972. Zur Morphologie und Phylogenie der Mollusca: Die Beziehungen der Caudofoveata und der Solenogastres als Aculifera, als Mollusca und als Spiralia. Z. Wiss. Zool. Abt. A 184:205–394.

Salvini-Plawen, L. von. 1977. On the evolution of photoreceptors and eyes. Evol. Biol. 10:207–263.

Salvini-Plawen, L. von. 1980. A reconsideration of systematics in the Mollusca. Phylogeny and higher classification. Malacologia 19:249–278.

Salvini-Plawen, L. von. 1981. On the origin and evolution of the Mollusca. Atti Accad. Naz. Lincei, Atti. Conv. Lincei 49:235–293.

Scheltema, A. H. 1978. Position of the class Aplacophora in the phylum Mollusca. Malacologia 17:99–109.

Scheltema, A. H. 1988. Ancestors and descendents: Relationships of the Aplacophora and Polyplacophora. Am. Malacol. Bull. 6: 57–68.

Vagvolgyi, J. 1967. On the origin of molluscs, the coelom, and coelomic segmentation. Syst. Zool. 16:153–168.

Wingstrand, K. G. 1985. On the anatomy and relationships of recent Monoplacophora. Galathea Rpt. 16:7–94.

Yochelson, E. L. 1978. An alternative approach to the interpretaton of the phylogeny of ancient mollusks. Malacologia 17: 165–191.

Yochelson, E. L. 1979. Early radiation of Mollusca and mollusc-like groups. *In* M. R. House (ed.), *The Origin of the Major Invertebrate Groups*. Academic Press, New York, pp. 323–358.

Yonge, C. M. 1957. *Neopilina*: Survival from the Paleozoic. Discovery (London), June 1957:255–256.

Yonge, C. M. 1957. Reflections on the monoplacophoran *Neopilina galatheae* Lemche. Nature 179:672–673.

Chapter Twenty-One

The Lophophorate Phyla: Phoronids, Ectoprocts, and Brachiopods

Anyone who starts looking at bryozoans will continue to do so, for their biology is full of interest and unsolved mysteries.

J. S. Ryland
Bryozoans, 1970

Introduction to the deuterostomes

Most of the rest of this book is devoted to seven phyla composing the group known as the deuterostomes: Echinodermata, Chaetognatha, Hemichordata, Chordata, and the three lophophorate phyla—Phoronida, Ectoprocta, and Brachiopoda. We regard the Deuterostomia to be a natural group, an evolutionary clade, and we elucidate this view in chapters 4 and 24. Some of the characteristics of the deuterostomes are primitive features and are shared by other taxa (such as radial cleavage in the cnidarians). However, the deuterostome clade is clearly separated from all other invertebrate phyla by the unique synapomorphies of enterocoelic coelom development, archenteric derived mesoderm, sheets of subepidermal muscles derived (at least in part) from archenteric mesoderm, and a tripartite *Bauplan*. Although the variations on this theme have not led to the extreme diversity of species seen among the protostomes, they have resulted in several fundamentally different and distinct groups of animals. In fact, there are greater differences among the major taxa of deuterostomes than among the major taxa of protostomes. Thus, at least at higher taxonomic levels, the deuterostome *Bauplan* has proved evolutionarily flexible and successful in a wide range of lifestyles.

The lophophorates: An overview

In his address introducing a symposium on the biology of lophophorates, Joel W. Hedgpeth referred to these creatures as an "aggregation of animals that possess a feeding structure known as a lophophore" (*American Zoologist* [1977] 17:3–4). Although the term *aggregation* is certainly not a valid taxonomic category, it is an appropriate description in this case and reflects the great disagreements about the relationships of the lophophorates to each other and to other phyla.

Traditionally included as lophophorates (Greek, "crest-bearers") are the taxa Phoronida, Brachiopoda, and Ectoprocta or Bryozoa. At first glance, the members of these groups may seem to have little in common (Figures 1, 5, 6, and 17). However, they do display a number of important similarities. They are allied with the deuterostomes and are built along a trimeric (= tripartite) *Bauplan* wherein the body is divided into an anterior **prosome**, a middle **mesosome**, and a posterior **metasome**. In most cases, at least developmentally, each of these regions contains a separate, often paired, coelomic compartment: the **protocoel**, **mesocoel**, and **metacoel**, respectively (see Chapter 4). Furthermore, the lophophorates all have a U-shaped gut and very simple, often transient reproductive systems. Nearly all secrete outer casings in the form of tubes, shells, or compartmented exoskeletons. Even though these features represent shared similarities among the lophophorates, none is unique to them. Their most significant common feature is the lophophore itself, which, simply put, is a ciliated, tentacular outgrowth containing coelomic extensions. It surrounds the mouth but not the anus.

A similar description might also apply to the condition in some other groups, such as sipunculans (Chapter 14). However, the tentacular crown of lophophorates is distinctive in that it arises specifically from the mesosome and its lumen is derived from the mesocoel. Sipunculans, of course, are not trimeric and thus have no mesosome–mesocoel homologues. Furthermore, sipunculan development is clearly protostomous.

The distinction between the lophophore and the tentaculate structures of pterobranchs (see Chapter 23) is less obvious. These odd little creatures possess mesosomal tentacles with a coelomic lumen, and some authors suggest a relationship to the lophophorates. The homology between pterobranch tentacles and a "true" lophophore is questionable, because in the former case the tentacles do not fully surround the mouth. Thus, we tentatively view the lophophore as a unique synapomorphy of members of the phyla Phoronida, Ectoprocta, and Brachiopoda and treat the three groups as a monophyletic clade. To be sure, there are some problems with this hypothesis, some of which are discussed later.

With the exception of a few freshwater ectoprocts, the lophophorates are exclusively marine. All are benthic, living either in tubes (phoronids) or in secreted shells, or casings. The phoronids form a small group, comprising only two genera and about 15 species of solitary or gregarious worms. Ectoprocts, on the other hand, make up a diverse taxon of about 4,500 species of colonial forms. The brachiopods, or lamp shells, include about 335 extant species. However, the brachiopods have left a record of over 12,000 fossil species as evidence of a greater past. They were well established by the early Cambrian, over 600 million years ago. These animals flourished in both abundance and diversity from the Ordovician through the Carboniferous, but they have declined in numbers and kinds ever since.

Taxonomic history

The earliest records of any lophophorate are of various ectoprocts reported in the sixteenth century. With few exceptions, the early zoologists treated them as plantlike and included them in the taxon Zoophyta, a misconception that persisted into the 1700s. Peyssonal (1729) finally established the animal nature of ectoprocts, and Jussieu (1742) noted the compartmentalized condition of the colonies and coined the term *polyps* to refer to the individual animals. Still, most well-known workers of the day (e.g.,

Linnaeus and Cuvier) insisted on allying them with the cnidarians in the group Zoophyta.

Eventually, de Blainville (1820) noted the complete gut of the ectoprocts and "raised" them above the cnidarians. By 1830, two names had been coined for these animals—Bryozoa (by the German zoologist Ehrenberg) and Polyzoa (by the Englishman Thompson). Just about the time the ectoprocts (under one or the other name) were being recognized as a separate group, concurrent events confused the issue further. The entoprocts were being described, and most workers tossed them in with the ectoprocts (as Bryozoa), while other workers were recognizing a relationship between ectoprocts and other lophophorates. All of this became terribly entangled in Milne-Edwards's (1843) concept of a taxon Molluscoides, which he established to include the ectoprocts and compound ascidians (see Chapter 23).

The brachiopods were known, at least from fossils, in the 1600s and were allied with the molluscs. This mistaken view was held until late in the nineteenth century.

Phoronids were first described from larvae by Müller (1846, 1847), who thought they were adults and named them *Actinotrocha brachiata*. But not long after, Gegenbaur (1854) recognized these animals as larval stages. The adults were found and described by Wright (1856), who named them *Phoronis*. Finally, the renowned embryologist Kowalevsky (1867) studied the metamorphosis of "*Actinotrocha*" and established the relationship between the two stages. The name "actinotroch" persists as a general term for the phoronid larva.

In 1857 Hancock recognized the relationship between brachiopods and bryozoans, but the whole matter was mixed up with confusion over the entoprocts and Milne-Edwards's Molluscoides. Nitsche (1869) made a valiant attempt to separate the entoprocts and ectoprocts, and in 1882 Caldwell first put forth the idea of a relationship between the brachiopods, phoronids, and bryozoans. This view was supported by Hatschek (1888), who suggested the establishment of a phylum Tentaculata to include the classes Phoronida, Brachiopoda, and Ectoprocta (but excluding the entoprocts). Since that time several attempts to unite some or all of these animals into a single taxon have been made, including Schneider's (1902) Lophophorata (for phoronids and bryozoans). Hyman (1959) rejected these ideas and retained separate phylum status for the three groups, an arrangement that remained popular for years.

Recently, however, some of the earlier views have reemerged with new evidence and new vigor.

On one front, Nielsen (1971, 1977) has revived the idea of an entoproct–ectoproct affinity and insists the two groups are closely related. Although this hypothesis has found its way into several recent texts, we find it untenable in light of the evidence. In support of lophophorate unity, some recent authors suggest that a single phylum (Emig 1982) or "superphylum" (Valentine 1973, 1975) should be formally established for the three groups. Classifications are given separately for each phylum.

The lophophorate *Bauplan*

In spite of the obvious diversity in external form among the lophophorates, there is a fundamental unity here. The diversity, great or small, in any monophyletic lineage is a reflection of evolutionary variations on a theme—the theme being the *Bauplan*. We have explored this idea in the protostomes, both in general and within each phylum. Among the deuterostomes, the general theme consists of those features of ontogeny that define the lineage and the resulting trimeric body plan with a tripartite coelomic system. Impose upon this plan the lophophore, and the result is a unique clade derived from the main deuterostome lineage. Couple this single unique synapomorphy with a number of other shared features, and the lophophorate *Bauplan* begins to emerge as a set of character states that are more easily explained as arising from common ancestry than as resulting from convergence.

All lophophorates are built for benthic life and suspension feeding, the latter being a primary function of the lophophore. The anterior body region, or prosome, is reduced to a small, flaplike **epistome** (or lost in some) associated with an overall reduction of the head, the elaboration of the mesosome as the lophophore, and the inactive lifestyle of these sessile organisms. As in most deuterostomes, the metasome houses the bulk of the viscera. The U-shaped gut is clearly advantageous for living encased in tubes, compartments, and shells; such animals do not "foul their nests," as it were. We have seen similar adaptations in some other groups with comparable habits, such as the recurved gut of sipunculans, and the ciliated fecal-removal grooves of some tube-dwelling polychaetes. In lophophorates, this condition not only prevents the unacceptable problem of fecal accumulation in the encasement but generally brings the anus close to rejection currents produced by the cilia on the lophophore.

The phoronids most clearly display the above traits, and retain the vermiform shape of the probable ancestral form (Box 1). Evolutionarily, the ectoprocts have exploited asexual reproduction, colonialism, and small size (Box 2). Thus relieved of long-distance internal transport problems, the ectoprocts have lost the circulatory and excretory systems. The brachiopods evolved a pair of valves or shells that encase and protect the body, including the lophophore (Box 3). Thus, instead of exposing the lophophore in the water, as phoronids and ectoprocts do, the brachiopods draw water into their **mantle cavity** for suspension feeding, an action analogous to that of bivalved molluscs. Sessile animals with soft parts (e.g., the lophophore), living on benthic surfaces, are exposed to potentially high levels of predation—thus the selective advantage of the brachiopod shell is clear. The phoronids are motile to the extent that their bodies can be retracted within the tubes, protecting the soft parts. Ectoprocts are entirely anchored in their casings, but the lophophore itself is retractable into and protrusible from the body, as the result of a unique arrangement of muscles and hydraulic mechanisms.

The phoronids

Phoronids are all tube dwellers; their chitinous tubes are usually either cemented (often in clusters) to hard substrata, or buried vertically in soft sediments (Box 1 and Figure 1). They are assigned to two

---

Box One

Characteristics of the Phylum Phoronida

1. Trimeric, modified enterocoelic, vermiform lophophorates
2. Body divided into flaplike epistome (prosome), lophophore-bearing mesosome, and elongate trunk (metasome), each with associated coelom
3. Gut U-shaped, anus close to mouth
4. One pair metanephridia in metasome
5. Closed circulatory system
6. Gonochoristic or hermaphroditic, with mixed or indirect life histories
7. Marine benthic tube dwellers

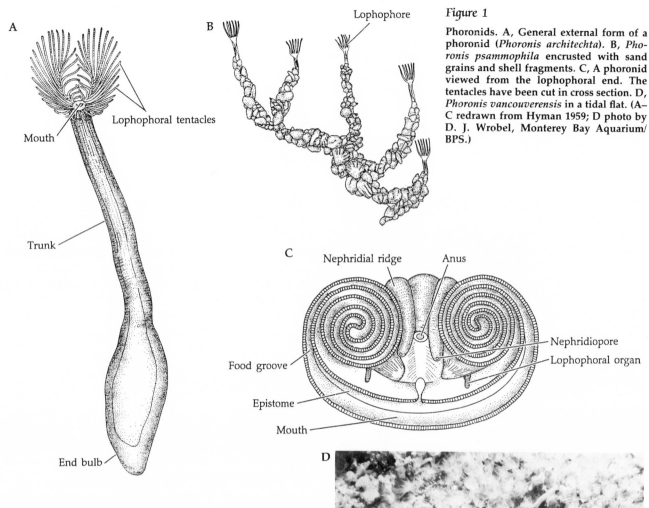

Figure 1

Phoronids. A, General external form of a phoronid (*Phoronis architechta*). B, *Phoronis psammophila* encrusted with sand grains and shell fragments. C, A phoronid viewed from the lophophoral end. The tentacles have been cut in cross section. D, *Phoronis vancouverensis* in a tidal flat. (A–C redrawn from Hyman 1959; D photo by D. J. Wrobel, Monterey Bay Aquarium/BPS.)

genera: *Phoronis* and *Phoronopsis*. The phylum name was apparently derived from the Latin term *Phoronis*, the surname of Io (who, according to mythology, was changed into a cow and roamed the Earth, eventually to be returned to her former body). Remember, these worms were first described from larval stages drifting in the sea, and only much later were the adults recognized as part of the same life cycle. Phoronids are known from intertidal mud flats to depths of about 400 meters. They range from about 5 to 25 cm in length. The vermiform body shows little regional specialization, except for the distinct lophophore and a modest inflation of the **end bulb**, which houses the stomach and aids in anchoring the animals in their tubes. The slitlike mouth is located between the tentacle-bearing lophophoral ridges and is overlaid by the flaplike epistome. The lateral aspects of the ridges are distinctly coiled and flank the dorsal anus and paired nephridiopores (Figure 1). Using the terms *anterior* and *posterior* when referring to phoronids can

be misleading. During metamorphosis the true anterior (mouth-bearing) and posterior (anus-bearing) ends are brought very close together by rapid enlargement of the ventral surface (see Figure 4). The dorsal surface is reduced to only the area between the adult mouth and anus. Because of these conditions, we refer to the "ends" of the adult worm as the lophophoral end and the stomachic end.

The phoronid *Bauplan*

Body wall, body cavity, and support

The phoronid body wall includes an epidermis of columnar cells overlaid by a very thin cuticle. Within the epidermis are sensory neurons and various gland cells, the latter responsible for the production of mucus and chitin (Figure 2A). The epidermis of the lophophore is densely ciliated. Internal to the epidermis is a well developed basement membrane of varying thickness, followed by a rather thin layer of circular muscle and a thicker layer of longitudinal muscle. A thin peritoneum lines the longitudinal muscles, forming the outer boundary of the coelomic spaces.

The coelom is clearly tripartite. The protocoel is reduced in all lophophorates, and in phoronids it is limited to a single small cavity within the epistome. The unpaired mesocoel comprises a coelomic ring in the lophophoral collar and extensions into each tentacle (Figure 2B). The protocoel and mesocoel are

Figure 2

Phoronid internal anatomy. A, The trunk of a phoronid (cross section). Note the body wall layers and coelomic partitioning. B, A tentacle of the lophophore (cross section). C, The digestive tract. D, The circulatory system. E, The major internal organs in the lophophoral and stomachic ends. (Redrawn from Hyman 1959.)

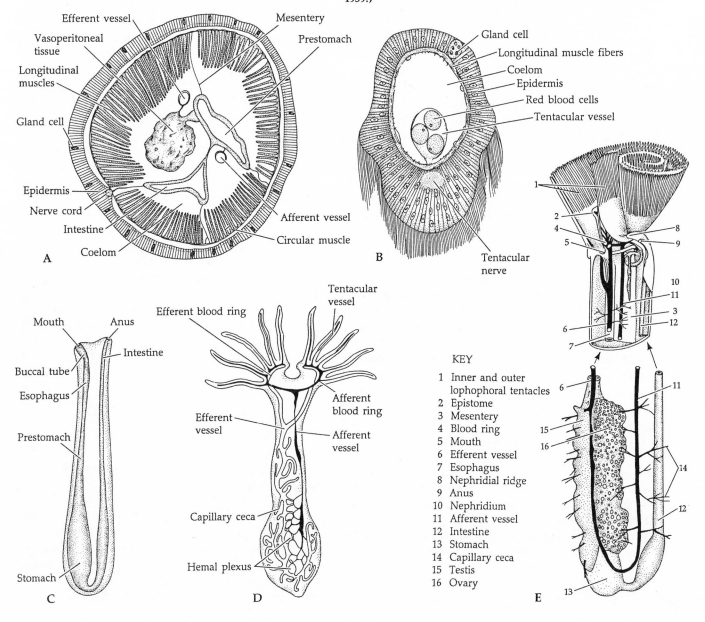

A — Efferent vessel, Vasoperitoneal tissue, Longitudinal muscles, Gland cell, Epidermis, Nerve cord, Intestine, Coelom, Mesentery, Prestomach, Afferent vessel, Circular muscle

B — Gland cell, Longitudinal muscle fibers, Coelom, Epidermis, Red blood cells, Tentacular vessel, Tentacular nerve

C — Mouth, Anus, Buccal tube, Intestine, Esophagus, Prestomach, Stomach

D — Efferent blood ring, Tentacular vessel, Afferent blood ring, Afferent vessel, Efferent vessel, Capillary ceca, Hemal plexus

E — KEY
1 Inner and outer lophophoral tentacles
2 Epistome
3 Mesentery
4 Blood ring
5 Mouth
6 Efferent vessel
7 Esophagus
8 Nephridial ridge
9 Anus
10 Nephridium
11 Afferent vessel
12 Intestine
13 Stomach
14 Capillary ceca
15 Testis
16 Ovary

connected to one another along the lateral aspects of the epistome. The metacoel forms the main trunk coelom, which is separated from the mesocoel by a transverse septum. Ontogenetically, the metacoel is an uninterrupted cavity with only one, midventral, mesentery. However, secondary mesenteries form later in development, yielding four longitudinal spaces (Figure 2A). The coelomic fluid contains several kinds of coelomocytes, including phagocytic amebocytes.

Body support is maintained primarily by the hydrostatic qualities of the coelomic chambers and by the tube. The muscles of the body wall are rather weak, particularly the circular layer, and once removed from their tubes phoronids are capable of only limited movement. Normally, however, the body wall of the end bulb is pressed against the tube, holding the worm in place. When disturbed, the animal simply contracts into the tube; the lophophore is not itself retractable.

The tube is secreted by epidermal gland cells. When first produced, the chitinous secretion is sticky, but upon contact with water it solidifies to a flexible parchment-like consistency. Sand grains and bits of other material adhere to the tube during the sticky phase of its formation in those phoronids that inhabit soft substrata (e.g., *Phoronopsis harmeri*). In some species the tubes intertwine with one another, with the whole tangled aggregation attached to a substratum or actually embedded in calcareous stone or shells (e.g., *Phoronis hippocrepia*).

The lophophore, feeding, and digestion

The tentacles of the lophophore are hollow, ciliated outgrowths of the mesosome, and each contains a blind-ended blood vessel and a coelomic extension (Figure 1B). The tentacles are in a double row, arising from two ridges. The ridges lie close to one another and form a narrow food groove in which the slitlike mouth is located (Figure 1C). Because the sides of the lophophoral ridges are coiled, many tentacles are compacted into a small area.

Phoronids are ciliary–mucus suspension feeders. The lophophoral cilia generate a water current that passes down between the two rows of tentacles and out between the tentacles. Food particles are trapped in mucus lining the food groove and then passed along the groove by cilia to the mouth. As the water current passes between the tentacles and out of the area of the food groove, some is directed over the anus and nephridiopores away from the animal (Figure 1C).

The digestive tract is U-shaped, but rather simple and not coiled (Figure 2C). The mouth is overlaid by the epistomal flap and leads inward to a short **buccal tube**, which is followed by an esophagus and a narrow **prestomach**. Within the end bulb the gut expands into a **stomach**, from which emerges the intestine. The intestine bends up toward the lophophore and leads to a short rectum and the anus. The gut is supported by peritoneal mesenteries (Figure 2A).

The entire digestive tract is apparently derived from entoderm. Some parts are muscular, but only weakly so, and much of the movement of food is by ciliary action. A middorsal strip of densely ciliated cells arises in the prestomach and extends into the stomach, and it is probably responsible for directing food along that portion of the gut. Gland cells occur in the esophagus, but their function remains uncertain. Transitory syncytial bulges in the stomach walls are the site of intracellular digestion in that organ.

Circulation, gas exchange, and excretion

Phoronids contain an extensive circulatory system comprising two major longitudinal vessels between which blood is exchanged in the lophophoral and stomachic ends of the body (Figures 2D and E). Various names have been applied to these vessels relative to their positions in the body. These terms are often confusing, because the positions of vessels vary along the length of the trunk, and they are not clearly dorsal, median, lateral, or ventral as the names imply. Some authors use the terms **afferent** and **efferent vessels** in reference to the direction of blood flow relative to the lophophore, and we prefer these clearer and more descriptive names.

The afferent vessel extends unbranched from the region of the end bulb to the base of the lophophore. For most of its length it lies more-or-less between the descending and ascending portions of the gut. In the mesosome the afferent vessel forks as an **afferent "ring" vessel** (U-shaped) at the base of the lophophoral tentacles. A series of **lophophoral vessels**, one in each tentacle, arises from the afferent ring. Each of these vessels joins with an **efferent "ring" vessel** (also U-shaped), which drains blood from the lophophore. Thus the afferent and efferent blood rings lie against one another, and generally share openings into the lophophoral tentacles within which blood moves back and forth, there being but a single vessel in each tentacle. Backflow into the afferent ring is largely prevented by tiny, one-way, flap valves.

The arms of the efferent ring unite to form the main efferent blood vessel, which extends through the trunk. This vessel gives off numerous branches or simple blind diverticula called **capillary ceca**, which bring blood close to the gut wall and other organs. In the end bulb, around the stomach and first part of the intestine, blood flows from efferent to afferent vessels through spaces composing the **hemal (stomachic) plexus** (Figures 2D and E). Blood actually leaves the true vessels here and flows through spaces between the organs and their bordering layers of peritoneum. Thus, technically speaking, the system is open at this point; however, blood flow is directed within the confines of these passages. Blood is moved through the circulatory system largely by muscular action of the blood vessel walls.

The intimate association of blood and the stomach wall suggests that nutrients are picked up from the stomach by the circulatory fluid and transported throughout the body. The tentacles of the lophophore are probably the most important site of gas exchange. Oxygenated blood flows from the lophophore into the efferent vessel and from there is distributed to all parts of the trunk. The blood contains nucleated red corpuscles, with hemoglobin as the respiratory pigment.

A pair of metanephridia lies in the trunk, and each bears two nephrostomes opening to the metacoel (Figures 2E and 3A). In each nephridium, the nephrostomes—one large and one small—join a curved nephridioduct, which leads to a nephridiopore adjacent to the anus. Although virtually nothing is known about excretory physiology in phoronids, particulate crystalline matter has been observed exiting the nephridiopores and probably represents precipitated nitrogenous waste products. The nephridia also function as pathways for the release of gametes. Osmoregulatory problems are presumably insignificant in most phoronids, being marine animals.

Nervous system

The nervous system of phoronids is rather diffuse and lacks a distinct cerebral ganglion. This condition is related to the sedentary lifestyle and overall reduction in cephalization in these worms. Most of the nervous system is intimately associated with the body wall, being either intraepidermal or immediately subepidermal. The body is everywhere supplied with a layer of nerve fibers between the epidermis and the circular muscle layer. Simple sensory neurons arise from this layer, either singly or in bundles, and extend to the body surface as the only receptor structures. Motor neurons extend inward to the muscle layers.

The central nervous system comprises a simple intraepidermal nerve ring, which lies at the base of the lophophore and is continuous with the subepidermal nerve layer. It is slightly swollen middorsally. The nerve ring supplies the tentacles with nerves as well as giving rise to motor nerves to some of the longitudinal muscles in the metasome. In addition, a bundle of sensory neurons extends from the nerve ring to each of the lophophoral organs (see below).

Phoronids possess one or two longitudinal giant motor fibers in the trunk (absent in the very small *Phoronis ovalis*). When only one longitudinal fiber is present, it lies on the left side. Actually, this fiber originates within the right side of the nerve ring, through which it passes to emerge on the left side.

Figure 3

A, A phoronid metanephridium (from *Phoronis australis*). Note the paired nephrostomes. B, The lophophore of *Phoronis vancouverensis*. Note the accessory lophophoral organs. C, Spermatophore of *Phoronopsis harmeri*. D, Spermatophore of *Phoronis vancouverensis*. (A after Hyman 1959; B–D after Zimmer 1967.)

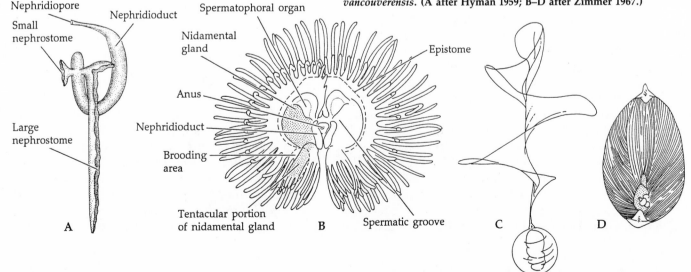

This nerve fiber is intraepidermal except where it extends inward along the left nephridium. In those species where two longitudinal fibers are present, the right one originates on the left side of the nerve ring and extends to the opposite side of the body.

Reproduction and development

Asexual reproduction by transverse fission or by a form of budding has been documented in *Phoronis ovalis* and *Phoronis psammophilia*. Phoronids are also capable of regenerating lost parts of the body.

Both gonochoristic and hermaphroditic species of phoronids occur, and in the latter case some are simultaneous hermaphrodites (e.g., *Phoronis vancouverensis*). The gonads are transient and form as thickened areas of the peritoneum around the hemal plexus. The resulting mass of gamete-forming tissue and blood sinuses is sometimes called **vasoperitoneal tissue** (Figure 2E). Gametes are proliferated into the metacoel and carried to the outside via the nephridia. Fertilization is probably internal in some species, and (at least in *Phoronopsis harmeri* and *Phoronis vancouverensis*) it occurs in the metacoel.

The developmental strategy differs among species (see Silén 1954; Zimmer 1967; Emig 1982). The particular pattern of any species depends in part on the size of the egg and the nature of the aforementioned **lophophoral organs**. These structures (Figure 3B) function as accessory sex glands (Zimmer 1967), serving to fashion spermatophores (Figures 3C and D) in males, and as brooding areas called **nidamental glands** in females of some species. In several species the females lack nidamental glands, and the eggs are shed directly into the water where they are fertilized (e.g., *Phoronis pallida*, *P. mulleri*, and *Phoronopsis harmeri*). The ova of these species contain little yolk, and they develop quickly to planktotrophic **actinotroch larvae** (Figure 4). In species that possess nidamental glands, fertilization is followed by brooding until release at the actinotroch stage (e.g., *Phoronis hippocrepia*, *P. vancouverensis*). The eggs of these species are moderately rich in yolk, providing nutrients for the embryos during the brooding period. *Phoronis ovalis* lacks nidamental glands, but the yolky eggs are shed into the maternal tube, where they are brooded. Development in *P. ovalis* does not include a typical actinotroch; instead, the embryos emerge as ciliated, sluglike larvae that have a short, planktonic life.

Internal fertilization in *Phoronopsis harmeri* (= *P. viridis*) has been elucidated by Zimmer (1972). Spermatophores on the tentacles of females rupture, releasing ameboid masses of sperm. The sperm enter the lophophoral coelom by lysis of the tentacular wall and then digest their way through the septum separating the mesocoel and metacoel into the trunk coelom, where fertilization occurs. Although fertilization has not actually been observed, Zimmer's experimental data, coupled with the fact that fertilized ova occur internally, suggest that this scenario is the only tenable explanation.

Despite continuing reference to phoronids as prostostome-like in terms of their early development, it had been clearly established descriptively and experimentally that such is not the case (see, for example, Zimmer 1964; Fairfax 1977). Early reports of spiral cleavage may have resulted from mechanical displacement of the blastomeres because of the tightness of the fertilization membrane. Otherwise, early cleavage is clearly radial and has been shown to be indeterminate. A coeloblastula forms and gastrulates by invagination. Mesoderm arises from the presumptive archenteron, and coelom formation is by a highly modified enterocoelous method. The only protostome-like feature is the formation of the mouth from the blastopore.

With the exception of *Phoronis ovalis*, all phoronids that have been studied produce distinctive **actinotroch larvae** (Figure 4). Earlier works alleging similarities between actinotrochs and trochophores are no longer given credence; the actinotroch is clearly a tripartite stage and lends additional support to the deuterostome affinity of the phoronids. The fully formed actinotroch bears a **preoral hood**, or lobe, over the mouth. The hood houses the protocoel and becomes the epistome. A partial ring of larval tentacles contains the mesocoel and eventually forms the lophophore. As the actinotroch develops, an inpocketing (called the **metasomal sac**) forms on the ventral surface. At settlement and metamorphosis this sac everts, extending the ventral surface such that the anus and mouth remain close to one another as the gut is drawn out into the characteristic U shape (Figure 4). It is during this metamorphic growth that the larval worms settle and begin secreting their tubes.

The ectoprocts

Members of the phylum Ectoprocta (Greek *ecto*, "outside"; *procta*, "anus") are sessile colonies of zooids living in marine and freshwater environments (Box 2; Figures 5 and 6). In most cases each colony is the product of asexual reproduction from a single, sexually produced individual called an **ancestrula**. The colony form differs greatly among species, but their general plantlike appearance earned these ani-

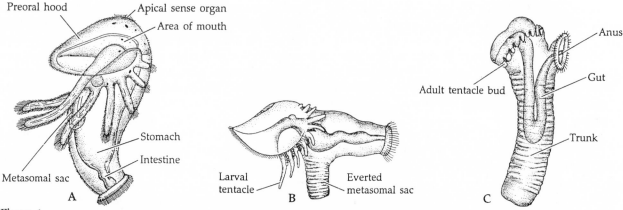

Figure 4

Phoronid larvae and metamorphosis. A, An actinotroch larva. B–C, Stages in the metamorphosis of an actinotroch. The gut is drawn into a U-shape, leaving the mouth and anus at the lophophoral end. (A after Hyman 1959; B–C after Dawydoff and Grassé 1959.)

mals the common name "moss animals," from which the old phylum name Bryozoa was coined. Marine ectoprocts are known from all depths and latitudes, mostly on solid substrata. One recently discovered Antarctic species forms gelatinous colonies on floating pieces of ice. Only a few species occur in fresh and brackish water. Littoral regions in most parts of the world harbor luxuriant growths of ectoprocts that often cover large areas of rock surfaces. Some species have corallike growth forms that can create miniature "reefs" in some shallow-water habitats. Others form dense bushlike colonies or gelatinous spaghetti-like masses. Many encrusting forms grow on the shells or exoskeletons of other invertebrates.

PHYLUM ECTOPROCTA

CLASS PHYLACTOLAEMATA: Freshwater ectoprocts; colonies with chitinous or gelatinous coverings; zooids cylindrical, large, and monomorphic; epistome with protocoel present; lophophore large and horseshoe-shaped; body wall muscles well developed; metacoel extensions interconnect zooids; a cord of tissue, the funiculus, extends from gut to body wall, but not between zooids; most produce asexual bodies called statoblasts. (e.g., *Cristatella, Hyalinella, Lophopodella, Plumatella*.)

CLASS STENOLAEMATA: Marine ectoprocts with zooids housed in tubular, calcified skeletal compartments; zooids cylindrical or trumpet-shaped, some polymorphic; epistome and protocoel absent; lophophore circular; body walls inflexible, lacking well developed musculature; without special coelomic extensions between zooids, but adjacent zooids connected by pores; funiculus does not extend between zooids; with a unique membranous sac housing the internal parts of the polypide; reproduction involves unique polyembryony, whereby single embryos reproduce asexually; one extant order, Cyclostomata. (e.g., *Actinopora, Crisia, Diaperoecia, Tubulipora*)

CLASS GYMNOLAEMATA: Highly diverse group of primarily marine ectoprocts; colony form extremely variable, soft or calcified, encrusting to arborescent; body wall lacks muscles; zooids variably modified from basic cylindrical form; zooids usually polymorphic; lophophore circular, epistome and protocoel absent; zooids joined by pores through which cords of tissue extend and join with each funiculus; two orders, Ctenostomata and Cheilostomata.

Box Two

Characteristics of the Phylum Ectoprocta

1. Trimeric, coelomate, colonial lophophorates
2. Epistome and protocoel absent in most species
3. Lophophore circular or U-shaped
4. Gut U-shaped, anus close to mouth
5. Typical circulatory and excretory structures absent
6. Colonies produced by asexual budding; zooids often polymorphic
7. Zooids usually gonochoristic, but both sexes often occur in a single colony
8. Most undergo mixed development
9. Sessile in marine and freshwater habitats

ORDER CTENOSTOMATA: colonies vary in shape; skeleton leathery, chitinous, or gelatinous, not calcified; openings through which zooids protrude lack opercula; without ovicells for brooding embryos; without avicularia. (e.g., *Aethozoon, Alcyonidium, Bowerbankia, Flustrellidra, Tubiporella*)

ORDER CHEILOSTOMATA: colony form varies, but generally of box-shaped zooids with calcareous walls; openings usually with opercula; zooids often polymorphic; embryos usually brooded in ovicells. (e.g., *Bugula, Callopora, Carbasea, Cellaria, Conopeum, Cornucopina, Cribrilaria, Eurystomella, Hippothoa, Membranipora, Microporella, Porella, Pyripora, Tricellaria*)

A special terminology has evolved among ectoproct specialists, especially concerning the morphology of the zooids. The colony itself is called a **zoarium** and the secreted exoskeleton the **zoecium**. Early workers mistakenly thought that ectoproct zooids were ac-

tually composed of two organisms, the exoskeletal compartment and the internal soft parts, which they named the cystid and polypide, respectively. These terms were redefined by Hyman (1959) and now have some meaning relative to the functional morphology of ectoprocts. The **cystid** comprises the outer casing or zoecium and the attached parts of the body wall, that is, the nonliving *and* living housing of each zooid. The **polypide** includes the lophophore and soft viscera that are movable within the housing (Figure 5). The opening in the cystid through which the lophophore extends is termed the **orifice** and often bears a flaplike covering, or **operculum**.

The nature of the exoskeleton differs among ectoprocts, as does the form of the colony. The outer covering may be gelatinous or chitinous, as it is in the Phylactolaemata and Ctenostomata, or calcified, as it is in the Stenolaemata and Cheilostomata. The

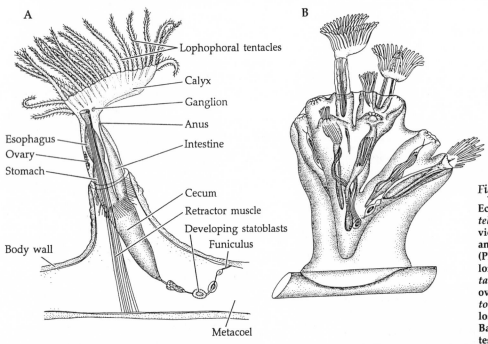

A

Lophophoral tentacles

Calyx

Ganglion

Anus

Esophagus

Ovary

Stomach

Intestine

Body wall

Cecum

Retractor muscle

Developing statoblasts

Funiculus

Metacoel

B

Figure 5

Ectoprocts. A, A single zooid of *Plumatella* (class Phylactolaemata) (cutaway view). Note the distinction between cystid and polypide. B, A colony of *Lophopus* (Phylactolaemata). Note the confluent coelomic cavities. C, A motile colony of *Cristatella mucedo* (Phylactolaemata) crawling over a plant stem. D, A colony of *Eurystomella* (class Gymnolaemata: order Cheilostomata). (A after Pennak 1978; B after Barnes 1980; C after Hyman 1959; D courtesy of G. McDonald.)

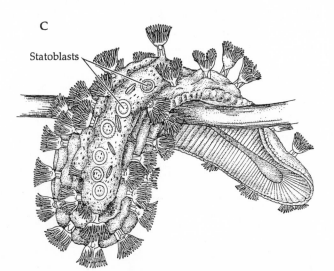

C

Statoblasts

D

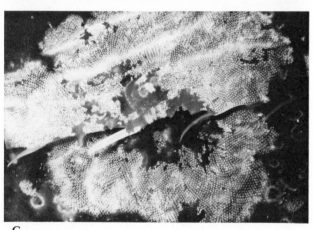

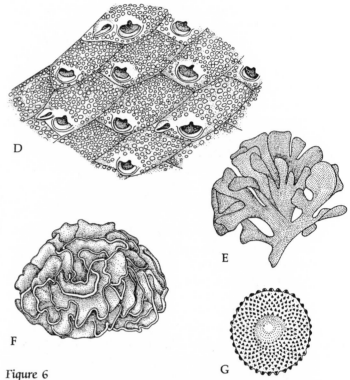

Figure 6

Ectoproct colonies. A, A stoloniferous colony of *Bowerbankia* (order Ctenostomata). B, Arborescent colonies of *Bugula* (order Cheilostomata); C, Patches of an encrusting ectoproct, *Membranipora* (Cheilostomata). D, Part of a colony of the encrusting ectoproct *Schizoporella* (Cheilostomata). E, Leaflike colony of *Flustra* (Cheilostomata). F, Cabbage-like colony of *Pentapora* (Cheilostomata). G, Discoidal colony of *Cupuladria* (Cheilostomata). (A, E–G after Ryland 1970; B photo by H. W. Pratt/BPS; C photo by K. Lucas/BPS; D after Barnes 1980.)

different growth patterns among ectoprocts result in a great variety of colony shapes (Figures 5 and 6). Most phylactolaemates display either **lophopodid** or **plumatellid** colony forms. In the former case, the gelatinous covering forms an irregular clump from which the zooids protrude, as seen in *Lophopus* (Figure 5B). Plumatellid colonies are usually erect or prostrate, and often are highly branched, like *Plumatella*. One remarkable phylactolaemate, *Cristatella*, grows in a distinct gelatinous strip, somewhat sluglike in form, and is capable of locomotion, creeping at rates of over 1 cm a day (Figure 5C).

The Stenolaemata and Gymnolaemata include a bewildering array of colony forms that may be gen-erally categorized as **stoloniferous** or **nonstoloniferous**. Stoloniferous colonies are characteristic of some members of the order Ctenostomata, in which the zooids arise separately from horizontal "runners" or stolons (e.g., *Bowerbankia*; Figure 6A). Nonstoloniferous colonies may be encrusting, arborescent, discoidal, and so on (Figure 6), but in all cases the zooids are compacted and adjacent to one another, rather than arising separately and at some distance from one another.

In addition to variation in overall form of the colonies, zooids of many gymnolaemates and some stenolaemates are polymorphic. Typical lophophore-bearing individuals are called **autozooids** and are responsible for feeding and digestion. All other individuals of a colony are collectively referred to as **heterozooids**, of which there are several types, all incapable of feeding. **Kenozooids** are reduced individuals modified for attachment to a substratum; various types of attachment discs, "holdfasts," and sto-lons are in this category. Many gymnolaemates possess **avicularia**, each of which bears an operculum modified as an articulating "jaw"; these zooids defend the colony against small organisms and keep the surface clean of debris (Figure 7A). The latter function is also facilitated by yet another type of heterozooid called a **vibraculum** (Figure 7B). These individuals have a flagellum-like operculum that sweeps over the colony surface, removing sediment particles and other material.

The body wall, coelom, muscles, and movement

The body wall comprises the outer secreted zoecium, the underlying epidermis, and the peritoneum. Sheets of circular and longitudinal muscles are present between the epidermis and peritoneum in phylactolaemates, but these muscle sheets are absent or greatly reduced in the other groups and are replaced by various muscle bands. Ectoprocts differ from other lophophorates in their ability to retract their lophophores into their zoecial casings, a clear protective device for these tiny sessile animals, whose soft parts would otherwise be continuously exposed to grazing predators.

Many ectoprocts possess ornate and species-specific surface sculpturing, the sculpted structures including spines, pits, and protuberances. Recent experiments by C. D. Harvell (1984) indicate that the cheilostomate *Membranipora* undergoes a rapid growth of new protective surface spines after grazing by predators (e.g., nudibranchs). Some ectoprocts also produce chemicals used as defense against would-be predators.

The mechanisms of lophophore retraction and protraction differ among ectoproct species. The specific mechanism depends largely upon the arrangement of muscles, the degree of rigidity of the zoecium, and the hydraulic qualities of the coelomic compartments. Recall the morphological distinction between the cystid and the polypide; extension and retraction of the lophophore basically involves movement of the latter relative to the former.

In all ectoprocts the main coeloms provide fluid-filled spaces on which muscles act directly or indirectly to increase hydraulic pressure for protraction of the lophophore. The epistome and protocoel (present only in the phylactolaemates) play no part in this process. The septum between the mesocoel and metacoel is perforated, so the fluid is continuous between the two chambers. Thus, when the metacoel is compressed the polypide is partially forced out of

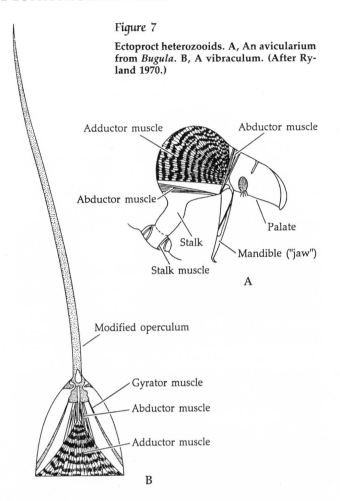

Figure 7

Ectoproct heterozooids. A, An avicularium from *Bugula*. B, A vibraculum. (After Ryland 1970.)

Adductor muscle
Abductor muscle
Abductor muscle
Palate
Stalk
Mandible ("jaw")
Stalk muscle
A

Modified operculum
Gyrator muscle
Abductor muscle
Adductor muscle
B

the cystid, thereby protracting the lophophore; at the same time coelomic fluid is moved into the mesocoel, thereby erecting the tentacles. Various retractor muscles serve to pull the polypide back within the cystid. Generally, these actions are common to all ectoprocts, but the mechanisms involved differ considerably. Below we describe a few examples of how these movements are accomplished and at the same time illustrate variations on the basic ectoproct *Bauplan*.

Members of the class Phylactolaemata protract their lophophores by contraction of the circular muscles of the flexible body wall around the metacoel. This action imposes pressure directly on the coelomic fluid and is similar to the mechanisms we have seen in many other coelomate animals. These ectoprocts possess a ring-shaped, muscular **diaphragm** just internal to the orifice through which the lophophore protrudes. The diaphragm dilates as the lophophore is protracted and serves to partially close off the orifice after the lophophore is withdrawn by retractor muscles, which extend from the body wall to the base of the lophophore (Figures 5A and 14).

Stenolaemate ectoprocts (Cyclostomata) have erect, tubular zooids surrounded by heavily calcified zoecia (Figures 8A and B). The inflexibility of the body wall and the absence of well developed sheets of muscles preclude use of the direct compression action of the phylactolaemates. Stenolaemates have evolved a mechanism of lophophore protraction unique among the ectoprocts. The structural features associated with this mechanism comprise a synapomorphy on which this group was established as a separate class. The key structure is a **membranous sac** attached by ligaments to the body wall and formed of a thin epithelial layer dividing the metacoel into an inner **entosaccal coelom** and an outer **exosaccal coelom** (Figures 8A and B). Distally, the exosaccal coelom lies between the outer body wall and a thick layer of **atrial dilator muscles**. When a zooid is re-

tracted, the atrial dilators are relaxed and a special **atrial sphincter** effectively closes the inner end of the atrium. The retracted polypide presses against the membranous sac, thereby forcing fluid into the distal region of the exosaccal coelom (Figure 8A). A reversal of these events causes protraction of the lophophore. As the retractor and atrial sphincter muscles relax, the atrial dilators contract. The contracted muscles press outward against the atrial wall and force coelomic fluid into the basal region of the exosaccal coelom, thereby pressing the polypide outward (Figure 8B).

Several methods of lophophore action have evolved among the gymnolaemates. Members of the order Ctenostomata possess an uncalcified, flexible zoecium composed of gelatinous, chitinous, or leathery material. Retraction of the lophophore is accomplished by the usual retractor muscle, which is aided by **longitudial parietal muscles** that pull in the atrial

Figure 8

Ectoproct anatomy and operation of the lophophore. See text for explanations. A–B, Two zooids of a stenolaemate, with the lophophore retracted (A) and protracted (B). C–D, The ctenostomate *Bowerbankia* with the lophophore protracted (C) and retracted (D). (After Ryland 1970.)

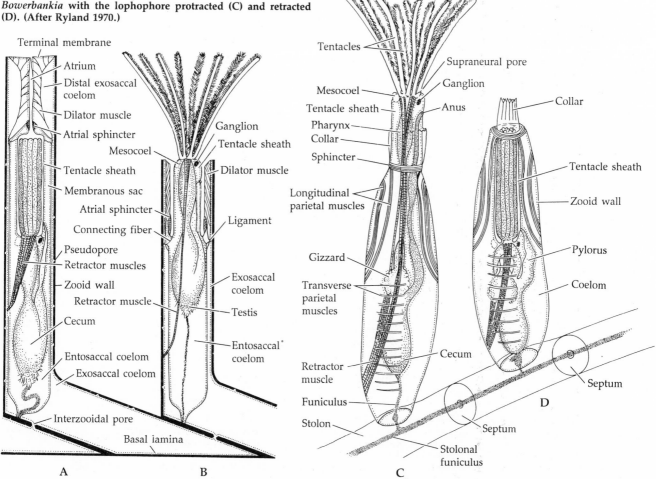

chamber. When the lophophore is fully retracted, a sphincter contracts, closing the orifice and, in some species, folding a pleated **collar** over the end of the zooid (Figure 8D). Contraction of **transverse parietal muscles** pulls the cystid walls inward, thereby causing an increase in coelomic pressure that then protracts the lophophore (Figure 8C).

Cheilostomate ectoprocts are housed in zoecia that have various amounts of calcium carbonate deposited between the epidermis and an outer chitinous zoecial layer (Figure 9). The problem of creating changes in coelomic pressure has been solved here by the retention of special uncalcified parts of the cystid wall upon which muscles can act. Each zoecium is more-or-less boxlike (rather than erect or tubular). The outer surface of the box that bears the orifice is called the **frontal surface**. In many cheilostomates the **frontal membrane** is uncalcified and flexible (Figure 9). Contraction of **parietal protractor muscles** pulls the frontal membrane inward, thus increasing coelomic pressure and pushing out the lophophore.

Nearly all cheilostomates possess a calcified operculum that closes over the orifice when the polypide is retracted, but the exposed frontal membrane presents a weakness in their defense against predation, and many species have evolved additional protective devices. Some forms, known as the **cribrimorph** ectoprocts, bear hard spines that project over the membrane and in some cases actually meet and fuse to form a cage above the vulnerable area (Figures 9B and C). In others a calcified partition, called the **cryptocyst**, lies beneath the frontal membrane, separating it from the soft parts within. The cryptocyst bears pores through which the protractor muscles extend (Figures 9F and G).

The most drastic modifications are seen in the so-called **ascophoran** cheilostomates, wherein the entire frontal surface is calcified except for a small opening. This **ascopore** leads inwards to a blind sac called the **compensation sac**, or **ascus**; this structure is an inwardly pouched, flexible portion of the body wall on which the protractor muscles insert (Figures 9H and I). Contraction of these muscles pulls the wall of the compensation sac inward as the ascopore allows water to enter the sac. Thus, pressure is exerted on the coelom and the lophophore is protracted. The ascophorans are the most diverse and successful group of ectoprocts.

Zooid interconnections

Before continuing, there are some aspects of ectoproct colony organization that must be addressed. As discussed in Chapter 3, clear definitions of the term *colony* are somewhat elusive. This difficulty arises because it is not always easy to tell where one individual ends and another begins (e.g., sponges)

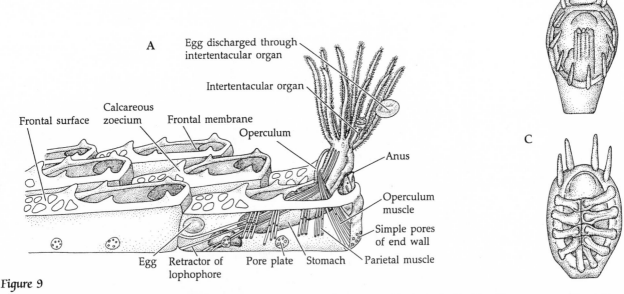

Figure 9

A, A portion of a colony of *Electra* (order Cheilostomata), with a cutaway view of one zooid. B–C, Zooids of two species of *Callopora* (Cheilostomata). Note the calcareous spines projecting over the frontal membrane. D–I, Parietal muscles and frontal membranes in cheilostomates. (D, F, and H are cross sections; E, G, and I are longitudinal sections.) D–E, Zooid with unprotected frontal membrane. F–G, Zooid with porous cryptocyst beneath frontal membrane. H–I, Ascophoran zooid with compensation sac and calcified frontal membrane. J, Internal anatomy of an ascophoran cheilostomate zooid. See text for explanation. (A after Barnes 1980; B–C, J after Ryland 1970; D–I redrawn from Meglitsch 1972.)

or because the degree of structural and functional communication among individuals is uncertain or variable. Ectoproct zooids, at least autozooids, are clearly demarcated by the elements of the polypide (lophophore, gut, and so on), but the way in which the zooids are interconnected differs among groups.

In phylactolaemates the metacoel is continuous among zooids, uninterrupted by septa (Figure 5A). Each zooid bears a tissue cord, which is called a **funiculus** and extends from the inner end of the curved gut to the body wall. All other ectoprocts lack extensive coelomic connections, and the zooids are separated by various sorts of structural components. The walls of adjacent zooids of stenolaemates bear interzooidal pores that allow communication of exosaccal coelomic fluid (Figures 8A and B). The funiculus is contained within the entosaccal space with the rest of the viscera and attaches the gut to the body wall.

Stoloniferous gymnolaemates (e.g., *Bowerbankia*; Figures 8C and D) have septa spaced along the stolons between the zooids. A cord of tissue passes along the stolons and through pores in each septum. This cord, called a **stolonal funiculus**, connects with the funiculus of each zooid arising from the stolon. In most nonstoloniferous gymnolaemates, the cystid walls of adjacent zooids are pressed tightly together, producing what are called **duplex walls** (Figures 9A and J). These double walls bear pores with tissue plugs, which, again, usually connect with the funiculus of adjacent zooids.

It is clear, then, that ectoproct zooids are interconnected structurally, either by direct sharing of coelomic spaces or by funicular tissue cords. Functionally, these connections provide a means of distributing materials through the colony, and perhaps other communal activities as well. Other special functions of the funiculus are discussed later.

The lophophore, feeding, and digestion

Ectoprocts are unique among lophophorates in that the lophophore is retractable, by mechanisms already explained. The lophophore is horseshoe-shaped in the phylactolaemates (except for the primitive *Fredericella*) and circular in the other two classes; the tentacular epidermis is ciliated. Ectoprocts are typically suspension feeders, although supplemental methods occur. The crescentic lophophore of phylactolaemates bears at its base a food groove that leads to the mouth and functions in a way similar to that described for phoronids. Feeding in other groups is somewhat different and has been more extensively studied.

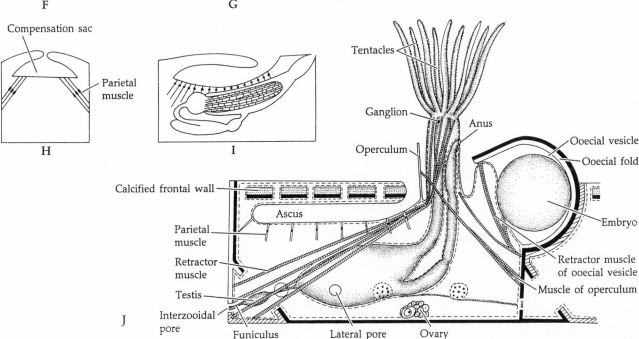

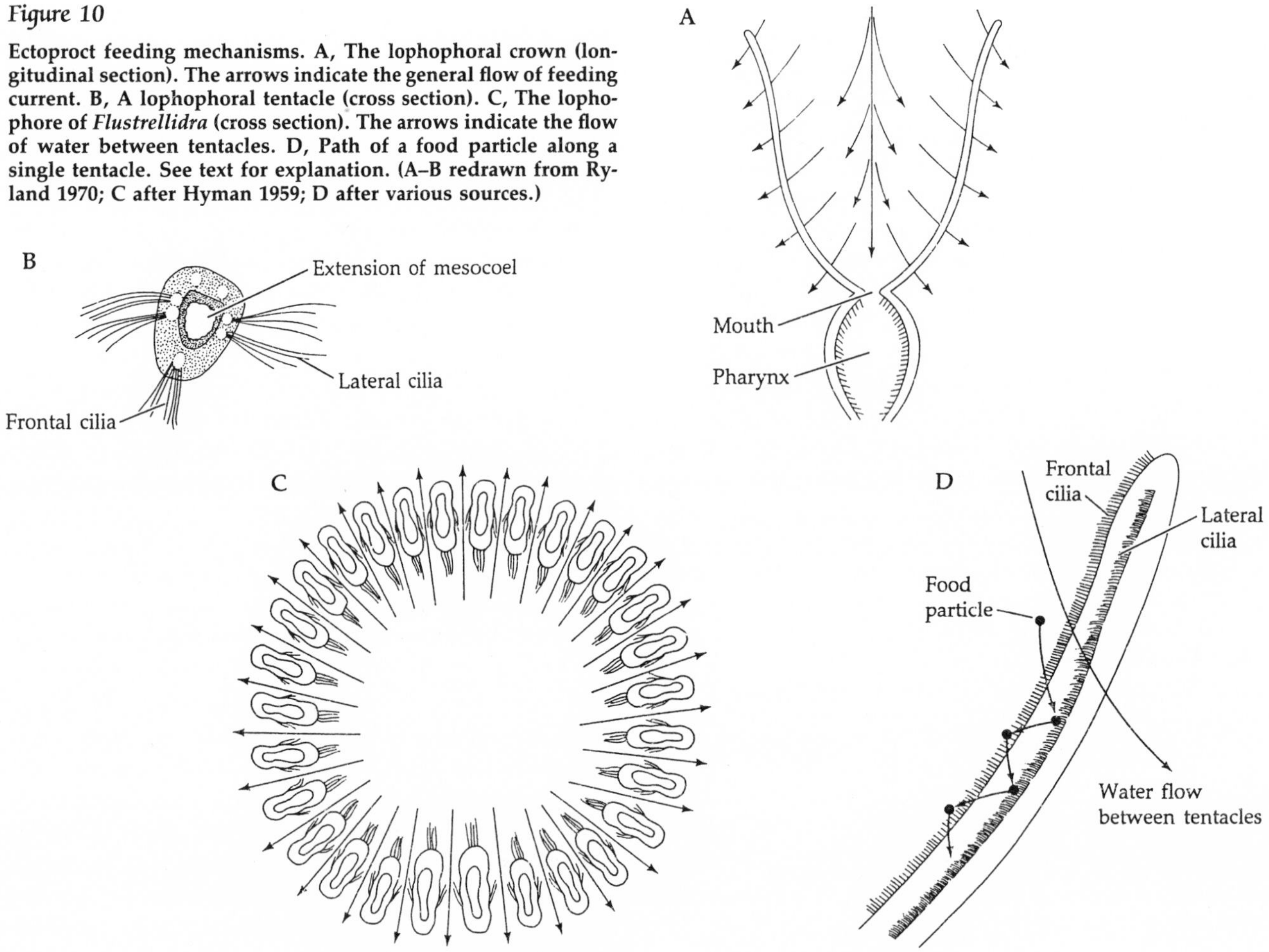

Figure 10

Ectoproct feeding mechanisms. A, The lophophoral crown (longitudinal section). The arrows indicate the general flow of feeding current. B, A lophophoral tentacle (cross section). C, The lophophore of *Flustrellidra* (cross section). The arrows indicate the flow of water between tentacles. D, Path of a food particle along a single tentacle. See text for explanation. (A–B redrawn from Ryland 1970; C after Hyman 1959; D after various sources.)

Upon protraction, the tentacles of the circular lophophores of stenolaemates and gymnolaemates are erected to form a funnel or bell-shaped arrangement around the mouth (Figure 10A). Each tentacle bears three ciliary tracts along its length, one **frontal tract** and two **lateral tracts** (Figure 10B). During normal suspension feeding, the lateral cilia create a current that enters the open end of the funnel, flows toward the mouth, and then out between the tentacles (Figure 10C). Some food particles are carried directly to the area of the mouth by the central flow of water. Other potential food, however, moves peripherally with the current toward the intertentacular spaces. When a particle contacts lateral cilia, a localized reversal of power stroke direction is initiated in those cilia, and the particle is tossed onto the frontal edge of the tentacle. The particle is repeatedly bounced in this fashion, from lateral to frontal, and is moved toward the mouth under the influence of a current generated by the frontal cilia (Figure 10D).

Many ectoprocts augment suspension feeding by various means that allow them to capture relatively large food particles, including live zooplankton. Winston (1978) has demonstrated that many species engage in flicking movements of individual tentacles with which a food particle has come in contact. By this means a single tentacle curls and strikes at the particle, moving it to the mouth. At least one species (*Bugula neritina*) is capable of trapping zooplankton by folding the tentacles over the prey and pulling it to the mouth. A number of ectoprocts rock or rotate the entire lophophore, apparently "sampling" reachable water for food material (Figure 11A).

In some ectoprocts the zooids of the colony function together in feeding and rejection of waste or nonfood materials (Winston 1979). In many genera (e.g., *Cauloramphus* and *Hippothoa*), groups of zooids "cooperate" to produce general currents that bring water to several clustered zooids and then flow away via "excurrent chimneys" between the clusters (Figures 11B and C). Such currents, which move larger amounts of water over the lophophores than could be moved by individual zooids, may be especially important to colonies inhabiting quiet water. The generation of strong excurrent water flow away from the colony surface helps to push nonfood material and feces far enough to reduce the possibility of recycling. In some, such as *Cauloramphus spiniferum*, large particles are actually passed from zooid to zooid and then dumped into an excurrent chimney (Figure 11D).

As in all lophophorates, the digestive tract of ectoprocts is U-shaped (Figures 5, 8, 9, and 12). The mouth lies within the lophophoral ring; in the Phylactolaemata, it is overlaid by an epistome. Ciliary tracts lead into the mouth from the surrounding **peristomial field**. Internal to the mouth is a muscular pharynx. A valve separates the lower end of the pharynx from the descending portion of the stomach, which is called the **cardia** and in some species is modified as a grinding gizzard. The cardia leads to a **central stomach** from which arises a large cecum; the funiculus attaches to the cecum. The ascending portion of the stomach, or **pylorus**, also arises from the central stomach and leads to a proctodeal rectum and the anus, which lies outside the lophophoral ring. The flow of material from the pylorus to the rectum is controlled by a sphincter. In phylactolaemates an esophagus precedes the stomach and the hindgut is elongated as an intestine.

Ingestion is accomplished by the sweeping action of the peristomial and oral cilia and by muscular contractions of the pharynx. Digestion begins extracellularly in the cardia and central stomach and is completed intracellularly in the cecum. Food is moved through the gut by peristalsis and cilia. Undigested material is rotated and formed into a spindle-shaped mass by the cilia of the pylorus and then passed to the rectum for expulsion.

Circulation, gas exchange, and excretion

Circulation of metabolites in single zooids is by diffusion, because there is no structural system for this purpose. Given the small size of these animals,

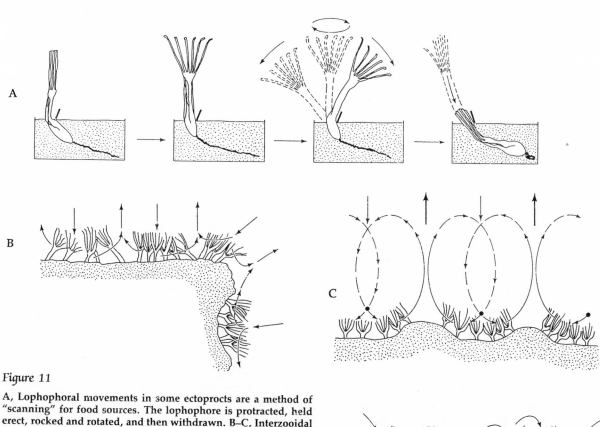

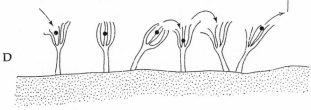

Figure 11

A, Lophophoral movements in some ectoprocts are a method of "scanning" for food sources. The lophophore is protracted, held erect, rocked and rotated, and then withdrawn. B–C, Interzooidal "cooperation" in the production of feeding currents, including "excurrent chimneys." B, *Hippothoa*. C, *Cauloramphus*. D, Cooperative rejection of a large particle. The particle is passed to an excurrent chimney (*Cauloramphus*). (A after Winston 1978; B–D after Winston 1979.)

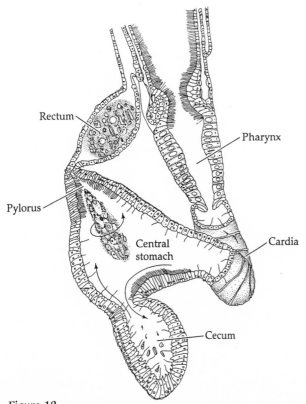

Figure 12

Digestive tract of the cheilostomate *Cryptosula pallasiana* (section). The arrows show food movement caused by ciliary action. (After Gordon 1975.)

intrazooid diffusion distances are small, and the coelomic fluid provides a medium for passive transport. Interzooid circulation is facilitated by the confluent coelom in phylactolaemates, the cystid pores in stenolaemates, and the funicular cords of most gymnolaemates. Gas exchange occurs across the walls of the protracted parts of the polypide, particularly the lophophore, the tentacles providing a very high surface area. Ectoprocts contain no respiratory pigments, and gases are carried in solution.

Metabolic waste products are accumulated and transported by phagocytic coelomocytes. The elimination of these wastes is not fully understood, but apparently it occurs in part by the formation of structures called **brown bodies**. The appearance of brown bodies is usually associated with the degeneration of polypides in adverse or stressful conditions; this degeneration is followed by reformation of a new polypide. In most gymnolaemates a brown body is left within the cystid following polypide degeneration, but in some cheilostomates the new polypide regenerates in such a way that the brown body is housed

within the gut of the new zooid and is then expelled via the anus. Note that the new polypide forms entirely from the tissue components of the cystid, that is, from the epidermis and peritoneum of the body wall. In most stoloniferous ctenostomates the old cystid with its brown body drops from the colony and an entire new zooid regenerates from the stolon. Cyclostomates and some other ectoprocts tend to form brown bodies within the coelom. In all cases, it is presumed that metabolic wastes are precipitated and concentrated in the brown bodies and thus eliminated or at least rendered inert.

Nervous system and sense organs

In concert with their sessile lifestyle and the general reduction of the anterior end, the ectoproct nervous system and sense organs are predictably reduced. A neuronal mass, or cerebral ganglion, lies dorsally in the mesosome near the pharynx. Arising from this structure is a circumenteric nerve ring. Nerves extend from the ring and ganglion to the viscera, and motor and sensory nerves extend into each tentacle. Interzooidal nerve fibers occur in some species, but their function remains obscure. The only known receptors are tactile cells on the lophophore and on avicularia. The planktonic larvae of at least some ectoprocts exhibit a marked negative geotaxis prior to settling. Experiments suggest that this geotaxis is a direct response to gravity, but the mechanism mediating this phenomenon is unknown. Larvae also usually have well developed ocelli and are positively phototactic while free swimming. Settlement is often accompanied by a shift to a negative phototaxis.

Reproduction and development

Asexual reproduction. As in most colonial animals, asexual reproduction is an indispensable part of the life history of ectoprocts and is responsible for colony growth and regeneration of zooids. Except for the unique cases of polyembryony in stenolaemates (see below), each colony begins from a single, sexually produced, primary zooid called the **ancestrula** (Figure 13A). The ancestrula undergoes budding to produce a group of daughter zooids, which themselves subsequently form more buds, and so on. The initial group of daughter zooids may arise in a chainlike series, a plate, or a disc; the budding pattern determines the growth form of the colony and is highly variable among species.

Budding involves only elements of the body wall. In most gymnolaemates a partition forms that isolates a small chamber, the developing bud, from the parent zooid. The bud initially includes only components of the cystid and an internal coelomic compartment. A new polypide is then generated from the living tissues of the bud, that is, the epidermis and the peritoneum. The epidermis and peritoneum invaginate, the former producing the lophophore and the gut. The peritoneum produces all of the new coelomic linings and the funiculus. Budding in phylactolaemates and stenolaemates is similar, except that the polypide develops first and is then encased by a new cystid wall.

In addition to budding, freshwater ectoprocts (Phylactolaemata) reproduce asexually by the formation of **statoblasts** (Figures 5, 13B, 13C, and 14). These structures are extremely resistant to drying and freezing, and are often produced in huge numbers during adverse environmental conditions. Statoblasts generally form on the funiculus of an autozooid and include peritoneal and epidermal cells plus a store of nutrient material. Each cellular mass secretes a pair of chitinous protective valves, differing among species in shape and ornamentation. The parent colony usually degenerates, freeing the statoblasts. Some statoblasts sink to the bottom, but others float by means of enclosed gas spaces. Some bear surface hooks or spines and are dispersed by passive attachment to aquatic animals or vegetation. With the return of favorable conditions, the cell mass generates a new zooid, which sheds its outer casing and attaches as a functional individual.

Sexual reproduction. Most ectoprocts are hermaphroditic, and each zooid is capable of producing sperm and eggs. The colonies of gonochoristic species may consist of zooids of one sex or, more commonly, may include both male and female individuals. Gametes usually arise from transient patches of germinal tissue developed from special areas of the metacoel peritoneum or the funiculus (Figures 8, 9, and 14). Only in the stenolaemates is any real "organ" present, that is, the testis is surrounded by a discrete cellular lining. Gametes are proliferated into the metacoel and migrate to the mesocoel prior to release. Sperm migrate into the coelomic lumina of the tentacles and, at least in some species, escape through special coelomopores on particular tentacles. A few cheilostomates (e.g., *Electra* and *Membranipora*) exhibit free spawning of eggs as well as of sperm, and fertilization and development are fully external. In all other ectoprocts thus far studied, the ova are retained by the parental zooids and brooded at least during early ontogeny.

In those gymnolaemates that release their ova to the sea water or to some external brooding area, the eggs are shed from the mesocoel through an opening called the **supraneural pore**, which is located between the bases of two tentacles. In some species this pore is elevated on a pedestal called the **intertentacular organ** (Figure 9A). A few ctenostomates (e.g., *Nolella* and *Victorella*) retain the ova within the coelom, where development takes place. Stenolaemates and phylactolaemates brood their embryos, the for-

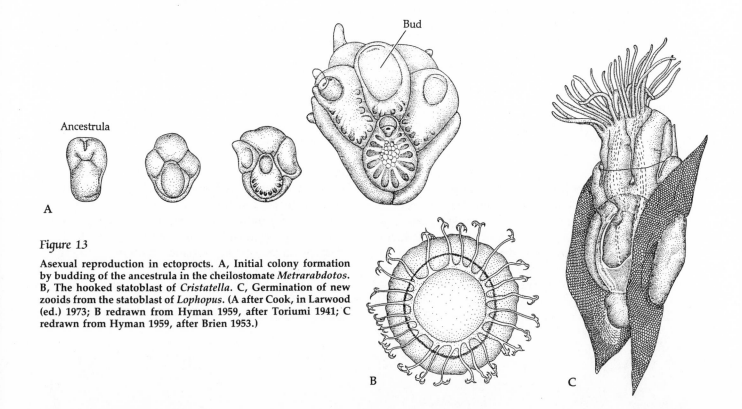

Figure 13

Asexual reproduction in ectoprocts. A, Initial colony formation by budding of the ancestrula in the cheilostomate *Metrarabdotos*. B, The hooked statoblast of *Cristatella*. C, Germination of new zooids from the statoblast of *Lophopus*. (A after Cook, in Larwood (ed.) 1973; B redrawn from Hyman 1959, after Toriumi 1941; C redrawn from Hyman 1959, after Brien 1953.)

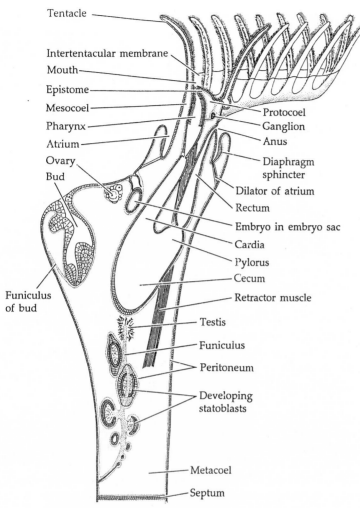

Tentacle

Intertentacular membrane

Mouth

Epistome

Mesocoel

Pharynx

Atrium

Ovary

Bud

Funiculus of bud

Protocoel

Ganglion

Anus

Diaphragm sphincter

Dilator of atrium

Rectum

Embryo in embryo sac

Cardia

Pylorus

Cecum

Retractor muscle

Testis

Funiculus

Peritoneum

Developing statoblasts

Metacoel

Septum

Figure 14

General anatomy of a phylactolaemate zooid. Note the positions of the bud, statoblasts, gonads, and brood pouch. (After Ryland 1970.)

mer in special individuals called **gonozooids** that are modified by loss of the polypide, and the latter in **embryo sacs** produced by invaginations of the body wall (Figure 14).

A variety of brooding methods occurs among gymnolaemates, usually involving the formation of an external brooding area called an **ovicell**, or **ooecium**. The most detailed and complete studies on the formation and functioning of these structures have been done by R. M. Woollacott and R. L. Zimmer on the cheilostome *Bugula neritina*. In this species, and probably many others, the ovicell develops from evaginations of the body wall of the parent autozooid. One of these evaginations is the **ooecial vesicle**, the lumen of which is confluent with the coelom of the

maternal zooid. The other evagination is called the **ooecial fold**; this structure forms a hoodlike covering over the ovicell. The embryo develops between the ooecial vesicle and fold (Figure 15). In many species the coelomic connection probably provides an avenue for nutrient transfer from parent to embryo. In *B. neritina* an actual tissue union develops between the epithelium of the ooecial vesicle and funicular extensions of the parent autozooid, producing a kind of placental system.

Ectoprocts undergo radial, holoblastic, nearly equal cleavage to form a coeloblastula. Subsequent development differs greatly among groups, but in all cases it involves a free-swimming dispersal form. Thus, development is either fully indirect (in those few species that free spawn) or mixed, with a planktonic stage following a period of brooding. Very little solid information exists on the derivation and fates of germ layers in ectoprocts. This is especially true for mesoderm and coelomic linings. It appears certain, however, that there is no indication of a 4d mesentoblast precursor for mesoderm, or any other convincing evidence of a protostome affinity.

In phylactolaemates the coeloblastula develops into a cystid-like stage lacking entoderm and then generates a polypide much like that seen in the usual bud formation. This zooid precursor is ciliated and escapes the embryo sac for a short swimming life before settling and attaching. The embryos of stenolaemates cleave to form a hollow ball, probably homologous to a coeloblastula. At this point, however, the embryo undergoes a budding process, forming secondary embryos, which in turn bud tertiary embryos. In some cases hundreds of small, solid, asexually produced embryos may result from a single primary ball of cells. This phenomenon of **polyembryony** is unique to these animals and may represent a heterochronic displacement of the usual asexual budding process of other ectoprocts. Each embryo develops cilia and escapes as a simple "larva," which settles and undergoes a metamorphosis similar to that described below for gymnolaemates.

The coeloblastulae of gymnolaemates undergo gastrulation by delamination; in this process four cells divide such that one of each pair of daughter cells is shunted to the blastocoel as presumptive entoderm and mesoderm. Free-swimming larvae are eventually produced. Many of the species that free spawn have a characteristic, flattened, triangular larva called a **cyphonautes** (Figure 16A). These larvae have a functional gut and may remain in the plankton for months, whereas the larvae of brooding species lack a digestive tract and lead very short, pelagic lives

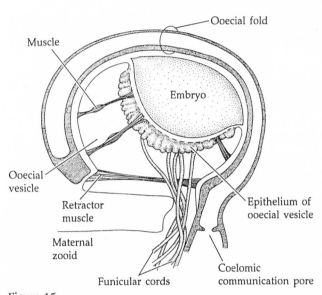

Figure 15

An ooecium (ovicell) of *Bugula neritina*. Note the tissue connection and coelomic communication with the parent zooid. Also see Figure 9J. (After Woollacott and Zimmer 1972.)

(Figures 16B and C). Despite these differences, gymnolaemate larvae have some fundamental similarities. For example, they characteristically possess a sensory **pyriform organ complex** and a pouchlike **adhesive sac**, both of which are important in settling and metamorphosis (Figure 16).

As mentioned earlier, ecotproct larvae are at first positively phototactic, and most possess pigment spots that are thought to be light sensitive. The pigment spots are ciliary in origin, supporting further the deuterostome alliance of the ectoprocts. Following a planktonic phase, the larvae usually become negatively phototactic and swim toward the bottom. Once in contact with the substratum, the pyriform organ complex is apparently used to test for chemical and tactile cues reflecting the suitability of the substratum for settling. Once a proper surface has been "selected," the adhesive sac everts and secretes sticky material for attachment. After attachment, there is a remarkable reorganization of tissue positions accompanied by histolysis of various larval structures. The metamorphosed larva then generates the primary zooid or ancestrula. The most detailed account of this process is, again, by Woollacott and Zimmer (1971) for *Bugula neritina*.

The brachiopods

Members of the phylum Brachiopoda (Greek *brachium*, "arm"; *poda*, "feet") are called the lamp shells (Box 3 and Figure 17). All are solitary, marine, benthic creatures. The body, including the lophophore, is enclosed between a pair of dorsoventrally oriented valves. Most brachiopods are attached to the substratum by a fleshy **pedicle** (Figure 17). Some

Figure 16

Ectoproct larvae. A, The cyphonautes larva of *Membranipora* (order Cheilostomata). The arrows indicate the direction of the feeding currents. B, Nonfeeding larva of *Alcyonidium* (order Ctenostomata). C, Nonfeeding larva of *Bugula* (Cheilostomata). (A after Ryland 1970; B after Hyman 1959; C after Woollacott and Zimmer 1971.)

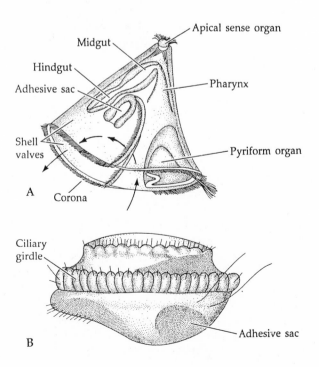

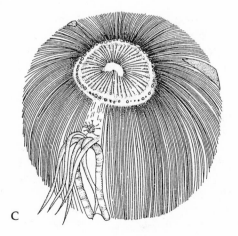

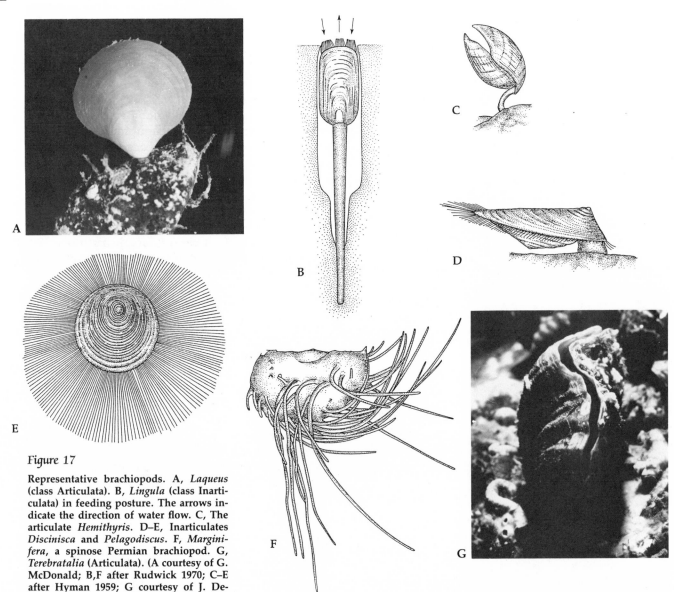

Figure 17

Representative brachiopods. A, *Laqueus* (class Articulata). B, *Lingula* (class Inarticulata) in feeding posture. The arrows indicate the direction of water flow. C, The articulate *Hemithyris*. D–E, Inarticulates *Discinisca* and *Pelagodiscus*. F, *Marginifera*, a spinose Permian brachiopod. G, *Terebratalia* (Articulata). (A courtesy of G. McDonald; B,F after Rudwick 1970; C–E after Hyman 1959; G courtesy of J. DeMartini.)

species lack a pedicle (e.g., *Crania*) and usually cement themselves directly to the substratum. Some forms that possess a pedicle do not form permanent attachments, such as *Magadina cumingi*, which lies free, and *Lingula*, which anchors in loose sand (Figure 17B). A few species possess both unattached and attached populations (e.g., *Neothyris lenticularis* and *Terebratella sanguinea*).

The shells are usually unequal, except in some inarticulates like *Lingula* and *Glottidia*, and are attached to one another posteriorly either by a tooth-and-socket hinge (Articulata) or simply by muscles (Inarticulata). The animals "sit" ventral side up, the pedicle arising from the ventral or **pedicle valve**; the dorsal shell is called the **brachial valve**.

Most brachiopods measure from 4 to 6 cm along the greatest shell dimension, but range from 1 mm to over 9 cm in extreme cases. Although they are known from nearly all ocean depths, they are most abundant on the continental shelf. The approximately 335 living species represent a small surviving fraction of the more than 12,000 species that once lived. Their rich fossil record dates back at least 600 million years. Brachiopods, especially articulates, were among the most abundant animals of the Paleozoic, but they declined in numbers and diversity after that time. C. W. Thayer (1985) has presented experimental evidence to support the idea that competition with epibenthic, bivalve molluscs was at least partly responsible for the reduction in brachiopod diversity following their Paleozoic success.

PHYLUM BRACHIOPODA

CLASS INARTICULATA: Valves not hinged, attached by muscles only; valves of organic composition, including chitin, or else calcareous; pedicle (absent in a few species) usually with intrinsic muscles and a coelomic lumen; epistome with coelomic channels confluent with lophophoral mesocoel; lophophore without internal skeletal support; anus present. Two orders, Lingulida and Acrotretida, comprising about 45 extant species. (e.g., *Crania, Discinisca, Glottidia, Lingula*)

CLASS ARTICULATA: Valves articulate by tooth-and-socket hinge; valves composed of scleroprotein and calcium carbonate; pedicle usually present, but lacking muscles and coelomic lumen; epistome small and tissue filled; lophophore generally with internal supportive elements; gut ends blindly, anus lacking. Three extant orders: Rhynchonellida, Terebratulida, and Thecideidina with just over 290 species. (e.g., *Argyrotheca, Dallina, Gryphus, Lacazella, Laqueus, Liothyrella, Magellania, Thecidellina, Terebratalia, Terebratella, Terebratulina, Tichosina*)

The body wall, coelom, and support

The shells of brachiopods comprise an outer organic **periostracum** and an inner structural layer or layers composed variably of calcium carbonate, calcium phosphate, scleroproteins, and chitinophosphate. Various spines are present in some species as outgrowths of the periostracum and serve to anchor the animals in place (Figure 17). In a fashion analagous to that of molluscs, brachiopod shells are secreted by **mantle lobes**, which are formed as outgrowths of the body wall (Figure 18). The periostracum is secreted by the mantle edges, and the inner layer by the general mantle surface. The shells of many brachiopods bear perforations, or **punctae**, extending from their inner surfaces nearly to the periostracum and containing tiny tissue extensions of the mantle (Figure 18B). The function of these mantle papillae is unknown, but some workers have suggested that they might serve as areas for food storage and gas exchange, or in some way deter the activities of borers. Shells that lack perforations are termed **impunctate**.

The soft mantle lobes line and are attached to the shells and form the water-filled **mantle cavity**, which houses the lophophore. The mantle edges often bear chitinous setae, which may protect the fleshy tissue and perhaps serve to prevent the entrance of large particles into the mantle cavity.

The epidermal cells of the mantle lobes and general body surface vary from cuboidal to columnar, and are densely ciliated on the lophophore. Beneath the epidermis is a connective tissue layer of varying thickness, which houses longitudinal muscle fibers where the body is not attached to the valves. The inner surface of the body wall is lined by peritoneum, which forms the outer boundary of the coelom. Being folds of the body wall, the mantle lobes contain extensions of the coelom, called **mantle canals** (Figure 18C).

The pedicle is an outgrowth of the body wall, arising from the posterior area of the ventral valve (Figures 17 and 18A). In inarticulates it contains all the usual layers beneath the epidermis, including connective tissue, muscles, and a coelomic lumen. However, the pedicle of articulates lacks muscles and a coelomic cavity. In the latter case the pedicle is operated by extrinsic muscle bands from the body wall itself. In brachiopods that attach firmly, the tip of the pedicle bears papillae or finger-like extensions that adhere tightly to the substratum.

The coelomic system of brachiopods includes the typical mesocoel and metacoel as the lophophoral and body coeloms, respectively. The epistome is solid in the articulates, but in inarticulates it contains a

---

Box Three

Characteristics of the Phylum Brachiopoda

1. Trimeric, coelomate lophophorates
2. Body enclosed between two shells (valves) oriented one dorsal and one ventral
3. Usually attached to the substratum by a stalk, or pedicle
4. Valves lined (and produced) by mantle lobes formed by outgrowths of the body wall and producing a water-filled mantle cavity
5. Trimeric condition partially obscured by modified body form
6. Epistome present, with or without coelomic lumen
7. Lophophore circular to variably coiled, with or without internal skeletal support
8. Gut U-shaped; anus present or absent
9. One or two pairs of metanephridia
10. Circulatory system reduced and open
11. Most are gonochoristic and undergo mixed or indirect life histories
12. Solitary, benthic, marine

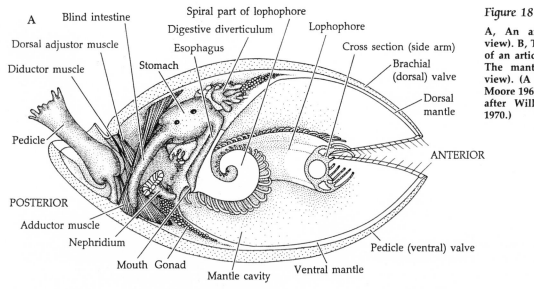

A   Blind intestine   Spiral part of lophophore
                        Digestive diverticulum   Lophophore
Dorsal adjustor muscle   Esophagus   Cross section (side arm)
Diductor muscle   Stomach   Brachial (dorsal) valve
                                Dorsal mantle
Pedicle   ANTERIOR
POSTERIOR
Adductor muscle   Pedicle (ventral) valve
    Nephridium   Ventral mantle
        Mouth Gonad   Mantle cavity

Figure 18

A, An articulate brachiopod (cutaway view). B, The edge of the shell and mantle of an articulate (longitudinal section). C, The mantle edge of *Notosaria* (inside view). (A after Williams and Rowell, in Moore 1965; B redrawn from Hyman 1959, after Williams 1956; C after Rudwick 1970.)

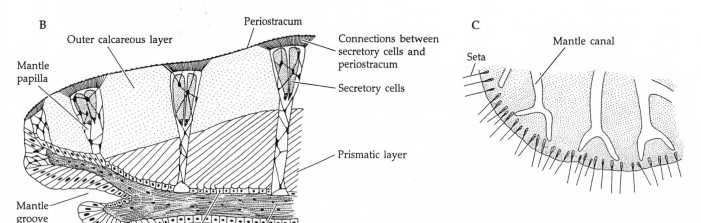

B   Periostracum
    Outer calcareous layer   Connections between secretory cells and periostracum
Mantle papilla   Secretory cells
    Prismatic layer
Mantle groove
    Epidermis   Connective tissue

C   Mantle canal
Seta

protocoel that is confluent with mesocoel. The coelomic fluid includes various coelomocytes, some of which contain hemerythrin.

The lophophore, feeding, and digestion

Like that of phoronids and ectoprocts, the lophophore of brachiopods comprises a ring of tentacles surrounding the mouth. In brachiopods however, the lophophore is produced as a pair of tentacle-bearing arms that extend anteriorly into the mantle cavity. The overall shape of the lophophore varies among taxa from a simple circular or U-shape to those with highly coiled arms (Figures 18 and 19). The brachiopod lophophore also differs in that it is always contained within the protection of the valves and is essentially immovable. In inarticulates the lophophore and tentacles are held in position by coelomic

pressure, whereas in articulates the tentacle-bearing ridge includes supportive skeletal elements. In addition, the dorsal valve often bears inwardly directed ridges and grooves that help support and position the lophophore.

In order to pass a water current through the mantle cavity, the two valves must be opened slightly. The mechanisms of valve operation differ between members of the two classes. Articulate brachiopods possess several sets of muscles including a pair of **diductor muscles**, which open the valves (Figure 20A). The tooth-and-socket hinges prevent a large gape. The adductor muscles include both striated and smooth fibers such that the valves can be quickly closed and then held together for long periods of time. Inarticulates lack a hinge and do not possess diductor muscles. Instead, the gape is produced by retraction of the body, an action that increases the internal pressure in the coelomic fluid and forces the valves apart. Adductor muscles are used to close the valves.

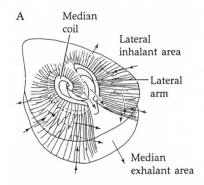

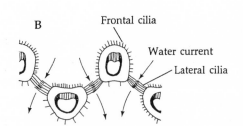

Figure 19

Feeding currents in brachiopods. A, Feeding currents (arrows) of *Waltonia*. B, Lophophoral tentacles (section). The water (arrows) passes over lateral ciliary bands. C, A portion of a lophophore. Food particles are transported along tentacles and the brachial food groove. (Redrawn from Rudwick 1970.)

Feeding currents are generated by the lophophoral cilia. Specific incurrent and excurrent flow patterns occur, varying with shell morphology and the shape and orientation of the lophophore. In any case, water is directed over and between the tentacles before passing out of the mantle cavity (Figure 19A). Each tentacle bears lateral and frontal ciliary tracts (Figure 19B). The lateral cilia of adjacent tentacles overlap and redirect food particles from the water to the frontal cilia by beat reversal. The frontal cilia beat toward the base of the tentacles, helping to direct trapped food. The lophophoral ridge, or **brachial axis**, bears a **brachial food groove** within which food material is moved to the mouth (Figure 19C). Brachiopods feed on nearly any appropriately small organic particles, especially phytoplankton.

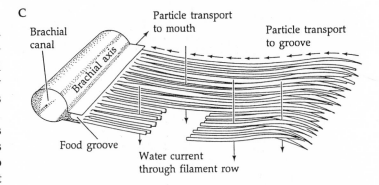

The digestive system is U-shaped (Figures 18, 20B, and 20C). The mouth is followed by a short esophagus, which extends dorsally and then posteriorly to the stomach. A digestive gland covers most of the stomach and connects to it via paired ducts. The intestine extends posteriorly, where it ends blindly in articulates or recurves as a rectum termi-

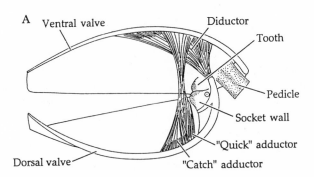

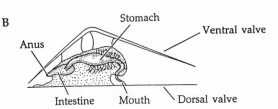

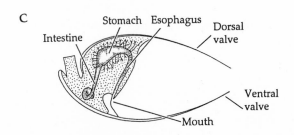

Figure 20

A, An articulate brachiopod (ventral side up; cutaway view). Note the major muscles that operate the valves. B, The complete gut of an inarticulate. C, The blind gut of an articulate. D, The nervous system of *Magellania*. Note the dorsal and ventral aspects on the left and right sides of the drawing, respectively. (Redrawn from Rudwick 1970.)

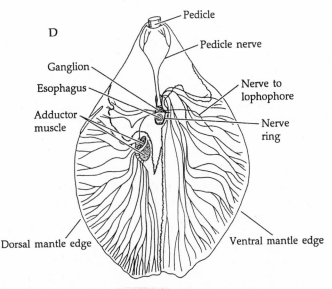

nating in an anal opening in inarticulates. In the latter case the anus opens either medially or on the right side of the mantle cavity. The absence of an anus is almost certainly a secondary loss in the articulates, and may be associated with the evolution of the articulate hinge, which restricts the posterior flow of water from the mantle cavity.

Little is known about digestion in brachiopods, but some work on *Lingula* (Chuang 1959) indicates that it occurs intracellularly in the digestive gland.

Circulation, gas exchange, and excretion

The brachiopod circulatory system is open, much reduced, and largely unstudied. A contractile heart lies in the dorsal mesentery just above the gut (*Crania* possesses several "hearts"). Leading anteriorly and posteriorly from the heart are channels bounded only by mesentery peritoneum, thus no true vessels are present. These channels branch to various parts of the body, but the pattern of circulation is not fully understood. It appears that the blood is separate from the coelomic fluid, although both contain certain similar cells. The function of the circulatory system is thought to be largely restricted to nutrient distribution.

Gas exchange probably occurs across the general body surface, especially the tentacles and mantle. These structures not only provide large surface areas but are also sites over which water moves and is brought close to underlying coelomic fluid. This general arrangement and the presence of hemerythrin in certain coelomocytes suggest that the coelomic fluid, not the blood, is the medium for oxygen transport.

Brachiopods possess one or two pairs of metanephridia, with the nephrostomes opening to the metacoel. The nephridioducts exit through pores into the mantle cavity. The nephridia function as gonoducts as well as discharging phagocytic coelomocytes that have accumulated metabolic wastes.

Nervous system and sense organs

The nervous system of brachiopods is somewhat reduced. A dorsal ganglion and a ventral ganglion lie against the esophagus and are connected by a circumenteric nerve ring. Nerves emerge from the ganglia and nerve ring and extend to various parts of the body, especially the muscles, mantle, and lophophore (Figure 20D).

As expected, the array of sense organs in these animals is compatible with their lifestyles. The mantle edges and setae are richly supplied with sensory neurons, probably tactile receptors. There is also evidence that brachiopods are sensitive to dissolved chemicals, perhaps through surface receptors on the tentacles or mantle edge. At least one species of *Lingula* possesses a pair of statocysts, which are associated in this burrowing form with orientation in the substratum.

Reproduction and development

Asexual reproduction does not occur in brachiopods. Most species are gonochoristic, with gametes developing from patches of transient, gonadal tissue derived from the metacoel peritoneum. Gametes are released into the metacoel and escape through the nephridia. In most cases both eggs and sperm are shed freely and fertilization is external. A few species, however, brood their embryos until the larval stage is reached. In these cases sperm are picked up in the water currents of females and the eggs are retained in a brooding area, where they are fertilized. *Argyrotheca*, for example, broods its embryos in the enlarged nephridia. Others retain their embryos on the arms of the lophophore, in special regions of the mantle cavity, or in modified depressions in a valve.

Cleavage is holoblastic, radial, and nearly equal; it leads to the formation of a coeloblastula. Gastrulation is by invagination, except in the brooding form *Lacazella*, where it apparently occurs by delamination. The blastopore closes and the mouth forms secondarily. The anus, when present, breaks through late as the gut grows and approaches the body wall. Mesoderm and coelom formation are enterocoelic. All of these developmental features bear witness to the deuterostome affinities of the Brachiopoda.

Whether the developmental pattern is mixed or fully indirect, all brachiopods eventually enter a free-swimming larval stage (Figure 21). These larvae are heretofore unnamed, but we suggest the term **lobate larvae**, in reference to the body regions visible as primordia at this stage and to the existing terminology traditionally used to describe these regions. The larvae of articulates and inarticulates differ morphologically and in the events at the time of settling. In inarticulates, such as *Lingula*, the larva is constructed much like the adult, except the pedicle is curled inside the mantle cavity and the body and lophophoral lobes are disproportionately large compared with the mantle lobes (Figures 21A and B). Thus the lophophore can be protruded out from between the mantle lobes and function to propel and feed the larva. The mantle lobes lie dorsoventrally on the body. Shell secretion commences early and, with added weight, the larva sinks, the pedicle is extended, and the juvenile brachiopod assumes benthic life. Thus, there is no drastic metamorphosis at the time of settling.

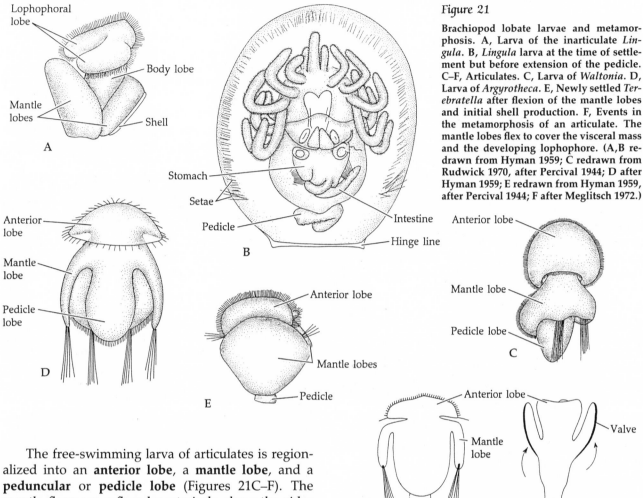

Figure 21

Brachiopod lobate larvae and metamorphosis. A, Larva of the inarticulate *Lingula*. B, *Lingula* larva at the time of settlement but before extension of the pedicle. C–F, Articulates. C, Larva of *Waltonia*. D, Larva of *Argyrotheca*. E, Newly settled *Terebratella* after flexion of the mantle lobes and initial shell production. F, Events in the metamorphosis of an articulate. The mantle lobes flex to cover the visceral mass and the developing lophophore. (A,B redrawn from Hyman 1959; C redrawn from Rudwick 1970, after Percival 1944; D after Hyman 1959; E redrawn from Hyman 1959, after Percival 1944; F after Meglitsch 1972.)

The free-swimming larva of articulates is regionalized into an **anterior lobe**, a **mantle lobe**, and a **peduncular** or **pedicle lobe** (Figures 21C–F). The mantle flaps are reflexed posteriorly along the sides of the presumptive pedicle, rather than anteriorly over the body as seen in inarticulates. After a short larval life of 1 or 2 days, the larva settles and metamorphoses. As the pedicle attaches to the substratum, the mantle lobes flex forward over the anterior lobe. The now exterior surfaces of the mantle lobes commence secretion of the valves, while the anterior lobe differentiates into the body and the lophophore.

Lophophorate phylogeny

Despite continuing arguments, evidence now seems overwhelming that the lophophorates are a monophyletic clade and that they are deuterostomes. The groups Phoronida, Ectoprocta, and Brachiopoda are united by their possession of lophophores, a clearly unique synapomorphy. In addition, they possess U-shaped digestive tracts, peritoneal gonads, metanephridia (absent in ectoprocts), and a tendency to secrete outer casings. We view these features as

homologous within this clade but convergent with similar conditions in many other taxa. As discussed later in this section, each of the three lophophorate groups displays enough derived character states to merit separate phylum status, but the idea of a "superphylum" (perhaps Lophophorata) may be warranted.

Zimmer (1973) has logically made the case for the deuterostome nature of the lophophorates. They all show radial cleavage, enterocoely, and (except for the phoronids), a mouth that is not derived from the blastopore. In addition, the body plan and coelomic arrangement is clearly trimerous or obviously derived therefrom.

An alliance between the groups Ectoprocta and Entoprocta, proposed by Nielsen (1971, 1977), is rejected on the basis of incompatibility with the idea of lophophorate unity and on direct comparative grounds. Entoprocts do not possess a lophophore as we have defined it, because the entoproct anus lies within the tentacular whorl. Furthermore, they lack any vestiges of a coelom. The feeding currents are virtually opposite in the two groups, and the methods of food capture and transport are entirely different. Entoprocts possess ducted gonads, ectoprocts do not. Cleavage in entoprocts is spiral, whereas it is radial in ectoprocts. Larval forms and particularly metamorphosis are clearly different in the two groups. More important, if the two groups are related, then they must share common (homologous) characteristics—that is, synapomorphies. The similarities pointed out by Nielsen are superficial and common to many colonial sessile animals (e.g., budding, metamorphosis, and life cycles). The U-shaped guts are convergent adaptations to zooid life in "boxes"—no other condition would function. Thus, in the absence of unifying synapomorphies and the presence of multiple and significant differences, we can only consider one conclusion: the two groups are unrelated.

The origin of the lophophorates is puzzling, largely because it is tied, in part, to the origin of the entire deuterostome line, which is itself very uncertain. Most workers agree that the phoronids show the least amount of change from the presumed ancestral form. That ancestor may have been a trimeric, coelomate, infaunal burrower. It probably evolved during the Precambrian as one evolutionary experiment at a coelomate *Bauplan*. Once this basic deuterostome plan was established, the stage was set for subsequent adaptive radiation.

The first lophophorate was probably phoronid-like and became adapted to tube dwelling and feeding above the substratum. Modern phoronids have probably changed little from this tube-dwelling protolophophorate.

The origin of the Ectoprocta clearly involved a reduction in body size and the development of colonial habits. The epidermal secretions became compartmentalized, with the exploitation of asexual budding as a means of colony formation. The acquisition of a retractable lophophore allowed protection of the soft tentacles. The absence of nephridia and circulatory structures provides space for the retraction of the polypide, and the short diffusion distances associated with small size allowed the disappearance of these systems. Without nephridia as a means of gamete release, other avenues of egg and sperm escape arose in the form of coelomopores from the mesocoel and communication between the metacoel and the mesocoel.

The origin of the brachiopods is marked by the appearance of several novel features largely associated with the evolution of mantle folds, their secretion of valves, and the enclosure of the lophophore and body proper in the mantle cavity. The lophophore lost most of its hydraulic qualities and became more-or-less stationary, held by various structural support mechanisms. The circulatory system was reduced. The origin of a pedicle allowed a means of attachment in these solitary animals, supporting the body off the substratum. The first brachiopods were probably lingulid types that used the pedicle for anchorage in soft substrata.

Valentine (1973, 1975) has attempted to support a polyphyletic origin for the brachiopods, but he does recognize a monophyletic lophophorate clade, somewhat as we have described here. However, Rowell's (1982) cladistic treatment of the brachiopods, living and extinct, presents a convincing case for monophyly, although his subgroups do not correspond exactly with the Articulata–Inarticulata division.

In summary, the major groups of lophophorates seen today are probably derived from common stock. But it is very difficult to define sister-groups within the lophophorate clade, and we have not attempted to produce a cladogram to depict the relationships among the three phyla (see Chapter 24). The origin of the lophophore allowed various avenues of escape from the infaunal life of their Precambrian ancestor and the exploitation of three different lifestyles, all involving suspension feeding. There remain, of course, many questions and alternative hypotheses on the matter of lophophorate evolution, especially their placement within the deuterostome lineage. Some workers (see Jeffries 1986) view the lophophorates as somewhat intermediate between the protostomes and deuterostomes, a hypothesis based largely on the fate of the blastopore (see Chapter 23).

Selected References

General References

Hyman, L. H. 1959. *The Invertebrates*, Vol. 5, Smaller Coelomate Groups. McGraw-Hill, New York.

Jefferies, R. P. S. 1986. *The Ancestry of the Vertebrates*. British Museum of Natural History, London.

McCammon, H. M. and W. A. Reynolds (organizers). 1977. Symposium: Biology of Lophophorates. Am. Zool. 17:3–150.

Moore, R. C. (ed.). 1965. *Treatise on Invertebrate Paleontology*. Pts. G and H (Vols. 1 and 2). Geological Society of America and University of Kansas Press, Lawrence.

Valentine, J. W. 1973. Coelomate superphyla. Syst. Zool. 22(2):97–102.

Valentine, J. W. 1975. Adaptive strategy and the origin of grades and ground-plans. Am. Zool. 15:391–404.

Zimmer, R. L. 1973. Morphological and developmental affinities of the lophophorates. *In* G. P. Larwood (ed.), *Living and Fossil Bryozoa*. Academic Press, London, pp. 593–600.

Phoronida

Emig, C. C. 1974. The systematics and evolution of the phylum Phoronida. Z. Zool. Syst. Evolutionsforsch. 12(2):128–151.

Emig, C. C. 1977. The embryology of Phoronida. Am. Zool. 17:21–38.

Emig, C. C. 1982. Phoronida. *In* S. P. Parker (ed.), *Synopsis and Classification of Living Organisms*. McGraw-Hill, New York, p. 741.

Fairfax, R. E. 1977. The embryology and reproductive biology of *Phoronopsis viridis*. Master's Thesis, Humboldt State University.

Silén, L. 1954. Developmental biology of the Phoronidea of the Gullmar Fiord area of the west coast of Sweden. Acta Zool. 35:215–257.

Zimmer, R. L. 1964. Reproductive biology and development of Phoronida. Ph.D. Dissertation, University of Washington. University Microfilms Inc., Michigan.

Zimmer, R. L. 1967. The morphology and function of accessory reproductive glands in the lophophores of *Phoronis vancouverensis* and *Phoronopsis harmeri*. J. Morphol. 121(2):159–178.

Zimmer, R. L. 1972. Structure and transfer of spermatozoa in *Phoronopsis viridis*. *In* C. J. Arceneaux (ed.), *30th Annual Proceedings of the Electron Microscopical Society of America*.

Ectoprocta

Baluk, N. and A. Radwanski. 1977. The colony regeneration and life habitat of free living bryozoans. Acta Geol. Pol. 27:143–156.

Boardman, R. S. and A. H. Cheetham. 1969. Skeletal growth, intracolony variation, and evolution in Bryozoa: A review. Paleontology 43:205–233.

Brien, P. 1953. Étude sur les phylactolemates. Ann. Soc. R. Zool. Belg. 84.

Buss, L. W. 1981. Group living, competition, and the evolution of cooperation in a sessile invertebrate. Science 213:1012–1014.

Chimonides, P. J. and P. L. Cook. 1981. Observations on living colonies of *Selenaria* (Bryozoa, Cheilostomata) II. Cah. Biol. Mar. 22:207–219.

Cook, P. L. 1977. Colony water currents in living Bryozoa. Cah. Biol. Mar. 18:31–47.

Cook, P. L. and P. J. Chimonides. 1981. Morphology and systematics of some rooted cheilostome Bryozoa. J. Nat. Hist. 15:97–134.

Cook, P. L. and P. J. Chimonides. 1983. A short history of the lunulite Bryozoa. Bull. Mar. Sci. 33(3):566–581.

Driscoll, E. C., J. W. Gibson and S. W. Mitchell. 1971. Larval selection of substrate by the bryozoans *Discoporella* and *Cupuladria*. Hydrobiologia 37:347–359.

Farmer, J. D., J. W. Valentine and R. Cowen. 1973. Adaptive strategies leading to the ectoproct groundplan. Syst. Zool. 22(3):233–239.

Gordon, D. P. 1975. Ultrastructure and function of the gut of a marine bryozoan. Cah. Biol. Mar. 16:367–382.

Harvell, C. D. 1984. Predator-induced defense in a marine bryozoan. Science 224:1357–1359.

Hughes, R. L. and R. M. Woollacott. 1980. Photoreceptors of bryozoan larvae. Zool. Scr. 9:129–138.

Jackson, J. B. C. 1979. Overgrowth competition between encrusting cheilostome ectoprocts in a Jamaican cryptic reef environment. J. Anim. Ecol. 48:805–823.

Jebram, D. 1980. Laboratory diets and qualitative nutritional requirements for bryozoans. Zool. Anz. 205:333–334.

Kaufmann, K. W. 1971. The form and functions of the avicularia of *Bugula*. Postilla 151:1–26.

Larwood, G. P. (ed.). 1973. *Living and Fossil Bryozoa*. Academic Press, London.

Larwood, G. P. and M. B. Abbott (eds.). 1979. *Advances in Bryozoology*. The Systematics Association Special Volume No. 13, Academic Press, New York.

Larwood, G. P. and C. Nielsen (eds.). 1981. *Recent and Fossil Bryozoa*. Olsen and Olsen, Fredensborg, Denmark.

Lidgard, S. 1986. Ontogeny in animal colonies: A persistent trend in the bryozoan fossil record. Science 232:230–232.

Moyano, H. I. 1982. Magellanic Bryozoa: Some ecological and zoogeographical aspects. Mar. Biol. 67:81–96.

Mukai, H. and S. Oda. 1980. Comparative studies on the statoblasts of higher phylactolaemate bryozoans. J. Morphol. 165:131–156.

Nielsen, C. 1971. Entoproct life cycles and the entoproct/ectoproct relationship. Ophelia 9:209–341.

Nielsen, C. 1977. The relationship of Entoprocta, Ectoprocta, and Phoronida. Am. Zool. 17(1):149–150.

Palumbi, S. R. and J. B. C. Jackson. 1983. Aging in modular organisms: Ecology of zooid senescence in *Steginoporella* sp. (Bryozoa; Cheilostomata). Biol. Bull. 164:267–278.

Pires, A. and R. M. Woollacott. 1982. A direct and active influence of gravity on the behavior of a marine invertebrate larva. Science 220:731–733.

Rider, J. and R. Cowen. 1977. Adaptive architectural trends in encrusting ectoprocts. Lethaia 10:29–41.

Rogick, M. D. 1959. Bryozoa. *In* W. T. Edmondson, H. B. Ward and G. C. Whipple (eds.), *Freshwater Biology*, 2nd Ed. Wiley, New York, pp. 495–507.

Ryland, J. S. 1970. *Bryozoans*. Hutchinson University Library, London.

Ryland, J. S. 1976. Physiology and ecology of marine bryozoans. Adv. Mar. Biol. 14:285–443.

Ryland, J. S. 1982. Bryozoa. *In* S. P. Parker (ed.), *Synopsis and Classification of Living Organisms*. McGraw-Hill, New York, pp. 743–769.

Silén, L. 1972. Fertilization in the Bryozoa. Ophelia 19(1):27–34.

Silén, L. 1980. Colony–substratum relations in Scrupocellariidae (Bryozoa, Cheilostomata). Zool. Scr. 9:211–217.

Soule, D. F. and J. D. Soule. 1964. The Ectoprocta (Bryozoa) of Scammon's Lagoon, Baja California, Mexico. Amer. Mus. (Nov.) 2199. [Includes key to bryozoan higher taxa.]

Thorpe, J. P., G. A. Shelton and M. S. Laverack. 1975. Colonial nervous control of lophophore retraction in Cheilostome Bryozoa. Science 189:60–61.

Toriumi, M. 1941. Studies on fresh-water Bryozoa of Japan. Sci. Rep. Tohoku Univ. Ser. 4.

Winston, J. E. 1978. Polypide morphology and feeding behavior in marine ectoprocts. Bull. Mar. Sci. 28(1):1–31.

Winston, J. E. 1979. Current-related morphology and behavior in some Pacific coast bryozoans. *In* G. P. Larwood and M. B. Abbott (eds.), *Advances in Bryozoology.* Academic Press, London, pp. 247–268.

Woollacott, R. M. and R. L. Zimmer. 1971. Attachment and metamorphosis of the cheiloctenostome bryozoan *Begula neritina* (Linné). J. Morphol. 134(3):351–382.

Woollacott, R. M. and R. L. Zimmer. 1972. Origin and structure of the brood chamber in *Bugula neritina* (Bryozoa). Mar. Biol. 16:165–170.

Woollacott, R. M. and R. L. Zimmer. 1972. Fine structure of a potential photoreceptor organ in the larva of *Bugula neritina* (Bryozoa). Z. Zellforsch. 123:458–469.

Woollacott, R. M. and R. L. Zimmer. (eds.). 1977. *Biology of Bryozoans.* Academic Press, New York.

Brachiopoda

Chuang, S. H. 1959. Structure and function of the alimentary canal in *Lingula unguis.* Proc. Zool. Soc. London 132:293–311.

Foster, M. W. 1982. Brachiopoda. *In* S. P. Parker (ed.), *Synopsis and Classification of Living Organisms.* McGraw-Hill, New York, pp 773–780.

Gutman, W. F., K. Vogel and H. Zorn. 1978. Brachiopods: Biochemical interdependencies governing their origin and phylogeny. Science 199:890–893.

MacKay, S. and R. A. Hewitt. 1978. Ultrastructure studies on the brachiopod pedicle. Lethaia 11:331–339.

Percival, E. 1944. A contribution to the life-history of the Brachiopod *Terebratella inconspicua.* Trans. R. Soc. N. Z. Zool. 74.

Richardson, J. R. 1981. Brachiopods in mud: Resolution of a dilemma. Science 211:1161–1163.

Rowell, A. J. 1982. The monophyletic origin of the Brachiopoda. Lethaia 15:299–307.

Rudwick, M. J. S. 1970. *Living and Fossil Brachiopods.* Hutchinson University Library, London.

Steele-Petrovic, H. M. 1976. Brachiopod food and feeding processes. Paleontology 19(3):417–436.

Thayer, C. W. 1985. Brachiopods versus mussels: Competition, predation, and palatability. Science 228:1527–1528.

Watabe, N. and C.-M. Pan. 1984. Phosphatic shell formation in atremate brachiopods. Am. Zool. 24:977–985.

Williams, A. 1956. The calcareous shell of the Brachiopoda and its importance to their classification. Biol. Rev. 31:243–287.

Williams, A. 1977. Differentiation and growth of the brachiopod mantle. Am. Zool. 17:107–120.

Chapter Twenty-Two

Phylum Echinodermata

Some of the most familiar seashore animals are members of the phylum Echinodermata (Greek *echinos,* "spiny"; *derma,* "skin"). There are about 7,000 living species, including the sea lilies, feather stars, sea stars, brittle stars, sea urchins, sand dollars, and sea cucumbers (Figures 1, 2, and 3). Another 13,000 or so species are known from a rich fossil record dating back at least to early Cambrian times.

Echinoderms range from tiny sea cucumbers and brittle stars less than 1 cm in size to sea stars that exceed 1 m in diameter and sea cucumbers that reach 2 m in length. Except for a few brackish-water forms, echinoderms are strictly marine; invasion of land and fresh water has been prevented by their cutaneous gas exchange methods and their lack of excretory–osmoregulatory structures. In the sea, however, they are widely distributed in all oceans at all depths. With the exception of a few odd pelagic sea cucumbers (Figures 1P and Q) and one (*Rynkatropa pawsoni*) that is commensal on deep-sea angler fish, all echinoderms are benthic. In the deep sea they may constitute more than 90 percent of the benthic biomass.

Echinoderms are deuterostomes, and their development is frequently cited as typical of that assemblage. With a few exceptions, living echinoderms possess a well developed coelom, an endoskeleton composed of unique calcareous **ossicles**, and pentamerous radial symmetry. This symmetry is second-arily derived, both evolutionarily and developmentally, and the larval forms are always bilateral. Among these and other defining characteristics (Box 1) is a uniquely echinoderm feature known as the **water vascular system** (Figure 5A), which is derived from the coelom as a complex system of channels and reservoirs that serve a variety of functions.

Taxonomic history and classification

Echinoderms have been known since ancient times; likenesses appear in 4,000-year-old frescoes of Crete. Jacob Klein is credited with coining the name Echinodermata in about 1734 in reference to sea urchins. Linnaeus placed the echinoderms in his taxon Mollusca, along with a mixed bag of other invertebrates. For nearly 100 years these animals were allied with various other groups, including the cnidarians, in Lamarck's Radiata. It was not until 1847 that Frey and Leukart recognized the echinoderms as a distinct taxon.

Since the middle of the nineteenth century, the controversies have centered on classification within the phylum, and arguments continue today. The abundant fossil record has been both a blessing and a burden, because authors have treated the fossil evidence in different ways. Some emphasize differences between morphological types and assign higher categorical ranks to nearly every fossil taxon discovered; consequently, certain schemes recognize as many as 25 separate classes of echinoderms. Others apply the evidence more parsimoniously, seeking to establish fundamental similarities; these schemes recognize fewer classes. A recent analysis of extant and fossil forms by Paul and Smith (1984) offers one view on echinoderm phylogeny and classification, and we basically follow their scheme. Their evidence

A

B

C

G

H

I

L

M

O

P

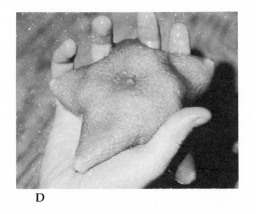

D

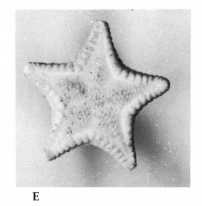

E

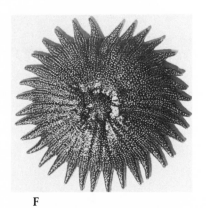

F

J

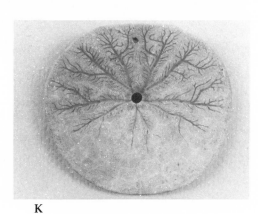

K

N

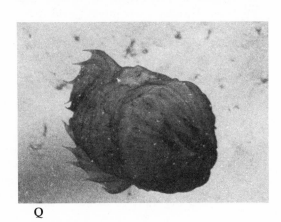

Q

Figure 1

Representative echinoderms. A, Three crinoids (class Crinoidea). B, *Luidia phragma* (class Asteroidea, order Platyasterida). C, *Astropecten armatus* (class Asteroidea, order Paxillosida). D, *Pteraster tesselatus* (class Asteroidea, order Spinulosida). E, *Odontaster crassus* (class Asteroidea, order Valvatida). F, *Heliaster microbranchius* (class Asteroidea, order Forcipulatida). G, The recently discovered "sea daisy" *Xyloplax medusiformis* (class Concentricycloidea). H, A brittle star, *Ophiopholis aculeata* (class Ophiuroidea). I, A basket star, *Astrocaneum* (Ophiuroidea). J, A young *Strongylocentrotus purpuratus* (class Echinoidea, superorder Echinacea) feeding on kelp. Note the elongate podia. K, Oral view of the test of *Dendraster excentricus*, a sand dollar (class Echinoidea, order Clypeasteroidea). L, An "irregular" sea urchin, *Lovenia* (class Echinoidea, superorder Atelostomata). M, A common sea cucumber, *Eupentacta* (class Holothuroidea, subclass Dendrochirotacea). The tentacles are retracted, but the rows of podia are clearly visible. N, The strange deep-sea sea cucumber *Scotoplanes*, which lacks podia on the "upper" surface (class Holothuroidea, subclass Aspidochirotacea). O, *Euapta* (class Holothuroidea, subclass Apodacea). P, A pelagic holothurian, *Pelagothuria* (class Holothuroidea, subclass Aspidochirotacea). Q, An epibenthic swimming holothurian, *Enypniastes* (class Holothuroidea, subclass Aspidochirotacea), photographed at 1,586 meters. (A courtesy of C. Arneson; B,F,I,J,K,L,O,P photos by the authors; C courtesy of A. Kirstitch; D,H,M courtesy of P. Fankboner; E,N courtesy of G. McDonald; G after Baker et al. 1986; Q courtesy of S. Ohta.)

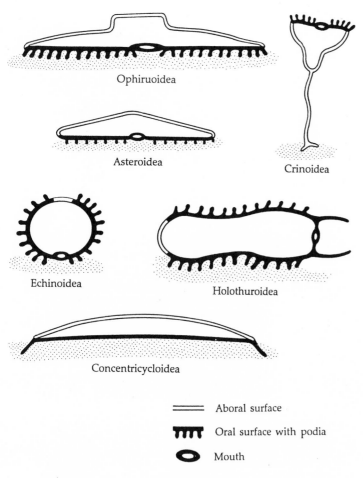

Aboral surface

Oral surface with podia

Mouth

Figure 2
Schematic sections of the six living classes of echinoderms, showing body orientations to the substratum and disposition of the oral and aboral surfaces. (Modified from Russell-Hunter 1979.)

suggests that the living echinoderms are divisible into two monophyletic clades, for which they have resurrected the subphyla Pelmatozoa and Eleutherozoa. This arrangement has been rejected by some specialists, and other schemes abound in the literature. The classification below includes mainly the living echinoderms, but we introduce some of the important fossil forms in the phylogeny section at the end of the chapter.

PHYLUM ECHINODERMATA

SUBPHYLUM PELMATOZOA: Sea lilies, feather stars, and extinct cystoids. Body form as a cup or calyx, with oral surface directed upward; aboral stalk arising from calyx; ambulacra on arms or brachioles; ambulacral grooves open; skeletal plates fused in calyx but articulated elsewhere; madreporite absent; mouth and anus on oral surface.

CLASS CRINOIDEA: Sea lilies and feather stars (Figures 1A, 3A, and 3B). Ambulacra may branch more than once, branches equal; arms bear pinnules. About 625 living species. (e.g., *Antedon*, *Asterometra*, *Comantheria*, *Comanthina*, *Florometra*, *Isometra*, *Metacrinus*, *Neometra*, *Oligometra*, *Phixometra*, *Zygometra*)

SUBPHYLUM ELEUTHEROZOA: Body form stellate, discoidal, globular, or sausage-shaped; oriented with oral side down or lying horizontally; without aboral stalk; body with or without arms, never with brachioles; ambulacral grooves open or closed; skeletal plates separate, or fused as a rigid test; madreporite present.

CLASS ASTEROIDEA: Sea stars (Figures 1B–F and 3C). Body stellate with five or more arms; arms not set off from central disc by distinct articulations; ambulacral grooves open; tube feet with internal ampullae, with or without suckers; madreporite aboral on CD interambulacrum. About 1,500 extant species in five orders (listed but not fully diagnosed below).

ORDER PLATYASTERIDA: Often considered most primitive asteroids; tube feet lack suckers, and animals restricted to soft substrata; anus absent. Living species confined to two genera: *Luidia* with about 60 species, and the monotypic *Platyasterias* (*P. latiradiata*).

ORDER PAXILLOSIDA: Upper surface with numerous umbrella-like arrangements of ossicles called paxillae; tube feet without suckers; anus present or absent; epibenthic or shallow burrowers. About 400 known species. (e.g., *Astropecten*, *Caymanostella*, *Ctenodiscus*, *Lethmaster*)

ORDER VALVATIDA: Tube feet with suckers; anus present. Widely distributed group with several hundred species. (e.g., *Amphiaster*, *Archaster*, *Asterodon*, *Chaetaster*, *Hoplaster*, *Linckia*, *Odontaster*, *Oreaster*)

ORDER SPINULOSIDA: With 5–18 arms; tube feet with suckers; anus present; generally lacking pedicellariae. With a few hundred species. (e.g., *Acanthaster*, *Dermasterias*, *Echinaster*, *Henricia*, *Patiria*, *Pteraster*, *Remaster*, *Solaster*)

ORDER FORCIPULATIDA: With 5–50 arms; tube feet with suckers; anus present; with pincer-like pedicellariae. Widely distributed, and including most intertidal forms. Several hundred extant species. (e.g., *Asterias*, *Astrometis*, *Evasterias*, *Heliaster*, *Leptasterias*, *Pisaster*, *Pycnopodia*, *Stylasterias*)

CLASS CONCENTRICYCLOIDEA: Sea daisies (Figures 1G, 3D, and 3E). Body discoidal, with ring of marginal spines but without radiating arms; skeletal plates arranged concentrically; suckerless podia arranged in a ring near body margin; two ring canals, one marginal, one submarginal, with hydropore on CD interambulacrum; five large plates on "dorsum" mark positions of ambulacra; gut present or absent; anus never present. Recently discovered in deep water associated with bacteria-rich sunken wood; less than 1 cm diameter; two species: *Xyloplax medusiformis* and *X. turnerae*.

CLASS OPHIUROIDEA: Brittle stars and basket stars (Figures 1H, 1I, and 3F). Body stellate with five unbranched or branched articulated ambulacral arms; arms clearly set off from central disc; ambulacral grooves closed; tube feet with internal ampullae but without suckers; anus lacking; madreporite on CD interambulacral plate on oral surface. About 2,000 extant species. (e.g., *Amphiura, Asteronyx, Astrocaneum, Astrophyton, Gorgonocephalus, Ophiacantha, Ophiocoma, Ophioderma, Ophiomusium, Ophionereis, Ophiopholis, Ophiothrix, Ophiura*)

Box One

Characteristics of the Phylum Echinodermata

1. Calcareous endoskeleton arising from mesodermal tissue and composed of separate plates or ossicles; each plate formed as a single calcite crystal and developed as an open meshwork structure called a stereom, the interstices of which are filled with living tissue (the stroma)
2. Adults with basic pentamerous radial symmetry derived from bilaterally symmetrical larvae (when present); body parts organized about an oral–aboral axis
3. Coelomic water vascular system composed of a complex series of fluid-filled canals, usually evident externally as muscular podia
4. Embryogeny primitively deuterostomous, with radial cleavage, entodermally derived mesoderm, enterocoely, and mouth not derived from the blastopore
5. Gut complete except where secondarily incomplete or lost
6. No excretory organs
7. Circulatory structures, when present, compose a hemal system derived from coelomic cavities and sinuses
8. Nervous system diffuse, decentralized, usually of a nerve net, nerve ring, and radial nerves
9. Mostly gonochoristic; development direct or indirect

CLASS ECHINOIDEA: Sea urchins and sand dollars (Figures 1J–L and 3G–I). Body globose or discoidal, often secondarily bilateral; skeletal plates fused as solid test; with movable spines; ambulacral grooves closed; two extant subclasses.

SUBCLASS PERISCHOECHINOIDEA: Test globular, each ambulacral plate bears one tube foot; spines large, pencil-like; anus at aboral pole; dermal gills absent; mostly extinct; often considered primitive in the class. Pencil urchins; about 140 surviving species. (e.g., *Cidaris, Eucidaris, Psychocidaris*).

SUBCLASS EUECHINOIDEA: Test globular or discoidal; numbers of tube feet and spines per plate vary; anal position varies from aboral to "posterior." "True" urchins; about 800 species in 4 superorders.

SUPERORDER DIADEMATACEA: Anus aboral; ambulacral plates compound (from fusion of several plates); gills usually present; primary skeletal tubercles bear terminal pits (perforate); radial symmetry. Includes various "regular" urchins. (e.g., *Aspidodiadema, Caenopedina, Diadema, Plesiodiadema*).

SUPERORDER ECHINACEA: Anus aboral; ambulacral plates compound; primary tubercles imperforate; radial symmetry. Includes most of the common "regular" urchins. (e.g., *Arbacia, Echinometra, Echinus, Heterocentrotus, Paracentrotus, Salenia, Strongylocentrotus, Toxopneustes, Tripneustes*)

SUPERORDER GNATHOSTOMATA: Anus variably shifted from aboral center; ambulacral plates simple; primary tubercles perforate when present; bilateral symmetry; two extant orders.

ORDER HOLECTYPOIDA: Mostly extinct, often viewed as intermediate between "regular" and "irregular" urchins; anus slightly posterior on aboral surface; no petaloids. Three extant species in two genera: *Echinoneus* and *Micropetalon*.

ORDER CLYPEASTEROIDA: True sand dollars. Body discoidal; anus shifted to posterior edge or onto oral surface; aboral ambulacra modified as petaloids, bearing flattened podia for gas exchange; disc margin sometimes with deep notches or perforations (lunules) (e.g., *Clypeaster, Dendraster, Echinarachnius, Echinocardium, Echinodiscus, Encope, Fibularia, Mellita*).

SUPERORDER ATELOSTOMATA: "Irregular" urchins. Anus moved posteriorly along CD interambulacrum; ambulacral plates simple; petaloids absent or present on aboral surface with podia modified for gas exchange; gills absent; primary tubercles perforate when present; test round to ovoid, variously bilateral, often flattened orally. Includes the spatangoid heart urchins and other "irregular" urchins. (e.g., *Cassidulus, Echinolampus, Lovenia, Maretia, Metalia, Urechinus*)

CLASS HOLOTHUROIDEA: Sea cucumbers (Figures 1M–Q; and 3J–K). Body fleshy, sausage-shaped, elongate on oral–aboral axis; skeleton usually reduced to isolated ossicles; symmetry pentamerous or secondarily modified by loss of "dorsal" (bivium) tube feet along ambulacra C and D; tube feet sometimes entirely absent; madreporite internal; ambulacral grooves closed; with circlet of feeding tentacles around mouth. About 1,150 extant species in three subclasses.

SUBCLASS DENDROCHIROTACEA: With 8–30 oral tentacles ranging from digitiform to highly branched; tentacles and oral region with retractor muscles; tube feet present. (e.g., *Cucumaria, Eupentacta, Psolus, Thyone*)

SUBCLASS ASPIDOCHIROTACEA: With 10–30 leaflike or shieldlike oral tentacles; oral region lacks retractor muscles; tube feet present. (e.g., *Enypniastes, Holothuria, Isostichopus, Parastichopus, Pelagothuria, Scotoplanes, Stichopus*)

SUBCLASS APODACEA: With up to 25 tentacles; tentacles vary from digitate to pinnate; tube feet highly reduced or absent. (e.g., *Caudina, Euapta, Leptosynapta, Molpadia, Synapta, Trochoderma*)

The echinoderm *Bauplan*

The key to the success of the echinoderm *Bauplan* may lie in the exploitation of radial symmetry imposed upon a relatively "advanced" coelomate architecture, including a mesodermally derived calcareous endoskeleton. We have seen the tendency among radially symmetrical animals to be either sessile or planktonic and to face their environments on all sides as suspension feeders or passive predators. This generalization applies not only to those creatures with primary radial symmetry (e.g., cnidarians) but also to those that have secondarily become "functionally radial" by way of a sessile lifestyle (e.g., tube-dwelling polychaetes, entoprocts, ectoprocts, phoronids, and others). Many echinoderms, on the other hand, have combined mobility with radial symmetry. Like other radially arranged animals, the echinoderms have a noncentralized nervous system, a feature allowing them to engage their environments equally from all sides.

Much of the biology of echinoderms is associated with their unique **water vascular system**, which is derived largely from specialized parts of the left mesocoelic portion of their tripartite coelom. The water vascular system is a complex of fluid-filled canals and reservoirs that aid in internal transport and hydraulically operate fleshy projections called **tube feet**. The external parts of the tube feet, or **podia**, serve a variety of functions including locomotion, gas exchange, feeding, attachment, and sensory reception.

These versatile structures have contributed greatly to the success of the echinoderm *Bauplan*.

Modern echinoderms are basically pentaradial creatures, although several secondarily derived conditions exist. In the general case, five sets of body parts are oriented about an oral–aboral axis. Except for the concentricycloids, the oral surface of echinoderms is defined by the presence of **ambulacral grooves** and their associated podia (Figure 2); the podia lie along radii called **ambulacra**. A radius bisecting adjacent ambulacra is called an interambulacrum. In a sea star, for example, the ambulacra are represented by the arms, and the interambulacra by the areas between the arms. In many echinoderms (e.g., ophiuroids and echinoids), the ambulacra are not marked by external furrows, in which case the animals are said to have "closed" ambulacral grooves.

The pentaradial symmetry of modern echinoderms is thought to have evolved from a triradiate plan; such a condition occurs in an extinct group called the helicoplacoids (Figure 19B). Although it may not be immediately obvious, the pentamerism of all echinoderms can be described in terms of reference to particular radii. When present externally, the position of the opening to the water vascular system (the **madreporite**) gives a clue to body orientation, because it lies on a particular interambulacrum. A system of lettering has been developed in which the ambulacrum opposite the madreporite is coded A; the others are then noted B through E in a counterclockwise fashion viewed from the aboral surface (Figure 3C). Thus, the madreporite lies between ambulacra C and D (i.e., on the CD interambulacrum). Radii C and D are said to compose the **bivium**, while radii A, B, and E compose the **trivium** (Figure 3C).

As we explore more details about echinoderms, keep these generalities in mind and treat echinoderm diversity as variations on this theme.

Figure 3

External anatomy of echinoderms. A, *Botryocrinus*, a stalked, fossil crinoid. **B,** *Neometra*, a 30-armed, nonstalked crinoid. **C,** Aboral view of *Ctenodiscus* (Asteroidea). The ambulacral radii are labeled according to convention. **D–E,** Aboral and oral views of *Xyloplax* (Concentricycloidea). **F,** The ophiuroid *Asteronyx* crawling on a gorgonian. Note the highly articulate arms. **G,** The sand dollar *Dendraster* (aboral view). Note the petaloids through which the gas exchange podia extend. **H,** Oral view of *Encope* (Echinoidea). Note the food grooves and lunules. **I,** The sea urchin *Plesiodiadema* has extremely long spines and podia. **J,** *Cucumaria planci*, a dendrochirotacean sea cucumber. **K,** The highly modified pelagic holothurian, *Pelagothuria*. (A modified from various sources; B,C,G,H,K after Hyman 1955; D,E after Baker et al. 1986; F after Barnes 1980; I,J after Cuénot 1948.)

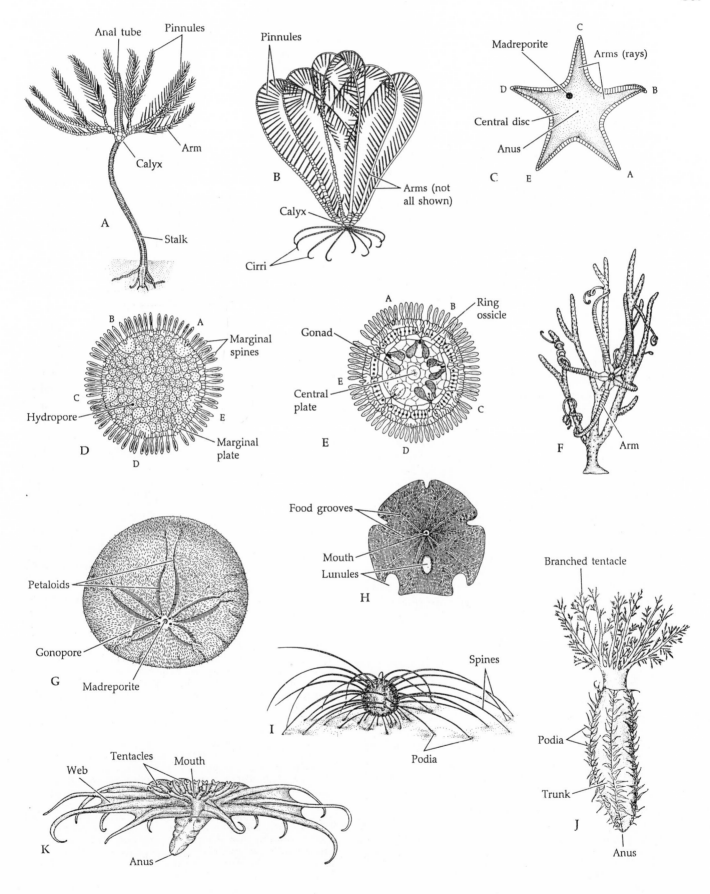

Body wall and coelom

An epidermis covers the body of all echinoderms and overlies a mesodermally derived dermis, which contains the skeletal elements, called **ossicles** (Figures 4A and D). Internal to the dermis and ossicles are muscle fibers or layers and the peritoneum of the coelom. The degree of development of the skeleton and muscles varies greatly among groups. In urchins and sand dollars, the ossicles are fused to form a rigid **test**, and the body wall muscles are weakly developed. In sea cucumbers, however, the ossicles are separate and lie scattered in the fleshy dermis (Figure 4D); here distinct muscle layers are present. Between these extreme conditions are cases where adjacent skeletal plates articulate to various degrees. In the arms of sea stars and brittle stars, for example, the body wall muscles are arranged in bands between the plates, providing various degrees of arm motion. In some groups the skeletal plates are developed to such a degree that they nearly obliterate internal cavities. In brittle stars, for example, each arm "segment" contains a central skeletal ossicle called a **vertebra** (Figures 9A and B) and the arm coeloms are reduced to small channels. Similarly, the arm coeloms in crinoids are greatly reduced by skeletal plates.

The endoskeleton is calcareous, mostly $CaCO_3$ in the form of calcite, with small amounts of $MgCO_3$ added. Developmentally, the skeleton of all echinoderms begins as numerous, separate, spicule-like elements, each behaving as a single calcite crystal. Additional material is deposited on these crystals in various amounts, depending on the ultimate condition of the skeleton. Each ossicle is porous, has an internal meshwork (**stereom**) of lattice-like or labyrinth-like spaces, and generally is filled with dermal cells and fibers (**stroma**). This structure (Figure 4D) is unique to members of the phylum Echinodermata.

During the formation of the skeleton, the plates may remain single (**simple**) or they may fuse to form **compound plates**. In addition, they frequently give rise to bumps and knobs called **tubercles**, to granules, and to various sorts of movable and fixed spines (Figures 4A and E). In some groups, especially the asteroids and echinoids, the skeleton also produces unique pincer-like structures called **pedicellariae** (Figures 4E–I). These structures respond to external stimuli independent of the main nervous system, and they possess their own neuromuscular reflex components. Pedicellariae were discovered by O. F. Müller in 1778, who described them as parasitic polyps and gave them the generic name *Pedicellaria*. He recorded three species of these "parasites" (*P. globifera*, *P. triphylla*, and *P. tridens*); forms of these names

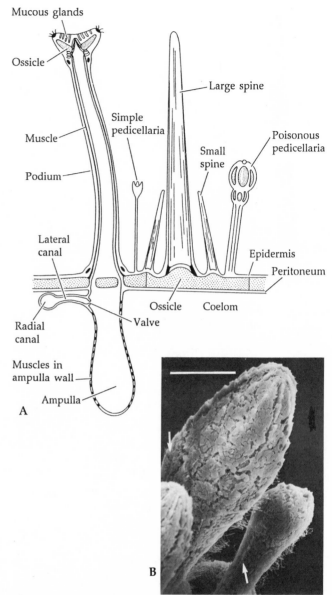

Figure 4

Structure of the echinoderm body wall and some skeletal elements. A, The body wall of an urchin (composite section). B, Spines on the sand dollar *Echinarachnius parma* (scanning electron micrograph). The arrows point to ciliary tracts. Scale bar represents 100 μm. C, Skeletal granules from the central discs of four species of brittle stars (Ophiuroidea), shown in top (top row), side (middle row), and basal (bottom row) views. Scale bar represents 0.05 mm. D, Skeletal ossicles from the holothurian *Psolus chitonoides*. The stereom structure is shown at two magnifications. Scale bar represents 100 μm. E, Four types of echinoid pedicellariae surrounding the base of a large spine. F–G, Elevated pedicellariae used for prey capture by the sea star *Stylasterias forreri*. F, Pedicellariae open and extended. G, Pedicellariae retracted. H, Details of a generalized pedicellaria. I, Two types of muscle systems in pedicellariae. J, A movable spine (section). (A after Barnes 1980, modified from Nichols 1962; B from Ellers and Telford 1984, courtesy of M. Telford; C from Turner 1984, photograph courtesy of R. Turner; D from Emlet 1982, photographs courtesy of R. Emlet; E after Campbell 1983; F,G after Chia and Amerongen 1975; H after Barnes 1980; I,J after Russell-Hunter 1979.)

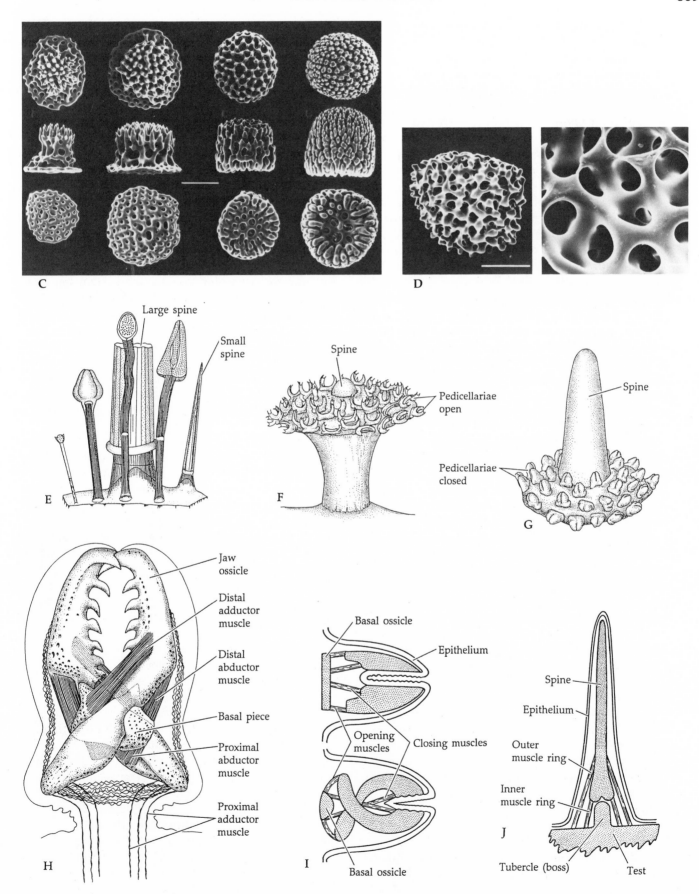

C

D

E

Large spine

Small spine

F

Spine

Pedicellariae open

G

Spine

Pedicellariae closed

H

Jaw ossicle

Distal adductor muscle

Distal abductor muscle

Basal piece

Proximal abductor muscle

Proximal adductor muscle

I

Basal ossicle

Epithelium

Opening muscles

Closing muscles

Basal ossicle

J

Spine

Epithelium

Outer muscle ring

Inner muscle ring

Tubercle (boss)

Test

are still used to describe different types of pedicellariae.

Nearly a century after Müller's discovery, it was realized that pedicellariae are actually produced by the echinoderms, but their exact nature remained elusive. Louis Agassiz believed they were the young of the animals on which they occurred. Even today there are competing opinions about the functions of pedicellariae (see Campbell, 1983, for a review). Pedicellariae differ not only in structural details, but in size and distribution on the body. Some are elevated on stalks, whereas others lie nestled directly on the body surface either singly or in clusters. Some help keep debris and settling larvae off the body, and others are used to defend against larger organisms. The sea urchin *Toxopneustes* bears toxin-producing pedicellariae with which it discourages would-be predators. In some urchins the pincers grasp and hold objects as camouflage and protection. Some sea stars (e.g., *Stylasterias forreri*) use their pedicellariae to capture and hold prey (Figures 4E and F).

Movable spines and pedicellariae contain muscles and other tissues that lie outside the main skeletal framework of the body wall (Figures 4H–J). This arrangement raises some interesting questions concerning the method of nutrient supply to these tissues, because they are isolated from the coelom and gut. Pedicellariae may absorb nutrients directly from the water, or they may actually trap and digest small organisms and then absorb the products (Stephens 1968; Pequignat 1966, 1970; Ferguson 1970).

The coelomic system of echinoderms usually develops as a tripartite series, the proto-, meso-, and metacoels, typical of deuterostomes. However, with the transformation to radial symmetry, these coelomic cavities do not come to lie in the three body regions usually associated with deuterostome *Baupläne*. The main body coeloms are derived from the embryonic metacoels and are well developed in most groups. Other coelomic derivatives include the water vascular system, gonadal linings, and certain neural sinuses.

The main body cavities, or **perivisceral coeloms**, are lined with ciliated peritoneum, and the coelomic fluid plays a major circulatory role. A variety of coelomocytes occurs in the body fluid and in the water vascular system. Many of these cells are phagocytic. Hemoglobin occurs in the coelomocytes of holothurians, but their respiratory physiology is largely unstudied.

Water vascular system

The water vascular system is intimately involved in several major aspects of echinoderm biology, and a discussion of its comparative anatomy among groups is a necessary preface to other considerations. It is perhaps easiest to begin with a detailed examination of the system in a sea star and then treat the other taxa comparatively.

Asteroidea. Figure 5A is a schematic representation of the water vascular system of a sea star. The system opens to the exterior through a special skeletal plate, the **madreporite**, or **sieve plate**, located off-center on the aboral surface on the CD interambulacrum (Figure 3C). The madreporite is deeply grooved, and the overlying epidermis is ciliated and porous where it lines the furrows. The function of the madreporite has been the subject of much controversy. The traditional view that it serves as an avenue for sea water to enter the system has been challenged in recent years. However, recent work using radioactive tracers has demonstrated that, in fact, water does enter through the madreporite (Ferguson 1984). The fluid in the water vascular system is similar to sea water, but it includes various coelomocytes, certain organic compounds such as proteins, and a relatively high concentration of potassium ions. This fluid is moved through the system largely by the action of cilia that line the canal epithelium.

Internally, the madreporite forms a cuplike depression, the lumen of which is called the **ampulla**, that communicates with other coelomic derivatives of the water vascular system and the **hemal system** (see below). From the lower end of the ampulla arises the **stone canal**, so named because of skeletal deposits in its wall. The stone canal descends orally and joins with a circular **ring** or **circumoral canal**, which extends around the central disc on a plane perpendicular to the oral-aboral axis. In addition to a **radial canal** extending into each arm, the ring canal gives rise to blind pouches called **Tiedemann's bodies** and **polian vesicles** (Figures 5A and B). There is some uncertainty about the functions of these pouches, but it is suspected that the former produce certain coelomocytes, and the latter help regulate internal pressure within the water vascular system.

In each arm the radial canal gives rise to numerous **lateral canals**, each of which passes through a pore between skeletal plates and terminates in a tube foot. In most asteroids each tube foot consists of a bulbous **ampulla** and a hollow, muscular, suckered **podium** (Figure 5B). Members of the primitive asteroid orders Platyasterida and Paxillosida lack suckers on their tube feet. The ampullae are internal and lie above the skeletal plates of the ambulacral groove. The podia extend to the outside and contain the usual body wall muscle layers around a coelomic lumen

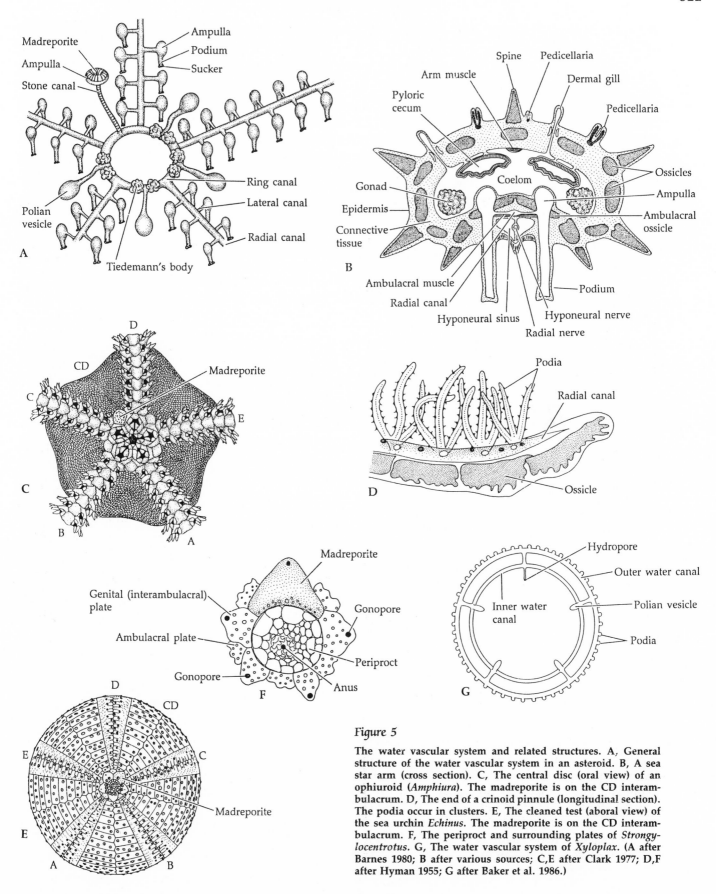

Figure 5

The water vascular system and related structures. A, General structure of the water vascular system in an asteroid. B, A sea star arm (cross section). C, The central disc (oral view) of an ophiuroid (*Amphiura*). The madreporite is on the CD interambulacrum. D, The end of a crinoid pinnule (longitudinal section). The podia occur in clusters. E, The cleaned test (aboral view) of the sea urchin *Echinus*. The madreporite is on the CD interambulacrum. F, The periproct and surrounding plates of *Strongylocentrotus*. G, The water vascular system of *Xyloplax*. (A after Barnes 1980; B after various sources; C,E after Clark 1977; D,F after Hyman 1955; G after Baker et al. 1986.)

and sometimes include supportive ossicles. At the tip of each radial canal is an unsuckered, tentacle-like, sensory **terminal tube foot**. In asteroids the tube feet serve primarily for locomotion and temporary attachment, and to hold prey during feeding. In addition, they are usually highly touch sensitive.

The operation of the tube feet depends on hydraulic pressure regulation, and on muscle action of the individual ampullae and podia. Fluid is supplied to each podium from the main canal system. The ampulla acts as a reservoir for fluid used to operate its associated podium. A valve in the lateral canal effectively isolates the tube foot from the rest of the system. Each podium contains longitudinal muscles, some of which insert on the center of the suction disc. When the ampulla is filled with fluid and the lateral canal valve is closed, the ampulla contracts and forces fluid into the podium to extend it. The sucker is then pressed against the substratum and held there by adhesive secretions of the epidermis. Next the longitudinal muscles of the podium contract; this action shortens the tube foot and forces the fluid back into the now relaxed ampulla. At the same time muscles raise the center of the sucker disc and create a vacuum, like that of a suction cup. Release of the sucker involves relaxation of the podial muscles and contraction of the ampulla; this action again forces fluid into the lumen of the podium and releases the suction. In addition to this attachment–detachment action, the podia are also capable of bending by differential contraction of the longitudinal muscles.

The above description may seem straightforward, but there are still many unresolved questions. The complete nature and source of the water vascular system fluid are unclear because of its coelomic-like contents. The precise function of the madreporite is puzzling, and overall system dynamics are not well understood (isolated arms crawl about normally without difficulty).

Concentricycloidea. The water vascular system of the sea daisies, *Xyloplax*, is unique among the echinoderms (Baker et al. 1986). A madreporite homologue, the **hydropore**, opens on the aboral surface on the CD interambulacrum (Figure 3D) and connects internally to a pair of concentric ring canals (Figure 5G). Polian vesicles lie on the other four interambulacra. The outer, marginal ring canal gives rise to peripherally located suckerless podia. Each podium bears an internal ampulla. This is the only echinoderm water vascular system in which the podia are not arranged along the ambulacra.

Ophiuroidea. The water vascular system of brittle stars is similar to that of asteroids. However, the madreporite is on the oral surface of the central disc, on the CD interambulacrum, and the internal plumbing is modified accordingly (Figure 5C). The ring canal bears polian vesicles, but apparently lacks Tiedemann's bodies. The ring canal gives off the usual five radial canals and also branches to a wreath of **buccal tube feet** around the mouth. The podia lack suckers. They are highly flexible, finger-like structures that secrete copious amounts of sticky mucus. They function primarily as feeding, digging, and sensory organs.

Crinoidea. The water vascular system of crinoids operates entirely on coelomic fluid. There is no external madreporite, rather a number of "stone canals" arise from the ring canal and open to coelomic channels. Some species possess hundreds of such stone canals. The main perivisceral coeloms bear ciliated funnels to the exterior; through these water enters the body cavities, perhaps as an indirect method of regulating hydraulic pressure in the water vascular system.

From the ring canal arise the main radial canals that extend into each arm and paired oral tube feet that appear at each interambulacrum. The number of arms in crinoids ranges from five to as many as 200, and in many cases the arms are branched. The number of radial canals corresponds to the arm number in each species, and they are branched in those with branched arms (as they are in ophiuroid basket stars). Furthermore, crinoid arms bear side branches called **pinnules** (Figure 3B) into which branches of the radial canals extend. Suckerless podia occur along the pinnules, often in clusters of three (Figure 5D). Each cluster is served by a branch of the water vascular system. The podia are highly mobile and usually bear adhesive papillae on their surfaces; they function primarily as feeding and sensory organs.

Echinoidea. The water vascular systems of sea urchins and sand dollars may be viewed as modifications of the asteroid plan. These animals bear a special set of skeletal plates around the aboral pole; one of these plates is the CD-interambulacral madreporite (Figures 5E and F). To understand the water vascular system of sea urchins, it is necessary to realize that the ambulacra, and thus the rows of podia and their internal plumbing, extend around the sides of the body (like five longitude lines on a globe) to the upper surface, where they converge toward the aboral pole. We may interpret this arrangement as the oral surface extending aborally around the sides and top of the animal (Figures 2 and 5E). Thus, the ancestral aboral surface is here represented only by the area around the anus on the animal's upper surface.

The madreporite of echinoids, like that of asteroids, leads to an ampulla and then to a stone canal (short in sand dollars and long in sea urchins) that extends orally to a ring canal surrounding a complex system of muscles and plates that comprise the feeding apparatus. The ring canal gives rise to five radial canals, one beneath each of the ambulacral areas. Each radial canal gives off lateral canals leading to tube feet and terminates in a sensory podium on the aboral surface. Unlike the plates in other echinoderms, the ambulacral plates of echinoids have holes in them through which the podia pass to the outside. The tube feet of echinoids may be suckered or unsuckered, and they serve a variety of functions including attachment, locomotion, feeding, and gas exchange.

Holothuroidea. In sea cucumbers the water vascular system contains the major elements seen in other taxa, but it is organized to accommodate the elongation of the body. In most holothurians the madreporite is internal and opens to the coelom. The madreporite lies beneath the pharynx in the CD-interambulacral position and gives rise to a short stone canal. A ring canal encircles the gut and bears from 1 to 50 polian vesicles. Five radial canals arise from the ring canal and give off extensions to the oral tentacles before extending aborally ("posteriorly") beneath closed ambulacral grooves. In those species that retain clear pentamerous symmetry, each radial canal gives rise to rows of ampullae and suckered podia. In some species the podia of the bivium (the "dorsal" or upper surface) are reduced or lost, and in the apodaceans all of the tube feet are greatly reduced or absent. The podia of holothurians serve in locomotion and attachment, and are touch sensitive.

Support and locomotion

Except for the holothurians, the general body shape and structural support of echinoderms are maintained by the skeletal elements. Particular structures, such as podia and gills, are supported mostly by hydrostatic pressure. In most sea cucumbers, where the skeletal plates are tiny separate ossicles, the body wall muscles form thick sheets, adding structural integrity to the body by working on the coelomic spaces to provide a hydrostatic skeleton.

Recent work by Motokawa (1984) indicates that many echinoderms possess certain connective tissues that contribute to body "tone" by rapid changes in their mechanical properties. In a matter of seconds or minutes the fibers of these tissues can become relatively rigid, thereby reducing body flexibility. This transformation appears to be under direct nervous control but does not involve muscular activity.

There are several locomotor methods among echinoderms, each determined by overall body configuration, the animals' habits, and the nature of the skeletal, muscular, and water vascular systems. Apart from the sessile sea lilies (e.g., *Ptilocrinus*), most extant crinoids are capable of crawling and swimming, both of which are done with the oral side directed away from the substratum. The well developed muscles of the arms account for these movements. The aboral **cirri** (Figure 3B) are used primarily for temporary attachment and for righting the animal if overturned.

During crawling, the arms are bent downward and used to lift the body off of the substratum; the animal then walks on its arm tips (Figures 6A and B). Swimming is accomplished by up and down sweeps of the arms, which are divided into functional sets that move alternately. For example, in 10-armed species, five arms move upward while the other five arms move downward. As any given arm is moving one way, its two neighboring arms are moving the opposite way. In animals with more arms (usually multiples of five), the arms are divided into functional sets of five.

Asteroids exemplify locomotion using podia. The action of a single podium involves power and recovery strokes, the process following the same fundamental mechanical principles we have seen in the appendages of many other invertebrates. The sea star's arms are held more-or-less stationary relative to the central disc, even in species with a flexible skeletal framework (e.g., *Pycnopodia*), and movement is accomplished by the thousands of podia on the oral surface. Overall movement is generally smooth, as a result of the high number of podia and the fact that at any given moment they are in different phases of the power and recovery strokes (Figure 6C). Although there is some coordination of the action of the tube feet to produce movement in a particular direction, there are no metachronal waves of podial motion as seen in many other "multilegged" creatures. Most sea stars move very slowly, but a few (e.g., *Pycnopodia*) are relative speedsters. Some asteroids that are usually rather sedentary become extraordinarily rapid "runners" upon encountering a potential predator (often another sea star). Some species that cannot escape by fast movement have evolved other defense mechanisms. The slow-moving Pacific "cushion star," *Pteraster tesselatus* (Figure 1D), secretes copious amounts of mucus, which serves to discourage predators such as *Solaster* and *Pycnopodia*.

If one can follow the action of a single podium during movement (not an easy assignment), the locomotor forces can be understood (Figure 6D). At the end of a recovery stroke, the podium extends in the direction of movement and attaches to the substratum. The sucker remains attached during the power stroke as the longitudinal muscles in the wall of the podium begin to contract, thereby shortening the podium and pulling the body forward. At the end of the power stroke, the podium lifts from the substratum and swings forward again. As illustrated, the ability to bend the podia is essential to the overall action.

Ophiuroids use their flexible articulated arms primarily for crawling or clinging (Figure 3F). The skeletal arrangement of the arms allows for extensive "lateral" movement on a plane perpendicular to the oral-aboral body axis, but the arms have almost no flexibility parallel to this axis. This feature, coupled with the fragile nature of these animals, causes them to break easily when lifted by an appendage—hence the common name "brittle stars." The tube feet lack suckers and ampullae but are equipped with a well developed lattice of muscles in their walls. They are capable of protraction and retraction and of swinging powerfully through arcs. These actions and the relative strength of the podia are important elements in the digging behavior of burrowing ophiuroids.

Sea urchins move by the use of podia and movable spines. Their long suckered podia are capable of the same types of movements described for the tube feet of sea stars. Many irregular urchins burrow in soft sediments, mostly clean sands, and some regular urchins excavate shallow depressions in hard rock. *Strongylocentrotus purpuratus* (Figure 1J), a common West Coast urchin of North America, forms such pockets in hard substrata, and members of this species often become trapped in their self-made homes. These urchins bore largely by the action of the teeth of their feeding apparatus. These excavations provide protection in areas of high wave and surge action.

Some of the irregular urchins burrow well below the sand surface and maintain an open chimney from their cavern to the overlying water (Figure 11G). Most of these soft-sediment burrowers have special spatulate spines along the sides of the body that aid in digging.

Sand dollars live in or on soft sediments. Some bury themselves completely, but most keep part of the body above the surface (Figure 11F). A few, such as *Clypeaster rosaceus*, do not burrow at all. Burrowing and crawling are accomplished largely by the action of short movable spines. There has been some controversy about the function of the deep marginal notches and holes (**lunules**) in the tests of some sand

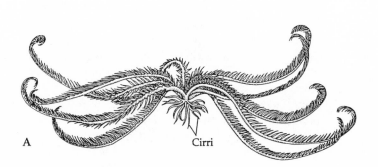

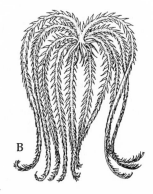

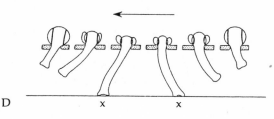

Figure 6

A, *Antedon* (Crinoidea) in a resting position. B, *Antedon* as it might appear walking on its arm tips. C, A sea star arm (side view) with tube feet in motion. D, Changes in position of an individual podium as the animal moves in the direction of the arrow. The podium executes its power stroke while in contact with the substratum (x), and its recovery stroke while lifted from the substratum. Note the changes in podium length and corresponding changes in volume of the ampulla. (A,B after Hyman 1955; C,D after Clark 1977.)

dollars (Figure 3H). Some workers (e.g., Alexander and Ghiold 1980) suggest that these openings play a role in feeding by facilitating the passage of food from the aboral to the oral surface. However, studies by Telford (1981, 1983) indicate that these structures help the animals maintain stability in strong currents. Excessive water pressure is eased by flow along surface channels from the center of the body to the lunules and notches and then away from the test margin.

Holothurians live on the surfaces of various substrata or else burrow into soft sediments. Crawling is accomplished by the podia (especially of the trivium) or by action of the body wall muscles. Many epibenthic species are cryptic and usually remain lodged in cracks and crevices or under rocks. In these forms the podia are used primarily for anchorage and to hold bits of shell and stone against the body for protection. In a few deep-sea forms (e.g., *Scotoplanes*; Figure 1N), some of the podia are elongate and used for walking. In some holothurians (e.g., *Psolus*), the trivium surface is modified as a creeping, footlike sole. A few sea cucumbers are pelagic and capable of weak swimming (Figures 1P and Q).

The apodaceans lack locomotor tube feet, and most burrow in sand or mud by means of peristaltic action of the body wall muscles. Some live completely buried, whereas others form U-shaped burrows.

Feeding and digestion

Echinoderms display a great variety of feeding strategies, and we present only a brief survey here. In addition, the structure of the digestive tract differs among groups, and these systems are described only briefly.

Crinoidea. Sea lilies and feather stars sit with their oral sides up and feed by removing suspended material from the surrounding water. The arms and pinnules are usually held outstretched on a plane perpendicular to the ambient water flow, thus presenting a large food-trapping surface. Crinoids are usually considered downstream feeders (i.e., the feeding surface faces away from the prevailing current, rather than into it). However, at least some species are now known to be upstream feeders, while some others apparently can feed in either direction. Many errant forms are negatively phototactic and emerge from their hiding places to feed only at night. Some deep-water species hold their arms upward and outward, forming a funnel with which they capture detrital rain.

The open ambulacral grooves extend onto the pinnules and are lined with cilia that beat toward the mouth. Food particles, including plankton and organic particulates, contact the podia, which then transport or flick the food into the grooves (Figure 7B). The food is carried to the oral area and ingested. The primitive nature of crinoids suggests that suspension feeding may have been the original function of the water vascular system (i.e., the podia and ambulacral grooves).

The mouth opens to a short esophagus that leads to a long intestine (Figures 7A and C). The intestine loops around the calyx and then straightens to a short rectum terminating in the anus, which is borne on an **anal cone** near the base of one of the arms. In most species the intestine bears diverticula, some of which are branched. Although the histology of the crinoid gut has been described, nothing is known about their digestive physiology.

Asteroidea. Most sea stars are opportunistic predators or scavengers. They feed on nearly any dead animal matter and prey on a variety of invertebrates. Many species are generalists in terms of their food preferences and often play important roles as high-level predators in intertidal and subtidal communities. Others are strict specialists. For example, *Solaster stimpsoni*, a large northeastern Pacific sea star, feeds exclusively on holothurians, while a related species (*S. dawsoni*) preys on *S. stimpsoni*!

Among the best known sea stars is the tropical "crown-of-thorns," *Acanthaster planci*. This animal feeds on coral polyps and has received great notoriety in recent years because of its implication in the destruction of Indo-West Pacific coral reefs. There is still disagreement concerning the reason for the increase in population size of *Acanthaster*, but some specialists think that it is a result of human interference in the predator–prey balance of the reef communities. Among the major predators of *Acanthaster* is the giant triton, *Charonia* (Gastropoda), which is collected in high numbers for its handsome shell.

Except for a few suspension feeders (discussed below), most sea stars depend on an eversible portion of their stomach to obtain food. Some forms, including *Acanthaster*, *Culcita* (the cushion star), and *Patiria* (the bat star), spread their stomachs over the surface of a food source, secrete primary enzymes, and suck in the partially digested soup. In the case of *Culcita*, the food may include encrusting sponges or algal mats, or organic detritus that has accumulated on the substratum. *Patiria* feeds in much the same manner, digesting organic matter under its spread everted

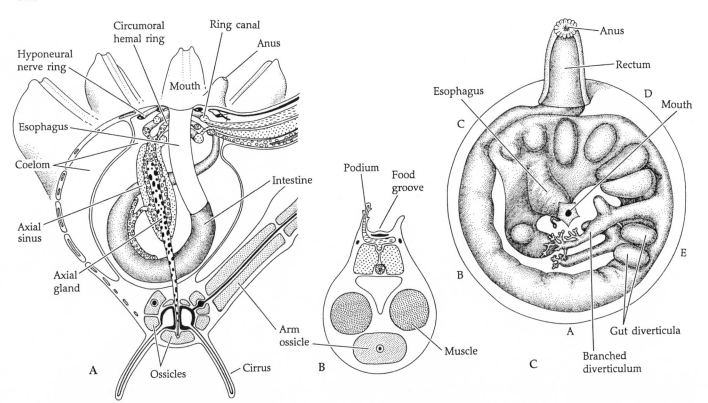

Figure 7

Internal anatomy of crinoids. A, Central disc and base of one arm (vertical section). B, An arm with open ambulacral (food) groove (cross section) C, The oral surface of *Antedon* (cutaway view). The positions of ambulacral radii are indicated by the letters around the periphery. (A,B after Nichols 1962; C after Cuénot 1948.)

stomach. *Oreaster* extrudes its stomach over sand, algae, or sea grass and ingests the associated microorganisms and particulate detritus. It can, however, switch to a predatory or scavenging mode when appropriate food sources are encountered.

Many sea stars that feed on large prey also utilize external digestion by everting their stomachs. Sedentary or sessile prey, such as gastropods, bivalves, and barnacles, are eaten by a host of asteroid predators. The sea star hunches over the prey with the oral area pressed against the potential victim, holding itself in position with its podia. It then everts the stomach and begins secreting digestive enzymes. The stomach is very thin and flexible; it can be slid between even the tightly clamped valves of mussels and clams, thus liquefying the prey's body inside its own shells. The fluid nutrients are drawn in with the retracting stomach.

Some sea stars are suspension feeders, consuming plankton and organic detritus. *Henricia* (Figure 8D), *Porania*, and a few others are typically full-time

suspension feeders, and some predatory types, such as *Astropecten*, are capable of periodic suspension feeding as a means of supplementing their usual diet. In most of these cases particulate food material that contacts the body surface is moved along with mucus to the ambulacral grooves and ultimately to the mouth. Food movement is by ciliary action. *Leptasterias tenera* is able to capture suspended food, such as phytoplankton and small crustaceans, with its pedicellariae and tube feet.

The digestive system of sea stars extends from the mouth in the center of the oral surface to the anus in the center of the aboral surface (Figure 8A). The mouth is surrounded by a leathery **peristomial membrane**. The membrane is flexible, allowing eversion of the stomach, and contains a sphincter muscle to close the mouth orifice. Internal to the mouth is a very short esophagus leading to the **cardiac stomach**, which is the portion that is everted during feeding. Radially arranged retractor muscles serve to pull the stomach back within the body. Aboral to the cardiac stomach is a flat **pyloric stomach**, from which arises a pair of **pyloric ducts** extending into each arm. These ducts lead to paired digestive glands or **pyloric ceca** in each arm (Figure 8). A short intestine, which leads from the pyloric stomach to the anus, often bears outpocketings called **rectal glands** or **rectal sacs**.

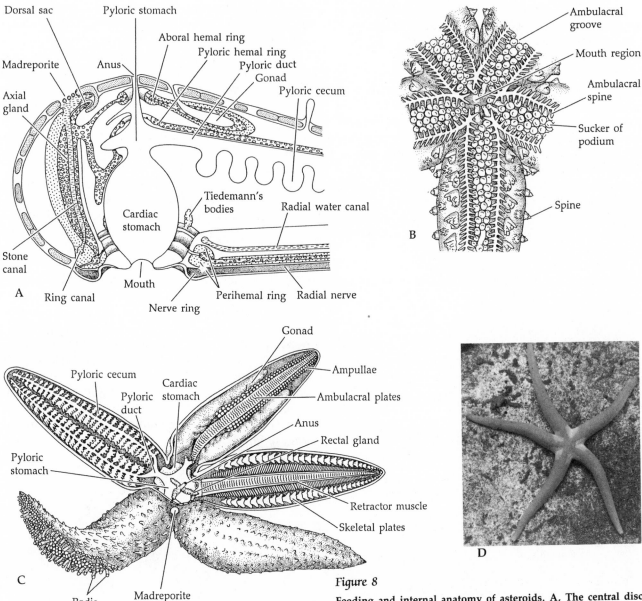

Figure 8

Feeding and internal anatomy of asteroids. A, The central disc and base of one arm of a sea star (vertical section). B, *Asterias* (oral view). The mouth is ringed by oral spines and podia. C, The internal organs in the central disc and arms of the trivium of *Asterias*. Each dissected arm has various organs removed. D, *Henricia*, a suspension-feeding sea star. (A after Nichols 1962; B,C after Barnes 1980; D photo by the authors.)

The digestive glands and cardiac stomach are the main sites of enzyme production. These enzymes, mostly proteases, are carried by ciliary action through the everted stomach and released onto the food material. Chemical digestion is completed internally, but extracellularly, after ingestion of the liquefied food. Digested products are moved through the pyloric ducts to the digestive glands, where they are absorbed and stored. The intestine apparently serves little purpose in the digestive process, but the rectal sacs are known to pick up nutrients from the intestine, probably salvaging them from potential loss through the anus.

Many sea stars harbor various commensals that derive their food from scraps of the host's meals. One well known relationship is that of a polynoid scale worm, *Arctonoe vittata*, and several species of asteroid hosts, including the Pacific leather star, *Dermasterias imbricata*. The worm is an obligate symbiont, spending most of its life cruising and feeding in the host's ambulacral grooves. Not only is the polychaete chemically attracted to its host, but recent studies indicate that *Dermasterias* is also attracted to *Arctonoe*; this observation suggests that the sea star also may derive some benefit from the association.

Concentricycloidea. *Xyloplax medusiformis* lacks a digestive system, but the oral surface is covered by a membranous **velum** that may have been derived from the gut (Baker et al. 1986). These animals may absorb dissolved organic matter across this velum. Perhaps the source of the nutrients is bacteria that live in the decomposing-wood habitat of these strange echinoderms.

Xyloplax turnerae has an incomplete gut. The mouth opens into a shallow, saclike stomach but intestine and anus are lacking.

Ophiuroidea. Brittle stars exhibit a variety of feeding methods, including predation, deposit feed-

ing, scavenging, and suspension feeding; some species are capable of feeding by more than one process. Some forms, such as the basket stars (Figure 9F), are really predators that utilize suspension feeding strategies to capture relatively large swimming prey (up to about 3 cm long).

Selective deposit feeding is accomplished by the podia and sometimes by the arm spines. The epidermis of the arms secretes mucus, to which organic material adheres. The podia roll the mucus and food into a clump, or food bolus. Near the base of each podium is a flaplike projection called a **tentacular scale** (Figures 9B and D). The food bolus is transferred from a podium onto its adjacent scale, picked up by the next podium, and so on, a process that transports the food along the arm to the mouth. Suspension feeding by brittle stars usually involves a similar method of transport once food is trapped. Food capture is sometimes accomplished by secreting mucous threads among the arm spines and waving the arms about to trap plankton and drifting organic detritus. Brittle stars that use this technique typically have very long arm spines (e.g., *Ophiocoma*, *Ophiothrix*, *Ophionereis*). The food is moved to the podia and then transported to the mouth. Other brittle stars suspension feed by using extended podia to form a trap; then the podia pass clumps of food to the mouth (Figure 9E).

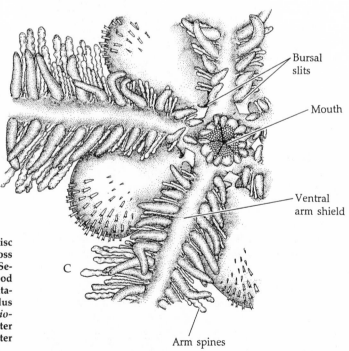

Figure 9

Feeding and internal anatomy of ophiuroids. A, The central disc and base of one arm (vertical section). B, An ophiuroid arm (cross section). C, The central disc of *Ophiothrix* (oral view). D, Sequence (1–5) of movements of a single podium as it passes food toward the mouth by scraping the podium on an adjacent tentacular scale (*Ophionereis fasciata*). E, Podia moving food bolus toward the mouth in the suspension-feeding ophiuroid *Ophiothrix fragilis*. F, The basket star, *Gorgonocephalus*. (A,B after Nichols 1962; C after Barnes 1980; D after Pentreath 1970; E after Warner and Woodley 1975; F courtesy of G. McDonald.)

Predatory suspension feeding by basket stars occurs mostly at night. At dusk the animals emerge from their hiding places and assume a feeding position, with their branched arms held fanlike into the prevailing current. *Astrophyton muricatum* changes position with the ebb and flow of the tide, always orienting its arms into the current; it stops feeding at slack tide (Hendler 1982b). When a small animal contacts an arm, the appendage curls to capture the prey. Ingestion is often postponed until darkness has passed, then the prey is transferred to the mouth by the flexible arm. These basket stars feed on a variety of invertebrates, such as swimming crustaceans and demersal polychaetes.

Some brittle stars are active predators, capturing benthic organisms by curling an arm into a loop around the prey, then pulling it to the mouth. Species that feed in this manner usually have short arm spines that lie flat against the arm itself (e.g., *Ophioderma*). Several species of brittle stars dig beneath the surface of the substratum and form semipermanent mucus-lined burrows. The arms extend to the surface and help maintain ventilation currents within the burrows. Such species are able to extract food from within the sediment, the substratum surface, and the overlying water (Woodley 1975).

The commensal brittle star *Ophiothrix lineata* lives in the atrium of the large sponge *Callyspongia vaginalis*, emerging to feed on detritus adhering to the host's outer surface. While keeping the sponge clean, the ophiuroid is supplied with food and afforded protection from predators (Hendler 1983).

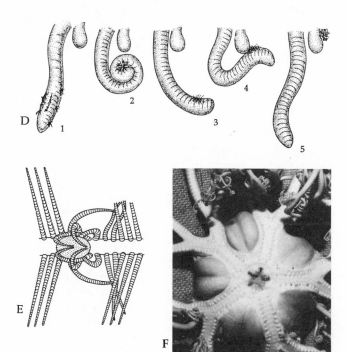

The digestive tract of ophiuroids is reduced relative to that of asteroids. The intestine and anus have been lost and the system is confined entirely to the central disc (Figure 9A). The mouth leads to a short esophagus and large folded stomach, which mostly fills the interior of the disc and reduces the coelom to a thin chamber. The stomach is presumably the site of digestion and absorption.

Echinoidea. Food-getting strategies among echinoids include various kinds of herbivory, suspension feeding, and detritivory and a few forms of predation. In radially symmetrical regular urchins, feeding depends in large part on the action of a complex masticatory apparatus that lies just inside the mouth and bears five protractable calcareous teeth. This apparatus is commonly called **Aristotle's lantern** (Figures 10 and 11A–D). It is a real architectural marvel: a complex of hard plates and muscles that control protraction, retraction, and grasping movements of the five teeth. In many species the entire apparatus can be rocked such that the teeth protrude at different angles. There is some variation in the lantern structure among the regular urchins, but the following brief description applies to most conditions (Figure 10).

The main structural elements of the apparatus are five vertically oriented triangular plates called **pyramids**. The calcareous pyramids are positioned in interambulacral spaces and are attached to one another by **comminator muscles**, which provide the rocking motion of the pyramids. The aboral edge of each pyramid is a thickened bar called an **epiphysis**. Each pyramid has a canal within which lies a tooth. The sharp end of the tooth extends out from the oral end of the pyramid into the mouth region. A soft **dental sac** of coelomic origin covers the unhardened aboral end of each tooth where it emerges from the top of the pyramid. As the teeth are worn down by use, more tooth material is produced within the dental sacs and becomes calcified as it grows through the pyramid canal. Measurements on some species indicate that, with normal wear, teeth grow about 1 mm each week. Lying atop the main structure of the lantern, on the oral surface, are five **compasses** and five **rotules**, one of each along each ambulacral radius. The compasses and their associated muscles regulate hydrostatic pressure within the gills (see below).

The teeth are protracted by pushing the entire lantern orally by contraction of sheetlike protractor muscles that originate around the mouth (on the interambulacral areas of the internal skeleton) and in-

Figure 10

The feeding complex (Aristotle's lantern) in sea urchins. A, The feeding complex in a regular urchin as seen from inside the test. B, The feeding apparatus of *Paracentrotus* (vertical section). C, The apparatus of *Cidaris* (aboral view). The compasses are removed to expose the rotules. (A after various sources; B,C after Hyman 1955.)

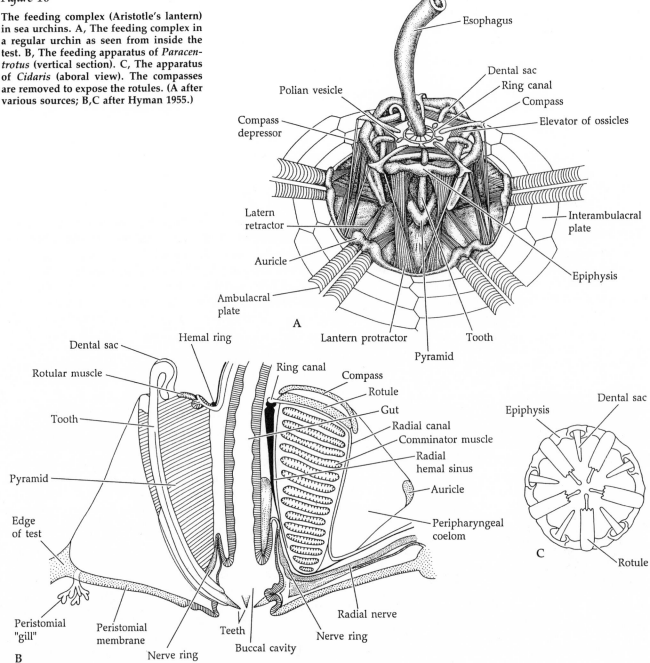

sert on the epiphyses, near the aboral end of each pyramid. Their action also serves to spread the teeth apart as protraction occurs. Retractor muscles originate on thick ambulacral plates called **auricles**, and they insert on the oral end of the lantern apparatus. Additional muscles associated with the pyramids and the rotules can produce a variety of teeth movements.

Most regular urchins use their teeth to scrape algal material from the substratum and to tear chunks of food into "bite-sized" pieces. Many species also feed on animal matter by similar actions. Some regular urchins excavate burrows in hard substrata and then feed on the algal film that develops on the burrow wall, or else they feed on suspended particles or drift algae that enter the chamber. Other burrowers establish a feeding position at the burrow entrance and catch floating debris with their podia and pedicellariae. Irregular urchins (sea biscuits and heart ur-

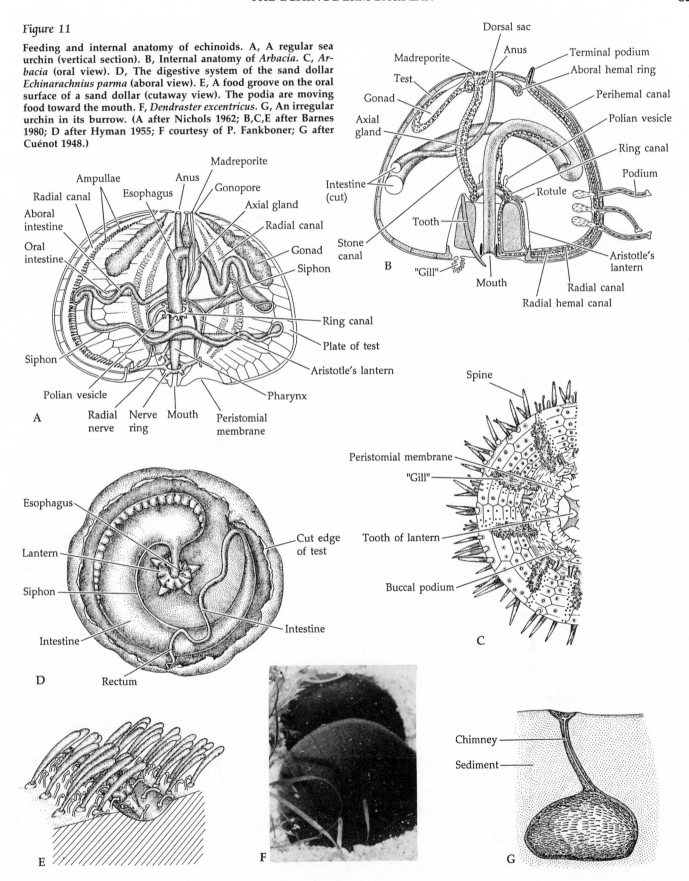

Figure 11

Feeding and internal anatomy of echinoids. A, A regular sea urchin (vertical section). B, Internal anatomy of *Arbacia*. C, *Arbacia* (oral view). D, The digestive system of the sand dollar *Echinarachnius parma* (aboral view). E, A food groove on the oral surface of a sand dollar (cutaway view). The podia are moving food toward the mouth. F, *Dendraster excentricus*. G, An irregular urchin in its burrow. (A after Nichols 1962; B,C,E after Barnes 1980; D after Hyman 1955; F courtesy of P. Fankboner; G after Cuénot 1948.)

Dorsal sac

Anus

Madreporite

Test

Gonad

Axial gland

Intestine (cut)

Stone canal

"Gill"

Tooth

Mouth

Terminal podium

Aboral hemal ring

Perihemal canal

Polian vesicle

Ring canal

Podium

Rotule

Aristotle's lantern

Radial canal

Radial hemal canal

B

Madreporite

Ampullae

Anus

Esophagus

Gonopore

Radial canal

Axial gland

Aboral intestine

Radial canal

Oral intestine

Gonad

Siphon

Siphon

Ring canal

Plate of test

Aristotle's lantern

Polian vesicle

Pharynx

Radial nerve

Nerve ring

Mouth

Peristomial membrane

A

Spine

Peristomial membrane

"Gill"

Tooth of lantern

Buccal podium

C

Esophagus

Lantern

Siphon

Intestine

Cut edge of test

Intestine

Rectum

D

E

F

Chimney

Sediment

G

chins) burrow into soft sediments and feed on organic particles (Figure 11G). These types of urchins usually use their podia to sort food material from the mud or sand and pass it to the mouth. A lantern is absent from most irregular urchins, and food is limited to small particles.

Most sand dollars (Clypeasteroida) are detritus and particulate feeders. They possess a highly modified lantern with nonprotractable teeth. Most of these animals burrow completely or partially in soft sediments and extract food particles from among the sand grains or from the overlying water. As most species plow along, a layer of sediment passes over the aboral surface. There is some controversy regarding exactly how sand dollars feed. Large particles are held by the tips of club-shaped spines and passed posteriorly off the body. Small particles drop between the spines and are trapped by the mucous coating of the body. The sheet of mucus containing the food is moved by ciliary action to the body edge. It was previously thought that these small particles collected by the mucous sheet were moved into food grooves on the oral surface (Figure 3H). However, Ellers and Telford (1984) and Telford et al. (1985) have shown that most of these particles also drop off the edge of the body. Apparently, at least some sand dollars (e.g., *Mellita quinquiesperforata*) feed on relatively large particles by selectively picking them out of the sediment with special podia. Other podia then pass the particles to food grooves where they are incorporated into cords of mucus. The mucous strands and the food are then moved to the mouth by podia lining the margins of the grooves.

A few species of sand dollars (e.g., *Dendraster excentricus*) burrow into the substratum but leave the posterior part of the body extended at an angle above the sediment (Figure 11F). *Dendraster* is a suspension feeder that traps diatoms and other particulate food in the water with its podia and then passes it to the mouth as described above. Larger prey, such as tiny crustaceans, are captured by the pedicellariae. Some young sand dollars eat high-density sand grains (especially those containing iron oxides), which they store in the gut as ballast to help stabilize their positions on the sea bottom.

Telford et al. (1983) described a unique feeding method by the clypeasteroid *Echinocyamus pusillus*. These sand dollars nestle among pebbles, which are brought to the mouth by podia and then rotated by the peristomial membrane while the teeth scrape off attached diatoms and organic detritus.

The digestive system of echinoids is basically a rather simple tube extending from the mouth to the anus. The mouth is located in the center of the oral surface or is shifted somewhat anteriorly in some irregular urchins. An esophagus extends aborally, through the center of the lantern (when present), and then joins an elongate intestine (Figures 11A, B, and D). In most echinoids a narrow duct, called the **siphon**, parallels the intestinal tract for part of its length. Both ends of the siphon open to the intestine, providing a shunt for excess water and helping to concentrate food material in the gut lumen. In many species, blind ceca arise from the gut near the junction of the esophagus and intestine. The intestine narrows into a short rectum leading to the anus, which is located either centrally on the aboral surface, on the posterior margin, or posteriorly on the oral surface. Digestive enzymes are produced by the intestinal and cecal walls, and breakdown is largely extracellular.

Holothuroidea. Most sea cucumbers are suspension or deposit feeders. Many of the sedentary epibenthic or nestling forms (e.g., *Eupentacta, Aslia, Selenkothuria, Psolus, Cucumaria*) extend their branched, mucus-covered tentacles (Figures 12D and E) into the water to trap suspended particles, including live plankton. The tentacles are then pushed into the mouth one at a time and the food is ingested (Figure 12F). A fresh supply of mucus is provided by secretory cells in the papillae of the tentacles and apparently also by gland cells of the foregut.

More errant epibenthic types (e.g., *Stichopus, Parastichopus*) crawl across the substratum and use their tentacles to ingest sediment and organic detritus (Figure 12C). Several studies indicate that some holothurians (e.g., *Stichopus, Holothuria*) are highly selective deposit feeders, preferentially ingesting sediments high in organic content. Sediment extracted from the gut of *Holothuria tubulosa* contains a much higher percentage of organic material than the general surrounding sediment. This animal is so adept at selective feeding that even its fecal pellets have a higher organic content than the environmental sediments (Massin 1980). Many apodacean holothurians burrow through the substratum by peristaltic movements and ingest the sediment as they move, like earthworms.

The mouth is located at the anterior end of the body and is surrounded by a whorl of buccal tentacles. The esophagus (or pharynx) leads inward and passes through a ring of calcareous plates that support the foregut and the ring canal of the water vascular system. The esophagus joins an elongate intestine, the anterior end of which is often enlarged as a

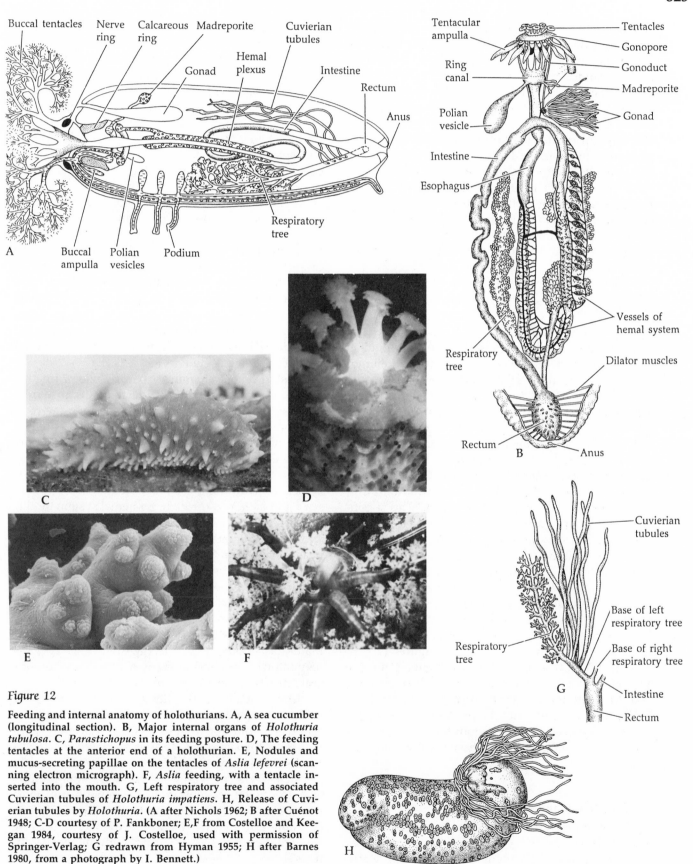

Figure 12

Feeding and internal anatomy of holothurians. A, A sea cucumber (longitudinal section). B, Major internal organs of *Holothuria tubulosa*. C, *Parastichopus* in its feeding posture. D, The feeding tentacles at the anterior end of a holothurian. E, Nodules and mucus-secreting papillae on the tentacles of *Aslia lefevrei* (scanning electron micrograph). F, *Aslia* feeding, with a tentacle inserted into the mouth. G, Left respiratory tree and associated Cuvierian tubules of *Holothuria impatiens*. H, Release of Cuvierian tubules by *Holothuria*. (A after Nichols 1962; B after Cuénot 1948; C–D courtesy of P. Fankboner; E,F from Costelloe and Keegan 1984, courtesy of J. Costelloe, used with permission of Springer-Verlag; G redrawn from Hyman 1955; H after Barnes 1980, from a photograph by I. Bennett.)

stomach. The intestine extends posteriorly, loops forward, and then extends posteriorly again; it may be coiled (Figure 12A and B). The intestine terminates in an expanded region leading to the posterior anus. The term cloaca has traditionally been applied to this enlarged hindgut, but it does not receive reproductive products and should probably be called a rectum. The hindgut is attached to the body wall by a series of suspensor muscles. In most species the hindgut bears highly branched outgrowths that extend anteriorly in the body cavity. These structures are the **respiratory trees** into which water is pumped via the anus for gas exchange (Figures 12A, B, and G). Little work has been done on the digestive physiology of holothurians, but digestion and absorption probably take place along the length of the intestine.

The digestive system of sea cucumbers is associated with two fascinating phenomena: (1) **evisceration**, and (2) the discharge of structures called **Cuvierian tubules** (Figure 12H). Evisceration occurs in many species. It is the expulsion by muscular action of part or all of the digestive tract and sometimes other organs, including the respiratory trees and gonads. In some forms (e.g., *Holothuria*), all of these structures are expelled following rupture of the hindgut region. In others (e.g., *Thyone* and *Eupentacta*), rupture occurs anteriorly and the tentacular crown and foregut are lost. Evisceration can be induced by a variety of experimental conditions (e.g., chemical stress, physical manipulation, and crowding), but it also occurs in nature in some species. The significance of this process is unclear. It is viewed by some zoologists as a seasonal event associated with adverse conditions and by others as a defense mechanism wherein the eviscerated parts serve as a decoy. In any case, the lost parts are usually regenerated.

Cuvierian tubules are defensive structures. These organs are clusters of sticky, blind tubules arising from the bases of the respiratory trees in certain genera (e.g., *Actinopyga* and *Holothuria*) (Figures 12A, B, and H). When threatened, these cucumbers aim the anus at the potential predator, contract the body wall, and discharge the tubules by rupturing the hindgut. The tubules are shot onto the predator, entangling it in the sticky mass. The Cuvierian tubules are regenerated along with any other tissue lost during discharge.

Circulation and gas exchange

Circulation. Internal transport in echinoderms is accomplished largely by the main perivisceral coeloms, augmented to various degrees by the water vascular system and the hemal system, all of which are derived from the coelom. Fluids are moved through these systems largely by ciliary action and, in some cases, by muscular pumping. In at least one species of sea urchin (*Lytechinus variegatus*) coelomic fluid is driven by movements of Aristotle's lantern (Hanson and Gust 1986).

The **hemal system** is a complex array of canals and spaces, mostly enclosed within coelomic channels called **perihemal sinuses** (Figure 13). The system is best developed in holothurians, where it is bilaterally arranged, and in the crinoids, where some of the channels form netlike plexi. In other groups the system is radially arranged and generally parallels the elements of the water vascular system. In these cases the hemal system consists of an oral and an aboral **hemal ring**, each with radial extensions. The two rings are connected to one another by an **axial sinus** lying against the stone canal (Figure 13A). Within the axial sinus is a core of spongy tissue called the **axial gland**, which is apparently responsible for producing some coelomocytes.

Radial hemal channels from the aboral ring extend to the gonads. Other radial channels arise from the oral hemal ring and are associated with the rows of tube feet; these channels are housed within a perihemal space called the **hyponeural sinus** (Figure 5B). A third hemal ring, the **gastric ring**, occurs in many echinoderms, including most asteroids, and is associated with the digestive system.

Fluid is moved through the hemal system by cilia and in some cases by muscular "pumps." In asteroids and most echinoids the axial sinus bears a **dorsal sac** near its junction with the aboral hemal ring. The dorsal sac pulsates, apparently aiding the movement of fluid within the hemal channels and spaces.

The well developed hemal system of holothurians comprises an elaborate set of longitudinal and connecting vessels (Figure 13B). This system is intimately associated with the digestive tract and the respiratory trees, and in many species includes several "hearts" or circulatory pumps.

The function of the hemal system is not fully understood, but it probably helps distribute nutrients absorbed from the digestive tract. Recent experiments on the sea star *Echinaster graminicolus* fed ^{14}C-labeled food show that absorbed nutrients appear in the hemal system within a few hours after feeding and eventually concentrate in the gonads and podia (Ferguson 1984). In sea cucumbers the hemal system probably also plays a role in gas exchange, because some of the vessels are in contact with the respiratory trees.

Figure 13

Hemal system. A, The central portion of the hemal system and some associated structures in an asteroid. B, The complex hemal system of *Isostichopus badionotus* (Holothuroidea), and its association with the gut and respiratory tree. (A after Ubaghs 1967; B after Herreid et al. 1976.)

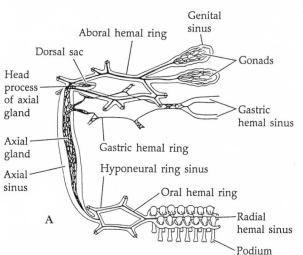

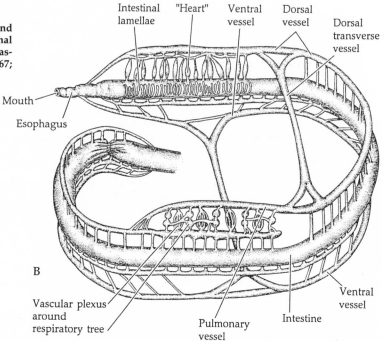

Gas exchange. Most echinoderms rely on thin-walled external processes as gas exchange surfaces. Only ophiuroids and holothurians have special internal organs for this purpose. Given the relatively large body sizes and volumes of many echinoderms, the fluid transport mechanisms discussed above are of major importance in moving dissolved gases between internal tissues and the body surface.

Crinoids have been inadequately studied in regards to gas exchange, but apparently oxygen and carbon dioxide are traded across all exposed thin parts of the body wall, especially the podia.

Gas exchange in asteroids occurs across the podia and special outpocketings of the body wall called **papulae** or **dermal gills** (Figures 14A and B). These structures are evaginations of the epidermis and peritoneum. Both tissues are ciliated, and currents are produced both in the coelomic fluid and in the overlying water. The two currents move in opposite directions, thus creating a countercurrent and maintaining maximum exchange gradients across the papulae surfaces.

Ophiuroids possess ten invaginations of the body wall called **bursae**, which open to the outside through ciliated slits (Figures 9A and C). Water is circulated through the bursae by the cilia and, in some species, by muscular pumping of the internal bursal sacs. Gases are exchanged between the flowing water and the body fluids.

Regular urchins possess five pairs of "gills" that are located in the peristomium (Figures 11A and C) and have long been viewed as the major gas exchange organs. However, recent work suggests that such is not the case (see Farmanfarmaian 1966; Kier 1974; Shick 1983). The pressure within these "gills" changes by manipulation of the compasses of the feeding apparatus. They probably function largely to accommodate pressure changes in the peripharyngeal coelom during feeding movements of the lantern complex and perhaps to provide an immediate oxygen supply to the associated muscles. The main gas exchange structures in these urchins are apparently thin-walled podia that operate on a countercurrent system similar to that associated with the papulae of asteroids (Figures 14C and D).

Irregular sea urchins and sand dollars bear highly modified podia on the aboral **petaloids** (Figure 3G). The external parts of these podia are flaplike and thin-walled, and serve as the main gas exchange surfaces. A countercurrent flow occurs between the water vascular system fluid in the podia and the sea water, and between the water vascular system fluid in the ampullae and the coelomic fluid (Figure 14E).

We have already described the respiratory trees of holothurians. Water is pumped in and out of the hindgut and branches of the respiratory trees, and gases are exchanged between the water and the coelom and hemal system. This device is augmented by

Figure 14

Gas exchange in echinoderms. A, A portion of the aboral surface of *Asterias*. Note the digitiform papulae and their surrounding structures. B, An asteroid papula (section). This structure is lined by the peritoneum and is filled with coelomic fluid. C, An ampulla and podium (longitudinal section) of *Strongylocentrotus purpuratus* (Echinoidea). The arrows represent the countercurrents between the ambient water, the fluid of the water vascular system, and the coelomic fluid. D, Three lamelliform ampullae from *Strongylocentrotus*. Gases are exchanged between the fluids of the water vascular system and the coelom. E, A "respiratory" tube foot and ampulla (section) of the irregular sea urchin *Echinocardium*. The arrows represent the countercurrents. (A after Barnes 1980; B after Cuénot 1948; C,D,E redrawn from Shick 1983, after the following original sources: C after Phelan 1977, and Smith 1978; D after Fenner 1973; E after Smith 1980.)

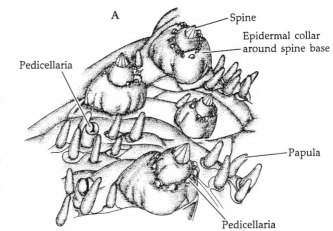

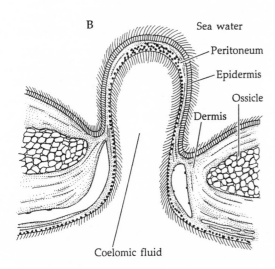

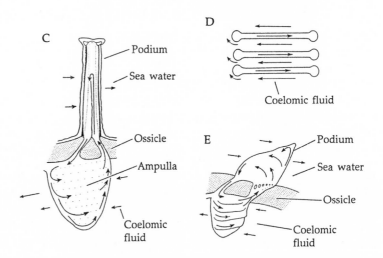

exchange across the podia, which is facilitated by a countercurrent system. Hemoglobin occurs in many holothurians, but its significance as a respiratory pigment in these animals is not well studied.

Excretion and osmoregulation

In some echinoderms dissolved nitrogenous wastes (ammonia) diffuse across body surfaces to the outside. This type of excretion occurs across the podia and papulae in asteroids, and is suspected to occur across the respiratory trees in holothurians. At least some excretion by simple diffusion probably takes place in most echinoderms. Precipitated nitrogenous material and other particulate wastes are phagocytosed by certain coelomocytes in the body fluids and then discharged by various methods.

In asteroids, waste-laden coelomocytes accumulate in the papulae, which then pinch off their distal ends, expelling the cells and waste material. Some studies indicate that the rectal ceca may also be involved in excretion. In ophiuroids it is suspected that coelomocytes deliver wastes to the bursae and are released there. Phagocytic coelomocytes in echinoids accumulate wastes and transport them to the podia and gills for release. Similarly, in holothurians particulate wastes are carried by coelomocytes to the respiratory trees, gut, and even the gonads, and released to the outside through the respective plumbing systems. Crinoid coelomocytes deposit wastes in tiny pockets along the sides of the ambulacral grooves, but discharge has not been observed.

Echinoderms are generally considered to be marine stenohaline creatures. Consequently, they do not have problems of osmotic and ionic regulation. However, a number of species have been reported from brackish water (see Binyon 1966, for a review). For examples, *Asterias rubens* (Asteroidea) has been collected from the Baltic Sea (8‰), *Ophiophragmus filagraneous* (Ophiuroidea) from Cedar Key, Florida (7.7‰), and various holothurians from the Black Sea (18‰). Obviously some mechanism allows them to survive in these low salinities.

The evidence to date suggests that echinoderms are osmoconformers. Both water and ions pass relatively freely across thin body surfaces, and the tonicity of the body fluids varies with environmental fluctuations. There appears to be some ionic regulation through active transport, but it is minimal.

Nervous system and sense organs

The secondarily derived radial *Bauplan* of echinoderms is clearly reflected in the anatomy of their nervous systems and the distribution of their sense organs. The nervous system is decentralized, somewhat diffuse, and without a cerebral ganglion. There are three main neuronal networks, developed and integrated with one another to various degrees among the classes. These networks are the **ectoneural (oral) system**, the **hyponeural (deep oral) system**, and the **entoneural (aboral) system**. The ectoneural system is predominately sensory, although motor fibers do occur; the hyponeural system, however, is largely motor in function. The entoneural system is absent from holothurians and reduced to different degrees in other groups—except the crinoids, where it is the primary nerve component and serves both motor and sensory functions.

The three nervous "systems" are interconnected by a **nerve net** derived primarily from the ectoneural and entoneural components. The nerve net is often described as a subepidermal plexus, but it gives rise to intraepidermal neurons and clearly has an intimate association with the epithelium. The nerve cords, as described below, may be viewed as thickenings of this plexus.

Except for the crinoids, where the entoneural component dominates, the most obvious nerves in echinoderms are derived from the ectoneural system. A circular or pentagonal **circumoral nerve ring** lies just beneath the oral epithelium and encircles the esophagus. From this ring arise **radial nerves** that extend along each ambulacrum. In sea stars, for example, these radial nerves appear as distinct V-shaped thickenings in the epidermis of each ambulacral groove (Figure 5B). In some cases, the entoneural components of the nerve plexus are also produced as radial cords, such as those along the lateral margins of the arms of asteroids. The hyponeural system generally parallels the nerves of the ectoneural system. Hyponeural neurons lie mostly within the radial perihemal sinuses and give rise to motor fibers and ganglia in the tube feet.

Sensory receptors are largely restricted to relatively simple epithelial structures innervated by a plexus of the ectoneural system. Sensory neurons in the epidermis respond to touch, dissolved chemicals, water currents, and light. They are frequently associated with outgrowths of the body wall such as spines and pedicellariae. Special photoreceptors occur in asteroids as eyespots, or **optic cushions**, each of which comprises a cluster of pigment-cup ocelli at the tip of an arm. Statocysts are known in some holothurians, and georeception is presumed to be the function of structures called **sphaeridia** in certain echinoids. Chemoreception has not been well studied in echinoderms, but there is some evidence that the buccal tentacles of holothurians and the oral podia of some echinoids may be sensitive to dissolved chemicals. Chemoreception in asteroids appears to depend largely on direct contact, although distance chemoreception is reported in some species.

In spite of the rather simple nervous system and the lack of specialized sense organs, many echinoderms engage in complex behaviors. As is so often the case in such matters, there is still much to be learned about the functional mediation between the circuitry of the nervous system and the observed behavioral responses. We have mentioned the coordination of the podia during locomotion and feeding in asteroids and ophiuroids, and the same phenomenon occurs in echinoids and holothurians. Most of these animals also exhibit distinct righting behaviors when overturned. These actions probably involve touch, georeception, and perhaps photoreception. Orientation to currents is known in some sand dollars and in many ophiuroids and crinoids. There is even evidence to support the contention that some degree of learning occurs in echinoderms (see Valentincic 1983).

Reproduction and development

Regeneration and asexual reproduction. Most echinoderms are capable of regenerating lost parts. Even the casual observer of tide-pool life will encounter a sea star regenerating a new arm, or notice the suckers of the podia left on a rock from which a sea star or urchin has been pulled free. Lost suckers are quickly replaced by regeneration. We have described earlier the dramatic processes of evisceration and expulsion of Cuvierian tubules—again, the lost organs are replaced. However, studies on regeneration in asteroids have put to rest the tales of oystermen who once claimed that chopping sea stars into small pieces resulted in the regeneration of an entire new animal from each part. While it is true that a damaged animal can grow new arms if a substantial portion of the central disc remains intact, an isolated arm soon dies. The exception to this generality is

Figure 15

Regeneration in *Linckia*. A, Initial regeneration from a single arm, here yielding a central disc with dual madreporites and five new rays. B, At a later stage, the animal has a single madreporite and the normal ray number. (After Cuénot 1948.)

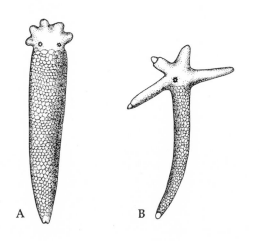

Linckia, which can regenerate an entire individual from a single arm, the regenerating stage being appropriately called a **comet** (Figure 15). Ophiuroids and crinoids frequently cast off arms or arm fragments when disturbed, and then regenerate the lost part.

Asexual reproduction occurs in some asteroids and ophiuroids by a process called **fissiparity**, wherein the central disc divides in two and each half forms a complete animal by regeneration. When the small, six-rayed brittle star *Ophiactis* divides, each half retains three arms. Asexual fission also occurs in some holothurians, but the process is not well understood.

Sexual reproduction. The majority of echinoderms are gonochoristic, but hermaphroditic species are known among the asteroids, holothuroids, and especially the ophiuroids. The reproductive systems are relatively simple and intimately associated with derivatives of the coelom. The gonads are usually housed within peritoneally lined **genital sinuses**. Holothurians are unique among echinoderms in possessing but a single gonad, which lies dorsally in the CD interambulacrum (Figure 12B). A single gonoduct opens between the bases of two dorsal buccal tentacles or just posterior to the tentacular whorl.

Crinoids lack distinct gonads. The gametes arise from the peritoneum of special coelomic extensions called **genital canals** in the pinnules on the proximal portion of each arm. There are no gonoducts. Ga-

metes are released by rupture of the pinnule walls. Ophiuroids possess from one to many gonads attached to the peritoneal side of each bursa adjacent to the bursal slits (Figure 9A). Gametes are released into the bursae and expelled through the slits.

Asteroids, concentricycloids, and echinoids possess multiple gonads with gonoducts leading to interambulacral gonopores between adjacent arms (Figures 8C and 11B). Sea daisies have a pair of gonads in each ambulacrum (Figure 3E). Regular sea urchins contain five gonads, one lying along the inside of each interambulacral radius. The gonopores are located on the five interambulacral **genital plates** surrounding the **periproct** (Figures 5F and 11B). The periproct and anus have migrated posteriorly in irregular urchins, but the genital plates remain more-or-less centrally located on the aboral surface. In all urchins one of the genital plates is perforated and doubles as the madreporite. In many irregular urchins and sand dollars there are only four (and sometimes fewer) gonads, one being lost along the line of migration of the anus. In such cases there is a corresponding reduction in the number of gonopores.

Life history strategies among echinoderms vary from free spawning followed by external fertilization and indirect development to various forms of brooding and direct development. Spawning has been observed in nature in only a few species of echinoderms. Some studies indicate that spawning is mostly a nocturnal event, wherein the animals assume characteristic postures with their bodies elevated off the substratum. Gametogenesis in at least some asteroids and echinoids is regulated by photoperiod (Pearse et al. 1986), which in turn ensures more-or-less synchronous spawning among members of the same population. Brooding is especially common in boreal and polar species among all groups of echinoderms and in certain deep-sea asteroids, whose environments are unfavorable for larval life. As expected, brooding species produce fewer but larger and more yolky eggs than do their free-spawning counterparts.

Brooding methods vary. Among the crinoids, *Antedon* and a few others cement their eggs to the epidermis of the pinnules from which they emerge (Figures 16A and B). Once fertilized by free sperm, the embryos are held by the parent until hatching. Most brooding asteroids hold their embryos on the body surface. One species (*Asterina gibbosa*) cements its eggs to the substratum, and another (*Leptasterias tenera*) broods its early embryos in the pyloric stomach before moving them to the outer body surface (Hendler and Franz 1982). Concentricycloids brood within the gonads and apparently release juveniles that may

Figure 16

Brooding in a crinoid and a holothurian. A, Portion of an arm of *Antedon*. The ova are housed within a pinnule (lower portion) and released to exterior (upper portion). B, Part of a pinnule of the crinoid *Phixometra* with developing young. C, *Cucumaria crocea*, brooding its young. (A,C after Cuénot 1948; B redrawn from Hyman 1955.)

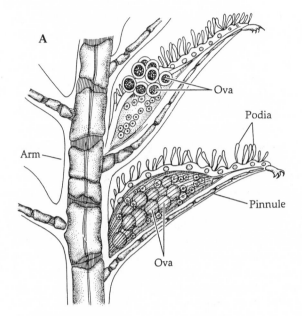

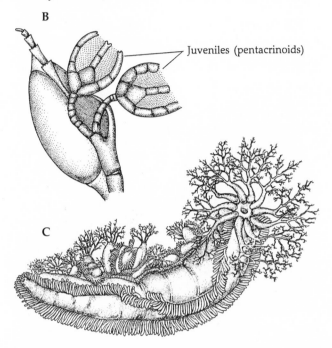

drift for some time before settling. Brooding is common in many ophiuroids. Sperm enter the bursae and fertilize the eggs, and the embryos are held within these sacs during development. Some echinoids brood their embryos among clusters of spines on the body or, in the case of sand dollars, on the petaloids. Brooding holothurians usually carry their embryos externally (Figure 16C), but some species of *Thyone* and *Leptosynapta* brood inside the coelom.

Development. The tremendous numbers of eggs produced by many echinoderms and the ease with which they can be reared in the laboratory have made these animals favorite objects of study by embryologists. Much of our information about the biology of animal fertilization and early development comes from over a century of work focusing particularly on urchins and sea stars. In addition, the early ontogeny of some echinoderms has served as a model of deuterostome development against which many other developmental patterns are measured. Except in brooding species, where development is modified by large amounts of yolk, the sequence of ontogenetic events is remarkably similar throughout the phylum.

We cannot cover the vast amount of information on this subject, and present an overview of indirect development, emphasizing urchins and asteroids and including some comparative comments on other taxa.

The ova of free-spawning echinoderms are usually isolecithal with relatively small amounts of yolk. Cleavage is radial, holoblastic, and initially equal and leads to a spacious coeloblastula. In some groups, such as urchins, cleavage preceding the blastula is unequal, resulting in blastomere tiers of mesomeres underlaid by slightly larger macromeres and a cluster of micromeres at the vegetal pole. These terms refer only to the relative sizes of the cells, and are not to be confused with the terms as they are used in describing spiral cleavage. The coeloblastula usually becomes ciliated and breaks free of the fertilization membrane as a swimming embryo.

The blastula flattens slightly at the vegetal pole, forming the **gastral plate** from which some cells proliferate into the blastocoel as **primary** or **larval mesenchyme**. In most echinoids these cells are the micromeres. The surrounding macromeres are the presumptive entoderm and adult mesoderm, and the mesomeres are the presumptive ectoderm. A coelogastrula is produced by invagination of the vegetal pole cells. The blastopore typically forms the anus; the archenteron grows to connect with a stomodeal inpocketing that forms the mouth. Before the gut is complete, however, the inner end of the archenteron proliferates **secondary mesenchyme** into the blastocoel, and one or two evaginations of mesoderm. Thus, coelom formation (enterocoely) is by archenteric pouching.

During the later stages of gastrulation and coelom development, the embryo assumes bilateral symmetry and eventually becomes a planktotrophic or lecithotrophic larva. The primary mesenchyme contributes to larval muscles and supportive structures, including in some cases calcareous spicules or ossicles (Figure 17G). The various larval stages of the major taxa are shown in Figure 17. The larvae of some echinoderms contain distinctive reddish pigment spots. Recent work suggests that these pigments may be involved in photochemical energy-producing reactions associated with the active life of these larvae (Ryberg 1980).

In order to understand the development of echinoderm larvae and their eventual metamorphosis to radially symmetrical adults, it is necessary to examine carefully the embryogeny and fates of the coelomic spaces. Although there are some differences in the details of these events among groups, they are similar enough to generalize for our purposes. The initial archenteric pouching typically occurs from the blind end of the developing gut, either as a pair of coeloms or as one cavity that divides into two. These coeloms pinch off another pair of cavities posteriorly, and then a third pair between the anterior and posterior ones (Figure 18A). From front to back, these pairs of coelomic spaces are called right and left **axocoels**, **hydrocoels**, and **somatocoels**. Clearly, these spaces correspond to the protocoels, mesocoels, and metacoels of other trimeric deuterostomes. As illustrated, the left axocoel and left hydrocoel do not fully separate, but remain connected to one another by the **stone canal**. From the left axocoel-hydrocoel complex arises a **hydrotube**, which grows to the dorsal surface of the larva and opens to the outside via a **hydropore**. In most cases the right axocoel disappears and the right hydrocoel becomes associated with the hydrotube as the **dorsal sac**.

As we explain metamorphosis, it will help you to note the fates of these various coelomic derivatives outlined in Table 1. As the time for metamorphosis approaches, the larva swims to the bottom and selects and attaches to an appropriate substratum. In

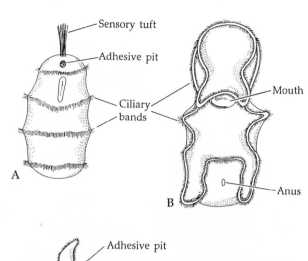

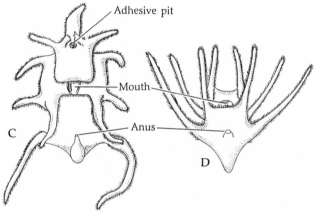

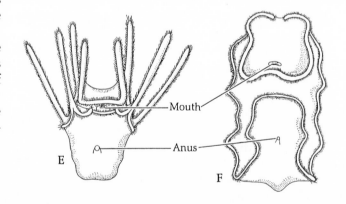

Figure 17

Echinoderm larval types. A, Vitellaria larva of a crinoid. B–C, Bipinnaria and later brachiolaria larvae of a sea star. D, Ophiopluteus larva of a brittle star. E, Echinopluteus of a sea urchin. F, Auricularia larva of a sea cucumber. G, Isolated larval spicule from the sand dollar *Dendraster*. H, Late pentacula stage (postlarva) of a sea cucumber. Skeletal ossicles and juvenile podia are present. (A–F adapted from several sources; G from Emlet 1982, photo courtesy of R. Emlet; H courtesy of and copyright by M. Apley.)

G   H

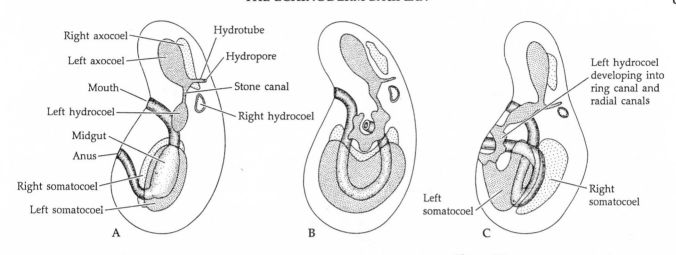

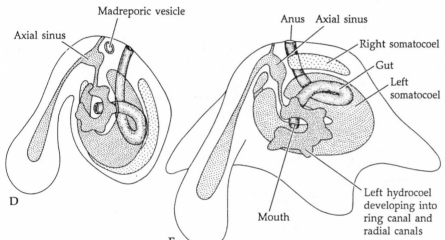

Figure 18

A–D, The development of the coelom and its derivatives in an asteroid. E, Metamorphosis in the same animal. See text for details. (Redrawn from Meglitsch 1972, after Dawydoff 1948.)

Table One

Adult fates of the major coelomic derivatives in generalized echinoderm development

EMBRYONIC COELOMIC STRUCTURE	ADULT FATE
Right somatocoel	Aboral perivisceral coelom
Left somatocoel	Oral perivisceral coelom; genital sinuses; most of the hyponeural sinus
Right axocoel	Largely lost
Hydropore	Incorporated into madreporite
Hydrotube and dorsal sac	Parts of madreporic vesicle and ampulla
Stone canal	Stone canal
Left hydrocoel	Ring canal; radial canals; lining of tube feet lumina, plus other components of the water vascular system including Tiedemann's bodies and polian vesicles

general, the larval left and right sides become the adult oral and aboral surfaces, respectively, although this often varies from a precise 90° reorientation. The remarkable change from bilateral to radial symmetry involves shifts in the positions of the mouth and anus. In many cases these embryonic openings disappear, and the stumps of the foregut and hindgut migrate beneath the body surface to their adult positions. The foregut swings from its larval anteroventral location to the left side, and the hindgut moves anteriorly and to the right (Figures 18B and C). As the foregut migrates, it presses into the wall of the left hydrocoel, which encircles the foregut as the precursor of the ring canal. Once affixed in the adult positions, the mouth and anus reopen. The ring canal grows radial extensions (Figures 18D and E) destined to become the radial canals and podial linings, and outgrowths of the left somatocoel contribute to the formation of the hyponeural sinus. Aborally, the madreporite complex arises from various parts of the left axocoel and its derivatives plus the dorsal sac and marks the position of the CD interambulacrum. The axial sinus arises from an outpocketing of the left axocoel.

As these transformations take place, most of the larval structures are lost and the juvenile assumes benthic life. Many echinoderm larvae appear to settle preferentially near conspecific adults. In at least some species (e.g., *Dendraster excentricus*) successful metamorphosis is triggered by pheromones that are released by the adults and act on the larval nervous system.

Echinoderm phylogeny

In spite of the rich fossil record and many decades of work, the origin and subsequent evolution of echinoderms are highly controversial issues. There have been a number of popular and competing ideas on these matters, as evidenced by the instability of the classification of the phylum in past years. A sampling of the important literature is in the references at the end of the chapter, including some recent work on the phylogeny of the Asteroidea (see Blake 1987; Gale 1987). C. R. C. Paul and A. B. Smith (1984) have subjected the echinoderms to cladistic analysis. The following ideas are based in part on their work, but we have made modifications for the sake of simplicity and additional treatment of extant classes.

The origin of the echinoderm lineage probably took place during the Precambrian invasion of epibenthic habitats from an ancestral burrowing deuter-

ostome. The line diversified rapidly, and most of the fundamental body plans within the phylum were probably established by the early Cambrian. Echinoderm diversity reached its zenith during the early-to-mid Paleozoic, but by the beginning of the Mesozoic had declined greatly in terms of higher taxa, leaving only five major groups (and the sea daisies) persisting to recent times. Some fossil forms are shown in Figure 19.

The origin of echinoderms probably involved the evolution of the endoskeleton composed of plates with the unique stereom structure. Evidence suggests that the origin of the skeleton occurred prior to the adoption of radial symmetry, the latter marking the appearance of the first "true" echinoderms as shown in the cladograms in Figure 20. The first dichotomy separates the main echinoderm clade from a now extinct group called the carpoids (Figure 19A). Many authors view the carpoids as echinoderms and place them in a separate subphylum, the Homalozoa. Although they possessed stereom ossicles, they were not radially symmetrical, and the nature of their water vascular system (if they had one) is uncertain. Because they lacked some of the fundamental defining characteristics of the echinoderms, they are best considered as an early pre-echinoderm group. In fact, some workers view certain carpoids as ancestral to both the modern echinoderms and the chordates (see Jefferies 1967, 1981, 1986); we explore this idea in more detail in Chapter 23. These early epibenthic creatures were probably suspension feeders, and bore a grooved "arm" or **brachiole** that apparently led to the mouth.

The first true echinoderms may have been the helicoplacoids (Figure 19B). These odd creatures appear in the fossil record in the early Cambrian but died out soon thereafter. They were spindle shaped, with spirally arranged skeletal plates and three ambulacra. The mouth was located on one side of the body rather than apically, so these animals were not constructed on an obvious oral–aboral axis as are modern echinoderms. It appears that the skeletal plates articulated somewhat, and the helicoplacoids may have been either surface plowers or attached with the oral side against the substratum. The ambulacra probably conveyed detrital or suspended food material to the mouth. Some authors speculate that the "lateral" placement of the mouth in helicoplacoids is indicative of the origin of metamorphic events during the conversion from bilateral to radial symmetry seen in extant echinoderms.

The next major dichotomy separates the pelmatozoan and eleutherozoan lineages (Figures 20 and

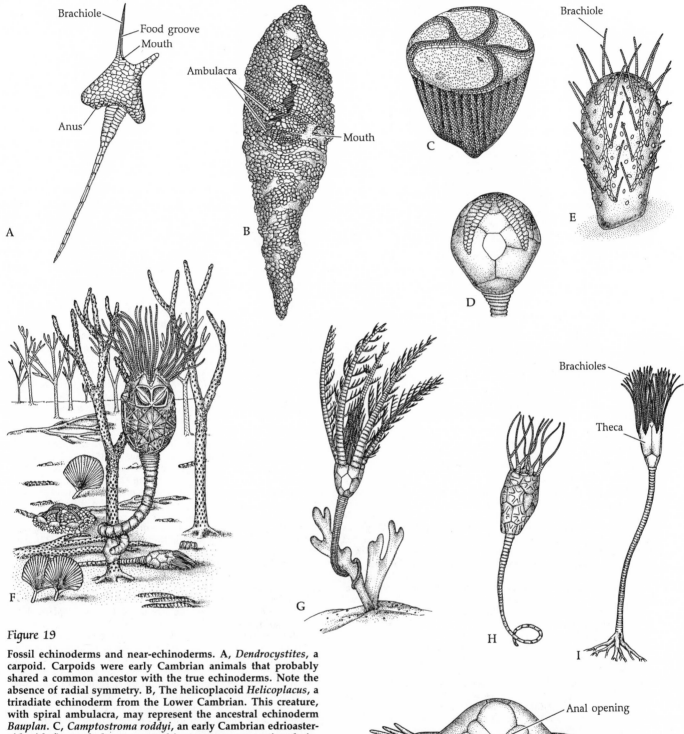

Brachiole
Food groove
Mouth
Anus
A

Ambulacra
Mouth
B

C

D

Brachiole
E

Brachioles
Theca
H
I

F

G

Anal opening
J
Podia
Possible combined
gonopore and hydropore

Figure 19

Fossil echinoderms and near-echinoderms. A, *Dendrocystites*, a carpoid. Carpoids were early Cambrian animals that probably shared a common ancestor with the true echinoderms. Note the absence of radial symmetry. B, The helicoplacoid *Helicoplacus*, a triradiate echinoderm from the Lower Cambrian. This creature, with spiral ambulacra, may represent the ancestral echinoderm *Bauplan*. C, *Camptostroma roddyi*, an early Cambrian edrioasteroid with five ambulacra arranged in a manner suggesting derivation from a triradiate form. D, *Steganoblastus*, a stalked edrioasteroid showing clear pentamery. E, A generalized cystoid. F, *Lepadocystis*, a stalked Ordovician cystoid. Note also the attached edrioasteroid. G, *Eifelocrinus*, an extinct crinoid. H, The eocrinoid *Macrocystella*. I, A generalized blastoid from the Carboniferous. J, *Volchovia*, a strange extinct echinoid. (A after Barnes 1980; B,C after Paul and Smith 1984; D,E,H,I after Nichols 1962; F redrawn from Barnes 1980, from Kesling in Moore (ed.) 1967; G redrawn from Cuénot 1948; J after Hyman 1955.)

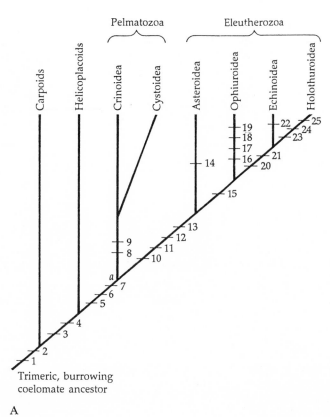

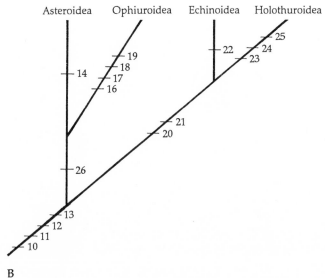

Figure 20

A, A cladogram depicting one set of hypotheses about the origins of some important synapomorphies among the major groups of echinoderms. B, A competing cladogram illustrating a slightly different view on the placement of the ophiuroids. The origin of the echinoderm lineage involved an escape from infaunal life with the evolution of a supportive system of endoskeletal plates with a stereom structure (1) and the use of external ciliary grooves for suspension feeding (2). This "preechinoderm" condition is represented in the fossil record by the carpoids. The first true echinoderms may have been the helicoplacoids, whose appearance was marked by the origin of triradial symmetry with three, spirally arranged, open ambulacral grooves (3), and a water vascular system (4), probably with the madreporite opening near the mouth. The immediate common ancestor of the pelmatozoan and eleutherozoan lineages (a) may have been similar to the extinct *Camptostroma*, with pentaradial symmetry (5) evidenced by five open ambulacral grooves, mouth and anus on the oral surface (6), and attachment to the substratum by the aboral surface (7). From this ancestral form, the pelmatozoans (Crinoidea and Cystoidea) diverged, with the evolution of arms or brachioles bearing open ciliated grooves used for suspension feeding (8) and the loss of the external madreporite (9). The origin of the eleutherozoan clade involved the movement of the anus to the aboral surface

(10), a change associated with the orientation of the body with the oral surface against the substratum (11). In this "new" position, eleutherozoans adopted alternative feeding modes and an errant lifestyle; the podia became suckered (12) and used for locomotion rather than feeding. The madreporite migrated along the CD interambulacrum to the aboral surface (13). Given this scenario, the asteroids arose with the evolution of five rays (arms) broadly connected to a central disc (14). The remaining three groups have closed ambulacral grooves (15) in common. The ophiuroids invaded soft substrata and lost the podial suckers (16). In addition, they evolved five highly articulated rays, with internal vertebral plates in each arm "segment" (17) and secondarily lost the anus (18). The madreporite migrated back to the oral surface along the CD interambulacrum (19). The echinoid–holothurian clade arose with the extension of the ambulacral grooves along the sides of the body from the oral to the aboral pole (20), thereby reducing the aboral surface to a small region around the anus (21). The echinoids evolved with the fusion of the skeletal plates, which formed a rigid globular or discoidal test (22). The origin of the holothurians involved a reduction of the skeletal plates to isolated ossicles (23), movement of the madreporite internally (24), and elongation of the fleshy body on the oral–aboral axis (25).

Cladogram B unites the asteroids and ophiuroids into a single clade on the basis of the five-rayed (26) body plan. Both cladograms include some unresolved problems. Both treat the evolution of ambulacral rays (arms) as convergent between the pelmatozoans (crinoids) and eleutherozoans (asteroids and ophiuroids), and cladogram A indicates that this condition is also convergent in the asteroids and ophiuroids. Cladogram A treats the oral madreporite position in ophiuroids as secondary, and thus convergent with the same condition in groups that diverged earlier (e.g., crinoids). In addition, cladogram B accepts as convergent the closed ambulacral grooves of ophiuroids and those of echinoids and holothurians.

21). Paul and Smith suggest that the extinct genus *Camptostroma* (Figure 19C) is similar to what may have been the common ancestor of these two monophyletic sister clades. It is the earliest known pentaradial echinoderm, with the five ambulacra devel-

oped in a 2-1-2 pattern, perhaps derived from the triradial pattern of the helicoplacoids. *Camptostroma* is usually assigned to a wholly extinct group called the Edrioasteroidea (Figure 19D). The group Pelmatozoa includes those taxa that primitively bore at-

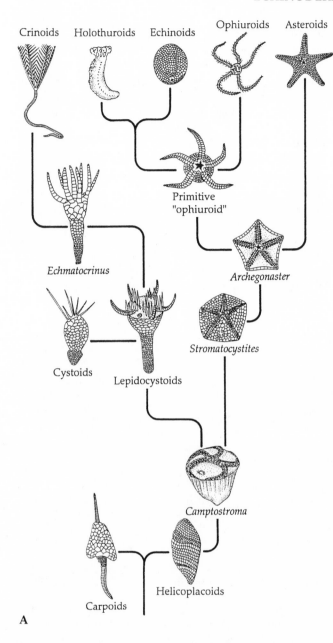

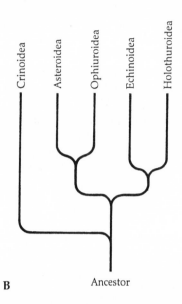

Figure 21

Two orthodox evolutionary trees of the echinoderms. A, A tree compatible with cladogram A in Figure 20. B, A "traditional" tree to the extant classes, compatible with cladogram B in Figure 20. (A after Paul and Smith 1984.)

tachment stalks arising from the aboral surface (e.g., extinct cystoids and crinoids; Figures 19E–H). These animals exploited suspension feeding by orienting with the oral side upward and using the open ambulacral grooves in the arms or brachioles for transporting food to the mouth. The taxon Eleutherozoa comprises the asteroids, ophiuroids, echinoids, and holothurians; as a group, they originally capitalized on other feeding modes and used the water vascular system largely for locomotion. These animals became more-or-less errant and, with the exception of the holothurians, oriented themselves with the oral side against the substratum.

The phylogeny of the eleutherozoans is controversial and as yet unsettled. Paul and Smith did not apply their cladistic analysis to the eleutherozoan classes directly but did include them in a suggested evolutionary tree (Figure 21A). On the basis of the temporal appearance in the fossil record, these authors placed the extinct genus *Stromatocystites* (Figure 21) at the base of the eleutherozoan line and indicated that the familiar benthic-feeding asteroids, ophiuroids, and echinoids did not appear until later (Ordovician), and the holothurians even more recently (early Mississippian). However, the emphasis on fossil chronology must be viewed with some caution, even in a group as well represented as the echinoderms. The evidence preserved in rocks is still fragmentary and may easily be misleading. The holothurians, for example, leave only isolated ossicles from which to draw inferences. Deep-sea ophiuroids and asteroids probably did not fossilize well, and the skeletons of others are known to disarticulate soon after death. Although we cannot ignore the fossils, we cannot base an entire evolutionary scenario on what they seem to tell us, nor can we rely heavily on their stratigraphy to date the origins of taxa.

Among the eleutherozoan classes, the close relationship between the echinoids and holothurians is accepted by most specialists, but there is much debate about whether the ophiuroids are closer to the asteroids or to the echinoid–holothurian clade. The position of the recently discovered concentricycloids is highly speculative. They may represent an offshoot of the early asteroid lineage that capitalized on small size, losing the gut and inhabiting specific bacteria-rich environments. They are not included in the cladograms. Two alternative cladograms are presented and discussed in Figure 20.

In summary, the evolution of a mesodermal skeleton of stereom-type ossicles was followed by the appearance of a water vascular system and pentaradial symmetry. These features presumably allowed escape from infaunal life to epibenthic surface dwelling. The water vascular system probably originally served for suspension and perhaps detritus feeding, facilitated first by simple ciliary tracts, as in the carpoid brachioles, and, later by the development of

ambulacral grooves in the helicoplacoids and pelmatozoans. Radial symmetry became the popular architecture among the echinoderms and enhanced their new lifestyle. The use of suckered podia for locomotion was a secondary event that provided a means of exploiting new habitats and food resources. Later, in many eleutherozoans, the ambulacral grooves closed, with a concomitant loss of feeding functions, and food-getting became the responsibility of other structures (e.g., buccal tentacles, teeth, and spines).

There is no doubting the success of the basic echinoderm *Bauplan*. The combined qualities of the supportive endoskeleton, coelomic water vascular system, and pentaradial symmetry are unique to these animals and have provided the basis for their diversification along several distinct lineages.

Selected References

General and Fossil Forms

Bather, F. A. 1900. The Echinodermata. *In* E. R. Lankester (ed.), *A Treatise on Zoology*, Pt. 3:1–216. A. and C. Black, London.

Bell, B. M. 1976. *A Study of North American Edrioasteroidea*. Mem. New York State Mus. and Sci. Services 21.

Binyon, J. 1966. Salinity tolerance and ionic regulation. *In* R. A. Boolootian (ed.), *Physiology of Echinodermata*. Wiley-Interscience, New York, pp. 359–378.

Binyon, J. 1972. *Physiology of Echinoderms*. Pergamon Press, Oxford.

Boolootian, R. A. (ed.). 1966. *Physiology of Echinodermata*. Wiley-Interscience, New York.

Breimer, A. 1978. General morphology: Recent crinoids. *In* R. C. Moore and C. Teichert (eds.), *Treatise on Invertebrate Paleontology*, Pt. 5, Echinodermata 2. Geological Society of America and University of Kansas Press.

Campbell, A. C. 1983. Form and function of pedicellariae. *In* M. Jangoux and J. M. Lawrence (eds.), *Echinoderm Studies*, Vol. 1. A. A. Balkema, Rotterdam, pp. 139–168.

Chia, F. S. and A. H. Whitely (eds.). 1975. Developmental biology of the echinoderms. Am. Zool. 15(3):483–775. [A collection of papers on various aspects of echinoderm reproduction and embryology from the 1973 ASZ meetings.]

Clark, A. M. 1977. *Starfishes and Related Echinoderms*. British Museum, London.

Cuénot, L. 1948. Anatomie, Éthologie, et Systématique des Échinodermes. *In* P. Grassé (ed.), *Traité de Zoologie*, Vol. 11. Masson et Cie, Paris.

Dawydoff, C. 1948. Embryologie des Échinodermes. *In* P. Grassé (ed.), *Traité de Zoologie*, Vol. 11. Masson et Cie, Paris.

Drestler, K. L. 1981. Morphological diversity of early Cambrian echinoderms. *In* M. E. Taylor (ed.), *Short Papers for the Second International Symposium on the Cambrian System*. United States Open File Report No. 81-743:71–75.

Durham, J. W. 1967. Notes on the Helicoplacoidea and early echinoderms. J. Paleontol. 41:97–102.

Durham, J. W. and K. E. Caster. 1963. Helicoplacoidea: A new class of echinoderms. Science 140:820–822.

Emlet, R. B. 1982. Echinoderm calcite: A mechanical analysis from larval spicules. Biol. Bull. 163:264–275.

Emlet, R. B. 1983. Locomotion, drag, and the rigid skeleton of larval echinoderms. Biol. Bull. 164:433–445.

Emlet, R. B. 1985. Crystal axes in recent and fossil adult echinoids indicate trophic mode in larval development. Science 230:937–940.

Farmanfarmaian, A. 1966. The respiratory physiology of echinoderms. *In* R. A. Boolootian (ed.), *Physiology of Echinodermata*. Wiley-Interscience, New York, pp. 245–266.

Fell, H. B. 1962. A classification of echinoderms. Tuatara 10:138–140.

Fell, H. B. 1966. Ancient echinoderms in modern seas. Annu. Rev. Oceanogr. Mar. Biol. 4:233–245.

Fell, H. B. 1982. Echinodermata. *In* S. P. Parker (ed.), *Synopsis and Classification of Living Organisms*. McGraw-Hill, New York, pp. 785–813.

Hammond, L. S. 1982. Patterns of feeding and activity in deposit-feeding holothurians and echinoids (Echinodermata) from a shallow back-reef lagoon, Discovery Bay, Jamaica. Bull. Mar. Sci. 32(2):549–571.

Hammond, L. S. 1983. Nutrition of deposit-feeding holothuroids and echinoids (Echinodermata) from a shallow reef lagoon, Discovery Bay, Jamaica. Mar. Ecol. Prog. Ser. 10:297–305.

Haugh, B. N. and B. M. Bell. 1980. Fossilized viscera in primitive echinoderms. Science 209:653–657.

Heddle, D. 1967. *Echinoderm Biology*. Academic Press, New York.

Hyman, L. H. 1955. *The Invertebrates*, Vol. 4, Echinodermata. The Coelomate Bilateria. McGraw-Hill, New York.

Jangoux, M. (ed.). 1980. *Echinoderms: Present and Past*. A. A. Balkema, Rotterdam.

Jangoux, M. and J. M. Lawrence (eds.). 1982. *Echinoderm Nutrition*. A. A. Balkema, Rotterdam.

Jangoux, M. and J. M. Lawrence (eds.). 1983. *Echinoderm Studies*, Vol. 1. A. A. Balkema, Rotterdam.

Jangoux, M. and J. M. Lawrence (eds.). 1987. *Echinoderm Studies*, Vol 2. A. A. Balkema, Rotterdam.

Jefferies, R. P. S. 1967. Some fossil chordates with echinoderm affinities. Symp. Zool. Soc. London 20:163–208.

Jefferies, R. P. S. 1980. The phylogenetic connection between Echinoderms and Chordates. *In* M. Jangoux (ed.), *Echinoderms: Present and Past*, A. A. Balkema, Rotterdam, pp. 29–30.

Jefferies, R. P. S. 1981. Fossil evidence for the origin of the chordates and echinoderms. Atti di Convengi Lincei, 49:487–561.

Jefferies, R. P. S. 1986. *The Ancestry of the Vertebrates*. British Museum of Natural History, London.

Lawrence, J. M. (ed.). 1982. *Echinoderms: Proceedings of the International Conference, Tampa Bay, 14–17 September 1981*. A. A. Balkema, Rotterdam. [Contains 119 papers and abstracts.]

Lawrence, J. M. 1987. *A Functional Biology of Echinoderms*. Johns Hopkins University Press, Baltimore.

Millott, N. (ed.). 1967. *Echinoderm Biology*. Symp. Zool. Soc. Lond. No. 20. Academic Press, London.

Moore, R. C. (ed.). 1966–1978. *Treatise on Invertebrate Paleontology*. Parts S–U, Echinodermata. Geological Society of America and University of of Kansas Press.

Motokawa, T. 1984. Connective tissue catch in echinoderms. Biol. Rev. 59:255–270.

Nichols, D. 1962. *Echinoderms*. Hutchinson University Library, London.

Nichols, D. 1972. The water-vascular system in living and fossil echinoderms. Paleontology 15:519–538.

Paul, C. R. C. 1979. Early echinoderm radiation. *In* M. R. House (ed.), *Origin of Major Invertebrate Groups*. Syst. Assoc. Spec. Vol. 12. Academic Press, London, pp. 415–434.

Paul, C. R. C. and A. B. Smith. 1984. The early radiation and phylogeny of echinoderms. Biol. Rev. 59:443–481. [A cladistic analysis.]

Pequignat, E. 1970. On the biology of *Echinocardium cordatum* (Pennant) of the Seine Estuary: New researches on skin-digestion and epidermal absorption in Echinoidea and Asteroidea. Forma et Functio 2:121–168.

Philip, G. M. 1979. Carpoids—echinoderms or chordates? Biol. Rev. 54:439–471.

Shick, J. M. 1983. Respiratory gas exchange in echinoderms. *In* M. Jangoux and J. M. Lawrence (eds.), *Echinoderm Studies*. Vol. 1. A. A. Balkema, Rotterdam, pp. 67–110.

Sprinkle, J. 1973. Morphology and evolution of blastozoan echinoderms. Spec. Publ. Mus. Comp. Zool. Harv. Univ.

Sprinkle, J. 1983. Patterns and problems in echinoderm evolution. *In* M. Jangoux and J. M. Lawrence (eds.), *Echinoderm Studies*, Vol 1. A. A. Balkema, Rotterdam, pp. 1–18.

Stephens, G. 1968. Dissolved organic matter as a potential source of nutrition for marine organisms. Am. Zool. 8:95–106.

Stephenson, D. G. 1974. Pentamerism and the ancestral echinoderm. Nature 250:82–83.

Ubaghs, G. 1967. *In* R. C. Moore (ed.), *Treatise on Invertebrate Paleontology*, Pt. 5, Vol. 1. Geological Society of America and University of Kansas Press.

Crinoidea

Bather, F. A. 1899. A phylogenetic classification of the Pelmatozoa. Br. Assoc. Adv. Science, Rpt. D (1898):916–923.

Breimer, A. and G. Ubaghs. 1974. A critical comment on the classification of the pelmatozoan echinoderms. Proc. K. Neder. Akad. Wet. B(78):398–417.

Clark, A. H. 1915, 1921, 1931, 1941, 1947, 1950. *A Monograph of the Existing Crinoids*, Vol. 1, Pts. 1, 2, 3, 4a, 4b, 4c. Bull. U.S. Nat. Mus. 82.

Grimmer, J. C., N. D. Holland and C. G. Messing. 1984. Fine structure of the stalk of the bourgueticrinid sea lily *Democrinus conifer* (Echinodermata: Crinoidea). Mar. Biol. 81:163–176.

Holland, N. D., A. B. Leonard and J. R. Strickler. 1987. Upstream and downstream capture during suspension feeding by *Oligometra serripinna* (Echinodermata: Crinoidea) under surge conditions. Biol. Bull. 173:552–556.

LaTouche, R. W. 1978. The feeding behavior of the feather star *Antedon bifida*. J. Mar. Biol. Assoc. U.K. 58:877–890.

Sprinkle, J. 1976. Classification and phylogeny of 'pelmatozoan' echinoderms. Syst. Zool. 25:83–91.

Stephenson, D. G. 1980. Symmetry and suspension-feeding in Pelmatozoan echinoderms. *In* M. Jangoux (ed.), *Echinoderms: Present and Past*. A. A. Balkema, Rotterdam, pp. 53–58.

Zmarzly, D. L. 1985. The shallow-water crinoid fauna of Kwajalein Atoll, Marshall Islands: ecological observations, interatoll comparisons, and zoogeographic affinities. Pac. Sci. 39:340–358.

Asteroidea

Blake, D. B. 1987. A classification and phylogeny of post-Paleozoic sea stars (Asteroidea: Echinodermata). J. Nat. Hist. 21:481–528.

Carey, A. G. 1972. Food sources of sublittoral, bathyal and abyssal asteroids in the northeast Pacific Ocean. Ophelia 10:35–47.

Chia, F. S. and H. Amerongen. 1975. On the prey-catching pedicellariae of a starfish, *Stylasterias forreri*. Can. J. Zool. 53:748–755.

Domanski, P. A. 1984. Giant larvae: Prolonged planktonic larval phase in the asteroid *Luidia sarsi*. Mar. Biol. 80:189–195.

Fell, H. B. 1963. The phylogeny of sea-stars. Philos. Trans. Roy. Soc. Lond. Ser. B. 246:381–435.

Ferguson, J. C. 1970. An autoradiographic study of the translocation and utilization of amino acids by starfish. Biol. Bull. 138:14–25.

Ferguson, J. C. 1984. Translocative functions of the enigmatic organs of starfish—the axial organ, hemal vessels, Tiedemann's bodies, and rectal caeca—and autoradiographic study. Biol. Bull. 166:140–155.

Gale, A. S. 1987. Phylogeny and classification of the Asteroidea (Echinodermata). Zool. J. Linn. Soc. 89:107–132.

Glynn, P. W. 1974. The impact of *Acanthaster* on corals and coral reefs in the Eastern Pacific. Environ. Conserv. 1(4):295–304.

Hendler, G. and D. Franz. 1982. The biology of a brooding seastar, *Leptasterias tenera*, in Block Island Sound. Biol. Bull. 162:273–289.

Menge, B. 1975. Brood or broadcast? The adaptive significance of different reproductive strategies in the two intertidal sea stars *Leptasterias hexactis* and *Pisaster ochraceus*. Mar. Biol. 31(1):87–100. [Big seastars free spawn; little ones brood.]

Nance, J. M. and L. F. Braithwaite. 1979. The function of mucous secretions in the cushion star *Pteraster tesselatus* Ives. J. Exp. Mar. Biol. Ecol. 40:259–266.

Pearse, J. S., D. J. Eernisse, V. B. Pearse and K. A. Beauchamp. 1986. Photoperiodic regulation of gametogenesis in sea stars, with evidence for an annual calendar independent of fixed daylength. Am. Zool. 26:417–431.

Rivkin, R. B., I. Bosch, J. S. Pearse and E. J. Lessard. 1986. Bacterivory: A novel feeding mode for asteroid larvae. Science 233:1311–1314.

Symposium on the biology and ecology of the crown-of-thorns starfish, *Acanthaster planci* (L.). 1973. Micronesica 9(2). [Contributions of numerous authors from a symposium convened in response to the population explosion of *A. planci* on Indo-West Pacific coral reefs.]

Valentincic, T. 1983. Innate and learned responses to external stimuli in asteroids. *In* M. Jangoux and J. M. Lawrence (eds.), *Echinoderm Studies*, Vol. 1. A. A. Balkema, Rotterdam, pp. 111–138.

Wagner, R. H., D. W. Phillips, J. D. Standing and C. Hand. 1979. Commensalism or mutualism: Attraction of a sea star towards its symbiotic polychaete. J. Exp. Mar. Biol. Ecol. 39:205–210.

Yamagata, A. 1982. Studies on reproduction in the hermaphroditic sea star, *Asterina minor*: The functional male gonads, "ovitestes." Biol. Bull. 162:449–456.

Yamaguchi, M. 1973. Early life histories of coral reef asteroids, with special reference to *Acanthaster planci* (L.). *In* O. A. Jones and R. Endean (eds.), *Biology and Geology of Coral Reefs*, Vol. 2, Biology 1. Academic Press, New York, pp. 369–387.

Concentricycloidea

Baker, A. N., F. W. E. Rowe and H. E. S. Clark. 1986. A new class of Echinodermata from New Zealand. Nature 321:862–864.

Nichols, D. 1986. A new class of echinoderms. Nature 321:808.

Rowe, F. W., A. N. Baker and H. E. S. Clark. 1988. The morphology, development, and taxonomic status of Xyloplax Baker, Rowe and Clark (1986) (Echinodermata: Concentricycloidea), with description of a new species. Proc. R. Soc. London Ser. B 233:431–459.

Ophiuroidea

Hendler, G. 1975. Adaptational significance of the patterns of ophiuroid development. Am. Zool. 15:692–715.

Hendler, G. 1978. Development of Amphioplus abditus (Verrill) (Echinodermata: Ophiuroidea). II. Description and discussion of ophiuroid skeletal ontogeny and homologies. Biol. Bull. 154:79–95.

Hendler, G. 1979. Sex-reversal and viviparity in Ophiolepis kieri, n. sp., with notes on viviparous brittlestars from the Caribbean (Echinodermata: Ophiuroidea). Proc. Biol. Soc. Wash. 92:783–795.

Hendler, G. 1982a. An echinoderm vitellaria with a bilateral larval skeleton: evidence for the evolution of ophiuroid vitellariae from ophiuroplutei. Biol. Bull. 163:431–437.

Hendler, G. 1982b. Slow flicks show star tricks: Elapsed-time analysis of basketstar (Astrophyton muricatum) feeding behavior. Bull. Mar. Sci. 32:909–918.

Hendler, G. 1983. The association of Ophiothrix lineata and Callyspongia vaginalis: A brittlestar–sponge cleaning symbiosis? Mar. Ecol. 5(1):9–27.

Hendler, G. 1984. Brittlestar color-change and phototaxis (Echinodermata: Ophiuroidea: Ophiocomidae). Mar. Ecol. 5(4):379–401.

Mladenov, P. V., R. H. Emerson, L. V. Colpit and I. C. Wilkie. 1983. Asexual reproduction in the West Indian brittle star Ophiocomella ophiactoides (H. L. Clark) (Echinodermata: Ophiuroidea). J. Exp. Mar. Biol. Ecol. 72:1–23.

Pentreath, R. J. 1970. Feeding mechanisms and the functional morphology of podia and spines in some New Zealand ophiuroids. J. Zool. 161:395–429.

Turner, R. L. 1984. Revision of Ophiopaepale Ljungman, 1872 (Echinodermata: Ophiuroidea), with a redescription of O. goesiana Ljungman, 1872, and notes on O. diplax (Nielsen, 1932), new combination. Proc. Biol. Soc. Wash. 97:927–941.

Warner, G. F. and J. D. Woodley. 1975. Suspension feeding in the brittle star Ophiothrix fragilis. J. Mar. Biol. Assoc. U.K. 55:199–210.

Woodley, J. D. 1975. The behavior of some amphiurid brittle-stars. J. Exp. Mar. Biol. Ecol. 18:29–46.

Woodley, J. D. 1980. The biomechanics of Ophiuroid tube-feet. In M. Jangoux (ed.), Echinoderms: Present and Past. A. A. Balkema, Rotterdam, pp. 293–299.

Echinoidea

Alexander, D. E. and J. Ghiold. 1980. The functional significance of the lunules in the sand dollar, Mellita quinquiesperforata. Biol. Bull. 159:561–570.

Angerer, R. C. and E. H. Davidson. 1984. Molecular indices of cell lineages specification in sea urchin embryos. Science 226:1153–1160.

Burke, R. D. 1980. Podial sensory receptors and the induction of metamorphosis in echinoids. J. Exp. Mar. Biol. Ecol. 47:223–234.

Burke, R. D. 1983. Neural control of metamorphosis in Dendraster excentricus. Biol. Bull. 164:176–188.

Burke, R. D. 1984. Pheromonal control of metamorphosis in the Pacific sand dollar, Dendraster excentricus. Science 225:223–224.

Chia, F. S. 1969. Some observations on the locomotion and feeding of the sand dollar, Dendraster excentricus. J. Exp. Mar. Biol. Ecol. 3(2):162–170.

Chia, F. S. 1973. Sand dollar: A weight belt for the juvenile. Science 181:73–74.

Ellers, O. and M. Telford. 1984. Collection of food by oral surface podia in the sand dollar, Echinarachnius parma (Lamarck). Biol. Bull. 166:574–582.

Fell, H. B. 1965. The early evolution of the Echinozoa. Breviora 219:1–17.

Fenner, D. H. 1973. The respiratory adaptations of the podia and ampullae of echinoids (Echinodermata). Biol. Bull. 145:323–339.

Ferguson, J. C. 1969. Feeding, digestion, and nutrition in Echinodermata. In M. Florkin and B. T. Scheer (eds.), Chemical Zoology, Vol. 3, Echinodermata. Academic Press, New York, pp. 71–100.

Ghiold, J. 1979. Spine morphology and its significance in feeding and burrowing in the sand dollar, Mellita quinquiesperforata (Echinodermata: Echinoidea). Bull. Mar. Sci. 29:481–490.

Hanson, J. L. and G. Gust. 1986. Circulation of perivisceral fluid in the sea urchin Lytechinus variegatus. Mar. Biol. 92:125–134.

Kier, P. M. 1974. Evolutionary trends and their functional significance in the post-Paleozoic echinoids. J. Paleontol. 48 (supplement to No. 3: Palontol. Soc. Mem. 5):1–95.

Lewis, J. B. 1968. The function of sphaeridia of sea urchins. Can. J. Zool. 46:1135–1138.

Mann, K. H. 1982. Kelp, sea urchins and predators: A review of strong interactions in rocky subtidal systems of eastern Canada, 1970–1980. Neth. J. Sea. Res. 16:414–423.

Markel, K. 1980. The lantern of Aristotle. In M. Jangoux (ed.), Echinoderms: Present and Past. A. A. Balkema, Rotterdam, pp. 91–92.

Millott, N. 1975. The photosensitivity of echinoids. Adv. Mar. Biol. 13:1–52.

Pawson, D. L. and J. E. Miller. 1983. Systematics and ecology of the sea-urchin genus Centrostephanus (Echinodermata: Echinoidea) from the Atlantic and eastern Pacific oceans. Smithsonian Contrib. Mar. Sci. 20:1–15.

Pennington, J. T. 1985. The ecology of fertilization of echinoid eggs: The consequences of sperm dilution, adult aggregation, and synchronous spawning. Biol. Bull. 169:417–430.

Pequignat, E. 1966. Skin digestion and epidermal absorption in irregular and regular urchins. Nature 210:396–399.

Phelan, T. F. 1977. Comments on the water vascular system, food grooves, and ancestry of the clypeasteroid echinoids. Bull. Mar. Sci. 27:400–422.

Raff, R. A. 1987. Constraint, flexibility, and phylogenetic history in the evolution of direct development in sea urchins. Dev. Biol. 119:6–19.

Ryberg, E. 1980. Development and possible function of the red pigment in sea urchin larvae. In M. Jangoux (ed.), Echinoderms: Present and Past. A. A. Balkema, Rotterdam, pp. 405–408.

Seilacher, A. 1979. Constructional morphology of sand dollars. Paleobiology 5:191–221.

Smith, A. B. 1978. A functional classification of the coronal pores of regular echinoids. Paleontology 21:759–789.

Smith, A. B. 1980. The structure, function, and evolution of tube feet and ambulacral pores in irregular echinoids. Paleontology 23:39–83.

Telford, M. 1981. A hydrodynamic interpretation of sand dollar morphology. Bull. Mar. Sci. 31:605–622.

Telford, M. 1983. An experimental analysis of lunule function in the sand dollar Mellita quinquiesperforata. Mar. Biol. 76:125–134.

Telford, M. 1985. Domes, arches, and urchins: The skeletal architecture of echinoids (Echinodermata). Zoomorphologie 105:114–124.

Telford, M., A. S. Harold and R. Mooi. 1983. Feeding structures, behavior, and microhabitat of Echinocyamus pusillus (Echinoidea: Clypeasteroida). Biol. Bull. 165:745–757.

Telford, M., R. Mooi and O. Ellers. 1985. A new model of podial deposit feeding in the sand dollar, *Mellita quinquiesperforata* (Leske): The sieve hypothesis challenged. Biol. Bull. 169:431–448.

Holothuroidea

Bakus, G. J. 1973. The biology and ecology of tropical holothurians. *In* O. A. Jones and R. Endean (eds.), *Biology and Geology of Coral Reefs*. Vol. 1, Biology 1. Academic Press, New York, pp. 326–368.

Cameron, J. L. and P. V. Fankboner. 1984. Tentacle structure and feeding processes in life stages of the commercial sea cucumber *Parastichopus californicus* (Stimpson). J. Exp. Mar. Biol. Ecol. 81:193–209.

Costelloe, J. and B. Keegan. 1984. Feeding and related morphological structures in the dendrochirote *Aslia lefevrei* (Holothuroidea: Echinodermata). Mar. Biol. 84:135–142.

Fish, J. D. 1967. Biology of *Cucumaria elongata*. J. Mar. Biol. Assoc. U.K. 47:129–143.

Hauksson, E. 1979. Feeding biology of *Stichopus tremulus*, a deposit-feeding holothurian. Sarsia 63(3):155–160.

Herreid, C. F., V. F. LaRussa and C. R. DeFesi. 1976. Blood vascular system of the sea cucumber *Stichopus moebii*. J. Morphol. 150(2):423–451.

Martin, W. E. 1969. *Rynkatropa pawsoni* n. sp. (Echinodermata: Holothuroidea) a commensal sea cucumber. Biol. Bull. 137:332–337.

Massin, C. 1980. The sediment ingested by *Holothuria tubulosa* Gmel (Holothuroidea: Echinodermata). *In* M. Jangoux (ed.), *Echinoderms: Present and Past*. A. A. Balkema, Rotterdam, pp. 205–208.

Moriarty, D. J. W. 1982. Feeding of *Holothuria atra* and *Stichopus chloronotus* on bacteria, organic carbon and organic nitrogen in sediments of the Great Barrier Reef. Aust. J. Mar. Freshwater Res. 33:255–263.

Pawson, D. L. 1982. Holothuroidea. *In* S. P. Parker (ed.), *Synopsis and Classification of Living Organisms*. McGraw-Hill, New York, pp. 813–818.

Pawson, D. L. and H. B. Fell. 1965. A revised classification of the dendrochirote holothurians. Breviora 214:1–7.

Roberts, D. and C. Bryce. 1982. Further observations on tentacular feeding mechanisms in Holothurians. J. Exp. Mar. Biol. Ecol. 59:151–163.

Rowe, F. W. E. 1969. A review of the family Holothuriidae (Holothuriodea: Aspidochirotida). Bull. Br. Mus. Nat. Hist. 18:119–170.

Swan, E. F. 1961. Seasonal variation in the sea cucumber *Parastichopus californicus*. Science 133:1078–1079.

Trott, L. B. 1981. A general review of the pearlfishes (Pisces, Carapidae). Bull. Mar. Sci. 31:623–629. [A brief review of the 20 or so species of these odd fishes, most of which are commensals with holothurians and asteroids.]

Vanden Spiegel, D. and M. Jangoux. 1987. Cuvierian tubules of the holothuroid *Holothuria ferskali* (Echinodermata): A morphofunctional study. Mar. Biol. 96:263–275.

Chapter Twenty-Three

Other Deuterostomes: Chaetognatha, Hemichordata, Chordata

In a flash it covers a distance of some five or six times its own length; and its great jaws, opened wide, snap upon its victim.

> Sir Alister Hardy, describing a chaetognath
> *The Open Sea*, 1965

In addition to the lophophorates and echinoderms (discussed in the previous two chapters), the deuterostome lineage includes three other phyla: Chaetognatha (Greek, "spine-jaws"), Hemichordata (Greek, "half-chordates"), and Chordata (= the chordates). These three phyla are discussed in this chapter, which concludes with an overview of ideas concerning deuterostome phylogeny and comments on the origin of the vertebrates.

The chaetognaths are called arrow worms and comprise about 100 species of marine, mostly planktonic creatures. The hemichordates include 85 or so species, most of which are benthic burrowers known as tongue worms or acorn worms. The phylum Chordata includes three subphyla: Vertebrata (fishes, amphibians, reptiles, birds, and mammals), Cephalochordata (= Acrania; the lancelets), and Urochordata (= Tunicata; the ascidians, larvaceans, and thaliaceans). There are about 47,000 species of vertebrates, 23 species of cephalochordates, and 3,000 species of urochordates. Although cephalochordates lack a backbone, they are also often included in courses dealing with the vertebrates.

Taxonomic history and classification

The first record of a chaetognath was made by the Dutch naturalist Martinus Slabber in 1775. For nearly 100 years, as more and more descriptive work was conducted, the systematic position of the group was hotly debated. The arrow worms were at times allied with molluscs, arthropods, and certain pseudocoelomates (particularly nematodes), generally within the catch-all taxon Vermes. Some of these arguments continued well into this century. Although the question of chaetognath phylogenetic affinities is still unsettled, embryological studies favor a deuterostome relationship, and several unique characteristics support its separate phylum status.

Hemichordates were discovered in 1825 by Eschscholtz, who thought his specimen was a holothurian. Additional early records allied these animals with nemerteans. Bateson (1885) conducted developmental studies on some hemichordates and coined the present phylum name after recognizing similarities with chordate embryogeny. The chordate nature of tunicates (Urochordata) had been recognized by this time, also on the basis of developmental studies. Until very recently, the hemichordates were considered by many workers to be a subphylum of Chordata; but although they are clearly related, the hemichordates lack a true notochord—a key defining synapomorphy of the chordates.

PHYLUM CHAETOGNATHA

Arrow worms (Figures 1A and 2)

ORDER PHRAGMOPHORA: With ventral transverse muscle bands (phragma) that appear whitish in living animals. Two families with about 30 species: Spadellidae (*Paraspadella, Spadella*), and Eukrohniidae (*Bathyspadella, Eukrohnia, Heterokrohnia, Kukrohnia*).

ORDER APHRAGMOPHORA: Without ventral transverse muscle bands. Five families with about 70 species: Sagittidae (e.g., *Parasagitta, Sagitta*), Pterosagittidae (monotypic: *Pterosagitta draco*), Bathybelidae (*Bathybelos*), Krohnittellidae (*Krohnittella*), and Krohnittidae (*Krohnitta*).

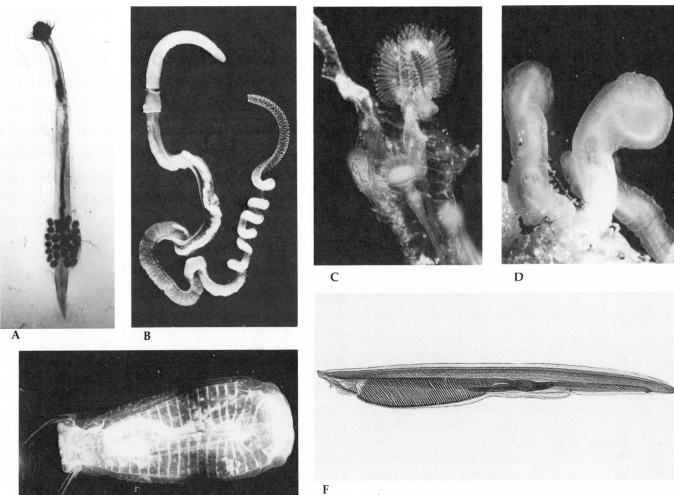

Figure 1

Representative deuterostomes. A, A chaetognath, *Eukrohnia bathypelagica* (order Phragmophora) carrying its fertilized eggs. B, *Saccoglossus*, an acorn worm (phylum Hemichordata, class Enteropneusta). C, Portion of a *Cephalodiscus* colony (phylum Hemichordata, class Pterobranchia). Note the large feeding individual with expanded tentacles. D, *Clavelina*, a social ascidian (phylum Urochordata). E, A pelagic tunicate (subphylum Urochordata, class Thaliacea). Note the circular bands of muscle in the body wall. F, A lancelet (subphylum Cephalochordata). (A courtesy of M. Terazaki and C. B. Miller; B photo by C. R. Wyttenbach, Univ. of Kansas/BPS; C from Lester 1985; D,E courtesy of G. McDonald; F courtesy Carolina Biological Supply.)

PHYLUM HEMICHORDATA

Acorn worms, pterobranchs, and the unusual *Planctosphaera pelagica* (Figures 1B, 1C, and 6)

CLASS ENTEROPNEUSTA: Acorn worms (Figures 1B, 6A, and 6B). Vermiform, with three body regions as proboscis, collar and trunk; coeloms reduced; gut elongate, straight; mouth ventral at anterior end of collar; without an endostyle; anus posterior, terminal; marine, burrow in soft sediments or nestle under rocks or in algal holdfasts; largely intertidal although a few deep-water species are known. About 75 species. (e.g., *Balanoglossus, Protoglossus, Saccoglossus, Saxipendium, Xenopleura*)

CLASS PTEROBRANCHIA: Pterobranchs (Figures 1C and 6E–H). Body sacciform; with three body regions as preoral disc (=cephalic shield), tentaculate mesosome, and metasome subdivided as trunk and stalk; neurochord lacking; gut U-shaped; marine; generally small (less than 1 cm), aggregating or colonial; three genera: *Atubaria, Cephalodiscus*, and *Rhabdopleura*.

CLASS PLANCTOSPHAEROIDEA: (Figure 6I) Body spherical but bilateral, jelly-like, with complexly arranged surface ciliary bands; gut U-shaped; coeloms poorly developed; monospecific (*Planctosphaera pelagica*); world-wide and considered to be probable larval stage of unknown adult hemichordate, perhaps an enteropneust.

Characteristics of the Phylum Chaetognatha

1. Bilateral deuterostomes, with streamlined, elongate, trimeric body comprising head, trunk, and postanal tail divided from one another by transverse septa; head with one "coelomic" space; trunk and tail with two "coelomic" spaces; body cavities not lined with mesodermal peritoneum
2. Body with lateral and caudal fins, supported by "rays" (apparently derived from epidermal basement membrane)
3. Head with a pair of uniquely arranged eyes (probably ciliary in origin) and, around the mouth, sets of grasping spines and teeth used in prey capture; mouth set in ventral vestibule; anterolateral fold of body wall forms hood that can enclose vestibule
4. Longitudinal muscles of unusual type, arranged in quadrants; no circular muscle
5. No discrete circulatory, gas exchange, or excretory systems
6. Complete gut; anus ventral, at trunk–tail junction
7. Large dorsal (cerebral) and ventral (subenteric) ganglia connected by circumenteric connectives; ciliary fans for detection of water-borne vibrations; anterior ciliary loop (= corona ciliata) of uncertain function
8. Hermaphroditic, with direct development
9. Strictly marine

PHYLUM CHORDATA

The chordates (Figures 1D–F)

SUBPHYLUM UROCHORDATA (= TUNICATA): The tunicates. Adult body form varies, but usually lacking obvious trimeric organization; body covered by thick or thin tunic (test) of a cellulose-like polysaccharide; without bony tissue; notochord restricted to tail and usually found only in larval stage (in adult appendicularians); gut U-shaped, pharynx (branchial chamber) typically with numerous gill slits (stigmata); coelom not developed; notochord present in larval stages; all marine; 4 classes.

CLASS ASCIDIACEA: Ascidians, sea squirts (Figures 1D and 13). Benthic, solitary or colonial, sessile tunicates; incurrent and excurrent siphons directed upwards, away from the substratum; without dorsal nerve cord in adult stages; occurring at all depths. About 13 families and many genera. (e.g., *Aplidium, Ascidia, Botryllus, Chelyosoma, Ciona, Clavelina, Corella, Diazona, Diplosoma, Lissoclinum, Molgula, Psammascidia, Pyura, Styela*)

CLASS THALIACEA: Pelagic tunicates or salps (Figures 1E and 14). Solitary or colonial; incurrent and excurrent siphons at opposite ends, providing locomotor current; adults without a tail; gill clefts not subdivided by gill bars; 3 orders: Pyrosomida, Salpida, and Doliolida. (e.g., *Dolioletta, Doliolum, Pyrosoma, Salpa, Thetys*)

CLASS APPENDICULARIA (= LARVACEA): Appendicularians or larvaceans (Figures 14F and G). Solitary planktonic tunicates; apparently neotenic, adults retain larval characteristics, including notochord and muscular tail; body enclosed in a complex gelatinous "house" involved in feeding. (e.g., *Fritillaria, Oikopleura*)

CLASS SORBERACEA: Benthic, abyssal, ascidian-like urochordates possessing dorsal nerve cords in adult stages; carnivorous, lacking perforated branchial sac. (e.g., *Octacnemus*)

SUBPHYLUM CEPHALOCHORDATA: Lancelets (amphioxus) (Figures 1F and 18). Small (to 7 cm), fishlike chordates with notochord, gill slits, dorsal nerve cord, and postanal tail present in adults, but without vertebral column or cranial skeleton structure; gonads numerous (25–38) and serially arranged. Marine and brackish water, usually associated with clean sand or gravel sediments in which they burrow. (e.g., *Branchiostoma, Asymmetron, Epigonichthyes*)

SUBPHYLUM VERTEBRATA: Vertebrates. Chordates usually possessing a vertebral column that forms the axis of the skeleton; most have paired appendages, a brain case, and (except for members of the class Agnatha) jaws. Seven classes are generally recognized, although not all are strictly monophyletic: Agnatha (lampreys and hagfishes), Chondrichthyes (sharks, skates, and rays), Osteichthyes (all bony fishes, e.g., trout, tunas, mackerals, perch), Amphibia (salamanders, frogs, toads, caecilians), Reptilia (turtles, snakes, lizards, crocodiles, and alligators), Aves (birds), Mammalia (mammals).

Phylum Chaetognatha

Arrow worms (Figure 2) comprise about 100 species of wholly marine, largely planktonic animals of moderate size, ranging from about 0.5 to 12 cm in length. With the exception of a few benthic species (e.g., *Spadella*), chaetognaths are adapted to life as

Box Two

Characteristics of the Phylum Hemichordata

1. Bilateral deuterostomes, body vermiform or saccate and fundamentally trimeric, with prosome, mesosome, and metasome, each with coelomic compartments; solitary or colonial
2. With ciliated, pharyngeal gill slits (or pores)
3. Well developed, open circulatory system
4. Unique excretory structures ("glomerulus")
5. Gonads extracoelic, in metasome
6. Complete gut
7. Short, dorsal, mesosomal, occasionally hollow nerve cord (neurochord)—homology with chordate nerve cord uncertain; without a notochord
8. Gonochoristic, with indirect or mixed development; asexual reproduction common
9. Strictly marine

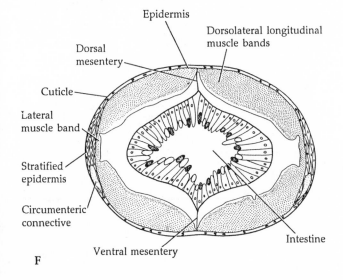

pelagic predators. They are distributed throughout the world's oceans and in some estuarine habitats. They often occur in very high numbers and may dominate the biomass in mid-water plankton tows. They are most abundant in neritic waters, but some occur at great depths. A few troglobitic species have been reported from anchialine caves in the Bahamas and France. Major characteristics of the group are listed in Box 1.

The chaetognath *Bauplan*

Externally, arrow worms are streamlined, with virtually perfect bilateral symmetry and transparent bodies, although some meso- and bathypelagic species have orange carotenoid pigmentation and some (phragmophorids) appear milky white because of the opaque ventral transverse musculature. The trimeric body includes a head, a trunk and a tail—corresponding to the prosome, mesosome, and metasome of the general deuterostome *Bauplan*. The trunk bears paired lateral fins and the tail bears a single tail fin, all of which are supported by rays of unknown composition. The body surface bears various sensory structures, but the functions of many are not well understood. The head has a ventrally placed mouth,

Figure 2

General anatomy of chaetognaths. A, *Sagitta elegans* **(ventral view). B,** The benthic chaetognath *Spadella* **(dorsal view). C,** *Krohnitta subtilis* **(dorsal view). D,** Outline of *Sagitta hispida*, **showing sensory bristles. E,** Anatomy of the head of *Sagitta*. **F,** The trunk of *Sagitta* **(cross section). G,** The nervous system of a **generalized chaetognath. H,** Arrangement of eye units in a chaetognath. **I,** Cerebral ganglion and major nerves (ventral view). **J,** Reproductive systems in *Sagitta*. (A,B after Barnes 1980; C,F,I after Hyman 1959; D after Feigenbaum 1978; E after Meglitsch 1972; G after various sources; H after Burfield 1927.)

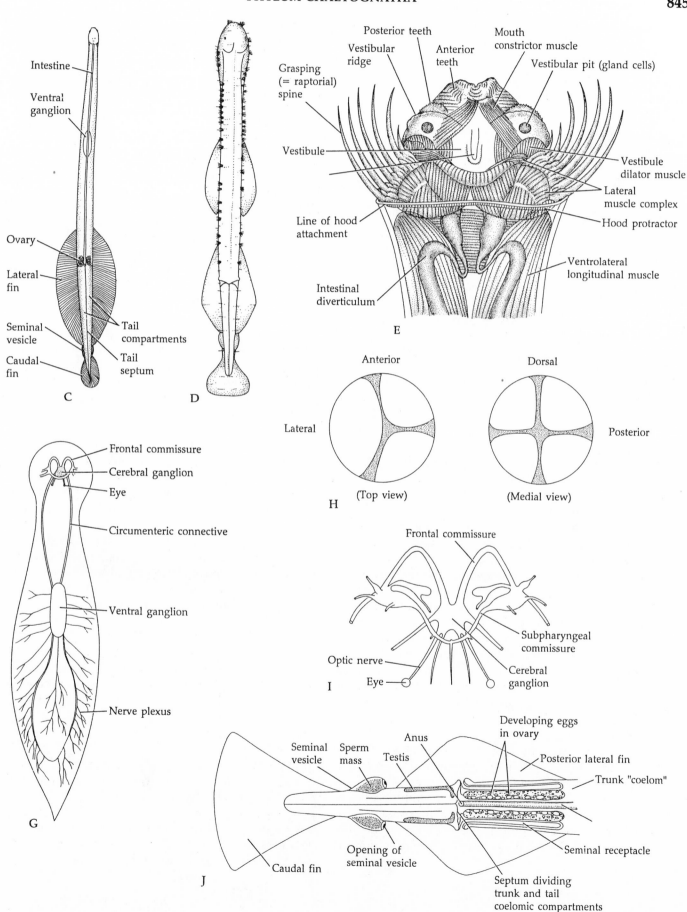

Intestine

Ventral ganglion

Ovary

Lateral fin

Seminal vesicle

Caudal fin

C

D

Posterior teeth

Vestibular ridge

Anterior teeth

Mouth constrictor muscle

Grasping (= raptorial) spine

Vestibular pit (gland cells)

Vestibule

Vestibule dilator muscle

Lateral muscle complex

Hood protractor

Line of hood attachment

Ventrolateral longitudinal muscle

Intestinal diverticulum

E

Tail compartments

Tail septum

Frontal commissure

Cerebral ganglion

Eye

Circumenteric connective

Ventral ganglion

Nerve plexus

G

Anterior

Lateral

Dorsal

Posterior

(Top view)

(Medial view)

H

Frontal commissure

Subpharyngeal commissure

Optic nerve

Eye

Cerebral ganglion

I

Developing eggs in ovary

Anus

Seminal vesicle

Sperm mass

Testis

Posterior lateral fin

Trunk "coelom"

Caudal fin

Opening of seminal vesicle

Seminal receptacle

Septum dividing trunk and tail coelomic compartments

J

Box Three

Characteristics of the Phylum Chordata

1. Bilaterally symmetrical, coelomate deuterostomes (coelom lost in some groups)
2. Pharyngeal gill slits present at some stage in development
3. Dorsal notochord present at some stage in development
4. Dorsal, hollow nerve cord
5. With an endostyle (Urochordata, Cephalochordata) or thyroid gland (Vertebrata)
6. Muscular, locomotor, postanal tail at some stage in development
7. Gut complete, usually regionally specialized
8. Ventral, contractile blood vessel (or heart)
9. Gonochoristic or hermaphroditic; development variable

---

set in a depression called the **vestibule**. Lateral to the mouth are heavy **grasping spines**, or "hooks," and in front of the mouth are smaller spines called **teeth** —all used in prey capture. Dorsally the head bears a pair of photoreceptors, usually called eyes, of unique structure (Figure 2H). All chaetognaths possess an anterolateral folding of the body wall called the **hood**, which can be drawn over the front and sides of the head enclosing the vestibule.

Other external features of note include a unique **ciliary loop** (or **corona ciliata**) located on the dorsal surface at the head-trunk junction (Figure 2B). This organ consists of two rings of ciliated epithelial cells and may be involved in chemoreception, or perhaps sperm transfer. Male and female gonopores are located laterally and posteriorly in the tail and trunk, respectively. The anus is ventral at the trunk–tail junction.

The overall body form of chaetognaths, coupled with their high degree of transparency and locomotor abilities have contributed to their great success as planktonic predators. They are frequently recovered in plankton hauls with their grasping spines firmly affixed to another animal (Figure 3C)—we are fortunate indeed that they do not attain a very large size!

Body wall, support, and locomotion. A very thin, flexible cuticle overlies the epidermis and helps maintain body shape. Over most of the body, the cells of the epidermis are squamous; they have sinuous interlocking margins and may be stratified. Columnar epithelial cells line the vestibule. In some areas the epidermis lacks a cuticle and bears abundant secretory cells. A well developed basement membrane lies beneath the epidermis and may form the supportive rays of the fins. The body wall musculature consists largely of four quadrants of well developed dorsolateral and ventrolateral longitudinal bands, plus the complex head musculature (Figures 2E and F). In the Phragmophora, there are additional transverse muscle bands along the ventral aspect of the trunk.

The nature of the adult body cavities is somewhat enigmatic in chaetognaths. Enterocoelic spaces do form during ontogeny, but these are apparently lost, and the precise origin of the adult body cavities remains uncertain. The adult body spaces are not lined by mesodermal peritoneum. Still, even in the adult, there is a clear 1:2:2, tripartite "coelomic" arrangement. The head cavity is greatly reduced by the complex cephalic musculature (Figure 2E). The paired trunk and tail cavities contain dorsal and ventral longitudinal mesenteries or septa. Internal transverse septa separate the three body regions. The mesenteries and septa are thought by some to be extensions of the epidermal basement membrane and not true peritoneum. The "coelomic" fluid contains various cells or cell-like inclusions, but their functions are unknown.

Body support in chaetognaths is provided by the hydrostatic qualities of the internal cavities and the well developed musculature, aided somewhat by the thin cuticle. The complex musculature of the head operates the spines and vestibule, and the closure of the hood. Locomotion in pelagic forms involves forward darting motions caused by rapid lateral body flexion, by alternately contracting the right- and left-side longitudinal muscles. Although these motions may alternate with brief quiescent periods when the animal slowly sinks, pelagic species seem to spend most of their time actively swimming about, presumably in search of prey. The fins are not used as propelling surfaces but are placed so that they slice through the water and serve as stabilizers. They also increase resistance to sinking between swimming bursts. Although chaetognaths are highly effective predators, watching a live arrow worm moving in the water gives one the impression that they are not very efficient swimmers. However, it has been suggested that the seemingly erratic movements may

Grasping spines | Vestibular pit | Papillae of vestibular ridge | Anterior teeth | Posterior teeth

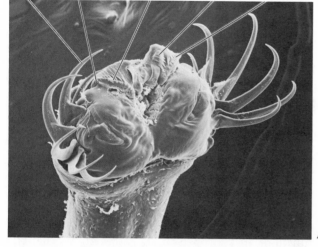

A

Figure 3

A–B, Heads of chaetognaths. A, *Zonosagitta pulchra*. Note the well developed raptorial structures. B, *Z. bedoti*, from the eastern Pacific. The hooks are clearly visible on either side of the head surrounding the exceptionally large number (17–20) of long, narrow, posterior teeth. The shorter anterior teeth lie just above the mouth. The vestibular ridge with its pores is partially visible behind the left set of posterior teeth. C, The chaetognath in this photo, *Flaccisagitta hexaptera*, has partially swallowed a fish larva (probably an anchovy). (Photos and information courtesy of E. Thuesen and R. Bieri.)

Vestibular pit

Posterior teeth

Anterior tooth

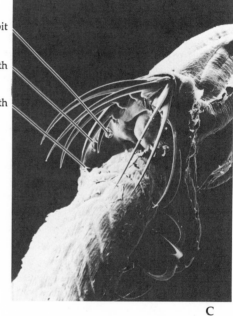

C

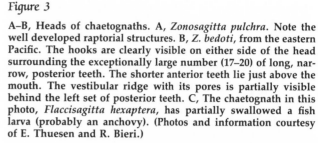

B

serve to confuse or elude predators. The ventral transverse muscles in benthic chaetognaths (e.g., *Spadella*) probably aid in crawling, and are generally considered to represent a primitive condition in the phylum. These benthic species are also capable of swimming over short distances.

Feeding and digestion. Chaetognaths are predatory carnivores that feed on a variety of other animals, including planktonic crustaceans, small fishes, and even other arrow worms. They seem to have a special fondness for copepods as food. Benthic forms, such as *Spadella*, are ambush predators. They affix themselves to a substratum by adhesive secretions, raise the head and protrude the mouth and vestibule, flaring the spines around the mouth. Prey swimming within reach are detected (probably by vibration receptors) and then captured by a quick strike of the

head while the rest of the body remains firmly attached to the substratum. The spines grasp and manipulate the prey, orienting it for ingestion. Prey is ingested whole.

When a planktonic chaetognath detects a nearby prey, it quickly darts forward to grab the victim with its grasping spines (Figure 3). The spines and/or teeth of some species bear serrations; the teeth are cuspidate, a shape that aids in the penetration of prey, especially crustacean exoskeletons. Many, if not most, species inject a poison into their prey when they attack. Erik Thuesen and several colleagues have recently shown that the poison contains a potent neurotoxin called tetrodotoxin that blocks sodium transport across cell membranes. Many marine bacteria synthesize tetrodotoxin, but only one other animal is known to produce a tetrodotoxin-based venom—the blue-ringed octopus. Recent studies suggest that in chaetognaths this toxin is produced by a commensal bacterium (*Vibrio alginolyticus*) inhabiting the head region of the arrow worm. Because neither the spines nor the teeth are hollow, the venom is probably secreted from minute pores or papillae that occur within the vestibule. Some large species of *Sagitta* and other chaetognaths are capable of capturing and devouring fishes as large as themselves! The bodies of chaetognaths bear arrays of "hairs" called **ciliary fans** (not to be confused with the ciliary loop mentioned earlier) that are sensitive to water-borne vibrations (Figure 2D). Because of the unique distribution of these structures a chaetognath can determine both the distance and direction to potential prey at close range.

As is typical of carnivores, the gut is a relatively simple straight tube running from the mouth in the vestibule to the ventral anus at the trunk–tail junction (Figure 2). The mouth leads to a short **pharynx**, which is equipped with mucus-secreting cells. Swallowing is accomplished by means of well developed pharyngeal muscles and is aided by lubricants from the mucous glands. When rigid prey such as small crustaceans are captured, the chaetognath positions the victim longitudinally for swallowing. The gut narrows to a long intestine, where it passes through the head–trunk septum. A short rectum joins the posterior intestine to the anus. Apparently most digestion occurs extracellularly in the posterior region of the intestine and can be extremely rapid.

Circulation, gas exchange, and excretion. No special organs are known to be associated with these functions. The body shape must yield a surface-to-volume ratio large enough and diffusion distances small enough that exchanges and internal transport can be accomplished largely by passive means. The spacious fluid-filled body cavities provide a medium in which diffusion can occur; this process is aided by the frequent body movements associated with swimming. We emphasize, however, that there is much to be learned about chaetognath physiology, and other mechanisms may eventually be shown to be involved.

Nervous system and sense organs. Of paramount importance to the success of chaetognaths as active predators are features of the nervous system and associated sensory receptors. A *Bauplan* that emphasizes cephalization is, as we have seen in other groups, frequently an integral factor in adapting to a predatory lifestyle. The central nervous system of chaetognaths includes a large **cerebral ganglion** in the head, dorsal to the pharynx. Several other ganglia arise from the cerebral ganglion and serve various muscles and sense organs of the head. A pair of circumenteric connectives emerges from the hind part of the cerebral ganglion and extends posteroventrally to meet in a large **ventral** or **subenteric ganglion** located in the trunk epidermis (Figure 2G). The ventral ganglion controls swimming. From it emerge a dozen or so pairs of nerves that extend to various parts of the body, many branching to form a dense subepidermal nerve plexus.

The body is covered with patches and tracts of bristle-like, cilia-derived ciliary fans, long thought to be tactile receptors but more recently shown to be sensors of water-borne vibrations or movements (Figure 2D). These structures operate in a fashion similar to that of the lateral line system of fishes, which functions in prey detection. Although specific chemoreceptors have yet to be positively identified in chaetognaths, they almost certainly occur. The aforementioned ciliary loop may have a chemoreceptive function, and many species have a set of pores called **transvestibular pores** that run more-or-less parallel with the vestibular ridge in the buccal area; these structures may also be chemosensory in function. Arrow worms possess a pair of eyes situated just below the epidermis on the dorsal surface of the head. The structure of these eyes is unusual in that each consists of five inverted pigment-cup ocelli, arranged with a large ocellus directed laterally and four small ones directed medially (Figure 2H). We may imply from this structure that chaetognaths have a more-or-less uninterrupted visual field enabling them to orient to light direction and intensity. Chaetognath eyes lack lenses, except in two deep-water species of *Eukrohnia* that reportedly bear hexagonal cuticular lenses. In most, however, the eyes probably do not form images but are used for orientation during vertical migration. There is disagreement about whether or not the ocelli are of the ciliary type like those found in most other deuterostomes.

Reproduction and development. Arrow worms are hermaphroditic, with paired ovaries in the trunk and paired testes in the tail (Figures 2A and B). Spermatogonia are released from the testes into the body cavity of the tail, where they mature. From there they are picked up by open ciliated funnels leading to sperm ducts, which open laterally at a pair of seminal vesicles. Sperm masses variously called "sperm balls" or "sperm clusters" form within the seminal vesicles. Chaetognaths apparently do not form true spermatophores (although the term is often used in the literature), because a definitive covering or "spermatophore membrane" has never been observed. Each ovary bears along its side an oviduct that leads to a genital pore just in front of the trunk–tail septum. Immature eggs are transferred to the oviducts, but the details of this process are unclear. In at least some species (e.g., *Spadella cephaloptera, Parasagitta hispida*), exchange of sperm may be mutual (reciprocal). *Parasagitta hispida* may undergo reciprocal or nonreciprocal fertilization, or even self-fertilization.

Transfer of the sperm mass has been most extensively studied in benthic species of the genus *Spadella*. After a rather elaborate mating "dance," the sperm are deposited as balls onto the mate's body. Rupture of the ball allows the sperm to stream posteriorly to enter the female gonopores and oviducts, where fertilization occurs. Benthic chaetognaths (e.g., *Spadella*) tend to deposit fertilized eggs on algae or other suitable substrata. Neritic species may secrete a jelly-like coating around each zygote and then shed the floating embryos to the sea (e.g., *Sagitta*), or they may

Figure 4

The chaetognath *Eukrohnia*, with temporary gelatinous marsupia housing the developing embryos. A–C, *E. bathypelagica* carrying fertilized eggs and young in the marsupium. D, *E. fowleri* carrying fertilized eggs in posterior marsupial sacs. E, Young of *E. fowleri* just after hatching. F, *E. fowleri* carrying the empty marsupial sacs from which the young have already escaped. (From Terazaki and Miller 1982.)

A

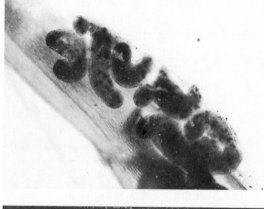

B

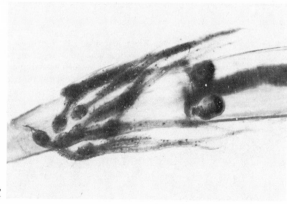

C

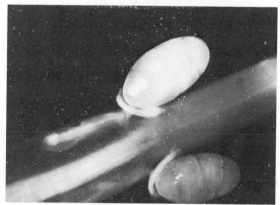

D

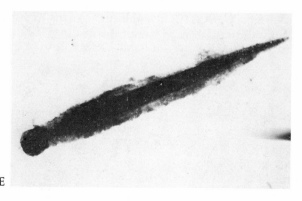

E

F

release developing embryos that sink to the bottom for development, or they may undertake downward migrations to lay eggs that sink to the bottom or attach to stationary objects. Species in the deep-water genus *Eukrohnia* carry the developing embryos in two temporary gelatinous marsupial pouches, one on either side of the body near the tail, until the young are ready to swim (Figure 4).

Development is direct, lacking any larval stage or metamorphosis. The transparent eggs contain very little yolk, and cleavage is radial, holoblastic, and equal, yielding a coeloblastula with a small blastocoel (Figure 5). The blastocoel enlarges during subsequent divisions, and gastrulation occurs by invagination of the presumptive entoderm. The blastopore marks the eventual posterior end of the animal, but it closes so that both the mouth and the anus form secondarily, the former by a stomodeal invagination of ectoderm. Entodermal pockets (Figure 5) eventually pinch off as first the head and then the trunk coeloms (the tail coeloms apparently form later than those in the trunk). Thus, embryonic coelom formation is clearly enterocoelous, although the relationship of these cavities to the adult body spaces is unclear. The embryo grows quickly, elongates, and hatches as a juvenile chaetognath. Development from egg release to hatching is rapid, on the order of about 48 hours. Strategically, the life history of most chaetognaths compares with indirect development, although no independent larval stage occurs. Still, parental investment is small, the eggs contain little yolk and are abandoned soon after fertilization, except in brooding forms. The rapid development to a feeding juvenile is essential to the success of this strategy.

The hemichordates

The small phylum Hemichordata contains the enteropneusts (acorn worms) and pterobranchs (Figures 1B and 6; Box 2). Also included here is *Planctosphaera pelagica*, which is assigned to the monotypic class Planctosphaeroidea (Figure 6I). This creature was discovered in 1932 and has since been collected in several localities in the Pacific and Atlantic oceans. It is viewed by most authorities as a hemichordate larva, but it has not yet been linked to a specific adult. In addition to the general traits associated with deuterostomes, hemichordates possess pharyngeal gill slits and most have a dorsal (sometimes hollow) nerve cord (Figure 7A). They were once thought to possess a notochord and therefore were placed as a subphylum of the Chordata. However, studies eventually showed that the suspected structure (an evagination of the anterior gut called the **buccal diverticulum**) is not at all homologous with a notochord.

As adults, all hemichordates are benthic marine animals. About 75 of the 85 or so living species belong to the class Enteropneusta, the acorn worms. These worms generally live buried in soft sediments, among algal holdfasts, or under rocks; they are largely intertidal. Of the four or five deep-water acorn worms

Figure 5

Chaetognath development. A, Early blastula. B, Gastrula. C, Later gastrula. D, Production of mesodermal folds from archenteron. E, Blastopore closure and secondary mouth opening with the formation of a stomodeum. F, Formation of coelomic pouches. (After Hyman 1959.)

KEY

1. Egg membrane
2. Blastomere
3. Blastocoel
4. Ectoderm
5. Entoderm
6. Archenteron
7. Blastopore
8. Primordial germ cell
9. Mesodermal fold
10. Mouth
11. Stomodeal pharynx
12. Gut
13. Anterior coelomic space
14. Developing trunk coelom

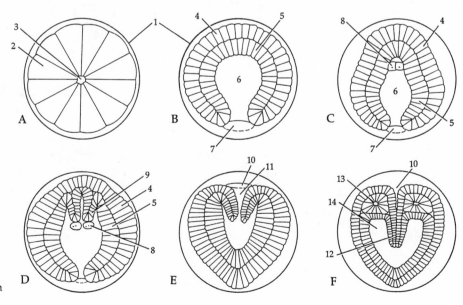

known, the most recently described species, *Saxipendium coronatum* (the "spaghetti worm"), is a member of the Galapagos geothermal vent community. Hemichordates range in length from a few centimeters to over 2 meters. Except for *Planctosphaera*, all of the remaining hemichordates belong to the class Pterobranchia. Pterobranchs are all colonial. The zooids are small, rarely exceeding 1 cm in length, but colonies may measure 10 cm or more across. They are very lophophorate-like in general structure (Figures 1C and 6E–H). Such similarities, however, as well as the differences between pterobranchs and enteropneusts, are probably the results of small size and colonial life.

The groups Enteropneusta and Pterobranchia represent two highly divergent clades within this phylum, each adapted to different ways of making a living. Both represent exploitations of the basic tripartite deuterostome architecture, attesting again to the evolutionary plasticity of this fundamental body plan.

The hemichordate *Bauplan*

Enteropneusts are solitary, elongate, vermiform animals, with bodies clearly divided into three regions. The **proboscis, collar,** and **trunk** (Figures 6A and B) are homologous to the prosome, mesosome, and metasome of other deuterostomes. The proboscis is short and often conical. A short, thin **proboscis stalk** connects the proboscis to the collar, the latter bearing the ventral mouth at its anterior end. The anus terminates at the posterior end of the long trunk. The trunk bears middorsal and midventral longitudinal ridges, which correspond to the location of certain longitudinal nerves and blood vessels. In addition, the trunk is differentiated regionally along its length, the amount of differentiation varying from species to species. Most species bear a distinct, anterior, **branchial region** of the trunk. This region is characterized by the presence of numerous **gill (= branchial) pores** flanking the middorsal ridge. Some, such as *Balanoglossus*, have a **genital region** housing the gonads and bearing external longitudinal **genital wings**. Also in *Balanoglossus*, the anterior portion of the intestine is thickened to such an extent that it is visible through the body wall as a distinctly colored area called the **hepatic region**. In a number of genera no clear body subdivisions occur, except for the region of the gill pores.

The bodies of pterobranchs are small, usually pyriform or globular, but still retain the ancestral tripartite regional division. Like the prosome of lophophorates, the pterobranch prosome is reduced to a small plate, called the **preoral disc** or **cephalic shield**, which generally folds ventrally over the mouth. The mesosome forms a collar that bears the anteroventral mouth and forms two to several **arms** on which tentacles are located. The arms are arranged dorsally and do not encircle the mouth. The gut is U-shaped and the anus lies anterodorsally (Figures 6E–H, and 9). The metasome is subdivided into a trunk and posterior stalk.

The three known genera are somewhat different from one another in habits and in some anatomical details. Most pterobranchs live in colonies (*Rhabdopleura*) or aggregations (*Cephalodiscus*) consisting of zooids housed with tubular secreted casings (Figures 6E–H). In colonies of *Rhabdopleura*, the zooids are connected to one another by tissue extensions called **stolons**, but no such interzooidal communication occurs in *Cephalodiscus*. The overall forms of the aggregations and colonies vary among species. In all cases the associated zooids are products of asexual reproduction initiated by a single sexually produced individual. The third genus is represented by a single species (*Atubaria heterolopha*), known only from 43 specimens collected in 1935 in Sagami Bay, Japan. These animals were recovered from dredge samples taken in 200–300 meters of water and were found clinging to hydroid colonies by their prehensile stalks. Although *Atubaria* is very similar anatomically to *Cephalodiscus*, it is a solitary form without any secreted casing (Figure 6H).

Body wall, coelom, support, and locomotion. Hemichordates in general possess a ciliated epidermis overlying a nerve plexus. The epidermis is usually richly supplied with gland cells, many of which are involved in mucus production, particularly on the proboscis and collar of enteropneusts and on the tentacles of pterobranchs. In some enteropneusts, epithelial mucocytes are known to produce noxious mucopolysaccharide compounds that may repel predators. Most of these secretions have a distinctive iodine-like odor. Both circular and longitudinal muscles are present in the wall of the proboscis and anterior collar of acorn worms, but elsewhere only longitudinal fibers exist. Apparently, pterobranchs possess only longitudinal fibers in their body walls, some of which are produced by the peritoneum. A basement membrane lies between the epidermis and the musculature, and in enteropneusts it is produced as thickened, rigid plates, called the **proboscis skeleton** (Figure 7A), and as supportive structures of the gill slits.

The peritoneum is variably reduced or trans-

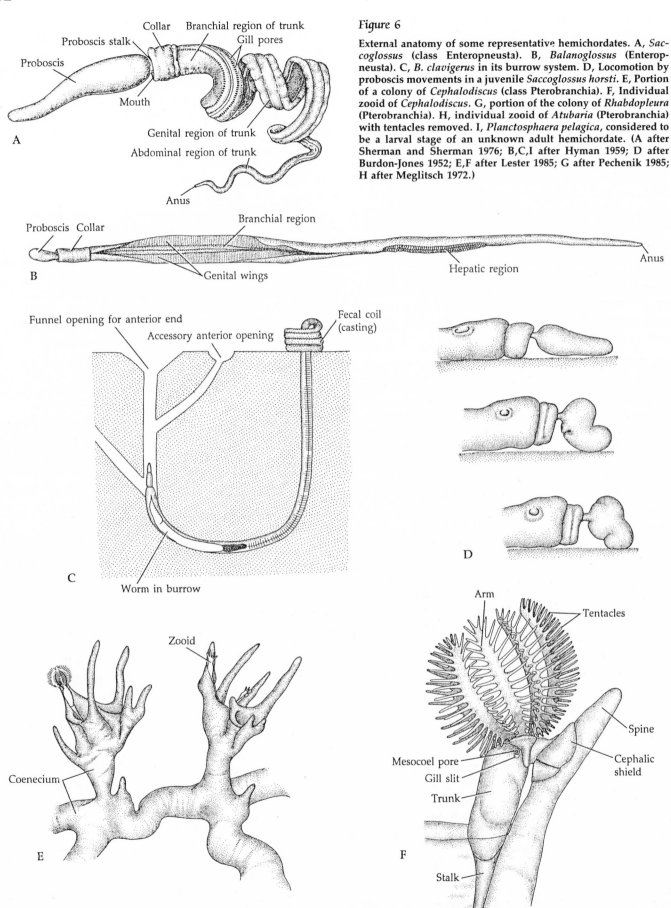

Figure 6

External anatomy of some representative hemichordates. A, *Saccoglossus* (class Enteropneusta). B, *Balanoglossus* (Enteropneusta). C, *B. clavigerus* in its burrow system. D, Locomotion by proboscis movements in a juvenile *Saccoglossus horsti*. E, Portion of a colony of *Cephalodiscus* (class Pterobranchia). F, Individual zooid of *Cephalodiscus*. G, portion of the colony of *Rhabdopleura* (Pterobranchia). H, individual zooid of *Atubaria* (Pterobranchia) with tentacles removed. I, *Planctosphaera pelagica*, considered to be a larval stage of an unknown adult hemichordate. (A after Sherman and Sherman 1976; B,C,I after Hyman 1959; D after Burdon-Jones 1952; E,F after Lester 1985; G after Pechenik 1985; H after Meglitsch 1972.)

formed into musculature in different parts of the bodies of hemichordates. Still, the tripartite coelomic arrangement is evident in all forms, even though the cavities are frequently reduced by the invasion of connective or supportive tissue. The usual three coelomic spaces, or their remnants, are present as a single protocoel, followed by paired mesocoels and metacoels. These spaces occur, in order, in the proboscis, collar, and trunk of enteropneusts, and in the cephalic shield, collar-arms-tentacles, and trunk of pterobranchs. As in lophophorates, mesocoelic extensions are present in the arms and tentacles of pterobranchs.

Body support is a function primarily of the hydrostatic nature of the body cavities, and secondarily of the structural integrity of the body wall, connective tissues, and supplemental structures such as the proboscis skeleton of acorn worms. In pterobranchs, except *Atubaria*, additional support and protection are provided by the secreted outer casing of the colony or aggregation.

Hemichordates are sessile or sedentary and capable of only limited movement at best. Enteropneusts crawl slowly or burrow by peristaltic action of the proboscis (Figure 6D), which contains necessary circular and longitudinal muscles. Although most are probably strictly benthic, and largely sedentary, at least one species of enteropneust (*Glandiceps hacksii*) has been reported swarming at the surface in shallow water, feeding on phytoplankton. The protraction and retraction of pterobranchs within their tubular houses are accomplished by hydrostatic pressure and contraction of longitudinal muscles, respectively. Some crawl within their tubes by using the muscular cephalic shield. The prehensile, tail-like stalk of *Atubaria* is quite mobile, probably as a result of a combination of hydraulics and muscle action.

Feeding and digestion. Enteropneusts that burrow through soft sediments are largely direct deposit feeders, ingesting the substratum and digesting organic material therein. Those that live in permanent burrows (Figure 6C) or among loose rubble or holdfasts tend to be suspension feeders, selectively trapping organic particulates from the water with the proboscis. Most species are probably capable of feeding by both methods. The details of suspension feeding have been examined in some acorn worms (e.g. *Protoglossus*; Figure 8). Food material, including detritus and live plankton, is trapped in mucus secreted over the surface of the proboscis and moved posteriorly by ciliary currents. The sorting that occurs at the posterior end of the proboscis and the stalk passes most large particles over the lip of the collar; these particles are then removed by special rejection currents (Figure 8). Most of the food is moved ventrally around the proboscis stalk, over a structure called the **preoral ciliary organ**, and condensed into a mucous cord that is then passed into the mouth. The preoral ciliary organ includes a concentration of sensory neurons and probably functions in chemoreception. Swallowing appears to be facilitated by a combination of ciliary action and the flow of water moving into the mouth and out the gill pores. The gill pores thus function to facilitate water flow—not as a filter-feeding device themselves. The cilia on the gill pores probably serve to keep the pores clean and unclogged.

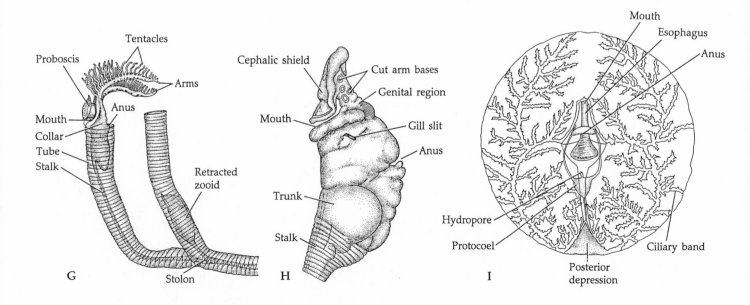

G — Proboscis, Tentacles, Arms, Anus, Mouth, Collar, Tube, Stalk, Retracted zooid, Stolon

H — Cephalic shield, Cut arm bases, Genital region, Mouth, Gill slit, Anus, Trunk, Stalk

I — Mouth, Esophagus, Anus, Hydropore, Protocoel, Posterior depression, Ciliary band

A

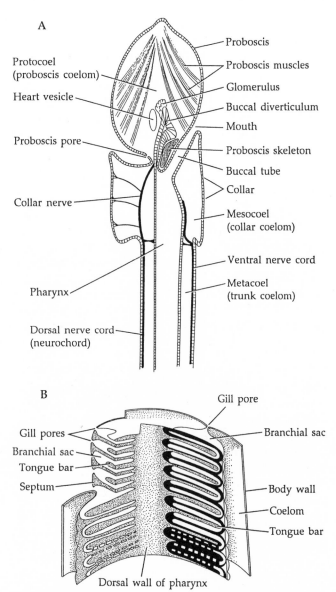

Protocoel (proboscis coelom)

Heart vesicle

Proboscis pore

Collar nerve

Pharynx

Dorsal nerve cord (neurochord)

Proboscis

Proboscis muscles

Glomerulus

Buccal diverticulum

Mouth

Proboscis skeleton

Buccal tube

Collar

Mesocoel (collar coelom)

Ventral nerve cord

Metacoel (trunk coelom)

B

Gill pores

Branchial sac

Tongue bar

Septum

Gill pore

Branchial sac

Body wall

Coelom

Tongue bar

Dorsal wall of pharynx

Figure 7

Internal anatomy of enteropneusts. A, The front end of an enteropneust (sagittal section). B, Cutaway view from inside the pharynx of an enteropneust, showing the arrangement of the gill slits. C, Enteropneust circulatory system. (A after various sources; B,C after Hyman 1959.)

C

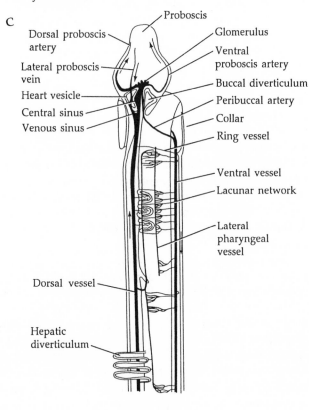

Dorsal proboscis artery

Lateral proboscis vein

Heart vesicle

Central sinus

Venous sinus

Dorsal vessel

Hepatic diverticulum

Proboscis

Glomerulus

Ventral proboscis artery

Buccal diverticulum

Peribuccal artery

Collar

Ring vessel

Ventral vessel

Lacunar network

Lateral pharyngeal vessel

The digestive tract of enteropneusts is a straight, regionally specialized tube, extending from the mouth to the anus (Figure 7A). Gut musculature is scant, and the food is moved along largely by cilia. The mouth leads to a **buccal tube**, which is housed within the collar and gives rise anterodorsally to the forwardly projecting **buccal diverticulum**. Behind the buccal tube and lying in the anterior part of the trunk is the **pharynx**. Both the buccal tube and the pharynx are derived from a stomodeal invagination of ectoderm. The digestive portion of the pharynx is restricted to a thin midventral **hypobranchial ridge**, while the lateral and dorsal parts bear the **gill slits** (Figure 7B). The gill slits number from a few to over 100 pairs. Each one is a U-shaped opening in the wall

Figure 8

Food sorting and rejection currents (arrows) on the proboscis and collar region of the enteropneust *Protoglossus kohleri*. (After Burdon-Jones 1956.)

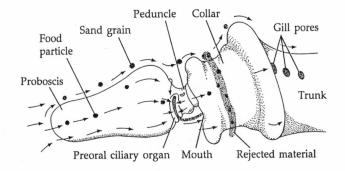

Food particle

Sand grain

Peduncle

Collar

Gill pores

Proboscis

Trunk

Preoral ciliary organ

Mouth

Rejected material

of the pharynx that leads to a **branchial sac** and then to a dorsolateral **gill pore** through which water exits to the outside. The septum between adjacent gill slits and the partition between the arms of the U of each slit (called a **tongue bar**) are supported by skeletal elements derived from the basement membrane of the gut lining.

Behind the pharynx is an esophagus, which at least in some forms (e.g., *Saccoglossus*) bears openings to the outside through the dorsal body wall. Unlike most of the gut, the middle region of the esophagus bears intrinsic muscles and moves the food into the intestine by peristalsis. The mechanical squeezing of the food material may press out excess water through the esophageal pores. In some species the wall of the anterior intestine contains dense green or brown inclusions that are visible externally and delimit the so-called **hepatic region** of the trunk. The intestine extends, more-or-less undifferentiated, to a short rectum terminating in the anus. Digestion is probably largely extracellular in the intestine, but details of this process are not fully known.

The major feeding structures of pterobranchs are the arms and tentacles derived from the mesosome. *Rhabdopleura* bears one pair of arms, each with numerous tentacles, whereas *Cephalodiscus* bears from five to nine pairs of arms, depending on the species. Pterobranchs are ciliary mucus suspension feeders. During feeding, they assume a position near an

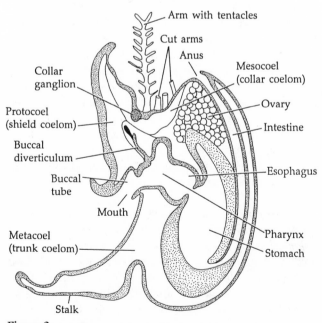

Figure 9
Internal anatomy of the pterobranch hemichordate *Cephalodiscus* (sagittal section). (After Hyman 1959.)

opening in their tubular cases and extend the arms and tentacles into the water (Figures 9 and 10). The tentacles on either side of the arms interdigitate to form a latticework, across which a mucous net is secreted. Food is trapped in the mucus and moved to the mouth by the action of cilia on the tentacles and arms, a pattern similar in some respects to that seen in lophophorates. At least in *Cephalodiscus*, the cilia over the general body surface may also move food to the mouth. Like enteropneusts, the pterobranch mouth is located under the anteroventral edge of the collar or mesosome (Figure 9). The gut is U-shaped, beginning with a **buccal tube** from which arises a buccal diverticulum of various forms and complexities. The pharynx bears one pair of gill slits in *Cephalodiscus* and *Atubaria*, but none in *Rhabdopleura*. When present, the gill apparatus is much simpler than that in enteropneusts, there being no supporting structures and less well defined branchial sacs. The slits open through pores on the exterior. An esophagus connects the pharynx to a sacciform **stomach** at the bottom of the U, and occupies most of the space within the trunk (Figure 9). The ascending portion of the gut consists of the intestine, which leads anteriorly to the dorsal anus. Digestion probably occurs in the stomach and intestine.

Circulation, gas exchange, and excretion. Enteropneusts possess a well developed open circulatory system comprising blood vessels, sinuses, and a contractile organ called the **heart vesicle** that is located in the proboscis (Figures 7A and C). A variety of names have been applied to parts of this system by different workers, and students should be alert to the possibility of encountering other terms in various publications on this group. Two main longitudinal vessels lie in the dorsal and ventral mesenteries along the length of the trunk and in the collar. Blood flows anteriorly in the dorsal vessel and posteriorly in the ventral vessel. The dorsal vessel expands in the collar as the **venous sinus**, which also receives blood anteriorly from a pair of **lateral proboscis veins**. The venous sinus leads to a larger, elongate, **central sinus** in the proboscis, lying between the buccal diverticulum and the dorsal heart vesicle. The heart vesicle has a muscular ventral wall that pulsates against the central sinus and aids the movement of blood. From the central sinus, blood moves into the **glomerulus**, which is an excretory organ unique to the hemichordates; it is formed of finger-like outpocketings of peritoneum associated with the blood sinuses. All of the blood leaving the central sinus passes through the glomerular sinuses, within which metabolic wastes

Rejection current

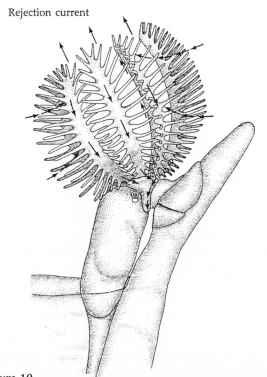

Figure 10

Feeding and rejection currents (arrows) of the pterobranch hemichordate *Cephalodiscus*. Water is drawn in between the tentacles and moves as a rejection current upward and away from the animal. Food particles are moved proximally along the tentacles to a food canal indicated by the arrowheads. (After Lester 1985.)

are presumably extracted. The glomerulus, buccal diverticulum, central sinus, and heart vesicle compose what is often referred to as the **proboscis complex** of enteropneusts (Figure 7C).

Blood leaves the glomerulus and passes through various vessels and sinuses supplying the anterior end of the worm, eventually reaching the ventral longitudinal vessel. Along the length of the trunk, blood leaves the ventral vessel to pass into networks of sinuses supplying the gut and the body wall; from these sinuses it then passes to the dorsal vessel.

Gas exchange occurs between the environment and the blood across the walls of the gill structures, especially the branchial sacs, all of which are richly supplied with blood from the ventral vessel. The gill septa and tongue bars bear cilia, which drive water into the mouth through the pharynx and out the gill slits, or pores. Actually, little work has been done on the matter of gas exchange, and it may be that other areas of the body surface are also involved in this activity. Enteropneust blood lacks pigments and contains very few cells, and gases are probably carried in solution.

The circulatory system of pterobranchs has not been fully studied. In general, it is weakly developed compared with that of entropneusts—a condition not unexpected in tiny animals. There is a central sinus and a heart vesicle near the buccal diverticulum, but no major vessels through the body. Rather, blood is carried in sinuses and lacunae that lack complete walls. A glomerulus is usually present, but it is weakly developed. The single pair of gill slits in *Cephalodiscus* and *Atubaria* may aid in gas exchange, but the small diffusion distances throughout the body of these tiny animals probably allows general cutaneous exchange, especially over the high surface areas of the tentacles.

Nervous system and sense organs. Most of the nervous system of all hemichordates consists of a netlike nerve plexus lying among the bases of the epithelial cells outside the basement membrane. A subepidermal dorsal nerve cord, or **neurochord**, is present in the collar of enteropneusts, but is reduced to a mere thickening of the plexus in pterobranchs. The plexus is thickened in enteropneusts as longitudinal tracts of neurons along the middorsal and midventral lines of the body. The evolutionary relationship between the dorsal hollow nerve cord of chordates and the mesosomal neurochord of enteropneusts is uncertain. The neurochord, however, is formed by an invagination of ectoderm and is actually hollow in some species, a condition indicating a probable homology with the nerve cord of the Chordata.

There are few types of sensory receptors in the hemichordates. Enteropneusts possess sensory cells over most of the body, probably serving as touch receptors that give these cryptic animals some information about their surroundings. As mentioned earlier, they also bear a preoral ciliary organ, presumed to be a chemoreceptor used during feeding. Little is known about the sensory apparatus of pterobranchs. Touch receptors are presumably present in the tentacles and perhaps on the cephalic shield and tip of the stalk in the noncolonial forms.

Reproduction and development. Asexual reproduction occurs in at least some enteropneusts (e.g., *Balanoglossus*) and in most pterobranchs. Acorn worms fragment small pieces from the trunk, and each one is able to grow into a new individual. They are very fragile worms and often break when handled; presumably they can regenerate missing parts.

As in most colonial invertebrates, asexual reproduction by budding is an integral part of the life history of aggregating and colonial pterobranchs. In

Cephalodiscus the buds arise from near the base of the stalk of adult individuals (Figure 11), and pass through a complex development cycle before they are released. Budding in *Rhabdopleura* occurs along the stolons that grow from the tips of the stalks of adult zooids. In both genera, the aggregations or colonies arise by budding after the formation of a single sexually produced individual. There is no evidence that *Atubaria* undergoes budding.

Hemichordates are gonochoristic but possess no outward evidence of sexual differences. Paired sacciform gonads lie in the trunk, outside the peritoneum; they are often very elongate in the acorn worms. Among the pterobranchs, *Rhabdopleura* possesses but a single gonad, which lies along the right side of the trunk coelom. Enteropneusts bear a pair of gonopores located dorsolaterally on the anterior trunk. The gonopores of *Cephalodiscus* are at the base of the arms; in *Rhabdopleura*, a single pore opens on the right side of the trunk.

Spawning in enteropneusts involves the release of mucoid egg masses by the females, followed by shedding of sperm by neighboring males. Once the eggs are fertilized, the mucous coating breaks down, thereby freeing the eggs into the sea water, where all subsequent development occurs. There are very few reports on spawning or fertilization in pterobranchs,

and most are quite incomplete. Apparently, the eggs are shed into the tubes of the colony or aggregation, where they are fertilized. Some species apparently brood the developing embryos within their tubes.

Early development in enteropneusts takes place wholly in the water. All features of this developmental period clearly attest to the deuterostome nature of acorn worms. Some species produce relatively yolky eggs, others produce eggs with very little yolk. In both cases, however, cleavage is holoblastic, radial, and more-or-less equal. A coeloblastula forms, which gastrulates by invagination. The blastopore is at the presumptive posterior end, but it closes and the anus and mouth form later. By late gastrula, the embryo has acquired cilia and breaks free from the egg membrane as a free-floating planker. Coelom formation is by archenteric pouching (typical enterocoely). Usually, a single protocoel arises from the inner end of the archenteron, and from it arise paired mesocoels and metacoels, establishing very early the tripartite body plan.

From this state, those species that produced yolky eggs develop directly to juvenile worms, without an intervening larval phase (e.g., *Saccoglossus*). In those that shed nonyolky eggs (e.g., *Balanoglossus*), the hatching stage develops quickly to a characteristic, planktotrophic, **tornaria larva** with ciliary bands reminiscent of certain echinoderm larvae. This larva soon elongates with the three body regions becoming externally apparent (Figure 12).

Among the pterobranchs, only *Cephalodiscus* has been studied embryologically, and even here the data are scanty. The large yolky eggs undergo radial, holoblastic, subequal cleavage. There is some argument about the form of the blastula and the precise nature of gastrulation. The embryo escapes the brooding area within the parental tube as a fully ciliated (but unnamed) larval stage. Coelom formation is by archenteric pouching, but the sequence of production is not clear. In terms of larval body orientation, the region between the mouth and anus apparently represents a much shortened dorsal surface and the "lower" side of the saclike trunk the ventral surface. These terms of reference are typically abandoned when describing the adult.

Thus, as we have seen in so many other benthic, sessile, and sedentary invertebrates, the hemichordates include a dispersal phase in their life history strategies. Even in those enteropneusts with technically direct development, the pattern is strategically "indirect," the embryos being at least planktonic free-living animals, even if not full-fledged larvae. The pterobranchs display a mixed life history pattern,

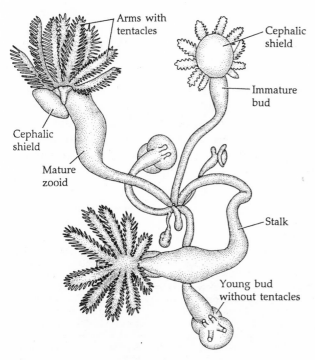

Figure 11

Budding in the pterobranch hemichordate *Cephalodiscus*, showing zooids at different stages of development. (After Lester 1985.)

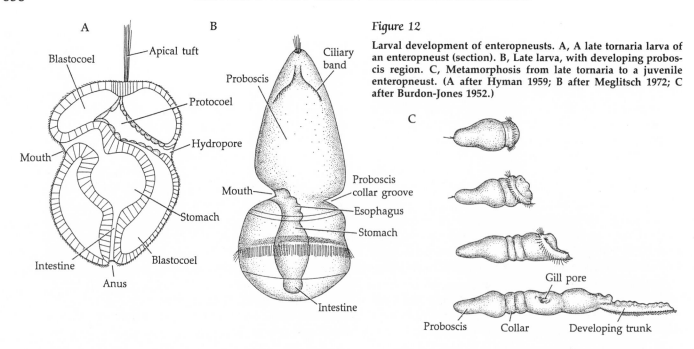

Figure 12

Larval development of enteropneusts. A, A late tornaria larva of an enteropneust (section). B, Late larva, with developing proboscis region. C, Metamorphosis from late tornaria to a juvenile enteropneust. (A after Hyman 1959; B after Meglitsch 1972; C after Burdon-Jones 1952.)

with a period of brooding followed by a free larval stage. This pattern is common among small, sessile animals, which cannot afford to produce huge numbers of eggs but depend on at least a short-lived dispersal phase.

The chordates

We are chordates. So are cats and dogs, horses and cows, birds and fishes, frogs and snakes, whales and elephants—all conspicuous by their size and familiarity. In addition to having a notochord, a dorsal, hollow nerve cord, and pharyngeal gill slits (Box 3), we and the rest of the these creatures also possess a skeletal "backbone," a vertebral column housing our dorsal nerve cord and defining us a members of the subphylum Vertebrata. But there are two other chordate groups, both lacking vertebrae. These are the invertebrate chordates, the subphyla Urochordata and Cephalochordata. The Cephalochordata comprise 20 or so species of small, fishlike animals called lancelets, or amphioxus (e.g., *Branchiostoma*) (Figure 1F). The Urochordata (= Tunicata), which is our main concern in this section, is composed of about 3,000 species, in four classes: the sessile filter-feeding sea squirts or ascidians (class Ascidiacea); the abyssal ascidian-like sorberaceans (class Sorberacea); the pelagic tunicates or salps (class Thaliacea); and planktonic larva-like tunicates called appendicularians or larvaceans (class Appendicularia) (Figures 13 and 14).

Members of these four classes are almost all marine suspension feeders, but they conduct their lives in very different ways. The ascidians include both solitary and colonial forms, with individuals ranging in size from less than 1 mm to 60 cm, and some colonies measuring several meters across. Ascidians are found worldwide and at all ocean depths, attached to nearly any substratum. They are most abundant and diverse in rocky littoral habitats and on deep-sea muds. Salps float about singly or in cylindrical or chainlike colonies sometimes several meters long. They are known from all oceans but are especially abundant in tropical and subtropical waters. Salps occur from the surface to depths of about 1,500 meters. The feces of salps and larvaceans, and the abandoned houses of the latter, constitute important sources of food and particulate organic

Figure 13

Representative ascidians. A–B, The solitary ascidians *Styela* and *Ciona*. C, A social ascidian, *Clavelina*. D–E, Zooid clusters of the compound ascidian *Botryllus*. F, A compound ascidian in which groups of zooids are clustered on stalks. G–H, The compound ascidian *Aplidium* sp. Note the external appearance of the colony. The colony is sliced open to expose the clusters of zooids. I, Portion of the colony of *Perophora*, with zooids arising from stolons. J, *Bolteniopsis*, a strange deep-sea ascidian. K, *Octacnemus*, a predatory deep-sea sorberacean. (A,B photos by H. W. Pratt/BPS; C photo by D. J. Wrobel, Monterey Bay Aquarium/BPS; D after Sherman and Sherman 1976; E,I photos by C. R. Wyttenbach, Univ. of Kansas/BPS; F after Romer 1956; G photo by authors; H courtesy of R. Vail; J,K after Barnes 1980, after photographs by Monniot and Monniot 1975.)

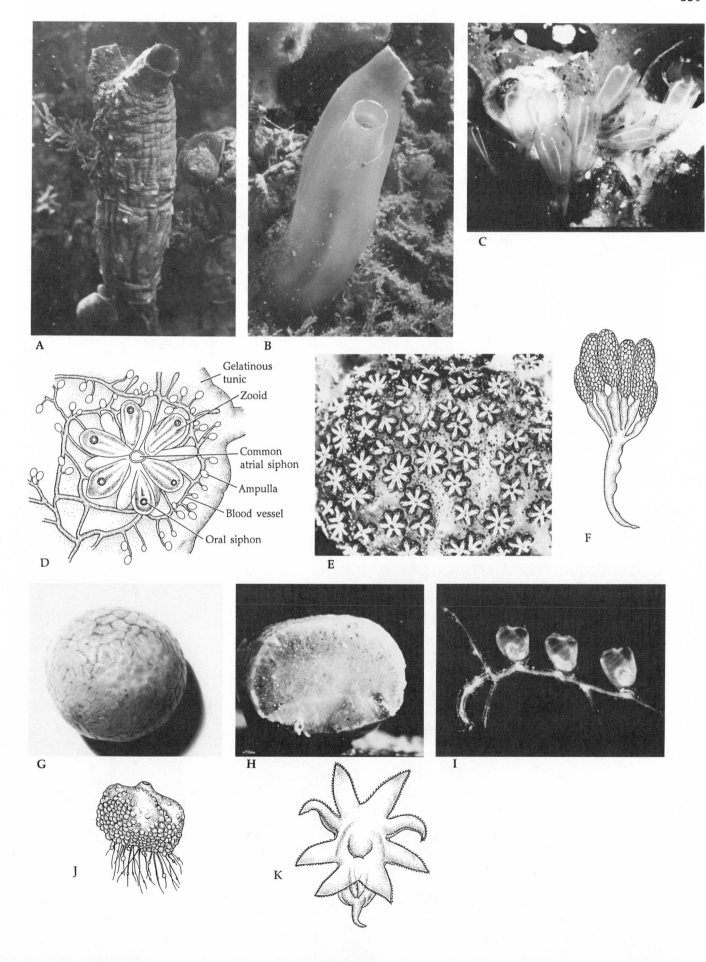

D labels: Gelatinous tunic; Zooid; Common atrial siphon; Ampulla; Blood vessel; Oral siphon

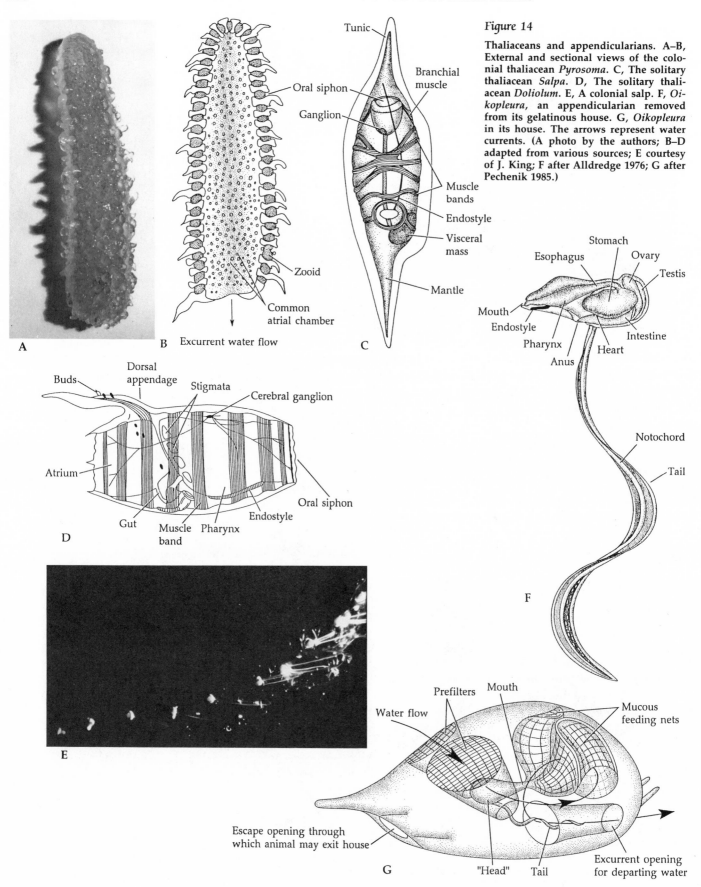

Tunic

Oral siphon

Ganglion

Branchial muscle

Muscle bands

Endostyle

Visceral mass

Mantle

Zooid

Common atrial chamber

B Excurrent water flow

C

Figure 14

Thaliaceans and appendicularians. A–B, External and sectional views of the colonial thaliacean *Pyrosoma*. C, The solitary thaliacean *Salpa*. D, The solitary thaliacean *Doliolum*. E, A colonial salp. F, *Oikopleura*, an appendicularian removed from its gelatinous house. G, *Oikopleura* in its house. The arrows represent water currents. (A photo by the authors; B–D adapted from various sources; E courtesy of J. King; F after Alldredge 1976; G after Pechenik 1985.)

Stomach

Esophagus

Ovary

Testis

Mouth

Endostyle

Pharynx

Intestine

Anus

Heart

Notochord

Tail

A

Buds

Dorsal appendage

Stigmata

Cerebral ganglion

Atrium

Oral siphon

Gut

Muscle band

Pharynx

Endostyle

D

Prefilters

Mouth

Mucous feeding nets

Water flow

Escape opening through which animal may exit house

Excurrent opening for departing water

"Head"

Tail

E

F

G

carbon in the sea. The larvaceans, or appendicularians, are solitary, luminescent planktonic creatures rarely more than about 5 mm long. They resemble in certain ways the larval stages of some other tunicates, hence the name "larvaceans." Their retention of larval features, including a notochord and a nerve cord, suggests that they arose by a process of paedomorphosis.

The tunicate *Bauplan*

Tunicates are bilaterally symmetrical, at least during early developmental stages. They utilize mucus-covered pharyngeal gill slits for suspension feeding. Although somewhat modified in the appendicularians, water flows into the mouth and pharynx by way of an incurrent siphon, or **oral siphon**, passes through the gill slits into a spacious water-filled atrium, and exits through an excurrent siphon, or **atrial siphon** (Figure 15). The gut is simple and U-shaped, with the anus emptying into the excurrent flow of water. Because of the drastic modification in body form relative to that of more familiar chordates, the general orientation of the bodies of urochordates is not immediately apparent and can only be fully understood by examining the events of metamorphosis, as described later. The oral siphon is generally anterior, and the atrial siphon is either anterodorsal (in ascidians) or posterior (in thaliaceans) (Figure 14C). In any case, the dorsoventral orientation of the body can be determined internally by the locations of the dorsal ganglion and a thickened ciliated groove, called the **endostyle**, that runs along the ventral side of the pharynx or branchial basket (Figures 13 and 14).

The Ascidiacea is the largest and most diverse class of tunicates. Some interstitial forms are known and a few live anchored in soft sediments, but the majority of ascidians are attached to hard substrata (Figure 1D). Three general types of ascidians are usually recognized, although these categories do not relate directly to formal taxa. Most of the large (up to 60 cm long) species are called **solitary ascidians** because they live singly and unattached to one another (e.g., *Styela*, *Molgula*, *Ciona*; Figures 13A and B), although many are highly gregarious. **Social ascidians** tend to live in clumps of individuals that are vascularly attached to one another at their bases (e.g., *Clavelina*; Figure 13C). Finally, a great number of species are **compound ascidians** and are characterized by many small individuals (zooids) living together in a common gelatinous matrix (e.g., *Aplidium*, *Botryllus*; Figures 13D–H). In extreme cases colonies may measure several meters across.

Ascidians have become so highly modified during their evolution as sessile suspension feeders that adults are not easily recognizable as chordates. The dorsal hollow nerve cord and the notochord are present only in the larval stage, but the pharyngeal gill slits persist in the adults. The oral and atrial siphons are generally directed away from the substratum and set at an angle to one another, thereby reducing the potential for recycling waste water. The bodies of some solitary and social forms are set on stalks and elevated above the substratum, where the compound forms typically grow as thin or thick sheets conforming to the topography of the surface on which they live. In some of these compound ascidians, zooids are arranged in regular rosettes and share a common atrial chamber formed within the tunic (Figures 13D, 13E, and 15D). The body wall, or **tunic**, varies from thick to thin and from smooth and slick to wrinkled and leathery. Many species are brightly colored.

Thaliaceans are pelagic ascidian-like urochordates. They are constructed much like their sessile counterparts except that the oral and atrial siphons are at opposite ends of the body, and in many forms the pharyngeal filtering basket is modified to accommodate the linear flow of water through the animal. The exiting water provides a means of "jet propulsion." Most are highly gelatinous and transparent.

The class Thaliacea comprises three orders. Members of the order Pyrosomida are considered the most primitive thaliaceans and most resemble their presumed ascidian ancestors. Pyrosomes are remarkable colonies of tiny ascidian-like zooids embedded in a matrix and arranged around a long central, tubular chamber called a **common cloaca** (Figures 14A and B). The cloaca receives exhalant water from the inwardly directed atrial siphons of all the zooids; the water then exits through a single large aperture, thereby propelling the barrel-shaped colony slowly through the surrounding sea water. As in the ascidians, water movement in pyrosomes is generated entirely by ciliary action of the individual zooids.

The orders Doliolida and Salpida include thaliaceans that alternate between solitary sexual forms and colonial asexual stages (Figure 14C–E). Doliolid individuals are generally small, less than 1 cm long, whereas single salps may be 15–20 cm long and form chainlike colonies several meters in length. The members of these two orders move water through their bodies and propel themselves partially (salps) or wholly (doliolids) by muscular action.

Thaliaceans are predominately warm-water creatures, although certain species are found in temperate and even polar seas. They are particularly abundant over the continental shelf and are frequently captured in surface waters or seen stranded on wave-swept sandy beaches after storms. Some, however, have been recorded from depths to 1,500 meters.

Appendicularians are among the strangest of all urochordates and are characterized by the retention of larval features. These solitary animals live in a gelatinous casing, or **house**, that they secrete around their body (Figures 14F and G). The bulbous trunk of the body contains the major organs, including the gut, and bears a muscular tail in which the notochord is retained. The dorsal nerve cord, though reduced, extends partway along the length of the tail. Thus, we see clearly the evidence of the chordate nature of these animals retained through paedomorphosis — evidence that is present only during developmental stages of other urochordates. (The phenomenon of paedomorphosis was discussed in Chapter 4.)

The pharynx of these animals is reduced and bears only two clefts. When positioned within its house, an appendicularian produces a complex water current by beating its tail. Filtering is accomplished by meshes in the house wall and by mucous nets secreted by the animal (Figure 14G). The exiting water provides the locomotor force. Appendicularians are found in the surface waters of all oceans and are sometimes extremely abundant.

Body wall, support, and movement. The body wall of tunicates includes a simple epithelium overlain by a secreted **tunic** of varying thickness and consistency. The tunic is most well developed in the ascidians and some thaliaceans. It varies from soft and gelatinous to tough and leathery, and it sometimes includes calcareous spicules. The matrix of the tunic contains fibers and is composed largely of cellulose-like carbohydrate called **tunicin**. The tunic is not a simple, secreted, nonliving cuticle, however, since it also contains amebocytes and, in some cases, blood cells and even blood vessels. The tunic may be viewed as an exoskeleton providing support and maintaining body shape.

Beneath the epidermis are muscle bands. In many species, especially ascidians, these muscles lie within an ectodermally produced mesenchyme called the **mantle** (Figures 15A and B). Ascidians possess longitudinal muscles extending along the body wall that serve to pull the flared siphons down against the body. Circular sphincter muscles close the siphonal openings. Doliolids and salps have well developed bands of circular muscles that pump water through the body for feeding and locomotion. When they contract, water within the body is forced out the

atrial siphon, thereby propelling the animal forward. When they relax, the body expands because of the resilience of the tunic, and water is drawn in. The tail muscles of appendicularians provide the action for moving water through the houses of these animals.

Tunicates do not have a coelom; the body cavity has been lost in concert with the evolution of a water chamber called the **atrium**, or **cloacal water chamber**, that functions in filter feeding. This chamber is a saclike, ectodermally derived structure continuous with the epidermis of the atrial siphon (Figures 15A and B). The inner wall lies against the pharynx and is perforated over the **gill slits**, or **stigmata**. Thus, water that enters the pharynx via the oral siphon flows through the stigmata, into the cloacal chamber, and out the atrial siphon.

Feeding and digestion. We have hinted at various aspects of the feeding biology of tunicates in our comments above. Most of these animals are suspension feeders and use various kinds of mucous nets to filter plankton and organic detritus from the sea water. A few ascidians live partially embedded in soft sediments and feed on organic material in the substratum, and certain bizarre deep-sea species actually prey on small invertebrates by grasping them with the lips of the oral siphon. Below we provide a description of the details of feeding and digestion in a suspension-feeding ascidian and then brief comparisons of the ascidian method with those of thaliaceans and appendicularians.

Water is moved through the body of an ascidian largely by the action of cilia lining the pharyngeal basket. Water enters the oral siphon and passes through a short **siphonal chamber** at the inner end of which is the mouth, surrounded by a ring of fleshy tentacles; the tentacles prevent the entrance of large particles (Figure 15A). Food-laden water then passes into the pharynx, which bears a ventral longitudinal groove called the **endostyle**. The bottom of the groove is lined with mucus-secreting cells and bears a longitudinal row of flagella; the sides of the groove bear cilia (Figure 15F). The mucus, a complex mucoprotein containing iodine, is moved to the sides of the endostyle by the basal flagella and then outward by the lateral cilia. Cells near the opening of the endostyle are responsible for binding environmental iodine and incorporating it into the mucus. Sheets of mucus then move dorsally along the inner wall of the pharynx and pass over the stigmata. The slitlike stigmata are arranged in rows and bear lateral cilia that drive water from the pharynx into the cloacal

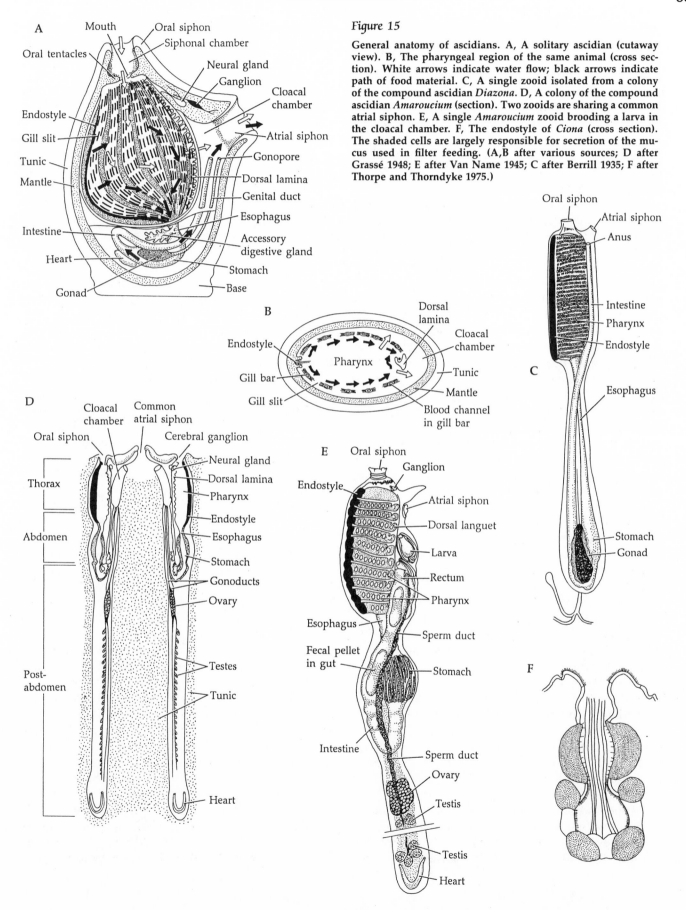

Figure 15

General anatomy of ascidians. A, A solitary ascidian (cutaway view). B, The pharyngeal region of the same animal (cross section). White arrows indicate water flow; black arrows indicate path of food material. C, A single zooid isolated from a colony of the compound ascidian *Diazona*. D, A colony of the compound ascidian *Amaroucium* (section). Two zooids are sharing a common atrial siphon. E, A single *Amaroucium* zooid brooding a larva in the cloacal chamber. F, The endostyle of *Ciona* (cross section). The shaded cells are largely responsible for secretion of the mucus used in filter feeding. (A,B after various sources; D after Grassé 1948; E after Van Name 1945; C after Berrill 1935; F after Thorpe and Thorndyke 1975.)

water chamber (Figure 15). Thus, water passing through the stigmata also passes through the mucous sheets on which food particles are retained. On the dorsal surface of the pharynx is a longitudinal curved ridge called the **dorsal lamina**, or a row of ciliated projections called **languets**, or both (Figures 15A and B). These structures serve to roll the mucous sheets into cords, which are then passed posteriorly to a short esophagus and then to a stomach. Attached to the stomach is a small **pyloric gland**, which extends around the intestine as a network of small tubes. Some species also bear an **accessory digestive gland** (Figure 15A). Digestive enzymes are secreted into the stomach lumen by secretory cells of the gut wall and perhaps by the associated glands, and digestion is largely extracellular. From the stomach, the gut loops forward as an intestine, through which undigested material passes to the anus and into the atrium near the excurrent siphon.

The unique ascidian family Didemnidae comprises colonial forms in which the cloacal systems are confluent, and the colonies are usually hardened with aragonitic "**spiculospheres.**" Among certain tropical genera (e.g., *Didemnum, Trididemnum, Lissoclinum, Diplosoma*) are species that maintain symbiotic algae in the test, the branchial basket, or the cloacal system. These algae are prokaryotic, resembling blue-green algae, but possess chlorophylls *a* and *b* (like those in the green algae, Chlorophyta); they are placed in the genus *Prochloron*. There is evidence that the host ascidian benefits from the association by feeding directly on the "algal" cells, perhaps by amebocytic phagocytosis. Didemnid ascidians housing such symbiotic algae are also remarkable for their powers of locomotion, limited though it is—colonies have been clocked at speeds of 4.7 mm per 12-hr period. Such movement may allow these ascidians to position themselves in light conditions favorable to their algal symbionts. A similar symbiosis has been reported between red cyanophyte algae and certain didemnid ascidians.

Thaliaceans feed in much the same way as ascidians do except that the siphons are at opposite ends of the body. The number of pharyngeal stigmata is usually reduced, especially in doliolids, and restricted to the posterior portion of the pharynx (Figures 14B–D). The intestine extends posteriorly from the pharynx and opens into the enlarged cloacal chamber.

Appendicularians (Figures 14F and G) secrete a hollow gelatinous (mucopolysaccharide) house in which they reside and upon which they depend for feeding. The tail is directed through a tube in the house structure toward an excurrent opening. Sinusoidal beating of the muscular tail generates a current that pulls water into the house through coarse, mesh-like, mucous filters that screen large particles; eventually the water leaves the house via the excurrent opening. The pharynx of appendicularians bears only two small gill slits, which open directly to the exterior. Mucous feeding nets are secreted through the mouth and lie within the house chamber. The water current is directed through these fine-mesh nets, where food particles are concentrated. The food, net and all, is periodically ingested by way of a short buccal tube. The houses of most appendicularians bear an additional opening that serves as an escape hole through which the animal can leave and reenter. The houses are fragile and easily damaged. Damaged or clogged houses are abandoned, and new ones are manufactured rapidly, in a matter of seconds or minutes. In some species a new or "spare" house may be found beneath the functional house; after escaping the clogged house, the appendicularian rapidly inflates the "spare".

The gut is U-shaped and the anus opens directly to the outside rather than into a cloacal chamber. Fecal material is released into the path of excurrent water leaving the filter nets. Appendicularians are primarily herbivorous, feeding on minute phytoplankton and bacteria down to a size of 0.1 μm. They sometimes constitute the dominant planktonic herbivore in waters over the continental shelf, reaching densities of many thousands per cubic meter.

Circulation, gas exchange, and excretion. The circulatory system is most well developed and best understood in the ascidians and is much reduced in the thaliaceans and appendicularians. Ascidians possess a short, tubular heart. It lies posteroventrally in the body near the stomach and behind the pharyngeal basket (Figure 15A). The heart is surrounded by a pericardial sac. Blood vessels extend anteriorly and posteriorly, opening into spaces around the internal organs and also providing the blood supply to the tunic. The heartbeat is by peristaltic action, and the direction of this motion is periodically reversed, flushing the blood first one way through the heart and then the other. Blood physiology and function are largely speculative. Ascidians are known to accumulate high concentrations of certain heavy metals in their blood, especially vanadium and iron. Preliminary evidence suggests that the presence of high vanadium levels in at least some species serves to deter would-be predators. In addition, the blood includes a large variety of cell types, including amebocytes that are thought to function in nutrient transport, tunic deposition, and accumulation of metabolic wastes. The blood also contains several vertebrate-like hormones, including thyroxine, oxytocins, and vasoconstrictors. Compound species, such as *Botryllus*, have special blood cells that play a vital role in rejecting conspecific, but nonclone-mate colonies (allogenic colonies).

Gas exchange occurs across the body wall, especially across the linings of the pharynx and the cloacal chamber. Little is known about respiratory physiology in these animals.

In most ascidians and some other tunicates, two evaginations arise from the posterior wall of the pharynx and lie along each side of the heart. These structures are called **epicardial sacs** and may represent coelomic remnants. In some species the epicardial sacs are involved in bud formation during asexual reproduction, and they may also function in the accumulation of nitrogenous waste products by forming storage capsules called **renal vesicles**. Other than these vesicles and certain blood cells (**nephrocytes**), it is likely that much of the metabolic waste is lost from the body by simple diffusion.

Nervous system and sense organs. The nervous system of tunicates is much reduced and reflects their relatively inactive sessile and floating planktonic lifestyles. A small **cerebral ganglion** lies just dorsal to the anterior end of the pharynx and gives rise to a few nerves to various parts of the body, especially the muscles and siphonal areas. As described later, a well developed dorsal nerve cord is present in the tails of tunicate larvae, but this structure is lost during metamorphosis. Most tunicates possess a **neural**, or **subneural, gland** located between the cerebral ganglion and the anterodorsal portion of the pharynx (Figure 15A). This gland opens to the pharynx through a small duct, but its function is unknown. Some workers have suggested that it may be the precursor of the pituitary gland of vertebrates. Cycles of cellular activity coinciding with tidal rhythms have been found in some *Ciona* species, but the function of this cellular rhythm remains enigmatic. Sensory receptors are poorly developed in tunicates, although touch-sensitive neurons are prevalent around the siphons.

Asexual reproduction. While the appendicularians are entirely sexual in their reproductive habits, thaliaceans and many ascidians include asexual processes in their life history strategies. In social and especially compound ascidians, asexual budding allows rapid exploitation of available substrata, as we have seen in other sessile colonial invertebrates.

Budding in tunicates occurs in a great variety of ways and from different organs and germinative tissues (Figure 16). In general, initial buds are formed by a sexually produced individual (**oozooid**), then the asexually produced individuals (**blastozooids**) produce additional buds. The simplest and perhaps most primitive budding process occurs in certain social colonial ascidians, including species of *Perophora* and *Clavelina*, where blastozooids arise from the body wall of stolons. In more complicated budding processes, the germinal tissues include various combinations of the epidermis, gonads, epicardial sacs, and gut. Doliolid thaliaceans often produce chains of buds. The chains are sometimes released intact, but eventually each blastozooid breaks loose as a separate individual. Budding in pyrosomes results in the floating colonies characteristic of this group (Figures 14A and B).

Much of the recent information about asexual reproduction in ascidians was reviewed and synthesized by M. Nakauchi (1982), who suggested that the various types of budding among ascidians can be

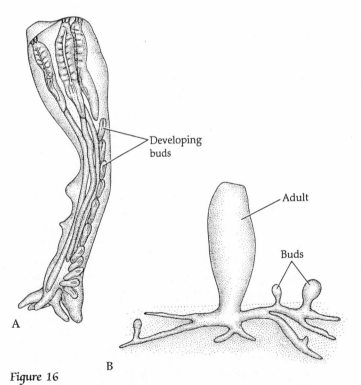

Figure 16

Asexual reproduction in ascidians. A, Formation of buds in the colonial ascidian *Circinalium*. B, General pattern of stoloniferous budding in an ascidian. (A after Brien, in Grassé 1948.)

divided into two basic categories on the basis of their functional significance. **Propagative budding** generally occurs during favorable environmental conditions and serves to increase colony size and exploit available resources. On the other hand, **survival budding** tends to take place during the onset of adverse conditions and may be viewed as an overwintering or other survival device. In this case colony size is generally reduced by the resorption of zooids, leaving potential buds of presumptive germinative tissues. With the return of more favorable growing conditions, these "pre-buds" quickly develop as new blastozooids.

Another fascinating aspect of asexual reproduction in colonial ascidians is that some colonies fuse with one another when they grow large enough to make contact. Both intraspecific and interspecific fusion are known to occur. This is a histocompatibility phenomenon that is based upon the genetic make up of the colonies involved; it is viewed as a self/nonself recognition phenomenon (see papers by Sabbadin, Scofield, Watanabe, and others in the 1982 Lambert-chaired symposium reports).

Sexual reproduction and development. Most tunicates are hermaphroditic, with relatively simple reproductive systems. Generally a single ovary and a single testis lie near the loop of the digestive tract in the posterior part of the body and, in most cases, connect through separate single sperm ducts and oviducts to the cloacal chamber near the anus (Figures 15D and E). Members of some families (e.g., Styelidae, Pyuridae) have multiple gonads.

There is a great deal of variation in the overall reproductive strategies among tunicates. Most large solitary ascidians produce high numbers of weakly yolked ova, which are shed to the sea coincidentally with the release of sperm from other individuals. External fertilization is followed by the development of a free-swimming **tadpole larva**, which eventually settles and metamorphoses to an oozooid (Figure 17). In contrast to this fully indirect life history pattern, many compound ascidians composed of tiny zooids produce relatively few eggs, but each egg has a high yolk content. These eggs are fertilized and subsequently brooded within the cloacal chamber; they are not released until the swimming tadpole larvae develop. Various degrees of larval suppression occur among ascidians with this mixed life history strategy, and some species actually display fully direct development.

Although all thaliaceans lack a free-swimming larval stage, they differ markedly in their approaches to direct development. In pyrosomes each zygote develops directly to an oozooid, with no evidence of a larval stage. The oozooid then buds to produce a colony. Doliolids produce tailed larvae, but each is encased in a cuticular capsule and does not swim. The larva metamorphoses to an oozooid. Salps undergo internal fertilization within the oviduct. The zygotes implant and form a placenta-like association with the parent in a uterine chamber in the oviduct. Here the embryos develop directly to the adult form.

Appendicularians free spawn, and fertilization occurs externally. They develop to a tadpole-like stage and then mature by protandry into the characteristic larva-like adults.

In most tunicate species studied, cleavage is holoblastic, radial, and slightly unequal; it leads to the formation of a coeloblastula, which undergoes gastrulation by invagination. The blastopore lies at the presumptive posterior end of the body but closes as development proceeds.

The development of the chordate features is most easily seen and understood in those species that form free tadpole larvae, such as most ascidians. As the embryo elongates, the gut proliferates three longitudinal strips of mesoderm—a middorsal strip that becomes the notochord and lateral strips that form the mesenchyme and body musculature. Thus, even though the mesoderm arises from the archenteron (entoderm), it does not pouch from the gut wall; in fact, no coelomic cavity is ever formed. A middorsal strip of ectoderm differentiates as a **neural plate**, which sinks inward and curls to produce the dorsal hollow nerve cord. The epidermis secretes a larval tunic, which often develops dorsal and ventral tail fins. The anterior part of the gut differentiates as the pharyngeal basket during larval life, and the rudiment of a cloacal water cavity forms by an ectodermal invagination producing the atrial siphon. However, these larvae are all lecithotrophic, and the gut and filtering devices do not become functional until after metamorphosis.

Ascidian larvae are short lived. When development is fully indirect, the larvae are planktonic for only about two days or less. In some forms with a mixed life history pattern (e.g., *Botryllus*), the free larval life lasts only a few minutes. Even though a short larval life allows dispersal over only small distances, the larvae are probably very important in the selection of suitable substrata. Gregarious settlement gives rise to large masses in many species. The events of settling and metamorphosis of ascidian larvae are complex and not fully understood. These phenomena are summarized briefly below, but the interested

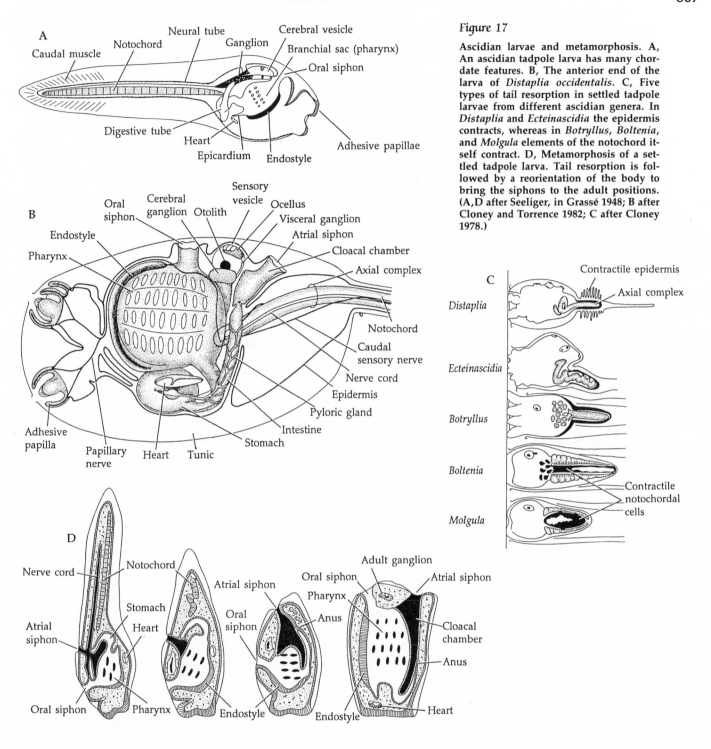

Figure 17

Ascidian larvae and metamorphosis. A, An ascidian tadpole larva has many chordate features. B, The anterior end of the larva of *Distaplia occidentalis*. C, Five types of tail resorption in settled tadpole larvae from different ascidian genera. In *Distaplia* and *Ecteinascidia* the epidermis contracts, whereas in *Botryllus*, *Boltenia*, and *Molgula* elements of the notochord itself contract. D, Metamorphosis of a settled tadpole larva. Tail resorption is followed by a reorientation of the body to bring the siphons to the adult positions. (A,D after Seeliger, in Grassé 1948; B after Cloney and Torrence 1982; C after Cloney 1978.)

reader is referred to the careful work of R. Cloney (1982) for additional details.

Ascidian larvae possess several sensory receptors that function in settling and probably substratum selection and are absent from the adults. A small **sensory vesicle** lies near the anterior end of the dorsal nerve cord adjacent to the developing cerebral ganglion (Figure 17B). This vesicle houses a light-sensitive ocellus and a statocyst (called an **otolith**). At the time of settlement, the larva becomes negatively phototactic and positively geotactic. The anterior end of the larva bears two or three **adhesive papillae**, each of which is supplied with nerves. There is variation in the details of the events of settlement and metamorphosis, and the following is a generalized account.

The settling larva contacts a substratum with its anterior end and secretes an adhesive from the papillae. In many compound ascidians the papillae evert during this process. The secretion of the adhesive apparently triggers an irreversible sequence of metamorphic events. Within minutes after attachment, resorption of the larval tail commences by one of several methods involving various contractile elements in the tail region (Figure 17C). The outer layer of the cuticle is shed, removing the larval fins from the settled juvenile. The animal then undergoes a remarkable 90° rotation of the viscera and siphons that brings these organs to their adult positions (Figure 17D). The pharynx enlarges and the filtering mechanisms become functional. During all of these processes, secondary attachment organs, called **ampullae**, extend from the body and permanently affix the animal to the substratum. Finally, various transient larval organs are lost, such as most of the larval nervous system and sense organs.

The cephalochordates

The subphylum Cephalochordata includes about two dozen species of small, fishlike creatures that rarely exceed 5 cm in length (Figures 1F and 18). They are commonly called lancelets or amphioxus, a name frequently applied to *Branchiostoma lanceolatum*, which is a familiar species to general zoology students. Lancelets are cosmopolitan in shallow marine and brackish waters, where they lie burrowed in clean sands with only the head protruding above the sediment. They can and do swim, however, and locomotion is important to their dispersal and mating habits.

The cephalochordate *Bauplan*

Cephalochordates are especially interesting animals whose *Bauplan* demonstrates several qualities intermediate between those of the invertebrates and the vertebrates. As we discuss later in this chapter, lancelets may represent living descendants of the ancestors of the vertebrates.

Body wall, coelom, support, and locomotion. The body is everywhere covered by an epidermis of simple columnar epithelium, underlain by a thin connective tissue dermis. The body wall muscles are distinctly vertebrate-like and occur as chevron-shaped blocks that are called **myotomes** and are arranged longitudinally along the dorsolateral aspects of the body (Figure 18). These muscle blocks are large and occupy much of the interior of the body, thereby reducing the coelom to relatively small spaces. The notochord persists in adults and provides the major structural support for the body.

The notochord also plays a major role in locomotion in lancelets. As a result of the action of segmental myotomes, the swimming action of cephalochordates is much like that of fishes, consisting basically of lateral body undulations that drive water posteriorly and provide a forward thrust. The propulsive action of these body movements is enhanced by a vertical **caudal fin**. Unlike the vertebral column and its articulating bones, however, the notochord is an elastic flexible rod. It prevents the body from shortening when the muscles contract, causing lateral bending instead. Its elasticity tends to straighten the body, and thus assists the antagonistic action of paired myotomes. The notochord extends beyond the myotomes both anteriorly and posteriorly, providing support beyond the muscles and apparently aiding in holding the body rigid during burrowing.

Although the notochord of cephalochordates is homologous with the same structure in other chordates, including the vertebrates, it displays some unique and rather remarkable structural and functional characteristics associated with its persistence in the adult. It is not a homogeneous structure of predominantly cartilage-like matrix material. Rather, the notochord of lancelets is built of discoidal lamellae that are stacked like so many poker chips along its length and are surrounded by a sheath of collagenous connective tissue. The lamellae are composed of muscle cells whose fibers are oriented transversely. Furthermore, a significant amount of extracellular fluid exists in spaces and channels around and between the lamellae within the collagenous sheath. These muscle cells are innervated by motor neurons from the dorsal nerve cord. Upon contraction, the hydrostatic pressure in the extracellular spaces increases, thereby resulting in an increase in stiffness of the whole notochord complex. It is suspected that this action may facilitate certain kinds of movement patterns, especially burrowing.

Feeding and digestion. Cephalochordates are ciliary-mucus suspension feeders, and they employ a food-gathering mechanism similar to that of tunicates. Water is driven into the mouth and pharynx and out through the pharyngeal gill slits into a sur-

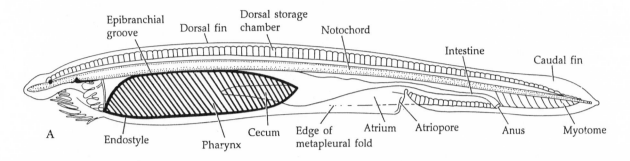

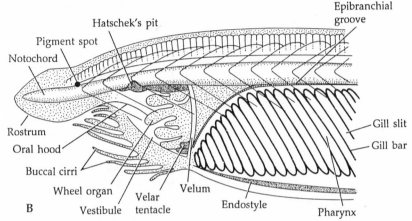

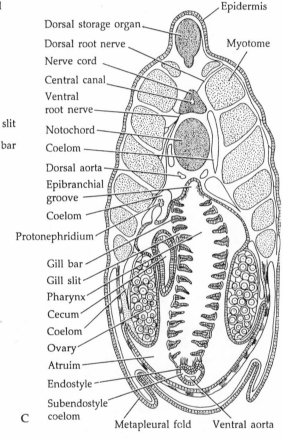

Figure 18

General anatomy of a cephalochordate (*Branchiostoma*). A, General external and internal anatomy. B, The anterior end. C, The region of the pharynx (cross section). D, The major blood vessels in the area of the gut cecum. (A,C after various sources; B after McFarland et al. 1985; D after Romer 1956.)

rounding **atrium**; it exits the body through a ventral **atriopore** (Figure 18A). Unlike the gill ventilation currents of aquatic vertebrates that are generated by muscular action, the feeding currents of lancelets are driven by pharyngeal cilia, a condition similar to that in tunicates. The gill slits are committed largely to feeding in cephalochordates and have little to do with gas exchange. There are up to 200 gill slits separated from one another by **gill bars**, which are supported by cartilaginous rods.

The trapping of food from the inflowing water involves sorting activities that actually occur before water enters the mouth. The mouth is housed within a depression called the **vestibule**, which is formed by an anterior extension of the body called the **oral hood** (Figures 18A and B). The oral hood is supported by the notochord and bears finger-like projections called **buccal cirri**. As water enters the vestibule, the cirri prevent sediments and other large particles from reaching the mouth. The mouth itself is a perforation in a membranous **velum**, which bears a set of **velar**

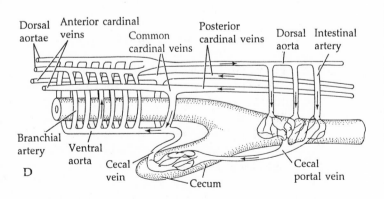

tentacles; this arrangement also screens large material from entering the mouth. The lateral walls of the vestibule bear complex ciliary bands that collectively constitute the **wheel organ**. The wheel organ appears as brown, thickened, folded epithelium in the roof and sides of the vestibule, visible through the skin of living animals. Cilia of the wheel organ drive food particles to the mouth and give the impression of a rotation, hence the name. Lying on the roof of the vestibule is a mucus-secreting structure called **Hatschek's pit**. Mucus from this pit flows over the wheel organ, and food particles trapped by the mucus are carried to the mouth along the ciliary tracts. This material enters the general water current moving through the mouth and into the pharynx.

The ventral surface of the pharynx bears the **endostyle**, or **hypobranchial groove** (Figure 18C). As in tunicates, the endostyle binds iodine, although the specific cells involved differ between the two groups. Many workers view the endostyle as a homologue and probable precursor to the vertebrate thyroid gland. The endostyle produces strings of mucus that trap food from the water as it passes through the gill slits and into the atrium. The food-laden mucus is then passed dorsally along the walls of the pharynx by ciliary tracts and into the **epibranchial groove**, which carries the material by cilia along the dorsal midline of the pharynx and into a short esophagus. The gut extends posteriorly as an elongate intestine and opens through the anus just in front of the caudal fin. From near the junction of the pharynx and esophagus arises an anteriorly projecting **digestive cecum** (sometimes called the **hepatic cecum**). Studies indicate that the cecum functions in lipid and glycogen storage and protein synthesis, and it is regarded by many workers as the evolutionary precursor of the vertebrate liver and perhaps the pancreas. Digestion initially is extracellular in the gut lumen and is completed intracellularly in the walls of the intestine and especially the cecum.

In addition to storage in the cecum, food reserves accumulate in longitudinally arranged **dorsal storage chambers** along the dorsal midline and along the ventral body margin posterior to the atriopore (Figure 18A). These structures were formerly called dorsal and ventral fins and fin rays, but recent studies indicate that their function is the storage of large quantities of food reserves, increased storage correlating with a great increase in gonadal mass before the onset of the reproductive season.

Circulation, gas exchange, and excretion. The circulatory system of lancelets comprises a set of closed vessels through which blood flows in a pattern similar to that in primitive vertebrates (e.g., fishes). There is no heart. Blood flows posteriorly along the pharyngeal region in a pair of **dorsal aortae**. Just posterior to the pharynx, these vessels merge into a single **median dorsal aorta** that extends into the region of the caudal fin (Figure 18D). Blood is supplied to the myotomes and notochord via a series of short **segmental arteries** and to the intestine through **intestinal arteries**. A capillary network in the intestinal wall collects the nutrient-laden blood and leads to a series of **intestinal veins** that join a large subintestinal vein called the **cecal portal vein**, which carries blood forward beneath the gut to another capillary bed in the digestive cecum. As in vertebrates, the vein that connects two capillary beds is called a portal vein (e.g., the hepatic portal vein and renal portal veins in fishes). The cecal portal vein of cephalochordates may be the homologue of the hepatic portal vein of vertebrates. In the digestive cecum, the nutrient and chemical composition of the blood is regulated before being distributed to the body tissues. (The vertebrate liver serves the same function via the hepatic portal system.)

Leaving the cecal capillaries is a **cecal vein**, which is joined by a pair of **common cardinal veins** formed by the union of paired **anterior** and **posterior cardinal veins** returning from the body tissues. These vessels merge to form the **ventral aorta** beneath the pharynx. From here blood is carried through the gill bars via **afferent** and **efferent branchial arteries** to the paired dorsal aortae, thus completing the circulatory cycle. Blood is moved through this system by peristaltic contractions of the major longitudinal vessels and by pulsating areas at the bases of the afferent branchial arteries.

The blood contains no pigments or cells and is thought to function largely in nutrient distribution rather than in gas exchange and transport. Although some diffusion of oxygen and carbon dioxide may occur across the gills, most of the gas exchange probably takes place across the walls of the **metapleural folds**, thin flaps off the body wall that lie just anterior to the atriopore (Figure 18C).

The excretory units in cephalochordates are protonephridia similar to the solenocytes of some other groups (e.g., primitive annelids). The numerous clusters of protonephridia accumulate nitrogenous wastes, which are carried by a nephridioduct to a pore in the atrium. Despite the structural similarities, the homology of lancelet protonephridia with those of other invertebrates is uncertain and most specialists regard this as a case of convergent evolution.

Nervous system and sense organs. The central nervous system of cephalochordates is very simple. A dorsal nerve cord extends most of the length of the body and may be expanded slightly as a **cerebral vesicle** in the base of the oral hood. Segmentally arranged nerves arise from the cord along the body in the typical vertebrate pattern of dorsal and ventral roots. The epidermis is rich in sensory nerve endings, most of which are probably tactile and important in burrowing. Some lancelets have a single simple eye spot near the anterior end of the dorsal nerve cord.

Reproduction and development. Cephalochordates are gonochoristic, but the sexes are structurally very similar. Rows of from 25 to 38 pairs of gonads are arranged serially along the body on each side of the atrium. The volume of gonadal tissue varies seasonally, and during the reproductive period it may occupy so much of the body that it interferes with feeding. Spawning typically occurs at dusk. The atrial wall ruptures, and eggs and sperm are released into the excurrent flow of water from the atrium; external fertilization follows.

The ova are isolecithal, with very little yolk. Cleavage is holoblastic, radial, and subequal, and leads to a coeloblastula that gastrulates by invagination (Figure 19). The roof of the archenteron eventually produces first a solid middorsal strip of mesoderm destined to become the notochord and then, sequentially, an anterior to posterior series of paired archenteric pouches along each side of the notochord. These enterocoelic pouches form the coelom and the other mesodermally derived structures such as the muscle bundles. The roof of the archenteron closes following mesoderm proliferation.

Dorsally, the ectoderm differentiates into a **neural plate**. The neural plate eventually rolls inward, separating from the bordering cells, and then sinks inward as a **neural tube**, which forms the dorsal hollow nerve cord. As this process occurs at the posterior end of the embryo, the developing nervous tissue contacts the blastopore, which remains open temporarily and connects the archenteron to the lumen of the nerve cord as a **neuropore** or **neurenteric canal**. Later the two structures separate, and the blastopore opens to the exterior as the anus. The mouth breaks through as a secondarily produced opening at the front end of the developing gut.

In the absence of abundant yolk reserves, development to a free-swimming larva takes place rapidly. As soon as they are able, the larvae swim upward in the water column, where they remain planktonic for 75–200 days. The larvae are planktotrophic. They al-

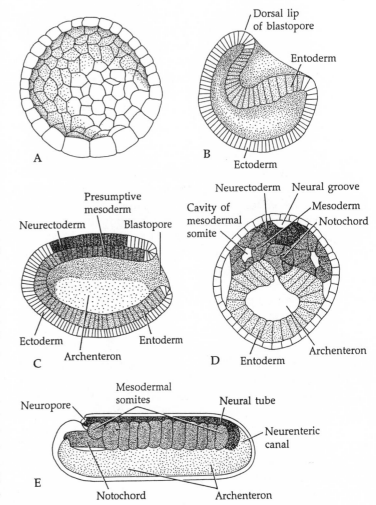

Figure 19

A, Development in a cephalochordate. A, Coeloblastula. B–C, Early and late gastrulae. D, The neural groove stage (cross section). Note the proliferation of mesoderm as central notochord and lateral coelomic cavities. E, Lateral view showing major structures and a temporary confluence of the gut and the neural tube (neurenteric canal). (Adapted from Romer 1956.)

ternately swim upward and then passively sink with the body held horizontally and the mouth directed downward, feeding on plankton and other suspended matter. Development to the juvenile is generally gradual.

Some phylogenetic considerations

In the two preceding chapters we discussed the phylogeny of the lophophorates and the echinoderms with reference to a hypothetical burrowing ancestor. Here we build on that foundation to speculate on the evolutionary history of the remaining deuterostomes, while summarizing current views of deuterostome phylogeny in general. Although there is strong evidence for the monophyly of the deuterostome clade (see Chapter 24), there is controversy about the relationships among the phyla and classes. We discuss below a set of hypotheses (summarized in Figure 20A) that we believe to be the most parsimonious and biologically reasonable. Then we briefly discuss an alternative scenario espoused largely by R. P. S. Jefferies.

The echinoderms capitalized on radial symmetry and on the various functional aspects of their ossicle-based skeleton and water vascular system. Their precise placement along the deuterostome line has long been debated, and our view of this group as an early offshoot of the main deuterostome line is a conservative one. There seem to be no solid synapomorphies linking the echinoderms to any other deuterostome phylum as a sister-group. Thus we believe it is most parsimonious to derive them directly from an ancestral form than from any existing group of deuterostomes. The rest of the deuterostomes may have originated as surface dwellers that employed ciliated mesosomal tentacles for epibenthic feeding. The lophophorates are characterized by their mesosomal lophophore, a synapomorphy that could have arisen by way of only minor modifications of the tentacular apparatus. The evolution of pharyngeal gill slits led on one hand to the hemichordates and on the other to the chordates (Figure 20).

Gill slits probably first evolved as a mechanism to facilitate water flow through the mouth and then out via the slits and pores, thereby increasing gas exchange. The gill slits function in this gas exchange mode in the hemichordates. The pterobranchs retain the mesosomal arms and tentacles as feeding devices, but the enteropneusts have lost these structures, no doubt in connection with their development of an infaunal lifestyle.

There is little doubt that the echinoderms, lophophorates, and hemichordates are closely related to one another. This affinity is evidenced by striking similarities in larval forms and by the mesocoelic origin of the water vascular canals in the echinoderms and the feeding tentacles in the other groups. However, there are various hypotheses about the arrangement of these three taxa along the general deuterostome line and about their relationship to the chordates. There is some question as to whether the gill slits of ascidians are homologous with those of hemichordates and cephalochordates. There is also some question whether the hemichordate neurochord is homologous with the chordate nerve cord.

However, we believe the data favor both homologies, and our placement of the hemichordates closest to the chordate clade is based on those shared synapomorphies.

The origin of the chaetognaths remains a great mystery, although they probably arose as an exploitation of a holoplanktonic lifestyle sometime before the origin of gill slits. Whether they evolved from a lophophorate-like ancestor or earlier in the history of the deuterostomes is quite unresolved, and we have not included them in this cladogram (but see Chapter 24).

The most primitive chordates may have arisen with a shift to internal feeding through the use of the pharyngeal gill slits for extracting suspended food from the water. In addition, chordate origin was marked by the appearance of a pharyngeal endostyle (associated with the feeding mechanism); a notochord; a somatic, muscular, locomotor, postanal tail; and other features. These events were probably accompanied by the appearance of a tadpole-like larva wherein other chordate features are developmentally manifested. Recent work indicates that the vertebrate thyroid gland may be homologous with the endostyle of tunicates, cephalochordates, and larval lampreys.

Figure 20

Deuterostome phylogeny. A cladogram (A) and an orthodox evolutionary tree (B) summarizing phylogeny in the deuterostomes. Defining synapomorphies of each monophyletic group are listed below (the numbers in the legend correspond to those in the cladogram), and are discussed in the text. Synapomorphies defining the deuterostomes as a monophyletic clade include (1) enterocoely; (2) mesoderm derived directly from archenteron; (3) mouth not derived from blastopore (except in phoronids); and (4) coelom tripartite (or derived therefrom).

The hypothesis depicted here suggests that the echinoderms arose early in the deuterostome line with the evolution of radial symmetry (5), an endoskeleton of plates, or ossicles, with a stereom structure (6); a hemal system (7); development of the mesocoel into the water vascular system (8), and the use of external ciliary grooves for suspension feeding (9). (See Chapter 22 for a discussion of echinoderm phylogeny.) The origin of the rest of the deuterostome line is defined by the development of a crown of ciliated feeding tentacles housing extensions of the mesocoel (10). The ancestral form to this line (indicated by *a* on this cladogram) could be viewed as a surface dwelling creature that capitalized on suspension feeding and exploited near-bottom food sources in early Cambrian seas.

The origin of the lophophorate line (Phoronida, Brachiopoda, Ectoprocta) was marked by modification of the ciliary crown into a lophophore (11); reduction of the prosome, as the lophophore became the dominant anterior-end structure (12); and development of a U-shaped gut (13) (see Chapter 21).

The hemichordate–chordate line and ancestor *b* can be defined by the evolution of ciliated pharyngeal gill slits, or clefts (14) and a dorsal hollow nerve cord (15). The gill slits may have first served to localize and increase the surface area for gas exchange, as water was passed through the foregut and feeding was

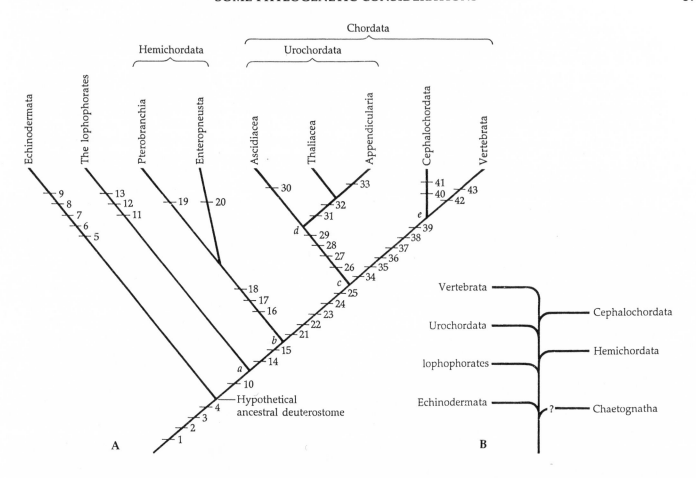

still accomplished by the tentacles. This characteristic is retained as a symplesiomorphy in the primitive hemichordates.

The Hemichordata is defined by the possession of a preoral gut diverticulum (16), the tripartite proboscis collar/trunk arrangement (17), and a unique excretory structure called the glomerulus (18). The latter is less apparent in pterobranchs than in enteropneusts. Pterobranchs evolved toward sessile, colonial lifestyles (19) and retained the tentacular mode of feeding, while the entropneusts lost the tentacles (20) in concert with adopting burrowing lifestyles, but they retained external food capture by ciliary action over the proboscis.

The origin of the chordates is marked by the appearance of several synapomorphies, including loss of the external mesosomai tentacles (21); evolution of a notochord (22); a somatic, post-anal tail region extending behind the visceral cavity, with lateral muscle bands derived from caudal mesoderm (23); an endostyle (in urochordates, cephalochordates, and larval lampreys) or a presumably homologous thyroid gland (in adult lampreys and other vertebrates) (24); and the tadpole larva (25).

The ancestral chordate (c on the cladogram) gave rise to two main lines. The tunicates (urochordates) arose with loss of the coelom as the atrium enlarged in association with their internal water filtering mechanisms (in this phylum, the caudal musculature and body mesenchyme develop directly from mesoderm) (26); development of an outer acellular covering, containing cellulose-like tunicin and capable of iodine fixation (27); evolution of incurrent and excurrent siphons (28); and a much enlarged pharynx, the branchial basket (29). The ancestral tunicate (d on the cladogram) was probably an ascidian-like creature from which the ascidian line evolved by capitalizing on a sessile, benthic life (30). The thaliaceans and appendicularians arose with a shift of

the atrial siphon to the posterior end (31), and the adoption of pelagic lifestyles (32). While the thaliaceans retained these primitive features, the appendicularians probably evolved by way of paedomorphosis from a thaliacean-like stock, retaining most of the tadpole larva features into adulthood (33).

The common ancestor of the cephalochordates and vertebrates (indicated by e on the cladogram) is defined by a large suite of synapomorphies, including: segmentally arranged body muscles, or myotomes (34); differentiation of the neural tube into an inner cell layer ("gray matter") and synaptic outer layer ("white matter"), with separate dorsal and ventral innervation (35); circulatory system with dorsal and ventral aortae, with branchial vessels linking them, and a host of other unique vascular modifications (36); a biosynthetic pathway for thyroid hormone production (nonfunctional in cephalochordates) (37); loss of chitin in the branchial skeleton (38); a large digestive cecum ("liver") (39); plus numerous other features. This ancestor, e, may have arisen from a tadpole larval-like form, in which case the myotomes were probably an adaption allowing greater movement and larger size.

The cephalochordates are defined by their unique buccal apparatuses (e.g., wheel organ, vestibule, oral hood, buccal cirri) (40), and the appearance of protonephridia as excretory structures (41). Cephalochordate protonephridia are thought to be convergent (nonhomologous) with the protonephridia of other invertebrates.

The vertebrates arose with the evolution of a full endoskeleton, including most notably a cranium (42), along with enlargement of the anterior end of the neural tube to form a true brain (43), and a host of other neuronal features.

The urochordates probably arose early from the chordate line. The coelom was lost as the main body cavity in conjunction with the formation of the large cloacal chamber through which water passes from the body after pharyngeal filtration. The earliest urochordates may have been ascidians, which, as a group, adopted a sessile lifestyle in association with suspension feeding and loss of adult locomotor musculature. Another line led to the thaliaceans, which exploited planktonic habits. The most primitive thaliaceans are thought to be the colonial pyrosomids (order Pyrosomida) because of their similarities to compound ascidians. They may have given rise to the doliolids (order Doliolida), and the doliolids to the salps (order Salpida)—both of which show increased zooid size and independence. The thaliaceans in general are characterized by the posterior placement of the excurrent siphon, a condition that provides them with locomotive powers in a pelagic environment. Recent study suggests that the appendicularians might have arisen from a doliolid-like ancestor. Many earlier hypotheses, however, supported an ascidian ancestry to the appendicularians. In any case, there is little doubt that the appendicularians arose by paedomorphic evolutionary events in which sexual maturation occurred in an animal that retained larval characteristics, as evidenced not only by the tadpole-larval form of the adults, but also by the persistence of chordate features (e.g., the notochord) present only in the larvae of other tunicates. Walter Garstang was of the opinion that the notochord originally evolved to give support to pelagic larvae as they developed longer and longer planktonic lives, a trend eventually leading to the neotenic appendicularians. As these forms grew larger in size, Garstang reasoned, the notochord provided a basis for muscular support and locomotion, supplementing previous reliance on ciliary locomotion.

Most workers feel that the paedomorphic tendencies of the tadpole larvae of the ancestral chordate

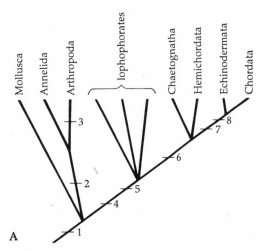

A

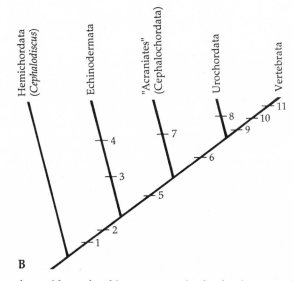

B

Figure 21

Some aspects of deuterostome phylogeny, adapted from the ideas of R. P. S. Jefferies (1986). Numbering on the cladograms corresponds to the synapomorphies noted below. A, Part of a cladogram of the animal kingdom, showing Jefferies' views on the origin of the deuterostomes from the protostomes and on phylogeny within the deuterstome line. The portion of the cladogram shown originates with the evolution of the coelom (1) as a monophyletic event within the protostomes. The annelid–arthropod clade is defined by metamerism (2), and the arthropods in particular by molting (3). The first appearance of deuterostome-like features included a switch from spiral to radial cleavage (4) and the origin of a tripartite body plan (5); the derivation of the anus (instead of the mouth) from the blastopore (6) occurred subsequent to the origin(s) of the lophophorates, placing the origin of those three phyla somewhere between the protostomes and the deuterostomes. The origin of the echinoderm–chordate clade is defined by the origin of the calcite, stereom endoskeleton (7), and an emphasis on left-side structures and reduction or loss of right-side structures, a phenomenon called dexiothetism (8). Jefferies cites evidence for this asymmetry in the development of echinoderms and cephalochordates, and in fossil calcichordates.

B, A cladogram summarizing Jefferies' arrangement of the echinoderm–chordate clade (with some details omitted) as it arose from a *Cephalodiscus*-like ancestor with the origin of the calcite skeleton (1) and dexiothetism (2). The echinoderms are defined by their radial symmetry (3) and loss of the primitive calcichordate gill slit (4). Jefferies details the events along the line to the chordates (5) by using a sequence of increasingly more chordate-like (and less echinoderm-like) calcichordate fossils. These events include the origin of the notochord and dorsal nerve cord. He employs both developmental and fossil evidence (6) to support placing the urochordates closest to the ancestral vertebrate. By this set of hypotheses, the calcite, stereom skeleton was lost three times during chordate evolution (7, 8, 9). The line to the vertebrates also includes a sequence of calcichordates and the acquisition of the vertebrate endoskeleton (10) and "segmentation" (11).

played a major role in the origin of the cephalochordates and vertebrates (Jefferies' theory, discussed below, being a notable exception to this opinion). The evolution of segmental muscle bundles (**myotomes**) marked the beginning of the cephalochordate–vertebrate clade and allowed greater locomotor facility than the tail thrashing movements of the tadpole-larval ancestor. The cephalochordates have retained the notochord, and the use of pharyngeal gill slits for feeding. The vertebrates are, of course, characterized by the development of an endoskeleton with cranium, vertebral column, and, ultimately, limbs. All of these features eventually provided increased body support and much more effective skeletomuscular mechanics and allowed great increases in locomotor abilities and body size.

We have already mentioned that arguments exist concerning the phylogeny of deuterostomes and that there are a number of alternatives to the hypotheses presented here. Perhaps the most unorthodox, but interesting, hypothesis is espoused by R. P. S. Jefferies. In his book *The Ancestry of the Vertebrates* (1986), Jefferies discusses in great detail his years of research on this matter. Here we present only a brief summary of his ideas.

Jefferies, a paleontologist, focuses on an extinct group of creatures that have traditionally been classified as carpoid echinoderms. These animals had an echinoderm-like skeleton composed of calcitic stereom ossicles, but, according to Jefferies, they also possessed a number of chordate features (including gill slits). Jefferies calls these animals **calcichordates** and interprets the inflated portion of their body and their appendage-like structure as homologous with the head and tail (respectively) of ascidian tadpole larvae and of chordates in general. He has "reconstructed" the soft parts of certain calcichordates to include a brain, a notochord, a dorsal nerve cord, and other features associated with the chordate *Bauplan*. If one accepts Jefferies' view of the calcichordates, there emerges a scenario of deuterostome phylogeny very different from the one we present above. These differences are triggered in part by treating the echinoderms and chordates as sister-groups, stemming from a calcichordate ancestor. Figure 21 depicts and explains some of the ideas embodied in this phylogeny.

The fact that we have given Jefferies' work only superficial coverage here should not be taken as a criticism of his efforts or hypotheses. His book and previous work are the products of years of very careful and detailed work not only on the problem of chordate origin, but also on deuterostomes in general, and they provide fascinating reading and topics for debate and further investigation. Clearly, the jury is still out on the question of chordate origin.

Selected References

General References

Barrington, E. J. W. 1965. *The Biology of Hemichordata and Protochordata*. W. H. Freeman, San Francisco.

Barrington, E. J. W. and R. P. S. Jefferies (eds.). 1975. *Protochordates*. Academic Press, New York. [A collection of papers generated from a symposium of the Zoological Society of London. Some of the papers are cited here individually.]

Delage, Y. and E. Herouard. 1898. *Traité de Zoologie Concrete*. Vol. 8. Les procordes.

Eaton, T. H. 1970. The stem–tail problem and the ancestry of the chordates. J. Paleontol. 44:969–979.

Gans, C. and R. G. Northcutt. 1983. Neural crest and the origin of vertebrates: A new head. Science 220:268–274.

Hamner, W. M., L. P. Madin, A. L. Alldredge, R. W. Gilmer and P. P. Hamner. 1975. Underwater observations of gelatinous zooplankton: Sampling problems, feeding biology, and behavior. Limnol. Oceanogr. 20:904–917.

Hyman, L. H. 1959. *The Invertebrates*, Vol. 5. Smaller Coelomate Groups. McGraw-Hill, New York. [This volume includes the Chaetognatha and Hemichordata.]

Jefferies, R. P. S. 1986. *The Ancestry of the Vertebrates*. British Museum (Natural History), London. [A detailed account of Jefferies' theory of the origin of vertebrates from a Paleozoic fossil echinoderm group; the book summarizes his previous studies and presents an excellent review of anatomy and embryology of living deuterostomes.]

Thorpe, A. and M. C. Thorndyke. 1975. The endostyle in relation to iodine. *In* E. J. W. Barrington and R. P. S. Jefferies (eds.), *Protochordates*. Acad. Press, New York, pp. 159–178.

Willmer, E. N. 1975. The possible contribution of the nemertines to the problem of the phylogeny of the protochordates. *In* E. J. W. Barrington and R. P. S. Jefferies (eds.), *Protochordates*. Academic Press, New York, pp. 319–345.

Chaetognatha

Alvarino, A. 1965. Chaetognaths. Annu. Rev. Oceanogr. Mar. Biol. 3:115–194.

Bieri, R. 1966. The function of the "wings" of *Pterosagitta draco* and the so-called tangoreceptors in other species of Chaetognatha. Publ. Seto Mar. Biol. Lab. 14:23–26.

Bieri, R. 1989. Krohnittellidae and Bathybelidae, new families in the phylum Chaetognatha; the rejection of the family Tokiokaispadellidae and the genera Tokiokaispadella, Zahonya, and Aberrospadella. Proc. Biol. Soc. Wash. 102:973–976.

Bowman, T. E. and R. Bieri. 1989. *Paraspadella anops*, n. sp., from Sagittarius Cave, Grand Bahama Island, the second troglobitic chaetognath. Proc. Biol. Soc. Wash. 102:586–589.

Burfield, S. 1927. *Sagitta*. Liverpool Mar. Biol. Comm., Mem. 28. *In* Proc. Trans. Liverpool Biol. Soc. 41:1–104.

Eakin, R. M. and J. A. Westfall. 1964. Fine structure of the eye of a chaetognath. J. Cell Biol. 21:115-132.

Feigenbaum, D. L. 1978. Hair-fan patterns in the Chaetognatha. Can. J. Zool. 56:536–546.

Feigenbaum, D. L. and R. C. Maris. 1984. Feeding in the Chaetognatha. Annu. Rev. Oceanogr. Mar. Biol. 22:343–392.

Ghirardelli, E. 1968. Some aspects of the biology of the chaetognaths. Adv. Mar. Biol. 6:271–375.

Goto, T. and M. Yoshida. 1985. The mating sequence of the benthic arrowworm Spadella schizoptera. Biol. Bull. 169:328–333.

Goto, T., N. Takasu and M. Yoshida. 1984. A unique photoreceptive structure in the arrowworms Sagitta crassa and Spadella schizoptera (Chaetognatha). Cell Tissue Res. 235:471–478.

Michel, H. B. 1982. Chaetognatha. In S. P. Parker (ed.), Synopsis and Classification of Living Organisms, Vol. 2. McGraw-Hill, New York, pp 781–783.

Michel, H. B. 1984. Chaetognatha of the Caribbean Sea and adjacent areas. NOAA Tech. Rpt. Nat. Mar. Fish. Serv. Circular 15:1–33.

Nagasawa, S. 1985. Ecological significance of deformed chaetognaths associated with bacteria. Bull. Mar. Sci. 37:707–715.

Pierrot-Bults, A. C. and K. C. Chidgey. 1988. Chaetognatha. Synopses of the British Fauna (New Series), No. 39. E. J. Brill, New York.

Salvini-Plawen, L. von. 1986. Systematic notes on Spadella and on the Chaetognatha in general. Z. Zool. Syst. Evolutionsforsch. 24:122–128.

Terazaki, M. and C. B. Miller. 1982. Reproduction of meso- and bathypelagic chaetognaths in the genus Eukrohnia. Mar. Biol. 71:193–196.

Terazaki, M., R. Marumo and Y. Fujita. 1977. Pigments of meso- and bathypelagic chaetognaths. Mar. Biol. 41:119–125.

Thuesen, E. V. and R. Bieri. 1987. Tooth structure and buccal pores in the chaetognath Flaccisagitta hexaptera and their relation to the capture of fish larvae and copepods. Can. J. Zool. 65:181–187.

Thuesen, E. V. and K. Kogure. 1989. Bacterial production of tetrodotoxin in four species of Chaetognatha. Biol. Bull. 176: 191–194.

Tokioka, T. 1965. The taxonomical outline of Chaetognatha. Publ. Seto Mar. Biol. Lab. 12:335–357.

Hemichordata

Armstrong, W. G., P. N. Dilly, and A. Urbanek. 1984. Collagen in the pterobranch coenecium and the problem of graptolite affinities. Lethaia 17:145–152.

Barrington, E. J. W. 1940. Observation of feeding and digestion in Glossobalanus. Q. J. Microsc. Sci. 82:227–260.

Benito, J. 1982. Hemichordata. In S. P. Parker (ed.), Synopsis and Classification of Living Organisms, Vol. 2. McGraw-Hill, New York, pp. 819–821.

Bullock, T. H. 1946. The anatomical organization of the nervous system of the Enteropneusta. Q. J. Microsc. Sci. 86:55–111.

Burdon-Jones, C. 1952. Development and biology of the larva of Saccoglossus horsti (Enteropneusta). Philos. Trans. R. Soc. London Ser. B 236:553–590.

Burdon-Jones, C. 1956. Observations on the enteropneust, Protoglossus kohleri. Proc. Zool. Soc. Lond. 127:35–58.

Dilly, P. N. 1985. The habitat and behavior of Cephalodiscus gracilis (Pterobranchia, Hemichordata) from Bermuda. J. Zool. 207:223–239.

Gilmour, T. H. J. 1979. Feeding in pterobranch hemichordates and the evolution of gill slits. Can. J. Zool. 57:1136–1142.

Gilmour, T. H. J. 1982. Feeding in tornaria larvae and the development of gill slits in enteropneust hemichordates. Can. J. Zool. 60:3010–3020.

Hadfield, M. G. 1975. Hemichordata. In A. C. Geise and J. S. Pearse (eds.), Reproduction of Marine Invertebrates, Vol. 2. Academic Press, New York, pp. 185–240.

Hadfield, M. G. and R. E. Young. 1983. Planctosphaera (Hemichordata: Enteropneusta) in the Pacific Ocean. Mar. Biol. 73:151–153.

Lester, S. M. 1985. Cephalodiscus sp. (Hemichordata: Pterobranchia): Observations of functional morphology, behavior and occurrence in shallow water around Bermuda. Mar. Biol. 85:263–268.

Spengel, J. 1932. Planctosphaera pelagica. Sci. Results Michael Sars North Atl. Deep Sea Exped. 5(5).

Stebbings, A. R. D. and P. N. Dilly. 1972. Some observations of living Rhabdopleura compacta (Hemichordata). J. Mar. Biol. Assoc. U. K. 52:443–448.

Strathmann, R. and D. Bonar. 1976. Ciliary feeding of tornaria larvae of Ptychodera flava. Mar. Biol. 34; 317–324.

Woodwick, K. H. and T. Sensenbaugh. 1985. Saxipendium coronatum, new genus, new species (Hemichordata: Enteropneusta): The unusual spaghetti worms of the Galapagos rift hydrothermal vents. Proc. Biol. Soc. Wash. 98:351–365.

Chordata

Abbott, D. P. and J. V. Johnson. 1972. The ascidians Styela barnharti, S. plicata, S. clava and S. montereyensis in Californian waters. Bull. South. Calif. Acad. Sci. 71:95–105.

Alexander, R. M. 1975. The Chordates. Cambridge University Press, New York.

Alldredge, A. 1976a. Appendicularians. Sci. Am. 235:94–102.

Alldredge, A. 1976b. Discarded appendicularian houses as sources of food, surface habitats, and particulate organic matter in planktonic environments. Limnol. Oceanogr. 21:14–23.

Alldredge, A. 1977. Morphology and mechanisms of feeding in the Oikopleuridae (Tunicata, Appendicularians). J. Zool. 181:175–188.

Azariah, J. 1982. Cephalochordata. In S. P. Parker (ed.), Synopsis and Classification of Living Organisms, Vol. 2. McGraw-Hill, New York, pp. 829–830.

Barham, E. 1979. Giant larvacean houses: Observations from deep submersibles. Science 205:1129–1131.

Berrill, N. J. 1935. Studies in tunicate development. IV. Asexual reproduction. Philos. Trans. R. Soc. London Ser. B 225:327–379.

Berrill, N. J. 1961. Salpa. Sci. Amer. 204: 150–160.

Berrill, N. J. 1975. Chordata: Tunicata. In A. C. Geise and J. S. Pearse (eds.), Reproduction of Marine Invertebrates, Vol. 2. Academic Press, New York, pp. 241–282.

Birkeland, C., L. Cheng and R. A. Lewis. 1981. Mobility of didemnid ascidean colonies. Bull. Mar. Sci. 31:170–173.

Cloney, R. A. 1978. Ascidian metamorphosis review and analysis. In F. S. Chia and M. E. Rice (eds.), Settlement and Metamorphosis of Marine Invertebrate Larvae. Elsevier North-Holland, New York, pp. 225–282.

Cloney, R. A. 1982. Ascidian larvae and the events of metamorphosis. In C. C. Lambert and G. Lambert (eds.), The developmental biology of the ascidians. Am. Zool. 22:817–826.

Cloney, R. A. and S. A. Torrence. 1982. Ascidian larvae: Structure and settlement. In J. D. Costlow (ed.), Biodeterioration. U. S. Naval Institute, Annapolis, Maryland.

Cox, G. 1983. Engulfment of Prochloron cells by cells of the ascidean, Lissoclinum. J. Mar. Biol. Assoc. U.K. 63:195–198.

Deibel, D., M. L. Dickson, and C. Powell. 1985. Ultrastructure of the mucus feeding filters of the house of the appendicularians Oikopleura vanhoeffeni. Mar. Ecol. Prog. Ser. 27:79–86.

Fenaux, R. 1985. Rhythm of secretion of Oikopleurid's houses. Bull. Mar. Sci. 37: 498–503.

Goodbody, I. 1974. The physiology of ascidians. Adv. Mar. Biol. 12:1–149.

Goodbody, I. 1982. Tunicata. *In* S. P. Parker (ed.), *Synopsis and Classification of Living Organisms Vol. 2*. McGraw-Hill, New York, pp. 823–829.

Hubbs, C. L. 1922. A list of the lancelets of the world with diagnoses of five new species of *Branchiostoma*. Occ. Pap. Mus. Zool. Univ. Mich. 105:1–16.

Katz, M. J. 1983. Comparative anatomy of the tunicate tadpole, *Ciona intentinalis*. Biol. Bull. 164:1–27.

Kott, P. 1982. Didemnid–algal symbioses: Host species in the western Pacific with notes on the symbiosis. Micronesica 18:95–127.

Koyama, H. and H. Watanabe. 1986. Studies on the fusion reaction in two species of *Perophora* (Ascidiacea). Mar. Biol. 92:267–275.

Lambert, C. C. and G. Lambert (eds.). 1982. The developmental biology of the ascidians. Am. Zool. 22:751–849. [Results of a 1981 symposium at the annual meeting of the American Society of Zoologists; nine papers, plus introductory remarks by C. Lambert; some papers cited separately here].

Lambert, G. 1968. The general ecology and growth of a solitary ascidian, *Corella willmeriana*. Biol. Bull. 135:296–307.

Levine, E. P. 1962. Studies on the structure, reproduction, development, and accumulation of metals in the colonial ascidian *Eudistoma ritteri* Van Name. J. Morphol. 111:105–137.

Løvtrup, S. 1977. *The Phylogeny of Vertebrata*. Wiley, London.

Maisey, J. G. 1986. Heads and tails: A chordate phylogeny. Cladistics 2:201–256.

McFarland, W. N., F. H. Pough, T. J. Cade, and J. B. Heiser. 1985. *Vertebrate Life*, 2nd Ed. Macmillan, New York.

Millar, R. H. 1962. Some ascidians from the Caribbean. Caribb. Mar. Biol. Inst. Curacao. Sel. Pap. 21:61–77. [Studies on the fauna of Curacao and other Caribbean islands, No. 59.]

Millar, R. H. 1966. Evolution in ascidians. *In* H. Barnes (ed.), *Some Contemporary Studies in Marine Science*. Allen & Unwin, London, pp. 519–534.

Millar, R. H. 1971. The biology of ascidians. Adv. Mar. Biol. 9:1–100.

Monniot, C. and F. Monniot. 1975. Abyssal tunicates: An ecological paradox. Ann. Inst. Oceanogr. 51(1):99–129.

Monniot, C. and F. Monniot. 1978. Recent work on the deep-sea tunicates. Annu. Rev. Oceanogr. Mar. Biol. 16:181–228.

Monniot, F. 1975. Microfiltres et ciliatures branchiales des ascidies littorales en microscopie electronique. Bull. Mus. Hist. Nat. Paris Ser. 4 1A(4):843–859.

Mukai, H., H. Koyama, and H. Watanabe. 1983. Studies on the reproduction of three species of *Perophora* (Ascidiacea). Biol. Bull. 164:251–266.

Nakauchi, M. 1982. Asexual development of ascidians:Its biological significance, diversity, and morphogenesis. *In* C. C. Lambert and G. Lambert (eds.), The developmental biology of the ascidians. Am. Zool. 22:753–764.

Parry, D. L. and P. Kott. 1988. Co-symbiosis in the Ascidiacea. Bull. Mar. Sci. 42:149–153.

Romer, A. S. 1956. *The Vertebrate Body*, 2nd Ed. Saunders, Philadelphia.

Romer, A. S. 1967. Major steps in vertebrate evolution. Science 158:1629–1637.

Schmidt, G. H. 1982. Aggregation and fusion between conspecifics of a solitary ascidean. Biol. Bull. 162:195–201.

Stoeker, D. 1980. Chemical defenses of ascidians against predators. Ecology 61:1327–1334.

Tokioka, T. 1963. The outline of Japanese ascidian fauna as compared with that of the Pacific coast of North America. Publ. Seto Mar. Biol. Lab. 11:131–156.

Torrence, S. A. and R. A. Cloney. 1982. The nervous system of ascidian larvae: Primary sensory neurons in the tail. Zoomorphologie 99:103–115.

Van Name, W. G. 1930. The ascidians of Porto Rico and the Virgin Islands. Ann. N.Y. Acad. Sci. 10:403–535.

Van Name, W. G. 1945. The North and South American ascidians. Bull. Am. Mus. Nat. Hist. 84:1–476.

Young, C. M. and L. F. Braithwaite. 1980. Orientation and current–induced flow in the stalked ascidian *Styela montereyensis*. Biol. Bull. 159:428–440.

Young, C. M. and L. F. Braithwaite. 1980. Larval behavior and post-settling morphology in the ascidian, *Chelyosoma productum* Stimpson. J. Exp. Mar. Biol. Ecol. 42:157–169.

Chapter Twenty-Four

Perspectives on Invertebrate Phylogeny

As a result of such speculations multitudes of phyloge-
netic trees sprang up in the thin soil of embryological
fact and developed a capacity of branching and produc-
ing hypothetical ancestors which was in inverse propor-
tion to their hold on solid ground.

    E. G. Conklin
      Embryology and Evolution, 1928

We shall not cease from exploration
And the end of all our exploring
Will be to arrive where we started
And know the place for the first time.
Through the unknown, remembered gate
When the last of earth left to discover
Is that which was the beginning

    T. S. Eliot
      Little Gidding, 1943

Ideas about invertebrate phylogeny are often presented as though they were widely agreed-upon theories or, worse yet, as though alternative ideas did
not even exist. Nothing could be farther from the
truth; we actually know very little about the relationships among animal phyla. The science of comparative biology can be viewed as a two-step process; the
first step is to recognize meaningful patterns in nature, and the second step is to propose testable hypotheses to explain those patterns. As for the origins
of and relationships among the invertebrate phyla,
we are still at an early stage of pattern analysis. In
this chapter we undertake such an analysis of the
invertebrates by developing a cladogram of the better
known phyla. We discuss this tree and some of its
implications, and then we discuss the more poorly
understood phyla. In doing so, we elucidate some of
the problems zoologists face in their efforts to unravel
the evolutionary history of the animal kingdom.

We are tempted at this point to ask you to return
to the introductory chapters of this book and reread
them, to examine your perceptions of invertebrate
zoology after several months of detailed study. If we
have all done our jobs and you have consulted those
chapters frequently, then you can now reduce the
multitude of details about invertebrates to some major concepts. We have explored how animals are put
together and how they work (the *Bauplan* concept),
how reproductive and developmental patterns relate
to adult structure and lifestyle (life history strategies),
and how the various groups of animals may be related to one another (phylogeny and evolutionary
history).

The paradigm of evolution is, of course, the foundation of modern biology. It is the only tenable scientific theory that explains the history of the living
world; no other general view of life is both compatible
with the available evidence and open to scientific
testing. Because of these qualities, evolution is not
only one of the major principles addressed in this
book, it is also the underlying framework supporting
all the other concepts and ideas we have discussed.

All animals must solve the same fundamental
problems of staying alive long enough to reproduce
themselves, and they must do so under the limitations imposed by environmental conditions on earth.
At the cellular level, organisms are very similar: their
activities are governed by the basic principles of
physics and chemistry and the ability of creatures to
convert energy from one form to another and put it
to use. Given this principle of unity among living
things, how is it that we see such enormous variety
and diversity of animals (and plants)? There are a
number of factors that contribute to answering this

question. For example, there is more than one way to "solve" any particular problem of staying alive. We have seen various anatomical and physiological methods of nutrition, excretion, support and locomotion, gas exchange, internal transport, and so on. Furthermore, the earth is an extremely heterogeneous place — what works in one habitat often will not work in another. So the impact of different environmental selection pressures is to favor and promote diversity. Mutation, of course, introduces variation at the most fundamental genetic level. Thus, while genetic events and sexual processes are unifying and cohesive attributes of biological systems, they are at the same time the foundation of variation and diversity.

We have seen that successful *Baupläne* occur at various levels of structural complexity, ranging from unicellular protozoa to multisystemed metazoa. The organisms that live and survive today, and those that survived in the past, are those with structural and functional body plans that work(ed) well enough to maintain the species. Given this condition of "success," perfection is irrelevant in the living world— adequate (i.e., "good enough") is sufficient. However, an adequate *Bauplan* is not guaranteed by haphazardly tossing together parts; the components must be compatible with one another if they are to produce a functional and successful animal. (Just because an open circulatory system works well as part of the arthropod *Bauplan* does not mean that it would support life in the body of a mammal.)

Throughout this book you have been exposed to various body architectures, the taxa that display them, and the advantages and constraints they impart in particular environmental settings. In this chapter we hope to tie much of this information together by way of a phylogenetic overview of major animal taxa and *Baupläne*. You have already read about many previously espoused ideas concerning invertebrate phylogeny, and countless others exist in the literature—far too many to review here. Recent studies concerning the evolution of the invertebrate phyla are cited in the reference section of Chapter 4 and elsewhere. Until fairly recently, analyses of the invertebrate phyla as a whole generally had been presented by way of narrative scenarios. Although such methods are perfectly legitimate, and we have used them ourselves, they are often ambiguous and very difficult, if not impossible, to test in any rigorous scientific fashion. Almost any kind of scenario can be concocted to explain how one group of organisms might have arisen from another. Furthermore, such narratives are often based on a priori assumptions

about *hypothetical* ancestors. Despite atonements to "workable ancestors," virtually any complicated evolutionary transition can be described on paper, given enough imagination.

So, rather than begin our discussion with a narrative of invertebrate evolution, we begin by first constructing a cladogram, based on real taxa and real characters (Figure 1). This cladogram of invertebrate phyla summarizes our own ideas, largely as discussed in previous chapters. Having developed a cladogram, we then briefly describe the sequence of evolutionary events implied by the tree. As new characters are elucidated, the cladogram can be tested and either accepted, rejected, or revised. The cladogram was produced by a computer program called PAUP (see Chapter 2 for details). It is the most parsimonious tree that exists given the characters and interpretations of the data that we have applied; that is, it is the tree with the fewest steps and the fewest hypothesized evolutionary convergences and reversals. We have tried to use embryological evidence as much as possible to evaluate homologies among the characters on the tree. Other interpretations of some of these characters exist, and of course many other possible trees exist.

You will notice that not all of the animal phyla are represented in the cladogram; groups whose major homologies are still not known or understood are not included, but they are discussed below. We felt it best to omit these groups from the final analysis to avoid distorting the branching sequences and giving suspect or false impressions. As evolutionary biology moves forward, we will undoubtedly see more studies of structure (especially ultrastructure), function, behavior, ecology, genetics, molecular data, and all of the other fields that ultimately shed light on the history and relationships among living creatures.

Our cladogram is only one of many possible sets of hypotheses and is certainly open to a host of challenges. Remember, always question these and other ideas about the evolutionary history of life. The late Donald P. Abbott left all of us some good advice— "Cultivate a suspicious attitude towards people who do phylogeny."

Origin of the metazoa

A large body of evidence accumulated since the 1960s supports the view that eukaryotic life evolved in concert with the evolution of mitochondria and plastids from certain endosymbiotic prokaryotes (Margulis 1981 provides a good summary). It seems likely that the eukaryotic condition arose more than once, perhaps in more than one way, to produce the polyphyletic group of single-celled organisms we call the Protista. Multicelled plants evolved from one or more protistan ancestors, and so did the multicelled animals (the metazoa). Opinion has long been split on whether the metazoa are monophyletic or polyphyletic. As explained below, we think that the available data best support a monophyletic view of the metazoa. (See Chapter 4; for recent studies espousing the polyphyletic view, see Anderson 1982, Ingils 1985, and Field et al. 1988).

From which protozoan group(s) did the metazoa arise, and what was the nature of the first metazoa? As we noted in Chapter 4, several competing hypotheses of metazoan origin have enjoyed popularity in the past. The best evidence to date supports a colonial flagellate ancestry for the metazoa. The ancestral form may have existed as a clump or a hollow ball of cells, something like a coeloblastula in basic structure, each cell with a single cilium/flagellum. Three very compelling lines of reasoning support this contention. First, the possession of monociliated (= monoflagellated) cells appears to be primitive among the metazoa; this condition occurs in poriferans, cnidarians, and the odd creature *Trichoplax* (as well as in some gastrotrichs, gnathostomulids, and generally in deuterostomes). Second, the flagellated collar cells of sponges and certain flagellate protozoa (the choanoflagellates) are unique and remarkably similar. Third, early stages in the origin of the metazoan condition must have included the formation of layered tissues, an event akin to the embryonic process of gastrulation. A blastula-like ancestor would have set the stage, so to speak, for such a gastrulation-like phenomenon. Once achieved, the advantages of a multilayered body plan were selected for and led to the rapid early eumetazoan diversification documented in the fossil record.

Despite the logical strength of the colonial flagellate hypothesis, the origin and relationships of the metazoa still remain essentially unknown and constitute one of the unanswered fundamental questions in modern biology. Three principal circumstances have contributed to the insolubility of this problem. First is the sheer age of the metazoan phyla, most or all of which probably arose in the Precambrian era and took part in the great "Cambrian explosion of life" that established virtually all major animal *Baupläne*. This age, coupled with a paucity of hard parts in the early metazoa, has resulted in the near absence of fossils from the period of major separations of the animal phyla. Almost no creatures have been iden-

tified in the fossil record that form good intermediary forms linking major taxa. Although the fossil record is limited, it does indicate that all animal phyla had their origin in the sea.

Second, it is clear from the fossil record of the Cambrian period that many major groups (phyla?) came and went long ago, leaving few clues and many mysteries as to their relationships. Some paleontologists have estimated that more than 100 animal phyla may have evolved in the Precambrian and the Cambrian period, with fewer than half of these surviving to the present. One of the great mysteries of life is why this burst of fundamental evolutionary invention occurred only once in earth history.

Third, most animal phyla are so distinct that very few good synapomorphies can be identified that define sister-groups among them. That is, structural homologies *among* phyla are difficult to positively identify. This problem is exacerbated by the enormous amount of evolutionary convergence that has occurred among the metazoan phyla over their 680+ million-year history. Thus, phyla seemingly related on some grounds may appear quite unrelated on others.

Because of these constraints, hypotheses concerning the origins of animal phyla carry a heavy burden of speculation, no matter what method of analysis is used or how carefully it is done. Nevertheless, comparative anatomy and especially embryology provide powerful clues to ancestry, and by seeking fundamental similarities that may constitute synapomorphies at the level of phyla, testable phylogenetic hypotheses can be created. Our cladogram represents one set of such hypotheses.

The cladogram in Figure 1 depicts a monophyletic origin for the metazoa and hypothesizes that certain key metazoan synapomorphies evolved only once—for example, multicellularity, the acetylcholine/choline esterase system, collagen, and characteristic intercellular features of metazoa (septate junctions, tight junctions, gap junctions*). The Porifera probably diverged early from the metazoan lineage, prior to the evolution of features such as an epidermal basement membrane, a synaptic nervous system,

---

*Specialized cell junctions are characteristic of multicelled animals (metazoa) and serve a variety of functions. For example, septate and tight junctions seal off the extracellular space in areas such as epithelial surfaces to prevent loss of intercellular fluids and molecules to the surroundings. Gap junctions are more specialized and complex, and play a variety of functional roles in cell–cell communication (e.g., nerve–muscle and nerve–nerve impulse transmission). Septate or tight junctions apparently occur in all metazoa, particularly along the distal cell contact areas of the epithelia; the more specialized gap junctions appear to be lacking in sponges.

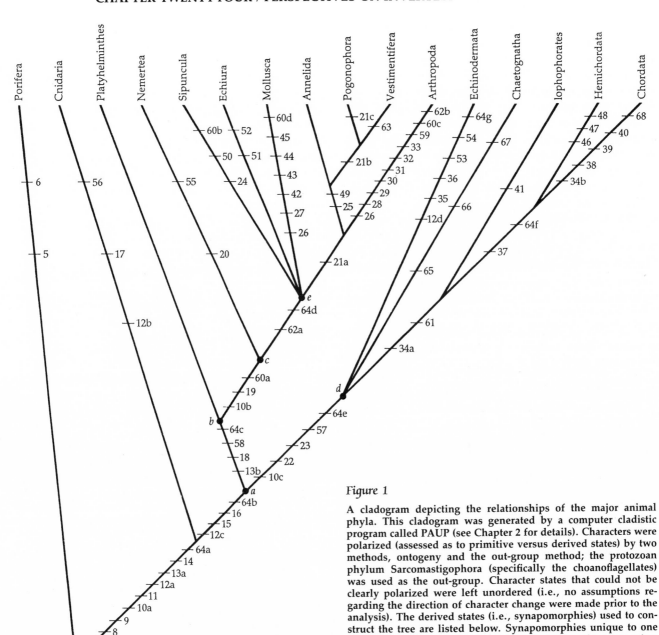

Figure 1

A cladogram depicting the relationships of the major animal phyla. This cladogram was generated by a computer cladistic program called PAUP (see Chapter 2 for details). Characters were polarized (assessed as to primitive versus derived states) by two methods, ontogeny and the out-group method; the protozoan phylum Sarcomastigophora (specifically the choanoflagellates) was used as the out-group. Character states that could not be clearly polarized were left unordered (i.e., no assumptions regarding the direction of character change were made prior to the analysis). The derived states (i.e., synapomorphies) used to construct the tree are listed below. Synapomorphies unique to one phylum are indicated in the list by including the appropriate phylum name in parentheses. Some character shifts that occur *within* a particular phylum are also noted in the list. For more detailed cladograms on individual phyla or groups of phyla, see the appropriate chapters.

1. Multicellularity, with a high degree of division of labor
2. Acetylcholine/cholinesterase system
3. Collagen
4. Septate/tight junctions between cells
5. Aquiferous system (Porifera)
6. Unique poriferan ontogeny of layered construction (Porifera)
7. Gap junctions between cells
8. Striated myofibrils
9. Loss of flagellated collar cells
10a. Gastrovascular cavity (i.e., incomplete gut) with mouth arising from blastopore

10b. Complete gut with mouth arising from blastopore
10c. Complete gut with mouth not arising from blastopore
11. Gastrulation (origin of true germ layers and diploblastic construction)
12a. Symmetrical body plan
12b. Fundamentally radially symmetrical (Cnidaria)
12c. Fundamentally bilaterally symmetrical
12d. Secondarily pentaradially symmetrical (Echinodermata)
13a. Typical radial cleavage
13b. Typical spiral cleavage
14. Basement membrane (= basal lamina) beneath epidermis
15. Multiciliate/multiflagellate cells

and gap junctions between cells, all characteristic features of the higher metazoa (the eumetazoa). Sponges retain a high degree of cellular totipotency and do not have the true tissues seen in higher metazoa; the latter conclusion is based on the observation that the embryonic processes leading to a layered construction in poriferans are very different from eumetazoan gastrulation. These differences have been taken by some biologists as evidence that the sponges arose separately from the main metazoan lineage, perhaps from a different group of colonial flagellates, or if from the same ancestral group, by way of a separate set of events.

The distribution of characters on the cladogram suggests that the first postporiferan metazoa were symmetrical (perhaps radially symmetrical) diploblastic creatures with synaptic nervous systems, a basement membrane beneath their epidermis, and perhaps a distinct gastrovascular cavity. They may or may not have been planuloid in appearance. This analysis suggests that the gastrulation-like event that led to their origin was probably akin to invagination, although a postgastrular hollowing of the gut remains a possibility. The cnidarian clade probably arose from this ancestral type, and its members retained the basic diploblastic condition.

16. Cephalization
17. Cnidae (Cnidaria)
18. Mesoderm arises from mesentoblast (primitively the 4d cell)
19. Body cavity (coelom) arises by schizocoely
20. Coelom reduced and restricted to rhynchocoel (Nemertea)
21a. True segmentation arising by teloblastic growth and resulting in serial repetition of body parts
21b. True segmentation restricted to opisthosoma
21c. Midsagittal mesenteries lost in opisthosoma (Pogonophora)
22. Mesoderm derived directly from archenteron
23. Body cavity (coelom) tripartite and derived by enterocoely
24. Unique anterior introvert (Sipuncula)
25. Serially arranged clusters of lateral epidermal setae
26. Schizocoelom reduced; body cavity a hemocoel
27. With schizocoelomic pericardial remnant around heart (Mollusca)
28. Cephalic ecdysial glands (Arthropoda)
29. Arthropodous ecdysis (Arthropoda)
30. Unique arthropodan compound eyes (Arthropoda)
31. Sclerotized, jointed, chitinous cuticle with segmental sclerites (Arthropoda)
32. Articulated appendages surrounded by chitinous cuticle and containing intrinsic musculature (Arthropoda)
33. Loss of motile cilia/flagella (except in sperm of some species) (Arthropoda)
34a. Ciliated feeding tentacles derived from mesosome and containing extensions of the mesocoel
34b. Loss of mesosomal tentacles (Chordata) (also in some Hemichordata)
35. Stereom calcite ossicle endoskeleton (Echinodermata)
36. Water vascular system derived from mesocoel (Echinodermata)
37. Pharyngeal gill slits
38. Notochord (Chordata)
39. Muscular, locomotor, postanal tail (Chordata)
40. Endostyle (Chordata)
41. Development of lophophore, accompanied by reduction of the prosome (lophophorates)
42. Radula (Mollusca)
43. Molluscan mantle (Mollusca)
44. Dorsal mantle shell gland produces spicules or shell (Mollusca)
45. Ventral, muscular, ciliated, solelike foot or its precursor (Mollusca)
46. Anterior (preoral) gut diverticulum (Hemichordata)
47. Tripartite body arranged as unique proboscis/collar/trunk (Hemichordata)
48. Unique hemichordate kidney "glomerulus" (Hemichordata)

49. Unique annelidan head of presegmental prostomium and peristomium
50. Compensation system (Sipuncula)
51. Unique preoral proboscis (Echiura)
52. Anal vesicles, with excretory funnels (Echiura)
53. Hemal system (Echinodermata)
54. Use of external ciliary grooves for suspension feeding (Echinodermata)
55. Unique proboscis apparatus (Nemertea)
56. Planula larva (Cnidaria)
57. Sheets of subepidermal muscles derived, at least in part, from archenteric mesoderm
58. Sheets of subepidermal muscles derived, at least in part, from 4d mesoderm
59. Muscles concentrated into isolated bands attached to internally directed skeletal apodemes (Arthropoda)
60a. Mesoderm (from mesentoblast) gives rise in part to closed circulatory system of vessels only
60b. Complete loss of closed circulatory system (Sipuncula)
60c. Reduction to open circulatory system with dorsal ostiate heart (Arthropoda)
60d. Reduction to open circulatory system with dorsal chambered heart (Mollusca)
61. Circulatory system derived, at least in part, from archenteric mesoderm (varies among taxa)
62a. Trochophore larva
62b. Trochophore lost (Arthropoda)
63. Vestimentum ("wings") (Vestimentifera)
64. Synaptic nervous system
64a. Nervous system arranged in noncentralized, netlike fashion
64b. Nervous system with anterior concentration of neurons (cerebral ganglion) and tending toward presence of longitudinal cords
64c. Cerebral ganglion issues pairs of longitudinal cords connected by transverse commissures (ladder-like), with tendency to emphasize ventral or ventrolateral cords (reduced to a single ventral cord in some taxa)
64d. With circumenteric nerve ring attached to one or more ventral nerve cords
64e. Longitudinal cords not ladder-like in arrangement and not emphasized ventrally
64f. Dorsal hollow nerve cord
64g. Nervous system pentaradially arranged, loss of cerebral ganglion (Echinodermata)
65. Unique chaetognathan fins (Chaetognatha)
66. Unique chaetognathan buccal apparatus (Chaetognatha)
67. Unique chaetognathan ciliary loop (Chaetognatha)
68. Tadpole larva (Chordata)

Radiation of the eumetazoa

Having established the origin of the metazoan condition from a protozoan ancestor, we are left to sort out the history of radiations that has resulted in the great diversity of animal life seen today and represented in the fossil record. We build our discussion by examining the origins and radiations of the two great metazoan lineages depicted in the cladogram, which correspond to long-recognized groups called the protostomes and the deuterostomes.* We then comment on the possible relationships of the more poorly known taxa not included on the tree.

Origins of the coelomate condition and true mesoderm

If the ancestor of the bilateral eumetazoa (node *a* on the cladogram) arose as depicted in Figure 1, then we may hypothesize that it possessed the following synapomorphic attributes: bilateral symmetry, rudimentary cephalization, and a nervous system that included some sort of cerebral ganglion and some arrangement of longitudinal nerve cords. It also retained the following symplesiomorphies (among others): radial cleavage, an embryogeny that included gastrulation (by ingression or invagination), and a blastopore that became the mouth. It should be evident to you that the changes one would have to impose on such an ancestor to derive a deuterostome are quite different from those necessary to derive a protostome. Our cladogram hypothesizes that the protostomes and deuterostomes do not share a common, *coelomate* ancestor; that is, the body cavity we call a coelom (and also the middle germ layer we call mesoderm) arose not once within the animal kingdom, but twice. Two views of the evolutionary origin of mesoderm and the coelomic condition—one monophyletic and one diphyletic—are depicted in Figure 2 and discussed below.

In a monophyletic theory of the origin of eucoelomates, we are faced with the following complicated and protracted scenarios. The ancestral coelomate presumably arose from a diploblastic ancestor by the addition of mesoderm and coelomic spaces by either "deuterostomous methods" (mesoderm from the archenteron and enterocoely), or "protostomous methods" (mesentoblast and schizocoely). The latter case

*Although different authors have proposed a variety of names for what is usually viewed as a fundamental dichotomy in the evolution of the higher metazoa (Eumetazoa), the most commonly used names are Protostomia and Deuterostomia. These names, and the initial consolidation of this "diphyletic theory" of eucoelomate evolution, are attributable to the great German morphologist K. Grobben (1908).

would also suggest that the ancestral coelomate displayed spiral cleavage. One or the other sets of these events would have occurred at step a in Figure 2A. If we assume that the ancestral coelomate was deuterostome-like in nature, then step a also involved switching the fate of the blastopore from the mouth to the anus, or loss of the blastopore and secondary formation of both gut openings. Following this hypothesis, step b is solved—we already have a deuterostome *Bauplan*. However, step c leading to the protostome line becomes highly complicated and would require the loss of entodermally derived mesoderm and of enterocoely, and reversal of the blastopore fate *back* to formation of the mouth, as well as the acquisition of spiral cleavage, a mesentoblast precursor to mesoderm, and schizocoely.

If, on the other hand, we assume that the ancestral coelomate in Figure 2A was protostome-like, then even more complications arise. Step a would include the origin of spiral cleavage, a mesentoblast, and schizocoely, leading neatly to the protostomes via step c. Step b, however, would involve changing the fate of the blastopore, a reversal to the radial cleavage of the ancestor, loss of the mesentoblast and schizo-

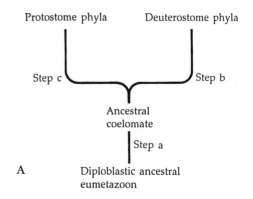

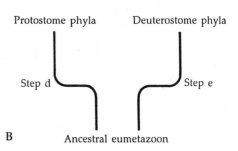

Figure 2

Monophyletic (A) and diphyletic (B) hypotheses for the origin of the coelomate condition. See the text for an explanation of steps a–e.

coely, and acquisition of archenteric mesoderm and enterocoely. Both schemes are very complicated, include major evolutionary reversals, and call out for simplification.

The diphyletic origin of the coelomate condition, as proposed by the cladogram and depicted in Figure 2B, avoids the problems inherent in the above scenarios and offers a more parsimonious explanation. From one (or more) kinds of diploblastic ancestors, two relatively direct sets of events can be outlined that led to the protostome and deuterostome lines. Step d involved the origin of spiral cleavage, accompanied by the mesentoblast and mesoderm (leading to node *b* on the cladogram in Figure 1) and eventually the formation of a coelom by schizocoely. The ancestral fate of the blastopore was retained. Step e included a change in the fate of the blastopore and the appearance of archenteric mesoderm and enterocoely; the primitive radial cleavage was retained in this lineage. Additional character changes are noted on the cladogram.

We believe that an a priori assumption that the coelom of all invertebrates is a homologous feature is unparsimonious and probably incorrect. Simply defining a "coelom" as a mesodermally derived body cavity is insufficient and does not establish it as a monophyletic (homologous) feature in all groups that possess such a cavity. Homology of a developmental feature should be based on the actual *process* by which that feature appears during embryogeny. That is, we must seek *developmental homology*. The embryonic development of the protostome and deuterostome coeloms (and their mesoderms) is very different, and we do not regard the schizocoelous coelom of the former to be homologous with the enterocoelic coelom of the latter. This idea was also suggested by Claus Nielsen (1985, 1987) based on an entirely different kind of analysis (a postulated series of hypothetical ancestral forms among the early metazoa).

The diphyletic theory of coelomate origins, as depicted in the cladogram, suggests that the first coelomate animals in each lineage differed profoundly from one another in their embryogeny and development (i.e., each line was initiated by the acquisition of numerous new developmental synapomorphies). However, having a common origin in a diploblastic ancestor, early representatives in both lines also would have retained some basic primitive similarities (i.e., symplesiomorphies) such as synaptic nervous systems and gastrulation events. Furthermore, the development of a body coelom, albeit in different ways, would have led to evolutionary convergences in various other attributes. For example, the appearance of mesodermally derived circulatory systems and sheets of body wall muscles must have been derived independently in the protostomes and the deuterostomes because the mesoderm itself arose independently. Primitive creatures early in the evolution of both lineages (nodes *c* and *d* on the cladogram) were probably bilaterally symmetrical and soft-bodied, and possessed a spacious trunk coelom that would have allowed them to burrow. Below we briefly discuss some ideas concerning the descent of such creatures.

The protostome and deuterostome clades

An examination of the cladogram in Figure 1 indicates that early protostomes (node *c* on the cladogram) not only possessed spiral cleavage, mesentoblast mesoderm formation, schizocoelous coelom formation, and a ladder-like nervous system, but also were probably unsegmented, wormlike creatures that produced free-swimming, trochophore-like larvae.* The Sipuncula, the Echiura, and to some extent the Nemertea bear a strong resemblance to this ancestral condition and appear to have diverged early within the protostome lineage. They may be among the most primitive living coelomate protostomes. In their retention of ancestral features, sipunculans and echiurans have remained as largely infaunal burrowers, using the large trunk coelom for peristalsis. The coelom in nemerteans is reduced to the rhyncocoel around the proboscis apparatus. The other major protostome groups escaped from infaunal life, perhaps in part by evolution of adaptations such as exoskeleton formation or tube building. The emergence of the molluscs (as well as the sipunculans, the echiurans, and the annelid–arthropod clade) from a common ancestral pool (node *e* on the cladogram) is especially interesting. If we continue our reasoning from the inference that the first protostome was a wormlike creature with a large body coelom, then it had a body architecture from which the primitive molluscan *Bauplan* can be easily and logically derived. Recall from Chapter 20 that our analysis of molluscan phylogeny depicts the caudofoveatans and aplacophorans as the primitive forms in that phylum. Thus we can conclude that the vermiform molluscs arose from an early wormlike protostome and eventually led to the shelled molluscs.

We have defined *segmentation* as the repetition of homologous body structures (i.e., serial homology)

---

*The planktotrophic Muller's larva (of polyclads) and pilidium larva (of nemerteans) are regarded by some workers as modified trochophore larvae.

derived by way of teloblastic growth (see Chapter 13). In spite of the fact that only this single synapomorphy defines the annelid–arthropod clade, we view it as a profound evolutionary step, not only because it is unique but also because of its far-reaching influence on the *Baupläne* of the taxa in which it occurs. Furthermore, serial repetition of some body structures occasionally occurs within presegmental members of the protostome line, where it may represent a developmental precursor, or preadaptation, for the true segmentation of the annelids and arthropods (e.g., serial repetition of reproductive structures occurs in some nemerteans and even flatworms).

According to the cladogram, the first organisms recognizable as deuterostomes (node *d* on the cladogram) retained a primitive radial cleavage but had evolved an enterocoelous coelom and archenteric mesoderm formation. The cladogram also hypothesizes that the first deuterostome possessed the tripartite coelomic condition. Like the earliest protostomes, the first deuterostome probably also was a bilaterally symmetrical wormlike burrower (perhaps producing free-swimming larvae). The main trunk coelom (the metacoel) probably functioned in peristaltic burrowing, leaving the small protocoel and mesocoel "free" for modification during early deuterostome radiation. Each of the major groups of deuterostomes has evolved very differently. In general, each group abandoned the ancestral burrowing lifestyle and assumed various strategies for other modes of existence.

The echinoderms probably appeared soon after the origin of the basic deuterostome *Bauplan*. The evolution of the echinoderm water vascular system (derived largely from parts of the mesocoel) and its suite of associated features, and the unique stereom endoskeleton, probably helped to solve some of the problems of exploiting epibenthic habitats. Their adoption of secondary pentaradial symmetry also met with obvious success. We treat the lophophorates here as a single monophyletic group within the deuterostome lineage because of the difficulties in defining sister-groups within the lophophorate clade. The lophophorate phyla (Brachiopoda, Phoronida, and Ectoprocta) evolved through elaboration of the mesosoma and mesocoel, producing the lophophore as an efficient means of suspension feeding. The phoronids retained the vermiform ancestral body, adopting a tube-dwelling epibenthic lifestyle; in this group the mouth develops near the blastopore, a reversal to the primitive condition of the premesodermal bilateral ancestor. The ectoprocts capitalized on small body size and colony formation, and the shell and

mantle cavity of brachiopods provided a protective housing for their nonretractile lophophore. The line leading to the hemichordates and chordates involved, in part, the origin of pharyngeal gill slits, serving as suspension feeding and/or gas exchange devices. More details of these hypotheses were discussed in Chapter 23.

A word about some characters

A discussion of all of the characters used to construct the cladogram would fill many pages. Most are discussed in various chapters of the text and can be located by using the index.

Several obvious examples of convergent evolution are not included on the cladogram: the evolution of U-shaped guts (in sipunculans, lophophorates, and tunicates); the appearance of feeding tentacles around the mouth (in cnidarians, sipunculans, lophophorates, and many annelids); an eversible buccal region or a proboscis (in nemerteans, priapulans, kinorhynchs, and many annelids); and giant neurons (in many different groups). Features such as these show high levels of convergence throughout the metazoa and must be viewed with great caution when assessing relationships among phyla. However, when we can convincingly establish, on embryological grounds, the nonhomologous nature of two or more convergent conditions, then those features can be used as valid (but separate) synapomorphies in their respective clades. For example, the origin and subsequent modifications of mesoderm and the coelom can be considered separately in the protostomes and deuterostomes *because* their independent developmental origins can be demonstrated. The same approach can be applied to structures of mesodermal origin in the two clades, again because they are developmentally independent of each other (e.g., circulatory systems).

One potentially important character that we have not used in the construction of our cladogram is nephridia. This omission is because of persisting uncertainties about homologies within and among such structures as protonephridia, metanephridia, and the excretory units of hemichordates and chordates (for some recent hypotheses, see Ruppert and Smith 1988 and Smith and Ruppert 1988). It is tempting to place the appearance of protonephridia just below node *a* on the cladogram along with the origin of bilateral symmetry. Protonephridia occur in the larvae and adults of many protostomes and at least in the larvae of phoronids among the deuterostomes. If all of these structures that we call protonephridia are homologous, then such a placement is justified. Furthermore, if protonephridia did arise early on, they have been subsequently lost in several phyla (e.g., Arthropoda, Echinodermata, Ectoprocta). Metanephridia may be the derived homologues of protonephridia; they would appear at least twice on the cladogram, once just above node *c* in the protostomes and again within the lophophorates (where they occur in adult phoronids and brachiopods). However, until the possible homologies of these and other invertebrate excretory structures are better understood, we are reluctant to speculate further by formally adding these characters to the cladogram.

Most of the other characters that are included on the tree are traceable without much difficulty within the phyla shown. However, their distribution on the tree is, of course, open to various interpretations, and we strongly encourage you to study and evaluate our cladogram. In doing so you will obtain a good review of invertebrate *Baupläne*, and at the same time gain an understanding of the kinds of questions and problems that one confronts when undertaking phylogenetic analyses.

Some difficult groups

Mesozoa. The odd creatures relegated to the mesozoan grade of complexity (i.e., *Trichoplax*, orthonectids, dicyemids, and, if it exists, *Salinella*) have puzzled many generations of biologists. As we noted in Chapter 6, there are three ways one might view the phylogenetic positions of these different enigmatic organisms: as one or more descendants that arose from protozoan stock independently of the metazoan line; as groups derived from higher metazoa through reduction and simplification; or as the descendants of lines emerging during the early period of transition from protozoa to metazoa (i.e., as phylogenetic "intermediates" between the protozoa and the higher metazoa). The first hypothesis is unparsimonious because it denies certain fundamental metazoan synapomorphies and seems highly unlikely. Furthermore, the mesozoan phyla share no unique synapomorphies, so almost certainly they do not represent a monophyletic clade themselves. Thus, it seems likely that they derived from as many as four separate lines that originated during the early stages of evolutionary "experimentation" with multicellularity, either from simple primitive ancestors or by way of reduction from more complex ancestors. There is growing evidence that a basement membrane is lacking in *Trichoplax* and that the entire epithelium consists of monociliate cells. Hence, this odd creature, and perhaps other mesozoa, might be very primitive indeed (fitting into the cladogram somewhere below node *a*).

Ctenophora. Since about 1925, the ctenophores have rested comfortably as close allies of the cnidarians. These two phyla are often depicted as sister-groups, largely on the basis of their similar gelatinous mesenchymes and radial symmetries. Harbison (1985) carefully analyzed the characteristics of ctenophores, cnidarians, and platyhelminths and concluded that there is no more justification for linking the ctenophores with the cnidarians than there is for linking them with the flatworms. As he pointed out, the radial symmetry and gelatinous nature of ctenophores and cnidarians may be superficial, convergent resemblances; we see similar adaptations to pelagic life in a variety of unrelated taxa. There are no unique synapomorphies to support a ctenophore–cnidarian clade. In addition, ctenophoran embryogeny is unique, and there is still the unresolved question of whether or not they possess true mesoderm. Also, in contrast to cnidarians, ctenophores have distinct subepidermal muscles. By the same token, none of the similarities between ctenophores and platyhelminths appear to be unique synapomorphies of a possible ctenophore–flatworm clade. Our current understanding of the Ctenophora (and a good deal of speculation about their development) leads us to consider them as a uniquely defined lineage arising near node *a* on our cladogram, or perhaps at the base of the deuterostome line (before node *d*).

Platyhelminthes. As discussed in earlier chapters, the phylogenetic position of the flatworms has long been uncertain. There is no question that the platyhelminths are closely allied to the protostome clade, with which they share several important developmental features (e.g., spiral cleavage and mesentoblast origin of mesoderm). The question is whether the acoelomate condition in flatworms is ancestral to the origin of schizocoely, as has traditionally been held and as we have depicted it in our cladogram, or whether the acoelomate condition was secondarily derived from a coelomate protostome ancestor by way of reduction of the coelom. We examined these two ideas in Chapter 10. If we posit the origin of the flatworms prior to the evolution of the schizocoelous coelom, then platyhelminths cannot be distinguished from the protostome ancestor itself. That is, the ancestor of the protostome line (node *b* on the cladogram) was perhaps, in fact, a flatworm. This conclusion is based on the observation that there appear to be no unique synapomorphies that define the flatworms (except perhaps the presence of rhabdites in a few groups). On the other hand, if we accept that the acoelomate condition arose secondarily from a coelomate ancestry, then the loss of the schizocoelous coelom (and perhaps also the circulatory system and anus) would be synapomorphies that could define the platyhelminth clade.

Suggesting loss of the coelom and other functional structures (e.g., anuses and circulatory systems) may seem to go against the grain of "progressive" evolution. However, we frequently see similar events in more solidly supported phylogenies: reduction of the coelom in arthropods, molluscs, and urochordates; loss of the gut in many parasitic taxa (and in pogonophorans and vestimentiferans); loss of the circulatory system in small annelids; and so on. These sorts of evolutionary reductions may be more commonplace than once thought and, while they may complicate issues, they are clearly legitimate biological hypotheses. On the other hand, many platyhelminths are quite large, free-living predators, with highly branched guts. It may seem biologically unreasonable to assume that such large and active creatures would derive through degeneration or loss of fundamentally "useful" features, such as a coelom and an anus. Thus, until convincing developmental evidence to the contrary is forthcoming, we favor the more parsimonious view that platyhelminths are the true precursors to the protostome line.

Nemertea. Ribbon worms find themselves today in much the same precarious situation as the platyhelminths. Historically, zoologists have usually viewed them as descended from, or at least closely related to, the flatworms. This view is based largely on the acoelomate nature of their bodies. Certainly the unique proboscis apparatus of the nemerteans allows us to define them more clearly than the flatworms, but their ancestry is a point of controversy. However, the strong evidence that the rhyncocoel is a true coelomic remnant (see Turbeville and Ruppert 1985) suggests a coelomate, protostomous ancestry, as depicted in our cladogram. The presence of a circulatory system and a complete digestive tube lends additional support to this hypothesis.

Pseudocoelomates, gnathostomulids, priapulans, and loriciferans. In Chapter 12 we discussed several enigmatic phyla lumped as "pseudocoelomates,"

along with the Gnathostomulida, Priapula, and Loricifera. Problems associated with evaluating the evolutionary histories of these groups stem from our understanding of the nature of the "pseudocoelom" itself, our inability to identify homologues among many of the taxa (that could be sister-groups), and a general lack of adequate data on many of the more obscure groups. A growing body of evidence suggests that the concept of a "pseudocoelom" may encompass the end product (adult condition) of several different developmental pathways and thus is not in itself a homologous or a unique feature. Furthermore, we now know that some of these groups actually lack a body cavity altogether and hence are really at an acoelomate grade of construction (e.g., some gastrotrichs and gnathostomulids), a feature probably associated with extreme reduction of body size; even some small annelids appear to be "acoelomate" in this regard. Since the welcome abandonment of the taxon "Aschelminthes," a great many hypotheses have been proposed regarding the origins and relationships of these puzzling animals. Perhaps no other group of phyla is such a phylogenetic mystery as the pseudocoelomates. (For some excellent reviews of the "pseudocoelomate problem," see the edited volume by Morris et al. 1985).

Very few generalizations can be made. We do know that these creatures are all triploblastic, and, where ontogenetic data are available, most seem to be more closely allied with the protostomes than with the deuterostomes. Beyond this, their roots in any particular protostomous ancestry lie hidden by time and change, particularly change associated with the adoption of small size seen in many of these animals. Taken together, these groups probably represent a polyphyletic assemblage that arose from various points within the flatworm–protostome clade, and we have elected to treat them as independent phyla in the text. We refer you to Chapter 12 and especially to the references cited therein for a sampling of ideas.

Pogonophora and Vestimentifera. Recall the historical confusion surrounding pogonophorans before and after the discovery of the segmented opisthosomal region of the body, and their varied alliance with both the deuterostomes and protostomes. The discovery of the vestimentiferans also led to controversy. While the final word on these two groups is not yet in, our cladogram supports the view that they are probably highly modified annelids. Among the most convincing pieces of evidence are the segmented opisthosoma and the recent discovery of an apparent transient gut in the vestimentiferans (see

Jones and Gardiner 1988 and Southward 1988) giving us a real clue as to the dorsoventral locations of the major blood vessel and the nerve cord—an annelid-like organization. Thus, the evidence suggests that these animals represent a line of segmented worms, perhaps derived from some ancient polychaete with extreme heteronomy.

Chaetognatha. Even though the chaetognaths can be defined by some unique synapomorphies (see Figure 1), they cannot be clearly allied as a sister-group to any other phylum. Their extreme modifications associated with a motile, pelagic, predatory life have masked clues to their relationships with other groups. Their development is convincingly deuterostomous (although somewhat modified), and on the cladogram they emerge as a separate line from node *d*, which also leads to the echinoderms and to the other deuterostomes.

Hemichordata. It seems clear that the Hemichordata and Chordata are sister-groups, united by possession of pharyngeal gill slits and a dorsal hollow nerve cord. We find it difficult, however, to adequately define the Hemichordata so as to fully encompass all of its members. Even ignoring the enigmatic *Planktosphaera*, the group still contains animals displaying two distinctly different *Baupläne* and the character states appearing on our cladogram as hemichordate synapomorphies are admittedly a bit shaky.

On the one hand are the enteropneusts (vermiform, infaunal burrowers), and on the other hand are the pterobranchs (tiny, zooid-like suspension feeders). The latter bear striking similarities to the lophophorates in the nature of the mesosomal tentacles. One might reasonably question whether the Hemichordata is a natural (monophyletic) taxon.

Where are we now?

Although we have derived a biologically sound phylogenetic tree, or cladogram, of the better known metazoa, there is no doubt other hypotheses of relationships will continue to be proposed as more data and analyses become available. Students interested in the history of life are encouraged to rigorously criticize our cladogram and compare it to those of other authors. We welcome these new ideas, for they will lead to increased discussion and insight on matters of invertebrate phylogenetics. As specialists provide more information on their favorite creatures, and the application of new methods of analysis continue to refine our views about the history of life on Earth, branches of evolutionary trees will be cut and grafted as evidence dictates. If we have stimulated you to think about such things and, more important, to appreciate invertebrates, then our task has been worthwhile.

Selected References

Hundreds of books and papers have been published on matters of invertebrate phylogeny. Many of these references are listed in Chapter 4 and in chapters pertaining to specific phyla. The list below includes only those references cited in this chapter.

Anderson, D. T. 1982. Origins and relationships among the animal phyla. Proc. Linn. Soc. N.S.W. 106:151–166.

Field, K. G., G. J. Olsen, D. J. Lane, S. J. Giovannoni, M. T. Ghiselin, E. C. Raff, N. R. Pace and R. A. Raff. 1988. Molecular phylogeny of the animal kingdom. Science 239:748–753. [Also see responses in Science 243:548–551.]

Grobben, K. 1908. Die systematische Einteilung des Tierreisches. Verh. Zool. Bot. Ges. Wien 58:491–511.

Harbison, G. R. 1985. On the classification and evolution of the Ctenophora. *In* C. Morris et al. (eds.), *The Origins and Relationships of Lower Invertebrates*. Clarendon Press, Oxford, pp. 78–100.

Inglis, W. G. 1985. Evolutionary waves: Patterns in the origins of animal phyla. Aust. J. Zool. 33:153–178.

Jones, M. L. and S. L. Gardiner. 1988. Evidence for a transient digestive tract in Vestimentifera. Proc. Biol. Soc. Wash. 101:423–433.

Margulis, L. 1981. *Symbiosis in Cell Evolution: Life and Its Environment on the Early Earth*. W. H. Freeman, San Francisco.

Morris, S. C., J. D. George, R. Gibson and H. M. Platt (eds.). 1985. *The Origins and Relationships of Lower Invertebrates*. Clarendon Press, Oxford.

Nielsen, C. 1985. Animal phylogeny in the light of the trochaea theory. Biol. J. Linn. Soc. 25:243–299.

Nielsen, C. 1987. Structure and function of metazoan ciliary bands and their phylogenetic significance. Acta Zool. 68:205–262.

Ruppert, E. E. and P. R. Smith. 1988. The functional organization of filtration nephridia. Biol. Rev. 63:231–258.

Smith, P. R. and E. E. Ruppert. 1988. Nephridia. *In* W. Westheide and C. O. Hermans (eds.), *Microfauna Marina*, Vol. 4. Gustav Fischer Verlag, Stuttgart, pp. 231–262.

Southward, E. C. 1988. Development of the gut and segmentation of newly settled stages of *Ridgeia* (Vestimentifera): Implications for relationship between Vestimentifera and Pogonophora. J. Mar. Biol. Assoc. U.K. 68:465–487.

Turbeville, J. M. and E. E. Ruppert. 1985. Comparative ultrastructure and the evolution of nemertines. Am. Zool. 25:53–71.

Index

Numbers in **boldface** signify definitions of the listed terms; numbers in *italic* indicate information in an illustration.